Stochastic Processes
From Applications to Theory

CHAPMAN & HALL/CRC
Texts in Statistical Science Series

Series Editors

Francesca Dominici, *Harvard School of Public Health, USA*
Julian J. Faraway, *University of Bath, UK*
Martin Tanner, *Northwestern University, USA*
Jim Zidek, *University of British Columbia, Canada*

Statistical Theory: A Concise Introduction
F. Abramovich and Y. Ritov

Practical Multivariate Analysis, Fifth Edition
A. Afifi, S. May, and V.A. Clark

Practical Statistics for Medical Research
D.G. Altman

**Interpreting Data: A First Course
in Statistics**
A.J.B. Anderson

Introduction to Probability with R
K. Baclawski

**Linear Algebra and Matrix Analysis for
Statistics**
S. Banerjee and A. Roy

**Mathematical Statistics: Basic Ideas and
Selected Topics, Volume I,
Second Edition**
P. J. Bickel and K. A. Doksum

**Mathematical Statistics: Basic Ideas and
Selected Topics, Volume II**
P. J. Bickel and K. A. Doksum

Analysis of Categorical Data with R
C. R. Bilder and T. M. Loughin

Statistical Methods for SPC and TQM
D. Bissell

Introduction to Probability
J. K. Blitzstein and J. Hwang

**Bayesian Methods for Data Analysis,
Third Edition**
B.P. Carlin and T.A. Louis

Second Edition
R. Caulcutt

**The Analysis of Time Series: An Introduction,
Sixth Edition**
C. Chatfield

Introduction to Multivariate Analysis
C. Chatfield and A.J. Collins

**Problem Solving: A Statistician's Guide,
Second Edition**
C. Chatfield

**Statistics for Technology: A Course in Applied
Statistics, Third Edition**
C. Chatfield

**Analysis of Variance, Design, and Regression :
Linear Modeling for Unbalanced Data,
Second Edition**
R. Christensen

**Bayesian Ideas and Data Analysis: An
Introduction for Scientists and Statisticians**
R. Christensen, W. Johnson, A. Branscum,
and T.E. Hanson

Modelling Binary Data, Second Edition
D. Collett

**Modelling Survival Data in Medical Research,
Third Edition**
D. Collett

**Introduction to Statistical Methods for
Clinical Trials**
T.D. Cook and D.L. DeMets

Applied Statistics: Principles and Examples
D.R. Cox and E.J. Snell

**Multivariate Survival Analysis and Competing
Risks**
M. Crowder

Statistical Analysis of Reliability Data
M.J. Crowder, A.C. Kimber,
T.J. Sweeting, and R.L. Smith

**An Introduction to Generalized
Linear Models, Third Edition**
A.J. Dobson and A.G. Barnett

**Nonlinear Time Series: Theory, Methods, and
Applications with R Examples**
R. Douc, E. Moulines, and D.S. Stoffer

**Introduction to Optimization Methods and
Their Applications in Statistics**
B.S. Everitt

**Extending the Linear Model with R:
Generalized Linear, Mixed Effects and
Nonparametric Regression Models, Second
Edition**
J.J. Faraway

Linear Models with R, Second Edition
J.J. Faraway

A Course in Large Sample Theory
T.S. Ferguson

Multivariate Statistics: A Practical Approach
B. Flury and H. Riedwyl

Readings in Decision Analysis
S. French

Discrete Data Analysis with R: Visualization and Modeling Techniques for Categorical and Count Data
M. Friendly and D. Meyer

Markov Chain Monte Carlo: Stochastic Simulation for Bayesian Inference, Second Edition
D. Gamerman and H.F. Lopes

Bayesian Data Analysis, Third Edition
A. Gelman, J.B. Carlin, H.S. Stern, D.B. Dunson, A. Vehtari, and D.B. Rubin

Multivariate Analysis of Variance and Repeated Measures: A Practical Approach for Behavioural Scientists
D.J. Hand and C.C. Taylor

Practical Longitudinal Data Analysis
D.J. Hand and M. Crowder

Logistic Regression Models
J.M. Hilbe

Richly Parameterized Linear Models: Additive, Time Series, and Spatial Models Using Random Effects
J.S. Hodges

Statistics for Epidemiology
N.P. Jewell

Stochastic Processes: An Introduction, Second Edition
P.W. Jones and P. Smith

The Theory of Linear Models
B. Jørgensen

Pragmatics of Uncertainty
J.B. Kadane

Principles of Uncertainty
J.B. Kadane

Graphics for Statistics and Data Analysis with R
K.J. Keen

Mathematical Statistics
K. Knight

Introduction to Multivariate Analysis: Linear and Nonlinear Modeling
S. Konishi

Nonparametric Methods in Statistics with SAS Applications
O. Korosteleva

Modeling and Analysis of Stochastic Systems, Second Edition
V.G. Kulkarni

Exercises and Solutions in Biostatistical Theory
L.L. Kupper, B.H. Neelon, and S.M. O'Brien

Exercises and Solutions in Statistical Theory
L.L. Kupper, B.H. Neelon, and S.M. O'Brien

Design and Analysis of Experiments with R
J. Lawson

Design and Analysis of Experiments with SAS
J. Lawson

A Course in Categorical Data Analysis
T. Leonard

Statistics for Accountants
S. Letchford

Introduction to the Theory of Statistical Inference
H. Liero and S. Zwanzig

Statistical Theory, Fourth Edition
B.W. Lindgren

Stationary Stochastic Processes: Theory and Applications
G. Lindgren

Statistics for Finance
E. Lindström, H. Madsen, and J. N. Nielsen

The BUGS Book: A Practical Introduction to Bayesian Analysis
D. Lunn, C. Jackson, N. Best, A. Thomas, and D. Spiegelhalter

Introduction to General and Generalized Linear Models
H. Madsen and P. Thyregod

Time Series Analysis
H. Madsen

Pólya Urn Models
H. Mahmoud

Randomization, Bootstrap and Monte Carlo Methods in Biology, Third Edition
B.F.J. Manly

Introduction to Randomized Controlled Clinical Trials, Second Edition
J.N.S. Matthews

Statistical Rethinking: A Bayesian Course with Examples in R and Stan
R. McElreath

Statistical Methods in Agriculture and
Experimental Biology, Second Edition
R. Mead, R.N. Curnow, and A.M. Hasted

Statistics in Engineering: A Practical Approach
A.V. Metcalfe

Statistical Inference: An Integrated Approach,
Second Edition
H. S. Migon, D. Gamerman, and
F. Louzada

Beyond ANOVA: Basics of Applied Statistics
R.G. Miller, Jr.

A Primer on Linear Models
J.F. Monahan

Stochastic Processes: From Applications to
Theory
P.D Moral and S. Penev

Applied Stochastic Modelling, Second Edition
B.J.T. Morgan

Elements of Simulation
B.J.T. Morgan

Probability: Methods and Measurement
A. O'Hagan

Introduction to Statistical Limit Theory
A.M. Polansky

Applied Bayesian Forecasting and Time Series
Analysis
A. Pole, M. West, and J. Harrison

Statistics in Research and Development,
Time Series: Modeling, Computation, and
Inference
R. Prado and M. West

Essentials of Probability Theory for
Statisticians
M.A. Proschan and P.A. Shaw

Introduction to Statistical Process Control
P. Qiu

Sampling Methodologies with Applications
P.S.R.S. Rao

A First Course in Linear Model Theory
N. Ravishanker and D.K. Dey

Essential Statistics, Fourth Edition
D.A.G. Rees

Stochastic Modeling and Mathematical
Statistics: A Text for Statisticians and
Quantitative Scientists
F.J. Samaniego

Statistical Methods for Spatial Data Analysis
O. Schabenberger and C.A. Gotway

Bayesian Networks: With Examples in R
M. Scutari and J.-B. Denis

Large Sample Methods in Statistics
P.K. Sen and J. da Motta Singer

Spatio-Temporal Methods in Environmental
Epidemiology
G. Shaddick and J.V. Zidek

Decision Analysis: A Bayesian Approach
J.Q. Smith

Analysis of Failure and Survival Data
P. J. Smith

Applied Statistics: Handbook of GENSTAT
Analyses
E.J. Snell and H. Simpson

Applied Nonparametric Statistical Methods,
Fourth Edition
P. Sprent and N.C. Smeeton

Data Driven Statistical Methods
P. Sprent

Generalized Linear Mixed Models:
Modern Concepts, Methods and Applications
W. W. Stroup

Survival Analysis Using S: Analysis of
Time-to-Event Data
M. Tableman and J.S. Kim

Applied Categorical and Count Data Analysis
W. Tang, H. He, and X.M. Tu

Elementary Applications of Probability Theory,
Second Edition
H.C. Tuckwell

Introduction to Statistical Inference and Its
Applications with R
M.W. Trosset

Understanding Advanced Statistical Methods
P.H. Westfall and K.S.S. Henning

Statistical Process Control: Theory and
Practice, Third Edition
G.B. Wetherill and D.W. Brown

Generalized Additive Models:
An Introduction with R
S. Wood

Epidemiology: Study Design and
Data Analysis, Third Edition
M. Woodward

Practical Data Analysis for Designed
Experiments
B.S. Yandell

Texts in Statistical Science

Stochastic Processes
From Applications to Theory

Pierre Del Moral

University of New South Wales
Sydney, Australia
and
INRIA Sud Ouest Research Center
Bordeaux, France

Spiridon Penev

University of New South Wales
Sydney, Australia

With illustrations by Timothée Del Moral

CRC Press
Taylor & Francis Group
Boca Raton   London   New York

CRC Press is an imprint of the
Taylor & Francis Group an **informa** business

A CHAPMAN & HALL BOOK

CRC Press
Taylor & Francis Group
6000 Broken Sound Parkway NW, Suite 300
Boca Raton, FL 33487-2742

© 2017 by Taylor & Francis Group, LLC
CRC Press is an imprint of Taylor & Francis Group, an Informa business

No claim to original U.S. Government works

Printed on acid-free paper by Ashford Colour Press Ltd.
Version Date: 20170119

International Standard Book Number-13: 978-1-4987-0183-9 (Hardback)

Visit the Taylor & Francis Web site at
http://www.taylorandfrancis.com

and the CRC Press Web site at
http://www.crcpress.com

To Laurence, Tiffany and Timothée;
to Tatiana, Iva and Alexander.

Contents

Introduction   **xxi**

I   An illustrated guide   **1**

1   Motivating examples   **3**
1.1   Lost in the Great Sloan Wall 3
1.2   Meeting Alice in Wonderland 6
1.3   The lucky MIT Blackjack Team 7
1.4   Kruskal's magic trap card 10
1.5   The magic fern from Daisetsuzan 12
1.6   The Kepler-22b Eve 15
1.7   Poisson's typos 17
1.8   Exercises 21

2   Selected topics   **25**
2.1   Stabilizing populations 25
2.2   The traps of reinforcement 28
2.3   Casino roulette 31
2.4   Surfing Google's waves 34
2.5   Pinging hackers 36
2.6   Exercises 37

3   Computational and theoretical aspects   **43**
3.1   From Monte Carlo to Los Alamos 43
3.2   Signal processing and population dynamics 45
3.3   The lost equation 49
3.4   Towards a general theory 56
3.5   The theory of speculation 60
3.6   Exercises 66

II   Stochastic simulation   **69**

4   Simulation toolbox   **71**
4.1   Inversion technique 71
4.2   Change of variables 74
4.3   Rejection techniques 75
4.4   Sampling probabilities 77
    4.4.1   Bayesian inference 77
    4.4.2   Laplace's rule of successions 79
    4.4.3   Fragmentation and coagulation 79
4.5   Conditional probabilities 80
    4.5.1   Bayes' formula 80
    4.5.2   The regression formula 81

        4.5.3   Gaussian updates   82
        4.5.4   Conjugate priors   84
    4.6   Spatial Poisson point processes   85
        4.6.1   Some preliminary results   85
        4.6.2   Conditioning principles   88
        4.6.3   Poisson-Gaussian clusters   91
    4.7   Exercises   92

5   **Monte Carlo integration**   **99**
    5.1   Law of large numbers   99
    5.2   Importance sampling 102
        5.2.1   Twisted distributions 102
        5.2.2   Sequential Monte Carlo 102
        5.2.3   Tails distributions 103
    5.3   Exercises 104

6   **Some illustrations**   **107**
    6.1   Stochastic processes 107
    6.2   Markov chain models 108
    6.3   Black-box type models 108
    6.4   Boltzmann-Gibbs measures 110
        6.4.1   Ising model 110
        6.4.2   Sherrington-Kirkpatrick model 111
        6.4.3   The traveling salesman model 111
    6.5   Filtering and statistical learning 113
        6.5.1   Bayes' formula 113
        6.5.2   Singer's radar model 114
    6.6   Exercises 115

III   **Discrete time processes**   **119**

7   **Markov chains**   **121**
    7.1   Description of the models 121
    7.2   Elementary transitions 122
    7.3   Markov integral operators 123
    7.4   Equilibrium measures 124
    7.5   Stochastic matrices 125
    7.6   Random dynamical systems 126
        7.6.1   Linear Markov chain model 126
        7.6.2   Two-states Markov models 127
    7.7   Transition diagrams 128
    7.8   The tree of outcomes 128
    7.9   General state space models 129
    7.10 Nonlinear Markov chains 132
        7.10.1 Self interacting processes 132
        7.10.2 Mean field particle models 134
        7.10.3 McKean-Vlasov diffusions 135
        7.10.4 Interacting jump processes 136
    7.11 Exercises 138

8   **Analysis toolbox**   **141**

  8.1   Linear algebra 141

      8.1.1   Diagonalisation type techniques 141

      8.1.2   Perron Frobenius theorem 143

  8.2   Functional analysis 145

      8.2.1   Spectral decompositions 145

      8.2.2   Total variation norms 149

      8.2.3   Contraction inequalities 152

      8.2.4   Poisson equation 156

      8.2.5   V-norms 156

      8.2.6   Geometric drift conditions 160

      8.2.7   *V*-norm contractions 164

  8.3   Stochastic analysis 166

      8.3.1   Coupling techniques 166

          8.3.1.1   The total variation distance 166

          8.3.1.2   Wasserstein metric 169

      8.3.2   Stopping times and coupling 172

      8.3.3   Strong stationary times 173

      8.3.4   Some illustrations 174

          8.3.4.1   Minorization condition and coupling 174

          8.3.4.2   Markov chains on complete graphs 176

          8.3.4.3   A Kruskal random walk 177

  8.4   Martingales 178

      8.4.1   Some preliminaries 178

      8.4.2   Applications to Markov chains 183

          8.4.2.1   Martingales with fixed terminal values 183

          8.4.2.2   Doeblin-Itō formula 184

          8.4.2.3   Occupation measures 185

      8.4.3   Optional stopping theorems 187

      8.4.4   A gambling model 191

          8.4.4.1   Fair games 192

          8.4.4.2   Unfair games 193

      8.4.5   Maximal inequalities 194

      8.4.6   Limit theorems 196

  8.5   Topological aspects 203

      8.5.1   Irreducibility and aperiodicity 203

      8.5.2   Recurrent and transient states 206

      8.5.3   Continuous state spaces 210

      8.5.4   Path space models 211

  8.6   Exercises 212

9   **Computational toolbox**   **221**

  9.1   A weak ergodic theorem 221

  9.2   Some illustrations 224

      9.2.1   Parameter estimation 224

      9.2.2   Gaussian subset shaker 225

      9.2.3   Exploration of the unit disk 226

  9.3   Markov Chain Monte Carlo methods 226

      9.3.1   Introduction 226

      9.3.2   Metropolis and Hastings models 227

      9.3.3   Gibbs-Glauber dynamics 229

	9.3.4	Propp and Wilson sampler	233
9.4		Time inhomogeneous MCMC models	236
	9.4.1	Simulated annealing algorithm	236
	9.4.2	A perfect sampling algorithm	237
9.5		Feynman-Kac path integration	239
	9.5.1	Weighted Markov chains	239
	9.5.2	Evolution equations	240
	9.5.3	Particle absorption models	242
	9.5.4	Doob h-processes	244
	9.5.5	Quasi-invariant measures	245
	9.5.6	Cauchy problems with terminal conditions	247
	9.5.7	Dirichlet-Poisson problems	248
	9.5.8	Cauchy-Dirichlet-Poisson problems	250
9.6		Feynman-Kac particle methodology	252
	9.6.1	Mean field genetic type particle models	252
	9.6.2	Path space models	254
	9.6.3	Backward integration	255
	9.6.4	A random particle matrix model	257
	9.6.5	A conditional formula for ancestral trees	258
9.7		Particle Markov chain Monte Carlo methods	260
	9.7.1	Many-body Feynman-Kac measures	260
	9.7.2	A particle Metropolis-Hastings model	261
	9.7.3	Duality formulae for many-body models	262
	9.7.4	A couple particle Gibbs samplers	266
9.8		Quenched and annealed measures	267
	9.8.1	Feynman-Kac models	267
	9.8.2	Particle Gibbs models	269
	9.8.3	Particle Metropolis-Hastings models	271
9.9		Some application domains	272
	9.9.1	Interacting MCMC algorithms	272
	9.9.2	Nonlinear filtering models	276
	9.9.3	Markov chain restrictions	276
	9.9.4	Self avoiding walks	277
	9.9.5	Twisted measure importance sampling	279
	9.9.6	Kalman-Bucy filters	280
		9.9.6.1 Forward filters	280
		9.9.6.2 Backward filters	281
		9.9.6.3 Ensemble Kalman filters	283
		9.9.6.4 Interacting Kalman filters	285
9.10		Exercises	286

IV Continuous time processes   297

10 Poisson processes   **299**
10.1	A counting process	299
10.2	Memoryless property	301
10.3	Uniform random times	302
10.4	Doeblin-Itō formula	303
10.5	Bernoulli process	304
10.6	Time inhomogeneous models	306
	10.6.1 Description of the models	306

10.6.2 Poisson thinning simulation 309
10.6.3 Geometric random clocks 310
10.7 Exercises 311

11 Markov chain embeddings **313**
11.1 Homogeneous embeddings 313
11.1.1 Description of the models 313
11.1.2 Semigroup evolution equations 314
11.2 Some illustrations 317
11.2.1 A two-state Markov process 317
11.2.2 Matrix valued equations 318
11.2.3 Discrete Laplacian 320
11.3 Spatially inhomogeneous models 322
11.3.1 Explosion phenomenon 324
11.3.2 Finite state space models 328
11.4 Time inhomogeneous models 329
11.4.1 Description of the models 329
11.4.2 Poisson thinning models 331
11.4.3 Exponential and geometric clocks 332
11.5 Exercises 332

12 Jump processes **337**
12.1 A class of pure jump models 337
12.2 Semigroup evolution equations 338
12.3 Approximation schemes 340
12.4 Sum of generators 342
12.5 Doob-Meyer decompositions 344
12.5.1 Discrete time models 344
12.5.2 Continuous time martingales 346
12.5.3 Optional stopping theorems 349
12.6 Doeblin-Itō-Taylor formulae 350
12.7 Stability properties 351
12.7.1 Invariant measures 351
12.7.2 Dobrushin contraction properties 353
12.8 Exercises 356

13 Piecewise deterministic processes **363**
13.1 Dynamical systems basics 363
13.1.1 Semigroup and flow maps 363
13.1.2 Time discretization schemes 366
13.2 Piecewise deterministic jump models 367
13.2.1 Excursion valued Markov chains 367
13.2.2 Evolution semigroups 369
13.2.3 Infinitesimal generators 371
13.2.4 Fokker-Planck equation 372
13.2.5 A time discretization scheme 373
13.2.6 Doeblin-Itō-Taylor formulae 376
13.3 Stability properties 377
13.3.1 Switching processes 377
13.3.2 Invariant measures 379
13.4 An application to Internet architectures 379

13.4.1 The transmission control protocol 379
13.4.2 Regularity and stability properties 381
13.4.3 The limiting distribution 383
13.5 Exercises 384

14 Diffusion processes **393**
14.1 Brownian motion 393
14.1.1 Discrete vs continuous time models 393
14.1.2 Evolution semigroups 395
14.1.3 The heat equation 397
14.1.4 Doeblin-Itō-Taylor formula 398
14.2 Stochastic differential equations 401
14.2.1 Diffusion processes 401
14.2.2 Doeblin-Itō differential calculus 402
14.3 Evolution equations 405
14.3.1 Fokker-Planck equation 405
14.3.2 Weak approximation processes 406
14.3.3 A backward stochastic differential equation 408
14.4 Multidimensional diffusions 409
14.4.1 Multidimensional stochastic differential equations 409
14.4.2 An integration by parts formula 411
14.4.3 Laplacian and orthogonal transformations 412
14.4.4 Fokker-Planck equation 413
14.5 Exercises 413

15 Jump diffusion processes **425**
15.1 Piecewise diffusion processes 425
15.2 Evolution semigroups 426
15.3 Doeblin-Itō formula 428
15.4 Fokker-Planck equation 433
15.5 An abstract class of stochastic processes 434
15.5.1 Generators and carré du champ operators 434
15.5.2 Perturbation formulae 437
15.6 Jump diffusion processes with killing 439
15.6.1 Feynman-Kac semigroups 439
15.6.2 Cauchy problems with terminal conditions 440
15.6.3 Dirichlet-Poisson problems 442
15.6.4 Cauchy-Dirichlet-Poisson problems 447
15.7 Some illustrations 450
15.7.1 One-dimensional Dirichlet-Poisson problems 450
15.7.2 A backward stochastic differential equation 451
15.8 Exercises 451

16 Nonlinear jump diffusion processes **463**
16.1 Nonlinear Markov processes 463
16.1.1 Pure diffusion models 463
16.1.2 Burgers equation 464
16.1.3 Feynman-Kac jump type models 466
16.1.4 A jump type Langevin model 467
16.2 Mean field particle models 468
16.3 Some application domains 470

16.3.1 Fouque-Sun systemic risk model 470
16.3.2 Burgers equation 471
16.3.3 Langevin-McKean-Vlasov model 472
16.3.4 Dyson equation 473
16.4 Exercises 474

17 Stochastic analysis toolbox **481**
17.1 Time changes 481
17.2 Stability properties 482
17.3 Some illustrations 483
17.3.1 Gradient flow processes 483
17.3.2 One-dimensional diffusions 484
17.4 Foster-Lyapunov techniques 485
17.4.1 Contraction inequalities 485
17.4.2 Minorization properties 486
17.5 Some applications 487
17.5.1 Ornstein-Uhlenbeck processes 487
17.5.2 Stochastic gradient processes 487
17.5.3 Langevin diffusions 488
17.6 Spectral analysis 490
17.6.1 Hilbert spaces and Schauder bases 490
17.6.2 Spectral decompositions 493
17.6.3 Poincaré inequality 494
17.7 Exercises 495

18 Path space measures **501**
18.1 Pure jump models 501
18.1.1 Likelihood functionals 504
18.1.2 Girsanov's transformations 505
18.1.3 Exponential martingales 506
18.2 Diffusion models 507
18.2.1 Wiener measure 507
18.2.2 Path space diffusions 508
18.2.3 Girsanov transformations 509
18.3 Exponential change twisted measures 512
18.3.1 Diffusion processes 513
18.3.2 Pure jump processes 514
18.4 Some illustrations 514
18.4.1 Risk neutral financial markets 514
18.4.1.1 Poisson markets 514
18.4.1.2 Diffusion markets 515
18.4.2 Elliptic diffusions 516
18.5 Nonlinear filtering 517
18.5.1 Diffusion observations 517
18.5.2 Duncan-Zakai equation 518
18.5.3 Kushner-Stratonovitch equation 520
18.5.4 Kalman-Bucy filters 521
18.5.5 Nonlinear diffusion and ensemble Kalman-Bucy filters 523
18.5.6 Robust filtering equations 524
18.5.7 Poisson observations 525
18.6 Exercises 527

V Processes on manifolds 533

19 A review of differential geometry 535
19.1 Projection operators 535
19.2 Covariant derivatives of vector fields 541
    19.2.1 First order derivatives 543
    19.2.2 Second order derivatives 546
19.3 Divergence and mean curvature 547
19.4 Lie brackets and commutation formulae 554
19.5 Inner product derivation formulae 556
19.6 Second order derivatives and some trace formulae 559
19.7 Laplacian operator 562
19.8 Ricci curvature 563
19.9 Bochner-Lichnerowicz formula 568
19.10 Exercises 576

20 Stochastic differential calculus on manifolds 579
20.1 Embedded manifolds 579
20.2 Brownian motion on manifolds 581
    20.2.1 A diffusion model in the ambient space 581
    20.2.2 The infinitesimal generator 583
    20.2.3 Monte Carlo simulation 584
20.3 Stratonovitch differential calculus 584
20.4 Projected diffusions on manifolds 586
20.5 Brownian motion on orbifolds 589
20.6 Exercises 591

21 Parametrizations and charts 593
21.1 Differentiable manifolds and charts 593
21.2 Orthogonal projection operators 596
21.3 Riemannian structures 599
21.4 First order covariant derivatives 602
    21.4.1 Pushed forward functions 602
    21.4.2 Pushed forward vector fields 604
    21.4.3 Directional derivatives 606
21.5 Second order covariant derivative 609
    21.5.1 Tangent basis functions 609
    21.5.2 Composition formulae 612
    21.5.3 Hessian operators 613
21.6 Bochner-Lichnerowicz formula 617
21.7 Exercises 623

22 Stochastic calculus in chart spaces 629
22.1 Brownian motion on Riemannian manifolds 629
22.2 Diffusions on chart spaces 631
22.3 Brownian motion on spheres 632
    22.3.1 The unit circle $S = \mathbb{S}^1 \subset \mathbb{R}^2$ 632
    22.3.2 The unit sphere $S = \mathbb{S}^2 \subset \mathbb{R}^3$ 633
22.4 Brownian motion on the torus 634
22.5 Diffusions on the simplex 635
22.6 Exercises 637

23 Some analytical aspects **639**
23.1 Geodesics and the exponential map 639
23.2 Taylor expansion 643
23.3 Integration on manifolds 645
    23.3.1 The volume measure on the manifold 645
    23.3.2 Wedge product and volume forms 648
    23.3.3 The divergence theorem 650
23.4 Gradient flow models 657
    23.4.1 Steepest descent model 657
    23.4.2 Euclidian state spaces 658
23.5 Drift changes and irreversible Langevin diffusions 659
    23.5.1 Langevin diffusions on closed manifolds 661
    23.5.2 Riemannian Langevin diffusions 662
23.6 Metropolis-adjusted Langevin models 665
23.7 Stability and some functional inequalities 666
23.8 Exercises 669

24 Some illustrations **673**
24.1 Prototype manifolds 673
    24.1.1 The circle 673
    24.1.2 The 2-sphere 674
    24.1.3 The torus 678
24.2 Information theory 681
    24.2.1 Nash embedding theorem 681
    24.2.2 Distribution manifolds 682
    24.2.3 Bayesian statistical manifolds 683
    24.2.4 Cramer-Rao lower bound 685
    24.2.5 Some illustrations 685
        24.2.5.1 Boltzmann-Gibbs measures 685
        24.2.5.2 Multivariate normal distributions 686

VI   Some application areas **691**

25 Simple random walks **693**
25.1 Random walk on lattices 693
    25.1.1 Description 693
    25.1.2 Dimension 1 693
    25.1.3 Dimension 2 694
    25.1.4 Dimension $d \geq 3$ 694
25.2 Random walks on graphs 694
25.3 Simple exclusion process 695
25.4 Random walks on the circle 695
    25.4.1 Markov chain on cycle 695
    25.4.2 Markov chain on circle 696
    25.4.3 Spectral decomposition 696
25.5 Random walk on hypercubes 697
    25.5.1 Description 697
    25.5.2 A macroscopic model 698
    25.5.3 A lazy random walk 698
25.6 Urn processes 699
    25.6.1 Ehrenfest model 699

        25.6.2 Pólya urn model   700
    25.7 Exercises   701

26 Iterated random functions   **705**
    26.1 Description   705
    26.2 A motivating example   707
    26.3 Uniform selection   708
        26.3.1 An ancestral type evolution model   708
        26.3.2 An absorbed Markov chain   709
    26.4 Shuffling cards   712
        26.4.1 Introduction   712
        26.4.2 The top-in-at-random shuffle   712
        26.4.3 The random transposition shuffle   713
        26.4.4 The riffle shuffle   716
    26.5 Fractal models   719
        26.5.1 Exploration of Cantor's discontinuum   720
        26.5.2 Some fractal images   723
    26.6 Exercises   726

27 Computational and statistical physics   **731**
    27.1 Molecular dynamics simulation   731
        27.1.1 Newton's second law of motion   731
        27.1.2 Langevin diffusion processes   734
    27.2 Schrödinger equation   737
        27.2.1 A physical derivation   737
        27.2.2 Feynman-Kac formulation   739
        27.2.3 Bra-kets and path integral formalism   742
        27.2.4 Spectral decompositions   743
        27.2.5 The harmonic oscillator   745
        27.2.6 Diffusion Monte Carlo models   748
    27.3 Interacting particle systems   749
        27.3.1 Introduction   749
        27.3.2 Contact process   751
        27.3.3 Voter process   751
        27.3.4 Exclusion process   752
    27.4 Exercises   753

28 Dynamic population models   **759**
    28.1 Discrete time birth and death models   759
    28.2 Continuous time models   762
        28.2.1 Birth and death generators   762
        28.2.2 Logistic processes   762
        28.2.3 Epidemic model with immunity   764
        28.2.4 Lotka-Volterra predator-prey stochastic model   765
        28.2.5 Moran genetic model   769
    28.3 Genetic evolution models   769
    28.4 Branching processes   770
        28.4.1 Birth and death models with linear rates   770
        28.4.2 Discrete time branching processes   772
        28.4.3 Continuous time branching processes   773
            28.4.3.1 Absorption-death process   775

28.4.3.2 Birth type branching process 776
28.4.3.3 Birth and death branching processes 777
28.4.3.4 Kolmogorov-Petrovskii-Piskunov equation 779
28.5 Exercises 781

29 Gambling, ranking and control **787**
29.1 Google page rank 787
29.2 Gambling betting systems 788
29.2.1 Martingale systems 788
29.2.2 St. Petersburg martingales 789
29.2.3 Conditional gains and losses 791
29.2.3.1 Conditional gains 791
29.2.3.2 Conditional losses 791
29.2.4 Bankroll management 792
29.2.5 Grand martingale 794
29.2.6 D'Alembert martingale 794
29.2.7 Whittacker martingale 796
29.3 Stochastic optimal control 797
29.3.1 Bellman equations 797
29.3.2 Control dependent value functions 802
29.3.3 Continuous time models 804
29.4 Optimal stopping 807
29.4.1 Games with fixed terminal condition 807
29.4.2 Snell envelope 809
29.4.3 Continuous time models 811
29.5 Exercises 812

30 Mathematical finance **821**
30.1 Stock price models 821
30.1.1 Up and down martingales 821
30.1.2 Cox-Ross-Rubinstein model 824
30.1.3 Black-Scholes-Merton model 825
30.2 European option pricing 826
30.2.1 Call and put options 826
30.2.2 Self-financing portfolios 827
30.2.3 Binomial pricing technique 828
30.2.4 Black-Scholes-Merton pricing model 830
30.2.5 Black-Scholes partial differential equation 831
30.2.6 Replicating portfolios 832
30.2.7 Option price and hedging computations 833
30.2.8 A numerical illustration 834
30.3 Exercises 835

Bibliography **839**

Index **855**

0

Introduction

A brief discussion on stochastic processes

These lectures deal with the foundations and the applications of stochastic processes also called random processes.

The term "stochastic" comes from the Greek word "stokhastikos" which means a skillful person capable of guessing and predicting. The first use of this term in probability theory can be traced back to the Russian economist and statistician Ladislaus Bortkiewicz (1868-1931). In his paper *Die Iterationen* published in 1917, he defines the term "stochastik" as follows: "The investigation of empirical varieties, which is based on probability theory, and, therefore, on the law of the large numbers, may be denoted as stochastic. But stochastic is not simply probability theory, but above all probability theory and its applications".

As their name indicates, stochastic processes are dynamic evolution models with random ingredients. Leaving aside the old well-known dilemma of determinism and freedom [26, 225] (solved by some old fashion scientific reasoning which cannot accommodate any piece of random "un-caused causations"), the complex structure of any "concrete" and sophisticated real life model is always better represented by stochastic mathematical models as a first level of approximation.

The sources of randomness reflect different sources of model uncertainties, including unknown initial conditions, model uncertainties such as misspecified kinetic parameters, as well as the external random effects on the system.

Stochastic modelling techniques are of great importance in many scientific areas, to name a few:

- Computer and engineering sciences: signal and image processing, filtering and inverse problems, stochastic control, game theory, mathematical finance, risk analysis and rare event simulation, operation research and global optimization, artificial intelligence and evolutionary computing, queueing and communication networks.

- Statistical machine learning: hidden Markov chain models, frequentist and Bayesian statistical inference.

- Biology and environmental sciences: branching processes, dynamic population models, genetic and genealogical tree-based evolution.

- Physics and chemistry: turbulent fluid models, disordered and quantum models, statistical physics and magnetic models, polymers in solvents, molecular dynamics and Schrödinger equations.

A fairly large body of the literature on stochastic processes is concerned with the probabilistic modelling and the convergence analysis of random style dynamical systems. As expected these developments are closely related to the theory of dynamical systems, to partial differential and integro-differential equations, but also to ergodic and chaos theory, differential geometry, as well as the classical linear algebra theory, combinatorics, topology, operator theory, and spectral analysis.

The theory of stochastic processes has also developed its own sophisticated probabilistic tools, including diffusion and jump type stochastic differential equations, Doeblin-Itō calculus, martingale theory, coupling techniques, and advanced stochastic simulation techniques.

One of the objectives of stochastic process theory is to derive explicit analytic type results for a variety of simplified stochastic processes, including birth and death models, simple random walks, spatially homogeneous branching processes, and many others. Nevertheless, most of the more realistic processes of interest in the real world of physics, biology, and engineering sciences are "unfortunately" highly nonlinear systems evolving in very high dimensions. As a result, it is generally impossible to come up with any type of closed form solutions.

In this connection, the theory of stochastic processes is also concerned with the modelling and with the convergence analysis of a variety of sophisticated stochastic algorithms. These stochastic processes are designed to solve numerically complex integration problems that arise in a variety of application areas. Their common feature is to mimic and to use repeated random samples of a given stochastic process to estimate some *averaging* type property using empirical approximations.

We emphasize that all of these stochastic methods are expressed in terms of a particular stochastic process, or a collection of stochastic processes, depending on the precision parameter being the time horizon, or the number of samples. The central idea is to approximate the expectation of a given random variable using the empirical averages (in space or in time) based on a sequence of random samples. In this context, the rigorous analysis of these complex numerical schemes also relies on advanced stochastic analysis techniques.

The interpretation of the random variables of interest depends on the application. In some instances, these variables represent the random states of a complex stochastic process, or some unknown kinetic or statistical parameters. In other situations, the random variables of interest are specified by a complex target probability measure. Stochastic simulation and Monte Carlo methods are used to predict the evolution of given random phenomena, as well as to solve several estimation problems using random searches. To name but a few: computing complex measures and multidimensional integrals, counting, ranking, and rating problems, spectral analysis, computation of eigenvalues and eigenvectors, as well as solving linear and nonlinear integro-differential equations.

This book is almost self-contained. There are no strict prerequisites but it is envisaged that students would have taken a course in elementary probability and that they have some knowledge of elementary linear algebra, functional analysis and geometry. It is not possible to define rigorously stochastic processes without some basic knowledge on measure theory and differential calculus. Readers who lack such background should instead consult some introductory textbook on probability and integration, and elementary differential calculus.

The book also contains around 500 exercises with detailed solutions on a variety of topics, with many explicit computations. Each chapter ends with a section containing a series of exercises ranging from simple calculations to more advanced technical questions. This book can serve as a reference book, as well as a textbook. The next section provides a brief description of the organization of the book.

On page xxxii we propose a series of lectures and research projects which can be developed using the material presented in this book. The descriptions of these course projects also provide detailed discussions on the connections between the different topics treated in this book. Thus, they also provide a reading guide to enter into a given specialized topic.

Organization of the book

The book is organized in six parts. The synthetic diagram below provides the connections betweens the parts and indicates the different possible ways of reading the book.

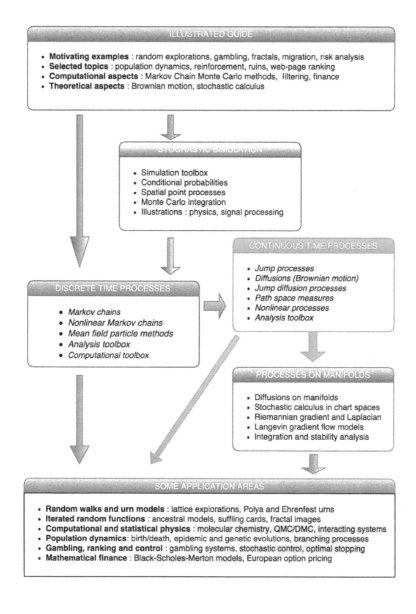

Part I provides an illustrative guide with a series of motivating examples. Each of them is related to a deep mathematical result on stochastic processes. The examples include the recurrence and the transience properties of simple random walks, stochastic coupling techniques and mixing properties of Markov chains, random iterated functional models, Poisson processes, dynamic population models, Markov chain Monte Carlo methods, and the Doeblin-Itō differential calculus. In each situation, we provide a brief discussion on the mathematical analysis of these models. We also provide precise pointers to the chapters and sections where these results are discussed in more details.

Part II is concerned with the notion of randomness, and with some techniques to "simulate perfectly" some traditional random variables (*abbreviated r.v.*). For a more thorough discussion on this subject, we refer to the encyclopedic and seminal reference book of Luc Devroye [92], dedicated to the simulation of nonuniform random variables. Several concrete applications are provided to illustrate many of the key ideas, as well as the usefulness of stochastic simulation methods in some scientific disciplines.

The next three parts are dedicated to discrete and continuous time stochastic processes. The synthetic diagram below provides some inclusion type links between the different classes of stochastic processes discussed in this book.

Part III is concerned with discrete time stochastic processes, and more particularly with Markov chain models.

Firstly, we provide a detailed discussion on the different descriptions and interpretations of these stochastic models. We also discuss nonlinear Markov chain models, including self-interacting Markov chains and mean field type particle models.

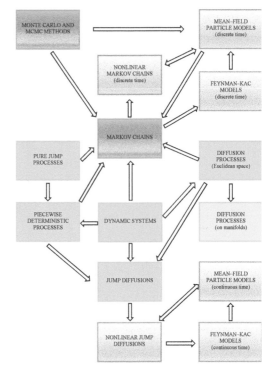

Then we present a panorama of analytical tools to study the convergence of Markov chain models when the time parameter tends to ∞. The first class of mathematical techniques includes linear algebraic tools, spectral and functional analysis. We also present some more probabilistic-type tools such as coupling techniques, strong stationary times, and martingale limit theorems.

Finally, we review the traditional and the more recent computation techniques of Markov chain models. These techniques include Markov chain Monte Carlo methodologies, perfect sampling algorithms, and time inhomogeneous models. We also provide a discussion of the more recent Feynman-Kac particle methodologies, with a series of application domains.

Part IV of the book is concerned with continuous time stochastic processes and stochastic differential calculus, a pretty vast subject at the core of applied mathematics. Nevertheless, most of the literature on these processes in probability textbooks is dauntingly mathematical. Here we provide a pedagogical introduction, sacrificing from time to time some technical mathematical aspects.

Continuous time stochastic processes can always be thought as a limit of a discrete generation process defined on some time mesh, as the time step tends to 0. This viewpoint which is at the foundation of of stochastic calculus and integration is often hidden in purely theoretical textbooks on stochastic processes. In the reverse angle, more applied presentations found in textbooks in engineering, economy, statistics and physics are often too far from recent advances in applied probability so that students are not really prepared to enter into deeper analysis nor pursue any research level type projects.

One of the main new points we have adopted here is to make a bridge between applications and advanced probability analysis. Continuous time stochastic processes, including nonlinear jump-diffusion processes, are introduced as a limit of a discrete generation process which can be easily simulated on a computer. In the same vein, any sophisticated formula in stochastic analysis and differential calculus arises as the limit of a discrete time mathematical model. It is of course not within the scope of this book to quantify precisely and in a systematic way all of these approximations. Precise reference pointers to articles and books dealing with these questions are provided in the text.

All approximations treated in this book are developed through a single and systematic mathematical basis transforming the elementary Markov transitions of practical discrete generation models into infinitesimal generators of continuous time processes. The stochastic version of the Taylor expansion, also called the Doeblin-Itō formula or simply the Itō formula, is described in terms of the generator of a Markov process and its corresponding carré du champ operator.

In this framework, the evolution of the law of the random states of a continuous time process resumes to a natural weak formulation of linear and nonlinear Fokker-Planck type equations (a.k.a. Kolmogorov equations). In this interpretation, the random paths of a given continuous time process provide a natural interpolation and coupling between the solutions of these integro-differential equations at different times. These path-space probability measures provide natural probabilistic interpretations of important Cauchy-Dirichlet-Poisson problems arising in analysis and integro-partial differential equation problems. The limiting behavior of these functional equations is also directly related to the long time behavior of a stochastic process.

All of these subjects are developed from chapter 10 to chapter 16. The presentation starts with elementary Poisson and jump type processes, including piecewise deterministic models, and develops forward to more advanced jump-diffusion processes and nonlinear Markov processes with their mean field particle interpretations. An abstract and universal class of continuous time stochastic process is discussed in section 15.5.

The last two chapters, chapter 17 and chapter 18 present respectively a panorama of analytical tools to study the long time behavior of continuous time stochastic processes, and path-space probability measures with some application domains.

Part V is dedicated to diffusion processes on manifolds. These stochastic processes arise when we need to explore state spaces associated with some constraints, such as the sphere, the torus, or the simplex. The literature on this subject is often geared towards purely mathematical aspects, with highly complex stochastic differential geometry methodologies. This book provides a self-contained and pedagogical treatment with a variety of examples and application domains. Special emphasis is given to the modelling, the stability analysis, and the numerical simulation of these processes on a computer.

The synthetic diagram below provides some inclusion type links between the different classes of stochastic computational techniques discussed in this book.

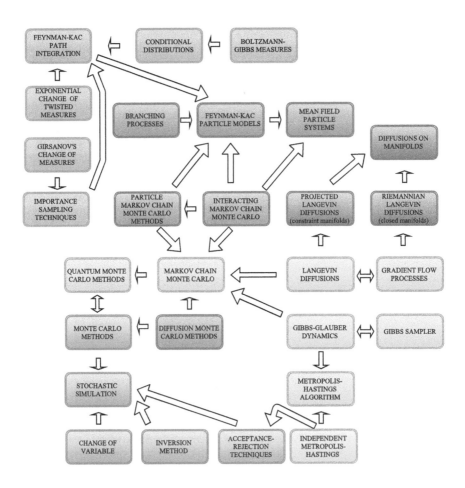

Part VI is dedicated to some application domains of stochastic processes, including random walks, iterated random functions, computational physics, dynamic population models, gambling and ranking, and mathematical finance.

Chapter 25 is mainly concerned with simple random walks. It starts with the recurrence or transience properties of these processes on integer lattices depending on their dimensions. We also discuss random walks on graphs, the exclusion process, as well as random walks on the circle and hypercubes and their spectral properties. The last part of the chapter is dedicated to the applications of Markov chains to analyze the behavior of urn type models such as the Polya and the Ehrenfest processes. We end this chapter with a series of exercises on these random walk models including diffusion approximation techniques.

Chapter 26 is dedicated to stochastic processes expressed in terms of iterated random functions. We examine three classes of models. The first one is related to branching processes and ancestral tree evolutions. These stochastic processes are expressed in terms of compositions of one to one mappings between finite sets of indices. The second one is concerned with shuffling cards. In this context, the processes are expressed in terms of random compositions of permutations. The last class of models are based on random compositions of linear transformations on the plane and their limiting fractal image compositions. We illustrate these models with the random construction of the Cantor's discontinuum, and provide some examples of fractal images such as a fractal leaf, a fractal tree, the Sierpinski carpet, and the Highways dragons. We analyze the stability and the limiting measures associated with each class of processes. The chapter ends with a series of exercises related to the stability properties of these models, including some techniques used to obtain quantitative estimates.

Chapter 27 explores some selected applications of stochastic processes in computational and statistical physics.

The first part provides a short introduction to molecular dynamics models and to their numerical approximations. We connect these deterministic models with Langevin diffusion processes and their reversible Boltzmann-Gibbs distributions.

The second part of the chapter is dedicated to the spectral analysis of the Schrödinger wave equation and its description in terms of Feynman-Kac semigroups. We provide an equivalent description of these models in terms of the Bra-kets and path integral formalisms commonly used in physics. We present a complete description of the spectral decomposition of the Schrödinger-Feynman-Kac generator, the classical harmonic oscillator in terms of the Hermite polynomials introduced in chapter 17. Last but not least we introduce the reader to the Monte Carlo approximation of these models based on the mean field particle algorithms presented in the third part of the book.

The third part of the chapter discusses different applications of interacting particle systems in physics, including the contact model, the voter and the exclusion process. The chapter ends with a series of exercises on the ground state of Schrödinger operators, quasi-invariant measures, twisted guiding waves, and variational principles.

Chapter 28 is dedicated to applications of stochastic processes in biology and epidemiology. We present several classes of deterministic and stochastic dynamic population models. We analyse logistic type and Lotka-Volterra processes, as well as branching and genetic processes. This chapter ends with a series of exercises related to logistic diffusions, bimodal growth models, facultative mutualism processes, infection and branching processes.

Chapter 29 is concerned with applications of stochastic processes in gambling and ranking. We start with the celebrated Google page rank algorithm. The second part of the chapter is dedicated to gambling betting systems. We review and analyze some famous martingales such as the St Petersburg martingale, the grand martingale, the D'Alembert and the Whittaker martingales. The third part of the chapter is mainly concerned with stochastic control and optimal stopping strategies. The chapter ends with a series of exercises on the Monty Hall game show, the Parrondo's game, the bold play strategy, the ballot and the secretary problems.

The last chapter, chapter 30, is dedicated to applications in mathematical finance. We discuss the discrete time Cox-Ross-Rubinstein models and the continuous Black-Scholes-Merton models. The second part of the chapter is concerned with European pricing options. We provide a discussion on the modelling of self-financing portfolios in terms of controlled martingales. We also describe a series of pricing and hedging techniques with some numerical illustrations. The chapter ends with a series of exercises related to neutralization techniques of financial markets, replicating portfolios, Wilkie inflation models, and life function martingales.

The synthetic diagram below provides some inclusion type links between the stochastic processes and the different application domains discussed in this book. The black arrows indicate the estimations and sampling problems arising in different application areas. The blue arrows indicate the stochastic tools that can be used to solve these problems. The red arrows emphasize the modelling, the design and the convergence analysis of stochastic methods. The green arrows emphasize the class of stochastic models arising in several scientific disciplines.

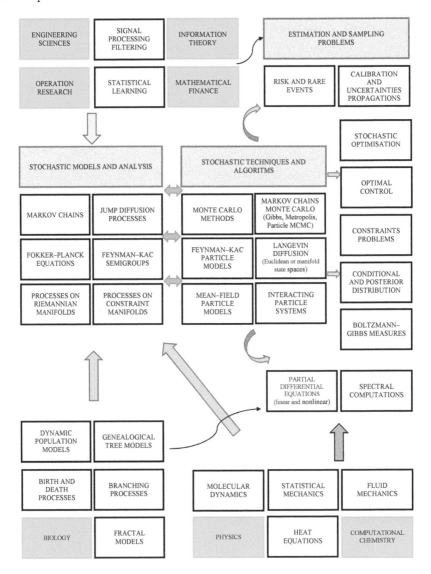

The book illustrations were hand-picked by the French artist Timothée Del Moral. He has been studying in the Fine Art School in Bordeaux and later joined several art studios in Toulouse, Nice and Bordeaux. Among his art works, he designed the logo of the Biips INRIA software, and posters for the French Society of Mathematics and Industry. He is now specialized in tattooing, developing new styles and emulating old school-type illustrations and tattoos.

Preview and objectives

This book provides an introduction to the probabilistic modelling of discrete and continuous time stochastic processes. The emphasis is put on the understanding of the main mathematical concepts and numerical tools across a range of illustrations presented through models with increasing complexity.

> *If you can't explain it simply, you don't understand it well enough.*
> Albert Einstein (1879-1955).

While exploring the analysis of stochastic processes, the reader will also encounter a variety of mathematical techniques related to matrix theory, spectral analysis, dynamical systems theory and differential geometry.

We discuss a large class of models ranging from finite space valued Markov chains to jump diffusion processes, nonlinear Markov processes, self-interacting processes, mean field particle models, branching and interacting particle systems, as well as diffusions on constraint and Riemannian manifolds.

Each of these stochastic processes is illustrated in a variety of applications. Of course, the detailed description of real-world models requires a deep understanding of the physical or the biological principles behind the problem at hand. These discussions are out of the scope of the book. We only discuss academic-type related situations, providing precise reference pointers for a more detailed discussion. The illustrations we have chosen are very often at the crossroads of several seemingly disconnected scientific disciplines, including biology, mathematical finance, gambling, physics, engineering sciences, operations research, as well as probability, and statistical inference.

The book also provides an introduction to stochastic analysis and stochastic differential calculus, including the analysis of probability measures on path spaces and the analysis of Feynman-Kac semigroups. We present a series of powerful tools to analyze the long time behavior of discrete and continuous time stochastic processes, including coupling, contraction inequalities, spectral decompositions, Lyapunov techniques, and martingale theory.

The book also provides a series of probabilistic interpretations of integro-partial differential equations, including stochastic partial differential equations (such as the nonlinear filtering equation), Fokker-Planck equations and Hamilton-Jacobi-Bellman equations, as well as Cauchy-Dirichlet-Poisson equations with boundary conditions.

Last but not least, the book also provides a rather detailed description of some traditional and more advanced computational techniques based on stochastic processes. These techniques include Markov chain Monte Carlo methods, coupling from the past techniques, as well as more advanced particle methods. Here again, each of these numerical techniques is illustrated in a variety of applications.

Having said that, let us explain some other important subjects which are not treated in this book. We provide some precise reference pointers to supplement this book on some of these subjects.

As any lecture on deterministic dynamical systems does not really need to start with a fully developed Lebesgue, or Riemann-Stieltjes integration theory, nor with a specialized course on differential calculus, we believe that a full and detailed description of stochastic integrals is not really needed to start a lecture on stochastic processes.

For instance, Markov chains are simply defined as a sequence of random variables indexed by integers. Thus, the mathematical foundations of Markov chain theory only rely on rather elementary algebra or Lebesgue integration theory.

It is always assumed implicitly that these (infinite) sequences of random variables are defined on a common probability space; otherwise it would be impossible to quantify, nor to define properly limiting events and objects. From a pure mathematical point of view, this innocent technical question is not so obvious even for sequences of independent random variables. The construction of these probability spaces relies on abstract projective limits, the well known Carathéodory extension theorem and the Kolmogorov extension theorem. The Ionescu-Tulcea theorem also provides a natural probabilistic solution on general measurable spaces. It is clearly not within the scope of this book to enter into these rather well known and sophisticated constructions.

As dynamical systems, continuous time stochastic processes can be defined as the limit of a discrete generation Markov process on some time mesh when the time step tends to 0. For the same reasons as above, these limiting objects are defined on a probability space constructed using projective limit techniques and extension type theorems to define in a unique way infinite dimensional distributions from finite dimensional ones defined on cylindrical type sets.

In addition, the weak convergence of any discrete time approximation process to the desired continuous time limiting process requires us to analyze the convergence of the finite dimensional distributions on any finite sequence of times. We will discuss these finite dimensional approximations in some details for jump as well as for diffusion processes.

Nevertheless, the weak convergence at the level of the random trajectories requires us to ensure that the laws of the random paths of the approximating processes are relatively compact in the sense that any sequence admits a convergent subsequence. For complete separable metric spaces, this condition is equivalent to a tightness condition that ensures that these sequences of probability measures are almost concentrated on a compact subset in the whole set of trajectories.

Therefore to check this compactness condition we first need to characterize the compact subsets of the set of continuous trajectories, or of the set of right continuous trajectories with left hand limits. By the Arzelà-Ascoli theorem, these compact sets are described in terms of equicontinuous trajectories. Besides the fact that the most of stochastic processes encountered in practice satisfy this tightness property, the proof of this condition relies on sophisticated probabilistic tools. A very useful and commonly used to check this tightness property is the Aldous criterion introduced by D. Aldous in his PhD dissertation and published in the *Annals of Probability* in 1978.

Here again, entering in some details into these rather well known and well developed subjects would cause digression. More details on these subjects can be found in the books

- R. Bass. *Stochastic Processes*. CUP (2011).

- S. Ethier, T. Kurtz. *Markov Processes: Characterization and Convergence*. Wiley (1986).

- J. Jacod, A. Shiryaev, *Limit Theorems for Stochastic Processes*, Springer (1987).

The seminal book by J. Jacod and A. N. Shiryaev is highly technical and presents difficult material but it contains almost all the limiting convergence theorems encountered in the theory of stochastic processes. This reference is dedicated to researchers in pure and applied probability. The book by S.N. Ethier and T. G. Kurtz is in spirit closer to our approach based on the description of general stochastic processes in terms of their generators. This book is recommended to anyone interested into the precise mathematical definition of generators and their regularity properties. The book of R. Bass is more accessible and presents a detailed mathematical construction of stochastic integrals and stochastic calculus.

The integral representations of these limiting stochastic differential equations (abbreviated SDE) are expressed in terms of stochastic integrals. These limiting mathematical models coincide with the definition of the stochastic integral. The convergence can be made precise in a variety of well known senses, such as in probability, almost sure, or w.r.t. some $\mathbb{L}_p$-norms.

Recall that integral description of any deterministic dynamical system is based on the so-called fundamental theorem of calculus. In the same vein, the integral description of stochastic processes is now based on the fundamental Doeblin-Itō differential calculus. This stochastic differential calculus can be interpreted as the natural extension of Taylor differential calculus to random dynamical systems, such as diffusions and random jump type processes. Here again the Doeblin-Itō differential calculus can also be expressed in terms of natural Taylor type expansions associated with a discrete time approximation of a given continuous type stochastic process.

In the further development of this book, we have chosen to describe in full details these stochastic expansions for the most fundamental classes of stochastic processes including jump diffusions and more abstract stochastic models in general state spaces, as well as stochastic processes on manifolds.

Precise reference pointers to textbooks dedicated to deeper mathematical foundations of stochastic integrals and the Doeblin-Itō differential calculus are given in the text.

This choice of presentation has many advantages.

Firstly, the time discretization of continuous time processes provides a natural way to simulate these processes on a computer.

In addition, any of these simulation techniques also improves the physical and the probabilistic understanding of the nature and the long time behavior of the limiting stochastic process.

Furthermore, this pedagogical presentation does not really require a strong background in sophisticated stochastic analysis techniques, as it is based only on discrete time approximation schemes. As stochastic integrals and the Doeblin-Itō differential calculus are themselves defined in terms of limiting formulae associated with a discrete generation process, our approach provides a way to introduce these somehow sophisticated mathematical objects in a simple form.

In this connection, most of the limiting objects as well as quantitative estimates are generally obtained by using standard technical tools such as the monotone or the dominated convergence theorems, Cauchy-Schwartz, Hölder or Minskowski inequalities. Some of these results are provided in the text and in the exercises. Further mathematical details can also be found in the textbooks cited in the text.

Last, but not least, this choice of presentation also lays a great foundation for research, as most of the research questions in probability are asymptotic in nature. In order to understand these questions, the basic non-asymptotic manipulations need to be clearly understood.

Rather than pepper the text with repeated citations, we mention here (in alphabetical order) some other classical and supplemental texts on the full mathematical construction of stochastic integrals, the characterization and the convergence of continuous time stochastic processes:

- R. Durrett, *Stochastic Calculus*. CRC Press (1996).

- I. Gikhman, A. Skorokhod. *Stochastic Differential Equations*. Springer (1972).

- F.C Klebaner. *Introduction to Stochastic Calculus with Applications*. World Scientific Publishing (2012).

- P. Protter. *Stochastic Integration and Differential Equations*. Springer (2004).

More specialized books on Brownian motion, martingales with continuous paths, diffusions processes and their discrete time approximations are:

- N. Ikeda, S. Watanabe. *Stochastic Differential Equations and Diffusion Processes.* North-Holland Publishing Co. (1989).

- I. Karatzas, S. E. Shreve. *Brownian Motion and Stochastic Calculus.* Springer (2004).

- P. E. Kloeden, E. Platen. *Numerical Solution of Stochastic Differential Equations.* Springer (2011).

- B. Øksendal. *Stochastic Differential Equations: An Introduction with Applications.* Springer (2005).

- L. C. G. Rogers, D. Williams. *Diffusions, Markov Processes, and Martingales.* CUP (2000).

The seminal book by P. E. Kloeden and E. Platen provides an extensive discussion on the convergence properties of discrete time approximations of stochastic differential equations.

Other specialized mathematical textbooks and lecture notes which can be useful for supplemental reading on stochastic processes and analysis on manifolds are:

- K.D. Elworthy. *Stochastic Differential Equations on Manifolds.* CUP(1982).

- M. Emery. *Stochastic Calculus on Manifolds.* Springer (1989).

- E.P. Hsu. *Stochastic analysis on manifolds.* Providence AMS (2002).

Some lecture projects

The material in this book can serve as a basis for different types of advanced undergraduate and graduate level courses as well as master-level and post-doctoral level courses on stochastic processes and their application domains. As we mentioned above, the book also contains around 500 exercises with detailed solutions. Each chapter ends with a section containing a series of exercises ranging from simple calculations to more advanced technical questions. These exercises can serve for training the students through tutorials or homework.

> *Running overtime is the one unforgivable error a lecturer can make.*
> *After fifty minutes (one microcentury as von Neumann used to say)*
> *everybody's attention will turn elsewhere.* Gian Carlo Rota (1932-1999).

Introduction to stochastic processes (with applications)

This first type of a lecture course would be a pedagogical introduction to elementary stochastic processes based on the detailed illustrations provided in the first three chapters, chapter 1, chapter 2, and chapter 3. The lectures can be completed with the simulation techniques presented in chapter 4, the Monte Carlo methods discussed in chapter 5, and the illustrations provided in chapter 6.

The lecture presents discrete as well as continuous time stochastic processes through a variety of motivating illustrations including dynamic population models, fractals, card shuffling, signal processing, Bayesian statistical inference, finance, gambling, ranking, and stochastic search, reinforcement learning and sampling techniques.

Through a variety of applications, the lecture provides a brief introduction to Markov chains, including the recurrence properties of simple random walks, Markov chains on permutation groups, and iterated random functions. This course also covers Poisson processes, and provides an introduction to Brownian motion, as well as piecewise deterministic Markov processes, diffusions and jump processes.

This lecture course also covers computational aspects with an introduction of Markov chain Monte Carlo methodologies and the celebrated Metropolis-Hastings algorithm. The text also provides a brief introduction to more advanced and sophisticated nonlinear stochastic techniques such as self-interacting processes particle filters, and mean field particle sampling methodologies.

The theoretical topics also include an accessible introduction to stochastic differential calculus. The lecture provides a natural and unified treatment of stochastic processes based on infinitesimal generators and their carré du champ operators.

This book is also designed to stimulate the development of research projects that offer the opportunity to apply the theory of stochastic processes discussed during the lectures to an application domain selected among the ones discussed in this introductory course: *card shuffling, fractal images, ancestral evolutions, molecular dynamics, branching and interacting processes, genetic models, gambling betting systems, financial option pricing, advanced signal processing, stochastic optimization, Bayesian statistical inference, and many others.*

These research projects provide opportunity to the lecturer to immerse the student in a favorite application area. They also require the student to do a background study and perform personal research by exploring one of the chapters in the last part of the book dedicated to application domains. The diagrams provided on page xxiii and on page xxviii guide the students on the reading order to enter into a specific application domain and explore the different links and inclusions between the stochastic models discussed during the lectures.

More theoretical research projects can also be covered. For instance, the recurrence questions of random walks discussed in the first section of chapter 1 can be further developed by using the topological aspects of Markov chains presented in section 8.5. Another research project based on the martingale theory developed in section 8.4 will complement the discussion on gambling and ruin processes discussed in chapter 2. The discussion on signal processing and particle filters presented in chapter 3 can be complemented by research projects on Kalman filters (section 9.9.6), mean field particle methodologies (section 7.10.2), and Feynman-Kac particle models (section 9.5 and section 9.6).

Other textbooks and lecture notes which can be useful for supplemental reading in probability and applications during this lecture are:

- P. Billingsley. *Convergence of Probability Measures*, Wiley (1999).

- L. Devroye. *Non-Uniform Random Variate Generation*, Springer (1986).

- W. Feller. *An Introduction to Probability Theory*, Wiley (1971).

- G.R. Grimmett, D.R. Stirzaker, *One Thousand Exercises in Probability*, OUP (2001).

- A. Shiryaev. *Probability*, Graduate Texts in Mathematics, Springer (2013).

- P. Del Moral, B. Remillard, S. Rubenthaler. *Introduction aux Probabilités*. Ellipses Edition [in French] (2006).

We also refer to some online resources such as the review of probability theory by T. Tao, the Wikipedia article on Brownian Motion, and a Java applet simulating Brownian motion.

Of course it normally takes more than an year to cover the full scope of this course with fully rigorous mathematical details. This course is designed for advanced undergraduate to master-level audiences with different educational backgrounds ranging from engineering sciences, economics, business, mathematical finance, physics, statistics to applied mathematics.

Giving a course for such a diverse audience with different backgrounds poses a significant challenge to the lecturer. To make the material accessible to all groups, the course begins with the series of illustrations in the first three chapters. This fairly gentle introduction is designed to cover a variety of subjects in a single semester course of around 40 hours, at a somewhat more superficial mathematical level from the pure and applied mathematician perspective. These introductory chapters contain precise pointers to specialized chapters with more advanced mathematical material. Students with deep backgrounds in applied mathematics are invited to explore deeper into the book, while students with limited mathematical backgrounds will concentrate on the more elementary material discussed in chapter 1, chapter 2, and chapter 3.

Discrete time stochastic processes

The second type of lecture course is a pedagogical introduction to discrete time processes and Markov chain theory, a rather vast subject at the core of applied probability and many other scientific disciplines that use stochastic models. This more theoretical type course is geared towards *discrete time stochastic processes and their stability analysis*. It would essentially cover chapter 7 and chapter 8.

As in the previous lecture course, this would start with some selected motivating illustrations provided in chapter 1 and chapter 2, as well as from the material provided in section 3.1, chapter 5 and chapter 6 dedicated to Markov chain Monte Carlo methodologies.

The second part of this course will center around some selected material presented in the third part of the book dedicated to discrete time stochastic processes. Chapter 7 offers a description of the main classes of Markov chain models including nonlinear Markov chain models and their mean field particle interpretations.

After this exploration of discrete generation random processes, at least two possible different directions can be taken.

The first one is based on chapter 8 with an emphasis on the topological aspects and the stability properties of Markov chains. This type of course presents a panorama of analytical tools to study the convergence of Markov chain models when the time parameter tends to ∞. These mathematical techniques include *matrix theory, spectral and functional analysis, as well as contraction inequalities, geometric drift conditions, coupling techniques and martingale limit theorems.*

The second one, based on chapter 9, will emphasize computational aspects ranging from traditional Markov chain Monte Carlo methods and perfect sampling techniques to more advanced Feynman-Kac particle methodologies. This part or the course will cover *Metropolis-Hastings algorithms, Gibbs-Glauber dynamics, as well as the Propp and Wilson coupling from the past sampler*. The lecture would also offer *a brief introduction to more advanced particle methods and related particle Markov chain Monte Carlo methodologies*.

The course could be complemented with some selected application domains described in chapter 25, such as random walks on lattices and graphs, or urn type random walks. Another strategy would be to illustrate the theory of Markov chains with chapter 26 dedicated to iterated random functions and their applications in biology, card shuffling or fractal imaging, or with the discrete time birth and death branching processes presented in chapter 28.

Several research projects could be based on previous topics as well as on the gambling betting systems and discrete time stochastic control problems discussed in chapter 29, or on the applications in mathematical finance discussed in chapter 30.

This course is designed for master-level audiences in engineering, physics, statistics and applied probability.

Other textbooks and lecture notes which can be useful for supplemental reading in Markov chains and their applications during this lecture are:

- D. Aldous, J. Fill. *Reversible Markov Chains and Random Walks on Graphs* (1999).

- S. Meyn, R. Tweedie. *Markov Chains and Stochastic Stability*, Springer (1993).

- J. Norris. *Markov Chains*. CUP (1998).

- S.I. Resnick. *Adventures in Stochastic Processes*. Springer Birkhauser (1992).

We also refer to some online resources such as *the Wikipedia article on Markov chains*, and to the *Markov Chains chapter* in American Mathematical Society's introductory probability book, the article by D. Saupe *Algorithms for random fractals*, and the seminal and very clear article by P. Diaconis titled *The Markov Chain Monte Carlo Revolution*.

Continuous time stochastic processes

This semester-long course is dedicated to continuous time stochastic processes and stochastic differential calculus. The lecture notes would essentially cover the series of chapters 10 to 18. The detailed description of this course follows essentially the presentation of the fourth part of the book provided on page xxiv. *This course is designed for master-level audiences in engineering, mathematics and physics with some background in probability theory and elementary Markov chain theory.*

To motivate the lectures, this course would start with some selected topics presented in the first three chapters, such as the Poisson's typos discussed in section 1.7 (Poisson processes), the pinging hacker story presented in section 2.5 (piecewise deterministic stochastic processes), the lost equation discussed in section 3.3 (Brownian motion), the formal stochastic calculus derivations discussed in section 3.4 (Doeblin-Itō differential calculus), and the theory of speculation presented in section 3.5 (backward equations and financial mathematics).

The second part of the lecture would cover elementary Poisson processes, Markov chain continuous time embedding techniques, pure jump processes and piecewise deterministic models. It would also cover diffusion processes and more general jump-diffusion processes, with detailed and systematic descriptions of the infinitesimal generators, the Doeblin-Itō differential formula and the natural derivation of the corresponding Fokker-Planck equations.

The third part could be dedicated to jump diffusion processes with killing and their use to solve Cauchy problems with terminal conditions, as well as Dirichlet-Poisson problems. The course can also provide an introduction to nonlinear stochastic processes and their mean field particle interpretations, with illustrations in the context of risk analysis, fluid mechanics and particle physics.

The last part of the course would cover path-space probability measures including Girsanov's type change of probability measures and exponential type martingales, with illustrations in the analysis of nonlinear filtering processes, in financial mathematics and in rare event analysis, including in quantum Monte Carlo methodologies.

Three different applications could be discussed, depending on the background and the scientific interests of the audience (or of the lecturer):

- *Computational and statistical physics (chapter 27):*

  The material on interacting particle systems presented in section 27.3 only requires some basic knowledge on generators of pure jump processes. This topic can also be used to illustrate pure jump processes before entering into the descriptions of diffusion type processes.

  The material on molecular dynamics simulation discussed in section 27.1 only requires some knowledge on pure diffusion processes and their invariant measures. We recommend studying chapter 17 before entering into these topics.

  Section 27.2 dedicated to the Schrödinger equation is based on more advanced stochastic models such as the Feynman-Kac formulae, jump diffusion processes and exponential changes of probability measures. We recommend studying chapter 15 and chapter 18 (and more particularly section 18.1.3) before entering into this more advanced application area. Section 28.4.3 also provides a detailed discussion on branching particle interpretations of Feynman-Kac formulae.

- *Dynamic population models (chapter 28):*

  All the material discussed in this chapter can be used to illustrate jump and diffusion stochastic processes. Section 28.4.3 requires some basic knowledge on Feynman-Kac semigroups. It is recommended to study chapter 15 before entering into this topic.

- *Stochastic optimal control (section 29.3.3 and section 29.4.3):*

  The material discussed in this chapter can also be used to illustrate jump and diffusion stochastic processes. It is recommended to start with discrete time stochastic control theory (section 29.3.1, section 29.3.2 and section 29.4.2) before entering into more sophisticated continuous time problems.

- *Mathematical finance (chapter 30):*

  The sections 30.2.4 to 30.2.7 are essentially dedicated to the Black-Scholes stochastic model. This topic can be covered with only some basic knowledge on diffusion processes and more particularly on geometric Brownian motion.

Depending on the selected application domains to illustrate the course, we also recommend for supplemental reading the following references:

- K.B. Athreya, P.E. Ney. *Branching Processes*. Springer (1972).

- D. P. Bertsekas. *Dynamic Programming and Optimal Control*, Athena Scientific (2012).

- M. Caffarel, R. Assaraf. A pedagogical introduction to quantum Monte Carlo. In *Mathematical Models and Methods for Ab Initio Quantum Chemistry*. Lecture Notes in Chemistry, eds. M. Defranceschi and C. Le Bris, Springer (2000).

- P. Del Moral. *Mean Field Simulation for Monte Carlo Integration*. Chapman & Hall/CRC Press (2013).

- P. Glasserman. *Monte Carlo methods in financial engineering*. Springer (2004).

- T. Harris. *The Theory of Branching Processes*, R-381, Report US Air Force (1964) and Dover Publication (2002).

- H. J. Kushner. *Introduction to Stochastic Control*, New York: Holt, Reinhart, and Winston (1971).

- J.M. Steele. *Stochastic Calculus and Financial Applications.* Springer (2001).

- A. Lasota, M.C. Mackey. *Chaos, Fractals, and Noise: Stochastic Aspects of Dynamics.* Springer (1994).

Stochastic processes on manifolds

This advanced theoretical style course covers the fifth part of the book on stochastic diffusions on (differentiable) manifolds. It essentially covers the chapters 19 to 23. *This course is intended for master-level students as well as post-doctoral audiences with some background in differential calculus and stochastic processes.* The purpose of these series of lectures is to introduce the students to the basic concepts of stochastic processes on manifolds. The course is illustrated with a series of concrete applications.

In the first part of the course we review some basics tools of differential geometry, including orthogonal projection techniques, and the related first and second order covariant derivatives (19). The lectures offer a pedagogical introduction to the main mathematical models used in differential geometry such as the notions of divergence, mean curvature vector, Lie brackets between vector fields, Laplacian operators, and the Ricci curvature on a manifold. One of the main objectives is to help the students understand the basic concepts of deterministic and stochastic differential calculus that are relevant to the theory of stochastic processes on manifolds. The other objective is to show how these mathematical techniques apply to designing and to analyzing random explorations on constraint manifolds.

One could end this part of the course with the detailed proof of the Bochner-Lichnerowicz formula. This formula is pivotal in the stability analysis of diffusion processes on manifolds. It connects the second order properties of the generator of these processes in terms of the Hessian operator and the Ricci curvature of the state space.

The second part of the lecture course could be dedicated to stochastic calculus on embedded manifolds (chapter 20) and the notion of charts (a.k.a. atlases) and paramtrization spaces (chapter 21):

In a first series of lectures we introduce the Brownian motion on a manifold defined in terms of the level sets of a smooth and regular constraint function. Then, we analyze the infinitesimal generator of these models and we present the corresponding Doeblin-Itō differential calculus. We illustrate these models with the detailed description of the Brownian motion on the two-dimensional sphere and on the cylinder. These stochastic processes do not really differ from the ones discussed in chapter 14. The only difference is that they evolve in constraint manifolds embedded in the ambient space.

In another series of lectures we describe how the geometry of the constraint manifold is lifted to the parameter space (a.k.a. chart space). The corresponding geometry on the parameter space is equipped with a Riemannian scalar product. This main objective is to explore the ways to link the differential operators on the manifold in the ambient space to the Riemannian parameter space manifold. We provide some tools to compute the Ricci curvature in local coordinates and we describe the expression of Bochner-Lichnerowicz formula in Riemannian manifolds.

Several detailed examples are provided in chapter 22 which is dedicated to stochastic calculus in chart spaces. We define Brownian motion and general diffusions on Riemannian manifolds. We illustrate these stochastic models with a detailed description of the Brownian motion on the circle, the two-dimensional sphere and on the torus. Chapter 24 also offers a series of illustrations, starting with a detailed discussion of prototype manifolds such as the circle, the sphere and the torus. In each situation, we describe the Riemannian metric, the corresponding Laplacian, as well as the `geodesics`, the `Christoffel symbols` and the Ricci curvature. In the second part of the chapter we present applications in information theory. We start with a discussion on distribution and Bayesian statistical manifolds and we analyze the corresponding Fisher information metric and the Cramer-Rao lower bounds. We also discuss the Riemannian metric associated with Bolzmann-Gibbs models and multivariate normal distributions.

The course could end with the presentation of material in chapter 23 dedicated to some important analytical aspects, including the construction of geodesics and the integration on a manifold. We illustrate the impact of these mathematical objects with the design of gradient type diffusions with a Boltzmann-Gibbs invariant measure on a constraint manifold. The time discretization of these models is also presented using Metropolis-adjusted Langevin algorithms (a.k.a. MALA in engineering sciences and information theory).

This chapter also presents some analytical tools for analyzing the stability of stochastic processes on a manifold in terms of the second order properties of their generators. The gradient estimates of the Markov semigroups are mainly based on the Bochner-Lichnerowicz formula presented in chapter 19 and chapter 21. The series of exercises at the end of this chapter also provide illustrations of the stability analysis of Langevin models including projected and Riemannian Langevin processes.

Other specialized mathematical textbooks and lecture notes which can be useful for supplemental reading during this course in stochastic processes and analysis on manifolds are presented on page xxxii.

We also refer to some online resources in differential geometry such as `the Wikipedia article on Riemannian manifolds`, the very useful `list of formulae in Riemannian geometry`, and the book by M. Spivak titled *A Comprehensive Introduction to Differential Geometry* (1999). Other online resources in stochastic geometry are:

The pioneering articles by K. Itō Stochastic Differential Equations in a Differentiable Manifold. [*Nagoya Math. J.* Volume 1, pp.35–47 (1950)], and The Brownian Motion and Tensor Fields on Riemannian Manifolds [*Proc. Int. Congr. Math.*, Stockholm (1962)], and the very nice article by T. Lelièvre, G. Ciccotti, E. Vanden-Eijnden. . Projection of diffusions on submanifolds: Application to mean force computation. [*Communi. Pure Appl. Math.*, 61(3), 371-408, (2008)].

Stochastic analysis and computational techniques

There is enough material in the book to support more specialized courses on stochastic analysis and computational methodologies. A synthetic description of some of these course projects is provided below.

Stability of stochastic processes

As its name indicates, this course is concerned with the long time behavior of stochastic processes. *This course is designed for master-level audiences in mathematics and physics with some background in Markov chains theory and/or in continuous time jump diffusion processes, depending on whether the lectures are dedicated to discrete generation and/or continuous time models.*

Here again the lectures could start with some selected topics presented in the first three chapters, such as the stabilization of population subject discussed in section 2.1 (coupling techniques), the pinging hackers subject presented in section 2.5 (long time behavior of piecewise deterministic processes), and the discussion on Markov chain Monte Carlo methods (a.k.a. MCMC algorithms) provided in section 3.1.

To illustrate the lectures on discrete generation stochastic processes, the course would also benefit from the material provided in chapter 5 and chapter 6 dedicated to Markov chain Monte Carlo methodologies. Another series of illustrations based on shuffling cards problems and fractals can be found in chapter 26.

We also recommend presenting the formal stochastic calculus derivations discussed in section 3.4 (Doeblin-Itō differential calculus) if the course is dedicated to continuous time processes.

The second part of the course could be centered around the analysis toolboxes presented in chapter 8 (discrete time models) and chapter 17 (continuous time models). The objective is to introduce the students to a class of traditional approaches to analyze the long time behavior of Markov chains, including *coupling, spectral analysis, Lyapunov techniques, and contraction inequalities.* The coupling techniques can be illustrated using the discrete generation birth and death process discussed in section 28.1. The stability of continuous time birth and death models (with linear rates) can also be discussed using the explicit calculations developed in section 28.4.1.

The lectures could be illustrated by the Hamiltonian and Lagrangian molecular dynamics simulation techniques discussed in section 27.1.

The last part of the course could focus on the stability of stochastic processes on manifolds if the audience has some background on manifold-valued stochastic processes. It would essentially cover the sections 23.4 to 23.7. Otherwise, this final part of the course offers a pedagogical introduction on these models based on the material provided in chapter 19, including the detailed proof of the `Bochner-Lichnerowicz formula` and followed by gradient estimates discussed in section 23.7.

Other lecture notes and articles which can be useful for supplemental reading on Markov chains and their applications during this course are:

- S. Meyn, R. Tweedie. *Markov Chains and Stochastic Stability*, Springer (1993).

- P. Diaconis. *Something we've learned (about the Metropolis-Hasting algorithm)*, Bernoulli (2013).

- R. Douc, E. Moulines, D. Stoffer. *Nonlinear Time Series Analysis: Theory, Methods and Applications with R Examples*, Chapman & Hall/CRC Press (2014).

- M. Hairer. *Convergence of Markov Processes.* Lecture notes, Warwick University (2010).

- A. Katok, B. Hasselblatt. *Introduction to the Modern Theory of Dynamical Systems.* CUP (1997).

- L. Saloff-Costes. *Lectures on Finite Markov Chains.* Springer (1997).

- P. Del Moral, M. Ledoux, L. Miclo. Contraction properties of Markov kernels. *Probability Theory and Related Fields*, vol. 126, pp. 395–420 (2003).

- F. Wang. *Functional Inequalities, Markov Semigroups and Spectral Theory.* Elsevier (2006).

Stochastic processes and partial differential equations

A one-quarter or one-semester course could be dedicated to probabilistic interpretations of integro-differential equations. *This course is designed for master-level audiences in applied probability and mathematical physics, with some background in continuous time stochastic processes and elementary integro-partial differential equation theory.*

The course could start with a review of continuous time processes, the description of the notion of generators, the Doeblin-Itō differential formula and the natural derivation of the corresponding Fokker-Planck equations. This part of the lecture could follow the formal stochastic calculus derivations discussed in section 3.4 (Doeblin-Itō differential calculus) and and the material presented in chapter 15 with an emphasis on the probabilistic interpretations of Cauchy problems with terminal conditions, as well as Dirichlet-Poisson problems (section 15.6).

The second part of the course could cover chapter 16 dedicated to nonlinear jump diffusions and the mean field particle interpretation of a class of nonlinear integro-partial differential equations (section 16.2), including Burger's equation, nonlinear Langevin diffusions, McKean-Vlasov models and Feynman-Kac semigroups.

The third part of the course could be dedicated to some selected application domains. The lectures could cover the Duncan-Zakai and the Kushner-Stratonovitch stochastic partial differential equations arising in nonlinear filtering theory (section 18.5), the Feynman-Kac description of the Schrödinger equation discussed in section 27.2, and the branching particle interpretations of the Kolmogorov-Petrovskii-Piskunov equations presented in section 28.4.3.4.

The last part of the lecture could cover the Hamilton-Jacobi-Bellman equations arising in stochastic control theory and presented in section 29.3.3.

Other textbooks and lecture notes (in alphabetical order) which can be useful for supplemental reading during this course are:

- M. Bossy, N. Champagnat. *Markov processes and parabolic partial differential equations.* Book section in R. Cont. *Encyclopedia of Quantitative Finance*, pp.1142-1159, John Wiley & Sons (2010).

- P. Del Moral. *Mean field simulation for Monte Carlo integration.* Chapman and Hall/CRC Press (2013).

- E.B. Dynkin. *Diffusions, Superdiffusions and PDEs.* AMS (2002).

- M.I. Freidlin. *Markov Processes and Differential Equations.* Springer (1996).

- H.M. Soner. Stochastic representations for nonlinear parabolic PDEs. In *Handbook of differential equations. Evolutionary Equations, volume 3.* Edited by C.M. Dafermos and E. Feireisl. Elsevier (2007).

- N. Touzi. *Optimal Stochastic Control, Stochastic Target Problems and Backward Stochastic Differential Equations.* Springer (2010).

Advanced Monte Carlo methodologies

A more applied course geared toward numerical probability and computational aspects would cover chapter 8 and the Monte Carlo techniques and the more advanced particle methodologies developed in chapter 9. Depending on the mathematical background of the audience, the course could also offer a review on the stability of Markov processes, following the description provided on page xxxviii.

Some application-oriented courses and research projects

There is also enough material in the book to support more applied courses on one or two selected application domains of stochastic processes. These lectures could cover random walk type models and urn processes (chapter 25), iterated random functions including shuffling cards, fractal models, and ancestral processes (chapter 26), computational physics and interacting particle systems (chapter 27), dynamic population models and branching processes (chapter 28), ranking and gambling betting martingale systems (chapter 29), and mathematical finance (chapter 30).

The detailed description of these course projects follows essentially the presentation of the sixth part of the book provided on page xxvi. The application domains discussed above can also be used to stimulate the development of research projects. The background requirements to enter into these topics are also discussed on page xxxvi.

We would like to thank John Kimmel for his editorial assistance, as well as for his immense support and encouragement during these last three years.

Some basic notation

We end this introduction with some probabilistic notation of current use in these lectures.

> *We could, of course, use any notation we want; do not laugh at notations; invent them, they are powerful. In fact, mathematics is, to a large extent, invention of better notations.* Richard P. Feynman (1918-1988).

We will use the symbol $a := b$ to define a mathematical object a in terms of b, or vice versa. We often use the letters m, n, p, q, k, l to denote integers and r, s, t, u to denote real numbers. We also use the capital letters $U, V, W X, Y, Z$ to denote random variables, and the letters u, v, w, x, y, z denote their possible outcomes.

Unless otherwise stated, S stands for some general state space model. These general state spaces and all the functions on S are assumed to be measurable; that is, they are equipped with some sigma field $\mathcal{S}$ so that the Lebesgue integral is well defined with respect to (w.r.t.) these functions (for instance $S = \mathbb{R}^d$ equipped with the sigma field generated by the open sets, as well as $\mathbb{N}^d$, $\mathbb{Z}^d$ or any other countable state space equipped with the discrete sigma field).

We also often use the letters f, g, h or F, G, H to denote functions on some state space, and μ, ν, η or $\mu(dx), \nu(dx), \eta(dx)$ for measures on some state space.

We let $\mathcal{M}(S)$ be the set of signed measures on some state space S, $\mathcal{P}(S) \subset \mathcal{M}(S)$ the subset of probability measures on the same state space S, and $\mathcal{B}(S)$ the set of bounded functions $f : x \in S \mapsto f(x) \in \mathbb{R}$.

We use the notation $dx(= dx_1 \times \ldots \times dx_k)$ to denote the Lebesgue measure on some Euclidian space $\mathbb{R}^k$, of some $k \geq 1$. For finite or countable state spaces, measures are identified to functions and we write $\mu(x), \nu(x), \eta(x)$ instead of $\mu(dx), \nu(dx), \eta(dx)$. The oscillations of a given bounded function f on some state space S are defined by $\mathrm{osc}(f) = \sup_{x,y \in S} |f(x) - f(y)|$.

We also use the proportionality sign $f \propto g$ between functions to state that $f = \lambda g$ for some $\lambda \in \mathbb{R}$.

Given a measure η on a state space S and a function f from S into $\mathbb{R}$ we set

$$\eta(f) = \int \eta(dx) f(x).$$

For multidimensional functions $f : x \in S \mapsto f(x) = (f_1(x), \ldots, f_r(x)) \in \mathbb{R}^r$, for some $r \geq 1$, we also set

$$\eta(f) = (\eta(f_1), \ldots, \eta(f_r)) \quad \text{and} \quad \eta(f^T) = (\eta(f_1), \ldots, \eta(f_r))^T$$

where a^T stands for the transpose of a vector $a \in \mathbb{R}^r$. For indicator functions $f = 1_A$, sometimes we slightly abuse notation and we set $\eta(A)$ instead of $\eta(1_A)$:

$$\eta(1_A) = \int \eta(dx) 1_A(x) = \int_A \eta(dx) = \eta(A).$$

We also consider the partial order relation between functions f_1, f_2 and measures μ_1, μ_2 given by

$$f_1 \le f_2 \Longleftrightarrow \forall x \in S \quad f_1(x) \le f_2(x)$$

and

$$\mu_1 \le \mu_2 \Longleftrightarrow \forall A \in \mathcal{S} \quad \mu_1(A) \le \mu_2(A).$$

The Dirac measure δ_a at some point $a \in S$ is defined by

$$\delta_a(f) = \int f(x) \delta_a(dx) = f(a).$$

When η is the distribution of some random variable X taking values in S, we have

$$\eta(dx) = \mathbb{P}(X \in dx) \quad \text{and} \quad \eta(f) = \mathbb{E}(f(X)).$$

For instance, the measure on $\mathbb{R}$ given by

$$\eta(dx) = \frac{1}{2} \left(\frac{1}{\sqrt{2\pi}} e^{-x^2/2} \, dx \right) + \frac{1}{2} \left(\frac{1}{2} \delta_0(dx) + \frac{1}{2} \delta_1(dx) \right)$$

represents the distribution of the random variable

$$X := \epsilon \, Y + (1 - \epsilon) Z$$

where (ϵ, Y, Z) are independent random variables with distribution

$$\mathbb{P}(\epsilon = 1) \quad = \quad 1 - \mathbb{P}(\epsilon = 0) = 1/2$$
$$\mathbb{P}(Z = 1) \quad = \quad 1 - \mathbb{P}(Z = 0) = 1/2 \quad \text{and} \quad \mathbb{P}(Y \in dy) = \frac{1}{\sqrt{2\pi}} e^{-y^2/2} \, dy.$$

For finite spaces of the form $S = \{e_1, \ldots, e_d\} \subset \mathbb{R}^d$, measures are defined by the weighted Dirac measures

$$\eta = \sum_{1 \le i \le d} w_i \, \delta_{e_i} \quad \text{with} \quad w_i = \eta(\{e_i\}) := \eta(e_i)$$

so that

$$\eta(f) = \int \eta(dx) f(x) = \sum_{1 \le i \le d} \eta(e_i) f(e_i).$$

Thus, if we identify measures and functions by the line and column vectors

$$\eta = [\eta(e_1), \ldots, \eta(e_d)] \quad \text{and} \quad f = \begin{pmatrix} f(e_1) \\ \vdots \\ f(e_d) \end{pmatrix} \tag{0.1}$$

we have

$$\eta f = [\eta(e_1), \ldots, \eta(e_d)] \begin{pmatrix} f(e_1) \\ \vdots \\ f(e_d) \end{pmatrix} = \sum_{1 \le i \le d} \eta(e_i) f(e_i) = \eta(f).$$

The Dirac measure δ_{e_i} is simply given by the line vector

$$\delta_{e_i} = \left[0, \ldots, 0, \underbrace{1}_{i-th}, 0, \ldots, 0 \right].$$

In this notation, probability measures on S can be interpreted as points $(\eta(e_i))_{1 \leq i \leq d}$ in the $(d-1)$-dimensional simplex $\Delta_{d-1} \subset [0,1]^d$ defined by

$$\Delta_{d-1} = \left\{ (p_1, \ldots, p_d) \in [0,1]^d \; : \; \sum_{1 \leq i \leq d} p_i = 1 \right\}. \tag{0.2}$$

We consider a couple of random variables (X_1, X_2) on a state space $(S_1 \times S_2)$, with marginal distributions

$$\eta_1(dx_1) = \mathbb{P}(X_1 \in dx_1) \quad \text{and} \quad \eta_2(dx_2) = \mathbb{P}(X_2 \in dx_2)$$

and conditional distribution

$$M(x_1, dx_2) = \mathbb{P}(X_2 \in dx_2 \mid X_1 = x_1).$$

For finite spaces of the form $S_1 = \{a_1, \ldots, a_{d_1}\}$ and $S_2 := \{b_1, \ldots, b_{d_2}\} \subset E = \mathbb{R}^d$, the above conditional distribution can be represented by a matrix

$$\begin{pmatrix} M(a_1, b_1) & M(a_1, b_2) & \ldots & M(a_1, b_{d_2}) \\ M(a_2, b_1) & M(a_2, b_2) & \ldots & M(a_2, b_{d_2}) \\ \vdots & \vdots & \vdots & \vdots \\ M(a_{d_1}, b_1) & M(a_2, b_2) & \ldots & M(a_{d_2}, b_{d_2}) \end{pmatrix}.$$

By construction, we have

$$\underbrace{\mathbb{P}(X_2 \in dx_2)}_{=\eta_2(dx_2)} = \int_{S_1} \underbrace{\mathbb{P}(X_1 \in dx_1)}_{\eta_1(dx_1)} \times \underbrace{\mathbb{P}(X_2 \in dx_2 \mid X_1 = x_1)}_{M(x_1, dx_2)}.$$

In other words, we have

$$\eta_2(dx_2) = \int_{S_1} \eta_1(dx_1) \, M(x_1, dx_2) := (\eta_1 M)(dx_2)$$

or in a more synthetic form $\eta_2 = \eta_1 M$.

Notice that for the finite state space model discussed above we have the matrix formulation

$$\begin{aligned} \eta_2 &= [\eta_2(b_1), \ldots, \eta_2(b_{d_2})] \\ &= [\eta_1(a_1), \ldots, \eta_1(a_{d_1})] \begin{pmatrix} M(a_1, b_1) & M(a_1, b_2) & \ldots & M(a_1, b_{d_2}) \\ M(a_2, b_1) & M(a_2, b_2) & \ldots & M(a_2, b_{d_2}) \\ \vdots & \vdots & \vdots & \vdots \\ M(a_{d_1}, b_1) & M(a_2, b_2) & \ldots & M(a_{d_2}, b_{d_2}) \end{pmatrix} \\ &= \eta_1 M. \end{aligned}$$

In this context, a matrix M with positive entries whose rows sum to 1 is also called a stochastic matrix.

Given a function f on S_2, we consider the function $M(f)$ on S_1 defined by

$$M(f)(x_1) = \int_{S_2} M(x_1, dx_2)\, f(x_2) = \mathbb{E}\left(f(X_2) \mid X_1 = x_1\right).$$

For functions $f : x \in S_2 \mapsto f(x) = (f_1(x), \ldots, f_r(x)) \in \mathbb{R}^r$ we also set

$$M(f)(x_1) = \int_{S_2} M(x_1, dx_2)\, f(x_2) = \mathbb{E}\left(f(X_2) \mid X_1 = x_1\right) = (M(f_r)(x_1), \ldots, M(f_r)(x_1))$$

or $M(f^T)(x_1) = (M(f_r)(x_1), \ldots, M(f_r)(x_1))^T$ for multidimensional functions defined in terms of column vectors.

Here again, for the finite state space model discussed above, these definitions resume to matrix operations

$$
\begin{aligned}
M(f) &= \begin{pmatrix} M(f)(a_1) \\ \vdots \\ M(f)(a_{d_1}) \end{pmatrix} \\
&= \begin{pmatrix} M(a_1, b_1) & M(a_1, b_2) & \cdots & M(a_1, b_{d_2}) \\ M(a_2, b_1) & M(a_2, b_2) & \cdots & M(a_2, b_{d_2}) \\ \vdots & \vdots & \vdots & \vdots \\ M(a_{d_1}, b_1) & M(a_2, b_2) & \cdots & M(a_{d_2}, b_{d_2}) \end{pmatrix} \begin{pmatrix} f(b_1) \\ \vdots \\ f(b_{d_2}) \end{pmatrix}.
\end{aligned}
$$

By construction,

$$\eta_1(M(f)) = (\eta_1 M)(f) = \eta_2(f) \iff \mathbb{E}\left(\mathbb{E}(f(X_2)|X_1)\right) = \mathbb{E}(f(X_2)).$$

Given some matrices M, M_1 and M_2, we denote by $M_1 M_2$ the composition of the matrices M_1 and M_2, and by $M^n = M^{n-1} M = M M^{n-1}$ the n iterates of M. For $n = 0$, we use the convention $M^0 = Id$, the identity matrix on S.

We use the same integral operations for any bounded integral operator. For instance, if $Q(x_1, dx_2) := G_1(x_1) M(x_1, dx_2) G_2(x_2)$ for some bounded functions G_1 and G_2 on S_1 and S_2 we set

$$(\eta Q)(dx_2) = \int_{S_1} \eta_1(dx_1)\, Q(x_1, dx_2) \quad \text{and} \quad Q(f)(x_1) = \int_{S_2} Q(x_1, dx_2)\, f(x_2)$$

for any measure η on S_1 and any function f on S_2.

When (S, d) is equipped with a metric d, we denote by $\mathrm{Lip}(S)$ the set of Lipschitz functions f such that

$$\mathrm{lip}(f) := \sup_{x \neq y} \frac{|f(x) - f(y)|}{d(x, y)} < \infty$$

and $\mathrm{BLip}(S) = \mathrm{Lip}(S) \cap \mathcal{B}(S)$ the subset of bounded Lipschitz functions equipped with the norm

$$\mathrm{blip}(f) := \|f\| + \mathrm{lip}(f).$$

In Leibniz notation, the partial derivative of a smooth function f on a product space $S = \mathbb{R}^r$ w.r.t. the i-th coordinate is denoted by

$$y \mapsto \frac{\partial f}{\partial x_i}(y) = \lim_{\epsilon \downarrow 0} \epsilon^{-1}\, [f(y_1, \ldots, y_{i-1}, y_i + \epsilon, y_{i+1}, \ldots, y_r) - f(y_1, \ldots, y_{i-1}, y_i, y_{i+1}, \ldots, y_r)]$$

with $1 \leq i \leq r$. We also denote by $\frac{\partial^2 f}{\partial x_i \partial x_j}$ the second order derivatives defined as above by replacing f by the function $\frac{\partial f}{\partial x_j}$, with $1 \leq i, j \leq r$. High order derivatives $\frac{\partial^n f}{\partial x_{i_1} \ldots \partial x_{i_n}}$ are defined in the same way. We also often use the following synthetic notation

$$\partial_{x_i} f := \frac{\partial f}{\partial x_i} \qquad \partial_{x_i, x_j} f := \frac{\partial^2 f}{\partial x_i \partial x_j} \quad \text{and} \quad \partial_{x_{i_1}, \ldots, x_{i_n}} f := \frac{\partial^n f}{\partial x_{i_1} \ldots \partial x_{i_n}}$$

as well as $\partial_{x_i}^2 = \partial_{x_i, x_i}$, and $\partial_{x_i}^3 = \partial_{x_i, x_i, x_i}$, and so on for any $1 \leq i \leq r$. When there is no confusion w.r.t. the coordinate system, we also use the shorthand

$$\partial_i = \partial_{x_i} \qquad \partial_{i,j} := \partial_{x_i, x_j} \qquad \partial_i^2 := \partial_{x_i, x_i} \quad \text{and} \quad \partial_{i_1, \ldots, i_n} = \partial_{x_{i_1}, \ldots, x_{i_n}}.$$

When $r = 1$ we often use the Lagrange notation

$$f' = \frac{\partial f}{\partial x} \qquad f'' = \frac{\partial^2 f}{\partial^2 x} \quad \text{and} \quad f^{(n)} = \frac{\partial^n f}{\partial^n x}$$

for any $n \geq 0$. For $n = 0$, we use the convention $f^{(0)} = f$. All of the above partial derivatives can be interpreted as operators on functional spaces mapping a smooth function f into another function with less regularity, for instance, if f is three times differentiable f' is only twice differentiable, and so on.

We also use integro-differential operators L defined by

$$L(f)(x) = \sum_{1 \leq i \leq r} a^i(x) \, \partial_i f(x) + \frac{1}{2} \sum_{1 \leq i, j \leq r} b^{i,j}(x) \, \partial_{i,j} f(x) + \lambda(x) \int (f(y) - f(x)) \, M(x, dy)$$

for some functions $a(x) = (a^i(x))_{1 \leq i \leq r}$, $b(x) = (b^{i,j}(x))_{1 \leq i, j \leq r}$, $\lambda(x)$ and some Markov transition M.

The operator L maps functions which are at least twice differentiable into functions with less regularity. For instance, L maps the set of twice continuously differentiable functions into the set of continuous functions, as soon as the functions a, b, λ are continuous. Notice that L also maps the set of infinitely differentiable functions into itself, as soon as the functions a, b, λ are infinitely differentiable functions. When $a = 0 = b$, the operator L resumes to an integral operator and it maps the set of bounded integrable functions into itself, as soon as the functions a, b, λ are bounded. To avoid repetition, these operators are assumed to be defined on sufficiently smooth functions. We often use the terms "sufficiently smooth" or "sufficiently regular" functions to avoid entering into unnecessary discussions on the domain of definition of these operators.

In theoretical and computational quantum physics, the inner product and more generally dual operators on vector spaces are often represented using a bra-ket formalism introduced at the end of the 1930s by P. Dirac [108] to avoid too sophisticated matrix operations (not so developed and of current use in the beginning of the 20th century).

For finite d-dimensional Euclidian vector spaces $\mathbb{R}^d$, the bras $\prec \alpha \mid$ and the kets $\mid \beta \succ$ are simply given for any $\alpha = (\alpha_i)_{1 \leq i \leq d} \in \mathbb{R}^d$ and $\beta = (\beta_i)_{1 \leq i \leq d} \in \mathbb{R}^d$ row and column

$$\prec \alpha \mid := [\alpha_1, \ldots, \alpha_d] \quad \text{and} \quad \mid \beta \succ := \begin{bmatrix} \beta_1 \\ \vdots \\ \beta_d \end{bmatrix} \Rightarrow \prec \alpha \mid\mid \beta \succ := \prec \alpha \mid \beta \succ := \sum_{1 \leq i \leq d} \alpha_i \beta_i.$$

In much the same way, the product of bra $\prec \alpha \mid$ with a linear (matrix) operator Q corresponds to the product of the row vector by the matrix

$$\prec \alpha \mid Q := [\alpha_1, \ldots, \alpha_d] \begin{bmatrix} Q(1,1) & Q(1,2) & \ldots & Q(1,d) \\ Q(2,1) & Q(2,2) & \ldots & Q(2,d) \\ \vdots & \vdots & \ldots & \vdots \\ Q(d,1) & Q(d,2) & \ldots & Q(d,d) \end{bmatrix}.$$

Likewise, the product of a linear (matrix) operator Q with a ket $\mid \beta \succ$ corresponds to the product of the matrix by the column vector

$$Q \mid \beta \succ := \begin{bmatrix} Q(1,1) & Q(1,2) & \ldots & Q(1,d) \\ Q(2,1) & Q(2,2) & \ldots & Q(2,d) \\ \vdots & \vdots & \ldots & \vdots \\ Q(d,1) & Q(d,2) & \ldots & Q(d,d) \end{bmatrix} \begin{bmatrix} \beta_1 \\ \beta_2 \\ \vdots \\ \beta_d \end{bmatrix}.$$

Combining these operations, we find that

$$\prec \alpha \mid Q \mid \beta \succ = [\alpha_1, \ldots, \alpha_d] \begin{bmatrix} Q(1,1) & Q(1,2) & \ldots & Q(1,d) \\ Q(2,1) & Q(2,2) & \ldots & Q(2,d) \\ \vdots & \vdots & \ldots & \vdots \\ Q(d,1) & Q(d,2) & \ldots & Q(d,d) \end{bmatrix} \begin{bmatrix} \beta_1 \\ \beta_2 \\ \vdots \\ \beta_d \end{bmatrix}.$$

Using the vector representation (0.1) of functions f and measures η on finite state spaces $S = \{e_1, \ldots, e_d\}$, the duality formula between functions and measures takes the following form

$$\prec \eta \mid f \succ = \eta f = \sum_{1 \leq i \leq d} \eta(e_i) f(e_i) := \eta(f).$$

Likewise, for any Markov transition M from $E_1 = \{a_1, \ldots, a_{d_1}\}$ into $E_2 := \{b_1, \ldots, b_{d_2}\}$, and function f on E_2 and any measure η_1 on E_1, we have

$$\prec \eta_1 \mid M \mid f \succ = \prec \eta_1 \mid Mf \succ = \eta_1(Mf) = (\eta_1 M) f = \prec \eta_1 M \mid f \succ.$$

The bra-ket formalism is extended to differential operations by setting

$$\prec g \mid L \mid f \succ = \int g(x) \, L(f)(x) \, dx := \langle g, L(f) \rangle.$$

Given a positive and bounded function G on some state space S, we denote by Ψ_G the Boltzmann-Gibbs mapping from $\mathcal{P}(S)$ into itself, defined for any $\mu \in \mathcal{P}(S)$ by

$$\Psi_G(\mu)(dx) = \frac{1}{\mu(G)} \, G(x) \, \mu(dx). \tag{0.3}$$

In other words, for any bounded function f on S we have

$$\Psi_G(\mu)(f) = \int \Psi_G(\mu)(dx) f(x) = \frac{\mu(Gf)}{\mu(G)}.$$

We let $\mu = \mathrm{Law}(X)$ be the distribution of a random variable X and $G = 1_A$ the indicator function of some subset $A \subset S$. In this situation, we have

$$\Psi_G(\mu)(f) = \frac{\mu(Gf)}{\mu(G)} = \frac{\mathbb{E}(f(X) 1_A(X))}{\mathbb{E}(1_A(X))} = \mathbb{E}\left(f(X) \mid X \in A\right).$$

In other words, we have
$$\Psi_{1_A}(\mu) = \text{Law}(X \mid X \in A).$$

Let (X, Y) be a couple of r.v. with probability density $p(x, y)$ on $\mathbb{R}^{d+d'}$. With a slight abuse of notation, we recall that the conditional density $p(x|y)$ of X given Y is given by the Bayes' formula

$$p(x|y) = \frac{1}{p(y)} \, p(y|x) \, p(x) \quad p(y) = \int p(y|x) \, p(x) \, dx.$$

In other words,

$$\mu(dx) := p(x)dx \quad \text{and} \quad G_y(x) := p(y|x) \Rightarrow \Psi_{G_y}(\mu)(dx) = p(x|y) \, dx.$$

Given a random matrix $A = (A_{i,j})_{i,j}$, we denote by $\mathbb{E}(A) = (\mathbb{E}(A_{i,j}))_{i,j}$ the matrix of the mean values of its entries. We slightly abuse notation and we denote by 0 the null real number and the null matrix. Given some $\mathbb{R}^d$-valued random variables X, Y we denote by $\text{Cov}(X, Y)$ the covariance matrix

$$\text{Cov}(X, Y) = \mathbb{E}\left((X - \mathbb{E}(X))(Y - \mathbb{E}(Y))'\right).$$

Sometimes, we will also use the conditional covariance w.r.t. some auxiliary random variable Z given by

$$\text{Cov}((X, Y) \mid Z) = \mathbb{E}\left((X - \mathbb{E}(X))(Y - \mathbb{E}(Y))' \mid Z\right).$$

For one-dimensional random variables, the variance of a random variable is given by $\text{Var}(X) = \text{Cov}(X, X)$, and

$$\text{Var}(X \mid Z) = \text{Cov}((X, X) \mid Z) = \frac{1}{2} \, \mathbb{E}((X - X')^2 \mid Z), \tag{0.4}$$

where X, X' are two independent copies of X given Z. We also use the notation $\sigma(X)$ to denote the σ-algebra generated by some possibly multi-dimensional r.v. X on some state space S. In this case, for any real valued random variable Y, we recall that

$$\mathbb{E}(Y \mid \sigma(X)) = \mathbb{E}(Y \mid X)$$

which has to be distinguished from the conditional expectations w.r.t. some event, say $X \in A$, for some subset $A \in \sigma(X)$

$$\mathbb{E}(Y \mid X \in A) = \mathbb{E}(Y \, 1_A(X))/\mathbb{E}(1_A(X)).$$

We also write $X \perp Y$ when a random variable X is independent of Y

$$X \text{ independent of } Y \implies_{def.} X \perp Y.$$

The maximum and minimum operations are denoted respectively by

$$a \vee b := \max\{a, b\} \quad a \wedge b := \min\{a, b\} \quad \text{as well as} \quad a_+ := a \vee 0 \quad \text{and} \quad a_- := -(a \wedge 0)$$

so that $a = a_+ - a_-$ and $|a| = a_+ + a_-$. We also denote by $\lfloor a \rfloor$ and $\{a\} = a - \lfloor a \rfloor$ the integer part, and respectively the fractional part, of some real number a.

We also use the Bachmann-Landau notation

$$f(\epsilon) = g(\epsilon) + \text{O}(\epsilon) \iff \limsup_{\epsilon \to 0} \frac{1}{\epsilon} \, |f(\epsilon) - g(\epsilon)| < \infty$$

and

$$f(\epsilon) = g(\epsilon) + o(\epsilon) \iff \limsup_{\epsilon \to 0} \frac{1}{\epsilon} |f(\epsilon) - g(\epsilon)| = 0.$$

When there is no confusion, sometimes we write $o(1)$ for a function that tends to 0 when the parameter $\epsilon \to 0$. We also denote by $O_P(\epsilon)$ some possibly random function such that

$$\mathbb{E}\left(|O_P(\epsilon)|\right) = O(\epsilon).$$

We also use the traditional conventions

$$\prod_{\emptyset} = 1 \qquad \sum_{\emptyset} = 0 \qquad \inf_{\emptyset} = \infty \quad \text{and} \quad \sup_{\emptyset} = -\infty.$$

Jacques Neveu, Poster Random Models and Statistics, MAS - SMAI Conferences, Prize of the Applied and Industrial Applied Mathematics Society (SMAI), France. *Graphical conception: Timothée Del Moral.*

Part I

An illustrated guide

1

Motivating examples

This chapter provides an illustrated guide to the topics to be discussed in this book, with a series of motivating examples. The seemingly simple examples are all related to certain non-trivial mathematical results on stochastic processes which will be discussed more thoroughly in a later chapter. The illustrations involve the recurrence and the transience properties of simple random walks, discussion of some gambling examples, fractals, ancestral evolution processes, as well as some simple examples of Poisson and Bernoulli processes. In each case, we refer to a later chapter or section where the related mathematical results will be discussed in more details.

> *Since the mathematicians have invaded the theory of relativity,*
> *I do not understand it myself anymore.* Albert Einstein (1879-1955).

1.1 Lost in the Great Sloan Wall

In 2214, the quantum field theory and the general relativity have been successfully combined within the eleven-dimensional M-string theory to explain most of the fundamental concepts of nature. The Tau Zero Foundation and the NASA's Breakthrough Propulsion Physics Project propose new interstellar travels on string lattices in any dimension. Light speed being too slow, to accelerate the interstellar flights in our 100,000 light years diameter universe, the central idea was to move the time-space itself instead of moving withinthe space-time.

A famous novel writer, Tony Gonzales, chooses the unlimited random traveling plan to avoid visiting twice the same places at these superluminal speeds. He starts his trip from Reykjavik, traveling back and forth randomly in a one-dimensional string stretched between the universes.

At the end of his "infinite travel", he was so happy to have visited all the places (infinitely many times) but he was also obliged to come back home an infinite number of times.

A little disappointed, the next summer holidays he chooses to travel on the two-dimensional string lattice, visiting randomly up/down/right/left every place in the two-dimensional universe. His trip has exactly the same characteristics and he returns to Reykjavik infinitely often.

To keep its customer happy, the Tau Zero Foundation offers him a life free trip voucher, with a special random tour on the three-dimensional Great Sloan Wall advertised in all the walls of the travel agency gallery.

Tony decides to start this nice trip immediately. After some finite number of returns to

Reykjavik, he wanders off into the infinite universe, visiting every new place only a finite number of times, but never returns back to his home.

This phenomenon is known as the recurrence of the simple random walks on lattices up to dimension 2, and the transience of the process in dimensions larger or equal than 3. These mathematical principles are discussed in section 8.5.2, as well as in 25.1. These elementary stochastic processes are also called "the drunkard walks". For a more recreational discussion, we refer the reader to the leisure book of L. Mlodinow [202].

We end this section with an intuitive and more formal derivation of these results. Tony's random exploration can be formalized by a simple random walk on the lattice $\mathbb{Z}^d$, with the dimension parameter $p = 1, 2, 3$. For instance, for $d = 1$ this random walk is given by

$$X_n = X_{n-1} + U_n \qquad (1.1)$$

with the starting state $X_0 = 0$, and a sequence U_n of independent and identically distributed (abbreviated i.i.d.) random variables (abbreviated r.v.) taking values in $\{-1, 1\}$ with a common law

$$\mathbb{P}(U_n = 1) = \mathbb{P}(U_n = -1) = 1/2.$$

The random walk on $\mathbb{Z}^2$ is also defined by
(1.1) with a starting state $X_0 = \begin{pmatrix} 0 \\ 0 \end{pmatrix}$, and
a sequence U_n of independent and identically
distributed random variables taking values in
the set of the four directional unit vectors

$$\mathcal{U} := \left\{ \begin{pmatrix} -1 \\ 0 \end{pmatrix}, \begin{pmatrix} 0 \\ -1 \end{pmatrix}, \begin{pmatrix} 1 \\ 0 \end{pmatrix}, \begin{pmatrix} 0 \\ 1 \end{pmatrix} \right\}$$

with common law

$$\forall u \in \mathcal{U} \qquad \mathbb{P}(U_n = u) = 1/4.$$

In the same vein, the random walk on $\mathbb{Z}^3$ is also defined by (1.1) with a starting state $X_0 = \begin{pmatrix} 0 \\ 0 \\ 0 \end{pmatrix}$, and a sequence U_n of i.i.d. random variables taking values in the set of the four directional unit vectors

$$\mathcal{U} := \left\{ \begin{pmatrix} -1 \\ 0 \\ 0 \end{pmatrix}, \begin{pmatrix} 0 \\ -1 \\ 0 \end{pmatrix}, \begin{pmatrix} 0 \\ 0 \\ -1 \end{pmatrix}, \begin{pmatrix} 1 \\ 0 \\ 0 \end{pmatrix}, \begin{pmatrix} 0 \\ 1 \\ 0 \end{pmatrix}, \begin{pmatrix} 0 \\ 0 \\ 1 \end{pmatrix} \right\}$$

with common law

$$\forall u \in \mathcal{U} \qquad \mathbb{P}(U_n = u) = 1/6.$$

The simulation of these random walks amounts to tossing a coin or a dice with four or six faces at each time step.

The excursions of the walker between consecutive returns to the origin are independent random trajectories with the same law. In particular, the durations $(T_i)_{i \geq 1}$ of consecutive excursions are independent copies of some random variable T that represents the length of a given random excursion. We let N be the number of returns to the origin. By construction, we have

$$\{N \geq m\} = \{(T_1 < \infty), \ldots, (T_m < \infty)\} \Rightarrow \mathbb{P}(N \geq m) = \mathbb{P}(T < \infty)^m.$$

Recalling that

$$\mathbb{E}(N) = \sum_{n \geq 1} \overbrace{n}^{\sum_{m=1}^{n} 1} \mathbb{P}(N = n)$$

$$= \sum_{1 \leq m \leq n} \mathbb{P}(N = n) = \sum_{1 \leq m} \overbrace{\sum_{m \leq n} \mathbb{P}(N = n)}^{\mathbb{P}(N \geq m)} = \sum_{1 \leq m} \mathbb{P}(N \geq m)$$

we conclude that

$$\mathbb{E}(N) = \sum_{1 \leq m} \mathbb{P}(T < \infty)^m < \infty \iff \mathbb{P}(T < \infty) < 1 \iff \mathbb{P}(T = \infty) > 0$$

$$\mathbb{E}(N) = \sum_{1 \leq m} \mathbb{P}(T < \infty)^m = \infty \iff \mathbb{P}(T < \infty) = 1.$$

On the other hand, if 0 stands for the origin state in $\mathbb{Z}^d$, we also have that

$$N = \sum_{n \geq 1} 1_0(X_n) \implies \mathbb{E}(N) = \sum_{n \geq 1} \mathbb{P}(X_n = 0 \mid X_0 = 0).$$

For $d = 1$, the walker can return to the origin after $2n$ steps with n up and n down one-step evolutions. Therefore, we have that

$$\mathbb{P}(X_{2n} = 0 \mid X_0 = 0) = \binom{2n}{n} \left(\frac{1}{2}\right)^n \left(\frac{1}{2}\right)^{2n-n} = \frac{(2n)!}{n!^2 \, 2^{2n}}.$$

Using Stirling's approximation $n! \simeq \sqrt{2\pi n} \, n^n e^{-n}$, we find that

$$\mathbb{P}(X_{2n} = 0 \mid X_0 = 0) \simeq \frac{2 \sqrt{\pi n} \, (2n)^{2n} e^{-2n} \, 2^{-2n}}{2\pi n \, n^{2n} e^{-2n}} = \frac{1}{\sqrt{\pi n}}$$

$$\implies \mathbb{E}(N) \sim \sum_{n \geq 1} \frac{1}{\sqrt{n}} = \infty.$$

In dimension $d = 2$, both coordinates $X_n^{(1)}$ and $X_n^{(2)}$ of the random state

$$X_{2n} = \begin{pmatrix} X_{2n}^{(1)} \\ X_{2n}^{(2)} \end{pmatrix} \in \mathbb{Z}^2$$

must return to the origin, so that

$$\mathbb{P}\left[X_{2n} = \begin{pmatrix} 0 \\ 0 \end{pmatrix} \mid X_0 = \begin{pmatrix} 0 \\ 0 \end{pmatrix}\right] \simeq \frac{1}{\sqrt{\pi n}} \times \frac{1}{\sqrt{\pi n}} \Rightarrow \mathbb{E}(N) \sim \sum_{n \geq 1} \frac{1}{n} = \infty.$$

In much the same vein, in $d = 3$ dimensions we have

$$\mathbb{P}\left[X_{2n} = \begin{pmatrix} 0 \\ 0 \\ 0 \end{pmatrix} \mid X_0 = \begin{pmatrix} 0 \\ 0 \\ 0 \end{pmatrix}\right] \simeq \frac{1}{\sqrt{\pi n}} \times \frac{1}{\sqrt{\pi n}} \times \frac{1}{\sqrt{\pi n}} \Rightarrow \mathbb{E}(N) \sim \sum_{n \geq 1} \frac{1}{n^{1+\frac{1}{2}}} < \infty.$$

1.2   Meeting Alice in Wonderland

To escape the Queen of Hearts' guards, Alice and the white rabbit run into the following polygonal labyrinth (with non-communicating edges) starting from doors 1 and 3:

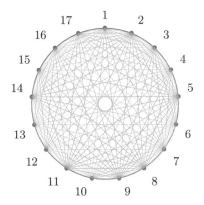

What is their chance to meet again (at some vertex)? After only 5 moves, they have more than a 25% chance to meet. This is increased to 51% after 12 steps, and to more than 90% after 40 steps.

These *come together* type principles were further developed by J. Lennon on Abbey Road, in the end of the 1960's, but their rigorous analysis certainly goes back to 1938, with the pioneering work of the French-German probabilist Wolfgang Doeblin [109].

Nowadays, these random walks on graphs arise in a variety of application domains, including in the analysis of the web-graph presented below for ranking websites, to offer personal advertisements and to improve search recommendations.

The basics of these coupling techniques are developed in section 8.3.1. The above assertions are a simple consequence of the coupling default estimates presented in Equation (8.39).

We end this section with an intuitive and more formal derivation of these results.

Firstly we observe that the random simulation of Alice's or the rabbit's evolution in the labyrinth amounts to choosing a random number I between 1 and $d = 17$ at every time step. If U is a $[0, 1]$-valued and uniform r.v. we can set

$$I := 1 + \lfloor d\, U \rfloor.$$

Notice that

$$I = i \iff \lfloor d\, U \rfloor = (i-1) \iff \frac{(i-1)}{d} \leq U < \frac{i}{d} \Rightarrow \mathbb{P}(I = i) = \mathbb{P}\left(U \in \left[\frac{(i-1)}{d}, \frac{i}{d}\right)\right) = \frac{1}{d}.$$

We let T be the first time Alice and the white rabbit meet. Since they have $1/17$ chances to meet at each time step, we readily check that

$$\forall n \geq 1 \qquad \mathbb{P}(T = n) = \left(1 - \frac{1}{d}\right)^{n-1} \frac{1}{d} \quad \text{with} \quad d = 17.$$

More formally, let I_n^a and I_n^r be the locations of Alice and the white rabbit (the indices of their doors). By construction, we have

$$\{T = n\} = \left\{(I_1^a \neq I_1^r), (I_2^a \neq I_2^r), \ldots, (I_{n-1}^a \neq I_{n-1}^r), (I_n^a \neq I_n^r)\right\}.$$

Since the r.v. $(I_k^a, I_k^r)_{k \geq 1}$ are independent, we have

$$\mathbb{P}(T = n) = \underbrace{\left[\mathbb{P}(I_1^a \neq I_1^r) \times \ldots \times \mathbb{P}(I_{n-1}^a \neq I_{n-1}^r)\right]}_{(1-1/d)^{n-1}} \times \underbrace{\mathbb{P}(I_n^a \neq I_n^r)}_{=1/d}.$$

In other words, T is a geometric r.v. with a success parameter $1/d$. This implies that

$$\mathbb{P}(T > n) = \sum_{k \geq n} \left(1 - \frac{1}{d}\right)^k \frac{1}{d} = \left(1 - \frac{1}{d}\right)^n \frac{1}{d} \sum_{k \geq 0} \left(1 - \frac{1}{d}\right)^k = \left(1 - \frac{1}{d}\right)^n \leq e^{-n/d}.$$

In the right hand side (r.h.s.) estimate we have used the fact that $\log(1 - x) \leq -x$, for any $x < 1$. For instance, we have

$$(n = 5 \quad \text{and} \quad d = 17) \Rightarrow \mathbb{P}(T > 5) = (16/17)^5 \leq .738 \leq 74\% \Longleftrightarrow \mathbb{P}(T \leq 5) \geq 25\%.$$

1.3 The lucky MIT Blackjack Team

Mr. M arrives in a private club in Atlantic City with his MIT blackjack team, intending to win against the dealer by using their brand new card shuffle tracking techniques [200].

Assuming that a perfectly shuffled deck is fully unpredictable, up to how many riffle shuffles are required for Mr. M to make some probabilistic predictions? If the number of random shuffles performed by the dealer is fewer than 5, the chance to make some predictions is larger than 90%, and larger than 40% after 6 riffle shuffles! The true story of the MIT blackjack team can be found in the Youtube 1.5h documentary.

These miraculous formulae discovered by the genius American mathematician Persi Warren Diaconis from Stanford University were reported on January 9th, 1990, in the *New York Times* [53] (see for instance the Youtube video on Coincidences and the one on Randomness). P. W. Diaconis is the Mary V. Sunseri Professor of Statistics and Mathematics at Stanford University. He is particularly well known for tackling mathematical problems involving randomness and randomization, such as coin flipping and shuffling playing cards. We also refer to his recent book with R. Graham on

Persi Warren Diaconis

magical mathematics [106]. He has also developed numerous important mathematical
results on Markov Chain Monte Carlo mod-
els [96, 97, 98, 99, 100, 101, 102, 103, 104], random walks on finite groups [263], and statistics
such as the the `Freedman-Diaconis` rule to select the sizes of the bins to be used in a
histogram.

To have a feeling about the main difficulty behind these problems, we notice that the number of different arrangements of a deck of 52
cards is given by $52! \simeq \sqrt{2\pi \times 52} \, 52^{52} \, e^{-52} \sim$
$8.053 \, 10^{67}$, and for the traditional French
tarot deck with 78 cards by $78! \simeq 1.13 \, 10^{115}$.

These large numbers can be related to the
estimated number 10^{80} of fundamental particles in the known universe, including subatomic particles. The reader should be convinced that brute force calculation of all the
possible outcomes to estimate probabilities of
some shuffling configurations is often hopeless. On the other hand, besides the fact that
a "theoretically" fully randomized deck is by essence unpredictable, a "normal" card deck
which is not perfectly mixed can be highly predictable.

The theories of card shuffling and more general Markov chains on finite groups arise in
several concrete problems, to name a few:

- Software security and design: simulation of unpredictable deck shuffles in online gaming
  or in casino card games (poker, blackjack, and others). Online card gambling sites
  often provide a description of their shuffling algorithms. The iPod also uses random
  style shuffling, reordering songs as one shuffles a deck of cards so that listeners hear
  everything, just once.

- Cryptography: random prime generation, random binary sequences, random ordering
  of integers for the design of secure unpredictable random key generators (encrypted
  messages must be seen as random from the adversary, but not random for those knowing
  the seed of the pseudo random number generator), protection against traffic analysis
  using fake random messages.

- Design of random search algorithms of complex combinatorial structures: matching
  problems, graph coloring, data association problems. For instance, solving of the traveling salesman problem requires us to explore the spaces of permutations. Each of them
  represents a circuit between the different cities.

- Computer sciences: design of random permutations for load balancing or load sharing
  protocols between computer workstations, random dynamic reallocation of resources in
  communication services (to protect attacks and faults caused by an adversary) [62].

In section 26.4, we introduce the reader to the modeling and the analysis of card shuffling
processes. In the final part, we provide an estimation formula (26.16) which gives an
indication on the predictions discussed above.

We end this section with an intuitive and more formal derivation of some of the results
discussed above. Shuffling cards is intimately related to sequence of random permutations.

For instance, the permutation

$$\sigma = \left(\begin{array}{ccccccccc} 1 & 2 & 3 & 4 & \ldots & 50 & 51 & 52 \\ 51 & 4 & 15 & 8 & \ldots & 1 & 7 & 2 \end{array} \right) \tag{1.2}$$

can be interpreted as a re-ordering of a deck of the 52 cards with labels 1, 2,..., 52. The first card is the one with label 51, the second one has label 4, and so on, up to the final card with label 2. In this context, an unpredictable shuffled deck is simply a random deck with uniform distribution on the set of permutations over 52 cards.

More precisely, the top card of an unpredictable deck can be one of the 52 cards, the second one (given the first) can be one of the 51 remaining cards, and so on. In other words, any (random) unpredictable deck occurs with a probability of 1/52!. In this interpretation, shuffling cards is a way of exploring the set of permutations. For instance, flipping the positions of two randomly chosen cards is equivalent to compose a given order-permutation by a random transposition $\tau_{I,J}$ of two randomly chosen labels I, J. The r.h.s. composition permutes the values of the cards at location I, J; and the l.h.s. composition permutes the location of the cards with values I, J. For instance, for the re-ordered deck (1.2) we have

$$\sigma \circ \tau_{1,2} = \left(\begin{array}{ccccccccc} 1 & 2 & 3 & 4 & \ldots & 50 & 51 & 52 \\ 4 & 51 & 15 & 8 & \ldots & 1 & 7 & 2 \end{array} \right)$$

and

$$\tau_{1,2} \circ \sigma = \left(\begin{array}{ccccccccc} 1 & 2 & 3 & 4 & \ldots & 50 & 51 & 52 \\ 51 & 4 & 15 & 8 & \ldots & 2 & 7 & 1 \end{array} \right)$$

Another strategy to shuffle randomly a deck of cards is to divide the deck of 52 into 2 parts, say $52 = k + (52 - k)$ where k is a binomial r.v. with parameters $(n, p) = (52, 1/2)$. Then, we assign a sequence of i.i.d. Bernoulli r.v. ϵ_l, $l = 1, \ldots, 52$ with parameter $1/2$ to each card. The random numbers $R = \sum_{1 \leq l \leq 52} \epsilon_l$, and respectively $L = \sum_{1 \leq l \leq 52} (1 - \epsilon_l)$, represent the number of cards that go to the r.h.s. stack, respectively to the l.h.s. stack before performing the riffle shuffle. For instance, the following schematic picture shows a riffle of $7 = 3 + 4$ cards cut into packs of 3 and 4 cards associated with the sequence $(0, 1, 1, 0, 1, 1, 0)$:

0	c_1		0	c_1
0	c_2		1	c_4
0	c_3		1	c_5
1	c_4	$\rightsquigarrow$	0	c_2
1	c_5		1	c_6
1	c_6		1	c_7
1	c_7		0	c_3

Starting from the bottom, first we drop the card c_3 from the left hand stack, then the card c_6, c_7 from the right hand stack, then c_2 from the left hand stack, then c_4, c_5 from the right hand stack, and finally the remaining card c_1 from the left hand stack.

The time reversed shuffle takes cards with 0's, 1's, in the top deck, respectively bottom deck, without changing their (relative) order. Note that cards with different $\{0, 1\}$-labels are in uniform random relative order, while cards with the same label have the same order before and after the time reverse shuffle.

Repeating inverse shuffles, we retain and mark the sequence of $0 - 1$ labels at the back of each card, so that after n inverse shuffles each card has an n-digit binary number. After some random number of steps, say T steps, all the cards have distinct labels so that the deck becomes uniform. These random times T are called strong stationary times and they are discussed in section 8.3.3. By construction, if $d = 52$ stands for the number of cards, we have

$$
\begin{aligned}
\mathbb{P}\left(T > n\right) &= 1 - \frac{2^n}{2^n} \times \frac{2^n - 1}{2^n} \times \frac{2^n - 2}{2^n} \cdots \times \frac{2^n - (d - 1)}{2^n} \\
&\leq 1 - \exp\left\{ -\frac{d(d + 1)}{2^{n+1}} \left[1 + \frac{1}{2^n} \left(\frac{2}{3} \, d + \frac{1}{3} \right) \right] \right\}.
\end{aligned}
\tag{1.3}
$$

The estimate stated above is proved in section 26.4.4 (it is based on the fact that $-x \geq \log\left(1 - x\right) \geq -x - x^2$, for any $0 \leq x \leq 1/2$). When $d = 52$ and $n = 12$, (1.3) gives the crude estimate .29, and .01 for $n = 14$. Using more sophisticated but more powerful techniques, D. Bayer and P. Diaconis obtained in 1992 the exact variation distances .95, .61, .33, respectively, .17, for $n = 5$, $n = 6$, $n = 7$, and $n = 8$ [13].

1.4   Kruskal's magic trap card

During his lectures, the mathematician and physicist Martin Kruskal proposes to his students to play one of his new card tricks. All of his students are aware of the work of D. Bayer and P. Diaconis [13] on riffle shuffles; therefore they ask their professor to shuffle the deck of 52 cards 8 times. In this way, they are pretty sure that the order of the cards will be unpredictable (up to some 17% error distance to the perfectly random sequence).

After this pretty long series of shuffles, he asks one of his students to pick secretly a number between 1 and 10. Then he starts displaying the cards one by one. He explains to his student that the card in the position he has secretly chosen will be his first secret card, and its value will be his second secret number. Counting again starting from this first secret card, the card in the position of the second secret number will be his second secret card, and so on. Repeating this procedure until he runs out of cards, the student will finally end on his last secret card.

Martin David Kruskal (1925-2006)

In this counting model, aces are worth 1; jacks, queens, and kings are worth 5, and the other cards takes their face values. Despite the initial (almost) perfectly random shuffled deck, and despite the secret initial number chosen by the student, the professor has more than 80% chance to know the last secret card. How it is possible? What is the the trick used by Martin Kruskal to find the last secret card? The answer is simpler than it appears: *Picking "secretly" any number between 1 and 10 and following the same procedure, we end up on the same "trap card" 80% of the time.*

Much more is true! Starting from 1, one has a chance of more than 85% to find the same last secret card. In addition, two decks of 52 cards will increase the percentage of success to 95%. See also a Youtube video explaining the Kruskal's count. Several heuristic-like arguments have been proposed in the literature to analyze these probabilities. We refer the reader to the series of articles [140, 176, 181, 207], and to the references therein. In these lectures, using Markov chain coupling techniques, some crude estimates are presented in section 8.3.4.3.

We end this section with the analysis of a simplified version of the Kruskal's magic card trick. The idea is to represent the sequential counting process as a *deterministic dynamical system* evolving in a *random environment*.

Assume that we have 10 possible values (the value of jack, queen and king is assumed to be 5). Arguing as in (1.2), the values of the cards are represented by a given random mapping $\tau : \{1, \ldots, 52\} \mapsto \{1, \ldots, 10\}$, with i.i.d. random states $(\tau(k))_{1 \leq k \leq 52}$ with some common distribution μ on $\{1, \ldots, 10\}$.

We let T_k be the k-th time the chain hits the value 1.

Suppose the initial state X_0 of the counting process starts at a jack, so that $X_0 = 4 + 1$ is the "secret" starting card at the origin. Then, the chain counts backward

$$X_0 = 4 + 1 \longrightarrow X_1 = 4 \longrightarrow X_2 = 3 \longrightarrow X_3 = 2 \longrightarrow X_4 = 1$$

so that $T_1 = 4$ and $X_{T_1+1} = X_5$ is the first "secret card", say $X_{T_1+1} = 9 + 1 = 10$.

More generally, starting at $X_0 := (d_0 + 1)$ for some $d_0 \in \{0, 1, \ldots, 9\}$, we have

$$X_0 := (d_0+1) \longrightarrow X_1 = d_0 \longrightarrow \ldots \longrightarrow X_k = (d_0+1) - k \longrightarrow \ldots \longrightarrow X_{d_0} = X_{T_1} = 1.$$

This shows that $T_1 = d_0$ and

$$X_{T_1+1} = X_{d_0+1} := d_1 \in \{1, \ldots, 10\}$$

is the first "secret card". By construction, we have

$$X_{T_1+1} = d_1 \longrightarrow X_{T_1+2} = d_1 - 1 \longrightarrow \ldots$$

$$\ldots \longrightarrow X_{T_1+k} = d_1 - (k-1) \longrightarrow \ldots \longrightarrow X_{T_1+d_1} = 1 = X_{T_2}$$

with $T_2 = T_1 + d_1$, and

$$X_{T_2+1} = d_2 \in \{1, \ldots, 10\}$$

is the second "secret card".

Iterating this process, we obtain a sequence of i.i.d. random states

$$X_{T_1+1}, X_{T_2+1}, \ldots, X_{T_k+1} \quad \text{with a common law} \quad \mu(y) = \mathbb{P}(X_{T_1+1} = y).$$

The worst case scenario for long excursions between visits of the unit state is to have $d_1 = d_2 = \ldots = d$, the maximum value of a card. This shows that on an interval $[0, nd]$ of time horizon nd, the chain $(X_l)_{l \geq 0}$ hits 1 at least n times.

To take the final step, we let $(X'_l)_{l \geq 0}$ be an independent copy of $(X_l)_{l \geq 0}$ starting at some possibly different value and we set

$$Y_k = X_{T_k+1} \quad \text{and} \quad Y'_k = X'_{T'_k+1}$$

the k-th secret card of each chain.

One of the chains represents the student's counting process, and the second one stands for the professor's counting process. We let T be the first time the chains agree and hit the same value. By construction, if we set $\mu_\star := \inf_{1 \leq k \leq d} \mu(k)$, then we have

$$\mathbb{P}(T > nd) \leq \mathbb{P}(Y_1 \neq Y_1', \ldots, Y_n \neq Y_n') = \mathbb{E}\left(\mathbb{P}\left[(Y_1 \neq Y_1', \ldots, Y_n \neq Y_n') \mid (Y_k')_{1 \leq k \leq n}\right]\right)$$

$$= \mathbb{E}\left(\prod_{1 \leq k \leq n} \mathbb{P}\left[Y_k \neq Y_k' \mid (Y_l')_{1 \leq l \leq n}\right]\right) = \mathbb{E}\left(\prod_{1 \leq k \leq n} [1 - \mu(Y_k')]\right) \leq (1 - \mu_\star)^n.$$

When $d = 10$ and $\mu(k) = 1/10$, for any $m = nd = 10n$ we have

$$\mathbb{P}(T > m) \leq p_m := \left(1 - \frac{1}{10}\right)^{m/10}.$$

The graphs of these rather crude estimates are given by the following diagram.

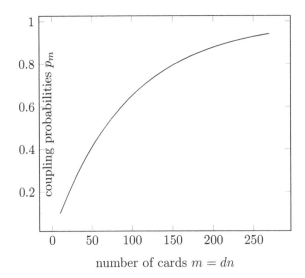

number of cards $m = dn$

1.5   The magic fern from Daisetsuzan

His Majesty the Emperor Akihito found a magical fern leaf in the mountains of Daisetsuzan National Park. With one of his powerful microscopes, he always sees the same pattern. One of his bright researchers in probability Kiyoshi Itō shows him that this mysterious leaf has been designed by someone using an elementary molecular-energy bidding only based on four distinct stippling type rules.

These fractal models were introduced in the beginning of the 1980's by the mathematician Benoit Mandelbrot [190]. We also refer the reader to the book of Michael Field and Martin Golubitsky [126] which has inspired most of the work in section 26.5, dedicated to

the modeling and the convergence analysis of fractal objects drawn by iterated random functions.

Since this period, many researchers have observed the fractal nature of some plants and vegetables such as the Romansco cabbage, or the nature of some physical phenomena such as the Von Koch snowflakes, the atmospheric turbulence, or the planet top layers structures. They are also used in algorithmically fractal arts zooming animations (see for instance YouTube videos on fractal zooming 1, 2, Fibonacci fractals 3, 4, or "pure" arts style videos 5, 6, and 3d-fractal animations 7). Nowadays, there is also considerable interest in analyzing the fractal nature of asset prices evolution in financial markets.

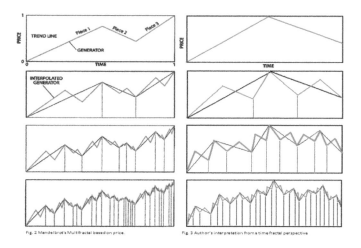

Fig. 2 Mandelbrot's Multifractal based on price.　Fig. 3 Author's interpretation from a time fractal perspective

We end this section with a simple illustration of iterated random functions. We consider a given point $A = \begin{pmatrix} x_A \\ y_A \end{pmatrix} \in \mathbb{R}^2$ and we set

$$S_A \ : \ P = \begin{pmatrix} x \\ y \end{pmatrix} \mapsto S_A(P) = S_A \begin{pmatrix} x \\ y \end{pmatrix} = \begin{pmatrix} x + \frac{2}{3}\left(x_A - x\right) \\ y + \frac{2}{3}\left(y_A - y\right) \end{pmatrix} = \begin{pmatrix} \frac{1}{3}\,x + \frac{2}{3}\,x_A \\ \frac{1}{3}\,y + \frac{2}{3}\,y_A \end{pmatrix}.$$
$$(1.4)$$

Given a couple of points

$$P_1 = \begin{pmatrix} x_1 \\ y_1 \end{pmatrix} \quad \text{and} \quad P_2 = \begin{pmatrix} x_2 \\ y_2 \end{pmatrix}$$

with the Euclidian distance

$$d(P_1, P_2) := \sqrt{(x_1 - x_2)^2 + (y_1 - y_2)^2}$$

we notice that

$$d(S_A(P_1), S_A(P_2)) \ = \ \frac{1}{3} \times d(P_1, P_2).$$

We consider the unit square square $\mathcal{C} = [0,1]^2$ with the A edges

$$A_1 = \begin{pmatrix} 1 \\ 1 \end{pmatrix}, \quad A_2 = \begin{pmatrix} 1 \\ 0 \end{pmatrix}, \quad A_3 = \begin{pmatrix} 0 \\ 0 \end{pmatrix}, \quad A_4 = \begin{pmatrix} 0 \\ 1 \end{pmatrix}. \qquad (1.5)$$

By construction, we have

$$\begin{array}{ll} S_{A_1}(\mathcal{C}) = \left[\frac{2}{3},1\right] \times \left[\frac{2}{3},1\right] & S_{A_2}(\mathcal{C}) = \left[\frac{2}{3},1\right] \times \left[0,\frac{1}{3}\right] \\ S_{A_3}(\mathcal{C}) = \left[0,\frac{1}{3}\right] \times \left[0,\frac{1}{3}\right] & S_{A_4}(\mathcal{C}) = \left[0,\frac{1}{3}\right] \times \left[\frac{2}{3},1\right]. \end{array}$$

A graphical description of the transformations S_{A_i} is provided in figure 1.1.

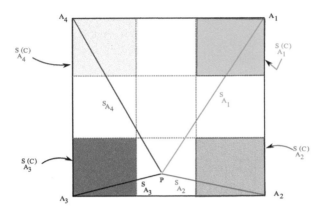

FIGURE 1.1: Graphical description of the transformations S_{A_i}

Applying S_{A_1} to the 4 squares $S_{A_i}(\mathcal{C})$, $i = 1, 2, 3, 4$, we obtain 4 new squares inside $S_{A_1}(\mathcal{C})$

$$\begin{array}{ll} S_{A_1}(S_{A_1}(\mathcal{C})) = \left[\frac{8}{9},1\right] \times \left[\frac{8}{9},1\right] & S_{A_1}(S_{A_2}(\mathcal{C})) = \left[\frac{8}{9},1\right] \times \left[\frac{2}{3},\frac{7}{9}\right] \\ S_{A_1}(S_{A_3}(\mathcal{C})) = \left[\frac{2}{3},\frac{7}{9}\right] \times \left[\frac{2}{3},\frac{7}{9}\right] & S_{A_1}(S_{A_4}(\mathcal{C})) = \left[\frac{2}{3},\frac{7}{9}\right] \times \left[\frac{8}{9},1\right]. \end{array}$$

A graphical description of the sets $S_{A_i}S_{A_j}(\mathcal{C})$ is provided in figure 1.2.

Iterating the set transformation

$$S \ : \ \mathcal{R} \subset \mathbb{R}^2 \mapsto S(\mathcal{R}) = S_{A_1}(\mathcal{R}) \cup S_{A_2}(\mathcal{R}) \cup S_{A_3}(\mathcal{R}) \cup S_{A_4}(\mathcal{R}) \subset \mathbb{R}^2$$

we obtain a sequence of decreasing sets

$$\mathcal{C}_n = S(\mathcal{C}_{n-1}) \subset \mathcal{C}_{n-1}$$

with 4^n squares with surface 9^{-n}. When $n \uparrow \infty$ this compact set-valued sequence converges to the Sierpinski fractal subset

$$\mathcal{C}_n = S^n(\mathcal{C}) \downarrow \mathcal{C}_\infty = \cap_{n \geq 1} \mathcal{C}_n \neq \emptyset.$$

It is also readily checked that S is a fixed point of the set transformation

$$S(\mathcal{C}_\infty) = \cap_{n \geq 1} S(\mathcal{C}_n) = \cap_{n \geq 1} \mathcal{C}_{n+1} = \cap_{n \geq 2} \mathcal{C}_n$$

$$\Downarrow$$

$$S(\mathcal{C}_\infty) = \mathcal{C}_\infty.$$

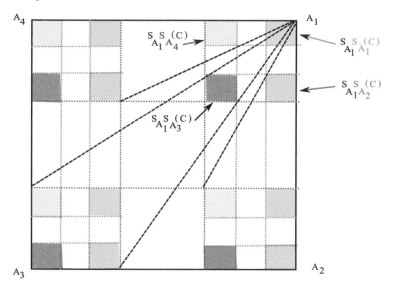

FIGURE 1.2: Graphical description of the sets $S_{A_i}S_{A_j}(\mathcal{C})$

1.6 The Kepler-22b Eve

On 5th December 2011, Brian Vastag from the *Washington Post* revealed that the NASA's Kepler mission discovered a new alien planet with just the right temperature for life. This similar-to-the-Earth planet is named Keppler-22b. It is the best candidate for a life-bearing new world. See the Youtube videos SkyNews on Earth twin, and the Kelpler-22B, first confirmed life-sustaining planet for a tour on this planet.

By 2457 this planet has been populated by a thousand selected humans from diverse countries and different cultures. The individuals have been selected rigorously after a long series of gene and chromosomes tests to avoid any common genomic ancestry.

These individuals have also been chosen so that they have approximatively the same reproduction rate. At each generation, say at every 20 years period, a pool of randomly selected individuals give birth to possibly many children. The descent lineage of the other ones is stopped. The reproduction process is designed so that the total size of the population remains almost constant during the ages. All the individuals have a specific genogram, i.e., a pictorial display of their family relationships and medical history.

Ship to Kepler-22b

After 5,500 years, more than a quarter of the population will belong to the same family! Furthermore, after every period of a thousand years this percentage grows from 68%, 86%, 94%, 97%, and finally to 99% after 10,000 years. In other words, 99% of the population are the children of a single ancestral Eve in the originally selected individuals. How could this happen? How does the ancestral line disappear during this evolution process?

In the field of human genetics, researchers are very concerned with the genealogical structure of the population. The Mitochondrial Eve is the matrilineal most recent common ancestor of all living humans on earth. She is estimated to have lived approximately 140,000 to 200,000 years ago.

We refer the reader to the pedagogical book "The Seven Daughters of Eve" of Bryan Sykes for more details on these hypothetical stories. See also the Youtube videos We are All Blacks of Geneticist Bryan Sykes, about the mitochondrial DNA of women, and the genetic diversity of Africans.

The rigorous analysis of these probabilities is developed in section 26.3, dedicated to genealogical tree evolution models, random iterated mappings, and coalescent processes. The numerical illustrations discussed above are direct consequences of the formulae (26.7).

We end this section with a more formal discussion on the simulation and the stochastic models associated with these ancestral evolution processes.

Assume that we have a population of d individuals with labels $S := \{1, \ldots, d\}$. At each time n, some individuals die while other individuals give birth to offsprings. Parents are selected uniformly in the pool. In other words, at each time step n we sample d i.i.d. r.v. $(A_n(i))_{1 \leq i \leq d}$ with uniform distribution on $\{1, \ldots, d\}$. For each $1 \leq i \leq d$, $A_n(i)$ stands for the label of the parent (at generation $(n-1)$) of the i-th individual at generation n. In other words, the range of the mapping A_n represents the successful parents with direct descendants. Notice that A_n is a random mapping from S into itself; for each state $i \in S$ the value $A_n(i)$ is chosen uniformly in the set S. For instance, we can set $A_n(i) = 1 + \lfloor d\, U_i \rfloor$, where $(U_i)_{1 \leq i \leq d}$ is a sequence of i.i.d. $[0, 1[$-valued uniform random variables.

We assume that the initial population is given by the set of d individuals with label $\{1, \ldots, d\}$. A schematic picture of the genealogical evolution process of $d = 5$ in a time horizon $n = 2$ is provided in the following picture.

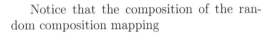

The Mitochondrial Eve

Notice that the composition of the random composition mapping

$$A_1 \circ A_2 \overset{\text{in law}}{=} A_2 \circ A_1$$

represents the ancestors at level 0 of the population of individual at generation $n = 2$. More generally, the number of ancestors at the origin is characterized by the cardinality

$$C_n = \text{Card}\left\{(A_n \circ \ldots \circ A_1)(S)\right\}$$

of the random sets $(A_n \circ \ldots \circ A_1)(S)$, where $(A_k)_{k \geq 1}$ stands for a sequence of i.i.d. random mappings from S into itself. The central idea is to observe that the Markov chain C_n is defined for any $1 \leq p \leq q \leq d$ by the elementary transitions

$$\mathbb{P}(C_n = p \mid C_{n-1} = q) = \frac{1}{d^q} \, S(q,p) \, (d)_p$$

where $S(q,p)$ stands for the number of ways of partitioning $\{1, \ldots, q\}$ into p sets (the Stirling numbers of the second kind), and $(d)_p = d!/(q-p)!$ is the number of one-to-one mappings from $\{1, \ldots, p\}$ into $\{1, \ldots, d\}$.

We let T be the first time all the individuals have the same ancestors; that is,

$$T = \inf \left\{ n \geq 0 \; : \; C_n = 1 \right\}.$$

In this notation, the estimates discussed above are consequences of the following formula

$$n_m := \left(m + \frac{7}{2} \right) d \quad \Longrightarrow \quad \mathbb{P}\left(T > n_m \right) \leq \left(m + \frac{7}{2} \right) e^{-m}.$$

In other words, after n_m evolution steps the chance for a population of d individuals to have a common ancestor is larger than $1 - \left(m + \frac{7}{2} \right) e^{-m}$. The proof of this formula is provided in section 26.3.2 and in exercise 431.

1.7 Poisson's typos

A young talented fisherman Siméon Denis Poisson was writing a series of essays on fishing techniques in the Loiret river in the spring season 1823.

There were no powerful spell-checking software products at that time, and he was very concerned with correcting all the misprints and spelling mistakes. He starts reading the first two pages, and he already finds four misprints: *A bab day of fishing is still better than a good day at the office. ...If people concentrated on the really importants things in life, there would be a shortage of fishing poles. ...Anyway, a woman who has never sen her husband fishing, doesn't known what a patient man she married.*

He assumes that all of these silly misprints were done at a unit rate per half a page during the typing. After four pages, he made the following rigorous predictions:

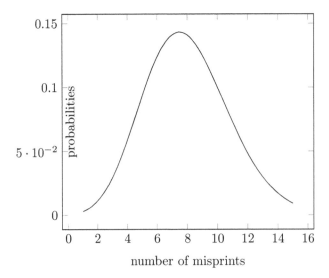

number of misprints

In addition, he also shows rigorously that for any given number of misprints he may have done on any given number of pages, all these typos will be uniformly placed in the text. He corrected the first typos, and after reading carefully, several times, the first 300 pages he could not find any misprints. He was very surprised and felt a little lucky. He really expected to find many typos later in the text. After some calculations, he saw that the chances to find misprints in the next four pages remain the same as the ones he already calculated for the very first pages. These intriguing memoryless properties opened his mind to swap fishing with creating a brand new theory of counting random events.

Siméon Denis Poisson was a famous French mathematician and physicist. He developed his probability theory in 1837 in his work "Research on the Probability of Judgments in Criminal and Civil Matters". He introduced a discrete probability distribution for counting random events arriving independently of one another in a given interval, or in some space-time window (cf. section 4.6). This distribution is known as the Poisson distribution, and the counting process is called the Poisson process.

These stochastic models were used to count the number of hits of V1 buzz bombs launched by the Nazi army on London [52], and the outbreaks of war from 1820 to 1950 [148]. We also mention the "celebrated" statistical analysis of Ladislaus Bortkiewicz. He was counting the number of cavalrymen deaths due to horse kicks in the Prussian army, covering the years 1875–1894 and 14 Prussian Cavalry Corps!

Nowadays, many random counting processes are represented by Poisson processes, including cellular telephone calls and/or wrong number connections [251], arrival of clients in a shop, flows of students entering a university building, the number of aircraft arriving at an airport, the number of claims to an insurance company, and so forth. More recently they have been used in mathematical finance to model trades in limit order books [32, 209].

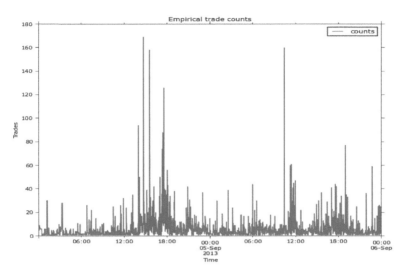

Hawkes process and trade counts

These stochastic models are also used to count and predict arrivals of trains, subways or buses, cyclones arrivals in the Arctic sea [181]. In physics, they are also used to count emissions of particles due to radioactive decay by an unstable substance [59], the stream of photons from an optical modulator, or photons landing on a photodiode in low light environments.

In engineering sciences, they are used to count the arrival of customers and jobs in communication queueing networks, internet packets at a router, and the number of requests in web servers [264]. They are also of current use in reliability and risk analysis to model the number of failures, degradation aging processes, including lifetime of systems. In geophysics, they model the random arrival of natural disasters or eco-catastrophes, such as avalanches, earthquakes, lightnings, wild fires, volcanic eruptions, floods, tsunamis, storms and hurricanes, and many others [10, 149, 211, 212].

In biology, they are used to count the number of cells chromosome interchanges subjected to X-ray radiation [44], as well as the number of bacteria and/or red blood cells in a drop of blood, birth and death counts, survival of endangered species in predator-prey systems, the location of harmless mutations in a genome (the time axis is here represented by the length of the genome).

In psychology, these processes are also used to count short or long responses from an internal clock/pacemaker to stimuli signals [172]. Self-exciting Poisson processes are used in criminology to model and analyze crimes and gang violence [115, 203, 243]. We refer the interested reader to the recent article of G. O. Mohlera, and his co-authors on the statistical analysis of point processes based on residential burglary data provided by the Los Angeles Police Department [203]. These stochastic models were also used to analyze the temporal dynamics of violence in Iraq based on civilian death reports data from 2003 to 2007 [192].

More recently, they have even been used as a statistical model for counting the number of goals or try scores in soccer or rugby games [152].

We refer the reader to section 4.6 and to chapter 10 for the modeling and the analysis of Poisson processes, as well as to chapter 28 for some workout examples of jump processes arising in biology and engineering sciences.

We end this section by a more formal brief discussion on Poisson and Bernoulli processes.

Suppose we are given a discretization $t_n = nh$ of some interval $[0, t]$, with a time mesh parameter $h \simeq 0$, with $0 \leq k \leq \lfloor t/h \rfloor$. These time steps can be interpreted as the letters or as the words of series of sentences in a book. We assume that misprints arise randomly according to a sequence of i.i.d. Bernoulli r.v. $\epsilon_{t_n}^h$ with common law

$$\mathbb{P}\left(\epsilon_{t_n}^h = 0\right) = 1 - \mathbb{P}\left(\epsilon_{t_n}^h = 1\right) = e^{-\lambda h}$$

for some parameter $\lambda > 0$. A misprint at the n-th location is characterized by the event $\{\epsilon_{t_n}^h = 1\}$. We let T_n^h indicate the location of the misprints in the text. More formally, we have that $T_0^h = 0$ and

$$T_{n+1}^h = T_n^h + \mathcal{E}_n^h \quad \text{with} \quad \mathcal{E}_n^h := h\left(1 + \lfloor \mathcal{E}_n/h \rfloor\right).$$

In the above display, the inter-occurrence of misprints $\mathcal{E}_n$ is a sequence of i.i.d. exponential r.v. with parameter λ, so that the misprints spacing rate is given by the geometric r.v.

$$\begin{aligned}
\mathbb{P}(\mathcal{E}_n^h = k\,h) &= \mathbb{P}(k-1 \leq \mathcal{E}_n/h < k) \\
&= \int_{(k-1)h}^{kh} \lambda e^{-\lambda t} dt = e^{-\lambda(k-1)h} - e^{-\lambda kh} = \left(e^{-\lambda h}\right)^{(k-1)} \left(1 - e^{-\lambda h}\right).
\end{aligned}$$

The number of misprints detected at time t is given by the Bernoulli process

$$N_t^h = \sum_{n \geq 0} n \, 1_{[T_n^h, T_{n+1}^h[}(t) = \sum_{k=1}^{\lfloor t/h \rfloor} \epsilon_{t_k}^h. \tag{1.6}$$

By construction, for any $n \leq \lfloor t/h \rfloor$ we have that

$$\begin{aligned}
\mathbb{P}\left(N_t^h = n\right) &= \binom{\lfloor t/h \rfloor}{n} \left(1 - e^{-\lambda h}\right)^n e^{-\lambda h(\lfloor t/h \rfloor - n)} \\
&= \underbrace{\left\{ \prod_{1 \leq p < n} \left(1 - \frac{p}{\lfloor t/h \rfloor}\right) \right\} \left(\frac{e^{\lambda h} - 1}{\lambda h} \frac{h \lfloor t/h \rfloor}{t}\right)^n \frac{(\lambda t)^n}{n!} e^{-\lambda h \lfloor t/h \rfloor}}_{\longrightarrow_{h \downarrow 0} 1} \\
&\longrightarrow_{h \downarrow 0} \quad \mathbb{P}\left(N_t = n\right) = \frac{(\lambda t)^n}{n!} e^{-\lambda t}. \tag{1.7}
\end{aligned}$$

This implies that

$$\mathbb{P}\left(T_n > t\right) = \mathbb{P}\left(N_t < n\right) = e^{-\lambda t} \sum_{0 \leq p < n} \frac{(\lambda t)^p}{p!}$$

from which we prove that

$$\mathbb{P}\left(T_n \in dt\right) = -\frac{d}{dt} \mathbb{P}\left(T_n > t\right) = \lambda e^{-\lambda t} \left[\sum_{0 \leq p < n} \frac{(\lambda t)^p}{p!} - \sum_{0 \leq p < (n-1)} \frac{(\lambda t)^p}{p!} \right] = \lambda e^{-\lambda t} \frac{(\lambda t)^{n-1}}{(n-1)!}.$$

We conclude that

$$\mathbb{P}\left((T_1, \ldots, T_n) \in d(t_1, \ldots, t_n)\right)$$

$$= \lambda\, e^{-\lambda(t_n - t_{n-1})} \times \lambda\, e^{-\lambda(t_1 - t_0)} \, 1_{t_0 = 0 \leq t_1 < \ldots < t_n} \, dt_1 \ldots dt_n$$

$$= \underbrace{\left(\lambda e^{-\lambda t_n} \frac{(\lambda t_n)^{n-1}}{(n-1)!}\right)}_{=\mathbb{P}(T_n \in dt_n)} \times \underbrace{\left[\frac{(n-1)!}{t_n^{n-1}} 1_{0 \leq t_1 < \ldots < t_{n-1} < t_n} dt_1 \ldots dt_{n-1}\right]}_{=\mathbb{P}((T_1,\ldots,T_{n-1}) \in d(t_1,\ldots,t_{n-1}) \mid T_n = t_n)}.$$

The above formula shows that the r.v. $(T_1, \ldots, T_{n-1})$ are uniformly distributed in the slot $[0, t_n]$ given the n-th value $T_n = t_n$.

In addition, using (1.6), we readily show that the number of misprints within the time slot $]s, r]$ is the same for any time slot of duration $(r - s)$. In addition, it is independent of the number of misprints in the time slot $]0, s]$. More formally, we have

$$N_r^h - N_s^h = \sum_{p=\lfloor s/h \rfloor + 1}^{\lfloor r/h \rfloor} \epsilon_{t_p}^h \overset{law}{=} N_{r-s}^h \qquad \text{and} \qquad N_r^h - N_s^h \perp N_s^h.$$

The same conclusions apply for the limiting Poisson process N_t and for the random times T_n. These are defined in the same way as N_t^h and T_n^h, just by replacing $\mathcal{E}_n^h$ by $\mathcal{E}_n$. Convergence rates can also be developed using the rather elementary estimates

$$\mathcal{E}_n^h \leq \mathcal{E}_n = h \lfloor \mathcal{E}_n / h \rfloor + h \{ \mathcal{E}_n / h \} = \mathcal{E}_n^h + h \{ \mathcal{E}_n / h \} \leq \mathcal{E}_n^h + h.$$

An important memoryless property of the geometric and the exponential random variables is given below

$$\begin{aligned} \mathbb{P} \left(\mathcal{E}_1^h > t_l + kh \mid \mathcal{E}_1^h > t_l \right) &= \mathbb{P} \left(\mathcal{E}_1^h > (k+l)h \mid \mathcal{E}_1^h > lh \right) \\ &= \frac{(e^{-\lambda h})^{k+l}}{(e^{-\lambda h})^l} = (e^{-\lambda h})^k = \mathbb{P} \left(\mathcal{E}_n^h > kh \right) \end{aligned}$$

for any $k, l \geq 1$, and and $s, t \geq 0$

$$\mathbb{P} \left(\mathcal{E}_1 > t + s \mid \mathcal{E}_1 \geq s \right) = \frac{e^{-\lambda(t+s)}}{e^{-\lambda s}} = e^{-\lambda t} = \mathbb{P} \left(\mathcal{E}_1 > t \right).$$

The r.v. $\mathcal{E}_1^h$ stands for the first time a misprint occurs in the text. Given there are no misprints at time t_l, the chance to see a misprint after k-text units (words or letters) is the same as if we were starting a new book from the scratch. Here the time axis is interpreted as the lengths of sentences with words or letter units. In other important instances, Poisson processes arise in the stochastic modeling of client arrivals or equipment failures.

The prototype of model is the exponential duration of a light bulb. Given the fact that the light bulb has not yet burned out at some time s, its future duration is the same as the one of a brand new light bulb. For both of them, their future duration is given by an exponential random variable.

The geometric and the exponential statistical nature of these random phenomenons depends on the problem at hand. For instance, given the fact that a standard light bulb has not yet burned after 3,000 years, it is difficult to imagine that its future duration is the same as that of a brand new one. This indicates that the exponential or the geometrical statistical nature of these light bulbs is questionable for very long time periods. Another critical illustration of exponential random duration sometimes used in the literature is the time spent by patients in a general practitioner physician's office. Given the fact that a certain patient is still in the office after two days should indicate that something wrong is going on. At least in this case these random waiting times cannot be considered as independent exponential random variables.

1.8   Exercises

Exercise 1 (Simple random walk - Mean and variance) *Consider the one-dimensional random walk X_n introduced in (1.1). Compute the quantities $\mathbb{E}(X_n)$ and $\mathrm{Var}(X_n)$.*

Exercise 2 (Simple random walk - Returns to the origin) *Consider the one-dimensional random walk introduced in (1.1). Check that for any $n \in \mathbb{N}$ we have*

$$\mathbb{P}\left(X_{2n+1} = 0\right) = 0 \quad and \quad \mathbb{P}\left(X_{2n} = 0\right) = \binom{2n}{n} 2^{-2n}.$$

Prove that for any $n \geq m$, we also have

$$\mathbb{P}\left(X_{2n} = 2m\right) = \binom{2n}{n+m} 2^{-2n}.$$

Exercise 3 (Poisson process) *Consider the Poisson process N_t with intensity λ introduced on page 20. Compute $\mathbb{E}(N_s N_t)$ and $\mathrm{Cov}(N_s, N_t)$, for any $0 \leq s \leq t$.*

Exercise 4 (Telegraph signal - Poisson process) *Consider a telegraph signal X_t taking values in $\{-1, +1\}$ starting at $X_0 = 1$ and switching at the arrival times of a Poisson process N_t with intensity λ introduced on page 20. Express X_t in terms of N_t.*

Exercise 5 (Telegraph signal - Correlation function) *Consider the random telegraph signal X_t presented in exercise 4. Assume that $\mathbb{E}(X_0) = 0$, $\mathbb{E}(X_0^2) = 1$, and X_0 and N_t are independent. Compute $\mathbb{E}(X_t)$ and $\mathbb{E}(X_t X_{t+s})$, for any $s \geq 0$.*

Exercise 6 (Reflection principle) *We consider the simple random walk X_n on the lattice $\mathbb{Z}$ (starting at the origin) given in (1.1). We consider the graph (n, X_n) of the random walk with the time axis $n \geq 0$. We take a couple of points $P_0 := (n_0, x_{n_0})$ and $P_1 := (n_1, x_{n_1})$ on the same side of the time axis, say P_0 and $P_1 \in (\mathbb{N} - \{0\})^2$. Using a graphical representation, prove that there is a one-to-one correspondence between the set of paths from P_0 to P_1 hitting the time axis at some time $n_0 < n < n_1$, and the set of all paths from $P_0^- := (n_0, -x_{n_0})$ to P_1.*

Exercise 7 (Coupling) ✍ *Let p_1, p_2 be a couple of probability densities on $\mathbb{R}$ (with respect to (w.r.t.) the Lebesgue measure dx) such that $p_1(x) \wedge p_2(x) \geq \rho \, q(x)$ for some probability density q and some parameter $\rho \in]0, 1]$. We let $\epsilon = (\epsilon_n)_{n \geq 1}$ be a sequence of i.i.d. Bernoulli random variables*

$$\mathbb{P}(\epsilon_1 = 1) = \rho = 1 - \mathbb{P}(\epsilon_1 = 0).$$

We also consider a sequence of i.i.d. random variables $X = (X_n)_{n \geq 0}$ with common probability density q, and a sequence of i.i.d. random variables $Y_i := (Y_{i,n})_{n \geq 0}$ with probability density

$$p_{\epsilon,i}(x) := (p_i(x) - \rho q(x))/(1 - \rho),$$

with $i \in \{1, 2\}$. We assume that (ϵ, X, Y_1, Y_2) are independent, and we set

$$\forall i \in \{1, 2\} \qquad Z_{i,n} := \epsilon_n \, X_n + (1 - \epsilon_n) \, Y_{i,n}.$$

Check that $(Z_{i,n})_{i \geq 0}$ are i.i.d. random variables with common probability density p_i, for $i = 1, 2$. Using the coupling model introduced above, show that $Z_{1,T} = Z_{2,T}$ for some, possibly random, finite time horizon T.

Exercise 8 (Shuffling cards) *We consider a random deck of n cards labelled from 1 to n so that the order of the cards is unpredictable. The game consists in guessing sequentially the cards in each position of the deck. We let X be the number of correct guesses. Without any information during the game, check that $\mathbb{E}(X) = 1$. Next, we assume that we are shown the card after each (correct or not) guess. Find a guessing strategy for which $\mathbb{E}(X) \simeq \log n$.*

Exercise 9 (Fisher and Yates shuffling) ✎ *Let $a = [a(1), \ldots, a(n)]$ be an array of n elements. The Fisher and Yates algorithm processes all elements one by one. At the i-th step (with $i \leq n$), generate a random integer $1 \leq j \leq i$ and switch elements $a(i)$ and $a(j)$. By induction w.r.t. the number of steps $m \in \{1, \ldots, n\}$, prove that all the elements $[a(1), \ldots, a(m)]$ are uniformly shuffled, that is,*

$$\forall 1 \leq i, j \leq m \qquad \mathbb{P}\left(\{a(i) \text{ is in the } j\text{-th location}\}\right) = 1/m.$$

Exercise 10 (Fractal images) *We consider the sequence of independent random variables*

$$\mathbb{P}(\epsilon_n = 1) = 0.01 \quad \mathbb{P}(\epsilon_n = 2) = 0.85 \quad \mathbb{P}(\epsilon_n = 3) = \mathbb{P}(\epsilon_n = 4) = 0.07$$

and the affine functions $f_i(x) = A_i.x + b_i$ with the matrices and the vectors defined below

$$A_1 = \begin{pmatrix} 0 & 0 \\ 0 & 0.16 \end{pmatrix} \quad b_1 = \begin{pmatrix} 0 \\ 0 \end{pmatrix} \quad A_2 = \begin{pmatrix} 0.85 & 0.04 \\ -0.04 & 0.85 \end{pmatrix} \quad b_2 = \begin{pmatrix} 0 \\ 1.6 \end{pmatrix}$$

and

$$A_3 = \begin{pmatrix} 0.2 & -0.26 \\ 0.23 & 0.22 \end{pmatrix} \quad b_3 = \begin{pmatrix} 0 \\ 1.6 \end{pmatrix} \quad A_4 = \begin{pmatrix} -0.15 & 0.28 \\ 0.26 & 0.24 \end{pmatrix} \quad b_4 = \begin{pmatrix} 0 \\ 0.44 \end{pmatrix}.$$

Run on a computer the Markov chain with 10^5 iterations to obtain the fractal image presented on page 19.

2

Selected topics

This chapter continues the presentation of guiding examples from the application areas of stochastic processes. The selected topics include a variety of effects in population dynamics, reinforcement learning and its unusual traps, and ruin models. Further on, the popular web-page ranking algorithms are interestingly linked to some stability properties of Markov chains. Finally, some internet traffic architectures are related to piecewise deterministic processes. We provide links to the chapters and sections where a more thorough discussion is accomplished.

The art of doing mathematics consists in finding that special case which contains all the germs of generality. David Hilbert (1862-1943).

2.1   Stabilizing populations

The last report of the Population Division of the United Nations in 2013 reveals 232 million international migrants living abroad worldwide: "Half of all international migrants lived in 10 countries, with the US hosting the largest number (45.8 million), followed by the Russian Federation (11 million); Germany (9.8 million); Saudi Arabia (9.1 million); United Arab Emirates (7.8 million); United Kingdom (7.8 million); France (7.4 million); Canada (7.3 million); Australia (6.5 million); and Spain (6.5 million)".

The migration flows between the 193 countries (recognized by the United Nations) could be represented by proportions $M_n(i,j)$ of residents in a country c_i that move every year, (or any other unit of time: hour, day, week, or month), to another country c_j, with the country indices i and j running from 1 to 193. The index parameter n represents the number of the time units. The quantities $M_n(i,j)$ can also be interpreted as the "chances" for an individual or the probability to go from country c_i to country c_j.

The population $q_n(j)$ of each country c_j at a given time n evolves according to the recursive equations

$$q_n(j) = \sum_{1 \leq i \leq 193} q_{n-1}(i) \, M_n(i,j). \qquad (2.1)$$

We can also work with the proportions of the population $p_n(j) = q_n(i)/N_n$ of each country c_j, where N_n is the (rapidly increasing) current world population given in real time by Worldometer.

To simplify the presentation, we assume that the number of individuals in the world is fixed as $N_n = N$. In other words, the number of individuals in country j at time n is the sum of all the individuals coming randomly from all the other countries i (including $i = j$ to take into account the non-migrants in country j).

In this case, the equation (2.1) can be derived from an individual-based evolution model in which $q_{n-1}(i)$ is the mean number of individuals in the country c_i.

At time $(n-1)$, each country c_i has $m_{n-1}(i)$ individuals denoted by $I_{i,n-1}^k$ with $1 \leq k \leq m_{n-1}(i)$. Each of these individuals $I_{i,n-1}^k$ moves from c_i to c_j with probability $M_n(i,j)$. In other words, the individual $I_{i,n-1}^k$ selects the index $\widehat{I}_{i,n-1}^k = j$ of country c_j with probability $M_n(i,j)$. In this notation, the number of individuals $m_n(i,j)$ in country c_i migrating to country c_j at time n is given by the random numbers

$$m_n(i,j) := \sum_{1 \leq k \leq m_{n-1}(i)} 1_j\left(\widehat{I}_{i,n-1}^k\right)$$

$$\Rightarrow \quad \mathbb{E}(m_n(i,j) \mid m_{n-1}(i)) \;=\; \mathbb{E}\left(\sum_{1 \leq k \leq m_{n-1}(i)} 1_j\left(\widehat{I}_{i,n-1}^k\right) \mid m_{n-1}(i)\right) \qquad (2.2)$$
$$= \quad m_{n-1}(i)\, M_n(i,j).$$

Under our assumption,

$$m_n(j) \;=\; \sum_{1 \leq i \leq 193} m_n(i,j)$$

$$\mathbb{E}\left(m_n(j)\right) \;=\; \mathbb{E}\left(\sum_{1 \leq i \leq 193} m_n(i,j)\right) = \sum_{1 \leq i \leq 193} \mathbb{E}(m_{n-1}(i))\, M_n(i,j).$$

The r.h.s. assertion is a direct consequence of (2.2).

To get one step further in our discussion, suppose that there is always a chance (even very small), say $\epsilon > 0$, to move from a country c_i to another country c_j after (perhaps very large) m units of time. In this situation, the initial population data $q_0(i)$ is not really important. More precisely, if we start from some wrong data, say $q_0'(i)$, after some number of generations n, the sum of all differences tends to 0 exponentially fast. More precisely, for certain positive constant B, we have that

$$\sum_{1 \leq i \leq 193} |q_n'(i) - q_n(i)| \leq B\, (1 - \epsilon)^{n/m}$$

where $q_n'(i)$ stands for the solution of (2.1) that starts at $q_0'(i)$. In the same vein, assuming that $M_n(i,j) = M(i,j)$ is roughly constant over time, the population $q_n(i)$ of each country stabilizes exponentially fast to some stationary population $\pi(i)$

$$\sum_{1 \leq i \leq 193} |q_n(i) - \pi(i)| \leq B\, (1 - \epsilon)^{n/m}.$$

For instance, suppose that the individuals have a 50% chance to migrate from one country to another every 5 years. In this case, the populations of *all* the countries in the world will have a 50% chance to stabilize after 10 years, and a 97% chance to stabilize after 30 years.

The stationary populations $\pi(i)$ are the unique solution of the fixed point equation

$$\pi(j) = \sum_{1 \leq j \leq 193} \pi(i)\, M(i,j)$$

with $\sum_{1 \leq i \leq 193} \pi(i) = N$.

These numbers can be approximated by sampling sequentially random variables $X_n \in \{c_1, \ldots, c_{193}\}$ mimicking the evolution of *a single* individual from one country to another

$$X_0 \rightsquigarrow X_1 \rightsquigarrow X_2 \rightsquigarrow \ldots \rightsquigarrow X_n \rightsquigarrow \ldots$$

with the transition probabilities

$$\mathbb{P}\left(X_n = c_j \mid X_{n-1} = c_i\right) = M(i,j).$$

The number $\pi(i)/N$ is roughly given by the proportion of times the individual visits the i-th country:

$$\pi^n(i) := \frac{1}{n} \sum_{1 \leq p \leq n} 1_{X_p=i} = \left(1 - \frac{1}{n}\right) \pi^{n-1}(i) + \frac{1}{n} 1_{X_n=i} \simeq_{n \to \infty} \pi(i)/N.$$

Of course, the individual X_n starting from the i-th country will return eventually to his initial country. Much more is true, the expected return time is precisely equal to $N/\pi(i)$.

Under the numerical assumptions described above, what are the chances for two individuals, say X_n and X'_n starting from different countries to meet in some country at the same period of time? The meeting time is again given by the estimate described above. They will have a 50% chance to meet after 10 years, and a 97% chance after 30 years.

The stochastic modeling techniques discussed above are described in chapter 7 dedicated to Markov chains and their different interpretations. We also used some analytic estimates presented in chapter 8. The coupling technology is housed in section 8.3.1.

It is important to notice that the interpretation $q_n(i)/N = \mathbb{P}(X_n = c_i)$ is only valid when the number of individuals in the whole world does not change. In a more general situation, each individual $I^k_{i,n-1}$ may die or may give birth to a random number of children at time $n-1$. One way to model this birth and death process is to consider the number of offsprings $I^{k,l}_{i,n-1}$, with $1 \leq l \leq N^k_{i,n-1}$, where $N^k_{i,n-1}$ stands for $\mathbb{N}$-valued r.v. with a given mean

$$\mathbb{E}\left(N^k_{i,n-1}\right) = G_{n-1}(i)$$

that depends on his or her country's birth rate.

In this simplified model, the individual $I^{k,l}_{i,n-1}$ dies when $N^k_{i,n-1} = 0$ or gives no birth when $N^k_{i,n-1} = 1$.

The resulting model is called a branching process .

population at time $(n-1)$ after migration

$$\xrightarrow{\text{branching}} \text{population at time } (n-1) \text{ after birth and death}$$

$$\xrightarrow{n\text{-th migration}} \text{population at time } n \text{ after migration.}$$

We let $\widehat{I}^{k,l}_{i,n-1}$ be the index of the country chosen by the individual $I^{k,l}_{i,n-1}$. In this notation, we have

$$m_n(j,i) = \sum_{1 \leq k \leq m_{n-1}(j)} \sum_{1 \leq l \leq N^k_{j,n-1}} 1_i\left(\widehat{I}^{k,l}_{j,n-1}\right)$$

$$\Rightarrow \mathbb{E}\left(m_n(j,i) \mid m_{n-1}(j), N^l_{j,n-1}, 1 \leq l \leq m_{n-1}(j)\right) = \sum_{1 \leq k \leq m_{n-1}(j)} N^k_{j,n-1} \, M_n(j,i)$$

$$\Rightarrow \mathbb{E}\left(m_n(j,i) \mid m_{n-1}(j)\right) = m_{n-1}(j) \, G_{n-1}(j) \, M_n(j,i).$$

This implies that

$$N_n = \sum_{1 \leq i \leq 193} m_n(i)$$

$$\Rightarrow \mathbb{E}\left(N_n \mid m_{n-1}(j), 1 \leq j \leq 193\right) = \sum_{1 \leq j \leq 193} m_{n-1}(j)\, G_{n-1}(j)$$

(2.3)

and

$$q_n(i) := \mathbb{E}\left(m_n(i)\right) = \sum_{1 \leq j \leq 193} q_{n-1}(j)\, G_{n-1}(j)\, M_n(j,i).$$

We conclude that

$$\mathbb{E}(N_n) = \sum_{1 \leq i \leq 193} q_n(i) = \sum_{1 \leq j \leq 193} q_{n-1}(j)\, G_{n-1}(j)$$

$$= \mathbb{E}(N_n) \times \sum_{1 \leq j \leq 193} \frac{q_{n-1}(j)}{\sum_{1 \leq k \leq 193} q_{n-1}(k)}\, G_{n-1}(j).$$

For constant birth rates $G_n(i) = g$, for any i and n we have three cases:

$$\begin{cases} g > 1 & \Leftrightarrow & \mathbb{E}(N_n) = g^n\, \mathbb{E}(N_0) \uparrow \infty & \text{super-critical} \\ g = 1 & \Leftrightarrow & \mathbb{E}(N_n) = \mathbb{E}(N_0) & \text{critical} \\ g < 1 & \Leftrightarrow & \mathbb{E}(N_n) = g^n\, \mathbb{E}(N_0) \downarrow 0 & \text{sub-critical} \end{cases}$$

The reader may have noticed that

$$(2.3) \implies \mathbb{E}\left(\frac{N_n}{g^n} \mid m_{n-1}(j), 1 \leq j \leq 193\right) = \frac{N_{n-1}}{g^{n-1}}$$

$$\Leftrightarrow \quad \overline{N}_n := \frac{N_n}{g^n} \text{ martingale.}$$

Using martingale theorems, we can show the almost sure convergence $\overline{N}_n \to_{n \uparrow \infty} \overline{N}_\infty$ with $\overline{N}_\infty = 0$ or $\overline{N}_\infty > 0$. Conditionally on non-extinction, $g > 1 \Rightarrow \overline{N}_\infty = \infty$. A detailed presentation of martingale processes and their limit theorems is provided in section 8.4.

2.2   The traps of reinforcement

The interpretation of reinforcement processes depends on the application domains they are used for.

As its name indicates, the reinforcement of some material is a reparation type process that strengthens and reinforces its structure.

In behavioral psychology, there are two types of reinforcement, namely the positive and the negative reinforcement. Both reinforcement processes increase the likelihood that some positive or negative behavior associated with a favorable or non-favorable outcome occurs more frequently [201].

The Hero of Waterloo pub in Sydney

In computer sciences reinforcement learning models are reward-based control algorithms. The "software" agent visiting a solution space receives a reward for any positive action aiming to maximize a given cumulative reward functional [249].

In all the situations discussed above, a reinforcement process is associated with some repeated experiment or some repeated outcome.

These reinforced experiments can be modeled by self-interacting processes. The characteristic of these stochastic processes is that the transitions from one state to another depend on the history and on the way the process has explored the state space.

New South Wales is the Australian state with the largest number of 2100 hotel pubs, taverns and bars among 3450 total pubs.

A French probabilist who recently arrived to Sydney wants to visit all of its total internationally known hotel pubs. Every evening he is tempted to look back in the past and to return to one of the pubs he visited the days before. To avoid visiting too many times the same place, from time to time he chooses randomly a new pub according to the distribution of the pubs presented above.

We can use a biased coin tossing to represent the reinforced behavior of the French tourist. The sequence of reinforced events can be associated to the sequence of head outcomes. In other words, every evening he flips a coin and when head occurs he looks back in the past events and returns to one of the previously visited pubs.

More formally, we assume that all the pubs have the same probability to be chosen. In other words, the distribution of the pubs is given by $\pi(x) = 1/d$, for any pub x, where d stands for the total number of pubs.

We also let $\epsilon \in [0, 1]$ be the probability of head, and we denote by X_k the name of the pub he visited the k-th day. On the n-th evening, the tourist flips the coin. If a head turns up he comes back randomly to one of the pubs $X_0, \ldots, X_{n-1}$ he has visited the days before. If he gets a tail he chooses a new pub randomly according to the distribution π of the pubs discussed above.

If we let $X_0 = x_0$ be the first visited pub, then the probability to visit some pub x on the n-th evening is given by the formula

$$\mathbb{P}\left(X_n = x\right) = \pi(x) + \epsilon \; \alpha_\epsilon(n) \; [1_{x=x_0} - \pi(x)]$$

with the error function $\alpha_\epsilon(n)$

$$\alpha_\epsilon(n) := \prod_{1 \leq k < n} \frac{k + \epsilon}{k + 1}.$$

When the reinforcement rate $\epsilon = 0$ is null, the tourist will visit the pubs uniformly at any time n. For a rather weak reinforcement rate, say $\epsilon = 1/10$, the remainder function $\alpha_\epsilon(n)$ lies between the red and the blue lines in the next figure. After one month, the tourist will visit the pubs almost uniformly.

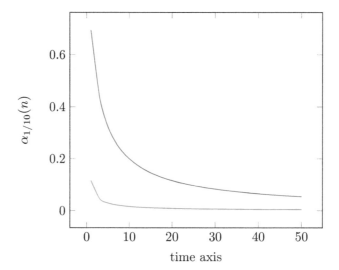

If the tourist looks back over the past 50% of the time, he will eventually visit the pubs almost uniformly only after three months.

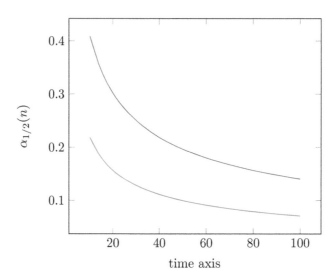

Finally, if he looks 90% of the time back in the past, then even after 2700 years the fluctuations around the desired uniform distribution are of the order of 25%.

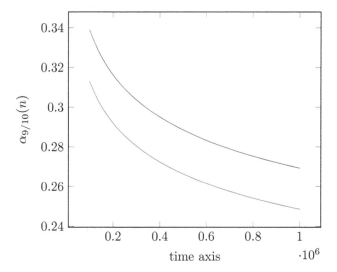

With this strong reinforcement rate, he will eventually succeed to visit the pubs almost uniformly after more than 2.6 million years.

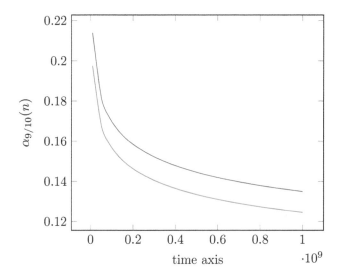

We leave the reader to find the moral of the story. The quote from the American poet Henry David Thoreau summarizes the situation in a few words: "Never look back unless you are planning to go that way." The detailed proofs of these results are provided in section 7.10.1 and in exercise 17.

2.3 Casino roulette

Asheyl Revell visited the Plaza Hotel & Casino, in Las Vegas in the summer of 2004. After having "some" beers in a London pub with some of his friends, he decided to sell all his worldly possessions, including his house, his clothes, his car, and his watch, and to risk everything on a single bet on the **black** color in a casino roulette wheel. This Friday night is a little special, and he suddenly decides to change the color at the last second, and to gamble all his   US\$135,300   on   a   single   spin   on   red. The event was filmed by Britain's Sky One television as a short reality series called $Double\ or\ Nothing$. When the ball landed on 7 red, Revell collected his winnings of \$270,600, and left a \$600 tip. The day after, he decided to play with more moderation the same strategy with a \$100 per bet, and to stop whenever he won \$100,000, or lost all of his gains.

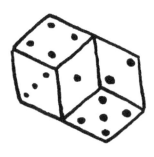

If the game was fair, he will have 1 chance over 100 to succeed to win \$100,000 before losing all of his gains. But the game is slightly biased, since his real chances in the USA roulette are smaller than 2.96×10^{-6} (and around 2.504×10^{-4} if he was playing in Europe).

To understand these numbers, we need to analyze very carefully the parameters of the problem. In the USA, the roulette wheel has 36 numbers with red and black colors, but there are also two green blocks with numbers 0 and 00. Thus, the chance q to win his bet at any time is not 1/2 but

$$q = 18/(36+2) < p = 20/(36+2).$$

The European roulette has the same 36 numbers with red and black colors, but there is a single green number 0.

The ratio $\delta = p/q = 1+2/18$ is of a particular importance, since the probabilities P_{US} to win $b = 100$ times an initial unit bet x before losing all gains are given by

$$P_{US}(x) = \frac{\delta^{-(100-x)} - \delta^{-100}}{1 - \delta^{-100}}.$$

In addition, the mean duration of the game is given by the formula

$$\text{Mean\_duration}_{US}(x) = \frac{\delta+1}{\delta-1}\left[100\ (1 - P_{US}(x)) - (100-x)\right].$$

If Revell was playing a fair game, these quantities would be given by

$$P_{US}(x) = x/100 \quad \text{and} \quad \text{Mean\_duration}(x) = x \times (100-x).$$

It is interesting to compare these formulae with their counterparts for the case of the European roulette. The chance q to win his bet at any time is still not 1/2 but is now equal to

$$q = 18/(36+1) < p = 19/(36+1).$$

The formulae for the $P_{Europe}(x)$ and the mean duration time $\text{Mean\_duration}_{Europe}(x)$ are the same as the ones for the US roulette wheel by replacing δ with $\delta = p/q = 1 + 1/18$.

The probability curves with respect to the starting bet are depicted in the following picture:

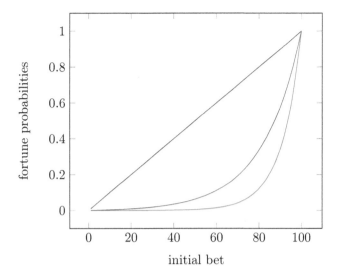

A zoom in the picture for a bet between 1 and 75 units shows that the US and the European roulette are not really the same.

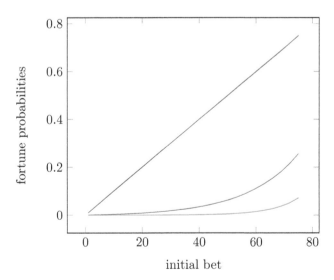

The mean duration time curves with respect to the starting bet are described below:

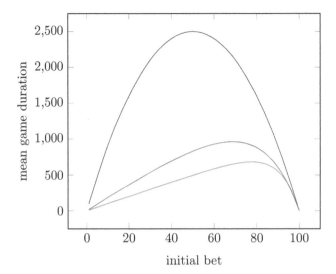

The proof of the formula described above relies on martingale theory. We refer the reader to section 8.4 for an introduction to this subject. The section 8.4.4 is dedicated to the applications of martingale techniques to the analysis of a gambling model of the same form as the one discussed above.

2.4 Surfing Google's waves

In 1996, Larry Page and Sergey Brin were working on a PhD research project on new web search engines at Stanford University. "Cramming their dormitory room full of cheap computers, they applied their method to web pages and found that they had hit upon a superior way to build a search engine. Their project grew quickly enough to cause problems for Stanford's computing infrastructure". (The Economist, *Enlightenment man*, Dec. (2008)). They invented a new page ranking algorithm called PageRank which is still the basis for all of Google's web search tools. According to Google:

PageRank works by counting the number and quality of links to a page to determine a rough estimate of how important the website is. The underlying assumption is that more important websites are likely to receive more links from other websites.

The PageRank U.S. Patent 6, 285, 999 is not assigned to Google but to Stanford University. The original PageRank algorithm is related to the so-called random surfer model on the $d = 15 \times 10^9$ estimated web pages (January 2014) visited by a virtual surfer. At each point in time, the surfer flips a biased coin with a probability of a head, say $\epsilon = .85 \in [0, 1[$. If head turns up, he jumps to one of the outgoing linked pages. Otherwise, he jumps randomly to a page that is uniformly chosen on the web.

We let X_n be the one of d web pages la-
belled by some index $i \in \{1, \dots, d\}$. We denote by $M(i, j)$ the Markov transition of the surfer

$$\mathbb{P}(X_n = j \mid X_{n-1} = i) = M(i, j).$$

After a long surfing period, the surfer is no more affected by the initial starting page

$$\mathbb{P}(X_n = j \mid X_0 = i) = M^n(i, j) \longrightarrow_{n \to \infty} \mathbb{P}(X_\infty = j) := \pi(j) \qquad (2.4)$$

and the proportions of visits of each page tend to stabilize to some values

$$\frac{1}{n} \times \mathrm{Card}\{0 \le p < n \;:\; X_p = i\} \longrightarrow_{n \to \infty} \mathbb{P}(X_\infty = i) := \pi(i). \qquad (2.5)$$

Thus, it is natural to define a new order relation by setting

$$i \prec j \iff \pi(i) \le \pi(j).$$

In addition, using Markov chain theory, we show that

$$\|\mathrm{Law}(X_n) - \mathrm{Law}(X_\infty)\|_{tv} := \frac{1}{2} \sum_{1 \le i \le d} |\mathbb{P}(X_n = i) - \pi(i)| \le .85^n.$$

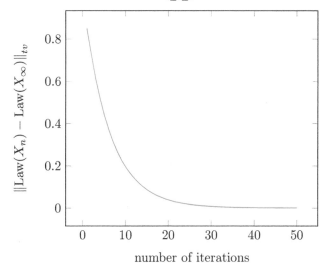

number of iterations

The formulae (2.4) and (2.5) provide two methods of computing the line vector π. The first one is called the power method. It consists of multiplying the matrices M up to some number of times n making sure that the column becomes almost stable. More precisely, we have

$$\forall i \in \{1, \dots, d\} \qquad M^n(i, j) \simeq_{n \uparrow \infty} \pi(j) \quad (\implies \pi M = \pi).$$

The second one is based on sampling the evolution of the surfer and counting the proportion of times the sites i are visited. Since the matrix M is very sparse, the power method is cheaper (the cost of sparse matrix operations only depends on the number of non null entries). One difficulty arises from the fact that the power matrix is rather slowly varying and the number of web pages changes with a speed of around $250,000$ new domain names every day. For any starting point $i \in S$, one can prove that

$$\|M^{n+1}(i, .) - \pi\|_{tv} \le .85 \times \|M^n(i, .) - \pi\|_{tv}.$$

The detailed proofs of these results and their connections with the stability properties of Markov chains are provided in section 29.1. How can Google store and update all the web data? Some answers can be found in the YouTube video on Google data center.

2.5   Pinging hackers

Kevin Mitnick, and his friend Tsutomu Shimomura sent a communication request to a Transmission Control Protocol (*abbreviated TCP*) in Illinois to communicate with a computer on internet. They know that TCP is simply a transmission layer with an architecture that is free of any security feature, but it includes several congestion algorithms to control the number of packets that can be transmitted safely.

Kevin is using his favorite packet sniffer to design hacking packets and to sniff the return traffic. He launches many trains of echo-reply packets with his brand new Ping program to some randomly chosen targets to select the best host with the smallest dormancy from his machine. They don't really want to steal or destroy data, but only to penetrate and to use several powerful host computers to run their own applications.   The main task for Tsutomu this afternoon is to check the status of the size of the TCP congestion window to avoid too many timeouts, with lost data segments and/or data receipt acknowledgements. This window gives the number of data units (also called segments).

Every time he detects a missing segment, he halves the congestion window. These timeouts result from their router buffer overflows, when the maximum capacity of the connection has been probed.

Otherwise, he tunes the system slowly and opens a window by the inverse of the current window size. After some statistical experiments, he checks that the congestion rate is around 3 timeouts per 4 seconds. Based on this statistical data, he designs a judicious piecewise deterministic process to predict analytically the behavior of his protocol. As their name indicates, these stochastic processes are deterministic between some jump times associated to the timeout periods. Between timeouts, the size of the congestion window evolves according to the traditional deterministic dynamical system

$$\frac{dW_t}{dt} = 1/W_t.$$

When a timeout occurs, say at some random time T, the size is halved

$$W_{T-} \quad \rightsquigarrow \quad W_T := W_{T-}/2.$$

The distance between two timeouts is an exponential random variable with parameter $4/3$. A more thorough discussion of the tools used by Tsutomu to model and analyze this TCP congestion window is provided in section 13.4. Further details on piecewise deterministic models can also be found in section 10.5.

In the early connection lifetime, the initial value of congestion window size is quite large, say in the critical order of 10^3 segments. This critical effect comes from the very large number of acknowledged data segments they received in the early stage of their hacking process. Right after a few tens of seconds, the size of the congestion window stabilizes extremely fast around these critical values. The limiting average of the square of the window size is given precisely by 2.

Kevin started analyzing the stochastic model designed by Tsutomu, and found that the probability to reach the equilibrium congestion size is given by the following exponential curve:

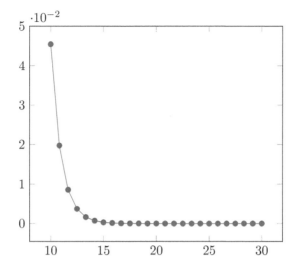

Working a little harder, he shows that the limiting congestion size is given by random variable

$$W_t \longrightarrow_{t \uparrow \infty} W_\infty := \sqrt{3/2} \sqrt{\sum_{n \geq 0} 4^{-n} \mathcal{E}_n}$$

where $\mathcal{E}_n$ represent independent standard exponential random variables. In addition, there is more than a 95% chance to have a size of the congestion window below 4 units; more precisely, it holds that

$$\mathbb{P}\left(W_\infty \leq 3.87\right) \geq .95.$$

The details of this probabilistic reasoning are given in section 13.4.

2.6 Exercises

Exercise 11 (Moment generating functions of offspring numbers) *We let $\varphi(s) := \mathbb{E}(s^X)$ be the moment generating function of a random number of offsprings X with some probability distribution $p(x) := \mathbb{P}(X = x)$, with $x \in \mathbb{N}$. Compute φ in the following situations*

- *The Bernoulli distribution $p(x) = p^x(1-p)^{1-x}$, with $p \in [0,1]$ and $x \in \{0,1\}$.*

- *The binomial distribution $p(x) = \binom{n}{x} p^x(1-p)^{n-x}$, with $p \in [0,1]$ and $x \in \{0, \ldots, n\}$ for some $n \in \mathbb{N}$.*

- *The Poisson distribution $p(x) = e^{-\lambda}\lambda^x/x!$, with $\lambda > 0$ and $x \in \mathbb{N}$.*

- *The geometric distribution $p(x) = (1-p)^{x-1}p$, with $\lambda > 0$ and $x \in \mathbb{N} - \{0\}$.*

Exercise 12 (Population size mean and variance) *Consider a branching process in which every individual in the population gives birth to a random number of offsprings*

*X. We let N_n be the total number of individuals in the population at the n-th genera-
tion. We also consider a sequence $(X_n^i)_{1 \leq i \leq N_n},\ _{n \geq 0}$ of independent copies of X, so that
$N_{n+1} = \sum_{1 \leq i \leq N_n} X_n^i$. Compute the mean and the variance of N_n in terms of the mean
$\mathbb{E}(X) = m$ and the variance $\mathrm{Var}(X) = \sigma^2$ of the random variable X. When $N_0 = 1$, show
that*

$$\mathrm{Var}(N_n) = \begin{cases} n\,\mathrm{Var}(X) & \text{when} \quad m = 1 \\ \mathrm{Var}(X)\,m^{n-1}\,\dfrac{m^n - 1}{m - 1} & \text{when} \quad m \neq 1. \end{cases}$$

Exercise 13 (Moment generating functions of population sizes) *We let $\varphi_n(s) := \mathbb{E}(s^{N_n})$ be the moment generating function of the branching process presented in exercise 12.
Check that $\varphi_n = \varphi_{n-1} \circ \varphi_1$, for any $n \geq 1$; and $\varphi_n(0) = \mathbb{P}(N_n = 0)$. Compute φ_n for the
Bernoulli offspring distribution introduced in exercise 11 for some $p \in [0, 1[$, and conclude
that*

$$\mathbb{P}(N_n = 0) = 1 - \mathbb{P}(N_n = 1) = p^n.$$

Exercise 14 (Spatial branching process) *Consider a positive function $i \in S \mapsto G(i) \in
]0, \infty[$ and a Markov transition $M = (M(i, j))_{i,j \in S}$ on a finite set S. We let $(g_n^i(j))_{i,j \in S, n \in \mathbb{N}}$
be a sequence of $\mathbb{N}$-valued random variables such that $\mathbb{E}(g_n^i(j)) = G(j)$, for each $i, j \in S$
and any $n \in \mathbb{N}$. We let $f : i \in S \mapsto f(i) \in \mathbb{R}$ be some function on S.*

- *We start with N_0 individuals $\xi_0 := (\xi_0^i)_{1 \leq i \leq N_0}$ with some common distribution η_0 on
  S, for some given integer $N_0 \in \mathbb{N} - \{0\}$. Given ξ_0, each individual ξ_0^i gives birth to
  $g_0^i(\xi_0^i)$ offsprings for each $1 \leq i \leq N_0$. At the end of this branching transition we obtain
  $N_1 = \sum_{1 \leq i \leq N_0} g_0^i(\xi_0^i)$ individuals $\widehat{\xi}_0 := (\widehat{\xi}_0^i)_{1 \leq i \leq N_1}$. Check that*

$$\mathbb{E}\left(\sum_{1 \leq i \leq N_1} f(\widehat{\xi}_0^i)\right) = N_0\,\eta_0(Gf).$$

- *Given $\widehat{\xi}_0$, each offspring $\widehat{\xi}_0^i$ moves to a new location ξ_1^i according to the Markov transition
  M. Check that*

$$\mathbb{E}\left(\sum_{1 \leq i \leq N_1} f(\xi_1^i) \mid N_1,\ \widehat{\xi}_0\right) = \sum_{1 \leq i \leq N_1} M(f)(\widehat{\xi}_0^i)$$

  and deduce that

$$\mathbb{E}\left(\sum_{1 \leq i \leq N_1} f(\xi_1^i)\right) = N_0\,\eta_0(Q(f))$$

  with the matrix Q with positive entries $Q(i, j) = G(i)M(i, j)$.

- *Given $\xi_1 := (\xi_1^i)_{1 \leq i \leq N_1}$, each individual ξ_1^i gives birth to $g_1^i(\xi_1^i)$ offsprings for each
  $1 \leq i \leq N_1$. At the end of this branching transition we obtain $N_2 = \sum_{1 \leq i \leq N_1} g_1^i(\xi_1^i)$
  individuals $\widehat{\xi}_1 := (\widehat{\xi}_1^i)_{1 \leq i \leq N_2}$. Check that*

$$\mathbb{E}\left(\sum_{1 \leq i \leq N_2} f(\widehat{\xi}_1^i)\right) = N_0\,\eta_0(Q(Gf)).$$

- *Given $\widehat{\xi}_1 := (\widehat{\xi}_1^i)_{1 \leq i \leq N_2}$, each offspring $\widehat{\xi}_1^i$ moves to a new location ξ_2^i according to the*

Markov transition M. Iterating these branching and exploration transitions we define a sequence of populations $\xi_n := (\xi_n^i)_{1 \leq i \leq N_n}$ and $\widehat{\xi}_n := (\widehat{\xi}_n^i)_{1 \leq i \leq N_{n+1}}$. Check that

$$\mathbb{E}\left(\sum_{1 \leq i \leq N_2} f(\xi_2^i) \mid N_2, \widehat{\xi}_1\right) = \sum_{1 \leq i \leq N_2} M(f)(\widehat{\xi}_1^i)$$

and deduce that

$$\mathbb{E}\left(\sum_{1 \leq i \leq N_2} f(\xi_2^i)\right) = N_0 \; \eta_0(Q^2(f)).$$

Iterating these calculations check that for any $n \in \mathbb{N}$ we have

$$\mathbb{E}\left(\sum_{1 \leq i \leq N_n} f(\xi_n^i)\right) = N_0 \; \eta_0(Q^n(f)).$$

Exercise 15 (Chinese restaurant process 1) ✍ *A Chinese restaurant has an infinite number of tables indexed by the integers $\mathbb{N} - \{0\}$ equipped with some probability distribution μ. Each table can seat an infinite number of customers. Let $\alpha > 0$ be a given parameter. We let X_n be the index of the table chosen by the n-th customer. The first customer enters and sits at one table, which he chooses randomly, with a distribution μ. With a probability $\frac{\alpha}{\alpha+n}$, the $(n+1)$-th customer chooses a new table randomly, according to the distribution μ; otherwise he chooses to join one of the occupied tables $i \in \{X_1, \ldots, X_n\}$ randomly chosen with the distribution $\frac{1}{n}\sum_{1 \leq p \leq n} 1_{X_p}(i)$.*

- *Write down the probability transition of X_{n+1} given $(X_1, \ldots, X_n)$.*

- *Find the expected number of different tables occupied by the first n customers.*

- *Check that*

$$\mathbb{P}\left(X_{n+1} = i \mid X_1, \ldots, X_n\right) = \frac{\alpha\mu(i) + V_n(i)}{\alpha + n} \quad \text{with} \quad V_n(i) = \sum_{1 \leq p \leq n} 1_{X_p}(i).$$

Exercise 16 (Chinese restaurant process 2) ✍ *Consider the Chinese restaurant process presented in exercise 15, with $\mu(S) = 1$ for some finite subset S.*

- *Check that*

$$\mathbb{P}\left(X_1 = x_1, \ldots, X_{n+1} = x_{n+1}\right) = \left[\prod_{0 \leq k \leq n} \frac{1}{\alpha + k}\right] \prod_{s \in S} \prod_{0 \leq k < v_{n+1}(s)} (\alpha\mu(s) + k) \quad (2.6)$$

with $v_{n+1}(s) = \sum_{1 \leq p \leq (n+1)} 1_{x_p}(s)$.

- *Compare the above formula with the moments of the Dirichlet distribution given in (4.9).*

- *Deduce that $(X_i)_{i \geq 1}$ can be interpreted as a sequence of independent random variables on the set $S := \{1, \ldots, d\}$ with probability distribution given by the following formula*

$$\forall s \in S \qquad \mathbb{P}\left(X_1 = s \mid U\right) = U_s \quad \text{with} \quad U = (U_1, \ldots, U_d) \sim D(\alpha\mu(1), \ldots, \alpha\mu(d)).$$
$$(2.7)$$

In this situation, given U check that $\frac{1}{n}\sum_{1\leq p\leq n} 1_{X_p}(i)$ converges almost surely to U_i, as $n\uparrow\infty$, and we have

$$\mathbb{E}\left(\frac{1}{n}\sum_{1\leq p\leq n} 1_{X_p}(i) \mid U\right) = U_i \quad and \quad \mathrm{Var}\left(\frac{1}{n}\sum_{1\leq p\leq n} 1_{X_p}(i) \mid U\right) = \frac{1}{n}U_i(1-U_i).$$

Exercise 17 (Self interacting processes) ✎ *We define sequentially a self-interacting process X_n evolving on some state space S equipped with a probability measure $\mu = \mathrm{Law}(X_0)$. Given $(X_0,\ldots,X_{n-1})$, we let X_n be a random variable with a conditional distribution*

$$\epsilon \frac{1}{n}\sum_{0\leq p<n} \delta_{X_p}(dx) + (1-\epsilon)\,\mu(dx)$$

and we set

$$\eta_n = \mathrm{Law}(X_n) \qquad S_n = \frac{1}{n}\sum_{k=0}^{n-1} \delta_{X_k} \qquad \overline{S}_n = \frac{1}{n}\sum_{k=0}^{n-1} \eta_k.$$

Check that for any bounded function f we have $\mathbb{E}(S_n(f)) = \overline{S}_n(f)$, and

$$S_n(f) = \frac{n}{n+1}S_{n-1}(f) + \frac{1}{n+1}f(X_n).$$

For any function f s.t. $\mu(f) = 0$, check that

$$\mathbb{E}(f(X_{n+1}) \mid X_0,\ldots,X_n) = \epsilon\, S_n(f) \quad and \quad \eta_{n+1}(f) = \epsilon\, \overline{S}_n(f) ,$$

and deduce from the above that

$$\overline{S}_n(f) = \alpha_\epsilon(n) \times \mathbb{E}(f(X_0)) \quad with \quad \alpha_\epsilon(n) := \prod_{k=1}^{n} \frac{k+\epsilon}{k+1}.$$

Prove that

$$\int_1^{n+1} \log\left(1 - \frac{(1-\epsilon)}{t}\right) dt \leq \log\alpha_\epsilon(n) \leq \int_2^{n+2} \log\left(1 - \frac{(1-\epsilon)}{t}\right) dt. \qquad (2.8)$$

Using the estimates

$$\forall x \in [0,1[\qquad -\frac{x}{1-x} \leq \log(1-x) \leq -x$$

check that

$$\frac{1}{(1+n/\epsilon)^{1-\epsilon}} \leq \alpha_\epsilon(n) \leq \frac{1}{(1+n/2)^{1-\epsilon}}.$$

Exercise 18 (Web surfer) *We consider the web surfer Markov chain X_n discussed in section 2.4. We let $\nu(i) = 1/d$, $i \in S = \{1,\ldots,d\}$ be the uniform probability measure on the web page indexes S, $\epsilon \in [0,1[$ be a given parameter, and we set*

$$M(i,j) = \epsilon\, K(i,j) + (1-\epsilon)\,\nu(i) \quad with \quad K(i,j) = \frac{1}{\mathrm{Card}(\mathrm{Link}(i))}\, 1_{\mathrm{Link}(i)}(j).$$

In the above display, $\mathrm{Link}(i)$ denotes the set of indexes of the outgoing pages linked to the page with index i. Using induction on $n \geq 0$, prove that $\mathrm{osc}(M^n(f)) \leq \epsilon^n\,\mathrm{osc}(f)$, for any function f on S, and deduce that

$$\forall(i,j,k) \in S^k \qquad \lim_{n\to\infty} |\mathbb{P}(X_n = k \mid X_0 = i) - \mathbb{P}(X_n = k \mid X_0 = j)| = 0.$$

Exercise 19 (Transmission protocol) *We consider the transmission protocol model W_t discussed in section 2.5. We let $\overline{W}_t = W_t^2/2$. Between timeouts, show that $\frac{d\overline{W}_t}{dt} = 1$. A discrete time version of $\overline{W}_t$ is given by the Markov chain*

$$X_n = (1 - \epsilon_n) \ (X_{n-1} + h) + \epsilon_n \ (X_{n-1}/4).$$

In the above display, $h > 0$ denotes a small time step, and ϵ_n is a sequence of i.i.d. $\{0, +1\}$-valued Bernoulli random variables with a common law

$$\mathbb{P}(\epsilon_1 = 1) = p = 1 - \mathbb{P}(\epsilon_1 = 0).$$

Show that

$$X_n - 4^{-\sum_{1 \le k \le n} \epsilon_k} \overset{law}{=} h \sum_{0 \le k < n} 4^{-\sum_{1 \le l \le k} \epsilon_l} \ (1 - \epsilon_{k+1}).$$

Exercise 20 (Gambler's ruin problem) *We consider the gambler's ruin problem discussed in section 2.3. During the game, the total fortune of the gambler X_n at each time n is described by the Markov chain*

$$X_n = X_{n-1} + \epsilon_n,$$

where ϵ_n stands for a sequence of i.i.d. $\{-1, +1\}$-valued Bernoulli random variables with a common law

$$\mathbb{P}(\epsilon_1 = 1) = p = 1 - \mathbb{P}(\epsilon_1 = -1).$$

We assume that the gambler starts with an initial fortune, say $X_0 = x \in \mathbb{N} - \{0\}$, and that the game ends when either $X_n = 0$ (running out of money) or $X_n = x_{max}$ (some maximal total fortune). We denote by $T = \inf \{n \ge 0 : X_n \in \{0, x_{max}\}\}$ the random time at which the game stops, and $P(x) = \mathbb{P}(X_T = x_{max} | X_0 = x)$. Note that $P(0) = 0$ and $P(x_{max}) = 1$. Using a conditioning argument w.r.t. X_1, check that

$$P(x) = p \ P(x + 1) + (1 - p) \ P(x - 1) \quad and \quad [P(x + 1) - P(x)] = \frac{p}{q} [P(x) - P(x - 1)]$$

with $q = 1 - p$ and for any $0 < x < x_{max}$. Deduce that

$$P(x + 1) = \begin{cases} P(1) \ \dfrac{1 - \left(\frac{q}{p}\right)^{x+1}}{1 - \left(\frac{q}{p}\right)} & \text{if} \quad p \neq q \\ P(1) \ (x + 1) & \text{if} \quad p = q \end{cases}$$

and for any $0 \le x \le x_{max}$

$$P(x) = \begin{cases} \dfrac{1 - \left(\frac{q}{p}\right)^{x}}{1 - \left(\frac{q}{p}\right)^{x_{max}}} & \text{if} \quad p \neq q \\ \dfrac{x}{x_{max}} & \text{if} \quad p = q. \end{cases}$$

3

Computational and theoretical aspects

In this chapter, we give a historical perspective to the origination of two of the main computational tools used in stochastic analysis: the Monte Carlo methods and the stochastic calculus. Particular emphasis is put to the Metropolis-Hastings simulation model. Next, the Kalman filter as a major stochastic algorithm used in engineering sciences and in information theory is presented in historical context. Furthermore, a moving story about the origination of the Doeblin-Itō stochastic calculus is told. One of the major application areas of stochastic calculus is finance. The financial applications are also put in historical perspective. Starting with Louis Bachelier's theory of speculation we move forward to give a first glimpse in option pricing. Much more details and derivations of these topics will be given in the subsequent chapters and sections and pointers to these are provided across this chapter.

> *If one disqualifies the Pythagorean Theorem from contention,*
> *it is hard to think of a mathematical result which is better known*
> *and more widely applied in the world today than "Itō's Lemma".*
> Citation from the US National Academy of Science.

3.1 From Monte Carlo to Los Alamos

Monte Carlo is a city in the tiny Principality of Monaco on the Mediterranean Sea, well known for its famous casino. It attracts a lot of celebrities and avid gamblers. One of them is the uncle of Stanislaw Ulam, a famous computational physicist from Los Alamos National Laboratories in the USA.

At the other side of the Atlantic ocean, in the beginning of the 1950's a small group of physicists and mathematicians from this laboratory, including Stanislaw Ulam, Nicholas Metropolis, and John von Neumann, spent their evenings playing poker with small sums. During one of these gambling evenings, Nick Metropolis, a Greek-American with a great personality, "described what a triumph it was when he won $10 to John (the author of the famous treatise on game theory). He bought this book with $5, and then pasted the other $5 inside the cover as a symbol of his victory" (S. Ulam, taken from Wikipedia). Nick also played the part of a scientist in the Woody Allen film "Husbands and Wives" in 1992.

Nicholas Metropolis

During his working days, Nick and his colleagues were focusing on nuclear weapon projects. In the 1940s, they invented a new stochastic simulation technique coined by John von Neumann the `Monte Carlo Method` due to the secrecy of their project and in "honor" of the uncle of Stanislaw Ulam who had a propensity to gamble [196].

Nowadays, `Monte Carlo methods` rather refer to a broader class of computer stochastic simulation methods. These methods include independent random sampling of a given random event with a prescribed probability distribution. However, without any doubt, the most famous stochastic algorithm is the *Metropolis-Hastings model*. This model was developed in the mid-1960s by Nicholas Metropolis, Arianna W. Rosenbluth, Marshall N. Rosenbluth, Augusta H. Teller, and Edward Teller in their seminal article [198].

Formally, we let $\pi(x)$ be any type of positive probability density on $\mathbb{R}^d$, for some $d \geq 1$.

Starting with a given state x, we propose randomly a new random state y with a positive probability density $P(x, y)$. Then, we accept the state y with probability

$$a(x, y) = \min\left(1, \frac{\pi(y)P(y, x)}{\pi(x)P(x, y)}\right).$$

Otherwise, we stay in x. In both cases, we propose another random state and accept or reject it following the same policy as before. This process continues in the same way. The main simplification of the above judicious choice of acceptance-rejection ratio comes from the fact that it does not depend on the often unknown normalizing constant of the so-called target density measure π. For instance the normalizing constant of some conditional distribution is often given by the unknown probability of the conditioning event; hence it is difficult to obtain this constant.

Other important examples of target measures are the Boltzmann-Gibbs measures described in chapter 6 (see for instance section 6.4).

Repeating these two operations, we obtain a sequence of random variables X_n, with $n \in \mathbb{N}$ such that

$$\text{Law}(X_n) \simeq_{n\uparrow\infty} \pi \quad \text{and} \quad \frac{1}{n} \sum_{0 \leq p < n} f(X_p) \simeq_{n\uparrow\infty} \int f(x) \, \pi(x) \, dx$$

for any bounded function f on $\mathbb{R}^d$.

The overall probability density $M(x, y)$ of the transition $x \rightsquigarrow y$ is given by the probability of the proposition $x \rightsquigarrow y$ <u>and</u> the chance to accept the state y. By simple conditioning argument, for any $x \neq y$ we find that

$$M(x, y) := P(x, y) \times \min\left(1, \frac{\pi(y)P(y, x)}{\pi(x)P(x, y)}\right)$$

from which we prove that

$$\pi(x)M(x, y) = \min\left(\pi(x)P(x, y), \pi(y)P(y, x)\right).$$

From the symmetry of this formula we obtain the so-called master equation

$$\pi(x)M(x, y) = \pi(y)M(y, x) \implies \int \pi(x) \, M(x, y) \, dx = \pi(y).$$

The right hand side formula says that if we start with a random variable with density $\pi(x)$ and perform the Metropolis-Hastings transition $x \rightsquigarrow y$, the probability of the resulting random variable is again $\pi(y)$.

In probability theory, this property is called the invariance property of π with respect to the Markov transition M, and the sequence of random states X_n is called a Markov chain with elementary transitions M.

The numerical analysis of these Markov Chain Monte Carlo methods *(abbreviated MCMC)* is directly related to the convergence analysis of Markov chain models towards their invariant measures, as the time parameter tends to ∞.

These questions are discussed in chapter 8. The reader will find a panorama of the main mathematical techniques involved in the stability analysis of Markov chain models, including linear algebraic methods, topological and functional methodologies, as well as more probabilistic approaches based on coupling techniques, strong stationary times, and martingale theory.

In the last century, stochastic simulation techniques have become some of the most important numerical techniques in applied sciences.

The rather elementary Metropolis-Hastings method is cited in *Computing in Science and Engineering* as one of the top-10 algorithms having the "greatest influence on the development and practice of science and engineering in the 20th century".

In chapter 9 we provide a review of the main stochastic simulation techniques in probability and statistics as well as in computational physics in the last decades.

3.2   Signal processing and population dynamics

Rudolf Emil Kalman is a member of the National Academy of Sciences (USA), the National Academy of Engineering (USA), and the American Academy of Arts and Sciences (USA). He is a foreign member of the Hungarian, French, and Russian Academies of Science.

On the 19th of February 2008, he received a US$500,000 Charles Stark Draper Prize in Washington from the US National Academy of Engineering "for the development and dissemination of the optimal digital technique (known as the Kalman Filter) that is pervasively used to control a vast array of consumer, health, commercial and defense products."

For instance, "the Kalman filters have been vital in the implementation of the navigation systems of US Navy nuclear ballistic missile submarines, and in the guidance and navigation systems of cruise missiles such as the Navy's Tomahawk missile and the US Air Force's Air Launched Cruise Missile.

It is also used in the guidance and navigation systems of the NASA space shuttle and the attitude control and navigation systems of the International Space Station" (taken from Wikipedia).

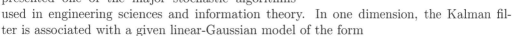

In a seminal paper published in 1960, Kalman presented one of the major stochastic algorithms used in engineering sciences and information theory. In one dimension, the Kalman filter is associated with a given linear-Gaussian model of the form

$$\begin{cases} X_n &= a\, X_{n-1} + W_n \\ Y_n &= b\, X_n + V_n \end{cases} \tag{3.1}$$

with some given parameters $(a, b) \in \mathbb{R}^2$ and some independent Gaussian r.v. (X_0, W_n, V_n)

with mean and variance parameters given respectively by $(m_0, 0, 0)$ and $(\sigma_0^2, \sigma^2, \tau^2)$. Apart from a few technicalities, the multi-dimensional and the time non-homogeneous cases follow the same line of arguments.

The two equations stated above constitute a filtering problem. The stochastic process X_n is called the signal. It may represent the evolution of some target such as a missile, a submarine, or a vehicle. The stochastic process Y_n is called the observation sequence. These random variables are delivered by some sensor such as a radar, sonar, or some global positioning system (a.k.a. GPS).

For a simplified but more sophisticated and concrete radar filtering model, we refer the reader to section 6.5.2.

Abusing notation, the objective is to compute *sequentially* the conditional probability densities

$$p(x_n \mid y_0, \ldots, y_{n-1}) \quad \text{and} \quad p(x_n \mid y_0, \ldots, y_n). \tag{3.2}$$

The states $(X_p, Y_p)_{p \leq n}$, being linear combinations of independent Gaussian variables, are necessarily Gaussian with some mean and variance parameters given by (m_n, σ_n^2) and $(\widehat{m}_n, \widehat{\sigma}_n^2)$, respectively.

The density on the l.h.s. in (3.2) is called the one step optimal predictor, while the r.h.s. is called the optimal filter. Using Bayes' rule, these probability densities are connected by the product formula

$$p(x_n \mid y_0, \ldots, y_n) \propto p(y_n \mid x_n) \times p(x_n \mid y_0, \ldots, y_{n-1}). \tag{3.3}$$

Using the celebrated regression formula that can be traced back to the German and the French mathematicians Carl Friedrich Gauss and Adrien Marie Legendre in the early 1800*s* (see [131, 132, 178] as well as section 4.5.2), we find that

$$\widehat{m}_n = \underbrace{\mathbb{E}_n(X_n)}_{=m_n} + \frac{\mathrm{Cov}_n(X_n, Y_n)}{\mathrm{Var}_n(Y_n)} \, (y_n - \underbrace{\mathbb{E}_n(Y_n)}_{=bm_n}) \tag{3.4}$$

where $\mathbb{E}_n(.)$ stands for the conditional expectation w.r.t. the observation sequence $Y_p = y_p$, with $p < n$.

In addition, we have

$$\mathrm{Cov}_n(X_n, Y_n) \quad := \quad \mathbb{E}_n \left((X_n - m_n) \overbrace{(Y_n - b \, m_n)}^{b \, (X_n - m_n)} \right) = b \, \sigma_n^2$$

$$\mathrm{Var}_n(Y_n) \quad := \quad \mathbb{E}_n \left((Y_n - \mathbb{E}_n(Y_n)^2) \right)$$
$$= \quad \mathbb{E}_n \left((b(X_n - m_n) + V_n)^2 \right) = b^2 \sigma_n^2 + \tau^2.$$

Using (3.4), we also check that

$$\widehat{\sigma}_n^2 \quad = \quad \mathbb{E} \left((X_n - \widehat{m}_n)^2 \right)$$
$$= \quad \mathrm{Var}_n(X_n) - \mathrm{Cov}_n(X_n, Y_n)^2 / \mathrm{Var}_n(Y_n).$$

This implies that

$$\widehat{\sigma}_n^2 = \sigma_n^2 - \frac{b^2\,\sigma_n^4}{b^2\sigma_n^2 + \tau^2} = \sigma_n^2\left(1 - \frac{b^2\sigma_n^2}{b^2\sigma_n^2 + \tau^2}\right) = (b^2\tau^{-2} + \sigma_n^{-2})^{-1}.$$

To obtain a recursive algorithm, it remains to find the reverse relation. To this end, we again apply the Bayes' rule to check that

$$p(x_{n+1} \mid y_0,\ldots,y_n) = \int \underbrace{p(x_{n+1} \mid x_n) \times p(x_n \mid y_0,\ldots,y_n)}_{=p(x_{n+1},\,x_n \mid y_0,\ldots,y_n)}\,dx_n \qquad (3.5)$$

from which we conclude that

$$\begin{aligned} m_{n+1} &= a\,\widehat{m}_n \\ \sigma_{n+1}^2 &= \mathbb{E}_n\left([X_{n+1} - m_{n+1}]^2\right) = \mathbb{E}_n\left([a(X_n - \widehat{m}_n) + W_n]^2\right) = a^2\widehat{\sigma}_n^2 + \sigma^2. \end{aligned} \qquad (3.6)$$

The so-called Kalman filter reduces to the resulting recursive equations

$$(m_n,\sigma_n) \xrightarrow{\text{updating}} (\widehat{m}_n,\widehat{\sigma}_n) \xrightarrow{\text{prediction}} (m_{n+1},\sigma_{n+1}).$$

The next picture illustrates one-dimensional Kalman filter associated with $\sigma^2 = 1 = \tau^2$.

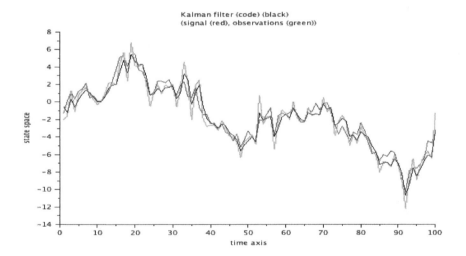

Unfortunately, most of the filtering problems arising in practice are nonlinear and/or non-Gaussian. In this context, the pivotal regression updating formula is clearly useless, and we need to find another strategy to solve these estimation problems.

To this end, an important observation is that formulae (3.3) and (3.5) are always satisfied for any type of filtering model. In probability theory and engineering sciences, these formulae are called the filtering equations.

One natural way to approximate a given distribution is to consider the empirical measure associated with a sequence of random variables. Following this idea, let us suppose that we have designed a sequence of N random particles $(\xi_n^1,\ldots,\xi_n^N)$ such that in some sense we have

$$p(x_n \mid y_0, \ldots, y_{n-1}) \, dx_n \ \simeq_{N\uparrow\infty} \ \frac{1}{N} \sum_{1 \leq i \leq N} \delta_{\xi_n^i}(dx_n). \tag{3.7}$$

Then, by (3.3), we have the following approximation

$$p(x_n \mid y_0, \ldots, y_n) \, dx_n \ \simeq_{N\uparrow\infty} \ \sum_{1 \leq i \leq N} \frac{p(y_n \mid \xi_n^i)}{\sum_{1 \leq j \leq N} p(y_n \mid \xi_n^j)} \, \delta_{\xi_n^i}(dx_n). \tag{3.8}$$

Thus, sampling N independent random variables with probability density $p(x_n \mid y_0, \ldots, y_n)$ is "almost equivalent" to sampling N independent random variables $(\widehat{\xi}_n^1, \ldots, \widehat{\xi}_n^N)$ with the discrete probability measure (3.8).

By the law of large numbers, we have

$$p(x_n \mid y_0, \ldots, y_n) \, dx_n \ \simeq_{N\uparrow\infty} \ \frac{1}{N} \sum_{1 \leq i \leq N} \delta_{\widehat{\xi}_n^i}(dx_n).$$

The construction of an empirical measure satisfying this approximation property is far from being unique. For instance, we can also replace the weights by any function $G_n(\xi_n^i) \propto p(y_n \mid \xi_n^i)$. Whenever G_n takes values in $[0,1]$, one can accept every particle ξ_n^i with the probability $G_n(\xi_n^i)$, and set $\widehat{\xi}_n^i = \xi_n^i$. The rejected particles are replaced by the same number of randomly chosen particles w.r.t. the weighted measure defined above.

The random transition $(\xi_n^1, \ldots, \xi_n^N) \rightsquigarrow (\widehat{\xi}_n^1, \ldots, \widehat{\xi}_n^N)$ can be interpreted in many different ways.

For instance, it can be interpreted as a selection of the particles ξ_n^i with high value of the likelihood $p(y_n \mid \xi_n^i)$.

It can also be seen as a branching process or a birth and death process.

Last, but not least, it can be seen as a acceptance-rejection scheme equipped with a recycling mechanism.

Now, using (3.5) we have

$$p(x_{n+1} \mid y_0, \ldots, y_n) \ \simeq_{N\uparrow\infty} \ \frac{1}{N} \sum_{1 \leq i \leq N} p\left(x_{n+1} \mid \widehat{\xi}_n^i\right).$$

Therefore, from each selected particle $\widehat{\xi}_n^i$, we sample a transition of the signal $X_n = \widehat{\xi}_n^i \rightsquigarrow X_{n+1} = \xi_{n+1}^i$ and we have the empirical approximation

$$p(x_{n+1} \mid y_0, \ldots, y_n) \, dx_{n+1} \ \simeq_{N\uparrow\infty} \ \frac{1}{N} \sum_{1 \leq i \leq N} \delta_{\xi_{n+1}^i}(dx_{n+1}).$$

The next animation illustrates the evolution of the genetic type population discussed above. We also quote at every time step the proportion of accepted individuals in the acceptance-rejection transition presented above.

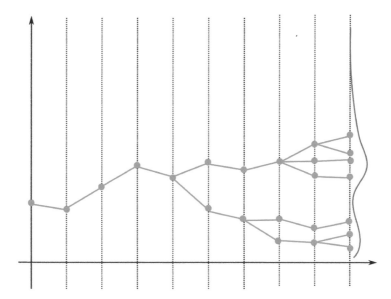

Running back in time, we can trace back the ancestral lines of every individual. The animation provides a graphical description of the genealogical tree associated with the dynamic population model.

It is out of the scope of these lectures to discuss the convergence of the occupation measures of these ancestral lines. We content ourselves by presenting the following important approximation result

$$\frac{1}{N} \sum_{1 \leq i \leq N} \delta_{i\text{-th ancestral line at time n}} \simeq_{N\uparrow\infty} \text{Law}\left((X_0, \ldots, X_n) \mid Y_p = y_p, \ p < n\right).$$

In addition, the product of the proportions of accepted particles converges as $N \uparrow \infty$ to the density of the observation sequence.

The genetic type particle model presented in (3.7) is often called a `particle filter`. These particle models can be interpreted as a mean field particle interpretation of Feynman-Kac models. These particle methodologies and their application domains are discussed in some details in section 9.6 and in section 9.9 (see for instance 9.9.2 for nonlinear filtering problems).

Tree of life

3.3 The lost equation

`Wolfgang Doeblin` was born in Berlin in March 17th 1915, and he spent his first three years in the German army during the World War I in Saargemünd with his father who volunteered as a physician. After the war, he moved back to Berlin, but the new Nazi army put him and his family on their black list. When he was 21 years old, he obtained the French nationality. After earning his PhD from the Sorbonne in 1938 with Maurice René Fréchet, he was enrolled with the French army in the World War II and was sent to the German front. During the next two wartime years, the young French-German mathematician wrote a series of works on probability theory. In the winter of 1940, he sent a sealed letter to M. Fréchet at the Académie of Sciences in Paris containing a new treatise on stochastic calculus. A few months later, in the little village Housseras, about 100 km from Sarreguemines, he decided to take his own life rather than be caught as a prisoner of war.

M. Fréchet forgot to open the letter, and died in June 1973 in Paris. The general rule of the French Académie of Sciences is that only the sender, his relatives, or the Academy of Sciences itself (after a period of 100 years), had the permission to open this sealed letter no. 11668.

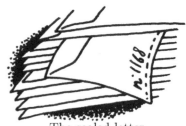

The sealed letter

In 1991, Bernard Bru, Professor of the History of Mathematics at the University René Descartes in Paris, was looking for information about Wolfgang's vitae. In the dusty archive he discovered the letter of W. Doeblin to M. Fréchet. Doeblin's brothers, Claude and Stephan, gave him permission to open the letter. Bernard found an exercise book with a handwritten article on a new stochastic calculus, rediscovered independently by the famous Japanese mathematician Kiyoshi Itō in 1944.

One of the central questions addressed by this stochastic calculus is to integrate random dynamical systems driven by the Brownian motion. This mysterious stochastic process was discovered incidentally in 1827 by the botanist Robert Brown. When analyzing the sexual relations of plants, he observed under his microscope the jittery motion of pollen grains of the plant *Clarkia pulchella*

when suspended in water. His first hypothesis was that these grains were the equivalent of a sperm and they were jiggling around just because they were alive. Then he made the same experiment with dead plants, and concluded that life exists at the microscopic level.

The earliest attempt to model Brownian motion can be traced to T. N. Thiele in 1880, and to Louis Bachelier in 1900 in his PhD thesis *The theory of speculation* [8], dedicated   to   the   stochastic   analysis   of   stock   option   markets   [255].

Some years later, in 1905 Albert Einstein [118] and Marian Smoluchowski [244] also used the fluctuation properties of this stochastic process to confirm in some way the existence of atoms and molecules in matter, gases or fluids. In their model, the chaotic behavior of these particles (such as pollen grains, coal dust, atoms, or gas molecules) results from successive collisions with particles of their environment.

The mathematical foundations of diffusion processes driven by Brownian motion

Kiyochi Itō

originates with the celebrated work of Kiyoshi
Itō published in Japanese in 1942. The citation from the US National Academy of Science
states:

*If one disqualifies the Pythagorean Theorem from contention, it is hard to think of a
mathematical result which is better known and more widely applied in the world today than
"Ito's Lemma". This result holds the same position in stochastic analysis that Newton's
fundamental theorem holds in classical analysis. That is, it is the sine qua non of the
subject.*

The following animation shows five Brownian particles colliding with a thousand of
smaller particles.

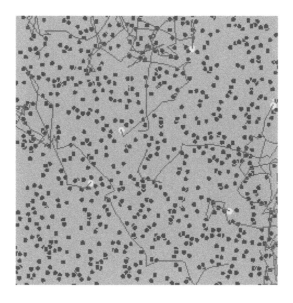

From a purely mathematical point of view, the Brownian process W_t is the continuous-
time analog of the simple random walk: a walker evolves in the real line performing randomly
local steps with amplitude $+dx = +\sqrt{dt}$ or $-dx = -\sqrt{dt}$ every dt units of time. When the
time step dt tends to 0, we obtain a random process W_t, indexed by the continuous time
parameter $t \in [0, \infty[$.

The next figure shows seven different and random realizations of this Brownian process.

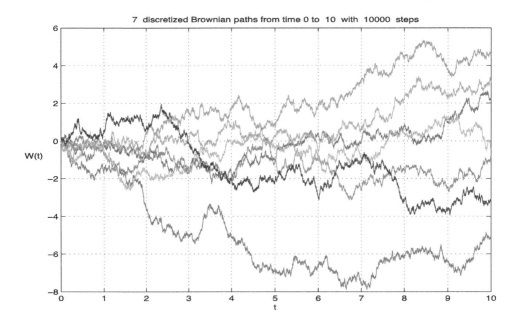

Extending these constructions, we define the following stochastic differential equations

$$dX_t := X_{t+dt} - X_t = b_t(X_t)\ dt + \ \sigma_t(X_t)\ dW_t \tag{3.9}$$

for some regular functions (b, σ).

Using the rules

$$
\begin{aligned}
dW_t \times dW_t &= \pm\sqrt{dt} \times \pm\sqrt{dt} = dt \\
dt \times dt &= 0 \quad \text{and} \quad dt \times dW_t = dt \ \times \pm\sqrt{dt} = 0
\end{aligned}
$$

for any smooth function f, we have the second order Taylor expansion

$$
\begin{aligned}
df(t, X_t) &= f(t + dt, X_t + dX_t) - f(t, X_t) \\
&= \partial_t f(t, X_t)\ dt + \partial_x f(t, X_t)\ dX_t + \frac{1}{2}\ \partial_x^2 f(t, X_t)\ dX_t\ dX_t \\
&= \Big[\partial_t f(t, X_t)\ + \partial_x f(t, X_t)\ b_t(X_t) \\
&\qquad + \frac{1}{2}\ \partial_x^2 f(t, X_t)\ \sigma_t^2(X_t)\ \Big]\ dt + \partial_x f(X_t)\ \sigma_t(X_t)\ dW_t.
\end{aligned}
$$

This yields the Doeblin-Itō differential formulae

$$df(t, X_t) = [\partial_t + L_t] f(t, X_t) \, dt + dM_t(f) \tag{3.10}$$

with the operator L_t and a remainder stochastic process $M_t(f)$ given by

$$L_t = b_t \, \partial_x + \frac{1}{2} \sigma_t^2 \, \partial_x^2 \quad \text{and} \quad dM_t(f) = \partial_x f(t, X_t) \, \sigma_t(X_t) \, dW_t. \tag{3.11}$$

For a more rigorous description of the Brownian diffusion processes, including a detailed presentation of the Doeblin-Itō differential calculus, we refer the reader to section 14.1.

By construction, we also have that

$$\mathbb{E}\left(dM_t(f) \mid X_t\right) = \partial_x f(X_t) \, \sigma_t(X_t) \overbrace{\mathbb{E}\left(dW_t \mid X_t\right) = 0}^{=0} \tag{3.12}$$

$$\mathbb{E}\left((dM_t(f))^2 \mid X_t\right) = (\partial_x f(X_t) \, \sigma_t(X_t))^2 \underbrace{\mathbb{E}\left((dW_t)^2 \mid X_t\right)}_{=1/2(\sqrt{dt})^2 + 1/2(-\sqrt{dt})^2 = dt}$$

$$= (\partial_x f(X_t) \, \sigma_t(X_t))^2 \, dt. \tag{3.13}$$

Without any doubt, this differential calculus is one of the most important tools of the modern theory of stochastic processes. Next, we present two more or less direct applications of this result.

When $\sigma_t = 0$, the process X_t reduces to the standard deterministic dynamical system

$$\frac{dX_t}{dt} = \dot{X}_t = b_t(X_t). \tag{3.14}$$

In this situation, for time homogeneous functions $f(t, x) = f(x)$, we have

$$df(X_t) = b_t(X_t) \, f'(X_t) \, dt \quad \text{and} \quad M_t(f) = 0.$$

When $b_t = 0$ and $\sigma_t = 1$, for time homogeneous functions $f(t, x) = f(x)$, we find that $X_t = W_t$ and

$$df(W_t) = \frac{1}{2} f''(W_t) \, dt + f'(W_t) \, dW_t.$$

This implies that

$$d\mathbb{E}\left(f(W_t)\right) = \mathbb{E}\left(f(W_{t+dt})\right) - \mathbb{E}\left(f(W_t)\right)$$

$$= \mathbb{E}\left(df(W_t)\right) = \frac{1}{2}\mathbb{E}\left(f''(W_t)\right) \, dt.$$

We let $p_t(w)$ be the probability density of the random variable W_t

$$\mathbb{P}\left(W_t \in dw\right) = p_t(w) \, dw \implies \mathbb{E}\left(f(W_t)\right) = \int f(w) \, p_t(w) \, dw.$$

From previous calculations, for any smooth function f with compact support, we have that

$$\int f(w) \, \frac{p_{t+dt}(w) - p_t(w)}{dt} \, dw = \frac{1}{2} \int f''(w) \, p_t(w) \, dw = \int f(w) \, \frac{1}{2} \, p_t''(w) \, dw.$$

The equality on the r.h.s. is proved using integration by parts.

We conclude that p_t satisfies the heat equation $\partial_t p_t = \frac{1}{2} \, \partial_w^2 p_t$, whose solution is given by the Gaussian density

$$p_t(w) = \frac{1}{\sqrt{2\pi t}} \, \exp\left(-\frac{w^2}{2t}\right).$$

Using the same line of arguments, the probability density $p_t(x)$ of the random states X_t of the general diffusion model presented in (3.9) satisfies the partial differential equation

$$\partial_t p_t = -\partial_x \left(b_t \, p_t\right) + \frac{1}{2} \, \partial_x^2 \left(\sigma_t^2 \, p_t\right). \tag{3.15}$$

The picture below illustrates the solution $p_t(x)$ of the Fokker-Planck equation (3.15) associated with the pure diffusion process starting at the origin

$$dX(t) = \sin(X(t))dt + \sigma \, dW(t).$$

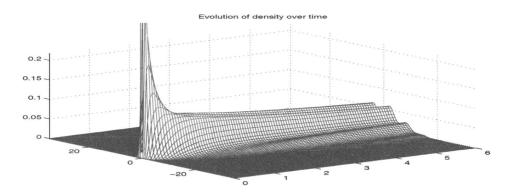

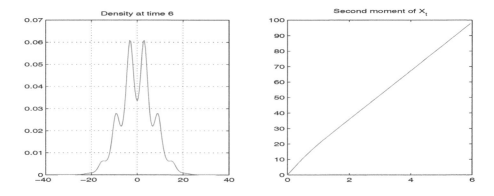

Sampling N independent copies of the diffusion model (3.9), we design the Monte Carlo approximation model

$$\frac{1}{N} \sum_{1 \le i \le N} f(X_t^i) \longrightarrow_{N\to\infty} \mathbb{E}\left(f(X_t)\right) = \int f(x) \, p_t(x) \, dx.$$

Next, we present another application of the second order differential formula. For spatially homogeneous functions $b_t(x) = b_t$ and $\sigma_t(x) = \sigma_t$, the diffusion model (3.9) takes the form

$$dX_t = b_t \, X_t \, dt + \sigma_t \, X_t \, dW_t. \tag{3.16}$$

If we take $f(x) = \log(x)$, we find that

$$d \log X_t = \frac{1}{X_t} \, dX_t - \frac{1}{2} \frac{1}{X_t^2} \, dX_t \, dX_t = b_t \, dt + \sigma_t \, dW_t - \frac{1}{2} \sigma_t^2 \, dt.$$

Integrating over the time interval $[0, t]$, we conclude that

$$\log X_t - \log X_0 = \int_0^t \left(b_s - \frac{1}{2} \sigma_s^2 \right) \, ds + \int_0^t \sigma_s dW.$$

Taking the exponential, we end up with the so-called geometric Brownian motion formula

$$X_t = X_0 \, \exp \left(\int_0^t \left(b_s - \frac{1}{2} \sigma_s^2 \right) \, ds + \int_0^t \sigma_s \, dW_s \right). \tag{3.17}$$

These non-negative stochastic processes are used in mathematical finance to represent the evolution of the stock price of a risky asset. The prototype financial model is defined in terms of some "reference" cash-flow process $\mathcal{S}_t^{(0)}$ with a given riskless return rate r_t, and a stochastic risky asset $\mathcal{S}_t$ with return rate b_t and volatility σ_t. These processes are often defined by the stochastic differential equations

$$\begin{cases} d\mathcal{S}_t^{(0)} &= \mathcal{S}_t^{(0)} \, r_t \, dt \\ d\mathcal{S}_t &= b_t \, \mathcal{S}_t \, dt + \sigma_t \, \mathcal{S}_t \, dW_t. \end{cases} \tag{3.18}$$

The deflated risky asset is given by

$$\begin{aligned} \overline{\mathcal{S}}_t &= \mathcal{S}_t / \mathcal{S}_t^{(0)} = e^{-\int_0^t r_s ds} \, \mathcal{S}_t / \mathcal{S}_t^{(0)} \\ &= \overline{\mathcal{S}}_0 \, \exp \left(\int_0^t \left([b_s - r_s] - \frac{1}{2} \sigma_s^2 \right) \, ds + \int_0^t \sigma_s \, dW_s \right). \end{aligned}$$

We notice that $\overline{\mathcal{S}}_t$ is the solution of the deflated stochastic differential equation

$$d\overline{\mathcal{S}}_t = [b_t - r_t] \, \overline{\mathcal{S}}_t \, dt + \sigma_t \, \overline{\mathcal{S}}_t \, dW_t. \tag{3.19}$$

The values $\mathcal{P}_t$ of a self-financing portfolio $\left(p_t^{(0)}, p_t \right)$ are defined by

$$\mathcal{P}_t = \underbrace{p_{t-dt}^{(0)} \, \mathcal{S}_t^{(0)} + p_{t-dt} \, \mathcal{S}_t}_{\text{value of the portfolio at time } t} = \underbrace{p_t^{(0)} \, \mathcal{S}_t^{(0)} + p_t \, \mathcal{S}_t}_{\text{choice of a (self financed) new strategy}}.$$

This implies that

$$\begin{aligned} d\mathcal{P}_t &= \mathcal{P}_{t+dt} - \mathcal{P}_t = \left[p_t^{(0)} \, \mathcal{S}_{t+dt}^{(0)} + p_t \, \mathcal{S}_{t+dt} \right] - \left[p_t^{(0)} \, \mathcal{S}_t^{(0)} + p_t \, \mathcal{S}_t \right] \\ &= p_t^{(0)} \left(\mathcal{S}_{t+dt}^{(0)} - \mathcal{S}_t^{(0)} \right) + p_t \, (\mathcal{S}_{t+dt} - \mathcal{S}_t) \\ &= p_t^{(0)} \, d\mathcal{S}_t^{(0)} + p_t \, d\mathcal{S}_t = \left[p_t^{(0)} \, r_t \, \mathcal{S}_t^{(0)} + p_t b_t \, \mathcal{S}_t \right] dt + p_t \sigma_t \, \mathcal{S}_t \, dW_t. \end{aligned}$$

When a security pays dividends or coupons with a return rate α_t, the self-financing portfolio is defined as above by replacing the term $d\mathcal{S}_t$ by

$$\begin{aligned} d\mathcal{S}_t^\alpha &:= d \left(e^{\int_0^t \alpha_s ds} \mathcal{S}_t \right) \\ &= \alpha_t \, \mathcal{S}_t^\alpha \, dt + e^{\int_0^t \alpha_s ds} \, d\mathcal{S}_t = [b_t + \alpha_t] \, \mathcal{S}_t^\alpha \, dt + \sigma_t \, \mathcal{S}_t^\alpha \, dW_t. \end{aligned}$$

We refer the reader to the end of section 18.2.3, as well as to chapter 30 for a more detailed discussion on the applications of these stochastic processes in pricing European financial options.

3.4   Towards a general theory

The theory of deterministic or stochastic differential equations (*abbreviated SDE*) plays a crucial role in pure and applied mathematics, as well as in physics, biology, finance, and engineering sciences. In (3.14) and (3.15) we have seen that standard deterministic dynamical systems and some classes of partial differential equations (*abbreviated PDE*) can be encapsulated in the Doeblin-Itō differential calculus.

The aim of this section is to extend this stochastic differential calculus to general stochastic processes, including jump type diffusions and integro-differential equations.

The square field vector (also called the "carré du champ operator") associated with some differential operator L is defined for any sufficiently smooth function f by the formula

$$f(x) \mapsto \Gamma_L(f,f)(x) = L([f - f(x)]^2)(x) = L(f^2)(x) - 2f(x)L(f)(x). \quad (3.20)$$

For instance, when $L = L_t$ is the operator defined in (3.11) we find that

$$\Gamma_{L_t}(f,f)(x) = b_t(x) \underbrace{\partial_y[f - f(x)]^2_{|\ y=x}}_{=0} + \sigma_t^2(x) \underbrace{\frac{1}{2} \ \partial_y^2[f - f(x)]^2_{|\ y=x}}_{=\partial_y([f-f(x)]\partial_y f)_{|\ y=x}=(\partial_x f(x))^2}$$

$$= [\sigma_t(x)\partial_x f(x)]^2. \quad (3.21)$$

If we set $\boldsymbol{X_t} := (X_s)_{s \le t}$ and $M_t := M_t(f)$, then by (3.13) we find that

$$\mathbb{E}(dM_t \mid \boldsymbol{X_t}) = 0 \quad (3.22)$$

$$\mathbb{E}\left((dM_t)^2 \mid \boldsymbol{X_t}\right) = \Gamma_{L_t}(f,f)(X_t) \ dt.$$

The first condition ensures that the stochastic process M_t is a martingale w.r.t. the information encapsulated in the increasing sequence of sigma-algebras $\mathcal{F}_t := \sigma(\boldsymbol{X_t}) := \sigma\left((X_s)_{s \le t}\right)$.

In addition, we have

$$\mathbb{E}\left((dM_t)^2 \mid \boldsymbol{X_t}\right) = \mathbb{E}\left((M_{t+dt} - M_t)^2 \mid \boldsymbol{X_t}\right)$$

$$= \mathbb{E}\left(M_{t+dt}^2 \mid \boldsymbol{X_t}\right) - 2M_t \, \mathbb{E}\left(M_{t+dt} \mid \boldsymbol{X_t}\right) + M_t^2$$

$$= \mathbb{E}\left(M_{t+dt}^2 \mid \boldsymbol{X_t}\right) - M_t^2 \quad \text{(by (3.12))}$$

$$= \mathbb{E}\left(M_{t+dt}^2 - M_t^2 \mid \boldsymbol{X_t}\right) = \mathbb{E}\left(dM_t^2 \mid \boldsymbol{X_t}\right)$$

with the increments of the square process $dM_t^2 := M_{t+dt}^2 - M_t^2$. This shows that the stochastic process

$$\overline{M}_t := M_t^2 - \int_0^t \Gamma_{L_t}(f, f)(X_t) \, dt$$

has conditionally centered increments

$$d\overline{M}_t = dM_t^2 - \mathbb{E}\left(dM_t^2 \mid \boldsymbol{X_t}\right) \implies \mathbb{E}\left(d\overline{M}_t \mid \boldsymbol{X_t}\right) = 0.$$

In this notation, the Doeblin-Itō differential formulae (3.10) can be rewritten as

$$df(t, X_t) = [\partial_t + L_t] f(t, X_t) \, dt + dM_t(f) \tag{3.23}$$

with the differential operator L_t defined in (3.11) and

$$\begin{aligned} \mathbb{E}\left(dM_t(f) \mid \boldsymbol{X_t}\right) &= 0 \\ \mathbb{E}\left(dM_t(f)^2 \mid \boldsymbol{X_t}\right) &= \Gamma_{L_t}(f, f)(X_t) \, dt. \end{aligned}$$

It may happen that the process X_t jumps at some random times $T = t$ to some locations $X_{t-dt} = x \rightsquigarrow X_t = y$, randomly chosen w.r.t. some probability measure $M_t(x, dy)$. We denote by ΔX_t the jump increment given by

$$\Delta X_t := X_t - X_{t-dt} \quad \text{or by the limiting object} \quad \Delta X_t := X_t - X_{t-}.$$

The jump times T_n, with $n \geq 1$, are defined sequentially for any $n \geq 1$ by some non-negative rate function $\lambda_t(x)$ with the formula

$$T_n := \inf\left\{t > T_{n-1} \text{ s.t. } \exp\left(-\int_{T_{n-1}}^t \lambda_s(X_s) \, ds\right) \leq U_n\right\}$$

with $T_0 = 0$, where U_n stands for a sequence of independent and uniform random numbers on $[0, 1]$. Formally, on a dt-time mesh, the n-th jump arrives between time t and $t + dt$ as soon as we have

$$\left(\prod_{T_{n-1} \leq s < t} e^{-\lambda_s(X_s) \, ds}\right) e^{-\lambda_t(X_t) \, dt} \leq U_n < \left(\prod_{T_{n-1} \leq s < t} e^{-\lambda_s(X_s) \, ds}\right).$$

Given the process evolution before the jump, the chance for this to happen is given by the geometric distribution

$$\left(\prod_{T_{n-1} \leq s < t} e^{-\lambda_s(X_s) \, ds}\right) \underbrace{\left(1 - e^{-\lambda_t(X_t) \, dt}\right)}_{\text{probability of a jump between } t \text{ and } t + dt \simeq \lambda_t(X_t) \, dt}$$

These stochastic processes are often called jump-diffusion processes.

In this situation, formula (3.23) remains valid with the integro-differential operator

$$L_t(f)(x) = L_t^{diff}(f)(x) + L_t^{jump}(f)(x),$$

with the diffusion operator

$$L_t^{diff} = b_t \, \partial_x \; + \frac{1}{2} \; \sigma_t^2 \; \partial_x^2,$$

and the jump integral operator

$$L_t^{jump}(f)(x) = \lambda_t(x) \int \; (f(y) - f(x)) \; M_t(x, dy).$$

Next, we provide a formal proof of this assertion and a more explicit description of the generator L_t and of the remainder term $M_t(f)$ in the Doeblin-Itō formula (3.23). We use the Taylor expansion

$$
\begin{aligned}
df(t, X_t) &= f(t + dt, X_t + dX_t) - f(t, X_t) \\
&= \partial_t f(t, X_t) \, dt + \partial_x f(t, X_t) \, dX_t + \frac{1}{2} \; \partial_x^2 f(t, X_t) \, dX_t \, dX_t \\
&\quad + \Delta f(t, X_t) - \left[\partial_x f(t, X_t) \, \Delta X_t + \frac{1}{2} \; \partial_x^2 f(t, X_t) \, \Delta X_t \, \Delta X_t \right].
\end{aligned}
$$

The r.h.s. term arises from the fact that

$$dX_t = dX_t^c + \Delta X_t$$

with the jump term ΔX_t, and the increment

$$dX_t^c := b_t(X_t) \, dt + \sigma_t(X_t) \, dW_t.$$

The above decomposition means that $dX_t = dX_t^c$ between the jumps and $dX_t = \Delta X_t$ as soon as a jump occurs.

Using the new set of rules

$$dt \times \Delta X_t = 0 \quad \text{and} \quad dW_t \times \Delta X_t = 0$$

we check that the Taylor expansion stated above is the same as the one discussed earlier up to an addition jump term

$$df(t, X_t) = \partial_t f(t, X_t) \, dt + \partial_x f(t, X_t) \, dX_t^c + \frac{1}{2} \; \partial_x^2 f(t, X_t) \, dX_t^c \, dX_t^c + \Delta f(t, X_t)$$

with the jump increment

$$\Delta f(t, X_t) := f(t, X_t) - f(t-, X_{t-})$$

of the function/process $f(t, X_t)$.

Finally, we observe that

$$\Delta f(t, X_t) = \mathbb{E} \left(\Delta f(t, X_t) \mid \mathbf{X_t} \right) + dM_t^d(f)$$

with a centered increment

$$dM_t^d(f) := \Delta f(t, X_t) - \mathbb{E}\left(\Delta f(t, X_t) \mid \boldsymbol{X_t}\right)$$

and

$$\begin{aligned}
\mathbb{E}\left(\Delta f(t, X_t) \mid \boldsymbol{X_t}\right) &= \mathbb{E}\left(\left[f(t, X_t + \Delta X_t) - f(t, X_t)\right] \mid \boldsymbol{X_t}\right) \\
&= \lambda_t(X_t) \, dt \int \left(f(t, y) - f(t, x)\right) M_t(x, dy).
\end{aligned}$$

The last assertion comes from the fact that the probability that the jump occurs between times t and $t + dt$ is given by $\lambda_t(X_t) \, dt$. We conclude that the remainder term $dM_t(f)$ in (3.23) is given by

$$dM_t(f) = dM_t^d(f) + \frac{\partial f}{\partial x}(t, X_t) \, \sigma_t(X_t) \, dW_t.$$

We end this section with a direct application of this formula for jump-diffusion processes of the form

$$dX_t = b_t \, X_t \, dt + \sigma_t \, X_t \, dW_t + c_t \, X_{t-} \, dN_t$$

where N_t stands for a unit jump process with $dN_t = \Delta N_t = 1$ at rate $\lambda_t \, dt$. These processes are called Poisson processes with unit rate.

If we take $f(x) = \log(x)$, we find that

$$d\log X_t = \frac{1}{X_t} \, dX_t^c - \frac{1}{2} \frac{1}{X_t^2} \, dX_t^c \, dX_t^c + \Delta \log X_t.$$

Now, we notice that

$$\Delta \log X_t = \log\left(X_{t-} + \Delta X_t\right) - \log X_{t-} = \log\left(1 + \frac{\Delta X_t}{X_{t-}}\right) = \log\left(1 + c_t\right) \, dN_t.$$

This implies that

$$d\log X_t = b_t \, dt + \sigma_t \, dW_t - \frac{1}{2} \sigma_t^2 \, dt + \log\left(1 + c_t\right) \, dN_t.$$

Integrating over the time interval $[0, t]$, we conclude that

$$\log X_t - \log X_0 = \int_0^t \left(b_s - \frac{1}{2} \sigma_s^2\right) \, ds + \int_0^t \sigma_s dW_s + \int_0^t \log\left(1 + c_s\right) \, dN_s.$$

Taking the exponential, we end up with the so-called geometric Brownian-Poisson motion formula

$$X_t = X_0 \, \exp\left(\int_0^t \left(b_s - \frac{1}{2} \sigma_s^2\right) \, ds + \int_0^t \sigma_s \, dW_s + \int_0^t \log\left(1 + c_s\right) \, dN_s\right).$$

3.5 The theory of speculation

Without any doubt, the French mathematician Louis Bachelier is the father of financial mathematics and the theory of speculation.

As we mentioned in *the lost equation* section, L. Bachelier is also one of the founders of the study of Brownian motion and of stochastic differential equations. In his PhD thesis in 1900, he already presented the link between Markov processes and partial differential equations (a.k.a. the Fokker-Planck

Louis Bachelier

equations), as well as the connections between the heat equation and the Brownian motion. His PhD thesis also discusses the option pricing problem and ends with a formula very close to the Nobel Prize winning solution of the option pricing problem by Fischer Black, Myron Scholes and Robert Merton in 1997 (a.k.a. the Black-Scholes, the Black-Scholes-Merton, or simply the Midas equation).

Unfortunately, the mathematical standards of PhDs in France and more particularly at La Sorbonne University were more focused on pure mathematics such as functional analysis and integration theory in the spirit of Emile Borel (1871-1956), René Louis Baire (1874-1932), and Henri Lebesgue (1875-1941). The PhD thesis of L. Bachelier only received the grade of *honorable*, but not the mark of *très honorable*. Probability theory became a mathematical discipline in France only after 1925, and it is still underestimated by several "pure" mathematicians. The rather positive report of Henri Poincaré (1854-1912) on Louis Bachelier's thesis illustrates the French viewpoint of probability and its applications:

"The subject chosen by Mr. Bachelier is somewhat removed from those which are normally dealt with by our applicants. His thesis is entitled "Theory of Speculation" and focuses on the application of probability to the stock market. First, one may fear that the author had exaggerated the applicability of probability as is often done. Fortunately, this is not the case. In his introduction and further in the paragraph entitled "Probability in Stock Exchange Operations", he strives to set limits within which one can legitimately apply this type of reasoning. He does not exaggerate the range of his results, and I do not think he is deceived by his formulas." (taken from [255])

Henri Poincaré

Nowadays, all of these results can be put in a better shape. As in most problems in probability theory, the key idea is to design a judicious martingale. To this end, we return to the Doeblin-Itō differential formulae (3.23). An alternative way to define the generator L_t is to check that for time homogeneous functions $f(t, x) = f(x)$,

$$\mathbb{E}\left(\, df(X_t) \mid X_t \right) = \mathbb{E}\left(\, f(X_{t+dt}) - f(X_t) \mid X_t \right) = L_t(f)(X_t)\, dt. \tag{3.24}$$

In terms of the conditional expectations

$$\forall s \leq t \qquad P_{s,t}(f)(x) = \mathbb{E}\left(f(X_t) \mid X_s = x\right)$$

we have

$$\mathbb{E}\left(\, f(X_{t+dt}) - f(X_t) \mid X_t = x\right) = P_{t,t+dt}(f)(x) - f(x) = L_t(f)(x)\, dt. \qquad (3.25)$$

We emphasize that the properties of conditional expectations

$$
\begin{aligned}
P_{r,t}(f)(X_r) \;&=\; \mathbb{E}\left(f(X_t) \mid X_r\right) \\[2mm]
&=\; \mathbb{E}\left(\underbrace{\mathbb{E}\left(f(X_t) \mid X_s\right)}_{=P_{s,t}(f)(X_s)} \;\Big|\; X_r \right) = P_{r,s}(P_{s,t}(f))(X_r)
\end{aligned}
$$

for any $0 \leq r \leq s \leq t$, translate into the the following property of the integral operators

$$P_{r,t} = P_{r,s}P_{s,t}.$$

In probability theory, this is called the semigroup property of the operators $P_{s,t}$.

In terms of these operators, we can rewrite the equation (3.25) in terms of the forward or backward formulae

$$\frac{1}{dt}\left[P_{t,t+dt} - Id\right] = L_t \quad \text{and/or} \quad \frac{1}{dt}\left[Id - P_{t-dt,t}\right] = L_t.$$

We fix a time horizon T, and some function $f_t(x)$ and we set

$$\forall t \in [0,T] \qquad g_t(X_t) := P_{t,T}(f_T)(X_t) = \mathbb{E}\left(f_T(X_T) \mid X_t\right).$$

Combining the Doeblin-Itō differential formulae with

$$
\begin{aligned}
\partial_t g(x) \;&=\; \frac{1}{dt}\left[P_{t,T}(f_T)(x) - P_{t-dt,T}(f_T)\right](x) \\[2mm]
&=\; \frac{1}{dt}\left[Id - P_{t-dt,t}\right](P_{t,T}(f_T))(x) = -L_t(g_t)
\end{aligned}
$$

we conclude that

$$dg(t,X_t) = dM_t(g) = \partial_x g(t,X_t)\, \sigma_t(X_t)\, dW_t.$$

In summary, we have proved the following result.

> For any fixed time horizon T, and for any function $f_t(x)$, the stochastic process $g(t,X_t)$ defined for any $t \in [0,T]$ by
>
> $$
> \begin{aligned}
> g_t(X_t) \;&:=\; P_{t,T}(f_T)(X_t) = \mathbb{E}\left(f_T(X_T) \mid X_t\right) \\[2mm]
> &=\; g_0(X_0) + \int_0^t \partial_x g(s,X_s)\, \sigma_s(X_s)\, dW_s \qquad (3.26)
> \end{aligned}
> $$
>
> is a martingale ending at $g_T(X_T) = f_T(X_T)$.

The fundamental theorem of financial mathematics states that the market is tradeable (i.e., absence of arbitrage opportunities) if and only if the deflated risky assets are martingales w.r.t. some probability measure. This is called the risk neutral measure.

More precisely, returning to the financial model discussed in (3.18), the deflated risky assets (3.19) are martingales if, and only if $b_t = r_t$. For time homogeneous models $(r_t, \sigma_t) = (r, \sigma)$, these deflated assets are given by the following SDE

$$d\overline{S}_t = \sigma\,\overline{S}_t\,dW_t \Rightarrow \mathbb{E}\left(d\overline{S}_t \mid \overline{S}_t\right) = 0.$$

The value of a self financing portfolio with the number $(\phi_{t-dt}, \psi_{t-dt})$ of assets $(S_t, S_t^{(0)})$ given by the SDE (3.18) is given by the equation

$$V(t, S_t) = \phi_{t-dt}\,S_t + \psi_{t-dt}\,S_t^{(0)}.$$

By the self financing property, before seeing the next value of the asset S_{t+dt} we choose a new number ϕ_{t-dt} of risky assets S_t

$$V(t, S_t) = \phi_t\,S_t + \psi_t\,S_t^{(0)} \quad \text{and} \quad \psi_t = e^{-rt}[V(t, S_t) - \psi_t\,S_t].$$

This implies that the wealth increments of the portfolio are defined by

$$\begin{aligned}
dV(t, S_t) &= V(t+dt, S_{t+dt}) - V(t, S_t) \\
&= \left[\phi_t\,S_{t+dt} + \psi_t\,S_{t+dt}^{(0)}\right] - \left[\phi_t\,S_t + \psi_t\,S_t^{(0)}\right] \\
&= \phi_t\,dS_t + \psi_t\,dS_t^{(0)}.
\end{aligned}$$

Assuming that the reference cash flow risk free asset starts at one unit, we have

$$e^{-rt}\,V(t, S_t) = \phi_{t-dt}\,\overline{S}_t + \psi_{t-dt} = \phi_t\,\overline{S}_t + \psi_t$$

and

$$e^{-r(t+dt)}\,V(t+dt, S_{t+dt}) = \phi_t\,\overline{S}_{t+dt} + \psi_t.$$

In other words, in terms of the deflated portfolio we have

$$\overline{V}(t, \overline{S}_t) := e^{-rt}\,V(t, e^{rt}\,S_t) = e^{-rt}\,V(t, \overline{S}_t) \Rightarrow d\overline{V}(t, \overline{S}_t) = \phi_t\,d\overline{S}_t.$$

An option is a security contract that allows the owner to trade the shares of some stock at a given fixed price (a.k.a. the strike) at a given terminal time (a.k.a. the maturity or expiration date). A call option gives the right to buy shares, while a put option gives the right to sell shares. At the expiration date, the owner of a Call can buy the shares at a lower price than the one given by market at that time. In the same way, the owner of a put can sell the shares at a higher price than the one proposed by the financial market at that time. These transactions are referred to as exercising the call or the put option.

The payoff function of a Call with strike K, with an expiration date T is given by

$$f_T(\overline{S}_T) := \left(\overline{S}_T - e^{-rT}K\right)^+ = e^{-rT}\left(S_T - K\right)^+. \tag{3.27}$$

In view of (3.26), a natural strategy is to use a martingale process

$$\begin{aligned}
\overline{V}(t, \overline{S}_t) &= \overline{P}_{t,T}(f_T)(\overline{S}_t) = \mathbb{E}\left(\left(\overline{S}_T - e^{-rT}K\right)^+ \mid \overline{S}_t\right) \\
&= \underbrace{\overline{P}_{0,T}(f_T)(\overline{S}_0)}_{\overline{V}(0, \overline{S}_0)} + \int_0^t \underbrace{\partial_x\overline{P}_{s,T}(f_T)(\overline{S}_s)}_{:=\phi_s}\,d\overline{S}_s
\end{aligned}$$

ending at $f_T(\overline{S}_T)$, with the semigroup $\overline{P}_{t,T}$ of the deflated risky asset price $\overline{S}_t$.

The arbitrage free option price is given by the initial value of the deflated portfolio

$$\overline{V}(0, \overline{S}_0) = \overline{P}_{0,T}(f_T)(\overline{S}_0) = \mathbb{E}\left(e^{-rT}\ (\mathcal{S}_T - K)^+ \mid \overline{S}_0\right) := C_T(\overline{S}_0, K)$$

and the number of risky assets (a.k.a. edging strategies) in the replicating portfolio is given by the derivative of the option prices

$$\phi_s = \partial_x \overline{P}_{s,T}(f_T)(\overline{S}_s).$$

In a final step, we recall that the risky asset $\overline{S}_t$ is given by the geometric Brownian motion

$$\overline{S}_t := \overline{S}_0 \, \exp\left(\sigma W_t - \frac{\sigma^2 t}{2}\right).$$

In addition, for any Gaussian r.v. V with mean m and variance τ^2, and any function φ we have the change of variable formula

$$\mathbb{E}\left(e^V \varphi(V)\right) = e^{m + \frac{\tau^2}{2}} \, \mathbb{E}\left(\varphi(V + \tau^2)\right). \tag{3.28}$$

We check this claim using the elementary formula

$$-\frac{1}{2\tau^2}\ (v - m)^2 + v = -\frac{1}{2\tau^2}\ \left(v - (m + \tau^2)\right)^2 \ + \ \left(m + \frac{\tau^2}{2}\right).$$

Using (3.28), we find the `Black-Scholes-Merton` formula

$$C_T(x, K) \quad = \quad x\, G\left(d_{T,K}^{(1)}(x)\right) - e^{-rT} K\, G\left(d_{T,K}^{(2)}(x)\right)$$

with $G(y) = \frac{1}{\sqrt{2\pi}} \int_{-\infty}^y e^{-x^2/2}\, dx$, and the functions

$$d_{T,K}^{(1)}(x) \quad = \quad \frac{1}{\sigma\sqrt{T}} \left[\log\left(\frac{x}{K}\right) + \left(r + \frac{\sigma^2}{2}\right) T\right]$$
$$d_{T,K}^{(2)}(x)) \quad = \quad d_{T,K}^{(1)}(x) - \sigma\sqrt{T}.$$

The hedging strategies are easily computed using the following differential formula

$$\frac{\partial}{\partial x} C_T(x, K) \quad = \quad G\left(d_{T,K}^{(1)}(x)\right) + \frac{1}{\sigma\sqrt{T}} \left[g\left(d_{T,K}^{(1)}(x)\right) - \frac{e^{-rT} K}{x} g\left(d_{T,K}^{(2)}(x)\right)\right]$$

with the Gaussian density $g(x) = G'(x) = \frac{1}{\sqrt{2\pi}} e^{-x^2/2}$.

We illustrate these formulae with a numerical example. We assume that the current price of shares of a company is $\mathcal{S}_0 = \$100$, and you would like to get a call option that allows you to purchase one share of this company stock for $K = \$90$. The standard deviation of the daily logarithmic stocks return is 1%, and the annual return of the risk-free stock is 4%.

The next graph provides a description of the call option prices for different values of the strike and the maturity.

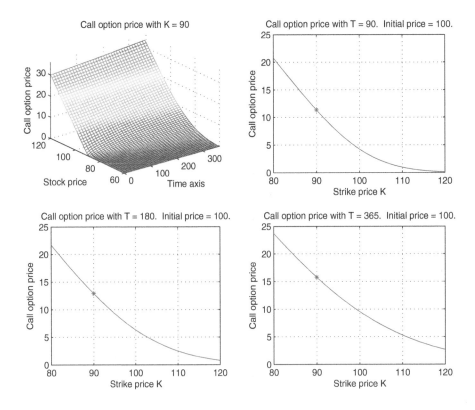

Using these graphs, we find that the price of the call is \$11.355 for $T = 90$ days, \$12.92 for $T = 180$ days, and \$15.76 for $T = 365$ days.

We can compare these values with the `Black-Scholes calculator of the Canadian Economic Research Institute`; \$11.3927 for $T = 90$ days, \$12.9899 for $T = 180$ days, and \$15.8084 for $T = 365$ days.

The next graph provides a description of a replicating portfolio.

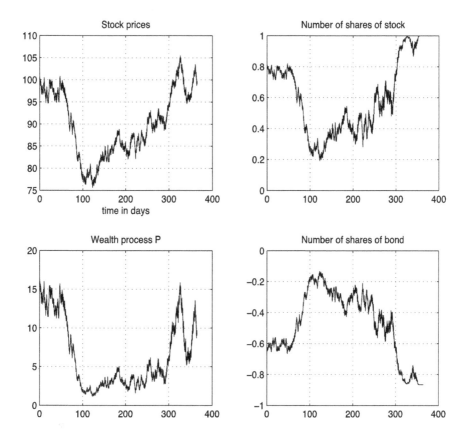

The wealth process represents the values of the portfolio starting with the price of the call option and terminating on the payoff function. When the price of the risky asset goes above the strike, we notice that the number of shares of the risky asset increases. In this situation, the value of the portfolio also increases to match the final payoff function. Conversely, when the market price of the risky asset goes below the strike, the value of the replicating portfolio decreases, and the number of risk free bounds increases. At the end of the year, the value of the portfolio $V(365, S_{365}) \simeq \$8$ coincides with the "realization" payoff function $\left(\$100 \; e^{\sigma W_{365} + 365\left(r - \frac{\sigma^2}{2}\right)} - \$90 \right)^{+} = \$(98 - 90)$.

More details on these derivations are provided in section 30.2.7, and in section 30.2.8. We also refer the interested reader to the NASDAQ Option Trading Guide, for more information on trading strategies.

3.6 Exercises

Exercise 21 (Gaussian reversible moves) *We consider a probability measure on $\mathbb{R}$ with a density $\pi(x)$ (w.r.t. the Lebesgue measure dx) of the form*

$$\pi(x) = \frac{1}{\mathcal{Z}} \; e^{-\beta V(x)} \; \lambda(x) \quad \text{with} \quad \lambda(x) = \frac{1}{\sqrt{2\pi}} \; e^{-x^2/2},$$

where $\beta > 0$ stands for a given parameter and V is a positive function on $\mathbb{R}$. We consider the Metropolis-Hastings algorithm discussed in section 3.1 with a proposal density

$$P(x, y) = \frac{1}{\sqrt{2\pi\epsilon}} \; \exp\left[-\frac{1}{2\epsilon} \; \left(y - \sqrt{(1-\epsilon)}x \right)^2 \right]$$

for some $\epsilon \in]0, 1]$. Check that

$$\lambda(x)P(x, y) = \lambda(y)P(y, x)$$

for any $x, y \in \mathbb{R}$ and describe the acceptance rate of the corresponding Metropolis-Hastings algorithm. Answer the same questions when $\lambda(x) = 1$ and $P(x, y) = \frac{1}{\sqrt{2\pi}} \exp\left[-\frac{1}{2} \; (y - x)^2 \right]$.

Exercise 22 (Metropolis-Hastings - Simple random walk) *We consider the Markov transition $P(x, x+1) = 1/2 = P(x, x-1)$ on the integer lattice $\mathbb{N} - \{0\}$, with $P(0, 0) = 1/2$. We let $\pi(x) = e^{-\lambda} \; \lambda^x / x!$ be the Poisson distribution on $x \in \mathbb{N}$ with parameter $\lambda > 0$. Describe the transitions of the Metropolis-Hastings algorithm with proposal transition P (also called the instrumental transition) and with target distribution π.*

Exercise 23 (Truncated Markov chain) *We consider a π-reversible Markov transition $M(i, j)$ on a finite set $S = \{1, \ldots, d\}$ w.r.t. some probability distribution π, for some given $d \geq 1$. We let $d' \leq d$ and we consider the truncated Markov transition $M'(i, j)$ on $S' = \{1, \ldots, d'\}$ defined by*

$$M'(i, j) = \begin{cases} M(i, j) & \text{for} & 0 \leq i \neq j \leq d' \\ M(i, j) + \sum_{k > d'} M(i, k) & \text{for} & 0 \leq i = j \leq d' \\ 0 & \text{otherwise} . \end{cases}$$

Check that M' is π' reversible w.r.t. the probability measure

$$\forall i \in S' \qquad \pi'(i) := \pi(i) / \sum_{j \in S'} \pi(j).$$

Exercise 24 (Doubly stochastic matrix) *We consider a Markov transition $M(x,y)$ on a finite set S such that $\sum_{z \in S} M(z,y) = 1 = \sum_{z \in S} M(x,z)$, for any $x,y \in S$. These are called doubly stochastic matrices. Check that the uniform measure on S is an invariant probability measure of M, that is, $\pi = \pi M$.*

Exercise 25 ✎ *We consider the filtering problem*

$$\begin{cases} X_n = X_{n-1} + W_n \\ Y_n = X_n + V_n \end{cases} \tag{3.29}$$

with some some independent Gaussian r.v. (X_0, W_n, V_n) with mean and variance parameters given by $(m_0, 0, 0)$ and $(\sigma_0, \sigma, \tau) \in]0, \infty[^3$, respectively, and with $\sigma_0 \geq \sigma$. We let m_n and m_n' be the solutions of the Kalman recursion (3.6) starting at m_0 and at some erroneous initial condition m_0'. Prove that

$$m_0 \geq m_0' \implies \forall n \geq 0 \quad m_n \geq m_n'$$

and

$$|m_n - m_n'| \leq \left(\frac{\tau^2}{\tau^2 + \sigma^2} \right)^n |m_0 - m_0'| \longrightarrow_{n \uparrow \infty} 0.$$

Exercise 26 *Describe the evolution of the particle filter (3.8) associated to the filtering problem (3.29).*

Exercise 27 *Applying the Doeblin-Itō formula to the function $f(X_t) = \log X_t$ prove that the solution of diffusion equation (3.16) is given by the exponential formula (3.17).*

Exercise 28 *We let W_t be the Brownian process (starting at the origin) introduced in section 3.3. Check that*

$$W_t^2 = 2 \int_0^t W_s \, dW_s + t \quad and \quad \mathbb{E}(W_t^2) = t.$$

Exercise 29 *We let W_t be the Brownian process (starting at the origin) introduced in section 3.3. Check that*

$$W_t^4 = 4 \int_0^t W_s^3 \, dW_s + 6 \int_0^t W_s^2 \, ds \quad and \quad \mathbb{E}(W_t^4) = 3t^2 = 6 \, \mathbb{E}\left(\int_0^t W_s^2 \, ds \right).$$

Exercise 30 *Consider the diffusion process X_t defined in (3.16). Using (3.17) check that for any $\alpha \in \mathbb{R}$ we have*

$$\mathbb{E}(X_t^\alpha) = \mathbb{E}(X_0^\alpha) \, \exp\left(\alpha \int_0^t b_s \, ds + \frac{\alpha(\alpha - 1)}{2} \int_0^t \sigma_s^2 \, ds \right).$$

Exercise 31 *Consider a jump-diffusion process X_t defined on page 57, and let W_t be a (standard) Brownian motion. Check that*

$$Y_t = \exp\left(\int_0^t \sigma_s(X_s) \, dW_s \right) \implies \frac{dY_t}{Y_t} = \frac{1}{2} \, \sigma_t^2(X_t) \, dt + \sigma_t(X_t) \, dW_t$$

and

$$Z_t = \exp\left(\int_0^t \sigma_s(X_s) \, dW_s - \frac{1}{2} \int_0^t \sigma_s^2(X_s) \, ds \right) \implies \frac{dZ_t}{Z_t} = \sigma_t(X_t) \, dW_t.$$

Exercise 32 ✎ *Let W_t be a (standard) Brownian motion. Consider the diffusion process defined by*

$$dX_t = (a_t + b_t \, X_t) \ dt + \ (\tau_t + \sigma_t \, X_t) \ dW_t$$

for some functions a_t, b_t, τ_t and σ_t. For any $n \in \mathbb{N}$, we set $m_t^n := \mathbb{E}(X_t^n)$. Check the recursive equations

$$\frac{dm_t^n}{dt} = n \, m_t^n \left(b_t + \frac{(n-1)}{2} \, \sigma_t^2 \right) + n \, m_t^{n-1} \, (a_t + (n-1)\tau_t\sigma_t) + \frac{n(n-1)}{2} \, m_t^{n-2} \, \tau_t^2$$

for any $n \geq 2$.

Exercise 33 *Check that the density $p_t(x)$ of the random states X_t of the general diffusion model presented in (3.9) satisfies the partial differential equation given in (3.17).*

Exercise 34 ✎✎ *A random variable Y is log-normal with parameters (μ, σ^2) if $Y = e^X$ with X being standard normal. Let $G(.)$ be the cumulative distribution function of the standard normal distribution. Show that for any constant $K > 0$ it holds:*

$$\mathbb{E}\left[(Y - K)^+ \right] = e^{\mu + \sigma^2/2} \, G\left(\frac{\mu + \sigma^2 - \log(K)}{\sigma} \right) - K \, G\left(\frac{\mu - \log(K)}{\sigma} \right).$$

Part II

Stochastic simulation

4

Simulation toolbox

Part II of the book is concerned with stochastic simulation. In this first, we start with the classical inversion technique for simulation. This is followed by other popular methods such as the change of variable and the rejection techniques, and several sampling techniques of probability measures over a finite set. Most of the contemporary simulation methods are employed to sample conditional distributions and for this reason, we review conditional probabilities next. The conditioning principles are then applied in formulating some basic properties of spatial Poisson point processes. We refer to later chapters for more detailed discussions.

The generation of random numbers is too important to be left to chance.
Robert R. Coveyou (1915-1996)

4.1   Inversion technique

The inversion simulation technique was introduced in 1947 by John Von Neumann [113]. We consider a real valued r.v. with a distribution function given by $F(x) = \mathbb{P}(X \leq x) \in [0, 1]$. The generalized inverse of F is given by

$$F^{-1}(u) = \inf \{x \ : \ F(x) \geq u\}.$$

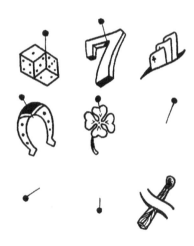

For strictly increasing functions F, F^{-1} coincides with the traditional inverse of the function F. The inversion technique is based on a single sample of a uniform r.v. U on $[0, 1]$, abbreviated $U \sim \text{Unif}([0, 1])$.

Theorem 4.1.1 *For any* $U \sim \text{Unif}([0,1])$, *we have*

$$\text{Law}(F^{-1}(U)) = \text{Law}(X) \quad \textit{and} \quad \text{Law}(F(X)) = \text{Law}(U).$$

Hint of proof :
We only prove the theorem for strictly increasing functions F. In this situation we have

$$\mathbb{P}\left(F^{-1}(U) \leq x\right) = \mathbb{P}\left(U \leq F(x)\right) = F(x) = \mathbb{P}\left(X \leq x\right).$$

Inversely, we have

$$\mathbb{P}\left(F(X) \leq u\right) = \mathbb{P}\left(X \leq F^{-1}(u)\right) = F(F^{-1}(u)) = u = \mathbb{P}\left(U \leq u\right).$$

This clearly ends the proof of the theorem.

One way to sample a sequence of independent copies $(X_p)_{1 \le i \le n}$ of the r.v. X is to set

$$X_1 = F^{-1}(U_1), \ldots, X_n = F^{-1}(U_n)$$

where $(U_p)_{1 \le i \le n}$ stands for a sequence of uniform and independent r.v. on $[0, 1]$. A schematic of this inversion sampling scheme is provided in figure 4.1. Inversely, if we have a sequence of independent copies $(X_p)_{1 \le i \le n}$ of the r.v. X, then the sequence

$$U_1 = F(X_1), \ldots, U_n = F(X_n)$$

forms a sequence of uniform and independent r.v. on $[0, 1]$. This property is sometimes used in statistics to check whether a sequence of data is distributed according to some hypothetical distribution.

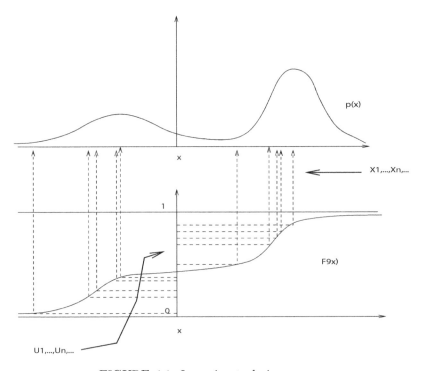

FIGURE 4.1: Inversion technique

- **Exponential distribution :** An exponential r.v. X with parameter $\lambda > 0$ is defined by

$$\mathbb{P}(X \in dx) = \lambda \, e^{-\lambda x} \, 1_{[0,\infty[}(x) \, dx.$$

We often write in a more synthetic way $X \sim \text{Exp}(\lambda)$. A simple calculation gives

$$\forall x \ge 0 \qquad F(x) = \mathbb{P}(X \le x) = -\int_0^x \frac{\partial}{\partial y}\left(e^{-\lambda y}\right) \, dy = 1 - e^{-\lambda x}.$$

This implies that

$$F^{-1}(U) = -\frac{1}{\lambda} \log\left(1 - U\right) \overset{law}{=} -\frac{1}{\lambda} \log U \tag{4.1}$$

where U stands for a uniform r.v. on $]0, 1[$.

- **Discrete distribution :** We consider r.v. X taking a finite number of values $x_i \in \mathbb{R}$, with probability p_i, with $1 \le i \le d$, for some $d \ge 1$. In other words, we have

$$\mathbb{P}\left(X \in dy\right) = \sum_{1 \le i \le d} p_i \; \delta_{x_i}(dy) \tag{4.2}$$

where dy stands for an infinitesimal neighborhood of the point $x \in \mathbb{R}$, and δ_a stands for the Dirac measure at the point $a \in \mathbb{R}$. In this situation, we have

$$F = \sum_{1 \le i \le d} p_i \; 1_{[x_i, \infty[}$$

and

$$F^{-1} = \sum_{1 \le i \le d} x_i \; 1_{\left[\sum_{1 \le j < i} p_j, \sum_{1 \le j \le i} p_j\right[}.$$

In particular, for a **Bernoulli** r.v. with distribution

$$\mathbb{P}\left(X \in dy\right) = p \; \delta_1(dy) + (1-p) \; \delta_0(dy)$$

associated with some (success) probability parameter $p \in [0, 1]$, we have $F^{-1} = 1_{[0,p[}$.

We also notice that the **uniform distribution on a finite set** $S = \{x_1, \dots, x_d\}$ is given by (4.2) with $p_i = 1/d$, for each $1 \le i \le d$. In this situation, we have

$$F^{-1}(u) = \sum_{1 \le i \le d} x_i \; 1_{\left[\frac{i-1}{d}, \frac{i}{d}\right[}(u) = x_{1 + \lfloor d \times u \rfloor}.$$

- **Counting variables :** We consider a sequence of independent Bernoulli r.v. X_n with parameter $p \in [0, 1]$. In this situation, the date of the first success

$$T = \inf\left\{n \ge 1 \; : \; X_n = 1\right\}$$

is a **geometric** r.v. with parameter p; that is,

$$\forall n \ge 1 \qquad \mathbb{P}\left(T = n\right) = (1-p)^{n-1} \; p.$$

We often write in a more synthetic way $T \sim \text{Geo}(p)$.

We consider a sequence $(U_m)_{1 \le m \le n}$ of n independent and uniform r.v. on $[0, 1]$. Then the counting variable

$$Y_n = \sum_{1 \le m \le n} 1_{[0,p[}(U_m)$$

is a **binomial** r.v. with parameters (n, p); that is, we have that

$$\forall 0 \le m \le n \qquad \mathbb{P}\left(Y_n = m\right) = \frac{n!}{m!(n-m)!} \; p^m \; (1-p)^{n-m}. \tag{4.3}$$

We often write the binomial distribution in a more synthetic way as $Y_n \sim \text{binomial}(n, p)$. More generally, for any sequence of parameters $(p_i)_{1 \le i \le d}$ s.t. $p_i \ge 0$ and $\sum_{1 \le i \le d} p_i = 1$, the sequence of counting numbers

$$\forall 1 \le k \le n \quad Y_n^k = \sum_{1 \le m \le n} 1_{[p_1 + \dots + p_{k-1}, p_1 + \dots + p_k[}(U_m)$$

forms a **multinomial** random variable

$$\forall (m_1, \dots, m_d) \in \mathbb{N}^d \ s.t. \ \sum_{1 \leq i \leq d} m_i = n$$

$$\mathbb{P}\left((Y_n^1, \dots, Y_n^d) = (m_1, \dots, m_d)\right) = \frac{n!}{m_1! \dots m_d!} \ p_1^{m_1} \dots p_d^{m_d}.$$

(4.4)

We often write in a more synthetic way $\text{multi}(n, p_1, \dots, p_d)$.

4.2 Change of variables

We recall the well known change of variable integration formula.

> For any pair of open subsets $\mathcal{D}$ and $\mathcal{D}' \subset \mathbb{R}^d$ and for any invertible function φ that maps $\mathcal{D}$ into $\mathcal{D}'$ and such that φ and φ^{-1} are smooth (a.k.a. diffeomorphism) we have
>
> $$\int_{\mathcal{D}} f(x) \ dx = \int_{\mathcal{D}'} f\left(\varphi^{-1}(y)\right) \ \left|\text{Jac}(\varphi^{-1})(y)\right| dy$$
>
> for any bounded function f on $\mathbb{R}^d$. In the above display, $\text{Jac}(\varphi)(y) = \text{Det}\left(\frac{\partial \varphi^j}{\partial x^j}\right)_{i,j}$ stands for the Jacobian of a diffeomorphism φ.

We illustrate this formula with the transformation

$$\varphi(u_1, u_2) = (a_1 + (b_1 - a_1)u_1, a_2 + (b_2 - a_2)u_2)$$

that maps $\mathcal{D} = [0,1]^2 \ni (u_1, u_2)$ into the cell $\mathcal{D}' = ([a_1, b_1] \times [a_2, b_2])$, with $a_1 < b_1$ and $a_2 < b_2$. For any pair of independent and uniform r.v. U_1 and U_2 on $[0,1]$, we have

$$\mathbb{E}\left(f(\varphi(U_1, U_2))\right) = \int_{\mathcal{D}} f(\varphi(u_1, u_2)) \ du_1 du_2.$$

On the other hand, we have

$$(x_1, x_2) = \varphi(u_1, u_2) \Longrightarrow (u_1, u_2) = \varphi^{-1}(x_1, x_2) = \left(\frac{x_1 - a_1}{b_1 - a_1}, \frac{x_2 - a_2}{b_2 - a_2}\right)$$

and

$$du_1 du_2 \ = \ \left|\frac{\partial u_1}{\partial x_1}\frac{\partial u_2}{\partial x_2} - \frac{\partial u_1}{\partial x_2}\frac{\partial u_2}{\partial x_1}\right| \times dx_1 dx_2 = \frac{1}{(b_1 - a_1)(b_2 - a_2)} \times dx_1 dx_2.$$

This implies that

$$(X_1, X_2) = (a_1 + (b_1 - a_1)U_1, a_2 + (b_2 - a_2)U_2)$$

is a uniform r.v. on $[a_1, b_1] \times [a_2, b_2]$; that is, for any bounded function f, we have that

$$\mathbb{E}\left(f(\varphi(U_1, U_2))\right) \ = \ \frac{1}{(b_1 - a_1)(b_2 - a_2)} \int_{[a_1, b_1] \times [a_2, b_2]} f(x_1, x_2) \ dx_1 dx_2$$

$$= \ \mathbb{E}\left(f(X_1, X_2)\right).$$

Now we consider the Box-Muller transform given by

$$\begin{cases} Y_1 := \sqrt{-2\log(U_1)} \ \cos(2\pi U_2) \\ Y_2 := \sqrt{-2\log(U_1)} \ \sin(2\pi U_2). \end{cases} \tag{4.5}$$

If we set $(Y_1, Y_2) = \varphi(U_1, U_2)$ we have

$$\mathbb{E}\left(f(\varphi(U_1, U_2))\right) = \int_{[0,1]^2} f(\varphi(u_1, u_2)) \ du_1 du_2.$$

On the other hand, we have

$$(u_1, u_2) = \varphi^{-1}(x_1, x_2) = \left(e^{-(y_1^2+y_2^2)/2}, \frac{1}{2\pi} \arctan\left(\frac{y_2}{y_1}\right) \right)$$

so that

$$\begin{aligned} du_1 du_2 &= \left| \frac{\partial u_1}{\partial y_1} \frac{\partial u_2}{\partial y_2} - \frac{\partial u_1}{\partial y_2} \frac{\partial u_2}{\partial y_1} \right| \times dy_1 dy_2 \\ &= \left| -y_1 \frac{1}{1+(y_2/y_1)^2} \frac{1}{y_1} - y_2 \frac{1}{1+(y_2/y_1)^2} \frac{y_2}{y_1^2} \right| \times \frac{1}{2\pi} e^{-(y_1^2+y_2^2)/2} \times dy_1 dy_2 \\ &= \frac{1}{2\pi} e^{-(y_1^2+y_2^2)/2} \times dy_1 dy_2. \end{aligned}$$

This implies that (Y_1, Y_2) is a pair of independent and centered Gaussian random variables with unit variance; that is, for any bounded function f,

$$\mathbb{E}\left(f(\varphi(U_1, U_2))\right) = \int_{\mathbb{R}^2} f(y_1, y_2) \frac{1}{2\pi} e^{-(y_1^2+y_2^2)/2} \times dy_1 dy_2.$$

Conversely, for any bounded functions f_1 and f_2 we have

$$\mathbb{E}\left(f_1\left(\sqrt{Y_1^2+Y_2^2}\right) f_2\left(\frac{Y_1}{\sqrt{Y_1^2+Y_2^2}}, \frac{Y_2}{\sqrt{Y_1^2+Y_2^2}}\right) \right)$$

$$= \mathbb{E}\left(f_1\left(\sqrt{-2\log(U_1)}\right) f_2\left(\cos(2\pi U_2), \sin(2\pi U_2)\right) \right)$$

$$= \mathbb{E}\left(f_1\left(\sqrt{-2\log(U_1)}\right) \right) \times \mathbb{E}\left(f_2\left(\cos(2\pi U_2), \sin(2\pi U_2)\right) \right).$$

This shows that the projection on the unit circle of a couple of Gaussian independent r.v. is uniform on the circle. This result is also true for projection of Gaussians on unit spheres of any dimension.

4.3 Rejection techniques

We let X be a uniform r.v. on the cell $[0, 1]^2$, and we consider the event $A = \{X \in B\}$ where $B \subset [0, 1]^2$ stands for some subset such that $p = \mathbb{P}(X \in B) > 0$. To fix ideas, we can consider $B = \{(u, v) \in [0, 1]^2 \ : \ u^2 + v^2 \le 1\}$. We consider a sequence of independent uniform random samples $(X_n)_{n \ge 1}$ on the unit square $[0, 1]^2$, and we set $A_n = \{X_n \in B\}$. In this situation, if we wait up to the first time n the sample X_n hits the set B we obtain a random variable with the conditional distribution of X given the fact that $\{X \in B\}$. More formally, we have that

$$T = \inf \{n \geq 1 \,:\, X_n \in B\} \quad \Rightarrow \quad \text{Law}(X_T) = \text{Law}\,(X \mid X \in B).$$

The extension of this result to general random events and its proof are provided in the following theorem.

> **Theorem 4.3.1** *We let X be some r.v. taking values in some state space E, and A some random event with probability $p = \mathbb{P}(A) > 0$. We consider a sequence of independent copies $(X_n, 1_{A_n})_{n \geq 1}$ of the r.v. $(X, 1_A)$. Then we have*
>
> $$T = \inf \{n \geq 1 \,:\, 1_{A_n} = 1\} \quad \Rightarrow \quad \text{Law}(X_T) = \text{Law}\,(X \mid A).$$
>
> *In addition T is a geometric random variable with parameter p.*

Proof :
By construction, we have

$$
\begin{aligned}
\mathbb{P}\,(X_T \in dx) &= \sum_{n \geq 1} \mathbb{P}\,(X_n \in dx \,,\, T = n) \\
&= \sum_{n \geq 1} \mathbb{P}\,(X_n \in dx \,,\, 1_{A_1} = 0, \ldots, 1_{A_{n-1}} = 0, 1_{A_n} = 1) \\
&= \sum_{n \geq 1} \mathbb{P}\,(X_n \in dx \,,\, 1_{A_n} = 1)\ \mathbb{P}\,(1_{A_1} = 0, \ldots, 1_{A_{n-1}} = 0).
\end{aligned}
$$

Using the fact that $(X_n, 1_{A_n}) \stackrel{law}{=} (X, 1_A)$ we conclude that

$$
\begin{aligned}
\mathbb{P}\,(X_T \in dx) &= \mathbb{P}\,(X \in dx \mid 1_A = 1) \sum_{n \geq 1} \mathbb{P}\,(1_A = 1) \times \mathbb{P}\,(1_{A_1} = 0, \ldots, 1_{A_{n-1}} = 0) \\
&= \mathbb{P}\,(X \in dx \mid A) \,\times\, \sum_{n \geq 1} p\,(1-p)^{n-1} = \mathbb{P}\,(X \in dx \mid A).
\end{aligned}
$$

This ends the proof of the theorem.   ∎

> **Lemma 4.3.2** *We let (X, Y) be a couple of $\mathbb{R}^d$-valued random vectors with probability densities (p, q) with respect to the Lebesgue measure dx on $\mathbb{R}^d$. We further assume that $p(x) \leq Cq(x)$, for any $x \in \mathbb{R}^d$, for some finite constant $C(\geq 1)$. For any uniform random variable U on $[0, 1]$ we have*
>
> $$A := \left\{ U \leq \frac{p(Y)}{Cq(Y)} \right\} \Rightarrow \mathbb{P}(A) = 1/C \quad and \quad \text{Law}\,(Y \mid A) = \text{Law}\,(X).$$

Proof :
For any bounded function f we have

$$
\begin{aligned}
\mathbb{E}\left(f(Y)\, 1_{U \leq \frac{p(Y)}{Cq(Y)}} \right) &= \mathbb{E}\left(f(Y)\, \mathbb{E}\left(1_{U \leq \frac{p(Y)}{Cq(Y)}} \mid Y \right) \right) \\
&= \mathbb{E}\left(f(Y)\, \frac{p(Y)}{Cq(Y)} \right) = \frac{1}{C}\, \mathbb{E}(f(X)).
\end{aligned}
$$

The end of the proof is clear now. ■

In the settings of the above lemma, we let Y_n be a sequence of independent copies of the r.v. Y, and for any sequence of independent and uniform r.v. U_n on $[0,1]$ we set

$$T = \inf \left\{ n \geq 1 \; : \; U_n \leq \frac{p(Y_n)}{Cq(Y_n)} \right\}$$

Combining the above lemma with theorem 4.3.1, we find that

$$\text{Law}\,(Y_T) = \text{Law}\,(Y \mid A) = \text{Law}\,(X)\,.$$

The dominating density $q(y)$ is sometimes called the instrumental distribution. A variety of examples illustrating this rejection technique are discussed in exercises 42 and 43 provided in section 4.7.

4.4 Sampling probabilities

This section is concerned with sampling techniques of probability measures over a finite set. These methodologies are used in Bayesian statistical machine learning, in physics and chemistry in the modeling of fragmentation (a.k.a aggregation) and coagulation processes, as well as in the modeling of dynamic population models using stochastic partial differential equations. We illustrate these stochastic models in the first two situations.

4.4.1 Bayesian inference

We notice that the density of the beta distribution

$$p(u) = \frac{\Gamma(a+b)}{\Gamma(a)\Gamma(b)} \; u^{a-1} \; (1-u)^{b-1} \, 1_{[0,1]}(u) \tag{4.6}$$

with parameters $a, b > 0$ can be interpreted as a distribution on the set of Bernoulli probability measures. In the above display, Γ stands for the gamma function

$$\Gamma(z) = \int_0^\infty t^{z-1} \; e^{-t} \; dt$$

with the factorial properties $\Gamma(z+1) = z\,\Gamma(z)$, and $\Gamma(1) = 1$. The normalizing constant in (4.6) comes from the fact that

$$
\begin{aligned}
\Gamma(a)\Gamma(b) &= \int_0^\infty s^{a-1} \; e^{-s} \left(\int_s^\infty (t-s)^{b-1} \; e^{-(t-s)} \; dt \right) ds \\
&= \int_0^\infty e^{-t} \left(\int_0^t s^{a-1} \; (t-s)^{b-1} \; ds \right) dt \\
&= \int_0^\infty e^{-t} \; t^{a+b-1} \; dt \left(\int_0^1 u^{a-1} \; (1-u)^{b-1} \; du \right) \\
&= \Gamma(a+b) \int_0^1 u^{a-1} \; (1-u)^{b-1} \; du.
\end{aligned}
$$

More formally, given a realization of some random variable $\Theta \sim \text{beta}(a, b)$, we can define the conditional Bernoulli variable

$$\mathbb{P}\left(X = 1 \mid \Theta = \theta\right) = 1 - \mathbb{P}\left(X = 0 \mid \Theta = \theta\right) = \theta. \tag{4.7}$$

This shows that given Θ, X is a Bernoulli $\{0, 1\}$-valued r.v. with parameter θ. In addition, using the fact that $\Gamma(z + 1) = z\Gamma(z)$, we readily check that the marginal distribution of X is again a Bernoulli random variable

$$\mathbb{P}\left(X = 1\right) = 1 - \mathbb{P}\left(X = 0\right) = \frac{a}{a + b}.$$

We notice that

$$\forall x \in \{0, 1\} \quad \mathbb{P}\left(X = x \mid \Theta = \theta\right) = \theta^x \times (1 - \theta)^{1-x}.$$

Using Bayes' rule, we also find that

$$\begin{aligned} \mathbb{P}\left(\Theta \in d\theta \mid X = x\right) &\propto \mathbb{P}\left(X = x \mid \Theta = \theta\right) \times \mathbb{P}\left(\Theta \in d\theta\right) \\ &\propto \theta^{(a+x)-1} \, (1 - \theta)^{(b+(1-x))-1} \, 1_{[0,1]}(\theta). \end{aligned}$$

This implies that the conditional distribution of Θ given a realization of the Bernoulli trial $X = x$ is given by

$$\text{Law}\left(\Theta \mid X = x\right) = \text{Beta}\left(a + x, b + (1 - x)\right).$$

The multivariate extension of the beta distribution is the Dirichlet distribution defined on the $(d - 1)$-dimensional simplex Δ_{d-1} (introduced in (0.2)) by the formula

$$p(u_1, \ldots, u_d) = \frac{\Gamma\left(\sum_{1 \leq i \leq d} a_i\right)}{\prod_{1 \leq i \leq d} \Gamma(a_i)} \left[\prod_{1 \leq i \leq d} u_i^{a_i - 1}\right] 1_{\Delta_{d-1}}(u_1, \ldots, u_d). \tag{4.8}$$

These distributions are often denoted by $D(a_1, \ldots, a_d)$, where $(a_i)_{1 \leq i \leq d} \in]0, \infty[$ are called the concentration parameters. Let $U = (U_1, \ldots, U_d) \sim D(a_1, \ldots, a_d)$, and $(m_i)_{1 \leq i \leq d} \in \mathbb{N}^d$. If we set $|a| = \sum_{1 \leq i \leq d} a_i$ and $|m| = \sum_{1 \leq i \leq d} m_i$, then we have

$$\begin{aligned} \mathbb{E}(U_1^{m_1} \ldots U_d^{m_d}) &= \frac{\Gamma(|a|)}{\prod_{1 \leq i \leq d} \Gamma(a_i)} \frac{\prod_{1 \leq i \leq d} \Gamma(a_i + m_i)}{\Gamma(|a| + |m|)} \\ &= \left[\prod_{1 \leq i \leq d} \frac{\Gamma(a_i + m_i)}{\Gamma(a_i)}\right] \left[\frac{\Gamma(|a|)}{\Gamma(|a| + |m|)}\right] \\ &= \left[\prod_{0 \leq l < |m|} \frac{1}{|a| + l}\right] \prod_{1 \leq i \leq d} \prod_{0 \leq k < m_i} (a_i + k). \end{aligned}$$

Given U we let $(X_i)_{i \geq 1}$ be a sequence of independent random variables on the set $S := \{1, \ldots, d\}$ with probability distribution given by the following formula

$$\forall s \in S \qquad \mathbb{P}\left(X_1 = s \mid U\right) = U_s.$$

By construction, we have

$$\mathbb{P}\left(X_1 = x_1, \ldots, X_{n+1} = x_{n+1} \mid U\right) = \prod_{s \in S} U_s^{v_{n+1}(s)}$$

with $v_{n+1}(s) = \sum_{1 \leq p \leq (n+1)} 1_{x_p}(s)$ (notice that $\sum_{s \in S} v_{n+1}(s) = n+1$).

$$\mathbb{P}(X_1 = x_1, \ldots, X_{n+1} = x_{n+1}) = \mathbb{E}\left(\prod_{s \in S} U_s^{v_{n+1}(s)}\right)$$

$$= \left[\prod_{0 \leq l \leq n} \frac{1}{|a| + l}\right] \prod_{s \in S} \prod_{0 \leq k < v_{n+1}(s)} (a_s + k). \quad (4.9)$$

When $a_1 = \ldots = a_d = 1$, the Dirichlet measure coincides with the uniform measure on the simplex Δ_{d-1}. For $d = 2$, we also have that $D(a_1, a_2) = beta(a_1, a_2)$.

These distributions can be sampled easily using gamma distributions and rejection techniques. We refer the reader to the exercises 43, 44, 45, 46 provided in section 4.7.

4.4.2 Laplace's rule of successions

The French mathematician and astronomer Pierre-Simon Laplace said in 1774 that an event that has already appeared k times during n experiments, has a chance of $(k+1)/(n+2)$ to occur in the $(n+1)$-th experiment.

To explain this statement, we start by a sequence of independent $\{0,1\}$-valued Bernoulli r.v. $(X_i)_{i \geq 1}$ with a given success parameter $\mathbb{P}(X_i = 1) = \Theta \in [0,1]$. In this case, the r.v. $\overline{X}_n := \sum_{1 \leq i \leq n} X_i$ represents the number of successes among n trials.

Of course, the success parameter Θ is unknown, so that we assume that it is is uniform on $[0,1]$. In this situation, we have

$$\mathbb{P}\left(X_{n+1} = 1 \mid \overline{X}_n = k\right) = \mathbb{E}\left(\mathbb{E}\left(X_{n+1} \mid \Theta, \overline{X}_n = k\right) \mid \overline{X}_n = k\right)$$
$$= \mathbb{E}\left(\Theta \mid \overline{X}_n = k\right).$$

Using Bayes' rule, we find that

$$\mathbb{P}\left(\Theta \in d\theta \mid \overline{X}_n = k\right) \propto \mathbb{P}\left(\overline{X}_n = k \mid \Theta = \theta\right) 1_{[0,1]}(\theta) \, d\theta$$
$$\propto \theta^k (1-\theta)^{n-k} 1_{[0,1]}(\theta) \, d\theta.$$

This shows that the conditional distribution of Θ given $\overline{X}_n = k$ is the beta distribution $beta(k+1, n-k+1)$. By (4.6), we conclude that

$$\mathbb{P}\left(X_{n+1} = 1 \mid \overline{X}_n = k\right)$$

$$= \frac{\Gamma(k+1+(n-k)+1)}{\Gamma(k+1)\Gamma((n-k)+1)} \int_0^1 \theta^{k+1} (1-\theta)^{n-k} \, d\theta$$

$$= \frac{\Gamma(k+1+(n-k)+1)}{\Gamma(k+1)\Gamma((n-k)+1)} \frac{\Gamma(k+2)\Gamma((n-k)+1)}{\Gamma(k+2+(n-k)+1)} = \frac{k+1}{k+2+(n-k)}$$

and therefore

$$\mathbb{P}\left(X_{n+1} = 1 \mid \overline{X}_n\right) = \mathbb{E}\left(\Theta \mid \overline{X}_n\right) = \left(\overline{X}_n + 1\right) / (n+2).$$

4.4.3 Fragmentation and coagulation

The uniform measure on the simplex Δ_d can be easily sampled using the ordered and uniform statistic $(V_1, \ldots, V_d)$ on $[0,1]$ discussed at the end of problem 41 from section 4.7.

As its name indicates, an ordered and uniform statistic $(V_1, \ldots, V_d)$ on $[0, 1]$ is obtained by ordering a sequence of d independent and uniform samples on $[0, 1]$. Then we have:

$$(V_1, (V_2 - V_1), \ldots, (V_d - V_{d-1}), 1 - V_d) \sim \text{Unif}(\Delta_d)$$

where $\text{Unif}(A)$ stands for the uniform measure on some subset $A \subset \mathbb{R}^d$, for some d.

These uniform distributions can also be interpreted in terms of elementary fragmentation and coagulation processes [20].

The fragmentation process is defined as follows: Suppose we have a uniform r.v. $X = (X_1, \ldots, X_d) \in \Delta_{d-1}$. Given this variable, we let I be the integer valued random variable defined by

$$\forall 1 \leq i \leq d \qquad \mathbb{P}(I = i \mid X) = X_i.$$

We sample independently an index I and some $U \sim \text{Unif}([0, 1])$, and we set

$$Y = (X_1, \ldots, X_{I-1}, \ U \times X_I, \ (1 - U) \times X_I, \ X_{I+1}, \ldots, X_d) \in \Delta_d.$$

The sequence Y can be seen as a fragmentation of a randomly chosen block of size X_I into two blocks of random sizes $U \times X_I$ and $(1 - U) \times X_I$.

The coagulation process can be seen as a time reversal of the fragmentation process. Given a uniform r.v. $X = (X_1, \ldots, X_{d+1}) \in \Delta_d$, we choose randomly a couple of integers (I_1, I_2) without replacement in the set $\{1, \ldots, d + 1\}$ and then set

$$Z = (X_1, \ldots, X_{J_1-1}, \ X_{J_1} + X_{J_2}, X_{J_1+1}, \ldots, X_{J_2-1}, X_{J_2+1}, \ldots, X_{d+1}) \in \Delta_{d-1}$$

with $J_1 = I_1 \wedge I_2$ and $J_2 = I_1 \vee I_2$. We end up with $Z \sim \text{Unif}(\Delta_{d-1})$.

4.5 Conditional probabilities

4.5.1 Bayes' formula

Sampling conditional probabilities is one of the most important problems in Bayesian statistics and in applied probability. For instance, suppose that we have a partial and noisy observation Y of some r.v. X. The estimation of the realization of X that produced a given observation $Y = y$ amounts to computing the conditional distribution of X given Y. The amount of information contained in the observation Y depends on the model at hand.

We further assume that (X, Y) is a $\mathbb{R}^2$-valued r.v. with density $p(x, y)$ w.r.t. the Lebesgue measure $dxdy$ on $\mathbb{R}^2$. Slightly abusing notation, we denote by $p(x)$, $p(y)$, the densities of the r.v. X and Y, and by $p(x|y)$ and $p(y|x)$ the conditional density of X given $Y = y$, and the one of Y given $X = x$. These notations are commonly used in the Bayesian literature.

> The density $p(x)$ is called the prior density, $p(y|x)$ the likelihood function, and $p(x|y)$ is the posterior density. The Bayes' formula states that
>
> $$p(x|y) = \frac{1}{p(y)} \, p(y|x) \, p(x) \quad \text{with} \quad p(y) = \int p(y|x) \, p(x) \, dx.$$

4.5.2 The regression formula

For some classes of prior densities and likelihood functions, the posterior density can be computed explicitly. For instance, suppose that (W, V) is a pair of centered independent Gaussian random variables with variances (σ^2, τ^2), respectively. We set

$$X = m + W \quad \text{and} \quad Y = aX + V \tag{4.10}$$

for some parameters $m(= \mathbb{E}(X))$ and $a \in \mathbb{R}$. By construction, we have

$$p(x) \propto \exp\left(-\frac{1}{2\sigma^2}(x - m)^2\right) \quad \text{and} \quad p(y|x) \propto \exp\left(-\frac{1}{2\tau^2}(y - ax)^2\right). \tag{4.11}$$

Expanding the squares we find that

$$p(x|y) \propto p(y|x)\, p(x) \propto \exp\left(-\frac{1}{2\rho}(x - (\alpha + \beta y))^2\right) \tag{4.12}$$

for some parameters (α, β, ρ) that depend on (m, a, σ, τ). This shows that the conditional distribution of X given Y is again Gaussian with

$$\mathbb{E}(X|Y) = \alpha + \beta Y$$

and

$$\mathbb{E}\left((X - \mathbb{E}(X|Y))^2 | Y\right) = \mathbb{E}\left((X - \mathbb{E}(X|Y))^2\right) = \rho.$$

Our aim is to find a couple of parameters (α', β') such that

$$\mathbb{E}(X|Y) = \alpha' + \beta'(Y - \mathbb{E}(Y)).$$

In this notation, we have $\alpha = \alpha' - \beta'\mathbb{E}(Y)$ and $\beta = \beta'$.

Taking the expectations, we readily observe that $\alpha' = \mathbb{E}(X)$. On the other hand, by the definition of the conditional expectations, we have

$$\mathbb{E}\left([X - \mathbb{E}(X|Y)][Y - \mathbb{E}(Y)]\right) = 0.$$

This yields that

$$\mathbb{E}\left([(X - \mathbb{E}(X)) - \beta'(Y - \mathbb{E}(Y))][Y - \mathbb{E}(Y)]\right) = 0.$$

Using the fact that

$$\begin{aligned}
\mathrm{Cov}(X, Y) &= \mathbb{E}\left((X - \mathbb{E}(X))(Y - \mathbb{E}(Y))\right) \\
&= \mathbb{E}\left((X - \mathbb{E}(X))(a(X - \mathbb{E}(X)) - V)\right) = a\sigma^2
\end{aligned}$$

and

$$\mathrm{Var}(Y) := \mathbb{E}\left((Y - \mathbb{E}(Y))^2\right) = \mathbb{E}\left((a(X - \mathbb{E}(X)) + V)^2\right) = a^2\sigma^2 + \tau^2$$

we conclude that $\beta' = a\,\sigma^2/(a^2\sigma^2 + \tau^2)$. Finally, we observe that

$$\begin{aligned}
\rho &= \mathbb{E}\left(((X - \mathbb{E}(X)) - \beta'(Y - \mathbb{E}(Y)))^2\right) \\
&= \sigma^2 - \beta'a\sigma^2 = \sigma^2\left(1 - \frac{a^2\sigma^2}{(a^2\sigma^2 + \tau^2)}\right) = (a^2\tau^{-2} + \sigma^{-2})^{-1}.
\end{aligned}$$

This shows that the random variable

$$m + \frac{a\sigma^2}{(a^2\sigma^2 + \tau^2)}(y - am) + W'$$

$$= \frac{\tau^2}{a^2\sigma^2 + \tau^2}\, m + \left(1 - \frac{\tau^2}{a^2\sigma^2 + \tau^2}\right)\, a^{-1}y + W'$$

with a centered Gaussian variable W' with variance ρ, is distributed according to the conditional distribution $p(x|y)$.

In statistics,

$$\mathbb{E}\left(X|Y\right) = \mathbb{E}(X) + \frac{\mathrm{Cov}(X, Y)}{\mathrm{Var}(Y)}(Y - \mathbb{E}(Y))$$

is called the regression formula. For nonlinear and/or non-Gaussian models the right hand side represents the best linear estimator of X based on the observation Y. This formula coincides with the conditional expectation for linear and Gaussian models.

In signal processing, the parameter

$$\boldsymbol{g} = a\sigma^2(a^2\sigma^2 + \tau^2)^{-1}$$

is called the gain. In terms of this parameter, the above equation takes the form

$$m + \boldsymbol{g}\,(y - am) + W' \quad \text{and} \quad \rho = (1 - \boldsymbol{g}a)\,\sigma^2.$$

In terms of the Boltzmann-Gibbs transformation (0.3), we have proved that

$$G(x) = e^{-\frac{1}{2\tau^2}(y - ax)^2} \quad \text{and} \quad \eta(dx) \propto e^{-\frac{1}{2\sigma^2}(x - m)^2}\, dx$$

$$\Downarrow$$

$$\Psi_G(\eta)(dx) \propto \exp\left\{-\frac{1}{2\rho}\,(x - [m + \boldsymbol{g}(y - am)])^2\right\}\, dx.$$

4.5.3   Gaussian updates

This short section is concerned with the extension of the regression formula to multi-dimensional Gaussian models.

Definition 4.5.1 *We denote by $\mathcal{N}(m, R)$ the Gaussian distribution on a d-dimensional space $\mathbb{R}^d$ with mean column vector $m \in \mathbb{R}^d$ and covariance matrix $R \in \mathbb{R}^{d \times d}$*

$$\mathcal{N}(m, R)(dx) = \frac{1}{(2\pi)^{d/2}\sqrt{|R|}}\, \exp\left[-2^{-1}(x - m)'R^{-1}(x - m)\right]\, dx. \qquad (4.13)$$

Slightly abusing notation, sometimes we denote by

$$\mathcal{N}\left[m, R\right](x) := \frac{1}{(2\pi)^{d/2}\sqrt{|R|}}\, \exp\left[-2^{-1}(x - m)'R^{-1}(x - m)\right] \qquad (4.14)$$

the density of a Gaussian distribution w.r.t. the Lebesgue measure dx.

In this notation, the multi-dimensional version of the Gaussian updating formula discussed above is given by the following proposition.

Proposition 4.5.2 *Given an observation state $y \in \mathbb{R}^q$, some matrix $A \in \mathbb{R}^{q \times p}$, and some covariance matrix $R_0 \in \mathbb{R}^{q \times p}$ and some point $a \in \mathbb{R}^q$, we have*

$$G(x) := \mathcal{N}\left[Ax + a; R_0\right](y) \implies \Psi_G\left(\mathcal{N}(m_1, R_1)\right) = \mathcal{N}(m_2, R_2) \qquad (4.15)$$

with

$$m_2 = m_1 + \mathbb{G}\left(y - (Am_1 + a)\right) \quad and \quad R_2 = (Id - \mathbb{G}A)R_1$$

with the gain matrix

$$\mathbb{G} = R_1 A'(AR_1 A' + R_0)^{-1}.$$

Proof :

There are several ways of proving formula (4.15). Next, we present the most elementary (but rather tedious) proof. There is no loss of generality (replacing y by $(y - a)$) to assume that $a = 0$.

Firstly, we observe that

$$(A'R_0^{-1}A + R_1^{-1})^{-1} = \left[I - R_1 A'(AR_1 A' + R_0)^{-1}A\right] R_1 = (I - \mathbb{G}A)R_1.$$

The r.h.s. is immediate. We check the l.h.s. using the fact that

$$\left[I - R_1 A'(AR_1 A' + R_0)^{-1}A\right] R_1 \left[R_1^{-1} + A'R_0^{-1}A\right] - I$$

$$= \left[I - R_1 A'(AR_1 A' + R_0)^{-1}A\right] \left[I + R_1 A'R_0^{-1}A\right] - I$$

$$= -\left\{R_1 A'(AR_1 A' + R_0)^{-1}A + R_1 A'(AR_1 A' + R_0)^{-1}AR_1 A'R_0^{-1}A\right\} + R_1 A'R_0^{-1}A$$

$$= -R_1 A'(AR_1 A' + R_0)^{-1}\underbrace{\left\{A + AR_1 A'R_0^{-1}A\right\}}_{\{R_0 + AR_1 A'\}R_0^{-1}A} + R_1 A'R_0^{-1}A = 0.$$

Thus, if we set $R_2 = (I - \mathbb{G}A)R_1$ then it clearly suffices to check that

$$(x - m_1)'R_1^{-1}(x - m_1) + (y - Ax)'R_0^{-1}(y - Ax)$$

$$= [x - \{m_1 + \mathbb{G}(y - Am_1)\}]'(A'R_0^{-1}A + R_1^{-1}) \times [x - \{m_1 + \mathbb{G}(y - Am_1)\}].$$

We notice that

$$(x - m_1)'R_1^{-1}(x - m_1) + (y - Ax)'R_0^{-1}(y - Ax)$$

$$= x'\left(R_1^{-1} + A'R_0^{-1}A\right)x - 2\left(m_1'R_1^{-1}x + y'R_0^{-1}Ax\right) + y'R_0^{-1}y + m_1'R_1^{-1}m_1$$

$$= x'\left(R_1^{-1} + A'R_0^{-1}A\right)x - 2\left(m_1'R_1^{-1}x + y'R_0^{-1}Ax\right) + \alpha(y)$$

with some function $\alpha(y)$ that only depends on the parameters y. In the same way, we have

$$[x - \{m_1 + \mathbb{G}(y - Am_1)\}]'(A'R_0^{-1}A + R_1^{-1})[x - \{m_1 + \mathbb{G}(y - Am_1)\}]$$

$$= x'\left(R_1^{-1} + A'R_0^{-1}A\right)x - 2\left(m_1' + (y' - m_1'A')\overbrace{(R_0 + AR_1 A')^{-1}AR_1}^{=\mathbb{G}'}\right)$$

$$\times \left(R_1^{-1} + A'R_0^{-1}A\right)x + \beta(y)$$

with some function $\beta(y)$ that only depends on the parameters y. Thus, it suffices to check that

$$m_1' R_1^{-1} x + y' R_0^{-1} A x$$

$$= \left(m_1' + (y' - m_1' A') \left(R_0 + A R_1 A' \right)^{-1} A R_1 \right) \left(R_1^{-1} + A' R_0^{-1} A \right) x.$$

To this end, we observe that

$$\left(R_0 + A R_1 A' \right)^{-1} A R_1 \left(R_1^{-1} + A' R_0^{-1} A \right)$$

$$= \left(R_0 + A R_1 A' \right)^{-1} \left(A + A R_1 A' R_0^{-1} A \right)$$

$$= \left(R_0 + A R_1 A' \right)^{-1} \left(R_0 + A R_1 A' \right) R_0^{-1} A = R_0^{-1} A.$$

This implies that

$$\left(m_1' + (y' - m_1' A') \left(R_0 + A R_1 A' \right)^{-1} A R_1 \right) \left(R_1^{-1} + A' R_0^{-1} A \right) x$$

$$= m_1' \left(R_1^{-1} + A' R_0^{-1} A \right) x + (y' - m_1' A') R_0^{-1} A x = m_1' R_1^{-1} x + y' R_0^{-1} A x.$$

This ends the proof of the proposition.   ∎

4.5.4   Conjugate priors

Definition 4.5.3 *We say that a class of prior distributions* $\mathbb{P}(X \in dx) \in \mathcal{P}$ *is conjugate to a class of likelihood distributions* $\mathbb{P}(Y \in dy \mid X = x) \in \mathcal{C}$ *whenever the conditional distributions given by the Bayes rule* $\mathbb{P}(X \in dx \mid Y = y) \in \mathcal{P}$. *The index parameters* x *of the likelihood distributions* $\mathbb{P}(Y \in dy \mid X = x)$ *are sometimes called the hyperparameters.*

For instance, the class of Gaussian prior distributions is conjugate with the class of Gaussian likelihood functions $p(y|x)$ of the form (4.11).

The beta distribution

$$\mathbb{P}(X \in dx) \propto x^{a-1} \left(1 - x \right)^{b-1} 1_{[0,1]}(x)$$

introduced in (4.6) is conjugate to the binomial likelihood function defined by

$$\forall y \in \{0, \ldots, n\} \quad \mathbb{P}(Y = y \mid X = x) = \binom{n}{y} x^y \left(1 - x \right)^{n-y}.$$

We check this claim using the fact that

$$\mathbb{P}(X \in dx \mid Y = y) \propto x^y \left(1 - x \right)^{n-y} x^{a-1} \left(1 - x \right)^{b-1} 1_{[0,1]}(x)$$

$$= x^{a+y-1} \left(1 - x \right)^{b+(n-y)-1} 1_{[0,1]}(x).$$

This shows that $\text{Law}(X \mid Y) = \text{beta}(a + y, b + (n - Y))$.

In much the same way, we find that the beta distribution is conjugate to the geometric distributions defined by

$$\forall y \geq 1 \quad \mathbb{P}(Y = y \mid X = x) = (1 - x)^{y-1} x$$

with

$$\mathbb{P}(X \in dx \mid Y = y) \propto x^{(a+1)-1} \left(1 - x \right)^{b+(y-1)-1} 1_{[0,1]}(x).$$

This shows that $\text{Law}(X \mid Y) = \text{beta}(a + 1, b + (Y - 1))$. For a more thorough discussion on conjugate distributions and their applications in Bayesian statistics, we refer to the research book [19].

4.6 Spatial Poisson point processes ////

4.6.1 Some preliminary results

We let $\mathcal{M}(S)$ be the set of non-negative measures on some state space S. We consider a non-negative measure γ on S with bounded and positive mass $\gamma(1) = \int \gamma(dx) > 0$. We let N be an integer valued Poisson random variable with parameter $\lambda = \gamma(1)$ with distribution given by the formula

$$\forall n \geq 0 \qquad \mathbb{P}\left(N = n\right) := \frac{\lambda^n}{n!}\, e^{-\lambda}.$$

We often write in a more synthetic way $N \sim \text{Po}(\lambda)$. We also denote by $X = (X^i)_{i \geq 1}$ a sequence of independent and identically distributed r.v. with common distribution $\eta(dx) := \gamma(dx)/\gamma(1)$. We also assume that N and X are independent.

A simple example illustrating these models is given by the one-dimensional Gaussian model on $S = \mathbb{R}$

$$\gamma(dx) = 10 \times \frac{1}{\sqrt{2\pi}\sigma}\, e^{-\frac{(x-m)^2}{2\sigma^2}}.$$

Hence $N \sim \text{Po}(10)$, and X^i are independent Gaussian r.v. with means $\mathbb{E}(X^i) = m$, and variance $\text{Var}(X^i) = \sigma^2$.

Definition 4.6.1 *For any probability distribution μ on S, and any natural number p, we let*

$$\mu^{\otimes p}(d(x^1, \ldots, x^p)) = \mu(dx^1) \times \ldots \times \mu(dx^p)$$

be the distribution of p independent r.v. $(Y^1, \ldots, Y^p) \in S^p$ with the common law μ on S. For any non-negative measure γ on some state space S, we also denote by $\gamma^{\otimes p}$ the non-negative measure on S^p defined by

$$\gamma^{\otimes p}(d(x^1, \ldots, x^p)) = \gamma(dx^1) \times \ldots \times \gamma(dx^p).$$

Definition 4.6.2 *The (spatial) Poisson point process $\mathcal{X}$ with intensity measure γ on some state space S is the random measure defined below*

$$\mathcal{X} := m_N(X) = \sum_{1 \leq i \leq N} \delta_{X^i}.$$

Remark :

Spatial Poisson processes are sometimes called complete spatial randomness (CSR) [22, 107]. They arise in statistical inference in social sciences, biology, pharmacology, as well as in astronomy. The random states X^i can be interpreted in many different ways, depending on the application model they describe: locations of trees or nests in a forest, ill individuals in a given population, invasive bacteria or other species in some environment, as well as craters in a planet, and multiple targets in advanced signal processing [41, 217].

These random measures are used to model any type of non-interacting events in some state space (not restricted to spatial patterns only). They can be used to estimate the intensity of events in some region (such as the presence of noisy and partially observed targets in tracking problems). They are also often used as a null hypothesis of total randomness to test the presence of some spatial dependence between events.

Chapter 10 is dedicated to Poisson point processes w.r.t. the time parameter. In this context, we are given some time horizon say t, and N will be some Poisson random variable with a given parameter $\int_0^t \lambda_s \, ds$, for some non-negative intensity parameter λ_t.

Given $N = n$, the non-ordered random times X^i are uniform and independent on $[0, t]$. The ordered random times $(T_1, \ldots, T_n)$ can be interpreted as the jump times of a stochastic process indexed by the time parameter $s \in [0, t]$. We refer the reader to chapter 10 for further details on these Poisson jump processes. The spatial version of these stochastic processes is discussed in section 28.4.3.2.

One of the main simplifications of Poisson point processes arises from the fact that their expectation measure coincides with their intensity measure:

$$\mathbb{E}\left(\mathcal{X}(f)\right) = \mathbb{E}\left(\mathbb{E}\left(\mathcal{X}(f) \mid N\right)\right) = \mathbb{E}\left(N\eta(f)\right) = \gamma(1)\eta(f) = \gamma(f).$$

Definition 4.6.3 *For every sequence of points $x = (x^i)_{i \geq 1}$ in S, any subset $A \subset S$, and every $p \geq 0$, we denote by $m_{p,A}(x)$ the restriction of the occupation measure $m_p(x)$ to the set A*

$$m_{p,A}(x)(dy) = m_p(x)(dy)1_A(y) = \sum_{1 \leq i \leq p} 1_A(x^i)\delta_{x^i}(dy).$$

Lemma 4.6.4 *Let $(\mathcal{X}_j)_{j \geq 1}$ be a sequence of independent Poisson point processes with intensity measure $(\gamma_i)_{i \geq 1}$ on some common state space S. For any $d \geq 1$, $\mathcal{X}$ is a Poisson point process with intensity measure $\sum_{1 \leq i \leq d} \gamma_i$ if, and only if, $\mathcal{X}$ is equal in law to the Poisson point process $\sum_{1 \leq i \leq d} \mathcal{X}_i$.*

Proof :
By symmetry arguments, for any bounded function F on $\mathcal{M}(S)$, and for any $d \geq 1$, we have

$$\mathbb{E}\left(F\left(\sum_{1 \leq i \leq d} \mathcal{X}_i\right)\right) = e^{-\sum_{1 \leq i \leq d} \gamma_i(1)} \sum_{p_1, \ldots, p_d \geq 0} \frac{\gamma_1(1)^{p_1} \ldots \gamma_d(1)^{p_d}}{p_1! \ldots p_d!}$$

$$\times \int F\left(\sum_{1 \leq i \leq d} m_{p_i}(x_i)\right) \prod_{1 \leq i \leq d} \eta_i^{\otimes p_i}(dx_i)$$

where $dx_i = d(x_i^1, \ldots, x_i^{p_i})$ stands for an infinitesimal neighborhood of a point $x_i = (x_i^1, \ldots, x_i^{p_i}) \in S^{p_i}$. This implies that

$$\mathbb{E}\left(F\left(\sum_{1 \leq i \leq d} \mathcal{X}_i\right)\right) = e^{-\sum_{1 \leq i \leq d} \gamma_i(1)} \sum_{q \geq 0} \frac{1}{q!} \sum_{p_1 + \ldots + p_d = q} \frac{q!}{p_1! \ldots p_d!} \gamma_1(1)^{p_1} \ldots \gamma_d(1)^{p_d}$$

$$\times \int F\left(m_q(y)\right) \left(\eta_1^{\otimes p_1} \otimes \ldots \otimes \eta_d^{\otimes p_d}\right)(dy).$$

In the above integral $dy = d(y^1, \ldots, y^q)$ stands for an infinitesimal neighborhood of the point $y = (y^i)_{1 \leq i \leq q} \in S^q$. This implies that

$$\mathbb{E}\left(F\left(\sum_{1 \leq i \leq d} \mathcal{X}_i\right)\right)$$

$$= e^{-\sum_{1 \leq i \leq d} \gamma_i(1)} \sum_{q \geq 0} \frac{\left(\sum_{i=1}^d \gamma_i(1)\right)^q}{q!} \int F\left(m_q(y)\right) \left(\frac{\sum_{i=1}^d \gamma_i(1)\, \eta_i}{\sum_{i=1}^d \gamma_i(1)}\right)^{\otimes q} (dy). \tag{4.16}$$

Therefore $\sum_{1 \leq i \leq d} \mathcal{X}_i$ is a Poisson point process with intensity measure $\sum_{1 \leq i \leq d} \gamma_i$. In addition, by (4.16), any Poisson point process with such an intensity measure has the same law as $\sum_{1 \leq i \leq d} \mathcal{X}_i$. ∎

The two main direct consequences of this lemma are the following important properties of Poisson processes:

> **• Superposition of Poisson processes:**
>
> $$\forall i \in I \quad \mathcal{X}_i \text{ independent Poisson with intensity } \gamma_i$$
>
> $$\implies \quad \sum_{i \in I} \mathcal{X}_i \text{ Poisson with intensity } \sum_{i \in I} \gamma_i.$$

> **• Thinning Poisson processes:**
> For any sum of independent Poisson processes $\mathcal{X} = \sum_{i \in I} \mathcal{X}_i$ with intensities γ_i, with $i \in I$, we have
>
> $$\forall i \in I \quad \mathcal{X}_i = \sum_{1 \leq j \leq N} 1_{\epsilon_j = i}\, \delta_{X^{\epsilon_i, i}} \quad \text{and} \quad \mathcal{X} = \sum_{1 \leq i \leq N} \delta_{X^{\epsilon_i, i}}$$
>
> with $N = \mathrm{Po}\left(\sum_{i \in I} \gamma_i(1)\right)$ and an independent sequence of independent r.v. ϵ_i with discrete distribution $\sum_{j \in I} \frac{\gamma_j(1)}{\sum_{i \in I} \gamma_i(1)}\, 1_j$. The last assertion can also be checked directly from the formula (4.16).

The next result is a direct consequence of Lemma 4.6.4.

Lemma 4.6.5 *Let $\mathcal{X} := \sum_{1 \leq i \leq N} \delta_{X^i}$ be a Poisson point process with intensity measure γ that is the random measure on S. We consider a measurable subset $A \subset S$, such that $\gamma(A) > 0$. Then, the restriction, or the trace, $\mathcal{X}_A = m_{N,A}(X)$ of $\mathcal{X}$ on the set A is again a Poisson point process with intensity measure $\gamma_A(dx) := 1_A(x)\gamma(dx)$.*

In addition, the conditional distribution of $\mathcal{X}$ given $\mathcal{X}_A$, can be calculated for any bounded function F on $\mathcal{M}(S)$, by the formula

$$\mathbb{E}\left(F(\mathcal{X}) \,|\, \mathcal{X}_A\right) = e^{-\gamma(A^c)} \sum_{p \geq 0} \frac{1}{p!} \int F\left(\mathcal{X}_A + m_p(x)\right) \gamma_{A^c}^{\otimes p}(dx).$$

In the above integral $dx = d(x^1, \ldots, x^p)$ stands for an infinitesimal neighborhood of the point $x = (x^i)_{1 \leq i \leq p} \in S^p$.

Proof :

Using the decomposition

$$\gamma(dx) = 1_A(x)\gamma(dx) + 1_{A^c}(x)\gamma(dx) \;\Rightarrow\; \gamma = \gamma_A + \gamma_{A^c}$$

we find that

$$\mathbb{E}\left(F\left(\mathscr{X}_A\right)\right) = e^{-\gamma(1)} \sum_{s \geq 0} \frac{1}{s!} \int F\left(m_{s,A}(x)\right) \; (\gamma_A + \gamma_{A^c})^{\otimes s}(dx).$$

By symmetry arguments, this implies that

$$\mathbb{E}\left(F\left(\mathscr{X}_A\right)\right)$$

$$= e^{-\gamma(1)} \sum_{s \geq 0} \frac{1}{s!} \sum_{p+q=s} \frac{s!}{p!q!} \int F\left(m_{s,A}(x)\right) \left[\gamma_A^{\otimes p} \otimes \gamma_{A^c}^{\otimes(s-p)}\right](dx),$$

from which we find that

$$\mathbb{E}\left(F\left(\mathscr{X}_A\right)\right) \;=\; e^{-\gamma(1)} \sum_{p \geq 0} \frac{1}{p!} \left(\sum_{s \geq p} \frac{\gamma(A^c)^{s-p}}{(s-p)!}\right) \int F\left(m_p(x)\right) \gamma_A^{\otimes p}(dx)$$

$$=\; e^{-(\gamma(E)-\gamma(A^c))} \sum_{p \geq 0} \frac{1}{p!} \int F\left(m_p(x)\right) \gamma_A^{\otimes p}(dx).$$

The last assertion is a direct consequence of lemma 4.6.4, applied to $d = 2$, replacing $(\mathscr{X}_1, \mathscr{X}_2)$ by $(\mathscr{X}_A, \mathscr{X}_{A^c})$. This ends the proof of the lemma. ∎

4.6.2 Conditioning principles

We consider a measure $\gamma \in \mathcal{M}(S)$ on some state space S and a Markov transition M from S into itself. We denote by $\eta(dx) = \gamma(dx)/\gamma(1)$ the normalized probability measure on S. We let

$$\mathscr{Z} := m_N(X, Y) = \sum_{1 \leq i \leq N} \delta_{(X^i, Y^i)} \tag{4.17}$$

be the Poisson point process on the product space S^2 with intensity measure Γ of the following form

$$\Gamma(d(x, y)) := \gamma(dx) \, M(x, dy) = \gamma(1) \, \eta(dx) \times M(x, dy).$$

In other words, $N \sim \text{Po}(\gamma(1))$ and given N, X^i, Y^i are independent random variables with common distribution

$$\mathbb{P}\left((X^i, Y^i) \in d(x, y)\right) = \eta(dx) \times M(x, dy).$$

We observe that

$$\mathbb{P}\left(X^i \in dx\right) = \eta(dx) \quad \text{and} \quad \mathbb{P}\left(Y^i \in dy \mid X^i = x\right) = M(x, dy)$$

from which we check that

$$\mathbb{P}\left(Y^i \in dy\right) \;=\; \int \mathbb{P}\left(Y^i \in dy \mid X^i = x\right) \mathbb{P}\left(X^i \in dx\right)$$

$$=\; \int \eta(dx) \, M(x, dy) = (\eta M)(dy).$$

We further assume that the conditional distribution of X^i given $Y^i = y$ exists and is given by a Markov transition

$$\mathbb{P}\left(X^i \in dx \mid Y^i = y\right) = \widehat{M}(y, dx).$$

The calculation of these conditional probabilities is based on the Bayes rule described by the synthetic formula

$$p(x|y) = \frac{1}{p(y)}\, p(y|x)\, p(x) \quad \text{with} \quad p(y) = \int p(y|x)\, p(x)\, dx.$$

In our context, the formula applies to

$$\eta(dx) = p(x)dx, \qquad M(x, dy) = p(y|x)\, dy \quad \text{and} \quad \widehat{M}(y, dx) = p(x|y)\, dx$$

when the measures η and the Markov transitions are absolutely continuous w.r.t. the Lebesgue measures dx and dy on $S = \mathbb{R}^d$.

Example 4.6.6 *For instance, for the Gaussian model discussed in (4.11) we have $S = \mathbb{R}$ and*

$$\eta(dx) = \frac{1}{\sqrt{2\pi\sigma^2}}\, \exp\left(-\frac{1}{2\sigma^2}\,(x - m)^2\right)\, dx$$

$$M(x, dy) = \frac{1}{\sqrt{2\pi\tau^2}}\, \exp\left(-\frac{1}{2\tau^2}\,(y - ax)^2\right)\, dy.$$

Hence the conditional distribution $\widehat{M}$ is given by the Gaussian density

$$\widehat{M}(y, dx) = \frac{1}{\sqrt{2\pi\widehat{\sigma}^2}}\, \exp\left(-\frac{1}{2\widehat{\sigma}^2}\,(x - \widehat{m}(y))^2\right)\, dx$$

with $\widehat{\sigma}^2 = \left(a^2\tau^{-2} + \sigma^{-2}\right)^{-1}$, and the linear regression parameter given by

$$\widehat{m}(y) = \frac{\tau^2}{a^2\sigma^2 + \tau^2}\, m + \left(1 - \frac{\tau^2}{a^2\sigma^2 + \tau^2}\right)\, a^{-1}y = \left(\frac{\widehat{\sigma}}{\sigma}\right)^2 m + \left(1 - \left(\frac{\widehat{\sigma}}{\sigma}\right)^2\right)\, a^{-1}y.$$

By construction, we have

$$\begin{aligned}
\mathbb{P}\left((X^i, Y^i) \in d(x, y)\right) &= \eta(dx) \times M(x, dy) \\
&= \mathbb{P}\left(X^i \in dx \mid Y^i = y\right) \times \mathbb{P}\left(Y^i \in dy\right) = (\eta M)(dy) \times \widehat{M}(y, dx)
\end{aligned}$$

so that

$$\eta(dx) \times M(x, dy) = (\eta M)(dy) \times \widehat{M}(y, dx). \tag{4.18}$$

Under our assumptions, using lemma 4.6.5 it is immediate to check that the marginal random measures given by

$$\mathcal{X} := m_N(X) = \sum_{1 \leq i \leq N} \delta_{X^i} \quad \text{and} \quad \mathcal{Y} := m_N(Y) = \sum_{1 \leq i \leq N} \delta_{Y^i}$$

are Poisson point processes on S, with intensity measures

$$\gamma_X(dx) := \gamma(1)\, \eta(dx) \quad \text{and} \quad \gamma_Y(dy) := \gamma(1)\, (\eta M)(dy).$$

Our next objective is to describe the conditional distributions of the random measures $\mathcal{X}$ w.r.t. $\mathcal{Y}$, and of $\mathcal{Y}$ w.r.t. $\mathcal{X}$. Using rather elementary manipulations, we prove the following lemma.

Lemma 4.6.7 *For any bounded function F on $\mathcal{M}(S)$, we have the conditioning formulae:*

$$\mathbb{E}\left(F(\mathcal{X}) \mid \mathcal{Y}\right) = \int F\left(m_N(x)\right) \prod_{1 \leq i \leq N} \widehat{M}(Y^i, dx^i)$$

and

$$\mathbb{E}\left(F(\mathcal{Y}) \mid \mathcal{X}\right) = \int F\left(m_N(y)\right) \prod_{1 \leq i \leq N} M(X^i, dy^i).$$

Proof :
Using the Bayes rule (4.18), we prove that for any bounded functions F and G on $\mathcal{M}(S)$

$$\mathbb{E}\left(F(\mathcal{Y}) \left\{\int G\left(m_N(x)\right) \prod_{1 \leq i \leq N} \widehat{M}(Y^i, dx^i)\right\}\right)$$

$$= e^{-\gamma(1)} \sum_{p \geq 0} \frac{\gamma(1)^p}{p!} \times \int F\left(m_p(y)\right) G\left(m_p(x)\right) \prod_{1 \leq i \leq p} \left[(\eta M)(dy^i)\widehat{M}(y^i, dx^i)\right]$$

$$= e^{-\gamma(1)} \sum_{p \geq 0} \frac{\gamma(1)^p}{p!} \times \int F\left(m_p(y)\right) G\left(m_p(x)\right) \prod_{1 \leq i \leq p} \left[\eta(dx) \times M(x, dy)\right]$$

$$= \mathbb{E}\left(F(\mathcal{Y})G(\mathcal{X})\right).$$

By symmetry arguments, the second assertion is a direct consequence of the first one. This ends the proof of the lemma. $\blacksquare$

Important remark : The conditional first moments of $\mathcal{X}$ given $\mathcal{Y}$ are given for any function f on S by

$$\widehat{\gamma}(f) := \mathbb{E}\left(\mathcal{X}(f) \mid \mathcal{Y}\right) = \sum_{1 \leq i \leq N} \int f(x^i) \prod_{1 \leq i \leq N} \widehat{M}(Y^i, dx^i)$$
$$= \sum_{1 \leq i \leq N} \widehat{M}(f)(y^i) = \mathcal{Y}\widehat{M}(f).$$

These conditional intensities can be used to estimate the location of "real" data points in some region, to evaluate their correlations, or to design some clusters, using the observation process $\mathcal{Y}$.

Take for example the linear Gaussian model presented in example 4.6.6. Assume that we have a Poisson observation process $\mathcal{Y} = \sum_{1 \leq i \leq N} Y^i$, with $N \neq 0$. Then we have

$$\mathbb{E}\left(\frac{1}{N} \sum_{1 \leq i \leq N} X^i \mid \mathcal{Y}\right) = \left(\frac{\widehat{\sigma}}{\sigma}\right)^2 m + \left(1 - \left(\frac{\widehat{\sigma}}{\sigma}\right)^2\right) \frac{1}{N} \sum_{1 \leq i \leq N} Y^i \qquad (4.19)$$

and

$$N\,\mathbb{E}\left(\left[\frac{1}{N} \sum_{1 \leq i \leq N} X^i - \mathbb{E}\left(\frac{1}{N} \sum_{1 \leq i \leq N} X^i \mid \mathcal{Y}\right)\right]^2 \mid \mathcal{Y}\right) = \widehat{\sigma}^2. \qquad (4.20)$$

We refer to exercise 53 for a detailed proof of these assertions.

4.6.3 Poisson-Gaussian clusters

We return to the Gaussian model discussed in example 4.6.6. We replace the prior distribution η by the mixed Gaussian model

$$\eta(dx) = \epsilon \, \eta_1(dx) + (1 - \epsilon) \, \eta_0(dx)$$

where $\epsilon \in [0, 1]$, and η_1 and η_0 are Gaussian distributions with mean and variance parameters (m_1, σ_1^2) and (m_0, σ_0^2). We let N be a Poisson random variable with some parameter, say $\lambda > 0$. We sample N, and independently N independent r.v. $(X^i)_{1 \leq i \leq N}$ with law η. Each of these points is observed with some noise

$$Y^i = a \, X^i + V^i$$

where a is a given real number, and V^i stands for a sequence of independent centered Gaussian r.v. with variance τ^2. Given the observation process $\mathcal{Y} = \sum_{1 \leq i \leq N} \delta_{Y^i}$ we want to estimate the conditional intensity

$$\widehat{\gamma}(f) := \mathbb{E}\left(\mathcal{X}(f) \mid \mathcal{Y}\right).$$

By construction, we have

$$
\begin{aligned}
\mathbb{P}\left(Y^i \in dy\right) &= \int \eta(dx) \, M(x, dy) \\
&= \epsilon \int \eta_1(dx) \, M(x, dy) + (1 - \epsilon) \int \eta_0(dx) \, M(x, dy) \\
&:= \epsilon \, \eta_1 M(dy) + (1 - \epsilon) \, \eta_0 M(dy).
\end{aligned}
$$

The distribution $\eta_i M$ clearly has a Gaussian density

$$p_i(y) = \frac{1}{\sqrt{2\pi(a^2\sigma_i^2 + \tau^2)}} \exp\left(-\frac{1}{2(a^2\sigma_i^2 + \tau^2)} \, (y - am_i)^2\right)$$

so that

$$\mathbb{P}\left(Y^i \in dy\right) = p(y)dy \quad \text{with} \quad p(y) := \epsilon \, p_1(y) + (1 - \epsilon) \, p_0(y).$$

In addition, using Bayes rule, we check that $\widehat{M}(y, dx)$ has the density

$$p(x|y) = \frac{1}{p(y)} \, p(y|x) \, p(x)$$

where $p(y|x)$ stands for the Gaussian density of $M(x, dy)$ and $p(x)$ denotes the mixture Gaussian density of $\eta(dx)$. A simple calculation shows that

$$p(x|y) = \frac{\epsilon \, p_1(y)}{\epsilon \, p_1(y) + (1 - \epsilon) \, p_0(y)} \, p_1(x|y) + \frac{(1 - \epsilon) \, p_0(y)}{\epsilon \, p_1(y) + (1 - \epsilon) \, p_0(y)} \, p_0(x|y)$$

with the Gaussian densities

$$p_i(x|y) := \frac{1}{\sqrt{2\pi\widehat{\sigma}_i^2}} \exp\left(-\frac{1}{2\widehat{\sigma}_i^2} \, (x - \widehat{m}_i(y))^2\right) \, dx$$

with $\widehat{\sigma}_i^2 = \left(a^2\tau^{-2} + \sigma_i^{-2}\right)^{-1}$, and the linear regression parameters given by

$$\widehat{m}_i(y) = \left(\frac{\widehat{\sigma}_i}{\sigma_i}\right)^2 m_i + \left(1 - \left(\frac{\widehat{\sigma}_i}{\sigma_i}\right)^2\right) a^{-1} y.$$

We conclude that

$$\widehat{\gamma}(dx) = \sum_{1 \le i \le N} \frac{\epsilon \, p_1(Y^i)}{\epsilon \, p_1(Y^i) + (1-\epsilon) \, p_0(Y^i)} \; p_1(x|Y^i) \, dx$$

$$+ \frac{(1-\epsilon) \, p_0(Y^i)}{\epsilon \, p_1(Y^i) + (1-\epsilon) \, p_0(Y^i)} \; p_0(x|Y^i) \, dx.$$

When m_1 are m_0 are far enough (and $a \ne 0$), this (weighted) conditional distribution allows us to separate or cluster the observations Y^i which are close to am_1 and the ones close to am_0. The number of observations in each cluster provides an estimate of the initial number of states X^i that were sampled according to each distribution $p_i(x)$. The conditional distributions allow us to estimate their locations in each cluster.

4.7   Exercises

Exercise 35 (Geometric) *We let X be an exponential r.v. with parameter $\lambda = -\log(1-p)$, with $p \in \,]0,1[$. Check that $Y = 1 + \lfloor X \rfloor$ is a geometric r.v. with parameter p.*

Exercise 36 (Cauchy) *We let X be a Cauchy r.v. with parameter $\sigma > 0$; that is,*

$$\mathbb{P}(X \in dx) = \frac{\sigma}{\pi} \, \frac{1}{\sigma^2 + x^2} \, dx.$$

Check that

$$\mathbb{P}(X \le x) = \frac{1}{2} + \frac{1}{\pi} \arctan\left(\frac{x}{\sigma}\right) \quad \text{and deduce that} \quad X \stackrel{law}{=} \sigma \, \tan\left(\pi \left(U - \frac{1}{2}\right)\right)$$

where U denotes a uniform r.v. on $[0,1]$.

Exercise 37 (Rayleigh) *We let X be a Rayleigh r.v. with parameter $\sigma > 0$; that is,*

$$\mathbb{P}(X \in dx) = \frac{x}{\sigma^2} \, e^{-\frac{x^2}{2\sigma^2}} \, 1_{[0,\infty[}(x) \, dx.$$

Check that

$$\mathbb{P}(X \le x) = 1 - e^{-\frac{x^2}{2\sigma^2}} \quad \text{and} \quad X \stackrel{law}{=} \sigma \, \sqrt{-2\log U}$$

where U stands for a uniform r.v. on $[0,1]$.

Exercise 38 (Pareto) *We let X be a Pareto r.v. with parameters $(a,b) \in (\mathbb{R} \times \,]0,\infty[)$; that is,*

$$\mathbb{P}(X \in dx) = \frac{ab^a}{x^{a+1}} \, 1_{[b,\infty[}(x) \, dx.$$

Check that

$$\mathbb{P}(X \le x) = 1 - \left(\frac{b}{x}\right)^a \quad \text{and} \quad X \stackrel{law}{=} b \, U^{-\frac{1}{a}}$$

where U stands for a uniform r.v. on $]0,1]$.

Exercise 39 (Triangular) *We let X be a triangular r.v. with parameter $a \in]0, \infty[$; that is,*

$$\mathbb{P}(X \in dx) = \frac{2}{a} \left(1 - \frac{x}{a}\right) 1_{[0,a]}(x) \, dx.$$

Check that

$$\mathbb{P}(X \le x) = \frac{2}{a} \left(x - \frac{x^2}{2a}\right) \quad and \quad X \overset{law}{=} a \left(1 - \sqrt{U}\right)$$

where U stands for a uniform r.v. on $]0,1]$.

Exercise 40 (Weibull) *We let X be a Weibull r.v. with parameter $(a, b) \in]0, \infty[^2$; that is,*

$$\mathbb{P}(X \in dx) = \frac{a}{b^a} \, x^{a-1} \, e^{-(x/b)^a} \, 1_{[0,\infty[}(x) \, dx.$$

Check that

$$\mathbb{P}(X \le x) = 1 - e^{-(x/b)^a} \quad and \quad X \overset{law}{=} b \, (-\log U)^{\frac{1}{a}}$$

where U stands for a uniform r.v. on $]0,1]$.

Exercise 41 (Uniform ordered statistics) *We consider a sequence of independent and exponential r.v. $(E_n)_{n \ge 1}$ with parameter $\lambda > 0$, and for any $n \ge 1$ we set*

$$T_n = T_{n-1} + E_n \quad with \quad T_0 = 0.$$

For any bounded function f on $[0, \infty[^n$, prove that

$$\mathbb{E}(f(T_1, \ldots, T_n)) = \int_{[0,\infty[^n} f(s_1, s_1 + s_2, \ldots, s_1 + \ldots + s_n) \, \lambda^n e^{-\lambda(s_1 + \ldots + s_n)} ds_1 \ldots ds_n.$$

- *Using the change of variable formula*

$$(\forall k \ge 1 \quad t_k = s_1 + \ldots + s_k) \Longleftrightarrow (\forall k \ge 0 \quad s_k = t_k - t_{k-1})$$

with the convention $t_0 = 0$, we have $ds_k = dt_k$, for any $1 \le k \le n$ and

$$\mathbb{E}(f(T_1, \ldots, T_n)) = \int_{t_1 < t_2 < \ldots < t_n} f(t_1, t_2, \ldots, t_n) \, \lambda^n e^{-\lambda t_n} dt_1 \ldots dt_n.$$

- *Using induction, prove that*

$$\int_{t_1 < t_2 < \ldots < t_n < t} dt_1 \ldots dt_n = \frac{t^n}{n!}$$

and conclude that $T_n = E_1 + \ldots + E_n$ is a gamma r.v. with parameter (n, λ).

$$\mathbb{P}(T_n \in dt) = \frac{t^{n-1}}{(n-1)!} \times \lambda^n \, e^{-\lambda t} \, 1_{[0,\infty[}(t) \, dt. \qquad (4.21)$$

This distribution is sometimes called the Erlang distribution with parameter (n, λ). We often write in a more synthetic way $T_n \sim \text{gamma}(n, \lambda)$.

- *When $\lambda = 1$, show that T_{n+1} is a gamma r.v. with parameter $((n+1), 1)$; that is,*

$$\mathbb{P}(T_{n+1} \in dt) = \frac{t^n}{n!} \times e^{-t} \, 1_{[0,\infty[}(t) \, dt. \qquad (4.22)$$

- *For any $\gamma > 0$ check that*

$$\frac{\gamma^n}{n!}\, e^{-\gamma} = \int_\gamma^\infty \frac{\partial}{\partial t}\left(-\frac{t^n}{n!}\, e^{-t}\right)\, dt \quad and \quad \frac{\partial}{\partial t}\left(-\frac{t^n}{n!}\, e^{-t}\right) = e^{-t}\frac{t^n}{n!} - e^{-t}\frac{t^{n-1}}{(n-1)!}.$$

Deduce from the above that

$$\frac{\gamma^n}{n!}\, e^{-\gamma} = \mathbb{P}\left(T_{n+1} > \gamma\right) - \mathbb{P}\left(T_n > \gamma\right) = \mathbb{P}\left(T_{n+1} > \gamma \ and \ T_n \le \gamma\right)$$

with the random times T_n defined in (4.22). Conclude that

$$\forall n \ge 0 \qquad \frac{\gamma^n}{n!}\, e^{-\gamma} = \mathbb{P}\left(N = n\right) \quad with \quad N := \inf\left\{n \ge 0 \ : \ T_{n+1} \ge \gamma\right\}.$$

- *Using the fact that*

$$1_{t_1 < \ldots < t_{n+1}} \lambda^{n+1}\, e^{-\lambda t_{n+1}} dt_1 \ldots dt_{n+1}$$

$$= \left(\frac{n!}{t_{n+1}^n} 1_{t_1 < \ldots < t_n < t_{n+1}} dt_1 \ldots dt_n\right) \times \left(\lambda e^{-\lambda t_{n+1}} \frac{(\lambda t_{n+1})^n}{n!}\, 1_{[0,\infty[}(t_{n+1})\, dt_{n+1}\right)$$

prove that

$$\mathbb{P}\left((T_1, \ldots, T_n) \in d(t_1, \ldots, t_n) \ \middle| \ T_{n+1} = t\right) = \frac{n!}{t^n}\, 1_{t_1 < \ldots < t_n < t}\, dt_1 \ldots dt_n$$

and conclude that

$$\mathbb{P}\left(\left(\frac{T_1}{T_{n+1}}, \ldots, \frac{T_n}{T_{n+1}}\right) \in d(v_1, \ldots, v_n) \ \middle| \ T_{n+1} = t\right) = n!\, 1_{0 \le v_1 < \ldots < v_n < 1}\, dv_1 \ldots dv_n.$$

- *For any sequence of uniform r.v. $(U_1, \ldots, U_n)$ there exists a permutation σ_U of the indices $\{1, \ldots n\}$ s.t.*

$$V_1 = U_{\sigma_U(1)} \le V_2 = U_{\sigma_U(2)} \le \ldots \le V_n = U_{\sigma_U(n)}.$$

Check that

$$\mathbb{P}\left((V_1, \ldots, V_n) \in d(v_1, \ldots, v_n) \ and \ \sigma_U = \tau\right) = 1_{0 \le v_1 < \ldots < v_n}\, dv_1 \ldots dv_n.$$

Conclude that

$$\left(\frac{T_1}{T_{n+1}}, \ldots, \frac{T_n}{T_{n+1}}\right) \stackrel{law}{=} (V_1, \ldots, V_n).$$

The sequence of r.v. $(V_1, \ldots, V_n)$ is sometimes called uniform order statistics on $[0,1]$.

Exercise 42 (Acceptance rejection sampling) *We consider the following cases*

$$1) \quad p(x) = \frac{2}{\pi}\, \sqrt{1-x^2}\, 1_{[-1,1]}(x) \qquad\qquad q(x) = \frac{1}{2}\, 1_{[-1,1]}(x)$$

$$2) \quad p(x) = \frac{1}{\pi}\, 1_{\{(y_1,y_2) \in \mathbb{R}^2 \ : \ y_1^2 + y_2^2 < 1\}}(x) \quad q(x) = \frac{1}{4}\, 1_{[-1,1]^2}(x)$$

$$3) \quad p(x) = \frac{1}{\sqrt{2\pi}}\, \exp\left(-\frac{x^2}{2}\right) \qquad\qquad q(x) = \frac{1}{\pi(x^2+1)}.$$

1. Prove that $C = \sup_{x \in [-1,1]} \frac{p(x)}{q(x)} = \frac{4}{\pi}$.

2. *Prove that* $C = \sup_{x \in [-1,1]^2} \frac{p(x)}{q(x)} = \frac{4}{\pi}$.

3. *In the third case, check that*

$$\frac{\partial}{\partial x} (p/q)(x) = \sqrt{\frac{\pi}{2}} \, x \, (1 - x) \, (1 + x) \, e^{-x^2/2}$$

and

$$C = \sup_{x \in \mathbb{R}} (p/q)(x) = (p/q)(1) = (p/q)(-1) = \sqrt{\frac{2\pi}{e}}.$$

Exercise 43 (Acceptance rejection sampling - gamma) *The density of a* gamma(α, λ)
r.v. X with parameters $\alpha, \lambda > 0$ is given by

$$p(x) = \frac{\lambda^\alpha}{\Gamma(\alpha)} \, x^{\alpha-1} \, e^{-\lambda x} \, 1_{]0,\infty[}(x).$$

*When $\alpha = n \in \mathbb{N}$ is an integer number, we have seen in (4.21) that X can be sampled by
summing n independent exponential r.v. with parameter $\lambda > 0$. When $\lambda > 1$, we can use
the rejection technique with the* gamma$(\lfloor \alpha \rfloor, \lambda - 1)$ *instrumental density*

$$q(x) = \frac{1}{\Gamma(\lfloor \alpha \rfloor)} \, x^{\lfloor \alpha \rfloor - 1} \, (\lambda - 1)^{\lfloor \alpha \rfloor} \, e^{-(\lambda-1)x} \, 1_{]0,\infty[}(x).$$

Prove that

$$\frac{p(x)}{q(x)} = \frac{\Gamma(\lfloor \alpha \rfloor)}{\Gamma(\lfloor \alpha \rfloor)} \, \frac{\lambda^\alpha}{(\lambda - 1)^{\lfloor \alpha \rfloor}} \, x^{\alpha - \lfloor \alpha \rfloor} \, e^{-x} \, 1_{]0,\infty[}(x)$$

and

$$C = \sup_{x \in]0,\infty[} \frac{p(x)}{q(x)} = \frac{p(\alpha - \lfloor \alpha \rfloor)}{q(\alpha - \lfloor \alpha \rfloor)}.$$

Prove that for any $\lambda > 0$, we have that gamma$(\alpha, \lambda) \overset{law}{=} \lambda \times$ gamma$(\alpha, 1)$. *Propose a way
of sampling the distribution* gamma(α, λ), *when $\lambda \leq 1$.*

Exercise 44 (Sums of Gamma) *If $X \sim$ gamma(a, c) and $Y \sim$ gamma(b, c) are independent r.v. with parameters $a, b, c > 0$, prove that*

$$X + Y \sim \text{gamma}(a + b, c).$$

Exercise 45 (Dirichlet beta gamma) *If $X \sim$ gamma(a, c) and $Y \sim$ gamma(b, c) are
independent r.v. with parameters $a, b, c > 0$, prove that*

$$\frac{X}{X + Y} \sim \text{beta}(a, b)$$

where beta(a, b) *stands for the Beta distribution with parameter (a, b) defined by the probability density*

$$p(u) = \frac{\Gamma(a + b)}{\Gamma(a)\Gamma(b)} \, u^{a-1} \, (1 - u)^{b-1} \, 1_{[0,1]}(u). \tag{4.23}$$

Notice that for $a = b = 1$ we have $p(u) = 1_{[0,1]}(u)$.

Exercise 46 (Dirichlet and gamma) *Let $X_i \sim \text{gamma}(a_i, c)$, $1 \leq i \leq d$ be a collection of independent r.v. with parameters $c > 0$, and $a_i > 0$, for each $1 \leq i \leq d$. Prove that*

$$\left(\frac{X_1}{\sum_{1 \leq i \leq d} X_i}, \ldots, \frac{X_d}{\sum_{1 \leq i \leq d} X_i} \right) \sim D(a_1, \ldots, a_d).$$

Combining this formula with the one presented in exercise 44 prove that for any $(U_1, \ldots, U_d) \sim D(a_1, \ldots, a_d)$, and any $1 \leq i < j \leq d$, we have the coagulation (a.k.a aggregation) formula

$$(U_1, \ldots, U_{i-1}, U_i + U_j, U_{i+1}, \ldots, U_j, U_{j+1}, \ldots, U_d)$$

$$\sim D(a_1, \ldots, a_{i-1}, a_i + a_j, a_{i+1}, \ldots, a_j, a_{j+1}, \ldots, a_d).$$

Exercise 47 (Dirichlet - Mean and variance) *Let $(U_1, \ldots, U_d) \sim D(a_1, \ldots, a_d)$, for some concentration parameters $(a_i)_{1 \leq i \leq d}$. Prove that*

$$\forall 1 \leq i \leq d \quad \mathbb{E}(U_i) = \frac{a_i}{\sum_{1 \leq i \leq d} a_i} \quad \text{and} \quad \mathbb{E}(U_i^2) = \frac{1 + a_i}{1 + \sum_{1 \leq i \leq d} a_i} \times \frac{a_i}{\sum_{1 \leq i \leq d} a_i}.$$

Exercise 48 (Bayes' formula) *We consider the partial observation model presented in (4.6) and (4.7). Suppose that we have n conditionally independent observations $(X_i)_{1 \leq i \leq n}$ of the parameter Θ. Prove that*

$$\mathbb{E}\left(\Theta \mid (X_1 \ldots, X_n) \right) \longrightarrow_{n \uparrow \infty} \Theta.$$

Exercise 49 (Conditional distribution - Gaussian variables) *We consider the partial observation model (4.10). Suppose that we have n conditionally independent observations*

$$\forall 1 \leq i \leq n \quad Y_i = aX + V_i$$

of the r.v. X, where $(V_i)_{1 \leq i \leq n}$ stands for n independent copies of V. Prove that the conditional density of X given $(Y_1, \ldots, Y_n)$ is the Gaussian density

$$p(x \mid y_1, \ldots, y_n) \propto \exp\left(-\frac{n}{2\rho_n} \left((x - m) - \beta_n \frac{1}{n} \sum_{1 \leq i \leq n} (y_i - am) \right)^2 \right)$$

with the parameters

$$\beta_n = a \, \sigma^2 n / (a^2 \sigma^2 n + \tau^2) \to a^{-1} \quad \text{and} \quad \rho_n = \left(a^2 \tau^{-2} + \sigma^{-2}/n \right)^{-1} \to \tau^2 / a^2.$$

Deduce that

$$\mathbb{E}\left(X \mid (Y_1, \ldots, Y_n) \right) \longrightarrow_{n \uparrow \infty} X.$$

Exercise 50 (Conditional distribution - Bernoulli variables) *We let X be a uniform random variable on $[0, 1]$. Given X, we let $Y = (Y_n)_{n \geq 1}$ be a sequence of i.i.d. Bernoulli random variables with common law*

$$\mathbb{P}(Y_1 = 1 \mid X) = X = 1 - \mathbb{P}(Y_1 = 0).$$

Compute the conditional distribution of X given a sequence of observations $(Y_k)_{0 \leq k \leq n}$.

Exercise 51 (Gamma and exponential) *Prove that the Gamma distributions*

$$\mathbb{P}(X \in dx) \propto x^{a-1}\, e^{-b\,x}\, 1_{]0,\infty[}(x)\, dx$$

associated with some parameters $a, b > 0$ are conjugate with the exponential densities $\mathbb{P}(Y \in y|x) \propto x\, e^{-xy} 1_{[0,\infty[}(y)\, dy$, with $c > 0$, as well as with the Poisson distributions

$$\forall y \in \mathbb{N} \quad \mathbb{P}\left(Y = y \mid X = x\right) = \frac{x^y}{y!}\, e^{-x}.$$

Exercise 52 (Dirichlet and multinomial) *Prove that the Dirichlet probability distributions $D(a_1, \ldots, a_d)$ defined in (4.8) are conjugate with the multinomial distributions $\mathrm{Multi}(n, p_1, \ldots, p_n)$ presented in (4.4).*

Exercise 53 *Check the formulae (4.19) and (4.20).*

Exercise 54 (Gaussian integration by part) *We let W be a centered Gaussian random variable with unit variance. For any differentiable function f such that $W f(W)$ and $f'(W)$ are integrable check the integration by part formula*

$$\mathbb{E}\left(f'(W)\right) = \mathbb{E}(W f(W)).$$

5

Monte Carlo integration

The rejection method from the previous chapter serves as a basis for a variety of simulation algorithms of the importance sampling class of algorithms. These algorithms can be used to evaluate expected values of functionals with high precision. They form the core of the sequential Monte Carlo methods and of the mean field particle integration theory. Applications of the techniques for precise evaluation of tail probabilities, or, similarly, in rare event simulation, are also presented at the end of the chapter.

> *A person who never made a mistake never tried anything new.*
> Albert Einstein (1879-1955).

5.1 Law of large numbers

Monte Carlo integration is a powerful numerical technique that uses statistical and probabilistic techniques for evaluating integrals that could not easily be evaluated analytically. Buffon's needle problem is one of the oldest Monte Carlo integration problems posed in the 18th century by Georges-Louis Leclerc, Comte de Buffon. A Youtube video by Dr Tony Padilla illustrates this experiment using 163 matches. Suppose we are given an integral of the form

$$\eta(f) := \int f(x)\,\eta(dx) = \mathbb{E}(f(X))$$

where $\eta(dx)$ is a probability measure on some state space S, and f is a function from S into $\mathbb{R}$ such that $\eta(|f|) = \mathbb{E}(|f(X)|) < \infty$.

> The central idea is to sample a sequence of independent random copies $(X^i)_{i \geq 1}$ of the r.v. X and to use the so-called empirical average estimates
>
> $$\eta^N(f) := \int f(x)\,\eta^N(dx) = \frac{1}{N}\sum_{1 \leq i \leq N} f(X^i) \quad \text{with} \quad \eta^N := \frac{1}{N}\sum_{1 \leq i \leq N} \delta_{X^i}. \quad (5.1)$$
>
> In (5.1) δ_a stands for the Dirac measure at some point $a \in S$.

We set

$$\sqrt{N}\left(\eta^N(f) - \eta(f)\right) := V^N(f) \Longleftrightarrow \eta^N(f) = \eta(f) + \frac{1}{\sqrt{N}}\,V^N(f). \qquad (5.2)$$

The r.h.s. formula in the above display can be interpreted as a first order type decomposition of the random deviations between the empirical measure η^N and its limiting value η.

In this notation, a simple calculation shows that

$$\mathbb{E}\left(V^N(f)\right) = 0 \quad \text{and} \quad \mathbb{E}\left(V^N(f)^2\right) = \sigma^2(f) \qquad (5.3)$$

for any function f such that $\eta(f^2) < \infty$ with

$$\begin{aligned}
\sigma^2(f) &:= \mathbb{E}\left(f(X)^2\right) - \mathbb{E}\left(f(X)\right)^2 = \eta(f^2) - \eta(f)^2 \\
&= \eta([f - \eta(f)]^2) = \frac{1}{2}\int (f(x) - f(y))^2\,\eta(dx)\eta(dy). \qquad (5.4)
\end{aligned}$$

For any couple of functions (f_1, f_2) s.t. $\max_{i=1,2} \eta(|f_i|) < \infty$, and any $N \geq 1$ we also have the formula

$$\mathbb{E}\left(V^N(f_1)V^N(f_2)\right) = C(f_1, f_2) \qquad (5.5)$$

with the covariance function

$$\begin{aligned}
C(f_1, f_2) &:= \eta([f_1 - \eta(f_1)] \times [f_2 - \eta(f_2)]) \\
&= \frac{1}{2}\int (f_1(x) - f_1(y))\,(f_2(x) - f_2(y))\,\eta(dx)\eta(dy). \qquad (5.6)
\end{aligned}$$

It is a simple exercise (cf. exercise 56) to prove that

$$\mathbb{E}\left(\eta^N(f)\right) = \eta(f) \quad \text{and} \quad \mathbb{E}\left(\left[\eta^N(f) - \eta(f)\right]^2\right) = \sigma^2(f)/N \downarrow_{N\uparrow\infty} 0 \qquad (5.7)$$

for any function f s.t. $\eta(f^2) < \infty$.

Working a little harder we prove the following theorem.

Theorem 5.1.1 (Law of large numbers)

$$\mathbb{E}(|f(X)|) < \infty \quad \Longrightarrow \quad \lim_{N\to\infty} \eta^N(f) = \eta(f) \quad \mathbb{P} - a.s.$$

This theorem and the fluctuation theorem given below are central in any first course on stochastic simulation and Monte Carlo methods.

Theorem 5.1.2 (Central limit theorem) *For any sequence of functions $(f_i)_{1 \leq i \leq d}$ s.t. $\mathbb{E}(f_i(X)^2) < \infty$ for any $1 \leq i \leq d$, as $N \to \infty$ we have the convergence in distribution*

$$\left(V^N(f_1), \ldots, V^N(f_d)\right) \implies (V(f_1), \ldots, V(f_d))$$

where $(V(f_1), \ldots, V(f_d))$ is a sequence of <u>centered Gaussian</u> r.v. with covariance function given for any $1 \leq i, j \leq d$ by

$$\mathbb{E}\left(V(f_i)V(f_j)\right) = \eta([f_i - \eta(f_i)]\,[f_j - \eta(f_j)]).$$

Hint of proof :

We assume without loss of generality that the functions f_i are centered, that is, $\eta(f_i) = 0$ for any $1 \leq i \leq d$. We consider a column vector $\lambda := (\lambda_1, \ldots, \lambda_d)' \in \mathbb{R}^d$, the column vector function $f = (f_1, \ldots, f_d)'$, the column random numbers $V^N(f) = (V^N(f_1), \ldots, V^N(f_d))'$, and we set $g = \lambda' f = \sum_{1 \leq j \leq d} \lambda_j f_j$.

$$\mathbb{E}\left(e^{i\lambda' V^N(f)}\right) = \mathbb{E}\left(e^{i \sum_{1 \leq j \leq d} \lambda_j V^N(f_j)}\right) = \mathbb{E}\left(e^{iV^N(g)}\right).$$

Recalling that $V^N(g) = \frac{1}{\sqrt{N}} \sum_{1 \leq j \leq N} [g(X^j) - \eta(g)]$ and using the fact that $\eta(g) = 0$ we check that

$$\begin{aligned}
\mathbb{E}\left(e^{iV^N(g)}\right) &= \left(\mathbb{E}\left(e^{i\frac{g(X)}{\sqrt{N}}}\right)\right)^N \\
&\simeq \left(\mathbb{E}\left(1 + i\frac{g(X)}{\sqrt{N}} - \frac{g(X)^2}{2N}\right)\right)^N = \left(1 - \frac{\eta(g^2)}{2N}\right)^N \simeq e^{-\eta(g^2)/2}.
\end{aligned}$$

The next step is to observe that

$$g^2 = \sum_{1 \leq j,k \leq d} \lambda_j \lambda_k f_j f_k \Rightarrow \eta(g^2) = \sum_{1 \leq j,k \leq d} \lambda_j \lambda_k C(f_j, f_k) = \lambda' C(f,f) \lambda$$

with the covariance matrix $C(f,f) := (C(f_i, f_j))_{1 \leq i,j \leq d}$. The end of the proof is left to the reader. ∎

Remark : Theorem 5.1.2 is often used in statistical inference to derive confidence intervals. Notice that $V(f)/\sigma(f) \sim N(0,1)$.

$$\left|\eta^N(f) - \eta(f)\right| = \frac{V^N(f)}{\sqrt{N}} \leq \lambda \frac{\sigma(f)}{\sqrt{N}} \quad \Leftrightarrow \quad \frac{V^N(f)}{\sigma(f)} \leq \lambda$$

so that

$$\mathbb{P}\left(\eta(f) \in \left[\eta^N(f) - \lambda \frac{\sigma(f)}{\sqrt{N}} , \eta^N(f) + \lambda \frac{\sigma(f)}{\sqrt{N}}\right]\right)$$

$$\simeq \mathbb{P}\left(|N(0,1)| \leq \lambda\right) = 2 \int_{-\infty}^{\lambda} \frac{1}{\sqrt{2\pi}} e^{-x^2/2} \, dx - 1 \simeq .95 \quad \text{for } \lambda = 1.96.$$

Remark :

In the settings of theorem 5.1.2, we also have Wick's formulae

$$\mathbb{E}\left(V(f_1) \ldots V(f_n)\right) = \sum_{1 \leq i < n} \mathbb{E}\left(\prod_{1 \leq j \neq i < n} V_N(f_j)\right) \times \mathbb{E}(V(f_i)V(f_n)) \tag{5.8}$$

and

$$\mathbb{E}\left(V(f_1) \ldots V(f_{2n})\right) = \sum_{P \in \mathcal{P}_n} \prod_{\{i,j\} \in P} \mathbb{E}\left(V(f_i)V(f_j)\right). \tag{5.9}$$

In the above display, $\mathcal{P}_n$ stands for the set of $(2n)!/(n!2^n)$ partitions of $\{1, \ldots, 2n\}$ into n partitions of 2 (ordered) indexes. A given partition $P \in \mathcal{P}_n$ is also called a pairing. Notice that

$$(5.8) \Rightarrow \mathbb{E}\left(V(f_1) \ldots V(f_{2n+1})\right) = 0$$

and for homogeneous functions $f_i = f$, for any $1 \leq i \leq n$ we have

$$\mathbb{E}\left(V(f)^{2n+1}\right) = 0 \quad \text{and} \quad \mathbb{E}\left(V(f)^{2n}\right) = \frac{(2n)!}{n!2^n} \mathbb{E}\left(V(f)^2\right). \tag{5.10}$$

The proof of Wick's formulae is discussed in exercise 61.

5.2 Importance sampling

5.2.1 Twisted distributions

The importance sampling technique is closely related to the rejection method discussed in section 4.3 (cf. lemma 4.3.2). We let (Y, X) be a couple of $\mathbb{R}^d$-valued r.v. with probability densities (p, q) with respect to the Lebesgue measure dx on $\mathbb{R}^d$. We further assume that the ratio $p(x)/q(x)$ exists for any $x \in \mathbb{R}^d$. We denote by $\mu(dy) = p(y)dy$ the distribution of Y and by $\eta(dx) = q(x)dx$ the law of X. We assume that $\eta((p/q)^2) < \infty$.

Our next objective is to estimate the quantities $\mathbb{E}(f(Y))$, for any function f on $\mathbb{R}^d$ s.t. $\eta(|f|^2) < \infty$, using a sequence of independent random copies $(X^i)_{i \geq 1}$ of the r.v. X. In importance sampling literature, the law of X is sometimes called the importance or the twisted probability.

To this end, we use the formula

$$p(y) = \frac{1}{\int g(y') \, q(y') \, dy'} \, g(y) \, q(y) \quad \text{with the function} \quad g = p/q$$

to check that the μ and η are related by the (non-linear) transformation

$$\mu(dy) = \Psi_g(\eta)(dy) := \frac{1}{\eta(g)} \, g(y) \, \eta(dy) \quad \text{with} \quad \eta(g) = \int g(y) \, \eta(dy).$$

The transformation Ψ_g is called the Boltzmann-Gibbs transformation associated with the potential function g. This implies that

$$\begin{aligned}
\mathbb{E}\left(f(Y)\right) &= \mu(f) = \int \Psi_g(\eta)(dy) \, f(y) = \frac{1}{\eta(g)} \int f(y) \, g(y) \, \eta(dy) \\
&= \eta(fg)/\eta(g) = \mathbb{E}\left(f(X)g(X)\right)/\mathbb{E}\left(g(X)\right).
\end{aligned}$$

Replacing the law η of X by its empirical approximation η^N discussed in section 5.1, we construct the following importance sampling approximation

$$\frac{\eta^N(fg)}{\eta^N(g)} = \sum_{1 \leq i \leq N} \frac{g(X^i)}{\sum_{1 \leq j \leq N} g(X^j)} \, f(X^i) \simeq_{N \uparrow \infty} \mathbb{E}\left(f(Y)\right).$$

5.2.2 Sequential Monte Carlo

In terms of probability measures, we have the weighted approximation

$$\Psi_g(\eta^N) = \sum_{1 \leq i \leq N} \frac{g(X^i)}{\sum_{1 \leq j \leq N} g(X^j)} \, \delta_{X^i} \simeq_{N \uparrow \infty} \mu = \text{Law}(Y). \tag{5.11}$$

The formula (5.11) shows that the sampling from the l.h.s. discrete probability measure is "almost" equivalent to that of sampling copies of the r.v. Y. This observation is at the core of sequential Monte Carlo methods (a.k.a. particle filtering methods) and of the mean field particle integration theory [39, 66, 67, 111].

We emphasize that these particle estimates do not depend on the normalizing constants of the densities p and q. This fact is important since the function $g = p/q$ is often known up to some normalizing constant. When the function g is explicitly known, we can use the estimate

$$\eta^N(fg) := \frac{1}{N} \sum_{1 \leq i \leq N} g(X^i) \, f(X^i) \simeq_{N \uparrow \infty} \eta(fg) = \mu(f). \tag{5.12}$$

5.2.3 Tails distributions

In this section we give a brief discussion on the use of importance sampling techniques in rare event simulation. We consider a standard Gaussian r.v. X. Suppose we want to evaluate the quantity

$$\mathbb{P}\left(X \in [a, \infty[\right) = \mathbb{E}\left(1_{[a,\infty[}(X)\right) = \eta(1_{[a,+\infty[})$$

with the Gaussian distribution η of X. We consider the Monte Carlo approximation (5.1) associated with independent copies $(X^i)_{i \geq 1}$ of X and defined by

$$\eta^N(1_{[a,+\infty[}) = \frac{1}{N} \sum_{1 \leq i \leq N} 1_{[a,+\infty[}(X^i).$$

We check easily that the relative variance of this estimator is given by

$$\mathbb{E}\left(\left[\frac{\eta^N\left(1_{[a,+\infty[}\right)}{\mathbb{P}(X \in [a,+\infty[)} - 1\right]^2\right) = \frac{1}{N} \ \mathbb{P}(X \geq a)^{-1} \ (1 - \mathbb{P}(X \geq a)).$$

Using Mill's inequalities (cf. exercise 64)

$$\forall a > 0 \qquad \frac{1}{a + a^{-1}} \ \frac{1}{\sqrt{2\pi}} \ e^{-\frac{a^2}{2}} \leq \mathbb{P}(X \geq a) \leq \frac{1}{a} \ \frac{1}{\sqrt{2\pi}} \ e^{-\frac{a^2}{2}} \tag{5.13}$$

for large values of a we find the equivalence relation

$$N \ \mathbb{E}\left(\left[\frac{\eta^N\left(1_{[a,+\infty[}\right)}{\mathbb{P}(X \geq a)} - 1\right]^2\right) \simeq a \ e^{\frac{a^2}{2}}. \tag{5.14}$$

We consider a r.v. Z with Gaussian density

$$q(x) \propto e^{-(x-a)^2/2} \quad \text{and we set} \quad p(x) \propto e^{-x^2/2}.$$

We let $\mu(dx) = q(x)dx$ be the twisted distribution. We observe that

$$g(x) := p(x)/q(x) = e^{\frac{1}{2}\left((x-a)^2 - x^2\right)} = e^{a\left(\frac{a}{2} - x\right)}.$$

We let $\mu^N = \frac{1}{N} \sum_{1 \leq i \leq N} \delta_{Z^i}$ be the empirical measure associated with N independent copies of Z. Using (5.12), we have

$$\mu^N\left(g \ 1_{[a,+\infty[}\right) \simeq_{N \uparrow \infty} \mu\left(1_{[a,+\infty[} \ g\right) = \mathbb{P}(X \geq a).$$

It is also easily checked that

$$N \ \mathbb{E}\left(\left[\mu^N\left(g \ 1_{[a,+\infty[}\right) - \mu\left(g \ 1_{[a,+\infty[}\right)\right]^2\right)$$

$$= \mu\left(g^2 \ 1_{[a,+\infty[}\right) - \mu\left(g \ 1_{[a,+\infty[}\right)^2 = \eta\left(g \ 1_{[a,+\infty[}\right) - \mathbb{P}(X \geq a)^2.$$

Now we observe that

$$\eta\left(g \ 1_{[a,+\infty[}\right) = \frac{\eta\left(g \ 1_{[a,+\infty[}\right)}{\eta\left(1_{[a,+\infty[}\right)} \times \eta\left(1_{[a,+\infty[}\right) = \mathbb{E}\left(\exp\left[a\left(\frac{a}{2} - X\right)\right] \mid X \geq a\right) \times \mathbb{P}(X \geq a)$$

$$\leq e^{-a^2/2} \times \mathbb{P}(X \geq a) \simeq a \ \mathbb{P}(X \geq a)^2.$$

In contrast to the exponential rate of the crude Monte Carlo method (5.14), this implies that the relative variance grows linearly with the parameter a

$$N \ \mathbb{E}\left(\left[\frac{\mu^N\left(g \ 1_{[a,+\infty[}\right)}{\mathbb{P}(X \geq a)} - 1\right]^2\right) \simeq a.$$

5.3   Exercises

Exercise 55 (Empirical random fields) *Check that the random and signed measures V^N satisfy the following property*

$$a_1 \, V^N(f_1) + a_2 \, V^N(f_2) = V^N \left(a_1 \, f_1 + a_2 \, f_2 \right)$$

for any pair $(a_1, a_2) \in \mathbb{R}^2$ and any couple of functions (f_1, f_2) s.t. $\max_{i=1,2} \eta(|f_i|) < \infty$.

Exercise 56 (Covariance functionals) *Prove formulae (5.5), (5.6), and deduce that*

$$[(5.5) \ \& \ (5.6)] \implies [(5.3) \ \& \ (5.4) \text{ with } C(f,f) = \sigma^2(f)] \ .$$

Using (5.2), prove (5.7).

Exercise 57 (Importance sampling 1) *Consider the couple of $\mathbb{R}^d$-valued random variables (Y, X) with probability densities (p, q) discussed in the beginning of section 5.2.1. We wish to estimate $\mathbb{E}(f(Y))$, for some bounded function f on $\mathbb{R}^d$ using a sequence of independent random copies $(X^i)_{i \geq 1}$ of the r.v. X. Prove that*

$$\mathbb{E}\left(f(Y) \right) = \mathbb{E}\left(f(X) \, \frac{p(X)}{q(X)} \right) = \mathbb{E}\left(\frac{1}{N} \sum_{1 \leq i \leq N} f(X^i) \, \frac{p(X^i)}{q(X^i)} \right)$$

and

$$N \, \mathrm{Var}\left(\frac{1}{N} \sum_{1 \leq i \leq N} f(X^i) \, \frac{p(X^i)}{q(X^i)} \right) = \mathbb{E}\left(f^2(Y) \, \frac{p(Y)}{q(Y)} \right) - \mathbb{E}\left(f(Y) \right)^2 .$$

Exercise 58 (Importance sampling 2) *We consider a couple of random walks starting at the origin and given by*

$$X_n = X_{n-1} + U_n \quad \text{and} \quad \overline{X}_n = \overline{X}_{n-1} + \overline{U}_n$$

where U_n and $\overline{U}_n$ stand for a sequence of independent Bernoulli random variables on $\{0, 1\}$ with

$$\mathbb{P}\left(U_n = 1 \right) = p \quad \text{and} \quad \mathbb{P}\left(\overline{U}_n = 1 \right) = \overline{p}.$$

Check that

$$\mathbb{P}\left(X_n \geq a \right) = \mathbb{E}\left(1_{[a,\infty[}\left(\sum_{1 \leq k \leq n} \overline{U}_k \right) \prod_{1 \leq k \leq n} G_k(\overline{U}_k) \right)$$

for any $a \in \mathbb{R}$ s.t. $a \leq n$, with the potential functions $G_k(u) = \frac{p}{\overline{p}} \, 1_{u=1} + \frac{1-p}{1-\overline{p}} \, 1_{u=0}$, with $k \geq 1$. Deduce that

$$\mathbb{P}\left(X_n \geq a \right) = \left(\frac{1-p}{1-\overline{p}} \right)^n \mathbb{E}\left(1_{[a,n]}\left(\overline{X}_n \right) \left(\frac{p}{1-p} \times \frac{1-\overline{p}}{\overline{p}} \right)^{\overline{X}_n} \right).$$

Exercise 59 (Chernov estimates)

- *Let U be an uniform random variable on $\{-1, +1\}$. Check that for any $\lambda > 0$ we have*

$$\mathbb{E}\left(e^{\lambda U} \right) \leq e^{\frac{\lambda^2}{2}}.$$

- Let X be an $[a, b]$-valued random variable for some $a < b$. Deduce from the above that for any $\lambda > 0$ we have
$$\mathbb{E}\left(e^{\lambda(X - \mathbb{E}(X))}\right) \leq e^{\frac{\lambda^2 (b-a)^2}{2}}.$$

- Prove that
$$\mathbb{P}\left(X \geq \mathbb{E}(X) + \rho\right) \leq \exp\left(-\frac{\rho^2}{2(b-a)^2}\right).$$

Exercise 60 (Bernoulli trials - Chernov estimates) ✐ *We let $(U_k)_{k \geq 1}$ be a sequence of independent Bernoulli random variables with*
$$\forall k \geq 1 \qquad \mathbb{P}\left(U_k = 1\right) = 1 - \mathbb{P}\left(U_k = 0\right) = p_k \in [0, 1].$$
We set $X_n = \sum_{1 \leq k \leq n} U_k$, and $m_n = \mathbb{E}(X_n)$.

- For any $\epsilon \geq 0$ and $\lambda > 0$ check that
$$\mathbb{P}(X_n \geq (1 + \epsilon)m_n) \leq e^{-\lambda(1+\epsilon)m_n} \mathbb{E}\left(e^{\lambda X_n}\right) \quad and \quad \mathbb{E}\left(e^{\lambda X_n}\right) \leq e^{m_n\left(e^\lambda - 1\right)}.$$

- Deduce that
$$\mathbb{P}(X_n \geq (1 + \epsilon)m_n) \leq \rho_\epsilon^{\epsilon m_n} \quad with \quad \rho_\epsilon := e\left(1 - \frac{\epsilon}{1 + \epsilon}\right)^{\frac{1+\epsilon}{\epsilon}} < 1.$$

Exercise 61 (Wick's formula) ✐ *Prove (5.8) and then (5.9) using an induction w.r.t. the parameter n.*

Exercise 62 (Novikov theorem) *Let $X = (X_i)_{1 \leq i \leq r}$ be an r-dimensional centered Gaussian $\mathcal{N}(0, R)$ with $(r \times r)$ with an invertible covariance matrix R (cf. for instance (4.13)). We let f be a smooth function on $\mathbb{R}^r$. Check that for any $1 \leq i \leq r$ we have*
$$x_j = \frac{1}{2} \sum_{1 \leq k \leq r} R_{j,k} \, \partial_{x_k}\left(x' R^{-1} x\right).$$

Deduce that
$$\mathbb{E}\left(X_i \, f(X)\right) = \sum_{1 \leq k \leq r} R_{j,k} \, \mathbb{E}\left(\partial_{x_k} f(X)\right).$$

Exercise 63 (An almost sure convergence result) *Let $a = (a_n)_{n \geq 0}$ and $b = (b_n)_{n \geq 0}$ be sequences of numbers s.t. $\lim_{n \uparrow \infty} b_n = b$. Let X_n be a sequence of independent random variables given by*
$$\mathbb{P}\left(X_n = a_n\right) = \frac{1}{n^{1+\epsilon}} \quad and \quad \mathbb{P}\left(X_n = b_n\right) = 1 - \frac{1}{n^{1+\epsilon}}$$
for some $\epsilon > 0$. Check that $\frac{1}{n} \sum_{1 \leq k \leq n} X_k \to_{n \uparrow \infty} b$ almost surely.

Exercise 64 (Mill's inequalities) ✐ *Check Mill's inequalities (5.13). For any Gaussian and centered r.v. X_n with unit variance prove that*
$$\mathbb{P}\left(|X_n| \geq \sqrt{2(1 + \alpha) \, \log n}\right) \leq \frac{1}{\sqrt{\pi \alpha}} \frac{1}{n^{1+\alpha}} \frac{1}{\sqrt{\log(n)}}$$
for any $\alpha > 0$, and any $n > 1$, and deduce that
$$\mathbb{P}\left(\exists n_\alpha \geq 1 \; : \; \forall n \geq n_\alpha \qquad |X_n| \leq \sqrt{2(1 + \alpha) \, \log n}\right) = 1.$$

Exercise 65 (Birnbaum's improvement [24]) ✎ *Show that the l.h.s of Mill's inequalities (5.13) can be sharpened. It holds:*

$$\forall a > 0 \qquad \frac{2}{\sqrt{4 + a^2} + a} \; \frac{1}{\sqrt{2\pi}} \; e^{-\frac{a^2}{2}} \leq \mathbb{P}(X \geq a).$$

Exercise 66 (A rare event problem) ✎ *We consider a pair of independent Gaussian and centered r.v. $X = (X_1, X_2)$. Suppose we want to evaluate the quantity*

$$\mathbb{P}\left(X \in A(a, \epsilon)\right) = \mathbb{E}\left(1_{A(a,\epsilon)}(X)\right) = \eta(1_{A(a,\epsilon)})$$

with the Gaussian distribution η of X, and the the indicator function $1_{A(a,\epsilon)}$ of the set

$$A(a, \epsilon) = \{(x_1, x_2) \in \mathbb{R}^2 \; : \; x_1^2 + x_2^2 \geq a \quad \text{and} \quad 0 \leq \arctan(x_2/x_1) \leq 2\pi e^{-b}\}.$$

Using the Box-Muller transformation $(X_1, X_2) = \varphi(U_1, U_2)$ presented in (4.5), check that

$$\mathbb{P}\left(X \in A(a, \epsilon)\right) = \int_{[0,1]^2} 1_{[0,e^{-a}] \times [0,e^{-b}]}(u_1, u_2) \; du_1 du_2 = e^{-(a+b)}.$$

We let $(U_1^i, U_2^i)_{i \geq 1}$ be a sequence of independent copies of the variable (U_1, U_2). We consider the Monte Carlo approximation (5.1) associated with the r.v. $X^i = (X_1^i, X_2^i) = \varphi(U_1^i, U_2^i)$ and defined by

$$\eta^N(1_{A(a,\epsilon)}) = \frac{1}{N} \sum_{1 \leq i \leq N} 1_{[0,e^{-a}] \times [0,e^{-b}]}(U_1^i, U_2^i).$$

Check that the relative variance is given by

$$N \; \mathbb{E}\left(\left[\frac{\eta^N\left(1_{A(a,\epsilon)}\right)}{\eta\left(1_{A(a,\epsilon)}\right)} - 1\right]^2\right) = e^{a+b} - 1.$$

6

Some illustrations

In this last chapter of Part II, we show several specific demonstrations of successful applications of stochastic simulations methods in some scientific disciplines. The simulation methods are needed and applied as tools to analyse, in a possibly simpler way, some complex stochastic phenomena modelled in physics, biology or in engineering. The discussion focuses on the description of the models only and the more rigorous mathematical treatment is dealt with in the further development of the course.

> *There is no branch of mathematics, however abstract,*
> *which may not some day be applied to phenomena of the real world.*
> *Nikolai Lobatchevsky (1792-1856)*

6.1 Stochastic processes

The random simulation techniques developed in this opening chapter are rather elementary but they are used as essential steps when sampling more complex random phenomena that evolve w.r.t. the time parameter.

The discrete time stochastic processes are defined in terms of a given sequence of random variables X_n, indexed by the time parameter $n \in \mathbb{N}$. These r.v. change sequentially and randomly according to some prescribed elementary transitions

$$X_0 \rightsquigarrow X_1 \rightsquigarrow X_2 \rightsquigarrow \ldots \rightsquigarrow X_{n-1} \rightsquigarrow X_n \rightsquigarrow \ldots \tag{6.1}$$

These processes are called a Markov chain if the random state X_n at one time n depends only on the state in the previous time X_{n-1}. More precisely, this means that the elementary transitions $X_{n-1} \rightsquigarrow X_n$ between integer time steps $(n-1) \rightsquigarrow n$ only depend on X_{n-1} as well as on a given finite number of elementary random variables, but not on the previous history.

More formally, we assume that X_n takes values in a state space S, and we let W_n be a sequence of random variables taking values in a space $\mathcal{W}$. In this notation, the elementary transitions of a Markov process (6.1) are given by an evolution equation of the form

$$X_n = F_n\left(X_{n-1}, W_n\right)$$

for a collection of functions F_n from $(S \times \mathcal{W})$ into S, and for an initial random state variable X_0. These elementary transitions can be sampled using the simulation toolbox presented in this chapter. Given a random state X_{n-1}, we compute the next random state

$X_n = F_n(X_{n-1}, W_n)$ by sampling sequentially the random variables W_n. The design of the continuous time version of these stochastic processes is technically more involved but using an appropriate time discretization scheme it can serve as a discrete generation model. These continuous time processes and their connections with partial differential equations are described in some details in the final chapter of these lecture notes.

In the rest of this section, we discuss some important questions related to the effective simulation and the convergence analysis of stochastic processes and complex probability measures. This discussion can serve as an introduction to the chapters in the further development of the course.

6.2   Markov chain models

> Most of the theory of Markov processes is concerned with computing the distributions of the random states, and with analysing the convergence of empirical time averages when the time parameter tends to ∞. More formally, we need to make precise the next two assertions:
>
> $$\lim_{n \to \infty} \mathrm{Law}(X_n) = \mathrm{Law}(X_\infty) \quad \text{and} \quad \lim_{n \to \infty} \frac{1}{n} \sum_{1 \leq p \leq n} \delta_{X_p} = \mathrm{Law}(X_\infty).$$

The first question we focus on is to identify (whenever it exists) the limiting distribution of the random states of a given stochastic process. The second important question is how fast the law of the random states converges to its limiting measure as n tends to ∞. Does a certain central limit theorem, similar to the one we have presented for independent random sequences, hold for Markov chain models?

In the next chapter, we provide a rather complete discussion on the theory of Monte Carlo methods, including a variety of illustrative examples borrowed from different scientific disciplines. For a more leisurely discussion at the undergraduate-level, the reader should look the recent lectures notes on Markovian modeling by N. Privault [228]. For a more detailed discussion the reader should consult the research monograph by S. P. Meyn and R.L. Tweedie [199].

6.3   Black-box type models

The fast developments of probability theory and computer science have presented us with more realistic and sophisticated stochastic processes to model complex random phenomena arising in physics, biology and other branches of engineering sciences. These stochastic models are often based on the simulation of high dimensional stochastic processes, including nonlinear stochastic partial differential equations. The source of randomness often comes from unknown initial conditions, and from the uncertainties of the models, including unknown kinetic parameters.

For instance, in reliability and risk analysis of offshore structures in extreme sea conditions, the random inputs of the processes depend on the wave spectra and forecasting data, such as temperature profiles, as well as on the model uncertainties. Given these data, the energy profiles on the offshore platform (boat, gas or petrol-offshore platform) can be treated as the outputs of a sophisticated partial differential equation based on hydrodynamic and mechanical principles.

These input-output systems can be interpreted as a black box type model

$$\text{Inputs} \; = \; X \; \longrightarrow \boxed{\text{Numerical codes } F} \longrightarrow \; \text{Outputs} \; = \; Y = F(X).$$

In this case, we are interested in analyzing the set A of some inputs X that make the ouput $Y = F(X) \in B$ hit some reference or critical event. More formally, the objective is to compute the following quantities

$$\mathbb{P}\left(X \in A\right) \quad \text{and} \quad \text{Law}(X \mid X \in A).$$

These problems are often termed "calibration of propagations of uncertainties" in numerical codes in computer and numerical sciences. In the example discussed above, the input random variables $X = (X_t)_{t \in [0,T]}$ represent the spectral characteristics of the wave as well as the atmosphere temperature profiles. For instance, a discrete generation perturbation process with a given temperature reference trajectory $(c_t)_{t \in [0,T]}$ can be represented by an autoregression model

$$(X_t - c_t) = \sqrt{1 - \epsilon} \; (X_{t-1} - c_{t-1}) + \sqrt{\epsilon} \; W_t \tag{6.2}$$

for a parameter $\epsilon \in [0,1]$, and a sequence of independent centered Gaussian r.v. $W = (W_t)_{t \in [0,T]}$ with variance σ^2. Assuming that $X_0 = c_0$, we prove that

$$\forall t \in [0,T] \quad \mathbb{E}\left(X_t\right) = c_t \quad \text{and} \quad \mathbb{E}\left((X_t - c_t)^2\right) = \sigma^2 \left(1 - (1 - \epsilon)^t\right). \tag{6.3}$$

We refer to exercise 67 for a proof of these formulae.

The evolution of the offshore structure w.r.t. the time parameter $t \in [0,T]$ is described by a random process

$$Y = (Y_t)_{t \in [0,T]} = (F_t(X))_{t \in [0,T]}.$$

In this context, the event of interest is given by $A := \left\{x \; : \; \sup_{t \in [0,T]} F_t(x) \geq a_\star\right\}$, where $a_\star$ denotes a suitably chosen critical threshold.

For rare event probabilities smaller than $\mathbb{P}(X \in A) \simeq 10^{-9}$ it is hopeless to use the crude Monte Carlo techniques presented in this chapter. In addition, the importance sampling methodology discussed in section 5.2 is too intrusive to be feasible, in the sense that it requires the user to change the evolution of the numerical code which is determined by precise physical laws and encoded sequentially using grid-type or projective-type numerical techniques.

Thus, the challenging problem is to design a robust and powerful stochastic search model that could be used to find the input regions involved in the rare event of interest. The central idea is to explore gradually the input domains leading to more and more critical values. These sophisticated stochastic search models belong to the class of Markov chain Monte Carlo methods (*abbreviated MCMC*), and particle simulation techniques also called sequential Monte Carlo methodologies (*abbreviated SMC*). Roughly speaking, the central idea of MCMC is to find a Markov chain model $(X_n)_{n \geq 0}$ of the form (6.1) with prescribed limiting distribution

$$\text{Law}(X_\infty) = \text{Law}(X \mid X \in A)$$

in the space of inputs, say S. In the example discussed in (6.2), the state of inputs is given by the set of trajectories $S = \mathbb{R}^{[0,T]}$.

In particle methodologies, we are given a sequence of decreasing gateways $A_n \downarrow A$ and we run a stochastic population type process $(X_n^1, \ldots, X_n^N) \in S^N$ with a sufficiently large number N of individuals such that *at any time step n*, we have

$$\lim_{N \to \infty} \frac{1}{N} \sum_{1 \leq i \leq N} \delta_{X_n^i} = \mathrm{Law}(X \mid X \in A).$$

Some of these advanced stochastic search processes are discussed in the next chapter. For a more thorough study, we refer to article [68], and research books [5, 39, 66, 67, 111, 134] .

6.4   Boltzmann-Gibbs measures

In computational physics, as well as in stochastic optimization, we are interested in computing Boltzmann-Gibbs distributions associated with an inverse temperature parameter β.

These distributions have the following form

$$\mu_\beta(dx) = \frac{1}{\mathcal{Z}_\beta} \, e^{-\beta V(x)} \, \lambda(dx) \tag{6.4}$$

where λ stands for a distribution on a state space S, and V is a function on S. The parameter β is interpreted as an inverse temperature parameter. It is also often important to compute the normalizing constants (a.k.a. partition functions in physics) $\mathcal{Z}_\beta = \lambda \left(e^{-\beta V} \right)$.

6.4.1   Ising model

The Ising model currently used in electromagnetism, statistical mechanics, as well as in image processing is associated with the state space

$$S = \{-1, +1\}^E \qquad E = \{1, \ldots, L\} \times \{1, \ldots, L\}$$

equipped with the uniform measure $\lambda(x) = 2^{-L^2}$. The lattice E is equipped with the following graph structure

$$j^1 = (i_1, i_2 + 1), \qquad j^2 = (i_1 + 1, i_2), \qquad j^3 = (i_1, i_2 - 1), \qquad j^4 = (i_1 - 1, i_2)$$

around some state $(i_1, i_2) \in E$. In other words, we have

$$
\begin{array}{c}
j^1 \\
\updownarrow \\
j^4 \leftrightarrow \quad i \quad \leftrightarrow j^2. \\
\updownarrow \\
j^3
\end{array}
$$

Two neighbors $i, j \in E$ are denoted by $i \sim i'$. The energy of a configuration $x \in S$ is given by the Hamiltonian function

$$V(x) = h \sum_{i \in E} x(i) - J \sum_{i \sim j} x(i)x(j). \tag{6.5}$$

The parameter $h \in \mathbb{R}$ represents the strength of an external magnetic field, and $J \in \mathbb{R}$ reflects the interaction degree between the sites.

6.4.2 Sherrington-Kirkpatrick model

In the Sherrington-Kirkpatrick model introduced in 1975 in their seminal article [241], the potential function is given by

$$V(\theta, x) := \sum_{1 \leq i \leq j \leq d} \theta_{i,j}\, x(i)\, x(j) + h \sum_{i=1}^{d} x(i)$$

where $\Theta_{i,j}$ are assumed to be i.i.d. centered Gaussian random variables. More general disordered models can be defined in terms of random mappings $\Theta : (i,j) \in \{1,\dots,d\}^2 \mapsto \theta_{i,j}$.

Directed polymer models arising in statistical physics are defined in much the same way. For instance, the micro state of a system consists of d particles $x_i = (p_i, r_i)$, with a momentum vector p_i and a position coordinate $r_i = (r_i^1, r_i^2, r_i^3)$, with $1 \leq i \leq d$. The energy of the system is given by some function

$$V(x) := \sum_{i=1}^{d} \left(\frac{1}{2m}\, \|p_i\|^2 + mgr_i^1 \right)$$

where m represents the mass of the particle, r_i^1 its height, and g is the gravitation constant. The probability distribution of the physical system at inverse temperature β_n is again given by the Boltzmann-Gibbs measures (6.4), with the Lebesgue measure λ. For a more thorough discussion on these models, we refer the reader to [54, 90, 91], and to the references therein.

6.4.3 The traveling salesman model

We now present another example of Boltzmann-Gibbs measure arising in operations research. We are given a finite state space $\mathcal{E}_m = \{e_1, \dots, e_m\}$ equipped with some metric d. We can think of a a finite number of cities and the distances between them. One typical problem known as the traveling salesman problem, consists in finding a way to visit all the cities by covering minimal traveling distance. We can model a given sequence of visits by a permutation σ on the index set $\{1, \dots, m\}$. In this situation, the state space S is given by the set of these permutations equipped with the uniform probability measure $\lambda(\sigma) = 1/m!$, and an energy function V defined by

$$V(\sigma) = \sum_{p=1}^{m} d(e_{\sigma(p)}, e_{\sigma(p+1)}).$$

It is not difficult to check that

$$\lim_{\beta \to \infty} \mu_\beta(\sigma) = \mu_\infty(\sigma) := \frac{1}{\mathrm{Card}(V^\star)}\, 1_{V^\star}(\sigma) \tag{6.6}$$

where $\mathrm{Card}(V^\star)$ stands for the cardinality of the set $V^\star = \{\sigma \in S : V(\sigma) = \inf_S V\}$ of the optimal traveling strategies (cf. Exercise 68). This result shows that at low temperature (i.e. $\beta \uparrow \infty$), the sampling of the distribution μ_β amounts to choosing randomly an unknown optimal solution of the problem.

As the reader may have noticed, the Monte Carlo methodologies described in this opening chapter cannot be used to sample exact random configurations according to the Boltzmann-Gibbs measure.

It is tempting to use the importance sampling strategies developed in section 5.2 to approximate the target distribution $p(\sigma) = \mu_\beta(\sigma)$ using certain twisted measure $q(\sigma)$ on the state space S. Besides the fact that these importance sampling measures are difficult to design, the empirical weight functions $W(\sigma) := p(\sigma)/q(\sigma)$ are evaluated at random states σ^i visited by sampling according to $q(\sigma)$. At low temperature, we observe that $p(\sigma) \simeq 0$ except for the optimal permutations. Thus, one expects to have a very poor estimation based on almost null weighted empirical averages using a twisted measure $q(\sigma)$ that is not concentrated around the unknown optimal permutations.

One way to solve these problems is to use the MCMC or the SMC methodologies discussed above. As before, the central idea of MCMC methods is to find a Markov chain model $(X_n)_{n \geq 0}$ of the form (6.1) with prescribed limiting distribution $\mathrm{Law}(X_\infty) = \mu_\beta$ on the set S.

The following pictures illustrate the initial circuit, the best historical circuit, and the evolution of the lengths of the circuits in the traveling salesman problem (TSP) with 30 cities.

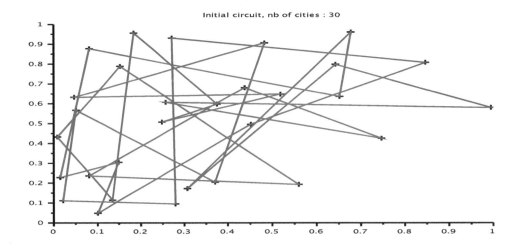

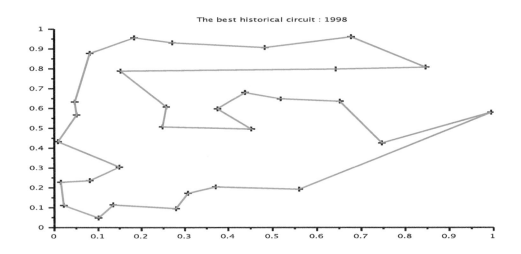

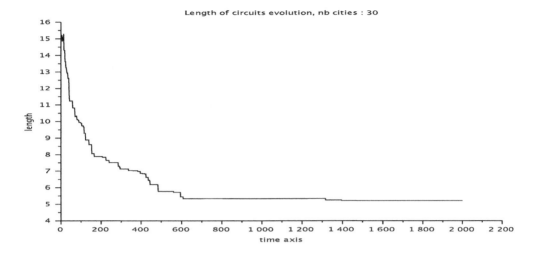

In particle methodologies, we are given a sequence of increasing parameters $\beta_n \uparrow \beta$ and we run a stochastic population type process $(X_n^1, \ldots, X_n^N) \in S^N$ with a sufficiently large number N of individuals s.t. *at any time step* n, we have

$$\lim_{N \to \infty} \frac{1}{N} \sum_{1 \leq i \leq N} \delta_{X_n^i} = \mu_{\beta_n}.$$

We refer to section 9.4.1 and section 9.9.1 for more detailed discussion on the simulation of these Boltzmann-Gibbs measures using simulated annealing algorithms and more advanced mean field type interacting MCMC methodologies.

6.5 Filtering and statistical learning

6.5.1 Bayes' formula

The filtering problem is defined as follows. A sensor (such as a radar, or a sonar) delivers at each time some partial and noisy observations Y_n of an evolving Markov chain X_n on some state space $\mathbb{R}^d$.

$$
\begin{array}{ccccccccl}
X_0 & \to & X_1 & \to & X_2 & \to & X_3 & \to & \ldots & \text{signal} \\
\downarrow & & \downarrow & & \downarrow & & \downarrow & & \ldots & \\
Y_0 & & Y_1 & & Y_2 & & Y_3 & & \ldots & \text{observation.}
\end{array}
$$

To fix ideas, we further assume that the observation process Y_n takes values in $\mathbb{R}$ and

$$Y_n = h_n(X_n) + V_n$$

holds for a function h_n from $\mathbb{R}^d$ into $\mathbb{R}$, and a sequence of independent and standard Gaussian r.v. V_n. Given a realization of the chain $(X_0, \ldots, X_n) = (x_0, \ldots, x_n)$ the random variables $(Y_0, \ldots, Y_n)$ are independent Gaussian r.v. with mean values $(h_0(x_0), \ldots, h_n(x_n))$. More formally, the density of the observations $(Y_0, \ldots, Y_n)$ given $(X_0, \ldots, X_n) = (x_0, \ldots, x_n)$

is given by

$$p_n((y_0, \ldots, y_n) \mid (x_0, \ldots, x_n)) \propto \prod_{p=0}^{n} \underbrace{\exp\left(-\frac{1}{2}(y_p - h_p(x_p))^2\right)}_{\propto p_n(y_n \mid x_n)}.$$

We denote by $p_n(x_0, \ldots, x_n)$ (whenever it exists) the density of the random states $(X_0, \ldots, X_n)$ w.r.t. some reference measure. To simplify the presentation, we also set $\boldsymbol{x_n} = (x_0, \ldots, x_n)$ and $\boldsymbol{y_n} = (y_0, \ldots, y_n)$. We also assume that the random variables $(X_0, \ldots, X_n)$ have a density $p_n(\boldsymbol{x_n})$ with respect to the product Lebesgue measure on $\mathbb{R}^d$.

In this notation, by applying Bayes' rule, we get the density of $(X_0, \ldots, X_n)$ given the observations $(Y_0, \ldots, Y_n)$:

$$p_n(\boldsymbol{x_n} \mid \boldsymbol{y_n}) = \frac{1}{p_n(\boldsymbol{y_n})} \; p_n(\boldsymbol{y_n} \mid \boldsymbol{x_n}) \; \times \; p_n(\boldsymbol{x_n}) \tag{6.7}$$

with the normalizing constant $p_n(\boldsymbol{y_n})$.

In some instances, the signal and the observation processes depend on an unknown and fixed parameter Θ (such as the variance of the perturbations, or the initial conditions or some kinetic parameters). In this context, another important question is to compute the conditional distribution of Θ given the sequence of observations $(Y_0, \ldots, Y_n)$.

When the r.v. Θ has a prior density $p(\theta)$, these conditional distributions are given with some obvious abusive notation by the formula

$$p_n(\theta \mid \boldsymbol{y_n}) \propto p_n(\boldsymbol{y_n} \mid \theta) \times p(\theta). \tag{6.8}$$

We notice that the likelihood functions $p_n(\boldsymbol{y_n} \mid \theta)$ coincide with the normalizing constants of the filtering problem discussed above when the parameter θ is given.

We also notice that the conditional distributions (6.7) can also be interpreted as Boltzmann-Gibbs distributions on the space of trajectories. More formally, we have that

$$p_n(\boldsymbol{x_n} \mid \boldsymbol{y_n}) = \frac{1}{\mathcal{Z}_{n,\boldsymbol{y_n}}} \; e^{-V_{n,\boldsymbol{y_n}}(\boldsymbol{x_n})} \; \lambda_n(\boldsymbol{x_n}) \tag{6.9}$$

with some normalizing constants $\mathcal{Z}_{n,\boldsymbol{y_n}}$, and $(\lambda_n, V_{n,\boldsymbol{y_n}})$ given by

$$\lambda_n(\boldsymbol{x_n}) = p_n(\boldsymbol{x_n}) \quad \text{and} \quad V_{n,\boldsymbol{y_n}}(\boldsymbol{x_n}) = -\log p_n(\boldsymbol{y_n} \mid \boldsymbol{x_n}).$$

6.5.2   Singer's radar model

We illustrate this stochastic model with the Singer radar filtering model [240]. In this context, $X_n = (X_n^{(i)})_{1 \le i \le 3}$ represents the evolution in $\mathbb{R}^3$ of a given target

$$\begin{cases} X_n^{(1)} - X_{n-1}^{(1)} &= \epsilon_n \, W_n \\ X_n^{(2)} - X_{n-1}^{(2)} &= -\alpha \, X_{n-1}^{(2)} \, \Delta + \beta \, \Delta \, X_n^{(1)} \\ X_n^{(3)} - X_{n-1}^{(3)} &= X_n^{(2)} \, \Delta. \end{cases}$$

The first coordinate $X_n^{(1)}$ represents the acceleration, the second $X_n^{(2)}$ the velocity, and the last one $X_n^{(3)}$ the position. We can suppose that the viscosity parameters are given by $\alpha = 1$ and $\beta = 18$, and the initial conditions are given by $X_0^{(1)} \sim \mathcal{N}(30, \sigma_0)$, $X_0^{(2)}$ and $X_0^{(3)} \sim \mathcal{N}(500, \sigma_0)$, with $\sigma_0 \in \{10, 100\}$. Here $\mathcal{N}(m, \sigma^2)$ denotes the Gaussian distribution with mean m and variance σ^2. The parameter $\Delta = 10^{-2}$ represents the radar sampling period. The acceleration periods $(\epsilon_n)_{n \geq 1}$ stand for a collection of independent Bernoulli r.v. with parameter Δ. Finally, $(W_n)_{n \geq 1}$ represent the acceleration amplitudes. For instance, we can suppose that these r.v. are independent and uniform on $[0, 60]$.

The observations delivered by the radar are given by the equations

$$\forall n \geq 0, \qquad Y_n = X_n^{(3)} + \Delta\, V_n.$$

Here $V = (V_n)_{n \geq 1}$ sequence of independent and centered Gaussian r.v. V_n with variance $\sigma_v^2 = 100$.

For linear-Gaussian models these conditional distributions can be computed recursively w.r.t. the time parameter using the filtering equations introduced by R. E. Kalman and R. S. Bucy in the beginning of the 1960's [160, 164] (see also [66, 67] for a more recent account on these filters and their applications). For nonlinear and/or non-Gaussian models, one solution is to find a judicious stochastic simulation technique to sample the Boltzmann-Gibbs distributions (6.9). By construction we also need to compute these conditional distributions (6.7) *recursively w.r.t. to time parameter.*

Since the target distributions change at every time step, the MCMC methodologies discussed above will require to change the Markov chain model with the prescribed limiting distribution at every time step. This important drawback, together with the so-called burning period needed for each chain to approximate its equilibrium limiting measure, shows that these MCMC methodologies fail to compute recursively the conditional distributions without a dramatic computational cost. In the reverse angle, the population-based design of particle and the SMC methodologies allow us to construct a stochastic process that estimates the conditional distributions (6.7) and (6.8) at every time step.

Some of these advanced stochastic particle processes are discussed in the next chapters. For a more detailed discussion on these advanced stochastic simulation techniques the reader could review article [4] and books [66, 67, 111].

6.6 Exercises

Exercise 67 (Autoregression model) *We consider the autoregression model presented in (6.2). Prove that*

$$(X_t - c_t) = \sqrt{\epsilon} \sum_{0 \leq s < t} (\sqrt{1 - \epsilon})^s\, W_{t-s}$$

and deduce (6.3).

Exercise 68 *Prove the convergence property (6.6).*

Exercise 69 (Random walk Metropolis-Hastings) *Let $p(x)$ and $q(x) = q(-x)$ be a couple of probability densities on $\mathbb{R}^d$, for some $d \geq 1$. We let $\pi(dx) = p(x)dx$ be a target probability distribution on $\mathbb{R}^d$. Write the acceptance ratio of the Metropolis-Hastings algorithm with proposal transition $P(x, dy) = \epsilon^{-d} q((y - x)/\epsilon)\, dy$.*

Exercise 70 (Random walk on a weighted graph) *Consider an undirected and connected graph $(\mathcal{V}, \mathcal{E})$ with the set of vertices $\mathcal{V}$ and the set of edges $\mathcal{E}$. The graph is equipped with a positive weight function $w(x, y)$ on each edge $(x, y) \in \mathcal{E}$. When the vertices x and y are not connected we set $w(x, y) = 0$. Assume that $0 < \sum_{y \in \mathcal{V}} w(x, y) < \infty$. We consider the Markov chain X_n on $\mathcal{V}$ with probability transitions*

$$M(x, y) = \frac{w(x, y)}{\sum_{z \in \mathcal{V}} w(x, z)}.$$

We assume that X_0 is distributed according to the probability distribution

$$\pi(x) = \sum_{z \in \mathcal{V}} w(x, z) / \sum_{u, v \in \mathcal{V}} w(u, v).$$

Check that

$$\pi(x)\, M(x, y) = \pi(y)\, M(x, y) \quad and \;\; \forall n \geq 1, \; \forall x \in \mathcal{V} \quad \mathbb{P}\left(X_n = x\right) = \pi(x).$$

Exercise 71 (Filtering model) *We consider the filtering problem presented in section 6.5. Using Bayesian notation, we let $p(x_{n+1}|x_n)$ the density of X_{n+1} given $X_n = x_n$ with respect to the Lebesgue measure dx_n. Prove that*

$$p(x_{n+1} \mid y_0, \ldots, y_n) = \int p(x_{n+1}|x_n)\, p(x_n \mid y_0, \ldots, y_n)\, dx_n$$

and

$$p(x_{n+1} \mid y_0, \ldots, y_n, y_{n+1}) = \frac{1}{p(y_{n+1}|y_0 \ldots, y_n)}\, p(y_{n+1}|x_{n+1})\, p(x_{n+1} \mid y_0, \ldots, y_n).$$

Check that

$$p(y_0 \ldots, y_n) = \prod_{0 \leq k \leq n} \int p(y_k|x_k)\, p(x_k \mid y_0, \ldots, y_{k-1})\, dx_k.$$

Exercise 72 (Ising model) 🏃 *Consider the Boltzmann-Gibbs measure*

$$\mu_\beta(x) = \mathcal{Z}_\beta^{-1}\, \exp\left(-\beta V(x)\right)$$

associated with the Ising model in one dimension

$$S = \{-1, +1\}^{\{1, \ldots, L\}} \ni x = (x(i))_{i=1,\ldots,L} \mapsto V(x) = -h \sum_{i=1}^{L} x(i) - J \sum_{i=1}^{L} x(i)x(i+1).$$

We use the periodic boundary condition $x(L+1) = x(1)$. Prove that

$$\mathcal{Z}_\beta \;\; = \;\; \mathrm{Trace}(T^L),$$

with the transfer (symmetric) matrix

$$T = \left(\begin{array}{cc} T(+1, +1) & T(+1, -1) \\ T(-1, +1) & T(-1, -1) \end{array} \right) = \left(\begin{array}{cc} e^{J_\beta + h_\beta} & e^{-J_\beta} \\ e^{-J_\beta} & e^{J_\beta - h_\beta} \end{array} \right)$$

with entries

$$T(x(i), x(i+1)) = \exp\left[J_\beta x(i)x(i+1) + \frac{h_\beta}{2}\left(x(i) + x(i+1)\right) \right] \quad and \;\; J_\beta = \beta J \qquad h_\beta = h\beta.$$

Check that

$$\mathcal{Z}_\beta = \lambda_{+,\beta}^L \left(1 + \left[\frac{\lambda_{-,\beta}}{\lambda_{+,\beta}}\right]^L\right) \quad and \quad \frac{1}{\beta L} \log \mathcal{Z}_\beta \to_{L \uparrow \infty} \frac{1}{\beta} \log \lambda_{+,\beta}$$

with

$$\lambda_{+,\beta} = e^{J_\beta} \cosh{(h_\beta)} + \sqrt{e^{2J_\beta} \cosh{(h_\beta)}^2 - 2\sinh(2J_\beta)}$$

$$\lambda_{-,\beta} = e^{J_\beta} \cosh{(h_\beta)} - \sqrt{e^{2J_\beta} \cosh{(h_\beta)}^2 - 2\sinh(2J_\beta)}.$$

Exercise 73 (Particle filter) *Write the particle filter algorithm (3.7) to solve the Singer filtering problem discussed in section 6.5.2.*

Exercise 74 (Random fields) *We let $(U_n)_{n \geq 0}$ be a sequence of centered and independent random variables with the same probability distribution on $\mathbb{R}$ and such that $\mathbb{E}(U_1^2) = 1$. Compute the covariance function $C(x,y) = \mathbb{E}(V(x)V(y))$ of the random fields defined below.*

- *Random polynomials: $V : x \in \mathbb{R} \mapsto V(x) = \sum_{0 \leq n \leq d} U_n \, x^n$.*

- *Cosine random field: $V : x \in \mathbb{R} \mapsto V(x) = U_1 \cos{(ax)} + U_2 \sin{(ax)}$.*

Exercise 75 (Reproducing Kernel Hilbert Space) *We let $(U_n)_{n \geq 0}$ be the sequence random variables introduced in exercise 74. Let S be some compact subset of $\mathbb{R}^d$ and $C(x,y)$ be a symmetric function on S^2. We further assume that the integral operator*

$$\mathcal{C} : f \in \mathbb{L}_2(T) \mapsto \mathcal{C}(f) \in \mathbb{L}_2(T) \quad with \quad \mathcal{C}(f)(x) = \int_T C(x,y) \, f(y) \, dy$$

has positive decreasing eigenvalues λ_n and respective normalized eigenvectors φ_n, with $n \geq 1$. In this situation, we have the spectral decomposition

$$C(x,y) = \sum_{n \geq 1} \lambda_n \, \varphi_n(x) \varphi_n(y).$$

Check that

$$V(x) = \sum_{n \geq 1} \sqrt{\lambda_n} \, U_n \, \varphi_n(x) \Longrightarrow C(x,y) = \mathbb{E}(V(x)V(y)).$$

Exercise 76 (Kriging interpolation - Black-Box metamodels) ✎ *We let V be a centered random field on a compact subset $S \subset \mathbb{R}^d$, for certain $d \geq 1$. We observe a sequence of n values $V(x_i)$ for some states $x_i \in S$, with $1 \leq i \leq n$. We estimate $V(x)$ at $x \in S$ by a linear interpolation*

$$\widehat{V}(x) = \sum_{1 \leq i \leq n} w_i(x) \, V(x_i)$$

with weight functions w_i that minimize the variance in the sense that

$$\mathbb{E}\left[(V(x) - \widehat{V}(x))^2\right] = \inf_{a_1,\ldots,a_m} \mathbb{E}\left[\left((V(x) - \sum_{1 \leq i \leq n} a_i \, V(x_i)\right)^2\right].$$

- *Check that*

$$\widehat{V}(x) = [V(x_1), \ldots, V(x_n)] \begin{pmatrix} C(x_1, x_1) & \ldots & C(x_1, x_n) \\ \vdots & \vdots & \vdots \\ C(x_n, x_1) & \ldots & C(x_n, x_n) \end{pmatrix}^{-1} \begin{pmatrix} C(x, x_1) \\ \vdots \\ C(x, x_n) \end{pmatrix}.$$

- *Prove that*

$$\mathbb{E}\left[(V(x) - \widehat{V}(x))^2\right]$$

$$= C(x, x) - (C(x, x_1), \ldots, C(x, x_n)) \begin{pmatrix} C(x_1, x_1) & \ldots & C(x_1, x_n) \\ \vdots & \vdots & \vdots \\ C(x_n, x_1) & \ldots & C(x_n, x_n) \end{pmatrix}^{-1} \begin{pmatrix} C(x, x_1) \\ \vdots \\ C(x, x_n) \end{pmatrix}.$$

The function $x \mapsto \widehat{V}(x)$ can be interpreted as the inputs and outputs of a metamodel representing a complex black-box model with a training set of inputs and outputs $x_i \mapsto V(x_i)$, with $1 \leq i \leq n$.

Part III

Discrete time processes

7

Markov chains

This chapter introduces the Markov chains as simplest examples of discrete time stochastic processes. The simplicity may be misleading since by utilizing long enough memory, Markov chains can be used as the epitome of almost every discrete time stochastic process. Starting with simple linear chain models, we move on to describe more advanced nonlinear Markov chain models, including self-interacting Markov chains, mean field particle models, McKean-Vlasov chains and interacting jump processes.

> *Goals transform a random walk into a chase.*
> Mihaly Csikszentmihalyi (1934-)

7.1 Description of the models

Markov chain models where introduced in the 1920s by A.A. Markov (*Calculus of Probabilities,*) 3rd ed., St. Petersburg, 1913). Informally a Markov chain is simply a sequence of random variables evolving with time. The random states are defined sequentially based on the current state and some additional random variables. The theory of Markov processes has led to rather intense activity in various scientific disciplines, providing natural probabilistic interpretations of various random evolution models arising in engineering, physics, biology and many other scientific disciplines. A more formal definition is given below.

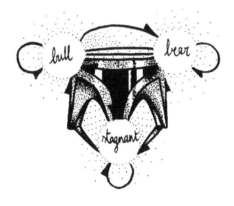

Definition 7.1.1 *A Markov chain is a sequence of random variables $(X_n)_{n \geq 0}$ indexed by the (integer) time parameter $n \in \mathbb{N}$, and taking values in some state space S for which the conditional distribution of the random state X_{n+1} w.r.t. to its past $(X_0, \ldots, X_n)$ only depends on the present state X_n. That is, we have*

$$\mathbb{P}\left(X_{n+1} \in dx \mid (X_0, \ldots, X_n)\right) = \mathbb{P}\left(X_{n+1} \in dx \mid X_n\right) \tag{7.1}$$

in the sense that for any bounded function f on S,

$$\mathbb{E}\left(f(X_{n+1}) \mid (X_0, \ldots, X_n)\right) = \mathbb{E}\left(f(X_{n+1}) \mid X_n\right)$$

holds. The property (7.1) is called the Markov property. The Markov chain is called homogeneous whenever the transition probabilities (7.1) do not depend on the time parameter n.

Of course, with some "large memory", every sequence of random variables Y_n taking values on a possibly different state space E_n at each time step, and with more sophisticated correlations, can be encapsulated in a Markov chain by considering the path space models

$$X_n = (n, (Y_0, \ldots, Y_n)) \in S = \cup_{n \geq 0} (\{n\} \times (S_0 \times \ldots \times S_n)). \tag{7.2}$$

Thus, the definition of a Markov chain is implicitly associated with some simple topological aspects of the state space. We shall return to these questions in section 8.5.4. The latter section is dedicated to historical and path space models associated with elementary Markov chains.

Sequences of independent real valued r.v. are clearly Markov chains. However, these random sequences that do not exhibit any memory, are clearly useless to represent random evolutions that involve any correlation between the random states. The simplest non-trivial example with some variability is the two-states $S = \{0, 1\}$ valued homogeneous Markov chain model. We will use this elementary example to illustrate the main probabilistic models used in the theory of Markov chains.

We emphasize that Markov chains can be introduced and interpreted in several equivalent ways, each being related to a specific mathematical model.

In the next sections, we discuss these equivalent formulations. Their extension to more general finite state space Markov chains follows the same line of arguments; thus their detailed description is omitted.

For finite space spaces $S := \{e_1, \ldots, e_d\}$, a probability measure η on S can be regarded as a a mapping

$$\eta \ : \ e_i \in S \ \mapsto \ \eta(e_i) := p_i \in [0, 1] \quad \text{s.t.} \quad \sum_{1 \leq i \leq d} p_i = 1.$$

In other words, a probability measure $(\eta(e_i))_{1 \leq i \leq d}$ can be seen as a point in the $(d-1)$-dimensional simplex Δ_{d-1} defined in (0.2), and $\mathcal{B}(S) = \mathbb{R}^S$.

When $S = \{0, 1\}$, these identifications are summarized by

$$\begin{aligned} \mathcal{P}(S) \quad &:= \quad \Delta_1 = \{(p_0, p_1) \in [0, 1] \ : \ p_0 + p_1 = 1\} \\ &= \quad \{(u, 1-u) \ : \ u \in [0, 1]\} \subset [0, 1]^2 \quad \text{and} \quad \mathcal{B}(S) = \mathbb{R}^{\{0,1\}}. \end{aligned}$$

7.2 Elementary transitions

The first way to introduce a non-trivial homogeneous Markov chain model on a finite or a countable state space S is to describe the law of the random state X_{n+1} given the present value X_n. These distributions are called Markov transitions, and they are defined by the following formula

$$\forall x, y \in S \qquad \mathbb{P}(X_{n+1} = y \mid X_n = x) = M(x, y) \tag{7.3}$$

for some given numbers $M(x, y) \in [0, 1]$ such that $\sum_y M(x, y) = 1$, for any $x \in S$; in other words, we have $M(x, .) \in \mathcal{P}(S)$, for any $x \in S$. Note that for any sequence of states $(x_0, \ldots, x_n) \in S^{n+1}$ we have

$$\mathbb{P}((X_0, \ldots, X_n) = (x_0, \ldots, x_n)) = M(x_0, x_1) \ldots M(x_{n-1}, x_n). \tag{7.4}$$

In the case $S = \{0, 1\}$, we notice that

$$
\begin{aligned}
\mathbb{P}\left(X_{n+1} = 0 \mid X_n = 0\right) &= M(0,0) = 1 - \mathbb{P}\left(X_{n+1} = 1 \mid X_n = 0\right) = 1 - M(0,1) \\
\mathbb{P}\left(X_{n+1} = 0 \mid X_n = 1\right) &= M(1,0) = 1 - \mathbb{P}\left(X_{n+1} = 1 \mid X_n = 1\right) = 1 - M(1,1).
\end{aligned}
$$
(7.5)

7.3 Markov integral operators

The Markov transition M defined in (7.3) is associated with two integral operators.

> The first one maps the set of functions $\mathcal{B}(S)$ into itself
> $$
> M : f \in \mathcal{B}(S) \mapsto M(f) \in \mathcal{B}(S)
> $$
> using the conditional expectation operator
> $$
> x \mapsto \mathbb{E}\left(f(X_{n+1}) \mid X_n = x\right) = \sum_{y \in S} M(x,y)\, f(y) := M(f)(x).
> $$

Recalling that

$$
\mathbb{P}\left(X_{n+1} = y\right) = \sum_{x \in S} \mathbb{P}\left(X_{n+1} = y \mid X_n = x\right)\, \mathbb{P}\left(X_n = x\right) = \sum_{x \in S} \mathbb{P}\left(X_n = x\right)\, M(x,y)
$$

we also check the transport formula:

> $$
> \eta_n(x) := \mathbb{P}\left(X_n = x\right) \Rightarrow \eta_{n+1}(y) = \sum_{x \in S} \eta_n(x)\, M(x,y) := \left(\eta_n M\right)(y).
> $$

> This shows that M also maps the set $\mathcal{P}(S)$ into itself
> $$
> M : \eta \in \mathcal{P}(S) \mapsto \eta M \in \mathcal{P}(S) \quad \text{with} \quad (\eta M)(y) := \sum_{x \in S} \eta(x)\, M(x,y).
> $$

Using (7.4) we clearly have that for any $0 \le m \le n$ and any pair of states $(x_m, x_n) \in S^2$

$$
\mathbb{P}\left(X_n = x_n \mid X_m = x_m\right)
$$

$$
= \sum_{(x_{m+1}, \ldots, x_n) \in S^{n-m}} M(x_m, x_{m+1}) \ldots M(x_{n-1}, x_n) := M^{(n-m)}(x_m, x_n)
$$

with the composition of the operators $M^n = M^{n-1}M = MM^{n-1}$ defined by induction using the formulae

$$M^n(x_0, x_n) = \sum_{x_{n-1} \in S} M^{n-1}(x_0, x_{n-1}) M(x_{n-1}, x_n) = \sum_{x_1 \in S} M(x_0, x_1) M^{n-1}(x_1, x_n).$$

The above formula is sometimes called the Chapman-Kolmogorov equation. The common link of Markov chain analysis with the theory of dynamical systems is provided by the following proposition.

> **Proposition 7.3.1** *The probability distributions η_n of the random states X_n of a Markov chain taking values in a finite set S are the solutions of a linear dynamical system in the simplex $\mathcal{P}(S)$*
>
> $$\forall n \geq 1 \qquad \eta_n = \eta_{n-1} M. \tag{7.6}$$
>
> *For any $0 \leq p \leq n$, we have*
>
> $$\eta_n = \eta_p M^{n-p} = \eta_0 M^n. \tag{7.7}$$

Important remark : Proposition 7.3.1 shows that we can reduce the analysis of Markov chain distributions to manipulations of operators on functions or on probability measures and their associated linear dynamical system.

7.4   Equilibrium measures

Proposition 7.3.1 shows that (whenever it exists) the limiting distribution

$$\lim_{n \to \infty} \eta_n = \lim_{n \to \infty} \mathrm{Law}(X_n) = \eta_\infty \tag{7.8}$$

satisfies the fixed point equation

$$\eta_\infty M = \eta_\infty. \tag{7.9}$$

We prove this claim taking the limits, as $n \uparrow \infty$ in the evolution equation

$$\underbrace{\eta_n}_{\to \eta_\infty} = \underbrace{\eta_{n-1}}_{\to \eta_\infty} M \Rightarrow (7.9).$$

We illustrate these equilibrium measures with the two-states Markov chain model (7.5). In this situation, the invariant measure is given by

$$\eta_\infty(0) \propto M(1,0) \quad \text{and} \quad \eta_\infty(1) \propto M(0,1).$$

Then we find that

$$
\begin{aligned}
(\eta_\infty M)(0) &= \eta_\infty(0) M(0,0) + \eta_\infty(1) M(1,0) \\
&\propto M(1,0)(1 - M(0,1)) + M(0,1) M(1,0) = M(1,0) \propto \eta_\infty(0).
\end{aligned}
$$

Definition 7.4.1 *Whenever it exists, the solution η_∞ of the fixed point equation (7.9) is called the invariant distribution or the limiting distribution of the Markov chain X_n with Markov transition M.*

Definition 7.4.2 *A Markov transition M satisfying the reversibility property*

$$\forall x, y \in S \qquad \pi(x)\, M(x, y) = \pi(y)\, M(y, x) \tag{7.10}$$

for some measure π is called a reversible transition w.r.t. the measure π.

For instance the Markov transition of the two-states Markov chain model (7.5) is reversible w.r.t. the uniform measure $\pi(0) = \pi(1) = 1/2$ on $\{0, 1\}$ as soon as $M(0, 1) = M(1, 0)$.

Proposition 7.4.3 *Suppose that a Markov transition M is a reversible transition w.r.t. some probability measure π. Then π is the invariant measure of the Markov chain with Markov transition M.*

Proof :
We have

$$(7.10) \Longrightarrow \forall y \in S \quad (\pi M)(y) = \sum_{x \in S} \pi(x)\, M(x, y) = \pi(y) \sum_{x \in S} M(y, x) = \pi(y).$$

This shows that $\pi = \pi M$ is a solution of the fixed point equation (7.9). ∎

7.5 Stochastic matrices

For finite and ordered state spaces $S = \{e_1, \ldots, e_d\}$, we can identify measures and functions using the vector notation defined in (0.1)

$$M_n = \begin{pmatrix} M_n(e_1, e_1) & \cdots & M_n(e_1, e_d) \\ \vdots & \vdots & \vdots \\ M_n(e_d, e_1) & \cdots & M_n(e_d, e_d) \end{pmatrix}.$$

Notice that the entries of each line are $[0, 1]$-valued and they sum to 1. These matrices are called stochastic matrices in the literature on finite state space valued Markov chain.

Using the identification of measures and functions with the line and column vectors discussed in (0.1), we check that

$$M_n(f) = \begin{pmatrix} M_n(f)(e_1) \\ \vdots \\ M_n(f)(e_d) \end{pmatrix} = \begin{pmatrix} M_n(e_1, e_1) & \cdots & M_n(e_1, e_d) \\ \vdots & \vdots & \vdots \\ M_n(e_d, e_1) & \cdots & M_n(e_d, e_d) \end{pmatrix} \begin{pmatrix} f(e_1) \\ \vdots \\ f(e_d) \end{pmatrix}$$

and

$$\begin{aligned} \eta M &= [(\eta M)(e_1), \ldots, (\eta M)(e_d)] \\ &= [\eta(e_1), \ldots, \eta(e_d)] \begin{pmatrix} M_n(e_1, e_1) & \cdots & M_n(e_1, e_d) \\ \vdots & \vdots & \vdots \\ M_n(e_d, e_1) & \cdots & M_n(e_d, e_d) \end{pmatrix}. \end{aligned}$$

Important remark : Using these matrix formulations of the Markov transition, the evolution equations (7.6) and (7.7) reduce to matrix operations. This shows that the analysis of a Markov chain on finite and ordered state spaces reduces to manipulations of finite-dimensional matrices.

For instance, the Markov transition (7.3) associated with the $S = \{0, 1\}$ valued Markov chain (7.5) is identified with the matrix

$$M = \left(\begin{array}{cc} M(0,0) & M(0,1) \\ M(1,0) & M(1,1) \end{array} \right).$$

If we set $M(0,1) = p$ and $M(1,0) = q$, this matrix can also be rewritten as follows:

$$M = \left(\begin{array}{cc} 1-p & p \\ q & 1-q \end{array} \right). \tag{7.11}$$

Using the fact that

$$[q,p] \left(\begin{array}{cc} 1-p & p \\ q & 1-q \end{array} \right) = [q,p]$$

we conclude that

$$\eta_n := \mathrm{Law}(X_n) \longrightarrow \mathrm{Law}(X_n) = [\eta_\infty(0), \eta_\infty(1)] = \left[\frac{q}{p+q}, \frac{p}{p+q} \right]. \tag{7.12}$$

We illustrate this model in the case where $p = .2$ and $q = .1$. The state 0 represents city dwellers, and 1 stands for suburbs dwellers. The transition matrix indicates that during every unit of time (years for instance) 20% of city dwellers move to suburbs, and 10% of suburbanites move to the city. In this context, a realization X_n of the Markov chain represents the random evolution of a given family from the suburbs to the city and vice versa. Sampling N independent chains X_n^i, will represent the evolution of N individuals randomly moving between suburbs and city.

7.6   Random dynamical systems

The dynamical system formulation of the Markov chain is related to the effective simulation of the chain. In principle, any Markov chain can be described inductively by a (*non unique*) recursion of the form

$$X_{n+1} = F_n(X_n, W_n) \tag{7.13}$$

where W_n stands for a sequence of independent random variables taking values on $\mathbb{R}^d$, for some $d \geq 1$.

Important remark : This formulation indicates that the analysis of Markov chain models of the form (7.13) is intimately related to the analysis of dynamical systems with random components. We illustrate this remark with a linear Markov chain model and the two-states Markov chain model presented in (7.5).

7.6.1   Linear Markov chain model

We consider the $\mathbb{R}^d$-valued linear model defined by

$$\forall n \geq 1 \quad X_n = A_n X_{n-1} + B_n W_n \tag{7.14}$$

where X_0, W_n are $\mathbb{R}^d$-valued independent random variables, and A_n, B_n are $(d \times d)$-matrices. We further assume that W_n are centered and we denote by $P_n = \mathrm{Cov}(W_n, W_n)$, $n \geq 1$ their covariance matrix; we also set $P_0 = \mathrm{Cov}(X_0, X_0)$. Using simple algebraic manipulations, we find that

$$X_n = (A_n \ldots A_1) X_0 + \sum_{1 \leq p \leq n} (A_n \ldots A_{p+1}) B_p W_p. \tag{7.15}$$

In (7.15) we have used the dots to indicate descending arrangement of indices. We also use the convention that $A_n \ldots A_{p+1} = Id$ for $p = n$. Now we conclude from (7.15) that $\mathbb{E}(X_n) = (A_n \ldots A_1) \mathbb{E}(X_0)$ and

$$\mathrm{Cov}(X_n, X_n)$$
$$= (A_n \ldots A_1) P_0 (A_n \ldots A_1)' + \sum_{1 \leq p \leq n} (A_n \ldots A_{p+1}) B_p P_p B_p' (A_n \ldots A_{p+1})'. \tag{7.16}$$

If we denote by X_n' the solution of the linear system (7.14) evolving with the same randomness but a different initial variable X_0' then

$$X_n - X_n' = (A_n \ldots A_1) (X_0 - X_0').$$

For one-dimensional and time homogeneous models $(A_n, B_n, \mathrm{Cov}(W_n, W_n)) = (a, b, \sigma^2)$ we find that

$$|X_n - X_n'| = |a|^n |X_0 - X_0'|$$
$$\mathbb{E}(X_n) = a^n \mathbb{E}(X_0) \quad \text{and} \quad \mathrm{Var}(X_n) = a^{2n} \mathrm{Var}(X_0) + (\sigma b)^2 \sum_{0 \leq p < n} a^{2p}. \tag{7.17}$$

For instance, when $a < 1$ and $\mathbb{E}(X_0) = 0$ we have $\mathbb{E}(X_n) = 0$ and

$$\mathrm{Var}(X_n) = \sigma^2 \frac{b^2}{1 - a^2} (1 - a^{2n}) \to_{n \to \infty} \sigma^2 \frac{b^2}{1 - a^2}.$$

We also easily check that the Markov chain forgets its initial condition exponentially fast in the sense that

$$|X_n - X_n'| = |a|^n |X_0 - X_0'| \to_{n \to \infty} 0.$$

7.6.2 Two-states Markov models

For the $\{0, 1\}$-valued Markov chain discussed in (7.5), we can choose independent r.v. $W_n = U_n$ uniformly distributed on $[0, 1]$, and set

$$X_{n+1} = F_n(X_n, U_n)$$
$$:= 1_{[0, M(X_n, 0)[}(U_n) \times 0 + 1_{[M(X_n, 0), 1]}(U_n) \times 1 = 1_{[M(X_n, 0), 1]}(U_n)$$

with the function

$$F : (x, u) \in S \times [0, 1] \mapsto F(x, u) := 1_{[M(x, 0), 1]}(u).$$

This dynamical formulation does not really help to analyze the time evolution of the Markov chain. In this context, the matrix interpretation appears to give a more adequate algebraic tool to analyze the convergence of this model when the time tends to infinity.

7.7   Transition diagrams

Markov chains on finite state spaces (with a reasonably large cardinality) are often defined by the a synthetic transition diagram that expresses all the possible transitions of the chain in a time step. The transition diagram of the $\{0,1\}$-valued Markov model discussed above is given by

$$M(0,0) \;\circlearrowright\; 0 \;\underset{M(1,0)}{\overset{M(0,1)}{\rightleftarrows}}\; 1 \;\circlearrowright\; M(1,1)\;\cdot$$

For instance, from state 0 we can move to state 1 with probability $M(0,1)$, or we can stay in 0 with probability $M(0,0) = 1 - M(0,1)$.

We extend easily these constructions to any reasonably large finite set. For instance the transition matrix

$$\begin{pmatrix} M(1,1) & M(1,2) & M(1,3) \\ M(2,1) & M(2,2) & M(2,3) \\ M(3,1) & M(3,2) & M(3,3) \end{pmatrix}$$

of a $\{1,2,3\}$-valued Markov chain is given by the following transition diagram

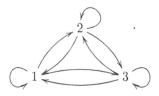

7.8   The tree of outcomes

One more way of introducing a Markov chain on a given state space, is to consider the tree of the possible outcomes w.r.t. the time parameter. The tree of outcomes of the $\{0,1\}$-valued Markov model starting at $X_0 = 0$ is given below

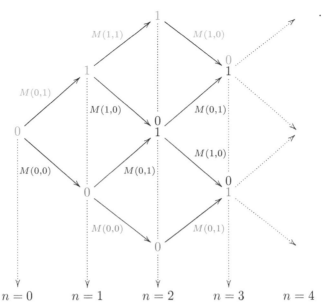

Time axis: $n = 0$ $n = 1$ $n = 2$ $n = 3$ $n = 4$

The conditional probabilities are computed using the product of the transition probability from the conditional starting point. For instance, we have that

$$\mathbb{P}\left((X_1, X_2, X_3) = (1, 1, 0) \mid X_0 = 0\right) = M(0, 1)M(1, 1)M(1, 0)$$

and

$$\mathbb{P}\left((X_2, X_3) = (0, 1) \mid X_1 = 0\right) = M(0, 0)M(0, 1).$$

7.9   General state space models

The choice of the formulation of a Markov chain depends on the problem at hand, but it is always preferable to keep in mind the different interpretations discussed in earlier sections to develop a certain physical or mathematical intuition about the evolution of a given stochastic process.

For general state space Markov chain models, only the first and the third formulations are generally used. In this context, the Markov chain is defined in terms of some not necessarily time homogeneous Markov transitions,

$$\forall x \in S \qquad \mathbb{P}\left(X_n = x_n \mid X_{n-1} = x_{n-1}\right) = M_n(x_{n-1}, dx_n)$$

where $M_n(x, .)$ stands for some probability measure over the set S. In this general framework, the Markov transitions M_n are associated with two integral operators

$$\mathbb{E}\left(f(X_n) \mid X_{n-1} = x_{n-1}\right) = \int M_n(x_{n-1}, dx_n)\, f(x_n) := M_n(f)(x_{n-1})$$

and if we set $\eta_n(dx_n) := \mathbb{P}\left(X_n \in dx_n\right)$, then

$$\eta_{n+1}(dx_{n+1}) = \int \eta_n(dx_n)\, M_{n+1}(x_n, dx_{n+1}) := (\eta_n M_{n+1})\,(dx_{n+1}).$$

This transport equation sometimes is written in the more synthetic form

$$\eta_{n+1} = \eta_n M_{n+1} = \eta_{n-1} M_n M_{n+1} = \ldots = \eta_0 M_1 \ldots M_{n+1}$$

with the composition of operators $M_1 \ldots M_n$ defined inductively using the formulae

$$
\begin{aligned}
(M_1 \ldots M_n)(x_0, dx_n) &= \int (M_1 \ldots M_{n-1})(x_0, dx_{n-1}) M_n(x_{n-1}, dx_n) \\
&= \int M_1(x_0, dx_1) \, (M_2 \ldots M_n)(x_1, dx_n).
\end{aligned}
$$

Definition 7.9.1 *The composition operators*

$$\forall 0 \le p \le n \qquad M_{p,n} = M_{p+1} M_{p+1} \ldots M_n$$

are called the semigroup (abbreviated sg) *of the Markov chain, with the convention $M_{n,n} = $ Id the identity operator, for $p = n$.*

Notice that for any $0 \le p \le q \le n$ we have

$$M_{p,n} = M_{p,q} M_{q,n} \quad \text{and} \quad \eta_n = \eta_p M_{p,n}.$$

We illustrate these abstract models with the Gaussian Markov transitions given by

$$M_n(x_{n-1}, dx_n) = \frac{1}{\sqrt{2\pi\sigma_n^2}} \, \exp\left\{-\frac{1}{2\sigma_n^2} \, (x_n - a_n \, x_{n-1})^2\right\} dx_n.$$

This collection of probability measures describes the elementary probability transitions of the Markov chain defined by the random evolution equation

$$X_n = a_n \, X_{n-1} + b_n \, W_n, \quad n \ge 1 \tag{7.18}$$

where $(a_n, \sigma_n) \in \mathbb{R}^2$, $b_n \in \mathbb{R} - \{0\}$, and $(W_n)_{n\ge 1}$ stands for a sequence of independent and centered Gaussian r.v. with variance $\sigma_n^2 > 0$.

We note that in this situation, the probability measures of the random trajectories are given by

$$
\begin{aligned}
\mathbb{P}^{(X_1,\ldots,X_n)}(d(x_1,\ldots,x_n) \mid X_0 = x_0) &= \prod_{p=1}^n M_p(x_{p-1}, dx_p) \\
&= \left[\prod_{p=1}^n \frac{1}{\sqrt{2\pi\sigma_p^2 b_p^2}} \, e^{-\frac{(x_p - a_p x_{p-1})^2}{2\sigma_p^2 b_p^2}} \, dx_p\right].
\end{aligned}
$$

This model has the same form as the one-dimensional version of the linear Markov chain discussed in (7.14). The only difference is that the random sequences W_n are (centered) *Gaussian* random variables with variance σ_n^2. We further assume that the initial state X_0 is a Gaussian random variable.

Since any linear combination of independent Gaussian variables is again a Gaussian random variable, we conclude that the random states X_n are Gaussian. In addition, if we take $a_n = a = \sqrt{1-\epsilon}$ (with $\epsilon \in]0,1]$), $b_n = b = \sqrt{\epsilon}$, and $\sigma_n = \sigma$ in (7.18) and in (7.17), then we find that

$$\lim_{n\to\infty} \mathbb{E}(X_n) = 0 \quad \text{and} \quad \lim_{n\to\infty} \mathrm{Var}(X_n) = \sigma^2.$$

Roughly speaking, this indicates that the limiting invariant measure of the chain is given by

$$\mathbb{P}(X_\infty \in dx) = \frac{1}{\sqrt{2\pi\sigma^2}} \exp\left\{-\frac{1}{2\sigma^2} x^2\right\} dx := \pi(dx).$$

Much more is true. We easily check that

$$M(x, dy) = \frac{1}{\sqrt{2\pi\sigma^2\epsilon}} \exp\left\{-\frac{1}{2\sigma^2\epsilon} (y - \sqrt{1-\epsilon}\, x)^2\right\} dx$$

is reversible w.r.t. the Gaussian measure π in the sense that

$$\pi(dx)M(x, dy) = \pi(dy)M(y, dx). \tag{7.19}$$

We end this section with a time homogeneous model representing the waiting time of the occurrence of some independent random events. We let $p(i)$ be a distribution on the set $\mathbb{N} - \{0\}$, such that $\sum_{i\geq 1} ip(i) < \infty$. We consider the Markov transitions on the set of integers $S = \mathbb{N}$ defined by

$$\begin{aligned} M(0, i) &= p(i) & \forall i \geq 1 \\ M(i, i-1) &= 1 & \forall i > 0 \end{aligned} \qquad \text{otherwise } M(i, j) = 0. \tag{7.20}$$

The transition diagram of the chain is given below:

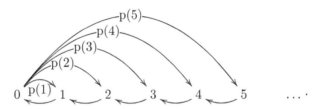

The parameters $p(i)$ represent the probability that the period between the occurrences of two events is i units of time. For instance $p(1)$ denotes the probability that the next event occurs after one unit of time, $p(2)$ denotes the probability that it occurs after 2 units of time.

The chain X_n represents the time until the occurrence of the next event.

For instance, $p(i)$ can be seen as the probability that a bus arrives after i minutes, and X_n represents the time till the next bus arrives. If we denote by I_n a sequence of independent r.v. with common law p, then X_n can be written in the following form

$$X_n = (X_{n-1} - 1)\, 1_{X_{n-1}\neq 0} + 1_{X_{n-1}=0}\, I_n.$$

After some elementary computations (cf. exercise 87), we prove that the invariant measure of X_n is given by

$$\forall i \geq 1 \qquad \pi(i) = \frac{\mathbb{P}(I_1 \geq i)}{1 + \mathbb{E}(I_1)} \quad \text{and} \quad \pi(0) = \frac{1}{1 + \mathbb{E}(I_1)}. \tag{7.21}$$

Important remark : The above observations show that the equilibrium measures discussed in section 7.4 and the reversible property presented in definition 7.4.2 are far from being restricted to finite state space models. The fixed point analysis (7.8) and the proposition 7.4.3 are also true for general state space models.

A more refined analysis of countable state space models is provided in section 8.5 dedicated to the topological properties of these models.

7.10   Nonlinear Markov chains

As their name indicates, interacting Markov chains are discrete time stochastic processes equipped with a particular spatial and/or temporal interaction structure. In this section, we discuss two different classes of interaction mechanisms, namely spatial and temporal interaction models.

The second class of models can be interpreted as the random evolution of a population of a fixed number of N interacting individuals. More precisely, the elementary transitions of each individual depend on the whole population, and more particularly on the occupation measure of the population of individuals. This type of interaction is called mean field interaction to reflect the fact that the individuals interact with the empirical distribution of the whole population.

7.10.1   Self interacting processes

The elementary transitions of self-interacting processes depend on the history of the process, and more particularly on the occupation measure of the chain from the origin up to the present time.

The resulting process can be interpreted as the motion of a single individual evolving with reinforced learning type strategies. More precisely, every elementary move of the individual at time n depends on the way it has explored the state space in the past up to the time horizon n. For instance, a site which has been visited many times can be more attractive (or repulsive) than other sites. This type of natural positive or negative reinforcement often arises when a tourist explores randomly some places in a city. In this context, the individual is often tempted to enter more often in a street, restaurant or pub that has been visited several times.

More formally, given the historical evolution of the process

$$\boldsymbol{X_n} := (X_0, \ldots, X_n)$$

up to current time n, the transition $X_n \rightsquigarrow X_{n+1}$ depends on the random occupation measure

$$m(\boldsymbol{X_n}) := \frac{1}{n+1} \sum_{0 \leq p \leq n} \delta_{X_p}.$$

We illustrate this class of models with a toy example. Let μ be a given probability measure on S, and $\epsilon \in [0, 1]$ be a reinforcement parameter. We associate with these objects the self-interacting process defined sequentially by the Markov transitions:

$$\mathbb{P}\left(X_{n+1} \in dx \mid \boldsymbol{X_n}\right) := \epsilon \; m(\boldsymbol{X_n})(dx) + (1 - \epsilon) \; \mu(dx). \tag{7.22}$$

In words, to sample the transition $X_n \rightsquigarrow X_{n+1}$, we flip a coin to determine whether or not we look back in the past. With a probability ϵ, the random state X_{n+1} chooses randomly and uniformly one of the values X_p, $0 \leq p \leq n$. With a probability $(1 - \epsilon)$, the random state X_{n+1} is a fresh new r.v. with distribution μ.

More generally, self-interacting processes are defined in terms of a collection of Markov transitions $K_\eta(x, dy)$ indexed by the set $\mathcal{P}(S)$ of all probability measures over some state space S.

The elementary transitions $X_n \rightsquigarrow X_{n+1}$ of the self interacting Markov chain associated with the collection of transitions M_η are given by the formula

$$\mathbb{P}(X_{n+1} \in dx \mid \boldsymbol{X_n}) = K_{m(\boldsymbol{X_n})}(X_n, dx) \qquad (7.23)$$

with the occupation measure $m(\boldsymbol{X_n})$ of the historical process $\boldsymbol{X_n}$ at time n given by

$$m(\boldsymbol{X_n}) := \frac{1}{n+1} \sum_{0 \le p \le n} \delta_{X_p}.$$

The reader has certainly noticed that X_n is not a Markov chain but the historical process

$$\boldsymbol{X_{n-1}} := (X_0, \ldots, X_{n-1}) \rightsquigarrow \boldsymbol{X_n} := (\boldsymbol{X_{n-1}}, X_n) = ((X_0, \ldots, X_{n-1}), X_n)$$

always has the Markov property.

For regular Markov chains M_η the occupation measures

$$S_n = m(\boldsymbol{X_{n-1}}) = \frac{1}{n} \sum_{k=0}^{n-1} \delta_{X_k}$$

converge when n tends to ∞ to the fixed point $\eta_\infty = \Phi(\eta_\infty)$ of the nonlinear transformation Φ. The fixed point is defined as

$$\Phi \ : \ \eta \in \mathcal{P}(S) \longrightarrow \Phi(\eta) := \eta K_\eta \in \mathcal{P}(S). \qquad (7.24)$$

The proof of this convergence result relies on the stability properties of nonlinear reinforced Markov chain models. It is far beyond the scope of these lecture notes to present these sophisticated tools. We refer the interested reader to [77, 78].

In the toy example (7.22) discussed above we have

$$\Phi(\eta) = \epsilon \, \eta + (1 - \epsilon) \, \mu \implies \eta_\infty = \mu.$$

We also notice that the reinforcement parameter reflects the contraction properties of the mapping Φ, in the sense that

$$\Phi(\eta_1) - \Phi(\eta_2) = \epsilon \, [\eta_1 - \eta_2].$$

We refer the reader to exercise 17 for an estimate of the convergence of the law of the random states X_n to μ, as n tends to ∞.

We end this section with a genetic type self-interacting Markov chain model associated with a Markov transition M and some $]0, 1]$-valued potential function G on some state space S. The transition of this chain is given by the formula

$$\mathbb{P}(X_n \in dy \mid X_0, \ldots, X_{n-1})$$

$$= G(X_{n-1}) \, M(X_{n-1}, dy) + (1 - G(X_{n-1})) \sum_{k=0}^{n-1} \frac{G(X_k)}{\sum_{l=0}^{n-1} G(X_l)} \, M(X_k, dy).$$

This evolutionary type model has the same form as (7.23) with the collection of transitions $K_\eta(x, dz)$ given by

$$K_\eta(x, dz) = G(x) \, M(x, dz) + (1 - G(x)) \int_S \eta(dy) \, \frac{G(y)}{\eta(G)} \, M(y, dz). \qquad (7.25)$$

It is a simple exercise to check that the mapping Φ defined in (7.24) is given for any bounded function f on S by

$$
\begin{aligned}
\eta(K_\eta(f)) &= \eta\left(\left[GM(f) + (1-G)\,\eta\left(\frac{GM(f)}{\eta(G)}\right)\right]\right) \\
&= \eta\left(GM(f)\right) + \eta\left(\frac{GM(f)}{\eta(G)}\right) - \eta(G)\eta\left(\frac{GM(f)}{\eta(G)}\right) \\
&= \eta\left(\frac{GM(f)}{\eta(G)}\right) := \Phi(\eta)(f).
\end{aligned}
$$

In other words, we have

$$
\Phi(\eta) = \Psi_G(\eta)M \quad \text{with} \quad \Psi_G(\eta)(dx) = \frac{1}{\eta(G)}\,G(x)\,\eta(dx). \tag{7.26}
$$

The r.h.s. transformation Ψ_G is called the Boltzmann-Gibbs transformation associated with the potential function G. In this notation, the Markov transition (7.25) takes the following form

$$
K_\eta(x, dz) = G(x)\,M(x, dz) + (1 - G(x))\,(\Psi_G(\eta)M)(dz).
$$

7.10.2   Mean field particle models

These processes are defined in terms of a Markov chain $\xi_n := (\xi_n^i)_{1 \leq i \leq N}$ on some product state space $S^N = (S \times \ldots \times S)$. The index $n \in \mathbb{N}$ represents the time parameter, and the number of dimensions N is interpreted as the number of individuals (a.k.a. particles). The random states in S^N are interpreted as a population of N individuals. The elementary transitions of each individual $\xi_n^i \rightsquigarrow \xi_{n+1}^i$ depend on the occupation measure of the whole population defined by

$$
m(\xi_n) := \frac{1}{N} \sum_{1 \leq i \leq N} \delta_{\xi_n^i}.
$$

More formally, given a collection of Markov transitions $K_{n+1,\eta}$ indexed by the time parameter, and the set $\mathcal{P}(S)$ of all probability measures over some state space S, we have

$$
\mathbb{P}\left(\xi_{n+1} \in dx \mid \xi_n\right) := \prod_{1 \leq j \leq N} K_{n+1,m(\xi_n)}\left(\xi_n^j, dx^j\right). \tag{7.27}
$$

In the above display, $dx = d(x^1, \ldots, x^N) := dx^1 \times \ldots \times dx^N$ stands for an infinitesimal neighborhood of the point $x = (x^1, \ldots, x^N)$ in S^N. The initial population $\xi_0 := (\xi_0^i)_{1 \leq i \leq N}$ consists of N independent r.v. with some given common law η_0.

We consider the sequence of random empirical measures

$$
\eta_n^N = m(\xi_n) := \frac{1}{N} \sum_{1 \leq i \leq N} \delta_{\xi_n^i}.
$$

Initially, by the law of large numbers discussed in section 5.1, for any function f we have

$$
\sqrt{N}\left(\eta_0^N(f) - \eta_0(f)\right) := V_0^N(f) \iff \eta_0^N(f) = \eta_0(f) + \frac{1}{\sqrt{N}}\,V_0^N(f)
$$

with the empirical random field V_0^N such that

$$\mathbb{E}\left(V_0^N(f)\right) = 0 \quad \text{and} \quad \mathbb{E}\left([V_0^N(f)]^2\right) \le 1$$

as soon as $\mathrm{osc}(f) \le 1$.

In much the same way, the local sampling errors induced by the mean field particle model (7.27) are expressed in terms of the empirical random field sequence V_n^N defined by

$$V_{n+1}^N = \sqrt{N}\ \left[\eta_{n+1}^N - \Phi_{n+1}\left(\eta_n^N\right)\right].$$

Here again, we notice that V_{n+1}^N is alternatively defined by the following stochastic perturbation formulae

$$\eta_{n+1}^N = \Phi_{n+1}\left(\eta_n^N\right) + \frac{1}{\sqrt{N}}\ V_{n+1}^N. \tag{7.28}$$

It is a matter of a rather elementary check to see that

$$
\begin{aligned}
\mathbb{E}\left(V_{n+1}^N(f) \mid \xi_n\right) &= 0 \\
\mathbb{E}\left(V_{n+1}^N(f)^2 \mid \xi_n\right) &= \int \eta_n^N(dx)\ K_{n+1,\eta_n^N}\left[\left(f - K_{n+1,\eta_n^N}(f)(x)\right)^2\right](x) \\
&\le \mathrm{osc}(f)^2. \tag{7.29}
\end{aligned}
$$

Here, the N-particle model can also be interpreted as a stochastic perturbation of the limiting system

$$\eta_{n+1} = \Phi_{n+1}\left(\eta_n\right) := \eta_n K_{n+1,\eta_n}. \tag{7.30}$$

In other words, for regular models one can prove that for any bounded function f we have the almost sure convergence result

$$\lim_{N\to\infty} \eta_n^N(f) = \eta_n(f).$$

It is important to observe that η_n can be interpreted as the law of the random states $\overline{X}_n$ of the nonlinear Markov chain model defined by

$$\mathbb{P}\left(\overline{X}_{n+1} \in dx \mid \overline{X}_n\right) = K_{n+1,\eta_n}\left(\overline{X}_n, dx\right) \quad \text{with} \quad \eta_n = \mathrm{Law}(\overline{X}_n). \tag{7.31}$$

We check this claim by a simple induction with the time parameter. The Markov chain $\overline{X}_n$ is called a McKean interpretation of the nonlinear measure valued equation (7.30).

7.10.3   McKean-Vlasov diffusions

Prototypes of discrete generation and nonlinear Markov-McKean models are given by McKean-Vlasov-Fokker-Planck diffusion type models arising in fluid mechanics, as well as in mean field game theory. In dimension $d = 1$, these non-homogeneous Markov models are given by an $\mathbb{R}$-valued stochastic process defined by the recursive equation

$$\overline{X}_n - \overline{X}_{n-1} = \mathbf{a_n}(\overline{X}_{n-1}, \eta_{n-1}) + \boldsymbol{\sigma_n}(\overline{X}_{n-1}, \eta_{n-1})\ W_n \tag{7.32}$$

with $\eta_{n-1} := \mathrm{Law}(\overline{X}_{n-1})$. In the above formula, $\overline{X}_0$ is a r.v. $(W_n)_{n \geq 0}$ is a collection of i.i.d. centered Gaussian random variables with unit variance, and the drift and diffusion functions are defined by

$$\mathbf{a_n}(\overline{X}_{n-1}, \eta_{n-1}) = \int a_n(\overline{X}_{n-1}, x_{n-1}) \, \eta_{n-1}(dx_{n-1})$$

$$\boldsymbol{\sigma_n}(\overline{X}_{n-1}, \eta_{n-1}) = \int \sigma_n(\overline{X}_{n-1}, x_{n-1}) \, \eta_{n-1}(dx_{n-1})$$

for some regular mappings a_n and σ_n. Whenever $\boldsymbol{\sigma_n}(x, \eta) \geq \epsilon$, for some $\epsilon > 0$, the law of the random states $\eta_n = \mathrm{Law}(\overline{X}_n)$ satisfies the evolution equation (7.30), with the McKean transitions given by

$$K_{n,\eta}(x, dy) = \frac{1}{\sqrt{2\pi\sigma_n^2(x,\eta)}} \, \exp\left\{ -\frac{1}{2} \left(\frac{(y-x) - \mathbf{a_n}(x,\eta)}{\boldsymbol{\sigma_n}(x,\eta)} \right)^2 \right\} \, dy.$$

The mean field particle model (7.27) associated with these McKean-Vlasov models is defined by the system of N interacting equations

$$\xi_n^i - \xi_{n-1}^i = \mathbf{a_n}(\xi_{n-1}^i, \eta_{n-1}^N) + \boldsymbol{\sigma_n}(\xi_{n-1}^i, \eta_{n-1}^N) \, W_n^i$$

$$= \frac{1}{N} \sum_{1 \leq j \leq N} a_n(\xi_{n-1}^i, \xi_{n-1}^j) + \frac{1}{N} \sum_{1 \leq j \leq N} \sigma_n(\xi_{n-1}^i, \xi_{n-1}^j) \, W_n^i$$

with $1 \leq i \leq N$. In the above displayed formulae, W_n^i stands for N independent copies of W_n.

A more general class of McKean-Markov chain models on some measurable state space S is given by the recursive formulae

$$\overline{X}_n = F_n(\overline{X}_{n-1}, \eta_{n-1}, W_n) \quad \text{with} \quad \eta_{n-1} := \mathrm{Law}(\overline{X}_{n-1}). \tag{7.33}$$

In the above display, W_n is a collection of independent, and independent of $(\overline{X}_p)_{0 \leq p < n}$ random variables taking values in some state space $\mathcal{W}$, and F_n is a measurable mapping from $(S \times \mathcal{P}(S) \times \mathcal{W})$ into S.

Here again, we easily check that the law of the random states $\eta_n = \mathrm{Law}(\overline{X}_n)$ satisfies the evolution equation (7.30) with the McKean transitions

$$K_{n,\eta}(f)(x) = \mathbb{E}\left[f(F_n(x, \eta, W_n)) \right].$$

7.10.4   Interacting jump processes

We illustrate the mean field particle models (7.27) with the non-homogeneous version of the Markov transitions discussed in (7.25). It is given by

$$K_{n+1,\eta}(u, dw) = G_n(u) \, M_{n+1}(u, dw) + (1 - G_n(u)) \, \Psi_{G_n}(\eta) M_{n+1}(dw) \tag{7.34}$$

for some Markov transitions M_{n+1} and some $]0, 1]$-valued potential functions G_n on S. In this context, arguing as in (7.26) we prove that

$$\eta K_{n+1,\eta} := \Phi_{n+1}(\eta) = \Psi_{G_n}(\eta) M_{n+1}. \tag{7.35}$$

By construction, it is readily checked that

$$\Psi_{G_n}(m(\xi_n)) := \sum_{1 \leq i \leq N} \frac{G_n(\xi_n^i)}{\sum_{1 \leq j \leq N} G_n(\xi_n^j)} \, \delta_{\xi_n^i}$$

and therefore

$$K_{n+1,m(\xi_n)}(u,dw)$$

$$= G_n(u)\, M_{n+1}(u,dw) + (1 - G_n(u)) \sum_{1 \le i \le N} \frac{G_n(\xi_n^i)}{\sum_{1 \le j \le N} G_n(\xi_n^j)}\, M_{n+1}(\xi_n^i, dw).$$

In this situation, the mean field model (7.27) can be interpreted as an interacting jump process transition.

Next, we illustrate the non-uniqueness of Markov transitions $K_{n+1,\eta}$ satisfying the compatibility condition (7.35). Let ϵ be any non-negative number such that $\epsilon G_n(x) \in [0,1]$, for any $x \in S$, and set

$$K_{n+1,\eta}(u,dw) := \epsilon G_n(u)\, M_{n+1}(u,dw) + (1 - \epsilon G_n(u))\, \Psi_{G_n}(\eta)M_{n+1}(dw)$$

It is not difficult to check that for any function f on S,

$$K_{n+1,\eta}(f)(u) = \epsilon G_n(u)\, M_{n+1}(f)(u) + (1 - \epsilon G_n(u))\, \Psi_{G_n}(\eta)M_{n+1}(f)$$

and

$$\begin{aligned} \eta K_{n+1,\eta}(f) &= \epsilon\eta(G_n M_{n+1}(f)) + (1 - \epsilon\eta(G_n))\, \Psi_{G_n}(\eta)M_{n+1}(f) \\ &= \Phi_{n+1}(\eta)(f) + \epsilon\eta(G_n M_{n+1}(f)) - \epsilon\eta(G_n)\frac{\eta(G_n M_{n+1}(f))}{\eta(G_n)} = \Phi_{n+1}(\eta)(f). \end{aligned}$$

When $\epsilon = 0$, the Markov transition $K_{n+1,\eta}$ reduces to

$$K_{n+1,\eta}(u,dw) := \Psi_{G_n}(\eta)M_{n+1}(dw).$$

In this particular situation, it is readily checked that

$$K_{n+1,m(\xi_n)}(u,dw) := \sum_{1 \le i \le N} \frac{G_n(\xi_n^i)}{\sum_{1 \le j \le N} G_n(\xi_n^j)}\, M_{n+1}(\xi_n^i, dw). \qquad (7.36)$$

The corresponding evolution of the mean field particle model (7.27) is given by the genetic type selection-mutation transitions

$$\mathbb{P}\left(\xi_{n+1} \in dx \mid \xi_n\right) := \prod_{1 \le i \le N} \sum_{1 \le j \le N} \frac{G_n(\xi_n^j)}{\sum_{1 \le k \le N} G_n(\xi_n^k)}\, M_{n+1}(\xi_n^j, dx^i). \qquad (7.37)$$

In the above display, $dx = d(x^1, \dots, x^N) := dx^1 \times \dots \times dx^N$ stands for an infinitesimal neighborhood of the point $x = (x^1, \dots, x^N)$ in S^N. In other words, given ξ_n the particles $\xi_{n+1} = (\xi_{n+1}^i)_{1 \le i \le N}$ are N independent random samples with common distribution

$$\Phi_{n+1}(m(\xi_n))(dw) := \sum_{1 \le i \le N} \frac{G_n(\xi_n^i)}{\sum_{1 \le j \le N} G_n(\xi_n^j)}\, M_{n+1}(\xi_n^i, dw).$$

We refer the reader to the end of section 9.6 for a more detailed description of these models in terms of evolutionary type processes.

7.11   Exercises

Exercise 77 (Reflected Markov chain) *Let $(W_n)_{n\geq 0}$ be a collection of independent random variables with probability distributions $(\mu_n)_{n\geq 0}$ on $S = \mathbb{R}^r$, for some $r \geq 1$. Let $A \subset S$ be a given subset. Consider the Markov chain starting on $A \ni X_0$ and defined by*

$$X_{n+1} - X_n = (b_n(X_n) + \sigma_n(X_n)\, W_{n+1})\, 1_A\, (X_n + b_n(X_n) + \sigma_n(X_n)\, W_{n+1})$$

for some collection of functions $b_n : x \in \mathbb{R}^r \mapsto \mathbb{R}^r$ and $\sigma_n : x \in \mathbb{R}^r \mapsto \mathbb{R}^{r\times r}$. Check that $X_n \in A$ for any $n \geq 0$. Compute the Markov transition M_n of the chain X_n in terms of the Markov transitions K_n of the chain Y_n defined by

$$Y_{n+1} - Y_n = b_n(Y_n) + \sigma_n(Y_n)\, W_{n+1}.$$

Exercise 78 (Random walks in a random environment) *Let Θ be a $[0, 1]$-valued random variable with distribution μ. Given Θ, we let W be a $\{-1, +1\}$-valued Bernoulli random variable*

$$\mathbb{P}(W = +1 \mid \Theta) = 1 - \mathbb{P}(W = -1 \mid \Theta) = \Theta.$$

Given Θ, we let W_n be a sequence of independent copies of W and we denote by X_n the random walk

$$X_n = X_{n-1} + W_n$$

starting at some given $X_0 = x_0$. Check that for any $\epsilon_i \in \{-1, +1\}$ with $1 \leq i \leq n$ we have

$$\mathbb{P}\left(W_1 = \epsilon_1, \ldots, W_n = \epsilon_n \mid \Theta\right) = \Theta^{\sum_{1\leq i\leq n} \frac{1+\epsilon_i}{2}}\, (1 - \Theta)^{n - \sum_{1\leq i\leq n} \frac{1+\epsilon_i}{2}}$$

and deduce that

$$\mathbb{P}\left(\Theta \in d\theta \mid W_1, \ldots, W_n\right) = \mathbb{P}\left(\Theta \in d\theta \mid \overline{W}_n\right) \propto \theta^{\overline{W}_n}\, (1 - \theta)^{n - \overline{W}_n}\, \mu(d\theta)$$

with $\overline{W}_n = \sum_{1\leq i\leq n} \frac{1+W_i}{2}$. Examine the situations where Θ is uniform on $[0, 1]$, and Θ follows a $\mathrm{beta}(a, b)$ distribution with parameters (a, b). Check that in both situations, we have

$$\mathbb{E}\left(\Theta \mid \overline{W}_n\right) \to_{n\to\infty} \Theta.$$

When Θ is a $\mathrm{beta}(a, b)$ distribution on $[0, 1]$ check that $\overline{W}_n = [(X_n - x_0) + n]/2$ and

$$\mathbb{P}\left(X_{n+1} = X_n + 1 \mid X_n\right) = \frac{n}{a+b+n}\, \frac{(X_n - x_0) + n}{2n} + \frac{a+b}{a+b+n}\, \frac{a}{a+b} \simeq_{n\uparrow\infty} \Theta.$$

Exercise 79 (Autoregressive models) *A q-th order real valued autoregressive model (abbreviated $AR(q)$), for some given integer $q \geq 1$, is defined by a recursion of the form*

$$Y_n = a + \sum_{1\leq p\leq q} b_p\, Y_{n-p} + V_n$$

for any $n \geq q$, with some given initial values $(Y_0, \ldots, Y_{q-1})$, some parameters a, b_k, c, and a sequence of independent random variables V_n. Check that the column vector $X_n = (Y_n, Y_{n+1} \ldots, Y_{n+q-1})'$ satisfies a linear equation of the form

$$X_{n+1} = c + BX_n + W_{n+1}$$

for some column vector $c \in \mathbb{R}^q$, some $(q \times q)$-matrix and some sequence of $\mathbb{R}^q$-valued independent random variables W_n.

Exercise 80 (Gun fight at OK Corral - Transition probabilities [237]) ✦

The Good (G), the Bad (B) and the Ugly (U) cowboys shoot with respective success probabilities $g > u > b \in [0,1]$. They draw and fire simultaneously, each of them firing at the better shot of his opponents. Assuming that there are at least two survivors, the second round of shooting starts and the gunfight continues, round after round, until there are fewer than two survivors. Design the transitions of the Markov chain on the space of the alive shooters at any round.

Exercise 81 (Gun fight at OK Corral - Random state distributions) ✦✦

We consider the OK Corral gun fight described in exercise 80. We denote by $p_1(n)$, $p_2(n)$, $p_3(n)$, and $p_4(n)$ the probabilities that the states $(\emptyset)$ (i.e., everybody has been killed), (G), (U), and respectively (B) are reached at time n. We also let $q_1(n)$, $q_2(n)$ and $q_3(n)$ the probabilities that the states (GB), (UB) and (GUB) are the result of the n-th round. Find recurrence relations between $p_i(n+1)$ and $(p_i(n), q_j(n))_{1 \leq j \leq 3}$ with $1 \leq i \leq 4$; and between $q_i(n+1)$ and $(q_j(n))_{1 \leq j \leq 3}$, with $i = 1, 2, 3$.

Exercise 82 *We consider the stochastic matrix*

$$M = \frac{1}{2} \begin{pmatrix} 1 & 1 & 0 \\ \frac{1}{2} & 1 & \frac{1}{2} \\ 0 & 1 & 1 \end{pmatrix}.$$

Check that for any $n \geq 1$ we have

$$M^n = \frac{1}{2^n} \begin{pmatrix} \frac{3}{2} + \left(2^{n-2} - 1\right) & 2^{n-1} & \frac{1}{2} + \left(2^{n-2} - 1\right) \\ 2^{n-2} & 2^{n-1} & 2^{n-2} \\ \frac{1}{2} + \left(2^{n-2} - 1\right) & 2^{n-1} & \frac{3}{2} + \left(2^{n-2} - 1\right) \end{pmatrix} \rightarrow_{n \uparrow \infty} \begin{pmatrix} \frac{1}{4} & \frac{1}{2} & \frac{1}{4} \\ \frac{1}{4} & \frac{1}{2} & \frac{1}{4} \\ \frac{1}{4} & \frac{1}{2} & \frac{1}{4} \end{pmatrix}.$$

Exercise 83 ✦✦ *Consider the simple random walk (1.1) discussed in section 1.1.*

- *Prove that*

$$\forall n \geq m \qquad \mathbb{P}(X_n = y \mid X_m = x) = 2^{-(n-m)} \begin{pmatrix} n - m \\ \frac{(n-m)+(y-x)}{2} \end{pmatrix}$$

if $(n - m) + (y - x)$ is even, and 0 otherwise.

- *Using the reflection principle presented in exercise 6, prove that for any $x, y > 0$ and $n \geq m$ we have*

$$\mathbb{P}(X_n = y, \ X_k > 0, \ 0 \leq k \leq n \mid X_m = x)$$

$$= 2^{-(n-m)} \left[\begin{pmatrix} n - m \\ \frac{(n-m)+(y-x)}{2} \end{pmatrix} - \begin{pmatrix} n - m \\ \frac{(n-m)-(x+y)}{2} \end{pmatrix} \right]$$

if $(n - m) + (y - x)$ is even, and 0 otherwise.

- *Deduce that for any $z < x \vee y$ we have*

$$\mathbb{P}(X_n = y, \ X_k > z, \ 0 \leq k \leq n \mid X_m = x)$$

$$= 2^{-(n-m)} \left[\begin{pmatrix} n - m \\ \frac{(n-m)+(y-x)}{2} \end{pmatrix} - \begin{pmatrix} n - m \\ \frac{(n-m)-(x+y)}{2} + z \end{pmatrix} \right] \tag{7.38}$$

if $(n - m) + (y - x)$ is even, and 0 otherwise.

- *Check that*

$$\mathbb{P}(X_n = x \ , \ X_k \geq 0, \ 0 \leq k \leq n \mid X_0 = 0)$$

$$= 2^{-n} \left[\binom{n}{\frac{n+x}{2}} - \binom{n}{\frac{n-x}{2} - 1} \right] \left(2^{-n} \frac{1}{n+1} \binom{n+1}{n/2} \ when \ x = 0 \right)$$

and deduce that for any $m \geq 0$ we have

$$\mathbb{P}(X_k \geq 0, \ 0 \leq k \leq 2m \mid X_0 = 0) = 2^{-2m} \binom{2m}{m}.$$

Exercise 84 *Describe the mean field particle model associated with the nonlinear $\mathbb{R}$-valued Markov chain defined by*

$$\overline{X}_n = \overline{X}_{n-1} + \mathbb{E}(\log\left(1 + \overline{X}_{n-1}^2\right)) + W_n$$

with $X_0 = x_0 > 0$, and with a sequence W_n of centered, independent and identically distributed Gaussian random variables s.t. $\mathbb{E}(W_n^2) = 1$.

Exercise 85 *Prove (7.15).*

Exercise 86 *Prove the reversible property (7.19).*

Exercise 87 *Prove formula (7.21).*

Exercise 88 *Prove the formulae (7.29).*

Exercise 89 *Consider the Markov transition on a state space S defined in (7.34). Prove that for any function f on S we have*

$$K_{n+1,\eta}(f) = \epsilon G_n \ M_{n+1}(f) + (1 - \epsilon G_n) \ \Psi_{G_n}(\eta) M_{n+1}(f).$$

Deduce that $\eta K_{n+1,\eta} = \Psi_{G_n}(\eta) M_{n+1}$.

8

Analysis toolbox

Following the introduction of some representative Markov chain models in the previous chapter, we are summarising now the main mathematical tools needed to analyse the convergence properties of Markov chains when the time parameter tends to infinity. There is a large variety of tools needed in the remaining chapters of the book. They range from simple tools from linear algebra and elementary probability theory to more advanced methods from functional analysis. We also present some more specific advanced probabilistic tools such as coupling, strong stationary times, martingales and martingale limit theorems.

> *Not everything that can be counted counts,*
> *and not everything that counts can be counted.*
> Albert Einstein (1879-1955).

8.1 Linear algebra

8.1.1 Diagonalisation type techniques

We return to the $\{0,1\}$-valued Markov chain model discussed in (7.11). We further assume that all entries of the Markov transition M are positive (i.e. $p \wedge q > 0$). We equip $\mathcal{B}(\{0,1\}) = \mathbb{R}^{\{0,1\}}$ with the scalar product

$$\langle f_1, f_2 \rangle := \sum_{x \in \{0,1\}} f_1(x) f_2(x)$$

for any $f_i = \begin{pmatrix} f_i(0) \\ f_i(1) \end{pmatrix}$, with $i = 1, 2$, and

the norm $|f_1|^2 = \langle f_1, f_1 \rangle$. Our next objective is to compute the diagonal form of M. To this end, we observe that the characteristic polynomial of M is given by

$$\mathrm{Det}(M - \lambda Id) = (\lambda - 1)\ (\lambda - (1 - (p + q))). \tag{8.1}$$

This shows that M has two real eigenvalues

$$\lambda_1 = 1 \quad \text{and} \quad \lambda_2 = 1 - (p+q) \quad (= (1-p) - q \le 1)$$

with the normalized eigenvectors $\overline{\varphi}_i := \begin{pmatrix} \overline{\varphi}_i(0) \\ \overline{\varphi}_i(1) \end{pmatrix}$, with $i = 1, 2$, given by

$$\overline{\varphi}_1 = \begin{pmatrix} 1/\sqrt{2} \\ 1/\sqrt{2} \end{pmatrix} \quad \text{and} \quad \overline{\varphi}_2 = \begin{pmatrix} p/\sqrt{p^2 + q^2} \\ -q/\sqrt{p^2 + q^2} \end{pmatrix}. \tag{8.2}$$

We associate with these vectors the change of variable formula

$$P := (\overline{\varphi}_1, \overline{\varphi}_2) = \begin{pmatrix} \overline{\varphi}_1(0) & \overline{\varphi}_2(0) \\ \overline{\varphi}_1(1) & \overline{\varphi}_2(1) \end{pmatrix} = \begin{pmatrix} 1/\sqrt{2} & p/\sqrt{p^2 + q^2} \\ 1/\sqrt{2} & -q/\sqrt{p^2 + q^2} \end{pmatrix}$$

and its inverse matrix

$$P^{-1} = \frac{1}{p+q} \begin{pmatrix} \dfrac{q\sqrt{2}}{\sqrt{p^2 + q^2}} & \dfrac{p\sqrt{2}}{-\sqrt{p^2 + q^2}} \end{pmatrix}. \tag{8.3}$$

By construction, we have that

$$\begin{aligned} PD &= \begin{pmatrix} \overline{\varphi}_1(0) & \overline{\varphi}_2(0) \\ \overline{\varphi}_1(1) & \overline{\varphi}_2(1) \end{pmatrix} \begin{pmatrix} \lambda_1 & 0 \\ 0 & \lambda_2 \end{pmatrix} \\ &= \begin{pmatrix} \lambda_1 \overline{\varphi}_1(0) & \lambda_2 \overline{\varphi}_2(0) \\ \lambda_1 \overline{\varphi}_1(1) & \lambda_2 \overline{\varphi}_2(1) \end{pmatrix} = M \begin{pmatrix} \overline{\varphi}_1(0) & \overline{\varphi}_2(0) \\ \overline{\varphi}_1(1) & \overline{\varphi}_2(1) \end{pmatrix} = MP \end{aligned}$$

from which we conclude that

$$M = PDP^{-1}$$

and therefore

$$M^2 = PDP^{-1}PDP^{-1} = PD^2P^{-1} \Rightarrow \ldots \Rightarrow M^n = PD^nP^{-1}.$$

After some elementary computations, we find that

$$M^n = \begin{pmatrix} \pi(0) & \pi(1) \\ \pi(0) & \pi(1) \end{pmatrix} + \lambda_2^n \, R \tag{8.4}$$

with the probability measure π and the matrix R given by

$$\begin{aligned} \pi &= [\pi(0), \pi(1)] = \left[\frac{q}{p+q}, \frac{p}{p+q} \right] \\ R &= \begin{pmatrix} R(0,0) & R(0,1) \\ R(1,0) & R(1,1) \end{pmatrix} = \begin{pmatrix} \pi(1) & -\pi(1) \\ -\pi(0) & \pi(0) \end{pmatrix}. \end{aligned} \tag{8.5}$$

In terms of operators acting on functions $f = \begin{pmatrix} f(0) \\ f(1) \end{pmatrix} \in \mathbb{R}^{\{0,1\}}$, we have proved the following proposition.

Proposition 8.1.1 *For any $f \in \mathcal{B}(\{0,1\})$ and and $x \in \{0,1\}$ we have*

$$M^n(f)(x) = \pi(f) + \lambda_2^n \, R(f)(x) \longrightarrow_{n \to \infty} \pi(f) \tag{8.6}$$

with the matrix R and the probability measure π defined in (8.5).

Taking indicator functions $f = 1_y$ of the states $y \in \{0,1\}$, we find that

$$\forall x, y \in \{0,1\} \qquad M^n(x,y) = \pi(y) + \lambda_2^n \, R(x,y) \longrightarrow_{n \to \infty} \pi(f).$$

In other words, the chain forgets its initial condition $X_0 = x$ and converges exponentially fast to the measure π, in the sense that

$$\sup_{x \in \{0,1\}} |\mathbb{P}\left(X_n = y \mid X_0 = x\right) - \pi(y)| \leq \lambda_2^n \longrightarrow_{n \to \infty} 0.$$

In the r.h.s. upper bound we have used the fact that $|R(x,y)| \leq \pi(0) \vee \pi(1) \leq 1$.

8.1.2 Perron Frobenius theorem

In section 8.1.1 we developed in some details the decomposition of a two-state Markov chain transition using its eigenvalue decomposition. The following theorem extends this result to general finite state space models.

Theorem 8.1.2 (Perron Frobenius) *Let* $M = (M(x,y))_{x,y \in S}$ *be a stochastic matrix on some finite (non-necessarily ordered) state space S such that M^m has all entries positive for some $m \geq 1$. Then, there exists a unique probability measure π on S such that $\wedge_{x \in S} \pi(x) > 0$ and*

$$\pi M = \pi \quad \text{with for any } x, y \in S \quad \lim_{n \to \infty} M^n(x,y) = \pi(y).$$

In addition, 1 is a simple root of the characteristic polynomial of M.

Remark : A detailed proof of this theorem can be found in the book of E. Seneta [239], and the lecture notes of L. Saloff Costes [236]. It is out of the scope of these lectures to enter into the details of the proof of this theorem.

To get some intuition, we provide a simple proof for $m = 1$.

We further assume that the following condition is met

$$M(x,y) \geq \epsilon \, \nu(y)$$

for some $\epsilon \in]0,1]$, and $x, y \in S$ and some probability measure ν on S such that $\nu(y) > 0$ for any $y \in S$.

Since the space space S is finite and all entries of M are positive, we can take $\epsilon = \mathrm{Card}(S) \times \inf_{x,y} M(x,y)$, and $\nu(x) = \frac{1}{\mathrm{Card}(S)}$, the uniform distribution on S.

To get one step further, we observe that

$$M(x,y) = (1 - \epsilon) \, M_\epsilon(x,y) + \epsilon \nu(y)$$

with the Markov transition

$$M_\epsilon(x,y) := \frac{M(x,y) - \epsilon \nu(y)}{1 - \epsilon}.$$

This implies that

$$[M(f)(x) - M(f)(y)] = (1 - \epsilon) \, [M_\epsilon(f)(x) - M_\epsilon(f)(y)].$$

On the other hand, for any Markov transition K on S, we have

$$K(f)(x) - K(f)(y) = \sum_{u,v} (f(u) - f(v)) \, K(x,u)K(y,v) \Rightarrow \mathrm{osc}\,(K(f)) \leq \mathrm{osc}(f).$$

This implies that

$$\operatorname{osc}(M(f)) = (1 - \epsilon) \operatorname{osc}(M_\epsilon(f)) \leq (1 - \epsilon) \operatorname{osc}(f).$$

We conclude that

$$\operatorname{osc}(M^n(f)) \leq (1 - \epsilon) \operatorname{osc}(M^{n-1}(f)) \leq \ldots \leq (1 - \epsilon)^n \operatorname{osc}(f). \qquad (8.7)$$

This implies that for any starting point $x \in S$ and for any bounded function f we have $\lim_{n \to \infty} M^n(f)(x) = \pi(f)$ for some probability measure π. Letting $n \uparrow \infty$, we find that

$$\pi(f) \longleftarrow \pi M^{n+1}(f) = \pi M^n(M(f)) \longrightarrow \pi(M(f))$$

from which we conclude that $\pi = \pi M$. We check that $\wedge_{x \in S} \pi(x) > 0$ using the fact that

$$\pi(y) = \sum_{x \in S} \pi(x) \, M(x, y) \geq \epsilon \sum_{x \in S} \pi(x) \, \nu(y) = \epsilon \, \nu(y) > 0.$$

This ends the proof of the theorem. ∎

Remark :
We end this section with some comments on the diagonalisation techniques developed above. Instead of computing the right action eigenfunctions $M(\overline{\varphi}_i) = \lambda_i \overline{\varphi}_i$ we can analyze the left action eigenmeasures $\pi_i M = \lambda_i \, \pi_i$, with $i = 1, 2$. We have already checked that the invariant measure $\pi_1 = \pi$ is an eigenmeasure $\pi M = \lambda_1 \, \pi$ associated with the eigenvalue $\lambda_1 = 1$. It is also not difficult to check that the signed measure

$$\pi_2 = [\pi_2(0), \pi_2(1)] = [1, -1]$$

is an eigenmeasure $\pi_2 M = \lambda_2 \, \pi_2$ associated with the eigenvalue $\lambda_2 = 1 - (p + q)$

$$[1, -1] \begin{pmatrix} 1 - p & p \\ q & 1 - q \end{pmatrix} = \begin{pmatrix} (1 - p) - q \\ p - (1 - q) \end{pmatrix} = (1 - (p + q)) \, [1, -1].$$

Consider the change of variable matrix

$$Q = \begin{pmatrix} \pi_1(0) & \pi_1(0) \\ \pi_2(0) & \pi_2(1) \end{pmatrix} = \begin{pmatrix} \frac{q}{p+q} & \frac{p}{p+q} \\ 1 & -1 \end{pmatrix} \implies Q^{-1} = \begin{pmatrix} 1 & \frac{p}{p+q} \\ 1 & -\frac{q}{p+q} \end{pmatrix}.$$

By construction, we have

$$\begin{pmatrix} \lambda_1 & 0 \\ 0 & \lambda_2 \end{pmatrix} Q = \begin{pmatrix} \lambda_1 \pi_1(0) & \lambda_1 \pi_1(1) \\ \lambda_2 \pi_2(0) & \lambda_2 \pi_2(1) \end{pmatrix} = \begin{pmatrix} \pi_1(0) & \pi_1(1) \\ \pi_2(0) & \pi_2(1) \end{pmatrix} M = QM.$$

This yields the alternative decompositions

$$M = Q^{-1} D Q \implies \forall n \geq 1 \quad M^n = Q^{-1} D^n Q.$$

8.2 Functional analysis

8.2.1 Spectral decompositions

We return to the $\{0,1\}$-valued Markov chain model (7.11) from section 8.1.1. We notice that

$$\langle \overline{\varphi}_1, \overline{\varphi}_2 \rangle = \frac{p-q}{\sqrt{2(p^2+q^2)}} = 0 \iff p = q$$

$$\iff M \text{ symmetric}$$

$$\implies \overline{\varphi}_2 = \frac{1}{\sqrt{2}} \begin{pmatrix} 1 \\ -1 \end{pmatrix} \quad \text{and} \quad \pi = \left[\frac{1}{2}, \frac{1}{2} \right].$$

In this situation, we also find that

$$P^{-1} = P' = \frac{1}{\sqrt{2}} \begin{pmatrix} 1 & 1 \\ 1 & -1 \end{pmatrix}.$$

Furthermore, the reversible property (7.10) is also met. By (8.6) we also find that

$$\forall n \geq 1 \quad \forall x \in S \qquad M^n(f)(x) = \pi(f) + \lambda_2^n \langle f, \overline{\varphi}_2 \rangle \overline{\varphi}_2(x). \tag{8.8}$$

Definition 8.2.1 *We let $l_2(\pi)$ be the Hilbert space of functions on some finite space S equipped with the scalar product*

$$\langle f_1, f_2 \rangle_\pi = \sum_{x \in S} \pi(x) f_1(x) f_2(x) \quad \text{and the norm} \quad |f_1|_\pi^2 = \langle f_1, f_1 \rangle_\pi.$$

In the two-states example discussed above, we readily check that

$$\langle f_1, f_2 \rangle_\pi = \frac{1}{2} \langle f_1, f_2 \rangle.$$

Using the fact that

$$|\sqrt{2}\, \overline{\varphi}_i|_\pi^2 = |\overline{\varphi}_i|^2 = 1$$

for any $i = 1, 2$, we show that the functions $\psi_i := \sqrt{2}\overline{\varphi}_i$ form an orthogonal basis of $l_2(\pi)$. Recalling that φ_1 is the unit function, we conclude that

$$M^n(f)(x) = \underbrace{\langle f, \psi_1 \rangle_\pi \, \psi_1(x)}_{=\pi(f)} + \lambda_2^n \langle f, \psi_2 \rangle_\pi \, \psi_2(x). \tag{8.9}$$

In other words

$$M^n(x,y)/\pi(y) = 1 + \lambda_2^n \, \psi_2(x)\psi_2(y) \left(= \lambda_1^n \, \psi_1(x)\psi_1(y) + \lambda_2^n \, \psi_2(x)\psi_2(y) \right).$$

This $\{0,1\} = \mathbb{Z}/2\mathbb{Z}$-valued Markov chain is a particular example of the $\mathbb{Z}/m\mathbb{Z}$-valued random walk models discussed in section 25.4. The following spectral decomposition theorem applies to any reversible Markov transition on some finite state space.

Theorem 8.2.2 *We let M be some reversible Markov transition w.r.t. some probability measure π on some finite space space S with cardinality d, such that M^m has all positive entries for some $m \geq 1$. In this situation M has a finite set of real valued eigenvalues $\lambda_1 = 1 \geq \lambda_2 \geq \ldots \geq \lambda_d > -1$, and there exists an orthonormal basis of $l_2(\pi)$ made of real valued eigenfunctions $(\psi_i)_{1 \leq i \leq d}$ of $(\lambda_i)_{1 \leq i \leq d}$, with $\psi_1 = 1$ being the unit function. Furthermore, for any $n \in \mathbb{N}$, we have the spectral decomposition*

$$M^n(x,y) = \pi(y) + \sum_{1 < i \leq d} \lambda_i^n \, \psi_i(x)\psi_i(y)\pi(y).$$

The difference $\lambda_2 - \lambda_1$ is called the spectral gap.

Proof :

We recall that for any symmetric matrix $K(x,y)$ on some finite space there exists some matrix P such that $PP' = Id$, and a diagonal matrix D such that $K = PDP'$ (this is a direct consequence of the spectral theorem for matrices). A proof of this result can be found in any textbook on elementary matrix theory. We let φ_i the column vectors of P. By construction, we have

$$PD = (\varphi_1, \ldots, \varphi_d) \, D = (\lambda_1\varphi_1, \ldots, \lambda_d\varphi_d) = K \, (\varphi_1, \ldots, \varphi_d) \, (= KP)$$

if and only if $K(\varphi_i) = \lambda_i\varphi_i$, for any $\forall 1 \leq i \leq d$. When $K(x,y) = \sqrt{\pi(x)} \, M(x,y) \, \frac{1}{\sqrt{\pi(y)}}$, we readily check that

$$\pi(x)M(x,y) = \pi(x)M(x,y)$$

$$\Leftrightarrow \sqrt{\pi(x)} \, M(x,y) \, \frac{1}{\sqrt{\pi(y)}} = \sqrt{\pi(y)} \, M(y,x) \, \frac{1}{\sqrt{\pi(x)}} \Leftrightarrow K(x,y) = K(y,x).$$

In addition, the eigenvalues of M are given by $\psi_i(x) = \frac{1}{\sqrt{\pi(x)}} \, \varphi_i(x)$. This follows from the fact that

$$\frac{1}{\sqrt{\pi(x)}}K(x,y)\sqrt{\pi(y)} = M(x,y) \ \Rightarrow M(\psi_i)(x) = \frac{1}{\sqrt{\pi(x)}}K(\varphi_i)(x) = \lambda_i \, \psi_i(x).$$

We also observe that these functions are orthonormal

$$\langle \psi_i, \psi_j \rangle_\pi = \langle \varphi_i, \varphi_j \rangle = \sum_{x \in S} \varphi_i(x)\varphi_j(x) = 1_{i=j}.$$

Finally, we use the decomposition of the indicator function

$$1_y = \sum_{1 \leq j \leq d} \langle 1_y, \psi_j \rangle_\pi \, \psi_i = \sum_{1 \leq j \leq d} \pi(y) \, \psi_i(y) \, \psi_i$$

to prove that

$$M^n(x,y) \ = \ M^n(1_y)(x) = \sum_{1 \leq j \leq d} \pi(y) \, \psi_i(y) \, M^n(\psi_i)(x) = \sum_{1 \leq j \leq d} \pi(y) \, \lambda_i^n\psi_i(x) \, \psi_i(y).$$

The fact that $\lambda_d > -1$ follows from observing that

$$\|\psi_d\| \geq \|M(\psi_d)\| = |\lambda_d| \ \|\psi_d\| \Rightarrow |\lambda_d| \ \leq 1.$$

In addition, if $\lambda_d = -1$ we would have the following contradiction

$$0 \overset{n\uparrow\infty}{\longleftarrow} M^n(\varphi_d)(x) = (-1)^n \varphi_d(x) \Rightarrow (-1)^n \to 0.$$

This ends the proof of the theorem. ∎

Corollary 8.2.3 *In the settings of theorem 8.2.2, we have*

$$\pi(x) \left| M^n(x,y) - \pi(y) \right| \leq \lambda_\star^n \sqrt{\pi(x)(1 - \pi(x))} \sqrt{\pi(y)(1 - \pi(y))} \leq e^{-\rho(M)n}/4$$

with

$$\lambda_\star := \sup_{1 < i \leq d} |\lambda_i| \quad and \quad \rho(M) := 1 - \lambda_\star \in]0,1[.$$

In addition, we have

$$\sum_{y \in S} |M^n(x,y) - \pi(y)| \leq \left[\frac{1}{\pi(x)} M^{2n}(x,x) - 1 \right]^{1/2} \leq \lambda_\star^n \sqrt{\frac{1 - \pi(x)}{\pi(x)}}$$

The quantity $\rho(M)$ is called the absolute spectral gap.

Proof :
By the Cauchy Schwartz inequality, we have

$$\left| \frac{M^n(x,y)}{\pi(y)} - 1 \right| \leq \lambda_\star^n \left[\sum_{1 < i \leq d} \psi_i(x)^2 \right]^{1/2} \left[\sum_{1 < i \leq d} \psi_i(y)^2 \right]^{1/2}$$

$$\pi(y) = \langle 1_y, 1_y \rangle_\pi = \langle \sum_{1 \leq j \leq d} \underbrace{\langle 1_y, \psi_j \rangle_\pi}_{=\pi(y)\psi_i(y)} \psi_i, \sum_{1 \leq j \leq d} \langle 1_y, \psi_j \rangle_\pi \psi_i \rangle_\pi = \pi(y)^2 \sum_{1 \leq j \leq d} \psi_i(y)^2.$$

This implies that

$$\sum_{1 \leq j \leq d} \psi_i(y)^2 = 1/\pi(y) \Rightarrow \sum_{1 < j \leq d} \psi_i(y)^2 = [1 - \pi(y)] /\pi(y).$$

Finally, we observe that $\lambda_\star^n = (1 - \rho(M))^n \leq e^{-\rho(M)\, n}$, and $x(1-x) \leq 1/4$ for any $x \in [0,1]$. This completes the proof of the first assertion. To check the last one, we use the variance inequality

$$\left[\sum_{y \in S} \pi(y)f(y) \right]^2 \leq \sum_{y \in S} \pi(y)f(y)^2 \Longleftrightarrow \sum_{y \in S} g(y) \leq \left[\sum_{y \in S} \pi(y)^{-1} g(y)^2 \right]^{1/2}$$

which is valid for any non negative functions $f(y) = g(y)/\pi(y)$ on S, to prove that

$$\sum_{y \in S} |M^n(x,y) - \pi(y)| \leq \left[\sum_{y \in S} \pi(y)^{-1} |M^n(x,y) - \pi(y)|^2 \right]^{1/2}$$

$$= \left[\sum_{y \in S} \pi(y)^{-1} M^n(x,y)^2 - 1 \right]^{1/2} = \left[\pi(x)^{-1} M^{2n}(x,x) - 1 \right]^{1/2}$$

In the last assertion we have used the reversibility property

$$\pi(x_0)M(x_0,x_1)\ldots M(x_{n-1},x_n) = M(x_0,x_1)\ldots M(x_{n-1},x_n)\pi(x_n)$$

$$\implies \pi(x_0)M^n(x_0,x_n) = M^n(x_0,x_n)\pi(x_n)$$

The second assertion is now a direct consequence of the first estimate. This ends the proof of the corollary. ∎

Corollary 8.2.4 *We equip $I = \{1,\ldots,d\}$ with some probability μ. We let M_i be a set of Markov transitions on a possibly different state spaces S_i, indexed by $i \in I$. We consider the Markov transition on the product space $\boldsymbol{S} = \prod_{1\leq i\leq d} S_i$ defined for any $\boldsymbol{x} = (x_i)_{1\leq i\leq d}$ and any $\boldsymbol{y} = (y_i)_{1\leq i\leq d} \in \boldsymbol{S}$ by*

$$\boldsymbol{M}(\boldsymbol{x},d\boldsymbol{y}) = \sum_{i\in I}\mu(i)\left\{\prod_{j\in I-\{i\}}\delta_{x_j}(dy_j)\right\}M_i(x_i,dy_i).$$

In essence, this means that the chain chooses randomly a coordinate, say i, with probability $\mu(i)$, and performs from the selected site a random move according to M_i (the other coordinates remain unchanged).

If M_i has an invariant probability measure π_i, for each $i \in I$, then $\boldsymbol{\pi}(d\boldsymbol{x}) = \prod_{i\in I}\pi_j(dx_j)$ is an invariant probability measure of $\boldsymbol{M}(\boldsymbol{x},d\boldsymbol{y})$.

In addition, if M_i has eigenfunctions φ_i with eigenvalues λ_i, for each $i \in I$, then $\boldsymbol{M}$ has eigenfunctions

$$\boldsymbol{\varphi}(\boldsymbol{x}) = \prod_{i\in I}\varphi_i(x_i)\quad\text{with eigenvalues}\quad\boldsymbol{\lambda} = \sum_{i\in I}\mu(i)\lambda_i.$$

Proof :
The first assertion follows from the fact that for any $\boldsymbol{x} = (x_i)_{1\leq i\leq d}$, we have

$$\int\prod_{j\in I}\pi_j(dx_j)\,\boldsymbol{M}(\boldsymbol{f})(\boldsymbol{x})$$

$$= \sum_{i\in I}\mu(i)\int\left\{\prod_{j\in I-\{i\}}\pi_j(dx_j)\right\}\underbrace{(\pi_iM)(dx_i)}_{=\pi_i}\boldsymbol{f}(\boldsymbol{x})$$

$$= \left[\sum_{i\in I}\mu(i)\right]\int\prod_{j\in I}\pi_j(dx_j)\boldsymbol{f}(\boldsymbol{x}) = \int\prod_{j\in I}\pi_j(dx_j)\boldsymbol{f}(\boldsymbol{x}).$$

The proof of the second assertion is a direct consequence of the following decomposition

$$\boldsymbol{M}(\boldsymbol{\varphi})(\boldsymbol{x}) = \sum_{i\in I}\mu(i)\left\{\prod_{i\in I-\{i\}}\varphi_j(x_j)\right\}\underbrace{M_i(\varphi_i)(x_i)}_{=\lambda_i\ \varphi_i(x_i)}$$

$$= \left(\sum_{i\in I}\mu(i)\,\lambda_i\right)\left\{\prod_{j\in I}\varphi_j(x_j)\right\} = \boldsymbol{\lambda}\,\boldsymbol{\varphi}(\boldsymbol{x}).$$

This ends the proof of the corollary. ∎

Example 8.2.5 *We illustrate this corollary with the Markov transitions on $S_i = \mathbb{R}$ given by*

$$M_i(x, dy) = \mu_i(dy)$$

where μ_i stands for the law of some random variable U_i with mean $\mathbb{E}(U_i) = m_i$. Then we have

$$M_i(\varphi_i) = \lambda_i \; \varphi_i$$

for $\lambda_i = 0$ and $\varphi_i(u) = (u - m_i)$, as well as for $\lambda_i = 1$ and $\varphi_i(u) = 1$. For any finite set $J \subset \{1, \ldots, d\}$ the functions

$$\varphi_J(x) := \prod_{j \in J} (x_i - m_i) \quad \left(= \left\{ \prod_{j \in J} (x_i - m_i) \right\} \times \left\{ \prod_{j \notin J} 1 \right\} \right)$$

are eigenfunctions associated with the eigenvalues

$$\lambda_J := \sum_{j \notin J} \mu(j) = 1 - \mu(J).$$

For more illustrations of this spectral decomposition theorem, we refer the reader to section 25.4, and section 25.5. A more detailed discussion on the consequences of these spectral decompositions can be found in the seminal Saint Flour summer school lectures of Laurent Saloff-Costes [236].

8.2.2 Total variation norms

Definition 8.2.6 *The total variation distance on the set $\mathcal{P}(S)$ of probability measures μ_1 and μ_2 on some finite state space S is given by*

$$\|\mu_1 - \mu_2\|_{tv} = \frac{1}{2} \sum_{x \in S} |\mu_1(x) - \mu_2(x)|.$$

The infimum measure $\mu_1 \wedge \mu_2$ is defined by

$$[\mu_1 \wedge \mu_2](x) := \mu_1(x) \wedge \mu_2(x).$$

Definition 8.2.7 ✎ *For a non-necessarily finite state space S, we have*

$$\|\mu_1 - \mu_2\|_{tv} = \frac{1}{2} \int \left| \frac{d\mu_1}{d\lambda}(x) - \frac{d\mu_2}{d\lambda}(x) \right| \lambda(dx)$$

where $p_i(x) = \frac{d\mu_i}{d\lambda}(x)$ stands for the density of the measure μ_i w.r.t. to some common (dominating) measure λ; that is, we have

$$\mu_i(dx) = p_i(x) \; \lambda(dx).$$

The infimum measure $\mu_1 \wedge \mu_2$ is defined by

$$[\mu_1 \wedge \mu_2](dx) := [p_1(x) \wedge p_2(x)] \; \lambda(dx).$$

For instance, when the measures μ_i have a density p_i w.r.t. the Lebesgue measure dx on $S = \mathbb{R}^r$, for some $r \geq 1$ we have

$$\|\mu_1 - \mu_2\|_{tv} = \frac{1}{2} \int |p_1(x) - p_2(x)| \ dx$$

$$[\mu_1 \wedge \mu_2](dx) = [p_1(x) \wedge p_2(x)] \ dx.$$

The choice of the reference measure is not unique. For instance, the distance can also be expressed in terms of the probability measure $\lambda(dx) = q(x) \, dx$ with $q(x) := \frac{1}{2} [p_1(x) + p_2(x)]$ by the formula

$$\|\mu_1 - \mu_2\|_{tv} = \frac{1}{2} \int \left| \frac{p_1(x)}{q(x)} - \frac{p_2(x)}{q(x)} \right| \ q(x) \ dx.$$

Example 8.2.8 *We let Z be a centered Gaussian r.v. with unit variance $\mathbb{E}(Z^2) = 1$. We consider the distributions $\mu_1 = \mathrm{Law}(X_1)$ and $\mu_2 = \mathrm{Law}(X_2)$ of the Gaussian r.v. $X_1 = m_1 + Z$ and $X_2 = m_2 + Z$, with some given parameters $m_1 \leq m_2$. We denote by p_1 and p_2 the Gaussian densities of μ_1 and μ_2. In this notation, the density $p_1 \wedge p_2$ of the infimum measure $\mu_1 \wedge \mu_2$ and its total mass $\int [p_1(x) \wedge p_2(x)] \, dx$ are illustrated in the following picture.*

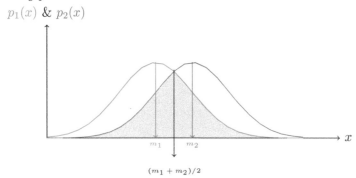

This shows that

$$\begin{aligned}
\|\mu_1 - \mu_2\|_{tv} &= 1 - \int [p_1(x) \wedge p_2(x)] \ dx \\
&= 1 - [\mathbb{P}(m_2 + Z \leq (m_1 + m_2)/2) + \mathbb{P}(m_1 + Z \geq (m_1 + m_2)/2)] \\
&= 1 - [\mathbb{P}(Z \leq -(m_1 - m_2)/2) + \mathbb{P}(Z \geq (m_2 - m_1)/2)] \\
&= 1 - \mathbb{P}[|Z| \geq (m_2 - m_1)/2] = \mathbb{P}[|Z| < (m_2 - m_1)/2] \\
&\leq (m_2 - m_1) \frac{1}{\sqrt{8\pi}} \ e^{-(m_2 - m_1)^2/8}.
\end{aligned}$$

Example 8.2.9 *When $S = \{0, 1\}$, we observe that*

$$\begin{aligned}
\|\mu_1 - \mu_2\|_{tv} &= \frac{1}{2} \left(|\mu_1(0) - \mu_2(0)| + |(1 - \mu_1(0)) - (1 - \mu_2(0))| \right) \\
&= |\mu_1(0) - \mu_2(0)| = (\mu_1(0) \vee \mu_2(0)) - (\mu_1(0) \wedge \mu_2(0)) \\
&= ([1 - \mu_1(1)] \vee [1 - \mu_2(1)]) - (\mu_1(0) \wedge \mu_2(0)) \\
&= 1 - [(\mu_1(0) \wedge \mu_2(0)) + (\mu_1(1) \wedge \mu_2(1))].
\end{aligned}$$

In the above two-state model illustration, we also notice that for any function $f \in \mathcal{B}(\{0, 1\})$, we have

$$\begin{aligned}
\mu_1(f) - \mu_2(f) &= [\mu_1(0) f(0) + (1 - \mu_1(0)) f(1)] - [\mu_2(0) f(0) + (1 - \mu_2(0)) f(1)] \\
&= [\mu_1(0) - \mu_2(0)] \times (f(1) - f(0)).
\end{aligned}$$

This proves the following proposition for two-state space models.

We note that the proposition also presents several equivalent definitions of the total variation distance between probability measures.

Proposition 8.2.10 *For any probability measures μ_1 and $\mu_2 \in \mathcal{P}(S)$ on some state space S, we have*

$$\|\mu_1 - \mu_2\|_{tv} = \sup\{|\mu_1(f) - \mu_2(f)| : f \text{ s.t. } \mathrm{osc}(f) \leq 1\} \tag{8.10}$$

$$= \frac{1}{2}\sup\{|\mu_1(f) - \mu_2(f)| : f \text{ s.t. } \|f\| \leq 1\} \tag{8.11}$$

$$= \sup\{|\mu_1(A) - \mu_2(A)| : A \subset S\}. \tag{8.12}$$

In addition, we have

$$\|\mu_1 - \mu_2\|_{tv} = 1 - [\mu_1 \wedge \mu_2](S). \tag{8.13}$$

Proof :
The proof of these equivalent descriptions is provided for finite state space models only. The analysis for general state spaces S equipped with some σ-field $\mathcal{S}$ can be found in [66].

We recall that for any subset $A \subset S$, and any measure ν on S we set $\nu(A) := \sum_{x \in S} \nu(x)$. In this notation, we notice that for any measure $\nu = \mu_1 - \mu_2$, with $\mu_1, \mu_2 \in \mathcal{P}(S)$,

$$\nu(S) = \mu_1(S) - \mu_2(S) = 1 - 1 = 0.$$

This implies that

$$\nu(S) = \nu(S^+) + \nu(S^-) = 0 \Rightarrow \nu(S^+) = -\nu(S^-) \tag{8.14}$$

with the subsets

$$S^+ := \{x \in S : \nu(x) = \mu_1(x) - \mu_2(x) \geq 0\}$$
$$S^- := \{x \in S : \nu(x) = \mu_1(x) - \mu_2(x) < 0\}.$$

In addition, for any $x \in S$ we have

$$\nu(x) = \nu^+(x) - \nu^-(x) \quad \text{with} \quad \begin{cases} \nu^+(x) = \nu(x)\,1_{S^+}(x)\ (\geq 0) \\ \nu^-(x) = -\nu(x)\,1_{S^-}(x)\ (> 0). \end{cases}$$

Notice that

$$(8.14) \Leftrightarrow \nu^+(S) = \nu^-(S).$$

Using this decomposition, we have $|\nu(x)| = \nu^+(x) + \nu^+(x)$ and

$$2\|\nu\|_{tv} = \sum_{x \in S} |\nu(x)| = \nu(S^+) + \nu(S^-) = 2\,\nu(S^+) = 2\nu^+(S) = 2\nu^-(S).$$

This also implies that

$$2\,\|\mu_1 - \mu_2\|_{tv} = \nu(S^+) - \nu(S^-) = \sum_{\mu_1 > \mu_2} (\mu_1(x) - \mu_2(x)) + \sum_{\mu_2 \geq \mu_1} (\mu_2(x) - \mu_1(x))$$

$$= \underbrace{\sum_{\mu_1 > \mu_2} \mu_1(x)}_{=1 - \sum_{\mu_2 \geq \mu_1} \mu_1(x)} - \sum_{\mu_2 \geq \mu_1} \mu_1(x) - \sum_{\mu_1 > \mu_2} \mu_2(x) + \underbrace{\sum_{\mu_2 \geq \mu_1} \mu_2(x)}_{=1 - \sum_{\mu_1 > \mu_2} \mu_2(x)} \quad .$$

Hence

$$\|\mu_1 - \mu_2\|_{tv} = 1 - \left[\sum_{\mu_2 \geq \mu_1} \mu_1(x) + \sum_{\mu_1 > \mu_2} \mu_2(x) \right] = 1 - \sum_{x \in S} \mu_1(x) \wedge \mu_2(x).$$

This ends the proof of (8.13). On the other hand, for any $f \in \mathcal{B}(S)$, with $\mathrm{osc}(f) \leq 1$,

$$
\begin{aligned}
|\nu(f)| &= |\nu^+(f) - \nu^-(f)| \\
&= \left| \sum_{x,y \in S} (f(x) - f(y)) \frac{\nu^+(x)}{\nu^+(S)} \frac{\nu^-(y)}{\nu^-(S)} \right| \|\nu\|_{tv} \leq \|\nu\|_{tv}.
\end{aligned}
$$

This implies that the l.h.s. of (8.10) is larger than the r.h.s. To prove the reverse inequality, we observe that $\mathrm{osc}(1_A) \leq 1$ so that

$$\|\nu\|_{tv} \geq \sup_{f \,:\, \mathrm{osc}(f) \leq 1} |\nu(f)| \geq \sup_{A \subset S} |\nu(A)|.$$

If we choose $A = S^+$, then we find that

$$\sup_{A \subset S} |\nu(A)| \geq \nu(S^+) = \|\nu\|_{tv}.$$

This ends the proof of (8.10). Now we turn to the proof of (8.12). If we take $f_A = \frac{1}{2}(1_A - 1_{S-A})$, we have $\mathrm{osc}(f_A) \leq 1$ and

$$
\begin{aligned}
\mu_1(f_A) - \mu_2(f_A) &= \frac{1}{2} \left[(\mu_1(A) - (1 - \mu_1(A))) - (\mu_2(A) - (1 - \mu_2(A))) \right] \\
&= \mu_1(A) - \mu_2(A).
\end{aligned}
$$

This shows that the r.h.s. of (8.12) is upper bounded by the r.h.s. of (8.10). In addition, the maxima in (8.12) and (8.10) (respectively in (8.11)) occurs for the functions f_A (respectively $2f_A$), with $A = S^+ = \{x \,:\, \mu_1(x) \geq \mu_2(x)\}$, since we have

$$(\mu_1 - \mu_2)(S^+) := \nu(S^+) = \|\mu_1 - \mu_2\|_{tv}.$$

This ends the proof of the proposition.   ∎

Important remark :
We mention without proof that the set of signed measures $\mathcal{M}(S)$ on some state space S equipped with the total variation norm

$$\|\mu\|_{tv} = \sup \{|\mu(f)| \,:\, f \text{ s.t. } \mathrm{osc}(f) \leq 1\} = \frac{1}{2} \sup \{|\mu(f)| \,:\, f \text{ s.t. } \|f\| \leq 1\}$$

is a complete Banach space.

8.2.3 Contraction inequalities

Definition 8.2.11 *The Dobrushin ergodic coefficient $\beta(M)$ of a Markov transition M on S is defined by*

$$\beta(M) = \sup_{x,y \in S} \|M(x, .) - M(y, .)\|_{tv}$$

We consider the $\{0,1\}$-valued Markov chain model discussed in (7.11). We further assume that all entries of the Markov transitions are positive (i.e., $M(x,y) > 0$), and we set $\mathcal{B}(\{0,1\}) = \mathbb{R}^{\{0,1\}}$, and $\mathcal{P}(\{0,1\})$ as the set of all probability measures on $\{0,1\}$.

$$M(f)(0) - M(f)(1)$$

$$= [M(0,0)f(0) + M(0,1)f(1)] - [M(1,0)f(0) + M(1,1)f(1)]$$

$$= [M(0,0)f(0) + (1 - M(0,0))f(1)] - [M(1,0)f(0) + (1 - M(1,0))f(1)]$$

$$= [M(0,0) - M(1,0)] [f(0) - f(1)].$$

Using the fact that

$$\beta(M) = \sup_{x,y \in S} \|M(x,.) - M(y,.)\|_{\mathrm{tv}} = |M(0,0) - M(1,0)|$$

we prove the following proposition for $\{0,1\}$-valued state space models.

Proposition 8.2.12 *For any Markov transition on some state space S we have*

$$\beta(M) = \sup_{f \in \mathcal{B}(S)} \frac{\mathrm{osc}(M(f))}{\mathrm{osc}(f)} = \sup_{f \,:\, \mathrm{osc}(f) \le 1} \mathrm{osc}(M(f)).$$

Proof :
We use the fact that

$$\sup_{f \,:\, \mathrm{osc}(f) \le 1} \mathrm{osc}(M(f)) = \sup_{f \,:\, \mathrm{osc}(f) \le 1} \sup_{x,y \in S} |M(f)(x) - M(f)(y)|$$

$$= \sup_{x,y \in S} \sup_{f \,:\, \mathrm{osc}(f) \le 1} |M(f)(x) - M(f)(y)|$$

$$= \sup_{x,y \in S} \|M(x,.) - M(y,.)\|_{tv}.$$

This ends the proof of the proposition. ∎

Extending the arguments we used in (8.7) to general state space models S, we prove that

$$[\, \exists\, \epsilon > 0 \,:\, \forall x \in S \ \ M(x,dy) \ge \epsilon\, \nu(dy)\,] \ \Rightarrow\ \beta(M) \le (1 - \epsilon). \tag{8.15}$$

The proof of this result is left as an exercise to the reader (cf. exercise 102).

We illustrate this condition with a series of examples (see also the hit-and-run samplers on bounded domains discussed in exercise 125):

- Let a be a bounded function on $S = \mathbb{R}$, and let M be the Markov transition associated with the evolution equation

$$X_n = a(X_{n-1}) + W_n$$

where W_n stands for a sequence of independent and absolutely continuous r.v. with common density $p(w) = \frac{1}{2\lambda}\, e^{-\lambda|w|}$. In this situation, if we fix a point $x_0 \in S$, we have

$$\nu(dy) := M(x_0, dy) \leq M(x, dy)\, e^{\lambda||y-a(x)|-|y-a(x_0)||} \leq M(x, dy)\, e^{\lambda \, \mathrm{osc}(a)}.$$

This implies that

$$M(x, dy) \geq \epsilon\, \nu(dy) \quad \text{with} \quad \epsilon = \exp\left(-\lambda_{\mathrm{osc}}(a)\right).$$

- We consider a compact set $S' \subset S = \mathbb{R}^d$. Let $p(x, y)$ be some continuous positive function on $(\mathbb{R}^d \times \mathbb{R}^d)$. The Markov transition $M(x, dy) \propto p(x, y)\, 1_{S'}(y)\, dy$ on S' satisfies (8.15). We check this claim using the fact that

$$\sqrt{\epsilon} \leq \frac{p(x', y)}{p(x, y)} \leq 1/\sqrt{\epsilon} \quad \text{with} \quad \sqrt{\epsilon} := \inf_{y \in S'} \frac{\inf_{x \in S'} p(x, y)}{\sup_{x' \in S'} p(x', y)} > 0. \tag{8.16}$$

These estimates imply (8.15) with $\nu(dx) \propto M(x_0, dy)$. This indicates that (8.15) is satisfied for any Markov transitions with a positive and continuous density on some compact space (equipped with some metric). For instance, any regular Markov chain (with transition densities) evolving in our galaxy satisfies (8.15).

- We assume that there exists a subset $A \subset S$ and a positive measure γ s.t. $\gamma(A) > 0$ and for any $x \in S$

$$M(x, dy)\, 1_A(y) \geq \gamma(dy)\, 1_A(y). \tag{8.17}$$

In this case, we have

$$M(x, dy) \geq M(x, dy)\, 1_A(y) \geq \gamma(dy)\, 1_A(y) = \epsilon\, \nu(dy)$$

with

$$\epsilon = \gamma(A) \quad \text{and} \quad \nu(dy) = \frac{\gamma(dy)1_A(y)}{\gamma(A)}.$$

For instance, the Gaussian transition on $S = \mathbb{R}$ defined by

$$M(x, dy) = \frac{1}{\sqrt{2\pi\sigma^2(x)}}\, \exp\left(-\frac{1}{2\sigma^2(x)}(y - a(x))^2\right)\, dy$$

satisfies (8.17) with $A = \mathbb{R}$ as soon as

$$0 < \sigma_{min}^2 \leq \sigma^2(x) \leq \sigma_{max}^2 < \infty \quad \text{and} \quad \|a\| := \sup_{x \in \mathbb{R}} |a(x)| < \infty.$$

We check this claim by using the fact that

$$y \geq 0 \quad \Rightarrow \quad \sup_{x \in S}(y - a(x))^2 \leq y^2 + 2y\|a\| + \|a\|^2 = (y + \|a\|)^2$$

$$y \leq 0 \quad \Rightarrow \quad \sup_{x \in S}(y - a(x))^2 \leq y^2 - 2y\|a\| + \|a\|^2 = (y - \|a\|)^2$$

and

$$M(x, dy) \geq \gamma(dy) := \gamma_1(dy) + \gamma_2(dy)$$

with

$$\gamma_1(dy) = \frac{1}{\sqrt{2\pi\sigma_{max}^2}}\, \exp\left(-\frac{1}{2\sigma_{min}^2}(y + \|a\|)^2\right)\, 1_{y \geq 0}\, dy$$

$$\gamma_2(dy) = \frac{1}{\sqrt{2\pi\sigma_{max}^2}}\, \exp\left(-\frac{1}{2\sigma_{min}^2}(y - \|a\|)^2\right)\, 1_{y < 0}\, dy.$$

Notice that in this case we have

$$\gamma_1(1) + \gamma_2(1) = \frac{\sigma_{min}}{\sigma_{max}} \left[\mathbb{P}\left(-\|a\| + \sigma_{min}Y \geq 0\right) + \mathbb{P}\left(\|a\| + \sigma_{min}Y \leq 0\right) \right]$$

$$= \frac{\sigma_{min}}{\sigma_{max}} \left[1 - \mathbb{P}\left(|Y| \leq \|a\|/\sigma_{min}\right) \right] > 0$$

where Y stands for a standard normal random variable.

We are now in a position to state and to prove the following theorem.

Theorem 8.2.13 *For any Markov transitions M, M_1, M_2 on some state space S, for any couple of measures $\mu_1, \mu_2 \in \mathcal{P}(S)$, for any function $f \in \mathcal{B}(S)$, and any $n \in \mathbb{N}$ we have*

$$\operatorname{osc}(M^n(f)) \leq \beta(M)^n \operatorname{osc}(f) \quad and \quad \beta(M_1 M_2) \leq \beta(M_1)\,\beta(M_2),$$

as well as

$$\|\mu_1 M^n - \mu_2 M^n\|_{tv} \leq \beta(M)^n \|\mu_1 - \mu_2\|_{tv}.$$

In addition when the regularity condition (8.15) is satisfied we have

$$\|\mu_1 M^n - \mu_2 M^n\|_{tv} \leq (1-\epsilon)^n \|\mu_1 - \mu_2\|_{tv} \rightarrow_{n \to \infty} 0. \tag{8.18}$$

In this situation, there exists a unique measure π such that $\pi = \pi M$.

Proof :
We observe that

$$\operatorname{osc}(M^n(f)) = \operatorname{osc}\left(M\left[\frac{M^{n-1}(f)}{\operatorname{osc}(M^{n-1}(f))} \right] \right) \times \operatorname{osc}(M^{n-1}(f)).$$

Since $\operatorname{osc}\left(\frac{M^{n-1}(f)}{\operatorname{osc}(M^{n-1}(f))} \right) \leq 1$, we conclude that

$$\operatorname{osc}(M^n(f)) \leq \left[\sup_{g\,:\,\operatorname{osc}(g) \leq 1} \operatorname{osc}\left(M(g)\right) \right] \times \operatorname{osc}(M^{n-1}(f)) = \beta(M) \times \operatorname{osc}(M^{n-1}(f)).$$

We prove in the same way that $\beta(M_1 M_2) \leq \beta(M_1)\,\beta(M_2)$. This ends the proof of the first assertion. The second assertion follows from the fact that

$$\|\mu_1 M - \mu_2 M\|_{tv} = \sup_{f\,:\,\operatorname{osc}(f) \leq 1} \left(\operatorname{osc}(M(f)) \times \left| (\mu_1 - \mu_2)\left[\frac{M(f)}{\operatorname{osc}(M(f))} \right] \right| \right)$$

$$= \beta(M) \times \sup_{g\,:\,\operatorname{osc}(g) \leq 1} |(\mu_1 - \mu_2)(g)| = \beta(M) \times \|\mu_1 - \mu_2\|_{tv}$$

for any Markov transition M. The last assertion is a direct consequence of the fixed point theorem. This clearly ends the proof of the proposition. ∎

8.2.4   Poisson equation

We let X_n be a Markov chain on some state space S. We further assume that the Dobrushin ergodic coefficient of the Markov transition M of the chain X_n is such that

$$\beta(M^n) \leq a \, e^{-b \, n}$$

for some parameters $a < \infty$, and $b > 0$, and for any $n \geq 0$. In this situation, the chain has an unique invariant measure $\pi = \pi M$ and we have

$$\mathrm{osc}\,(M^n(f)) \leq a \, e^{-b \, n}$$

for any f such that $\mathrm{osc}(f) \leq 1$. Using the fact that

$$\|M^n(f) - \pi(f)\| = \|M^n(f) - \pi M^n(f)\| \leq \mathrm{osc}\,(M^n(f))$$

we check that the functional series

$$P(f)(x) = \sum_{n \geq 0} M^n(f)(x)$$

are well defined bounded functions for any f such that $\pi(f) = 0$, and $\mathrm{osc}\,(f) \leq 1$. We check this claim using the fact that

$$\|P(f)\| \leq \sum_{n \geq 0} \|M^n(f) - \pi M^n(f)\| \leq \sum_{n \geq 0} \mathrm{osc}\,(M^n(f)) \leq a/(1 - e^{-b}).$$

In addition, the functional series solve the Poisson equation

$$g = P(f) \;\Rightarrow\; [Id - M](g) = f \qquad (8.19)$$

for any given function f s.t. $\pi(f) = 0$.

We check this claim using the fact that

$$[Id - M]P(f)(x) = \sum_{n \geq 0} M^n(f)(x) - \sum_{n \geq 1} M^n(f)(x) = f(x).$$

Sometimes we write

$$P = [Id - M]^{-1} = \sum_{n \geq 0} M^n.$$

8.2.5   *V*-norms ////

Definition 8.2.14 *The V-norm on the set of signed measures $\mathcal{M}(S)$ (on some state space S equipped with some σ-field $\mathcal{S}$) associated with some non-negative function V is defined for any $\mu \in \mathcal{M}(S)$ by*

$$\|\mu\|_V := \|\mu\|_{tv} + |\mu|\,(V). \qquad (8.20)$$

In the above display, $|\mu| = \mu^+ + \mu^-$ stands for the total variation of the measure μ, defined in terms of the Hahn-Jordan decomposition $\mu = \mu^+ - \mu^-$ of μ.

As seen in (8.14), these measures are prescribed by a partition of the state $S = S^+ \cup S^-$, with the positive μ^+ and the negative part μ^- of a signed measure μ given by

$$\mu^+(dx) = \mu(dx)1_{S^+}(x) \quad \text{and} \quad \mu^-(dx) = -\mu(dx)1_{S^-}(x).$$

We also mention that these measures are alternatively defined by

$$
\begin{aligned}
\mu^+(A) &= \sup\{\mu(B) : B \subset A, \ B \in \mathcal{S}\} \\
\mu^-(A) &= -\inf\{\mu(B) : B \subset A, \ B \in \mathcal{S}\} \\
|\mu|(A) &= \sup\left\{\sum_{i \in I} |\mu(A_i)| : A = \cup_{i \in I} A_i \text{ finite partition}\right\}.
\end{aligned}
\tag{8.21}
$$

For more details on these measures, we refer the reader to the seminal book of Paul Richard Halmos [144]. Using (8.21) for any signed measures μ_1 and μ_2 we have the triangle inequality

$$|\mu_1 + \mu_2| \le |\mu_1| + |\mu_2| \tag{8.22}$$

from which we readily check that $\|.\|_V$ is a well defined norm on $\mathcal{M}(S)$. When $V = 0$ the V-norm coincides with the total variation norm discussed in 8.2.10. In this connection, we recall that

$$\|\mu\|_{tv} = \mu(S^+) = \mu^+(S) = \mu(S^-) = \mu^-(S) = \frac{1}{2}\mu\left((1_{S^+} - 1_{S^-})\right)$$

so that

$$
\begin{aligned}
\|\mu\|_V &= \|\mu\|_{tv} + \mu^+(V) + \mu^-(V) \\
&= \|\mu\|_{tv} + \mu((1_{S^+} - 1_{S^-})V) = \mu\left((1_{S^+} - 1_{S^-})\left[V + \frac{1}{2}\right]\right).
\end{aligned}
$$

Definition 8.2.15 *We define the V-norm and the V-oscillation of a given function f by*

$$\|f\|_V := \left\|\frac{f}{V + 1/2}\right\| = \sup_{x \in S}\left(\frac{|f(x)|}{V(x) + 1/2}\right)$$

and

$$\operatorname{osc}_V(f) := \sup_{x,y \in S}\left(\frac{|f(x) - f(y)|}{[V(x) + V(y) + 1]}\right).$$

Remark :
Notice that

$$
\begin{aligned}
\frac{|f(x) - f(y)|}{[V(x) + V(y) + 1]} &\le \frac{|f(x)| + |f(y)|}{[V(x) + V(y) + 1]} \\
&\le \frac{|f(x)|}{[V(x) + 1/2]}\frac{V(x) + 1/2}{[V(x) + V(y) + 1]} + \frac{|f(y)|}{[V(y) + 1/2]}\frac{V(y) + 1/2}{[V(x) + V(y) + 1]} \\
&\le \|f\|_V
\end{aligned}
$$

from which we conclude that

$$\mathrm{osc}_V(f) \leq \|f\|_V. \quad (8.23)$$

When $V = 0$ we have

$$\mathrm{osc}_0(f) = \mathrm{osc}(f) \quad \text{and} \quad \|f\|_0 = 2\|f\|.$$

Sometimes these V-norm-type quantities are expressed in terms of the functions $W = 1/2 + V(x)$ with

$$\|f\|_{\mathbf{W}} := \sup_{x \in S} \frac{|f(x)|}{W(x)} \quad \text{and} \quad \mathbf{osc_W}(f) := \sup_{x,y \in S} \frac{|f(x) - f(y)|}{W(x) + W(y)}.$$

We mention without proof that the set of signed measures $\mathcal{M}(S)$ on some state space S equipped with some V-norm is a complete metric space.

The following result is an extension of proposition 8.2.10 to V-norms.

Proposition 8.2.16 *For any $\mu \in \mathcal{M}(S)$ on some state space S, we have*

$$\begin{aligned}
\|\mu\|_V &= \sup\{|\mu(f)| \; : \; f \text{ s.t. } \mathrm{osc}_V(f) \leq 1\} & (8.24) \\
&= \sup\{|\mu(f)| \; : \; f \text{ s.t. } \|f\|_V \leq 1\}. & (8.25)
\end{aligned}$$

Proof :

For any function f s.t. $\mathrm{osc}_V(f) \leq 1$ we have the decomposition

$$f(x) - f(y) = \overbrace{\frac{1}{1 + [V(x) + V(y)]} (f(x) - f(y))}^{:= f_1(x,y)}$$
$$+ \underbrace{[V(x) + V(y)] \frac{f(x) - f(y)}{1 + [V(x) + V(y)]}}_{= f_2(x,y)}$$

with $\|f_1\| \leq 1$, and $|f_2(x,y)| \leq [V(x) + V(y)]$. This implies that

$$\begin{aligned}
\mu(f) &= \mu^+(f) - \mu^-(f) \\
&= \int (f(x) - f(y)) \frac{\mu^+(dx)}{\mu^+(S)} \frac{\mu^-(dy)}{\mu^-(S)} \|\mu\|_{tv} \\
&= \int f_1(x,y) \frac{\mu^+(dx)}{\mu^+(S)} \frac{\mu^-(dy)}{\mu^-(S)} \|\mu\|_{tv} + \int f_2(x,y) \frac{\mu^+(dx)}{\mu^+(S)} \frac{\mu^-(dy)}{\mu^-(S)} \|\mu\|_{tv}
\end{aligned}$$

from which we prove that

$$\begin{aligned}
|\mu(f)| &\leq \|\mu\|_{tv} + \int [V(x) + V(y)] \frac{\mu^+(dx)}{\mu^+(S)} \frac{\mu^-(dy)}{\mu^-(S)} \|\mu\|_{tv} \\
&= \|\mu\|_{tv} + \mu^+(V) + \mu^-(V) = \|\mu\|_V.
\end{aligned}$$

Taking the supremum over all the functions f such that $\mathrm{osc}_V(f) \le 1$, we check that

$$\sup\{|\mu(f)| \ : \ f \text{ s.t. } \mathrm{osc}_V(f) \le 1\} \le \|\mu\|_V.$$

Furthermore we have

$$\|\mu\|_V = \|\mu\|_{tv} + \mu((1_{S^+} - 1_{S^-})V) = \mu\left(\underbrace{1_{S^+} + (1_{S^+} - 1_{S^-})V}_{:=W}\right)$$

with the function W such that

$$\mathrm{osc}_V(W) \le 1.$$

We check this claim using the decomposition

$$\begin{aligned} W(x) - W(y) \ &:= \ [1_{S^+}(x) - 1_{S^+}(y)] \\ &\quad + [1_{S^+}(x) - 1_{S^-}(x)] \ V(x) + [1_{S^-}(y) - 1_{S^+}(y)] \ V(y). \end{aligned}$$

This implies that

$$|W(x) - W(y)| \ := \ 1 + V(x) + V(y) \quad \Rightarrow \quad \mathrm{osc}_V(W) \le 1.$$

We conclude that

$$\|\mu\|_V = \mu(W) \quad \text{with} \quad \mathrm{osc}_V(W) \le 1$$

and therefore

$$\|\mu\|_V \le \sup\{|\mu(f)| \ : \ f \text{ s.t. } \mathrm{osc}_V(f) \le 1\}.$$

This ends the proof of (8.24).

Now we turn to the proof of (8.25). Firstly, we observe that

$$(8.23) \ \Rightarrow \ \|f\|_V \le 1 \Rightarrow \mathrm{osc}_V(f) \le 1. \tag{8.26}$$

This readily implies that

$$\sup\{|\mu(f)| \ : \ f \text{ s.t. } \|f\|_V \le 1\} \le \sup\{|\mu(f)| \ : \ f \text{ s.t. } \mathrm{osc}_V(f) \le 1\} = \|\mu\|_V.$$

To end the proof, we consider the function

$$U = \frac{1}{2}\left(1_{S^+} - 1_{S^-}\right) + V\left(1_{S^+} - 1_{S^-}\right).$$

We check that

$$|U(x)| \ \le \ V(x) + 1/2 \quad \Rightarrow \quad \|U\|_V \le 1$$

and

$$\mu(U) \ = \ \frac{\mu^+(S) + \mu^-(S)}{2} + \mu^+(V) + \mu^-(V) = \mu^+(S) + \mu^+(V) + \mu^-(V) = \|\mu\|_V.$$

This implies that

$$\|\mu\|_V = \mu(U) \le \sup\{|\mu(f)| \ : \ f \text{ s.t. } \|f\|_V \le 1\}.$$

∎

8.2.6   Geometric drift conditions ⫻

We let M be a Markov transition on some state space S. We further assume that S is a topological vector space equipped with the Borel σ-field $\mathcal{S}$. We recall that a subset $B \subset S$ is called bounded when every neighborhood A of the null vector state can be inflated to include the set B; that is,

$$\exists a \in \mathbb{R}_+ \qquad B \subset a\,A = \{a\,x \,:\, x \in A\}.$$

The reader who does not like too much abstraction can restrict the forthcoming discussion to metric state spaces or to the Euclidian space $\mathbb{R}^d$, with $d \geq 1$. In this situation, all the compact subsets of S are bounded and closed.

The aim of this section is to discuss the following conditions.

Dobrushin local contraction condition:
For any compact subset $C \subset S$, we have

$$\beta(C; M) := \sup_{(x,y) \in C^2} \|M(x, .) - M(y, .)\|_{tv} < 1. \tag{8.27}$$

The quantities $\beta(C; M)$ are called the Dobrushin local contraction coefficients.

Foster-Lyapunov condition:
There exists some non-negative function W on S with compact subset levels, such that
$$M(W) \leq \epsilon\,W + c \tag{8.28}$$
for some $\epsilon \in [0, 1[$ and some finite constant $c < \infty$. The function W is called a Lyapunov function.

Definition 8.2.17 *When the Dobrushin local contraction and the Foster-Lyapunov condition are satisfied, for any $R \in \mathbb{R}_+$, we set*

$$\beta^{(R)}(M) \;=\; \beta\left(\{W \leq R\}, M\right) := \sup_{(x,y)\,:\,W(x)\vee W(y)\leq R} \|M(x, .) - M(y, .)\|_{tv} < 1.$$

A simple way to check the Dobrushin local contraction condition is to prove that for any compact subset $C \subset S$, there exists some $\epsilon_C \in]0, 1]$ and some probability measure ν_C on S such that
$$\forall x \in C \quad M(x, dy) \geq \epsilon_C\,\nu_C(dy). \tag{8.29}$$
In this situation, using the same arguments as the ones we used in (8.7), we prove that

$$\beta(C; M) \leq (1 - \epsilon_C).$$

In the literature on Markov chain stability, the subsets C satisfying the minorization condition (8.29) are often called "small" sets.

This local contraction condition is satisfied for most of the Markov chains encountered in practice. For instance, for the Gaussian transition

$$M(x, dy) = \frac{1}{\sqrt{2\pi\sigma^2(x)}} \, \exp\left(-\frac{1}{2\sigma^2(x)}(y - a(x))^2\right) \, dy$$

associated with some *locally bounded* drift and variance functions a and σ^2 on $S = \mathbb{R}$ we have

$$y \geq 0 \quad \Rightarrow \quad \sup_{x \in A}(y - a(x))^2 \leq y^2 + 2y\|a\|_A + \|a\|_A^2 = (y + \|a\|_A)^2$$

$$y \leq 0 \quad \Rightarrow \quad \sup_{x \in A}(y - a(x))^2 \leq y^2 - 2y\|a\|_A + \|a\|_A^2 = (y - \|a\|_A)^2$$

for any bounded subset $A \subset S$, with $\|a\|_A := \sup_{x \in A} |a(x)|$. We further assume that

$$\forall x \in A \qquad 0 < \sigma^2_{min,A} \leq \sigma^2(x) \leq \sigma^2_{max,A} < \infty.$$

This implies that

$$
\begin{aligned}
M(x, dy) \quad &\geq \quad \gamma(dy) \\
&:= \quad 1_{y \geq 0} \left(\sqrt{2\pi\sigma^2_{max,A}}\right)^{-1} \exp\left(-\frac{1}{2\sigma^2_{min,A}}(y + \|a\|_A)^2\right) dy \\
&\quad + 1_{y < 0} \left(\sqrt{2\pi\sigma^2_{max,A}}\right)^{-1} \exp\left(-\frac{1}{2\sigma^2_{min,A}}(y - \|a\|_A)^2\right) dy \geq \tau_A \, \nu(dy)
\end{aligned}
$$

with the probability measure $\nu(dy) = \gamma(dy)/\gamma(1)$ and the $]0,1[$ valued constant

$$\tau_A = \frac{\sigma_{min,A}}{\sigma_{max,A}} \left[1 - \int_{-\|a\|_A/\sigma_{min,A}}^{\|a\|_A/\sigma_{min,A}} \frac{1}{\sqrt{2\pi}} e^{-y^2/2} \, dy\right].$$

Using the same lines of arguments as the ones we used in (8.7), this implies that

$$\sup_{(x,y) \in A^2} \|M(x, .) - M(y, .)\|_{tv} \leq 1 - \tau_A.$$

More generally, we have the following result.

Proposition 8.2.18 *We consider a Markov transition M on some complete separable metric space S such that*

$$M(x, dy) \geq m(x, y) \, \lambda(dy) \tag{8.30}$$

for some strictly positive Radon measure $\lambda(dy)$ (i.e., nonempty open balls have positive measure). We further assume that for any compact set A there exists some positive measurable function q_A such that

$$\inf_{x \in A} m(x, y) \geq q_A(y). \tag{8.31}$$

In this situation, the Dobrushin local contraction condition (8.29) is satisfied. For instance, the minorization condition (8.31) is met when the functions $m(x, y)$ are lower semicontinuous w.r.t. the first variable, and upper semicontinuous w.r.t. the second.

Proof :
We recall that a characteristic of Radon measures is that the measure of a Borel set B is the supremum of the measures $\lambda(A)$ of the compact sets $A \subset B$. Thus, if λ is a strictly positive Radon measure on S, one can always find for every open set $B \subset S$ a compact $A \subset B$ such that

$$\lambda(A) \geq (1/2) \, \lambda(B) \, (> 0).$$

This ensures that λ charges all the compact sets.

It is of course tempting to set $\inf_{x \in A} m(x, y) = q_A(y)$ but it is well known that the infimum of an uncountable collection of functions may fail to be measurable. Under the condition (8.31), we clearly have that

$$\forall x \in A \qquad M(x, dy) \geq q_A(y) \, \lambda(dy) := \gamma_A(dy) \quad \text{with} \quad \gamma_A(1) = \lambda(q_A) > 0.$$

In this situation, the Dobrushin condition (8.27) is clearly met with

$$\forall x \in A \qquad M(x, dy) \geq \epsilon_A \, \nu_A(dy)$$

with

$$\epsilon_A = \gamma_A(1) > 0 \quad \text{and} \quad \nu_A(dy) := \gamma_A(dy)/\gamma_A(1).$$

When the density function $m(x, y)$ is lower semicontinuous w.r.t. the first variable and upper semicontinuous w.r.t. the second variable, there exists some measurable function $h_A : y \mapsto h_A(y)$ such that

$$\inf_{x \in A} m(x, y) = m(h_A(y), y) := q_A(y) > 0.$$

A proof of this result can be found in [42]. It this situation, the minorization condition (8.31) is clearly satisfied. This ends the proof of the proposition. ∎

Remark : For instance, the Markov transitions

$$P_t(x, dy) = e^{-t} \sum_{n \geq 0} \frac{t^n}{n!} \, M^n(x, dy)$$

on some metric space S of the continuous time embedding of a Markov chain with transitions M discussed in section 11.1.2 may not have a density w.r.t. some measure $\lambda(dy)$, but they satisfy the condition (8.30) for any $t > 0$ as soon as M^n satisfies (8.30) for some integer $n \geq 1$.

> The Foster-Lyapunov condition ensures that the Markov chain X_n with transition probabilities M has little chance to escape from the level sets $\{W \leq w\}$ of the function W.

Indeed, we have

$$(8.28) \implies M^n(W) \leq \epsilon^n \, W + c \, (1 + \epsilon + \ldots + \epsilon^{n-1}) \leq \epsilon^n \, W + c/(1 - \epsilon)$$

from which we prove the uniform estimate

$$\sup_{n \geq 0} \mathbb{E}\left(W(X_n)\right) \leq \epsilon^n \, \mathbb{E}(W(X_0)) + c/(1 - \epsilon) \leq C := \mathbb{E}(W(X_0)) + c/(1 - \epsilon).$$

> Using Markov inequality, when the level sets of W are compact we have
>
> $$\forall \rho > 0 \qquad \exists \{W \leq C/\rho\} := A_\rho \text{ compact} \quad \text{s.t.} \quad \sup_{n \geq 0} \mathbb{P}\left(W(X_n) \notin A_\rho\right) \leq \rho.$$

In the Gaussian model discussed above, we notice that

$$|a(x)| \leq \epsilon \, |x| \Rightarrow \mathbb{E}\left[\, |X_n| \, \mid X_{n-1} = x \right] \leq \epsilon \, |x| + \sigma^2.$$

In this case, $W(x) = |x|$ satisfies (8.28) with $c = \sigma^2$.

More generally, suppose that X_n and X_n' are two not necessarily independent copies of the transition of the chain starting at x and x_0; that is,

$$
\begin{aligned}
M(x, dy) &:= \mathbb{P}\left(X_n \in dy \mid X_{n-1} = x\right) \\
M(x_0, dy) &:= \mathbb{P}\left(X_n' \in dy \mid X_{n-1}' = x_0\right) = \mathbb{P}\left(X_n \in dy \mid X_{n-1} = x_0\right).
\end{aligned}
$$

We suppose that the state space S is equipped with a metric d and we have the local contraction inequality

$$\mathbb{E}\left(d(X_n, X_n') \mid (X_{n-1}, X_{n-1}') = (x, x_0)\right) \leq \epsilon \, d(x, x_0).$$

Returning to the Gaussian model discussed above, we can take

$$X_n = a(x) + \sigma \, Y \quad \text{and} \quad X_n' = a(x_0) + \sigma \, Y$$

where Y stands for a centered Gaussian r.v. with unit variance. In this situation, the local contraction stated above is met for the Euclidian distance $d(x, y) = |x - y|$ provided that

$$|a(x) - a(x_0)| \leq \epsilon \, |x - x_0|.$$

We set $W(x) := d(x, x_0)$ for some fixed state $x_0 \in S$. Using the triangle inequality

$$d(X_n, x_0) - d(X_n', x_0) \leq d(X_n, X_n')$$

we prove that

$$\mathbb{E}\left(d(X_n, x_0) - d(X_n', x_0) \mid (X_{n-1}, X_{n-1}') = (x, x_0)\right)$$

$$= M(W)(x) - M(W)(x_0) \leq \epsilon \, d(x, x_0) = \epsilon \, W(x).$$

This implies that the Foster-Lyapunov condition is satisfied with

$$M(W)(x) \leq \epsilon \, W(x) + c \quad \text{where} \quad c := M(W)(x_0).$$

We end this section with a sufficient condition for the Foster-Lyapunov condition. Suppose there exists a subset $A \subset S$ s.t.

$$
\begin{cases}
M(W)(x) \leq \epsilon \, W(x) & \text{for any} \quad x \in S - A \\
M(W)(x) \leq c & \text{for any} \quad x \in A
\end{cases}
\tag{8.32}
$$

then we have

$$\forall x \in S \quad M(W)(x) \leq \epsilon \, W(x) \, 1_{S-A}(x) \, + \, c \, 1_A(x) \leq \epsilon W \, + \, c.$$

Whenever $M(W)$ is continuous w.r.t. some metric the condition (8.32) is satisfied as soon as we can find some compact set A such that

$$\forall x \notin A \qquad M(W)(x) \leq \epsilon \, W(x)$$

for some $\epsilon \in [0, 1[$. In this case (8.32) is satisfied with $c = \sup_{x \in A} |W(x)|$.

8.2.7   *V*-norm contractions ////

We let M be a Markov transition on some state space S. In the further development of this section, we assume that M satisfies the Foster-Lyapunov condition (8.28) and the local Dobrushin contraction inequality (8.27).

Replacing W by W/c in (8.28) there is no loss of generality to assume that $c = 1$. In addition, replacing W by $W_\epsilon = 1 + \epsilon\, W \geq 1$ we have

$$M\,(W_\epsilon) = \epsilon\, M(W) + 1 \leq \epsilon\, W_\epsilon + 1.$$

Therefore, there is no loss of generality to replace (8.28) by

$$M(W) \leq \epsilon\, W + 1 \quad \text{for some function } W \geq 1. \tag{8.33}$$

Definition 8.2.19 *We let V be a non-negative function V such that*

$$M(V) \leq c_1\, V + c_2$$

for some $c_1, c_2 \geq 0$. We equip the set of probability measures $\mathcal{P}(S)$ on some state space S with the V-norm. The V-Dobrushin ergodic coefficient $\beta_V(M)$ is defined by

$$
\begin{aligned}
\beta_V(M) &= \sup\left\{\operatorname{osc}_V(M(f))\,,\ f\ :\ \operatorname{osc}_V(f) \leq 1\right\} \\
&= \sup_{(x,y)\in S^2} \frac{\|M(x,\,.\,) - M(y,\,.\,)\|_V}{1 + [V(x) + V(y)]} \leq c_1 \vee (1 + 2c_2).
\end{aligned}
$$

The second formulation in the above display is readily checked using the fact that

$$\beta_V(M) = \sup_{(x,y)\in S^2} \sup_{f:\operatorname{osc}_V(f)\leq 1} \frac{|M(f)(x) - M(f)(y)|}{1 + [V(x) + V(y)]}.$$

Using the same arguments as in the proof of theorem 8.2.13 we have the following contraction inequalities.

Theorem 8.2.20 *For any couple of measures $\mu_1, \mu_2 \in \mathcal{P}(S)$, any Markov transitions M, M_1, M_2 on S, any function f s.t. $\operatorname{osc}_V(f) < \infty$, and any $n \in \mathbb{N}$ we have*

$$\beta_V(M^n) \leq \beta_V(M)^n \quad \text{and} \quad \beta_V(M_1 M_2) \leq \beta_V(M_1)\, \beta_V(M_2)$$

as well as

$$\operatorname{osc}_V(M(f)) \leq \beta_V(M)\, \operatorname{osc}_V(f) \quad \text{and} \quad \|\mu_1 M - \mu_2 M\|_V \leq \beta_V(M)\, \|\mu_1 - \mu_2\|_V.$$

In the further development of this section, we assume that the condition (8.33) is satisfied and we set $V_\rho = \rho\, W$, for some $\rho \in\,]0,1]$. Notice that

$$(8.33) \ \Rightarrow\ M(V_\rho) \leq \epsilon\, V_\rho + \rho.$$

In addition, we have the uniform estimate

$$(V_\rho = \rho\, W \quad \text{and} \quad W \geq 1) \;\Rightarrow\; V_\rho^{-1} M(V_\rho) \;\leq\; \epsilon + \frac{\rho}{\rho W} \leq 1 + \epsilon.$$

We also notice that for any $R \geq 1$ we have

$$W(x) \geq R \;\Rightarrow\; \left\{ \begin{array}{rcl} V_\rho(x)^{-1}\, M(V_\rho)(x) & \leq & \epsilon + \frac{1}{W(x)} \leq \epsilon + \frac{1}{R} \\ V_\rho(x) & = & \rho\, W(x) \geq \rho R \end{array} \right. \tag{8.34}$$

and

$$W(x) \leq R \;\Rightarrow\; \left\{ \begin{array}{rcl} V_\rho(x)^{-1}\, M(V_\rho)(x) & \leq & 1 + \epsilon \\ \rho \leq V_\rho(x) & = & \rho\, W(x) \leq \rho R. \end{array} \right. \tag{8.35}$$

We set

$$\Delta_\rho(x, y) := \frac{\| M(x, \cdot) - M(y, \cdot) \|_{V_\rho}}{1 + V_\rho(x) + V_\rho(y)}.$$

By definition of the V-norm, using the triangle inequality (8.22) we prove the following decomposition

$$
\begin{aligned}
\Delta_\rho(x, y) \;=\;& \frac{1}{1 + V_\rho(x) + V_\rho(y)} \, \| M(x, \cdot) - M(y, \cdot) \|_{tv} \\
&+ \frac{V_\rho(x)}{1 + V_\rho(x) + V_\rho(y)} \frac{M(V_\rho)(x)}{V_\rho(x)} + \frac{V_\rho(y)}{1 + V_\rho(x) + V_\rho(y)} \frac{M(V_\rho)(y)}{V_\rho(y)} \\
\leq\;& \frac{1}{1 + V_\rho(x) + V_\rho(y)} \, \| M(x, \cdot) - M(y, \cdot) \|_{tv} \\
&+ \frac{V_\rho(x) + V_\rho(y)}{1 + V_\rho(x) + V_\rho(y)} \left(\frac{M(V_\rho)(x)}{V_\rho(x)} \vee \frac{M(V_\rho)(y)}{V_\rho(y)} \right).
\end{aligned}
$$

When $W(x) \wedge W(y) \geq R$, using (8.34) we find that

$$
\begin{aligned}
\Delta_\rho(x, y) \;\leq\;& \frac{1}{1 + V_\rho(x) + V_\rho(y)} + \frac{V_\rho(x) + V_\rho(y)}{1 + V_\rho(x) + V_\rho(y)} \left(\epsilon + \frac{1}{R} \right) \\
=\;& 1 - \left(1 - \frac{1}{1 + V_\rho(x) + V_\rho(y)} \right) \left(1 - \left(\epsilon + \frac{1}{R} \right) \right)
\end{aligned}
$$

from which we conclude that

$$\sup_{W(x) \wedge W(y) \geq R} \Delta_\rho(x, y) \leq 1 - \left(1 - \frac{1}{1 + 2\rho R} \right) \left(1 - \left(\epsilon + \frac{1}{R} \right) \right) < 1$$

for any $\rho \in\,]0, 1]$. Using (8.35) we also find that

$$\sup_{W(x) \vee W(y) \leq R} \Delta_\rho(x, y) \;\leq\; \frac{1}{1 + 2\rho} \beta^{(R)}(M) + 2\, \frac{\rho R}{1 + 2\rho} (1 + \epsilon) \leq \beta^{(R)}(M) + 4\rho R < 1$$

for any $\rho < (1 - \beta^{(R)}(M))/(4R)$.

Combining these estimates with theorem 8.2.20 we readily prove the following theorem.

Theorem 8.2.21 *When the drift condition (8.28) and the local Dobrushin condition (8.27) are satisfied for some function W and some parameter $\epsilon \in [0, 1[$, there exist a positive function V such that $\beta_V(M) < 1$. In this case there exists an unique invariant measure $\pi = \pi M$ and we have the exponential contraction inequality*

$$\| \mu_1 M^n - \mu_2 M^n \|_V \leq \beta_V(M)^n \, \| \mu_1 - \mu_2 \|_V \longrightarrow_{n \uparrow \infty} 0.$$

In exercise 97 we design a function V such that

$$\beta_V(M) \leq 1 - \frac{1 - \beta^{(R_\epsilon)}(M)}{R_\epsilon(1 + 2\sqrt{3})} \quad \text{with} \quad R_\epsilon := 2/(1 - \epsilon).$$

Remark : By definition of $\beta_V(M^n)$, for any function f such that

$$|f(x)| \leq 1/2 + V(x) \quad (\Rightarrow |f(x) - f(y)| \leq 1 + V(x) + V(y))$$

and for any $(x, y) \in S$ we have

$$|M^n(f)(x) - M^n(f)(y)| \leq \beta_V(M^n) \ (1 + V(x) + V(y)).$$

This implies that

$$
\begin{aligned}
|M^n(f)(x) - \pi(f)| &\leq \int \pi(dy) \ |M^n(f)(x) - M^n(f)(y)| \\
&\leq \beta_V(M^n) \ (1 + V(x) + \pi(V)).
\end{aligned}
$$

8.3   Stochastic analysis

8.3.1   Coupling techniques

8.3.1.1   The total variation distance

Lemma 8.3.1 *For any couple of random variables (X, Y) with law (μ_1, μ_2) on some state space S we have*

$$\|\mu_1 - \mu_2\|_{tv} \leq \mathbb{P}(X \neq Y).$$

Proof :
The proof is a direct consequence of the following assertions

$$
\begin{aligned}
\mu_1(A) - \mu_2(A) &= \mathbb{P}(X \in A) - \mathbb{P}(Y \in A) \\
&= \mathbb{P}(X = Y \in A, \ X = Y) + \mathbb{P}(X \in A, \ X \neq Y) \\
&\quad -\mathbb{P}(Y = X \in A, \ Y = X) - \mathbb{P}(Y \in A, \ X \neq Y) \\
&= \mathbb{P}(X \in A, \ X \neq Y) - \mathbb{P}(Y \in A, \ X \neq Y) \\
&= [\mathbb{P}(X \in A \mid X \neq Y) - \mathbb{P}(Y \in A \mid X \neq Y)] \times \mathbb{P}(X \neq Y).
\end{aligned}
$$

∎

Much more is true. The following theorem provides an interpretation of the total variation distance in terms of the chances of coupling two random variables

Theorem 8.3.2 *For any probability measures μ_1, μ_2 on some state space S we have*

$$\|\mu_1 - \mu_2\|_{tv} = \inf \{\mathbb{P}(X \neq Y) \ : \ (X, Y) \text{ s.t. } \mathrm{Law}(X) = \mu_1 \ \& \ \mathrm{Law}(Y) = \mu_2\}.$$

Proof :

To avoid unnecessary technicalities, we only prove this theorem for finite state spaces. It clearly suffices to prove the existence of a pair of r.v. such that $\mathbb{P}(X \neq Y) = \|\mu_1 - \mu_2\|_{tv}$. We consider the pair of coupled random variables (X, Y) defined for any $x \neq y$ by

$$\mathbb{P}\left(X = x \, , \, Y = y\right)$$

$$= \frac{1}{\|\mu_1 - \mu_2\|_{tv}} \left(\mu_1(x) - \mu_1(x) \wedge \mu_2(x)\right) \left(\mu_2(y) - \mu_1(y) \wedge \mu_2(y)\right)$$

and for $x = y$ we set

$$\mathbb{P}\left(X = Y = x\right) = \mu_1(x) \wedge \mu_2(x).$$

By construction, it is readily checked that $\mathbb{P}(X \neq Y) = \|\mu_1 - \mu_2\|_{tv}$. On the other hand, using (8.13) we have

$$\sum_x \left(\mu_1(x) - \mu_1(x) \wedge \mu_2(x)\right) = 1 - \sum_x \mu_1(x) \wedge \mu_2(x)$$

$$= \sum_y \left(\mu_2(y) - \mu_1(y) \wedge \mu_2(y)\right) = \|\mu_1 - \mu_2\|_{tv}$$

and obviously

$$\left(\mu_1(x) - \mu_1(x) \wedge \mu_2(x)\right) \left(\mu_2(x) - \mu_1(x) \wedge \mu_2(x)\right) = 0.$$

From these observations, we prove that

$$\mathbb{P}\left(X = x\right) = \sum_{y \neq x} \mathbb{P}\left(X = x \, , \, Y = y\right) + \mathbb{P}\left(X = x, \, Y = x\right)$$

$$= \left(\mu_1(x) - \mu_1(x) \wedge \mu_2(x)\right) \times \left(1 - \frac{\left(\mu_2(x) - \mu_1(x) \wedge \mu_2(x)\right)}{\|\mu_1 - \mu_2\|_{tv}}\right)$$

$$+ \mu_1(x) \wedge \mu_2(x)$$

$$= \mu_1(x)$$

and by symmetry arguments $\mathbb{P}\left(Y = y\right) = \mu_2(y)$.

∎

Definition 8.3.3 *The coupling we have constructed in the proof of the theorem 8.3.2 for finite state spaces is called a maximal coupling between the distribution* $\mathrm{Law}(X) = \mu_1$ *and* $\mathrm{Law}(Y) = \mu_2$. *In a more simulation-based formulation, it is defined by the equation*

$$\mathbb{P}\left((X, Y) = (x, y)\right)$$

$$= \alpha_{(\mu_1, \mu_2)} \, p_{(\mu_1, \mu_2)}(x) \, 1_x(y) + \left(1 - \alpha_{(\mu_1, \mu_2)}\right) \, q_{(\mu_1, \mu_2)}(x) \, q_{(\mu_2, \mu_1)}(y)$$

with the coupling probability $\alpha_{(\mu_1, \mu_2)} = [\mu_1 \wedge \mu_2](S) \in [0, 1[$ *and the probability measures*

$$p_{(\mu_1, \mu_2)} := \frac{[\mu_1 \wedge \mu_2]}{\alpha(\mu_1, \mu_2)} \quad and \quad q_{(\mu_1, \mu_2)} := \frac{\mu_1 - [\mu_1 \wedge \mu_2]}{1 - \alpha(\mu_1, \mu_2)}.$$

Definition 8.3.4 *The maximal coupling for general state space models is defined in the same way in terms of the infimum measures presented in definition 8.2.7. In this situation, the maximal coupling can also be rewritten as*

$$\begin{cases} X &= (1-\epsilon)\, X' + \epsilon\, Z \\ Y &= (1-\epsilon)\, Y' + \epsilon\, Z \end{cases}$$

with independent random variables $(\epsilon, X, Y, X', Y', Z)$ with distributions

$$\mathbb{P}(\epsilon = 1) = 1 - \mathbb{P}(\epsilon = 0) = \int (p_1(x) \wedge p_2(x))\, \lambda(dx) := c \quad (8.36)$$

$$\mathbb{P}(X' \in dx) = [p_1(x) - (p_1(x) \wedge p_2(x))]\, \lambda(dx)/(1-c)$$
$$\mathbb{P}(Y' \in dy) = [p_2(y) - (p_1(y) \wedge p_2(y))]\, \lambda(dy)/(1-c)$$
$$\mathbb{P}(Z \in dz) = (p_1(z) \wedge p_2(z))\, \lambda(dz)/c$$

as soon as the coupling probability $c > 0$ (otherwise the distributions of X and Y are absolutely continuous so that $\|\mu_1 - \mu_2\|_{tv} = 1$) and $c < 1$ (otherwise $\mu_1 = \mu_2$).

Example 8.3.5 *Consider a couple of exponential random variables $(X, Y) = (X_1, X_2)$ with distribution*

$$\mathbb{P}(X_i \in dx) = \mu_i(dx) = \underbrace{\lambda\ e^{-\lambda(x-a_i)}\ 1_{[a_i,\infty[}(x)}_{:= p_i(x)}\ \underbrace{dx}_{:= \lambda(dx)}$$

for some given parameters $0 \le a_1 < a_2$. In this situation, the coupling probability c in (8.36) is given by

$$p_1(x) \wedge p_2(x) = e^{\lambda(a_2-a_1)}\, p_2(x) \ \Rightarrow\ c = e^{-\lambda(a_2-a_1)} = \mathbb{P}\left(X_1 \in [a_2, \infty[\right).$$

In addition, we have

$$\mathbb{P}(X' \in dx) = 1_{[a_1,a_2]}(x)\, p_1(x)\, dx/(1-c) = \mathbb{P}\left(X_1 \in dx \mid X_1 \in [a_1, a_2]\right)$$
$$\mathbb{P}(Y' \in dx) = p_2(x)\, dx = \mathbb{P}(Z \in dx) \quad and \quad \mathbb{P}(X_1 = X_2) = c = e^{-\lambda(a_2-a_1)}.$$

Since $p_2(x) \le e^{\lambda(a_2-a_1)} p_1(x)$, the same result can be proved using the acceptance-rejection technique.

Example 8.3.6 *Consider a couple of Bernoulli $\{0,1\}$-valued r.v. $(X, Y) = (X_1, X_2)$ with parameters $(p_1, p_2) \in [0,1]^2$ such that $p_2 \ge p_1$. We let $\mathrm{Law}(X_i) = \mu_i$, with $i = 1, 2$. We check that*

$$\|\mu_1 - \mu_2\|_{tv} = |p_1 - p_2|.$$

We let (Y_1, Y_2) be the r.v. on $\{0,1\}^2$ defined by the maximal coupling

$$\mathbb{P}\left((Y_1, Y_2) = (0,0)\right) = 1 - p_2 \ (= \inf(1-p_1, 1-p_2))$$
$$\mathbb{P}\left((Y_1, Y_2) = (0,1)\right) = p_2 - p_1$$
$$\left(= \frac{1}{p_2 - p_1}\left((1-p_1) - (1-p_1) \wedge (1-p_2)\right) \times (p_2 - p_1 \wedge p_2)\right)$$
$$\mathbb{P}\left((Y_1, Y_2) = (1,0)\right) = 0$$
$$\left(= \frac{1}{p_2 - p_1}\,(p_1 - p_1 \wedge p_2) \times \left((1-p_2) - (1-p_1) \wedge (1-p_2)\right)\right)$$
$$\mathbb{P}\left((Y_1, Y_2) = (1,1)\right) = p_1 \ (= \inf(p_1, p_2)).$$

We readily check that $\mathbb{P}(Y_1 = 0) = 1 - p_1 = 1 - \mathbb{P}(Y_1 = 1)$ *and* $\mathbb{P}(Y_2 = 0) = 1 - p_2 = 1 - \mathbb{P}(Y_2 = 1)$ *and* $\mathbb{P}(Y_1 \neq Y_2) = p_2 - p_1 = \|\mu_1 - \mu_2\|_{tv}$.

Example 8.3.7 *We let* (X_1, X_2) *be two random variables on some state space* S *with distributions* (μ_1, μ_2), *and consider a random variable* U *on some state space* $\mathcal{U}$. *We then consider a couple of random variables*

$$X_1' = F(X_1, U_1) \quad and \quad X_2' = F(X_2, U_2)$$

where F *stands for some function from* $(\mathcal{U} \times S)$ *into some possibly different state space* S', *and* (U_1, U_2) *are two (non-necessarily independent) copies of of* U. *Using the coupling* $U = U_1 = U_2$, *we have*

$$X_2' = 1_{X_1 = X_2} X_1' + 1_{X_1 \neq X_2} F(X_2, U) = X_1' + 1_{X_1 \neq X_2} [F(X_2, U) - F(X_1, U)]$$

from which we conclude that

$$\mathbb{P}(X_1' \neq X_2') \leq \mathbb{P}(X_1 \neq X_2).$$

In other words, if (μ_1', μ_2') *stands for the distributions of* (X_1', X_2') *we have*

$$\|\mu_1' - \mu_2'\|_{tv} \leq \|\mu_1 - \mu_2\|_{tv}.$$

8.3.1.2 Wasserstein metric

The following metric extends the coupling interpretation of the total variation distance presented in theorem 8.3.2

Definition 8.3.8 *The Wasserstein distance between two probability measures* μ_1, μ_2 *on some metric space* (S, d) *is defined by*

$$\mathbb{W}(\mu_1, \mu_2) = \inf \{\mathbb{E}(d(X, Y)) \ : \ (X, Y) \text{ s.t. } \mathrm{Law}(X) = \mu_1 \ \& \ \mathrm{Law}(Y) = \mu_2\}.$$

This metric is also known as the Vasershtein distance and/or the Kantorovich-Monge-Rubinstein metric.

If we choose the Hamming distance $d(x, y) = 1_{x \neq y}$, then by theorem 8.3.2 we have $\mathbb{W}(\mu_1, \mu_2) = \|\mu_1 - \mu_2\|_{tv}$. Notice that in this situation, a function f is 1-Lipschitz w.r.t. the trivial distance $d(x, y) = 1_{x \neq y}$ if, and only if, $\mathrm{osc}(f) \leq 1$.

For more general (separable) metric spaces (S, d), the celebrated Kantorovich-Rubinstein duality theorem [165] states that

$$\mathbb{W}(\mu_1, \mu_2) = \sup \{|\mu_1(f) - \mu_2(f)| \ : \ f \in \mathrm{Lip}(S) \text{ s.t. } \mathrm{lip}(f) \leq 1\}. \tag{8.37}$$

For any f such that $\mathrm{lip}(f) \leq 1$ we clearly have that

$$|\mu_1(f) - \mu_2(f)| \leq \mathbb{E}(|f(X) - f(Y)|) \leq \mathbb{E}(d(X, Y))$$

for any r.v. (X, Y) with distributions (μ_1, μ_2). This implies that the r.h.s. in (8.37) is upper bounded by $\mathbb{W}(\mu_1, \mu_2)$. The proof of this reversal inequality is out of the scope of this chapter. We refer to the book of C. Villani for a more thorough discussion on this theorem and related optimal transport problems [258].

Proposition 8.3.9 *We have*

$$\delta(S) \; \|\mu_1 - \mu_2\|_{tv} \leq \mathbb{W}(\mu_1, \mu_2) \leq \text{diam}(S) \; \|\mu_1 - \mu_2\|_{tv}$$

with

$$0 \leq \delta(S) := \inf_{x \neq y} d(x, y) \leq \text{diam}(S) := \sup_{x,y} d(x, y) \leq \infty.$$

Proof :
We clearly have that

$$\delta(S) \; 1_{X \neq Y} \; \leq d(X, Y) \leq \text{diam}(S) \; 1_{X \neq Y}.$$

The proposition is now a consequence of the coupling descriptions of the total variation distance and the Wasserstein metric. This completes the proof of the proposition. ∎

Important remark : When the state space S is finite, a given discrete probability measure $\mu = \sum_{1 \leq i \leq d} p_i \; \delta_{x_i} \in \mathcal{P}(S)$ can be interpreted as the number of stones in some pile (with p_i stones of some type x_i). In this interpretation, a pair of distributions $\mu = \sum_{1 \leq i \leq d} p_i \; \delta_{x_i}$ and $\mu' = \sum_{1 \leq i \leq d'} p_i' \; \delta_{x_i}$ represents two piles of stones. The balanced control problem is to find a flow of stones $P(i,j)$ from the pile μ to the pile μ' that minimizes the average transportation cost defined by the Wasserstein distance

$$\mathbb{W}(\mu_1, \mu_2) = \min \left\{ \sum_{i,j=1}^{d,d'} d(x_i, x_j) \; P(i,j) \right\},$$

with the marginal conditions

$$p(i) := \sum_{j=1}^{d'} P(i,j) \quad \text{and} \quad p'(i) := \sum_{i=1}^{d} P(i,j).$$

In computer sciences, this metric is known as the earth mover's distance. In fractal image processing, and in computational biology, the Wasserstein metric is also called the Hutchinson metric and it is used to measure the similarity between two images or two DNA sequences [163].

Remark : The *bounded Lipschitz distance* between the probability measures μ_1 and μ_2 is defined by replacing in the r.h.s. of (8.37) the condition $\text{lip}(f) \leq 1$ by $\text{blip}(f) := \|f\| + \text{lip}(f) \leq 1$ We also mention that this distance metrizes the convergence in distribution (a.k.a. the convergene in law).

Example 8.3.10 *We consider the couple of Gaussian distributions on $S = \mathbb{R}$ discussed in example 8.2.8. We equip $\mathbb{R}$ with the distance $d(x,y) = |x - y|$. In this situation, we have*

$$\mathbb{W}(\mu_1, \mu_2) \leq \mathbb{E}\left(|(m_1 + Z) - (m_2 + Z)|\right) = |m_2 - m_1|.$$

More generally, if we consider the distributions $\mu_1 = \text{Law}(X_1)$ and $\mu_2 = \text{Law}(X_2)$ of the Gaussian r.v. $X_1 = m_1 + \sigma_1 \, Z$ and $X_2 = m_2 + \sigma_2 \, Z$, with given positive parameters σ_1, σ_2, then we have the crude estimate

$$\mathbb{W}(\mu_1, \mu_2) = \mathbb{E}\left(|(m_1 - m_2) + (\sigma_1 - \sigma_2) \, Z)|\right) \leq |m_1 - m_2| + |\sigma_1 - \sigma_2|.$$

Example 8.3.11 *We consider the Bernoulli distributions discussed in example 8.2.9. By symmetry, there is no loss of generality to assume that $\mu_1(1) \leq \mu_2(1)$. In this case the r.v. X_1 and X_2 with distribution μ_1 and μ_2 can be coupled using a single uniform random variable U on $[0,1]$ with the following formula*

$$X_1 = 1_{[0,\mu_1(1)]}(U) \quad and \quad X_2 = 1_{[0,\mu_2(1)]}(U).$$

Using the fact that

$$1_{[0,\mu_2(1)]}(U) - 1_{[0,\mu_1(1)]}(U) = 1_{[\mu_1(1),\mu_2(1)]}(U)$$

we check that

$$\mathbb{W}(\mu_1,\mu_2) \leq \mathbb{P}\left(U \in [\mu_1(1),\mu_2(1)]\right) = \mu_2(1) - \mu_1(1) = \|\mu_1 - \mu_2\|_{tv}.$$

Example 8.3.12 *We return to the model discussed in example 8.3.7. We further assume that $S = S' \subset \mathbb{R}^d$ equipped with some norm $\|.\|$, and the function F is such that*

$$\int \|F(x_2,u) - F(x_1,u)\| \, du \leq a \, \|x_1 - x_2\|.$$

In this situation, using the fact that

$$\|X_1' - X_2'\| \leq \|F(X_2,U) - F(X_1,U)\|$$

we prove that

$$\mathbb{W}(\mu_1',\mu_2') \leq a \, \mathbb{W}(\mu_1,\mu_2).$$

We end this section with a conditioning property of the Wasserstein distance.

Proposition 8.3.13 *We consider a Markov transition M on some state space S. We assume that*

$$\forall (x,y) \in S^2 \quad \mathbb{W}(\delta_x M, \delta_y M) \leq w(x,y)$$

for some function w from S^2 into $\mathbb{R}_+$. In this situation, we have

$$\forall (\mu_1,\mu_2) \in \mathcal{P}(S)^2 \quad \mathbb{W}(\mu_1 M, \mu_2 M) \leq \int \mu_1(dx) \, \mu_2(dy) \, w(x,y).$$

Proof :
We let (X,Y) be a couple of r.v. with distributions (μ_1,μ_2). Given $(X,Y) = (x,y)$, we let (X',Y') be a couple of r.v. with distributions $(\delta_x M, \delta_y M)$. By construction, we have

$$
\begin{aligned}
|\mathbb{E}\left(f(X')\right) - \mathbb{E}\left(f(Y')\right)| &= |\mathbb{E}\left[\mathbb{E}\left(f(X') - f(Y') \mid (X,Y)\right)\right]| \\
&\leq \mathbb{E}\left[\mathbb{E}\left[|f(X') - f(Y')| \mid (X,Y)\right]\right] \\
&\leq \int \mathbb{P}((X,Y) \in d(x,y)) \, \mathbb{E}\left(d(X',Y') \mid (X,Y) = (x,y)\right).
\end{aligned}
$$

Taking the infimum of all possible couplings of (X',Y') given $(X,Y) = (x,y)$ we find

$$\mathbb{W}(\mu_1 M, \mu_2 M) \leq \int \mu_1(dx) \, \mu_2(dy) \, \mathbb{W}(\delta_x M, \delta_y M).$$

This ends the proof of the proposition. ∎

8.3.2   Stopping times and coupling

Definition 8.3.14 *Let X_n be a Markov chain taking values in some state space S. A stopping time T is a random variable taking values in $\mathbb{N} \cup \{\infty\}$ such that the events $\{T = n\}$ depend only on the $X_0, \ldots, X_n$.*

Informally, a random time is a stopping time if it does not depend on the future of the Markov chain. We often use the following immediate consequence of the Markov property without saying so.

Proposition 8.3.15 (Strong Markov property) *For any Markov chain taking values in some state space S, and for any stopping time T, we have*

$$\mathbb{P}\left(X_{n+1} \in dx_{n+1} \mid T = n,\ (X_0, \ldots, X_n) = (x_0, \ldots, x_n)\right) = \mathbb{P}\left(X_{n+1} \in dx_{n+1} \mid X_n = x_n\right).$$

In other words, the Markov chain $(X_{T+n})_{n \geq 0}$ is again a Markov chain but its initial condition is given by X_T.

For instance, the first time a Markov chain on a finite set S hits some subset A,

$$T = \inf\{n \geq 0\ :\ X_n \in A\}$$

is a stopping time since $\{T = n\} = \{X_0 \notin A,\ \ldots, X_{n-1} \notin A,\ X_n \in A\}$. Notice that T can also be interpreted as the first exit time of the set $B = S - A$. The first time a given chain reaches its maximum $X_n^\star := \max_{0 \leq p \leq n} X_p$ on some interval $[0, n]$ defined by

$$T' = \inf\{n \geq 0\ :\ X_p = X_n^\star\}$$

is not a stopping time since the event

$$\{T' = k\} = \cap_{p<k} \cup_{0 \leq q \leq n} \cap_{0 \leq r \leq n} \{X_k \geq X_r,\ X_p < X_q\}$$

depends on the whole sequence of states from the origin up to the terminal time n. Also notice that last exit or hitting times of the form

$$T'' = \max\{n \geq 0\ :\ X_n \in A\}$$

for some $A \subset S$ are *not* stopping times since the events $\{T'' = n\}$ also depend on the future of the chain

$$\{T'' = n\} = \cap_{p \geq 1}\{X_n \in A,\ X_{n+p} \notin A\}.$$

The following proposition is a direct consequence of the coupling lemma stated above.

Proposition 8.3.16 *We let X_n and Y_n be a couple of Markov chains such that $X_n = Y_n$ for any n after some stopping time T. In this case, we have*

$$\|\mathrm{Law}(X_n) - \mathrm{Law}(Y_n)\|_{tv} \leq \mathbb{P}\left(X_n \neq Y_n\right) \leq \mathbb{P}\left(T > n\right) \leq \mathbb{E}(T)/n.$$

In particular, we have

$$\mathbb{P}(T < \infty) = 1 \Longrightarrow \mathbb{P}\left(T > n\right) \downarrow_{n\uparrow\infty} 0 \Longrightarrow \|\mathrm{Law}(X_n) - \mathrm{Law}(Y_n)\|_{tv} \rightarrow_{n\uparrow\infty} 0.$$

In particular, if (X_n, Y_n) are two (non-necessarily independent) copies of the same Markov chain with Markov transition M and stationary distribution $\pi M = \pi$, we have

$$\text{Law}(Y_0) = \pi \implies \|\text{Law}(X_n) - \pi\|_{tv} \leq \mathbb{P}(T > n)$$

where T stands for the first time the chains merge (i.e., $T = \inf\{n \geq 0 \ : \ X_n = Y_n\}$). We can also use the inequality

$$\|\text{Law}(X_n) - \text{Law}(Y_n)\|_{tv}$$

$$\leq \int \mathbb{P}(X_0 \in dx)\mathbb{P}(Y_0 \in dy)\, \|\text{Law}(X_n \mid X_0 = x) - \text{Law}(Y_n \mid Y_0 = y))\|_{tv}$$

$$\leq \sup_{(x,y)\in S^2} \mathbb{P}(T_{x,y} > n)$$

where $T_{x,y}$ stands for the coupling time of X_n and Y_n starting at $X_0 = x$ and $Y_0 = y$.

8.3.3 Strong stationary times

This section is concerned with two main probabilistic techniques to quantify the convergence rate to equilibrium of a regular Markov chain, namely the coupling technique and the notion of strong stationary times introduced by D. Aldous and P. Diaconis in their seminal article [2].

Definition 8.3.17 *Let X_n be a Markov chain taking values in some state space S with an invariant measure π. A strong stationary time T (a.k.a strong uniform time) is a stopping time such that X_T and T are independent and $\text{Law}(X_T) = \pi$.*

Notice that the construction of strong stationary times can be seen as a perfect sampling technique of the invariant stationary measure π of a regular Markov chain.

We illustrate these strong stationary times with the top-in-at-random card shuffle introduced by D. Aldous and P. Diaconis in [2]. This is a rather silly way of shuffling cards but it is a preliminary toy model to analyze more complex shuffling techniques. In this context, shuffling cards processes are modelled by a Markov chain X_n taking values in the group $S = \mathcal{G}_{52}$ of all the permutations σ of the 52 cards. For a more thorough discussion on these shuffles, we refer the reader to section 26.4. Each transition of the chain consists of taking the first card and inserting it back at a random position. The invariant measure of this chain is the uniform distribution $\pi(\sigma) = 1/52!$ on $\mathcal{G}_{52}$. We check this claim by considering a uniform random deck with law π. After the top-in-at-random card shuffle the deck is still uniform.

We let τ be the first time the bottom card reaches the top of the deck, and $T = \tau + 1$. Clearly T is a stopping time. Furthermore, all the cards below the bottom cards at some given time are equally likely. In other words, we have $\text{Law}(X_T) = \pi$. An inductive proof of this claim is provided in exercise 101.

Theorem 8.3.18 *Let X_n be a Markov chain taking values in some state space S with an invariant measure π. For any strong uniform time T we have*

$$\|\text{Law}(X_n) - \pi\|_{tv} \leq \mathbb{P}(T > n).$$

Proof :
Using the Markov property, and recalling that $\{T = p\}$ depends on the r.v. $X_0, \ldots, X_p$ only, for any $0 \leq p \leq n$, we have

$$\mathbb{P}(X_n \in A \mid X_p = x_p, \ T = p) = \mathbb{P}(X_n \in A \mid X_p = x_p) = M^{n-p}(x_p, A)$$

and

$$\mathbb{E}\left(\mathbb{P}\left(X_n \in A \mid X_p\right) \mid T = p\right) \quad = \quad \mathbb{E}\left(M^{n-p}(1_A)(X_T) \mid T = p\right) = \pi\left(M^{n-p}(1_A)\right) = \pi(A).$$

This implies that

$$
\begin{aligned}
\mathbb{P}\left(X_n \in A \,,\, T \leq n\right) &= \sum_{0 \leq p \leq n} \mathbb{P}\left(X_n \in A \mid T = p\right) \mathbb{P}(T = p) \\
&= \sum_{0 \leq p \leq n} \mathbb{E}\left(\mathbb{P}\left(X_n \in A \mid X_p\right) \mid T = p\right) \mathbb{P}(T = p) \\
&= \pi(A)\, \mathbb{P}\left(T \leq n\right) = \pi(A)\, (1 - \mathbb{P}(T > n)).
\end{aligned}
$$

Hence we conclude that

$$
\begin{aligned}
\mathbb{P}\left(X_n \in A\right) &= \mathbb{P}\left(X_n \in A \,,\, T \leq n\right) + \mathbb{P}\left(X_n \in A \,,\, T > n\right) \\
&= \pi(A)\, (1 - \mathbb{P}(T > n)) + \mathbb{P}\left(X_n \in A \mid T > n\right)\, \mathbb{P}(T > n)
\end{aligned}
$$

and

$$\mathbb{P}\left(X_n \in A\right) - \pi(A) = \left(\mathbb{P}\left(X_n \in A \mid T > n\right) - \pi(A)\right)\, \mathbb{P}(T > n).$$

This ends the proof of the theorem. ∎

8.3.4 Some illustrations

8.3.4.1 Minorization condition and coupling

In this section we present a (more probabilistic in spirit) proof of the convergence result (8.18) presented in the contraction theorem 8.2.13. We let M be a Markov transition on some state space S satisfying the minorization condition (8.15). We let M_ϵ represent the Markov transition defined by

$$M_\epsilon(x, dy) = \frac{M(x, dy) - \epsilon \nu(dy)}{1 - \epsilon} \Leftrightarrow M(x, dy) = \epsilon\, \nu(dy) + (1 - \epsilon)M_\epsilon(x, dy).$$

We let $\boldsymbol{X_n} := (X_n, X'_n)$ be the Markov transition on $\boldsymbol{S} := S^2$ defined for any $\boldsymbol{x} = (x, x')$ by

$$\boldsymbol{M(x, dy)} = \epsilon \int \nu(dz)\delta_{(z,z)}(\boldsymbol{dy}) + (1 - \epsilon)\, M_\epsilon(x, dy)M_\epsilon(x', dy')$$

where $\boldsymbol{dy}$ stands for an infinitesimal neighborhood of the state $\boldsymbol{y} = (y, y')$.

By construction, we have

$$\boldsymbol{M(x}, A \times S) = \epsilon\, \nu(A) + (1 - \epsilon)\, M_\epsilon(x, A) = M(x, A)$$

and by symmetry $\boldsymbol{M(x}, S \times A) = M(y, A)$. This shows that each chain marginally follows the same Markov transition M.

Notice that at any time step $(n - 1) \rightsquigarrow n$, we flip a coin with heads probability ϵ. If the outcome is a head then we set $X_n = X'_n = U_n$ where U_n is a r.v. with distribution ν. Otherwise each of the chains follows its evolution independently according to the common Markov transition M_ϵ.

Therefore, the coupling time $T = \inf\{n \geq 0 \,:\, X_n = X'_n\}$ of the the couple of chains (X_n, X'_n) starting at any initial state (x, x') is such that

$$\mathbb{P}(T > n) \leq (1 - \epsilon)^n.$$

Now we only suppose that M^m satisfies the minorization condition (8.15) for some $m \geq 1$, and we set by

$$M_\epsilon^{(m)}(x_0, dx_m) = \frac{M^m(x_0, dx_m) - \epsilon\nu(dx_m)}{1 - \epsilon}.$$

Arguing as above, we can define a Markov chain $\boldsymbol{X_{km}} := (X_{km}, X'_{km})$ on $\boldsymbol{S} := S^2$ with transition

$$\boldsymbol{M^{(m)}}((x_0, x'_0), d(x_m, x'_m))$$

$$:= \epsilon \int \nu(dz)\delta_{(z,z)}(d(x_m, x'_m)) + (1 - \epsilon)\ M_\epsilon^{(m)}(x_0, dx_m)M_\epsilon^{(m)}(x'_0, dx'_m).$$

To design a coupling on the random trajectories the main difficulty here comes from the fact that $M_\epsilon^{(m)}$ is not the m-step transition of some Markov chain. A natural idea is to sample random bridges between the coupled random states (X_{km}, X'_{km}). To this end, we observe that

$$\begin{aligned} M(x_0, d(x_1, \ldots, x_m)) &:= M(x_0, dx_1)M(x_1, dx_2) \times \ldots \times M(x_{m-1}, dx_m) \\ &= M^m(x_0, dx_m) \times B((x_0, x_m), d(x_1, \ldots, x_{m-1})) \end{aligned}$$

with the conditional bridge distribution $B((x_0, x_m), d(x_1, \ldots, x_{m-1}))$ of the bridge path sequence $(X_1, \ldots, X_{m-1})$ given the initial and the terminal states $X_0 = x_0$ and $X_m = x_m$. With some abusive notation this bridge distribution is given by the formula

$$B((x_0, x_m), d(x_1, \ldots, x_{m-1})) = M(x_0, d(x_1, \ldots, x_m))/M^m(x_0, dx_m).$$

In this notation, the coupled bridge model between the random states (X_{km}, X'_{km}) defined above is defined by the Markov transition

$$\boldsymbol{B}((x_0, x'_0), d((x_1, \ldots, x_m), (x'_1, \ldots, x'_m)))$$

$$:= \boldsymbol{M^{(m)}}((x_0, x'_0), d(x_m, x'_m))\ B((x_0, x_m), d(x_1, \ldots, x_{m-1}))\ B((x'_0, x'_m), d(x'_1, \ldots, x'_{m-1})).$$

In this situation, we have

$$T := \inf\{n \geq 0 \,:\, X_n = X'_n\} \leq \inf\{n = km \geq 0 \,:\, X_{km} = X'_{km}\} := T_m$$

so that

$$\mathbb{P}(T > n) \leq \mathbb{P}(T_m > m\lfloor n/m \rfloor) \leq (1 - \epsilon)^{\lfloor n/m \rfloor} \leq (1 - \epsilon)^{-1}\ e^{-n\epsilon/m}. \qquad (8.38)$$

8.3.4.2 Markov chains on complete graphs

We consider a Markov chain with transition $M(x, y) = 1/d$ on a finite and complete graph with d vertices $S := \{1, \ldots, d\}$. This chain is reversible with the unique invariant uniform measure $\pi(x) = 1/d$, for any $x \in S$.

The case where $d = 5$ is described by the following polygon

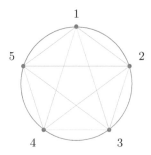

which is equivalent to the following transition diagram

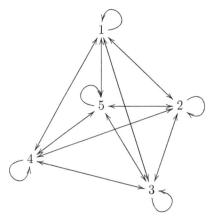

We let (X_n, Y_n) be a couple of independent Markov chains, up to the first time T they meet. After that time $n \geq T$, we set $X_n = Y_n$. The Markov transition of this chain is defined by

$$\mathbb{P}\left((X_n, Y_n) = (x', y') \mid (X_{n-1}, Y_{n-1}) = (x, y)\right)$$

$$= 1_{x=y} \, \mathbb{P}\left(X_n = x' \mid X_{n-1} = x\right) + 1_{x \neq y} \, \mathbb{P}\left(X_n = x' \mid X_{n-1} = x\right) \mathbb{P}\left(Y_n = y' \mid Y_{n-1} = y\right).$$

By construction, T is a geometric r.v. with success parameter $1/d$

$$\{T = n\} = \{X_1 \neq Y_1, \ \ldots, X_{n-1} \neq Y_{n-1}, \ X_n = Y_n\}$$

$$\Downarrow$$

$$\mathbb{P}(T = n) = \left[\prod_{1 \leq p < n} \mathbb{P}(X_p \neq Y_p \mid X_{p-1} \neq Y_{p-1})\right] \times \mathbb{P}(X_n = Y_n \mid X_{n-1} \neq Y_{n-1})$$

$$= \left(1 - \frac{1}{d}\right)^{n-1} \frac{1}{d}.$$

This implies that

$$\mathbb{P}\left(T > n\right) = \left(1 - \frac{1}{d}\right)^n \sum_{p \geq n} \left(1 - \frac{1}{d}\right)^{p-n} \frac{1}{d} = \left(1 - \frac{1}{d}\right)^n \leq e^{-n/d} \qquad (8.39)$$

from which we conclude that

$$\|\mathrm{Law}(X_n) - \pi\|_{tv} \leq e^{-n/d}.$$

8.3.4.3 A Kruskal random walk

One natural but simplified way to analyze the Kruskal magic trick discussed in section 1.4 is to consider the Markov chain X_n on the set of integers $S = \{1, \ldots, d\}$, for some $d \geq 1$, and defined by the elementary transition

$$M(i, j) = 1_{S - \{1\}} \, 1_{i-1}(j) + 1_1(i) \, \mu(j)$$

where μ stands for some distribution on S. The initial condition X_0 is a r.v. with distribution μ.

The set S represents the set of values of the cards. In the card counting model discussed in section 1.4 we have $d = 10$. The initial value X_0 represents the value of the first secret card, say $X_0 = 5$ if the card is a jack. Starting from this value, the chain counts backward $X_1 = 4$, $X_2 = 3$, $X_3 = 2$ up to $X_4 = 1$. Then X_5 represents the value of the second magic card, say $X_5 = 4$. Starting from this value, the chain counts backwards $X_6 = 3$, $X_7 = 2$ up to $X_8 = 1$. Then X_9 stands for the third magic card, and so on.

Any realization of the chain $(X_k)_{0 \leq k \leq nd}$ on the time interval $[0, nd]$ hits the state 1 at least n times, say $T_1, \ldots, T_n (< nd)$. We check this claim using the fact that the minimal number of hits is associated with the realization of the chain given by

$$X_0 = d \rightsquigarrow X_{1 \times d - 1} = 1 \to X_d = d \rightsquigarrow \ldots \rightsquigarrow X_{n \times d - 1} = 1 \to X_{nd}.$$

We consider a pair $(X_n^1, X_n^2)_{n \geq 0}$ of independent copies of $(X_n)_{n \geq 0}$, and we let T be the first time n these chains meet; that is

$$T = \inf\left\{n \geq 0 \; : \; X_n^1 = X_n^2\right\}.$$

After the time T, the chains are coupled; that is, we set $X_n^1 = X_n^2$ for any $n \geq T$.

We further assume that $X_0^1 = d_0 + 1$ for some d_0 and we let $T_1, \ldots, T_n$ be the first n times the chain X_k^1, hits the state 1. By construction, we have

$$T_1 = X_0^1 \quad \text{and} \quad \forall k \geq 0 \quad T_{k+1} - T_k = X_{T_k + 1}^1.$$

Thus, $\left(X_{T_k + 1}^1\right)_{1 \leq k \leq n}$ are independent r.v. with distribution μ, and these r.v. are independent of $X^2 := \left(X_k^2\right)_{k \geq 0}$.

Furthermore, it is easily checked that

$$T > nd \implies \forall 1 \leq k \leq n \quad X_{T_k + 1}^1 \neq X_{T_k + 1}^2.$$

This implies that

$$\begin{aligned}
\mathbb{P}\left(T > nd\right) &\leq \mathbb{E}\left(\mathbb{P}\left(\forall 1 \leq k \leq n \quad X_{T_k + 1}^1 \neq X_{T_k + 1}^2 \mid X^2\right)\right) \\
&= \mathbb{E}\left(\prod_{1 \leq k \leq n} \left(1 - \mathbb{P}\left(X_{T_k + 1}^1 = X_{T_k + 1}^2 \mid X^2\right)\right)\right)
\end{aligned}$$

from which we conclude that

$$\mathbb{P}\left(T > nd\right) \leq \left(1 - \mu_\star\right)^n \quad \text{with} \quad \mu_\star := \inf_{1 \leq i \leq d} \mu(i).$$

When μ is the uniform distribution on S, we find that

$$\forall m \geq d \qquad \mathbb{P}\left(T > m\right) \leq \left(1 - \frac{1}{d}\right)^{m/d}.$$

The graph of these rather crude estimates when $d = 10$ is shown below.

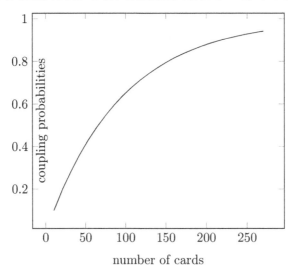

8.4   Martingales

8.4.1   Some preliminaries

Definition 8.4.1 *For any $\mathbb{R}$-valued stochastic process $Y = (Y_n)_{n \geq 0}$, we denote by $\Delta Y_n = Y_n - Y_{n-1}$ its increments, so that*

$$\forall 0 \leq p \leq q \qquad Y_n = Y_p + \sum_{p < q \leq n} \Delta Y_q.$$

To fix ideas, we consider the gambling ruin model

$$Y_n = Y_0 + X_1 + \ldots + X_n \Rightarrow \Delta Y_n = X_n \tag{8.40}$$

where X_n are independent $\{-1, 0, 1\}$-valued r.v. with common distribution

$$\mathbb{P}\left(X_n = -1\right) = p \quad \mathbb{P}\left(X_n = 0\right) = q \quad \mathbb{P}\left(X_n = 1\right) = r$$

where $p, q, r \in [0, 1]$ are s.t. $p + q + r = 1$. The initial random variable Y_0 represents the initial fortune of the gambler. When $Y_0 = 0$, the process Y_n can also be seen as the cumulative sum of the outcomes of the game.

Definition 8.4.2 *The filtration* $\mathcal{F} = (\mathcal{F}_n)_{n \geq 0}$ *generated by some Markov chain* X_n *is the increasing sequence of of* σ-*algebras*

$$\mathcal{F}_n = \sigma(X_0, X_1, \ldots, X_n) \subset \mathcal{F}_{n+1}$$

generated by the random state variables from the origin up to the different times horizon. A stochastic process $Y = (Y_n)_{n \geq 0}$ *on some possibly different state space* S' *is said to be adapted to the filtration* $\mathcal{F}$ *if there exist some functions* h_n *from* S^{n+1} *into* S' *s.t.*

$$Y_n = h_n(X_0, \ldots, X_n).$$

In this situation, we use the synthetic notation $Y_n \subseteq \mathcal{F}_n$.

Remark : Suppose a process Y_n is not adapted to filtration $\mathcal{F} = (\mathcal{F}_n)_{n \geq 0}$ generated by some Markov chain X. We define by $\widehat{Y}_n$ the sequence of random variables

$$\widehat{Y}_n = \mathbb{E}(Y_n \mid \mathcal{F}_n) = \mathbb{E}(Y_n \mid X_0, \ldots, X_n).$$

Recalling that $\mathbb{E}(Y_n \mid X_0, \ldots, X_n) = h_n(X_0, \ldots, X_n)$ for some deterministic function h_n, we check that $\widehat{Y}_n \subseteq \mathcal{F}_n$. This process is called the projection of Y on the filtration $\mathcal{F}$.

In the gambling ruin model (8.40), $\mathcal{F}_n$ represents all the information we have collected with observing the outcomes of the game X_p, from the origin up to the current time n. In this situation, the cumulative gains $Y_n - Y_0$ of the gambler are adapted to the filtration $\mathcal{F}_n$; that is,

$$Y_n - Y_0 = h_n(X_1, \ldots, X_n) = X_1 + \ldots + X_n \subseteq \mathcal{F}_n.$$

In the further development of this chapter we always assume that $\mathcal{F}_n$ is the filtration associated with some Markov chain $X = (X_n)_{n \geq 0}$ evolving on some state space S.

Definition 8.4.3 *We say that a* $\mathbb{R}$-*valued stochastic process* $M = (M_n)_{n \geq 0}$ *is a martingale w.r.t. the filtration* $\mathcal{F} = (\mathcal{F}_n)_{n \geq 0}$, *or an* $\mathcal{F}$-*martingale, if we have*

$$M_n \subseteq \mathcal{F}_n \quad and \quad \mathbb{E}(\Delta M_n \mid \mathcal{F}_{n-1}) = 0.$$

The process M *is also called a super-martingale, and respectively a sub-martingale (w.r.t. the filtration* $\mathcal{F}$), *if we can replace the r.h.s. in the above display by*

$$\mathbb{E}(\Delta M_n \mid \mathcal{F}_{n-1}) \leq 0, \quad and \ resp. \quad \mathbb{E}(\Delta M_n \mid \mathcal{F}_{n-1}) \geq 0.$$

Important remark : For martingales, we have

$$\mathbb{E}(M_n) = \mathbb{E}(M_{n-1}) + \mathbb{E}(\Delta M_n) = \mathbb{E}(M_{n-1}) = \ldots = \mathbb{E}(M_0).$$

In addition, for any $p < n$, we have

$$\mathcal{F}_p \subset \mathcal{F}_{n-1} \Longrightarrow \mathbb{E}(M_n \mid \mathcal{F}_p) = \mathbb{E}\left(\overbrace{\mathbb{E}(M_n \mid \mathcal{F}_{n-1})}^{=M_{n-1}} \mid \mathcal{F}_p \right)$$

$$= \mathbb{E}(M_{n-1} \mid \mathcal{F}_p).$$

Iterating the argument, we easily prove that

$$\mathbb{E}\left(M_n \mid \mathcal{F}_p\right) = \mathbb{E}\left(M_{n-1} \mid \mathcal{F}_p\right) = \ldots = \mathbb{E}\left(M_{p+1} \mid \mathcal{F}_p\right) = M_p.$$

For super-martingales, we have

$$\mathbb{E}(M_n) = \mathbb{E}(M_{n-1}) + \mathbb{E}(\Delta M_n) \leq \mathbb{E}(M_{n-1}) \leq \ldots \leq \mathbb{E}(M_0)$$

and $\mathbb{E}\left(M_n \mid \mathcal{F}_p\right) \leq M_p$, for any $p \leq n$. Finally, for sub-martingales

$$\mathbb{E}(M_n) = \mathbb{E}(M_{n-1}) + \mathbb{E}(\Delta M_n) \geq \mathbb{E}(M_{n-1}) \geq \ldots \geq \mathbb{E}(M_0)$$

and $\mathbb{E}\left(M_n \mid \mathcal{F}_p\right) \geq M_p$, for any $p \leq n$.

In the gambling ruin model (8.40), the process Y_n is clearly $\mathcal{F}$-adapted

$$Y_n = Y_0 + X_1 + \ldots + X_n \Rightarrow \Delta Y_n = X_n$$

and we have

$$\mathbb{E}\left(\Delta Y_n \mid \mathcal{F}_{n-1}\right) = \mathbb{E}\left(X_n \mid X_1, \ldots, X_n\right) = \mathbb{E}(X_n) = r - p. \tag{8.41}$$

This shows that Y_n is a martingale, resp. super-martingale, resp. sub-martingale, when $r = p$ (fair game), resp $r \leq p$ (unfair), resp. $r \geq p$ (superfair).

Lemma 8.4.4 *For any $\mathcal{F}$-adapted real process Y, the process M defined by*

$$M_n = Y_0 + \sum_{0 < p \leq n}\left(\Delta Y_p - \mathbb{E}\left(\Delta Y_p \mid \mathcal{F}_{p-1}\right)\right) \tag{8.42}$$

is a $\mathcal{F}$-martingale starting at $M_0 = Y_0$. In addition, for any $\mathcal{F}$-martingale M, and for any $\mathcal{F}$-adapted process H

$$(H \bullet M)_n := \sum_{0 < p \leq n} H_{k-1}\,\Delta M_k \tag{8.43}$$

is a $\mathcal{F}$-martingale starting at the origin $(H \bullet M)_0 = 0$.

Proof :
We simply check that

$$\Delta M_n = \Delta Y_n - \mathbb{E}\left(\Delta Y_n \mid \mathcal{F}_{n-1}\right) \Rightarrow \mathbb{E}\left(\Delta M_n \mid \mathcal{F}_{n-1}\right) = 0$$

and

$$\Delta(H \bullet M)_n = H_{n-1}\,\Delta M_n$$

$$\Rightarrow \mathbb{E}\left(\Delta(H \bullet M)_n \mid \mathcal{F}_{n-1}\right) = H_{n-1} \times \mathbb{E}\left(\Delta M_n \mid \mathcal{F}_{n-1}\right) = 0.$$

This ends the proof of the lemma.   ∎

In the gambling ruin model (8.40), using (8.41) we check that the martingale (8.42) is given by

$$M_n = Y_0 + \sum_{0 < k \leq n}\left(\Delta Y_k - (r - p)\right) = Y_n - n(r - p). \tag{8.44}$$

In the fair game case $r = p$, and $Y_0 = 0$, this martingale is simply given by

$$M_n = Y_n = X_1 + \ldots + X_n \Rightarrow \Delta M_n = X_n.$$

Notice that any bet size H_{n-1} of the gambler at the n-th coup is a deterministic function h_{n-1} of the outcomes X_p of the game up to time $p < n$; that is,

$$H_{n-1} = h_{n-1}(X_1, \ldots, X_{n-1}).$$

In the fair game case $p = r$, the evolution of the fortune of the gambler using this betting strategy is given by the martingale

$$(H \bullet Y)_n := \sum_{0 < p \leq n} H_{k-1} \, \Delta Y_k = \sum_{0 < p \leq n} H_{k-1} \, X_k. \tag{8.45}$$

We quote a technical lemma of independent interest that allows us to quantify the fluctuations of a martingale around the origin.

Lemma 8.4.5 *For any martingale M_n w.r.t. the filtration $\mathcal{F}$, and null at the origin the stochastic processes*

$$\widetilde{M}_n := M_n^2 - [M]_n \quad \text{and} \quad \overline{M}_n := M_n^2 - \langle M \rangle_n$$

with

$$[M]_n := \sum_{0 < k \leq n} (\Delta M_k)^2 \quad \text{and} \quad \langle M \rangle_n := \sum_{0 < k \leq n} \mathbb{E}\left((\Delta M_k)^2 \mid \mathcal{F}_{k-1}\right)$$

are martingales (null at the origin) w.r.t. the same filtration.
The processes $[M]_n$ and $\langle M \rangle_n$ are called the quadratic variation, and the predictable quadratic variation of M_n. The process $\langle M \rangle_n$ is often called the angle bracket of M_n.

Proof :
For a given a martingale M_n, with increments $\Delta M_n = M_n - M_{n-1}$, we have

$$\begin{aligned}
\Delta(M^2)_n &= M_n^2 - M_{n-1}^2 = (M_{n-1} + \Delta M_n)^2 - M_{n-1}^2 \\
&= (\Delta M_n)^2 - 2M_{n-1}\,\Delta M_n \\
&= (\Delta M_n)^2 + \Delta\widetilde{M}_n = \mathbb{E}\left((\Delta M_n)^2 \mid \mathcal{F}_{n-1}\right) + \Delta\overline{M}_n
\end{aligned}$$

with the conditionally centered random variables

$$\Delta\widetilde{M}_n = -2M_{n-1}\Delta M_n$$

and

$$\Delta\overline{M}_n := \left[(\Delta M_n)^2 - \mathbb{E}\left((\Delta M_n)^2 \mid \mathcal{F}_{n-1}\right)\right] - 2M_{n-1}\Delta M_n.$$

This shows that

$$M_n^2 - M_0^2 \quad := \quad \sum_{0 < k \leq n} \mathbb{E}\left((\Delta M_k)^2 \mid \mathcal{F}_{k-1}\right) + \overline{M}_n = \sum_{0 < k \leq n} (\Delta M_k)^2 + \widetilde{M}_n$$

with the martingales

$$\overline{M}_n := \sum_{0 < k \leq n} \Delta\overline{M}_k \quad \text{and} \quad \widetilde{M}_n := \sum_{0 < k \leq n} \Delta\widetilde{M}_k.$$

This ends the proof of the lemma.   ■

We next extend lemma 8.4.5 to products of martingales.

Lemma 8.4.6 *For any couple of martingales $M_n^{(1)}$ and $M_n^{(2)}$ w.r.t. the filtration $\mathcal{F}$, and null at the origin the stochastic processes*

$$\widetilde{M}_n := M_n^{(1)} M_n^{(2)} - \left[M^{(1)}, M^{(2)} \right]_n \quad \text{and} \quad \overline{M}_n := M_n^{(1)} M_n^{(2)} - \langle M^{(1)}, M^{(2)} \rangle_n$$

are martingales (null at the origin) w.r.t. the same filtration, with the quadratic covariation, and the predictable quadratic covariation of $M_n^{(1)}$ and $M_n^{(2)}$ defined by

$$\left[M^{(1)}, M^{(2)} \right]_n := \sum_{0 < k \le n} \Delta M_k^{(1)} \, \Delta M_k^{(2)}$$

and

$$\langle M^{(1)}, M^{(2)} \rangle_n := \sum_{0 < k \le n} \mathbb{E} \left(\Delta M_k^{(1)} \, \Delta M_k^{(2)} \mid \mathcal{F}_{k-1} \right).$$

When $M^{(1)} = M^{(2)} = M$ we often write $[M]$ and $\langle M \rangle$ instead of $[M, M]$ and $\langle M, M \rangle$.

Proof :
We have the increment decompositions

$$
\begin{aligned}
\Delta(M^{(1)} M^{(2)})_n \; &:= \; M_n^{(1)} M_n^{(2)} - M_{n-1}^{(1)} M_{n-1}^{(2)} \\
&= \; \left(M_{n-1}^{(1)} + \Delta M_n^{(1)} \right) \left(M_{n-1}^{(2)} + \Delta M_n^{(2)} \right) - M_{n-1}^{(1)} \, M_{n-1}^{(2)} \\
&= \; \Delta M_n^{(1)} \Delta M_n^{(2)} - M_{n-1}^{(1)} \, \Delta M_n^{(2)} - M_{n-1}^{(2)} \, \Delta M_n^{(1)} \\
&= \; \underbrace{\Delta M_n^{(1)} \Delta M_n^{(2)}}_{=\Delta\left[M^{(1)}, M^{(2)} \right]_n} + \Delta \widetilde{M}_n = \underbrace{\mathbb{E} \left(\Delta M_n^{(1)} \Delta M_n^{(2)} \mid \mathcal{F}_{n-1} \right)}_{=\Delta \langle M^{(1)}, M^{(2)} \rangle_n} + \Delta \overline{M}_n
\end{aligned}
$$

with the conditionally centered random variables

$$\Delta \widetilde{M}_n = -M_{n-1}^{(1)} \, \Delta M_n^{(2)} - M_{n-1}^{(2)} \, \Delta M_n^{(1)}$$

and

$$\Delta \overline{M}_n := \left[\Delta M_n^{(1)} \Delta M_n^{(2)} - \mathbb{E} \left(\Delta M_n^{(1)} \Delta M_n^{(2)} \mid \mathcal{F}_{n-1} \right) \right] - M_{n-1}^{(1)} \, \Delta M_n^{(2)} - M_{n-1}^{(2)} \, \Delta M_n^{(1)}.$$

This shows that

$$
\begin{aligned}
M_n^{(1)} M_n^{(2)} - M_0^{(1)} M_0^{(2)} \; &:= \; \sum_{0 < k \le n} \Delta(M^{(1)} M^{(2)})_k \\
&= \; \sum_{0 < k \le n} \mathbb{E} \left(\Delta M_k^{(1)} \Delta M_k^{(2)} \mid \mathcal{F}_{k-1} \right) + \overline{M}_n \\
&= \; \sum_{0 < k \le n} \Delta M_k^{(1)} \Delta M_k^{(2)} + \widetilde{M}_n
\end{aligned}
$$

with the martingales

$$\overline{M}_n := \sum_{0<k\leq n} \Delta\overline{M}_k \quad \text{and} \quad \widetilde{M}_n := \sum_{0<k\leq n} \Delta\widetilde{M}_k.$$

This ends the proof of the lemma.

∎

In the gambling ruin model (8.40), the increments of the martingale (8.44) are given by

$$\Delta M_n = (\Delta Y_n - (r-p)) = (X_n - \rho) \quad \text{with} \quad \rho := \mathbb{E}(X_n) = (r-p).$$

This yields

$$\begin{aligned}\Delta\langle M\rangle_n &:= \langle M\rangle_n - \langle M\rangle_{n-1} = \mathbb{E}\left((\Delta M_n)^2 \mid \mathcal{F}_{n-1}\right)\\ &= \mathrm{Var}(X_n) = (r+p) - (r-p)^2 = (2p+\rho) - \rho^2,\end{aligned}$$

from which we conclude that

$$\overline{M}_n := M_n^2 - \langle M\rangle_n = [Y_n - n\rho]^2 - \left[(2p+\rho) - \rho^2\right]n. \tag{8.46}$$

Notice that

$$\begin{aligned}q = 0 \quad &\Rightarrow \quad r = 1-p \Rightarrow \rho = r - p = 1 - 2p\\ &\Rightarrow \quad \left[(2p+\rho) - \rho^2\right] = 1 - (1-2p)^2 = 4p(1-p) = 4pr.\end{aligned}$$

One direct consequence of the martingale property is an explicit variance formula for the fortune of the player

$$\mathbb{E}\left(\left[\frac{Y_n}{n} - \rho\right]^2\right) = \frac{1}{n}\left[(2p+\rho) - \rho^2\right] \overset{if\ \underline{q=0}}{=} \frac{4pr}{n}.$$

8.4.2 Applications to Markov chains

8.4.2.1 Martingales with fixed terminal values

Let X_n be a Markov chain on a state space S with Markov transitions M_n. We let $\mathcal{F}_n$ be the filtration generated by the Markov chain X_n, and we denote by $M_{p,n}$, $p \leq n$, the semigroup of the Markov chain defined for any function f by the formulae

$$M_{p,n}(f)(x_p) = \mathbb{E}\left(f(X_n) \mid X_p = x_p\right).$$

Theorem 8.4.7 *Given some terminal time horizon n, and some bounded function f_n, the process*

$$0 \leq p \leq n \mapsto \mathcal{M}_p^{(n)} := M_{p,n}(f_n)(X_p)$$

is the <u>unique</u> martingale w.r.t. the filtration $\mathcal{F}_p$, with $p \leq n$, with terminal value $\mathcal{M}_n^{(n)} = f_n(X_n)$.

Proof :
The proof of the last assertion is obvious. The martingale property follows from the semi-group definition. Indeed, for any $p < n$ we have

$$
\mathbb{E}\left(\mathcal{M}_{p+1}^{(n)} \mid \mathcal{F}_p\right) = \mathbb{E}\left(M_{p+1,n}(f_n)(X_{p+1}) \mid X_p\right)
$$
$$
= M_p\left(M_{p+1,n}(f)\right)(X_p) = \left(M_p M_{p+1,n}\right)(f)(X_p) = M_{p,n}(f)(X_p) = \mathcal{M}_p^{(n)}.
$$

The uniqueness follows from the fact that the values of the martingale are pre-defined by the terminal value; that is, we have that

$$
\mathbb{E}\left(f_n(X_n) \mid \mathcal{F}_p\right) = \mathbb{E}\left(\mathcal{M}_n^{(n)} \mid \mathcal{F}_p\right) = \mathcal{M}_p^{(n)}.
$$

This ends the proof of the theorem.   ∎

Important remark : The above construction of martingales with a terminal end point is of current use in mathematical finance, and more precisely in pricing European options. In this situation, the terminal time condition is called the payoff function, and the martingale represents the evolution of a given portfolio that starts at the "real" price of an option and ends at the terminal time to the payoff function associated with the option contract. These covering portfolios are also called replicating portfolios to emphasize that they "mimic" the evolution of a given risky stock price to cover its random terminal values. We refer the reader to section 30.2 for a more thorough discussion on these application domains.

8.4.2.2   Doeblin-Itō formula

For any function f on S we have

$$
f(X_n) := f(X_0) + \sum_{0 < p \leq n} \Delta f(X_p) \quad \text{with} \quad \Delta f(X_p) = f(X_p) - f(X_{p-1}).
$$

We denote by $\mathcal{M}_n(f)$ the stochastic the process given by

$$
\mathcal{M}_n(f) := \sum_{0 < p \leq n} \left[\Delta f(X_p) - \mathbb{E}\left(\Delta f(X_p) \mid \mathcal{F}_{p-1}\right)\right]
$$
$$
= \sum_{0 < p \leq n} \left[f(X_p) - \mathbb{E}\left(f(X_p) \mid \mathcal{F}_{p-1}\right)\right].
$$

It should be clear that this process is a martingale, and the increments are given by

$$
\Delta \mathcal{M}_n(f) := \mathcal{M}_n(f) - \mathcal{M}_{n-1}(f) = f(X_n) - M_n(f)(X_{n-1}).
$$

In addition, we have

$$
\mathbb{E}\left((\Delta \mathcal{M}_n(f))^2 \mid \mathcal{F}_{n-1}\right) = \mathbb{E}\left(([f(X_n) - \mathbb{E}(f(X_n) \mid \mathcal{F}_{n-1})])^2 \mid \mathcal{F}_{n-1}\right)
$$
$$
= M_n(f^2)(X_{n-1}) - (M_n(f)(X_{n-1}))^2 \left(\leq \operatorname{osc}(f)^2\right).
$$

This shows that the predictable angle bracket of the martingale $\mathcal{M}_n(f)$ is given by

$$
\langle \mathcal{M}(f) \rangle_n := \sum_{0 < p \leq n} \left[M_p(f^2)(X_{p-1}) - (M_p(f)(X_{p-1}))^2\right].
$$

Definition 8.4.8 *The decomposition*

$$
\begin{aligned}
f(X_n) \quad &:= \quad f(X_0) + \sum_{0 < p \le n} \mathbb{E}\left(\Delta f(X_p)\middle|\ \mathcal{F}_{p-1}\right) + \mathcal{M}_n(f) \\
&= \quad f(X_0) + \sum_{0 < p \le n} \left(M_p(f) - f\right)(X_{p-1}) + \mathcal{M}_n(f) \quad (8.47)
\end{aligned}
$$

is called the martingale decomposition of the stochastic process $f(X_n)$.

Definition 8.4.9 *The generator* L_n *of the Markov chain* X_n *with Markov transition and the carré du champ operator* Γ_{L_n} *are defined by*

$$
L_n := M_n - Id \ : \ f \mapsto L_n(f) := M_n(f) - f
$$

and

$$
\Gamma_{L_n} \ : \ (f, g) \mapsto \Gamma_{L_n}(f, g) := L_n(fg) - fL_n(g) - gL_n(f).
$$

Using the fact that $M_n = L_n + Id$ we readily check that

$$
M_n(f^2) - (M_n(f))^2 \ = \ L_n(f^2) + f^2 - (L_n(f) + f)^2 = \Gamma_{L_n}(f, f) - (L_n(f))^2.
$$

The martingale decomposition and the predictable angle bracket of the martingale $\mathcal{M}_n(f)$ computed above are expressed in a natural way in terms of these operators. From previous considerations we readily prove the following theorem.

Theorem 8.4.10 *The Doeblin-Itō formula associated with the Markov chain* X_n *is given by the formula*

$$
f(X_n) \ = \ f(X_0) + \sum_{0 \le p \le n} L_p(f)(X_{p-1}) + \mathcal{M}_n(f) \quad (8.48)
$$

with a martingale $\mathcal{M}_n(f)$ *with predictable angle bracket*

$$
\langle \mathcal{M}(f) \rangle_n := \sum_{0 < p \le n} \Gamma_{L_p}(f, f)(X_{p-1}) - \sum_{0 < p \le n} \left(L_p(f)(X_{p-1})\right)^2.
$$

8.4.2.3 Occupation measures

In the further development of this section, we consider a time homogeneous Markov chain X_n with Markov transition M on some state space S. The next technical lemma provides a martingale decomposition of the occupation measure of the chain

$$
\pi^n := \frac{1}{n+1} \sum_{0 \le p \le n} \delta_{X_p} \quad (8.49)
$$

in terms of the solution of the Poisson equation presented in section 8.2.4.

Lemma 8.4.11 *We let X_n be a Markov chain on some state space S satisfying the regularity properties discussed in section 8.2.4. We denote by π the unique invariant measure of the chain. For any bounded function f on S we have*

$$\frac{1}{n+1}\left(g(X_{n+1}) - g(X_0)\right) = [\pi(f) - \pi^n(f)] + \frac{1}{n+1}\,\mathcal{M}_{n+1}(g) \tag{8.50}$$

with the (bounded) solution of the Poisson equation (8.19) given by

$$g = \sum_{n\geq 0}\left(M^n(f) - \pi(f)\right). \tag{8.51}$$

Proof :
Recalling that

$$g = \sum_{n\geq 0}\left(M^n(f) - \pi(f)\right) \Rightarrow [Id - M](g) = f - \pi(f),$$

we prove that for any $p \geq 0$, we have

$$
\begin{aligned}
f(X_p) - \pi(f) &= [Id - M](g)(X_p) = g(X_p) - M(g)(X_p) = g(X_p) - \mathbb{E}\left(g(X_{p+1}) \mid X_p\right)\\
&= -\underbrace{\left(g(X_{p+1}) - g(X_p)\right)}_{\Delta g(X_{p+1})} + \underbrace{\left(g(X_{p+1}) - \mathbb{E}\left(g(X_{p+1}) \mid X_p\right)\right)}_{=\Delta \mathcal{M}_{p+1}(g)}.
\end{aligned}
$$

Taking the sum of the index p, we find the desired decomposition. This ends the proof of the lemma. ∎

As a direct consequence of the lemma, we have

$$|\mathbb{E}\left[\pi^n(f)\right] - \pi(f)| = \frac{1}{n+1}\,|\mathbb{E}\left(g(X_{n+1}) - g(X_0)\right)| \leq \mathrm{osc}(g)/(n+1) \longrightarrow_{n\to\infty} 0.$$

We also notice that

$$
\begin{aligned}
\sqrt{n+1}\,[\pi(f) - \pi^n(f)] &= -\frac{\mathcal{M}_{n+1}(g)}{\sqrt{n+1}} - \left(g(X_{n+1}) - g(X_0)\right)/\sqrt{n+1}\\
&= -\frac{\mathcal{M}_{n+1}(g)}{\sqrt{n+1}} + \mathrm{O}(1/\sqrt{n}).
\end{aligned}
$$

This implies that

$$(n+1)\,\mathbb{E}\left(\left[\pi(f) - \pi^n(f)\right]^2\right) = \frac{1}{n+1}\,\underbrace{\mathbb{E}\left(\langle\mathcal{M}(g)\rangle_{n+1}\right)}_{\leq (n+1)\,\mathrm{osc}(g)^2} + o(1/n)$$

from which we conclude that

$$\mathbb{E}\left(\left[\pi(f) - \pi^n(f)\right]^2\right) = \mathrm{O}(1/n). \tag{8.52}$$

In addition, using the fact that

$$n^{-1}\,\mathbb{E}\left(\langle\mathcal{M}(g)\rangle_n\right)$$

$$
\begin{aligned}
&= \mathbb{E}\left(n^{-1}\sum_{0<p\leq n}\left[M(g^2) - M(g)^2\right](X_{p-1})\right)\\
&= \mathbb{E}\left(\pi^{n-1}(M(g^2) - M(g)^2)\right) \longrightarrow_{n\to\infty} \pi\left(M(g^2) - M(g)^2\right)
\end{aligned}
$$

and recalling that $\pi M = \pi$, we also have the asymptotic result

$$(n+1)\,\mathbb{E}\left(\left[[\pi(f) - \pi^n(f)]^2\right]\right) \longrightarrow_{n\to\infty} \sigma^2(f) := \pi\left(g^2\right) - \pi\left(M(g)^2\right).$$

Since g satisfies the Poisson equation we have

$$g = M(g) + (f - \pi(f)) \Rightarrow \pi(g^2) = \pi(M(g)^2) + 2\,\pi((f - \pi(f))M(g)) + \pi((f - \pi(f))^2).$$

This implies that

$$\begin{aligned}
\sigma^2(f) &= 2\,\pi((f - \pi(f))M(g)) + \pi((f - \pi(f))^2) \\
&= 2\sum_{p\geq 1}\pi([f - \pi(f)]M^p[f - \pi(f)]) + \pi([f - \pi(f)]^2).
\end{aligned}$$

More precise estimates are provided in section 9.1 which is dedicated to a simple proof of a weak ergodic theorem without using (implicitly) the Poisson equation and sophisticated martingale tools.

8.4.3 Optional stopping theorems

In practice a player chooses to stop gambling at some possibly large but random time. Unfortunately, the theorem states that the evolution of a fortune in a fair game up to any stopping time remains a martingale null at the origin.

In the further development of this section c, c_1, c_2 stands for some finite constants.

Theorem 8.4.12 (Optional stopping theorem 1) *We let M be a martingale, and T be a stopping time w.r.t. the filtration $\mathcal{F}$ generated by some Markov chain. The stopped process*

$$M_{n\wedge T} := M_n\,1_{[n,\infty[}(T) + M_T\,1_{[0,n[}(T) = M_n\,1_{]n,\infty[}(T) + M_T\,1_{[0,n]}(T) \quad (8.53)$$

is a martingale w.r.t. $\mathcal{F}$. If M_n is a super-martingale then $M_{n\wedge T}$ is also a super-martingale.

Proof :
We recall that a stopping time is such that $\{T = n\} \in \mathcal{F}_n$, for any $n \geq 0$. Since $\mathcal{F}_n \subset \mathcal{F}_{n+1}$, we have

$$\{T \leq n\} = \cup_{m\leq n}\{T = m\} \in \mathcal{F}_n \quad \text{and} \quad \{T > n\} = \Omega - \{T \leq n\} \in \mathcal{F}_n$$

and

$$M_T\,1_{T\leq n} = \sum_{0\leq m\leq n} M_m\,1_{T=m\in\mathcal{F}_n}.$$

We check this claim by combining the l.h.s. expression in (8.53) with the following conditioning formulas

$$\begin{aligned}
\mathbb{E}\left(M_{(n+1)\wedge T}\mid\mathcal{F}_n\right) &= \mathbb{E}\left(M_{n+1}\,1_{T\geq(n+1)} + M_T\,1_{T<(n+1)}\mid\mathcal{F}_n\right) \\
&= \mathbb{E}\left(M_{n+1}\,1_{T>n} + M_T\,1_{T\leq n}\mid\mathcal{F}_n\right) \\
&= \mathbb{E}\left(M_{n+1}\mid\mathcal{F}_n\right)\,1_{T>n} + M_T\,1_{T\leq n}.
\end{aligned}$$

By the martingale property, for the r.h.s. expression in (8.53) we conclude that

$$\mathbb{E}\left(M_{(n+1)\wedge T} \mid \mathcal{F}_n\right) \;=\; M_n \, 1_{T>n} + M_T \, 1_{T\leq n} = M_{n\wedge T}.$$

When M_n is a super-martingale we have $\mathbb{E}\left(M_{n+1} \mid \mathcal{F}_n\right) \leq M_n$ so that

$$\mathbb{E}\left(M_{(n+1)\wedge T} \mid \mathcal{F}_n\right) \leq M_n \, 1_{T>n} + M_T \, 1_{T\leq n} = M_{n\wedge T}.$$

This shows that $M_{n\wedge T}$ is also a super-martingale. This ends the proof of the theorem. ∎

Lemma 8.4.13 *Let M_n be a martingale w.r.t. some filtration $\mathcal{F}_n$. For any stopping time T, any $0 \leq n \leq m$, and any $A_n \subset \mathcal{F}_n$ we have*

$$\mathbb{E}(M_n \, 1_{A_n} \, 1_{T\geq n}) = \mathbb{E}\left(M_T \, 1_{A_n} \, 1_{n\leq T\leq m}\right) + \mathbb{E}\left(M_m \, 1_{A_n} \, 1_{T>m}\right).$$

Proof :
We fix $n \geq 0$ and we use induction w.r.t. the parameter $m \geq n$. For $m = n$, the result follows from the fact that

$$M_n \, 1_{T\geq n} = M_n \, 1_{T=n} + M_n \, 1_{T>n} = M_T \, 1_{T=n} + M_n \, 1_{T>n}. \tag{8.54}$$

Suppose we have proved the formula for some rank $m \geq n$. We use the decomposition

$$\mathbb{E}\left(M_{m+1} \mid \mathcal{F}_m\right) = M_m \quad \text{and} \quad (T > m) \in \mathcal{F}_m \ni 1_{A_n}$$

$$\Rightarrow \; M_m \, 1_{A_n} \, 1_{T>m} \;=\; \mathbb{E}\left(M_{m+1} \mid \mathcal{F}_m\right) \, 1_{T>m} \, 1_{A_n}$$
$$=\; \mathbb{E}\left(M_{m+1} \, 1_{A_n} \, 1_{T>m} \mid \mathcal{F}_m\right) = \mathbb{E}\left(M_{m+1} \, 1_{A_n} \, 1_{T\geq (m+1)} \mid \mathcal{F}_m\right).$$

Under the induction hypothesis, this implies that

$$\mathbb{E}(M_n \, 1_{A_n} \, 1_{T\geq n})$$

$$= \mathbb{E}\left(M_T \, 1_{A_n} \, 1_{n\leq T\leq m}\right) + \mathbb{E}\left(\mathbb{E}\left(M_{m+1} \, 1_{A_n} \, 1_{T\geq (m+1)} \mid \mathcal{F}_m\right)\right)$$

$$= \mathbb{E}\left(M_T \, 1_{A_n} \, 1_{n\leq T\leq m}\right) + \underbrace{\mathbb{E}\left(M_{m+1} \, 1_{A_n} \, 1_{T\geq (m+1)}\right)}_{\parallel \, \left(\text{by } (8.54)\right)}$$

$$= \mathbb{E}\left(M_T \, 1_{A_n} \, 1_{n\leq T\leq m}\right) + \mathbb{E}\left(M_T \, 1_{A_n} \, 1_{T=(m+1)}\right) + \mathbb{E}\left(M_{m+1} \, 1_{A_n} \, 1_{T>(m+1)}\right).$$

The end of the proof follows from the fact

$$\mathbb{E}\left(M_T \, 1_{A_n} \, 1_{n\leq T\leq m}\right) + \mathbb{E}\left(M_T \, 1_{A_n} \, 1_{T=(m+1)}\right) = \mathbb{E}\left(M_T \, 1_{A_n} \, 1_{n\leq T\leq (m+1)}\right).$$

This ends the proof of the induction step and the proof of the lemma is now completed. ∎

Theorem 8.4.14 (Optional stopping theorem 2) *Whenever a stopped martingale $M_{n \wedge T}$ is s.t. $|M_{n \wedge T}| \leq c$ we have*

$$\mathbb{E}(M_0) = \mathbb{E}(M_{n \wedge T}) \longrightarrow_{n \uparrow \infty} \mathbb{E}\left(\lim_{n \uparrow \infty} M_{n \wedge T} \right) = \mathbb{E}(M_T). \qquad (8.55)$$

In the same way, we also have

$$(|M_{n \wedge T}| \leq c_1 + c_2 \, T \quad with \quad \mathbb{E}(T) < \infty)$$

$$\Longrightarrow \mathbb{E}(M_0) = \mathbb{E}(M_{n \wedge T}) \longrightarrow_{n \uparrow \infty} \mathbb{E}\left(\lim_{n \uparrow \infty} M_{n \wedge T} \right) = \mathbb{E}(M_T). \qquad (8.56)$$

The proofs of (8.55) and (8.56) are based on the dominated convergence theorem. Observe that

$$|M_{n \wedge T}| \leq \sqrt{c_1} \quad and \quad \langle M \rangle_{n \wedge T} \leq c_2 \, T \; \Rightarrow \; \left| M_{n \wedge T}^2 - \langle M \rangle_{n \wedge T} \right| \leq c_1 + c_2 \, T.$$

Theorem 8.4.15 (A Wald's type indentity) *For any stopped martingale we have*

$$\left. \begin{array}{rcl} |M_{n \wedge T}| & \leq & c \\[2mm] \langle M \rangle_{n \wedge T} & \leq & c \, T \quad with \quad \mathbb{E}(T) < \infty \end{array} \right\} \; \Longrightarrow \; \mathbb{E}\left(M_T^2 \right) = \mathbb{E}\left(\langle M \rangle_T \right) \qquad (8.57)$$

as soon as $M_0 = 0 = \langle M \rangle_0$. In addition, we have

$$|M_{n \wedge T}| \leq c \; \Longrightarrow \; \mathbb{E}\left(\langle M \rangle_T \right) \leq c^2 \qquad (8.58)$$

Proof :
The assertion (8.57) is proved applying (8.56) to the martingale $N_n := M_n^2 - \langle M \rangle_n$. The second one follows from the fact that the stopped process is a martingale so that

$$\mathbb{E}(M_{n \wedge T}^2) = \mathbb{E}\left(\langle M \rangle_{T \wedge n} \right) \leq c^2.$$

Applying Fatou's lemma we have

$$\mathbb{E}\left(\langle M \rangle_T \right) = \mathbb{E}(\lim_{n \to \infty} \langle M \rangle_{T \wedge n}) \leq \liminf_{n \to \infty} \mathbb{E}\left(\langle M \rangle_{T \wedge n} \right) \leq c^2.$$

This ends the proof of (8.58). ∎

For instance, the condition $|M_{n \wedge T}| \leq c$ is clearly met when the martingale starts $M_0 \in [a, b]$ in some interval $[a, b] \subset \mathbb{R}$ and T is the first exit time of that interval.

More generally, we have the following theorem.

Theorem 8.4.16 (Optional stopping theorem) *Let* M_n *be a martingale w.r.t. a filtration* $\mathcal{F}_n$. *For any finite stopping time* T *(i.e.* $\mathbb{P}(T < \infty) = 1$*), such that* $\mathbb{E}(|M_T|) < \infty$, *we have*

$$\mathbb{E}\left(M_m \, 1_{T>m}\right) \to_{m\uparrow\infty} 0 \quad \Rightarrow \quad \mathbb{E}\left(M_n \, 1_{T\geq n}\right) = \mathbb{E}\left(M_T \, 1_{T\geq n}\right)$$

as well as

$$\mathbb{E}(M_T) = \mathbb{E}(M_0) \quad and \quad \mathbb{E}\left(M_T \mid \mathcal{F}_n\right) \, 1_{T<n} = M_n \, 1_{T<n}.$$

In addition we have

$$\mathbb{E}\left(|M_n| \, 1_{T>n}\right) \to_{n\uparrow\infty} 0 \quad \Rightarrow \quad \mathbb{E}(M_T \mid \mathcal{F}_n) = M_n.$$

Proof :

The first assertion is a convergence of the monotone convergence theorem and lemma 8.4.13 applied to $A_n = \Omega$. In this situation, we also notice that

$$\mathbb{E}(M_0) = \mathbb{E}\left(M_{n\wedge T}\right) = \overbrace{\mathbb{E}\left(M_n \, 1_{T\geq n}\right)}^{=\mathbb{E}\left(M_T \, 1_{T\geq n}\right)} + \mathbb{E}\left(M_T \, 1_{T<n}\right) = \mathbb{E}(M_T).$$

Furthermore, using the decomposition

$$M_{n\wedge T} := M_n \, 1_{T\geq n} + M_T \, 1_{T<n}$$

we check that

$$M_n = M_{n\wedge n} = \mathbb{E}\left(M_{n\wedge T} \mid \mathcal{F}_n\right) = M_n \, 1_{T\geq n} + \mathbb{E}\left(M_T \mid \mathcal{F}_n\right) \, 1_{T<n}$$

$$\Rightarrow \mathbb{E}\left(M_T \mid \mathcal{F}_n\right) \, 1_{T<n} = M_n \, 1_{T<n}.$$

Now, we move on to the proof of the second assertion. For any $A_n \in \mathcal{F}_n$ we have

$$\mathbb{E}\left(\mathbb{E}\left(M_T \mid \mathcal{F}_n\right) \, 1_{A_n} \, 1_{n\leq T}\right) = \mathbb{E}\left(M_T \, 1_{A_n} \, 1_{n\leq T}\right).$$

It remains to prove that

$$\mathbb{E}\left(M_T \, 1_{T\geq n} \, 1_{A_n}\right) = \mathbb{E}\left(M_n \, 1_{T\geq n} \, 1_{A_n}\right).$$

Arguing as above, this result is proved by using the monotone convergence theorem and lemma 8.4.13. We conclude that for any $A_n \in \mathcal{F}_n$ we have

$$\mathbb{E}\left(\mathbb{E}\left(M_T \mid \mathcal{F}_n\right) \, 1_{A_n} \, 1_{n\leq T}\right) = \mathbb{E}\left(M_n \, 1_{A_n} \, 1_{n\leq T}\right).$$

This implies that

$$\mathbb{E}\left(M_T \mid \mathcal{F}_n\right) \, 1_{n\leq T} = M_n \, 1_{n\leq T}.$$

This ends the proof of the theorem.   ∎

Important remark : Using the martingale property, it is tempting to deduce that $\mathbb{E}(M_T \mid \mathcal{F}_n) = M_n$ on the event $T > n$. The theorem 8.4.16 indicates that this result may fail when $\liminf_{n\to\infty} \mathbb{E}\left(|M_n| \, 1_{T>n}\right) > 0$. This obstacle in applying the conditioning formulae arises in the analysis of most of the martingale betting systems discussed in section 29.2.

8.4.4   A gambling model

We consider the gambling model defined in (8.40). The bettor's profit per unit of time is represented by a sequence of independent $\{-1, 1\}$-valued random variables X_n with common distribution

$$\mathbb{P}(X_n = -1) = p \quad \text{and} \quad \mathbb{P}(X_n = 1) = q = 1 - p \in [0, 1]$$

and we set $\mathbb{E}(X_n) = q - p = \rho$.

Definition 8.4.17 *We fix an initial fortune $Y_0 = y \in]a, b[$ on some maximal loss or gain interval $]a, b[\in \mathbb{N}$, and we let*

$$T_{a,b} = \inf\{n \geq 0 \;:\; Y_n \in]a, b[\} = \inf\{n \geq 0 \;:\; Y_n = a \text{ or } Y_n = b\}$$

to be the first time the process Y_n exits the open interval $]a, b[$ from the left or from the right.

The first result is that the player will reach fortune or ruin in finite time.

Lemma 8.4.18 *The exit time $T_{a,b} < \infty$ is almost surely finite. Moreover, there exist some constants $\alpha < \infty$, and $\beta > 0$ such that*

$$\sup_{y \in]a,b[} \mathbb{P}(T_{a,b} > n \mid Y_0 = y) \leq \alpha \, e^{-\beta n} \text{ and } \sup_{y \in]a,b[} \mathbb{E}(T_{a,b} \mid Y_0 = y) \leq \alpha \, (1 - e^{-\beta})^{-1}. \tag{8.59}$$

Proof :
The r.h.s. of (8.59) comes from the well known formula $\mathbb{E}(T) = \sum_{n \geq 0} \mathbb{P}(T > n)$ which is valid for any integer valued random variable T. When $\rho \neq 0$, we can deduce this result using the strong law of large numbers

$$\epsilon(n) := \frac{1}{n}[Y_n - \mathbb{E}(Y_n)] = \frac{1}{n} \sum_{0 < k \leq n} (X_k - \rho) \to_{n \uparrow \infty} 0$$

almost surely. This implies that

$$Y_n = \mathbb{E}(Y_n) + n \; \epsilon(n) = n(\rho + \epsilon(n)) \to_{n \uparrow \infty} \text{sign}(\rho) \; \infty$$

from which we conclude that $T_{a,b} < \infty$, when $\rho \neq 0$.

To prove that the exit time is finite even when $\rho = 0$, another more general strategy is to choose a large enough exit time period $m \geq 1$ so that

$$\inf_{y \in]a,b[} \mathbb{P}(Y_m \notin]a, b[\mid Y_0 = y) := \epsilon > 0. \tag{8.60}$$

Using the fact that

$$\forall y \in]a, b[\;\; \exists \, m(y) \geq 1 \; s.t. \; \mathbb{P}(Y_{m(y)} \in \{a, b\} \mid Y_0 = y) > 0$$

we check that condition (8.60) is satisfied for any $m \geq \max_{y \in]a,b[} m(y)$. In this case, we have

$$\begin{aligned} \mathbb{P}(T_{a,b} > m \mid Y_0 = y) &= 1 - \mathbb{P}(\exists n \in [1, m] \; Y_n \in \{a, b\} \mid Y_0 = y) \\ &\leq 1 - \mathbb{P}(Y_m \in \{a, b\} \mid Y_0 = y) \leq 1 - \epsilon \end{aligned}$$

and more generally

$$\mathbb{P}\left(T_{a,b} > nm \mid T_{a,b} > (n-1)m, \, Y_0 = y\right)$$

$$= \mathbb{P}\left(T_{a,b} > nm \mid Y_{(n-1)m} \in]a, b[, \ldots, \, Y_1 \in]a, b[, \, Y_0 = y\right)$$

$$= \mathbb{E}\left[\mathbb{P}\left(T_{a,b} > nm \mid Y_{(n-1)m}, \ldots, Y_1, \, Y_0\right) \mid Y_{(n-1)m} \in]a, b[, \ldots, Y_1 \in]a, b[\, Y_0 = y\right] \leq (1 - \epsilon).$$

This implies that for any $l = nm + k$, $0 \leq k < m$

$$
\begin{aligned}
\mathbb{P}\left(T_{a,b} > l\right) &\leq \mathbb{P}\left(T_{a,b} > nm \mid Y_0 = y\right) \\
&\leq \mathbb{P}\left(T_{a,b} > nm \mid T_{a,b} > (n-1)m, \, Y_0 = y\right) \times \mathbb{P}\left(T_{a,b} > (n-1)m \mid Y_0 = y\right) \\
&\leq (1 - \epsilon) \times \mathbb{P}\left(T_{a,b} > (n-1)m \mid Y_0 = y\right) \\
&\leq (1 - \epsilon)^n = (1 - \epsilon)^{-k}\left((1 - \epsilon)^{1/m}\right)^{nm+k} \leq (1 - \epsilon)^{-(1-1/m)}\left((1 - \epsilon)^{1/m}\right)^l.
\end{aligned}
$$

This clearly ends the proof of the lemma.   ∎

Definition 8.4.19 *We call the lucky player event*

$$\Omega_{a,b} := \left\{Y_{T_{a,b}} = b\right\}.$$

We have seen in (8.44) that it can be directly confirmed that the centered process

$$M_n = y + \sum_{0 < k \leq n} (X_k - \rho) = Y_n - \rho n \tag{8.61}$$

is a martingale.

8.4.4.1 Fair games

Firstly, we examine the case $\rho = q - p = 0$; that is $p = q = 1/2$. In this situation, using (8.46) we have the following martingales

$$
\begin{aligned}
M_n &= Y_n \\
\overline{M}_n &= M_n^2 - \langle M \rangle_n = Y_n^2 - n.
\end{aligned}
$$

By the optional stopping theorem 8.4.12, we have

$$M_0 = y = \mathbb{E}\left(M_{T_{a,b} \wedge n}\right) \to_{n \uparrow \infty} \mathbb{E}\left(M_{T_{a,b}}\right) \stackrel{\rho=0}{=} \mathbb{E}\left(Y_{T_{a,b}}\right).$$

This implies that

$$\mathbb{E}\left(M_{T_{a,b}}\right) = b \, \mathbb{P}\left(Y_{T_{a,b}} = b\right) + a \left(1 - \mathbb{P}\left(Y_{T_{a,b}} = b\right)\right) = y$$

from which we conclude that

$$\mathbb{P}\left(Y_{T_{a,b}} = b\right) = (y - a)/(b - a).$$

Having found these probabilities, another valuable bit of information for our player is to find the expected duration of the game. Arguing as above, we have

$$\overline{M}_0 = y^2 - 0 = \mathbb{E}\left(\overline{M}_{T_{a,b} \wedge n}\right) \to_{n \uparrow \infty} \mathbb{E}\left(\overline{M}_{T_{a,b}}\right) = \mathbb{E}\left(Y_{T_{a,b}}^2\right) - \mathbb{E}\left(T_{a,b}\right).$$

This implies that

$$
\begin{aligned}
\mathbb{E}\left(T_{a,b}\right) &= (b^2 - y^2)\,\mathbb{P}\left(Y_{T_{a,b}} = b\right) + (a^2 - y^2)\,\left(1 - \mathbb{P}\left(Y_{T_{a,b}} = b\right)\right) \\
&= (b^2 - y^2)\,\frac{(y - a)}{(b - a)} + (a^2 - y^2)\,\frac{(b - y)}{(b - a)} \\
&= (b - y)(y - a)\left[\frac{(b + y)}{(b - a)} - \frac{(a + y)}{(b - a)}\right] = (b - y)(y - a).
\end{aligned}
$$

8.4.4.2 Unfair games

For unfair games $\rho = q - p \neq 0$, we use the martingale (8.61) and the following exponential type martingale

$$
\mathcal{E}_n = (p/q)^{Y_n} \Rightarrow \Delta\mathcal{E}_n =
\begin{aligned}[t]
& (p/q)^{Y_{n-1} + \Delta Y_n} - (p/q)^{Y_{n-1}} \\
& = (p/q)^{Y_{n-1}}\left((p/q)^{X_n} - 1\right).
\end{aligned}
$$

It is readily checked that

$$
\mathbb{E}\left((p/q)^{X_n}\right) = (q/p)\,p + (p/q)\,q = p + q = 1
$$

from which we conclude that

$$
\mathbb{E}\left(\Delta\mathcal{E}_n \mid \mathcal{F}_{n-1}\right) = (p/q)^{Y_{n-1}}\left[\mathbb{E}\left((p/q)^{X_n}\right) - 1\right] = 0.
$$

This proves that $\mathcal{E}_n$ is a martingale. Arguing as above, we have

$$
\mathcal{E}_0 = (p/q)^y = \mathbb{E}\left((p/q)^{X_{T_{a,b} \wedge n}}\right) \to_{n \uparrow \infty} \mathbb{E}\left((p/q)^{X_{T_{a,b}}}\right).
$$

This implies that

$$
\begin{aligned}
(p/q)^y &= \mathbb{E}\left((p/q)^{X_{T_{a,b}}}\right) \\
&= (p/q)^b\,\mathbb{P}\left(Y_{T_{a,b}} = b\right) + (p/q)^a\,\left(1 - \mathbb{P}\left(Y_{T_{a,b}} = b\right)\right) \\
&= (p/q)^a + \left[(p/q)^b - (p/q)^a\right]\,\mathbb{P}\left(Y_{T_{a,b}} = b\right)
\end{aligned}
$$

from which we conclude that

$$
\mathbb{P}\left(Y_{T_{a,b}} = b\right) = \frac{(p/q)^y - (p/q)^a}{(p/q)^b - (p/q)^a}.
$$

To compute the mean duration of the unfair game, we use the martingale $M_n = Y_n - \rho n$ discussed in (8.61). Using once again the same line of arguments as above, we find that

$$
M_0 = y - 0 = \mathbb{E}\left(M_{T_{a,b} \wedge n}\right) \to_{n \uparrow \infty} \mathbb{E}\left(M_{T_{a,b}}\right) = \mathbb{E}\left(Y_{T_{a,b}}\right) - \rho\,\mathbb{E}\left(T_{a,b}\right).
$$

This implies that

$$
\begin{aligned}
\rho\,\mathbb{E}\left(T_{a,b}\right) &= (b - y)\,\mathbb{P}\left(Y_{T_{a,b}} = b\right) + (a - y)\,\left(1 - \mathbb{P}\left(Y_{T_{a,b}} = b\right)\right) \\
&= (b - y)\left(1 - \frac{(p/q)^b - (p/q)^y}{(p/q)^b - (p/q)^a}\right) - (y - a)\,\frac{(p/q)^b - (p/q)^y}{(p/q)^b - (p/q)^a} \\
&= (b - y) - (b - a)\,\frac{(p/q)^b - (p/q)^y}{(p/q)^b - (p/q)^a}.
\end{aligned}
$$

Hence we get

$$\mathbb{E}\left(T_{a,b}\right) = \frac{1}{q-p}\left[(b-y)-(b-a)\frac{(p/q)^b-(p/q)^y}{(p/q)^b-(p/q)^a}\right].$$

For unfair games $\delta := p/q > 1$ holds and we have

$$p = \delta q = 1 - q \Rightarrow q = 1/(1+\delta) \quad \text{and} \quad p = \delta/(1+\delta).$$

In this case, we find that

$$\mathbb{P}\left(Y_{T_{a,b}}=b\right) = \frac{\delta^y-\delta^a}{\delta^b-\delta^a} = \frac{\delta^{-(b-y)}-\delta^{-(b-a)}}{1-\delta^{-(b-a)}} \simeq_{b-a\uparrow\infty} \delta^{-(b-y)}$$

and

$$\mathbb{E}\left(T_{a,b}\right) = \frac{\delta+1}{\delta-1}\left[(b-a)\left(1-\frac{\delta^{-(b-y)}-\delta^{-(b-a)}}{1-\delta^{-(b-a)}}\right)-(b-y)\right]$$

$$\simeq_{b-a\uparrow\infty} \frac{\delta+1}{\delta-1}(y-a)\left[1-\frac{b-a}{y-a}\delta^{-(b-y)}\right] \simeq_{b-a\uparrow\infty} \frac{\delta+1}{\delta-1}(y-a).$$

8.4.5 Maximal inequalities

In the further development of this section M_n stands for a martingale w.r.t. a filtration $\mathcal{F}_n$.

Lemma 8.4.20 *For any $x > 0$, we have*

$$\mathbb{P}\left(\sup_{0\leq p\leq n}|M_p|\geq x\right) \leq x^{-1}\,\mathbb{E}\left(|M_n|\,1_{\sup_{0\leq p\leq n}|M_p|\geq x)}\right)$$

or equivalently

$$\mathbb{E}\left(|M_n|\,\Big|\,\sup_{0\leq p\leq n}|M_p|\geq x\right) \geq x.$$

Proof :
Let T_a be the first time a martingale exceeds certain level $a > 0$:

$$T_a = \inf\{n\geq 0\,:\,M_n\geq a\}\quad\left(\leq n \Leftrightarrow \sup_{0\leq p\leq n}M_p\geq a\right).$$

By construction, we have

$$\underbrace{\mathbb{E}\left(M_{T_a\wedge n}\,1_{T_a\leq n}\right)}_{=M_{T_a}\,1_{T_a\leq n}} \geq a\,\mathbb{P}\left(T_a\geq n\right) = a\,\mathbb{P}\left(\sup_{0\leq p\leq n}M_p\geq a\right).$$

These inequalities are valid for any processes. The martingale property enters here to help us to prove that

$$\mathbb{E}\left(M_{n\wedge T_a}\,1_{T_a\leq n}\right) = \mathbb{E}\left(M_{n\wedge T_a}\,(1-1_{T_a>n})\right) = \underbrace{\mathbb{E}\left(M_{n\wedge T_a}\right)}_{\mathbb{E}(M_n)} - \underbrace{\mathbb{E}\left(M_{n\wedge T_a}\,1_{T_a>n}\right)}_{=\mathbb{E}(M_n\,1_{T_a>n})}$$

from which we conclude that

$$\mathbb{E}\left(M_{n\wedge T_a}\,1_{T_a\leq n}\right) = \mathbb{E}\left(M_n\,(1-1_{T_a>n})\right) = \mathbb{E}\left(M_n\,1_{T_a\leq n}\right) \leq \mathbb{E}\left(M_n^+\,1_{T_a\leq n}\right)$$

with $M_n^+ := \max(0, M_n)$. Summarizing, we have proved that for any martingale M_n

$$\mathbb{P}\left(\sup_{0 \le p \le n} M_p \ge a\right) \le a^{-1}\, \mathbb{E}\left(M_n^+\, 1_{T_a \le n}\right)$$

$$\mathbb{P}\left(\sup_{0 \le p \le n} (-M_p) \ge a\right) \le a^{-1}\, \mathbb{E}\left((-M_n)^+\, 1_{T_a^- \le n}\right) = a^{-1}\, \mathbb{E}\left(M_n^-\, 1_{T_a^- \le n}\right)$$

with

$$T_a^- = \inf\left\{n \ge 0 : -M_n \ge a\right\} \le n \quad \Leftrightarrow \quad \sup_{0 \le p \le n}(-M_p) \ge a \Leftrightarrow \inf_{0 \le p \le n} M_p \le -a.$$

The second assertion is deduced from the first using the fact that the inverse $(-M_n)$ of a martingale remains a martingale, and $(-M_n)^+ := \max(0, -M_n) = -\inf(0, M_n) = M_n^-$. Adding the two l.h.s. lines, we find that

$$\mathbb{P}\left(\sup_{0 \le p \le n} M_p \ge a\right) + \mathbb{P}\left(\inf_{0 \le p \le n} M_p \le -a\right) = \mathbb{P}\left(\sup_{0 \le p \le n} |M_p| \ge a\right).$$

Recalling that $|a| = a^+ + a^- \ge \max(a^+, a^-)$, for any $a \in \mathbb{R}$, adding the two r.h.s. lines, up to the multiplicative factor a^{-1} we get

$$\mathbb{E}\left(M_n^+\, 1_{T_a \le n}\right) + \mathbb{E}\left(M_n^-\, 1_{T_a^- \le n}\right) \le \mathbb{E}\left(|M_n|\, 1_{(T_a \le n) \cap (T_a^- \le n)}\right)$$

$$= \mathbb{E}\left(|M_n|\, 1_{(\sup_{0 \le p \le n} M_p \ge a) \cap (\inf_{0 \le p \le n} M_p \le -a)}\right)$$

$$= \mathbb{E}\left(|M_n|\, 1_{\sup_{0 \le p \le n} |M_p| \ge a}\right).$$

This ends the proof of the lemma. ∎

Lemma 8.4.21 *For any conjugate numbers $\frac{1}{p} + \frac{1}{q} = 1$, with $p > 1$, we have*

$$\mathbb{E}\left(\sup_{0 \le k \le n} |M_k|^p\right) \le q^p\, \mathbb{E}\left(|M_n|^p\right).$$

Proof :
For any non negative real valued r.v. X, and any $p > 1$, using Fubini's theorem we have

$$\mathbb{E}(X^p)$$

$$= \int_0^\infty \overbrace{\left[\int_0^y p\, x^{p-1}\, dx\right]}^{=y^p} \mathbb{P}(X \in dy) = \int_{0 \le x \le y < \infty} p\, x^{p-1}\, dx \mathbb{P}(X \in dy)$$

$$= \int_0^\infty p\, x^{p-1}\left[\int_x^\infty \mathbb{P}(X \in dy)\right] dx = \int p\, x^{p-1}\, \mathbb{P}(X \ge x)\, dx.$$

Applying this formula to $X = \sup_{0 \le p \le n} |M_p|$, and using lemma 8.4.20 we find that

$$\mathbb{E}\left(\sup_{0 \le k \le n} |M_k|^p\right) \le \frac{p}{p-1}\, \mathbb{E}\left(|M_n| \int_0^{\sup_{0 \le k \le n} |M_k|} (p-1)\, x^{p-1}\, x^{-1}\, dx\right)$$

$$= \frac{1}{1 - 1/p}\, \mathbb{E}\left(|M_n| \sup_{0 \le p \le n} |M_p|^{p-1}\right). \tag{8.62}$$

For any conjugate numbers $\frac{1}{p} + \frac{1}{q} = 1$ with $p > 1$, Hölder's inequality implies that

$$
\mathbb{E}\left(|M_n| \sup_{0 \leq k \leq n} |M_k|^{p-1}\right) \leq \mathbb{E}\left(|M_n|^p\right)^{1/p} \mathbb{E}\left(\sup_{0 \leq k \leq n} |M_k|^{qp(1-1/p)}\right)^{1/q}
$$

$$
= \mathbb{E}\left(|M_n|^p\right)^{1/p} \mathbb{E}\left(\sup_{0 \leq k \leq n} |M_k|^p\right)^{1/q}.
$$

Combining this estimate with (8.62) we prove that

$$
\mathbb{E}\left(\sup_{0 \leq k \leq n} |M_k|^p\right)^{1-1/q} = \mathbb{E}\left(\sup_{0 \leq k \leq n} |M_k|^p\right)^{1/p} \leq q\, \mathbb{E}\left(|M_n|^p\right)^{1/p}.
$$

This ends the proof of the lemma. ∎

8.4.6 Limit theorems

We recall that the upper and lower limits of a sequence of random variables M_n are given by the formulae

$$
\liminf_{n \to \infty} M_n = \inf_{n \geq 0} \sup_{m \geq n} M_m \qquad \text{et} \qquad \limsup_{n \to \infty} M_n = \sup_{n \geq 0} \inf_{m \geq n} M_m.
$$

The existence of the limit is defined by the event

$$
\{\omega \in \Omega \; : \; \lim_{n \to \infty} M_n(\omega) \text{ exists }\} = \Omega - \{\omega \in \Omega \; : \; \limsup_{n \to \infty} M_n(\omega) > \liminf_{n \to \infty} X_n(\omega)\}.
$$

with

$$
\{\limsup_{n \to \infty} M_n > \liminf_{n \to \infty} M_n\} = \bigcup_{\substack{a < b \\ a, b \in \mathbb{Q}}} \{\limsup_{n \to \infty} M_n > b > a \liminf_{n \to \infty} M_n\}.
$$

This shows that the sequence M_n will have a limit as $n \uparrow \infty$ with probability 1 if and only if the number of oscillations between two rational numbers $a < b$ is finite with probability 1.

Let $\{T_n \; ; \; n \geq 0\}$ the times the process M_n goes above and below some given parameters $a < b$; that is,

$$
\begin{aligned}
T_0 &= 0 \\
T_1 &= \min\{n > T_0 \; : \; M_n \leq a\} \\
T_2 &= \min\{n > T_1 \; : \; M_n \geq b\} \\
\ldots &= \ldots \\
T_{2m} &= \min\{n > T_{2m-1} \; : \; M_n \geq b\} \\
T_{2m+1} &= \min\{n > T_{2m} \; : \; M_n \leq a\} \\
\ldots &= \ldots
\end{aligned}
$$

with the convention $\inf_\emptyset = +\infty$. For each $n \geq 1$ the number of crossing $\beta_n^{(M)}([a, b])$ is defined by

$$
\beta_n^{(M)}([a, b]) = \begin{cases} 0 & \text{if } T_2 > n \\ \max\{m \geq 0 \; : \; T_{2m} \leq n\} & \text{if } T_2 \leq n. \end{cases}
$$

Lemma 8.4.22 (Doob's upcrossing lemma) *For any sub-martingale (respectively super-martingale) M_n w.r.t. some $\mathcal{F}_n$, and for any $a, b \in \mathbb{R}$, with $a < b$, we have*

$$\mathbb{E}\left(\beta_n^{(M)}(a, b) \right) \leq \mathbb{E}((M_n - a)^+)/(b - a) \qquad (\text{resp. } \mathbb{E}((b - M_n)^+)/(b - a)).$$

Proof :
Since M is a super-martingale if and only if $(-M_n)$ is a sub-martingale, and

$$\beta_n^{(M)}(a, b) = \beta_n^{(-M)}(-b, -a),$$

it clearly suffices to check the lemma for sub-martingales. We also notice that the number of crossing of the interval $[a, b]$ is the same as the number of crossing the interval $[0, b - a]$ by the sequence $(M_n - a)^+$. Therefore, we can assume that $a = 0$ and we need to check that

$$\mathbb{E}\left(\beta_n^{(M)}(0, b) \right) \leq \mathbb{E}(M_n)/b.$$

We assume that $X_0 = 0$ (and $\mathcal{F}_0 = \{\emptyset, \Omega\}$). For each $k \geq 1$ we set

$$H_k = \begin{cases} 1 & \text{si} \quad T_m < k \leq T_{m+1} \text{ for } m \text{ odd} \\ 0 & \text{si} \quad T_m < k \leq T_{m+1} \text{ for } m \text{ even}. \end{cases}$$

By construction, we have

$$b \, \beta_n^{(M)}(0, b) \leq \sum_{1 \leq k \leq n} H_k \, (M_k - M_{k-1})$$

and

$$\{H_k = 1\} = \cup_{m \text{ odd}} (\{T_m < k\} - \{T_{m+1} < k\}) \in \mathcal{F}_{k-1}.$$

Finally we have

$$\begin{aligned} b \, \mathbb{E}\left(\beta_n^{(M)}(0, b) \right) &\leq \sum_{1 \leq k \leq n} \mathbb{E}\left(1_{H_k=1} \, \mathbb{E}\left(M_k - M_{k-1} | \mathcal{F}_{k-1} \right) \right) \\ &\leq \sum_{1 \leq k \leq n} \mathbb{E}\left(\mathbb{E}\left(M_k - M_{k-1} | \mathcal{F}_{k-1} \right) \right) = \mathbb{E}\left(M_n \right). \end{aligned}$$

This ends the proof of the lemma. ∎

Theorem 8.4.23 (Doob's convergence theorem) *For any martingale, super-martingale or sub-martingale M_n s.t. $\sup_{n \geq 0} \mathbb{E}(|M_n|) < \infty$, the almost sure limit $\lim_{n \to \infty} M_n = M_\infty$ exists and we have $\mathbb{E}(|M_\infty|) < \infty$.*

Proof :
By the monotone convergence theorem, for any $a < b$ we have

$$\begin{aligned} \mathbb{E}\left(\lim_{n \to \infty} \beta_n^{(M)}(a, b) \right) &= \lim_{n \to \infty} \mathbb{E}\left(\beta_n^{(M)}(a, b) \right) \\ &\leq \frac{1}{b - a} \left(\sup_{n \geq 0} \mathbb{E}(|M_n|) + |a| + |b| \right) < +\infty. \end{aligned}$$

This implies that

$$\mathbb{P}\left(\limsup_{n\to\infty} M_n > b > a \liminf_{n\to\infty} M_n\right) = 0$$

so that $\lim_{n\to\infty} M_n = M_\infty$ exists. In addition, by Fatou's lemma we have $\mathbb{E}(|M_\infty|) < \infty$. ∎

In the further development of this section M_n stands for a martingale w.r.t. a filtration $\mathcal{F}_n$.

Theorem 8.4.24 *Any martingale M_n such that $\sup_{n\geq 0} \mathbb{E}(M_n^2) < \infty$ converges almost surely to some random variable M_∞, as $n \to \infty$, and*

$$\lim_{n\to\infty} \mathbb{E}\left[|M_n - M_\infty|^2\right] = 0.$$

Proof :
The existence of the limit is granted by Doob's convergence theorem. Next, we provide an alternative analysis based on the completeness of the Hilbert space of square integrable random variables. Without loss of generality we assume that the martingale is null at the origin. In this situation, from lemma 8.4.5, we readily check that $\mathbb{E}(M_n^2)$ is an increasing sequence and

$$\sup_{n\geq 0}\mathbb{E}(M_n^2) = \sup_{n\geq 0}\mathbb{E}([M]_n) = \lim_{n\uparrow\infty}\mathbb{E}\left([M]_n\right) = \sum_{n\geq 0}\mathbb{E}\left((\Delta M_n)^2\right) < \infty.$$

This implies that

$$\mathbb{E}\left([M]_{m+n} - [M]_m\right) = \sum_{m<k\leq m+n}\mathbb{E}\left((\Delta M_k)^2\right) \leq \sum_{m<n}\mathbb{E}\left((\Delta M_n)^2\right)\downarrow_{m\uparrow\infty} 0.$$

Hence

$$\begin{aligned}
\mathbb{E}\left((M_{m+n} - M_m)^2\right) &= \mathbb{E}(M_{m+n}^2 - M_m^2) \\
&= \mathbb{E}\left([M]_{m+n} - [M]_m\right)\downarrow_{m\uparrow\infty} 0.
\end{aligned}$$

This shows that M_n is a Cauchy sequence on the Hilbert space of square integrable random variables and hence converges to some random variable M_∞.

We recall that a sequence of random variables M_n converges almost surely to some random variable M_∞ if and only if $\sup_{m\geq n}|M_m - M_\infty|$ tends to 0 in probability, as $n \to \infty$. The proof of this classical result in probability theory is a consequence of the Borel Cantelli lemma. To check that this condition is satisfied we use Markov's inequality to prove that for any $\epsilon > 0$, and any fixed n, we have

$$\epsilon^2\,\mathbb{P}\left(\sup_{m\geq n}|M_m - M_\infty| \geq \epsilon\right) \leq \mathbb{E}\left(\sup_{m\geq n}|M_m - M_\infty|^2\right).$$

Our next objective is to prove that the r.h.s. term converges to 0 as $n \uparrow \infty$. Since

$$|M_m - M_\infty|^2 = |[M_m - M_n] + [M_n - M_\infty]|^2 \leq 2\left[|M_m - M_n|^2 + |M_n - M_\infty|^2\right]$$

we find that

$$2^{-1}\mathbb{E}\left(\sup_{m\geq n}|M_m - M_\infty|^2\right) \leq \mathbb{E}\left(\sup_{m\geq n}|M_m - M_n|^2\right) + \underbrace{\mathbb{E}\left(|M_n - M_\infty|^2\right)}_{\to_{n\uparrow\infty}0}.$$

It remains to treat the first summand in the r.h.s. To this end, we notice that by the monotone convergence theorem

$$\lim_{q\uparrow\infty}\mathbb{E}\left(\sup_{n\leq m\leq q}|M_m - M_n|^2\right) = \mathbb{E}\left(\sup_{n\leq m}|M_m - M_n|^2\right)$$

and

$$\sup_{n\leq m\leq q}|M_m - M_n| = \sup_{0\leq p\leq q-n}|M_{n+p} - M_n| = \sup_{0\leq p\leq q-n}|M_p^{(n)}|$$

with the martingale $M_p^{(n)} := M_{n+p} - M_n$, null at the origin. By lemma 8.4.21, we have

$$\mathbb{E}\left(\sup_{n\leq m\leq q}|M_m - M_n|^2\right) = \mathbb{E}\left(\sup_{0\leq p\leq q-n}|M_p^{(n)}|^2\right)$$
$$\leq 2^2\,\mathbb{E}\left(|M_{q-n}^{(n)}|^2\right) = 2^2\,\underbrace{\mathbb{E}\left(|M_q - M_n|^2\right)}_{\to_{q\uparrow\infty}\,\mathbb{E}(|M_\infty - M_n|^2)}.$$

This implies that

$$\mathbb{E}\left(\sup_{n\leq m}|M_m - M_n|^2\right) \leq 2^2\,\mathbb{E}\left(|M_\infty - M_n|^2\right) \to_{q\uparrow\infty} 0.$$

This ends the proof of the theorem. ■

Theorem 8.4.25 *For any square integrable martingale (i.e. $\mathbb{E}(M_n^2) < \infty$, for any $n \geq 0$), we have the almost sure convergence*

$$\lim_{n\to\infty}\langle M\rangle_n = \infty \implies \lim_{n\to\infty} M_n/\langle M\rangle_n = 0.$$

In addition, we have

$$\sum_{n\geq 1} n^{-2}\,\mathbb{E}([\Delta M_n]^2) < \infty \implies \lim_{n\to\infty} M_n/n = 0$$

as well as $\lim_{n\to\infty}\mathbb{E}\left((M_n/n)^2\right) = 0$.

Important remarks : Before getting into the details of the proof, we quote two direct but important consequences of this theorem.

• The first one applies to the convergence analysis of the occupation measures (8.49) of regular Markov chains. Under the assumptions of lemma 8.4.11, the martingale $\mathcal{M}_n(g)$ associated with the (bounded) solution g of the Poisson equation satisfies the hypothesis of the theorem 8.4.25. We conclude that $\pi^n(f)$ converges almost surely to $\pi(f)$, as $n \to \infty$, for any bounded function f. In section 9.1, we provide an elementary proof of a somehow weaker result without using these sophisticated martingale limit theorems.

- The second one provides a useful equivalence principle. We consider a random series $\sum_{1 \leq k \leq n} \varphi_{k-1} V_k$ associated with some random variables $V_k \subseteq \mathcal{F}_k$ with unit mean, and some possibly random functions $\varphi_k \subseteq \mathcal{F}_k$ such that $\mathbb{E}\left(\varphi_k^2\right) < \infty$, for any $k \geq 0$. In this situation, we have

$$\frac{1}{n} \sum_{1 \leq k \leq n} \varphi_{k-1} V_k = \frac{1}{n} \sum_{1 \leq k \leq n} \varphi_{k-1} + \epsilon_n$$

for some remainder term ϵ_n that tends to 0, almost surely as $n \uparrow \infty$, as well as $\mathbb{E}\left(\epsilon_n^2\right) \to_{n \uparrow \infty} 0$. The proof follows by applying the theorem to the martingale $M_n := \sum_{1 \leq p \leq n} \varphi_{k-1} (V_k - 1)$. For instance, if W_k denotes a sequence of independent and centered Gaussian random variables with unit variance, we have

$$\frac{1}{n} \sum_{1 \leq k \leq n} \varphi_{k-1} \left[W_k^2 - 1\right] = \epsilon_n \longrightarrow_{n \uparrow \infty} 0. \tag{8.63}$$

Now, we turn back to the proof of the theorem.

Proof of theorem 8.4.25:

There is no loss of generality to assume that $M_0 = 0$. We consider the martingale M_n^ϵ, null at the origin, with increments given by

$$\Delta M_n^\epsilon := \frac{\Delta M_n}{\epsilon + \langle M \rangle_n} \Rightarrow \Delta \langle M^\epsilon \rangle_n = \mathbb{E}\left((\Delta M_n^\epsilon)^2 \mid \mathcal{F}_{n-1}\right) = \frac{\Delta \langle M \rangle_n}{(\epsilon + \langle M \rangle_n)^2}.$$

We notice that

$$\langle M \rangle_1 = \Delta \langle M \rangle_1$$

$$\Rightarrow \Delta \langle M^\epsilon \rangle_1 = \frac{\Delta \langle M \rangle_1}{(\epsilon + \Delta \langle M \rangle_1)^2} = \underbrace{\frac{\Delta \langle M \rangle_1}{(\epsilon + \Delta \langle M \rangle_1)}}_{\leq 1} \frac{1}{(\epsilon + \Delta \langle M \rangle_1) \ (\geq \epsilon)} \leq 1/\epsilon$$

and

$$\frac{1}{\epsilon + \langle M \rangle_{n-1}} - \frac{1}{\epsilon + \langle M \rangle_n} = \frac{\Delta \langle M \rangle_n}{(\epsilon + \langle M \rangle_{n-1})(\epsilon + \langle M \rangle_n)} = \frac{\Delta \langle M \rangle_n}{(\epsilon + \langle M \rangle_n)^2} \frac{(\epsilon + \langle M \rangle_n)}{(\epsilon + \langle M \rangle_{n-1})}$$

$$\geq \frac{\Delta \langle M \rangle_n}{(\epsilon + \langle M \rangle_n)^2} = \Delta \langle M^\epsilon \rangle_n.$$

This yields the estimate

$$\begin{aligned}
\langle M^\epsilon \rangle_n &= \Delta \langle M^\epsilon \rangle_1 + \sum_{1 < p \leq n} \Delta \langle M^\epsilon \rangle_p \\
&\leq \Delta \langle M^\epsilon \rangle_1 + \sum_{1 < p \leq n} \left(\frac{1}{\epsilon + \langle M \rangle_{p-1}} - \frac{1}{\epsilon + \langle M \rangle_p} \right) \\
&= \Delta \langle M^\epsilon \rangle_1 + \frac{1}{\epsilon + \langle M \rangle_1} - \frac{1}{\epsilon + \langle M \rangle_n} \leq 2/\epsilon.
\end{aligned}$$

We conclude that

$$\lim_{n \to \infty} \langle M^\epsilon \rangle_n = \sup_{n \geq 1} \langle M^\epsilon \rangle_n := \langle M^\epsilon \rangle_\infty \leq 2/\epsilon$$

and therefore

$$\lim_{n \to \infty} \mathbb{E}\left((M_n^\epsilon)^2\right) = \sup_{n \geq 1} \mathbb{E}\left((M_n^\epsilon)^2\right) = \sup_{n \geq 1} \mathbb{E}\left(\langle M^\epsilon \rangle_n\right) \leq \mathbb{E}\left(\langle M^\epsilon \rangle_\infty\right) \leq 2/\epsilon.$$

By theorem 8.4.24 this implies that M_n^ϵ converges almost surely to some random variable M_∞^ϵ, as $n \to \infty$. Invoking the Toeplitz-Kronecker lemma we conclude that

$$\frac{M_n}{\langle M \rangle_n} = \underbrace{\frac{\epsilon + \langle M \rangle_n}{\langle M \rangle_n}}_{n \to n \to \infty 1} \underbrace{\frac{1}{\epsilon + \langle M \rangle_n} \sum_{1 \leq p \leq n} (\epsilon + \langle M \rangle_p) \, \Delta M_p^\epsilon}_{n \to n \to \infty 0}.$$

This ends the proof of the first assertion. The proof of the second one follows the same line of arguments. Firstly, we use the fact that $\overline{M}_n = \sum_{1 \leq p \leq n} p^{-1} \, \Delta M_p$ is a square integrable martingale. By theorem 8.4.24 it also converges almost surely to some random variable $\overline{M}_n$, and

$$\lim_{n \to \infty} \mathbb{E} \left([\overline{M}_n - \overline{M}_\infty]^2 \right) = 0.$$

It remains to observe that

$$\begin{aligned}
\frac{1}{n} M_n &= \frac{1}{n} \sum_{1 \leq p \leq n} p \, (\overline{M}_p - \overline{M}_{p-1}) = \frac{1}{n} \sum_{1 \leq p \leq n} [p \, \overline{M}_p - (p-1) \, \overline{M}_{p-1} - \overline{M}_{p-1}] \\
&= \frac{1}{n} \, n \, \overline{M}_n - \frac{1}{n} \sum_{0 \leq p < n} \overline{M}_p = \overline{M}_n - \frac{1}{n} \sum_{0 \leq p < n} \overline{M}_p \to_{n \to \infty} \overline{M}_\infty - \overline{M}_\infty = 0
\end{aligned}$$

almost surely. This ends the proof of the theorem. ∎

We end this section with a fluctuation theorem for martingales with bounded increments.

Theorem 8.4.26 *For any martingale M_n such that $|M_0| \vee |\Delta M_n| < c$, for some finite constant $c < \infty$, we have*

$$\lim_{n \to \infty} \frac{1}{n} \langle M \rangle_n = \sigma^2 \implies \lim_{n \to \infty} \frac{M_n}{\sqrt{n}} = N(0, \sigma^2).$$

The convergence in the l.h.s. is a convergence in probability, whereas the convergence in the r.h.s. is a convergence in law to a centered Gaussian random variable.

Hint of proof :
It clearly suffices to prove the result for $M_0 = 0$. We set

$$\phi(t) = \exp \left(-\frac{t^2 \sigma^2}{2} \right) \quad \text{and} \quad \phi_n(t) = \prod_{1 \leq p \leq n} \mathbb{E} \left(\exp \left(-it \frac{\Delta M_p}{\sqrt{n}} \right) \mid \mathcal{F}_{p-1} \right).$$

Using the approximation formula $\left| e^{ix} - (1 + ix - \frac{x^2}{2}) \right| \leq \frac{x^3}{3!}$ (which is valid for any $x \in \mathbb{R}$) we find that

$$\begin{aligned}
\mathbb{E} \left(\exp \left(-i \frac{t}{\sqrt{n}} \Delta M_p \right) \mid \mathcal{F}_{p-1} \right) &= 1 - \frac{t^2}{n} \Delta \langle M \rangle_p + \mathrm{O}(n^{-(1+\frac{1}{2})}) \\
&= e^{-\frac{t^2}{n} \Delta \langle M \rangle_p} + \mathrm{O}(n^{-(1+\frac{1}{2})}) \\
&= e^{-\frac{t^2}{n} \Delta \langle M \rangle_p} (1 + \mathrm{O}(n^{-(1+\frac{1}{2})})).
\end{aligned}$$

In the second line of the formula we use the fact that $e^{-\epsilon} = 1 - \epsilon + O(\epsilon^2)$ when $\epsilon \simeq 0$. This implies that

$$\phi_n(t) = e^{-\frac{t^2}{n} \sum_{1 \le p \le n} \Delta \langle M \rangle_p} \underbrace{(1 + O(n^{-(1+\frac{1}{2})}))^n}_{\to_{n \to \infty} 1} \longrightarrow_{n \to \infty} \phi(t).$$

To take the final step, we observe that

$$\mathcal{E}_p = \mathcal{E}_{p-1} \times \frac{\exp\left(-\frac{it}{\sqrt{n}} \Delta M_p\right)}{\mathbb{E}\left(\exp\left(-\frac{it}{\sqrt{n}} \Delta M_p\right) \mid \mathcal{F}_{p-1}\right)} = \frac{\exp\left(-it\frac{M_p}{\sqrt{n}}\right)}{\phi_p(t)}$$

is a martingale starting at 1, for $0 \le p \le n$. This implies that $\mathbb{E}(\mathcal{E}_n) = \mathbb{E}\left(\frac{e^{-it\frac{M_n}{\sqrt{n}}}}{\phi_n(t)}\right) = 1$ and therefore

$$\mathbb{E}\left(e^{-it\frac{M_n}{\sqrt{n}}}\right) - \phi(t) = \mathbb{E}\left(\frac{e^{-it\frac{M_n}{\sqrt{n}}}}{\phi_n(t)} \underbrace{(\phi_n(t) - \phi(t))}_{\to_{n \to \infty} 0}\right) \longrightarrow_{n \to \infty} 0.$$

This ends the proof of the theorem. ∎

As an illustration, we derive a central limit theorem for the occupation measures

$$\pi^n := \frac{1}{n+1} \sum_{0 \le p \le n} \delta_{X_p}$$

of time homogeneous Markov chain X_n discussed in section 8.4.2.3. We use the Poisson decomposition presented in lemma 8.4.11. We recall that

$$\frac{1}{n+1} \left(g(X_{n+1}) - g(X_0)\right) = [\pi(f) - \pi^n(f)] + \frac{1}{n+1} \mathcal{N}_{n+1}(g)$$

where $g := P(f)$ stands for the (bounded) solution of the Poisson equation (8.19) given in (8.51), and

$$\mathcal{N}_n(g) := \sum_{0 < p \le n} [g(X_p) - \mathbb{E}(g(X_p) \mid \mathcal{F}_{p-1})]$$

is the martingale with angle bracket given by

$$\langle \mathcal{N}(g) \rangle_n := \sum_{0 < p \le n} \left[M(g^2)(X_{p-1}) - (M(g)(X_{p-1}))^2 \right]$$

where $M(x, dy)$ stands for the Markov transition of the chain X_n. Using (8.52) we have the $\mathbb{L}_1$-convergence

$$n^{-1} \langle \mathcal{N}(g) \rangle_n = \pi^n(M(g^2)) - \pi^n(M(g^2))$$

$$\longrightarrow_{n \to \infty} \pi(M(g^2)) - \pi(M(g^2)) := \sigma^2(g) = \sigma^2(P(f))$$

with the invariant measure $\pi M = \pi$ of the chain.

We conclude that

$$\pi^n = \pi + \frac{1}{\sqrt{n+1}} \, V^n$$

with a sequence of random fields V^n s.t.

$$V^n(f) \longrightarrow_{n \to \infty} N(0, \sigma^2(P(f))).$$

We show the last assertion by using the fact that

$$V^n(f) = \sqrt{n+1} \, [\pi^n(f) - \pi(f)] = \underbrace{\frac{\mathcal{N}_{n+1}(P(f))}{\sqrt{n+1}}}_{\to_{n \to \infty} N(0, \sigma^2(P(f)))} + O\left(\frac{1}{\sqrt{n+1}}\right).$$

8.5 Topological aspects

8.5.1 Irreducibility and aperiodicity

When some state regions are not accessible for initialization, the chain cannot forget its initial condition. In this context, the invariant measure may depend on the initial value of the chain. For instance the $\{0, 1, 2, 3, 4\}$ valued Markov chain given by the following transition diagram cannot access the set $A = \{2, 3, 4\}$ when starting from $B = \{0, 1\}$, and vice versa.

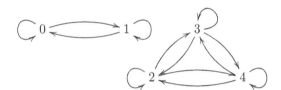

In this situation, any iteration of the Markov transition M^n will have the null entries $M^n(i, j) = 0$, for any $(i, j) \in (A \times B) \cup (B \times A)$. The chain starting at A will have the same invariant measure as the chain with transition diagram with vertices A; and the one starting at B will have the have the same invariant measure as the chain with transition diagram with vertices B. In summary, an elementary obstruction to uniqueness of invariant measures is when the state space contains regions for which the transition from one into the other occurs with zero probability.

It may happen that all states communicate while it is impossible to find some n such that all entries of M^n are positive. For instance, the stochastic matrix associated with the transition diagram

$$0 \rightleftarrows 1$$

given by $M = \begin{pmatrix} 0 & 1 \\ 1 & 0 \end{pmatrix}$ is such that $M^{2n+1} = M$ and $M^{2n} = Id$. The invariant measure $\pi = [.5, .5]$ but $M^n(x, y)$ does not converge to $1/2$.

Definition 8.5.1 *When all the states x and y communicate, in the sense that $M^n(x,y) > 0$ for some n, the chain is said to be irreducible. It is called aperiodic when*

$$\text{GCD}\{n \ : \ M^n(x,x) > 0\} = 1$$

for any x (where GCD stands for greatest common divisor) . Otherwise the chain is said to be periodic. The chain can be irreducible but periodic. For instance, in the two-state example discussed above, the chain is irreducible but we have $\text{GCD}\{n \ : \ M^n(x,x) > 0\} = 2$.

Periodicity is clearly not an obstacle to the existence to the invariant measure, but it prevents convergence to the equilibrium since the state can be partitioned in classes that we visit only at periodic times.

The aperiodicity *combined* with the irreducibility implies the existence of an integer m such that M^m has positive entries.

To check this claim, we use the fact that the $\{n \ : \ M^n(x,x) > 0\} \ni n_1, n_2$ is closed by addition. Indeed,

$$M^{n_1+n_2}(x,x) = \sum_{y \in S} M^{n_1}(x,y) M^{n_2}(y,x) \geq M^{n_1}(x,x) \times M^{n_2}(x,x) > 0$$

for any x. This implies the existence of $n(x)$ s.t. $M^k(x,x) > 0$ for any $k \geq n(x)$. Now, the irreducibility implies the existence of some $l(x,y)$ such that for any $l \geq l(x,y)$

$$M^{k+l}(x,y) \geq M^{k+(l-l(x,y))}(x,x) \, M^{l(x,y)}(x,y) > 0.$$

This shows that for any $m \geq \sup_{x,y}(n(x) + l(x,y))$, the matrix $M^m > 0$ has positive entries.

A natural and very simple way of turning a given Markov chain with transition M into an aperiodic chain is to consider the chain associated with the elementary transitions

$$M_\epsilon(x,y) = \epsilon 1_x(y) + (1 - \epsilon) \, M(x,y)$$

for some $\epsilon \in]0, 1[$. We leave the reader to check that the resulting chain is aperiodic and it has the same invariant measure as M

$$\pi M = \pi \implies \pi M_\epsilon = \pi.$$

Definition 8.5.2 *The chain with Markov transition M_ϵ is called the ϵ-lazy version of the Markov chain associated with M.*

Notice that the $1/2$-lazy version of the periodic chain discussed above is equivalent to sampling independent Bernoulli r.v., and it is immediately at equilibrium in one step.

We end this section with a more probabilistic description of the invariant measure of a Markov chain in terms of the mean return time to a given state.

Theorem 8.5.3 (Kac's formula) *The invariant measure of an irreducible and aperiodic Markov chain X_n on a finite space S is given by*

$$\pi(x) = 1/\mathbb{E}\left(T \mid X_0 = x\right) \tag{8.64}$$

where T stands for the first return time to x

$$T = \inf\{n \geq 1 \ : \ X_n = x\}.$$

Proof :

Since S is finite and X_n irreducible we have $\mathbb{P}_x(T < \infty) = 1$ and

$$\mathbb{E}\left(\sum_{0 \le n < T} 1_y(X_n) \mid X_0 = x\right) = \sum_{m > 0} \mathbb{E}\left(\sum_{0 \le n < m} 1_y(X_n) \, 1_{T=m} \mid X_0 = x\right)$$

$$= \sum_{0 \le n < m} \mathbb{P}\left(X_n = y, \, T = m \mid X_0 = x\right)$$

$$= \sum_{n \ge 0} \underbrace{\sum_{m > n} \mathbb{P}\left(X_n = y, \, T = m \mid X_0 = x\right)}_{=\mathbb{P}(X_n = y, \, T > n \mid X_0 = x)}.$$

This implies that

$$\gamma_x(y) := \mathbb{E}\left(\sum_{0 \le n < T} 1_y(X_n) \mid X_0 = x\right)$$

$$= \sum_{n \ge 0} \mathbb{P}\left(X_n = y \mid T > n, \, X_0 = x\right) \, \mathbb{P}\left(T > n \mid X_0 = x\right).$$

This alos implies that

$$\gamma_x(S) := \sum_{y \in S} \gamma_x(y) = \sum_{n \ge 0} \mathbb{P}\left(T > n \mid X_0 = x\right) = \mathbb{E}\left(T \mid X_0 = x\right) < \infty$$

and

$$(\gamma_x M)(z) = \sum_{y \in S} \gamma_x(y) \, M(y, z)$$

$$= \sum_{n \ge 0} \underbrace{\sum_{y \in S} \mathbb{P}\left(X_n = y \mid T > n, \, X_0 = x\right) M(y, z)}_{=\mathbb{P}(X_{n+1} = z \mid T > n, \, X_0 = x)} \mathbb{P}\left(T > n \mid X_0 = x\right)$$

and therefore

$$(\gamma_x M)(z) = \sum_{n \ge 0} \underbrace{\mathbb{P}\left(X_{n+1} = z, \, T > n, \, X_0 = x\right)}_{\sum_{m > n} \mathbb{P}(X_{n+1} = z, \, T = m \mid X_0 = x)}$$

$$= \sum_{m > 0} \sum_{0 \le n < m} \mathbb{P}\left(X_{n+1} = z, \, T = m \mid X_0 = x\right)$$

$$= \sum_{m > 0} \mathbb{E}\left(\sum_{0 \le n < m} 1_z(X_{n+1}) \, 1_{T=m} \mid X_0 = x\right).$$

Hence

$$(\gamma_x M)(z) = \mathbb{E}\left(\sum_{1 \le n \le T} 1_z(X_n) \mid X_0 = x\right)$$

$$= \mathbb{E}\left(\sum_{0 \le n < T} 1_z(X_n) \mid X_0 = x\right) = \gamma_x(z).$$

The l.h.s. formula follows from the fact that $X_T = x = X_0$. We conclude that

$$\forall x \in S \qquad \eta_x M = \eta_x$$

with the probability measures

$$\eta_x(y) = \mathbb{E}\left(\sum_{0 \leq n < T} 1_y(X_n) \mid X_0 = x\right) / \mathbb{E}\left(T \mid X_0 = x\right).$$

Notice that for $x = y$, we have $\gamma_x(x) = 1$ so that

$$\eta_x(x) = 1/\mathbb{E}\left(T \mid X_0 = x\right).$$

Since the invariant measure is unique, we have $\pi(x) = \eta_x(x)$. This ends the proof of the theorem. ∎

8.5.2   Recurrent and transient states

〽 We consider a Markov chain X_n on some countable state space S with Markov transition M.

Definition 8.5.4 *We denote by N_x the number of returns to x:*

$$N_x = \sum_{n \geq 1} 1_x(X_n) = \sum_{n \geq 1} 1_{X_n}(x).$$

The potential matrix is the matrix $G(x, y)$ with entries

$$
\begin{aligned}
G(x, y) &:= \mathbb{E}\left(\sum_{n \geq 0} 1_y(X_n) \mid X_0 = x\right) \\
&= 1_{x=y} + \mathbb{E}(N_y \mid X_0 = x) = \sum_{n \geq 0} M^n(x, y).
\end{aligned}
$$

Definition 8.5.5 *A state x is said to be recurrent as soon as the return probability to x starting from x is 1. Otherwise, the state is said to be transient. A Markov chain is said to be recurrent when all the states are recurrent.*

We notice that the excursions from a given state x to x are independent of each other. The Markov chain restarts its evolution from that state every time it comes back to that site. In particular, the duration $T_{x,n}$ of the n-th excursion, with $n \geq 1$, forms a sequence of independent random variables with common law. This implies that

$$\mathbb{P}(N_x \geq n \mid X_0 = x)$$

$$= \mathbb{P}\left(T_{x,1} < \infty,\ T_{x,2} < \infty, \ldots, T_{x,n} < \infty \mid X_0 = x\right) = \mathbb{P}\left(T_{x,1} < \infty \mid X_0 = x\right)^n.$$

Now

$$
\begin{aligned}
\mathbb{E}(N_x \mid X_0 = x) &= \sum_{p \geq 1} p\, \mathbb{P}(N_x = p \mid X_0 = x) = \sum_{1 \leq p} \sum_{1 \leq q \leq p} \mathbb{P}(N_x = p \mid X_0 = x) \\
&= \sum_{q \geq 1} \sum_{p \geq q} \mathbb{P}(N_x = p \mid X_0 = x) = \sum_{q \geq 1} \mathbb{P}(N_x \geq q \mid X_0 = x) \\
&= \sum_{q \geq 1} \mathbb{P}\left(T_{x,1} < \infty \mid X_0 = x\right)^q.
\end{aligned}
$$

Summarizing, we have proved the following proposition.

Proposition 8.5.6

$$\mathbb{E}(N_x \mid X_0 = x) = \infty \iff \mathbb{P}(T_{x,1} < \infty \mid X_0 = x) = 1$$
$$\iff G(x,x) = \infty \iff x \text{ is recurrent.}$$

For irreducible Markov chains, all the states have the same nature.

To check the last assertion we use the Chapman-Kolmogorov equation to prove that

$$M^{n_1+n_2+n_3}(x,x) \geq M^{n_1}(x,y)M^{n_2}(y,y)M^{n_3}(y,x)$$

from which we conclude that all series $\sum_{n\geq 0} M^n(x,x)$, with $x \in S$ have the same nature.

We let $T_{x\to y}$ the duration of an excursion from x to y. By construction, we have

$$\mathbb{P}(N_y \geq n \mid X_0 = x)$$

$$= \mathbb{P}(T_{x\to y} < \infty, \ T_{y,1} < \infty, \ T_{y,2} < \infty, \ldots, T_{y,n-1} < \infty \mid X_0 = x)$$

$$= \mathbb{P}(T_{x\to y} < \infty) \times \mathbb{P}(T_{y,1} < \infty \mid X_0 = y)^{n-1}.$$

In this situation, we have

$$\begin{aligned}
\mathbb{E}(N_y \mid X_0 = x) &= \sum_{q\geq 1} \mathbb{P}(N_y \geq q \mid X_0 = x) \\
&= \mathbb{P}(T_{x\to y} < \infty) \times \sum_{q\geq 0} \mathbb{P}(T_{y,1} < \infty \mid X_0 = y)^q \\
&= \mathbb{P}(T_{x\to y} < \infty) \times [1 + \mathbb{E}(N_y \mid X_0 = y)].
\end{aligned}$$

Summarizing, we have proved the following proposition.

Proposition 8.5.7

$$y \text{ transient} \iff (\forall x \quad \mathbb{E}(N_y \mid X_0 = x) < \infty) \iff (\forall x \quad G(x,y) < \infty).$$

Proposition 8.5.8 *For irreducible and recurrent Markov chains we have*

$$\forall x, y \in S \qquad \mathbb{P}(T_{x\to y} < \infty) = 1.$$

Proof :
We let T_x be the hitting time of the state x. We observe that

$$\begin{aligned}
\mathbb{P}(T_{x\to y} < \infty) &= \mathbb{P}(T_y < \infty \mid X_0 = x) \\
&= \mathbb{P}(T_y < \infty, \ T_x \leq T_y \mid X_0 = x) \\
&\qquad\qquad\qquad\qquad + \mathbb{P}[T_y < T_x(< \infty) \mid X_0 = x] \\
&= \mathbb{E}\left(\underbrace{\mathbb{P}(T_y < \infty, \mid T_x, \ X_0 = x)}_{=\mathbb{P}(T_{x\to y} < \infty)} \times 1_{T_x \leq T_y}\right) \\
&\qquad\qquad\qquad\qquad + \mathbb{P}[T_y < T_x(< \infty) \mid X_0 = x].
\end{aligned}$$

This implies that

$$\mathbb{P}(T_{x\to y} < \infty) = \mathbb{P}(T_{x\to y} < \infty)(1 - \mathbb{P}[T_y < T_x \mid X_0 = x]) + \mathbb{P}[T_y < T_x \mid X_0 = x]$$

from which we conclude that

$$\mathbb{P}\left[T_y < T_x \mid X_0 = x\right] (1 - \mathbb{P}\left(T_{x \to y} < \infty\right)) = 0.$$

It remains to check that $\mathbb{P}\left[T_y < T_x \mid X_0 = x\right] > 0$. We denote by R_n the successive epochs of return visits of the state x. Since $R_n \geq R_{n-1}$, we have

$$\mathbb{P}\left[T_y \geq R_n \mid X_0 = x\right] \downarrow_{n \uparrow \infty} \mathbb{P}\left[T_y = \infty \mid X_0 = x\right] = 0$$

$$= \underbrace{\mathbb{P}\left[T_y \geq R_n \mid T_y \geq R_{n-1}, X_0 = x\right]}_{=\mathbb{P}[T_y \geq T_x \mid X_0 = x]} \times \mathbb{P}\left[T_y \geq R_{n-1} \mid X_0 = x\right]$$

$$= \mathbb{P}\left[T_y \geq T_x \mid X_0 = x\right] \times \mathbb{P}\left[T_y \geq R_{n-1} \mid X_0 = x\right]$$
$$= \mathbb{P}\left[T_y \geq T_x \mid X_0 = x\right]^{n-1} \mathbb{P}\left[T_y \geq R_1 = T_x \mid X_0 = x\right] = \mathbb{P}\left[T_y \geq T_x \mid X_0 = x\right]^n.$$

This clearly implies that $\mathbb{P}\left[T_y \geq T_x \mid X_0 = x\right] < 1$. This ends the proof of the proposition. ∎

Rephrasing the proof of theorem 8.5.3 we have the following theorem.

Theorem 8.5.9 (Kac's theorem) *Let M be the transitions of an irreducible and recurrent Markov chain X_n. We let T_x be the first <u>return</u> time to a state x, with $x \in S$. For any state x, the expected time spent in state y between visits to x*

$$\gamma_x(y) := \mathbb{E}\left(\sum_{0 \leq n < T_x} 1_y(X_n) \mid X_0 = x\right) \tag{8.65}$$

is an invariant measure of the Markov chain $\gamma_x M = \gamma_x$. When the expected return times are finite $\mathbb{E}(T_x \mid X_0 = x) = \sum_y \gamma_x(y) < \infty$ the unique invariant measure of the chain is given by

$$\pi(x) = 1/\mathbb{E}\left(T_x \mid X_0 = x\right).$$

Important remark : By construction, we notice that

$$\gamma_x(x) = \mathbb{E}\left(\sum_{0 \leq n < T_x} 1_x(X_n) \mid X_0 = x\right) = 1.$$

In addition, using the Chapman-Kolmogorov equation and the irreducibility property, we have

$$0 < M^{n(x,y)}(x,y) = \gamma_x(x) \; M^{n(x,y)}(x,y) \leq \gamma_x M^{n(x,y)}(y) = \gamma_x(y)$$

and

$$\gamma_x(y) \; \underbrace{M^{n(y,x)}(y,x)}_{>0} \leq \gamma_x M^{n(y,x)}(x) = \gamma_x(x) = 1$$

for some integers $n(x,y) \geq 1$ and $n(y,x) \geq 1$. This implies that

$$\forall x, y \in S \qquad 0 < \gamma_x(y) < \infty.$$

Definition 8.5.10 *The (recurrent) states x s.t.*

$$\mathbb{E}(T_x \mid X_0 = x) = \sum_{n \geq 1} \mathbb{P}(T_x \geq n \mid X_0 = x) < \infty$$

are called null recurrent states.

Important remark : Since

$$\mathbb{P}\left(T_x \geq n \mid X_0 = x\right) \downarrow_{n\uparrow\infty} 0 \Rightarrow \mathbb{P}\left(T_x < \infty \mid X_0 = x\right) = 1,$$

the null recurrent states are clearly recurrent states. The null recurrent property depends on the decay rates.

One can also check that the existence of the invariant probability measure is a sufficient a necessary condition for all the states to be recurrent positive. To check this claim, we need the following technical lemma.

Lemma 8.5.11 *Let M be the transitions of an irreducible and recurrent Markov chain X_n. The invariant measure γ_x defined in (8.65) is the unique invariant measure of M s.t. $\gamma_x(x) = 1$.*

Proof :
For any $x \in S$, we denote by Q_x the matrix defined by

$$Q_x(u,v) = 1_{\neq x}(u)M(u,v) \Longleftrightarrow \forall f \in \mathcal{B}(S) \quad Q_x(f)(u) = 1_{\neq x}(u)M(f)(u).$$

By the fixed point equation, for any invariant measure ν_x s.t. $\nu_x(x) = \nu_x(1_x) = 1$ we have

$$\nu_x(f) \quad = \quad \nu_x(1_{\neq x}M(f)) + \nu(x)M(f)(x) = (\nu_x Q_x)(f) + \delta_x M(f).$$

In a more synthetic formulation we have

$$\begin{aligned}
\nu_x \quad &= \quad \nu_x Q_x + \delta_x M \\
&= \quad (\nu_x Q_x + \delta_x M)Q_x + \delta_x M = \nu_x Q_x^2 + \delta_x M\left(Q_x + Id\right) \\
&= \quad (\nu_x Q_x + \delta_x M)Q_x^2 + \delta_x M Q_x + \delta_x M = \nu_x Q_x^3 + \delta_x M\left(Q_x^2 + Q_x + Id\right).
\end{aligned}$$

An elementary induction yields that

$$\nu_x = \nu_x Q_x^n + \sum_{0 \leq p < n} \delta_x M Q_x^p.$$

Notice that for any $p \geq 0$ we have

$$Q_x^p(f)(y) = 1_{\neq x}(y)\, \mathbb{E}\left(f(X_{p+1})\, 1_{T_x \geq p+1} \mid X_1 = y\right)$$

and

$$\begin{aligned}
\nu_x^n(f) \quad &:= \quad \sum_{0 \leq p < n} \delta_x M Q_x^p(f) = \sum_{0 \leq p < n} \mathbb{E}\left(f(X_{p+1})\, 1_{T_x \geq p+1} \mid X_0 = x\right) \\
&= \quad \mathbb{E}\left(\sum_{1 \leq p \leq n} f(X_p)\, 1_{T_x \geq p} \mid X_0 = x\right) = \mathbb{E}\left(\sum_{1 \leq p \leq n \wedge T_x} f(X_p) \mid X_0 = x\right).
\end{aligned}$$

This implies that

$$\begin{aligned}
\nu_x^n(x) \quad &= \quad \mathbb{E}\left(\sum_{1 \leq p \leq T_x} 1_x(X_p)\, 1_{T_x \leq n} \mid X_0 = x\right) + \mathbb{E}\left(\sum_{1 \leq p \leq n} 1_x(X_p)\, 1_{T_x > n} \mid X_0 = x\right) \\
&= \quad \mathbb{P}\left(T_x \leq n \mid X_0 = x\right) \uparrow_{n\uparrow\infty} \mathbb{P}\left(T_x < \infty \mid X_0 = x\right) = 1.
\end{aligned}$$

In addition, for any $y \neq x$ we have

$$\nu_x(y) \;\geq\; \nu_x^n(y) = \mathbb{E}\left(\sum_{0 \leq p \leq n} 1_y(X_p)\, 1_{T_x > p} \mid X_0 = x \right) \uparrow_{n \uparrow \infty} \gamma_x(y).$$

To take the final step, we notice that $\mu_x(y) := \nu_x(y) - \gamma_x(y) \geq 0$ is also an invariant measure of M, with a null mass at x. Thus, invoking the Chapman-Kolmogorov equation we have

$$0 = \mu_x(x) \geq \mu_x(y) \underbrace{M^{n(y,x)}(y,x)}_{>0} \quad \Longrightarrow \quad \mu_x(y) = 0$$

for some $n(y,x) \geq 1$. This proves that $\nu_x = \gamma_x$. ∎

Proposition 8.5.12 *An irreducible and recurrent Markov chain X_n has an invariant probability measure if and only if one of the states is null recurrent. In this situation, all the states of the chain are null recurrent.*

Proof :

When one of the states x is null recurrent, theorem 8.5.9 implies that X_n has an invariant measure $\pi_x(y) = \gamma_x(y)/\sum_y \gamma_x(y)$. Inversely, suppose that X_n has an invariant measure, say $\pi(x) > 0$, for all x (by the irreducibility property). In this case, the measure $\nu_x(y) = \pi(y)/\pi(x)$ is an invariant measure such that $\pi(x) = 1$, and lemma 8.5.11 implies that $\nu_x = \gamma_x$ so that

$$\mathbb{E}(T_x \mid X_0 = x) = \sum_y \gamma_x(y) = \sum_y \pi(y)/\pi(x) = 1/\pi(x) \Longrightarrow x \text{ is null recurrent.}$$

This shows that all the states all null recurrent. This ends the proof of the proposition. ∎

8.5.3 Continuous state spaces

An example of a real valued Markov chain forgetting its initial condition exponentially fast is provided by the reversible Gaussian transition (7.19). In this case, the chain is given by a recursion of the form

$$X_n \;=\; \sqrt{1-\epsilon}\, X_{n-1} + \sqrt{\epsilon}\, W_n = (\sqrt{1-\epsilon})^n\, X_0 + \sqrt{\epsilon} \sum_{0 \leq p < n} (\sqrt{1-\epsilon})^p\, W_{n-p}$$

for some $\epsilon \in]0,1]$ and some independent centered Gaussian random variables W_n with unit variance. The r.h.s. formula has been proven in (7.15). In this situation, the invariant reversible measure is given by the standard normal distribution. For any Lipschitz function f on $\mathbb{R}$ with Lipschitz constant $\mathrm{lip}(f) \leq 1$ we find that

$$|M^n(f)(x) - M^n(f)(y)| \;\leq\; (\sqrt{1-\epsilon})^n\, |x - y|.$$

This implies that

$$\sup_{(x,y)\in[a,b]^2} |M^n(f)(x) - M^n(f)(y)| \leq (\sqrt{1-\epsilon})^n\, |b - a|$$

for any compact interval $[a, b] \subset \mathbb{R}$. In addition, we have that

$$
\begin{aligned}
\|M^n(f) - \pi(f)\|^2_{\mathbb{L}_2(\pi)} &:= \int \pi(dx)\, |M^n(f)(x) - \pi(f)|^2 \\
&\leq (\sqrt{1-\epsilon})^{2n} \int \pi(dx)\pi(dy)\, |x-y|^2 = 2\, (\sqrt{1-\epsilon})^{2n} \to_{n\to\infty} 0.
\end{aligned}
$$

Important remark : We notice that the Gaussian Markov transition

$$
M(x, dy) \propto \exp\left(-\frac{1}{2}(x-y)^2\right) dy
$$

is clearly reversible w.r.t. the Lebesgue $\lambda(dx) = dx$ on $\mathbb{R}$ but it cannot be normalized. The question whether a given reversible measure can be turned into a reversible (probability) distribution depends on the "size" of the state space and/or on the integrability properties of the measure at hand. We also notice that the transition restricted to some compact interval $S = [a, b]$

$$
M(x, dy) \propto \exp\left(-\frac{1}{2}(x-y)^2\right) 1_{[a,b]}(y)\, dy
$$

satisfies condition (8.16) and (8.15) with $\epsilon = \exp\left(-(b-a)^2\right)$, for any $a < b$.

8.5.4 Path space models

Definition 8.5.13 *In the further development of this section, we fix some parameter $p \geq 0$ and we set*

$$
\boldsymbol{X}_n = (X_n, X_{n+1}, \ldots, X_{n+p}).
$$

We notice that $\boldsymbol{X}_n$ is a Markov chain on S^{p+1} with Markov transitions given by

$$
\boldsymbol{M}((x_0, \ldots, x_p), d(y_1, \ldots, y_{p+1}))
$$

$$
= \delta_{(x_1, \ldots, x_p)}(d(y_1, \ldots, y_p)) \times M(y_p, dy_{p+1}).
$$

Lemma 8.5.14 *The probability measure*

$$
\boldsymbol{\pi}(d(y_1, \ldots, y_{p+1})) = \pi(dy_1)M(y_1, dy_2)\ldots M(y_p, dy_{p+1})
$$

is the unique invariant probability measure of the chain $\boldsymbol{X}_n$; that is,

$$
\boldsymbol{\pi}\boldsymbol{M} = \boldsymbol{\pi}.
$$

Proof :
Suppose that the initial state of the chain $\boldsymbol{X}_0 = (X_0, X_1, \ldots, X_p)$ is distributed according to $\boldsymbol{\pi}$. In this situation, the marginal distributions of each random state X_q, with $0 \leq q \leq p$, coincide with π. On the other hand, the elementary transition of the chain is given by

$$
\boldsymbol{X}_0 = (X_0, X_1, \ldots, X_p) \rightsquigarrow \boldsymbol{X}_1 = (X_1, \ldots, X_p, X_{p+1}).
$$

Since $\mathrm{Law}(X_1) = \pi$, we conclude that $\mathrm{Law}(\boldsymbol{X}_1) = \boldsymbol{\pi}$.

∎

Lemma 8.5.15 *The Markov transition* $\boldsymbol{M}^p$ *satisfies condition (8.15) with*

$$\boldsymbol{\nu}(d(y_1,\ldots,y_{p+1})) = \nu(dy_1)M(y_1,dy_2)\ldots M(y_p,dy_{p+1})$$

if and only if M *satisfies (8.15) for some probability measure* ν.

Proof :
Using the fact that $\boldsymbol{X}_{p+1} = (X_{p+1}, X_{p+2}, \ldots, X_{p+p})$, we prove that

$$\boldsymbol{M}^p((x_0,\ldots,x_p), d(y_1,\ldots,y_{p+1})) = M(x_p, dy_1)\ldots M(y_p, dy_{p+1}).$$

The end of the proof is now clear.   ∎

8.6   Exercises

Exercise 90 (Invariant measure - 1) *Let* X_n *be the cyclic random walk on* $\{0, 1, 2, 3\}$
with Markov transitions given by the matrix

$$M = \begin{pmatrix} 0 & 1/2 & 0 & 1/2 \\ 1/2 & 0 & 1/2 & 0 \\ 0 & 1/2 & 0 & 1/2 \\ 1/2 & 0 & 1/2 & 0 \end{pmatrix}.$$

Find the invariant measure of the chain.

Exercise 91 (Invariant measure - 2) *Let* X_n *be a random walk on* $\{0, 1, 2, 3\}$ *with Markov*
transitions given by the matrix

$$M = \begin{pmatrix} 0 & 1 & 0 & 0 \\ 1/2 & 0 & 1/2 & 0 \\ 0 & 1/2 & 0 & 1/2 \\ 0 & 0 & 1 & 0 \end{pmatrix}.$$

Find the invariant measure of the chain.

Exercise 92 (Invariant measure - 3) *Let* X_n *be a random walk on* $\{0, 1, 2, 3\}$ *with Markov*
transitions given by the matrix

$$M = \begin{pmatrix} 0 & 1 & 0 & 0 \\ 1/3 & 1/3 & 1/3 & 0 \\ 0 & 1/3 & 1/3 & 1/3 \\ 0 & 0 & 1 & 0 \end{pmatrix}.$$

Find the invariant measure of the chain.

Exercise 93 (Invariant measure - Reflecting random walk) *Let* $u, v \geq 0$ *be a couple*
of parameters s.t. $u+v = 1$. *Let* X_n *be the reflecting random walk on* $\{0, 1, 2, 3\}$ *with Markov*
transitions given by the matrix

$$M = \begin{pmatrix} u & v & 0 & 0 \\ u/2 & v & u/2 & 0 \\ 0 & u/2 & v & u/2 \\ 0 & 0 & v & u \end{pmatrix}.$$

Find the invariant measure of the chain.

Exercise 94 (Invariant measure - Birth and death process) *Let X_n be a random walk on $\mathbb{N}$ with Markov transitions given for any $x \geq 1$ by $M(x-1, x) = p = 1 - M(x, x-1) \in [0, 1]$, and $M(0, 0) = 1 - p$. Find the invariant measure of the chain.*

Exercise 95 (Invariant measure - Birth and death process - Finite space) *Let X_n be a random walk on $\{0, 1, \ldots, d\}$ with Markov transitions given for any $1 \leq x \leq d$ by $M(x-1, x) = p = 1 - M(x, x-1) \in [0, 1]$, $M(d, d) = p$ and $M(0, 0) = 1 - p$. Find the invariant measure of the chain.*

Exercise 96 *Prove the formulae (8.1), (8.2), (8.3), (8.4), and (8.8).*

Exercise 97 (Geometric drift contractions) ⫸ *The aim of this exercise is to quantify more explicitly the geometric drift contraction inequalities discussed in section 8.2.7. We set $R_\epsilon := 2/(1 - \epsilon)$, $\alpha_\epsilon := 1 - \beta^{(R_\epsilon)}(M)$, and $\delta := (1 - \epsilon)$.*

- *Choosing $R = R_\epsilon$ in (8.34) and (8.35), check that*

$$\forall \rho \in \,]0, 1] \qquad \sup_{W(x) \wedge W(y) \geq R_\epsilon} \Delta_\rho(x, y) \leq 1 - \frac{1}{2} \frac{4\rho(1 - \epsilon)}{(1 - \epsilon) + 4\rho} < 1$$

and

$$\forall \rho \in \,]0, \alpha_\epsilon \delta/8[\qquad \sup_{W(x) \vee W(y) \leq R_\epsilon} \Delta_\rho(x, y) \leq 1 - \left(\alpha_\epsilon - \frac{8\rho}{1 - \epsilon} \right) < 1.$$

- *We set $u := 4\rho/\delta$, and*

$$g(u) \quad := \quad \frac{1}{2} \frac{4\rho(1 - \epsilon)}{(1 - \epsilon) + 4\rho} = \frac{\delta}{2} \left(1 - \frac{1}{1 + u} \right)$$

$$h(u) \quad := \quad \left(\alpha_\epsilon - \frac{8\rho}{1 - \epsilon} \right) = (\alpha_\epsilon - 2u).$$

Check that these two functions intersect at the point

$$u = \sqrt{a^2 + b} - a \in [0, b]$$

with

$$a := \frac{1}{2} \left(1 - b + \frac{\delta}{4} \right) \leq \frac{1}{2} \quad \text{and} \quad b := \frac{\alpha_\epsilon}{2}.$$

Prove that for any $v \geq 0$ we have

$$\sqrt{1 + v} \geq 1 + \frac{v}{2\sqrt{1 + v}}$$

and deduce that

$$g(u) = h(u) \geq \frac{\delta b}{1 + 2\sqrt{3}}.$$

- *Deduce from the above that for $\rho = u\delta/4$ we have*

$$\beta_{V_\rho}(M) \leq 1 - \frac{(1 - \epsilon)(1 - \beta^{(R_\epsilon)}(M))}{2(1 + 2\sqrt{3})}.$$

Exercise 98 (Invariant measure - Graph formulation) *We consider a stochastic matrix with positive entries*

$$M = \begin{pmatrix} p_{11} & p_{12} & p_{13} \\ p_{21} & p_{22} & p_{23} \\ p_{31} & p_{32} & p_{33} \end{pmatrix}.$$

For any $x \in S = \{1,2,3\}$, we denote by $\mathcal{G}(i)$ the set of all the i-graphs. An i-graph is a set of directed edges without any loops connecting all the states $j \not\!{=} i$ without cycles to i; and with a single edge starting from the states $j \not\!{=} i$. For instance, the set of 1-graphs $\mathcal{G}(1) = \{g_1, g_2, g_3\}$ with the directed graphs is defined below

$$g_1 \qquad\qquad g_2 \qquad\qquad g_3$$

Prove that the unique invariant measure of the chain is given by

$$\pi(i) = \gamma(i)/\sum_{1 \le j \le 3} \gamma(j) \quad \text{with} \quad \gamma(i) = \sum_{g \in \mathcal{G}(i)} \prod_{(k,l) \in g} p_{k,l}.$$

For instance, for $i = 1$ we have

$$\gamma(1) = \sum_{g \in \mathcal{G}(1)} \prod_{(i,j) \in g} p_{i,j} = (p_{21}p_{31} + p_{32}p_{21} + p_{23}p_{31}).$$

Exercise 99 (Invariant measure - Graph formulation - General case) *We consider a Markov transition on some finite state space such that $M^m(x,y) > 0$, for any $x, y \in S$, for some $m \ge 1$. Prove that the unique invariant measure of the chain is given by*

$$\pi(x) = \gamma(x)/\sum_{y \in S} \gamma(y) \quad \text{with} \quad \gamma(x) = \sum_{g \in \mathcal{G}(x)} \prod_{(y,z) \in g} M(y,z)$$

where $\mathcal{G}(x)$ stands for the set of x-graphs defined in exercise 98.

Exercise 100 (Eigenvalues formula) *We consider the stochastic matrix presented in exercise 98.*

- *Prove that the characteristic polynomial $P(\lambda) = \mathrm{Det}(M - \lambda I)$ is given by*

$$P(\lambda) = (1 - \lambda)\ \left(\lambda^2 + (1 - A)\lambda + C\right)$$

  with $C = 1 - (A + B)$, and

$$\begin{aligned} A &= p_{11} + p_{22} + p_{33} \\ B &= p_{23}p_{32} + p_{12}p_{21} + p_{13}p_{31} - (p_{11}p_{22} + p_{11}p_{33} + p_{22}p_{33}). \end{aligned}$$

- *Check that*

$$-\frac{1 - A}{2} = 1 - (q_{12} + q_{13} + q_{23})$$

  with the parameters $q_{i,j} = (p_{ij} + p_{ji})/2$.

- *Check that*

$$B = -3 + 4 \ (q_{12} + q_{13} + q_{23}) - D$$

  with

$$D = 4 \ (q_{12}q_{13} + q_{12}q_{23} + q_{13}q_{23}) - (p_{21}p_{23} + p_{12}p_{13} + p_{31}p_{32}) \ .$$

- *Deduce that*

$$\left(\frac{1-A}{2} \right)^2 - C = \Delta(q) + \delta(p)$$

  with the parameters

$$\begin{aligned}
\Delta(q) &= \frac{1}{2} \left[(q_{12} - q_{13})^2 + (q_{12} - q_{23})^2 + (q_{13} - q_{23})^2 \right] \\
\delta(p) &= [p_{12}p_{13} - q_{12}q_{13}] + [p_{21}p_{23} - q_{21}q_{23}] + [p_{31}p_{32} - q_{31}q_{32}] \ .
\end{aligned}$$

- *Conclude that the eigenvalues of M are given by $\lambda_1 = 1$, and*

$$\begin{aligned}
\lambda_2 &= (1 - (q_{12} + q_{13} + q_{23})) + \sqrt{\Delta(q) + \delta(p)} \\
\lambda_3 &= (1 - (q_{12} + q_{13} + q_{23})) - \sqrt{\Delta(q) + \delta(p)}
\end{aligned}$$

  with the convention $\sqrt{-a} = i\sqrt{a}$, for any $a \geq 0$.

- *In the reversible case, check that $\delta(p) = 0$, and*

$$\lambda_3 \leq \lambda_2 \leq \lambda_1 = 1.$$

Exercise 101 (Top-in-at-random card shuffle) ✐ *We consider the top-in-at-random card shuffle discussed on page 173 and in section 26.4. Use an inductive proof to check that when there are k cards below the original bottom card, all $k!$ orderings of these cards are equally likely.*

Exercise 102 ✐ *Prove the formulae (8.15) for general state space models.*

Exercise 103 (Returns to the origin) *We consider a Markov chain X_n on the set of integers $S = \mathbb{N}$, defined by the elementary transition*

$$M(i,j) = 1_{S-\{0\}} \ 1_{i-1}(j) + 1_0(i) \ \mu(j)$$

where μ stands for some distribution on S. We also assume that $X_0 = 0$. This Markov model is a slight modification of the Markov chain discussed on page 177. We let X_n^1 and X_n^2 be two independent copies of X_n, and we let T be the first time n these chains return to the origin; that is

$$T = \inf \left\{ n \geq 1 \ : \ X_n^1 = 0 = X_n^2 \right\}.$$

- *Check that*

$$\mathbb{P}(T = 1) = 0 \quad \mathbb{P}(T = 2) = \mu(1)^2 \quad \text{and} \quad \mathbb{P}(T = 3) = \mu(2)^2.$$

- *Prove that*

$$(T = 4) = \{X_1^1 = 3 = X_1^2\}$$

$$\cup \ \{X_1^1 = 3 \ \& \ X_1^2 = 1 = X_3^2\} \ \cup \ \{X_1^1 = 1 = X_3^1 \ \& \ X_1^2 = 3\}$$

  and deduce that

$$\mathbb{P}(T = 4) = \mu(3)^2 + 2\mu(3)\mu(1)^2.$$

- *Using similar arguments, check that*

$$\mathbb{P}(T=5) = 2\mu(1)^2\mu(2)^2$$
$$\mathbb{P}(T=6) = 2\mu(1)\left(\mu(1)^2\mu(2)^2 + \mu(1)\mu(3)^2 + 2\mu(3)\mu(2)^2\right).$$

- *Compute the above probabilities for the geometric distribution with success parameter $p \in]0,1[$.*

Exercise 104 (Random sums) *Let $(X_n)_{n\geq 0}$ be a sequence of real valued i.i.d. copies of a random variable X with finite mean $m := \mathbb{E}(X) < \infty$. We let T be a stopping time (i.e., the event $\{T=n\}$ only depends on $(X_0,\ldots,X_n)$) with $\mathbb{E}(T) < \infty$. Prove that*

$$\mathbb{E}\left[\sum_{1\leq n\leq T} X_k\right] = \mathbb{E}(T)\,\mathbb{E}(X).$$

We further assume that $\mathbb{E}(X) = 0$ and $\sigma^2 := \mathrm{Var}(X) < \infty$. In this case, check that

$$\mathbb{E}\left[\left(\sum_{1\leq n\leq T} X_k\right)^2\right] = \mathbb{E}(T)\,\mathbb{E}(X^2).$$

Exercise 105 (Stochastic approximation - 1) 〃
   We let $\epsilon = (\epsilon^i)_{1\leq i\leq r}$ be some $\mathbb{R}^r$-valued random variable s.t. $\mathbb{E}(\epsilon^i) = 0$, and $\mathbb{E}((\epsilon^i)^2) < \infty$ for any $i \in \{1,\ldots,r\}$, $\tau = (\tau_n)_{n\geq 0} \in]0,1[^{\mathbb{N}}$ a sequence of parameters and $V : x \in \mathbb{R}^r$ some function such that

$$\|V(x) - V(y)\| \leq (1-\rho)\,\|x-y\|$$

with the Euclidian norm $\|v\|^2 = \langle v,v \rangle$ on $\mathbb{R}^r$, and some $\rho \in]0,1[$. We let $x^\star = V(x^\star)$ the unique fixed point of V. Consider the Markov chain X_n defined by the induction

$$X_{n+1} = X_n + \tau_n\,[(V(X_n) + \epsilon_n) - X_n]$$

with a sequence of independent and copies of ϵ, and some initial condition. We set $\mathcal{F}_n = \sigma(X_0,\ldots,X_n)$.

- *Check that*

$$\forall i \in \{1,\ldots,r\}\quad \epsilon_n^i = 0\quad and\quad \sum_{n\geq 0}\tau_n = \infty \implies \lim_{n\to\infty} X_n = x^\star.$$

- *We set $W(x) = \frac{1}{2}\,\|x - x^\star\|$. Check that*

$$\partial W(x^\star) = 0\quad and\quad (\partial W(x))^T\,(V(x) - x) < 0\ \forall x \neq x^\star$$

and deduce that

$$\langle x^\star - x, V(x) - x \rangle \geq \rho\,\|x^\star - x\|^2$$

and

$$\mathbb{E}\left[\|(V(x) + \epsilon) - x\|^2\right] \leq c\,\left(1 + \|x^\star - x\|^2\right)\quad for\ some\quad c < \infty.$$

- *We further assume that* $\tau_n \downarrow_{n\uparrow} 0$ *with*

$$\sum_{n\geq 0} \tau_n = \infty \quad and \quad \tau^2 := \sum_{n\geq 0} \tau_n^2 < \infty.$$

We set

$$I_n = \|X_n - x^\star\|^2 \quad and \quad M_n := I_n + c\,\tau^2 - c\sum_{0\leq k < n} \tau_k^2.$$

Check that

$$\mathbb{E}\left(I_{n+1} \mid \mathcal{F}_n\right) \leq \left(1 - \tau_n\left(2\rho - \tau_n\,c\right)\right) I_n + \tau_n^2\,c$$

and deduce that M_n is a non negative super-martingale such that $\sup_{n\geq 0} \mathbb{E}\left(M_n\right) < \infty$. Prove the almost sure convergences $\lim_{n\to\infty} I_n = 0 = \lim_{n\to\infty} M_n$.

Exercise 106 (Stochastic approximation - 2) ✎
  We let ϵ be some random variable on some state space S, $\tau = (\tau_n)_{n\geq 0} \in]0,1[^{\mathbb{N}}$ a sequence of parameters and $\mathcal{U} : (x,\epsilon) \in \mathbb{R}^r \times S \mapsto \mathbb{R}^r$ some function. We also set $U(x) = \mathbb{E}(\mathcal{U}(x,\epsilon))$ and we assume that

$$\langle x^\star - x, U(x)\rangle \geq \rho\,\|x^\star - x\|^2$$

and

$$\mathbb{E}\left[\|\mathcal{U}(x,\epsilon)\|^2\right] \leq c\,\left(1 + \|x^\star - x\|^2\right)$$

for some state $x^\star \in \mathbb{R}^r$ s.t. $U(x^\star) = 0$ and some finite constant $c < \infty$. Consider the Markov chain X_n defined by the induction

$$X_{n+1} = X_n + \tau_n\,\mathcal{U}\left(X_n, \epsilon_n\right)$$

with $\tau_n \downarrow_{n\uparrow} 0$. Check that

$$\mathcal{U}(x,\epsilon) = U(x) \quad and \quad \sum_{n\geq 0} \tau_n = \infty \implies \lim_{n\to\infty} X_n = x^\star.$$

We further assume that $\sum_{n\geq 0} \tau_n = \infty$ and $\tau^2 := \sum_{n\geq 0} \tau_n^2 < \infty$.. Check the estimates and the convergence results stated in the second part of exercise 105.
  Let W be some function given by $W(x) = \mathbb{E}(\mathcal{W}(x,\epsilon))$, for some $\mathcal{W} : (x,\epsilon) \in \mathbb{R}^r \times S \mapsto \mathbb{R}^r$. We let $a \in \mathbb{R}^r$ and x_a such that $W(x_a) = a$. We further assume that

$$\langle x_a - x, W(x_a) - W(x)\rangle \geq \rho\,\|x_a - x\|^2$$

$$\mathbb{E}\left[\|\mathcal{W}(x_a,\epsilon) - \mathcal{W}(x,\epsilon)\|^2\right] \leq c\,\left(1 + \|x_a - x\|^2\right).$$

Check that

$$X_{n+1} = X_n + \tau_n\,\left(a - \mathcal{W}(X_n, \epsilon_n)\right) \longrightarrow_{n\to\infty} x_a.$$

Exercise 107 (Symmetric random walk - Exponential martingale) *We consider the symmetric random walk $X_n = \sum_{1\leq k\leq n} U_n$ associated with a sequence of independent and identically distributed Bernoulli random variables U_n s.t. $\mathbb{P}(U_n = 1) = 1/2 = \mathbb{P}(U_n = 0)$. Find parameters (a,b) such that $Y_n = \exp\left(aX_n + bn\right)$ is a martingale w.r.t. the filtration $\mathcal{F}_n := \sigma(U_p,\ p \leq n)$.*

Exercise 108 (Symmetric random walk - Exit time) *We consider the symmetric random walk $X_n = X_0 + \sum_{1 \leq k \leq n} U_n$ associated with a sequence of independent and identically distributed Bernoulli random variables U_n s.t. $\mathbb{P}(U_n = 1) = 1/2 = \mathbb{P}(U_n = 0)$. Let T be the first time X_n exits the set $[-a, a] \ni X_0$, for given $a \in \mathbb{N} - \{0\}$. Prove that*

$$\mathbb{P}(T > 2a + n \mid T > n) \leq 1 - \frac{1}{2^{2a}}.$$

Deduce that $\mathbb{P}(T < \infty) = 1$ and $\mathbb{E}(T) < \infty$.

Exercise 109 (Martingales) *We let X be a real valued random variable s.t. $\mathbb{E}(|X|) < \infty$, and let $Y = (Y_n)_{n \geq 0}$ be a given Markov chain on some state space S. We let $\mathcal{F}_n = \sigma(Y_0, \ldots, Y_n)$ be the filtration associated with the chain Y. Prove that the sequence $Z_n := \mathbb{E}(X|\mathcal{F}_n)$ is a martingale w.r.t. the filtration $\mathcal{F}_n$.*

Exercise 110 (Product martingales) *We let $X = (X_n)_{n \geq 0}$ be a sequence of i.i.d. random variables s.t. $\mathbb{E}(X_1) = 1$ We let $\mathcal{F}_n = \sigma(X_0, \ldots, X_n)$ be the filtration associated with the sequence X. Prove that the sequence $Y_n := \prod_{1 \leq p \leq n} X_p$ is a martingale w.r.t. the filtration $\mathcal{F}_n$.*

Exercise 111 (Backward martingale models) *Let $X = (X_n)_{n \geq 0}$ be a given Markov chain on some state space S, and let f_n be some function on S s.t. $\mathbb{E}(|f_n(X_n)|) < \infty$ We fix some time horizon $n \geq 1$. Prove that the sequence $Y_k := \mathbb{E}(f_n(X_n)|X_k)$ is a martingale w.r.t. the filtration $(\mathcal{F}_k)_{0 \leq k \leq n}$ ending at $Y_n = f_n(X_n)$ for $k = n$.*

Exercise 112 (Lyapunov functions - 1) *Let M be the Markov transition of the $\mathbb{R}^r$-dimensional Markov chain defined by*

$$X_{n+1} - X_n = b(X_n) + \sigma(X_n) W_n$$

for some functions b from $\mathbb{R}^r$ into itself, some functions σ from $\mathbb{R}^r$ into the set of square $(r \times r)$-matrices, and a sequence of centered and independent variables W_n with unit covariance matrix. We set $V(x) = \|x\|^2$. Check that

$$\limsup_{\|x\| \to \infty} 2 \langle x, b(x) \rangle + \|b(x)\|^2 + \mathrm{tr}(\sigma(x)'\sigma(x)) < 0$$

$$\implies (\exists R > 0 \text{ s.t. } \forall \|x\| \geq R \quad \text{we have} \quad M(V)(x) - V(x) \leq 0)$$

and

$$\limsup_{\|x\| \to \infty} 2 \langle x, b(x) \rangle + \|b(x)\|^2 + \mathrm{tr}(\sigma(x)'\sigma(x)) < -1$$

$$\implies (\exists R > 0 \text{ s.t. } \forall \|x\| \geq R \quad \text{we have} \quad M(V)(x) - V(x) \leq -1).$$

Exercise 113 (Lyapunov functions - 2) *We consider the Markov chain model and the Lyapunov function discussed in exercise 112. We further assume that for any sufficiently large $R > 0$ we have*

$$\langle x, b(x) \rangle \leq -\rho_0 \|x\|^2 \qquad \|b(x)\|^2 \leq \rho_1 \|x\|^2 \quad \text{and} \quad \mathrm{tr}(\sigma(x)'\sigma(x)) \leq \rho_2 \|x\|^2$$

for any $x \notin B(0, R) = \{y \in \mathbb{R}^r : \|y\| < R\}$ and for some parameters ρ_i whose values do not depend on R. Find the best constants ρ_0, and ρ_1 for linear drift functions $b(x) = Ax$ associated with a symmetric matrix A. For constant diffusion matrices $\sigma(x) = \sigma$ check that we can choose $\rho_2 = \epsilon r$, where ϵ stands for the maximal eigenvalue of the symmetric matrix $\sigma'\sigma$, as soon as $R \geq 1$. Check that for any $x \notin B(0, R)$ we have

$$\rho_0 > (\rho_1 + \rho_2)/2 \implies [M(V)(x) - V(x)] \leq -\left(\rho_0 - \frac{\rho_1 + \rho_2}{2}\right) R^2.$$

Exercise 114 (Super-martingale design) *Consider a time homogeneous Markov chain X_n with Markov transitions M on some state space S. We let T_A be the first time X_n hits some set $A \subset S$. Assume that there exists some non-negative function V s.t. $(M(V) - V)(x) \le -1$ for any $x \in S - A$. We set $\mathcal{N}_n := V(X_{T_A \wedge n}) - V(X_0) + (T_A \wedge n)$.*
Using the martingale decomposition (8.47) of the process $V(X_n)$, check that

$$\mathcal{N}_n - \mathcal{N}_{n-1} \le \mathcal{M}_{T_A \wedge n}(V) - \mathcal{M}_{T_A \wedge (n-1)}(V).$$

Deduce that $\mathcal{N}_n$ is a super-martingale.

Exercise 115 (Return times - Foster-Lyapunov functions) *Consider the time homogeneous Markov chain X_n discussed in exercise 114. Assume that there exists some non negative function V s.t. $(M(V) - V)(x) \le 0$ for any $x \in S - A$, and $V(x_0) \le \inf_{x \in A} V(x)$. Using exercise 114, check that*

$$\mathbb{E}\left(V(X_{T_A}) + T_A \mid X_0 = x_0\right) - V(x_0) \le \mathbb{E}\left(\mathcal{N}_n \mid X_0 = x_0\right) \le \mathbb{E}\left(\mathcal{N}_0 \mid X_0 = x_0\right) = 0.$$

Deduce that

$$\mathbb{E}\left(T_A \mid X_0 = x_0\right) \le V(x_0) / \inf_{x \in A} V(x) \le 1.$$

Exercise 116 (Harris recurrent sets) *Consider the time homogeneous Markov chain X_n discussed in exercise 114. Assume that there exists some non negative function V s.t. $(M(V) - V)(x) \le 0$ for any $x \in S - A$, and the hitting times T_C of the sets $C = \{V > c\}$ are finite (for any starting state $x \in S$). Using the same reasoning as in exercise 114, check that $\mathcal{N}_n := V(X_{T_A \wedge n})$ is a super-martingale. Using the optional stopping theorem (theorem 8.4.12) check that $\mathcal{N}_{T_C \wedge n}$ is also a super-martingale. Prove that*

$$V(x) \ge \mathbb{E}\left(V(X_{T_C \wedge T_A}) \mathbb{1}_{T_A = \infty} \mid X_0 = x\right) \ge c\, \mathbb{P}\left(T_A = \infty \mid X_0 = x\right).$$

Deduce that $\mathbb{P}(T_A < \infty \mid X_0 = x) = 1$ for any $x \in S$. The sets A satisfying this property are called Harris recurrent sets.

Exercise 117 (Positive recurrent set) *Consider the time homogeneous Markov chain X_n discussed in exercise 114. Assume that there exists some non negative function V s.t. $(M(V) - V)(x) \le -1$ for any $x \in S - A$. Check that $\mathbb{E}(T_A \mid X_0 = x) \le V(x)$ for any $x \in S - A$. We let $T'_A = \inf\{n \ge 1 : X_n \in A\}$. Check that*

$$\forall x \in A \qquad \mathbb{E}\left(T'_A \mid X_0 = x\right) \le M(V)(x).$$

The sets A satisfying the above property are called positive recurrent sets.

9

Computational toolbox

The toolbox from the previous chapter is extended in this chapter. Whereas in the previous chapter we were focusing on analytic type tools, the focus in this chapter is on computational tools. We present a review of traditional Markov chain Monte Carlo methodology, perfect sampling, and time inhomogeneous Markov chain Monte Carlo models. We also provide an extensive introduction to the more recent Feynman-Kac particle methodolgy, by giving a unified treatment of a large set of models with different names and guises scattered around in a variety of application domains.

> *Computers are useless.*
> *They can only give you answers.*
> Pablo Picasso (1881-1973).

9.1 A weak ergodic theorem

We consider a Markov chain X_n on some state space S with a Markov transition that satisfies condition (8.15). By the contraction theorem 8.2.13, there exists a unique invariant measure $\pi = \pi M$.

Definition 9.1.1 *The occupation measures of the Markov chain X_n are the random measures π_n defined for any $n \geq 0$ by*

$$\pi_n = \frac{1}{n+1} \sum_{0 \leq p \leq n} \delta_{X_p}.$$

When the distribution of the Markov chain X_n converges to some invariant measure π, the sampling errors are defined by the empirical random field V^n defined for any function f on S by

$$V^n(f) = \sqrt{(n+1)} \, [\pi_n - \pi] \, (f)$$

or in an equivalent perturbation formulation

$$\pi_n = \pi + \frac{1}{\sqrt{(n+1)}} \, V^n.$$

Theorem 9.1.2 (Weak ergodic theorem) *We assume that the Markov transition M satisfies (8.15). For any $f \in \mathcal{B}(S)$, with $\mathrm{osc}(f) \leq 1$, and any initial distribution $\eta_0 = \mathrm{Law}(X_0)$, we have the bias and the variance estimates*

$$|\mathbb{E}\left(V^n(f)\right)| \leq \frac{1}{\sqrt{n+1}} \, \frac{1}{1 - \beta(M)} \tag{9.1}$$

221

and

$$\mathbb{E}\left([V^n(f)]^2\right) \le \left(1 + \frac{2}{1 - \beta(M)}\right). \tag{9.2}$$

In addition, we have the first order estimate

$$\left| \mathbb{E}\left(V^n(f)^2\right) - \left[\pi([f - \pi(f)]^2) + 2\sum_{p \ge 1} \pi\left((f - \pi(f))M^p(f - \pi(f))\right)\right]\right|$$

$$\le \frac{1}{(n+1)}\frac{5}{(1 - \beta(M))^2}$$

with the convention that $M^0 = Id$, the identity integral operator.

Important remark :
We emphasize that the weak ergodic theorem can be extended to any Markov chain which forgets its initial condition sufficiently fast. Condition (8.15) is a rather crude sufficient condition ensuring that the chain forgets its initial condition exponentially fast. For instance, the weak ergodic theorem is also true when M^p satisfies condition (8.15), for some $p \ge 1$ only. In this case, the constants in the r.h.s. of (9.1) differ and they depend on $\beta(M^p)$ instead of $\beta(M)$.
Proof :
We use the decomposition

$$\mathbb{E}\left(\pi_n(f)\right) - \pi(f) \quad = \quad \frac{1}{n+1}\sum_{0 \le p \le n} \left([\eta_0 - \pi]M^p\right)(f)$$

to prove that

$$|\mathbb{E}\left(\pi_n(f)\right) - \pi(f)| \quad = \quad \frac{1}{n+1}\sum_{0 \le p \le n} \|[\eta_0 - \pi]M^p\|_{tv}$$

$$= \quad \frac{1}{n+1}\sum_{0 \le p \le n} \beta(M)^p \le \frac{1}{n+1}\frac{1}{1 - \beta(M)}.$$

To prove the second assertion, we first notice that for any $p < q$ and any bounded function f it holds:

$$\mathbb{E}\left(f(X_p)\, f(X_q)\right) \quad = \quad \mathbb{E}\left(f(X_p)\,\mathbb{E}\left(f(X_q) \mid X_p\right)\right)$$

$$= \quad \mathbb{E}\left(f(X_p)\, M^{q-p}(f)(X_p)\right) = (\eta_0 M^p)\left(f M^{q-p}(f)\right).$$

Using these relations, we readily check that

$$(n+1)^2\mathbb{E}\left([\pi_n(f) - \pi(f)]^2\right)$$

$$= \mathbb{E}\left(\left[\sum_{0 \le p \le n}(f(X_p) - \pi(f))\right]^2\right)$$

$$= \sum_{0 \le p \le n} (\eta_0 M^p)\left[(f - \pi(f))^2\right]$$

$$+2\sum_{0 \le p < q \le n} (\eta_0 M^p)\left((f - \pi(f))M^{q-p}(f - \pi(f))\right).$$

On the other hand, for any function f s.t. $\operatorname{osc}(f) \leq 1$ we have

$$M^{q-p}(f - \pi(f))(x) = M^{q-p}(f)(x) - \pi M^{q-p}(f).$$

This implies that

$$M^{q-p}(f - \pi(f))(x) = [\delta_x - \eta_0] M^{(q-p)}(f)$$

$$\implies \|M^{q-p}(f - \pi(f))\| \leq \left\|[\delta_x - \eta_0] M^{(q-p)}\right\|_{tv} \leq \beta(M)^{q-p}.$$

Therefore

$$\left|\sum_{0 \leq p < q \leq n} (\eta_0 M^p)\left((f - \pi(f))M^{q-p}(f - \pi(f))\right)\right|$$

$$\leq \sum_{0 \leq p \leq n} \sum_{p < q \leq n} \beta(M)^{q-p} \leq (n+1)/(1 - \beta(M)).$$

The end of the proof of the variance estimate is now clear.

To prove the final assertion, for any function g s.t. $\operatorname{osc}(g) \leq 1$ we observe that

$$|(\eta_0 M^p)(g) - \pi(g)| \leq \beta(M)^p.$$

This yields that for g of the form $g = f - \pi(f)$, we have

$$\left|\frac{1}{n+1} \sum_{0 \leq p \leq n} [(\eta_0 M^p)(g) - \pi(g)]\right| \leq \frac{1}{n+1} \frac{1}{1 - \beta(M)}.$$

In much the same way,

$$\left|\sum_{q=1}^{(n-p)} M^q(f - \pi(f)) - \sum_{q \geq 1} M^q(f - \pi(f))\right|$$

$$= \left|\sum_{q > (n-p)} M^q(f - \pi(f))\right| \leq \sum_{q > (n-p)} \beta(M)^q = \frac{\beta(M)^{(n-p)+1}}{1 - \beta(M)}.$$

Thus, if we set

$$I_{n-p}(f) := \sum_{q=1}^{(n-p)} M^q(f - \pi(f)) \quad \text{and} \quad I_\infty(f) := \sum_{q \geq 1} M^q(f - \pi(f))$$

we have

$$\left|\sum_{0 \leq p \leq n} (\eta_0 M^p)\left((f - \pi(f))I_{n-p}(f)\right) - \sum_{0 \leq p \leq n} (\eta_0 M^p)\left((f - \pi(f))I_\infty(f)\right)\right|$$

$$\leq \frac{\beta(M)}{1 - \beta(M)} \sum_{0 \leq p \leq n} \beta(M)^{(n-p)} \leq \frac{\beta(M)}{[1 - \beta(M)]^2}.$$

On the other hand, we also have the estimate

$$\left|\left[\frac{1}{n+1} \sum_{0 \leq p \leq n} (\eta_0 M^p)\left((f - \pi(f))I_\infty(f)\right)\right] - \pi\left((f - \pi(f))I_\infty(f)\right)\right|$$

$$= \frac{1}{n+1} \left|\sum_{0 \leq p \leq n} [\eta_0 M^p - \pi M^p]\left((f - \pi(f))I_\infty(f)\right)\right|$$

$$\leq \frac{2}{n+1} \|I_\infty(f)\| \sum_{0 \leq p \leq n} \beta(M)^p \leq \frac{1}{n+1} \frac{2}{(1 - \beta(M))^2}.$$

The last assertion comes from the fact that

$$I_\infty(f) := \sum_{q \geq 1} [M^q - \pi M^q](f) \Rightarrow \|I_\infty(f)\| \leq 1/(1 - \beta(M)).$$

We conclude that

$$(n+1) \left| \mathbb{E}\left([\pi_n(f) - \pi(f)]^2\right) - [\pi([f - \pi(f)]^2) + 2\pi((f - \pi(f))I_\infty(f))]\right|$$

$$\leq \frac{1}{1 - \beta(M)} + \frac{\beta(M)}{[1 - \beta(M)]^2} + \frac{4}{(1 - \beta(M))^2} = \frac{5}{(1 - \beta(M))^2}.$$

This ends the proof of the theorem. ∎

We consider the path space model $\boldsymbol{X}_n = (X_n, X_{n+1}, \ldots, X_{n+p})$, with a fixed $p \geq 0$ presented in (8.5.13).

Using lemma 8.5.15 we readily prove that the ergodic theorem 9.1 is also true for the Markov chain $\boldsymbol{X}_n$.

Corollary 9.1.3 *We assume that M satisfies (8.15). In this situation, ss $n \uparrow \infty$, we have the convergence*

$$\boldsymbol{\pi_n} := \frac{1}{n+1} \sum_{0 \leq q \leq n} \delta_{\boldsymbol{X}_q} \longrightarrow_{n \to \infty} \boldsymbol{\pi}$$

in the sense that for any $f \in \mathcal{B}(S^{p+1})$:

$$|\mathbb{E}(\boldsymbol{\pi_n}(f)) - \boldsymbol{\pi}(f)| \vee \mathbb{E}\left([\boldsymbol{\pi_n}(f) - \boldsymbol{\pi}(f)]^2\right) = \mathrm{O}(1/n).$$

9.2 Some illustrations

9.2.1 Parameter estimation

Suppose that we are observing the random states X_n of the $\{0, 1\}$-valued Markov chain presented in (7.11), and we want to estimate the parameters p, q.

In (7.12) we have shown that the invariant measure of the chain X_n is given by

$$\pi = [\pi(0), \pi(1)] = \left[\frac{q}{p+q}, \frac{p}{p+q}\right].$$

By the ergodic theorem 9.1 we know that

$$\pi_n(1_0) := \frac{1}{n+1} \sum_{0 \leq p \leq n} 1_0(X_n) \longrightarrow \pi(1_0) = \pi(0) = \frac{q}{p+q}.$$

We also observe that the invariant measure of the chain $\boldsymbol{X}_n = (X_n, X_{n+1})$ is given by

$$\boldsymbol{\pi}(x, y) = \pi(x) \, M(x, y).$$

By corollary 9.1.3 we have that

$$\boldsymbol{\pi_n}(1_{(0,1)}) = \frac{1}{n+1} \mathrm{Card}\{0 \leq p \leq n \; : \; (X_p, X_{p+1}) = (0, 1)\} \simeq_{n \uparrow \infty} p \frac{q}{p+q}.$$

This implies that

$$\frac{\pi_n(1_{(0,1)})}{\pi_n(1_0)} \simeq_{n\uparrow\infty} p \, \frac{q}{p+q} \times \frac{p+q}{q} = p.$$

If we interpret π_n as the (random) distribution of a couple of random variables $(Y_1^{(n)}, Y_2^{(n)})$, then we have

$$\mathbb{P}\left(Y_2^{(n)} = 1 \,\Big|\, Y_1^{(n)} = 0 \,, (X_0, \dots, X_n)\right) = \frac{\pi_n(1_{(0,1)})}{\pi_n(1_0)} \simeq_{n\uparrow\infty} p.$$

9.2.2 Gaussian subset shaker

Our next objective is to sample a centered Gaussian random variable Z with unit variance restricted to some set A, say $A =]a, b[$, for some $-\infty \leq a < b \leq \infty$. We let λ be the distribution of Z, and we set

$$\pi(dz) := \frac{1}{\lambda(A)} \, 1_A(z) \, \lambda(dz) = \mathbb{P}\left(Z \in dz \mid Z \in A\right).$$

Of course, we can use the distribution function $F(z) = \mathbb{P}(Z \leq z)$ of the Gaussian random variable and set

$$Z_{a,b} = F^{-1}\left(F(a) + U\left(F(b) - F(a)\right)\right). \tag{9.3}$$

It is an elementary exercise to check that $\mathrm{Law}(Z_{a,b}) = \pi$ (cf. exercise 119). Nevertheless, the function F requires us to integrate the Gaussian density from $-\infty$ up to any state z, using some kind of numerical approximation scheme.

Another strategy is to use the rejection simulation technique described in section 4.3. In this case, we sample a sequence of independent copies of Z and we accept the ones that hit the desired set A.

Next, we describe an alternative approach based on Markov chain simulation.

We let Z_n be a sequence of independent copies of Z. We design a Markov chain with invariant measure π as follows.

Suppose, a chain with transition K is chosen and that the chain $X_n \in A$ at some time step $n \geq 0$. Starting from this point, we set

$$Y_{n+1} = \sqrt{1 - \epsilon_n} \, X_n + \sqrt{\epsilon_n} \, Z_n.$$

If $Y_{n+1} \in A$ we accept the move and we set $X_{n+1} = Y_{n+1}$. Otherwise, we stay in the same place: $X_{n+1} = X_n$. The Markov transition of the chain $X_n \rightsquigarrow X_{n+1}$ is now given by

$$M(x, dy) = K(x, dy) \, 1_A(y) + (1 - K(1_A)(x)) \, \delta_x(dy)$$

with $K(x, dy) = \mathbb{P}(Y_{n+1} \in dy \mid X_n = x)$. We claim that

$$\frac{1}{n+1} \sum_{0 \leq p \leq n} \delta_{X_p} \longrightarrow_\infty \pi.$$

We prove that $\pi M = \pi$ as follows. Recalling that the transition $K(x, dy)$ is a reversible w.r.t. to the Gaussian distribution λ, for any bounded function f on $\mathbb{R}$,

$$\begin{aligned}
\pi M(f) &\propto \lambda(1_A M(f)) \\
&= \lambda(1_A K(1_A \, f)) + \lambda(1_A(1 - K(1_A) \, f) \\
&= \lambda(K(1_A)1_A \, f) + \lambda(1_A f) - \lambda(1_A \, K(1_A) \, f) = \lambda(1_A f) \propto \pi(f).
\end{aligned}$$

This clearly implies that $\pi = \pi M$.

9.2.3 Exploration of the unit disk

Suppose we want to select uniformly a point $Z = (X, Y)$ in the unit disk

$$A := \{(x, y) \in [-1, 1]^2 \; : \; x^2 + y^2 \leq 1\}.$$

Here again, we can use the rejection simulation technique described in section 4.3. In this case, we sample a sequence of independent random variables on $[-1, 1]^2$ and we accept the ones that hit the desired set A. Notice that the invariant measure π on A is defined by

$$
\begin{aligned}
\pi(d(x, y)) \quad &\propto \quad 1_A(x, y) \, dx dy \\
&= \quad 1_{[-1,1]}(x) \, dx \; 1_{[-\sqrt{1-x^2}, +\sqrt{1-x^2}]}(y) \, dy \\
&= \quad 1_{[-1,1]}(y) \, dy \; 1_{[-\sqrt{1-y^2}, +\sqrt{1-y^2}]}(x) \, dx.
\end{aligned}
\tag{9.4}
$$

This implies that

$$
\begin{aligned}
\mathbb{P}\left(Y \in dy \mid X = x\right) \quad &\propto \quad 1_{[-\sqrt{1-x^2}, +\sqrt{1-x^2}]}(y) \, dy \\
\mathbb{P}\left(X \in dx \mid Y = y\right) \quad &\propto \quad 1_{[-\sqrt{1-y^2}, +\sqrt{1-y^2}]}(x) \, dx.
\end{aligned}
$$

Pick any initial point (X_0, Y_0) in A. The next state of the chain (X_1, Y_1) is defined as follows. Firstly, we choose uniformly a point X_1 on $\left[-\sqrt{1 - Y_0^2}, +\sqrt{1 - Y_0^2}\right]$. Then, we choose uniformly a state Y_1 on $\left[-\sqrt{1 - X_1^2}, +\sqrt{1 - X_1^2}\right]$. Iterating these transitions, we construct a Markov chain evolving inside the the unit disk A

$$
Z_n := \begin{pmatrix} X_n \\ Y_n \end{pmatrix} \rightsquigarrow \begin{pmatrix} X_{n+1} \\ Y_n \end{pmatrix} \rightsquigarrow Z_{n+1} = \begin{pmatrix} X_{n+1} \\ Y_{n+1} \end{pmatrix}.
\tag{9.5}
$$

After some elementary computations, we prove that π is the invariant measure of the chain Z_n (see exercise 121). Applying the ergodic theorem, we find that

$$
\frac{1}{n+1} \sum_{0 \leq p \leq n} \delta_{Z_p} \longrightarrow_{n \to \infty} \pi.
$$

9.3 Markov Chain Monte Carlo methods

9.3.1 Introduction

Markov chain Monte Carlo algorithms are rather standard stochastic simulation methods for sampling from a given target distribution, say π, on some state space S. The prototype of target a measure is given by Boltzmann-Gibbs measures of the following form

$$
\pi(dx) = \Psi_G(\lambda)(dx) = \frac{1}{\lambda(G)} \, G(x) \, \lambda(dx)
\tag{9.6}
$$

where λ denotes a reference probability measure and G is a potential function on some state space S.

Several examples have been discussed in section 6.4, section 9.2.2, and section 9.2.3.

The central idea behind MCMC methodologies is to design a judicious Markov transition $M(x, dy)$, with adequate stability properties, that has the target probability measure $\pi = \pi M$ as its invariant measure. After a rather large number of runs, and when the chain is sufficiently stable, the ergodic theorem tells us that the occupation measures of the random states X_n of the chain with Markov transition M approximate π.

9.3.2 Metropolis and Hastings models

The Metropolis-Hastings algorithm is the most famous MCMC model of current use in practice.

Firstly, we choose a Markov transition K to explore randomly the state S. We further assume that $K(x, dy)$ and the target measure $\pi(dy)$ have a density with respect to some reference probability measure $\lambda(dy)$, that is,

$$K(x, dy) = k(x, y)\,\lambda(dy) \quad \text{and} \quad \pi(dy) = h(y)\,\lambda(dy)$$

with some density functions $k(x, y)$ and $h(y)$ such that

$$h(y)k(y, x) = 0 \implies h(x)k(x, y) = 0.$$

We set

$$G(x, y) := \frac{h(y)k(y, x)}{h(x)k(x, y)}$$

with the convention $0/0 = 0$. For more general models, we take

$$G(x, y) = \frac{\pi(dy)K(y, dx)}{\pi(dx)K(x, dy)}. \tag{9.7}$$

For Boltzmann-Gibbs measures π of the form (9.6), it is readily checked that the function G does not depend on the normalizing constant $\lambda(G)$, and is given by the formula

$$G(x, y) = \frac{G(y)}{G(x)} \times \frac{\lambda(dy)K(y, dx)}{\lambda(dx)K(x, dy)}.$$

In addition, when the proposal transition K is reversible w.r.t. the measure λ, the function G takes the simpler form

$$G(x, y) = G(y)/G(x). \tag{9.8}$$

Definition 9.3.1 *The Metropolis-Hastings model is a Markov chain with μ-reversible acceptance-rejection style transitions of the following form*

$$M(x, dy) = K(x, dy)\,a(x, y) + \left(1 - \int K(x, dz)\,a(x, z)\right)\delta_x(dy). \tag{9.9}$$

To guarantee the reversibility property, we often chose one of the following acceptance rates

$$a = G/(1 + G) \qquad \text{or} \qquad a = 1 \wedge G. \tag{9.10}$$

The Markov chain with acceptance ratio $a = G/(1 + G)$ is sometimes called the heat-bath Markov chain sampler. When the proposal transition $K(x, .) = \nu$ is given by some probability measure ν that does not depend on the current state x, the resulting MCMC sampler is sometimes called an independent Metropolis-Hastings sampler.

> **Theorem 9.3.2** *The Metropolis-Hastings transition (9.9) associated with one of the acceptance rates a given in (9.10) is reversible w.r.t. the target measure π; that is,*
>
> $$\pi(dx)M(x,dy) = \pi(dy)M(y,dx).$$

Proof :

When $a = 1 \wedge G$, for any $x \neq y$ we have

$$
\begin{aligned}
\pi(dx)M(x,dy) &= \pi(dx)K(x,dy)\, a(x,y) \\
&= \lambda(dx)\, h(x)\, k(x,y)\lambda(dy) \left\{ 1 \wedge \frac{(h(y)k(y,x))}{(h(x)k(x,y))} \right\} \\
&= \lambda(dx)\lambda(dy)\ \{(h(x)k(x,y)) \wedge (h(y)k(y,x))\}.
\end{aligned}
$$

This formula is clearly symmetric w.r.t. x and y.

When $a = G/(1+G)$, for any $x \neq y$ we have

$$
a(x,y) = \frac{\frac{h(y)k(y,x)}{h(x)k(x,y)}}{1 + \frac{h(y)k(y,x)}{h(x)k(x,y)}} = \frac{h(y)k(y,x)}{h(x)k(x,y) + h(y)k(y,x)}
$$

and

$$
\begin{aligned}
\pi(dx)M(x,dy) &= \pi(dx)K(x,dy)\, a(x,y) \\
&= \lambda(dx)\lambda(dy)\ \frac{(h(y)k(y,x))(h(x)k(x,y))}{h(x)k(x,y) + h(y)k(y,x)}.
\end{aligned}
$$

This formula is again symmetric w.r.t. x and y.   ∎

For a detailed discussion on this model, we refer the reader to the pioneering article by N. Metropolis, A. Rosenbluth, M. Rosenbluth, A. Teller, and E. Teller [198], the more recent review article by N. Metropolis [197], and the series of articles by P. Diaconis [99, 100, 102].

The mathematical analysis of this Markov chain model is also well developed. We refer the reader to the series of seminal articles by P. Diaconis and his co-authors [94, 95, 96, 97, 101, 103]. These works reveal fascinating connections between the design and the performance analysis of MCMC models with powerful pure and applied mathematical techniques, ranging from representation theory, micro-local analysis, log-Sobolev inequalities, and spectral analysis.

Because these techniques provide very sharp rates of convergence, it is clearly beyond the scope of this book to review these methods. In the end of this section, we content ourselves with presenting one of the simplest ways to analyze the convergence of an MCMC algorithm.

Suppose that K satisfies the minorization condition

$$K^m(x,dy) \geq \epsilon\, \nu(dy)$$

for some integer $m \geq 1$, some $\epsilon \in\,]0,1]$ and some probability measure ν. In this situation,

$$M^m(x,dy) \geq \epsilon'\, \nu(dy)$$

with

$$\epsilon' = \epsilon \inf_{x_0 \rightsquigarrow \ldots \rightsquigarrow x_m} \left[\prod_{0 \leq p < m} a(x_p, x_{p+1}) \right].$$

The r.h.s. infimum is taken over all sequences $(x_0, \ldots, x_m)$ of states in S of length m and such that

$$k(x_p, x_{p+1}) > 0.$$

Whenever $\epsilon > 0$, the Dobrushin contraction coefficient of M^m is s.t. $\beta(M^m) < 1$. For a more detailed discussion on MCMC models, and their stochastic analysis we refer the reader to the review articles by P. Diaconis [93, 98, 103], and the references therein.

9.3.3 Gibbs-Glauber dynamics

We let π be some target measure defined on some product state space $S = (S_1 \times S_2)$.

> We assume that the following disintegration property
>
> $$\pi(d(x_1, x_2)) = \pi_1(dx_1)\, L_{1,2}(x_1, dx_2) = \pi_2(dx_2)\, L_{2,1}(x_2, dx_1)$$
>
> is satisfied with the first, and second marginals, π_1 and π_2, and the corresponding conditional probability measures $L_{1,2}$ and $L_{2,1}$.

The `disintegration theorem` ensures that any Borel probability measure π on Radon spaces can be disintegrated. From a probabilistic view, the disintegration property is equivalent to finding the conditional distributions of one coordinate given the second. In Bayesian notation, the above disintegration formulae are often expressed as probability densities

$$p(x_1, x_2) = p(x_1)\, p(x_2|x_1) = p(x_2)\, p(x_1|x_2).$$

It is important to notice that these disintegration formulae depend on the geometry of the state space, that is, on the coordinate system on which we express a given probability measure. Several examples of disintegration formulae are discussed in the further development of this section. We also refer the reader to section 23.3.1 dedicated to uniform measures on geometric surfaces, as well as to the series of exercises in the final section of this chapter.

> **Definition 9.3.3** *The Gibbs sampler is the Markov chain with the elementary transition*
>
> $$M = K_1 K_2$$
>
> *with the transitions K_i given for any $i \in \{1, 2\}$ by*
>
> $$\begin{aligned} K_1((x_1, x_2), d(y_1, y_2)) &:= \delta_{x_1}(dy_1)L_{1,2}(y_1, dy_2) \\ K_2((x_1, x_2), d(y_1, y_2)) &:= \delta_{x_2}(dy_2)L_{2,1}(y_2, dy_1). \end{aligned}$$
>
> *We can alternatively choose the Markov transitions*
>
> $$M = K_2 K_1 \quad \text{or} \quad M = \frac{1}{2}\, K_1 + \frac{1}{2}\, K_2.$$

Theorem 9.3.4 *The transitions K_1 and K_2 are reversible w.r.t. the measure π. In addition, the Metropolis-Hastings transitions M_1, and respectively M_2, with proposal transition K_1, and respectively K_2, and acceptance rate $a = 1 \wedge G$ with G given by (9.7) have unit acceptance rate.*

Proof :
We check this claim using the fact that

$$\pi(d(y_1, y_2)) \times K_1((y_1, y_2), d(x_1, x_2))$$

$$= \pi_1(dy_1) L_{1,2}(y_1, dy_2) \times \delta_{y_1}(dx_1) L_{1,2}(x_1, dx_2)$$

$$= \underbrace{\pi_1(dy_1) \delta_{y_1}(dx_1)}_{= \pi_1(dx_1) \delta_{x_1}(dy_1)} \times (L_{1,2}(y_1, dy_2) L_{1,2}(x_1, dx_2)).$$

This formula is clearly symmetric w.r.t. $x = (x_1, x_2)$ and $y = (y_1, y_2)$. Thus, we have

$$G((x_1, x_2), (y_1, y_2)) = \frac{\pi(d(y_1, y_2)) \times K_1((y_1, y_2), d(x_1, x_2))}{\pi(d(x_1, x_2)) \times K_1((x_1, x_2), d(y_1, y_2))} = 1.$$

$\blacksquare$

The resulting Markov chain model is often called the Gibbs sampler or the Glauber dynamics.

Example 9.3.5 *The random exploration of the unit disk discussed in section 9.2.3 is the Gibbs model associated with the disintegration formulae (9.4).*

These constructions can be extended to product state spaces of any dimension. More formally, we let π be some target measure on some state space $S = E^I$, where I stands for some finite set.

Example 9.3.6 (Graph coloring model) *For instance $E = \{1, \ldots, d\}$ can be the set of colors on the vertices of some graph $I = (\mathcal{V}, \mathcal{E})$. When $E = \{0, 1\}$, the color 0 can be interpreted as an empty site, and the color 1 as an occupied site. In this situation, a given configuration $x = (x(i))_{i \in I}$ can be interpreted as a collection of particles placed on the vertices $i \in I$ s.t. $x(i) = 1$.*

We let $X = (X_i)_{i \in I}$ be some random variable with distribution π on E^I. For any fixed $i \in I$, and any $x \in E^I$ we set

$$x_{I-i} = (x_j)_{j \in I - \{i\}} \quad \text{with} \quad I - i := \{j \in I \, : \, j \neq i\}.$$

We assume that the following disintegration property is satisfied

$$\pi(dx) = \pi_{I-i}(dx_{I-i}) L_{I-i,i}(x_{I-i}, dx_i)$$

with the i-th marginals π_i of π and the conditional probability measure

$$L_{I-i,i}(x_{I-i}, dx_i) = \mathbb{P}(X_i \in dx_i \mid X_{I-i} = x_{I-i}).$$

In the above displayed formulae, dx, resp. dx_i and dx_{I-i}, stand for an infinitesimal neighborhood of the point $x \in E^I$, resp. $x_i \in E$ and $x_{I-i} \in E^{I-i}$.

We associate with these models the Markov transitions K_i given for any $i \in I$ by

$$K_i(x, dy) \quad := \quad \delta_{x_{I-i}}(dy_{I-i}) \, L_{I-i,i}(y_{I-i}, dy_i). \tag{9.11}$$

The corresponding Gibbs sampler is the Markov chain with the elementary transition

$$M = \prod_{i \in I} K_i \quad \text{in any order.} \tag{9.12}$$

We can alternatively choose the Markov transitions

$$M = \frac{1}{\text{Card}(I)} \sum_{i \in I} K_i. \tag{9.13}$$

Arguing as above, one checks that M is reversible w.r.t. π, so that $\pi M = \pi$.

Example 9.3.7 (Subset sampling) *We let λ be a reference probability measure on the set $S = E^I$ discussed above. We assume that λ satisfies the disintegration property*

$$\lambda(dx) = \lambda_{I-i}(dx_{I-i}) \, P_{I-i,i}(x_{I-i}, dx_i)$$

for some Markov transitions $P_{I-i,i}$ from E^{I-i} into E. For finite state spaces E, we can consider the product counting measures

$$\lambda(x) = \prod_{i \in I} \lambda_i(x_i) \quad \text{with} \quad \lambda_i(x_i) = \frac{1}{\text{Card}(E)}.$$

In this situation, we have

$$\lambda_{I-i}(dx_{I-i}) = \prod_{j \in I} \lambda_j(x_j) \quad \text{and} \quad P_{I-i,i}(x_{I-i}, dx_i) = \lambda_i(x_i).$$

We let π be the Boltzmann-Gibbs measure associated with some subset $A \subset S = E^I$ and defined by

$$\pi(dx) = \frac{1}{\lambda(A)} \, 1_A(x) \, \lambda(dx).$$

For each $i \in I$, we let A_{I-i} be the projection of the set A into the set E^{I-i} defined by the set of mappings $x_{I-i} \in E^{I-i}$ that can be extended to some mapping $x \in A$ by choosing some $k \in E$ and setting $x(i) = k$. In this slightly abusive notation, we let

$$A_i(x_{I-i}) = \{k \in E \ : \ x \in A\}.$$

By construction, we have

$$1_A(x) = 1_{A_{I-i}}(x_{I-i}) \times 1_{A_{I-i}(x_{I-i})}(x_i)$$

and therefore

$$\pi(dx) \propto \underbrace{1_{A_{I-i}}(x_{I-i}) \, \lambda_{I-i}(dx_{I-i})}_{\propto \, \pi_{I-i}(dx_{I-i})} \times \underbrace{P_{I-i,i}(x_{I-i}, dx_i) \, 1_{A_{I-i}(x_{I-i})}(x_i)}_{\propto \, L_{I-i,i}(x_{I-i}, dx_i)}.$$

The Gibbs sampler associated with the Markov transition (9.12) is defined by a Markov chain $X_n = (X_n(i))_{i \in I} \in S = E^I$.

At time n, we choose randomly a vertex $i \in I$ and we set $X_{n+1}(j) = X_n(j)$ for any $j \in I-i$. Finally $X_n(i)$ is a random variable with the distribution $P_{I-i,i}(x_{I-i}, dx_i)$ restricted to the set $A_{I-i}(x_{I-i})$.

Example 9.3.8 (Graph coloring and hard core models) *In the graph coloring model discussed above, we let A be the set of graph colorings such that two neighbor vertices have different colors.*

When $E = \{0, 1\}$ the set A can be chosen as to represent the configurations where no two occupied sites are adjacent (that is, $i \sim j \Rightarrow x(i)x(j) \neq 1$). In statistical physics, this model is often referred to as the hard-core model.

In this context, $I - i$ represents the set of all vertices that differ from the vertex $i \in I$. Given some coloring x_{I-i} of these vertices $I - i$, the set $A_i(x_{I-i})$ coincides with the set of colors $k \in E$ that do not appear in the neighborhood of the vertex i.

When the reference measure λ is given by the product counting measure on the set of colors, the Markov transition $L_{I-i,i}(x_{I-i}, dx_i)$ amounts to choosing uniformly at random some color k that does not appear in the neighborhood of the vertex i and setting $x_i = k$.

The Gibbs sampler associated with the Markov transition (9.12) is defined by a Markov chain $X_n = (X_n(i))_{i \in I} \in S = E^I$: At time n, we choose randomly a vertex $i \in I$ and we set $X_{n+1}(j) = X_n(j)$ for any $j \in I - i$. Finally $X_n(i)$ is an uniform r.v. on the set $E - X_n(N(i))$, where $N(i)$ denotes the set of all neighbors j of i (that is the vertices j s.t. $(i, j) \in \mathcal{E}$).

Example 9.3.9 (Ising model) *We consider the Ising model discussed in section 6.4. In this context, the target measure is given by the Boltzmann-Gibbs measure on*

$$S = \{-1, +1\}^I \qquad I = \{1, \ldots, L\} \times \{1, \ldots, L\}$$

defined by

$$\pi(x) = \frac{1}{\mathcal{Z}_\beta} \, e^{-\beta V(x)} \, \lambda(x)$$

with the uniform measure $\lambda(x) = 2^{-L^2}$ and the potential function

$$V(x) = h \sum_{i \in I} x(i) - J \sum_{i \sim j} x(i)x(j).$$

In this situation, for any fixed $i \in I$ we have

$$\pi(x) \quad \propto \quad e^{-\beta h \, x(i) - \beta J \, x(i) \sum_{j \sim i} x(j)}$$
$$\times \, e^{-\beta h \sum_{j \in I-i} x(j) - \beta J \sum_{j \sim k, \, j,k \in I-i} x(j)x(k)}.$$

This implies that

$$L_{I-i,i}(x_{I-i}; x_i) \propto \exp\left(-\beta h \, x(i) - \beta J \, x(i) \sum_{j \sim i} x(j)\right).$$

More precisely, we have the following spin-site updates

$$
\begin{aligned}
L_{I-i,i}(x_{I-i}; \{1\}) &= 1 - L_{I-i,i}(x_{I-i}; \{-1\}) \\
&= \frac{e^{-\beta h - \beta J \sum_{j \sim i} x(j)}}{e^{-\beta h - \beta J \sum_{j \sim i} x(j)} + e^{+\beta h + \beta J \sum_{j \sim i} x(j)}} \\
&= 1 / \left(1 + e^{2\beta[h + J \sum_{j \sim i} x(j)]}\right).
\end{aligned}
$$

To sample the transition $x \rightsquigarrow y$ w.r.t. the Markov transition K_i given in (9.11), we sample a uniform r.v. U on $[0, 1]$ and we set

$$y = F^{(i)}(x)$$

with the random function $F^{(i)} : x \in S \rightarrow F^{(i)}(x) \in S$ *defined by*

$$\forall j \in I - i \qquad F^{(i)}(x)(j) = x(j)$$

and

$$F^{(i)}(x)(i) := 1_{[0, p_i(x)[}(U) - 1_{[p_i(x), 1]}(U) \qquad (9.14)$$

with

$$p_i(x) := 1/\left(1 + e^{2\beta[h + J \sum_{j \sim i} x(j)]}\right). \qquad (9.15)$$

When $J < 0$, the chance to pick the spin $+1$ increases as the number $j \sim i$ that have the spin $+1$. This model is called an attractive spin system.

An illustration of the Gibbs sampler can be found in the YouTube video with varying temperatures.

9.3.4 Propp and Wilson sampler

In the further development of this section $M(x, y)$ stands for an aperiodic and irreducible Markov transition on some finite set S.

Definition 9.3.10 *A random mapping F is said to be M-compatible as soon as we have for any $(x, y) \in S^2$*

$$\mathbb{P}(F(x) = y) = M(x, y).$$

The existence of M-compatible mappings is proved as follows:

Up to a change of label, there is no loss of generality to assume that the state space $S = \{1, \ldots, d\}$, with $d = \text{Card}(S)$. In this notation, a mapping F is characterized by a column random vector $F = (F(1), \ldots, F(d))'$.

We let $(U_i)_{1 \leq i \leq d}$ be a sequence of independent and uniform random variables on $[0, 1[$, and we set

$$F(i) = \sum_{1 \leq j \leq d} j \, 1_{[\sum_{1 \leq k < j} M(i, k), \, \sum_{1 \leq k \leq j} M(i, k)[}(U_i). \qquad (9.16)$$

By construction, the random states $(F(i))_{1 \leq i \leq d}$ are *independent r.v.* and we have

$$\mathbb{P}(F(i) = j) = \mathbb{P}\left(\sum_{1 \leq k < j} M(i, k) \leq U_i \leq \sum_{1 \leq k \leq j} M(i, k)\right) = M(i, j).$$

From the above construction, we notice that F is not necessarily a one-to-one mapping.

In this notation, the Markov chain with elementary transition M is defined for any $n \geq 0$ by the recursion

$$X_{n+1} = F_n(X_n) = F_n(F_{n-1}(X_{n-1})) = \ldots = (F_n \circ \ldots \circ F_1 \circ F_0)(X_0)$$

where F_n, with $n \in \mathbb{N}$, stands for a sequence of independent copies of the mapping F.

Definition 9.3.11 *Given a sequence of independent copies $(F_n)_{n \geq 0}$ of the mapping F, we let*

$$\overleftarrow{\boldsymbol{F_n}} := F_0 \circ F_1 \circ \ldots \circ F_n \overset{law}{=} F_n \circ \ldots \circ F_1 \circ F_0 := \overrightarrow{\boldsymbol{F_n}}.$$

We also let $\overleftarrow{T}$ and $\overrightarrow{T}$ be the forward and backward coalescent times

$$\overleftarrow{T} = \inf\left\{n : \text{Card}\left(\overleftarrow{\boldsymbol{F_n}}(S)\right) = 1\right\}$$

$$\overrightarrow{T} = \inf\left\{n : \text{Card}\left(\overrightarrow{\boldsymbol{F_n}}(S)\right) = 1\right\} \overset{law}{=} \overleftarrow{T}.$$

The backward mapping $\overleftarrow{F_n}$ is better interpreted as running the chain forward from some random state X_{-n} up to the state X_1

$$X_1 = F_0(X_0) = F_0(F_{-1}(X_{-1})) = \ldots = (F_0 \circ F_{-1} \circ \ldots \circ F_{-n})(X_{-n}) \tag{9.17}$$

where F_n, with $n \in \mathbb{Z}$, stands for independent copies of F. In this situation, the initial condition is X_{-n} and X_1 is the terminal state of the chain after $(n+1)$ forward interactions.

Theorem 9.3.12 *Assume that the M-compatible mapping F is chosen so that*

$$\mathbb{P}\left(\overleftarrow{T} < \infty\right) = 1 = \mathbb{P}\left(\overrightarrow{T} < \infty\right). \tag{9.18}$$

In this situation, the value of $\overleftarrow{F_{\overleftarrow{T}}}(x) := Y$ does not depend on the state variable x and it is distributed according to the invariant measure of the chain $\pi = \pi M$.

Proof :
By construction the value $\overleftarrow{F_{\overleftarrow{T}}}$, and a fortiori the one of $\overleftarrow{F_{\overleftarrow{T}}}(x) := Y$ do not depend on the variable x. This implies that

$$T \leq n$$

$$\Rightarrow (F_0 \circ F_1 \circ \ldots \circ F_n)(x) = (F_0 \circ F_1 \circ \ldots \circ F_T) \circ (F_{T+1} \circ \ldots \circ F_n(x)) = Y.$$

We conclude that

$$\mathbb{P}(Y = y) \overset{\infty \leftarrow n}{\longleftarrow} \quad \mathbb{P}((F_0 \circ F_1 \circ \ldots \circ F_n)(x) = y)$$

$$= \quad \mathbb{P}((F_n \circ \ldots \circ F_0)(x) = y) \overset{n \to \infty}{\longrightarrow} \pi(y).$$

This ends the proof of the theorem.   ∎

The coalescent condition is not satisfied for some M-compatible mappings. For instance when $S = \{1, 2\}$ and $M(i, j) = 1/2$ for any $i, j \in S$ it is readily checked that the mapping F defined by

$$\mathbb{P}((F(1), F(2)) = (1, 2)) = 1/2 = \mathbb{P}((F(1), F(2)) = (2, 1))$$

is M-compatible but the above condition is not satisfied. Indeed,

$$\mathbb{P}(F(1) = 1) = 1/2 = \mathbb{P}(F(1) = 2) \quad \text{and} \quad \mathbb{P}(F(2) = 1) = 1/2 = \mathbb{P}(F(2) = 2)$$

but $\overleftarrow{F_n}$ and $\overrightarrow{F_n}$ are random permutations of the states $\{1, 2\}$.

Nevertheless the mappings defined in (9.16) satisfy the desired condition. To check this claim, we notice that

$$\begin{aligned}
\mathbb{P}((F(1), F(2)) = (1, 2)) &= \mathbb{P}((F(1), F(2)) = (1, 1)) \\
&= \mathbb{P}((F(1), F(2)) = (2, 1)) \\
&= \mathbb{P}((F(1), F(2)) = (2, 2)) = 1/4.
\end{aligned}$$

Hence

$$\mathbb{P}(F(2) = 1) = \mathbb{P}(F(1) = 1) = 1/4 + 1/4 = 1/2 = \mathbb{P}(F(1) = 2) = \mathbb{P}(F(2) = 2).$$

In addition, we have

$$\mathbb{P}\left(\overleftarrow{T} \leq 1\right) \geq \mathbb{P}\left(\mathrm{Card}(F(\{1,2\})) = 1\right) = 1/2 > 0 \Rightarrow \mathbb{P}\left(\overleftarrow{T} > 1\right) \leq 1/2 < 1.$$

Recall that

$$\mathbb{P}\left(\overleftarrow{T} > n \mid \overleftarrow{T} > (n-1)\right)$$

$$\propto \mathbb{E}\left(\mathbb{P}\left(\mathrm{Card}\left(\overleftarrow{F_{n-1}} \circ F_n(S)\right) > 1 \mid \overleftarrow{F_{n-1}}\right) 1_{\mathrm{Card}\left(\overleftarrow{F_{n-1}}(S)\right) > 1}\right)$$

$$= \mathbb{E}\left(\underbrace{\mathbb{P}\left(\mathrm{Card}\left(\overleftarrow{F_1}(S)\right) > 1 \mid \overleftarrow{F_0}\right)_{\overleftarrow{F_0} = \overleftarrow{F_{n-1}}}}_{\leq 1/2} 1_{\mathrm{Card}\left(\overleftarrow{F_{n-1}}(S)\right) > 1}\right).$$

This implies that

$$\mathbb{P}\left(\overleftarrow{T} > n\right) = \mathbb{P}\left(\overleftarrow{T} > n \mid \overleftarrow{T} > (n-1)\right) \times \mathbb{P}\left(\overleftarrow{T} > (n-1)\right) \leq 1/2^n$$

from which we conclude that $\mathbb{P}\left(\overleftarrow{T} < \infty\right) = 1$.

Important remarks :

Notice that the Propp and Wilson scheme requires us to store all the values of the functions F_n. This drawback reflects the main limitation of applying the Propp and Wilson sampler in large state spaces.

Nevertheless, we can overcome this difficulty when the state space S is equipped with a partial order with a minimal and a maximal state, $x_{min} \leq x \leq x_{max}$, for any $x \in S$. In this case, the strategy is to find a judicious monotone M-compatible mapping F. Combining the interpretation (9.17) with the fact that

$$\mathrm{Card}\left(F_0 \circ F_{-1} \circ \ldots \circ F_{-n}\right)(S) = 1$$
$$\Leftrightarrow$$
$$\left(F_0 \circ F_{-1} \circ \ldots \circ F_{-n}\right)(x_{min}) = \left(F_0 \circ F_{-1} \circ \ldots \circ F_{-n}\right)(x_{max}),$$

we only need to store the values of two chains starting at x_{min} and x_{max}. This also shows that the coalescence property (9.18) of *monotone* mapping F is granted as soon as the chain is ergodic.

The drawback is that the desired coalescence may not appear after just some initially chosen number of n steps. In this case, we need to restart the simulation with a larger number of steps. In practice we often choose these numbers of the form 2^k, with $k \geq 1$.

For a more detailed discussion on this simulation technique, we refer the reader to the book of S. Asmussen, P. W. Glynn [5]. The website of D.B. Wilson on perfect sampling with Markov chains also contains a rather complete list of references on this subject.

We end this section with some examples of monotone M-compatible mappings.

Example 9.3.13 (The ladder chain) *We consider the ladder Markov chain X_n defined by the following transition diagram*

We also consider the couple of monotone mappings

$$F_+(x) = \begin{cases} x+1 & \text{for} \quad x \in \{1, \ldots, d-1\} \\ d & \text{for} \quad x = d \end{cases}$$

and

$$F_-(x) = \begin{cases} 1 & \text{for} \quad x = 1 \\ x-1 & \text{for} \quad x \in \{2, \ldots, d-1\}. \end{cases}$$

Given some uniform r.v. U on $[0,1]$, we set

$$F = 1_{[0,1/2[}(U) \; F_- + 1_{[1/2,1]}(U) \; F_+.$$

It is a simple exercise to check that this random mapping is monotone and compatible w.r.t. the Markov transition M of the ladder chain.

Example 9.3.14 (Ising model) *We return to the Ising model discussed in example 9.3.9 and in section 6.4. We equip the state space $S = \{-1, +1\}^I$, with $I = \{1, \ldots, L\}^2$ with the partial order*

$$x \leq y \implies \forall i \in I \qquad x(i) \leq y(i).$$

The minimal and maximal states x_{min} and x_{max} are clearly given by

$$\forall i \in I \qquad x_{min}(i) = -1 \quad \text{and} \quad x_{max}(i) = +1.$$

We also observe that

$$x \leq y \;\Rightarrow\; \forall i \in I \qquad \sum_{j \sim i} x(j) \leq \sum_{j \sim i} y(j).$$

Thus, when $J < 0$ the functions $F^{(i)}$ and $p_i(x)$ defined in (9.14) and (9.15) are such that

$$x \leq y \quad \Rightarrow \quad \forall i \in I \qquad p_i(x) := 1/\left(1 + e^{2\beta[h - |J| \sum_{j \sim i} x(j)]}\right) \leq p_i(y)$$
$$\Rightarrow \quad \forall i \in I \qquad F^{(i)}(x) \leq F^{(i)}(y).$$

Given a sequence of independent functions $F^{(i)}$, with $i \in I$, the functions

$$F = \circ_{i \in I} F^{(i)} \quad \text{(in any order)} \quad \text{and} \quad F = \frac{1}{\text{Card}(I)} \sum_{i \in I} F^{(i)}$$

are monotone and compatible w.r.t. the Gibbs Markov transitions.

9.4   Time inhomogeneous MCMC models

9.4.1   Simulated annealing algorithm

We suppose that we are given a sequence of target measures π_n defined in terms of a sequence of Boltzmann-Gibbs measures

$$\pi_n(dx) = \mu_{\beta_n}(dx) = \frac{1}{\mathcal{Z}_{\beta_n}} \; e^{-\beta_n V(x)} \; \lambda(dx) \qquad (9.19)$$

associated with some inverse temperature parameter $\beta_n \uparrow \infty$, some non negative potential function V and some reference measure λ on some state space S. Several examples of Boltzmann-Gibbs measures are discussed in section 6.4, including the Ising model and the traveling salesman problem.

For finite state spaces equipped with the counting measure λ, we have seen in (6.6) that these measures converge to the uniform measure on the subset of all global minima of the potential function V, as β_n tends to ∞. This shows that the sampling of these measures at low temperature is equivalent to that of sampling uniformly a state with minimal energy. Since most of the time these minimal energy states are unknown, it is impossible to sample Boltzmann-Gibbs measures at low temperature.

One strategy is to consider a sequence of Metropolis-Hastings transitions M_n such that for any time n we have

$$\mu_{\beta_n} M_n = \mu_{\beta_n} \iff \pi_n M_n = \pi_n.$$

We recall that the Markov transition M_n associated with a λ-reversible proposition transition K_n is given by

$$M_n(x, dy) = K_n(x, dy) \, a_n(x, y) + \left[1 - \int K_n(x, dz) \, a_n(x, z)\right] \delta_x(dy) \tag{9.20}$$

with the acceptance rate

$$a_n(x, y) = 1 \wedge e^{-\beta_n \ (V(y) - V(x))} = e^{-\beta_n \ (V(y) - V(x))_+}.$$

To simplify the presentation, we start with a null inverse temperature parameter $\beta_0 = 0$, and a r.v. X_0 with distribution $\eta_0 = \mu_{\beta_0} = \lambda$. We run a series of m_1 MCMC moves with Markov transition M_1

$$X_0 \xrightarrow{\ M_1^{m_1}\ } X_{m_1}.$$

If m_1 is sufficiently large, we expect X_{m_1} to be approximately distributed according to the invariant measure $\pi_1 = \mu_{\beta_1}$ of the transition M_1. Nevertheless, when β_1 is too large, the acceptance rate $a_1(x, y) = e^{-\beta_1(V(y) - V(x))}$ is almost null for any $V(x) < V(y)$. In other words, the sequence of M_1-MCMC moves is almost equivalent to a series of gradient-descent-type transitions. Thus, for large values of β_1 we cannot expect to have $\mathrm{Law}(X_{m_1}) \simeq \pi_1$ but for very large values of the parameter m_1.

Thus, the natural idea is to find a judicious schedule (β_n, m_n) such that the time inhomogeneous model

$$X_0 \xrightarrow{\ M_1^{m_1}\ } X_{m_1} \xrightarrow{\ M_2^{m_2}\ } X_{m_1 + m_2} \xrightarrow{\ M_3^{m_3}\ } X_{m_1 + m_2 + m_3} \xrightarrow{\quad} \cdots$$

explores randomly the state space with

$$\forall n \in \mathbb{N} \qquad \mathrm{Law}(X_{m_1 + \ldots + m_n}) \simeq \pi_n.$$

Some illustrations of the evolution of the simulated annealing in the context of the traveling salesman model discussed in section 6.4.3 can be found in the YouTube video, including comparisons with greedy style algorithms.

9.4.2 A perfect sampling algorithm

One idea is to introduce an intermediate acceptance-rejection mechanism every time we change the temperature parameter. For instance, initially we set

$$\widehat{X}_0 = \begin{cases} X_0 & \text{with probability} \quad e^{-(\beta_1 - \beta_0)V(X_0)} \\ c & \text{with probability} \quad 1 - e^{-(\beta_1 - \beta_0)V(X_0)} \end{cases}$$

where c stands for some auxiliary cemetery state. Notice that for any function f on S we have

$$
\begin{aligned}
\mathbb{E}\left(f(\widehat{X}_0) \mid \widehat{X}_0 \neq c\right) \quad &\propto \quad \mathbb{E}\left(\mathbb{E}\left(f(\widehat{X}_0)\, 1_{\widehat{X}_0 \neq c} \mid X_0\right)\right) \\
&= \quad \mathbb{E}\left(f(X_0)\, e^{-(\beta_1 - \beta_0)V(X_0)}\right) \\
&\propto \quad \int f(x)\, e^{-(\beta_1 - \beta_0)V(x)}\, e^{-\beta_0 V(x)}\, \lambda(dx).
\end{aligned}
$$

This implies that

$$
\mathbb{E}\left(f(\widehat{X}_0) \mid \widehat{X}_0 \neq c\right) \propto \int f(x)\, e^{-\beta_1 V(x)}\, \lambda(dx) \propto \pi_1(f).
$$

In much the same way, recalling that $\beta_0 = 0$ we prove that

$$
\mathbb{P}(\widehat{X}_0 \neq c) = \lambda\left(e^{-\beta_1 V(x)}\right) / \lambda\left(e^{-\beta_0 V(x)}\right) = \lambda\left(e^{-\beta_1 V(x)}\right).
$$

If $\widehat{X}_0 = c$ then the algorithm stops. Otherwise, by starting from $\widehat{X}_0 = X_0$, as before, we run m_1 transitions M_1 up to some random state X_{m_1}. Notice that

$$
\text{Law}(\widehat{X}_0 \mid \widehat{X}_0 \neq c) = \pi_1 \quad \Longrightarrow \quad \forall m_1 \geq 1 \quad \text{Law}(X_{m_1}) = \pi_1.
$$

Then, we accept or reject this state as follows

$$
\widehat{X}_{m_1} = \begin{cases} X_{m_1} & \text{with probability} \quad e^{-(\beta_2 - \beta_1)V(X_{m_1})} \\ c & \text{with probability} \quad 1 - e^{-(\beta_2 - \beta_1)V(X_{m_1})}. \end{cases}
$$

Arguing as above, we find that

$$
\text{Law}(\widehat{X}_{m_1} \mid \widehat{X}_{m_1} \neq c,\ \widehat{X}_0 \neq c) = \pi_2.
$$

Similarly, we also prove that

$$
\mathbb{P}(\widehat{X}_{m_1} \neq c \mid \widehat{X}_0 \neq c) = \lambda\left(e^{-\beta_2 V(x)}\right) / \lambda\left(e^{-\beta_1 V(x)}\right) = \pi_1\left(e^{-(\beta_2 - \beta_1)V(x)}\right)
$$

$$
\Longrightarrow \quad \mathbb{P}(\widehat{X}_{m_1} \neq c) = \lambda\left(e^{-\beta_2 V(x)}\right)
$$

as well as

$$
\forall m_2 \geq 1 \qquad \text{Law}(X_{m_1 + m_2}) = \pi_2,
$$

where $X_{m_1 + m_2}$ stands for the random states of the model after m_2 iterations of the transition M_2, starting from $\widehat{X}_{m_1} = X_{m_1}(\neq c)$. As before, when $\widehat{X}_{m_1} = c$ the algorithm is stopped. Iterating this algorithm we obtain a sequence of perfect random samples w.r.t. the target measures π_n, as soon as the states are accepted at every acceptance-rejection transition. The main drawback of this algorithm is related to the fact that the acceptance rate decreases exponentially fast to 0; that is,

$$
\mathbb{P}(\widehat{X}_{m_1 + \ldots + m_n} \neq c) = \lambda\left(e^{-\beta_n V(x)}\right) \downarrow_{n \uparrow \infty} 0.
$$

9.5 Feynman-Kac path integration

9.5.1 Weighted Markov chains

We let $\mathbb{P}_n$ be the distribution of the random paths

$$\boldsymbol{X_n} := (X_0, \ldots, X_n) \in \boldsymbol{S_n} := (S_0 \times \ldots \times S_n)$$

of a given reference Markov process X_n, taking values in some state spaces S_n whose values may depend on the time parameter n. More precisely, if $\eta_0 = \mathrm{Law}(X_0)$ is the distribution of the initial random state, then we have

$$\mathbb{P}_n(dx) \quad := \quad \mathbb{P}(\boldsymbol{X_n} \in dx) = \eta_0(dx_0)\, M_1(x_0, dx_1) \times \ldots \times M_n(x_{n-1}, dx_n).$$

In the above formula, $dx = d(x_0, \ldots, x_n) := dx_0 \times \ldots \times dx_n$ denotes an infinitesimal neighborhood of some path sequence $x = (x_0, \ldots, x_n) \in \boldsymbol{S_n}$. Feynman-Kac measures represent the distributions $\mathbb{P}_n$ weighted by a collection of non-negative potential functions G_n, up to some normalizing constant $\mathcal{Z}_n$.

More formally, Feynman-Kac models are defined by the formulae

$$\mathbb{Q}_n(dx) := \frac{1}{\mathcal{Z}_n}\, Z_n(x)\, \mathbb{P}_n(dx) \quad \text{with} \quad Z_n(x) := \prod_{0 \le p < n} G_p(x_p). \qquad (9.21)$$

In other words, for any function $\boldsymbol{f_n}$ on the path space $\boldsymbol{S_n}$,

$$\mathbb{Q}_n(\boldsymbol{f_n}) \quad = \quad \int \mathbb{Q}_n(dx)\, \boldsymbol{f_n}(x) = \mathbb{E}\left(\boldsymbol{f_n}(\boldsymbol{X_n})\, Z_n(X)\right)$$

$$\propto \quad \int \boldsymbol{f_n}(x)\, Z_n(x)\, \mathbb{P}(\boldsymbol{X_n} \in dx).$$

We note that the normalizing constant is also defined by

$$\mathcal{Z}_n := \int Z_n(x)\, \mathbb{P}(\boldsymbol{X_n} \in dx) = \mathbb{E}\left(Z_n(X)\right).$$

The measures $\mathbb{Q}_n$ are well defined on $\boldsymbol{S_n}$, as soon as $\mathcal{Z}_n \neq 0$. We refer the reader to section 9.9 for a series of illustrations of Feynman-Kac models in rare event simulation, in nonlinear filtering, and in global optimization problems.

To get a step further in our discussion, we also denote by Γ_n the unnormalized measures defined by

$$\Gamma_n(dx) = \mathcal{Z}_n \times \mathbb{Q}_n(dx) = Z_n(x) \times \mathbb{P}_n(dx). \qquad (9.22)$$

In other words, for any function $\boldsymbol{f_n}$ on the path space $\boldsymbol{S_n}$,

$$\Gamma_n(\boldsymbol{f_n}) \quad := \quad \int \boldsymbol{f_n}(x)\, Z_n(x)\, \mathbb{P}(\boldsymbol{X_n} \in dx) = \mathbb{E}\left(\boldsymbol{f_n}(\boldsymbol{X_n})\, Z_n(X)\right).$$

Finally, we let γ_n and η_n be the n-th marginal measures of Γ_n and $\mathbb{Q}_n$. By construction, for any function f_n on S_n we have

$$\eta_n(f_n) = \gamma_n(f_n)/\gamma_n(1) \quad \text{with} \quad \gamma_n(f_n) = \mathbb{E}\left(f_n(X_n)\, Z_n(X)\right), \qquad (9.23)$$

as well as the path space formulae

$$\mathbb{Q}_n(\boldsymbol{f_n}) = \Gamma_n(\boldsymbol{f_n})/\Gamma_n(1) \quad \text{with} \quad \Gamma_n(\boldsymbol{f_n}) = \mathbb{E}\left(\boldsymbol{f_n}(\boldsymbol{X_n})\, Z_n(X)\right) \qquad (9.24)$$

for any $\boldsymbol{f_n}$ on $\boldsymbol{S_n}$.

We let X'_n be a Markov process with elementary transitions M'_n on state spaces S'_n. We also denote by G'_n some non-negative potential functions on S'_n.

We further assume that X_n is given by the historical process

$$X_n = \boldsymbol{X'_n} := (X'_0, \ldots, X'_n) \in S_n := \boldsymbol{S'_n} := (S'_0 \times \ldots \times S'_n) \qquad (9.25)$$

and the potential functions only depend on the terminal state of the trajectory; that is,

$$G_n(X_n) = G_n\left(X'_0, \ldots, X'_n\right) = G'_n(X'_n). \qquad (9.26)$$

for some function G'_n on S'_n.

We let $\mathbb{Q}'_n$ be Feynman-Kac measures on path space S_n given by

$$\mathbb{Q}'_n(dx') := \frac{1}{\mathcal{Z}'_n}\, Z'_n(x')\, \mathbb{P}'_n(dx') \quad \text{with} \quad Z'_n(x') := \prod_{0 \leq p < n} G'_p(x'_p). \qquad (9.27)$$

In the above display $\mathbb{P}'_n$ stands for the distribution of the random paths $(X'_0, \ldots, X'_n)$. and $dx' = d(x'_0, \ldots, x'_n) := dx'_0 \times \ldots \times dx'_n$ denotes an infinitesimal neighborhood of some path sequence $x = (x'_0, \ldots, x'_n) \in S_n$. We let $\Gamma'_n(dx) = \mathcal{Z}'_n \times \mathbb{Q}'_n(dx)$ the unnormalized measures.

Using the fact that

$$Z_n(X) := \prod_{0 \leq p < n} G_p(X_p) = \prod_{0 \leq p < n} G'_p(X'_p) =: Z'_n(X') \quad \text{and} \quad \mathcal{Z}_n = \mathcal{Z}'_n$$

we readily check the formulae

$$\gamma_n = \Gamma'_n \quad \text{and} \quad \eta_n = \mathbb{Q}'_n. \qquad (9.28)$$

9.5.2   Evolution equations

Using the Markov property, we prove that

$$\gamma_n(f_n) = \mathbb{E}\left(G_{n-1}(X_{n-1}) \times \overbrace{\mathbb{E}\left(f_n(X_n) \mid X_{n-1}\right)}^{=M_n(f_n)(X_{n-1})} Z_{n-1}(X)\right)$$

$$= \gamma_{n-1}\left(G_{n-1}M_n(f_n)\right) = \gamma_{n-1}\left(Q_n(f_n)\right) = [\gamma_{n-1}Q_n](f_n)$$

with the integral operator

$$G_{n-1}(x_{n-1})M_n(x_{n-1}, dx_n) = Q_n(x_{n-1}, dx_n)$$

$$\Downarrow$$

$$Q_n(f_n)(x_{n-1}) = G_{n-1}(x_{n-1})M_n(f_n)(x_{n-1}) \quad \text{and} \quad \gamma_n(dx_n) = \int \gamma_{n-1}(dx_{n-1})Q_n(x_{n-1}, dx_n).$$
$$(9.29)$$

This clearly implies that

$$
\begin{aligned}
\eta_n(f_n) &= \frac{\eta_{n-1}(Q_n(f_n))}{\eta_{n-1}(Q_n(1))} = \frac{\gamma_{n-1}\left(G_{n-1}M_n(f_n)\right)/\gamma_{n-1}(1)}{\gamma_{n-1}\left(G_{n-1}\right)/\gamma_{n-1}(1)} \\
&= \frac{\eta_{n-1}\left(G_{n-1}M_n(f_n)\right)}{\eta_{n-1}\left(G_{n-1}\right)} = \Psi_{G_{n-1}}(\eta_{n-1})M_n(f_n).
\end{aligned}
$$

We conclude that γ_n and η_n satisfy the measure-valued equations

$$\gamma_n = \gamma_{n-1}Q_n \quad \text{and} \quad \eta_n = \Psi_{G_{n-1}}(\eta_{n-1})M_n \qquad (9.30)$$

with the Boltzmann-Gibbs transformations $\Psi_{G_{n-1}}$ defined in (7.26). Conversely, the solution of the above equation is given by a Feynman-Kac measure of the form discussed above.

By construction, for any $0 \le p \le n$ we have the evolution formula

$$\gamma_n = \gamma_p Q_{p,n} \Longleftrightarrow \gamma_n(dx_n) = \int \gamma_p(dx_p)\, Q_{p,n}(x_p, dx_n)$$

with linear semigroup $Q_{p,n}$ defined by the induction formula $Q_{p,n} = Q_{p+1}Q_{p+1,n}$, that is,

$$
\begin{aligned}
Q_{p,n}(x_p, dx_n) &= \int Q_{p+1}(x_p, dx_{p+1})\, Q_{p+1,n}(x_{p+1}, dx_n) \\
&= \int \dots \int Q_{p+1}(x_p, dx_{p+1})\, Q_{p+2}(x_{p+1}, dx_{p+2}) \dots Q_n(x_{n-1}, dx_n)
\end{aligned}
$$

or equivalently, for any bounded measurable function f_n on S_n

$$Q_{p,n}(f_n)(x_p) = \int Q_{p,n}(x_p, dx_n)\, f_n(x_n) = \mathbb{E}\left(f_n(X_n)\, Z_{p,n}(X) \mid X_p = x_p\right)$$

with

$$Z_{p,n}(X) := \prod_{p \le q < n} G(X_q).$$

It is also important to notice that

$$
\begin{aligned}
\gamma_n(1) &= \mathbb{E}\left(G_{n-1}(X_{n-1})\, Z_{n-1}(X)\right) \\
&= \gamma_{n-1}(G_{n-1}) = \underbrace{\frac{\gamma_{n-1}(G_{n-1})}{\gamma_{n-1}(1)}}_{=\eta_{n-1}(G_{n-1})} \times\ \gamma_{n-1}(1).
\end{aligned}
$$

In summary, we have proved the product formula

$$\gamma_n(1) = \mathcal{Z}_n = \mathbb{E}\left(Z_n(X)\right) = \prod_{0 \le p < n} \eta_p(G_p), \qquad (9.31)$$

and therefore

$$\gamma_n(f_n) = \eta_n(f_n) \times \gamma_n(1) = \eta_n(f_n) \times \prod_{0 \le p < n} \eta_p(G_p).$$

9.5.3 Particle absorption models

In probability theory, particle absorption models are represented by Markov chains evolving in a deterministic or random environment associated with some absorption rate functions.

The interpretation of the absorption event clearly depends on their application models. In optical ray propagation problems, the event of interest is related to photon absorptions [229]. In particle physics or in chemistry, the absorption rate is dictated by the energy of an electronic or macro-molecular configuration.

Absorption and critical type events can also be thought of as network overflows in complex queueing systems [114] and production systems [223]. Absorbed Markov chains are also used in web engineering [224] and biochemistry [180], as well as in environmental analysis [264], and in many other scientific disciplines.

This rather extraordinary variety of application domains is not really surprising, since all of these absorption models can be represented by a Feynman-Kac model. Inversely, we emphasize that *any* Feynman-Kac model (9.21) can be interpreted as the distribution of the random trajectories of a Markov chain evolving in an absorbing environment.

We consider a collection of measurable state spaces S_n and an auxiliary coffin, or cemetery, state c. We set $S_{n,c} = S_n \cup \{c\}$. We also denote by G_n some $[0,1]$-valued potential functions on S_n, and M_{n+1} some Markov transitions from S_n, into S_{n+1}. We define an $S_{n,c}$-valued Markov chain X_n^c with two separate killing/exploration transitions:

$$X_n^c \xrightarrow{\quad \text{killing} \quad} \widehat{X}_n^c \xrightarrow{\quad \text{exploration} \quad} X_{n+1}^c. \qquad (9.32)$$

These killing/exploration mechanisms are defined as follows:

- **Killing:** If $X_n^c = c$, we set $\widehat{X}_n^c = c$. Otherwise the particle X_n^c is still alive. In this case, with a probability $G_n(X_n^c)$, it remains in the same site, so that $\widehat{X}_n^c = X_n^c$; and with a probability $1 - G_n(X_n^c)$, it is killed, and we set $\widehat{X}_n^c = c$.

- **Exploration:** Once a particle has been killed, it cannot be brought back to life; so if $\widehat{X}_n^c = c$, then we set $\widehat{X}_p^c = X_p = c$, for any $p > n$. Otherwise, the particle $\widehat{X}_n^c \in E_n$ evolves to a new location X_{n+1}^c in S_{n+1}, randomly chosen according to the distribution $M_{n+1}(X_n^c, dx_{n+1})$.

Definition 9.5.1 *The Markov chain X_n^c defined above is called a Markov chain with the absorption rates $(1 - G_n)$, and the free exploration transitions M_n, on the state spaces S_n.*

Notice that the Markov chain X_n^c on the augmented state spaces $S_{n,c}$ can be interpreted as a conventional Markov chain, with a single absorbing state $\{c\}$, as soon as $M_n(x_n, \{x_n\}) \ne 1$ for any $x_n \in S_n$. Inversely, any Markov chain with a single absorbing state can be represented in this form.

In branching processes and population dynamics literature, the model X_n^c often represents the number of individuals of a given species [125, 138, 247]. Each individual can die or reproduce. The state $0 \in S_n = \mathbb{N}$ is interpreted as a trap, or as a hard obstacle, in the sense that the species disappears as soon as X_n^c hits 0.

Next, we present a Feynman-Kac interpretation of particle absorption models. We denote by X_n the Markov chain on S_n, with elementary transitions M_n. In this notation, the Feynman-Kac measures $\mathbb{Q}_n$ associated with the parameters (G_n, M_n), and defined in (9.21), represent the conditional distributions of the random paths of a nonabsorbed Markov particle. To see this claim, we let T be the killing time; that is, the first time at which the particle enters in the cemetery state

$$T = \inf \{n \geq 0 \; ; \; \widehat{X}_n^c = c\}.$$

By construction, we have

$$\begin{aligned}
\mathbb{P}(T \geq n) &= \mathbb{P}(\widehat{X}_0^c \in S_0, \ldots, \widehat{X}_{n-1}^c \in S_{n-1}) \\
&= \int_{S_0 \times \ldots \times S_{n-1}} \eta_0(dx_0) \, G_0(x_0) \prod_{1 \leq p < n} (M_p(x_{p-1}, dx_p) G_p(x_p)).
\end{aligned}$$

This shows that the normalizing constants $\mathcal{Z}_n$ of the Feynman-Kac measures $\mathbb{Q}_n$ represent the probability for the particle to be alive at time $n - 1$; that is,

$$\mathcal{Z}_n = \mathbb{P}(T \geq n) = \mathbb{E}\left(Z_n(X)\right).$$

In the above display, X_n stands for a Markov chain on S_n, with initial distribution η_0 and elementary Markov transitions M_n.

In the same vein, in terms of the n-th time marginal Feynman-Kac models we have

$$\mathbb{E}(f(X_n^c) \, 1_{T \geq n}) = \gamma_n(f_n) := \mathbb{E}\left[f_n(X_n) \, Z_n(X)\right] \qquad (9.33)$$
$$\mathbb{E}(f(X_n^c) \mid T \geq n) = \eta_n(f_n) := \gamma_n(f_n)/\gamma_n(1). \qquad (9.34)$$

Using these formulae, we also find that

$$\mathcal{Z}_n = \mathbb{P}(T \geq n) = \gamma_n(1) = \mathbb{E}\left(Z_n(X)\right) = \prod_{0 \leq p < n} \eta_p(G_p).$$

More generally, similar arguments yield that it is the distribution of a particle conditional upon being alive at time $n - 1$ that is defined by the Feynman-Kac model introduced in (9.21), that is,

$$\mathbb{Q}_n(dx) = \mathbb{P}\left(\boldsymbol{X}_n^c \in dx \mid T \geq n\right) \qquad (9.35)$$

with the historical process

$$\boldsymbol{X}_n^c := (X_0^c, \ldots, X_n^c).$$

Inversely, any Feynman-Kac model of the form (9.21) associated with some bounded potential functions G_n can be interpreted in terms of a particle absorption model. To prove this claim, we further assume that $\|G_n\| \le c_n$ for some finite constant $c_n < \infty$. We let X_n^c be the Markov chain on $S_{n,c}$ defined in (9.32) with absorption rate $(1 - G_n(x_n)/c_n)$. By construction, we readily check that

$$\mathbb{Q}_n := \mathrm{Law}\left(\boldsymbol{X_n^c} \mid T \ge n\right).$$

9.5.4   Doob h-processes

We consider the time homogeneous Feynman-Kac model $(\Gamma_n, \mathbb{Q}_n)$, associated with the parameters $(S_n, G_n, M_n) = (S, G, M)$ on some measurable state space S, defined in (9.21). We also set

$$Q(x, dy) := G(x)M(x, dy).$$

We also assume that G is uniformly bounded above and below by some positive constant, and the Markov transition M is reversible w.r.t. some probability measure μ on S, with $M(x, .) \simeq \mu$ and $dM(x, .)/d\mu \in \mathbb{L}_2(\mu)$. We denote by λ the largest eigenvalue of the integral operator Q on $\mathbb{L}_2(\mu)$, and by $h(x)$ a positive eigenvector

$$Q(h) = \lambda h.$$

Under some regularity conditions on (G, M), there exists a constant $\rho \ge 1$ such that

$$1/\rho \le h(x)/h(y) \le \rho \tag{9.36}$$

for any $x, y \in S$. For instance, let us suppose that

$$M(x, dz) \ge \epsilon M(y, dz) \quad \text{and} \quad G(x) \le g G(y)$$

for some $\epsilon \in]0, 1]$ and some $g < \infty$. In this situation, we have

$$Q(h)(x)/Q(h)(y) = h(x)/h(y) \le \rho \quad \text{with} \quad \rho \le g/\epsilon.$$

> The Doob h-process, corresponding to the ground state eigenfunction h defined above, is a Markov chain X_n^h on S, with initial distribution $\eta_0^h = \Psi_h(\eta_0)$, and the Markov transition
>
> $$M^h(x, dy) := \frac{1}{\lambda} \times h^{-1}(x)Q(x, dy)h(y) = \frac{M(x, dy)h(y)}{M(h)(x)}.$$

We also denote by η_n^h the distribution of the random state X_n^h starting with initial distribution η_0^h, that is,

$$\mathrm{Law}(X_n^h) = \eta_n^h = \eta_0^h (M^h)^n.$$

Our next objective is to connect the distribution of the paths of the h-process

$$\mathbb{P}\left(\boldsymbol{X_n^h} \in dx\right) = \eta_0^h(dx_0)M^h(x_0, dx_1) \dots M^h(x_{n-1}, dx_n)$$

with the historical process

$$\boldsymbol{X_n^h} := (X_0^h, \dots, X_n^h)$$

with the Feynman-Kac measures Γ_n and $\mathbb{Q}_n$ introduced in (9.21). In the above formula, $dx = d(x_0, \dots, x_n) := dx_0 \times \dots \times dx_n$ denotes an infinitesimal neighborhood of some path sequence $x = (x_0, \dots, x_n) \in \boldsymbol{S_n} = (S_0 \times \dots \times S_n)$.

Firstly, by construction we have

$$G = \lambda \times h / M(h)$$

and therefore

$$
\begin{aligned}
\Gamma_n(dx) &= \eta_0(dx_0) \, Z_n(x) \prod_{1 \leq p \leq n} M(x_{p-1}, dx_p) \\
&= \lambda^n \, \eta_0(dx_0) \, h(x_0) \left\{ \prod_{1 \leq p \leq n} \frac{M(x_{p-1}, dx_p) h(x_p)}{M(h)(x_{p-1})} \right\} \frac{1}{h(x_n)}.
\end{aligned}
$$

We conclude that

$$\Gamma_n(dx) = \lambda^n \, \eta_0(h) \, \mathbb{P}_n^h(dx) \, \frac{1}{h(x_n)}$$

where $\mathbb{P}_n^h$ stands for the law of the historical process

$$\mathbf{X_n^h} = (X_0^h, \ldots, X_n^h).$$

This clearly implies that

$$\mathbb{Q}_n(dx) = \frac{1}{\mathbb{E}(h^{-1}(X_n^h))} \, h^{-1}(x_n) \, \mathbb{P}_n^h(dx)$$

with the normalizing constants

$$\mathcal{Z}_n = \lambda^n \, \eta_0(h) \, \mathbb{E}(h^{-1}(X_n^h)).$$

The above formula shows that the sampling of the Feynman-Kac measure $\mathbb{Q}_n$ reduces to sampling the h-process X_n^h.

9.5.5 Quasi-invariant measures

Under condition (9.36), using the multiplicative formula (9.31),

$$\frac{1}{n} \log \mathcal{Z}_n = \frac{1}{n} \sum_{0 \leq p < n} \log \eta_p(G) = \log \lambda + \frac{1}{n} \log \left(\eta_0(h) \, \mathbb{E}(h^{-1}(X_n^h)) \right)$$

and therefore

$$\log \lambda - \frac{1}{n} \log \rho \leq \frac{1}{n} \log \mathcal{Z}_n = \frac{1}{n} \sum_{0 \leq p < n} \log \eta_p(G) \leq \log \lambda + \frac{1}{n} \log \rho \qquad (9.37)$$

from which we conclude that

$$\lim_{n \to \infty} \frac{1}{n} \sum_{0 \leq p < n} \log \eta_p(G) = \log \lambda.$$

In terms of the h-process, the n-th time marginal γ_n of the Feynman-Kac measures Γ_n takes the following form:

$$\gamma_n(f) = \lambda^n \, \eta_0(h) \, \eta_0^h (M^h)^n(f/h) = \lambda^n \, \eta_0(h) \, \eta_n^h(f/h).$$

In terms of particle absorption models we have

$$\text{Law}(\boldsymbol{X_n^c} \mid T^c \geq n) = \frac{1}{\mathbb{E}(h^{-1}(X_n^h))} \; h^{-1}(X_n^h) \; d\mathbb{P}_n^h$$

and

$$\mathcal{Z}_n = \mathbb{P}\left(T^c \geq n\right) = \lambda^n \; \eta_0(h) \; \mathbb{E}(h^{-1}(X_n^h)) \longrightarrow_{n\uparrow\infty} 0. \qquad (9.38)$$

Whenever it exists, the Yaglom limit of the measure η_0 is defined as the limiting measure

$$\eta_n \longrightarrow_{n\uparrow\infty} \eta_\infty \qquad (9.39)$$

of the Feynman-Kac flow η_n, when n tends to infinity. We also say that η_0 is a quasi-invariant measure as we have $\eta_0 = \eta_n$, for any time step. When the Feynman-Kac flow η_n is asymptotically stable, in the sense that it forgets its initial conditions, we also say that the quasi-invariant measure η_∞ is the Yaglom measure.

Quantitative convergence estimates of the limiting formulae (9.39) can be derived using the stability properties of the Feynman-Kac models. For a more thorough discussion on these particle absorption models, we refer the reader to the series of articles of the author with A. Guionnet [74, 75], L. Miclo [81, 82] and A. Doucet [69], as well as the monograph [66].

We end this section with a more precise description of the measures η_∞. To simplify the presentation and avoid some unnecessary technical discussion of the integrability of the potential function w.r.t. reference measure μ, we further assume that $\epsilon \leq G \leq \epsilon^{-1}$, for some $\epsilon > 0$, and that $M(x, dy) = m(x, y)\mu(dy)$ for some density function $m \in \mathbb{L}_2(\mu \otimes \mu)$. In this situation, Q is a compact self-adjoint operator on $\mathbb{L}_2(\mu) = \mathbb{L}_2(\Psi_{G^{-1}}(\mu))$ (with $G^{-1} = 1/G$). The $\Psi_{G^{-1}}(\mu)$-reversibility of Q follows from the fact that for any functions $f_1, f_2 \in \mathbb{L}_2(\mu)$ we have

$$\Psi_{G^{-1}}(\mu)\left(f_1 Q(f_2)\right) \propto \mu(f_1 M(f_2)) = \mu(M(f_1) \; f_2) \propto \Psi_{G^{-1}}(\mu)\left(Q(f_1) \; f_2\right).$$

The spectral theorem for compact, self adjoint operators (preserving positivity) allows us to rewrite any power Q^n in terms of a countable (by the compactness property) orthonormal basis $(f_i)_{i\geq 0} \in \mathbb{L}_2(\Psi_{G^{-1}}(\mu))$ of eigenfunctions associated with a sequence of real eigenvalues $\lambda_0 > \lambda_1 \geq \lambda_2 \geq \ldots$ of Q

$$Q^n(x, dy) = \sum_{i\geq 0} \lambda_i^n \; f_i(x) f_i(y) \; \Psi_{G^{-1}}(\mu)(dy) \quad \text{with} \quad (\lambda_0, f_0) = (\lambda, h). \qquad (9.40)$$

When the number of eigenvalues λ_i is infinite they tends to 0 as $i \to \infty$. For finite spaces S with cardinality d the d eigenvalues $(\lambda_i)_{0\leq i < d}$ are such that $\lambda_0 > \lambda_1 \geq \ldots \geq \lambda_{d-1} > -\lambda_0$. It is also readily checked that M^h is reversible w.r.t. $\pi^h := \Psi_{hM(h)}(\mu)$

$$\Psi_{hM(h)}(\mu)\left(f_1 M^h(f_2)\right) \propto \mu((hf_1)M(hf_2)) = \mu(M(hf_1) \; (hf_2)) \propto \Psi_{hM(h)}(\mu)\left(M^h(f_1) \; f_2\right).$$

it is also readily checked that

$$(M^h)^n(x, dx') = \pi^h(dx') + \sum_{i\geq 1} \overline{\lambda}_i^n \; \overline{f}_i(x) \; \overline{f}_i(x') \; \pi^h(dx')$$

with the $\mathbb{L}_2(\pi^h)$-orthonormal basis functions $\overline{f}_i := f_i/f_0$ and the eigenvalues

$$\overline{\lambda}_i := \lambda_i/\lambda_0 \quad \text{s.t.} \quad M^h(\overline{f}_i) = \overline{\lambda}_i \; \overline{f}_i$$

Last but not least, we observe that

$$
\begin{aligned}
\eta_n(f) \quad &= \quad \frac{\mathbb{E}\left(h^{-1}(X_n^h)f(X_n^h)\right)}{\mathbb{E}\left(h^{-1}(X_n^h)\right)} \\
&\simeq_{n\uparrow\infty} \quad \frac{\Psi_{hM(h)}(\mu)(h^{-1}f)}{\Psi_{hM(h)}(\mu)(h^{-1})} = \frac{\mu(M(h)f)}{\mu(M(h))} = \Psi_{M(h)}(\mu)(f) := \eta_\infty(f).
\end{aligned}
$$

Using the fact that M is μ-reversible, we also find that

$$
\mu(M(h)f) = \mu(hM(f))
$$

$$
\Rightarrow \quad \eta_\infty(f) = \Psi_{M(h)}(\mu)(f) = \Psi_h(\mu)M(f) = \frac{\Psi_{G^{-1}}(\mu)(hf)}{\Psi_{G^{-1}}(\mu)(h)} \qquad (\Leftarrow h/G \propto M(h)).
$$

We also notice that

$$
\Psi_G(\eta_\infty) = \Psi_h(\mu).
$$

Finally using (9.40) we have

$$
\lambda_0 > \lambda_1 \Longrightarrow \eta_n(f) = \eta_\infty(f) + \mathrm{O}\left((\lambda_1/\lambda_0)^n\right)
$$

and by (9.31) we find that

$$
\frac{1}{n}\log \mathcal{Z}_n = \frac{1}{n}\log\gamma_n(1) = \frac{1}{n}\sum_{0\le p<n}\log\eta_n(G) = \log\eta_\infty(G) + \mathrm{O}\left(1/n\right) = \log\lambda + \mathrm{O}\left(1/n\right).
$$

The r.h.s. formula is a consequence of the fact that

$$
\eta_\infty(G) = \Psi_{M(h)}(\mu)(G) = \frac{\mu(GM(h))}{\mu(M(h))} = \lambda \qquad (\Leftarrow G\,M(h) = \lambda h).
$$

9.5.6 Cauchy problems with terminal conditions

We return to the Feynman-Kac semigroups introduced in section 9.5.2.

> We fix a time horizon n and some function f_n on the state space S_n. For any $0 \le p \le n$ we set
> $$
> x \in S_p \mapsto u_p(x) := Q_{p,n}(f_n)(x) = \mathbb{E}\left(f_n(X_n)\,Z_n(X) \mid X_p = x\right).
> $$

By construction, we have

$$
Q_{p,n}(f_n) = Q_{p+1}\left(Q_{p+1,n}(f_n)\right) = G_p\,M_{p+1}(Q_{p+1,n}(f_n)) \quad \text{and} \quad Q_{n,n}(f_n) = f_n.
$$

> This shows that u_p solves the discrete time Cauchy problem with terminal condition
> $$
> \begin{cases} u_p &= G_p\,M_{p+1}(u_{p+1}) \quad \text{for any} \quad p<n \\ u_n &= f_n. \end{cases} \tag{9.41}
> $$

We consider a collection of functions g_p on S_p, with $p \leq n$ and we set

$$v_p = Q_{p,n}(f_n) + \sum_{p \leq q < n} Q_{p,q}(g_q) \quad (\Rightarrow v_n = Q_{n,n}(f_n) = f_n).$$

Using the fact that

$$q = p \Rightarrow Q_{p,q}(g_q) = g_p \Rightarrow \sum_{p \leq q < n} Q_{p,q}(g_q) = g_p + \sum_{(p+1) \leq q < n} Q_{p+1}Q_{p+1,q}(g_q)$$

we readily prove that

$$
\begin{aligned}
v_p &= Q_{p+1}(Q_{p+1,n}(f_n)) + g_p + Q_{p+1}\left(\sum_{(p+1) \leq q \leq n} Q_{p+1,q}(g_q) \right) \\
&= Q_{p+1}(v_{p+1}) + g_p.
\end{aligned}
$$

This shows that v_p solves the discrete time Cauchy problem with terminal condition

$$
\begin{cases}
v_p &= G_p\, M_{p+1}(v_{p+1}) + g_p \quad \text{for any} \quad p < n \\
v_n &= f_n.
\end{cases}
\tag{9.42}
$$

9.5.7 Dirichlet-Poisson problems

We return to the Feynman-Kac semigroups introduced in section 9.5.2. We consider a time homogeneous model $(G_n, M_n, S_n) = (G, M, S)$, and we let D be some subset of some state space S.

The Dirichlet-Poisson problem consists with finding a function v on D satisfying the following equations:

$$
\begin{cases}
v(x) &= G(x)M(v)(x) + g(x) \quad \text{for any} \quad x \in D \\
v(x) &= h(x) \qquad\qquad\qquad\quad \text{for any} \quad x \notin D.
\end{cases}
\tag{9.43}
$$

These equations are also called `Fredholm integral equations` of the second kind.

When v solves (9.43) the stochastic process

$$\mathcal{N}_n = v(X_n)\, Z_n(X) + \sum_{0 \leq p < n} g(X_p)\, Z_p(X) \quad (\Longrightarrow \mathcal{N}_0 = v(X_0))$$

is a martingale w.r.t. the σ-fields $\mathcal{F}_n = \sigma(X_p\,,\ 0 \leq p \leq n)$. To check this claim we use the fact that

$$G(X_{n-1})\, M(v)(X_{n-1}) = v(X_{n-1}) - g(X_{n-1})$$

to prove that

$$
\begin{aligned}
\mathbb{E}\left(\mathcal{N}_n \mid \mathcal{F}_{n-1}\right) &= G(X_{n-1})M(v)(X_{n-1})\, Z_{n-1}(X) + \sum_{0 \leq p < n} g(X_p)\, Z_p(X) \\
&= v(X_{n-1})\, Z_{n-1}(X) \\
&\quad -g(X_{n-1})\, Z_{n-1}(X) + \sum_{0 \leq p < n} g(X_p)\, Z_p(X) = \mathcal{N}_{n-1}.
\end{aligned}
$$

Let T_D be the exit time of D. Whenever (9.43) is satisfied we have

$$v(x) = \mathbb{E}\left(h(X_{T_D})\, Z_{T_D}(X) + \sum_{0 \le p < T_D} g(X_p)\, Z_p(X) \mid X_0 = x \right) \qquad (9.44)$$

as soon as $\mathbb{E}(\mathcal{N}_{T_D} \mid \mathcal{F}_0) = \mathcal{N}_0 (= f(X_0))$. This formula provides an explicit description of the solution of (9.43) in terms of the functions (g, h) and T_D. For instance if $(V, g) = (0, 0)$ and $h = 1_A$ for some $A \subset D$ we have

$$v(x) = \mathbb{P}(X_{T_D} \in A \mid X_0 = x).$$

For null boundary conditions $h = 0$ the solution

$$v(x) = \mathbb{E}\left(\sum_{0 \le p < T_D} g(X_p)\, Z_p(X) \mid X_0 = x \right)$$

can be approximated using the sequence of finite time horizon functions

$$
\begin{aligned}
v_n(x) &= \mathbb{E}\left(\sum_{0 \le p < T_D \wedge n} g(X_p)\, Z_p(X) \mid X_0 = x \right) \longrightarrow_{n \uparrow \infty} v(x) \\
&= \sum_{0 \le p < n} \mathbb{E}\left(1_{T_D > p}\, g(X_p)\, Z_p(X) \mid X_0 = x \right).
\end{aligned}
$$

In other words, for null boundary conditions $h = 0$ we have the approximation

$$v_n(x) = \sum_{0 \le p < n} Q_{0,p}(1_D g)(x) \longrightarrow_{n \uparrow \infty} v(x)$$

with the Feynman-Kac semigroup $Q_{p,n}$ defined for any $p \le n$ by

$$Q_{p,n}(f)(x) := \mathbb{E}\left(f(X_n)\, Z_{p,n}(X) \mid X_p = x \right).$$

with

$$Z_{p,n}(X) := \prod_{p \le q < n} (1_D(X_q) G(X_q)).$$

We also notice that

$$G \le 1 \;\Rightarrow\; |\mathcal{N}_{T_D}| \le C(g, h)\, (1 + T_D) \quad \text{with} \quad C(g, h) = \sup_{x \notin D} |h(x)| \vee \sup_{x \in D} |g(x)|. \qquad (9.45)$$

In the above discussion, we have implicitly assumed that Doob's stopping theorem 8.4.12 applies so that $\mathbb{E}(\mathcal{N}_{T_D} \mid \mathcal{F}_0) = \mathcal{N}_0 (= f(X_0))$. According to (9.45) the assumptions of theorem 8.4.12 are satisfied as soon as $G \le 1$ and $\sup_{x \in D} \mathbb{E}(T_D \mid X_0 = x) < \infty$. Several examples of random walks models satisfying this last condition are discussed in section 8.4.4. For a more thorough discussion on these stochastic models and their numerical approximation using particle methods, we refer the reader to section 12.2 of the research monograph [66].

Observe that

$$G = 1 \quad \text{and} \quad h = 0 \;\Rightarrow\; \forall x \in D \quad v(x) = g(x) + M(f)(x) = \mathbb{E}\left(\sum_{0 \leq p < T_D} g(X_p) \mid X_0 = x\right).$$

In addition, if $D = S$ then we have $T_D = \infty$. In this case the solution of the Poisson equation is given by

$$v(x) = \sum_{n \geq 0} M^n(g)(x) = \mathbb{E}\left(\sum_{n \geq 0} g(X_n) \mid X_0 = x\right).$$

It is clearly of interest to find conditions on the Markov transition M that ensures that v is bounded. For $M = Id$ we clearly have $v(x) = \text{sign}(g(x)) \times \infty$, where $\text{sign}(a)$ stands for the sign of $a \in \mathbb{R}$. Let V be some non-negative function s.t. $\beta_V(M) < 1$, where $\beta_V(M)$ stands for the V-Dobrushin ergodic coefficient introduced in definition 8.2.19. Natural conditions under which $\beta_V(M) < 1$ are provided in theorem 8.2.21. In this situation, using theorem 8.2.20 we have the estimate

$$\text{osc}_V(g) < \infty \;\Rightarrow\; \text{osc}_V(v) \leq \left[\sum_{n \geq 0} \beta_V(M)^n\right] \text{osc}_V(g) = \text{osc}_V(g)/(1 - \beta_V(M)) < \infty.$$

9.5.8 Cauchy-Dirichlet-Poisson problems

We consider a collection of functions (g_n, h_n) on S_n and we let D_n be subsets of some state spaces S_n, with $n \geq 0$. We consider the Feynman-Kac semigroup $Q_{p,n}$ defined for any $p \leq n$ by

$$\begin{aligned} Q_{p,n}(f)(x) &:= \mathbb{E}\left(f(X_n) \, Z_{p,n}(X) \mid X_p = x\right) \\ &= G_p(x) \, 1_{D_p}(x) \, \mathbb{E}\left(f(X_n) \, Z_{p+1,n}(X) \mid X_p = x\right), \end{aligned} \tag{9.46}$$

with

$$Z_{p,n}(X) := \prod_{p \leq q < n} (G_q(X_q) \, 1_{D_q}(X_q)).$$

Notice that this Feynman-Kac model coincides with the one introduced in section 9.5.2, by replacing the functions G_n by the product potential function $1_{D_n} G_n$. We let $T_D^{(n)}$ be the first time after n the process $(X_p)_{p \geq n}$ exits one the sets $(D_p)_{p \geq n}$.

We fix the time horizon n, and for any $p \leq n$. We consider the collection of functions $(v_p)_{p \leq n}$ defined for any $p \leq n$ and any $x \in S_p$ by

$$\begin{aligned} v_p(x) &= Q_{p,n}(f_n 1_{D_n})(x) + \sum_{p \leq q < n} Q_{p,q}(g_q 1_{D_q})(x) \\ &\quad + \mathbb{E}\left(1_{T_D^{(p)} \leq n} \, h_{T_D^{(p)}}(X_{T_D^{(p)}}) \, Z_{p,T_D^{(p)}}(X) \mid X_p = x\right). \end{aligned}$$

By construction, we have

$$X_n \in D_n \Rightarrow T_D^{(n)} > n.$$

This yields

$$\mathbb{E}\left(1_{T_D^{(n)} \leq n} \, h_{T_D^{(n)}}(X_{T_D^{(n)}}) \, Z_{n,T_D^{(n)}}(X) \mid X_n\right)$$

$$= G_n(X_n) \, 1_{D_n}(X_n) \, \mathbb{E}\left(1_{T_D^{(n)} \leq n} \, h_{T_D^{(n)}}(X_{T_D^{(n)}}) \, Z_{n+1,T_D^{(n)}}(X) \mid X_n\right) = 0$$

and therefore

$$v_n = Q_{n,n}(1_{D_n} f_n) = 1_{D_n} f_n \quad \Rightarrow \quad \forall x \in D_n \quad v_n(x) = f_n(x).$$

We also observe that

$$q = p \Rightarrow Q_{p,q}(1_{D_q} g_q)(x) = 1_{D_p} g_p$$

$$\Rightarrow \sum_{p \leq q < n} Q_{p,q}(1_{D_q} g_q) = 1_{D_p} g_p + Q_{p+1}\left(\sum_{p+1 \leq q < n} Q_{p+1,q}(1_{D_q} g_q)\right).$$

This implies that for any $x \in D_p$

$$Q_{p,n}(1_{D_n} f_n)(x) + \sum_{p \leq q < n} Q_{p,q}(1_{D_q} g_q)(x)$$

$$= Q_{p+1}(Q_{p+1,n}(1_{D_n} f_n)) + g_p + Q_{p+1}\left(\sum_{(p+1) \leq q \leq n} Q_{p+1,q}(1_{D_q} g_q)\right). \tag{9.47}$$

In this situation, it is also readily checked that

$$\mathbb{E}\left(1_{T_D^{(p)} \leq n} \, h_{T_D^{(p)}}(X_{T_D^{(p)}}) \, Z_{p,T_D^{(p)}}(X) \mid X_p = x\right)$$

$$= G_p(x) \, 1_{D_p}(x) \int M_{p+1}(x, dy)$$

$$\times \mathbb{E}\left(1_{T_D^{(p+1)} \leq n} \, h_{T_D^{(p+1)}}(X_{T_D^{(p+1)}}) \, Z_{p+1,T_D^{(p+1)}}(X) \mid X_{p+1} = y\right).$$

In the last assertion we have used the fact that

$$X_p \in D_p \Longrightarrow T_D^{(p)} \geq (p+1) \Longrightarrow T_D^{(p)} = T_D^{(p+1)}.$$

Combining this result with (9.47) we conclude that

$$v_p(x) = Q_{p+1}(v_{p+1})(x) + g_p(x) \quad \text{for any} \quad 0 \leq p < n \quad \text{and any} \quad x \in D_p.$$

In much the same way, we have

$$(9.46) \Longrightarrow \forall x \notin D_p \quad \forall p \leq q \leq n \quad Q_{p,n}(f_n 1_{D_n})(x) = 0 = Q_{p,q}(1_{D_q} g_q)(x).$$

On the other hand, we have

$$\forall x \notin D_p \quad \Longrightarrow \quad T_D^{(p)} = p \, (\leq n)$$

$$\Longrightarrow \quad \mathbb{E}\left(1_{T_D^{(p)} \leq n} \, h_{T_D^{(p)}}(X_{T_D^{(p)}}) \, Z_{p,T_D^{(p)}}(X) \mid X_p = x\right) = h_p(x).$$

We conclude that $(v_p)_{p \leq n}$ satisfies the Cauchy-Dirichlet-Poisson problem

$$\begin{cases} v_p(x) & = & G_p M_{p+1}(v_{p+1})(x) + g_p(x) & \text{for any} \quad p < n \text{ and } x \in D_p \\ v_p(x) & = & h_p(x) & \text{for any} \quad p \leq n \text{ and } x \notin D_p \\ v_n(x) & = & f_n(x) & \text{for any} \quad x \in D_n. \end{cases} \tag{9.48}$$

9.6   Feynman-Kac particle methodology

9.6.1   Mean field genetic type particle models

The nonlinear evolution equation (9.30) coincides with the one discussed in (7.35). Thus, we can solve the equation (9.30) using the genetic type mean field particle model presented in the end of section 7.10.2.

At each step, each particle ξ_{n-1}^i evaluates its potential value $G_{n-1}(\xi_{n-1}^i)$. With a probability $G_{n-1}(\xi_{n-1}^i)$ it remains in the same location. Otherwise, it jumps to a fresh new location ξ_{n-1}^j, chosen randomly, with a probability proportional to $G_{n-1}(\xi_{n-1}^j)$. In a second stage, each particle evolves randomly according to the Markov transition M_n. In the context of Monte Carlo sampling methods this algorithm can be interpreted as a (biased) rejection-free Monte Carlo sampler. For a more detailed discussion on these samplers in the context of MCMC methods we refer to section 9.9.1 dedicated to interacting MCMC algorithms.

This mean field stochastic algorithm can also be interpreted as a population of individuals mimicking natural evolution mechanisms:

- During a mutation stage, the particles evolve independently of one another, according to the same probability transitions M_n.

- During the selection stage, each particle evaluates the potential value of its location. The ones with small relative values are killed, while the ones with high relative values are multiplied.

From the statistical or stochastic view, these interacting particle systems can be interpreted as a sophisticated acceptance-rejection sampling technique, equipped with an interacting recycling mechanism.

For a more thorough discussion on the interpretations of Feynman-Kac mean field models we refer the reader to books [66, 67].

We let

$$\eta_n^N = \frac{1}{N} \sum_{1 \le i \le N} \delta_{\xi_n^i} \longrightarrow_{N \uparrow \infty} \eta_n \qquad (9.49)$$

be the empirical approximation of η_n associated with the mean field particle model (7.27) with the Markov transitions (7.34).

In this notation, mimicking (9.31) an unbiased particle approximation of the normalizing constant $\gamma_n(1)$ and the unnormalized measures γ_n are given by the formulae

$$\gamma_n^N(1) := \prod_{0 \le p < n} \eta_p^N(G_p) \quad \text{and} \quad \gamma_n^N(f_n) := \gamma_n^N(1) \times \eta_n^N(f_n) \longrightarrow_{N \uparrow \infty} \gamma_n(f_n).$$

$$(9.50)$$

These genetic type interacting particle systems have been used with success in a variety of application domains as heuristic like Monte Carlo schemes since the end of the 1940s. We quote the pioneering articles by T.E. Harris and H. Kahn [147], published in 1951, and the one by Enrico Fermi and R.D. Richtmyer in 1948 on resampled type quantum Monte Carlo methodologies. To the best of our knowledge, the first rigorous mathematical foundations of these models were published in 1996 in [64] (see also [65]).

Depending on their application domains the genetic type selection-mutation transitions discussed above are also known under different guises, with a variety of different names and terminologies. For instance the r.v. ξ_n^i are called *samples, particles, individuals, or replica*. To guide the reader in these interdisciplinary literature, in the following table we have tried to summarize some more or less equivalent formulations of the two-step transitions of the algorithm discussed above.

Sequential Monte Carlo	Sampling	Resampling
Particle Filters	Prediction	Updating
Data Assimilation	Forecasting	Analysis
Genetic Algorithms	Mutation	Selection
Evolutionary Population	Exploration	Branching-selection
Diffusion Monte Carlo	Free evolutions	Absorption
Quantum Monte Carlo	Walker motions	Reconfiguration
Sampling Algorithms	Transition Proposals	Accept-reject-recycle

The selection transition in the r.h.s. column is also termed bootstrapping, spawning, cloning, pruning, replenish, multi-level splitting, enrichment, go with the winner, quantum teleportation, etc.

We end this section with a simple proof of the unbiasedness properties of the unnormalized particle measures γ_n^N.

By construction, we have

$$\mathbb{E}\left(\gamma_n^N(f_n) \mid \xi_0, \ldots, \xi_{n-1}\right) \;=\; \Phi_n\left(\eta_{n-1}^N\right)(f_n) \prod_{0 \leq p < n} \eta_p^N(G_p).$$

Notice that

$$\begin{aligned}
\Phi_n\left(\eta_{n-1}^N\right)(f_n) &= \Psi_n\left(\eta_{n-1}^N\right)(M_n(f_n)) \\
&= \frac{\eta_{n-1}^N(G_{n-1}M_n(f_n))}{\eta_{n-1}^N(G_{n-1})} = \frac{1}{\eta_{n-1}^N(G_{n-1})}\, \eta_{n-1}^N(Q_n(f_n))
\end{aligned}$$

with the integral operator Q_n defined in (9.29) and given by

$$Q_n(f_n) = G_{n-1}M_n(f_n).$$

This shows that

$$\begin{aligned}
\mathbb{E}\left(\gamma_n^N(f_n) \mid \xi_0, \ldots, \xi_{n-1}\right) &= \eta_{n-1}^N(Q_n(f_n)) \prod_{0 \leq p < (n-1)} \eta_p^N(G_p) \\
&= \gamma_{n-1}^N(Q_n(f_n)). \tag{9.51}
\end{aligned}$$

Iterating the argument, we find that

$$\left[\; \forall p \leq n \qquad \mathbb{E}\left(\gamma_n^N(f_n) \mid \xi_0, \ldots, \xi_p\right) = \gamma_p^N(Q_{p,n}(f_n)) \;\right] \Rightarrow \mathbb{E}\left(\gamma_n^N(f_n)\right) = \gamma_n(f_n). \tag{9.52}$$

9.6.2   Path space models

We let X'_n be a Markov process with elementary transitions M'_n on state spaces S'_n. We also denote by G'_n some non-negative potential functions on S'_n. We consider the Feynman-Kac model (9.23) and (9.28) discussed in section 9.6.

As shown in section 9.5.2 η_n satisfies the nonlinear measure valued equation

$$\eta_n = \Psi_{G_{n-1}}(\eta_{n-1})M_n \tag{9.53}$$

where M_n stands for the Markov transition of the historical process given by

$$M_n(X_{n-1}, dx')$$

$$= \delta_{X_{n-1}}\left(d(x'_0, \ldots, x'_{n-1})\right) \ M'_n(x'_{n-1}, dx'_n).$$

In the above display, $dx' = dx'_0 \times \ldots dx'_n$ stands for an infinitesimal neighborhood of the historical path $x' = (x'_0, \ldots, x'_n)$.

The N-mean field particle model associated with these Feynman-Kac models is defined in terms of path particles

$$\xi^i_n := \left(\xi^i_{0,n}, \xi^i_{1,n}, \ldots, \xi^i_{n,n}\right) \in S_n. \tag{9.54}$$

At each step, each path particle

$$\xi^i_{n-1} = \left(\xi^i_{0,n-1}, \xi^i_{1,n-1}, \ldots, \xi^i_{n-1,n-1}\right)$$

evaluates its potential value

$$G_{n-1}(\xi^j_{n-1}) = G'_{n-1}(\xi^j_{n-1,n-1}).$$

With a probability $G_{n-1}(\xi^i_{n-1})$ it remains in the same location, and we set $\widehat{\xi}^i_{n-1} = \xi^i_{n-1}$. Otherwise, it jumps to a fresh new path location $\widehat{\xi}^i_{n-1} = \xi^j_{n-1}$ randomly chosen with a probability proportional to $G_{n-1}(\xi^j_{n-1})$. In a second stage, each selected path particle

$$\widehat{\xi}^i_{n-1} = \left(\widehat{\xi}^i_{0,n-1}, \widehat{\xi}^i_{1,n-1}, \ldots, \widehat{\xi}^i_{n-1,n-1}\right)$$

evolves randomly according to the Markov transition of the historical process M_n. In other words, we set

$$\xi^i_n = \left(\underbrace{\left(\xi^i_{0,n}, \xi^i_{1,n}, \ldots, \xi^i_{n-1,n}\right)}_{=\widehat{\xi}^i_{n-1}}, \xi^i_{n,n}\right)$$

where $\xi^i_{n,n}$ is a r.v. with distribution $M'_n\left(\xi^i_{n-1,n}, dx'_n\right)$.

By construction, this N-interacting path-particle model coincides with the genealogical tree evolution of the N-mean field particle model associated with a Feynman-Kac model with potential function G'_n and the reference Markov chain X'_n.

Using the same arguments as above and by formula (9.28) we have

$$\eta^N_n := \frac{1}{N}\sum_{1 \le i \le N}\delta_{\xi^i_n} = \frac{1}{N}\sum_{1 \le i \le N}\delta_{\left(\xi^i_{0,n}, \xi^i_{1,n}, \ldots, \xi^i_{n,n}\right)} \xrightarrow{N\to\infty} \eta_n = \mathbb{Q}'_n. \tag{9.55}$$

9.6.3 Backward integration

Next, we present a description of the Feynman-Kac measure $\mathbb{Q}_n$ on path space defined in (9.21) in terms of $(\eta_p)_{0 \leq p \leq n}$.

> We further assume that the Markov transitions M_n are absolutely continuous with respect to some measures λ_n on S_n, and for any $(x_{n-1}, x_n) \in (S_{n-1} \times S_n)$ we have
>
> $$\begin{aligned} Q_n(x_{n-1}, dx_n) &:= G_{n-1}(x_{n-1}) \, M_n(x_{n-1}, dx_n) \\ &= H_n(x_{n-1}, x_n) \, \lambda_n(dx_n) \end{aligned} \tag{9.56}$$
>
> for some density function H_n.

In this situation, for any f on S_{n+1}, and for any $x_n \in S_n$ we have

$$Q_{n+1}(f)(x_n) = \int H_{n+1}(x_n, x_{n+1}) \, f(x_{n+1}) \, \lambda_{n+1}(dx_{n+1})$$

$$\begin{aligned} \Rightarrow \eta_{n+1}(f) &= \Psi_{G_n}(\eta_n)\,(M_{n+1}(f)) \\ &= \int_{S_{n+1}} \left[\int_{S_n} \eta_n(dx_n) \, \frac{1}{\eta_n(G_n)} \, H_{n+1}(x_n, x_{n+1}) \right] f(x_{n+1}) \, \lambda_{n+1}(dx_{n+1}). \end{aligned}$$

This shows that

$$\eta_{n+1}(dx_{n+1}) = \frac{1}{\eta_n(G_n)} \, \eta_n \left(H_{n+1}(., x_{n+1}) \right) \lambda_{n+1}(dx_{n+1})$$

from which we prove that

$$\begin{aligned} \mathbb{Q}_n(dx) &= \frac{1}{\mathcal{Z}_n} \, \eta_0(dx_0) \, Q_1(x_0, dx_1) \ldots Q_n(x_{n-1}, dx_n) \\ &= \frac{1}{\prod_{0 \leq p < n} \eta_p(G_p)} \, \eta_0(dx_0) \prod_{1 \leq p \leq n} (H_p(x_{p-1}, x_p) \, \lambda_p(dx_p)) \\ &= \eta_0(dx_0) \prod_{1 \leq p \leq n} \left(\frac{H_p(x_{p-1}, x_p)}{\eta_{p-1}(H_p(., x_p))} \, \eta_p(dx_p) \right). \end{aligned}$$

In the above formula, $dx = d(x_0, \ldots, x_n) := dx_0 \times \ldots \times dx_n$ denotes an infinitesimal neighborhood of some path sequence

$$x = (x_0, \ldots, x_n) \in \boldsymbol{S_n} = (S_0 \times \ldots \times S_n).$$

> This implies that
>
> $$\mathbb{Q}_n(dx) = \eta_n(dx_n) \prod_{q=1}^{n} \mathbb{M}_{q, \eta_{q-1}}(x_q, dx_{q-1}) \tag{9.57}$$
>
> with the collection of backward Markov transitions
>
> $$\mathbb{M}_{n+1, \eta_n}(x_{n+1}, dx_n) = \frac{\eta_n(dx_n) \, H_{n+1}(x_n, x_{n+1})}{\eta_n(H_{n+1}(., x_{n+1}))}.$$

If we take the unit potential functions $G_n = 1$, the backward formula (9.57) reduces to the conventional backward representation of conditional distribution of the random paths $(X_0, \ldots, X_{n-1})$ given the terminal time X_n; that is,

$$\mathbb{P}\left((X_0, X_1, \ldots, X_{n-1}) \in d(x_0, x_1, \ldots, x_{n-1}) \mid X_n = x_n\right)$$

$$= \mathbb{M}_{n,\eta_{n-1}}(x_n, dx_{n-1}) \cdots \mathbb{M}_{2,\eta_1}(x_2, dx_1)\mathbb{M}_{1,\eta_0}(x_1, dx_0).$$

To the best of our knowledge, these forward-backward representations of Feynman-Kac measures were introduced by Ruslan L. Stratonovitch in the early 1960s in the context of nonlinear filtering [246]. For a more thorough discussion on these backward Markov chain models and their application in advanced signal processing and in hidden Markov chain problems, we also refer the reader to [110] and to a series of articles of P. Del Moral, A. Doucet and S.S. Singh [71, 72, 73].

Mimicking the backward Markov chain formula (9.57), the measure $\mathbb{Q}_n$ can alternatively be approximated (under the assumption (9.56)) using the backward particle measures

$$\mathbb{Q}_n^N(dx) = \eta_n^N(dx_n) \prod_{q=1}^{n} \mathbb{M}_{q,\eta_{q-1}^N}(x_q, dx_{q-1}) \longrightarrow_{N\uparrow\infty} \mathbb{Q}_n(dx) \tag{9.58}$$

with the collection of Markov transitions

$$\begin{aligned}
\mathbb{M}_{n+1,\eta_n^N}(x_{n+1}, dx_n) &= \frac{\eta_n^N(dx_n) \, H_{n+1}(x_n, x_{n+1})}{\eta_n^N(H_{n+1}(., x_{n+1}))} \\
&= \sum_{1 \leq i \leq N} \frac{H_{n+1}(\xi_n^i, x_{n+1})}{\sum_{1 \leq j \leq N} H_{n+1}(\xi_n^j, x_{n+1})} \, \delta_{\xi_n^i}(dx_n).
\end{aligned} \tag{9.59}$$

For any function $\boldsymbol{f_n}$ on the path space $\boldsymbol{S_n}$ we have the unbiasedness property

$$\mathbb{E}\left(\mathbb{Q}_n^N(\boldsymbol{f_n}) \prod_{0 \leq p < n} \eta_p^N(G_p)\right) = \mathbb{E}\left(\boldsymbol{f_n}(\boldsymbol{X_n}) \, Z_n(X)\right) := \Gamma_n(\boldsymbol{f_n}).$$

We set

$$\Gamma_n^N(\boldsymbol{f_n}) := \mathbb{Q}_n^N(\boldsymbol{f_n}) \prod_{0 \leq p < n} \eta_p^N(G_p). \tag{9.60}$$

To check the unbiasedness property stated above, we notice that

$$\mathbb{E}\left(\Gamma_n^N(\boldsymbol{f_n}) \mid \xi_0, \ldots, \xi_{n-1}\right)$$

$$= \gamma_n^N(1) \int \Phi_n\left(\eta_{n-1}^N\right)(dx_n) \left\{\prod_{q=1}^{n} \mathbb{M}_{q,\eta_{q-1}^N}(x_q, dx_{q-1})\right\} \boldsymbol{f_n}(x_0, \ldots, x_n)$$

$$= \gamma_{n-1}^N(1) \int (\eta_{n-1}^N Q_n)(dx_n) \, \mathbb{M}_{n,\eta_{n-1}^N}(x_n, dx_{n-1}) \left\{\prod_{q=1}^{n-1} \mathbb{M}_{q,\eta_{q-1}^N}(x_q, dx_{q-1})\right\} \boldsymbol{f_n}(x_0, \ldots, x_n).$$

Observe that

$$(\eta_{n-1}^N Q_n)(dx_n) = \eta_{n-1}^N(H_n(., x_n)) \, \lambda_n(dx_n)$$

$$\Rightarrow (\eta_{n-1}^N Q_n)(dx_n) \, \mathbb{M}_{n,\eta_{n-1}^N}(x_n, dx_{n-1}) = \eta_{n-1}^N(dx_{n-1})Q_n(x_{n-1}, dx_n).$$

This implies that

$$\mathbb{E}\left(\Gamma_n^N(\boldsymbol{f_n}) \mid \xi_0, \ldots, \xi_{n-1}\right)$$

$$= \gamma_n^N(1) \int \Phi_n\left(\eta_{n-1}^N\right)(dx_n) \left\{\prod_{q=1}^{n} \mathbb{M}_{q,\eta_{q-1}^N}(x_q, dx_{q-1})\right\} \boldsymbol{f_n}(x_0, \ldots, x_n)$$

$$= \gamma_{n-1}^N(1) \int \eta_{n-1}^N(dx_{n-1}) \left\{\prod_{q=1}^{n-1} \mathbb{M}_{q,\eta_{q-1}^N}(x_q, dx_{q-1})\right\} \boldsymbol{f_{n-1,n}}(x_0, \ldots, x_{n-1}) = \Gamma_{n-1}^N(F_{n-1,n})$$

with

$$\boldsymbol{f_{n-1,n}}(x_0, \ldots, x_{n-1}) = \int_{S_n} Q_n(x_{n-1}, dx_n) \, \boldsymbol{f_n}(x_0, \ldots, x_n).$$

Iterating the argument, we prove that

$$\forall p \leq n \quad \mathbb{E}\left(\Gamma_n^N(\boldsymbol{f_n}) \mid \xi_0, \ldots, \xi_p\right) = \Gamma_p^N(\boldsymbol{f_{p,n}}) \implies \mathbb{E}(\Gamma_n^N(F_n)) = \Gamma_n(\boldsymbol{f_n})$$

with the collection of functions

$$\boldsymbol{f_{p,n}}(x_0, \ldots, x_p) = \int_{S_{p+1} \times \ldots \times S_n} Q_{p+1}(x_p, dx_{p+1}) \ldots Q_n(x_{n-1}, dx_n) \, \boldsymbol{f_n}(x_0, \ldots, x_n).$$

9.6.4 A random particle matrix model

The computation of integrals w.r.t. the particle measures $\mathbb{Q}_n^N$ is reduced to summation over the particle locations ξ_n^i. It is therefore natural to identify a population of individuals $(\xi_n^1, \ldots, \xi_n^N)$ at time n to the ordered set of indexes $\{1, \ldots, N\}$. In this framework, the occupation measures and the functions are identified with the following line and column vectors

$$\eta_n^N := \left[\frac{1}{N}, \ldots, \frac{1}{N}\right] \quad \text{and} \quad f_n := \begin{pmatrix} f_n(\xi_n^1) \\ \vdots \\ f_n(\xi_n^N) \end{pmatrix}$$

and the transitions $\mathbb{M}_{n,\eta_{n-1}^N}$ are identified by the $(N \times N)$ matrices

$$\mathbb{M}_{n,\eta_{n-1}^N} := \begin{pmatrix} \mathbb{M}_{n,\eta_{n-1}^N}(\xi_n^1, \xi_{n-1}^1) & \cdots & \mathbb{M}_{n,\eta_{n-1}^N}(\xi_n^1, \xi_{n-1}^N) \\ \vdots & \vdots & \vdots \\ \mathbb{M}_{n,\eta_{n-1}^N}(\xi_n^N, \xi_{n-1}^1) & \cdots & \mathbb{M}_{n,\eta_{n-1}^N}(\xi_n^N, \xi_{n-1}^N) \end{pmatrix} \qquad (9.61)$$

with the (i, j)-entries

$$\mathbb{M}_{n,\eta_{n-1}^N}(\xi_n^i, \xi_{n-1}^j) = \frac{H_n(\xi_{n-1}^j, \xi_n^i)}{\sum_{k=1}^{N} H_n(\xi_{n-1}^k, \xi_n^i)}.$$

For instance, the $\mathbb{Q}_n$-integration of normalized additive linear functionals of the form

$$\boldsymbol{f_n}(x_0, \ldots, x_n) = \frac{1}{n+1} \sum_{0 \leq p \leq n} f_p(x_p) \qquad (9.62)$$

is given by the particle matrix approximation model

$$\mathbb{Q}_n^N(\mathbf{f_n}) = \frac{1}{n+1} \sum_{0 \leq p \leq n} \eta_n^N \mathbb{M}_{n,\eta_{n-1}^N} \mathbb{M}_{n-1,\eta_{n-2}^N} \cdots \mathbb{M}_{p+1,\eta_p^N}(f_p).$$

This Markov interpretation allows computing complex Feynman-Kac path integrals using simple random matrix operations on finite sets. Roughly speaking, this methodology allows reducing Feynman-Kac path integration problems on general state spaces to Markov path integration on *finite state spaces*, with cardinality N.

9.6.5   A conditional formula for ancestral trees

We let X'_n be a Markov process with elementary transitions M'_n on state spaces S'_n. We also denote by G'_n some non-negative potential functions on S'_n. We consider the Feynman-Kac model (9.23) and (9.28) discussed in section 9.6.

The mean field particle model $\xi_n = (\xi^i_n)_{1 \leq i \leq N}$ associated with these Feynman-Kac measures is defined as in section 9.6.1. Following the arguments presented in section 9.6.2, the N path-valued particles $\xi_n = (\xi^i_n)_{1 \leq i \leq N}$ given by

$$\forall 1 \leq i \leq N \qquad \xi^i_n = (\xi^i_{0,n}, \xi^i_{1,n}, \ldots, \xi^i_{n-1,n}, \xi^i_{n,n}) \in S_n = (S'_0 \times \ldots \times S'_n) \quad (9.63)$$

represent the ancestral lines of the genetic particle model

$$\xi'_n := (\xi'^{,i}_n)_{1 \leq i \leq N} = (\xi^i_{n,n})_{1 \leq i \leq N} \in (S'_n)^N$$

defined in section 9.6.1.

We recall that ξ'_n is the mean field particle model associated with the Feynman-Kac measures (γ'_n, η'_n) on the path spaces S'_n defined for any function f' on S'_n by

$$\eta'_n(f') := \gamma'_n(f') / \gamma'_n(1) : \quad \text{with} \quad \gamma'_n(f') := \mathbb{E}\left(f'(X'_n) \, Z'_n(X')\right).$$

By construction, the particle model $\xi'_p = \xi_{p,p}$ represents the population of ancestors at level p of the genealogical tree $\xi_k = (\xi^i_k)_{1 \leq i \leq N}$ at every level $k = p, \ldots, n$.

Definition 9.6.1 *The set of these ancestors* $(\xi'_p)_{0 \leq p \leq n}$ *is called the complete ancestral tree (without the genealogical structure) associated with the sequence of genealogical trees* $(\xi_p)_{0 \leq p \leq n}$. *For any* $n \geq 0$ *we also denote by* $m(\xi_n)$ *and* $m(\xi'_n)$ *the empirical measures*

$$m(\xi'_n) = \frac{1}{N} \sum_{1 \leq i \leq N} \delta_{\xi'^{,i}_n} \quad and \quad m(\xi_n) = \frac{1}{N} \sum_{1 \leq i \leq N} \delta_{\xi^i_n} = \frac{1}{N} \sum_{1 \leq i \leq N} \delta_{(\xi^i_{0,n}, \xi^i_{1,n}, \ldots, \xi^i_{n-1,n}, \xi^i_{n,n})}.$$

Notice that $m(\xi'_n)$ is the n-th time marginal of $m(\xi_n)$ on the state space S'_n.

We further assume that the Markov transitions M'_n are absolutely continuous with respect to some measures λ'_n on S'_n, and for any $x \in S'_{n-1}$ we have

$$Q'_n(x, dy) := G'_{n-1}(x) \, M'_n(x, dy) = H'_n(x, y) \, \lambda'_n(dy) \quad (9.64)$$

for some density function H'_n.

We let

$$(\mathbb{Q}'_n, \mathbb{M}'_{n, \eta'_{n-1}}, \eta'^{,N}_n, \mathbb{Q}'^{,N}_n)$$

the mathematical model defined as the models $(\mathbb{Q}_n, \mathbb{M}_{n,\eta_{n-1}}, \eta_n^N, \mathbb{Q}_n^N)$ discussed in section 9.6.3 by replacing

$$(G_n, M_n, \xi_n, H_n, S_n) \qquad \text{by} \qquad (G'_n, M'_n, \xi'_n, H'_n, S'_n).$$

> **Definition 9.6.2** *Given the complete ancestral tree* $(\xi'_p)_{0 \le p \le n}$, *we let* $\mathbb{X}_n^\flat$ *be a random path with the backward particle measure* $\mathbb{Q}_n^{',N}$. *Given the genealogical tree* ξ_n, *we let* $\mathbb{X}_n$ *be a random ancestral line with uniform distribution* $m(\xi_n)$.

To simplify the forthcoming analysis, we further assume that the mean field particle models associated with the Feynman-Kac measures described above are given by the genetic type particle models (7.37). In this context, we have the following result.

> **Theorem 9.6.3** *Given the complete ancestral tree* $(\xi'_p)_{0 \le p \le n}$, *the sequence of genealogical trees* $(\xi_p)_{0 \le p \le n}$ *forms a Markov chain starting at* $\xi_0 = \xi'_0$. *The elementary transitions of the ancestral lines* $\xi_p \rightsquigarrow \xi_{p+1}$ *given the population of ancestors* ξ'_{p+1} *are given for any function f on* S_{p+1} *by*
>
> $$\mathbb{E}\left(f(\xi_{p+1}) \mid \xi_p, \xi'_{p+1} \right)$$
>
> $$= \int \left\{ \prod_{1 \le i \le N} \frac{m(\xi_p)(dx_p^i) \, H'_{p+1}(x_{p,p}^i, \xi'^{,i}_{p+1})}{m(\xi'_p)(H'_{p+1}(\cdot, \xi'^{,i}_{p+1}))} \right\} f\left(\left(x_p^j, \xi'^{,j}_{p+1} \right)_{1 \le j \le N} \right).$$
>
> *In the above display* $dx_p^i = dx_{0,p}^i \times dx_{1,p}^i \times \ldots \times dx_{p,p}^i$ *stands for an infinitesimal neighborhood of the path* $x_p^i = (x_{q,p}^i)_{0 \le q \le p} \in S_p$. *In particular, this implies that*
>
> $$\text{Law}\left(\mathbb{X}_n \mid (\xi'_p)_{0 \le p \le n} \right) = \mathbb{Q}_n^{',N} = \text{Law}\left(\mathbb{X}_n^\flat \mid (\xi'_p)_{0 \le p \le n} \right). \qquad (9.65)$$

Proof :

For any couple of functions f_1, f_2 on S_{p+1}, recalling that $\xi'_k = \xi_{k,k}$ we have

$$\mathbb{E}\left(f_1(\xi_{p+1}) \, f_2(\xi_p, \xi'_{p+1}) \mid \xi_p \right) = \int \left\{ \prod_{1 \le i \le N} \frac{m(\xi_p)(dx_p^i) \, Q'_{p+1}(x_{p,p}^i, dx'^{,i}_{p+1})}{m(\xi'_p)(G'_p)} \right\}$$

$$\times f_1\left(\left(x_p^j, x'^{,j}_{p+1} \right)_{1 \le j \le N} \right) f_2\left(\xi_p, \left(x'^{,i}_{p+1} \right)_{1 \le i \le N} \right).$$

To get one step further, we recall that

$$m(\xi_p)(dx_p^i) \, Q'_{p+1}(x_{p,p}^i, dx'^{,i}_{p+1})$$

$$= m(\xi_p)(dx_p^i) \, H'_{p+1}(x_{p,p}^i, x'^{,i}_{p+1}) \, \lambda'_{p+1}(dx'^{,i}_{p+1})$$

$$= \frac{m(\xi_p)(dx_p^i) \, H'_{p+1}(x_{p,p}^i, x'^{,i}_{p+1})}{m(\xi'_p)(H'_{p+1}(\cdot, x'^{,i}_{p+1}))} \, m(\xi'_p)(H'_{p+1}(\cdot, x'^{,i}_{p+1})) \, \lambda'_{p+1}(dx'^{,i}_{p+1})$$

and

$$\frac{m(\xi_p')(H_{p+1}'(.,x_{p+1}'^{,i}))}{m(\xi_p')(G_p')} \, \lambda_{p+1}'(dx_{p+1}'^{,i}) := \Phi_{p+1}'(m(\xi_p'))(dx_{p+1}'^{,i}).$$

This implies that

$$\mathbb{E}\left(f_1(\xi_{p+1}) \, f_2(\xi_p, \xi_{p+1}') \mid \xi_p\right)$$

$$= \int \left\{ \prod_{1\leq i\leq N} \Phi_{p+1}'(m(\xi_p'))(dx_{p+1}'^{,i}) \right\} f_2\left(\xi_p, \left(x_{p+1}'^{,i}\right)_{1\leq i\leq N}\right)$$

$$\int \left\{ \prod_{1\leq i\leq N} \frac{m(\xi_p)(dx_p^i) \, H_{p+1}'(x_{p,p}^i, x_{p+1}'^{,i})}{m(\xi_p')(H_{p+1}'(.,x_{p+1}'^{,i}))} \right\} f_1\left(\left(x_p^j, x_{p+1}'^{,j}\right)_{1\leq j\leq N}\right).$$

This ends the proof of the theorem.   ∎

9.7 Particle Markov chain Monte Carlo methods

9.7.1 Many-body Feynman-Kac measures

We return to the Feynman-Kac models and their particle interpretations presented in section 9.5.1 and section 9.6.1. Using the unbiased property (9.52), for any function f_n on S_n we have

$$\mathbb{E}\left(\gamma_n^N(f_n)\right) = \gamma_n(f_n) = \mathbb{E}\left(f_n(X_n) \, Z_n(X)\right) \quad \text{with} \quad \gamma_n^N(f_n) := \eta_n^N(f_n) \times \prod_{0\leq p<n} \eta_p^N(G_p).$$

This yields the following result.

For any $N \geq 1$ and $n \geq 0$, we have the unbiased formula

$$\mathbb{E}\left(\eta_n^N(f_n) \prod_{0\leq p<n} \eta_p^N(G_p)\right) = \mathbb{E}\left(f_n(X_n) \prod_{0\leq p<n} G_p(X_p)\right). \tag{9.66}$$

Rewritten in a slightly different way, for any function f_n on S_n we have

$$\overline{\gamma}_n(\overline{f}_n) := \mathbb{E}\left(\overline{f}_n(\overline{X}_n)\,\overline{Z}_n(\overline{X})\right) = \mathbb{E}\left(f_n(X_n)\,Z_n(X)\right) \qquad (9.67)$$

with the Markov chain and the Radon-Nikodym derivatives

$$\overline{X}_n := \left(\xi_n^i\right)_{1 \leq i \leq N} \in \overline{S}_n := S_n^N \quad \text{and} \quad \overline{Z}_n(\overline{X}) := \prod_{0 \leq p < n} \overline{G}_p(\overline{X}_p)$$

and the collection of (symmetric) functions $\overline{f}_n$ and $\overline{G}_n$ on S_n^N defined by

$$\overline{f}_n(\overline{X}_n) := \frac{1}{N} \sum_{1 \leq i \leq N} f_n(\xi_n^i) \quad \text{and} \quad \overline{G}_n(\overline{X}_n) := \frac{1}{N} \sum_{1 \leq i \leq N} G_n(\xi_n^i).$$

Notice that the Feynman-Kac measures $\overline{\gamma}_n$ and their normalized version $\overline{\eta}_n$ on $\overline{S}_n$ are defined for any functions $\overline{f}_n$ on $\overline{S}_n$ by the formulae

$$\overline{\eta}_n(\overline{f}_n) := \overline{\gamma}_n(\overline{f}_n)/\overline{\gamma}_n(1) \quad \text{with} \quad \overline{\gamma}_n(\overline{f}_n) := \mathbb{E}\left(\overline{f}_n(\overline{X}_n)\,\overline{Z}_n(\overline{X})\right). \qquad (9.68)$$

These models are defined as the Feynman-Kac measures (γ_n, η_n) by replacing (X_n, G_n) by $(\overline{X}_n, \overline{G}_n)$.

Definition 9.7.1 *The Feynman-Kac measures $(\overline{\gamma}_n, \overline{\eta}_n)$ on the product spaces $\overline{S}_n$ defined in (9.68) are called the many-body measures associated with the mean field particle interpretation $\overline{X}_n = \xi_n$ of the Feynman-Kac measures (γ_n, η_n) on S_n.*

From the pure mathematical point of view, the Feynman-Kac measures $(\overline{\gamma}_n, \overline{\eta}_n)$ have exactly the same form as the Feynman-Kac measures (γ_n, η_n). This observation allows us to define without further work their particle interpretations, denoted by $\overline{\xi}_n = \left(\overline{\xi}_n^i\right)_{1 \leq i \leq \overline{N}} \in \overline{S}_n^{\overline{N}}$ for some population size $\overline{N}$, using the mean field particle models discussed in section 9.6.1. Notice that in this situation, each particle $\overline{\xi}_n^i \in \overline{S}_n = S_n^N$ is itself a population of individuals, with $1 \leq i \leq \overline{N}$. These population-based particles are called islands. In this context, the particle model $\overline{\xi}_n$ can be interpreted as a mean field interacting evolution of islands. For a more detailed discussion on this class of island type particle models, we refer the reader to the article [256]. It is instructive to notice that for $N = 1$ the measures $(\overline{\gamma}_n, \overline{\eta}_n)$ coincide with (γ_n, η_n).

We consider the Feynman-Kac measures (γ_n, η_n) on the path spaces $S_n := (S_0' \times \ldots \times S_n')$ discussed in section 9.6.2 and section 9.6.5. In this situation, the many-body Feynman-Lac measures $(\overline{\gamma}_n, \overline{\eta}_n)$ associated with the mean field particle interpretation of the Feynman-Kac measures (γ_n, η_n) are defined in terms of path-valued particles $\overline{X}_n = \xi_n = (\xi_n^i)_{1 \leq i \leq N} \in S_n^N$. For each $1 \leq i \leq N$, ξ_n^i is given by the ancestral line $(\xi_{p,n}^i)_{0 \leq p \leq n}$ defined in (9.54). In addition, by (9.26) the potential functions $\overline{G}_n$ are now given by

$$\overline{G}_n(\overline{X}_n) = \frac{1}{N} \sum_{1 \leq i \leq N} G_n(\xi_n^i) = \frac{1}{N} \sum_{1 \leq i \leq N} G_n'(\xi_n'^i).$$

9.7.2 A particle Metropolis-Hastings model

Our next objective is to define an independent Metropolis-Hastings type model with the target measure $\overline{\eta}_n$ defined in (9.68), for *some given and fixed time horizon* $n \geq 0$.

To this end, it is convenient to extend these measures $(\overline{\gamma}_n, \overline{\eta}_n)$ to the path space measures $(\overline{\Gamma}_n, \overline{\mathbb{Q}}_n)$ on $\overline{S}_n = (\overline{S}_0 \times \ldots \times \overline{S}_n)$ defined for any function $\overline{f}_n$ on $\overline{S}_n$ by the formulae

$$\overline{\mathbb{Q}}_n(\overline{f}_n) := \overline{\Gamma}_n(\overline{f}_n)/\overline{\Gamma}_n(1) \quad \text{with} \quad \overline{\Gamma}_n(\overline{f}_n) := \mathbb{E}\left(\overline{f}_n(\overline{X}_n)\, \overline{Z}_n(\overline{X})\right)$$

with the historical process

$$\overline{X}_n := (\overline{X}_0, \ldots, \overline{X}_n) \in \overline{S}_n = (\overline{S}_0 \times \ldots \times \overline{S}_n).$$

In the further development of this section, we fix the time horizon n and we use the notation $\overline{x} = (\overline{x}_0, \ldots, \overline{x}_n)$ and $\overline{y} = (\overline{y}_0, \ldots, \overline{y}_n)$ to denote trajectories in the path space $\overline{S}_n$. Rewritten in a Boltzmann-Gibbs form, the measure $\overline{\mathbb{Q}}_n$ is defined by

$$\overline{\mathbb{Q}}_n(d\overline{x}) \propto \overline{Z}_n(\overline{x})\, \overline{\mathbb{P}}_n(d\overline{x}) \quad \text{with} \quad \overline{\mathbb{P}}_n = \text{Law}\left(\overline{X}_n\right)$$

and Radon-Nikodym derivatives $\overline{Z}_n$ defined by

$$\overline{Z}_n(\overline{x}) := \prod_{0 \le p < n} \overline{G}_p(\overline{x}_p).$$

We consider the Metropolis-Hastings Markov chain $(\mathcal{X}_k)_{k \ge 0}$ on $\mathcal{S}_n := \overline{S}_n$ with proposal transition

$$\mathcal{K}(\overline{x}, d\overline{y}) = \overline{\mathbb{P}}_n(d\overline{y})$$

and acceptance ratio

$$a\left(\overline{x}, \overline{y}\right) \quad = \quad 1 \wedge \frac{\overline{\mathbb{Q}}_n(d\overline{y})\mathcal{K}(\overline{y}, d\overline{x})}{\overline{\mathbb{Q}}_n(d\overline{x})\mathcal{K}(\overline{x}, d\overline{y})}.$$

By construction, $\mathcal{X}_k$ is a Markov chain with invariant measure $\overline{\mathbb{Q}}_n$. In addition, an elementary manipulation shows that

$$a\left(\overline{x}, \overline{y}\right) = 1 \wedge \frac{\overline{Z}_n(\overline{y})}{\overline{Z}_n(\overline{x})}.$$

The main advantage of this particle MCMC model comes from the fact that

$$\overline{G}_p(\overline{X}_p) \to_{N \uparrow \infty} \eta_p(G_p) \implies \overline{Z}_n(\overline{X}) \to_{N \uparrow \infty} \gamma_n(1)$$

so that the acceptance rate of the resulting particle Metropolis-Hastings model converges to 1 as the size of the particle model N tends to ∞.

9.7.3 Duality formulae for many-body models

We let X'_n be a Markov process with elementary transitions M'_n on state spaces S'_n. We also denote by G'_n some non-negative potential functions on S'_n. We consider the Feynman-Kac model (γ_n, η_n) defined in (9.23) and (9.28), and their mean field particle interpretations ξ_n discussed in section 9.6.2, section 9.6.3 and in section 9.6.5. The many-body Feynman-Kac measures $(\overline{\gamma}_n, \overline{\eta}_n)$ on $\overline{S}_n = S_n^N$ associated with the mean field particle interpretation $\overline{X}_n = \xi_n$ of the measures (γ_n, η_n) are defined in (9.68).

With a slight abuse of notation we denote by $\overline{S}_n$ the N-symmetric product spaces S_n^N / Σ_N, where Σ_N stands for the N-symmetric group. In other words, $\overline{S}_n$ stands for the set of non-ordered sequences $x = (x^i)_{1 \leq i \leq N} \in S_n^N$.

In this notation, the Markov transitions of the chain $\overline{X}_n$ are given by

$$\overline{M}_{n+1}(\overline{X}_n, dx) := \mathbb{P}\left(\overline{X}_{n+1} \in dx \mid \overline{X}_n\right) = \prod_{1 \leq i \leq N} \Phi_{n+1}(m(\overline{X}_n))(dx^i).$$

In the above display, $dx = dx^1 \times \ldots \times dx^N$ stands for an infinitesimal neighborhood of $x = (x^i)_{1 \leq i \leq N} \in \overline{S}_n$. The initial condition $\overline{X}_0 = (\xi_0^i)_{1 \leq i \leq N}$ is given by N i.i.d. samples with common law

$$\eta_0 := \mathrm{Law}(X_0) \implies \overline{\eta}_0 := \mathrm{Law}(\overline{X}_0) = \eta_0^{\otimes N}.$$

By exchangeability arguments, the chain $\overline{X}_n$ is well defined in the state spaces $\overline{S}_n$.

The following technical lemma is pivotal.

Lemma 9.7.2 *We have the duality formulae*

$$\overline{\eta}_0(dx) \ m(x)(dy) = \eta_0(dy) \ \overline{\eta}_{y,0}^{\#}(dx) \tag{9.69}$$

with the collection of probability measures $\overline{\eta}_{y,0}^{\#}$ on $\overline{S}_0$ defined by

$$\overline{\eta}_{y,0}^{\#}(dx) = \frac{1}{N} \sum_{1 \leq i \leq N} \left\{ \prod_{1 \leq j \neq i \leq N} \eta_0(dx^j) \right\} \delta_y\left(dx^i\right). \tag{9.70}$$

In the above displayed formulae, $dx = dx^1 \times \ldots \times dx^N$ stands for an infinitesimal neighborhood of $x = (x^i)_{1 \leq i \leq N} \in \overline{S}_0$ and dy an infinitesimal neighborhood of a given state $y \in S_0$.

Proof :

For any function f on $(S_0 \times \overline{S}_0)$ and for any $1 \leq i \leq N$ we have

$$\int \eta_0^{\otimes N}(dx) \ m(x)(dy) f(y, x) = \int \eta_0^{\otimes N}(dx) f(x^i, x)$$

$$= \int \eta_0(dy) \int \left[\left\{ \prod_{1 \leq j \neq i \leq N} \eta_0(dx^j) \right\} \delta_y\left(dx^i\right) \right] f(y, x) = \int \eta_0(dy) \int \overline{\eta}_{y,0}^{\#}(dx) \ f(y, x).$$

This ends the proof of the lemma. ∎

To get one step further, we consider the integral operators

$$\overline{Q}_n(\overline{x}_{n-1}, d\overline{x}_n) = \overline{G}_{n-1}(\overline{x}_{n-1}) \ \overline{M}_n(\overline{x}_{n-1}, d\overline{x}_n).$$

In the above displayed formulae, $d\overline{x}_n = dx_n^1 \times \ldots \times dx_n^N$ stands for an infinitesimal neighborhood of $\overline{x}_n = (x_n^i)_{1 \leq i \leq N} \in \overline{S}_n$ and $\overline{x}_{n-1} = (x_{n-1}^i)_{1 \leq i \leq N}$ is a given state in $\overline{S}_{n-1}$.

By construction, we have

$$\overline{G}_{n-1}(\overline{x}_{n-1}) \ \times \ \Phi_n\left(m(\overline{x}_{n-1})\right)(dy_n) = \left(m(\overline{x}_{n-1})Q_n\right)(dy_n)$$

where dy_n an infinitesimal neighborhood of a given state $y_n \in S_n$. We also recall that

$$\overline{M}_n(\overline{x}_{n-1}, d\overline{x}_n) = \Phi_n(m(\overline{x}_{n-1}))^{\otimes N}(d\overline{x}_n).$$

Arguing as in the proof of lemma 9.7.2 we prove that

$$\Phi_n\left(m(\overline{x}_{n-1})\right)^{\otimes N}(d\overline{x}_n) \, m(\overline{x}_n)(dy_n) = \Phi_n\left(m(\overline{x}_{n-1})\right)(dy_n) \, \overline{M}^{\#}_{y_n,n}(\overline{x}_{n-1}, d\overline{x}_n)$$

with the Markov transitions $\overline{M}^{\#}_{y_n,n}$ from $\overline{S}_{n-1}$ into $\overline{S}_n$ defined by

$$\overline{M}^{\#}_{y_n,n}(\overline{x}_{n-1}, d\overline{x}_n) = \frac{1}{N} \sum_{1 \le i \le N} \left\{ \prod_{1 \le j \neq i \le N} \Phi_n\left(m(\overline{x}_{n-1})\right)(dx_n^j) \right\} \delta_{y_n}\left(dx_n^i\right).$$

$$(9.71)$$

In summary, we have proved the following result.

Lemma 9.7.3 *For any $n \ge 1$, we have the duality formula*

$$\overline{Q}_n(\overline{x}_{n-1}, d\overline{x}_n) \, m(\overline{x}_n)(dy_n) = \left(m(\overline{x}_{n-1})Q_n\right)(dy_n) \, \overline{M}^{\#}_{y_n,n}(\overline{x}_{n-1}, d\overline{x}_n).$$ (9.72)

Definition 9.7.4 (Dual mean field model) *Given some realization of the historical chain $(X_n)_{n \ge 0}$, we let $\overline{X}^{\#}_n$ be the Markov chain with conditional initial distribution $\overline{\eta}^{\#}_{X_0,0}$ and elementary transitions $\overline{M}^{\#}_{X_n,n}$. The process $\overline{X}^{\#}_n$ is called the dual mean field model associated with the Feynman-Kac particle model $\overline{X}_n$ and the frozen path $(X_n)_{n \ge 0}$.*

Definition 9.7.5 *The historical processes of $\overline{X}_n$ and $\overline{X}^{\#}_n$ are defined by*

$$\overline{\boldsymbol{X}_n} := (\overline{X}_0, \ldots, \overline{X}_n) \quad and \quad \overline{\boldsymbol{X}^{\#}_n} := \left(\overline{X}^{\#}_0, \ldots, \overline{X}^{\#}_n\right) \in \overline{\boldsymbol{S}_n} := (\overline{S}_0 \times \ldots \times \overline{S}_n).$$

Given the genealogical tree $\overline{X}_n$, we also let $\mathbb{X}_n$ be a randomly chosen ancestral line with (conditional) distribution $m(\overline{X}_n)$.

We are now in position to state and prove the following duality theorem.

Theorem 9.7.6 (Duality theorem) *For any $n \ge 0$ and any function f_n on the product space $(\overline{\boldsymbol{S}_n} \times S_n)$*

$$\mathbb{E}\left(f_n(\overline{\boldsymbol{X}_n}, \mathbb{X}_n) \, \overline{Z}_n(\overline{X})\right) = \mathbb{E}\left(f_n(\overline{\boldsymbol{X}^{\#}_n}, X_n) \, Z_n(X)\right).$$ (9.73)

Proof :

By (9.72), we have

$$\overline{Q}_n(\overline{x}_{n-1}, d\overline{x}_n)\, m(\overline{x}_n)(dy_n) = \int m(\overline{x}_{n-1})(dy_{n-1}) Q_n(y_{n-1}, dy_n)\, \overline{M}^{\#}_{y_n,n}(\overline{x}_{n-1}, d\overline{x}_n).$$

This implies that

$$\overline{Q}_{n-1}(\overline{x}_{n-2}, d\overline{x}_{n-1})\overline{Q}_n(\overline{x}_{n-1}, d\overline{x}_n)\, m(\overline{x}_n)(dy_n)$$

$$= \int \overline{Q}_{n-1}(\overline{x}_{n-2}, d\overline{x}_{n-1})\, m(\overline{x}_{n-1})(dy_{n-1}) Q_n(y_{n-1}, dy_n)\, \overline{M}^{\#}_{y_n,n}(\overline{x}_{n-1}, d\overline{x}_n).$$

Using (9.72), we deduce that

$$\overline{Q}_{n-1}(\overline{x}_{n-2}, d\overline{x}_{n-1})\overline{Q}_n(\overline{x}_{n-1}, d\overline{x}_n)\, m(\overline{x}_n)(dy_n)$$

$$= \int m(\overline{x}_{n-2})(dy_{n-2}) Q_{n-1}(y_{n-2}, dy_{n-1})\, \overline{M}^{\#}_{n-1,y_{n-1}}(\overline{x}_{n-2}, d\overline{x}_{n-1})$$

$$\times Q_n(y_{n-1}, dy_n)\, \overline{M}^{\#}_{y_n}(\overline{x}_{n-1}, d\overline{x}_n).$$

Iterating backward in time we prove that

$$\overline{\eta}_0(d\overline{x}_0)\left\{\prod_{1\leq p\leq n}\overline{Q}_p(\overline{x}_{p-1}, d\overline{x}_p)\right\}\, m(\overline{x}_n)(dy_n)$$

$$= \int \eta_0(dy_0)\left\{\prod_{1\leq p\leq n} Q_p(y_{p-1}, dy_p)\right\}\, \overline{P}^{\#}_{[y_0,\ldots,y_n],n}(d(\overline{x}_0,\ldots,\overline{x}_n))$$

with the conditional distribution

$$\overline{P}^{\#}_{[y_0,\ldots,y_n],n}(d(\overline{x}_0,\ldots,\overline{x}_n)) := \overline{\eta}^{\#}_{y_0,0}(d\overline{x}_0)\prod_{1\leq p\leq n}\overline{M}^{\#}_{y_p,p}(\overline{x}_{p-1}, d\overline{x}_p).$$

This ends the proof of the theorem. ∎

We consider the marginal models

$$\overline{X}'_n := \left(\xi^i_{n,n}\right)_{1\leq i\leq N} = \left(\xi'^{,i}_n\right)_{1\leq i\leq N} = \xi'_n \in \overline{S}'_n = (S'_n)^N$$

and the historical process

$$\overline{X}'_n = \left(\overline{X}'_0,\ldots,\overline{X}'_n\right) \in \overline{S}'_n = \left(\overline{S}'_0 \times \ldots \times \overline{S}'_n\right).$$

In this context, given the historical path $X_n = (X'_0,\ldots,X'_0)$, the dual process $\overline{X}^{\#}_n$ is also given by a path-valued Markov chain

$$\overline{X}^{\#}_n = \left(\overline{X}^{\#}_{0,n},\ldots,\overline{X}^{\#}_{n,n}\right) \in \overline{S}_n = S^N_n = (S'_0 \times \ldots \times S'_n)^N.$$

In addition, the n-th time marginal process $\overline{X}_n^{',\#} = \overline{X}_{n,n}^{\#}$ is also a Markov chain on $\overline{S}_n' = (S_n')^N$ with initial distribution $\overline{\eta}_{X_0,0}^{\#} = \overline{\eta}_{X_0',0}^{\#}$ and the Markov transitions

$$\overline{M}_{X_n',n}^{',\#}(\overline{x}_{n-1}', d\overline{x}_n') = \frac{1}{N} \sum_{1 \leq i \leq N} \left\{ \prod_{1 \leq j \neq i \leq N} \Phi_n'\left(m(\overline{x}_{n-1}')\right)(dx_n'^j) \right\} \delta_{X_n'}\left(dx_n'^i\right).$$
(9.74)

In the above displayed formulae, $d\overline{x}_n' = dx_n'^1 \times \ldots \times dx_n'^N$ stands for an infinitesimal neighborhood of $\overline{x}_n' = (x_n'^i)_{1 \leq i \leq N} \in \overline{S}_n'$ and $\overline{x}_{n-1}' = (x_{n-1}'^i)_{1 \leq i \leq N}$ a given state in $\overline{S}_{n-1}'$. In addition, Φ_n' stands for the one-step Feynman-Kac mapping defined as Φ_n by replacing (G_{n-1}, M_n, S_n) by (G_{n-1}', M_n', S_n').

In this situation, we have

$$\overline{G}_n(\overline{X}_n) = \overline{G}_n'(\overline{X}_n') := m(\xi_n')(G_n') \quad \text{and} \quad G_n(X_n) = G_n'(X_n')$$

and using the conditioning formula (9.65) we readily prove the following corollary of theorem 9.7.6.

Corollary 9.7.7 *Under the regularity condition (9.64), for any $n \geq 0$ and any function f_n on $(\overline{S}_n' \times S_n)$*

$$\begin{aligned}
\mathbb{E}\left(f_n(\overline{X}_n', \mathbb{X}_n) \, \overline{Z}_n(\overline{X})\right) &= \mathbb{E}\left(f_n(\overline{X}_n', \mathbb{X}_n^\flat) \, \overline{Z}_n(\overline{X})\right) \\
&= \mathbb{E}\left(f_n(\overline{X}_n'^{,\#}, X_n) \, Z_n(X)\right)
\end{aligned}$$
(9.75)

with the random paths $(\mathbb{X}_n, \mathbb{X}_n^\flat)$ introduced in definition 9.6.2, and the historical process $\overline{X}_n'^{,\#} := \left(\overline{X}_0'^{,\#}, \ldots, \overline{X}_n'^{,\#}\right)$.

9.7.4   A couple particle Gibbs samplers

We let π_n be the probability measures on $(\overline{S}_n \times S_n)$ defined for any function f_n on $(\overline{S}_n \times S_n)$ by the Feynman-Kac measures

$$\pi_n(f_n) \propto \mathbb{E}\left(f_n(\overline{X}_n, \mathbb{X}_n) \, \overline{Z}_n(\overline{X})\right) = \mathbb{E}\left(f_n(\overline{X}_n^{\#}, X_n) \, Z_n(X)\right).$$
(9.76)

The transition probabilities of the Gibbs sampler

$$\left(\mathbb{X}_n^{(k)}, \overline{X}_n^{(k)}\right) \rightsquigarrow \left(\mathbb{X}_n^{(k+1)}, \overline{X}_n^{(k+1)}\right)$$

of the target multivariate distribution π_n on the product space $(\overline{S}_n \times S_n)$ are described by the synthetic diagram

$$
\left.\begin{array}{rcl}
\mathbb{X}_n^{(k)} & = & x \\
\overline{\boldsymbol{X}}_{\boldsymbol{n}}^{(k)} & = & x
\end{array}\right\}
$$

$$
\overset{(1)}{\rightarrow}\left\{\begin{array}{rcl}
\mathbb{X}_n^{(k+1)} & = & \overline{x} \sim \left(\mathbb{X}_n \mid \overline{\boldsymbol{X}}_{\boldsymbol{n}} = x\right) \\
\overline{\boldsymbol{X}}_{\boldsymbol{n}}^{(k)} & = & x
\end{array}\right\}
\overset{(2)}{\rightarrow}\left\{\begin{array}{rcl}
\mathbb{X}_n^{(k+1)} & = & \overline{x} \\
\overline{\boldsymbol{X}}_{\boldsymbol{n}}^{(k+1)} & = & \overline{x} \sim \left(\overline{\boldsymbol{X}}_{\boldsymbol{n}} \mid \mathbb{X}_n = \overline{x}\right).
\end{array}\right.
\tag{9.77}
$$

In the above display, $\left(\mathbb{X}_n \mid \overline{\boldsymbol{X}}_{\boldsymbol{n}}\right)$ and $\left(\overline{\boldsymbol{X}}_{\boldsymbol{n}} \mid \mathbb{X}_n\right)$ is a shorthand notation for the $\boldsymbol{\pi_n}$-conditional distributions of $\mathbb{X}_n$ given $\overline{\boldsymbol{X}}_{\boldsymbol{n}}$, and $\overline{\boldsymbol{X}}_{\boldsymbol{n}}$ given $\mathbb{X}_n$. Notice that the first transition of the Gibbs sampler reduces to the uniform sampling of an ancestral line. In addition, by (9.76), the second transition amounts to sampling a genetic particle model with a frozen ancestral line:

$$
\overline{\boldsymbol{X}}_{\boldsymbol{n}}^{(k+1)} = \overline{x} \sim \left(\overline{\boldsymbol{X}}_{\boldsymbol{n}} \mid \mathbb{X}_n = \overline{x}\right) = \left(\overline{\boldsymbol{X}}_{\boldsymbol{n}}^{\#} \mid X_n = \overline{x}\right).
$$

By construction $(\mathbb{X}_n^{(k)})_{k \geq 0}$ is a Markov chain on S_n with reversible Feynman-Kac probability measure η_n.

We let π_n' be the probability measures on $\left(\overline{\boldsymbol{S}}_{\boldsymbol{n}}' \times S_n\right)$ defined for any function f_n on $\left(\overline{\boldsymbol{S}}_{\boldsymbol{n}}' \times S_n\right)$ by the Feynman-Kac measures

$$
\pi_n'(f_n) \propto \mathbb{E}\left(f_n(\overline{\boldsymbol{X}}_{\boldsymbol{n}}', \mathbb{X}_n^{\flat})\, \overline{Z}_n(\overline{X})\right) = \mathbb{E}\left(f_n(\overline{\boldsymbol{X}}_{\boldsymbol{n}}'^{,\#}, X_n)\, Z_n(X)\right).
\tag{9.78}
$$

The transition probabilities of the Gibbs sampler

$$
\left(\overline{\mathbb{X}}_n^{\flat,(k)}, \overline{\boldsymbol{X}}_{\boldsymbol{n}}'^{,(k)}\right) \rightsquigarrow \left(\overline{\mathbb{X}}_n^{\flat,(k+1)}, \overline{\boldsymbol{X}}_{\boldsymbol{n}}'^{,(k+1)}\right)
\tag{9.79}
$$

of the target multivariate distribution $\overline{\pi}_n'$ are defined as above by replacing the uniform ancestral line sampling by a randomly chosen backward ancestral line. By construction $(\overline{\mathbb{X}}_n^{\flat,(k)})_{k \geq 0}$ is a Markov chain on S_n with reversible Feynman-Kac probability measure η_n.

9.8 Quenched and annealed measures

9.8.1 Feynman-Kac models

We let Θ be some parameter with distribution $\lambda(d\theta)$ on some state space Ξ. Given Θ, we consider a Markov chain X_n' evolving in some state spaces S_n' with an initial condition $\eta_{\Theta,0}'$, and some elementary transitions $M_{\Theta,n}'(x_{n-1}', dx_n')$ that depend on the parameter Θ. We also consider a collection of potential functions $G_{\theta,n}'(x_n')$ indexed by $\theta \in \Xi$. We let

$$
X_n = (X_0', \ldots, X_n') \in S_n = (S_0' \times \ldots \times S_n')
$$

be the historical process of the chain X_n' and we set

$$G_{\Theta,n}(X_n) := G_{\Theta,n}'(X_n').$$

Definition 9.8.1 *We let μ_n be the Feynman-Kac measures defined for any function f_n on $(\Xi \times S_n)$ by*

$$\mu_n(f_n) \propto \mathbb{E}(f_n(\Theta, X_n)\, Z_{\Theta,n}(X)) \quad \text{with} \quad Z_{\Theta,n}(X) = \prod_{0 \le p < n} G_{\Theta,n}(X_p).$$

We also consider the quenched measures defined for any function f_n on S_n by the Feynman-Kac formulae

$$\eta_{\theta,n}(f_n) = \gamma_{\theta,n}(f_n)/\gamma_{\theta,n}(1) \quad \text{with} \quad \gamma_{\theta,n}(f_n) := \mathbb{E}(f_n(X_n)\, Z_{\Theta,n}(X) \mid \Theta = \theta). \tag{9.80}$$

In a more synthetic form, we have

$$\mu_n(d(\theta, x)) = \frac{\lambda(d\theta)\, \gamma_{\theta,n}(dx)}{\int \lambda(d\theta')\, \gamma_{\theta',n}(1)} = \lambda_n(d\theta)\, \eta_{\theta,n}(dx) \tag{9.81}$$

with the probability measures $(\lambda_n, \eta_{\theta,n})$ defined by

$$\lambda_n(d\theta) := \frac{\lambda(d\theta)\, \gamma_{\theta,n}(1)}{\int \lambda(d\theta')\, \gamma_{\theta',n}(1)}. \tag{9.82}$$

We illustrate this rather abstract model with the nonlinear filtering models discussed in section 6.5.1. In this situation, the potential functions are defined in terms of the the likelihood functions of the observations Y_n' given the value of the parameter $\Theta = \theta$ with a prior distribution $\lambda(d\theta)$, and the random state of the signal $X_n' = x_n'$, that is,

$$G_{\theta,n}'(x_n') \propto p_n(y_n' \mid x_n', \theta).$$

In this context, we have

$$
\begin{aligned}
\mu_n &= \text{Law}\left((\Theta, X_n) \mid Y_p' = y_p', 0 \le p < n\right) \\
\lambda_n &= \text{Law}\left(\Theta \mid Y_p' = y_p', 0 \le p < n\right) \\
\eta_{\theta,n} &= \text{Law}\left(X_n \mid \Theta = \theta,\, Y_p' = y_p', 0 \le p < n\right) \\
\gamma_{\theta,n+1}(1) &= p(y_0', \ldots, y_n' \mid \theta).
\end{aligned}
\tag{9.83}
$$

In Bayesian notation, the formulae (9.81) and (9.82) take the form

$$p(x, \theta \mid y) = \frac{p(x, y \mid \theta)}{p(y)}\, p(\theta) = p(\theta \mid y)\, p(x \mid \theta, y) \quad \text{and} \quad p(\theta \mid y) = \frac{p(y \mid \theta)}{p(y)}\, p(\theta)$$

with the density $p(x, \theta \mid y)$ of the conditional distribution μ_n of (X_n, Θ) given

$$Y_{n-1} := (Y_k')_{0 \le k < n} = (y_k')_{0 \le k < n} = y,$$

the density $p(x, y|\theta)$ of the conditional distribution $\gamma_{\theta,n}$ of (X_n, Y_{n-1}) given $\Theta = \theta$, the density $p(\theta|y)$ of the conditional distribution λ_n of Θ given $Y_{n-1} = y$, the density $p(\theta)$ of the prior distribution λ of Θ, the density $p(x|\theta, y)$ of the conditional distribution $\eta_{\theta,n}$ of X_n given $(\Theta, Y_{n-1}) = (\theta, y)$, and the density $p(y)$ of the distribution of Y_{n-1}. In this notation, we also have the Bayes' formula

$$p(\theta|x, y) = \frac{p(y|\theta, x)}{p(y|x, \theta)} \; p(x|\theta)$$

with the density $p(\theta|x, y)$ of the conditional distribution $\lambda_{x,n}$ of Θ given $(X_n, Y_{n-1}) = (x, y)$, and the density $p(x|\theta)$ of the conditional distribution of Θ given $X_n = x$.

9.8.2 Particle Gibbs models

We further assume that $\gamma_{\theta,n} \sim \gamma_{\theta',n}$ for any couple of parameters $\theta, \theta' \in \Xi$.

In this situation, we have

$$\mu_n(d(\theta, x)) = \eta_n(dx) \times \lambda_{x,n}(d\theta)$$

with

$$\lambda_{x,n}(d\theta) := \lambda_n(d\theta) \; \times \; \frac{d\eta_{\theta,n}(dx)}{d\eta_n(dx)} \quad \text{and} \quad \eta_n(dx) := \int \lambda_n(d\theta) \; \eta_{\theta,n}(dx).$$

In the context of the nonlinear filtering model (9.83), we have

$$\lambda_{x,n} = \text{Law}\left(\Theta \mid X_n = x, \; Y'_p = y'_p, 0 \leq p < n\right)$$

and

$$\eta_n = \text{Law}\left(X_n \mid Y'_p = y'_p, 0 \leq p < n\right).$$

Suppose we are given a collection of Markov transitions $\mathbb{K}^{[1]}_{\theta,n}$ and $\mathbb{K}^{[2]}_{x,n}$ with invariant measures $\eta_{\theta,n}$ and $\lambda_{x,n}$ on S_n and Ξ, that is, we have the fixed point equations

$$\eta_{\theta,n} = \eta_{\theta,n}\mathbb{K}^{[1]}_{\theta,n} \quad \text{and} \quad \lambda_{x,n} = \lambda_{x,n}\mathbb{K}^{[2]}_{x,n}.$$

Given the parameter $\Theta = \theta$, the Feynman-Kac measures on path space $\eta_{\theta,n}$ have exactly the same forms as the ones discussed in section 9.7.1 and in section 9.7.3. Thus, we can choose the particle Metropolis-Hastings model discussed in section 9.7.2 or the couple of particle Gibbs samplers presented in section 9.7.4.

In Bayesian literature, the choice of the Markov transition $\mathbb{K}^{[2]}_{x,n}$ is often dictated by the model at hand. For hidden Markov chain models with judiciously chosen conjugate priors, it is also possible to sample directly from $\lambda_{x,n}$. In other words, we can choose $\mathbb{K}^{[2]}_{x,n}(\theta, d\theta') = \lambda_{x,n}(d\theta')$. We illustrate this assertion with an elementary $\mathbb{R}$-valued Markov chain model

$$X'_n := b'_n(X'_{n-1}) + \sqrt{\Theta} \; W'_n$$

with a function b'_n on $\mathbb{R}$, a sequence of i.i.d. Gaussian centered random variables W'_n with unit variance, and some given $X'_0 = x'_0$. We also assume that the observation process Y'_n has the form

$$Y'_n = h'_n(X'_n) + V'_n$$

for some sensor function h'_n on $\mathbb{R}$, and a sequence of i.i.d. Gaussian centered random variables V'_n. We consider the historical processes $X_n = (X'_k)_{0 \leq k \leq n}$ and $Y_n := (Y'_k)_{0 \leq k \leq n}$.

We further assume that the unknown parameter Θ takes values in $\mathbb{R}_+ := [0, \infty[$ according to an inverse gamma distribution

$$\lambda(d\theta) := \frac{\beta^\alpha}{\Gamma(\alpha)} \ \theta^{-(\alpha+1)} \ e^{-\beta/\theta} \ 1_{\mathbb{R}_+(\theta)} \ d\theta$$

with a shape parameter α and a scale parameter β; here $\Gamma(\alpha)$ stands for the gamma function. In this context, we have

$$\mathrm{Law}\,(\Theta \mid X_n = x) = \mathrm{Law}\,(\Theta \mid X_n = x, \ Y_{n-1} = y \,).$$

for any observation sequence $y = (y'_k)_{0 \leq k < n}$.

Using Bayesian notation, for any given sequence $x = (x'_k)_{0 \leq k \leq n}$ we have

$$
\begin{aligned}
\lambda_{x,n}(d\theta) \quad &\propto \quad p((x'_0, \ldots, x'_n)|\theta) \ p(\theta) \ d\theta \\
&= \quad \left\{ \prod_{1 \leq k \leq n} p(x'_k \mid x'_{k-1}, \ \theta) \right\} \ p(\theta) \ d\theta \\
&\propto \quad \frac{1}{\theta^{n/2}} \ \exp\left(-\frac{1}{2\theta} \sum_{1 \leq k \leq n} (x'_k - b_k(x'_{k-1}))^2 \right) \ \theta^{-(\alpha+1)} \ e^{-\beta/\theta} \ 1_{\mathbb{R}_+}(\theta) \ d\theta.
\end{aligned}
$$

This implies that $\lambda_{x,n}$ is an inverse gamma distribution

$$\lambda_{x,n}(d\theta) \propto \theta^{-(\alpha_n+1)} \ e^{-\beta_n(x)/\theta} \ 1_{\mathbb{R}_+}(\theta) \ d\theta$$

with shape and scale parameters

$$\alpha_n := \alpha + n/2 \quad \text{and} \quad \beta_n(x) := \beta + \frac{1}{2} \sum_{1 \leq k \leq n} (x'_k - b'_k(x'_{k-1}))^2.$$

Proposition 9.8.2 *The measure $\mu_n = \mu_n \mathbb{K}_n$ is an invariant probability measure of the Markov transition $\mathbb{K}_n$ defined by*

$$\mathbb{K}_n\left((\theta, x), d(\theta', x')\right) := \mathbb{K}^{[2]}_{x,n}(\theta, d\theta') \ \mathbb{K}^{[1]}_{\theta', n}(x, dx').$$

Proof :
Notice that

$$
\begin{aligned}
\int \mu_n(d(\theta, x)) \ \mathbb{K}_n\left((\theta, x'), d(\theta', x')\right) \ &= \ \int \eta_n(dx) \left[\int \lambda_{x,n}(d\theta) \mathbb{K}^{[2]}_{x,n}(\theta, d\theta') \right] \mathbb{K}^{[1]}_{\theta', n}(x, dx') \\
&= \ \int \underbrace{\eta_n(dx) \times \lambda_{x,n}(d\theta')}_{=\lambda_n(d\theta') \times \eta_{\theta', n}(dx)} \ \mathbb{K}^{[1]}_{\theta', n}(x, dx').
\end{aligned}
$$

This implies that

$$
\begin{aligned}
\int \mu_n(d(\theta, x)) \, \mathbb{K}_n\left((\theta, x'), d(\theta', x')\right) &= \lambda_n(d\theta') \int \eta_{\theta',n}(dx) \, \mathbb{K}_{\theta',n}^{[1]}(x, dx') \\
&= \lambda_n(d\theta') \, \eta_{\theta',n}(dx') = \mu_n(d(\theta', x')).
\end{aligned}
$$

This ends the proof of the proposition. ∎

9.8.3 Particle Metropolis-Hastings models

We return to the many-body Feynman-Kac measures discussed in section 9.7.1. Given the parameter Θ, we let $\overline{X}_n := (\xi_n^i)_{1 \le i \le N} \in S_n^N$ be the mean field particle approximation of the quenched Feynman-Kac measures $\eta_{\Theta,n}$ defined in (9.80).

Using the unbiasedness properties (9.67) we have

$$
\mathbb{E}\left(\overline{Z}_{\Theta,n}(\overline{X}) \mid \Theta\right) = \mathbb{E}\left(Z_{\Theta,n}(X) \mid \Theta\right) = \gamma_{\Theta,n}(1) \tag{9.84}
$$

with the Radon-Nikodym derivatives

$$
\overline{Z}_{\Theta,n}(\overline{X}) := \prod_{0 \le p < n} \overline{G}_{\Theta,p}(\overline{X}_p) \quad \text{and} \quad \overline{G}_{\Theta,n}(\overline{X}_n) := \frac{1}{N} \sum_{1 \le i \le N} G_{\Theta,n}(\xi_n^i).
$$

Definition 9.8.3 *We fix the time horizon $n \ge 0$, and we let $P(\theta, d\overline{x})$ be the conditional probability distribution of the historical process $\overline{X}_n = (\overline{X}_p)_{0 \le p \le n} \in \overline{S}_n := \prod_{0 \le p \le n} \overline{S}_p$, given the value of the parameter $\Theta = \theta$. We also set*

$$
\overline{\Theta} := (\Theta, \overline{X}_n) \in \overline{\Xi} := (\Xi \times \overline{S}_n) \quad \text{and} \quad \overline{\lambda}(d\overline{\theta}) := \lambda(d\theta) \, P(\theta, d\overline{x}).
$$

In the above displayed formula, $\overline{\theta} = (\theta, \overline{x})$ stands for a given state in $\overline{\Xi} := (\Xi \times \overline{S}_n)$.

Definition 9.8.4 *We let $(\overline{h}_p)_{0 \le p \le n}$ be the collection of non-negative functions on $\overline{\Xi}$ defined for any $\overline{\theta} = (\theta, \overline{x})$ with $\overline{x} = (\overline{x}_p)_{0 \le p \le n}$ by the formulae*

$$
\overline{h}_p(\overline{\theta}) = \overline{G}_{\theta,p}(\overline{x}_p) \implies \prod_{0 \le p < n} \overline{h}_p(\overline{\theta}) = \overline{Z}_{\theta,n}(\overline{x}).
$$

Definition 9.8.5 *We associate with these objects the Boltzmann-Gibbs measures*

$$
\overline{\lambda}_n(d\overline{\theta}) = \frac{1}{\mathcal{Z}_n} \left\{ \prod_{0 \le p \le n} \overline{h}_p(\overline{\theta}) \right\} \overline{\lambda}(d\overline{\theta})
$$

with some normalizing constant $\mathcal{Z}_n$.

The unbiasedness property (9.84) implies that the θ-marginal of $\overline{\lambda}_n$ coincides with the probability measure λ_n defined in (9.82).

Our next objective is to design a particle Metropolis-Hastings Markov chain with invariant distribution $\overline{\lambda}_n$. To this end, we let $K(\theta, d\theta')$ be a Markov transition on Ξ satisfying the regularity condition

$$\lambda(d\theta)K(\theta, d\theta') \sim \lambda(d\theta')K(\theta', d\theta).$$

We let $\overline{K}(\overline{\theta}, d\overline{\theta}')$ be the Markov transition on $\overline{\Xi}$ defined by

$$\overline{K}(\overline{\theta}, d\overline{\theta}') = K(\theta, d\theta') \times P(\theta', d\overline{x}').$$

In the above displayed formula, $\overline{\theta} = (\theta, \overline{x})$ and $\overline{\theta}' = (\theta', \overline{x}')$ stands for a couple of states in the product space $\overline{\overline{\Xi}} := (\Xi \times \overline{\boldsymbol{S}_n})$.

To take the final step, we notice that

$$\frac{\overline{\lambda}_n(d\overline{\theta}')\overline{K}(\overline{\theta}', d\overline{\theta})}{\overline{\lambda}_n(d\overline{\theta})\overline{K}(\overline{\theta}, d\overline{\theta}')} = \frac{\prod_{0 \leq p \leq n} \overline{h}_p(\overline{\theta}')}{\prod_{0 \leq p \leq n} \overline{h}_p(\overline{\theta})} \times \frac{\lambda(d\theta')K(\theta', d\theta)}{\lambda(d\theta)K(\theta, d\theta')}.$$

We readily define a Metropolis-Hastings transition with reversible measure $\overline{\lambda}_n$, proposal transition $\overline{K}$ and acceptance rate

$$a(\overline{\theta}, \overline{\theta}') = 1 \wedge \left(\left[\prod_{0 \leq p \leq n} \frac{\overline{h}_p(\overline{\theta}')}{\overline{h}_p(\overline{\theta})} \right] \times \frac{\lambda(d\theta')K(\theta', d\theta)}{\lambda(d\theta)K(\theta, d\theta')} \right).$$

9.9 Some application domains

Feynman-Kac methodologies and their mean field particle interpretations constitute a universal class of simulation-based stochastic algorithms to sample approximately from any sequence of probability distributions η_n, $n \in \mathbb{N}$, with an increasing complexity.

In section 9.6 we discussed a mean field particle interpretation of any sequence of measures given by a Feynman-Kac model associated with a potential function G_n and some reference Markov chain X_n.

In this section, we illustrate these Feynman-Kac formulations with a series of examples taken from diverse application domains. Further application areas with detailed and workout examples can be found in the literature [66, 67].

9.9.1 Interacting MCMC algorithms

We let $\eta_n = \pi_n$ be the sequence of Boltzmann-Gibbs measures discussed in (9.19), and M_n the sequence of MCMC transitions defined in (9.20). By construction, we have the fixed point equations

$$\eta_n = \eta_n M_n$$

and the updating formulae

$$\eta_n(dx) \propto e^{-(\beta_n - \beta_{n-1})V(x)} e^{-\beta_{n-1}V(x)} \lambda(dx) \propto e^{-(\beta_n - \beta_{n-1})V(x)} \eta_{n-1}(dx).$$

In terms of the Boltzmann-Gibbs transformations defined in (0.3) this formula can be rewritten as follows:

$$\eta_n = \Psi_{G_{n-1}}(\eta_{n-1}) \quad \text{with} \quad G_{n-1} := e^{-(\beta_n - \beta_{n-1})V}.$$

Combining this formula with the fixed point equation we prove that

$$\eta_n = \Psi_{G_{n-1}}(\eta_{n-1}) M_n. \tag{9.85}$$

This nonlinear evolution equation coincides with the Feynman-Kac model (9.30) discussed in section 9.6.

The N-particle approximation of the nonlinear evolution equation (9.85) is defined in terms of a sequence of N particles evolving with a two genetic type selection-mutation transitions:

$$\left(X_n^i\right)_{1\le i\le N} \xrightarrow{selection} \left(\widehat{X}_n^i\right)_{1\le i\le N} \xrightarrow{mutation} \left(X_{n+1}^i\right)_{1\le i\le N}.$$

During the selection stage, for each $1 \le i \le N$, we set

$$\widehat{X}_n^i = \begin{cases} X_n^i & \text{with probability} \quad e^{-(\beta_{n+1}-\beta_n)V(X_n^i)} \\ \widetilde{X}_n^i & \text{with probability} \quad 1 - e^{-(\beta_{n+1}-\beta_n)V(X_n^i)} \end{cases}$$

where $\widetilde{X}_n^i$ is a r.v. with distribution

$$\Psi_{G_n}\left(\frac{1}{N}\sum_{1\le i\le N}\delta_{X_n^i}\right) = \sum_{1\le i\le N}\frac{e^{-(\beta_{n+1}-\beta_n)V(X_n^i)}}{\sum_{1\le j\le N}e^{-(\beta_{n+1}-\beta_n)V(X_n^j)}}\,\delta_{X_n^i}.$$

During the mutation transition, each selected particle $\widehat{X}_n^i$ evolves to a new random state X_{n+1}^i chosen with the distribution $M_{n+1}\left(\widehat{X}_n^i, dx\right)$.

Since $\eta_n = \eta_n M_n = \eta_n M^{m_n}$, for any $m_n \ge 1$,

$$\eta_n = \Psi_{G_{n-1}}(\eta_{n-1}) M_n^{m_n}. \tag{9.86}$$

The corresponding N-particle model is defined as above, replacing the mutation transition M_n by $M_n^{m_n}$. In all the situations, we have

$$\forall n \in \mathbb{N} \qquad \eta_n^N := \frac{1}{N}\sum_{1\le i\le N}\delta_{X_n^i} \longrightarrow_{N\to} \eta_n. \tag{9.87}$$

More generally, suppose we are given a sequence of target measures π_n on some state space S of the following form

$$\eta_n(dx) = \frac{1}{\mathcal{Z}_n}\left\{\prod_{0\le p\le n} h_p(x)\right\}\lambda(dx) \tag{9.88}$$

for some positive functions h_n and some reference measure λ. Choosing $h_n = e^{-(\beta_n - \beta_{n-1})V}$ and $\beta_0 = 0 = \beta_{-1}$ the measure π_n coincides with the Boltzmann-Gibbs measure discussed above. Arguing as above we have

$$\eta_n = \Psi_{G_{n-1}}(\eta_{n-1}) M_n \quad \text{with} \quad G_{n-1} = h_n$$

and for any Markov transition M_n s.t. $\eta_n = \eta_n M_n$.

These interacting MCMC samplers belong to the class of biased rejection-free Monte Carlo samplers.

We illustrate these interacting MCMC models with two examples from operations research and statistical physics.

- **Global optimization:** From probabilistic view, one natural way to compute the global minima of a given non-negative potential function V on some state space is to sample r.v. with a Boltzmann-Gibbs measure associated with some inverse temperature parameter $\beta_n \uparrow \infty$. These measures are defined by

$$\eta_n(dx) = \frac{1}{\mathcal{Z}_{\beta_n}} \ e^{-\beta_n V(x)} \ \lambda(dx)$$

where λ denotes a reference measure on S (such as the Lebesgue measure on $\mathbb{R}^d$, or the counting measure on some finite space S, or the law of some random variable).

- **Partition functions:**

Boltzmann-Gibbs measures are also of current use in statistical physics (cf. the Ising model discussed in section 6.4). In this context, another important quantity to estimate is the normalizing constant $\mathcal{Z}_n$. To this end, we observe that

$$
\begin{aligned}
\mathcal{Z}_n \ &:= \ \lambda\left(e^{-\beta_n V}\right) \\
&= \ \frac{\lambda\left(e^{-(\beta_n - \beta_{n-1})V} \ e^{-\beta_{n-1} V}\right)}{\lambda\left(e^{-\beta_{n-1} V}\right)} \times \lambda\left(e^{-\beta_{n-1} V}\right) \\
&= \ \eta_{n-1}\left(e^{-(\beta_n - \beta_{n-1})V}\right) \times \lambda\left(e^{-\beta_{n-1} V}\right).
\end{aligned}
$$

This implies that

$$\mathcal{Z}_n / \mathcal{Z}_0 = \prod_{0 \le p < n} \eta_p\left(G_p\right)$$

with the functions G_n defined for any $n \ge 0$ by

$$G_n := \exp\left\{-(\beta_{n+1} - \beta_n)V\right\}.$$

An unbiased estimate of the partitions function ratio is given by the product formulae

$$\prod_{0 \le p < n} \eta_p^N\left(G_p\right)$$

with the N-empirical measures η_n^N defined in (9.87).

- **Rare event simulation:** Most of the rare event simulation problems, including the black-box type models discussed in section 6.3, can be reduced to studying the probability that some r.v. X enters in a subset A of some state space S

$$\mathbb{P}\left(X \in A\right) \quad \text{and} \quad \mathrm{Law}(X \mid X \in A).$$

If we consider an decreasing sequence of subsets $A_n \downarrow$ the above problem can be described by a sequence of Boltzmann-Gibbs measures

$$\eta_n(dx) = \frac{1}{\mathcal{Z}_n} \ 1_{A_n}(x) \ \lambda(dx)$$

where λ stands for the distribution of X. Notice that in this notation the rare event probabilities coincide with the normalizing constants; that is, we have that

$$\mathcal{Z}_n = \lambda(1_{A_n}) = \mathbb{P}\left(X \in A_n\right).$$

Arguing as above, using the fact that

$$1_{A_n} \times 1_{A_{n-1}} = 1_{A_n \cap A_{n-1}} = 1_{A_n}$$

we also have that

$$
\begin{aligned}
\mathcal{Z}_n &:= \lambda\left(1_{A_n}\right) \\
&= \frac{\lambda\left(1_{A_n} \, 1_{A_{n-1}}\right)}{\lambda\left(1_{A_{n-1}}\right)} \times \lambda\left(1_{A_{n-1}}\right) \\
&= \eta_{n-1}\left(1_{A_n}\right) \times \mathcal{Z}_{n-1}.
\end{aligned}
$$

This implies that

$$\mathcal{Z}_n / \mathcal{Z}_0 = \prod_{0 \le p < n} \eta_p\left(G_p\right) \quad \text{with } \forall n \ge 0 \quad G_n := 1_{A_n}.$$

Here again, an unbiased estimate of the partitions function ratio is given by the product formulae $\prod_{0 \le p < n} \eta_p^N\left(G_p\right)$, with the N-empirical measures η_n^N defined in (9.87).

The following picture illustrates the empirical histogram of a genetic type particle scheme with $N = 2000$ particles in the case $\lambda = \mathcal{N}(0,1)$ and a terminal level set $A_n = [5, \infty[$. The mutation transition of the particles is given by the Gaussian shakers discussed in section 9.2.2.

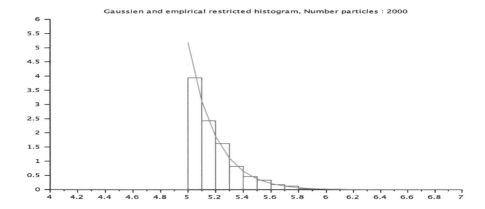

The picture below illustrates the particle estimation of the corresponding normalizing constants.

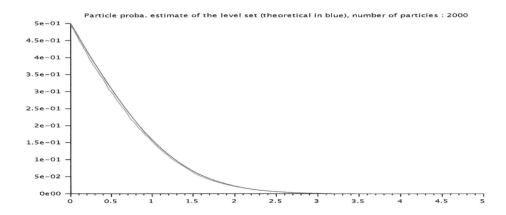

Particle proba. estimate of the level set (theoretical in blue), number of particles : 2000

9.9.2 Nonlinear filtering models

The filtering model discussed in (6.7) is a Feynman-Kac type (9.21) with

$$\mathbb{P}_n(d(x_0,\dots,x_n)) = p_n(x_0,\dots,x_n)\ dx_0\dots dx_n$$

and

$$\forall n \geq 0 \qquad G_n(x_n) := \frac{1}{\sqrt{2\pi}}\ \exp\left(-\frac{1}{2}\ (y_n - h_n(x_n))^2\right).$$

In the above display, $(y_n)_{n\geq 0}$ stands for a given and fixed sequence of observations. In this case,

$$\mathbb{Q}_n = \text{Law}\left((X_0,\dots,X_n) \mid Y_p = y_p,\ 0 \leq p < n\right).$$

In addition, with some obvious abusive notation we have

$$\mathcal{Z}_n = p_{n-1}(y_0,\dots,y_{n-1})$$

with the probability density of the observation sequence

$$\mathbb{P}\left((Y_0,\dots,Y_{n-1}) \in d(y_0,\dots,y_{n-1})\right) = p_{n-1}(y_0,\dots,y_{n-1})\ dy_0\dots dy_{n-1}.$$

9.9.3 Markov chain restrictions

The Feynman-Kac models (9.21) can always be interpreted as the conditional distribution of a Markov chain w.r.t. some collection of events. For instance, if we consider the indicator potential functions $G_n = 1_{A_n}$ we can readily check that

$$\prod_{0 \leq p < n} G_p(X_p) = 1_{(A_0 \times \dots \times A_{n-1})}(X_0,\dots,X_{n-1}).$$

This implies that

$$\mathcal{Z}_n = \mathbb{P}\left(X_p \in A_p,\ \forall 0 \leq p < n\right)$$

and

$$\mathbb{Q}_n = \text{Law}\left((X_0,\dots,X_n) \mid X_p \in A_p,\ \forall 0 \leq p < n\right).$$

9.9.4 Self avoiding walks

We assume that $X_n = (X_0', \ldots, X_n') \in S_n = E^{n+1}$ is the historical process associated with a simple random walk (*abbreviated SRW*) evolving in a d-dimensional lattice $E = \mathbb{Z}^d$. In this situation, if we set $G_n(X_n) = 1_{\mathbb{Z}^d - \{X_0', \ldots, X_{n-1}'\}}(X_n')$ in (9.21), then we find that

$$
\begin{aligned}
\mathcal{Z}_n &= \mathbb{P}\left(X_p' \neq X_q' , \ \forall 0 \leq p < q < n\right) \\
\mathbb{Q}_n &= \mathrm{Law}\left((X_0', \ldots, X_n') \mid X_p' \neq X_q' , \ \forall 0 \leq p < q < n\right).
\end{aligned}
$$

For $d = 2$, we notice that $\mathcal{Z}_{n+1} = 4^{-n} \, \mathrm{Card}(\mathcal{A}_n)$ with

$$
\mathcal{A}_n := \left\{(x_0, \ldots, x_n) \in (\mathbb{Z}^2)^{n+1} \mid \forall 0 \leq k \neq l \leq n \quad |x_k - x_{k-1}| = 1 \quad x_k \neq x_l\right\}.
$$

The following picture illustrates a sample of the SRW on $\mathbb{Z}^2$ on the time horizon $[0, 1000]$

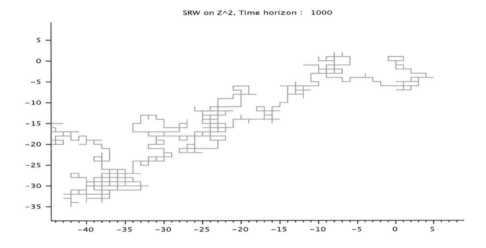

The following pictures illustrate an ancestral line and a genealogical tree-based particle model with $N = 100$ non-intersection on $\mathbb{Z}^2$ on the time horizon $[0, 1000]$.

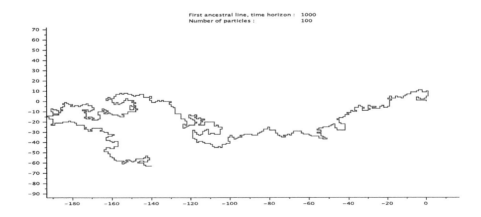

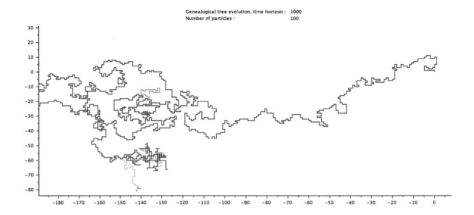

These particle models have been simulated using the mean field genetic type particle models presented in section $9.6.1$ with the free evolution or mutation associated with the SRW, and the selection indicator potential functions $G_n(X_n) = 1_{\mathbb{Z}^d - \{X'_0, \ldots, X'_{n-1}\}}(X'_n)$.

Using the fact that $\mathrm{Card}(\mathcal{A}_{p+q}) \leq \mathrm{Card}(\mathcal{A}_p) \times \mathrm{Card}(\mathcal{A}_q)$ and $2^n \leq \mathrm{Card}(\mathcal{A}_n) \leq 4 \times 3^n$, using sub-additivity arguments we find that

$$c := \lim_{n \to \infty} \mathrm{Card}(\mathcal{A}_n)^{1/n} \in [2, 3].$$

To estimate the so-called connectivity constant c (a.k.a. the critical fugacy), we use

$$\eta_n(G_n) \;=\; \mathbb{P}\left(X'_n \notin \{X'_0, \ldots, X'_{n-1}\} \mid \forall 0 \leq p < q < n \; X'_p \neq X'_q\right) = \mathrm{Card}(\mathcal{A}_n)/\left(4 \, \mathrm{Card}(\mathcal{A}_{n-1})\right)$$

and therefore

$$\frac{1}{n} \log \gamma_n(G_n) = \frac{1}{n} \sum_{0 \leq k \leq n} \log \eta_k(G_k) \simeq_{n \uparrow \infty} \log(c/4).$$

The following picture provides an estimate of the connectivity constant with $N = 100$ particles.

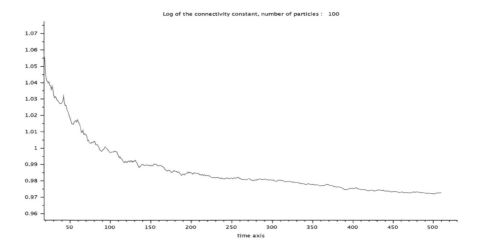

Self avoiding random walks (*abbreviated SAWs*) are used in physics to model the evolution of linear and directed polymers. These polymers represent the formation of long molecules consisting of monomers linked together in a chemical solvent. The location of the monomers is encoded in the random walk evolution, and the time horizon represents the length of the molecule. These models can be extended with a little extra work to analyze polymer models in a confined geometry.

9.9.5 Twisted measure importance sampling

Computing the probability of some events of the form $\{V_n(X_n) \geq a\}$, for some energy-like function V_n and a given threshold a is often performed using the importance sampling distribution of the state variable X_n with some multiplicative Boltzmann weight function $\exp(\beta V_n(X_n))$, associated with an inverse temperature parameter β. These twisted measures can be described by a Feynman-Kac model (9.21) in transition space by setting

$$G_n(X_{n-1}, X_n) = \exp\{\beta[V_n(X_n) - V_{n-1}(X_{n-1})]\}.$$

For instance, it is easily checked that

$$\mathbb{P}\left(V_n(X_n) \geq a\right) \;=\; \mathbb{E}\left(\mathbf{f_n}(\mathbf{X}_n) \prod_{0 \leq p < n} G_p(\mathbf{X}_p)\right)$$

with the function $f_n(\mathbf{X}_n) = 1_{V_n(X_n) \geq a}\, e^{-\beta V_n(X_n)}$, and the potential function and the reference Markov chain

$$\mathbf{X}_n = (X_n, X_{n+1}) \quad \text{and} \quad G_n(\mathbf{X}_n) = \exp\{\beta(V_{n+1}(X_{n+1}) - V_n(X_n))\}.$$

We let $\mathbb{Q}_n$ be the Feynman-Kac model (9.21) associated with the reference Markov chain $\mathbf{X}_n$ and the potential function G_n In the same vein, we have the Feynman-Kac formulae

$$\mathbb{E}\left(f_n(X_0, \ldots, X_n) \mid V_n(X_n) \geq a\right) = \mathbb{Q}_n(F_{n,f_n})/\mathbb{Q}_n(F_{n,1})$$

with the function

$$F_{n,f_n}(X_0, \ldots, X_n) = f_n(X_0, \ldots, X_n)\, 1_{V_n(X_n) \geq a}\, e^{-\beta V_n(X_n)}.$$

9.9.6   Kalman-Bucy filters

9.9.6.1   Forward filters

We consider a $\mathbb{R}^{p+q}$-valued Markov chain (X_n, Y_n) defined by the recursive relations

$$\begin{cases} X_n &= A_n\, X_{n-1} + a_n + W_n\,, & n \geq 1 \\ Y_n &= B_n\, X_n + b_n + V_n\,, & n \geq 0 \end{cases} \tag{9.89}$$

for some $\mathbb{R}^p$ and $\mathbb{R}^q$-valued independent random sequences W_n and V_n, independent of X_0, some matrices A_n, B_n with appropriate dimensions and finally some $(p+q)$-dimensional vector (a_n, b_n). We further assume that W_n and V_n centered Gaussian random sequences with covariance matrices R_n^v, R_n^w and X_0 is a Gaussian random variable in $\mathbb{R}^p$ with a mean and covariance matrix denoted by

$$\widehat{X}_0^- \;=\; \mathbb{E}(X_0) \quad \text{and} \quad \widehat{P}_0^- = \mathbb{E}((X_0 - \mathbb{E}(X_0))\,(X_0 - \mathbb{E}(X_0))').$$

Theorem 9.9.1 *The one-step predictors and the optimal filters are given by*

$$\begin{aligned} \eta_n &= \mathrm{Law}(X_n \mid (Y_0, \dots, Y_{n-1})) = \mathcal{N}(\widehat{X}_n^-, P_n^-) \\ \widehat{\eta}_n &= \mathrm{Law}(X_n \mid (Y_0, \dots, Y_{n-1}, Y_n)) = \mathcal{N}(\widehat{X}_n, P_n). \end{aligned} \tag{9.90}$$

In the above display, $\mathcal{N}(.,.)$ stands for the Gaussian distributions discussed in (4.13). The synthesis of the conditional mean and covariance matrices is carried out using the traditional Kalman-Bucy recursive updating-prediction equations

$$\left(\widehat{X}_n^-, P_n^-\right) \xrightarrow{\;updating\;} \left(\widehat{X}_n, P_n\right) \xrightarrow{\;prediction\;} \left(\widehat{X}_{n+1}^-, P_{n+1}^-\right). \tag{9.91}$$

The updating transition is given by

$$\widehat{X}_n = \widehat{X}_n^- + \mathbb{G}_n\,(Y_n - (B_n \widehat{X}_n^- + b_n)) \quad and \quad P_n = (Id - \mathbb{G}_n B_n) P_n^- \tag{9.92}$$

with the gain matrix

$$\mathbb{G}_n = P_n^- B_n'(B_n P_n^- B_n' + R_n^v)^{-1}.$$

The prediction transition is given by

$$\widehat{X}_{n+1}^- = A_{n+1}\widehat{X}_n + a_{n+1} \quad and \quad P_{n+1}^- = A_{n+1}' P_n A_{n+1} + R_{n+1}^w. \tag{9.93}$$

In addition, the density $p(y_0, \dots, y_n)$ of the observation sequence $(Y_0, \dots, Y_n)$ (w.r.t. the Lebesgue measure $dy_0 \times \dots \times dy_n$) evaluated on the observation path $p(Y_0, \dots, Y_n)$ is given by

$$p(Y_0, \dots, Y_n) = \prod_{0 \leq k \leq n} \mathcal{N}(B_k\,\widehat{X}_k^- + b_k, B_k P_k^- B_k' + R_k^v)(Y_k) \tag{9.94}$$

with the density $\mathcal{N}\,[m, R]\,(y)$ a Gaussian distribution w.r.t. the Lebesgue measure dy defined in (4.14).

Proof :

We use induction w.r.t. the time parameter. For $n = 0$, we have $\eta_0 = \text{Law}(X_0) = \mathcal{N}(\widehat{X}_0^-, P_0^-)$. In addition, recalling that

$$p(y_0|x_0) = \mathcal{N}\left[B_0 x + b_0; R^v\right](y_0) \quad \text{and} \quad p(x_0|y_0) = \frac{1}{p(y_0)} \, p(y_0|x_0) \, p(x_0)$$

using the Gaussian update formula (4.15), we have

$$\text{Law}(X_0 \mid Y_0) = \mathcal{N}(\widehat{X}_0, P_0)$$

with the parameters $(\widehat{X}_0, P_0)$ given in (9.92) for $n = 0$. By construction, we have the linear Gaussian equation

$$X_1 = A_1 \, X_0 + a_1 + W_1$$

and

$$p(x_1 \mid y_0) = \int p(x_1|x_0) \, p(x_0|y_0) \, dx_0 \quad \text{with} \quad p(x_0|y_0) \, dx_0 = \mathcal{N}(\widehat{X}_0, P_0)(dx_0).$$

This clearly implies that

$$p(x_1 \mid y_0) \, dx_1 = \mathcal{N}(\widehat{X}_1^-, P_1^-)(dx_1)$$

with the parameters $(\widehat{X}_1^-, P_1^-)$ given in (9.93) for $n = 0$. Assuming the result is true at some rank n, we check (4.15) and (9.93) using the same proof as above. This ends the proof of the first assertion.

The proof of (9.94) follows from the fact that

$$p(y_0, \ldots, y_n) = p(y_n|y_0, \ldots, y_{n-1})p(y_0, \ldots, y_{n-1}) = \prod_{0 \leq k \leq n} p(y_k|y_0, \ldots, y_{k-1})$$

with the conditional density $p(y_n|y_0, \ldots, y_{n-1})$ of Y_n given $(Y_0, \ldots, Y_{n-1}) = (y_0, \ldots, y_{n-1})$. On the other hand, we have

$$p(y_n|y_0, \ldots, y_{n-1}) = \int p(y_n|x_n) \, p(x_n|y_0, \ldots, y_{n-1}) \, dx_n \quad \text{and} \quad Y_n = B_n \, X_n + b_n + V_n.$$

This implies that

$$p(y_n|Y_0, \ldots, Y_{n-1})dy_n = \mathcal{N}(B_n \, \widehat{X}_n^- + b_n, B_n P_n^- B_n' + R_n^v)(dy_n).$$

This ends the proof of the theorem. ∎

9.9.6.2 Backward filters

In Bayesian notation, we have the conditional density formulae

$$p((x_0, \ldots, x_n) \mid (y_0, \ldots, y_{n-1}))$$

$$= p(x_n \mid (y_0, \ldots, y_{n-1})) \, p(x_{n-1} \mid x_n, (y_0, \ldots, y_{n-1}))$$

$$\times p(x_{n-2} \mid x_{n-1}, (y_0, \ldots, y_{n-2})) \ldots p(x_1 \mid x_2, (y_0, y_1)) \, p(x_0 \mid x_1, y_0).$$
$$(9.95)$$

This shows that

$$\mathbb{P}\left((X_0, \ldots, X_n) \in d(x_0, \ldots, x_n) \mid Y_p = y_p, \ p < n\right)$$

$$= \eta_n(dx_n) \prod_{1 \leq k \leq n} \widehat{\mathbb{M}}_{k,\widehat{\eta}_{k-1}}(x_k, dx_{k-1}) \tag{9.96}$$

with the (backward) Markov transitions

$$\widehat{\mathbb{M}}_{k,\widehat{\eta}_{k-1}}(x_k, dx_{k-1}) \ = \ p(x_{k-1} \mid x_k, (y_0, \ldots, y_{k-1})) \, dx_{k-1}.$$

Using Bayes' rule, we also have

$$\begin{aligned}
\widehat{\mathbb{M}}_{k,\widehat{\eta}_{k-1}}(x_k, dx_{k-1}) \ &= \ p(x_{k-1} \mid x_k, (y_0, \ldots, y_{k-1})) \, dx_{k-1} \\
&\propto \ p(x_k \mid x_{k-1}) \, \underbrace{p(x_{k-1} \mid (y_0, \ldots, y_{k-1})) \, dx_{k-1}}_{=\widehat{\eta}_{k-1}(dx_{k-1})}.
\end{aligned} \tag{9.97}$$

Recalling

$$p(x_{k-1} \mid (y_0, \ldots, y_{k-1})) dx_{k-1} \propto p(y_{k-1}|x_{k-1}) \, \underbrace{p(x_{k-1} \mid (y_0, \ldots, y_{k-2})) dx_{k-1}}_{=\eta_{k-1}(dx_{k-1})}$$

shows that

$$\widehat{\mathbb{M}}_{k,\widehat{\eta}_{k-1}}(x_k, dx_{k-1}) = \mathbb{M}_{k,\eta_{k-1}}(x_k, dx_{k-1}) \propto p(y_{k-1}|x_{k-1}) \, p(x_k \mid x_{k-1}) \, \eta_{k-1}(dx_{k-1}).$$

The backward transitions $\mathbb{M}_{k,\eta_{k-1}}$ coincide with the ones discussed in (9.57).

Our next objective is to provide an analytic expression of these Markov transitions. To this end, we observe that

$$\widehat{\eta}_{k-1}(dx_{k-1}) = \mathcal{N}\left(\widehat{X}_{k-1}, P_{k-1}\right)(dx_{k-1}) \quad \text{and} \quad X_k = A_k \, X_{k-1} + a_k + W_k.$$

Applying the updating formula (4.15) to the Bayes' rule (9.97), we readily check that

$$\widehat{\mathbb{M}}_{k,\widehat{\eta}_{k-1}}(x_k, dx_{k-1}) = \mathcal{N}\left(\widetilde{m}_{k-1}(x_k), \widetilde{P}_{k-1}\right)(dx_{k-1})$$

with

$$\widetilde{m}_{k-1}(x_k) := \widehat{X}_{k-1} + \widetilde{\mathbb{G}}_{k-1}\left(x_k - (A_k\widehat{X}_{k-1} + a_k)\right) \quad \text{and} \quad \widetilde{P}_{k-1} = (Id - \widetilde{\mathbb{G}}_{k-1}A_k)P_{k-1} \tag{9.98}$$

with the gain matrix

$$\widetilde{\mathbb{G}}_{k-1} = P_{k-1}A_k'\left(A_k P_{k-1} A_k' + R_k^v\right)^{-1} = P_{k-1}A_k'\left(P_k^-\right)^{-1}.$$

In other words, (9.96) is the distribution of the backward random trajectories

$$\widetilde{X}_n^{(n)} \to \widetilde{X}_{n-1}^{(n)} \to \ldots \to \widetilde{X}_1^{(n)} \to \widetilde{X}_0^{(n)}$$

defined by the backward equations

$$\widetilde{X}_{k-1}^{(n)} = \widetilde{m}_{k-1}\left(\widetilde{X}_k^{(n)}\right) + \widetilde{W}_{k-1}$$

with $\widetilde{X}_n^{(n)} \sim \mathcal{N}\left(\widehat{X}_n^-, P_n^-\right)$, and a sequence $\widetilde{W}_k$ of i.i.d. centered Gaussian variables with covariance matrices $\widetilde{P}_k$.

This also implies that the conditional mean and covariance matrices of this Gaussian linear model

$$\overline{X}_k^{(n)} = \mathbb{E}\left(\widetilde{X}_k^{(n)} \mid (Y_0, \ldots, Y_{k-1})\right)$$

$$\Sigma_k^{(n)} = \mathbb{E}\left(\left(\widetilde{X}_k^{(n)} - \overline{X}_k^{(n)}\right)\left(\widetilde{X}_k^{(n)} - \overline{X}_k^{(n)}\right)' \mid (Y_0, \ldots, Y_{n-1})\right)$$

satisfy the backward recursive formula

$$\begin{cases} \overline{X}_k^{(n)} = \widetilde{m}_k\left(\overline{X}_{k+1}^{(n)}\right) \\ \Sigma_k^{(n)} = \widetilde{P}_k + \widetilde{\mathbb{G}}_k \, \Sigma_{k+1}^{(n)} \widetilde{\mathbb{G}}_k' = P_k + \widetilde{\mathbb{G}}_k \, \left(\Sigma_{p+1}^{(n)} - P_{p+1}^-\right) \widetilde{\mathbb{G}}_k' \end{cases}$$

with final time horizon condition $\left(\overline{X}_n^{(n)}, \Sigma_n^{(n)}\right) = \left(\widehat{X}_n^-, P_n^-\right)$.

The covariance formula follows from the fact that

$$\widetilde{X}_k^{(n)} - \overline{X}_k^{(n)} = \left[\widetilde{m}_k\left(\widetilde{X}_{k+1}^{(n)}\right) - \widetilde{m}_k\left(\overline{X}_{k+1}^{(n)}\right)\right] + \widetilde{W}_k = \widetilde{\mathbb{G}}_k \, (\widetilde{X}_{k+1}^{(n)} - \overline{X}_k^{(n)}) + \widetilde{W}_k$$

and

$$P_k - \widetilde{\mathbb{G}}_k \, P_{k+1}^- \underbrace{\left(P_{k+1}^-\right)^{-1} A_{k+1} P_k}_{:=\widetilde{\mathbb{G}}_k'} = \left(Id - \widetilde{\mathbb{G}}_k A_{k+1}\right) P_k = \widetilde{P}_{k-1}.$$

9.9.6.3 Ensemble Kalman filters

In the further development of this section, the observation sequence $Y_n = y_n$ is assumed to be fixed, with $n \geq 0$. In this situation, the Kalman filter is a deterministic sequence of variables defined by the recursions (9.92) and (9.93).

Definition 9.9.2 *We let $\widetilde{X}_n^-$ be a Gaussian random variable with distribution $\eta_n = \mathcal{N}(\widehat{X}_n^-, P_n^-)$ and we set*

$$\widetilde{X}_n := \widetilde{X}_n^- + \mathbb{G}_n \left(y_n - (B_n \widetilde{X}_n^- + b_n + V_n)\right) \tag{9.99}$$

with some collection of independent (and independent of $\widetilde{X}_n^-$) Gaussian random variables V_n with distribution $\mathcal{N}(0, R_n^v)$.

Lemma 9.9.3 *The random variable $\widetilde{X}_n$ defined in (9.99) is a Gaussian random variable with mean $\widehat{X}_n$ and covariance matrix P_n.*

Proof :

By construction, $\widetilde{X}_n$ is a Gaussian random variable with mean

$$\mathbb{E}(\widetilde{X}_n) = \widehat{X}_n^- + \mathbb{G}_n \left(y_n - (B_n\widehat{X}_n^- + b_n)\right) = \widehat{X}_n$$

and covariance matrix

$$\mathbb{E}\left((\widetilde{X}_n - \mathbb{E}(\widetilde{X}_n))(\widetilde{X}_n - \mathbb{E}(\widetilde{X}_n))'\right) = (Id - \mathbb{G}_nB_n)P_n^-(Id - \mathbb{G}_nB_n)' + \mathbb{G}_nR_n^v\mathbb{G}_n'$$
$$= (Id - \mathbb{G}_nB_n)P_n^- = P_n.$$

The covariance formula follows from

$$\widetilde{X}_n - \mathbb{E}(\widetilde{X}_n) = (Id - \mathbb{G}_nB_n)\left(\widetilde{X}_n^- - \widehat{X}_n^-\right) - \mathbb{G}_nV_n$$

and

$$\mathbb{G}_nR_n^v\mathbb{G}_n' - (Id - \mathbb{G}_nB_n)P_n^-B_n'\mathbb{G}_n' = \mathbb{G}_n\left(R_n^v + B_nP_n^-B_n'\right)\mathbb{G}_n' - P_n^-B_n'\mathbb{G}_n' = 0.$$

This ends the proof of the lemma. ■

Using the prediction formula (9.93), we also check the following technical lemma.

Lemma 9.9.4 *We have*

$$\widetilde{X}_n \sim \mathcal{N}(\widehat{X}_n, P_n) \Rightarrow \widetilde{X}_{n+1}^- = A_{n+1}\,\widetilde{X}_n + a_{n+1} + W_{n+1} \sim \mathcal{N}(\widehat{X}_{n+1}^-, P_{n+1}^-)$$

with some Gaussian random variables W_{n+1} (independent of $\widetilde{X}_n$) with distribution $\mathcal{N}(0, R_{n+1}^w)$.

Given some probability measure η on $\mathbb{R}^d$, we denote by $\widetilde{P}_\eta$ the covariance matrix defined by

$$\widetilde{P}_\eta := \eta\left([\varphi - \eta(\varphi)]\ [\varphi - \eta(\varphi)]'\right)$$

with the column identity vector $\varphi(x) = x \in \mathbb{R}^p$. In this notation, the gain matrix $\mathbb{G}_n$ can be rewritten as follows:

$$\mathbb{G}_n = \widetilde{P}_{\eta_n}B_n'(B_n\widetilde{P}_{\eta_n}B_n' + R_n^v)^{-1} := \widetilde{\mathbb{G}}_{n,\eta_n}.$$

The Kalman filter recursion (9.91) can be interpreted as the evolution of the mean and the covariance matrix of the nonlinear Markov chain model

$$\widetilde{X}_n^- \xrightarrow{\text{updating/analysis}} \widetilde{X}_n := \widetilde{X}_n^- + \widetilde{\mathbb{G}}_{n,\eta_n}\left(Y_n - (B_n\widetilde{X}_n^- + b_n) - V_n\right)$$

$$\xrightarrow{\text{prediction/forecast}} \widetilde{X}_{n+1}^- = A_{n+1}\,\widetilde{X}_n + a_{n+1} + W_{n+1}$$

starting at some Gaussian random variable $\widetilde{X}_0^- \sim \mathcal{N}(\widehat{X}_0^-, P_0^-)$. In data assimilation and computer science literature the updating and prediction transitions are often called the analysis-forecast transitions.

Note that these Markov chain models belong to the class of nonlinear Markov chain models (7.31) discussed in section 7.10.2. Their N-mean field particle interpretations (7.27) are defined in terms of a Markov chain evolving in the product space $(\mathbb{R}^p)^N$

$$(\widetilde{X}_n^{-,i})_{1 \leq i \leq N} \xrightarrow{\text{analysis}} (\widetilde{X}_n^i)_{1 \leq i \leq N} \xrightarrow{\text{forecast}} \left(\widehat{X}_{n+1}^{-,i}\right)_{1 \leq i \leq N}. \tag{9.100}$$

The parameter N stands for the number of particles and the precision of the stochastic algorithm. We set $\eta_n^N := \frac{1}{N} \sum_{1 \leq i \leq N} \delta_{\widetilde{X}_n^{-,i}}$, we consider a sequence $(V_n^i, W_n^i)_{1 \leq i \leq N}$ of independent copies of the random variables (V_n, W_n), and we let $(\widetilde{X}_0^{-,i})_{1 \leq i \leq N}$ be N i.i.d. copies of $\widetilde{X}_0^-$.

The evolution of the N-mean field particle model (a.k.a. ensemble Kalman filter) is defined for any $1 \leq i \leq N$ by the synthetic diagram

$$\widetilde{X}_n^{-,i} \xrightarrow{\text{updating/analysis}} \widetilde{X}_n^i := \widetilde{X}_n^{-,i} + \widetilde{\mathbb{G}}_{n,\eta_n^N} \left(Y_n - (B_n \widetilde{X}_n^{-,i} + b_n) - V_n^i\right)$$

$$\xrightarrow{\text{prediction/forecast}} \widetilde{X}_{n+1}^{-,i} = A_{n+1} \widetilde{X}_n^i + a_{n+1} + W_{n+1}^i.$$

The continuous time version of the ensemble Kalman filters discussed above is presented in some details in section 18.5.5.

9.9.6.4 Interacting Kalman filters

Suppose that at every time step the state of a Markov chain with two coordinates (Θ_n, X_n) is partially observed according to the following schematic picture

$$
\begin{array}{ccccccc}
\Theta_0 & \longrightarrow & \Theta_1 & \longrightarrow & \Theta_2 & \longrightarrow & \cdots \\
\downarrow & & \downarrow & & \downarrow & & \\
X_0 & \longrightarrow & X_1 & \longrightarrow & X_2 & \longrightarrow & \cdots \\
\downarrow & & \downarrow & & \downarrow & & \\
Y_0 & & Y_1 & & Y_2 & & \cdots
\end{array}
\tag{9.101}
$$

We assume that Θ_n is a Markov chain evolving in some state spaces Ξ_n. Given a realization of the chain Θ_n, we assume that the pair signal observation (X_n, Y_n) is given by

$$\begin{cases} X_n = A_n(\Theta_n) X_{n-1} + a_n(\Theta_n) + W_n, & n \geq 1 \\ Y_n = B_n(\Theta_n) X_n + b_n(\Theta_n) + V_n, & n \geq 0 \end{cases} \tag{9.102}$$

for some collection of matrices and vectors $A_n(\theta), B_n(\theta), a_n(\theta), b_n(\theta)$, indexed by $\theta \in \Xi_n$, of the same dimension as the matrices and vectors (A_n, B_n, a_n, b_n) introduced in (9.89). Using Bayesian notation, by (9.94) we have

$$p(y_0, \ldots, y_n | \theta_0, \ldots, \theta_n) = \prod_{0 \leq k \leq n} \overline{G}_k(\overline{\theta}_k)$$

with

$$\overline{\theta}_k := \left(\theta_k, \widehat{X}_{k,\theta}^-, P_{k,\theta}^-\right)$$

$$\overline{G}_k(\overline{\Theta}_k) := \mathcal{N}(B_k(\theta_k) \widehat{X}_{k,\theta}^- + b_k(\theta_k), B_k(\theta_k) P_{k,\theta}^- B_k'(\theta_k) + R_k^v)(y_k)$$

where $(\widehat{X}_{k,\theta}^-, P_{k,\theta}^-)$ stands for the conditional mean and covariance matrix of the forward Kalman filters. Given $(\Theta_0, \ldots, \Theta_k) = (\theta_0, \ldots, \theta_k)$, the sequence $(\widehat{X}_{k,\theta}^-, P_{k,\theta}^-)$ is given by (the deterministic) forward Kalman filter recursion. Thus, the sequence of variables $\overline{\Theta}_n$ forms a Markov chain with some Markov transition $\overline{M}_n$ and some initial distribution $\overline{\eta}_0$. On the other hand, the posterior distribution

$$p(d(\theta_0, \ldots, \theta_n)|y_0, \ldots, y_n) \propto p(y_0, \ldots, y_n|\theta_0, \ldots, \theta_n)\, p(d(\theta_0, \ldots, \theta_n))$$

is the $(\theta_0, \ldots, \theta_n)$-marginal of the Feynman-Kac measures

$$\left\{ \prod_{0 \le k \le n} \overline{G}_k(\overline{\theta}_k) \right\} \times p(d(\overline{\theta}_0, \ldots, \overline{\theta}_n)).$$

The N-mean field particle interpretations of the Feynman-Kac measures defined above can be interpreted as a sequence of N interacting Kalman filters.

For fixed parameter $\Theta_n = \Theta_{n-1} := \Theta$, the posterior distribution of the unknown parameter Θ is given by

$$p(d\theta|y_0, \ldots, y_n) \propto p(y_0, \ldots, y_n|\theta)\, p(d\theta) \quad \text{with} \quad p(y_0, \ldots, y_n|\theta) = \prod_{0 \le k \le n} h_k(\theta)$$

and the positive functions

$$h_n(\theta) := \mathcal{N}(B_n(\theta)\, \widehat{X}_{n,\theta}^- + b_n(\theta),\, B_n(\theta)P_{k,\theta}^- B_k'(\theta) + R_k^v)(y_k).$$

These target probability measures have the same form as the ones discussed in (9.88). They can be approximated using the interacting MCMC methodologies discussed in section 9.9.1.

9.10 Exercises

Exercise 118 (Parameter inference - Direct observation) ✎ *Suppose that we are observing the random states X_n of the $\{0,1\}$-valued Markov chain presented in (7.11). Propose an estimate of the parameter q.*

Exercise 119 (Gaussian restrictions) ✎ *Check that the r.v. defined in (9.3) is distributed according to the Gaussian distribution restricted to the set $[a,b]$.*

Exercise 120 (A random direction Monte Carlo sampler) *Consider a Boltzmann-Gibbs probability measure $\nu(dx) \propto e^{-U(x)}dx$ on $\mathbb{R}$ associated with some potential function $U : \mathbb{R} \mapsto \mathbb{R}$ (s.t. the Gibbs measure is well defined). We let $\mu(dv)$ be any symmetric probability measure on $\mathbb{R}$ (in the sense that $V \sim \mu \Rightarrow (-V) \sim \mu$).*

- *We consider the Markov chain $\mathcal{X}_n = (V_n, X_n)$ on $S = (\mathbb{R} \times \mathbb{R})$ with elementary transition defined for any $(x,v) \in \mathbb{R}^2$ and any bounded function f on $\mathbb{R}^2$ by*

$$M_h(f)(v,x) = f(v, x + vh)\, e^{-(U(x+hv)-U(x))_+} + f(-v,x)\left(1 - e^{-(U(x+hv)-U(x))_+}\right)$$

  with some parameter $h \in \mathbb{R}$. Check that $\pi(d(v,x)) = \mu(dv) \times \nu(dx)$ is an invariant measure of $\mathcal{X}_n$. Discuss the non-uniqueness property of the invariant measure.

- We denote by $\overline{\mathcal{X}}_n = (\overline{V}_n, \overline{X}_n)$ the Markov chain on $S = (\mathbb{R} \times \mathbb{R})$ with elementary transition defined by

$$\overline{M}_h(f)(v, x)$$
$$= \int f(w, x + vh) \ e^{-(U(x+hv)-U(x))_+} \mu(dw) \ + \int f(w, x) \ \left(1 - e^{-(U(x+hv)-U(x))_+}\right) \mu(dw).$$

Check that

$$\overline{M}_h = M_h K \quad \text{with the transition} \quad K(f)(v, x) = \int f(w, x) \ \mu(dw).$$

This shows that the chain $\overline{\mathcal{X}}_n$ is defined as $\mathcal{X}_n$ but at each time we regenerate the velocity component according to the distribution μ. In this situation, the velocity coordinates $\overline{V}_n$ are independent copies of a random variable W with distribution μ (assuming implicitly that $\overline{V}_0 \sim \mu$). Prove that the second coordinate $\overline{X}_n$ resumes to a Metropolis-Hastings sampler with transitions

$$\overline{P}(x, dy) \ := \ \mathbb{P}\left(\overline{X}_{n+1} \in dy \mid \overline{X}_n = x\right)$$
$$= \ P(x, dy) \ a(x, y) \ + \left(1 - \int P(x, dz) \ a(x, z)\right) \ \delta_x(dy),$$

with the proposal transition

$$P(x, dy) = \mathbb{P}\left(x + W \in dy\right),$$

and the acceptance rate

$$a(x, y) = e^{-(U(y)-U(x))_+} = \min\left(1, \frac{\nu(dy)P(y, dx)}{\nu(dx)P(x, dy)}\right).$$

The continuous time version of this sampler is discussed in exercise 217 (see also exercise 229).

Exercise 121 (Gibbs sampler on a disk and its boundary) ✎ *We let M be the Markov transition of the chain Z_n presented in (9.5). Prove that $\pi M = \pi$, where π is the uniform measure on the unit disk. Describe the Gibbs sampler associated with the uniform target measure η on the circle $\{(x, y) \ : \ x^2 + y^2 = 1\}$ given by*

$$\eta(d(x, y)) = \frac{1}{\pi} \ \frac{1}{\sqrt{1-x^2}} \ 1_{]-1,1[}(x) \ dx \ \times \ \frac{1}{2} \left[\delta_{-\sqrt{1-x^2}} + \delta_{\sqrt{1-x^2}}\right](dy).$$

Check that the sampler starting at $\begin{pmatrix} x \\ y \end{pmatrix}$ gets stuck on the four states

$$\left\{\begin{pmatrix} x \\ y \end{pmatrix}, \begin{pmatrix} x \\ -y \end{pmatrix}, \begin{pmatrix} -x \\ -y \end{pmatrix}, \begin{pmatrix} -x \\ y \end{pmatrix}\right\}.$$

This example shows that Cartesian coordinates only offer two possible directions for the exploration of the boundary. We refer to exercises 125, 132 and 133 for the design of more flexible direction free Gibbs samplers and related stochastic billiards processes. We also refer to section 23.3.1 for other examples of uniform measures on boundary surfaces of embedded manifolds.

Exercise 122 (Metropolis-Hastings Poisson sampling) *We consider the Poisson distribution π on $\mathbb{N}$ given by*

$$\forall x \in \mathbb{N} \qquad \pi(x) = e^{-\lambda} \frac{\lambda^x}{x!}$$

for some given $\lambda > 0$. We let K be the Markov transition of the simple random walk on $\mathbb{N}$ given by

$$\forall x \in \mathbb{N} - \{0\} \qquad K(x,y) = \frac{1}{2} 1_{x-1}(y) + \frac{1}{2} 1_{x+1}(y).$$

For $x = 0$, we let $K(0,1) = 1$. Describe the acceptance ratio (9.10) of the Metropolis-Hastings model with proposal transition K.

Exercise 123 (Metropolis-Hastings Gaussian sampling) *We let $\pi(x)$ be the Gaussian density with mean m and variance σ^2. We let $K(x,y)$ be the Gaussian transition*

$$K(x,y) = \frac{1}{\sqrt{2\pi\tau^2}} \exp\left(-\frac{1}{2\tau^2} (y-x)^2\right).$$

Describe the acceptance ratio (9.10) of the Metropolis-Hastings model with proposal transition K.

Exercise 124 (Ball walk Metropolis-Hastings sampler) *For any given $x \in \mathbb{R}^r$ we let $K_\epsilon(x,dy)$ be the uniform distribution on the ball $B(x,\epsilon) := \{y \in \mathbb{R}^r : \|x-y\| \leq \epsilon\}$ centered at x with radius ϵ. Describe a Metropolis-Hastings sampler with proposal transition K_ϵ and a prescribed target measure $\pi(dx) = p(x)\ dx$ with a density $p(x)$.*

Exercise 125 (Hit-and-run and direction-free Gibbs samplers [50])

Let $X = (X_j)_{1 \leq j \leq r}$ be a random variable with a probability distribution $\eta(dx) = p(x)dx$ having a positive density $p(x)$ w.r.t. the Lebesgue measure dx on $\mathbb{R}^r$, for some $r \geq 1$. Let $\mathcal{A}$ be the set of lines $A(x,u) := \{x+tu : \|u\| = 1\}$ indexed by $x \in \mathbb{R}^r$ and unit vectors u. We let $\nu(du)$ be the uniform distribution on the unit sphere $\mathbb{S}^{r-1} := \{u \in \mathbb{R}^r : \|u\| = 1\}$. Let $\eta_{x,u}(dz)$ be the restriction of η to the line $A(x,u)$ defined by

$$\eta_{x,u}(dz) = \frac{\int p(x+tu)\ dt}{\int p(x+su)\ ds} \delta_{x+tu}(dz).$$

We consider the Markov transition $M(x,dy)$ defined by

$$M(x,dy) = \int \nu(du)\ \eta_{x,u}(dy).$$

Check that

$$\forall u \in \mathbb{S}^{r-1} \qquad \int \eta(dx)\ \eta_{x,u}(dz) = \eta(dz) \quad and \quad \eta M = \eta.$$

- *Design a Markov transition with a target measure η when the density $p(x)$ is supported by an open bounded subset $S \subset \mathbb{R}^r$. For any $x \neq y$ we set $v_{x,y} := (y-x)/\|y-x\| \in \mathbb{S}^1$. When $r = 2$ check that $\overline{M}(x,dy) = \overline{m}(x,y)\ dy$, with the Lebesgue measure dy on $\mathbb{R}^2$ and the probability density*

$$\overline{m}(x,y) = \frac{1}{\pi} \frac{p(y)}{\|x-y\|} \Big/ \int_{\mathcal{S}(x,v_{x,y})} p(x+tv_{x,y})\ dt$$

with $\mathcal{S}(x,u) := \{t \in \mathbb{R} : x+tu \in S\}$. When p is bounded, check that $\overline{m}(x,y) \geq \epsilon\ p(y)$, for some parameter $\epsilon > 0$ whose values do not depend on (x,y). Deduce that

$$\left\|\mu\overline{M}^n - \eta\right\|_{tv} \leq (1-\epsilon)^n\ \|\mu - \eta\|_{tv}$$

for any initial distribution μ on S.

- *Discuss the situation when ν is not necessarily uniform.*

- *Consider the case $\nu = \frac{1}{r}\sum_{1 \le i \le r}\delta_{e_i}$, where $e_i = (1_i(j))_{1 \le j \le r}$ stands for the r unit vectors of $\mathbb{R}^r$. Check that*

$$\forall 1 \le i \le r \quad \eta_{X,e_i} = \mathrm{Law}\,(X_i \mid X_{-i}) \quad with \quad X_{-i} := (X_1, \ldots, X_{i-1}, X_{i+1}, \ldots, X_r).$$

Compare the resulting sampler with the Gibbs sampler associated with (9.13).

Exercise 126 (Hit-and-run vs Gibbs samplers) *Consider some random variable X with a target measure $\eta(dx)$ on some state space S^X. Let $K(x, dy)$ be a Markov transition from S^X into a possibly different state space S^Y. We denote by (X, Y) a random variable with distribution $\pi(d(x,y)) = \eta(dx)K(x, dy)$. Assume that the following reversibility property (a.k.a. Bayes' rule) is satisfied*

$$\eta(dx)\ K(x, dy) = (\eta K)(dy)\ M(y, dx)$$

for some Markov transition $M(y, dx)$ from S^Y into S^X. Design a Gibbs sampler with target measure $\pi = \mathrm{Law}(X, Y)$. Illustrations of these hit-and-run samplers are discussed in exercises 127 and 128, see also the survey article [3] for applications to contingency tables, discrete exponential families, single move Metropolis-Hastings samplers, slice sampling and burnside processes.

Exercise 127 (Hit-and-run and Gibbs-Glauber samplers) *Let X be a real-valued random variable with a probability density $p(x) > 0$ w.r.t. the Lebesgue measure dx. Also let U be a uniform random variable on $\{-1, +1\}$ independent of X. We consider the observation $Y = X + U$. Check that*

$$\mathbb{P}(Y \in dy) = \frac{1}{2}\left[p(y+1) + p(y-1)\right]\ dy \quad and \quad \mathbb{P}\,(Y \in dy \mid X) = \frac{1}{2}\,\left[\delta_{X-1} + \delta_{X+1}\right](dy)$$

as well as

$$\mathbb{P}\,(X \in dx \mid Y) = \frac{p(Y-1)}{p(Y+1) + p(Y-1)}\ \delta_{Y-1}(dx) + \frac{p(Y+1)}{p(Y+1) + p(Y-1)}\ \delta_{Y+1}(dx).$$

Design a Gibbs sampler with target measure $\pi = \mathrm{Law}(X, Y)$. Discuss the situation when U is a uniform random variable on $\{-h, +h\}$, for some $h > 0$. A continuous time version of this sampler is discussed in exercise 179.

Exercise 128 (Hit-and-run - Conditional distributions [15]) *Let X be an $\mathbb{R}^r$-valued random variable with a probability density $p(x) > 0$ w.r.t. the Lebesgue measure dx. We also denote by $\mathbb{S} = \{x \in \mathbb{R}^r : \|x\| \le 1\}$ the unit sphere and $\partial\mathbb{S} = \{u \in \mathbb{R}^r : \|u\| = 1\}$ its boundary equipped with the uniform distribution $\nu(du)$. Also let U be a uniform random variable on $\partial\mathbb{S}$ and let T be an $\mathbb{R}$-valued random variable with distribution μ. Assume that (X, U, T) are independent. We consider the observation $Y = (U, Z)$ with $Z := X + TU$. Check the conditional distribution formulae*

$$\mathbb{P}\,((U, Z) \in d(u, z) \mid X) = \nu(du) \int \mu(dt)\ \delta_{X+tu}(dz)$$

and

$$\mathbb{P}\,(Z \in dz \mid U) = \left[\int\ p(z - tU)\ \mu(dt)\right]\ dz$$

as well as

$$\mathbb{P}\,(X \in dx \mid (U, Z)) = \int \frac{p\,(Z - tU)\ \mu(dt)}{\int\ p\,(Z - sU)\ \mu(ds)}\ \delta_{Z-tU}(dx).$$

Design a Gibbs sampler with target measure $\pi = \mathrm{Law}(X, Y)$.

Exercise 129 (Transformation group MCMC 1) *Let X be a random variable with the uniform distribution $\eta(dx)$ on a bounded open subset $S \subset \mathbb{R}^r$. There is no loss of generality to assume that $0 \in S$ and we equip $\mathbb{R}^r$ with the Cartesian coordinates associated with r unit vectors $e_i = (1_i(j))_{1 \leq j \leq r}$ of $\mathbb{R}^r$. Let μ be some probability distribution on the* special *orthogonal group $G = SO(r)$ (a.k.a. the rotation group). For any matrix $g \in G$ we set $g(S) := \{g(x) : x \in S\}$, and $\eta_g(dx) \propto 1_{g(S)}(x)\, dx$, and we let $M_g(x, dy)$ be a Markov transition on $g(S) \subset \mathbb{R}^r$ (equipped with the cartesian coordinates) such that $\eta_g = \eta_g M_g$. We consider the Markov transitions on S defined by*

$$
\overline{M}_g(x, dy) = \int_{g(S)} \delta_{g(x)}(dx') \int_{g(S)} M_g(x', dy')\, \delta_{g^{-1}(y')}(dy)
$$

$$
\overline{M}(x, dy) = \int_G \mu(dg)\, \overline{M}_g(x, dy).
$$

For any $x \in g(S)$ we let $T_g.x$ be a random variable with distribution $M_g(x, dy)$ on $g(S)$. Check that $\eta_g = \mathrm{Law}(g(X))$ and $g^{-1}T_g.g(X) \overset{law}{=} X$. Deduce that $\eta = \eta \overline{M}_g$ and $\eta = \eta \overline{M}$. Discuss the situation where $\eta(dx) \propto p(x)\, 1_S(x)\, dx$, for some density function $p(x)$ w.r.t. the Lebesgue measure dx. Discuss the choice of the MCMC transition M_g and extend these samplers to any target distribution and any group of transformations.

Illustrations of these transformation group MCMC samplers are discussed in exercises 131 and 132.

Exercise 130 (Transformation group MCMC 2) *We consider the MCMC sampler discussed in exercise 129. We further assume that $\eta(dx)$ has a density $p(x)$ and (G, ν) are chosen so that $g(S) = S$ for any $g \in G$ and*

$$
H \sim \nu \Rightarrow H^{-1} \sim \nu \quad \text{and} \quad \forall g \in G \quad H \circ g \sim \nu.
$$

Let $K(x, dg)$, resp. $\overline{K}(x, dy)$, be the Markov transition from S into G, resp. S, defined by

$$
K(x, dg) := \frac{p(g(x))\, |\partial g(x)/\partial x|\, \nu(dg)}{\int_G p(h(x))\, |\partial h(x)/\partial x|\, \nu(dh)}
$$

$$
\overline{K}(x, dy) = \int_G K(x, dg)\, \delta_{g(x)}(dy)
$$

where $|\partial g(x)/\partial x|$ stands for the Jacobian of $x \mapsto g(x)$. Check that $\eta \overline{K} = \eta$.

Exercise 131 (Hit-and-run vs rotation group MCMC) *We consider the rotation group MCMC sampler discussed in exercise 129 when $r = 2$ and $S = [-1, 1] \times [-1, 1]$. We let μ be the uniform distribution on $SO(2)$ associated with the uniform distribution $\nu(d\theta)$ on the set of angles $[0, 2\pi]$. Following exercise 121, describe some Gibbs samplers with the target uniform measures η_{g_θ} where g_θ stands for the rotation with angle θ. For any $i = 1, 2$, $g \in SO(2)$ and any $x' \in g(S)$ we let*

$$
M_g^{(i)}(x', dy') \propto \int_{\mathcal{T}_{i,g}(x')} dt\, \delta_{x' + t e_i}(dy') \quad \text{with} \quad \mathcal{T}_{i,g}(x') := \{t \in \mathbb{R} : x' + t e_i \in g(S)\}.
$$

Check that for any $x \in S$, $g \in SO(2)$ and any $i = 1, 2$ we have

$$
T_g(x) = x + t e_i \implies g^{-1}T_g.g(x) = x + t\, g^{-1}(e_i).
$$

Describe the rotation group MCMC on S associated with the collection of Gibbs samplers on $g(S)$ with probability transition $M_g := \frac{1}{2}\left(M_g^{(1)} + M_g^{(2)}\right)$, with $g \in SO(2)$. Discuss the connections between this rotation group MCMC and the hit-and-run sampler discussed in exercise 125.

Exercise 132 (Direction free Gibbs sampling on boundary surfaces) *We consider the direction free MCMC sampler discussed in exercise 129 and exercise 131. We assume that $r = 2$ and the set S is given by the boundary of the cell discussed in exercise 131, that is,*

$$S \;=\; \partial\left([-1,1] \times [-1,1]\right) := \left(\{-1,1\} \times [-1,1]\right) \cup \left([-1,1] \times \{-1,1\}\right).$$

Describe different ways of sampling the uniform probability measures

$$\eta(d(x_1,x_2)) \;\propto\; \left(\delta_{-1}(dx_1) + \delta_1(dx_1)\right) \, 1_{[-1,1]}(x_2) \, dx_2$$
$$+ 1_{[-1,1]}(x_1) \, dx_1 \left(\delta_{-1}(dx_2) + \delta_1(dx_2)\right)$$

$$\eta_{g_{\pi/4}}(d(x_1,x_2)) \;\propto\; 1_{[0,\sqrt{2}]}(x_1) \, dx_1 \left(\delta_{-(\sqrt{2}-x_1)}(dx_2) + \delta_{\sqrt{2}-x_1}(dx_2)\right)$$
$$+ 1_{[-\sqrt{2},0]}(x_1) \, dx_1 \left(\delta_{-(\sqrt{2}+x_1)}(dx_2) + \delta_{\sqrt{2}+x_1}(dx_2)\right).$$

Check that for any subset $C \subset S$ of length c we have $\mathbb{P}(X \in C) = c/8$ where X stands for a random variable with distribution η of η_{g_θ}, for some $\theta \in [0, 2\pi]$. Following exercises 121 and 131, design some Gibbs samplers with these target uniform measures. Describe the rotation group MCMC with transition $\overline{M}$ associated with these objects.

Exercise 133 (Shake-and-bake/Stochastic billiards) ⫼ *When S is given by the boundary of bounded polyhedra the direction free sampler discussed in exercise 132 is equivalent to the shake-and-bake samplers discussed in [27]. The direction free MCMC discussed above provides an alternative and simple way of designing stochastic billiards processes as the ones discussed in [55] and [124]. To illustrate these models, let us suppose that $D \subset \mathbb{R}^2$ is an open and smooth convex surface s.t. $S = \partial D$ is the null level set $\partial D = \varphi^{-1}(\{0\})$ of a continuously differentiable function s.t. $\partial_{y_2}\varphi(y_1,y_2) \neq 0$ on ∂D. We let $\nu(du)$ be the uniform distribution on the unit sphere $\mathbb{S}^1 := \{u \in \mathbb{R}^2 : \|u\| = 1\}$. For any $x \in S$ we let $n(x)$ be the outward pointing unit normal (column) vector to the curve ∂D and we set*

$$r_x \;:\; u \in \mathbb{S}^1 \mapsto r_x(u) = 1_{\langle u, n(x)\rangle > 0} \, (Id - 2n(x)n(x)')\,(u) + u \, 1_{\langle u, n(x)\rangle \leq 0}.$$

In the above display, $(.)'$ stands for the transpose operator. Check that r_x is the reflection w.r.t. the tangent line $T_x(D)$ at the surface at $x \in \partial D$. We let $\mathbb{S}^1_x$ be the set of admissible directions starting from x in the sense that $x + \epsilon u \in D$ for some $\epsilon \geq 0$; more formally, we have

$$\mathbb{S}^1_x := \{u \in \mathbb{S}^1 : \langle u, n(x)\rangle \leq 0\}.$$

We let U_x be a random variable with distribution $\nu_x(du) \propto \nu(du) \, 1_{\mathbb{S}^1_x}(u)$. Consider the Markov chain on $S = \partial D$ with elementary transition $\overline{M}$ defined by

$$\overline{M}(f)(x) := \mathbb{E}\left(f(x + t(x, U_x)U_x)\right) = \frac{1}{\nu(\mathbb{S}^1_x)} \int_{\mathbb{S}^1_x} f(x + t(x,u)u) \, 1_{\mathbb{S}^1_x}(u) \, \nu(du)$$

with the hitting time

$$t(x,u) := \inf\{t \geq 0 : x + tu \in \partial D\}.$$

Check that

$$\overline{M}(x,dz) \propto \frac{1}{\|z - x\|} \left\langle \frac{z - x}{\|z - x\|}, n(z) \right\rangle \sigma(dz)$$

with the surface measure $\sigma(dz)$ on S. (hint: using the implicit function theorem around any hitting point $z = (z_1, z_2)$, the set S can be seen locally as a graph $(z_1, h(z_1))$ of some height function h in Cartesian coordinates. The surface measure $\sigma(dz)$ expressed in this parametrization is discussed on page 647.) Discuss the situation $S = \partial D = \{x \in \mathbb{R}^2 : \|x\| = R\}$.

Exercise 134 (Knudsen random walk) *We consider the stochastic billiard model discussed in exercise 133. We examine the situation where ν_x is replaced by the measure*

$$\nu_{x,\kappa}(du) \propto \kappa_x(u)\nu_x(du) \quad \text{with} \quad \kappa_x(u) = -\langle u, n(x)\rangle = |\langle u, n(x)\rangle|.$$

Check that

$$\overline{M}(x,dz) \quad \propto \quad \frac{1}{\|z-x\|} \left\langle \frac{x-z}{\|x-z\|}, n(x)\right\rangle \left\langle \frac{z-x}{\|z-x\|}, n(z)\right\rangle \sigma(dz).$$

Prove that $\overline{M}$ is reversible w.r.t. σ. Deduce that the uniform measure on S is an invariant probability measure. The Markov chain associated with the above cosine law of reflection is often called the Knudsen random walk or the Knudsen stochastic billiard. Further details on the long time behavior of these stochastic models can be found in [55]. Discuss the situation $S = \partial D = \{x \in \mathbb{R}^2 : \|x\| = R\}$.

Exercise 135 (Slice sampling) *Let X be an $\mathbb{R}^r$-valued random variable with distribution $\eta(dx) \propto p(x)dx$ for some non-necessarily normalized density $p(x) > 0$ w.r.t. the Lebesgue measure dx. Let $Y = p(X)U$ for some uniform random variable U on $[0,1]$, independent of X. Check that*

$$\mathbb{P}(X \in dx \mid Y) = \frac{1}{\int \mathbb{1}_{p^{-1}([y,\infty[)}(x')\, dx'} \; \mathbb{1}_{p^{-1}([y,\infty[)}(x)\, dx$$

and design a Gibbs sampler with target measure $\pi = \text{Law}(X, Y)$.

Exercise 136 (Hit-and-run and MCMC within Gibbs) *Consider the hit-and-run sampler discussed in exercise 126. For each given $y \in S^Y$ let $M'_y(x, dx')$ be a Markov transition on S^X s.t.*

$$\int_{S^X} M(y, dx)\, M'_y(x, dx') = M(y, dx').$$

Check that η is an invariant measure of the Markov transition K' given

$$K'(x, dx') := \int K(x, dy)\, M'_y(x, dx').$$

Exercise 137 (Auxiliary variable - MCMC within Gibbs [3]) *Let η be the Boltzmann-Gibbs measure on some state space S given by*

$$\eta(dx) = \frac{1}{\mathcal{Z}} \exp\left\{\sum_{1\le i \le r} V_i(x)\right\} \lambda(dx)$$

where λ stands for some reference measure on S and $(V_i)_{1\le i \le r}$ a collection of potential functions s.t. $0 < \mathcal{Z} < \infty$. We let $(U_i)_{1\le i \le r}$ be a sequence of independent uniform random variables on $]0,1[$ and X an independent random variable with distribution η. Consider the observation sequence $Y = (Y_i)_{1\le i \le r}$ with

$$\forall 1 \le i \le r \qquad Y_i = e^{V_i(X)}\, U_i.$$

Check that the conditional distribution of Y given X is the uniform distribution on the r-cell $\prod_{1\le i \le r}\left[0, e^{V_i(X)}\right]$. Deduce that

$$M(y, dx) := \mathbb{P}(X \in dx \mid Y) \propto \left\{\prod_{1\le i \le r} \mathbb{1}_{\left[0, e^{V_i(x)}\right]}(Y_i)\right\} \lambda(dx).$$

For each given y let $M'_y(x, dx')$ be a Markov transition on S with invariant measure $M(y, dx)$. Design a MCMC within Gibbs sampler with target measure η.

Exercise 138 (Gibbs sampling - Conjugate priors) *✍ We let I and J be some finite sets and (V_1, V_2) be a couple of independent random variables with inverse Gamma distributions with shape parameters $(a_i)_{i=1,2}$ and scale parameters $(b_i)_{i=1,2}$, that is, for $i \in \{1,2\}$,*

$$\mathbb{P}(V_i \in dx) = \frac{b_i^{a_i}}{\Gamma(a_i)} \frac{1}{x^{a_i+1}} \exp\left(-\frac{b_i}{x}\right) 1_{]0,\infty[}(x) \, dx.$$

Given V_1, we let Z be a Gaussian random variable with some given mean m and variance v. Given (V_1, Z), we let $(X_i)_{i \in I}$ be a sequence of i.i.d. Gaussian random variables with mean Z and variance V_1, and $(W_{i,j})_{(i,j) \in (I \times J)}$ be a sequence of i.i.d. centered Gaussian random variables with variance V_2. We associate with these objects the variance component model

$$\forall (i,j) \in (I \times J) \qquad Y_{i,j} = X_i + W_{i,j}.$$

- *Given $Y = (Y_{i,j})_{(i,j)\in(I\times J)} = (y_{i,j})_{(i,j)\in(I\times J)}$, $Z = z$ and $(V_1, V_2) = (v_1, v_2)$ show that the $X = (X_i)_{i \in I}$ are independent Gaussian random variables with mean and variance parameters $\left(\alpha\left((y_{i,j})_{j\in J}, v_1, v_2\right), \sigma^{-2}(v)\right)_{i \in I}$ given by*

$$\forall i \in I \qquad \sigma^{-2}(v_1, v_2) = \left(\frac{1}{v_1} + \frac{|J|}{v_2}\right) \quad \text{and} \quad \alpha\left((y_{i,j})_{j\in J}, v_1, v_2\right) = \frac{\frac{z}{v_1} + \frac{1}{v_2}\sum_{j\in J} y_{i,j}}{\frac{1}{v_1} + \frac{|J|}{v_2}}.$$

- *Check that the conditional distribution of V_1 given $X = (X_i)_{i \in I} = x$, $Y = (Y_{i,j})_{(i,j)\in(I\times J)} = y$ and $(Z, V_2) = (z, v_2)$ coincides with the conditional distribution of V_1 given $X = x$ and $Z = z$; and it is given by an inverse Gamma distribution with shape parameter $A_1(a_1)$ and scale parameter $B_1(b_1, z)$ given by*

$$A_1 = \frac{|I|}{2} + a_1 \quad \text{and} \quad B_1(z) = b_1 + \frac{1}{2}\sum_{i\in I}(x_i - z)^2.$$

- *Check that the conditional distribution of V_2 given $X = (X_i)_{i \in I} = x$, $Y = (Y_{i,j})_{(i,j)\in(I\times J)} = y$ and $(Z, V_1) = (z, v_1)$ coincides with the conditional distribution of V_2 given $Y = y$ and $X = x$; and it is given by an inverse gamma distribution with shape parameter $A_2(a_2)$ and scale parameter $B_2(b_2, x, y)$ given by*

$$A_2 = \frac{|I|}{2} + a_2 \quad \text{and} \quad B_2(x, y) = b_2 + \frac{1}{2}\sum_{(i,j)\in(I\times J)}(y_{i,j} - x_i)^2.$$

- *Prove that the conditional distribution of Z given $X = (X_i)_{i \in I} = x$, $Y = (Y_{i,j})_{(i,j)\in(I\times J)} = y$ and $(V_1, V_2) = (v_1, v_2)$ coincides with the conditional distribution of Z given $X = y$ and $V_1 = v_1$; and it is given by a Gaussian probability with mean and variance $(\beta(x, v_1), \tau^2(v_1))$ defined by*

$$\tau^{-2}(v_1) = \left(\frac{|I|}{v_1} + \frac{1}{v}\right) \quad \text{and} \quad \beta(x, v_1) = \left(\frac{m}{v} + \frac{1}{v_1}\sum_{i\in I} x_i\right) \Big/ \left(\frac{|I|}{v_1} + \frac{1}{v}\right).$$

- *Design a Gibbs sampler targeting the conditional distribution of the random variables (X, Z, V_1, V_2) given the sequence of observations $Y = (Y_{i,j})_{(i,j)\in(I\times J)}$.*

Exercise 139 (Filtering one-dimensional signals) *We consider the 1d-nonlinear filtering problem defined by*

$$\begin{cases} X_n &= a_n(X_{n-1}) + W_n \\ Y_n &= b_n(X_n) + V_n. \end{cases} \tag{9.103}$$

We assume that X_0, W_n and V_n are i.i.d. centered Gaussian random variables with unit variance, and (a_n, b_n) are some possibly nonlinear functions on $\mathbb{R}$.

- *Describe the posterior distributions of X_n and $(X_k)_{0 \leq k \leq n}$ given the sequence of observations $(Y_k)_{0 \leq k < n}$ in terms of Feynman-Kac measures $(\mathbb{Q}_n, \gamma_n, \eta_n)$ defined in (9.21) and (9.23).*

- *Describe the density of the observations $(Y_k)_{0 \leq k \leq n}$ evaluated at some observation path sequence $(y_k)_{0 \leq k \leq n}$ in terms of the normalizing constants of Feynman-Kac measures (9.23).*

- *Describe the time evolution of the posterior distributions of X_n given the sequence of observations $(Y_k)_{0 \leq k < n}$.*

Exercise 140 (Particle filters for one-dimensional signals) *We consider the 1d-nonlinear filtering problem defined in (9.103).*

- *Design a particle methodology to estimate the conditional probabilities of the random states X_n given the sequence of observations $(Y_k)_{0 \leq k < n}$, and propose an unbiased estimate of the density of the observations $(Y_k)_{0 \leq k \leq n}$.*

- *Propose a couple of estimates of the the conditional probabilities of the random trajectories $(X_k)_{0 \leq k \leq n}$ given the sequence of observations $(Y_k)_{0 \leq k < n}$.*

Exercise 141 (Particle filters - Likelihood estimation) ✎ *We consider the 1d-nonlinear filtering problem defined in (9.103).*

- *We further assume that $b_n(X_n) = c_n(X_n) + \theta \, d_n(X_n)$ for some parameter $\theta \in \mathbb{R}$ and for given functions (c_n, d_n). We let $p_\theta(y_0, \ldots, y_n)$ be the density of the observations $(Y_k)_{0 \leq k \leq n}$ associated with a fixed parameter θ. Show that*

$$
\frac{\partial}{\partial \theta} \log p_\theta(y_0, \ldots, y_n) = \frac{\mathbb{E}\left(L_{n,\theta}(X_0, \ldots, X_n) \prod_{0 \leq k \leq n} G_{\theta,k}(X_k) \right)}{\mathbb{E}\left(\prod_{0 \leq k \leq n} G_{\theta,k}(X_k) \right)}
$$

with the likelihood functions

$$
G_{\theta,k}(x_k) \propto \exp\left(-\frac{1}{2} \left(y_k - c_k(X_k) - \theta d_k(X_k) \right)^2 \right) \tag{9.104}
$$

and the additive functional

$$
L_{n,\theta}(X_0, \ldots, X_n) = \sum_{0 \leq k \leq n} l_{\theta,k}(X_k) \quad \text{with} \quad l_{\theta,k}(x_k) = \frac{\partial}{\partial \theta} \log G_{\theta,k}(x_k).
$$

- *Propose a couple of particle estimates of $\frac{\partial}{\partial \theta} \log p_\theta(y_0, \ldots, y_n)$.*

Exercise 142 (Particle filters - Many body particle models) ✐ *We consider the 1d-nonlinear filtering problem defined in (9.103).*

- *Describe the posterior distributions of X_n given the sequence of observations $(Y_k)_{0 \leq k < n}$ in terms of Feynman-Kac measures (γ_n, η_n) defined in (9.23).*

- *Describe the many-body Feynman-Kac measures associated with the mean field particle interpretation of the measures (γ_n, η_n).*

Exercise 143 (Particle filters and smoothing) ✐ *We consider the 1d-nonlinear filtering problem defined by*

$$\begin{cases} X'_n &= a_n(X'_{n-1}) + W_n \\ Y_n &= b_n(X'_n) + V_n. \end{cases} \tag{9.105}$$

We assume that X'_0, W_n and V_n are i.i.d. centered Gaussian random variables with unit variance, and (a_n, b_n) are some possibly nonlinear functions on $\mathbb{R}$.

- *Describe the posterior distributions of $X_n = (X'_0, \ldots, X'_n) \in \mathbb{R}^{n+1}$ given the sequence of observations $(Y_k)_{0 \le k < n}$ in terms of the Feynman-Kac measures (γ_n, η_n) defined in (9.23).*

- *Compute (γ_n, η_n) in terms of the occupation measures of the genealogical tree model associated with a genetic type process.*

- *Describe the many-body Feynman-Kac measures associated with the mean field particle interpretation of the measures (γ_n, η_n).*

- *Design a particle Metropolis-Hastings algorithm to approximate the posterior distribution of the signal trajectories $X_n = (X'_0, \ldots, X'_n) \in \mathbb{R}^{n+1}$ given the sequence of observations $(Y_k)_{0 \le k < n}$.*

Exercise 144 (Particle Gibbs-Glauber samplers - Smoothing problems) ⫻ *We consider the 1d-nonlinear filtering problem defined in (9.105).*

- *Describe the posterior distributions of $X_n = (X'_0, \ldots, X'_n) \in \mathbb{R}^{n+1}$ given the sequence of observations $(Y'_k)_{0 \le k < n}$ in terms of Feynman-Kac measures on the path space (γ_n, η_n).*

- *Compute (γ_n, η_n) in terms of the occupation measures of the genealogical tree model associated with a genetic type process.*

- *Describe the many-body Feynman-Kac measures associated with the mean field particle interpretation of the measures (γ_n, η_n).*

- *Describe the dual mean field model with frozen trajectory X_n on some given time horizon n.*

- *Design a couple of particle Gibbs-Glauber algorithms to approximate the posterior distribution of the signal trajectories $X_n \in \mathbb{R}^{n+1}$ given the sequence of observations $(Y_k)_{0 \le k < n}$.*

Part IV

Continuous time processes

10

Poisson processes

Perhaps the simplest examples of continuous time stochastic processes are the counting processes. By including a bit more structure into these, we obtain the (homogeneous or inhomogeneous) Poisson processes and the Bernoulli process. We discuss important properties of Poisson processes in this chapter, such as the memoryless property and the Doeblin-Itō formula. Poisson processes and their generalizations have found a variety of applications and it is important to be able to sample such processes. We briefly discuss the Poisson thinning simulation technique.

> *The probability of an event is the reason we have to believe that it has taken place, or that it will take place.*
> Siméon Denis Poisson (1781-1840).

10.1 A counting process

We let $(\mathcal{E}_n)_{n\geq 1}$ be a sequence of independent exponential random variables with parameter $\lambda > 0$, and we consider the random times

$$T_n = T_{n-1} + \mathcal{E}_n := \sum_{1\leq p\leq n} \mathcal{E}_p \quad \text{with} \quad T_0 = \sum_{\emptyset} = 0.$$

By construction, we have

$$\mathbb{P}\left((T_1,\ldots,T_n) \in d(t_1,\ldots,t_n)\right)$$

$$= [1_{[0,\infty)}(t_1)\ \lambda\ e^{-\lambda t_1}\ dt_1][1_{[t_1,\infty)}(t_2)\ \lambda\ e^{-\lambda(t_2-t_1)}\ dt_2]$$

$$\ldots \times [1_{[t_{n-1},\infty)}(t_n)\ \lambda\ e^{-\lambda(t_n-t_{n-1})}\ dt_n].$$
(10.1)

The random variables can be interpreted as the random arrivals of customers. In this interpretation T_n is the sum of n first inter-arrival times, so that it corresponds to the arrival time of the n-th customer.

Definition 10.1.1 *The $\mathbb{N}$-valued counting process N_t defined below*

$$t \in \mathbb{R}_+ \mapsto N_t \ = \ \sum_{n\geq 0} 1_{[T_n,\infty)}(t) = \sum_{n\geq 0} 1_{T_n\leq t} \tag{10.2}$$

$$= \ \sup\{n\geq 1\ :\ T_n \leq t\} = \inf\{n\geq 1\ :\ T_{n+1} > t\} = \sum_{n\geq 0} 1_{[T_n,T_{n+1}[}(t)\ n$$

$$= \ N_0 + \sum_{0<T_n\leq t}(N_{T_n} - N_{T_n-}) = N_0 + \sum_{0<n\leq N_t}(N_{T_n} - N_{T_n-})$$

with unit jumps $N_{T_n} - N_{T_n-} = 1$, is called a Poisson process with intensity λ. A randomly sampled trajectory of this process is provided in figure *10.1*.

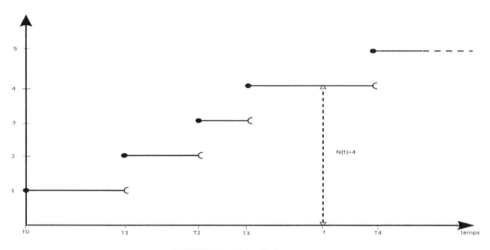

FIGURE 10.1: Poisson process

Remark : These counting processes can be extended to not necessarily exponential inter-arrival times. These general counting processes are sometimes called renewal processes.

10.2 Memoryless property

One of the most important properties of the Poisson process is its memoryless property which is inherited by the exponential inter-times variables.

$$
\begin{aligned}
\mathbb{P}\left(\mathcal{E}_1 > t + s \mid \mathcal{E}_1 \geq s\right) &= \frac{\mathbb{P}\left(\mathcal{E}_1 > t + s\right)}{\mathbb{P}\left(\mathcal{E}_1 > s\right)} \\
&= e^{-\lambda(t+s)} / e^{-\lambda s} = e^{-\lambda t} = \mathbb{P}\left(\mathcal{E}_1 > t\right).
\end{aligned}
$$

Remark :
The memoryless property of a random variable X implies that the function $t \mapsto \mathbb{P}\left(X > t\right)$ is decreasing, satisfying the functional equation

$$
\mathbb{P}\left(X > t + s\right) = \mathbb{P}\left(X > t\right)\,\mathbb{P}\left(X \geq s\right).
$$

This shows that X is an exponential type random variable. In discrete time settings, these random variables are given by the geometric random variable $\mathcal{G}$ with success parameter $p \in [0, 1]$

$$
\begin{aligned}
\mathbb{P}\left(\mathcal{G} > n + m \mid \mathcal{G} > m\right) &= \mathbb{P}\left(\mathcal{G} > n + m\right) / \mathbb{P}\left(\mathcal{G} > m\right) \\
&= (1 - p)^{n+m} / (1 - p)^m = (1 - p)^n = \mathbb{P}\left(\mathcal{G} > n\right).
\end{aligned}
$$
(10.3)

Proposition 10.2.1 *The distance R_t from t to the next arrival (after t) after some arrival epoch does not depend on the Poisson process up to to time t, and R_t is an exponential random variable with parameter λ.*

Proof :
Using the memoryless property of the first exponential $\mathcal{E}_1$ (and the fact that $\{N_t = 0\} = \{\mathcal{E}_1 > t\}$) we first observe that

$$
\mathbb{P}\left(R_t > r \mid N_t = 0\right) = \mathbb{P}\left(\mathcal{E}_1 > t + r \mid \mathcal{E}_1 > t\right) = \mathbb{P}\left(\mathcal{E}_1 > r\right) = e^{-\lambda r}.
$$

Given $N_t = n$, and $T_n = s(\leq t)$, the next arrival after t is given by

$$
\begin{aligned}
T_{n+1} &= T_n + \mathcal{E}_{n+1} \\
&= s + \mathcal{E}_{n+1} = s + \underbrace{[R_t + (t - s)]}_{=\mathcal{E}_{n+1}} = t + R_t.
\end{aligned}
$$

Also notice that
$$
\{N_t = n\,, T_n = s\} = \{\mathcal{E}_{n+1} > (t - s)\,, T_n = s\}.
$$

Using the memoryless property of the first exponential $\mathcal{E}_{n+1}$, we conclude that

$$
\mathbb{P}\left(R_t > r \mid N_t = n\,, T_n = s\right)
$$

$$
= \mathbb{P}\left(\mathcal{E}_{n+1} > r + (t - s) \mid N_t = n\,, T_n = s\right)
$$

$$
= \mathbb{P}\left(\mathcal{E}_{n+1} > r + (t - s) \mid \mathcal{E}_{n+1} > (t - s)\,, T_n = s\right)
$$

$$
= \mathbb{P}\left(\mathcal{E}_{n+1} > r + (t - s) \mid \mathcal{E}_{n+1} > (t - s)\right) = \mathbb{P}\left(\mathcal{E}_{n+1} > r\right) = e^{-\lambda r}.
$$

This ends the proof of the proposition ∎

> Proposition 10.2.1 shows that the Poisson process $(N_{t+s})_{s\geq 0}$ given the value $N_t = n$ is a perfect replica of the process $(N_s)_{s\geq 0}$ starting at $N_0 = n$; in other words $(N_{t+s} - N_t)$ is independent of $(N_t - N_0)$ and it has the same distribution as N_s. This shows that *the Poisson process N_t has independent Poisson increments*. More precisely, each of them $(N_t - N_s) \overset{law}{=} N_{t-s}$ is a Poisson r.v. with parameter $\lambda(t-s)$.

10.3   Uniform random times

The Poisson process can be interpreted as counting the number of clients arriving at exponential inter-arrival times. In this interpretation, the epochs T_n and the Poisson process can be easily related using the following properties

$$\{T_n \leq t\} = \{N_t \geq n\} \quad \text{and/or} \quad \{T_n > t\} = \{N_t < n\}.$$

For instance, the l.h.s. states that the n-arrival event is less than t if and only if the number of arrivals at time t is greater than n. We also have the following equivalence

$$\{N_t = n\} = \{T_n \leq t < T_{n+1}\} = \{t < T_{n+1}\} - \{t < T_n\}.$$

On the other hand, we have

$$-\frac{\partial}{\partial t}\left(\frac{(\lambda t)^n}{n!} e^{-\lambda t}\right) = \frac{(\lambda t)^n}{n!}\, \lambda\, e^{-\lambda t} - \frac{(\lambda t)^{n-1}}{(n-1)!}\, \lambda\, e^{-\lambda t}.$$

Recalling that $T_n \sim \text{gamma}(n,\lambda)$, cf. (4.22) on page 93), we prove that

$$\begin{aligned}
\mathbb{P}(N_t = n) &= \mathbb{P}(T_n \leq t < T_{n+1}) = \mathbb{P}(t < T_{n+1}) - \mathbb{P}(t < T_n) \\
&= -\int_t^\infty \frac{\partial}{\partial s}\left(\frac{(\lambda s)^n}{n!} e^{-\lambda s}\right) ds = \frac{(\lambda t)^n}{n!} e^{-\lambda t}.
\end{aligned}$$

This clearly implies that

$$\begin{aligned}
\mathbb{P}(T_n > t) &= \mathbb{P}(N_t < n) = e^{-\lambda t} \sum_{0 \leq p < n} \frac{(\lambda t)^p}{p!} \\
\mathbb{E}(N_t) &= \lambda t.
\end{aligned}$$

On the other hand, using (10.1) we find that

$$\mathbb{P}((T_1, \ldots, T_n) \in d(t_0, \ldots, t_n)\,;\, N_t = n)$$

$$= \int_t^\infty \mathbb{P}((T_1, \ldots, T_n, T_{n+1}) \in d(t_0, \ldots, t_n, t_{n+1}))$$

$$= \frac{(\lambda t)^n}{n!} \int_t^\infty \lambda e^{-\lambda t_{n+1}} dt_{n+1} \times \left[\frac{n!}{t^n}\, 1_{0 \leq t_1 \leq \ldots \leq t_n \leq t}\, dt_1 \ldots dt_n\right]$$

$$= \frac{(\lambda t)^n}{n!} e^{-\lambda t} \times \left[\frac{n!}{t^n}\, 1_{0 \leq t_1 \leq \ldots \leq t_n \leq t}\, dt_1 \ldots dt_n\right].$$

We summarize this discussion with the following proposition.

Proposition 10.3.1 *Given $N_t = n$, the jump times $T_0, \ldots, T_n$ are uniformly distributed on $[0, t]$*

$$\mathbb{P}\left((T_1, \ldots, T_n) \in d(t_0, \ldots, t_n) \mid N_t = n\right) = \frac{n!}{t^n} \, 1_{0 \leq t_1 \leq \ldots \leq t_n \leq t} \, dt_1 \ldots dt_n.$$

10.4 Doeblin-Itō formula

Using the remark at the end of section 10.2 we prove the following result.

> For any time sequence $0 \leq t_0 \leq t_1 \leq \ldots \leq t_n$ the random variables
>
> $$\left(N_{t_1} - N_{t_0}, N_{t_2} - N_{t_1}, \ldots, N_{t_n} - N_{t_{n-1}}\right)$$
>
> and independent Poisson random variables with parameters
>
> $$(\lambda(t_1 - t_0), \lambda(t_2 - t_1), \ldots, \lambda(t_n - t_{n-1})).$$
>
> In addition, the process $(N_t - \lambda t)_{t \geq 0}$ is a martingale w.r.t. the filtration $\mathcal{F}_t = \sigma(N_s, \, s \leq t)$.

The Doeblin-Itō formula for N_t is easily deduced from the following telescoping formula

$$f(N_t)$$

$$= f(N_0) + \sum_{0 < T_k \leq t} (f(N_{T_k}) - f(N_{T_k -}))$$

$$= f(N_0) + \sum_{0 \leq k \leq N_t} (f(k) - f(k-1))$$

$$= f(N_0) + \int_0^t (f(N_s) - f(N_{s-})) \, dN_s$$

$$= f(N_0) + \int_0^t (f(N_{s-} + 1) - f(N_{s-})) \, \lambda \, ds + \int_0^t (f(N_{s-} + 1) - f(N_{s-})) \, (dN_s - \lambda \, ds)$$

for any function f on $\mathbb{R}$.

This shows that
$$df(N_t) = L(f)(N_t)dt + d\mathcal{M}_t(f)$$
with the infinitesimal generator
$$L(f)(x) = \lambda \left[f(x+1) - f(x) \right]$$
and the martingale
$$d\mathcal{M}_t(f) = (f(N_{t-} + 1) - f(N_{t-})) \; (dN_t - \lambda dt)$$
with predictable angle bracket
$$\langle \mathcal{M}(f) \rangle_t = \int_0^t \left(f(N_{s-} + 1) - f(N_{s-}) \right)^2 \lambda ds.$$

Roughly speaking, the last assertion is due to

$$dN_t dt = 0 = (dt)^2 \quad \text{and} \quad (dN_t)^2 = dN_t$$

$$\Rightarrow \mathbb{E}((dN_t - \lambda dt)^2 \mid \mathcal{F}_t) = \mathbb{E}(dN_t \mid \mathcal{F}_t) = \lambda dt$$

$$\Rightarrow \mathbb{E} \left(d\mathcal{M}_t^2(f) \mid \mathcal{F}_t \right) = \mathbb{E} \left((d\mathcal{M}_t(f))^2 \mid \mathcal{F}_t \right) = (f(N_{t-} + 1) - f(N_{t-}))^2 \; \lambda dt.$$

10.5 Bernoulli process

We let $(\mathcal{E}_n)_{n \geq 1}$ be a sequence of independent exponential random variables with parameter $\lambda > 0$, and we let $h > 0$ be some time mesh parameter associated with discretization of an interval $[0, t]$

$$t_0 = 0 < t_1 = h < \ldots < t_n = nh < \ldots < h\lfloor t/h \rfloor \leq t \;\; (= h\lfloor t/h \rfloor + h\{t/h\}).$$

We let $\epsilon_{t_n}^h$ be a sequence of independent Bernoulli random variables with common law
$$\mathbb{P}\left(\epsilon_{t_n}^h = 0 \right) = 1 - \mathbb{P}\left(\epsilon_{t_n}^h = 1 \right) = e^{-\lambda h}.$$
We associate with these objects the geometric random variables
$$T_{n+1}^h \;\; = \;\; \inf\left\{ t_k > T_n^h \; : \; \epsilon_{t_k}^h = 1 \right\} = T_n^h + \mathcal{E}_n^h$$
with the geometric random variable with success parameter $1 - e^{-\lambda h}$ given by
$$\mathcal{E}_n^h := h \, \inf\left\{ k > 0 \; : \; \epsilon_{T_n^h + hk}^h = 1 \right\} \stackrel{law}{=} h\left(1 + \lfloor \mathcal{E}_n/h \rfloor \right). \tag{10.4}$$

We refer the reader to exercise 4.7 for the description of the geometric random variable with success parameter $p_{h,\lambda} = 1 - e^{-\lambda h}$ in terms of exponential random variables with parameter $-\log\left(1 - p_{h,\lambda} \right) = \lambda$. We also mention that the random times T_{n+1}^h can also be defined sequentially in terms of the decreasing function

$$\gamma_{T_n^h} \; : \; t_k \in [T_n^h, \infty[\; \mapsto \gamma_{T_n^h}(t_k) := \prod_{T_n^h < t_p \leq t_k} e^{-\lambda h} = e^{-\lambda(t_k - T_n^h)},$$

starting at 0 at the time $t_k = T_n^h$ and going to 0 as $t_k \uparrow \infty$. Given a sequence of uniform random variables U_n on $[0,1]$, we define T_{n+1}^h as the first time t_k s.t. $\gamma_{T_n^h}(t_k) \leq U_{n+1}$.

$$
\begin{aligned}
T_{n+1}^h &= \inf \left\{ t_k > T_n^h \; : \; \prod_{T_n^h < t_p \leq t_k} e^{-\lambda h} \leq U_{n+1} \right\} \\
&= \inf \left\{ t_k > T_n^h \; : \; \sum_{T_n^h < t_p \leq t_k} \lambda h \geq - \log U_{n+1} \right\}.
\end{aligned}
\tag{10.5}
$$

Recalling that $\mathcal{E}_{n+1}^h \overset{law}{=} h \left\lfloor -\frac{1}{\lambda h} \log U_{n+1} \right\rfloor$, we find that

$$T_{n+1}^h = \inf \left\{ t_k > T_n^h \; : \; (t_k - T_n^h) \geq -\frac{1}{\lambda h} \log U_{n+1} \right\} = T_n^h + \mathcal{E}_{n+1}^h.$$

We can also check this claim using the fact that

$$
\begin{aligned}
\mathbb{P}\left(T_{n+1}^h = t_k \mid T_n^h = t_l\right) &= \mathbb{P}\left(\prod_{t_l < t_p \leq t_k} e^{-\lambda h} \leq U_{n+1} < \prod_{t_l < t_p \leq t_{k-1}} e^{-\lambda h} \mid T_n^h \right) \\
&= \left\{ \prod_{t_l < t_p \leq t_{k-1}} e^{-\lambda h} \right\} \left(1 - e^{-\lambda h}\right) \\
&= e^{-\lambda(t_l - t_{k-1})} \left(1 - e^{-\lambda h}\right) = \left(e^{-\lambda h}\right)^{(l-k)-1} \left(1 - e^{-\lambda h}\right).
\end{aligned}
$$

Definition 10.5.1 *The $\mathbb{N}$-valued counting process N_t^h defined below*

$$t \in \mathbb{R}_+ \mapsto N_t^h = \sum_{n \geq 0} n \, 1_{[T_n^h, T_{n+1}^h[}(t) = \sum_{p=1}^{\lfloor t/h \rfloor} \epsilon_{t_p}^h \tag{10.6}$$

is called a Bernoulli counting process with success probability $1 - e^{-\lambda h}$. A randomly sampled trajectory of this process is provided in figure 10.1.

The interpretation of the geometric in terms of exponential random variables (10.4) allows coupling of the Bernoulli and the Poisson counting processes. More precisely, we have

$$T_n^h = T_{n-1}^h + \mathcal{E}_n^h := \sum_{1 \leq p \leq n} \mathcal{E}_p^h \quad \text{with} \quad T_0^h = \sum_{\emptyset} = 0.$$

Using the fact that

$$\mathcal{E}_n = h \lfloor \mathcal{E}_n/h \rfloor + h \left\{ \mathcal{E}_n/h \right\}$$

we find that

$$T_n \leq T_n^h = T_n + h \underbrace{\sum_{1 \leq p \leq n} (1 - \{\mathcal{E}_n/h\})}_{:=\tau_n^h} \leq T_n + nh.$$

This shows that the Bernoulli process has a τ_n^h-counting delay that tends to 0 as $h \downarrow 0$, but it is not an obstacle to the convergence in law of the Bernoulli process N_t^h to the Poisson process N_t, as $h \downarrow 0$. We also recall from (1.7) that

$$
\mathbb{P}\left(N_t^h = n\right) = \binom{\lfloor t/h \rfloor}{n} \left(1 - e^{-\lambda h}\right)^n e^{-\lambda h(\lfloor t/h \rfloor - n)}
$$

$$
= \underbrace{\left\{ \prod_{1 \leq p < n} \left(1 - \frac{p}{\lfloor t/h \rfloor}\right) \right\} \left(\frac{e^{\lambda h} - 1}{\lambda h} \frac{h \lfloor t/h \rfloor}{t}\right)^n}_{\longrightarrow_{h \downarrow 0} 1} \frac{(\lambda t)^n}{n!} e^{-\lambda h \lfloor t/h \rfloor}
$$

$$
\longrightarrow_{h \downarrow 0} \mathbb{P}\left(N_t = n\right).
$$

For any $s \leq r$ we have

$$
N_r^h - N_s^h = \sum_{p = \lfloor s/h \rfloor + 1}^{\lfloor r/h \rfloor} \epsilon_{t_p}^h \perp N_s^h
$$

and for any $t_p < t_q$

$$
N_{t_p}^h - N_{t_q}^h = \sum_{k=1}^{ph - qh} \epsilon_{qh + kh}^h \stackrel{law}{=} \sum_{p=1}^{(p-q)h} \epsilon_{kh}^h = N_{t_q - t_p}^h
$$

so that $N_r^h - N_s^h \stackrel{law}{=} N_{r-s}^h$.

> Using the memoryless property of the Bernoulli and the Poisson processes, we also prove that
> $$
> \lim_{h \to 0} \mathrm{Law}\left(N_{s_1}^h, \ldots, N_{s_n}^h\right) = \mathrm{Law}\left(N_{s_1}, \ldots, N_{s_n}\right)
> $$
> for any sequence of times $0 \leq s_1 \leq \ldots \leq s_n$, and for any $n \geq 1$.

10.6 Time inhomogeneous models

10.6.1 Description of the models

We start with an elementary observation. We let $\mathcal{E}$ be an exponential random variable with unit parameter. We can choose $\mathcal{E} = -\log U$, where U is uniform on $[0, 1]$. For any $t \in \mathbb{R}_+$ we have

$$
T = \inf\{t \geq 0 \ : \ t\lambda \geq \mathcal{E}\} = \frac{\mathcal{E}}{\lambda}
$$

$$
= \inf\{t \geq 0 \ : \ e^{-t\lambda} \leq U\} = -\frac{1}{\lambda} \log U \stackrel{law}{=} \mathrm{Exp}(\lambda).
$$

Definition 10.6.1 *We consider a bounded continuous function* $\lambda \ : \ t :\in \mathbb{R}_+ \mapsto \lambda(t) =: \lambda_t \in \mathbb{R}_+$, *and for any* $s \leq t$, *with* $s, t \in \mathbb{R}_+$ *we set*

$$
\varpi_s(t) = \int_s^t \lambda_r \ dr.
$$

We notice that

$$\forall s \leq t \leq r \qquad \varpi_s(t) + \varpi_t(r) = \varpi_s(r).$$

We let T be the random variable defined by

$$
\begin{aligned}
T & := \inf\left\{t \geq 0 \; : \; \int_0^t \lambda_s \, ds \geq \mathcal{E}\right\} \\
& = \inf\left\{t \geq 0 \; : \; e^{-\int_0^t \lambda_s ds} \leq U\right\} = \inf\left\{t \geq 0 \; : \; \varpi_0(t) \geq \mathcal{E}\right\} = \varpi_0^{-1}(\mathcal{E}).
\end{aligned}
$$

By construction, T is random time with distribution given by

$$
\begin{aligned}
\mathbb{P}\left(T \geq t\right) & = \mathbb{P}\left(U \leq e^{-\int_0^t \lambda_s ds}\right) \\
& = \mathbb{P}(\mathcal{E} \leq \varpi_0(t)) = e^{-\varpi_0(t)} = e^{-\int_0^t \lambda_s ds}.
\end{aligned}
$$

For any $s \leq t$ we have the memoryless property

$$\mathbb{P}\left(T \geq t \mid T \geq s\right) = \mathbb{P}\left(T \geq t\right)/\mathbb{P}\left(T \geq s\right) = e^{-\int_s^t \lambda_r dr}. \tag{10.7}$$

In addition, it is easily seen that the distribution of T is given by the formula

$$\mathbb{P}\left(T \in dt\right) = \mathbb{P}\left(\varpi^{-1}(\mathcal{E}) \in dt\right) = \lambda_t \, e^{-\int_s^t \lambda_r dr} \, dt.$$

Definition 10.6.2 *We let T_n be a sequence of random times defined by $T_0 = 0$ and*

$$
\begin{aligned}
T_{n+1} & = \inf\left\{t \geq T_n \; : \; \int_{T_n}^t \lambda_s \, ds \geq \mathcal{E}_{n+1}\right\} \\
& = \inf\left\{t \geq T_n \; : \; \exp\left[-\int_{T_n}^t \lambda_s ds\right] \leq U_{n+1}\right\} = \varpi_{T_n}^{-1}(\mathcal{E}_{n+1})
\end{aligned} \tag{10.8}
$$

where $\mathcal{E}_{n+1} = -\log U_{n+1}$ is a sequence of independent exponential random variables with unit parameters ($\Rightarrow U_n$ uniform on $[0,1]$). The counting process

$$N_t = \sum_{n \geq 0} 1_{[T_n, T_{n+1}[}(t) \, n \tag{10.9}$$

is called an inhomogeneous Poisson process (or non-homogeneous Poisson process) with intensity (or rate) parameter λ_t.

Important remark : When the intensity function λ_t is given by an auxiliary independent random process, this is sometimes called the Cox process [58], or a doubly stochastic Poisson process. There exist also some self-exiting Poisson models in which the intensity depends on the process itself. For instance λ_t can be given by some function of the form

$$\lambda_t = \alpha_t + \sum_{n \geq 0} 1_{T_n \leq t} \, \beta_{t-T_n}$$

for some positive and deterministic functions α_t, β_t. The function α_t is a deterministic base intensity function, and β_t is an auxiliary positive and deterministic function that expresses the influence β_{t-T_n} of the past events of the Poisson process on the current intensity λ_t. When $\alpha_t = \alpha_0$ is constant and $\beta_t = \beta_0 \, e^{-\gamma \, t}$ we have

$$\lambda_t = \alpha_0 + \beta_0 \sum_{n \geq 0} 1_{T_n \leq t} \, e^{-\gamma \, (t-T_n)}.$$

These jump models are often called the Hawkes processes [149]. These stochastic models are of current use in seismology, where they are sometimes called epidemic type aftershock sequence (ETAS) models [10]. A form of the clustering density used to model earthquake aftershocks is the so-called Omori-Utsu power law $\beta_t = a/b^t$ for some given parameters $a > 0$ and $b > 1$ [211, 212].

Remark : We observe that

$$
\varpi_0(T_n) = \int_0^{T_n} \lambda_s \, ds
$$

$$
= \sum_{0 \leq p < n} \int_{T_p}^{T_{p+1}} \lambda_s \, ds = \sum_{0 \leq p < n} \varpi_{T_n}(T_{n+1}) = \sum_{1 \leq p \leq n} \mathcal{E}_n := T'_n.
$$

This implies that

$$
N_t = \sum_{n \geq 0} 1_{[T_n, T_{n+1}[}(t) \, n
$$

$$
= \sum_{n \geq 0} 1_{[\varpi_0(T_n), \varpi_0(T_{n+1})[}(\varpi_0(t)) \, n = \sum_{n \geq 0} 1_{[T'_n, T'_{n+1}[}(\varpi_0(t)) \, n = N'_{\varpi_0(t)}
$$

with a Poisson process $N'_{\varpi_0(t)}$ with unit intensity $\lambda_t = 1$.

We summarize this discussion with the following transfer theorem.

Theorem 10.6.3 *For any Poisson processes N_t with intensity λ_t we have the time re-scaling properties*

$$
N_t = N'_{\varpi_0(t)} \quad \text{and} \quad N_{\varpi_0^{-1}(t)} = N'_t \tag{10.10}
$$

where N'_t stands for a Poisson process with unit intensity.

Proposition 10.6.4 *For any $n \geq 1$ we have*

$$
\mathbb{P}\left((T_1, \ldots, T_n) \in d(t_1, \ldots, t_n)\right) = \prod_{1 \leq p \leq n} 1_{[t_{p-1}, \infty[}(t_p) \, \lambda_{t_p} \, \exp\left(-\int_{t_{p-1}}^{t_p} \lambda_s \, ds\right) dt_p
$$

with the convention $t_{-1} = 0$, for $p = 0$. In addition, for any $n \geq 0$ we have

$$
\mathbb{P}\left(T_{n+1} \in dt\right) = \frac{\left(\int_0^t \lambda_s ds\right)^n}{n!} \, \lambda_t \, \exp\left(-\int_0^t \lambda_s \, ds\right) dt. \tag{10.11}
$$

Proof :
Using the fact that

$$
T_{n+1} = T_n + \inf\left\{t \geq 0 \; : \; \int_{T_n}^{T_n+t} \lambda_s \, ds \geq \mathcal{E}_n\right\}
$$

we prove that for any $t_{n+1} \geq t_n$

$$
\mathbb{P}\left(T_{n+1} \geq t_{n+1} \mid T_n = t_n\right) = \mathbb{P}\left(\mathcal{E}_n \geq \int_{t_n}^{t_{n+1}} \lambda_s \, ds\right) = \exp\left(-\int_{t_n}^{t_{n+1}} \lambda_s \, ds\right)
$$

from which we conclude that

$$\mathbb{P}\left(T_{n+1} \in dt_{n+1} \mid T_n = t_n\right) = 1_{[t_n, \infty[}(t_{n+1}) \; \lambda_{t_{n+1}} \; \exp\left(-\int_{t_n}^{t_{n+1}} \lambda_s \; ds\right) dt_{n+1}.$$

This ends the proof of the first assertion.

We prove (10.11) by induction w.r.t. the parameter n. For $n = 0$, the result is immediate. We assume that the result is true at rank n. In this case, we have

$$
\begin{aligned}
\mathbb{P}\left(T_{n+1} \in dt\right) &= \int_0^t \mathbb{P}\left(T_{n+1} \in dt \mid T_n = s\right) \mathbb{P}\left(T_n \in ds\right) \\
&= \lambda_t \; \exp\left(-\int_0^t \lambda_u \; du\right) dt \int_0^t \frac{\partial}{\partial s} \frac{\left(\int_0^s \lambda_u du\right)^n}{n!} \; ds.
\end{aligned}
$$

The proof of the proposition is now completed. ∎

10.6.2   Poisson thinning simulation

The random occupation measures $\sum_{1 \leq i \leq N_t} \delta_{T_i}$ associated with the counting processes (10.9) and (10.2). can be interpreted in terms of the spatial style point processes discussed in section 4.6. These measures can be thought as spatial Poisson point processes on the state space $S = \mathbb{R}_+$, with respective intensity functions ν_t and ν_t' defined by

$$
\begin{aligned}
\nu_t(ds) &= \lambda_s \; 1_{[0,t]}(s) \; ds & \overline{\nu}_t(ds) &= \frac{1}{\int_0^t \lambda_s ds} \; \lambda_s \; 1_{[0,t]}(s) \; ds \\
\nu_t'(ds) &= \lambda \; 1_{[0,t]}(s) \; ds & \overline{\nu}_t'(ds) &= \frac{1}{t} \; 1_{[0,t]}(s) \; ds.
\end{aligned}
$$

We refer the reader to section 4.6, definition 4.6.2 for a more detailed discussion on these models. We further assume that the intensity function λ_s is continuous and upper bounded by λ, for any $s \in [0, t]$. In this situation, the picture below illustrates the traditional acceptance-rejection technique for sampling the random times T_n from $\overline{\nu}_t$ using random samples T_n' from $\overline{\nu}_t'$. As the picture shows, the acceptance probability of a random sample, say $T_n = s$, from $\overline{\nu}_t$ is equal to

$$\mathbb{P}\left(U_n \leq \lambda_s\right) = \lambda_s / \lambda$$

where U_n are independent uniform random variables on $[0, \lambda]$.

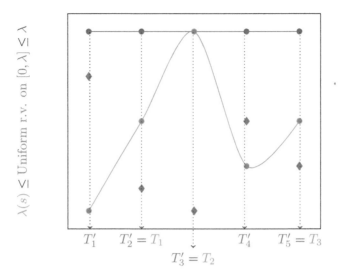

State space $S \ni s \in$ time axis

Important remark : The Poisson thinning simulation technique allows us to sample the jump times of a non-homogeneous Poisson process on any interval $[0, t]$, as soon as the intensity is bounded by some homogeneous finite constant. The price to pay for this simplification is that the time to obtain a random sample from the non-homogeneous model can be very large. We refer the reader to the exercise 166 for an example of a non-homogeneous model for which the average first acceptance time on $[0, t]$ tends to infinity when $t \uparrow \infty$.

10.6.3   Geometric random clocks

Using the re-scaling properties (10.10) we have

$$N_{t+h} - N_t = N'_{\varpi_0(t+h)} - N'_{\varpi_0(t)} \stackrel{law}{=} N'_{\varpi_0(t+h) - \varpi_0(t)} = N'_{\varpi_t(t+h)}.$$

This implies that

$$\mathbb{P}\left(N_{t+h} - N_t = 1\right) = \varpi_t(t+h) \, e^{-\varpi_t(t+h)} \simeq_{h \to 0} \lambda_t \, h$$

and for any $n \geq 2$

$$\mathbb{P}\left(N_{t+h} - N_t = n\right) = \frac{\varpi_t(t+h)^n}{n!} \, e^{-\varpi_t(t+h)} = \mathrm{O}(h^n).$$

Recalling the memoryless property of these processes, we see that the Poisson process can be simulated by a time inhomogeneous Bernoulli process on some time mesh t_n with $(t_n - t_{n-1}) = h$ with a success probability $p_{h, t_n} \simeq_{h \to 0} \lambda_{t_n} \, h$.

To formalize this claim, we let $h > 0$ be some time mesh parameter associated with some discretization of some interval $[0, t]$

$$t_0 = 0 < t_1 = h < \ldots < t_n = nh < \ldots < h\lfloor t/h \rfloor \leq t \ \left(= h\lfloor t/h \rfloor + h\{t/h\}\right).$$

Replacing in (10.8) the integral of the intensity function by its discrete time approximation

$$\varpi_{t_l}^h(t_n) := \sum_{t_l < t_m \leq t_n} \lambda_{t_m} \, h \simeq_{h \downarrow 0} \varpi_{t_l}(t_n) := \int_{t_l}^{t_n} \lambda_s \, ds,$$

we define a sequence of random times

$$T_{n+1}^h = \inf \left\{ t_k \geq T_n^h \; : \; \sum_{T_n^h < t_l \leq t_k} \lambda_{t_l} h \geq \mathcal{E}_{n+1} \right\}. \tag{10.12}$$

> These random times can be interpreted as the jump times of a time inhomogeneous Bernoulli counting process defined as in (10.6) with the sequence of independent Bernoulli random variables
>
> $$\mathbb{P}\left(\epsilon_{t_n}^h = 0 \right) = 1 - \mathbb{P}\left(\epsilon_{t_n}^h = 1 \right) = e^{-\lambda_{t_n} h}.$$

More precisely, using the same arguments as in (10.5), we find that

$$
\begin{aligned}
T_{n+1}^h &= \inf \left\{ t_k > T_n^h \; : \; \epsilon_{t_k}^h = 1 \right\} \\
&= \inf \left\{ t_k \geq T_n^h \; : \; \prod_{T_n^h < t_p \leq t_k} e^{-\lambda_{t_p} h} \leq U_{n+1} \right\}. \tag{10.13}
\end{aligned}
$$

> It follows that $N_{t_n}^h$ is an $\mathbb{N}$-valued Markov chain with transition probabilities
>
> $$
> \begin{aligned}
> \mathbb{P}\left(N_{t_{n+1}}^h \in dy \mid N_{t_n}^h = x \right) &= M_{t_n, t_{n+1}}^h(x, dy) \\
> &= e^{-\lambda_{t_n} h} \, \delta_x(dy) + \left(1 - e^{-\lambda_{t_n} h} \right) \, \delta_{x+1}(dy).
> \end{aligned}
> $$

10.7   Exercises

Exercise 145 (Poisson intensity) *Let N_t be a Poisson processes with intensity $\lambda > 0$. Check that $\mathbb{E}(N_t) = \lambda t = \mathrm{Var}(N_t)$ holds.*

Exercise 146 (Poisson conditional distributions) *Let N_t be a Poisson processes with intensity $\lambda > 0$, and T_1, resp. T_2 be the first time, resp. the second time of a jump of N_t.*

- *Find the conditional distribution of T_1 given $N_t = 1$.*

- *Find the conditional distribution of T_1 given $N_t = 2$.*

- *Find the conditional distribution of T_2 given $N_t = 2$.*

Exercise 147 (Conditioning principles) *We let $N_1(t)$ and $N_2(t)$ two independent Poisson processes with respective intensities λ_1 and λ_2. Compute the probability distribution of $N(t) = N_1(t) + N_2(t)$. Find the conditional distribution of $N_1(t)$ given $N(t)$.*

Exercise 148 (Arrival time distribution) *We consider a Bernoulli sequence of i.i.d. random variables $(\mathcal{E}_n)_{n \geq 1}$ with $\mathbb{P}(\mathcal{E} = 1) = 1 - \mathbb{P}(\mathcal{E}_0 = 0) = p \in]0,1[$. We interpret $\mathcal{E}_n = 1$ as the arrival of some individual at time n. We let $N_n := \sum_{1 \leq k \leq n} \mathcal{E}_k$ be the number of arrivals at time n. Find the distribution of the random variables N_n and $(N_n - N_m)$ for $m \leq n$. Describe the distribution of the random vector $(N_{n_1}, N_{n_2} - N_{n_1}, \ldots, N_{n_k} - N_{n_{k-1}})$, for a sequence $1 \leq n_1 < \ldots < n_k$. Find the distribution of the n-th arrival times T_n.*

Exercise 149 (Superposition property) *We let $N_1(t)$ and $N_2(t)$ two independent Poisson processes with respective intensities λ_1 and λ_2. Compute the probability that N_1 jumps n times before N_2 jumps m times.*

Exercise 150 (Non-homogeneous Poisson formula) 🖋 *We let $N_1(t)$ be a Poisson process with unit intensity $\lambda_1 = 1$, and let $n \in \mathbb{N} \mapsto \lambda(n)$ be a bounded function. We associate with these objects the jump process*

$$N_2(t) = N_2(0) + N_1 \left(\int_0^t \lambda(N_2(s)) \, ds \right).$$

Describe the jump times $T_n^{(2)}$ of N_2 in terms of the jump times $T_n^{(1)}$ of N_1.

Exercise 151 (Ordered uniform statistics) 🖋 *We consider the random times T_n defined in (10.8). Given $T_{n+1} = t$, check that the sequence $(T_1, \ldots, T_n)$ forms a re-ordered sequence of independent random variables with common distribution $\propto 1_{[0,t]}(s) \, \lambda_s \, ds$.*

Exercise 152 (Poisson stochastic differential equation - 1) *We let $a \in \mathbb{R}$ be some given parameter. We consider the stochastic process*

$$dX_t = a \, X_{t-} \, dN_t$$

starting at some $X_0 \neq 0$, where N_t stands for a Poisson process with intensity $\lambda > 0$. Check that

$$X_t = (1 + a)^{N_t} \, X_0.$$

Exercise 153 (Poisson stochastic differential equation - 2) *We let $t \mapsto a_t \in \mathbb{R}$ be a given function. We consider the stochastic process*

$$dX_t = a_t \, X_{t-} \, dN_t$$

starting at some $X_0 \neq 0$. Here N_t is a Poisson process with intensity $\lambda > 0$. Check that

$$X_t = X_0 \prod_{0 \leq s \leq t, \, dN_s = 1} (1 + a_s).$$

Exercise 154 (Poisson stochastic differential equation - 3) 🖋 *We let $t \mapsto a_t \in \mathbb{R}$ be some given function We consider the stochastic process*

$$dX_t = b_t \, X_{t-} \, dt + a_t \, X_{t-} \, (dN_t - \lambda dt)$$

starting at some $X_0 \neq 0$. Here N_t is a Poisson process with intensity $\lambda > 0$. Check that

$$X_t = X_0 \, \exp \left(\int_0^t (b_s - \lambda a_s) ds \right) \prod_{0 \leq s \leq t, \, dN_s = 1} (1 + a_s).$$

11

Markov chain embeddings

Embedding techniques allow us to calculate continuous time versions of Markov chains. Such embeddings allow us to study properties of a chain via properties of the related infinitesimal generators. In this chapter, we provide a detailed theoretical discussion of the embedding techniques, as well as series of illustrative examples that include spatially inhomogeneous models and time inhomogeneous models.

> *Only two things are infinite, the universe and human stupidity, and I'm not sure about the former.*
> Albert Einstein (1879-1955).

11.1 Homogeneous embeddings

11.1.1 Description of the models

This section is concerned with the continuous time version of a Markov chain.

We consider a Markov chain Y_n on some general state space S with a Markov transition $K(x, dy)$. We let $(\mathcal{E}_n)_{n \geq 1}$ be a sequence of independent exponential r.v. with unit parameter, and we consider the random times

$$T_n = \sum_{1 \leq p \leq n} \mathcal{E}_p$$

with $T_0 = \sum_\emptyset = 0$. Notice that

$$\mathbb{R}_+ := [0, \infty[= \cup_{n \geq 0} [T_n, T_{n+1}[.$$

We let X_t be the Markov process indexed by the continuous time parameter $t \in \mathbb{R}_+$ and defined by

$$
\begin{aligned}
X_t &= \sum_{n \geq 0} 1_{[T_n, T_{n+1}[}(t) \ Y_n = Y_{N_t} \qquad\qquad (11.1)\\
&= X_0 + \sum_{0 < T_n \leq t} (X_{T_n} - X_{T_n-}) = X_0 + \sum_{0 < n \leq N_t} (X_{T_n} - X_{T_n-})
\end{aligned}
$$

with jumps $X_{T_n} - X_{T_n-} = Y_n$. In the above display N_t stands for a Poisson process N_t with unit intensity. Notice that X_t is a jump process, with the jump times T_n, and the random positions are prescribed by the Markov chain sequence Y_n.

Definition 11.1.1 *The continuous time process (11.1) is right-continuous and left-limited (abbreviated r.c.l.l.) and it is called the continuous time embedding or version of the Markov chain Y_n. The r.c.l.l. property is often referred as the càdlàg property in reference to the French "continue à droite et limité à gauche". Inversely, the Markov chain Y_n is called the embedded Markov chain (abbreviated EMC) of the continuous time process X_t. The process X_t is sometimes called the uniform Markov jump process with Poisson clock N_t and subordinated chain Y_n.*

11.1.2   Semigroup evolution equations

We consider the operator P_t defined for any bounded function f on S by the expectation operator

$$P_t(f)(x) := \mathbb{E}(f(X_t) \mid X_0 = x). \tag{11.2}$$

Notice that for $t = 0$, we have $P_0(f)(x) = f(x)$. In addition, recalling that

$$\mathbb{E}(f(X_{t+s}) \mid X_s = x) = \mathbb{E}(f(X_t) \mid X_0 = x) = P_t(f)(x)$$

we find that

$$P_{t+s}(f)(x) = \mathbb{E}\left(f(X_{t+s})|X_0 = x\right) = \mathbb{E}\left(\underbrace{\mathbb{E}(f(X_{t+s}) \mid X_s)}_{P_t(f)(X_s)}|X_0 = x\right) = P_s(P_t(f))(x).$$

> These properties are often written in terms of the composition of integral operators:
>
> $$P_{s+t} = P_s P_t \quad \text{with} \quad P_0 = Id. \tag{11.3}$$

> **Theorem 11.1.2** *For any $t \in \mathbb{R}_+$, $x \in S$, and any bounded function f on S, we have the integral formula*
>
> $$P_t(f)(x) = e^{-t} f(x) + \int_0^t e^{-(t-s)} K(P_s(f))(x) \, ds. \tag{11.4}$$
>
> *In addition, we have the differential formula*
>
> $$\partial_t P_t(f)(x) = L(P_t(f))(x) = P_t(L(f))(x) \tag{11.5}$$
>
> *with the initial condition $P_0(f)(x) = f(x)$, for $t = 0$, and with the integral operator L defined by*
>
> $$L(f)(x) := K(f)(x) - f(x) = \int [f(y) - f(x)] \, K(x, dy)$$
>
> $$= \partial_t P_t(f)(x)_{|t=0} = \lim_{\epsilon \downarrow 0} \frac{P_\epsilon(f)(x) - f(x)}{\epsilon}. \tag{11.6}$$
>
> *In addition, the distribution η_t of the random states X_t satisfies the equation*
>
> $$\partial_t \eta_t(f) = \eta_t(L(f)). \tag{11.7}$$

Proof :

The proof is based on the following decomposition w.r.t. to the first jump time of the process

$$
\begin{aligned}
P_t(f)(x) &= \mathbb{E}(f(X_t)\, 1_{T_1>t} \mid X_0 = x) + \mathbb{E}(f(X_t)\, 1_{T_1\leq t} \mid X_0 = x) \\
&= f(x)\, \mathbb{P}\,(T_1 > t) + \mathbb{E}(\mathbb{E}\,(f(X_t) \mid T_1,\, X_{T_1})\, 1_{T_1\leq t} \mid X_0 = x).
\end{aligned}
\tag{11.8}
$$

By construction, the random variables $X_{T_1} = Y_1$ and T_1 are independent; hence

$$
\begin{aligned}
\mathbb{E}(g(T_1, X_{T_1})\, 1_{T_1\leq t} \mid X_0 = x) &= \int K(x,dy)\, \mathbb{E}(g(T_1,y)\, 1_{T_1\leq t}) \\
&= \int K(x,dy) \int_0^t g(s,y)\, e^{-s}\, ds
\end{aligned}
$$

for any function g on $\mathbb{R}_+ \times S$. On the other hand, we have

$$
\begin{aligned}
g_t(s,y) &:= \mathbb{E}\,(f(X_t) \mid T_1 = s,\, X_s = y) \\
&= \mathbb{E}\,(f(X_t) \mid X_s = y) = \mathbb{E}\,(f(X_{t-s}) \mid X_0 = y) = P_{t-s}(f)(y).
\end{aligned}
$$

This shows that

$$
\mathbb{E}(\overbrace{\mathbb{E}\,(f(X_t) \mid T_1,\, X_{T_1})}^{:=g_t(T_1,X_{T_1})}\, 1_{T_1\leq t} \mid X_0 = x) = \int K(x,dy) \int_0^t g_t(s,y)\, e^{-s}\, ds
$$

$$
= \int_0^t \underbrace{\left[\int K(x,dy) P_{t-s}(f)(y)\right]}_{=K(P_{t-s}(f))(x)} e^{-s}\, ds = \int_0^t K(P_{t-s}(f))(x)\, e^{-s}\, ds.
$$

Notice that

$$
\int_0^t e^{-s}\, K(P_{t-s}(f))(x)\, ds \overset{r=t-s}{=} \int_0^t e^{-(t-r)}\, K(P_r(f))(x)\, dr.
$$

The end of the proof of (11.4) is now a direct consequence of the decomposition (11.8).

$$
\begin{aligned}
\partial_t P_t(f)(x) &= -e^{-t}\, f(x) - \int_0^t e^{-(t-s)}\, K(P_s(f))(x)\, ds + K(P_t(f))(x) \\
&= K(P_t(f))(x) - P_t(f)(x) = L(P_t(f))(x).
\end{aligned}
$$

This ends the proof of the l.h.s. assertion in (11.5). Taking $t = 0$ in the l.h.s. of (11.5) we prove the differential formula (11.6).

The r.h.s. of (11.5) is easily checked using the semigroup (a.k.a. *sg*) property (11.3), that is,

$$
\frac{1}{\epsilon}\,[P_{t+\epsilon}(f) - P_t(f)] = \frac{1}{\epsilon}\,\underbrace{[P_\epsilon\,(P_t(f)) - P_t(f)]}_{\to_{\epsilon\downarrow 0}L(P_t(f))} = P_t\left(\underbrace{\frac{1}{\epsilon}\,[P_\epsilon(f) - f]}_{\to_{\epsilon\downarrow 0}L(f)}\right).
$$

The last assertion is a consequence of the fact that

$$
\eta_t(f) = \mathbb{E}(f(X_t)) = \mathbb{E}(\mathbb{E}(f(X_t) \mid X_0)) = \eta_0(P_t(f)) = \int \eta_0(dx)\, P_t(f)(x).
$$

We conclude that for any bounded function f on S we have

$$\partial_t \eta_t(f) = \eta_0 \left(\partial_t P_t(f)\right) = \eta_0 \left(P_t(L(f))\right) = \eta_t(L(f)).$$

∎

Remark :

For finite state space, $S = \{1, \ldots, d\}$, the Markov transitions $K(x, y)$ and generator $L(x, y)$ are given by matrices

$$L(f)(x) \;=\; \sum_{y \in S} [f(y) - f(x)] \; K(x, y) := \sum_{y \in S} \underbrace{[K(x, y) - 1_x(y)]}_{:= L(x, y)} \; f(y).$$

If we set $u_t(x) := P_t(f)(x)$, using (11.5) we readily find that $u_t(x)$ satisfies the integro-differential equation

$$\partial_t u_t(x) = \int [u_t(y) - u_t(x)] \; K(x, dy)$$

with the initial condition $u_0(x) = f(x)$.

Definition 11.1.3 *The operator L defined in (11.6) is called the (infinitesimal) generator of the process X_t defined in (11.1), and P_t is called its semigroup of integral operators acting on the set of bounded functions. The equations (11.4), (11.5), (11.6), and (11.7) are often written in the more synthetic forms*

$$P_t \;=\; e^{-t} \; Id \;+\; \int_0^t e^{-(t-s)} \; P_s \; ds \quad and \quad \partial_t \eta_t = \eta_t L \qquad (11.9)$$

as well as

$$\partial_t P_t \;=\; L P_t = P_t L \quad and \quad L = K - Id = \lim_{\epsilon \downarrow 0} \epsilon^{-1} [P_\epsilon - Id]. \qquad (11.10)$$

These integral operator formulations are sometimes called Gelfand-Pettis weak sense integral equations [34].

The same analysis works if we replace in (11.1) the standard exponential random variable by an exponential random variable $\mathcal{E}_n$ with some parameter $\lambda > 0$. In this situation, the continuous time embedding of Y_n is given by

$$X_t = \sum_{n \geq 0} 1_{[T_n, T_{n+1}[}(t) \; Y_n = Y_{N_t}$$

with the Poisson process N_t with intensity λ, and arrival times T_n.

Notice that the integral formulae (11.4) and (11.9) take the form

$$P_t \;=\; e^{-\lambda t} \; Id \;+\; \int_0^t \lambda \, e^{-\lambda(t-s)} \; K P_s \; ds.$$

In this case, the differential equations (11.5) meet with the integral generator

$$L(f)(x) := \lambda \int [f(y) - f(x)] \; K(x, dy) = \lim_{\epsilon \downarrow 0} \epsilon^{-1} [P_\epsilon - Id](f)(x). \qquad (11.11)$$

In this situation, the function

$$u_t(x) := P_t(f)(x) = \mathbb{E}\left(f(X_t) \mid X_0 = x\right)$$

satisfies the integro-differential equation

$$\partial_t u_t(x) = \lambda \int [u_t(y) - u_t(x)] K(x, dy)$$

with the initial condition $u_0(x) = f(x)$.

In addition, the semigroup of X_t is given by

$$
\begin{aligned}
P_t(f)(x) &= \sum_{n \geq 0} \mathbb{E}\left(f(Y_{N_t}) \mid N_t = n,\ X_0 = x\ \right)\ \mathbb{P}\left(N_t = n\right) \\
&= e^{-\lambda t} \sum_{n \geq 0} \frac{(\lambda t)^n}{n!}\ K^n(f)(x).
\end{aligned}
\tag{11.12}
$$

For finite state space models, we have the matrix formulae

$$P_t = e^{-\lambda t} \sum_{n \geq 0} \frac{(\lambda t)^n}{n!}\ K^n = e^{-\lambda t}\ e^{\lambda K t}.$$

We also notice that $P_0 = Id$ and

$$\partial_t P_t = P_t L = L P_t \Rightarrow P_t = e^{tL} Id = \sum_{n \geq 0} \frac{(\lambda t)^n}{n!}\ L^n.$$

11.2 Some illustrations

11.2.1 A two-state Markov process

We illustrate these models with the continuous embedding of the two-state Markov chain

$$.5\ \circlearrowleft\ 1 \underset{.5}{\overset{1}{\rightleftarrows}} 2.$$

For indicator functions we have

$$P_t(1_i)(j) = \mathbb{P}(X_t = i \mid X_0 = j) = P_t(j, i)$$

and

$$
\begin{aligned}
L(1_1)(1) &= \sum_{y \in S} K(1, y)\ 1_1(y)\ - 1_1(1) = .5 - 1 = -.5 \\
L(1_1)(2) &= \sum_{y \in S} K(2, y)\ 1_1(y)\ - 1_1(2) = 1 - 0 = 1.
\end{aligned}
$$

This implies that

$$
\begin{aligned}
\partial_t P_t(i,1) &= \partial_t P_t(1_1)(i) = P_t(L(1_1))(i) \\
&= P_t(i,1)L(1_1)(1) + P_t(i,2)L(1_1)(2) = -\frac{1}{2}\,P_t(i,1) + P_t(i,2).
\end{aligned}
$$

If we choose $i = 1,2$, we find that

$$
\partial_t P_t(1,1) = -\frac{1}{2}\,P_t(1,1) + (1 - P_t(1,1)) = 1 - \frac{3}{2}P_t(1,1)
$$

and

$$
\partial_t P_t(2,1) = -\frac{1}{2}\,P_t(2,1) + (1 - P_t(2,1)) = 1 - \frac{3}{2}P_t(2,1).
$$

Therefore

$$
\begin{aligned}
P_t(1,1) &:= e^{-3t/2}P_0(1,1) + \int_0^t e^{-3(t-s)/2}\,ds = e^{-3t/2} + \frac{2}{3}\,(1 - e^{-3t/2}) \\
&= 1 - P_t(1,2)
\end{aligned}
$$

and

$$
P_t(2,1) = \frac{2}{3}\,(1 - e^{-3t/2}) = 1 - P_t(2,2).
$$

This clearly implies that

$$
P_t = \begin{pmatrix} P_t(1,1) & P_t(1,2) \\ P_t(2,1) & P_t(2,2) \end{pmatrix} \xrightarrow{t\uparrow\infty} \begin{pmatrix} 2/3 & 1/3 \\ 2/3 & 1/3 \end{pmatrix} = \begin{pmatrix} \pi(1) & \pi(2) \\ \pi(1) & \pi(2) \end{pmatrix}.
$$

We observe that $\pi = (\pi(1), \pi(2))$ is the invariant measure of the embedded Markov chain with transition matrix K, that is,

$$
\begin{aligned}
\pi K &= (\pi(1), \pi(2)) \begin{pmatrix} K(1,1) & K(1,2) \\ K(2,1) & K(2,2) \end{pmatrix} \\
&= (2/3, 1/3) \begin{pmatrix} 1/2 & 1/2 \\ 1 & 0 \end{pmatrix} = (2/3, 1/3) = \pi.
\end{aligned}
$$

11.2.2 Matrix valued equations

Let us assume that $S = \{1, \ldots, d\}$ and let L be the operator

$$
L(f)(j) = \sum_{k \in S} (f(k) - f(j))\, Q(j,k)
$$

for some matrix Q with positive entries. The positive entries $Q(j,k)$ represent the rates of the transitions $j \rightsquigarrow k$. For a more detailed discussion on these models we refer to section 11.3 and section 11.3.2.

Next we rewrite this generator in terms of the homogeneous Markov chain embeddings discussed above.

Firstly, we observe that

$$
L(f)(j) = \lambda(j) \sum_{k \in S} (f(k) - f(j))\, K(j,k) \tag{11.13}
$$

with

$$
K(j,k) = \frac{Q(j,k)}{\sum_{l \in S} Q(j,l)} \quad \text{and} \quad \lambda(j) := \sum_{k \in S} Q(j,k).
$$

In addition, we have

$$L(f)(j) \;=\; \lambda \sum_{k \in S} (f(k) - f(j))\, K(j,k)$$

for any

$$\lambda \geq \sup_{i \in S} \lambda(i) \quad \text{and} \quad K(i,j) = \frac{\lambda(i)}{\lambda}\, K(i,j) + \left(1 - \frac{\lambda(i)}{\lambda}\right)\, 1_i(j).$$

This generator has exactly the same form as the one discussed in (11.11).

The Markov process X_t with generator L is a jump type process. The jump times T_n arrive at rate λ. At these jump times times, say T_n, the process is at a given location, say $X_{T_n-} = i$. With a probability $\frac{\lambda(i)}{\lambda}$, the process jumps to a new state $X_{T_n} = j$ randomly chosen with the probability $K(i,j)$. Otherwise, it stays in the same location $X_{T_n} = i$.

The self-loops transition represents the probability $\left(1 - \frac{\lambda(i)}{\lambda}\right)$ of fictitious jumps. The form of the generator (11.13) suggests a less time consuming sampling technique:

Starting from $X_0 = i_0$, we wait an exponential time with parameter $\lambda(i_0)$ and we jump from i_0 to i_1 according to the probability $K(i_0, i_1)$. Then we wait an exponential time with parameter $\lambda(i_1)$ and we jump from i_1 to i_2 according to the probability $K(i_1, i_2)$, and so on.

Further details on these models are provided in section 11.3 dedicated to general spatially inhomogeneous models and in section 11.3.2 covering finite state space models.

In this situation, for any indicator function $f_i = 1_i$, we have

$$\eta_t(f_i) = \eta_t(i) = \mathbb{P}(X_t = i).$$

As usual, we can identify the measures η_t and the functions f by the row and column vectors

$$\eta_t = [\eta_t(1), \ldots, \eta_t(d)] \quad \text{and} \quad f = \begin{pmatrix} f(1) \\ \vdots \\ f(d) \end{pmatrix}.$$

In the same way the Markov probability transitions $P_t(i,j) = \mathbb{P}(X_t = j \mid X_0 = i)$ are identified with the transition matrix

$$P_t = \begin{pmatrix} P_t(1,1) & P_t(1,2) & \ldots & P_t(1,d) \\ P_t(2,1) & P_t(2,2) & \ldots & P_t(1,d) \\ \vdots & \vdots & \ldots & \\ P_t(d,1) & P_t(d,2) & \ldots & P_t(d,d) \end{pmatrix}.$$

In this vector notation, the evolution equations (11.5) and (11.7) can be rewritten in terms of ordinary differential equations (*abbreviated ODE*)

$$\partial_t P_t = L P_t = P_t L \quad \text{and} \quad \partial_t \eta_t = \eta_t L$$

with the matrix

$$L = \begin{pmatrix} L(1,1) & L(1,2) & \ldots & L(1,d) \\ L(2,1) & L(2,2) & \ldots & L(1,d) \\ \vdots & \vdots & \ldots & \\ L(d,1) & L(d,2) & \ldots & L(d,d) \end{pmatrix}$$

defined by

$$\begin{aligned} L(j,i) &= L(f_i)(j) = \sum_{k \in S} (f_i(k) - f_i(j)) \, Q(j,k) \\ &= Q(j,i) - 1_i(j) \sum_k Q(j,k) = Q(j,i) - \lambda(j) 1_j(i) \\ &= \lambda(j) \, [K(j,i) - 1_j(i)]. \end{aligned}$$

We also notice that

$$L(f)(j) = \sum_{i \in S} L(j,i) f(i) = \lambda(j) \sum_{i \in S} [K(j,i) - 1_j(i)] \, (f)(i) = \lambda(j) \, [K(f)(j) - f(j)].$$

Example 11.2.1 *We consider a mathematician wandering every day between three libraries with the following transition rate diagram*

$$Q(1,1) \, \circlearrowright \, 1 \xrightleftharpoons[\; Q(1,2) \;]{\; Q(2,1) \;} 2 \xrightleftharpoons[\; Q(2,3) \;]{\; Q(3,2) \;} 3. \, \circlearrowright \, Q(3,3)$$

The transition matrix of the embedded Markov chain is given by

$$K = \begin{pmatrix} \frac{Q(1,1)}{Q(1,1)+Q(1,2)} & \frac{Q(1,2)}{Q(1,1)+Q(1,2)} & 0 \\ \frac{Q(2,1)}{Q(2,1)+Q(2,3)} & 0 & \frac{Q(2,3)}{Q(2,1)+Q(2,3)} \\ 0 & \frac{Q(3,2)}{Q(3,2)+Q(3,3)} & \frac{Q(3,3)}{Q(3,2)+Q(3,3)} \end{pmatrix}.$$

The infinitesimal generator L of X_t is given by

$$L = \begin{pmatrix} \lambda(1) & 0 & 0 \\ 0 & \lambda(2) & 0 \\ 0 & 0 & \lambda(3) \end{pmatrix}$$

$$\times \left[\begin{pmatrix} \frac{Q(1,1)}{Q(1,1)+Q(1,2)} & \frac{Q(1,2)}{Q(1,1)+Q(1,2)} & 0 \\ \frac{Q(2,1)}{Q(2,1)+Q(2,3)} & 0 & \frac{Q(2,3)}{Q(2,1)+Q(2,3)} \\ 0 & \frac{Q(3,2)}{Q(3,2)+Q(3,3)} & \frac{Q(3,3)}{Q(3,2)+Q(3,3)} \end{pmatrix} - \begin{pmatrix} 1 & 0 & 0 \\ 0 & 1 & 0 \\ 0 & 0 & 1 \end{pmatrix} \right]$$

$$= \begin{pmatrix} Q(1,1) & Q(1,2) & 0 \\ Q(2,1) & 0 & Q(2,3) \\ 0 & Q(3,2) & Q(3,3) \end{pmatrix} - \begin{pmatrix} \lambda(1) & 0 & 0 \\ 0 & \lambda(2) & 0 \\ 0 & 0 & \lambda(3) \end{pmatrix}.$$

11.2.3 Discrete Laplacian

We consider the unit vectors on the integer d-dimensional lattice $\mathbb{Z}^d$

$$\forall 1 \leq i \leq d \quad e_i = \Big(0, \ldots, 0, \underbrace{1}_{i\text{-th}}, 0, \ldots, 0 \Big) \quad \text{and} \quad e_{d+i} = -e_i.$$

The simple random walk on $\mathbb{Z}^d$ is the Markov chain with transitions

$$K(x,y) = \frac{1}{2d} \sum_{1 \leq i \leq 2d} 1_{x+e_i}(y).$$

It can be interpreted as a Markov chain on $S = \mathbb{R}^d$ with transitions given for any $f \in \mathcal{B}(S)$ by the following formulae

$$K(f)(x) = \frac{1}{2d} \sum_{1 \leq i \leq d} [f(x+e_i) + f(x-e_i)].$$

In this situation, the generator L of the continuous time embedding process is given by

$$\begin{aligned} L(f)(x) &= \frac{1}{2d} \sum_{1 \leq i \leq d} [f(x+e_i) + f(x-e_i) - 2f(x)] \\ &= [K - Id](f)(x) := \frac{1}{2} \Delta_{\mathbb{Z}^d}(f)(x). \end{aligned}$$

The generator $\Delta_{\mathbb{Z}^d}$ is called the discrete Laplacian on the lattice $\mathbb{Z}^d$.

Our next objective is to increase the jump rates and to decrease the amplitude of the moves on $\mathbb{R}^d$ with some proportional factor $h \downarrow 0$. Given some $h > 0$, we consider the Markov chain Y_n^h with transitions

$$M^h(f)(x) = \frac{1}{2d} \sum_{1 \leq i \leq d} \left[f(x + \sqrt{d}he_i) + f(x - \sqrt{d}he_i) \right]$$

and a Poisson process with intensity $\lambda^h := h^{-2}$. The generator of the continuous time embedding process is given by the generator

$$\begin{aligned} L^h(f)(x) &= \frac{1}{2dh^2} \sum_{1 \leq i \leq d} [f(x+he_i) + f(x-he_i) - 2f(x)] \\ &= \lambda^h [K^h - Id](f)(x). \end{aligned}$$

We end this section with a brief discussion on the connection between the discrete and the continuous Laplacian. To this end, we further assume that f is twice differentiable. In this situation, for any $\epsilon \in \{-1, +1\}$, and any $i \in \{1, \ldots, d\}$ we have

$$[f(x + \epsilon \sqrt{d}h\, e_i) - f(x)] = \frac{\partial f}{\partial x_i}(x)\, \sqrt{d}h + \frac{d}{2} \frac{\partial^2 f}{\partial x_i^2}(x)\, h^2 + O(h^3).$$

This implies that

$$f(x+he_i) + f(x-he_i) - 2f(x) = d\, \frac{\partial^2 f}{\partial x_i^2}(x)\, h^2 + O(h^3)$$

from which we find that

$$L^h(f)(x) = \frac{1}{2} \sum_{1 \leq i \leq d} \frac{\partial^2 f}{\partial x_i^2}(x) + O(h).$$

11.3   Spatially inhomogeneous models

We start with a rather elementary proposition.

Proposition 11.3.1 *For any positive operator $Q(x, dy)$ on some general state space S, such that $\lambda(x) := Q(x, S) \le \lambda$ for any x, and some $\lambda < \infty$, we have*

$$\int [f(y) - f(x)]\, Q(x, dy) = \lambda(x) \int [f(y) - f(x)]\, K(x, dy)$$

$$= \lambda \int [f(y) - f(x)]\, K'(x, dy) \qquad (11.14)$$

with the Markov transitions

$$K(x, dy) = Q(x, dy)/Q(x, S)$$

$$K'(x, y) = \frac{\lambda(x)}{\lambda}\, K(x, dy) + \left(1 - \frac{\lambda(x)}{\lambda}\right)\, \delta_x(dy).$$

The function $x \mapsto \lambda(x)$ is often called the jump rate or the holding time of the state x. The stochastic matrix K is called the Markov transition of the embedded Markov chain model.

Example 11.3.2 *For instance, if we take $S = \mathbb{R}$ and $Q(x, dy) = 10 \times \frac{1}{\sqrt{2\pi}}\, e^{-\frac{1}{2}(x-y)^2}\, dy$, we have $\lambda(x) = 10$ and K reduces to the Gaussian transition $K(x, dy) = \frac{1}{\sqrt{2\pi}}\, e^{-\frac{1}{2}(x-y)^2}\, dy$.*

Definition 11.3.3 *Take any positive operator $Q(x, dy)$ on some general state space S, such that $\lambda(x) := Q(x, S) \le \lambda$ for any x, and for some $\lambda < \infty$. Let Y_n be a Markov chain with transition probabilities $K(x, dy) = Q(x, dy)/Q(x, S)$.*

- *The spatially homogeneous embedding X_t of the Markov chain Y_n w.r.t the constant intensity function $\lambda(x) = \lambda$ is a process X_t, similar to the one defined in (11.1) but replacing the exponential random variable with a unit parameter thereby yields some random variable $\mathcal{E}_n$ with a parameter $\lambda > 0$.*

- *The spatially inhomogeneous embedding of the Markov chain Y_n w.r.t the intensity function $\lambda(x)$ is the Markov process X_t' defined sequentially as follows:*

    1. *At time T_n we set $X_{T_n}' := Y_n$.*
    2. *For a given $\mathcal{E}_{n+1}' \sim \mathrm{Exp}\,(\lambda(Y_n))$ we set*

    $$X_t' := Y_n \qquad \forall T_n \le t < T_{n+1} := T_n + \mathcal{E}_{n+1}'. \qquad (11.15)$$

Theorem 11.3.4 *The spatially inhomogeneous embedding X_t' w.r.t the intensity function $\lambda(x)$ coincides with the spatially homogeneous embedding X_t w.r.t. the constant intensity λ as soon as $\lambda(x) \le \lambda$.*

In addition, the semigroup operator (11.2) and the distributions of this process satisfy (11.5) and (11.7) with L given by (11.14).

Proof :

Starting from some value, say $X_0 = x$, the first time T_x a "real" jump occurs

$$X_0 = x \rightsquigarrow X_{T_x} = Y \sim K(x, dy)$$

in the spatially homogeneous model is clearly given by the formula

$$T_x \; := \; \mathcal{E}_1 + \ldots + \mathcal{E}_{N_x} \quad \text{with} \quad N_x = \inf \{n \geq 1 \, : \, \epsilon_n = 1\}$$

where ϵ_n is a sequence of $\{0, 1\}$-valued Bernoulli r.v. with success parameter $\lambda(x)/\lambda$, and $\mathcal{E}_n$ is a sequence of (independent of ϵ_n) independent exponential random variables with parameter λ.

Recalling that N_x is a geometric random variable with parameter $\lambda(x)/\lambda$, and the sum of independent exponential random variables is gamma distributed, we prove that for any bounded function f on $\mathbb{R}_+$ we have

$$\mathbb{E}(f(T_x)) \;\; = \;\; \mathbb{E}(\mathbb{E}(f(T_x) \,|\, N_x)) = \sum_{n \geq 1} \mathbb{E}(f(\mathcal{E}_1 + \ldots + \mathcal{E}_n) \,|\, N_x = n) \, \mathbb{P}(N_x = n).$$

This implies that

$$\begin{aligned}
\mathbb{E}(f(T_x)) \;\; &= \;\; \sum_{n \geq 1} \left(\int_0^\infty f(t) \frac{(\lambda t)^{n-1}}{(n-1)!} \, \lambda \, e^{-\lambda t} dt \right) \left(1 - \frac{\lambda(x)}{\lambda} \right)^{n-1} \frac{\lambda(x)}{\lambda} \\
&= \;\; \int_0^\infty f(t) \underbrace{\left[\sum_{n \geq 1} \frac{1}{(n-1)!} \left(\lambda t \left(1 - \frac{\lambda(x)}{\lambda} \right) \right)^{n-1} \right]}_{= \exp([\lambda - \lambda(x)]t)} \lambda(x) \, e^{-\lambda t} dt.
\end{aligned}$$

This also implies that T_x is an exponential random variable with parameter $\lambda(x)$

$$T_x \sim \text{Exp}(\lambda(x)). \tag{11.16}$$

Using these observations, we define X_t and X_t' in the same way. This ends the proof of the theorem. ∎

Important remark : In biology and chemistry, the algorithmic formulation of the process X_t' is usually called the Gillespie algorithm.

Inversely, any Markov process X_t' defined as in theorem 11.3.4 for some rate functions $\lambda(x)$ and some Markov transitions K can be interpreted as a continuous embedding of a discrete Markov chain with Markov transition K or K' depending on the choice of jump rate $\lambda(x)$ of λ. The times T_x and T_x' the processes X_t and X_t' transition away from some state x are exponential r.v. with rates $\lambda(x)$ and $\lambda(x) := \lambda$, thus we have

$$\mathbb{E}(T_x) = \frac{1}{\lambda(x)} \; \geq \; \mathbb{E}(T_x') = \frac{1}{\lambda'(x)} = \frac{1}{\lambda}.$$

This shows that the process X_t' jumps more often than X_t, but the chance $\lambda(x)/\lambda$ to perform a real jump (i.e., change its value) according to the embedded Markov transition K can be drastically small if λ is chosen too large.

Important remark : In this situation, the function

$$u_t(x) := P_t(f)(x) = \mathbb{E}(f(X_t) \,|\, X_0 = x)$$

satisfies the integro-differential equation

$$\partial_t u_t(x) = \lambda(x) \int [u_t(y) - u_t(x)] \, K(x, dy) = \int [u_t(y) - u_t(x)] \, Q(x, dy)$$

with the initial condition $u_t(x) = f(x)$.

In addition, given an auxiliary bounded function g, the function

$$
\begin{aligned}
v_t(x) &= \mathbb{E}(f(X_t) \mid X_0 = x) + \int_0^t \mathbb{E}(g(X_t) \mid X_s = x) \, ds \\
&= P_t(f)(x) + \int_0^t P_{t-s}(g)(x) \, ds
\end{aligned}
$$

satisfies the integro-differential equation

$$\partial_t v_t(x) = g(x) + \int [v_t(y) - v_t(x)] \, Q(x, dy)$$

with the initial condition $u_t(x)|_{t=0} = f(x)$.

11.3.1   Explosion phenomenon

The spatially homogeneous model described in definition 11.3.3 can be extended to not necessarily bounded intensity functions $\lambda \, : \, x \in S \mapsto \lambda(x) = Q(x, S) < \infty$. The generator of the corresponding process X_t is given by

$$L(f)(x) = \lambda(x) \int (f(y) - f(x)) \, K(x, dy) \tag{11.17}$$

with the Karkov transition $K(x, dy) = Q(x, dy)/Q(x, S)$. Caution is necessary to have a well defined process. For instance, suppose that $S = \mathbb{N}$ and $K(x, dy) = \delta_{x+1}(dy)$. Then for any integer $n \in \mathbb{N}$ we have

$$L(f)(n) = \lambda(n) \, [f(n+1) - f(n)]. \tag{11.18}$$

In this context, the embedded Markov chain in (11.15) is simply given by $Y_n = n$, and the random times $\mathcal{E}_n := (T_{n+1} - T_n)$ are independent exponential random variables with parameter $\lambda(n)$. We consider the explosion time

$$\lim_{n \to \infty} T_n = T_\infty = \sup_{n \geq 0} T_n = \sum_{n \geq 1} \mathcal{E}_n = \lim_{n \to \infty} \underbrace{\sum_{1 \leq p \leq n} \mathcal{E}_p}_{=T_n}. \tag{11.19}$$

Definition 11.3.5 *Whenever T_∞ takes some finite value, we have an infinite number of jumps accumulating right before that time. In this situation, we say that we have an explosion of the jump process.*

Lemma 11.3.6 *We have the equivalence*

$$T_\infty = \infty \iff \sum_{n \geq 1} \frac{1}{\lambda(n)} = \infty.$$

Proof :

Invoking the monotone convergence theorem, we have

$$\mathbb{E}(T_\infty) = \lim_{n\to\infty} \mathbb{E}(T_n) = \lim_{n\to\infty} \sum_{1\leq p\leq n} \frac{1}{\lambda(p)} = \sum_{n\geq 1} \frac{1}{\lambda(n)} \tag{11.20}$$

and

$$\begin{aligned} \mathbb{E}\left(e^{-T_\infty}\right) &= \mathbb{E}\left(e^{-T_\infty} 1_{T_\infty<\infty}\right) + \mathbb{E}\left(e^{-T_\infty} 1_{T_\infty=\infty}\right) \\ &= \prod_{n\geq 0} \frac{1}{1+\frac{1}{\lambda(n)}} = \frac{1}{\prod_{n\geq 0}\left[1+\frac{1}{\lambda(n)}\right]} \leq \frac{1}{1+\sum_{n\geq 1}\frac{1}{\lambda(n)}}. \end{aligned} \tag{11.21}$$

The last assertion follows from the fact that $\mathbb{E}(e^{-\mathcal{E}_n}) = \frac{1}{1+\frac{1}{\lambda(n)}}$, for any $n \geq 1$, and $\prod_{n\geq 1}(1+a_p) \geq \sum_{n\geq 1} a_n$, for any non-negative numbers a_n. We conclude that

$$\sum_{n\geq 1} \frac{1}{\lambda(n)} = \infty \overset{11.21}{\Rightarrow} \mathbb{P}\left(T_\infty = \infty\right) = 1 \overset{11.20}{\Rightarrow} \sum_{n\geq 1} \frac{1}{\lambda(n)} = \infty.$$

This ends the proof of the lemma. ∎

The above lemma tells us that if we choose $\lambda(n) = n^2$ in (11.18), the explosion time is given by

$$T_\infty = \sum_{n\geq 1} \mathcal{E}'_n/\lambda_n$$

where $\mathcal{E}'_n$ are independent exponential random variables with unit parameters. By construction, this random time is almost surely finite so that $X_t \uparrow \infty$ as $t \uparrow \infty$.

More generally, for the jump times T_n of a general jump process X_t with generator (11.17), rephrasing the proof of the lemma, we readily prove that

$$\mathbb{E}\left(T_\infty \mid X_{T_n}, \, n\geq 0\right) = \sum_{n\geq 1} \frac{1}{\lambda(X_{T_n})}$$

$$\mathbb{E}\left(e^{-T_\infty} \mid X_{T_n}, \, n\geq 0\right) = \left[1 + \sum_{n\geq 1} \frac{1}{\lambda(X_{T_n})}\right]^{-1}.$$

This yields the following almost sure equivalence principle

$$T_\infty = \infty \iff \sum_{n\geq 1} \frac{1}{\lambda(X_{T_n})} = \infty.$$

The following theorem allows us to quantify the deviation of these random times w.r.t. the time horizon and their mean values.

Theorem 11.3.7 *We associate with a sequence of independent exponential random variables $\mathcal{E}_n$ with unit parameter and a sequence a_n of $[0,1]$-valued real numbers, the random times*

$$T_n^a := \sum_{1\leq p\leq n} a_p \, \mathcal{E}_p.$$

In this situation, for any $t \geq 0$ we have the exponential deviation estimates

$$\mathbb{P}\left(\frac{T_n^a - \mathbb{E}\left(T_n^a\right)}{\mathrm{Var}(T_n^a)} \geq t\right) \leq \exp\left(-\mathrm{Var}(T_n^a)\left(\sqrt{t+1} - 1\right)^2\right) \tag{11.22}$$

with the mean and variance parameters

$$\mathbb{E}\left(T_n^a\right) = \sum_{1 \leq p \leq n} a_p \quad and \quad \mathrm{Var}(T_n^a) = \sum_{1 \leq p \leq n} a_p^2.$$

In other words, the probability of the event

$$\sum_{1 \leq p \leq n} a_p \, \mathcal{E}_p \leq |a|_{n,1} + t + 2|a|_{n,2}\sqrt{t} \tag{11.23}$$

is larger than $1 - e^{-t}$, with $|a|_{n,1} := \sum_{1 \leq p \leq n} a_p$ and $|a|_{n,2}^2 := \sum_{1 \leq p \leq n} a_p^2$.

Proof :
We start with the proof of this result for $n = 1$, and $a_1 = 1$. We observe that

$$\forall t \in [0, 1[\quad \varphi(t) := \mathbb{E}\left(e^{t(\mathcal{E}_1 - 1)}\right) = \frac{e^{-t}}{1-t} = e^{L_0(t)} \leq e^{L_1(t)}$$

with

$$L_0(t) := -t - \log\left(1 - t\right) = t^2 \left(\sum_{p \geq 0} \frac{t^p}{p+2}\right) \leq L_1(t) := t^2 \sum_{p \geq 0} t^p = \frac{t^2}{1-t}.$$

This implies that for any $x > 0$ and any $t \in [0, 1[$, we have

$$\mathbb{P}\left(\mathcal{E}_1 - 1 \geq x\right) = \mathbb{P}\left(e^{t(\mathcal{E}_1 - 1)} \geq e^{tx}\right) \leq e^{-tx} \varphi(t) = e^{-(tx - L_0(t))}.$$

Taking the infimum over $t \in [0, 1[$, we prove that

$$\mathbb{P}\left(\mathcal{E}_1 \geq 1 + x\right) \leq e^{-L_0^\star(x)} \leq e^{-L_1^\star(x)}$$

with

$$L_0^\star(x) := \sup_{t \in [0,1[} \left(tx - L_0(t)\right) \geq L_1^\star(x) := \sup_{t \in [0,1[} \left(tx - L_1(t)\right).$$

Our next objective is to compute the function $L_1^\star$. Using the fact that

$$L_1(t) = -(t+1) + \frac{1}{1-t} \Rightarrow L_1'(t) = -1 + \frac{1}{(1-t)^2}$$

we readily check that

$$\partial_t \left(tx - L_1(t)\right) = (x+1) - \frac{1}{(1-t)^2}$$

$$= 0 \Longleftrightarrow t = t(x) := 1 - \frac{1}{\sqrt{1+x}} \in [0, 1[.$$

Notice that

$$\frac{t(x)^2}{1 - t(x)} = \frac{\left(\sqrt{x+1} - 1\right)^2}{x+1} \sqrt{1+x}$$

$$= \frac{\left(\sqrt{x+1} - 1\right)^2}{\sqrt{x+1}} = \left(\sqrt{x+1} - 1\right) \left(1 - \frac{1}{\sqrt{x+1}}\right)$$

and

$$x\, t(x) = x\left(1 - \frac{1}{\sqrt{1+x}}\right)$$
$$= (\sqrt{1+x} - 1)\, \frac{x}{\sqrt{1+x}} = (\sqrt{1+x} - 1)\, \left(\sqrt{1+x} - \frac{1}{\sqrt{1+x}}\right).$$

This clearly implies that

$$L_1^\star(x) = t(x)x - L_1(t(x)) = \left(\sqrt{1+x} - 1\right)^2$$

and therefore

$$\mathbb{P}\left(\mathcal{E}_1 \geq 1 + x\right) \leq e^{-\left(\sqrt{1+x} - 1\right)^2}.$$

By simple manipulations, we find that

$$y = L_1^\star(x) = \left(\sqrt{1+x} - 1\right)^2 \Rightarrow x = (1 + \sqrt{y})^2 - 1 = y + 2\sqrt{y} = (L_1^\star)^{-1}(y)$$

from which we conclude that

$$\mathbb{P}\left(\mathcal{E}_1 \geq 1 + y + 2\sqrt{y}\right) \leq e^{-y}.$$

Now, we turn to the proof of (11.23). We set

$$T_n^a := \sum_{1 \leq p \leq n} a_p\, \mathcal{E}_p \quad \text{and} \quad \overline{T}_n^a = T_n^a - \mathbb{E}(T_n^a) = \sum_{1 \leq p \leq n} a_p\, (\mathcal{E}_p - 1).$$

Using the fact that $0 \leq ta_p \leq t < 1$, as soon as $t \in [0, 1[$, we prove that

$$\phi_n(t) := \mathbb{E}\left(e^{t\overline{T}_n^a}\right) = \prod_{1 \leq p \leq n} \mathbb{E}\left(e^{ta_p\, (\mathcal{E}_p - 1)}\right)$$
$$= \exp\left(\sum_{1 \leq p \leq n} L_0(a_p t)\right) \leq \exp\left(\sum_{1 \leq p \leq n} L_1(a_p t)\right)$$
$$\leq \exp\left(\sigma_n(a)^2\, t^2/(1 - t)\right)$$

with $\sigma_n(a)^2 := \mathrm{Var}(T_n^a) = |a|_{n,2}^2$. This implies that

$$\mathbb{P}\left(\overline{T}_n^a \geq x\right) = \mathbb{P}\left(e^{t\overline{T}_n^a} \geq e^{tx}\right)$$
$$\leq \exp\left(-tx + \sigma_n(a)^2\, \frac{t^2}{1-t}\right) = \exp\left(-\sigma_n(a)^2\left[\frac{tx}{\sigma_n(a)^2} - L_1(t)\right]\right).$$

Taking the infimum over $t \in [0, 1[$, we prove that

$$\mathbb{P}\left(T_n^a - \mathbb{E}(T_n^a) \geq x\right) \leq \exp\left(-\sigma_n(a)^2\, L_1^\star\left(\frac{x}{\sigma_n(a)^2}\right)\right)$$
$$= \exp\left(-\sigma_n(a)^2\left(\sqrt{1 + \frac{x}{\sigma_n(a)^2}} - 1\right)^2\right).$$

This ends the proof of (11.22). Now we turn to the proof of (11.23). We first notice that

$$\frac{y}{\sigma_n(a)^2} = L_1^\star\left(\frac{x}{\sigma_n(a)^2}\right) \iff x = \sigma_n(a)^2\, (L_1^\star)^{-1}\left(\frac{y}{\sigma_n(a)^2}\right)$$

and

$$\sigma_n(a)^2 \left(\frac{y}{\sigma_n(a)^2} + 2\sqrt{\frac{y}{\sigma_n(a)^2}} \right) = y + 2\sigma_n(a)\sqrt{y}.$$

Hence we conclude that

$$\mathbb{P}\left(A_n \geq m_n(a) + y + 2\sigma_n(a)\sqrt{y} \right) \leq e^{-y}.$$

This ends the proof of the theorem. ■

11.3.2   Finite state space models

It is also important to observe that for any matrix with positive entries $Q(x,y)$ indexed by a finite set $x, y \in S = \{1, \ldots, d\}$ with bounded mass $\lambda(x) := \sum_y Q(x,y) \leq \lambda < \infty$, we have the decompositions

$$\sum_{y \in S} [f(y) - f(x)] \, Q(x,y) \;=\; \lambda(x) \sum_{y \in S} [f(y) - f(x)] \, K(x,dy)$$

$$=\; \lambda \sum_{y \in S} [f(y) - f(x)] \, K'(x,dy)$$

with the stochastic matrices $K(x,y) := Q(x,y)/\sum_{y \in S} Q(x,y)$ and

$$K'(x,y) := \frac{\lambda(x)}{\lambda} \, K(x,y) + \left(1 - \frac{\lambda(x)}{\lambda} \right) \, 1_x(y).$$

Notice that the jump rate or the holding time of the state x is now given by the sum $\lambda(x) := Q(x,S) = \sum_{y \in S} Q(x,y) \leq \lambda$. In this context, $Q(x,y)$ may be called the local rate of the transition $x \rightsquigarrow y$.

These local rates induce another interpretation of the continuous time process in terms of "alarm type" jump times:

- When the process $X_{T_n} = x$ enters in some state x at some jump time T_n, all the states $y \in S$ (including $y = x$) start an exponential alarm $\mathcal{E}_{n,x,y} \sim \text{Exp}(Q(x,y))$.

- When the first alarm goes off

$$T_{n+1} := T_n + \inf_{y \in S} \mathcal{E}_{n,x,y} = T_n + \mathcal{E}_{n,x,Y_{n,x}}$$

  for some state $Y_{n,x}$, then the process X_t jumps to this state $X_{T_{n+1}} = Y_{n,x}$.

This algorithmic description coincides with the one discussed above since

$$\inf_y \mathcal{E}_{n,x,y} \stackrel{law}{=} \text{Exp}(\sum_y Q(x,y)) = \text{Exp}(\lambda(x)).$$

It remains to check that the state $Y_{n,x} = y$ is distributed according to $K(x,y)$. We check this claim using the decomposition

$$\mathbb{P}\left(\inf_{y \in S} \mathcal{E}_{n,x,y} = \mathcal{E}_{n,x,z} \; ; \; \inf_{y \in S} \mathcal{E}_{n,x,y} > t \right)$$

$$= \mathbb{P}\left(\forall y \in S \quad \mathcal{E}_{n,x,y} \geq \mathcal{E}_{n,x,z} > t \right)$$

$$= \int_t^\infty Q(x,z) e^{-\lambda(z)s} \underbrace{\left[\int_s^\infty \cdots \int_s^\infty \prod_{y \neq x} Q(x,y) e^{-Q(x,y)s_y} \, ds_y \right]}_{= \mathbb{P}(\forall y \in S - \{x\} \quad \mathcal{E}_{n,x,y} \geq s)} \, ds. \qquad (11.24)$$

Recalling that

$$
\mathbb{P}\left(\forall y \in S - \{x\} \quad \mathcal{E}_{n,x,y} \geq s\right) = \prod_{y \in S - \{x\}} \mathbb{P}\left(\mathcal{E}_{n,x,y} \geq s\right)
$$

$$
= \prod_{y \in S - \{x\}} e^{-Q(x,y)s} = e^{-\sum_{y \in S - \{x\}} Q(x,y)s}
$$

we conclude that

$$
(11.24) \quad = \quad Q(x,z) \int_t^\infty e^{-\sum_y Q(x,y)s}\, ds = \frac{Q(x,z)}{\sum_y Q(x,y)} \times \mathbb{P}\left(\inf_{y \in S} \mathcal{E}_{n,x,y} > t\right).
$$

We see that $Y_{n,x}$ does not depend on the first alarm time, and it is distributed with the desired Markov transition

$$
\mathbb{P}\left(\inf_{y \in S} \mathcal{E}_{n,x,y} = \mathcal{E}_{n,x,z} \mid \inf_{y \in S} \mathcal{E}_{n,x,y} > t\right) = \frac{Q(x,z)}{\sum_y Q(x,y)} = K(x,z).
$$

Important remark : The same analysis works for countable state space models, but obviously it is impossible to store a countable number of exponential random clocks.

11.4 Time inhomogeneous models

11.4.1 Description of the models

Definition 11.4.1 *We let T_n be the random jump times (10.8) of a time inhomogeneous Poisson process N_t with intensity λ_t. We consider a Markov chain Y_n on some general state space S with a Markov transition $K(x, dy)$, and we let X_t be the Markov process indexed by the continuous time parameter $t \in \mathbb{R}_+$ and defined by*

$$
X_t = \sum_{n \geq 0} 1_{[T_n, T_{n+1}[}(t)\, Y_n = Y_{N_t}. \tag{11.25}
$$

The Markov process X_t is the continuous time embedding of the Markov chain Y_n associated with the intensity function λ_t. Inversely, the Markov chain Y_n is the embedded Markov chain (abbreviated EMC) of the continuous time process X_t.

Important remark : In section 10.6, specifically dedicated to spatial Poisson point processes, we discuss effective simulation techniques of these time inhomogeneous models for bounded and continuous intensity functions λ_t (cf. proposition 11.4.4).

Definition 11.4.2 *The semigroup of the Markov process X_t is defined for any $s \leq t$ and any $f \in \mathcal{B}(S)$ by the conditional expectations*

$$
P_{s,t}(f)(x) = \mathbb{E}(f(X_t) \mid X_s = x).
$$

We also denote by η_t the distribution of the random states X_t.

Theorem 11.4.3 *The sg of X_t satisfies the integral equation*

$$P_{s,t} = e^{-\int_s^t \lambda_r \, dr} \, Id \, + \int_s^t \lambda_r \, e^{-\int_s^r \lambda_u \, du} \, KP_{r,t} \, dr. \qquad (11.26)$$

In addition, we have

$$\partial_t P_{s,t} \;=\; P_{s,t} L_t \qquad \partial_s P_{s,t} = -L_s P_{s,t} \quad and \quad \partial_t \eta_t = \eta_t L_t \qquad (11.27)$$

with the initial condition $P_{s,s} = Id$, for $s = t$, and the integral operator L_t defined for any $f \in \mathcal{B}(S)$ by

$$
\begin{aligned}
L_t(f)(x) \;:=\;& \lambda_t \int [f(y) - f(x)] \, K(x, dy) \\
\;=\;& \lim_{s \to t} \frac{1}{t-s} [P_{s,t}(f)(x) - f(x)]. \qquad (11.28)
\end{aligned}
$$

Proof :

The proof follows essentially the same lines of arguments as the proof of theorem 11.1.2; thus, it is just sketched. For any $s \in \mathbb{R}_+$, we let $T_{s,1}$ be the random time defined by

$$T_{s,1} = \inf \left\{ t \geq s \; : \; \int_s^t \lambda_s \, ds \geq \mathcal{E} \right\}.$$

We have

$$
\begin{aligned}
P_{s,t}(f)(x) \;=\;& \mathbb{E}(f(X_t) \mid X_s = x) \\
\;=\;& \mathbb{E}(f(X_t) \, 1_{T_{s,1} > t} \mid X_s = x) + \mathbb{E}(f(X_t) \, 1_{T_{s,1} \leq t} \mid X_s = x) \\
\;=\;& f(x) \, e^{-\int_s^t \lambda_r dr} + \mathbb{E}(\mathbb{E}\left(f(X_t) \mid T_{s,1}, \, X_{T_{s,1}}\right) 1_{T_{s,1} \leq t} \mid X_0 = x).
\end{aligned}
$$

In this situation, we have

$$g_{s,t}(r, y) \;:=\; \mathbb{E}\left(f(X_t) \mid T_{s,1} = r, \, X_r = y\right) = P_{r,t}(f)(y)$$

so that

$$\overbrace{\mathbb{E}(\mathbb{E}\left(f(X_t) \mid T_{s,1}, \, X_{T_{s,1}}\right)}^{=g_{s,t}(T_{s,1}, X_{T_{1,s}})} 1_{T_{s,1} \leq t} \mid X_0 = x)$$

$$= \int K(x, dy) \int_s^t P_{r,t}(f)(y) \, \lambda_r \, e^{-\int_s^r \lambda_u \, du} \, dr.$$

This ends the proof of the first assertion.

On the other hand, we have

$$\partial_s P_{s,t}(f)(x) \;=\; \lambda_s \, (P_{s,t}(f)(x) - KP_{s,t}(f)(x)) = -L_s(P_{s,t}(f))(x).$$

This implies that

$$-\frac{1}{t-s} [P_{s,t}(f)(x) - f(x)] \;=\; \frac{1}{t-s} \int_s^t \partial_r P_{r,t}(f)(x) dr = -\frac{1}{t-s} \int_s^t L_r(P_{r,t}(f))(x) \, dr$$

from which we conclude that

$$\lim_{s \to t} \frac{1}{t-s} \left[P_{s,t}(f)(x) - f(x) \right] = L_t(f)(x).$$

Using the decomposition

$$\frac{P_{s,t+\epsilon}(f)(x) - P_{s,t}(f)(x)}{\epsilon} = P_{s,t} \left[\frac{P_{t,t+\epsilon}(f) - f}{\epsilon} \right](x)$$

we also check that

$$\partial_t P_{s,t}(f)(x) = P_{s,t}(L_t(f))(x).$$

The proof of the last assertion follows the one for (11.7). This ends the proof of the theorem. ∎

Important remark : We fix a time horizon t and we set

$$u_s(x) := P_{s,t}(f)(x)$$

for any $s \le t$. Then using (11.27) we find that $u_s(x)$ satisfies the integro-differential equation

$$\partial_s u_s(x) + \lambda_s \int \left[u_s(y) - u_s(x) \right] K(x, dy) = 0$$

with the terminal condition $u_t(x) = f(x)$.

Important remark : In terms of integral operators, the limit (11.28) is sometimes rewritten in the following form

$$\lim_{s \to t} \frac{1}{t-s} \left[P_{s,t} - Id \right] = L_t := \lambda_t \, (K - Id). \tag{11.29}$$

11.4.2 Poisson thinning models

We associate with the Poisson point process T'_n defined on page 309 the continuous time Markov process X'_n defined as follows:

At the jump times $T'_n = t$, with a probability λ_t/λ the process $X'_t = x$ jumps to a random state $x \rightsquigarrow Y \sim K(x, dy)$; otherwise it stays in the same location. Arguing as in the proof of theorem 11.4.3, we can check that its semigroup $P'_{s,t}$ satisfies the equation

$$P'_{s,t} = e^{-\lambda(t-s)} \, Id + \int_s^t \lambda \, e^{-\lambda(r-s)} \, K_r P_{r,t} \, dr$$

with the Markov transitions

$$K_t(x, dy) := \frac{\lambda_t}{\lambda} \, K(x, dy) + \left(1 - \frac{\lambda_t}{\lambda} \right) \delta_x(dy).$$

In addition, the laws of the random states $\eta'_t = \text{Law}(X'_t)$, and the sg $P'_{s,t}$ satisfy the forward and backward equations (11.27) with the generator

$$\begin{aligned} L_t(f)(x) &:= \lambda \int [f(y) - f(x)] \left(\frac{\lambda_t}{\lambda} \, K(x, dy) + \left(1 - \frac{\lambda_t}{\lambda} \right) \delta_x(dy) \right) \\ &= \lambda_t \int [f(y) - f(x)] \, K(x, dy). \end{aligned}$$

The following proposition is now a direct consequence of this thinning property of the Poisson processes discussed above.

Proposition 11.4.4 *The Markov process X'_t coincides with the Markov process X_t defined in (11.25).*

Important remark : We emphasize that this (exact) sampling technique is based on the Poisson thinning properties discussed on page 309. Following the important remark we made in section 10.6.2, these sampling strategies can be really inefficient when $\frac{1}{t}\int_0^t (\lambda_s/\lambda)$ tends to 0, as $t \uparrow \infty$ (cf. exercise 166).

11.4.3   Exponential and geometric clocks

We let $h > 0$ be some time mesh parameter associated with some discretization of some interval $[0, t]$

$$t_0 = 0 < t_1 = h < \ldots < t_n = nh < \ldots < h\lfloor t/h \rfloor \leq t. \tag{11.30}$$

We let X_t^h be the Markov process

$$X_t^h = \sum_{n \geq 0} 1_{[T_n^h, T_{n+1}^h[}(t)\ Y_n$$

defined as in (11.25) by replacing the random jump times T_n defined in (10.8) by the discrete time approximation times T_n^h defined in (10.12). We recall that the random jump times T_n^h can be seen as the jump times of a time inhomogeneous Bernoulli counting process (10.13), that is,

$$T_{n+1}^h = \inf\left\{ t_k > T_n^h\ :\ \epsilon_{t_k}^h = 1 \right\}$$

where $\epsilon_{t_n}^h$ is a sequence of independent Bernoulli r.v. with a common law

$$\mathbb{P}\left(\epsilon_{t_n}^h = 0\right) = 1 - \mathbb{P}\left(\epsilon_{t_n}^h = 1\right) = e^{-\lambda_{t_n} h}.$$

This shows that $X_{t_n}^h$ is an S-valued Markov chain with transition probabilities

$$
\begin{aligned}
\mathbb{P}\left(X_{t_{n+1}}^h \in dy \mid X_{t_n}^h = x\right) &= P_{t_n, t_{n+1}}^h(x, dy) \\
&= e^{-\lambda_{t_n} h}\, \delta_x(dy) + \left(1 - e^{-\lambda_{t_n} h}\right) K(x, dy).
\end{aligned}
$$

In terms of integral operators, we have the equivalent synthetic formulation

$$P_{t_n, t_{n+1}}^h = e^{-\lambda_{t_n} h}\, Id + \left(1 - e^{-\lambda_{t_n} h}\right)\ K.$$

We observe that for any time t, we have

$$\frac{1}{h}\left[P_{t, t+h}^h - Id\right] = \frac{1}{h}\left(1 - e^{-\lambda_t h}\right)\ [K - Id] \to_{h \to 0} L_t = \lambda_t\ (K - Id).$$

This equation provides another interpretation of the formula (11.29). The convergence of these discrete time models to the limiting continuous time processes is discussed in chapter 12, dedicated to general pure jump models (compare theorem 12.3.1).

11.5   Exercises

Exercise 155 (Feynman-Kac measures - Embedded Markov chains) *Let X_t be a Markov process on some state space S with infinitesimal generator L given by*

$$L(f)(x) = \lambda\ \int\ (f(y) - f(x))\ K(x, dy)$$

for some parameter $\lambda > 0$ and some Markov transition $K(x, dy)$ on S. Let V : $x \in S \mapsto V(x) \in S$ be some bounded potential function. We let γ_t be the Feynman-Kac measure on S defined for any bounded function f on S by

$$\gamma_t(f) = \mathbb{E}\left(f(X_t) \, \exp\left(\int_0^t V(X_s) \, ds \right) \right).$$

We let $(T_n)_{n \geq 0}$ be the jump times of a Poisson process N_t with intensity λ. For any $n \geq 0$, check that

$$\mathbb{E}\left(f(X_{T_n}) \, \exp\left(\int_0^{T_n} V(X_s) \, ds \right) \mid (T_0, \dots, T_n) \right)$$

$$= \mathbb{E}\left(f(Y_n) \prod_{0 \leq k < n} e^{(T_{k+1} - T_k)V(Y_k)} \mid (T_0, \dots, T_n) \right)$$

as well as

$$\mathbb{E}\left(f(X_{T_n}) \, \exp\left(\int_0^{T_n} V(X_s) \, ds \right) \right) = \mathbb{E}\left(f(Y_n) \prod_{0 \leq k < n} e^{E_k V(Y_k)} \right)$$

where $Y = (Y_n)_{n \geq 0}$ is a Markov chain with transition probability K starting at $Y_0 = X_0$, and $E = (E_n)_{n \geq 0}$ is a sequence of independent exponential random variables with parameter λ; we assume that Y and E are independent. Prove that

$$\mathbb{E}\left(f(X_t) \, \exp\left(\int_0^t V(X_s) \, ds \right) \mid N_t = n \right)$$

$$= \mathbb{E}\left(f(Y_n) \exp\left\{ \left(t - \sum_{0 \leq k < N_t} E_k \right) V(Y_n) \right\} \prod_{0 \leq k < N_t} e^{E_k V(Y_k)} \right)$$

with a Poisson random variable N_t with parameter (λt) independent of the sequence E.

Exercise 156 (A soliton-like jump process [154]) *We let $q(u)$ be a probability density of a random variable U on the positive axis $[0, \infty[$. We consider the solution p_t : $x \in \mathbb{R} \mapsto p_t(x) \in \mathbb{R}$ of the evolution equation*

$$\partial_t p_t(x) = -p_t(x) + \int_{-\infty}^x q(x - y) \, p_t(y) \, dy$$

with some initial condition $p_0(x)$ given by some probability density on $\mathbb{R}$. We set $\eta_t(dx) = p_t(x)dx$. Check that

$$\partial_t \eta_t(f) = \eta_t(L(f))$$

with the generator

$$L = K - Id \quad (with \quad K(x, dy) = 1_{[x, \infty[}(y) \, q(y - x) \, dy)$$

of a jump process X_t. Describe the embedded Markov chain Y_n. We consider (whenever they exist) the Laplace transforms $\phi_t(\lambda) := \mathbb{E}(e^{\lambda X_t})$ and $\varphi(\lambda) := \mathbb{E}(e^{\lambda U})$. Find an explicit expression of ϕ_t in terms of φ. When U is an exponential random variable with some parameter $\alpha > 0$, describe the law of Y_n and compute the Markov transitions of X_t.

Exercise 157 (Telegraph process) *We consider a Markov process X_t taking values in $S = \{1, 2\}$. It switches from 1 to 2 at rate $\lambda(1) > 0$ and from 2 to 1 at rate $\lambda(2) > 0$. Describe the infinitesimal generator of the process X_t and compute the probability distributions $\eta(i) = \mathbb{P}(X_t = i)$, for any $i \in S$. Furthermore, we let π be the probability distribution on S defined by $\pi(1) = 1 - \pi(2) = \frac{\lambda(2)}{\lambda(1)+\lambda(2)}$, with $\lambda := \lambda(1) + \lambda(2)$. Check that*

$$\sup_{i \in S} |\eta_t(i) - \pi(i)| = \lambda^{-1} \exp(-\lambda t) \, |\lambda(1)\eta_0(1) - \lambda(2)(1 - \eta_0(1))| \,.$$

Exercise 158 (Birth and death process) *We consider a Markov process X_t taking values in $S = \mathbb{N}$ with infinitesimal generator L defined for any function f on $\mathbb{N}$ and any $x \in \mathbb{N} - \{0\}$ by*

$$L(f)(x) = \lambda_+(x) \, (f(x+1) - f(x)) + \lambda_-(x) \, (f(x-1) - f(x))$$

for some functions $\lambda_+(x), \lambda_-(x) \geq 0$ s.t. $\lambda_-(0) = 0$ and $\sum_{x \geq 0} \prod_{0 \leq y < x} \frac{\lambda_+(y)}{\lambda_-(y+1)} < \infty$. Find a probability distribution π on S such that $\pi L = 0$.

Exercise 159 (Poisson Process) *The formula (10.2) shows that N_t can be interpreted as a continuous time embedding of the $\mathbb{N}$-valued Markov chain X_n with the one-step transitions*

$$K(n, m) = 1_{n+1}(m).$$

Describe the generator L of the process N_t. We let $\eta_t = \mathrm{Law}(N_t)$. Using the evolution equation of η_t in terms of the generator L, check that $\eta_t(n) = \frac{(\lambda t)^n}{n!} \, e^{-\lambda t}$.

Exercise 160 (Poisson Processes - Birth and death process) *Let N_t and N' be two independent Poisson processes with respective intensity λ and λ'. Describe the generator of the process $X_t = N_t - N'_t$. Check that $X_t = Y_{N''_t}$ where N''_t is a Poisson process with intensity $\lambda'' = (\lambda + \lambda')$ and Y_n a random walk with Markov transitions*

$$\mathbb{P}(Y_{n+1} = Y_n + 1 \mid Y_n) = \frac{\lambda}{\lambda + \lambda'} = 1 - \mathbb{P}(Y_{n+1} = Y_n - 1 \mid Y_n).$$

We assume that $N_0 = N'_0$ and we let $\eta_t(x) = \mathbb{P}(X_t = x)$, with $x \in \mathbb{Z}$. Check that

$$\partial_t \eta_t(x) = \lambda \, (\eta_t(x-1) - \eta_t(x)) + \lambda' \, (\eta_t(x+1) - \eta_t(x)).$$

We let $g_t(z) = \mathbb{E}(z^{X_t})$ be the moment generating function of X_t. Check that

$$g_t(z) = e^{-(\lambda+\lambda')t} \, e^{(\lambda z + \lambda' z^{-1})t}.$$

Exercise 161 (Compound Poisson process) *We let $\lambda > 0$ be a parameter and Y be a random variable on $\mathbb{R}$ with some probability measure $\mu(dy)$, s.t. $\mathbb{E}(Y^2) < \infty$. A compound Poisson process is defined by*

$$X_t = \sum_{1 \leq n \leq N_t} Y_n$$

where N_t is a Poisson process with intensity λ and Y_n is a sequence of independent copies of Y, independent of N_t. Find the infinitesimal generator of X_t. Next, assume that Y_n are centered Gaussian random variables. Describe the cumulative function $\mathbb{P}(X_t \leq x)$ in terms of the cumulative function $\mathbb{P}(Y \leq y)$.

Exercise 162 (Yule process) ✎

Consider a branching process in which each individual splits independently into two offsprings after independent exponentially distributed clocks with common parameter $\lambda > 0$. We let X_t be the number of individuals in a branching process starting with one individual, and T_1 be the first time the first individual splits, and we let T_n represent the time interval between the $(n-1)$-th split and the n-th one.

- Describe the probability distribution of the sequence of variables $(T_n)_{n \geq 1}$.

- Describe the probability distribution of the n-th splitting time $\overline{T}_n = \sum_{1 \leq k \leq n} T_k$.

- Describe the probability distribution of X_t in terms of $\overline{T}_n$, deduce $\mathbb{E}(X_t)$ and

$$\forall n \geq 1 \qquad \mathbb{P}(X_t = n) = e^{-\lambda t} \times \left(1 - e^{-\lambda t}\right)^{n-1}.$$

Exercise 163 (Poisson martingales) ✎

We let $a \in \mathbb{R}$, $b > -1$, $\lambda > 0$ be some parameters, N_t be a Poisson process with intensity λ, and $\mathcal{F}_t = \sigma(N_s, \ s \leq t)$. Check that the following stochastic processes are martingales w.r.t. the sigma-fields $\mathcal{F}_t$.

$$
\begin{aligned}
M_t^{(1)} &:= N_t - \lambda t & M_t^{(2)} &:= \left(M_t^{(1)}\right)^2 - \lambda t \\
M_t^{(3)} &:= \exp\left(a N_t - \lambda t \left(e^a - 1\right)\right) & M_t^{(4)} &:= (1+b)^{N_t} \, e^{-\lambda b t}.
\end{aligned}
$$

Exercise 164 (First-in-first-out queueing process)

We consider a single one line server queue with Poisson arrival rate λ_1. The service times for each customer in the order of arrival are represented by i.i.d. exponential random variables S_n with parameter λ_2. We assume that these service times S_n are independent of the Poisson arrival process. We let X_t be the number of customers in the system (i.e., in the arrival queue and in the service area). Describe the evolution of the stochastic process X_t in terms of a Markov chain embedding.

Exercise 165 (Multi-server queueing processes)

We return to the exercise *164* and we assume that we have a ≥ 1 servers working in parallel (and a single arrival queue). Describe the evolution of the stochastic process X_t in terms of a Markov chain embedding.

Exercise 166 (Poisson thinning)

We return to the Poisson thinning techniques presented on page *309*. We consider a sequence of independent random variables U_n on $[0, \lambda]$, and we let A_t be the first integer k such that

$$A_t = \inf\left\{k \geq 1 \ : \ U_k \leq \lambda_{T'_k}\right\}.$$

Check that A_t is a geometric random variable with success probability given by the area ratio $\int_0^t \lambda_s \, ds / (\lambda t)$. When $\lambda_t = e^{-t}$, check that for any choice of $\lambda \geq \lambda_t$, we have $\lim_{t \to \infty} \mathbb{E}(A_t) = \infty$.

Exercise 167 (Explosion times)

We consider a jump process $(X_t)_{t \in [0,1[}$ with generator

$$\forall t \in [0, 1[\qquad L_t := \lambda_t \, (K - Id) \quad \text{with the intensity} \quad \lambda_t = 1/(1-t)$$

for some Markov transitions $K(x, dy)$ on $\mathbb{R}$. Check that the jump times T_n of X_t are defined by the recursion

$$\forall n \geq 1 \qquad T_n = (1 - U_n) + U_n \, T_{n-1} \in [0, 1]$$

with $T_0 = 0$, and a sequence U_n of independent copies of some uniform random variable U on $[0, 1[$. Deduce that

$$\forall p \geq 0 \qquad \mathbb{E}\left((1 - T_n)^p\right) = (p+1)^{-n} \quad \text{and} \quad T_n \to_{n \to \infty} T_\infty = 1.$$

The exercise *208* analyses non explosive and time homogeneous regenerative jump processes associated with the Markov transition $K(x, dy) = \delta_0(dy)$.

Exercise 168 (Explosion time concentration inequalities) ⚡

We consider the random times defined in *(11.19)*, for some $\lambda_n \geq 1$. Using theorem *11.3.7*, check that

$$\mathbb{P}\left(T_n \leq \sum_{1 \leq p \leq n} \lambda_p^{-1} + t + 2\sqrt{\sum_{1 \leq p \leq n} \lambda_p^{-2}} \sqrt{t}\right) \geq 1 - e^{-t}.$$

- In the explosive case (i.e. $\sum_{p \geq 1} \lambda_p^{-1} < \infty$), check that

$$\mathbb{P}\left(T_n \leq |\lambda|_1 + t + 2|\lambda|_2 \sqrt{t}\right) \geq 1 - e^{-t}$$

for any $t \geq 0$, with

$$|\lambda|_2^2 := \sum_{p \geq 1} \lambda_p^{-2} \leq \mathbb{E}(T_\infty) := |\lambda|_1 := \sum_{p \geq 1} \lambda_p^{-1} < \infty.$$

- Deduce from the above that

$$\mathbb{P}\left(T_\infty \leq t + \mathbb{E}(T_\infty)(1 + 2\sqrt{t})\right) \geq 1 - e^{-t} \qquad (11.31)$$

for any $t \geq 0$. Show that $T_\infty \leq 10.35$ with a probability larger than 95%. Compare this estimate with the one obtained using Markov's inequality

$$\mathbb{P}(T_\infty \geq t) \leq t^{-1} \, \mathbb{E}(T_\infty).$$

- We consider the non-explosive case (i.e., $\sum_{p \geq 1} \lambda_p^{-1} = \infty$). When $\lambda_n = 1$ check that

$$\mathbb{P}\left(T_n \leq n + t + 2\sqrt{nt}\right) = \mathbb{P}\left(\sqrt{T_n} \leq \sqrt{n} + \sqrt{t}\right) \geq 1 - e^{-t}$$

for any $t \geq 0$. Prove that the seventh jump time occurs before 20 units of time, with a probability 95%; and the 10^3-th jump occurs before $1.1113 \; 10^3$ units of time, with a probability larger than 95%.

12

Jump processes

Pure jump processes generalize Poisson processes with jumps. The semigroup evolutions for these pure jump processes have analytic form and possess simple discrete time approximations. We formulate basic statements about the error of these approximations. Furthermore, we study in details the Doob-Meyer decompositions for both discrete and continuous time models, and present the main optional stopping theorems. These results are used to formulate the all important Doeblin-Itō formula (a.k.a. Itō lemma) for smooth transformations of general continuous time pure jump models. At the end of the chapter, we investigate the stability properties of the time homogeneous jump processes.

> *Jump, and you will find out how to unfold your wings as you fall.*
> Ray Bradbury (1920-2012).

12.1 A class of pure jump models

We consider a collection of Markov transitions $K_t(x, dy)$ on a state space S, and a collection of bounded intensity functions $\lambda_t(x)$, indexed by $t \in \mathbb{R}_+$. We also assume that the mappings $t \mapsto K_t(f)(x)$ and $t \mapsto \lambda_t(x)$ are Lipschitz continuous for any x and for any function $f \in \mathcal{B}(S)$.

We associate with these objects the càdlàg Markov process X_t defined sequentially as follows. We let $\mathcal{E}_n = -\log U_n$ be a sequence of independent exponential random variables with unit parameters ($\Rightarrow U_n$ uniform on $[0, 1]$).

At the origin, the process starts at some initial state $X_0 = x_0$, and we set $T_0 = 0$.

1. We assume that the process is defined on the interval $[0, T_n]$. After the time T_n the process remains constant

$$\forall T_n \leq s < T_{n+1} \qquad X_s := X_{T_n} \qquad (12.1)$$

up to the random jump time T_{n+1} defined by

$$T_{n+1} = \inf \left\{ t \geq T_n \ : \ \int_{T_n}^t \lambda_s (X_s) \ ds \geq \mathcal{E}_{n+1} \right\}.$$

2. At the time T_{n+1}, the jump of the process $X_{T_{n+1}-} \rightsquigarrow X_{T_{n+1}}$ is defined by choosing a random state with the Markov transition at that time,

$$X_{T_{n+1}} \sim K_{T_{n+1}}\left(X_{T_{n+1}-}, dx\right).$$

When $K_t(x, dy) = \delta_{x+1}(dy)$ only positive and unit jumps occur. The corresponding X_t reduces to a Poisson process with jump intensity $\lambda_t(X_t)$.

We also have the telescoping sum

$$X_t \;=\; X_0 + \sum_{0 < T_n \le t} (X_{T_n} - X_{T_n-}) \qquad (12.2)$$

with jumps $X_{T_n} - X_{T_n-}$ defined in terms of a random variables X_{T_n} with conditional distribution $K_{T_n}(X_{T_n-}, dx)$ given the value of the process X_{T_n-} at the jump time T_n.

When the intensity function $\lambda_t(x) \le \lambda$ is upper bounded by some parameter λ, using the Poisson thinning techniques, we can replace $\lambda_t(x)$ by the parameter λ in the above description. Likewise, we can replace the Markov transitions K_t by

$$K_t^\lambda(x, dy) := \frac{\lambda_t(x)}{\lambda}\, K_t(x, dy) + \left(1 - \frac{\lambda_t(x)}{\lambda}\right) \delta_x(dy).$$

Important remark : This reformation of the process is based on the Poisson thinning properties discussed on page 309. As we mention in section 10.6.2, in practical situations these sampling strategies can be really inefficient (compare with exercise 166).

12.2   Semigroup evolution equations

By construction, for any $s \ge t$ we have

$$\mathbb{P}\left(T_{n+1} \in dt \,, X_{T_{n+1}} \in dy \mid T_n = s,\ X_{T_n} = x\right)$$

$$= \underbrace{\lambda_t(x)\ \exp\left(-\int_s^t \lambda_r(x)\ dr\right)\ dt}_{\mathbb{P}(T_{n+1} \in dt \ \mid\ T_n = s, X_{T_n} = x)} \quad \underbrace{K_t(x, dy)}_{\mathbb{P}(X_{T_{n+1}} \in dy \ \mid\ T_{n+1} = t,\ X_{T_n} = x).}$$

In the same vein, if $T^{(s)}$ is the first jump time after time s, we have

$$\mathbb{P}\left(T^{(s)} \in dt \,, X_{T^{(s)}} \in dy \mid X_s = x\right)$$

$$= \underbrace{\lambda_t(x)\ \exp\left(-\int_s^t \lambda_r(x)\ dr\right)\ 1_{s \le t}\ dt}_{\mathbb{P}(T^{(s)} \in dt \ \mid\ X_s = x)} \times \quad \underbrace{K_t(x, dy)}_{\mathbb{P}(X_{T^{(s)}} \in dy \ \mid\ T^{(s)} = t,\ X_s = x).}$$

This yields the formula

$$\mathbb{P}\left(T^{(t)} \in dt \,, X_{t+dt} \in dy \mid X_t = x\right) = \lambda_t(x) \, dt \, K_t(x, dy). \qquad (12.3)$$

Last but not least, we also observe that for any $t \geq s$ we have

$$\mathbb{P}\left(T^{(s)} > t \mid X_s = x\right) = \exp\left(-\int_s^t \lambda_r(x) \, dr\right) 1_{s \leq t}. \qquad (12.4)$$

We check this claim by using the fact that

$$\mathbb{P}\left(T^{(s)} > t \mid X_s = x\right) = \int_t^\infty \mathbb{P}\left(T^{(s)} \in du \mid X_s = x\right) \, du$$

$$= \int_t^\infty \lambda_u(x) \, \exp\left(-\int_s^u \lambda_r(x) \, dr\right) \, du = -\int_t^\infty \frac{\partial}{\partial u} \, \exp\left(-\int_s^u \lambda_r(x) \, dr\right) \, du.$$

We let $P_{s,t}$ and η_t be the semigroup and and the law of random states of the process X_t introduced in definition 11.4.2.

Arguing as in the proof of theorem 11.4.3, we have the integral formula

$$P_{s,t}(f)(x)$$

$$= \exp\left[-\int_s^t \lambda_r(x) dr\right] f(x)$$

$$+ \int_s^t \lambda_r(x) \, \exp\left[-\int_s^r \lambda_u(x) \, du\right] (K_r P_{r,t})(f)(x) \, dr.$$

In addition, we have

$$\partial_t P_{s,t} = P_{s,t} L_t \qquad \partial_s P_{s,t} = -L_s P_{s,t} \quad \text{and} \quad \partial_t \eta_t(f) = \eta_t L_t(f) \quad (12.5)$$

with the initial condition $P_{s,s} = Id$ for $s = t$, and with the integral operator L_t defined for any $f \in \mathcal{B}(S)$ by

$$L_t(f)(x) := \lambda_t(x) \int [f(y) - f(x)] \, K_t(x, dy)$$

$$= \lim_{s \to t} \frac{1}{t - s} [P_{s,t}(f)(x) - f(x)]. \qquad (12.6)$$

By (12.2), for any function f and any $s < t$ we also have the telescoping sum

$$f(X_t) \quad = \quad f(X_s) + \sum_{s < T_n \leq t} \left(f(X_{T_n}) - f(X_{T_n-}) \right). \qquad (12.7)$$

Roughly speaking this formula implies that

$$f(X_t) = f(X_0) + \int_0^t \left(f(X_{s+ds}) - f(X_s) \right).$$

Taking the expectations we have

$$(12.3) \Rightarrow \quad \mathbb{E}\left[\left(f(X_{s+ds}) - f(X_s) \right) \mid X_s \right] \quad = \quad \lambda_s(X_s) \, ds \int \left(f(y) - f(X_s) \right) K_s(X_s, dy)$$
$$= \quad L_s(f)(X_s) \, ds. \quad \Rightarrow (12.5)$$

12.3   Approximation schemes

This section is concerned with the discrete time approximation of these models.

Using the modeling techniques developed in section 11.4.3, a natural way to approximate these models on a discrete time mesh (11.30) is to consider the Markov chain

$$\mathbb{P}\left(X^h_{t_{n+1}} \in dy \mid X^h_{t_n} = x \right) \quad = \quad P^h_{t_n, t_{n+1}}(x, dy)$$
$$= \quad e^{-\lambda_{t_n}(x)h} \, \delta_x(dy) + \left(1 - e^{-\lambda_{t_n}(x)h} \right) K_{t_n}(x, dy). \qquad (12.8)$$

We have

$$P^h_{t_n, t_{n+1}} - Id \quad = \quad \left(1 - e^{-\lambda_{t_n} h} \right) \left[K_{t_n} - Id \right] = h \, L_{t_n} + \mathrm{O}(h^2).$$

On the other hand, using elementary Taylor first order expansion, we get

$$P_{s, s+h}(f) \quad = \quad P_{s,s}(f) + \partial_t P_{s,t}(f)_{|t=s} \, h + \mathrm{O}\left(h^2 \right) \qquad (12.9)$$
$$= \quad P_{s,s}(f) + P_{s,s}(L_s(f)) + \mathrm{O}\left(h^2 \right). \qquad (12.10)$$

Recalling that $P_{t,t} = Id$, we also have that

$$P_{t,t+h} = Id + h \, L_t + \mathrm{O}(h^2). \qquad (12.11)$$

Extending the Taylor expansion at the n-th order, we find that

$$P_{t,t+h} = Id + h \, L_t + \frac{h^2}{2!} \, L_t^2 + \ldots + \frac{h^n}{n!} \, L_t^n + \mathrm{O}(h^{n+1}) \qquad (12.12)$$

where L_t^n denotes the n-th iterate of the operator L_t.

It is out of the scope of these lectures to provide a detailed review on the convergence analysis of these discretized models to their limiting continuous time version. We simply quote the following general local perturbation theorem.

Theorem 12.3.1 *We let $P^h_{t_p,t_n}$ be the semigroup of a discrete time model defined at a time mesh (11.30), with $t_p \leq t_n$. We also consider the semigroup $P_{s,t}$ of a continuous time model, with $s \leq t$. We assume that for any $n \in \mathbb{N}$ we have*

$$P^h_{t_n,t_{n+1}} = Id + h\, L_{t_n} + O(h^{1+\epsilon}) = P_{t_n,t_{n+1}} \tag{12.13}$$

for some $\epsilon > 0$, and some infinitesimal generators L_t, with $t \in \mathbb{R}_+$. For any time steps $t_p = h\lfloor s/h \rfloor \leq t_n = h\lfloor t/h \rfloor$, we have the first order approximation formulae

$$P^h_{t_p,t_n} = P_{t_p,t_n} + O(h^\epsilon) \quad and \quad \eta^h_{t_n} = \eta_{t_n} + O(h^\epsilon)$$

as soon as $\eta^h_0 = \eta_0 + O(h^\epsilon)$.

Proof :

Under our assumptions, we have

$$\left[Id - P_{t_p,t_{p+1}} \right] = -L_{t_p}\, h + O\left(h^2 \right).$$

This implies that

$$\left[P^h_{t_p,t_{p+1}} - P_{t_p,t_{p+1}} \right] = \left[P^h_{t_p,t_{p+1}} - Id \right] - \left[P_{t_p,t_{p+1}} - Id \right] = O(h^2).$$

To take the final step, we consider the interpolating path

$$p \in [0,n] \mapsto P^h_{0,t_p} P_{t_p,t_n}$$

starting at P_{0,t_n} and ending at P^h_{0,t_n}, with the corresponding telescoping sum

$$P^h_{0,t_n} - P_{0,t_n} = \sum_{0 \leq t_p < t_n} \left[P^h_{0,t_{p+1}} P_{t_{p+1},t_n} - P^h_{0,t_p} P_{t_p,t_n} \right].$$

We observe that

$$P^h_{0,t_{p+1}} P_{t_{p+1},t_n} - P^h_{0,t_p} P_{t_p,t_n} = P^h_{0,t_p} \left[P^h_{t_p,t_{p+1}} - P_{t_p,t_{p+1}} \right] P_{t_{p+1},t_n} = O\left(h^{1+\epsilon} \right).$$

This clearly implies that

$$P^h_{0,t_n} - P_{0,t_n} = nh\, O\left(h^\epsilon \right) = t_n\, O\left(h^\epsilon \right) = O\left(h^\epsilon \right).$$

The same analysis applies to the interpolating path

$$q \in [p,n] \mapsto P^h_{t_p,t_q} P_{t_q,t_n}$$

starting at P_{t_p,t_n} and ending at $P^h_{t_p,t_n}$,. The last assertion is a direct consequence of the fact that $\eta_0 = \eta^h_0$ and of the semigroup property

$$\eta^h_{t_n} = \eta_0 P^h_{0,t_n} \quad and \quad \eta_{t_n} = \eta_0 P^h_{0,t_n}.$$

This ends the proof of the theorem. ∎

We extend the discrete generation process $X_{t_n}^h$ to a continuous time process $(X_t^h)_{t \geq 0}$ with càdlàg random trajectories by setting

$$\forall n \geq 0 \quad \forall t \in [t_n, t_{n+1}[\qquad X_t^h = X_{t_n}^h. \tag{12.14}$$

We denote by $P_{s,t}^h$ the corresponding semigroup defined for any $s \leq t$.

For any $s < t$ with $(t - s) > h$ we have

$$0 \leq t - h\lfloor t/h \rfloor < h \quad \Rightarrow \quad h\lfloor s/h \rfloor \leq s < h\lfloor t/h \rfloor \leq t$$
$$\Rightarrow \quad P_{h\lfloor s/h \rfloor, s}^h = Id = P_{h\lfloor t/h \rfloor, t}^h \Rightarrow P_{s,t}^h = P_{h\lfloor s/h \rfloor, h\lfloor t/h \rfloor}^h.$$

On the other hand, using (12.11), for any $s \leq t$ we readily check that

$$P_{h\lfloor s/h \rfloor, h\lfloor t/h \rfloor} = P_{s,t} + O(h).$$

We summarize the above discussion with the following direct corollary of theorem 12.3.1.

Corollary 12.3.2 *For any $s_1 < s_2$ with $(s_2 - s_1) > h$ we have*

$$P_{s_1, s_2}^h = P_{s_1, s_2} + O(h).$$

More generally, for any finite sequence of times $s_1 < \ldots < s_n$ with a sufficiently small h we have the finite dimensional approximation

$$\prod_{1 \leq i < n} P_{s_i, s_{i+1}}^h(x_i, dx_{i+1}) = \prod_{1 \leq i < n} P_{s_i, s_{i+1}}(x_i, dx_{i+1}) + O(h).$$

We end this section with a series of important observations.

Firstly, we mention that the approximation theorem 12.3.1 is not restricted to discrete time approximation schemes. It also applies to analyze the convergence of sg $P_{t_n, t_{n+1}}^h$ satisfying condition (12.13) on any time mesh sequence.

Secondly, the reader should be convinced that the theorem can be extended to approximations of any order. We refer to exercice 190 for high order discrete time schemes for pure jump models.

12.4   Sum of generators

In various situations, a Markov process is often introduced as a stochastic process with generator of the form

$$L_t(f)(x) := \int [f(y) - f(x)] \, Q_t(x, dy)$$

for some positive integral operator Q_t. Assuming that $\lambda_t(x) := Q_t(x, S) > 0$ this generator has exactly the same form as the one discussed in (12.6) with $K_t(x, dy) = Q_t(x, dy)/Q_t(x, S)$. Thus, we can apply the discrete time approximation techniques developed in this section to sample the process X_t.

In some important practical situations, the operator Q_t is often given by the sum of positive operators $Q_t^{(i)}$, indexed by some finite set I. Here again, we have the equivalent formulations

$$
\begin{aligned}
L_t(f)(x) &:= \sum_{i \in I} \int [f(y) - f(x)] \, Q_t^{(i)}(x, dy) \\
&= \sum_{i \in I} \lambda_t(i, x) \int [f(y) - f(x)] \, K_t^{(i)}(x, dy) \qquad (12.15) \\
&= \lambda(x) \int [f(y) - f(x)] \, K_t(x, dy)
\end{aligned}
$$

with the intensity functions

$$
\lambda_t(i, x) = Q_t^{(i)}(x, S) > 0 \quad \text{and} \quad \lambda(x) = \sum_{i \in I} \lambda_t(i, x),
$$

with the Markov transitions

$$
K_t^{(i)}(x, dy) = Q_t^{(i)}(x, dy) / Q_t^{(i)}(x, S)
$$

and with

$$
K_t(x, dy) = \sum_{i \in I} \frac{\lambda_t(i, x)}{\sum_{j \in I} \lambda_t(j, x)} \, K_t^{(i)}(x, dy).
$$

The final step consists in noting that the decomposition formula (12.15) gives a definition of L_t in terms of the sum of generators

$$
L_t = \sum_{i \in I} L_t^{(i)} \quad \text{with} \quad L_t^{(i)}(f)(x) := \lambda_t(i, x) \int [f(y) - f(x)] \, K_t^{(i)}(x, dy).
$$

Proposition 12.4.1 *For any collection of discrete time approximation sg $P_{t_n, t_{n+1}}^{(h,i)}$ satisfying the condition (12.13)*

$$
P_{t_n, t_{n+1}}^{(h,i)} = Id + h \, L_{t_n}^{(i)} + \mathrm{O}(h^2),
$$

the composition sg defined below also satisfies (12.13):

$$
P_{t_n, t_{n+1}}^h = \prod_{i \in I} P_{t_n, t_{n+1}}^{(h,i)} \quad \underline{\textit{in any order.}} \qquad (12.16)
$$

Proof :
We check this claim using the fact that

$$
\prod_{i \in I} P_{t_n, t_{n+1}}^{(h,i)} = \prod_{i \in I} \left(Id + h \, L_{t_n}^{(i)} + \mathrm{O}(h^2) \right) = Id + h \sum_{i \in I} L_{t_n}^{(i)} + \mathrm{O}(h^2).
$$

This ends the proof of the proposition. ∎

12.5   Doob-Meyer decompositions

12.5.1   Discrete time models

In the further development of this section, $X_{t_n}^h$ stands for the discrete time model with transitions (12.8).

Definition 12.5.1 *We let $\mathcal{F}_{t_n}^h = \sigma\left(X_{t_p}^h, \ 0 \leq t_p \leq t_n\right)$ be the increasing sequence of σ-fields generated by the Markov chain $X_{t_n}^h$.*

$$\mathcal{F}_{t_0}^h \subset \mathcal{F}_{t_1}^h \subset \ldots \subset \mathcal{F}_{t_n}^h \subset \mathcal{F}_{t_{n+1}}^h \subset \ldots$$

For any function $f \in \mathcal{B}(S)$, we let $\Delta f(X^h)_{t_n}$ be the increments given for any $n \geq 0$ by

$$\Delta f(X^h)_{t_n} := f(X_{t_n}^h) - f(X_{t_{n-1}}^h).$$

For $n = 0$, we use the convention $\Delta f(X^h)_{t_0} := f(X_{t_0}^h)$.

Definition 12.5.2 *We let $L_{t_{n-1}}^h$ be the integral operators*

$$L_{t_{n-1}}^h = \frac{1}{h}\left[P_{t_{n-1}, t_n}^h - Id\right] = L_{t_{n-1}} + \mathrm{O}(h).$$

Definition 12.5.3 *The square field vector (also called the "carré du champ operator") associated with the operators $L = L_{t_{n-1}}^h$ or $L = L_{t_{n-1}}$ is*

$$\begin{aligned}
\Gamma_L(f,g)(x) &= L((f - f(x))(g - g(x)))(x) \\
&= L(fg)(x) - f(x)L(g)(x) - g(x)L(f)(x). \quad (12.17)
\end{aligned}$$

Theorem 12.5.4 *For any $t_n > t_0$ we have*

$$\Delta f(X^h)_{t_n} = L_{t_{n-1}}^h(f)(X_{t_{n-1}}^h)\, h + \Delta M_{t_n}(f)$$

with a random variable $\Delta M_{t_n}(f)$ such that $\mathbb{E}\left(\Delta M_{t_n}(f) \mid X_{t_{n-1}}^h\right) = 0$, and

$$\mathrm{Var}\left(\Delta M_{t_n}(f)\mid X_{t_{n-1}}^h = x\right) = h\, \Gamma_{L_{t_{n-1}}^h}(f,f)(x) - \left(h\, L_{t_{n-1}}^h(f)(x)\right)^2.$$

More generally, for any couple of functions f, g we have

$$\mathbb{E}\left(\Delta M_{t_n}(f)\, \Delta M_{t_n}(g) \ \Big|\ X_{t_{n-1}}^h = x\right) = h\, \Gamma_{L_{t_{n-1}}^h}(f,g)(x) - \left(h\, L_{t_{n-1}}^h(f)(x)\right)\left(h\, L_{t_{n-1}}^h(g)(x)\right)$$

Proof :

Firstly, we observe that

$$\begin{aligned}
\Delta f(X^h)_{t_n} &= \mathbb{E}\left(\Delta f(X^h)_{t_n} \mid \mathcal{F}_{t_{n-1}}^h\right) + \left[\Delta f(X^h)_{t_n} - \mathbb{E}\left(\Delta f(X^h)_{t_n} \mid \mathcal{F}_{t_{n-1}}^h\right)\right] \\
&= [P_{t_{n-1}, t_n}^h - Id](f)(X_{t_{n-1}}^h) + \Delta M_{t_n}(f)
\end{aligned}$$

with the random variables

$$\Delta M_{t_n}(f) \;:=\; \left[\Delta f(X^h)_{t_n} - \mathbb{E}\left(\Delta f(X^h)_{t_n} \mid X^h_{t_{n-1}} \right) \right] = f(X^h_{t_n}) - \mathbb{E}\left(f(X^h_{t_n}) \mid X^h_{t_{n-1}} \right).$$

These random variables are clearly conditionally centered given $X^h_{t_{n-1}}$. In addition, it is also clear that

$$\mathbb{E}\left(\Delta M_{t_n}(f)^2 \mid \mathcal{F}^h_{t_{n-1}} \right) \;=\; \mathrm{Var}\left(f(X^h_{t_n}) \mid X^h_{t_{n-1}} \right) = \Gamma^h_{L^h_{t_{n-1}}}(f,f)(X^h_{t_{n-1}})$$

with

$$
\begin{aligned}
\Gamma^h_{L^h_{t_{n-1}}}(f,f)(x) \;&:=\; P^h_{t_{n-1},t_n}(f^2)(x) - P^h_{t_{n-1},t_n}(f)(x)^2 \\
&=\; f^2(x) + h\, L^h_{t_{n-1}}(f^2)(x) - \left(f(x) + h\, L^h_{t_{n-1}}(f)(x) \right)^2 \\
&=\; h\left[L^h_{t_{n-1}}(f^2) - 2f\, L^h_{t_{n-1}}(f) \right](x) - \left(h\, L^h_{t_{n-1}}(f)(x) \right)^2.
\end{aligned}
$$

This ends the proof of the first assertion. The second formula can be proved by polarization. Next, we provide an alternative proof. Arguing as above for any couple of functions f, g we have

$$\mathbb{E}\left(\Delta M_{t_n}(f)\, \Delta M_{t_n}(g) \mid \mathcal{F}^h_{t_{n-1}} \right) \;=\; \Gamma^h_{L^h_{t_{n-1}}}(f,g)(X^h_{t_{n-1}})$$

with

$$
\begin{aligned}
\Gamma^h_{L^h_{t_{n-1}}}(f,g)(x) \;&:=\; P^h_{t_{n-1},t_n}(fg)(x) - P^h_{t_{n-1},t_n}(f)(x)\, P^h_{t_{n-1},t_n}(g)(x) \\
&=\; (fg)(x) + h\, L^h_{t_{n-1}}(fg)(x) \\
&\qquad - \left(f(x) + h\, L^h_{t_{n-1}}(f)(x) \right)\left(g(x) + h\, L^h_{t_{n-1}}(g)(x) \right) \\
&=\; h\left[L^h_{t_{n-1}}(fg) - f\, L^h_{t_{n-1}}(g) - g\, L^h_{t_{n-1}}(f) \right](x) \\
&\qquad - \left(h\, L^h_{t_{n-1}}(f)(x) \right)\left(h\, L^h_{t_{n-1}}(g)(x) \right).
\end{aligned}
$$

This ends the proof of the theorem. ∎

Definition 12.5.5 *We say that A_{t_n}, respectively M_{t_n}, is a predictable process, respectively a martingale process w.r.t. the filtration $\mathcal{F}^h_{t_n}$, $n \geq 0$, when we have*

$$\mathbb{E}\left(A_{t_n} \mid \mathcal{F}^h_{t_{n-1}} \right) = A_{t_n} \quad \text{and respectively} \quad \mathbb{E}\left(M_{t_n} \mid \mathcal{F}^h_{t_{n-1}} \right) = M_{t_{n-1}}.$$

The following decomposition is more or less a direct consequence of the above theorem.

Corollary 12.5.6 *For any $n \geq 0$, we have the decomposition*

$$f(X^h_{t_n}) = f(X^h_{t_0}) + A^h_{t_n}(f) + M^h_{t_n}(f) \tag{12.18}$$

with the stochastic processes

$$A^h_{t_n}(f) = \sum_{t_0 < t_p \leq t_n} L^h_{t_{n-1}}(f)(X^h_{t_{n-1}})\, h \quad \text{and} \quad M^h_{t_n}(f) = \sum_{t_0 < t_p \leq t_n} \Delta M^h_{t_p}(f).$$

The stochastic processes $A_{t_n}^h(f)$ and $M_{t_n}^h(f)$ are called the predictable and the martingale parts of $X_{t_n}^h$. The decomposition (12.18) is called the Doob-Meyer decomposition of the process $f(X_{t_n}^h)$.

Proof :

For any function $f \in \mathcal{B}(S)$, we clearly have

$$f(X_{t_n}^h) = f(X_{t_0}^h) + \sum_{t_0 < t_p \leq t_n} \left(f(X_{t_p}^h) - f(X_{t_{p-1}}^h) \right) = \sum_{t_0 \leq t_p \leq t_n} \Delta f(X^h)_{t_p}.$$

This implies (12.18) with the stochastic processes $A_{t_n}^h(f)$ and $M_{t_n}^h(f)$ defined by

$$A_{t_n}^h(f) = \sum_{t_0 < t_p \leq t_n} \overbrace{\mathbb{E}\left(\Delta f(X^h)_{t_p} \mid X_{t_{p-1}}^h \right)}^{=\Delta A_{t_p}^h(f)}$$

$$M_{t_n}^h(f) = \sum_{t_0 < t_p \leq t_n} \underbrace{\left[\Delta f(X^h)_{t_p} - \mathbb{E}\left(\Delta f(X^h)_{t_p} \mid X_{t_{p-1}}^h \right) \right]}_{=\Delta M_{t_p}^h(f)}.$$

The end of the proof is now a direct consequence of theorem 12.5.4. This ends the proof of the corollary. ∎

12.5.2   Continuous time martingales

It is convenient to extend the class of martingale processes discussed in section 8.4 to continuous time processes. When studying continuous time processes X_t defined on some probability space $(\Omega, \mathbb{P}, \mathcal{G})$, we need to consider filtrations $\mathcal{F} = (\mathcal{F}_t)_{t \geq 0}$ indexed by the continuous time parameter $t \in [0, \infty[$. For instance $\mathcal{F}_t = \sigma(X_s, \ s \leq t)$ contains the information on the process up to a given time horizon. Jump type processes X_t and related martingales have càdlàg trajectories.

Roughly speaking, for càdlàg processes X_t, the σ-field $\mathcal{F}_{t-} = \sigma(X_s, \ s < t)$ contains information on the process from the origin up to the left limit X_{t-}. In the reverse angle, $\mathcal{F}_t = \mathcal{F}_{t+} = \sigma(X_s, \ s \leq t)$ contains information on the process from the origin up to the right hand side limit $X_t = X_{t+}$.

The more rigorous construction and the continuity analysis of these filtrations for general càdlàg stochastic processes is rather technical. For technical reasons it is commonly assumed that the filtration is right continuous and complete, in the sense that $\mathcal{F}_t = \mathcal{F}_{t+} := \cap_{h \geq 0} \mathcal{F}_{t+h}$, and all subsets of $\mathbb{P}$-null sets are in $\mathcal{F}_0$. The left limits are defined by $\mathcal{F}_{t-} := \vee_{s < t} \mathcal{F}_s$ (the smallest σ-field containing $\mathcal{F}_s$ for all $s < t$).

Analogous to the discrete time case (cf. section 8.3.2) stopping times w.r.t. a right continuous filtration $\mathcal{F}_t$ are random times T s.t. $\{T < t\} \in \mathcal{F}_t$, for all $t \geq 0$. For instance, let X_t be a càdlàg stochastic processes taking values in $\mathbb{R}^r$, and let $\mathcal{F}_t$ the associated canonical right continuous and complete filtration. For a given (measurable) subset $B \subset \mathbb{R}^r$, the random hitting time $T = \inf\{t \geq 0 \ : \ X_t \in B\}$ is an $\mathcal{F}$-stopping time as soon as B is an open set. The result is also true for closed sets but the process needs to have continuous trajectories. Inversely any stopping time can be seen as a hitting time. These results are based on the Début theorem; the proof of this result is quite technical, thus it is omitted.

In this context, the extension of definition 12.5.5 to continuous time processes takes the following form.

Definition 12.5.7 *We say that a process A_t, respectively M_t, is $\mathcal{F}$-predictable, respectively an $\mathcal{F}$-martingale, if we have*

$$\mathbb{E}(A_t \mid \mathcal{F}_{t-}) = A_t \quad \text{and respectively} \quad \forall s \leq t \quad \mathbb{E}(M_t \mid \mathcal{F}_s) = M_s.$$

As in the discrete time case (cf. definition 8.4.3), M is also called a super-martingale, and respectively a sub-martingale (w.r.t. the filtration $\mathcal{F}$), if we can replace the r.h.s. in the above display by

$$\mathbb{E}(M_t \mid \mathcal{F}_s) \geq M_s, \quad \text{and resp.} \quad \mathbb{E}(M_t \mid \mathcal{F}_s) \leq M_s.$$

Definition 12.5.8 *We let $\mathcal{F}_t$ be an increasing sequence of σ-fields. An $\mathcal{F}_t$-martingale M_t is a real valued stochastic process such that for any $s \leq t$*

$$\mathbb{E}(M_t \mid \mathcal{F}_s) = M_s.$$

The angle bracket $\langle M^{(1)}, M^{(2)} \rangle_t$ of a couple of martingales $M_t^{(1)}$ and $M_t^{(2)}$ is an $\mathcal{F}$-predictable process such that

$$M_t^{(1)} M_t^{(2)} - \langle M^{(1)}, M^{(2)} \rangle_t \quad \text{is an } \mathcal{F}_t\text{-martingale.}$$

When $M_t^{(1)} = M_t^{(2)} = M_t$ we often write $\langle M \rangle_t$ instead of $\langle M, M \rangle_t$.

Important remark : For any martingale M_t we have

$$
\begin{aligned}
\mathbb{E}\left((M_{t+dt} - M_t)^2 \mid \mathcal{F}_t\right) &= \mathbb{E}(M_{t+dt}^2 \mid \mathcal{F}_t) - 2M_t \, \mathbb{E}(M_{t+dt} \mid \mathcal{F}_t) + M_t^2 \\
&= \mathbb{E}(M_{t+dt}^2 \mid \mathcal{F}_t) - M_t^2 \\
&= \mathbb{E}(M_{t+dt}^2 - M_t^2 \mid \mathcal{F}_t) = \mathbb{E}(dM_t^2 \mid \mathcal{F}_t).
\end{aligned}
$$

In much the same way, for any couple of martingales $M_t^{(1)}$ and $M_t^{(2)}$ we have

$$\mathbb{E}\left(\left(M_{t+dt}^{(1)} - M_t^{(1)}\right)\left(M_{t+dt}^{(2)} - M_t^{(2)}\right) \mid \mathcal{F}_t\right)$$

$$= \mathbb{E}\left(M_{t+dt}^{(1)} M_{t+dt}^{(2)} \mid \mathcal{F}_t\right) - M_t^{(1)} \mathbb{E}\left(M_{t+dt}^{(2)} \mid \mathcal{F}_t\right) - M_t^{(2)} \mathbb{E}\left(M_{t+dt}^{(1)} \mid \mathcal{F}_t\right) + M_t^{(1)} M_t^{(2)}$$

$$= \mathbb{E}\left(M_{t+dt}^{(1)} M_{t+dt}^{(2)} \mid \mathcal{F}_t\right) - M_t^{(1)} M_t^{(2)} = \mathbb{E}\left(M_{t+dt}^{(1)} M_{t+dt}^{(2)} - M_t^{(1)} M_t^{(2)} \mid \mathcal{F}_t\right)$$

with the product increments

$$d(M^{(1)} M^{(2)})_t := M_{t+dt}^{(1)} M_{t+dt}^{(2)} - M_t^{(1)} M_t^{(2)} \quad \text{and} \quad dM_t^2 := M_{t+dt}^2 - M_t^2.$$

This yields the decomposition

$$M_{t+dt}^{(1)} M_{t+dt}^{(2)} - M_t^{(1)} M_t^{(2)}$$

$$= \underbrace{M_t^{(1)} \left(M_{t+dt}^{(2)} - M_t^{(2)} \right)}_{\text{martingale increment}} + \underbrace{M_t^{(2)} \left(M_{t+dt}^{(1)} - M_t^{(1)} \right))}_{\text{martingale increment}} + \left(M_{t+dt}^{(1)} - M_t^{(1)} \right) \left(M_{t+dt}^{(2)} - M_t^{(2)} \right)$$

$$= \mathbb{E} \left(\left(M_{t+dt}^{(1)} - M_t^{(1)} \right) \left(M_{t+dt}^{(2)} - M_t^{(2)} \right) \mid \mathcal{F}_t \right) + \text{Martingale increments},$$

with the predictable quadratic rules,

$$\mathbb{E} \left(d(M_t^2) \mid \mathcal{F}_t \right) = \mathbb{E} \left((dM_t)^2 \mid \mathcal{F}_t \right)$$

$$\mathbb{E} \left(d(M^{(1)} M^{(2)})_t \mid \mathcal{F}_t \right) = \mathbb{E} \left(dM_t^{(1)} dM_t^{(2)} \mid \mathcal{F}_t \right). \qquad (12.19)$$

Last but not least, suppose the martingales $M_t^{(i)} = M_t^{(i),c} + M_t^{(i),d}$ are decomposed into martingales $M_t^{(i),c}$ with continuous trajectories and martingales $M_t^{(i),d}$ with discontinuous trajectories such that

$$\left(M_{t+dt}^{(i),c} - M_t^{(i),c} \right) \left(M_{t+dt}^{(j),d} - M_t^{(j),d} \right) = 0$$

for any $i, j \in \{1, 2\}$. In this situation, we have

$$\left(M_{t+dt}^{(1)} - M_t^{(1)} \right) \left(M_{t+dt}^{(2)} - M_t^{(2)} \right)$$

$$= dM_t^{(1),c} dM_t^{(2),c} + \left(M_{t+dt}^{(1),d} - M_t^{(1),d} \right) \left(M_{t+dt}^{(2),d} - M_t^{(2),d} \right)$$

$$= d\langle M^{(1),c}, M_t^{(2),c} \rangle_t + \Delta \left[M^{(1),d}, M^{(2),d} \right]_t + \underbrace{dM_t^{(1),c} dM_t^{(2),c} - \mathbb{E} \left(dM_t^{(1),c} dM_t^{(2),c} \mid \mathcal{F}_t \right)}_{\text{martingale increment}}$$

with the square bracket pure jump process

$$\left[M^{(1),d}, M^{(2),d} \right]_t := \sum_{s \leq t} \Delta M_s^{(1)} \Delta M_s^{(2)}.$$

If we set

$$\left[M^{(1)}, M^{(2)} \right]_t := \langle M^{(1),c}, M^{(2),c} \rangle_t + \sum_{s \leq t} \Delta M_s^{(1)} \Delta M_s^{(2)}, \qquad (12.20)$$

we conclude that

$$M_t^{(1)} M_t^{(2)} - \left[M^{(1)}, M^{(2)} \right]_t \quad \text{is a martingale.}$$

12.5.3 Optional stopping theorems

This section provides a brief discussion on the extension to continuous time of the stopped martingales with respect to some stopping times discussed in section 8.4.3. Random times T are called stopping times with respect to some filtration $\mathcal{F} := (\mathcal{F}_t)_{t \geq 0}$ when we have $\{T \leq t\} \in \mathcal{F}_t$ for any $t \in [0, \infty[$. In the further development of this section c, c_1 and c_2 stand for some finite constants. The continuous time version of theorem 8.4.12 is given below.

Theorem 12.5.9 (Optional stopping theorem 1)

Every stopped martingale $M_{t \wedge T}$ w.r.t. some $\mathcal{F}$-stopping time is again an $\mathcal{F}$-martingale.

As in (8.55), applying the dominated convergence theorem we also prove the following theorem.

Theorem 12.5.10 (Optional stopping theorem 2) *Whenever the stopped martingale $M_{t \wedge T}$ is s.t. $|M_{t \wedge T}| \leq c$ we have*

$$\mathbb{E}(M_0) = \mathbb{E}(M_{t \wedge T}) \longrightarrow_{t \uparrow \infty} \mathbb{E}\left(\lim_{t \uparrow \infty} M_{t \wedge T}\right) = \mathbb{E}(M_T). \qquad (12.21)$$

We also have

$$|M_{t \wedge T}| \leq c_1 + c_2 \, T \quad with \quad \mathbb{E}(T) < \infty$$

$$\Longrightarrow \mathbb{E}(M_0) = \mathbb{E}(M_{t \wedge T}) \longrightarrow_{t \uparrow \infty} \mathbb{E}\left(\lim_{n \uparrow \infty} M_{t \wedge T}\right) = \mathbb{E}(M_T). \qquad (12.22)$$

Last but not least, applying (12.22) to the martingale $N_t := M_t^2 - \langle M \rangle_t$, we prove the theorem.

Theorem 12.5.11 (A Wald's type identity) *For any stopped martingale we have*

$$\left. \begin{array}{rcl} |M_{t \wedge T}| & \leq & c \\[2mm] \langle M \rangle_{t \wedge T} & \leq & c \, T \quad with \quad \mathbb{E}(T) < \infty \end{array} \right\} \Longrightarrow \mathbb{E}\left(M_T^2\right) = \mathbb{E}\left(\langle M \rangle_T\right)$$

as soon as $M_0 = 0 = \langle M \rangle_0$. In addition, we have

$$|M_{t \wedge T}| \leq c \implies \mathbb{E}\left(\langle M \rangle_T\right) \leq c^2. \qquad (12.23)$$

12.6   Doeblin-Itō-Taylor formulae

In the further development of this section, X_t stands for the continuous time jump model introduced in section 12.1, and $X_{t_n}^h$ its discrete time version on a time mesh (11.30) with the Markov transitions (12.8).

At time $t_n = h \langle t/h \rangle \uparrow t$, as $h \downarrow$ we have the increment approximation

$$
\begin{aligned}
df(X_t) \simeq \Delta f(X^h)_{t_n} &= L_{t_{n-1}}^h(f)(X_{t_{n-1}}^h)\, h + \Delta M_{t_n}^h(f) \quad \text{(by theorem 12.5.4)} \\
&\simeq L_t(f)(X_t)\, dt + dM_t(f)
\end{aligned}
$$

with a martingale $M_t(f)$ w.r.t. the σ-field

$$
\mathcal{F}_t = \sigma\left(X_s, s \leq t\right) \;\simeq_{h\downarrow 0}\; \mathcal{F}_{t_n}^h
$$

generated by the stochastic process X_t.

By lemma 8.4.6, for any couple of functions (f, g) we also have the martingale approximations

$$
M_{t_n}^h(f)\, M_{t_n}^h(g) - \langle M^h(f), M^h(g) \rangle_{t_n} \simeq M_t(f)\, M_t(g) - \langle M(f), M(g) \rangle_t
$$

with the angle bracket

$$
\langle M^h(f), M^h(g) \rangle_{t_n} - \langle M^h(f), M^h(g) \rangle_{t_{n-1}}
$$

$$
= \mathbb{E}\left(\Delta M_{t_n}^h(f) \Delta M_{t_n}^h(g) \mid \mathcal{F}_{t_{n-1}}^h\right) = \operatorname{Cov}\left(\Delta M_{t_n}(f), \Delta M_{t_n}(g) \mid X_{t_{n-1}}^h\right)
$$

$$
= h\, \Gamma_{L_{t_{n-1}}^h}(f, g)(X_{t_{n-1}}^h) - \left(h\, L_{t_{n-1}}^h(f)(X_{t_{n-1}}^h)\right)\left(h\, L_{t_{n-1}}^h(g)(X_{t_{n-1}}^h)\right) \quad \text{(by th. 12.5.4)}
$$

$$
\simeq \Gamma_{L_t}(f, g)(X_t)\, dt := d\langle M(f), M(g) \rangle_t.
$$

In other words the stochastic process

$$
t \mapsto M_t(f) M_t(g) - \langle M(f), M(g) \rangle_t \tag{12.24}
$$

is a martingale with the angle bracket $\langle M(f), M(g) \rangle_t$ defined by the integral formula

$$
\langle M(f), M(g) \rangle_t = \int_0^t d\langle M(f), M(g) \rangle_s = \int_0^t \Gamma_{L_s}(f, g)(X_s)\, ds.
$$

When $f = g$, we often write $\langle M(f) \rangle_t$ instead of $\langle M(f), M(f) \rangle_t$.

Working a little harder, we can prove the following theorem (a.k.a. Doeblin-Itō formula or Itō lemma).

Theorem 12.6.1 *For any smooth functional $t \mapsto f_t = f(t, .) \in \mathcal{B}(S)$, we have*

$$df(t, X_t) = (\partial_t + L_t) f(t, X_t) \, dt + dM_t(f) \qquad (12.25)$$

with a martingale $M_t(f)$ null at the origin and such that

$$M_t^2(f) - \langle M(f) \rangle_t := M_t^2(f) - \int_0^t \Gamma_{L_s}(f_s, f_s)(X_s) \, ds$$

is a martingale, with the square field operator given by

$$\Gamma_{L_t}(f, f)(x) = \lambda_t(x) \int [f(y) - f(x)]^2 \, M_t(x, dy).$$

We observe that a Poisson process $X_t = N_t$ with intensity λ_t is a pure jump process with generator

$$L_t(f)(x) = \lambda_t \, [f(x+1) - f(x)].$$

Applying theorem 12.6.1 to the identity function $f(N_t) = N_t$, we find that

---

The stochastic process

$$M_t := N_t - \int_0^t \lambda_s \, ds$$

is a martingale with angle bracket given by $d\langle M \rangle_t = \lambda_t \, dt$.

We can alternatively use

$$\mathbb{E}\left(dN_t \mid \mathcal{F}_t\right) \;=\; \mathbb{P}\left(N_{t+dt} - N_t = 1 \mid \mathcal{F}_t\right) = \lambda_t \, dt$$

to check that

$$\mathbb{E}\left(dM_t \mid \mathcal{F}_t\right) = \mathbb{E}\left(dN_t - \lambda_t dt \mid \mathcal{F}_t\right) = 0.$$

In addition, using (12.19), we have

$$\mathbb{E}\left(d(M_t^2) \mid \mathcal{F}_t\right) \;=\; \mathbb{E}\left((dM_t)^2 \mid \mathcal{F}_t\right) = \mathbb{E}\left((dN_t)^2 \mid \mathcal{F}_t\right) = \mathbb{E}\left(dN_t \mid \mathcal{F}_t\right) = \lambda_t \, dt.$$

12.7 Stability properties

12.7.1 Invariant measures

We consider a time homogeneous jump process X_t on some state space S with generator

$$L(f)(x) = \lambda(x) \int [f(y) - f(x)] \, K(x, dy) \qquad (12.26)$$

with a bounded intensity function $\lambda(x) \leq \lambda_{max}$ for some finite $\lambda_{max} < \infty$, and some Markov transition K. We set

$$K_\lambda(x, dy) = \frac{\lambda(x)}{\lambda_{max}} \, K(x, dy) + \left(1 - \frac{\lambda(x)}{\lambda_{max}}\right) \, \delta_x(dy) \Rightarrow L = \lambda_{max} \, [K_\lambda - Id]. \qquad (12.27)$$

By construction, the sg of X_t is given by

$$
\begin{aligned}
P_t(f)(x) &= \mathbb{E}\left(f(X_t) \mid X_0 = x\right) \\
&= \sum_{n \geq 0} \mathbb{E}\left(f(X_t) \, 1_{N_t = n} \mid X_0 = x\right) = e^{-\lambda_{max} t} \sum_{n \geq 0} \frac{(t\lambda_{max})^n}{n!} \, K_\lambda^n(f)(x)
\end{aligned}
$$

where N_t is a Poisson process with intensity λ_{max}.

Lemma 12.7.1 *Whenever they exist (and they are unique), the invariant probability measures π, respectively π_λ, of the Markov transitions K, respectively K_λ, are connected for any $f \in \mathcal{B}(S)$ by the formula*

$$\pi_\lambda(f) = \pi(f/\lambda)/\pi(1/\lambda) \quad \text{and we have} \quad \pi_\lambda L(f) = 0. \qquad (12.28)$$

Proof :
The l.h.s. follows from the following observations:

$$
\begin{aligned}
\pi K = \pi \Rightarrow \pi_\lambda(K_\lambda(f)) \propto \pi\left(\frac{1}{\lambda}(\lambda K(f) + (1 - \lambda)f)\right) &= \pi K(f) + \pi(f/\lambda) - \pi(f) \\
&= \pi(f/\lambda) \propto \pi_\lambda(f).
\end{aligned}
$$

We also notice that

$$
\begin{aligned}
\pi_\lambda K_\lambda = \pi_\lambda \Rightarrow \pi_\lambda(\lambda K(f)) &= \pi_\lambda(K_\lambda(f)) - \pi_\lambda((1 - \lambda)(f)) \\
&= \pi_\lambda(\lambda f) \Rightarrow \pi(f) \propto \pi_\lambda(\lambda f).
\end{aligned}
$$

The r.h.s. of (12.28) is now immediate. This ends the proof of the lemma. ∎

Definition 12.7.2 *A probability measure π on some state space S is invariant w.r.t. some time homogeneous Markov semigroup $P_t = P_{0,t}$ if we have $\pi P_t = \pi$, for any $t \geq 0$.*

Proposition 12.7.3 *The invariance property is also characterized in terms of the generator L of the semigroup:*

$$\pi \text{ is } P_t\text{-invariant} \iff (\forall t \geq 0 \;\; \pi P_t = \pi) \iff \pi L = 0.$$

Proof :
We check this claim using

$$\pi\left(\frac{P_t - Id}{t}\right) \to_{t \downarrow 0} \pi L \quad \text{and} \quad P_t = Id + \int_0^t L P_s \, ds.$$

Definition 12.7.4 *The measure π is reversible w.r.t. P_t if we have*

$$\pi(dx) \, P_t(x, dy) = \pi(dy) \, P_t(y, dx)$$

or equivalently, for any pair of functions $(f, g) \in \mathcal{B}(S)^2$

$$\forall t \geq 0 \quad \pi(f P_t(g)) = \pi(P_t(f) g) \iff \pi(f L(g)) = \pi(L(f) g).$$

By lemma 12.7.1 the invariant probability measures of jump processes with generator L given in (12.26) are directly connected to those of the Markov chain with probability transitions K_λ described in (12.27).

We summarize this result with the series of equivalences

$$\pi K = \pi \iff \pi_\lambda = \pi_\lambda K_\lambda \iff \pi_\lambda L = 0. \tag{12.29}$$

In addition, π_λ is reversible w.r.t. P_t if and only if π_λ is reversible w.r.t. K_λ. This reversibility property is also equivalent to the fact that π is K-reversible.

12.7.2 Dobrushin contraction properties

There are several ways to transfer the stability properties of the embedded discrete Markov chain X_n^λ with transition K_λ to the stability of the continuous time model X_t.

Next, we present three possible routes. The first one is expressed in terms of the Dobrushin ergodic coefficient $\beta(K)$ of a Markov transition K introduced in definition 8.2.11. The second one is expressed in terms of the V-Dobrushin local contraction coefficient $\beta_V(K_\lambda)$ presented in definition 8.2.19. The third one is related to coupling techniques.

Theorem 12.7.5 • *We assume that there exists some $m \geq 1$ s.t. $\beta(K_\lambda^m) < 1$. In this situation, for any $t \geq 0$, we have the exponential estimate*

$$\beta(P_t) \leq \frac{1}{\beta(K_\lambda^m)^{1-1/m}} \, \exp\left[-t\lambda_{max}\left(1 - \beta(K_\lambda^m)^{\frac{1}{m}}\right)\right]. \tag{12.30}$$

• *We assume that $\beta_V(K_\lambda^m) < 1$ for some $m \geq 1$ and some function $V \geq 0$. In this situation, for any $t \geq 0$, we have the exponential estimate*

$$\beta_V(P_t) \leq \frac{1}{\beta_V(K_\lambda^m)^{1-1/m}} \, \exp\left[-t\lambda_{max}\left(1 - \beta_V(K_\lambda^m)^{\frac{1}{m}}\right)\right]. \tag{12.31}$$

We further assume that $0 < \lambda_{min} \leq \lambda \leq \lambda_{max}$, and that K satisfies the Foster-Lyapunov condition (8.28) for some $\epsilon \in [0, 1[$, some finite $c < \infty$, and some function $W \geq 0$. In this situation, K_λ satisfies the Foster-Lyapunov condition (8.28) with

$$K_\lambda(W) \leq \left(1 - \frac{\lambda_{min}}{\lambda_{max}} \, (1 - \epsilon)\right) \, W + c. \tag{12.32}$$

In addition, if K_λ satisfies the Dobrushin local contraction condition (8.27), there exists some function V s.t. $\beta_V(K_\lambda) < 1$.

Proof :

Firstly, we prove (12.30). For any function f such that $\operatorname{osc}(f) \leq 1$, using theorem 8.2.13 we have

$$\operatorname{osc}\left(K_\lambda^{nm+p}(f)\right) \leq \beta(K_\lambda^m)^n \operatorname{osc}(K_\lambda^p(f)) \leq \beta(K_\lambda^m)^n.$$

$$\begin{aligned}
\operatorname{osc}\left(P_t(f)\right) &\leq e^{-\lambda_{max}t} \sum_{n\geq 0} \frac{(t\lambda_{max})^n}{n!} \operatorname{osc}\left(K_\lambda^n(f)\right) \\
&= e^{-t\lambda_{max}} \sum_{n\geq 0}\sum_{0\leq p<m} \frac{(t\lambda_{max})^{nm+p}}{(nm+p)!} \operatorname{osc}\left(K_\lambda^{nm+p}(f)\right) \\
&\leq e^{-t\lambda_{max}} \sum_{n\geq 0}\sum_{0\leq p<m} \frac{(t\lambda_{max})^{nm+p}}{(nm+p)!} \beta(K_\lambda^m)^n.
\end{aligned}$$

When $\beta(K_\lambda^m) = 0$, the result is obvious. When $\beta(K_\lambda^m) > 0$, we observe that

$$\begin{aligned}
\frac{(t\lambda_{max})^{nm+p}}{(nm+p)!} \beta(K_\lambda^m)^n &= \frac{(t\lambda_{max}\beta(K_\lambda^m)^{\frac{1}{m}})^{nm+p}}{(nm+p)!} \frac{1}{\beta(K_\lambda^m)^{p/m}} \\
&\leq \frac{(t\lambda_{max}\beta(K_\lambda^m)^{\frac{1}{m}})^{nm+p}}{(nm+p)!} \frac{1}{\beta(K_\lambda^m)^{1-1/m}}.
\end{aligned}$$

This implies that

$$\begin{aligned}
\operatorname{osc}\left(P_t(f)\right) &\leq \frac{e^{-t\lambda_{max}}}{\beta(K_\lambda^m)^{1-1/m}} \sum_{n\geq 0} \frac{1}{n!}\left(t\lambda_{max}\beta(K_\lambda^m)^{\frac{1}{m}}\right)^n \\
&= \frac{1}{\beta(K_\lambda^m)^{1-1/m}} \exp\left[-t\lambda_{max}\left(1 - \beta(K_\lambda^m)^{\frac{1}{m}}\right)\right].
\end{aligned}$$

This ends the proof of (12.30).

To prove (12.31) we use theorem 8.2.21 to check that

$$\begin{aligned}
\|P_t(x,.) - P_t(y,.)\|_V &\leq e^{-\lambda_{max}t} \sum_{n\geq 0} \frac{(t\lambda_{max})^n}{n!} \|K_\lambda^n(x,.) - K_\lambda^n(y,.)\|_V \\
&\leq e^{-\lambda_{max}t} \sum_{n\geq 0} \frac{(t\lambda_{max})^n}{n!} \beta_V(K_\lambda^n) \|\delta_x - \delta_x\|_V \\
&= e^{-\lambda_{max}t} \sum_{n\geq 0} \frac{(t\lambda_{max})^n}{n!} \beta_V(K_\lambda^n) \left(1 + V(x) + V(y)\right).
\end{aligned}$$

This implies that

$$\beta_V(P_t) \leq e^{-\lambda_{max}t} \sum_{n\geq 0} \frac{(t\lambda_{max})^n}{n!} \beta_V(K_\lambda^n).$$

The end of the proof of (12.31) follows the same arguments as the ones we used in the proof of (12.30). The proof of (12.32) is a consequence of

$$\begin{aligned}
K_\lambda(W) &= \frac{\lambda}{\lambda_{max}} K(W) + \left(1 - \frac{\lambda}{\lambda_{max}}\right) W \\
&\leq \left[\epsilon \frac{\lambda}{\lambda_{max}} + \left(1 - \frac{\lambda}{\lambda_{max}}\right)\right] W + \frac{\lambda}{\lambda_{max}} c \leq \left[1 - \frac{\lambda_{min}}{\lambda_{max}}(1-\epsilon)\right] W + c.
\end{aligned}$$

The last assertion is a consequence of theorem 8.2.21.

The proof of (12.33) is a direct consequence of the fact that

$$\|\mathrm{Law}(X_t \mid X_0 = x) - \mathrm{Law}(X_t \mid X_0 = y)\|_{tv} = \sup_{\mathrm{osc}(f) \leq 1} |P_t(f)(x) - P_t(f)(y)|$$

and

$$|P_t(f)(x) - P_t(f)(y)|$$

$$\leq e^{-\lambda_{max}t} \sum_{n \geq 0} \frac{(t\lambda_{max})^n}{n!} |K_\lambda^n(f)(x) - K_\lambda^n(f)(y)|$$

$$= e^{-\lambda_{max}t} \sum_{n \geq 0} \frac{(t\lambda_{max})^n}{n!} \|\mathrm{Law}(X_n^\lambda \mid X_0 = x) - \mathrm{Law}(Y_n^\lambda \mid Y_0 = y)\|_{tv}.$$

This ends the proof of the theorem. ∎

Theorem 12.7.6 • *We let $T_{x,y}^\lambda$ be a coupling time of two copies X_n^λ and Y_n^λ of the Markov chain with Markov transition K_λ starting at $X_n^\lambda = x$ and $Y_n^\lambda = y$. We assume that*

$$\mathbb{P}\left(T_{x,y}^\lambda \geq n\right) \leq a_\lambda(x,y) \, \exp\left(-b_\lambda n\right)$$

for some finite function $a_\lambda(x,y) < \infty$ and some positive constant $b_\lambda \in]0,1[$. In this situation, we have

$$\|\mathrm{Law}(X_t \mid X_0 = x) - \mathrm{Law}(X_t \mid X_0 = y)\|_{tv}$$

$$\tag{12.33}$$

$$\leq a_\lambda(x,y) \, \exp\left(-\lambda_{max}t(1 - e^{-b_\lambda})\right).$$

• *Assume that K^m satisfies the minimization condition (8.15) for some integer $m \geq 1$ and some parameter ϵ. Also assume that $0 < \lambda_{min} \leq \lambda(x) \leq \lambda_{max} < \infty$ for some parameters λ_{min} and λ_{max}. In this situation, there exists a coupling of two copies X_t and X_t' of the process with generator L with a coupling time T such that*

$$\mathbb{P}(T > t) \leq (1 - \epsilon_m)^{-1} \, e^{-\rho_m t}$$

$$\tag{12.34}$$

with $\epsilon_m := \epsilon \times (\lambda_{min}/\lambda_{max})^m$ and $\rho_m := \lambda_{max}\left(1 - e^{-\epsilon_m/m}\right)$.

Proof :

The first result is now a direct consequence of the coupling proposition 8.3.16. If K^m satisfies the minimization condition (8.15) for some integer $m \geq 1$ and some ϵ, then K_λ^m also satisfies this condition with ϵ replaced by ϵ_m. We couple the stochastic processes X_t and X_t' starting at different states $X_0 = x$ and $X_0' = x'$ using the jump times of a common Poisson process N_t with intensity λ_{max}. At these random times we use the coupled bridge chains (Y_n, Y_n') with transitions K_λ (starting at $Y_0 = x$ and $Y_0' = x'$) discussed in the end of section 8.3.4.1. We let T_c the coupling time of (X_t, X_t') and T_d the coupling time of (Y_n, Y_n'). By (8.38) we have

$$\mathbb{P}(T > t) \leq \mathbb{P}(T_d > N_t) \leq (1 - \epsilon_m)^{-1} \, \mathbb{E}\left(e^{-N_t \epsilon_m/m}\right) = (1 - \epsilon_m)^{-1} \, e^{-\lambda_{max}t\left(1 - e^{-\epsilon_m/m}\right)}.$$

This ends the proof of the theorem. ∎

We illustrate the continuous coupling inequality (12.33) with the Markov transition M on a finite and complete graph with d vertices $S := \{1, \ldots, d\}$ discussed in (8.39). We assume that $\lambda_t = \lambda_{max} = \lambda$ (so that $M_\lambda = M$). Combining (12.33) with (8.39), we readily check that

$$\|\text{Law}(X_t \mid X_0 = x) - \text{Law}(X_t \mid X_0 = y)\|_{tv} \le e^{-\lambda t(1 - e^{-1/d})} \simeq_{d \uparrow \infty} e^{-t\lambda/d}.$$

12.8 Exercises

Exercise 169 *Let N_t be a Poisson process with intensity $\lambda > 0$ (starting at the origin $N_0 = 0$), and X_t be the solution of the stochastic differential equation*

$$\begin{cases} dX_t &= -2X_t \, dN_t \\ X_0 &= 1. \end{cases}$$

Describe the generator of the process X_t. Check that $X_t = (-1)^{N_t}$.

Exercise 170 *We consider a pure jump process $X_t \in S = \{0, 1\}$ with generator L defined by*

$$L(f)(0) = \lambda(0) \, (f(1) - f(0)) \quad and \quad L(f)(1) = \lambda(1) \, (f(0) - f(1))$$

for some positive intensities $\lambda(0), \lambda(1) > 0$. Check that

$$\eta_t(0) = \frac{\lambda(1)}{\lambda(0) + \lambda(1)} + e^{-(\lambda(0) + \lambda(1))t} \left(\eta_0(0) - \frac{\lambda(1)}{\lambda(0) + \lambda(1)} \right).$$

Exercise 171 (Compound Poisson process - Mean and variance) *We consider the compound Poisson process X_t discussed in exercise 161. Applying the Doeblin-Itō formula (12.25) to $f(x) = x$ check that*

$$X_t = \lambda \, \mathbb{E}(Y) \, t + M_t$$

with a martingale M_t w.r.t. $\mathcal{F}_t = \sigma(X_s, \ s \le t)$, with angle bracket $\langle M \rangle_t = \lambda \, \mathbb{E}(Y^2) \, t..$ Deduce that

$$\mathbb{E}(X_t) = -\lambda \, \mathbb{E}(Y) \, t \quad and \quad \text{Var}(X_t) = \lambda \, \mathbb{E}(Y^2) \, t.$$

We further assume that $\mathbb{E}(Y) = 0 = \mathbb{E}(Y^3)$. Applying the Doeblin-Itō formula (12.25) to $f(x) = x^2$ compute $\mathbb{E}(X_t^2)$ and $\text{Var}(X_t^2)$.

Exercise 172 (Jump processes - Unit vector representation) *Let $S := \{e_i \ , \ 1 \le i \le r\}$ be an orthonormal basis of some vector space $\mathcal{V}$ equipped with some inner product $\langle v, w \rangle$. For instance if $\mathcal{V} = \mathbb{R}^r$ is equipped with the Euclidian inner product $\langle x, y \rangle = \sum_{1 \le i \le r} x_i y_i$ we can choose the r unit column vectors $e_i = (1_i(j))_{1 \le j \le r}$. Let $\left(N_t^{i,j} \right)_{1 \le i \ne j \le r}$ be $r(r-1)$ Poisson processes with intensity $\lambda(i, j) \ge 0$. Let X_t be given by the stochastic differential equation*

$$\begin{cases} dX_t &= \sum_{1 \le i \ne j \le r} (e_i - e_j) \, \langle e_j, X_t \rangle \, dN_t^{(i,j)} \\ X_0 &\in \{e_1, \ldots, e_r\}. \end{cases}$$

Check that $X_t \in S$ and describe its generator. We let η_t be the row vector $\eta_t = [\eta_t(1), \ldots, \eta_t(r)]$ with entries $\eta_t(i) = \mathbb{P}(X_t = e_i)$. Describe the evolution of η_t in matrix form.

Exercise 173 (Classical coupling - Jump processes) *Consider a jump process X_t on some state space S with generator L given by*

$$L(f)(x) := \int [f(y) - f(x)] \, Q(x, dy)$$

for some bounded and positive integral operator Q We let $\mathcal{Z}_t = (\mathcal{X}_t, \mathcal{Y}_t)$ be the jump process on $\mathcal{S} := (S \times S)$ with generator $\mathcal{L}$ defined for any bounded F on $\mathcal{S}$

$$\mathcal{L}(F)(x, y) = 1_{x \neq y} \, [L(F(x, .))(y) + L(F(., y))(x)] + 1_{x=y} \int [F(y, y) - F(x, x)] \, Q(x, dy).$$

Check that $\mathcal{X}_t$ and $\mathcal{Y}_t$ have the same law as X_t.

Exercise 174 (Basic coupling - Jump processes) *Consider the jump process X_t discussed in exercise 173. We further assume that $Q(x, dy) = q(x, y) \, \lambda(dy)$ has a density w.r.t. some reference measure λ on S. We let $\mathcal{Z}_t = (\mathcal{X}_t, \mathcal{Y}_t)$ be the jump process on $\mathcal{S} := (S \times S)$ with generator $\mathcal{L}$ defined for any bounded F on $\mathcal{S}$*

$$
\begin{aligned}
\mathcal{L}(F)(x, y) &= \int [F(z, z) - F(x, y)] \, (q(x, z) \wedge q(y, z)) \, \lambda(dz) \\
&\quad + \int [F(z, y) - F(x, y)] \, (q(x, z) - q(y, z))_+ \, \lambda(dz) \\
&\quad + \int [F(x, z) - F(x, y)] \, (q(y, z) - q(x, z))_+ \, \lambda(dz).
\end{aligned}
$$

Check that $\mathcal{X}_t$ and $\mathcal{Y}_t$ have the same law as X_t.

Exercise 175 (Jump process - Invariant measure) *Let X_t be a jump type process evolving in some state space S with an infinitesimal generator of the form*

$$L(f)(x) = \lambda \int (f(x') - f(x)) \, K(x, dx')$$

for some intensity parameter $\lambda > 0$ and some Markov transition K on S satisfying the minorization condition $K^m(x, dx') \geq \epsilon \, \nu(dx')$ for some probability measure ν and some parameter $\epsilon > 0$ and some integer $m \geq 1$. Check that X_t has an unique invariant measure π and

$$\sup_{x \in S} \|\mathrm{Law}(X_t \mid X_0 = x) - \pi\|_{tv} \leq c_1 \, e^{-c_2 \, t}$$

for some non negative constants (c_1, c_2).

Exercise 176 (Pure jump Markov chain samplers) *Let $\lambda(dx)$ be some probability measure on some state space S and let V be some non-negative function on S s.t. $\lambda(e^{-V}) \in \,]0, \infty[$. Let X_t be a jump type process evolving in some state space S with an infinitesimal generator of the form*

$$L(f)(x) = \lambda(x) \int (f(x') - f(x)) \, K(x, dx')$$

for some intensity rate function $\lambda(x) \propto e^{V(x)}$ and some Markov transition K s.t. $\lambda K = \lambda$. Check that $\pi(dx) \propto e^{-V(x)} \, \lambda(dx)$ is L-invariant. When K is λ-reversible, check that π is L-reversible.

Exercise 177 (Jump process - Feynman-Kac formulae) *We consider the pure jump process X_t discussed in exercise 175. Let V be some non-negative energy type potential function on S. For any $\beta > 1$ we let $\gamma_t^{[\beta]}$ be the Feynman-Kac measures defined for any function f on S by*

$$\eta_t^{[\beta]}(f) = \gamma_t^{[\beta]}(f)/\gamma_t^{[\beta]}(1) \quad \text{with} \quad \gamma_t^{[\beta]}(f) = \mathbb{E}\left[f(X_t) \exp\left(-\beta \int_0^t V(X_s)\, ds\right)\right].$$

Compute the derivative $\partial_t \log \gamma_t^{[\beta]}(1)$ and check that

$$\frac{1}{t} \log\left[\gamma_t^{[\beta]}(1)\right]^{1/\beta} = \frac{1}{t} \int_0^t \eta_s^{[\beta]}(V)\, ds.$$

Assume that $\eta_t^{[\beta]}$ converges to some limiting probability measure $\eta_\infty^{[\beta]}$ such that

$$\left|\frac{1}{t}\int_0^t \eta_s^{[\beta]}(V)\, ds - \eta_\infty^{[\beta]}(V)\right| \le \frac{c_\beta}{t}$$

for some finite constant c_β (see for instance (27.21), exercises 308, 446, and 447 for more a more thorough discussion on this condition as well as workout examples). Check that

$$C_\beta^{-1}\, e^{-t\eta_\infty^{[\beta]}(V)} \le \left[\gamma_t^{[\beta]}(1)\right]^{1/\beta} \le C_\beta\, e^{-t\eta_\infty^{[\beta]}(V)} \quad \text{with} \quad C_\beta = e^{c_\beta}.$$

Exercise 178 (1-dimensional Kramer-Moyal expansion) *Consider an $\mathbb{R}$-valued pure jump process X_t with generator*

$$L_t(f)(x) \;=\; \int\, (f(x+u) - f(x))\, g_t(x,u)\, du$$

for some regular rate function g_t. Assume that $\eta_t = \mathrm{Law}(X_t)$ has a density $\eta_t(dx) = p_t(x)\, dx$ w.r.t. the Lebesgue measure dx.

- *Check that*

$$L_t(f)(x) \;=\; \int\, (f(y) - f(x))\, q_t(x,y)\, dy \quad \text{with} \quad q_t(x,y) := g_t(x, y-x).$$

- *Prove that*

$$\partial_t p_t(x) \;=\; \int\, [p_t(y)\, q_t(y,x) - p_t(x)\, q_t(x,y)]\, dy.$$

- *Using the Taylor's expansion,*

$$p_t(x-z)\, g_t(x-z, z) = p_t(x)\, g_t(x, z) + \sum_{n \ge 1} \frac{(-1)^n}{n!}\, z^n\, \partial_x^n\, (p_t(x) g_t(x, z))$$

check the Kramer-Moyal expansion

$$\partial_t p_t = \sum_{n \ge 1} \frac{(-1)^n}{n!}\, \partial_x^n\, (\alpha_t^n\, p_t)$$

with the collection of conditional jump moment functionals

$$\alpha_t^n(x) := \int\, z^n\, g_t(x,z)\, dz = \mathbb{E}\left[\, (\Delta X_t)^n \mid X_{t_-} = x\right]/dt.$$

In the above display ΔX_t denotes the jump amplitude increment $\Delta X_t = X_t - X_{t_-}$.

Exercise 179 *Consider a probability distribution $\eta(dx) = p(x)dx$ having a density $p(x) > 0$ w.r.t. the Lebesgue measure dx on $\mathbb{R}$. Let X_t be the jump process with generator*

$$L_h(f)(x) = \lambda \left[(f(x+h) - f(x)) \, \frac{p(x+h)}{p(x) + p(x+h)} + (f(x-h) - f(x)) \, \frac{p(x-h)}{p(x) + p(x-h)} \right]$$

for some $h \neq 0$. Check that η is an L_h-invariant measure (i.e. $\eta L_h = 0$). A discrete generation Gibbs-type version of this sampler is discussed in exercise 127.

Exercise 180 (Multidimensional Kramer-Moyal expansion) *Consider an $\mathbb{R}^r$-valued pure jump process X_t with generator*

$$L_t(f)(x) \;=\; \int \, (f(x+u) - f(x)) \, g_t(x, u) \, du$$

for some regular rate function g_t on $\mathbb{R}^{r+r}$. Assume that $\eta_t = \mathrm{Law}(X_t)$ has a density $\eta_t(dx) = p_t(x) \, dx$ w.r.t. the Lebesgue measure $dx = dx_1 \times \ldots \times dx_r$ on $\mathbb{R}^r$. In this situation, check that the Kramer-Moyal expansion

$$\partial_t p_t = \sum_{n \geq 1} \frac{(-1)^n}{n!} \sum_{1 \leq i_1, \ldots, i_n \leq r} \partial_{x_{i_1}, \ldots, x_{i_n}} \left(\alpha_t^{i_1, \ldots, i_n} \, p_t \right)$$

with the collection of conditional jump moment functionals

$$\alpha_t^{i_1, \ldots, i_n}(x) := \int \, z_{i_1} \ldots z_{i_n} \, g_t(x, z) \, dz = \mathbb{E} \left[\prod_{1 \leq k \leq r} \Delta X_t^{i_k} \mid X_{t-} = x \right] / dt.$$

In the above display ΔX_t^i denotes the jump amplitude increment $\Delta X_t^i = X_t^i - X_{t-}^i$ of the i-th coordinate. When these moments are null (or negligible) for $m > 2$, check that the Kramer-Moyal expansion resumes to the Fokker-Planck equation

$$\partial_t p_t = - \sum_{1 \leq i \leq r} \partial_{x_i} \left(\alpha_t^i \, p_t \right) + \frac{1}{2} \sum_{1 \leq i, j \leq r} \partial_{x_i, x_j} \left(\alpha_t^{i,j} \, p_t \right).$$

Exercise 181 (Compound Poisson process - Exit times) *We consider the compound Poisson process X_t discussed in exercise 161 and in exercise 171. We further assume that $\mathbb{E}(Y) = 0$ and we denote by T_D the first time X_t exits the set $D = [-a, a] (\ni 0 = X_0)$, for some $a > 0$. Applying theorem 12.5.9 to the martingale presented in exercise 171 check that*

$$\mathbb{E}(T_D) \leq a^2 / (\lambda \mathbb{E}(Y^2)).$$

We let $X_t^x = x + X_t$ be the compound Poisson process starting at some $x \in D$, and T_D^x be its exit time from D. Check that

$$\mathbb{E}(T_D^x) \leq (a^2 - x^2) / (\lambda \mathbb{E}(Y^2)).$$

Exercise 182 (Queueing system) *Consider a queueing system with d servers. Customers arrive at rate $\lambda_1 > 0$ and the servers' service times are independent identically distributed exponential random variables with parameter $\lambda_2 > 0$. Customers wait in line before accessing the first free server.*

- *Describe the evolution of the Markov process X_t representing the number of customers being served in the system.*

- *Describe the infinitesimal generator of X_t and the invariant measure π of X_t.*

Exercise 183 (M/M/1 Queue) *Consider a queueing system with a single server. Customers arrive at rate $\lambda_1 > 0$ and the server's service times are independent and identically distributed exponential random variables with parameter $\lambda_2 > 0$. Customers wait in line in a buffer of infinite size before accessing the server. The server serves the customers one at a time according to a first-come/first-served rule.*

- *Describe the evolution of the Markov process X_t representing the number of customers in the system.*

- *Describe the infinitesimal generator of X_t and the invariant measure π of X_t.*

Exercise 184 (M/M/m Queue) *Customers arrive at rate $\lambda_1 > 0$ in a queueing system with m servers with a common service rate $\lambda_2 > 0$. Customers wait in a single buffer before accessing in the first free server.*

- *Describe the evolution of the Markov process X_t representing the number of customers being served in the system.*

- *Describe the infinitesimal generator of X_t and the invariant measure π of X_t.*

Exercise 185 (Kac's model of gases) 🖊 *We consider the velocities $X_t = \left(X_t^i\right)_{1 \le i \le N}$ of N interacting particles on $S = \mathbb{R}$. At rate $N\lambda$, a pair of particles (i,j) is chosen at random and their velocities (X_{t-}^i, X_{t-}^j) are changed to new values*

$$(X_t^i, X_t^j) = \left(\cos(\theta)\ X_t^i + \sin(\theta)\ X_t^j, -\sin(\theta)\ X_t^i + \cos(\theta)\ X_t^j\right)$$

with a randomly chosen angle θ on $[0, 2\pi[$. Write the infinitesimal generator of X_t in terms of an embedded discrete time Markov chain model. Show that the total kinetic energy $\sum_{1 \le i \le N}(X_t^i)^2$ is preserved.

Exercise 186 (Maximum principle) *We let L be the infinitesimal generator of a pure jump process on some state space S. For any bounded function f on S prove that*

$$f(x^\star) = \sup_{x \in S} f(x) \Longrightarrow L(f)(x^\star) \le 0.$$

Exercise 187 (Dirichlet forms) *We consider a jump process X_t on a finite state space S, with infinitesimal generator L defined in (12.26), and a semigroup P_t. We assume that X_t has an unique invariant probability measure π and we set*

$$\mathrm{Var}_\pi(f) = \pi((f - \pi(f))^2) \quad and \quad \mathcal{E}(f, g) := -\pi(fL(g)).$$

The functional $\mathcal{E}$ is called the Dirichlet form associated with L and π. Prove that

$$-\partial_t \pi(f P_t(g))_{|t=0} = \mathcal{E}(f, g) \quad and \quad -\frac{1}{2}\partial_t \mathrm{Var}_\pi(P_t(f)) = \mathcal{E}(P_t(f), P_t(f))$$

and

$$P_t(f) \to_{t \to \infty} \pi(f) \Longrightarrow \mathrm{Var}_\pi(f) = -2 \int_0^\infty \mathcal{E}(P_t(f), P_t(f))\ dt.$$

Exercise 188 (Poincaré inequality) ✎ *We return to the exercise 187. We say that π satisfies a Poincaré inequality for some parameter $a > 0$ if we have*

$$\frac{a}{2} \operatorname{Var}_\pi(f) \leq \mathcal{E}(f, f)$$

for any function f on S (see also exercise 258 in the context of diffusion processes). Check that this condition is equivalent to the exponential variance decay property of the semigroup

$$\operatorname{Var}_\pi(P_t(f)) \leq e^{-at} \operatorname{Var}_\pi(f).$$

Exercise 189 (Ising model) ✎✎ *We let $S = \{-1, +1\}^{\mathcal{V}}$ with some finite undirected graph $(\mathcal{V}, \mathcal{E})$. The set $\mathcal{V}$ denotes the set of vertices and $\mathcal{E}$ is the set of edges. For any pair of vertices $v_1, v_2 \in \mathcal{V}$, we set $v_1 \sim v_2$ when $(v_1, v_2) \in \mathcal{E}$ (and $(v, v) \notin \mathcal{E}$ and $v_1 \sim v_2 \Rightarrow (v_1, v_2) = (v_2, v_1)$). For any $x = (x(v))_{v \in \mathcal{V}} \in S$, $u \in \mathcal{V}$ and $\epsilon \in \{-1, +1\}$ we set*

$$x^{v,\epsilon} = (x^{v,\epsilon}(u))_{u \in \mathcal{V}} \quad \text{with} \quad x^{v,\epsilon}(u) := \begin{cases} x(u) & \text{for} \quad u \neq v \\ \epsilon & \text{for} \quad u = v. \end{cases}$$

We consider the S-valued process X_t that jumps from x to $y \in S$ at a rate

$$Q(x, y) = \sum_{v \in \mathcal{V}, \epsilon \in \{-1, +1\}} q_\epsilon(v, x) \, 1_{x^{v,\epsilon}}(y) \quad \text{with} \quad q_\epsilon(v, x) = 1 \wedge \exp\left(2\epsilon\beta \sum_{u \sim v} x(u)\right).$$

We also let π_β be the Boltzmann-Gibbs measure given by

$$\pi_\beta(x) = \frac{1}{\mathcal{Z}_\beta} \exp\left(-\beta H(x)\right) \quad \text{with the Hamiltonian function} \quad H(x) = \sum_{(v_1, v_2) \in \mathcal{E}} x(v_1) x(v_2)$$

and the normalizing constant $\mathcal{Z}_\beta := \sum_{x \in S} \exp\left(-\beta H(x)\right)$.

- *Prove that*

$$H(x) = |E| - \frac{1}{2} \sum_{(v_1, v_2) \in \mathcal{E}} (x(v_1) - x(v_2))^2.$$

- *Check that $\pi_\beta(x) Q(x, y) = \pi_\beta(y) Q(y, x)$ and deduce that π_β is an invariant measure of the pure jump process X_t described above (in the sense that $\operatorname{Law}(X_0) = \pi_\beta \Rightarrow \operatorname{Law}(X_t) = \pi_\beta$).*

Exercise 190 (High order discrete time schemes) ✎

The approximation theorem 12.3.1 can be extended to any order by considering discrete time sg $P^h_{t_n, t_{n+1}}$ of the form

$$P^h_{t_n, t_{n+1}} := \sum_{0 \leq p \leq m} \frac{1}{p!} \, h^p \, L^p_{t_n} \tag{12.35}$$

for some integer $m \geq 1$. For pure jump processes with generator (12.6), we need to find some sufficiently small h s.t. $\lambda_t(x) \leq 1/h$, for any $x \in S$. In this situation, check that

$$P^h_{t_p, t_n} = P_{t_p, t_n} + \mathrm{O}(h^m).$$

Describe the sg (12.35) when $m = 1$ and $m = 2$. In the general case, prove that

$$P^h_{t_n, t_{n+1}} = \sum_{0 \leq p \leq m} \alpha^h_{t_n}(p) \, M^p_{t_n}$$

with some probability measure $\alpha^h_{t_n}$ on the set of integers $\{0, \ldots, m\}$.

Exercise 191 ✒ *We let N_t be a Poisson process with a continuous intensity function λ_t. We denote by $\mathcal{F}_t = \sigma(N_s \ , \ s \leq t)$ the σ-field generated by the Poisson process N_s, up to time t. We also consider a parameter $\epsilon > -1$, and a càdlàg process φ_t adapted to the filtration $\mathcal{F}_t$, in the sense that φ_t is $\mathcal{F}_t$-measurable for any $t \geq 0$. Check that the following processes are martingales:*

$$
\begin{aligned}
dM_t &= dN_t - \lambda_t dt \qquad d\mathcal{M}_t = dM_t^2 - dN_t \\
d\overline{M}_t &:= \varphi_{t-} \, dM_t \qquad d\widetilde{M}_t = d\overline{M}_t^2 - \varphi_{t-}^2 \, \lambda_t^2 \, dt \\
\mathcal{E}_t &= \exp\left(\int_0^t \varphi_{s-} dN_s - \int_0^t \lambda_s \, [e^{\varphi_{s-}} - 1] \, ds \right) \\
\mathcal{E}_t^\epsilon &= (1+\epsilon)^{N_t} \, \exp\left(-\epsilon \int_0^t \lambda_s \, ds \right).
\end{aligned}
$$

Hint: Use the decomposition $\mathcal{E}_{t+dt}/\mathcal{E}_t = e^{\varphi_t - dN_t - \lambda_t \, (e^{\varphi_t} - 1)dt}$ and the fact that $dN_t = N_{t+dt} - N_t$ is a Poisson random variable with intensity $\lambda_t dt$.

13

Piecewise deterministic processes

The piecewise deterministic processes studied in this chapter are jump type stochastic processes whose trajectories between the jumps are deterministic. They may arise when approximating the solution of a dynamical system via time discretization. They also arise naturally when the constant evolution model between jumps of a pure jump process is replaced by some deterministic flow of a dynamical system. Application areas of such processes include ruin theory, communication networks, queuing theory, biochemistry and bacterial population growth. We describe the evolution semigroups and the infinitesimal generators, as well as the Doeblin-Itō formulae for these processes. At the end of the chapter, we provide a thorough discussion of applications in the transmission control protocol (TCP) used to control the information transmission in internet applications.

> *Anyone who attempts to generate random numbers*
> *by deterministic means is, of course, living in a state of sin.*
> John von Neumann (1903-1957).

13.1   Dynamical systems basics

13.1.1   Semigroup and flow maps

We consider a real valued (and deterministic) dynamical system

$$\frac{dx_t}{dt} = a_t(x_t) \qquad (13.1)$$

associated with some well behaving Lipschitz function $a : (t, x) \in \mathbb{R} \mapsto a_t(x) \in \mathbb{R}$ so that (13.1) has a unique solution over the entire time axis.

Definition 13.1.1 *The flow map of (13.1) is given by the family of mappings $x \mapsto \varphi_{s,t}(x)$, with $s \leq t$, where $\varphi_{s,t}(x)$ is the solution of (13.1) starting from some state $x_s = x$ at time s. The evolution semigroup of (13.1) is defined by the operators $P_{s,t}$ acting on the space of bounded functions by the formula*

$$f \mapsto P_{s,t}(f) = f \circ \varphi_{s,t}.$$

By construction, for any $s \leq r \leq t$ and $x \in \mathbb{R}$ we have the semigroup (sg) properties

$$\varphi_{s,t} = \varphi_{r,t} \circ \varphi_{s,r} \quad \text{and} \quad P_{s,r} P_{r,t} = P_{s,t}.$$

The r.h.s. formula is a consequence of the fact that

$$P_{s,r}(P_{r,t}(f)) = P_{s,r}(f \circ \varphi_{r,t}) = f \circ \varphi_{r,t} \circ \varphi_{s,r} = f \circ \varphi_{s,t} = P_{s,t}(f).$$

We further assume that

$$\varphi_{s,t+\epsilon}(x) = \varphi_{s,t}(x) + \mathrm{O}(\epsilon) \quad \text{and} \quad \varphi_{s+\epsilon,t}(x) = \varphi_{s,t}(x) + \mathrm{O}(\epsilon) \tag{13.2}$$

for some function $\mathrm{O}(\epsilon)$ that may depend on the parameters s, t, x, and that

$$\varphi_{s,t}(x) = \varphi_{s,t}(y) + \mathrm{O}\left(|x - y|\right)$$

for some function $\mathrm{O}(\epsilon)$ that may depend on the parameters t, x.

These conditions ensure the continuity of the evolution sg; for instance, we have

$$P_{s,t+\epsilon}(f)(x) \longrightarrow_{\epsilon \to 0} P_{s,t}(f)(x) \longleftarrow_{\epsilon \to 0} P_{s+\epsilon,t}(f)(x).$$

A detailed discussion of the conditions and hypotheses under which these properties hold would require an excursion to the theory of differential equations beyond what is appropriate for these lecture notes. We can make these hypotheses plausible by considering the traditional flow maps of linear dynamical systems. Given a couple of regular bounded Lipschitz functions α_t and β_t, the function

$$\begin{aligned}
x \mapsto \varphi_{s,t}(x) &:= \exp\left(\int_s^t \alpha_u \, du\right) x + \int_s^t \exp\left(\int_r^t \alpha_u \, du\right) \beta_r \, dr \\
&= \exp\left(\int_s^t \alpha_u \, du\right) \left[x + \int_s^t \exp\left(-\int_s^r \alpha_u \, du\right) \beta_r \, dr\right]
\end{aligned}$$

is the general solution of the linear system

$$\forall s \leq t \qquad \frac{dx_t}{dt} = a_t(x_t) = \alpha_t x_t + \beta_t$$

with the initial condition $x_s = x$. In this situation, conditions (13.2) are satisfied as soon as the functions α_t and β_t are Lipschitz continuous w.r.t. the time parameter. We notice that

$$|\varphi_{s,t}(x) - \varphi_{s,t}(y)| \leq e^{\|\alpha\|(t-s)} |x - y|.$$

In addition, we also have

$$\varphi_{s,t+\epsilon}(x) = \exp\left(\int_t^{t+\epsilon} \alpha_u \, du\right)$$

$$\times \left[\varphi_{s,t}(x) + \exp\left(\int_s^t \alpha_u \, du\right) \int_t^{t+\epsilon} \exp\left(-\int_s^r \alpha_u \, du\right) \beta_r \, dr\right]$$

from which we conclude that

$$\varphi_{s,t+\epsilon}(x) - \varphi_{s,t}(x) = \left[\exp\left(\int_t^{t+\epsilon} \alpha_u \, du\right) - 1\right] \varphi_{s,t}(x)$$

$$+ \exp\left(\int_s^{t+\epsilon} \alpha_u \, du\right) \int_t^{t+\epsilon} \exp\left(-\int_s^r \alpha_u \, du\right) \beta_r \, dr.$$

Using the rather crude upper bound $|e^a - e^b| \leq |a - b| \, e^{|a| \wedge |b|}$ which is valid for any $a, b \in \mathbb{R}$, we prove that

$$\epsilon^{-1} |\varphi_{s,t+\epsilon}(x) - \varphi_{s,t}(x)| \leq \|\alpha\| e^{1 \wedge (\|\alpha\|)} |\varphi_{s,t}(x)| + e^{\|\alpha\|((t-s)+1)} \, \exp\left(\|\alpha\|((t-s)+1)\right) \|\beta\| .$$

In further developments, we always assume that the flow maps are well defined $\varphi_{s,t}(x)$ for any $s \leq t$ and any x.

For strongly nonlinear systems, it may happen that the solution is not defined at any time or on any initial states. For instance, the solution of the quadratic system

$$\forall s \leq t \qquad \frac{dx_t}{dt} = a_t(x_t) = x_t^2 \tag{13.3}$$

starting at $x_s = x$ at time $t = s$ is given by

$$\varphi_{s,t}(x) = \frac{x}{1 - x \ (t - s)}.$$

When $x > 0$, the solution is defined for any $t \in \left[s, s + \frac{1}{x}\right[$ with an explosion $\lim_{t \to t_s(x)} \varphi_{s,t}(x) = \infty$ at time $t_s(x) = s + \frac{1}{x}$. If $x < 0$, the solution is defined for any $t \geq s$ and converges to 0 as $t \uparrow \infty$.

For smooth functions f we have

$$\begin{aligned}
\partial_t P_{s,t}(f)(x) &= \partial_t f(\varphi_{s,t}(x)) \\
&= \partial_t f\left(x + \int_s^t a_r(\varphi_{s,r}(x)) \ dr\right) = a_t(\varphi_{s,t}(x)) \ \partial_x f(\varphi_{s,t}(x)).
\end{aligned}$$

In terms of the first order differential operators

$$f \mapsto L_t(f) = a_t \ \partial_x f.$$

we have proved the forward evolution equation

$$\partial_t P_{s,t}(f) = P_{s,t}(L_t(f)). \tag{13.4}$$

In much the same way, we have the backward evolution equation

$$\partial_s P_{s,t}(f) = -L_s(P_{s,t}(f)). \tag{13.5}$$

To prove these claims we can use the following decompositions

$$\frac{1}{\epsilon} [P_{s+\epsilon,t} - P_{s,t}] = -\frac{1}{\epsilon} [P_{s,s+\epsilon} - Id] \, P_{s+\epsilon,t} \tag{13.6}$$

and

$$-\frac{1}{\epsilon} [P_{s-\epsilon,t} - P_{s,t}] = -\frac{1}{\epsilon} [P_{s-\epsilon,s} - Id] \, P_{s,t}, \tag{13.7}$$

as well as

$$\frac{1}{\epsilon} [P_{s,t+\epsilon} - P_{s,t}] = P_{s,t} \left(\frac{1}{\epsilon} [P_{t,t+\epsilon} - Id]\right). \tag{13.8}$$

For instance, we have

$$
\begin{aligned}
P_{s-\epsilon,s}(f)(x) - f(x) &= f\left(x + \int_{s-\epsilon}^{s} a_u(\varphi_{s-\epsilon,u}(x))du\right) - f(x) \\
&= \partial_x f(\varphi_{s-\epsilon,s}(x)) \int_{s-\epsilon}^{s} a_u(\varphi_{s-\epsilon,u}(x))du + O(\epsilon^2).
\end{aligned}
$$

Using the fact that

$$
a_u(\varphi_{s-\epsilon,u}(x)) = a_{s-\epsilon}(\varphi_{s-\epsilon,s-\epsilon}(x)) + O(\epsilon) = a_{s-\epsilon}(x) + O(\epsilon)
$$

we conclude that

$$
\epsilon^{-1}\left[P_{s-\epsilon,s} - Id\right](f)(x) = a_{s-\epsilon}(x)\,\partial_x f(\varphi_{s-\epsilon,s}(x)) + O(\epsilon) \longrightarrow_{\epsilon\to 0} L_s(f)(x).
$$

13.1.2   Time discretization schemes

To get some feasible solution of the evolution equation (13.1), we often need to resort to a discrete time approximation scheme on some time mesh sequence on some interval $[0,t]$

$$
t_0 = 0 < t_1 = h < \ldots < t_n = nh < \ldots < h\lfloor t/h\rfloor \le t
$$

with some time mesh parameter $h > 0$.

It is clearly beyond the scope of these lectures to review all the basic time discretization techniques of dynamical systems. Trying not to obscure the main ideas, we assume for simplicity that there exists a discrete time approximation $\varphi_{s,t}^h(x)$ of the flow $\varphi_{s,t}(x)$

$$
\varphi_{t_n,t_{n+1}}^h(x) = \mathcal{F}_{t_n,t_{n+1}}^h\left(\varphi_{t_{n-1},t_n}^h(x)\right) \tag{13.9}
$$

for functions $F_{s,t}^h$, such that for any s we have

$$
\varphi_{s,s+h}^h(x) = \varphi_{s,s+h}(x) + O(h^2) \tag{13.10}
$$

for a function $o(h^2)$ that may depend on x.

We also assume that $\sup_{h>0} \sup_{0\le t_n\le t} \left|\varphi_{0,t_n}^h(x)\right| < \infty$ for any $x \in \mathbb{R}$. For instance, we can choose the traditional Euler approximation of the equation (13.1)

$$
\varphi_{s,s+h}^h(x) = x + a_s(x)\,h := \mathcal{F}_{s,s+h}^h(x).
$$

In this situation, we have

$$
\varphi_{s,s+h}(x) - \varphi_{s,s+h}^h(x) = \int_s^{s+h} \left[a_r\left(\varphi_{s,r}(x)\right) - a_s(x)\right]\,dr.
$$

For bounded and Lipschitz functions $(t,x) \mapsto a_t(x)$, we readily prove that

$$
\begin{aligned}
\left|\varphi_{s,s+h}^h(x) - \varphi_{s,s+h}(x)\right| &\le l(a)\left[h^2 + \int_s^{s+h} \left|\varphi_{s,r}(x) - x\right|\,dr\right] \\
&\le l(a)[1 + \|a\|]\,h^2.
\end{aligned}
$$

This inequality follows from the fact that

$$
\left|\varphi_{s,r}(x) - x\right| = \left|\int_s^r a_u(\varphi_{s,u}(x))\,du\right| \le \|a\|\,(r-s).
$$

To convince the reader that these rather crude assumptions suffice to approximate the flow map on any interval, we end this section with a brief discussion around this theme. Using the telescoping sum

$$\varphi^h_{0,t_m}(x) - \varphi_{0,t_m}(x) = \sum_{0 \leq n < m} \left[\varphi_{t_{n+1},t_m}(\varphi^h_{0,t_{n+1}}(x)) - \varphi_{t_n,t_m}(\varphi^h_{0,t_n}(x)) \right]$$

and the sg property $\varphi_{t_n,t_m} = \varphi_{t_{n+1},t_m} \circ \varphi_{t_n,t_{n+1}}$, we readily prove that

$$\left| \varphi^h_{0,t_m}(x) - \varphi_{0,t_m}(x) \right|$$

$$\leq \sum_{0 \leq n < m} l(\varphi) \underbrace{\left| \varphi^h_{t_n,t_{n+1}}(\varphi^h_{0,t_n}(x)) - \varphi_{t_n,t_{n+1}}(\varphi^h_{0,t_n}(x)) \right|}_{=h\, \mathrm{O}(h)}.$$

This implies that

$$\left| \varphi^h_{0,t_m}(x) - \varphi_{0,t_m}(x) \right| \leq \mathrm{O}(h)\, l(\varphi)\, t_m.$$

13.2 Piecewise deterministic jump models

13.2.1 Excursion valued Markov chains

We use the framework and the notation introduced in section 12 and in section 13.1 dedicated to pure jump processes and dynamical systems.

> **Definition 13.2.1** *A piecewise deterministic Markov process (abbreviated PDMP) on $S = \mathbb{R}$ is defined as a pure jump process introduced in chapter 12 by replacing the constant evolution model between the jumps (12.1) by the deterministic flow of the dynamical system*
>
> $$\forall T_n \leq s < T_{n+1} \qquad X_s := \varphi_{T_n,s}(X_{T_n}). \qquad (13.11)$$
>
> *The extension of this definition to more general state space models S is immediate.*

Remark : Piecewise deterministic Markov processes were introduced by M.H.A. Davis [63] in 1984. Since then, these stochastic models have been used in several application domains, including ruin theory [117], communication networks and queueing theory [35], biochemistry, neuronal function models, bacterial population growth [43], and geosciences [211].

The piecewise deterministic jump model can be defined in terms of a simple *discrete generation Markov chain* in the space of excursions.

We start the process $X_0 = x_0$ at an initial location at time $T_0 = 0$, and we wait up to the time

$$T_1 = \inf \left\{ t \geq T_0 \ : \ \int_{T_0}^t \lambda_s(\varphi_{T_0,s}(X_{T_0}))\, ds \geq \mathcal{E}_1 \right\}.$$

At that time we have defined the random excursion

$$\mathcal{X}_0 = (\, T_0, \varphi_{T_0,s}(X_{T_0}), \ s \in [T_0, T_1[\,)$$

up to the time T_1 (*excluded*). Then, we sample a random variable

$$X_{T_1} \simeq K_{T_1} \left(\varphi_{T_1,s}(X_{T_0}), dx \right).$$

We wait up to the first time

$$T_2 = \inf \left\{ t \geq T_1 \ : \ \int_{T_1}^t \lambda_s(\varphi_{T_1,s}(X_{T_1})) \, ds \geq \mathcal{E}_2 \right\}.$$

At that time we have defined the random excursion

$$\mathcal{X}_1 = \left(T_1, \varphi_{T_1,s}(X_{T_1}), \ s \in [T_1, T_2[\ \right)$$

and so on. A synthetic description of these two excursions is given in the following picture.

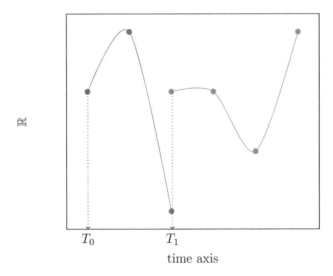

time axis

In summary, given the excursion

$$\mathcal{X}_n = \left(T_n, \varphi_{T_n,s}(X_{T_n}), \ s \in [T_n, T_{n+1}[\ \right)$$

up to the terminal time

$$T_{n+1} = \inf \left\{ t \geq T_n \ : \ \int_{T_n}^t \lambda_s(\varphi_{T_n,s}(X_{T_n})) \, ds \geq \mathcal{E}_n \right\}$$

we randomly sample a jump

$$X_{T_{n+1}-} \ \leadsto \ X_{T_{n+1}} \simeq K_{T_{n+1}} \left(\varphi_{T_n,s}(Y_n), dx \right).$$

Starting from this location, we consider the next excursion up to time T_{n+2}

$$\mathcal{X}_{n+1} = \left(T_{n+1}, \varphi_{T_{n+1},s}(X_{T_{n+1}}), \ s \in [T_{n+1}, T_{n+2}[\ \right).$$

The resulting stochastic process $\mathcal{X}_n$ is a Markov chain in excursion spaces.

By construction, for any $s \geq t$ we have

$$\mathbb{P}\left(T_{n+1} \in dt, X_{T_{n+1}} \in dy \mid T_n = s, X_{T_n} = x\right)$$

$$= \underbrace{\lambda_t(\varphi_{s,t}(x)) \, \exp\left(-\int_s^t \lambda_r(\varphi_{s,r}(x)) \, dr\right) \, dt}_{\mathbb{P}(T_{n+1} \in dt \mid T_n = s, X_{T_n} = x)} \qquad \underbrace{K_t(\varphi_{s,t}(x), dy)}_{\mathbb{P}(X_{T_{n+1}} \in dy \mid T_{n+1} = t, T_n = s, X_s = x)}.$$

$$(13.12)$$

For PDMP, the telescoping sums (12.2) and (12.2) take the form

$$X_t = X_s + \int_s^t a_r(X_r) \, dr + \sum_{s < T_n \leq t} (X_{T_n} - X_{T_n -}) \qquad (13.13)$$

$$f(X_t) = f(X_s) + \int_s^t L_r(f)(X_r) \, dr + \sum_{s < T_n \leq t} (f(X_{T_n}) - f(X_{T_n -})).$$

13.2.2 Evolution semigroups

Unless otherwise stated we always assume that the mappings

$$(t, x) \mapsto \lambda_t(x) \qquad (t, x) \mapsto \varphi_{s,t}(x) \quad \text{and} \quad (t, x) \mapsto K_t(f)(x) \qquad (13.14)$$

are smooth functions, for any $s \in [0, \infty[$ and $f \in \mathcal{B}(\mathbb{R})$. The term "smooth" is used to describe functions which have bounded derivatives of arbitrary order. These regularity conditions can be relaxed using appropriate domains of definitions. We have chosen these strong conditions to clarify the presentation and to concentrate on the stochastic modeling rather than on sophisticated analytical considerations.

The following proposition is the extended version of theorem 11.4.3 to PDMP. In contrast to the pure jump model discussed in section (11.4) we already mention that the forthcoming result cannot be used to derive directly the backward or the forward evolution equations of the sg of the PDMP.

Proposition 13.2.2 *The sg of the PDMP process, defined for any $s \leq t$ by*

$$P_{s,t}(f)(x) = \mathbb{E}\left(f(X_t) \mid X_s = x\right),$$

maps smooth functions into smooth functions, and it satisfies the equation

$$P_{s,t}(f)(x) = f(\varphi_{s,t}(x)) \, e^{-\int_s^t \lambda_u(\varphi_{s,u}(x))du}$$

$$+ \int_s^t \lambda_u(\varphi_{s,u}(x)) \, e^{-\int_s^u \lambda_r(\varphi_{s,r}(x))dr} \, (K_u P_{u,t})(f)(\varphi_{s,u}(x)) \, du. \qquad (13.15)$$

Proof :
We have the decomposition

$$P_{s,t}(f)(x) = \mathbb{E}\left(f(X_t) \, 1_{T^s > t} \mid X_s = x\right) + \mathbb{E}\left(f(X_t) \, 1_{T^s \leq t} \mid X_s = x\right)$$

with the first time of the jump after time s, and starting at x at that time

$$T^s = \inf \left\{ t \geq s \; : \; \int_s^t \lambda_r(\varphi_{s,r}(x)) \, dr \geq \mathcal{E}_n \right\}$$

$$
\begin{aligned}
\mathbb{E}\left(f(X_t) \, 1_{T^s > t} \mid X_s = x\right) &= f(\varphi_{s,t}(x)) \, \mathbb{P}\left(T^s > t \mid X_s = x\right) \\
&= f(\varphi_{s,t}(x)) \, \exp\left(-\int_s^t \lambda_r(\varphi_{s,r}(x)) \, dr\right).
\end{aligned}
$$

On the other hand, we have

$$
\begin{aligned}
\mathbb{E}\left(f(X_t) \, 1_{T^s \leq t} \mid X_s = x\right) &= \mathbb{E}\left(\underbrace{\mathbb{E}\left(f(X_t) \mid T^s, X_{T^s}\right)}_{\parallel} \, 1_{T^s \leq t} \mid X_s = x\right) \\
&= \mathbb{E}\left(\overbrace{P_{T^s,t}(f)(X_{T^s})} \, 1_{T^s \leq t} \mid X_s = x\right).
\end{aligned}
$$

Using (13.12) we prove that

$$\mathbb{E}\left(f(X_t) \, 1_{T^s \leq t} \mid X_s = x\right)$$

$$= \int_s^t \left[\int_S K_u(\varphi_{s,u}(x), dy) P_{u,t}(f)(y) \right] \lambda_u(\varphi_{s,u}(x)) \, e^{-\int_s^u \lambda_r(\varphi_{s,r}(x)) dr} \, du.$$

The fact that smooth functions are stable under the action of the sg is a consequence of the decomposition (13.15) and of the hypothesis (13.14). This ends the proof of the proposition. ∎

Definition 13.2.3 *For any intensity function λ, we let $P_{s,t}^\lambda$ with $s \leq t$ be defined by*

$$P_{s,t}^\lambda(f)(x) := f(\varphi_{s,t}(x)) \, \exp\left(-\int_s^t \lambda_u(\varphi_{s,u}(x)) du\right). \qquad (13.16)$$

For the null intensity function $\lambda_t = 0$, we notice that

$$P_{s,t}^0(f)(x) := f(\varphi_{s,t}(x)) = P_{s,r}^0(P_{r,t}^0(f))(x) := P_{r,t}^0(f)(\varphi_{s,r}(x))$$

so that

$$P_{s,t}^\lambda(f) = P_{s,t}^0(f) \, \exp\left[-\int_s^t P_{s,u}^0(\lambda_u) \, du\right]. \qquad (13.17)$$

Proposition 13.2.4 *For any $s \leq t$, and for any bounded function f, we have*

$$P_{s,t}(f) = P_{s,t}^\lambda(f) + \int_s^t P_{s,u}^\lambda(\lambda_u K_u P_{u,t}(f)) \, du. \qquad (13.18)$$

In addition, we have the integral formula

$$P_{s,t}^\lambda(f) = P_{s,t}^0(f) - \int_s^t P_{s,r}^\lambda(\lambda_r P_{r,t}^0(f)) \, dr.$$

Proof :

Formula (13.18) is a direct consequence of (13.15). To check the second claim, we use the fact that

$$\partial_t e^{-\int_s^t \lambda_u(\varphi_{s,u}(x))du} = -\lambda_t(\varphi_{s,t}(x)) \ e^{-\int_s^t \lambda_u(\varphi_{s,u}(x))du}$$

from which we find the integral formula

$$e^{-\int_s^t \lambda_u(\varphi_{s,u}(x))du} = 1 - \int_s^t \lambda_r(\varphi_{s,r}(x)) \ e^{-\int_s^r \lambda_u(\varphi_{s,u}(x))du} \ dr.$$

This implies that

$$P_{s,t}^\lambda(f)(x) = P_{s,t}^0(f)(x) - \int_s^t \lambda_r(\varphi_{s,r}(x)) \ P_{r,t}^0(f)(\varphi_{s,r}(x)) \ e^{-\int_s^r \lambda_u(\varphi_{s,u}(x))du} \ dr.$$

This ends the proof of the proposition. ∎

13.2.3   Infinitesimal generators

We let $\eta_t = \text{Law}(X_t)$ be the distribution of the random states X_t of the PDMP. By construction, we have

$$\forall s \leq t \qquad \eta_t = \eta_s P_{s,t}. \tag{13.19}$$

Theorem 13.2.5 *For any $s \leq t$, and for any smooth function f, we have*

$$\partial_t P_{s,t}^\lambda(f) = P_{s,t}^\lambda(L_t^\lambda(f)) \quad and \quad \partial_s P_{s,t}^\lambda(f) = -L_s^\lambda(P_{s,t}^\lambda(f)) \tag{13.20}$$

with the first order differential operator

$$L_t^\lambda(f) = a_t \ \partial_x f - \lambda_t \ f \tag{13.21}$$

and the drift function defined in (13.1). In addition, we have

$$\begin{aligned} \partial_t \eta_t(f) &= \eta_t(L_t(f)) \\ \partial_t P_{s,t}(f) &= P_{s,t}(L_t(f)) \quad and \quad \partial_s P_{s,t}(f) = -L_s(P_{s,t}(f)) \end{aligned} \tag{13.22}$$

with the infinitesimal generator

$$L_t = L_t^0 + \lambda_t \ [K_t - Id]. \tag{13.23}$$

Proof :

The proof of the first assertion in (13.20) is a direct consequence of the deterministic forward evolution equation (13.4) combined with (13.19). Indeed, combining these two results we have

$$\partial_t P_{s,t}^\lambda(f)(x) = [a_t(\varphi_{s,t}(x)) \ \partial_x f(\varphi_{s,t}(x)) - f(\varphi_{s,t}(x)) \ \lambda_t(\varphi_{s,t}(x)) \ ] e^{-\int_s^t \lambda_u(\varphi_{s,u}(x))du}.$$

To check the r.h.s. of (13.20), by (13.17) firstly we notice that

$$
\begin{aligned}
L_s^0\left(P_{s,t}^\lambda(f)\right) &= a_s\, \frac{\partial P_{s,t}^\lambda(f)}{\partial x} \\
&= \exp\left[-\int_s^t P_{s,u}^0\left(\lambda_u\right)(x)\,du\right] \\
&\quad \times \left[L_s^0\left(P_{s,t}^0(f)\right) - P_{s,t}^0(f)(x)\int_s^t L_s^0(P_{s,u}^0\left(\lambda_u\right))(x)\,du\right].
\end{aligned}
$$

By (13.17) and the deterministic backward evolution equation (13.5) we also have

$$
\begin{aligned}
\partial_s P_{s,t}^\lambda(f) &= e^{-\int_s^t P_{s,u}^0(\lambda_u)(x)\,du}\left[\partial_s P_{s,t}^0(f) - P_{s,t}^0(f)\,\partial_s \int_s^t P_{s,u}^0\left(\lambda_u\right)\right] \\
&= e^{-\int_s^t P_{s,u}^0(\lambda_u)\,du}\left[-L_s^0(P_{s,t}^0(f)) + P_{s,t}^0(f)\int_s^t L_s^0(P_{s,u}^0\left(\lambda_u\right))\right] \\
&\qquad\qquad\qquad + \lambda_s\,\exp\left[-\int_s^t P_{s,u}^0\left(\lambda_u\right)(x)\,du\right]P_{s,t}^0(f).
\end{aligned}
$$

This shows that

$$
\partial_s P_{s,t}^\lambda(f) = -\left[L_s^0\left(P_{s,t}^\lambda(f)\right) - \lambda_s(x)P_{s,t}^\lambda(f)\right].
$$

Now we derive the proof of the r.h.s. of (13.22). Using (13.18), we find that

$$
\partial_s P_{s,t}(f) = \underbrace{\partial_s P_{s,t}^\lambda(f) + \int_s^t \partial_s P_{s,u}^\lambda(K_u P_{u,t}(f))\,du}_{=-L_s^\lambda(P_{s,t}(f))} - K_s P_{s,t}(f).
$$

This implies that

$$
\partial_s P_{s,t}(f) = -\left[L_s^0(P_{s,t}(f)) - \lambda_s P_{s,t}(f) + K_s P_{s,t}(f)\right] = -L_s(P_{s,t}(f)).
$$

The proof of the l.h.s. of (13.22) follows from the following arguments

$$
\begin{aligned}
\partial_t P_{s,t}(f) &= \lim_{\epsilon\to 0} -\frac{1}{\epsilon}\left[P_{s,t-\epsilon} - P_{s,t}\right](f) \\
&= \lim_{\epsilon\to 0} P_{s,t-\epsilon}\left(-\frac{1}{-\epsilon}\left[P_{t-\epsilon,t} - Id\right]\right)(f) = P_{s,t}(L_t(f)).
\end{aligned}
$$

This ends the proof of the theorem. ∎

13.2.4 Fokker-Planck equation

We let X_t be a PDMP associated with a deterministic flow

$$
dx_t = a_t(x_t)\,dt
$$

on $\mathbb{R}^r$, for some $r \geq 1$. Here $a_t = (a_t^i)_{1\leq i\leq r}$ stands for some smooth vector field from $S = \mathbb{R}^r$ into itself. In this situation, the first order generator L_t^0 takes the form

$$
L_t^0(f) = \sum_{1\leq i\leq r} a_t^i\,\partial_{x_i}(f).
$$

We further assume that $K_t(y, dx) = m_t(y, x)\, dx$ and $\eta_t(dx) = p_t(x)dx$ have a density w.r.t. the Lebesgue measure dx on $\mathbb{R}^r$. In this situation, using (13.22) we have

$\int\ f(x)\ \partial_t p_t(x)\ dx$

$= \sum_{1 \leq i \leq r}\ \int\ a_t^i(x)\ \partial_{x_i}(f)(x)\ p_t(x)\ dx + \int\ \lambda_t(x)\ (f(y) - f(x))\ m_t(x, y)\ p_t(x)\ dx\ dy.$

Notice that

$$\int\ \lambda_t(x)\ (f(y) - f(x))\ m_t(x, y)\ p_t(x)\ dx\ dy$$

$$= \int\ f(x)\ \left\{ \left[\int\ m_t(y, x)\ \lambda_t(y)\ p_t(y)\ dy \right] - \lambda_t(x)\ p_t(x) \right\}\ dx.$$

For compactly supported functions f we also have the the integration by part formula

$$\int\ a_t^i(x)\ \partial_{x_i}(f)(x)\ p_t(x)\ dx = -\int\ f(x)\ \partial_{x_i}(a_t^i p_t)(x)\ dx.$$

Combining these two formulae we obtain

$\int\ f(x)\ \partial_t p_t(x)\ dx$

$$= \int\ f(x)\ \left[-\sum_{1 \leq i \leq r} \partial_{x_i}(a_t^i p_t)(x) + \left\{ \left[\int\ m_t(y, x)\ \lambda_t(y)\ p_t(y)\ dy \right] - \lambda_t(x)\ p_t(x) \right\} \right]\ dx$$

for any smooth and compactly supported function f.

This yields the Fokker-Planck equation

$$\partial_t p_t(x) = L_t^\star(p_t)(x) \qquad\qquad (13.24)$$

$$:= -\sum_{1 \leq i \leq r} \partial_{x_i}(a_t^i p_t)(x) + \left[\int\ m_t(y, x)\ \lambda_t(y)\ p_t(y)\ dy \right] - \lambda_t(x)\ p_t(x).$$

The dual $\star$-type notation is due to

$$\langle p_t, L_t(f) \rangle = \langle L_t^\star(p_t), f \rangle \quad \text{with the inner product} \quad \langle f, g \rangle = \int\ f(x)\ g(x)\ dx$$

associated with some vector spaces of functions satisfying some regularity properties.

The invariant and reversible measures of time homogeneous PDMP (with $(a_t, \lambda_t, L_t) = (a, \lambda, L)$) are defined as in definition 12.7.2 and definition 12.7.4. Proposition 12.7.3 is also valid for PDMP processes. In particular, if an invariant measure $\pi(dx) = p(x)dx$ has a density $p(x)$ w.r.t. the Lebesgue measure dx on $\mathbb{R}^r$, we have

$$(13.24) \Rightarrow L^\star(p) = 0.$$

13.2.5   A time discretization scheme

We let $h > 0$ be a time mesh parameter associated with some discrete time discretization of an interval $[0, t]$

$$t_0 = 0 < t_1 = h < \ldots < t_n = nh < \ldots < h\lfloor t/h \rfloor \leq t. \qquad (13.25)$$

We also consider a discrete time approximation $\varphi^h_{t_p,t_n}(x)$ of the flow $\varphi_{t_p,t_n}(x)$, $t_p \leq t_n$ (we refer the reader to section 13.1.2, formula (13.9) for a discussion on these discrete generation models). The semigroup $P^0_{t_p,t_n}$ associated with this discrete generation deterministic model is given by the one-step Dirac transitions

$$P^{0,h}_{t_n,t_{n+1}}(x,dy) = \delta_{\varphi^h_{t_n,t_{n+1}}(x)}(dy).$$

Using (13.10), we have

$$
\begin{aligned}
\frac{1}{h}\left[P^{0,h}_{t,t+h} - Id\right](f)(x) &= \frac{1}{h}\left[f\left(\varphi^h_{t,t+h}(x)\right) - f(x)\right] \\
&= \underbrace{\frac{1}{h}\left[\varphi_{t,t+h}(x) - x\right]\,\partial_x f(x)}_{\longrightarrow_{h\to 0} L^0_t(f)(x)} + O(h)
\end{aligned}
$$

with the first order infinitesimal generators L^0_t defined in section 13.2.3. This implies that

$$P^{0,h}_{t,t+h} = Id + h\, L^0_t + + O(h^2). \tag{13.26}$$

The jump terms of the PDMP are defined in terms of sequence of independent exponential r.v. $\mathcal{E}_n = -\log U_n$ with unit parameters ($\Rightarrow U_n$ uniform on $[0,1]$).

The discrete time approximation PDMP $X^h_{t_n}$, $n \in \mathbb{N}$, of the PDMP process X_t defined in (13.11) can be interpreted as jump type Markov chain on time steps t_n.

- Between the jumps times T^h_n and T^h_{n+1}, the process evolves according to the deterministic flow of the discrete time model

$$\forall T^h_n \leq t_k < T^h_{n+1} \qquad X^h_{t_k} := \varphi^h_{T^h_n,t_k}\left(X^h_{T^h_n}\right).$$

- Mimicking the formula (10.12), the $(n+1)$-th jump time is defined by

$$T^h_{n+1} = \inf\left\{t_k \geq T^h_n \;:\; \sum_{T^h_n < t_l \leq t_k} \lambda_{t_l}\left(\varphi^h_{T^h_n,t_l}(X^h_{T^h_n})\right)\, h \geq \mathcal{E}_{n+1}\right\}. \tag{13.27}$$

- At the time T^h_{n+1}, we sample a random variable

$$X^h_{T_{n+1}} \sim K_{T^h_{n+1}}\left(\varphi^h_{T^h_n,T^h_{n+1}}\left(X^h_{T^h_n}\right), dx\right).$$

Definition 13.2.6 *The jump process* $X^h_{t_n}$, $n \in \mathbb{N}$, *is called the discrete time version of the PDMP (abbreviated h-PDMP) associated with the time mesh (13.25). We let* $\eta^h_{t_n} = \mathrm{Law}(X^h_{t_n})$ *be the distribution of the random states* $X^h_{t_n}$ *of the PDMP. By construction, we have*

$$\forall t_p \leq t_n \qquad \eta^h_{t_n} = \eta^h_{t_p} P^h_{t_p,t_n} \tag{13.28}$$

with the semigroup $P^h_{t_p,t_n}$ *of the h-PDMP defined for any bounded function f by the conditional expectation operators*

$$P^h_{t_p,t_n}(f)(x) = \mathbb{E}\left(f(X^h_{t_n}) \;\Big|\; X^h_{t_p} = x\right).$$

Once again, these random times can be interpreted as the jump times of a stochastic time inhomogeneous Bernoulli process with the sequence of independent Bernoulli random variables

$$
\begin{aligned}
\mathbb{P}\left(\epsilon^h_{t_{n+1}} = 0 \ \big| X^h_{t_0}, \ldots, X^h_{t_n}\right) &= 1 - \mathbb{P}\left(\epsilon^h_{t_{n+1}} = 1 \ \big| X^h_{t_0}, \ldots, X^h_{t_n}\right) \\
&= \exp\left(-\lambda_{t_{n+1}}\left(\varphi^h_{t_n,t_{n+1}}(X^h_{t_n})\right)h\right).
\end{aligned}
$$

More precisely, using the same arguments as in (10.5), we find that

$$
\begin{aligned}
T^h_{n+1} &= \inf\left\{t_k > T^h_n \ : \ \epsilon^h_{t_k} = 1\right\} \\
&= \inf\left\{t_k \geq T^h_n \ : \ \prod_{T^h_n < t_l \leq t_k} e^{-\lambda_{t_l}\left(\varphi^h_{T^h_n,t_l}(X^h_{T^h_n})\right)h} \leq U_{n+1}\right\}.
\end{aligned} \tag{13.29}
$$

We see that the $X^h_{t_n}$ is a Markov chain. Given $X^h_{t_n} = x$, we sample a $\{0,1\}$-valued Bernoulli r.v. $\epsilon^h_{t_{n+1}}$ with success probability

$$
1 - \exp\left(-\lambda_{t_{n+1}}\left(\varphi^h_{t_n,t_{n+1}}(x)\right)h\right)
$$

and a random variable $Y^h_{t_{n+1}} \sim K_{t_{n+1}}\left(\varphi^h_{t_n,t_{n+1}}(x), dy\right)$. Then we set

$$
X^h_{t_{n+1}} = \left(1 - \epsilon^h_{t_{n+1}}\right)\varphi^h_{t_n,t_{n+1}}(x) + \epsilon^h_{t_{n+1}}\, Y^h_{t_{n+1}}.
$$

In summary, we have proved the following proposition.

Proposition 13.2.7 *The elementary Markov transitions of the h-PDMP are given by*
$$
P^h_{t_n,t_{n+1}} = P^{0,h}_{t_n,t_{n+1}} S^h_{t_n,t_{n+1}},
$$
that is, as the composition of the Dirac transition $P^{0,h}_{t_n,t_{n+1}}$ with the jump type transition
$$
S^h_{t_n,t_{n+1}}(x,dy) = e^{-\lambda_{t_{n+1}}(x)h}\,\delta_x(dy) + \left(1 - e^{-\lambda_{t_{n+1}}(x)h}\right)K_{t_{n+1}}(x,dy).
$$

We end this section with the version of theorem 12.3.1 and corollary 12.3.2 for PDMP processes.

Theorem 13.2.8 *For any time $t \in \mathbb{R}_+$, we have*

$$P^h_{t,t+h} = Id + h\, L_t + O(h^2) = P_{t,t+h} \qquad (13.30)$$

on the set of smooth functions. In addition, for any time steps $t_p = h\lfloor s/h \rfloor \leq t_n = h\lfloor t/h \rfloor$, we have the first order approximation formulae

$$P^h_{t_p,t_n} = P_{t_p,t_n} + O(h).$$

For any finite sequence of times $s_1 < \ldots < s_n$ and a sufficiently small h we have the finite dimensional approximation

$$\prod_{1 \leq i < n} P^h_{s_i,s_{i+1}}(x_i, dx_{i+1}) = \prod_{1 \leq i < n} P_{s_i,s_{i+1}}(x_i, dx_{i+1}) + O(h)$$

where $P^h_{s_1,s_2}$ stands for the continuous time semigroup of the process X^h_t defined in (12.14).

Proof :
The proof of the first assertion follows from the following decompositions

$$h^{-1}\left[S^h_{t,t+h} - Id\right] = h^{-1}\left(1 - e^{-\lambda_{t+h}(x)h}\right)[K_{t+h} - Id] = \lambda_t[K_t - Id] + O(h).$$

Using (13.26), we conclude that

$$h^{-1}\left[P^h_{t,t+h} - Id\right] = h^{-1}\left[P^{0,h}_{t,t+h} - Id\right]S^h_{t,t+h} + h^{-1}\left[S^h_{t,t+h} - Id\right] = L_t(f) + O(h)$$

with $L_t := L^0_t + \lambda_t[K_t - Id]$. This shows that the condition (12.13) in theorem 12.3.1 is satisfied for the PDMP models discussed in this section. The l.h.s. expansion in (13.30) has been checked in (13.30). The r.h.s. expansion in (13.30) is proved by combining (13.22) with the first order Taylor expansion presented in (12.10). The end of the proof of the theorem is then a direct consequence of theorem 12.3.1 and the proof of corollary 12.3.2. ∎

Remark : In contrast to the pure jump situation discussed in chapter 12, we observe that the generator L_t in (13.30) is a first order differential operator so that the approximation formulae discussed above such as (13.30) are only valid on sufficiently smooth functions. Following the proof of theorem 12.3.1 we need to ensure that the semigroup $P_{s,t}$ maps smooth functions into smooth functions. One strategy to check this condition is to use the perturbation techniques and the semigroup series expansions discussed in section 15.5.2 and exercises 285 through 287.

13.2.6 Doeblin-Itō-Taylor formulae

Using the same notation as in section 12.5.1, we have the discrete time Doeblin-Itō-Taylor formula

$$\Delta f(X^h)_{t_n} = f(X^h_{t_n}) - f(X^h_{t_{n-1}}) = L^h_{t_{n-1}}(f)(X^h_{t_{n-1}})\, h + \Delta M_{t_n}(f)$$

with the integral operators

$$L^h_{t_{n-1}} = h^{-1}\left[P^h_{t_{n-1},t_n} - Id\right] = L_{t_{n-1}} + O(h) \quad (\Longleftarrow (13.30))$$

and a martingale increment $\Delta M_{t_n}(f)$ (w.r.t. the σ-fields $\mathcal{F}_{t_n}^h = \sigma\left(X_{t_m}^h \ : \ t_m \leq t_n\right)$) such that

$$\mathbb{E}\left(\Delta M_{t_n}(f) \mid X_{t_{n-1}}^h\right) = 0$$

and

$$\text{Var}\left(\Delta M_{t_n}(f) \mid X_{t_{n-1}}^h = x\right) = h\,\Gamma_{L_{t_{n-1}}^h}(f,f)(x) - \left(h\,L_{t_{n-1}}^h(f)(x)\right)^2.$$

In the above display, $\Gamma_{L_{t_{n-1}}^h}$ denotes the square field vector that we defined in 12.5.4. The continuous time version is defined using the same lines of argument as the ones we used in section 12.5.2. In the context of real valued PDMP, we have the following theorem.

The definitions of continuous time martingales and their angle brackets are provided in section 12.5.2.

Theorem 13.2.9 *For any smooth function f, we have*

$$df(X_t) = L_t(f)(X_t)\,dt + dM_t(f)$$

with a martingale $M_t(f)$ (w.r.t. the σ-fields $\mathcal{F}_t = \sigma\left(X_s \ : \ s \leq t\right)$) null at the origin and such that

$$M_t(f)^2 - \langle M(f)\rangle_t \quad \text{with} \quad \langle M(f)\rangle_t := \int_0^t \Gamma_{L_s}(f,f)(X_s)\,ds$$

is a martingale.

13.3 Stability properties

13.3.1 Switching processes

We let $X_t = (X_t^1, X_t^2) \in S := \mathbb{R}^{r=r_1+r_2} = (\mathbb{R}^{r_1} \times \mathbb{R}^{r_2}) := S_1 \times S_2$ be a PDMP associated with a deterministic flow $x_t = \varphi_{s,t}(x_s)$, defined for any $t \geq s$ by the dynamical equations

$$\begin{cases} dx_t^1 &= a_t(x_t)\,dt \\ x_t^2 &= x_s^2 \end{cases}$$

on $\mathbb{R}^{r_1+r_2}$, for some $r_1, r_2 \geq 1$ and some smooth vector field $a_t = (a_t^i)_{1 \leq i \leq r_1}$. In this situation, the second component of the deterministic flow does not change between the jump times T_n. In the reverse angle, let us assume that the Markov transition M_t is defined by

$$K_t((x^1, x^2), d(y^1, y^2)) = \delta_{x^1}(dy^1)\,K_t^{(2)}((x^1, x^2), dy^2) \tag{13.31}$$

where $d(y^1, y^2)$ stands for an infinitesimal neighborhood of the state variable $(y^1, y^2) \in (\mathbb{R}^{r_1} \times \mathbb{R}^{r_2})$. In this situation, the first component of the PDMP does not change during the jump. Only the second switching coordinate $x^2 \rightsquigarrow y^2$ changes according to some Markov transition $K_t^{(2)}((x^1, x^2), dy^2)$ from $\mathbb{R}^r$ into $\mathbb{R}^{r_2}$. By construction the infinitesimal generator of X_t is given for any $x = (x^1, x^2) \in (S_1 \times S_2)$ with $x^1 = (x_i^1)_{1 \leq i \leq r_1} \in S_1 := \mathbb{R}^{r_1}$ by

$$L(f)(x) = \sum_{1 \leq i \leq r_1} a_t^i(x)\,(\partial_{x_i^1}f)(x) + \lambda_t(x)\int (f(x^1, y^2) - f(x^1, x^2))\,K_t^{(2)}(x, dy^2).$$

We further assume that

$$K_t^{(2)}(x, dy^2) = m_t^{(2)}(x, y^2)\, \nu(dy^2) \quad \text{and} \quad \eta_t(dx) = p_t(x^1, x^2)\, dx^1 \nu(dx^2)$$

have a density w.r.t. the Lebesgue measure dx^1 on $\mathbb{R}^{r_1}$ and some measure ν on S_2.

In this situation, using (13.22) we have

$$\int f(x)\, \partial_t p_t(x)\, dx^1 \nu(dx^2)$$

$$= \int \left[\sum_{1 \le i \le r_1} \int a_t^i(x)\, \partial_{x_i^1}(f)(x^1, x^2)\, p_t(x^1, x^2)\, dx^1 \right] \nu(dx^2)$$

$$+ \int \lambda_t(x) \int (f(x^1, y^2) - f(x))\, m_t^{(2)}(x, y^2)\, \nu(dy^2)\, p_t(x)\, dx^1 \nu(dx^2).$$

Notice that

$$\int f(x^1, y^2) \left[\int m_t^{(2)}((x^1, x^2), y^2)\, \lambda_t(x^1, x^2)\, p_t(x^1, x^2)\, \nu(dx^2) \right] dx^1\, \nu(dy^2)$$

$$= \int f(x^1, x^2) \left[\int m_t^{(2)}((x^1, y^2), x^2)\, \lambda_t(x^1, y^2)\, p_t(x^1, y^2)\, \nu(dy^2) \right] dx^1\, \nu(dx^2).$$

For compactly supported functions $x^1 \mapsto f(x^1, x^2)$ we also have the the integration by part formula

$$\sum_{1 \le i \le r_1} \int a_t^i(x)\, \partial_{x_i^1}(f)(x^1, x^2)\, p_t(x^1, x^2)\, dx^1 = - \sum_{1 \le i \le r} \int f(x^1, x^2)\, \partial_{x_i^1}(a_t^i p_t)(x^1, x^2)\, dx^1.$$

Combining these two formulae we obtain

$$\int f(x)\, \partial_t p_t(x)\, dx^1 \nu(dx^2)$$

$$= - \int f(x) \left[\sum_{1 \le i \le r_1} \partial_{x_i^1}(a_t^i p_t)(x^1, x^2) \right] dx^1 \nu(dx^2)$$

$$+ \int f(x) \left[\int m_t^{(2)}((x^1, y^2), x^2)\, \lambda_t(x^1, y^2)\, p_t(x^1, y^2)\, \nu(dy^2) - \lambda_t(x)\, p_t(x) \right] dx^1\, \nu(dx^2).$$

This yields the Fokker-Planck equation

$$\partial_t p_t(x) \quad = \quad L_t^\star(p_t)(x) \tag{13.32}$$
$$:= \quad - \sum_{1 \le i \le r_1} \partial_{x_i^1}(a_t^i p_t)(x)$$
$$+ \int m_t^{(2)}((x^1, y^2), x^2)\, \lambda_t(x^1, y^2)\, p_t(x^1, y^2)\, \nu(dy^2) - \lambda_t(x)\, p_t(x).$$

We can derive the same analysis replacing $\mathbb{R}^{r_2}$ by some abstract state space S_2 equipped with some measure ν. We further assume that ν is the uniform distribution on some state space S_2 and

$$K_t^{(2)}((x^1, x^2), dy^2) = \nu(dy^2) \implies m_t^{(2)}((x^1, y^2), x^2) = 1.$$

In this situation we have

$$(13.32) \iff \partial_t p_t(x) = -\sum_{1 \le i \le r} \partial_{x_i^1}(a_t^i p_t)(x) + \int \lambda_t(x^1, y^2)\, p_t(x^1, y^2)\, \nu(dy^2) - \lambda_t(x)\, p_t(x).$$

13.3.2 Invariant measures

We consider a time homogeneous switching model $(a_t, \lambda_t) = (a, \lambda)$ with some drift function $a_t(x^1, x^2) = a(x^2)$ that does not depend on the first coordinate and such that

$$\forall 1 \le i \le r_1 \qquad \int a^i(x^2)\, \nu(dx^2) = 0.$$

We set $\pi(dx) = q(x^1, x^2)\, dx^1\, \nu(dx^2)$ with some probability density function $q(x^1, x^2) = q(x^1)$ that only depends on the first coordinate and we choose a jump intensity λ such that

$$\lambda(x^1, x^2)\, q(x^1, x^2) = \lambda^\star(x^1) - \sum_{1 \le i \le r_1} a^i(x^2)\, \partial_{x_i^1} q(x^1) \ge 0$$

for some sufficiently large function

$$\lambda^\star(x^1) \ge \max\left(\sum_{1 \le i \le r_1} a^i(x^2)\, \partial_{x_i^1} q(x^1), 0\right). \qquad \cdot$$

We refer to exercise 219 for a more concrete illustration of these intensity rate functions. In this case, we have

$$\int \left[\lambda(x^1, y^2)\, q(x^1, y^2) - \lambda(x^1, x^2)\, q(x^1, x^2)\right]\, \nu(dy^2)$$

$$= \int \left[\lambda^\star(x^1) - \sum_{1 \le i \le r_1} a^i(y^2)\, \partial_{x_i^1}(q)(x^1) - \lambda^\star(x^1) + \sum_{1 \le i \le r_1} a^i(x^2)\, \partial_{x_i^1}(q)(x^1)\right]\, \nu(dy^2)$$

$$= \sum_{1 \le i \le r_1} a^i(x^2)\, \partial_{x_i^1}(q)(x^1) = \sum_{1 \le i \le r_1} \partial_{x_i^1}(a^i q)(x^1, x^2) \Rightarrow \pi L = 0.$$

We conclude that π is an invariant probability measure of the PDMP with infinitesimal generator L. We refer to exercises 213 to 219 for worked-out examples of switching processes with given target distributions.

13.4 An application to Internet architectures

13.4.1 The transmission control protocol

As its name indicates, the transmission control protocol (*abbreviated TCP*) is a sophisticated reliability protocol used by most Internet applications to control the transmission of information. More than 50% of Internet traffic uses TCP. Loosely speaking, computer applications often break long messages into very small pieces called datagrams, and wrap them into a collection of packets called IP transmission units. TCP defines the packaging of the messages and the rules of sending and reading them. Each host maintains a congestion window that measures the number of data units (also called segments). A data unit is set as the maximum segment size allowed on that connection [155, 227].

When the receiver TCP detects a missing segment, the size of the congestion window is decreased by some proportionality factor $(1 + \alpha) \in]0, 1[$, for some $\alpha \in] - 1, 0[$ (say half of the size $(1 + \alpha) = .5$ for $\alpha = -.5$). Otherwise, every time a segment is well received (and the corresponding acknowledgement receipt reaches the sender) the window is slowly opened by a proportion $\beta \in]0, 1[$ of the inverse of the current window size. We assume that all the packets are equal to the maximal system size. We refer the reader to the pioneering articles of T. J. Ott and his co-authors [215, 216] for a more thorough discussion on these models.

We consider a time mesh sequence $t_n = nh$, with $n \in \mathbb{N}$, and a given time step $h \in]0, 1[$. The timeout probabilities are supposed to be small, so that we further assume that the probabilities of a segment loss at every time step have the form

$$p_h = 1 - e^{-\lambda h} = \lambda h + O(h^2)$$

for some given rate $\lambda > 0$.

We let ϵ_{t_n} be a sequence of independent Bernoulli random variables with a common law

$$\mathbb{P}\left(\epsilon_{t_n}^h = 0\right) = 1 - \mathbb{P}\left(\epsilon_{t_n} = 1\right) = e^{-\lambda h}.$$

The congestion window size $W_{t_n}^h$ at time t_n is given by the Markov chain:

$$W_{t_n}^h = (1 - \epsilon_{t_n}) \left(W_{t_{n-1}}^h + h\beta/W_{t_{n-1}}^h\right) + \epsilon_{t_n} \left((1 + \alpha)W_{t_{n-1}}^h\right).$$

The initial condition $W_{t_0}^h = w_0$ is usually set to an unit of the maximal system size.

$$
\begin{aligned}
\Delta W_{t_n}^h &:= W_{t_n}^h - W_{t_{n-1}}^h \\
&= (1 - \epsilon_{t_n}) \ \beta h/W_{t_{n-1}}^h + \epsilon_{t_n} \ \alpha W_{t_{n-1}}^h.
\end{aligned}
$$

We let T_n^h be the jump times (10.4) of the Bernoulli process associated with the $\epsilon_{t_n}^h$. Between the jumps times, we have

$$T_n^h < t_k < T_{n+1}^h \qquad \frac{W_{t_k}^h - W_{t_{k-1}}^h}{t_k - t_{k-1}} = \frac{\beta}{W_{t_{k-1}}^h}.$$

At the jump time T_{n+1}^h, we have

$$\Delta W_{T_{n+1}^h}^h = W_{T_{n+1}^h}^h - W_{T_{n+1}^h - h}^h = \alpha W_{T_{n+1}^h - h}^h.$$

Taking the limits when $h \downarrow 0$, we have seen in section 10.5 that the random times T_n^h converge to the random times T_n of a Poisson process with intensity $\lambda_t = \lambda$. From previous considerations, $W_{t_n}^h$ converge (in law) to the continuous time jump process given by

$$T_n < t < T_{n+1} \qquad \frac{dW_t}{dt} = \frac{\beta}{W_t}.$$

At the jump time T_{n+1}, we have

$$\Delta W_{T_{n+1}} = W_{T_{n+1}} - W_{T_{n+1}-} = \alpha W_{T_{n+1}-} \Rightarrow W_{T_{n+1}} = (1 + \alpha) \ W_{T_{n+1}-}.$$

We let X_t be the process defined by

$$X_t := \frac{1}{2\beta} W_t^2.$$

Between the jumps times, we have

$$\frac{dX_t}{dt} = \frac{1}{\beta} W_t \frac{dW_t}{dt} = 1$$

and at the jump times

$$
\begin{aligned}
X_{T_{n+1}} - X_{T_{n+1}-} &= \frac{1}{2\beta} \left[W_{T_{n+1}}^2 - W_{T_{n+1}-}^2 \right] \\
&= \rho\, X_{T_{n+1}-} \quad \text{with} \quad \rho := (1+\alpha)^2.
\end{aligned}
\tag{13.33}
$$

Using the same reasoning as above, a discrete time approximation of this process is given by the Markov chain

$$X_{t_n}^h = (1 - \epsilon_{t_n}) \left(X_{t_{n-1}}^h + h \right) + \epsilon_{t_n} \left(\rho X_{t_{n-1}}^h \right).$$

13.4.2 Regularity and stability properties

We now turn to studying the long time behavior of these TCP models. Firstly, we observe that

$$
\begin{aligned}
X_{t_n}^h &= a_n\, X_{t_{n-1}}^h + b_n \\
&= a_n a_{n-1}\, X_{t_{n-2}}^h + a_n b_{n-1} + b_n = \ldots = A_{t_n}\, X_{t_0}^h + B_{t_n}
\end{aligned}
$$

with

$$A_{t_n} := \left[\prod_{p=1}^n a_p \right] \quad \text{and} \quad B_{t_n} := \sum_{1 \le p \le n} \left[\prod_{n \ge q > p} a_q \right] b_p$$

and the collection of random variables

$$a_n := (\epsilon_{t_n} \rho + (1 - \epsilon_{t_n})) = \rho^{\epsilon_{t_n}} \quad \text{and} \quad b_n := (1 - \epsilon_{t_n})\, h.$$

We notice that

$$\mathbb{E}(a_n) = \rho \left(1 - e^{-\lambda h}\right) + e^{-\lambda h} < 1 \quad \text{and} \quad \mathbb{E}(b_n) = h e^{-\lambda h}.$$

This implies that A_{t_n} converges almost surely exponentially fast to 0, as $n \uparrow \infty$ in the sense that

$$\mathbb{P}\left(A_{t_n} > \epsilon\right) \le \epsilon^{-1} \mathbb{E}(A_{t_n}) = \epsilon^{-1} \left[\rho \left(1 - e^{-\lambda h}\right) + e^{-\lambda h} \right]^n \longrightarrow_{n \uparrow \infty} 0$$

for any $\epsilon > 0$.

Note that we also have

$$
\begin{aligned}
&\text{Law}\left((a_1, \ldots, a_{p+1}, \ldots, a_n), (b_1, \ldots, b_p, \ldots, b_n)\right) \\
&= \\
&\text{Law}\left((a_n, \ldots, a_{n-p}, \ldots, a_1), (b_n, \ldots, b_{n-p+1}, \ldots, b_1)\right).
\end{aligned}
$$

Therefore, we have

$$
\begin{aligned}
B_{t_n} = \sum_{1 \le p \le n} b_p\, [a_{p+1} \ldots a_n] \quad &\overset{law}{=} \quad \sum_{1 \le p \le n} [a_1 \ldots a_{n-p}]\, b_{(n-p)+1} \\
&= \sum_{0 \le q < n} [a_1 \ldots a_q]\, b_{q+1} := \overline{B}_{t_n}.
\end{aligned}
\tag{13.34}
$$

We also have the almost sure convergence

$$\overline{B}_{t_n} \longrightarrow_{n\uparrow\infty} \overline{B}_\infty := \lim_{n\to\infty} \sum_{0\le q<n} [a_1 \ldots a_q] \, b_{q+1} = \sum_{q\ge 0} [a_1 \ldots a_q] \, b_{q+1}.$$

We check this claim using the fact that $\overline{B}_\infty \ge \overline{B}_{t_n}$, and for any $\epsilon > 0$, we have

$$\mathbb{P}\left(\overline{B}_\infty - \overline{B}_{t_n} > \epsilon\right) \le \epsilon^{-1} \, \mathbb{E}\left(\overline{B}_\infty - \overline{B}_{t_n}\right)$$

and

$$\mathbb{E}\left(\overline{B}_\infty - \overline{B}_{t_n}\right)$$

$$= he^{-\lambda h} \sum_{q\ge n} \left[\rho\left(1 - e^{-\lambda h}\right) + e^{-\lambda h}\right]^q$$

$$= he^{-\lambda h} \left[\rho\left(1 - e^{-\lambda h}\right) + e^{-\lambda h}\right]^n / \left[(1 - e^{-\lambda h})(1 - \rho)\right] \longrightarrow_{n\uparrow\infty} 0.$$

We conclude that

$$\mathrm{Law}\left(X_{t_n}^h\right) = \mathrm{Law}\left(A_{t_n} \, X_{t_n}^h + \overline{B}_{t_n}\right) \longrightarrow_{n\uparrow\infty} \mathrm{Law}\left(\overline{B}_\infty\right).$$

In addition, if $X_{t_n}^h$ and $Y_{t_n}^h$ denote two solutions that start from two different positive initial values $X_0^h = x$ and $Y_0^h = y$, we have the almost sure exponential forgetting property

$$\left[X_{t_n}^h - Y_{t_n}^h\right] = A_{t_n} \, [x - y] \longrightarrow_{n\uparrow\infty} 0.$$

In addition, in terms of the Wasserstein distance introduced in section 8.3.1.2, we have

$$\mathbb{W}\left(\mathrm{Law}\left(X_{t_n}^h\right), \mathrm{Law}\left(Y_{t_n}^h\right)\right) \le \left[\rho\left(1 - e^{-\lambda h}\right) + e^{-\lambda h}\right]^n |x - y|.$$

We notice that

$$\rho\left(1 - e^{-\lambda h}\right) + e^{-\lambda h} \simeq_{h\simeq 0} \rho\lambda h + (1 - \lambda h) = 1 - \lambda h(1 - \rho).$$

Recalling that $\log(1 - x) \le -x$, for any $x \in [0, 1]$, and $t_n = hn$ we conclude that

$$\left[\rho\left(1 - e^{-\lambda h}\right) + e^{-\lambda h}\right]^n \le \exp\left(-\lambda t_n(1 - \rho)\right).$$

To take the final step, we observe that

$$\mathbb{E}\left(\overline{B}_\infty\right) = he^{-\lambda h} \sum_{q\ge 0} \left(\rho\left(1 - e^{-\lambda h}\right) + e^{-\lambda h}\right)^q$$

$$= \left[\frac{e^{\lambda h} - 1}{h}(1 - \rho)\right]^{-1} \left(\longrightarrow_{h\uparrow 0} [\lambda(1 - \rho)]^{-1}\right).$$

Using proposition 8.3.13, we also have

$$\mathbb{W}\left(\mathrm{Law}\left(X_{t_n}^h\right), \mathrm{Law}\left(\overline{B}_\infty\right)\right) \le \left[\rho\left(1 - e^{-\lambda h}\right) + e^{-\lambda h}\right]^n \mathbb{E}(|x - \overline{B}_\infty|)$$

$$\le \left[\rho\left(1 - e^{-\lambda h}\right) + e^{-\lambda h}\right]^n \left[x + \left[\frac{e^{\lambda h} - 1}{h}(1 - \rho)\right]^{-1}\right].$$

This implies that

$$\mathbb{W}\left(\mathrm{Law}\left(X_{t_n}^h\right), \mathrm{Law}\left(\overline{B}_\infty\right)\right) \le \exp\left(-\lambda t_n(1 - \rho)\right) \left[x + [\lambda(1 - \rho)]^{-1}\right].$$

In the last r.h.s. estimate, we have used the fact that $\frac{e^{\lambda h}-1}{\lambda h} \geq 1$. If we choose the parameters $(\lambda, \alpha,) = (4/3, -1/2)$, we find that $\rho = (1+\alpha)^2 = 1/4$ and $\lambda(1-\rho) = 1$. In this case we have $\mathbb{E}\left(\overline{B}_\infty\right) = 1$, and

$$\mathbb{W}\left(\text{Law}\left(X_{t_n}^h\right), \text{Law}\left(\overline{B}_\infty\right)\right) \leq \exp\left(-t_n\right)(x+1).$$

When the time step unit h is expressed in milliseconds, say $h = 10^{-3}$, the jump times of the Bernoulli process are associated with geometric random variables $\mathcal{E}_n^h = T_n^h - T_{n-1}^h = h\left[1 + \lfloor \mathcal{E}_n/h \rfloor\right]$, where $\mathcal{E}_n$ denotes an exponential random variable with parameter $\lambda = 4/3$. In this case, the timeouts arrive in the average every

$$\mathbb{E}\left(\mathcal{E}_n^h\right) = h/p_h = \left(\frac{1-e^{-\lambda h}}{h}\right)^{-1} \simeq_{h\downarrow} \lambda^{-1} = 3/4 \text{ s}.$$

13.4.3   The limiting distribution

Rewriting the random variables $\overline{B}_{t_n}$ from (13.34) in terms of the Bernoulli sequence, we have

$$\overline{B}_{t_n} := \sum_{t_0=0\leq t_q < t_n} \rho^{\sum_{t_k=t_1}^{t_q} \epsilon_{t_k}}\left(1 - \epsilon_{t_{q+1}}\right) h.$$

With a slight abuse of notation (dropping the h index), we let N_{t_k} and T_k be the Bernoulli process given by

$$N_{t_k} = \sum_{l=1}^{k} \epsilon_{t_l}$$

and its jump times defined by

$$T_{k+1} = \inf\left\{t_p > T_k \; : \; \epsilon_{t_p} = 1\right\}$$

with $T_0 = 0$. In this notation, we have

$$\overline{B}_{t_n} = \sum_{l=1}^{k_n} \sum_{T_{l-1}\leq t_q < T_l} \rho^{\sum_{t_k=t_1}^{t_q} \epsilon_{t_k}}\left(1 - \epsilon_{t_{q+1}}\right) h + \sum_{T_{k_n} < t_q < t_n} \rho^{\sum_{t_k=t_1}^{t_q} \epsilon_{t_k}}\left(1 - \epsilon_{t_{q+1}}\right) h$$

where k_n denotes the index of the last jump time that happened earlier than t_n. This implies that

$$\overline{B}_{t_n} = \sum_{l=1}^{k_n} \rho^{l-1} \sum_{T_{l-1}\leq t_q < T_l} h + \rho^{k_n} \sum_{T_{k_n} < t_q < t_n} h = \sum_{l=1}^{k_n} \rho^{l-1}\left(T_l - T_{l-1}\right) + \rho^{k_n}\left(t_n - T_{k_n}\right).$$

To take the final step, we observe that k_n tends to infinity as $t_n \uparrow \infty$

$$k_n = N_{t_n-h} = \sum_{t_p=t_0=0}^{t_n-h} \epsilon_{t_p} \longrightarrow_{n\uparrow\infty} \infty.$$

Using this observation, we conclude that

$$\overline{B}_{t_n} \longrightarrow_{n\uparrow\infty} \overline{B}_\infty = \sum_{l\geq 0} \rho^l\left(T_{l+1} - T_l\right).$$

Recalling that the Bernoulli inter-arrival times $(T_{l+1} - T_l)$ are proportional to independent geometric random variables (cf. (10.4)) we have

$$\overline{B}_\infty = \sum_{l \geq 0} \rho^l \left(1 + \lfloor \mathcal{E}_l/h \rfloor\right) h \longrightarrow_{h\uparrow 0} \sum_{l \geq 0} \rho^l \mathcal{E}_l$$

where $\mathcal{E}_l$ are independent exponential r.v. with parameter λ. Using the formula (11.31) obtained in exercise 168, we have

$$\mathbb{P}\left(\overline{B}_\infty \leq 3 + (1 + 2\sqrt{3})/(\lambda(1-\rho))\right) \geq 1 - e^{-3} \geq 0.95. \tag{13.35}$$

When $\lambda = 4/3$ and $\rho = 1/4$ we have $\lambda(1-\rho) = 1$ and

$$\mathbb{P}\left(\overline{B}_\infty \leq 2(2 + \sqrt{3})\right) \geq 0.95.$$

If we set $W_\infty^2/2 = \overline{B}_\infty$ then we find that

$$\mathbb{P}\left(W_\infty \leq 2\sqrt{2 + \sqrt{3}}\right) \geq 0.95.$$

13.5 Exercises

Exercise 192 (Shot noise processes) *We consider the compound Poisson process X_t discussed in exercise 161. We consider a sequence of time steps $t_i \leq t_{i+1}$ and we set*

$$C_t = \sum_{1 \leq n \leq N_t} e^{a(t-t_n)} Y_n.$$

Describe the stochastic differential equation of C_t in terms of X_t. Find the infinitesimal generator of C_t.

Exercise 193 (PDMP stochastic differential equations) *We consider the ordinary differential equation*

$$dx_t = a(x_t) \, dt \tag{13.36}$$

for a regular function a on $S = \mathbb{R}$. We let b be a bounded function on $\mathbb{R}$ and N_t be a Poisson process with intensity $\lambda > 0$. We associate with these objects the stochastic differential equation

$$dX_t = a(X_t) \, dt + b(X_t) \, dN_t. \tag{13.37}$$

We assume that the flow map $\varphi_{s,t}(x)$ (cf. definition 13.1.1) of the deterministic system (13.40) is explicitly known. Propose an algorithm to sample the stochastic process (13.37). Compute the infinitesimal generator of X_t.

Exercise 194 *We consider the stochastic differential equation*

$$dX_t = a \, X_t \, dt + b \, X_t \, dN_t. \tag{13.38}$$

with a Poisson process N_t with intensity $\lambda > 0$ and some parameters $(a, b) \in \mathbb{R}$. Describe the infinitesimal generator of X_t. Compute the solution of X_t in terms of the jump times T_n of the Poisson process N_t.

Exercise 195 (Poisson intensity formulations) *We consider the ordinary differential equation (13.40) and we denote by $N(t)$ a Poisson process with unit intensity $\lambda = 1 > 0$. Let b denote a bounded positive function on $\mathbb{R}$.*

We associate with these objects the following stochastic process:

$$X_t = X_0 + \int_0^t a(X_s) \, ds + N\left(\int_0^t b(X_s) \, ds\right). \tag{13.39}$$

We assume that the flow map $\varphi_{s,t}(x)$ (cf. definition 13.1.1) of the deterministic system (13.40) and the integrals $\int_s^t b(\varphi_{s,r}(x))dr$ are explicitly known. Propose an algorithm to sample the stochastic process (13.39). Compute the infinitesimal generator of X_t.

Exercise 196 (Growth-fragmentation models) *The size X_t of a cell grows continuously at a rate $a(X_t)$, for some positive function a on $\mathbb{R}$. Its size $X_t = x$ is decreased to y according to some Markov transition $K(x, dy)$ at a rate $\lambda(X_t)$. Describe the infinitesimal generator of X_t.*

Exercise 197 (Degrowth-production models) *The quantity X_t of a chemical product in the body is degraded continuously at a rate $a(X_t)$ for some positive function a on $\mathbb{R}$. A random amount y is added to $X_t = x$ according to a Markov transition $K(x, dy)$, at a rate $\lambda(X_t)$. Describe the infinitesimal generator of X_t.*

Exercise 198 (Storage process) *Let X_t be a real valued PDMP with generator*

$$L(f)(x) = -ax \, \partial_x f(x) + \lambda \int [f(x+y) - f(x)] \, \nu(dy)$$

for some positive parameter $a, \lambda > 0$ and some probability measure ν on $\mathbb{R}$. Describe the evolution of X_t in terms of a sequence $(Y_n)_{n \geq 1}$ of independent random variables with common law ν and a sequence $(Z_n)_{n \geq 1}$ of independent exponential random variables with parameter λ.

Exercise 199 (Storage process - Mean values) *Consider the PDMP X_t discussed in exercise 198, and let $m := \int y \, \nu(dy)$. Check that*

$$\mathbb{E}(X_t) - (\lambda m/a) = e^{-at} \, (\mathbb{E}(X_0) - \lambda m/a) \longrightarrow_{t \to \infty} 0.$$

Exercise 200 (Storage process - Laplace transform) ✐ *Consider the PDMP X_t discussed in exercise 198 and set $g_t(u) = \mathbb{E}\left(e^{uX_t}\right)$. Describe the evolution equation of g_t in terms of the Laplace transform $h(u) = \mathbb{E}\left(e^{uY}\right)$ of a random variable Y with distribution ν. We assume that $h(u)$ is defined for any $u \leq u_0$ for some $u_0 > 0$. Check that*

$$\partial_t g_t(u) = -au \, \partial_u g_t(u) + g_t(u) \, V(u) \quad with \quad V(u) = \lambda \, [h(u) - 1].$$

Prove that the moment generating function $g_t(u)$ is given by the Feynman-Kac formula

$$g_t(u) = \exp\left(\int_0^t V\left[e^{-as}u\right] \, ds\right).$$

Let $\overline{V}$ be such that $\partial_u \log \overline{V}(u) = V(u)/u$. Check that

$$g_t(u) = \left(\overline{V}(u)/\overline{V}\left(e^{-at}u\right)\right)^{1/a}.$$

Examine the situation $\nu(dy) = b \, e^{-by} \, 1_{[0,\infty[}(y) \, dy$, for some $b > 0$. Prove that

$$g_t(u) \longrightarrow_{t \uparrow \infty} g_\infty(u) = (1 - u/b)^{-\lambda/a}.$$

Deduce that the invariant probability measure of X_t is given by the Laplace distribution with shape (λ/a) and rate b.

Exercise 201 (Storage process - Stability properties - Coupling) *Consider the pair of PDMP X_t^x and X_t^y discussed in exercise 198 and starting at x and $y \in [0, \infty[$. Consider the Wasserstein metric $\mathbb{W}(\mu_1, \mu_2)$ associated with the distance $d(x, y) = |x - y|$ discussed in section 8.3.1.2. Prove that*

$$\mathbb{W}\left(\text{Law}(X_t^x), \text{Law}(X_t^y)\right) \leq e^{-at}\,|x - y|.$$

Exercise 202 (Storage process - Stability properties - Total variation [186]) 🖋
Consider the pair of PDMP X_t^x and X_t^y discussed in exercise 201. Assume that the processes share the same jump times T_n and the jump amplitudes are defined in terms of a sequence of independent random variables with common exponential law $\nu(dy) = b\,e^{-by}\,1_{[0,\infty[}(y)\,dy$, for some given $b > 0$. Using the maximal coupling technique discussed in example 8.3.5, describe a coupling such that

$$\mathbb{P}\left(X_{T_n}^x = X_{T_n}^y \mid \left(X_{T_n-}^x, X_{T_n-}^y\right)\right) = \exp\left(-b\left|X_{T_n-}^x - X_{T_n-}^y\right|\right).$$

Deduce that at any time t s.t. $N_t > 0$ the chance of coupling is given by

$$\mathbb{P}\left(X_{T_{N_t}}^x = X_{T_{N_t}}^y \mid \left(X_{T_{N_t}-}^x, X_{T_{N_t}-}^y\right)\right) \geq 1 - b\,|x - y|\,e^{-aT_{N_t}-}.$$

Using theorem 8.3.2 check that

$$\|\text{Law}(X_t^x) - \text{Law}(X_t^y)\|_{tv} \leq e^{-\lambda t} + b\,|x - y|\,\mathbb{E}\left(e^{-aT_{N_t}}\,1_{N_t>0}\right).$$

Given $N_t = n$, we recall that T_n/t has the same law as the maximum $\max_{1 \leq i \leq n} U_i$ of n uniform random variables U_i on $[0, 1]$ (cf. exercise 41). Check that

$$\mathbb{E}\left(e^{-aT_{N_t}}\,1_{N_t>0}\right) = \int_0^1 e^{-atu}\,\mathbb{E}\left(N_t\,u^{N_t-1}\right)\,du = \frac{\lambda}{(\lambda - a)}\left[e^{-at} - e^{-\lambda t}\right]$$

as soon as $\lambda \neq a$. Conclude that

$$\|\text{Law}(X_t^x) - \text{Law}(X_t^y)\|_{tv} \leq e^{-\lambda t} + \frac{\lambda b}{(\lambda - a)}\left[e^{-at} - e^{-\lambda t}\right]\,|x - y|.$$

Exercise 203 (Minorization condition) *We let X_t be a PDMP on $S = \mathbb{R}^r$ associated with some regular deterministic flow maps $\varphi_{s,t}(x)$, a jump rate $\lambda_t(x)$ and Markov transitions $K_t(x, dy)$ describing the jumps from x to y. We assume that*

$$\forall x \in S \qquad \lambda_\star \leq \lambda_t(x) \leq \lambda^\star \quad \text{and} \quad K_t(x, dy) \geq \epsilon\,\nu_t(dy)$$

for some positive and finite parameters $\lambda^\star \geq \lambda_\star > 0$ and $\epsilon > 0$, and some probability measure $\nu_t(dy)$ on $\mathbb{R}^r$. Using the integral formula (13.15) check that the semigroup $P_{s,t}$, $s \leq t$, satisfies the minimization condition

$$\forall x \in S \quad \forall s < t \qquad P_{s,t}(x, dy) \geq \epsilon_{s,t}\,\nu_{s,t}(dy)$$

for some probability measure $\nu_{s,t}$ and some $\epsilon_{s,t} > 0$ s.t.

$$(t - s) \geq (\log 2)/\lambda^\star \Rightarrow \epsilon_{s,t} \geq \frac{\epsilon}{2}\,\frac{\lambda_\star}{\lambda^\star}.$$

Using theorem 8.2.13 check the contraction inequality

$$\|\mu_1 P_{t,t+nh} - \mu_2 P_{t,t+nh}\|_{tv} \leq \left(1 - \frac{\epsilon}{2}\,\frac{\lambda_\star}{\lambda^\star}\right)^n \|\mu_1 - \mu_2\|_{tv}$$

for any $t \geq 0$ and $n \geq 0$, and any probability measures μ_1, μ_2 on S, as soon as $h \geq (\log 2)/\lambda^\star$. Deduce that X_t has an unique invariant measure π.

Exercise 204 (Switching processes - Generator) *We let $S_j = \mathbb{R}^{r_j}$, with $r_j \geq 1$, and $a_j = (a_j^i)_{1 \leq j \leq r_j}$ be a collection of product state spaces and some regular functions $a_j \; S_j \mapsto \mathbb{R}$ indexed by a finite set $J \ni j$. Between the jumps, a process $\mathcal{X}_s = (I_s, X_s)$ follows the ordinary differential equation*

$$dX_s = a_{I_s}(X_s) \; ds \quad and \quad I_s = I_0 \tag{13.40}$$

for $s \geq 0$. At a rate $\lambda(I_s, X_s)$ it jumps to a new state, chosen randomly according to some Markov transition $K((I_s, X_s), d(i, y))$ on $S := \cup_{j \in J}(\{j\} \times S_j)$. Describe the infinitesimal generator of X_t.

Exercise 205 (Switching processes - Feynman-Kac formulae) *The switching process discussed in this exercise and the next two is a slight extension of the model presented in [17]. We consider the pure jump process X_t discussed in exercise 175 and the Feynman-Kac measures $(\gamma_t^{[\beta]}, \eta_t^{[\beta]})$ associated with some potential function V discussed in exercise 177. Let $\mathcal{X}_t = (X_t, Y_t) \in \mathcal{S} = (S \times \mathbb{R}^r)$, for some $r \geq 1$, be a switching process with generator*

$$\mathcal{L}(F)(x, y) = \sum_{1 \leq i \leq r} a^i(x, y) \; \partial_{y_i} F(x, y) + L(F(\,.\,, y))(x)$$

for some given drift functions $a^i(x, y)$. We assume that X_0 and Y_0 are independent. We equip $\mathbb{R}^r$ with the scalar product $\langle x, y \rangle = \sum_{1 \leq j \leq r} x_j y_j$ and the Euclidian norm $\|x\|^2 := \langle x, x \rangle$, and we assume that the drift functions are chosen so that

$$\forall (x, y, z) \in (S \times \mathbb{R}^r \times \mathbb{R}^r) \qquad \langle a(x, y) - a(x, z), y - z \rangle \leq -V(x) \; \|y - z\|^2.$$

Check that

$$\|Y_t\| \leq e^{-\int_0^t V(X_s)ds} \; \|Y_0\| + \int_0^t e^{-\int_s^t V(X_u)du} \; \|a(X_s, 0)\| \; ds.$$

Using the generalized Minkowski inequality (cf. for instance lemma 5.1 in [70]) check that for any $\beta \geq 1$ we have

$$\mathbb{E}\left[\|Y_t\|^\beta\right]^{1/\beta} \; \leq \; C_\beta \left[e^{-t\eta_\infty^{[\beta]}(V)} \; \mathbb{E}\left(\|Y_0\|^\beta\right)^{1/\beta} + C_\beta \int_0^t e^{-(t-s)\eta_\infty^{[\beta]}(V)} \; \left(\eta_s^{[\beta]}(f^\beta)\right)^{1/\beta} \; ds \right]$$

with the function $f(x) = \|a(x, 0)\|$ and some finite constants C_β. When f is uniformly bounded by some constant $a^\star$, deduce that

$$\mathbb{E}\left[\|Y_t\|^\beta\right]^{1/\beta} \; \leq \; C_\beta \; e^{-t\eta_\infty^{[\beta]}(V)} \; \mathbb{E}\left(\|Y_0\|^\beta\right)^{1/\beta} + C_\beta^2 \; a^\star \; \left(1 - e^{-t\eta_\infty^{[\beta]}(V)}\right).$$

In addition, when $\inf_{x \in \mathbb{R}^r} V(x) = V_\star > 0$ prove the almost sure estimate

$$\|Y_t\| \leq \; e^{-\int_0^t V(X_s)ds} \; \|Y_0\| + (a^\star/V_\star) \; \left(1 - e^{-\int_0^t V(X_u)du}\right) \leq \|Y_0\| \vee (a^\star/V_\star).$$

Exercise 206 (Switching processes - Coupling 1) *Consider the switching process discussed in exercise 205. Assume that $a^\star := \sup_{x \in S} \|a(x, 0)\| < \infty$ and $\inf_{x \in \mathbb{R}^r} V(x) = V_\star > 0$. We couple the stochastic processes $\mathcal{X}_t = (X_t, Y_t)$ and $\mathcal{X}_t' = (X_t, Y_t')$ starting at different states $\mathcal{X}_0 = (x, y)$ and $\mathcal{X}_0' = (x, y')$ (using with the same first coordinate process X_t). Check that for any parameter $\beta > 0$ we have*

$$\mathbb{E}\left(\|Y_t - Y_t'\|^\beta \mid (Y_0, Y_0') = (y, y'), X_0 = x\right) \leq \mathcal{Z}_{0,t}(x) \; \|y - y'\|^\beta$$

with the Feynman-Kac normalizing partition function $\mathcal{Z}_{s,t}(x)$ defined for any $0 \le s \le t$ by

$$\mathcal{Z}_{s,t}^{(\beta)}(x) := \mathbb{E}\left(\exp\left(-\beta \int_s^t V(X_s)\ ds\right) \mid X_0 = x\right).$$

Exercise 207 (Switching processes - Coupling 2) ✎ *Consider the switching process discussed in exercise 205. Assume that $a^\star := \sup_{x \in S} \|a(x, 0)\| < \infty$ and $\inf_{x \in \mathbb{R}^r} V(x) = V_\star > 0$. We couple the stochastic processes $\mathcal{X}_t = (X_t, Y_t)$ and $\mathcal{X}'_t = (X'_t, Y'_t)$ starting at different states $\mathcal{X}_0 = (x, y)$ and $\mathcal{X}'_0 = (x', y')$ using the coupling described in the second statement of theorem 12.7.6. To be more precise, we couple the first components (X_t, X'_t) until their coupling time T, and we set $X_t = X'_t$ for any $t \ge T$. Prove that for any parameter $\beta > 0$ there exist some constants c and $\delta_\beta > 0$ such that*

$$\mathbb{E}\left(\|Y_t - Y'_t\|^\beta \mid (Y_0, Y'_0) = (y, y'), (X_0, X'_0) = (x, x')\right)$$

$$\le c\ \exp\left(-\delta_\beta t\right)\left[\|y\|^\beta \vee \|y'\|^\beta \vee (a^\star/V_\star)^\beta\right].$$

Exercise 208 (Regenerative processes) *We consider a real valued PDMP X_t with generator*

$$L(f)(x) = \lambda\ (f(0) - f(x))\ +\ f'(x)$$

for some parameter $\lambda > 0$. Using the integral formula (13.15), find an explicit description of the conditional probability of X_t given it starts at the origin. Extend this formula to PDMP X_t with generator

$$L_t(f)(x) = \lambda\ (f(0) - f(x))\ +\ b_t(x)\ f'(x)$$

for some some smooth and bounded drift function b_t. This exercise can be complemented with exercise 167 on the analysis of time inhomogeneous regenerative processes with explosions.

Exercise 209 (A random 2-velocity process [128, 159]) ✎ *We let X_t be a homogeneous pure jump process taking values in $\{-1, 1\}$. At some rate $\lambda(X_t)$, the process X_t changes its sign. Given some smooth functions a and $b > |a|$ on $\mathbb{R}$ we set*

$$\frac{dY_t}{dt} = a(Y_t) + X_t\ b(Y_t).$$

Describe the generator of the process $Z_t = (X_t, Y_t)$. Assume that (X_t, Y_t) has a density given by

$$\forall x \in \{-1, 1\} \qquad \mathbb{P}(X_t = x\ ,\ Y_t \in dy) = p_t(x, y)\ dy.$$

We set

$$q_t^+(y)\ =\ p_t(1, y) + p_t(-1, y) \qquad q_t^-(y)\ =\ p_t(1, y) - p_t(-1, y).$$

Describe the evolution equations of the functions $(p_t(1, y), p_t(-1, y))$ and $(q_t^+(y), q_t^-(y))$. Discuss the fixed points of the equation governing the evolution of the density $q_t^+(y)$ of the random state Y_t. Discuss the situation $\lambda(1) = \lambda(-1)$, $a = 0$ and $0 < \int_c^\infty b^{-1}(y)dy < \infty$ for some constant c.

Exercise 210 (Switching processes - Marginal distributions) *We consider the switching process $X_t = (X_t^1, X_t^2)$ discussed in section 13.3.1. Find the evolution equations of the marginal probability measures*

$$p_t^1(x^1) = \int\ p_t(x^1, x^2)\ \nu(dx^2) \quad \text{and} \quad p_t^2(x^2) = \int\ p_t(x^1, x^2)\ dx^1.$$

Exercise 211 (Biochemical reaction network) ✎ *We consider a biochemical reaction network with n_c chemical reactions among n_s species $(S_i)_{1 \leq i \leq n_s}$*

$$\left\{ \begin{array}{rcl} \sum_{1 \leq i \leq n_s} a_{i,j} \, S_i & \rightarrow & \sum_{1 \leq i \leq n_s} b_{i,j} \, S_j \\ j & = & 1, \ldots, n_c \end{array} \right.$$

where $a_{i,j}$ and $b_{i,j}$ are non-negative integers. These parameters are often called the stoichiometric coefficients of the reaction; $a_{i,j}$ represents the number of molecules of species S_i consumed in the j-th reaction, and $b_{i,j}$ represents the number of molecules of species S_i produced in the j-th reaction. We denote by $X_t = (X_t^i)_{1 \leq i \leq n_s}$ the number of molecules of the species $(S_i)_{1 \leq i \leq n_s}$ at time t. We assume that the j-th reaction occurs according to a counting Poisson process with reaction rate $\lambda_j(X_t) = \kappa_j \prod_{1 \leq i \leq n_s} \begin{pmatrix} a_{i,j} \\ X_t^i \end{pmatrix}$, for a certain reaction rate parameter κ_j, with $1 \leq j \leq n_c$. Describe the process X_t in terms of n_c independent Poisson processes with unit intensity.

Exercise 212 (Gene expression) *We let X_t^1 be a $\{0,1\}$ valued random variable representing the state ("off" or "on") of a given gene at some time t. We let X_t^2 be the quantity of protein or mRNA produced by the gene at time t. We assume that X_t^1 switches between 0 and 1 at rates $\lambda_1(X_t^2)$ (for transitions $1 \rightsquigarrow 0$) and $\lambda_2(X_t^2)$ (for transitions $0 \rightsquigarrow 1$). When the gene expression is "on", i.e., when $X_t^1 = 1$, the gene produces protein at a rate $\lambda_3(X_t^2)$, and it is degraded at a rate $\lambda_4(X_t^2)$. Describe the process $X_t = (X_t^1, X_t^2)$ in terms of independent Poisson processes with unit intensity.*

Exercise 213 (Mesquita bacterial chemotaxis process) *In his PhD dissertation [195], A. R. Mesquita proposes a PDMP model to analyze the chemotactic activity of the bacteria* Escherichia coli*. The process $X_t = (U_t, V_t) \in S := (\mathbb{R}^2 \times \mathbb{S})$ represents the location U_t of the cell and its velocity/orientation $V_t \in \mathbb{S}$ in the unit one-dimensional sphere $\mathbb{S} = \{v = (v_1, v_2) \in \mathbb{R}^2 : \|v\|_2^2 := v_1^2 + v_2^2 = 1\}$ (thus assuming constant unit velocity variations). The deterministic flow $dU_t/dt = a(U_t, V_t)$ represents the exploration of the cell in terms of its velocity, for some drift function $a : (\mathbb{R}^2 \times \mathbb{S}) \rightarrow \mathbb{R}^2$. The velocity component V_t changes with some rate $\lambda(X_t)$ that may depend on the location and the speed of the cell. When the system $X_t = (U_t, V_t) = (u, v)$ jumps only the velocity component is changed $(u, v) \rightsquigarrow (u, w)$. The new velocity is chosen according to some Markov transition $K((u, v), dw)$ from S into $\mathbb{S}$. Describe the infinitesimal generator of the process X_t on S.*

Exercise 214 (PDMP process - Invariant measures) *We return to the Mesquita bacterial chemotaxis process discussed in exercise 213. We further assume that the drift function $a(u, v) = a(v)$ only depends on the velocity of the cell. We also assume that $K((u, v), dw) = \nu(dw)$ with the uniform measure ν on $\mathbb{S}$ with $\nu(a) = 0$. Find some jump intensity such that X_t has an invariant measure of the form $\pi(d(u, v)) \propto q(u) \, du \, \nu(dv)$ where $q(u)$ stands for some given smooth density of nutrients in the environment of the cell.*

Exercise 215 (PDMP with bi-Laplace invariant measure [129]) *We consider the stochastic process $\mathcal{X}_t = (X_t, Y_t)$ on $S = (\{-1, +1\} \times \mathbb{R})$ with generator L defined for any sufficiently regular function f on S (smooth w.r.t. the second coordinate) and for any $(\epsilon, y) \in S$ by*

$$L(f)(\epsilon, y) = \epsilon \, \partial_y f(\epsilon, y) + (a + (b - a) \, 1_{\epsilon y > 0}) \, (f(-\epsilon, y) - f(\epsilon, y))$$

for some parameters $0 < a < b$. Check that $\pi(d(x, y)) = \frac{1}{2}(\delta_{-1} + \delta_{+1})(dx) \times \nu(dy)$ is an invariant measure of $\mathcal{X}_t$.

Exercise 216 (Double telegraph - Bouncy particle samplers [230, 30]) *Consider a Boltzmann-Gibbs probability measure $\nu(dy) \propto e^{-V(y)}dy$ on $\mathbb{R}$ associated with some non-negative potential function V. We consider the stochastic process $\mathcal{X}_t = (X_t, Y_t)$ taking values in $S = (\{-1, +1\} \times \mathbb{R})$ with generator L defined for any sufficiently regular function f on S (smooth w.r.t. the second coordinate) and for any $(\epsilon, y) \in S$ by*

$$L(f)(\epsilon, y) = \epsilon \, \partial_y f(\epsilon, y) + (\epsilon \partial_y V(y))_+ \, (f(-\epsilon, y) - f(\epsilon, y)).$$

Check that $\pi(d(x, y)) = \frac{1}{2}(\delta_{-1} + \delta_{+1})(dx) \times \nu(dy)$ is an invariant measure of $\mathcal{X}_t$. The multidimensional case is discussed in exercise 219. The non-uniqueness property of the invariant measure is discussed in exercise 216.

Exercise 217 (Bouncy particle samplers [230, 30]) *Consider a Boltzmann-Gibbs probability measure $\nu(dx) \propto e^{-U(x)}dx$ on $\mathbb{R}$ associated with some potential function $U : \mathbb{R} \mapsto \mathbb{R}$ (s.t. the Gibbs measure is well defined). We let $\mu(dv)$ be any symmetric probability measure on $\mathbb{R}$ (in the sense that $V \sim \mu \Rightarrow (-V) \sim \mu$).*

We consider the stochastic process $\mathcal{X}_t = (V_t, X_t)$ taking values in $S = (\mathbb{R} \times \mathbb{R})$ with generator L defined for any sufficiently regular function f on S (smooth w.r.t. the second coordinate) and for any $(v, x) \in S$ by

$$L(f)(v, x) = a \, v \, \partial_x f(v, x) + \left(a \, v \, e^{U(x)} \partial_x e^{-U(x)} \right)_- \, (f(-v, x) - f(v, x))$$

for some given parameter $a \in \mathbb{R}$. Check that $\pi(d(v, x)) = \mu(dv) \times \nu(dx)$ is an invariant measure of $\mathcal{X}_t$. Compare this stochastic process with the one discussed in exercise 216. Discuss the non-uniqueness property of the invariant measure. The discrete time version of this sampler is discussed in exercise 120 (see also exercise 229).

Exercise 218 (Mesquita MCMC samplers) *Consider a Boltzmann-Gibbs probability measure $\nu(dx) \propto e^{-U(x)}dx$ on $\mathbb{R}$ associated with some potential function $U : \mathbb{R} \mapsto \mathbb{R}$ (s.t. the Gibbs measure is well defined). Let $\mu(dv)$ be any probability measure on $\mathbb{R}$ and $a : v \in \mathbb{R} \mapsto a(v) \in \mathbb{R}$ be some given drift function s.t. $\mu(a) = 0$. We also let $\alpha : \mathbb{R} \mapsto [0, \infty[$ be some intensity function.*

We consider the stochastic process $\mathcal{X}_t = (V_t, X_t)$ taking values in $S = (\mathbb{R} \times \mathbb{R})$ with generator L defined for any sufficiently regular function f on S (smooth w.r.t. the second coordinate) and for any $(v, x) \in S$ by

$$L(f)(v, x) = a(v) \, \partial_x f(v, x) + \lambda(v, x) \int \, (f(w, x) - f(v, x)) \, \mu(dw),$$

with the jump intensity

$$\lambda(v, x) = \alpha(x) + \sup_{w \in \mathbb{R}} \left(a(w) \, e^{U(x)} \partial_x e^{-U(x)} \right) - \left(a(v) \, e^{U(x)} \partial_x e^{-U(x)} \right).$$

Check that $\pi(d(v, x)) = \mu(dv) \times \nu(dx)$ is an invariant measure of $\mathcal{X}_t$. Discuss the choice of the tuning intensity function α.

Exercise 219 (Rejection-free Monte Carlo and bouncy particle samplers [30, 222]) *We equip $\mathbb{R}^r$ with the scalar product $\langle x, y \rangle = \sum_{1 \leq i \leq r} x_i y_i$ and the Euclidian norm $\|x\| = \langle x, x \rangle^{1/2}$. Let $\partial V = (\partial_{x_i} V)'_{1 \leq i \leq r}$ be the gradient column vector field associated with some non-negative smooth function V on $\mathbb{R}^r$ s.t. $\mathcal{Z} := \int e^{-V(x)} \, dx \in]0, \infty[$ (recall that v' stands for the transpose of a vector v). We set $U = \partial V / \|\partial V\|$ and $A = (I - 2UU')$, where I denotes the identity matrix on $\mathbb{R}^r$.*

Check that A is an $(r \times r)$-orthogonal matrix field ($A(x) = A(x)'$ and $A(x)^2 = I$) and for any $x, y \in \mathbb{R}^r$ we have

$$\langle U(x), A(x)y \rangle = -\langle U(x), y \rangle \quad \text{and} \quad \langle U(x), A(x)y \rangle_+ - \langle U(x), y \rangle_+ = -\langle U(x), y \rangle.$$

Deduce that

$$\langle \partial V(x), A(x)y \rangle_+ - \langle \partial V(x), y \rangle_+ = -\langle \partial V(x), y \rangle.$$

We denote by ν some spherically symmetric probability distribution on $\mathbb{R}^r$ (for instance a Gaussian centered distribution with unit covariance matrix). Let π be the Boltzmann-Gibbs distribution on $\mathbb{R}^2$ given by

$$\pi(d(x^1, x^2)) = \frac{1}{\mathcal{Z}} \ e^{-V(x^1)} \ dx^1 \ \nu(dx^2).$$

Consider the switching model $X_t = \left(X_t^1, X_t^2\right)$. discussed in section 13.3.1 and section 13.3.2 with $r_1 = r_2 = r$. Assume that the drift function, the jump intensity and the jump transition in (13.31) are given by

$$a_t(x^1, x^2) = x^2 \qquad \lambda(x^1, x^2) = \langle \partial V(x^1), x^2 \rangle_+ \quad \text{and} \quad K_t^{(2)}((x^1, x^2), dy^2) = \delta_{A(x^1)x^2}(dy^2).$$

Describe the infinitesimal generator L of the process X_t.

Check that

$$\int \pi(dx) \ \lambda(x) \ \left[f\left(x^1, A(x^1)x^2\right) - f(x) \right] = - \int \pi(dx) \ \langle \partial V(x^1), x^2 \rangle \ f(x)$$

and

$$\int \pi(dx) \sum_{1 \leq i \leq r} x_i^2 \ \partial_{x_i^1} f(x) = \int \pi(dx) \ \langle \partial V(x^1), x^2 \rangle \ f(x).$$

Deduce that $\pi L = 0$. Discuss the non-uniqueness property of the invariant distribution of X_t.

14

Diffusion processes

This chapter is devoted to the classical theory of diffusion processes and of stochastic differential equations driven by Brownian motion. It starts with the introduction of the Brownian process as a limiting stochastic process on a time mesh sequence of independent Rademacher random variables. The connection to the heat equation is discussed next. The Doeblin-Itō formula and its various applications in stochastic differential calculus are discussed in great details and is used to derive the Fokker-Planck equation for the density of the random states. The univariate results are extended to the case of multivariate diffusions at the end of the chapter.

> *These motions were such as to satisfy me, after frequently repeated observation,*
> *that they arose neither from currents in the fluid, nor from its gradual evaporation,*
> *but belonged to the particle itself.*
> Robert Brown (1773-1858).

14.1 Brownian motion

14.1.1 Discrete vs continuous time models

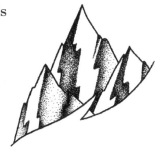

It is beyond the scope of this section to describe in full details the construction of the Brownian motion and the stochastic integrals w.r.t. to the induced random measure on the time axis. We have chosen to present these continuous time stochastic processes from the practitioner's point of view, using simple arguments based on their discrete approximations. The definitions of continuous time filtrations, predictable processes, martingales and their angle brackets are provided in section 12.5.2.

For a more thorough and rigorous discussion on these probabilistic models, we refer the reader to the seminal books by Daniel Revuz and Marc Yor [231], Ioannis Karatzas and Steven Shreve [161], as well as the book by Stewart Ethier and Thomas Kurtz [122], and the one by Bernt Øksendal [213].

The simplest way to introduce the Brownian process is to consider a virtual individual evolving on the real line performing randomly local steps with amplitude $+\Delta X$ or $-\Delta X$ every Δt units of time, in such a way that $[\Delta X]^2 = \Delta t$. For instance, he moves to the right with step $+\Delta X = +10^{-p}$ or to the left with step $-\Delta X = +10^{-p}$, every $\Delta t = 10^{-2p}$ units of time. As a result, his local speed $\Delta X / \Delta t = 10^p$ tends to infinity as $p \uparrow \infty$.

393

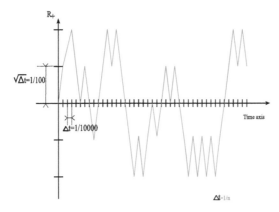

FIGURE 14.1: Discrete Brownian motion, with $\Delta t = 10^{-4}$.

To formalize these stochastic processes, we consider a time mesh sequence $t_n = nh$, with $n \in \mathbb{N}$, and a given time step $h \in]0, 1[$.

$$W_0^h = 0 \quad \text{and} \quad \Delta W_{t_n}^h = W_{t_n}^h - W_{t_{n-1}}^h = \epsilon_{t_n} \sqrt{h} \,,$$

with a sequence of independent random variables with a common law

$$\mathbb{P}(\epsilon_{t_n} = +1) = \mathbb{P}(\epsilon_{t_n} = -1) = \frac{1}{2} \,.$$

By construction, we have

$$
\begin{aligned}
\Delta W_{t_n}^h \times \Delta W_{t_n}^h &= h \\
\mathbb{E}(\Delta W_{t_n}^h \mid W_{t_0}^h, \ldots, W_{t_{n-1}}^h) &= 0.
\end{aligned}
$$

To get one step further in the analysis of the limiting model, we observe that for any $s \leq t$, and for any bounded and uniformly continuous function $r \mapsto \varphi_r$ on $\mathbb{R}_+$,

$$
\begin{aligned}
\sum_{0 < p \leq \lfloor t/h \rfloor} \varphi_{t_{p-1}} \left(\Delta W_{t_p}^h \right)^2 &= \sum_{0 < p \leq \lfloor t/h \rfloor} \varphi_{t_{p-1}} (t_p - t_{p-1}) \\
&= \int_0^{h \lfloor t/h \rfloor} \varphi_s \, ds + \sum_{0 < p \leq \lfloor t/h \rfloor} \int_{t_{p-1}}^{t_p} \left[\varphi_{t_{p-1}} - \varphi_r \right] \, dr \\
&\longrightarrow_{h \to 0} \int_0^t \varphi_s \, ds.
\end{aligned}
$$

In addition, we have

$$
\begin{aligned}
W_{h \lfloor t/h \rfloor}^h - W_{h \lfloor s/h \rfloor}^h &= \sum_{\lfloor s/h \rfloor < p \leq \lfloor t/h \rfloor} \Delta W_{t_p}^h \quad \left(\perp W_{h \lfloor s/h \rfloor}^h \right) \\
&= \underbrace{\sqrt{h(\lfloor t/h \rfloor - \lfloor s/h \rfloor)}}_{\to_{h \downarrow 0} \sqrt{t-s}} \left[\frac{1}{\sqrt{\lfloor t/h \rfloor - \lfloor s/h \rfloor}} \sum_{\lfloor s/h \rfloor < p \leq \lfloor t/h \rfloor} \epsilon_{t_p} \right].
\end{aligned}
$$

Invoking the central limit theorem, we prove the following theorem.

Theorem 14.1.1 *As the time step h tends to 0, we have the following convergence in law*

$$\left(W^h_{h\lfloor s/h \rfloor} \, , \, W^h_{h\lfloor t/h \rfloor} - W^h_{h\lfloor s/h \rfloor} \right) \, \to_{h \to 0} \, (W_s, (W_t - W_s)) \tag{14.1}$$

where W_s and $(W_t - W_s)$ are independent and centered Gaussian r.v. with variance s, and $t - s$.

Arguing as in the proof of theorem 14.1.1 we can show that the sequence of increments

$$\left(W_{t_1} - W_{t_0}, \ldots, W_{t_n} - W_{t_{n-1}} \right)$$

and independent centered Gaussian random variables with variances $(t_1 - t_0, \ldots, t_n - t_{n-1})$, for any sequence of time steps $t_0 \leq \ldots \leq t_n$. It can also be shown that the random paths $t \mapsto W_t$ are almost surely continuous.

Definition 14.1.2 *The limiting stochastic process W_t defined above is called the (standard) Brownian motion, or the Wiener process. The Brownian motion W_t is characterized by the continuity of the paths and the Gaussian properties stated above.*

In addition, for any bounded and uniformly continuous function g on $(\mathbb{R}_+ \times \mathbb{R})$ we have the almost sure convergence results

$$\sum_{0 < p \leq \lfloor t/h \rfloor} g\left(t_{p-1}, W_{t_{p-1}} \right) \left(W_{t_p} - W_{t_{p-1}} \right) \longrightarrow_{h \to 0} \int_0^t g(s, W_s) \, dW_s$$

and

$$\sum_{0 < p \leq \lfloor t/h \rfloor} g\left(t_{p-1}, W_{t_{p-1}} \right) \left(W_{t_p} - W_{t_{p-1}} \right)^2 \longrightarrow_{h \to 0} \int_0^t g(s, W_s) \, ds. \tag{14.2}$$

The proof of the last assertion (14.2) is a consequence of the equivalence principle (8.63), combined with the almost sure convergence result

$$\sum_{0 < p \leq \lfloor t/h \rfloor} g\left(t_{p-1}, W_{t_{p-1}} \right) h \longrightarrow_{h \to 0} \int_0^t g(s, W_s) \, ds.$$

14.1.2 Evolution semigroups

The evolution semigroup (sg)

$$P^h_{t_p, t_n}(f)(x) = \mathbb{E}\left(f(W^h_{t_n}) \mid W^h_{t_p} = x \right)$$

of the Markov chain $W_{t_n}^h$ is defined in terms of the one step transitions

$$
\begin{aligned}
P_{t_n,t_{n+1}}^h(f)(x) &= \mathbb{E}\left(f(W_{t_{n+1}}^h) \mid W_{t_n}^h = x\right) = \mathbb{E}\left(f(x + \epsilon_{t_{n+1}}\sqrt{h})\right) \\
&= \frac{1}{2}\left(f(x - \sqrt{h}) + f(x + \sqrt{h})\right).
\end{aligned}
$$

For smooth functions, using a second order Taylor expansion we find that

$$
P_{t_n,t_{n+1}}^h(f)(x) = f(x) + \frac{h}{2}\,\partial_x^2 f(x) + \mathrm{O}(h^{3/2}).
$$

By theorem 14.1.1,

$$
W_t - W_s \stackrel{law}{=} \sqrt{t-s}\,W_1.
$$

This shows that the evolution sg of the Brownian motion is given by

$$
\begin{aligned}
P_{s,t}(f)(x) &= \mathbb{E}\left(f(W_t) \mid W_s = x\right) \\
&= \mathbb{E}\left(f(W_s + (W_t - W_s)) \mid W_s = x\right) = \mathbb{E}\left(f(x + \sqrt{t-s}\,W_1)\right).
\end{aligned}
$$

For smooth and bounded functions, using a second order Taylor expansion we find that

$$
f(x + \sqrt{h}\,W_1) = f(x) + \sqrt{h}\,W_1\,f'(x) + \frac{h}{2}\,W_1^2\,f''(x) + \mathrm{O}_P(h^{1+1/2}).
$$

This also implies that

$$
P_{t,t+h}(f)(x) = \mathbb{E}\left(f(x + \sqrt{h}\,W_1)\right) = f(x) + \frac{h}{2}\,\partial_x^2 f(x) + \mathrm{O}(h^{1+1/2}).
$$

Combining these expansions with theorem 12.3.1 we prove the following theorem.

Theorem 14.1.3 *For any $t \geq 0$, we have the first order operator expansion*

$$
P_{t,t+h}^h = Id + h\,L + \mathrm{O}(h^{1+1/2}) = P_{t,t+h} \quad with \quad L = \frac{1}{2}\,\partial_x^2. \tag{14.3}
$$

In addition, for any time steps $t_p = h\lfloor s/h \rfloor \leq t_n = h\lfloor t/h \rfloor$, we have the first order approximation formulae

$$
P_{t_p,t_n}^h = P_{t_p,t_n} + \mathrm{O}\left(h^{1/2}\right).
$$

Important remark : In view of the Gaussian nature of the limiting process (14.1), an alternative approximation of W_t on a given time mesh is given by

$$
\Delta\overline{W}_{t_n}^h = \overline{W}_{t_n}^h - \overline{W}_{t_{n-1}}^h = \sqrt{t_n - t_{n-1}} \times \overline{W}_n = \sqrt{h}\,\overline{W}_n \tag{14.4}
$$

where $\overline{W}_n$ stands for a sequence of independent and centered Gaussian random variables. Notice that $\Delta\overline{W}_{t_n}^h \stackrel{law}{=} W_{t_n} - W_{t_{n-1}}$.

14.1.3 The heat equation

The connection between the heat equation and the Brownian motion is described by the following proposition.

> **Proposition 14.1.4** *The laws of the random states of the Brownian motion are given for any function f on $\mathbb{R}$ by*
>
> $$\mathbb{E}(f(W_t)) = \int_{\mathbb{R}} f(x) \underbrace{p_t(x)\ dx}_{\mathbb{P}(W_t \in dx)} \quad with \quad p_t(x) = \frac{1}{\sqrt{2\pi t}}\ \exp\left(-\frac{x^2}{2t}\right).$$
>
> *The measure $p_t(x)dx$ is the weak solution (in the sense of distributions) of the heat equation starting at δ_0 given by*
>
> $$\partial_t p_t = \frac{1}{2}\ \partial_x^2 p_t.$$

Proof :

We need to check that

$$\int_{\mathbb{R}} f(x)\ \partial_t p_t(x)\ dx = \frac{1}{2} \int_{\mathbb{R}} f(x)\ \partial_x^2 p_t(x)\ dx$$

for any smooth function with compact support, and $\int_{\mathbb{R}} f(x)\ p_0(x)\ dx = f(0)$. We check this claim by using

$$\partial_t p_t(x) = \frac{1}{\sqrt{2\pi t}} \left(-\frac{1}{2t} + \frac{x^2}{2t^2}\right)\ e^{-\frac{x^2}{2t}} = \frac{1}{2}\left(\frac{x^2}{t^2} - \frac{1}{t}\right)\ p_t(x)$$

and

$$\frac{1}{2}\partial_x^2 p_t(x) = -\frac{1}{2}\partial_x\left(\frac{x}{t}\ p_t(x)\right) = \frac{1}{2}\left(\frac{x^2}{t^2} - \frac{1}{t}\right) p_t(x).$$

This ends the proof of the proposition. ∎

We end this section with an alternative proof of the heat equation based on a second order Taylor expansion. To clarify the presentation, we fix a smooth bounded function f vanishing at the infinity and we set $m(t) = \mathbb{E}(f(W_t))$. By construction, we have

$$\partial_t\ \mathbb{E}(f(W_t)) \simeq \frac{m(t + \Delta t) - m(t)}{\Delta t} = \frac{\mathbb{E}[f(W_t + \Delta W_t) - f(W_t)]}{\Delta t}$$

with

$$\Delta W_t = (W_{t+\Delta t} - W_t) \quad \text{for some time step} \quad \Delta t \simeq 0.$$

Using a second order expansion of the function f at the random state W_t, we find that

$$
\begin{aligned}
\Delta f(W_t) &:= f(W_t + \Delta W_t) - f(W_t) \\
&= \partial_x f(W_t)\ \Delta W_t + \frac{1}{2}\ \partial_x^2 f(W_t)\ [\Delta W_t]^2 + \mathrm{O}((\Delta W_t)^3)
\end{aligned}
\tag{14.5}
$$

where $\mathrm{O}((\Delta W_t)^3)$ denotes a random function such that

$$\mathbb{E}\left(\left|\mathrm{O}((\Delta W_t)^3)\right|\right) \le \mathrm{C}\ \mathbb{E}((\sqrt{\Delta t})^3) \le \mathrm{C}\ \Delta t\sqrt{\Delta t}$$

for some finite constant $C < \infty$ whose values only depend on the function f. Using the fact that

$$\mathbb{E}\left(\partial_x f(W_t)\,\Delta W_t \mid W_t\right) = 0$$

and

$$\mathbb{E}\left(\partial_x^2 f(W_t)\,[\Delta W_t]^2 \mid W_t\right) = \partial_x^2 f(W_t) \times \Delta t$$

we conclude that

$$\frac{1}{\Delta t}\,\mathbb{E}[f(W_t + \Delta W_t) - f(W_t)] = \mathbb{E}\left(\frac{1}{2}\,\partial_x^2 f(W_t)\right) + \mathrm{o}(1)$$

with a certain deterministic function o(1) that converges to 0, as Δt tends to 0.

14.1.4   Doeblin-Itō-Taylor formula

Our next objective is to further develop in a rather informal way the Taylor expansion (14.5). We assume that all the derivatives of f are bounded at any order. From the previous analysis, we have the decomposition

$$\Delta f(W_t) \;=\; \frac{1}{2}\,\partial_x^2 f(W_t)\,\Delta t \;+\Delta M_t(f) + \;\epsilon_t(\Delta t)\,\Delta t + \mathrm{O}((\Delta W_t)^3)$$

with the martingale increment

$$\Delta M_t(f) = \partial_x f(W_t)\,\Delta W_t \quad \text{s.t.} \quad \mathbb{E}\left(\Delta M_t(f) \mid W_t\right) = 0$$

and the remainder term

$$\epsilon_t(\Delta) = \frac{1}{2}\,\partial_x^2 f(W_t)\,\left(\frac{[\Delta W_t]^2}{\Delta t} - 1\right)$$

such that

$$\sum_{0 \le t_n \le \Delta t \lfloor t/\Delta t \rfloor} \epsilon_{t_n}(\Delta)\,\Delta t \longrightarrow_{\Delta t \uparrow \infty} 0.$$

The proof of this assertion is a consequence of the equivalence principle (8.63). We also observe that $M_t(f)^2 - \Delta\langle M(f)\rangle_t$ is a martingale with the increments of the angle bracket process given by

$$\mathbb{E}\left((\Delta M_t(f))^2 \mid W_t\right) \;=\; (\partial_x f(W_t))^2\,\Delta t$$
$$:= \;\langle M(f)\rangle_{t+\Delta t} - \langle M(f)\rangle_t := \Delta\langle M(f)\rangle_t.$$

Finally, we observe that

$$L = \frac{1}{2}\partial_x^2 \;\Rightarrow\; \Gamma_L(f,f)(x) = L((f - f(x))^2)(x) = (\partial_x f)^2$$

where Γ_L stands for the "carré du champ operator" introduced in definition 12.5.3. Working a little harder, we prove the Doeblin-Itō lemma for Brownian motion.

Lemma 14.1.5 *For any smooth and bounded function f, we have*

$$df(W_t) = L(f)(W_t)\,dt + dM_t(f)$$

with the 1d-Laplacian generator $L = \frac{1}{2} \partial_x^2$, and the martingale

$$M_t(f) = \int_0^t \partial_x f(W_s) \, dW_s$$

with angle bracket given by

$$\langle M(f) \rangle_t := \int_0^t \Gamma_L(f, f)(X_s) \, ds = \int_0^t (\partial_x f(W_s))^2 \, ds.$$

When the function $f(t, x)$ depends on the time parameter, we have

$$df(t, W_t) = [\partial_t + L](f)(t, W_t) \, dt + dM_t(f).$$

In this case, the martingale part is defined by

$$M_t(f) = \int_0^t \partial_x f_s(W_s) \, dW_s$$

and its angle bracket given by

$$\langle M(f) \rangle_t := \int_0^t \Gamma_L(f_s, f_s)(X_s) \, ds = \int_0^t (\partial_x f_s(W_s))^2 \, ds.$$

We also mention that proposition 14.1.4 can be deduced directly from the Doeblin-Itō lemma. If we set $\eta_t = \mathrm{Law}(W_t)$, taking the expectations in the formulas of lemma 14.1.5, we have

$$d\eta_t(f) = d\mathbb{E}(f(W_t)) = \mathbb{E}(L(f)(W_t)) \, dt = \eta_t(L(f)) \, dt.$$

This implies that

$$\partial_t \eta_t(f) = \eta_t(L(f)).$$

Since $\eta_t(dx) = p_t(x) \, dx$, this implies that

$$\int f(x) \, \partial_t p_t(x) \, dx = \frac{1}{2} \int \partial_x^2 f(x) \, p_t(x) \, dx = \frac{1}{2} \int f(x) \, \partial_x^2 p_t(x) \, dx$$

for any smooth functions f with bounded derivatives and compact support. The last equation is proved by a double integration by parts formula.

We end this section with a couple of applications of Doeblin-Itō differential calculus.

Example 14.1.6 *If we choose $f(x) = x^2/2$, we have*

$$f'(x) = x \quad \text{and} \quad f''(x) = 1 \;\Rightarrow\; \frac{1}{2} \, d(W_t^2) = W_t \, dW_t + \frac{1}{2} \, dt.$$

This implies that

$$\frac{1}{2} \, W_t^2 = \frac{1}{2} \left[W_t^2 - W_0^2 \right] = \frac{1}{2} \int_0^t d(W_s^2) = \int_0^t W_s \, dW_s + \frac{1}{2} \, (t - 0).$$

Hence

$$\frac{1}{2} \int_0^t d(W_s^2) \neq \int_0^t W_s \, dW_s = \frac{1}{2} \, (W_t^2 - t).$$

Example 14.1.7 *If we choose the exponential functions $f(t,x) = e^{\sigma x - \frac{1}{2} \sigma^2 t}$, we have*

$$\partial_x f = \sigma\, f \quad and \quad \frac{1}{2}\, \partial_x^2 f = \frac{\sigma^2}{2}\, f = -\, \partial_t f$$

from which we conclude that

$$df(t, W_t) = \sigma\, f(t, W_t)\, dW_t.$$

We can rewrite in slightly different terms:

$$X_t := X_0\, f(t, W_t) = X_0\, e^{\sigma W_t - \frac{\sigma^2}{2} t} \quad\Longleftrightarrow\quad dX_t = \sigma\, X_t\, dW_t.$$

In the literature on stochastic processes, this formula is sometimes called the Doléan-Dade exponential formula for the Brownian motion. In mathematical finance, this process is also called the geometric Brownian motion. This formula can also be rewritten in the following form:

$$e^{\sigma W_t - \frac{\sigma^2}{2} t} = X_0 + \sigma \int_0^t e^{\sigma W_s - \frac{\sigma^2}{2} s}\, dW_s. \tag{14.6}$$

In mathematical finance, this geometric Brownian motion is often used to model the random evolution of asset prices.

We let $\mathcal{F}_t = \sigma(W_s\ :\ 0 \le s \le t)$ the σ-field generated by the Brownian motion. By construction, for any $s \le t$ we have

$$\mathbb{E}(W_t \mid \mathcal{F}_s) = W_s \quad and \quad \mathbb{E}((W_t - W_s)^2 \mid \mathcal{F}_s) = (t - s).$$

This shows that W_t is a martingale w.r.t. $\mathcal{F}_t$. In addition, the stochastic process

$$M_t := W_t^2 - \langle W \rangle_t \quad \text{with the angle bracket} \quad \langle W \rangle_t = t$$

is a martingale w.r.t. $\mathcal{F}_t$. The last assertion is a consequence of

$$W_t^2 - t = (W_s + (W_t - W_s))^2 - (t - s) - s \Rightarrow M_t - M_s = (W_t - W_s)^2 - (t - s) + 2 W_s (W_t - W_s)$$

and

$$\mathbb{E}\left((W_t - W_s)^2 - (t - s) + 2 W_s (W_t - W_s) \mid \mathcal{F}_s \right)$$

$$= \mathbb{E}\left((W_t - W_s)^2 \mid \mathcal{F}_s \right) - (t - s) + 2 W_s \mathbb{E}\left((W_t - W_s) \mid \mathcal{F}_s \right) = 0.$$

Inversely, we have the following important characterization property.

Theorem 14.1.8 (Levy's characterization of Brownian motion) *Any martingale W_t w.r.t. some filtration $\mathcal{F}_t$ with continuous paths and angle bracket $\langle W \rangle_t = t$ is a Brownian motion.*

Proof :
The proof follows the arguments provided in example 14.1.7. Applying the Doeblin-Itō lemma to the function $f(t, W_t) = e^{a W_t - a^2 t / 2}$ we have

$$df(t, W_t) = -\frac{a^2}{2}\, e^{a W_t - a^2 t / 2}\, dt + a\, e^{a W_t - a^2 t / 2}\, dW_t + \frac{1}{2}\, a^2\, e^{a W_t - a^2 t / 2}\, dt = a\, f(t, W_t)\, dW_t.$$

This shows that $X_t = f(t, W_t)$ is a martingale w.r.t. some filtration $\mathcal{F}_t$. Therefore we have

$$\mathbb{E}(f(t, W_t) \mid \mathcal{F}_s) = f(s, W_s) \iff \mathbb{E}\left(e^{a(W_t - W_s)} \mid \mathcal{F}_s\right) = e^{-a^2(t-s)/2}.$$

This shows that $(W_t - W_s)$ is independent of W_s and it is a centered Gaussian random variable with variance $(t - s)$. This ends the proof of the theorem. ∎

14.2 Stochastic differential equations

14.2.1 Diffusion processes

We consider a time mesh sequence $t_n = nh$, with $n \in \mathbb{N}$, and a given time step $h \in]0, 1[$. In view of theorem 14.1.1, The increments $\Delta W_{t_{n+1}} = W_{t_{n+1}} - W_{t_n}$ of the Brownian motion W_t on this time mesh form a sequence of independent and centered Gaussian random variables with variance h.

Given a couple of bounded smooth functions b_t and σ_t, we let $X_{t_n}^h$ be the real valued Markov chain given by the recursive equations

$$\Delta X_{t_{n+1}}^h := X_{t_{n+1}}^h - X_{t_n}^h = b_{t_n}(X_{t_n}^h)\, h + \sigma_{t_n}(X_{t_n}^h)\, \Delta W_{t_{n+1}}. \qquad (14.7)$$

We denote by $\mathcal{F}_{t_n}^h$ the filtration generated by the Markov chain

$$\mathcal{F}_{t_n}^h = \sigma\left(X_{t_p}^h, t_p \leq t_n\right) = \sigma\left(X_0^h,\, \Delta W_{t_p},\, t_p \leq t_n\right).$$

Under rather weak regularity conditions on the drift and the diffusion functions b_t and σ_t, the Markov chain $X_{t_n}^h$ converges, as h tends to 0 (in the weak and the strong sense) to a continuous time diffusion process

$$dX_t = b_t(X_t)\, dt + \sigma_t(X_t)\, dW_t, \qquad (14.8)$$

with $X_0^h = X_0$.

These continuous time processes are called diffusion models. As for deterministic dynamical systems discussed in section 13.1, the existence and the uniqueness of the solution of the stochastic differential equation (14.8) at any time and for any initial condition require some regularity property on the drift and the diffusion functions b_t and σ_t. As usual when the functions (b_t, σ_t) are Lipschitz, the system (14.8) has a unique solution at least for short time periods. The existence of the solution for all time horizon is ensured for instance under linear type growth conditions uniformly w.r.t. the time horizon. The exercise 236 provides an example of a diffusion equation with a non-unique solution.

Besides the fact that the discrete time model (14.7) is well defined for any h, even for $h = 100^{-100^{100^{100}}}$, the continuous time stochastic model (14.8) involves the random measures dW_t that cannot be handled using classical integrations w.r.t. the usual measure of time dt. The meaning of (14.8) is that the discrete time process $(X_t^h)_{t \geq 0}$ converges to some limiting process $(X_t)_{t \geq 0}$ as $h \downarrow 0$ such that

$$X_t = X_0 + \int_0^t b_s(X_s) \, ds + \int_0^t \sigma_s(X_s) \, dW_s.$$

Most of the continuous time diffusion processes encountered in applications are well defined for any starting point and for any time horizon. As for deterministic dynamical systems, (13.3) let us provide a strongly nonlinear example where things may go wrong. Applying the Doeblin-Itō formula to the function

$$\forall w \neq x_0 \quad f(w) = x_0/(1 - wx_0) \Rightarrow f'(w) = f(w)^2 \quad \text{and} \quad \frac{1}{2} f''(w) = f(w)f'(w) = f(w)^3$$

we readily check that $X_t = f(W_t) = X_0/(1 - W_t X_0)$ and $X_t = 0$ satisfy the equation

$$dX_t = X_t^3 \, dt + X_t^2 \, dW_t$$

starting at X_0. When $X_0 \neq 0$, the solution is defined up to the first time T the Brownian motion starting at the origin hits the set $\left[\frac{1}{X_0}, +\infty\right[$ (when $X_0 > 0$) or $\left]-\infty, \frac{1}{X_0}\right]$ when $X_0 < 0$. The explosion time T can be analyzed but the solution is not defined after this random time horizon.

For a more thorough discussion on these regularity conditions we refer the reader to books referenced in the beginning of section 14.1. For more details on the convergence of discrete time models to continuous time models, we refer the reader to the pioneering and seminal articles of P. E. Kloeden and E. Platen [169, 170, 171].

14.2.2   Doeblin-Itō differential calculus

We denote by $\mathcal{F}_t$ the filtration generated by the Brownian motion

$$\mathcal{F}_t = \sigma\left(X_s, s \leq t\right) = \sigma\left(X_0, \, W_s - W_r, \, r \leq s \leq t\right).$$

Using the rules

$$dt \times dt = 0 \qquad dW_t \times dW_t = dt \quad \text{and} \quad dt \times dW_t = 0$$

for any smooth function f we have

$$\begin{aligned} df(t, X_t) &= \partial_t f(t, X_t) \, dt + \partial_x f(t, X_t) \, dX_t + \frac{1}{2} \partial_x^2 f(t, X_t) \, dX_t dX_t \\ &= \left[\partial_t f(t, X_t) + \partial_x f(t, X_t) \, b_t(X_t) \right. \\ &\left. \quad + \frac{1}{2} \partial_x^2 f(t, X_t) \, \sigma_t^2(X_t)\right] dt + \partial_x f(X_t) \, \sigma_t(X_t) \, dW_t. \end{aligned}$$

This yields an informal proof of the extended version of lemma 14.1.5 to general diffusion models.

Theorem 14.2.1 *For any smooth and bounded function $f(t, x)$, we have*

$$df(t, X_t) = [\partial_t + L_t] (f)(t, X_t) \; dt + dM_t(f) \tag{14.9}$$

with the infinitesimal generator

$$L_t = b_t \; \partial_x \; + \frac{1}{2} \; \sigma_t^2 \; \partial_x^2 \tag{14.10}$$

and the $\mathcal{F}_t$ martingale

$$M_t(f) = \int_0^t \partial_x f(s, X_s) \; \sigma_s(X_s) \; dW_s. \tag{14.11}$$

In addition, the angle bracket of $M_t(f)$ is given by

$$\langle M(f) \rangle_t := \int_0^t \Gamma_{L_s}(f(s, .), f(s, .))(X_s) \; ds = \int_0^t \left(\sigma_s(X_s) \; \partial_x f(s, X_s) \right)^2 ds \tag{14.12}$$

where Γ_{L_t} stands for "carré du champ operator" introduced in definition 12.5.3.

Important remark : As a direct consequence of theorem 14.2.1, we have

$$\begin{aligned}
\mathbb{E}(M_t(f)^2) &= \mathbb{E}\left(\left[\int_0^t \sigma_s(X_s) \partial_x f(s, X_s) \; dW_s \right]^2 \right) \\
&= \mathbb{E}(\langle M(f) \rangle_t) = \mathbb{E}\left(\int_0^t \left(\sigma_s(X_s) \; \partial_x f(s, X_s) \right)^2 ds \right).
\end{aligned}$$

More generally, we have the following proposition.

Proposition 14.2.2 *For any regular function $g(s, x)$ the stochastic process*

$$(g \cdot W)_t := \int_0^t g(s, X_s) \; dW_s$$

is an $\mathcal{F}_t$-martingale with angle bracket

$$\langle (g \cdot W) \rangle_t = \int_0^t g(s, X_s)^2 \; ds.$$

In addition, if $g(s, x) = g(s)$ is a deterministic homogeneous function, then $(g \cdot W)_t$ is a centered Gaussian r.v. with variance function $\langle (g \cdot W) \rangle_t = \int_0^t g(s)^2 \; ds.$

Hint of proof :
The martingale property follows from the fact that

$$d (g \cdot W)_t = (g \cdot W)_{t+dt} - (g \cdot W)_t = g(t, X_t) \; dW_t$$

and

$$\mathbb{E} (g(t, X_t) \; dW_t \mid \mathcal{F}_t) = g(t, X_t) \; \mathbb{E} ((W_{t+dt} - W_t) \mid \mathcal{F}_t) = 0.$$

Applying (12.19) to the martingale

$$M_t = (g \cdot W)_t \quad \Longrightarrow \quad dM_t = g(t, X_t) \, dW_t$$
$$\Longrightarrow \quad \mathbb{E}\left((dM_t)^2 \mid \mathcal{F}_t\right) = g(t, X_t)^2 \, dt$$

we find that

$$\mathbb{E}\left(d\left(M_t^2\right) - g(t, X_t)^2 \, dt \mid \mathcal{F}_t\right) = \mathbb{E}\left(dN_t \mid \mathcal{F}_t\right) = 0$$

with

$$N_t := M_t^2 - \int_0^t g(s, X_s)^2 \, ds.$$

This ends the proof of the first assertion.

When $g(s, x) = g(s)$ is a deterministic homogeneous function, using the Doeblin-Itō differential rule we have

$$e^{-u \, (g \cdot W)_t} \, d e^{u \, (g \cdot W)_t} = u \, g(t, X_t) \, dW_t + \frac{u^2}{2} \, g^2(t) \, dt$$

for any $u \in \mathbb{R}$. This implies that

$$\partial_t \mathbb{E}\left(e^{u \, (g \cdot W)_t}\right) = \frac{u^2}{2} \, g^2(t) \, \mathbb{E}\left(e^{-u \, (g \cdot W)_t}\right).$$

Hence we conclude that

$$\mathbb{E}\left(e^{u \, (g \cdot W)_t}\right) = e^{\frac{u^2}{2} \int_0^t g^2(s) ds}.$$

This ends the proof of the proposition.   ∎

By proposition 14.2.2,

$$\mathbb{E}\left(\left[\int_0^t g(s, X_s) \, dW_s\right]^2\right) = \mathbb{E}\left(\int_0^t g(s, X_s)^2 \, ds\right).$$

Next we provide an alternative but informal proof. For any $r \leq s$ we have

$$\mathbb{E}\left(g(r, X_r) g(s, X_s) \, dW_r \, dW_s \mid \mathcal{F}_s\right)$$
$$= g(s, X_r) g(s, X_s) \, dW_r \times \mathbb{E}\left((W_{s+ds} - W_s) \mid \mathcal{F}_s\right) = 0$$

and for $r = s$

$$\mathbb{E}\left(g(s, X_s)^2 \, (dW_s)^2\right) = \mathbb{E}\left(g(s, X_s)^2\right) \, ds.$$

This implies that

$$\mathbb{E}\left(\left[\int_0^t g(s, X_s) \, dW_s\right]^2\right) = \int_0^t \int_0^t \mathbb{E}\left(g(s, X_r) g(s, X_s) \, dW_r \, dW_s\right)$$
$$= \int_0^t \mathbb{E}\left(g(s, X_s)^2\right) \, ds.$$

We end this section with an alternative proof of the theorem based on the discrete time approximation model. To simplify the presentation, we further assume that f is homogeneous w.r.t. the time parameter. In this situation, arguing as in (14.5), for any smooth and bounded function f we have

$$
\begin{aligned}
\Delta f(X^h_{t_{n+1}}) & := f(X^h_{t_n} + \Delta X^h_{t_{n+1}}) - f(X^h_{t_n}) \\
& = \partial_x f(X^h_{t_n})\, \Delta X^h_{t_{n+1}} + \frac{1}{2}\, \partial^2_x f(X^h_{t_n}) \left[\Delta X^h_{t_{n+1}}\right]^2 + \mathrm{O}_P(h^{1+1/2}),
\end{aligned} \tag{14.13}
$$

where $\mathrm{O}_P(h^{1+1/2})$ denotes a random function such that

$$
\mathbb{E}\left(\left|\mathrm{O}_P(h^{1+1/2})\right|\right) = \mathrm{O}(h^{1+1/2}).
$$

Developing the square in (14.13), we have

$$
\Delta f(X^h_{t_{n+1}}) = L_{t_n}(f)(X^h_{t_n})\, h + \Delta M^h_{t_{n+1}}(f) + \mathrm{O}_P(h^{1+1/2})
$$

with the $\mathcal{F}^h_{t_n}$-martingale increments

$$
\Delta M^h_{t_{n+1}}(f) := M^h_{t_{n+1}}(f) - M^h_{t_n}(f) = \partial_x f(X^h_{t_n})\, \sigma_{t_n}(X^h_{t_n})\, \Delta W_{t_{n+1}}.
$$

This implies that

$$
f(X^h_{t_n}) = f(X^h_0) + \int_0^{t_n} L_{\tau_h(s)}(f)(X^h_{\tau_h(s)})\, ds + M^h_{t_{n+1}}(f) + \mathrm{O}_P(h^{1/2})
$$

with $\tau_h(s) = t_n$, for any $t_n \le s < t_{n+1}$, with $n \ge 0$.

We also observe that the stochastic process

$$
t_n \mapsto M^h_{t_n}(f)^2 - \langle M^h(f)\rangle_{t_n}
$$

is a martingale with the increments of the angle bracket process given by

$$
\begin{aligned}
\Delta \langle M^h(f)\rangle_{t_{n+1}} & = \langle M^h(f)\rangle_{t_{n+1}} - \langle M^h(f)\rangle_{t_n} = \mathbb{E}\left(\left(\Delta M^h_{t_{n+1}}(f)\right)^2 \mid \mathcal{F}^h_{t_n}\right) \\
& = \left(\partial_x f(X^h_{t_n})\right)^2 \sigma^2_{t_n}(X^h_{t_n})\, h = \Gamma_{L_{t_n}}(f, f)(X^h_{t_n})\, h.
\end{aligned}
$$

The proof of the last assertion follows the same lines as in (3.21).

Letting h tends to 0, we prove the Doeblin-Itō differential formula (14.9).

14.3 Evolution equations

14.3.1 Fokker-Planck equation

We let p_t be the probability density of the random states X_t; that is,

$$
\mathbb{P}(X_t \in dx) = p_t(x)\, dx := \eta_t(dx).
$$

Using (14.9), we have

$$\partial_t \eta_t(f) \quad = \quad \eta_t(L_t(f)).$$

Rewritten in terms of the density $p_t(x)$, for any smooth function f with compact support we have

$$\int f(x) \ \partial_t p_t(x) \ dx \quad = \quad \int \left(b_t(x) \ f'(x) + \frac{1}{2} \ \sigma_t^2(x) f''(x) \right) \ p_t(x) \ dx$$

$$= \quad \int f(x) \ \left[-\partial_x \left(b_t(x) \ p_t(x) \right) + \frac{1}{2} \ \partial_x^2 \left(\sigma_t^2(x) p_t(x) \right) \right] \ dx.$$

The last assertion is proved using an elementary integration by parts technique. We summarize the above discussion with the following theorem.

Theorem 14.3.1 *The density of the random states satisfies the Fokker-Planck equation*

$$\partial_t p_t = L_t^\star (p_t) \quad \text{with} \quad L_t^\star (p_t) := -\partial_x \left(b_t \ p_t \right) + \frac{1}{2} \ \partial_x^2 \left(\sigma_t^2 \ p_t \right).$$

14.3.2 Weak approximation processes

The evolution semigroup

$$P_{t_p, t_n}^h (f)(x) = \mathbb{E} \left(f(X_{t_n}^h) \mid X_{t_p}^h = x \right)$$

of the Markov chain $X_{t_n}^h$ defined in (14.7) is defined in terms of the one-step transitions

$$P_{t,t+h}^h(f)(x) \quad = \quad \mathbb{E} \left(f(x + b_t(x) \ h + \sigma_t(x) \ \sqrt{h} \ W_1) \right).$$

For smooth functions, using a second order Taylor expansion we find that

$$f(x + b_t(x)h + \sigma_t(x)\sqrt{h} \ W_1)$$

$$= f(x) + f'(x) \ (b_t(x) \ h + \sigma_t(x) \ \sqrt{h} \ W_1)$$

$$+ \frac{1}{2} \ (b_t(x) \ h + \sigma_t(x) \ \sqrt{h} \ W_1)^2 \ f''(x) + O_P(h^{1+1/2})$$

$$= f(x) + f'(x) \ b_t(x) \ h + \frac{1}{2} \ \sigma_t^2(x) \ h \ W_1^2 \ f''(x) + f'(x) \ \sigma_t(x)\sqrt{h} \ W_1 + O_P(h^{1+1/2}).$$

This implies that

$$P_{t,t+h}^h(f)(x) = f(x) + L_t(f)(x) \ h + O(h^{1+1/2})$$

with the infinitesimal generator L_t defined in (14.10).

On the other hand, by theorem 14.2.1 the sg $P_{s,t}$ of X_t satisfies

$$P_{t,t+h}(f)(x) = f(x) + \int_t^{t+h} \mathbb{E}\left(L_s(f)(X_s) \mid X_t = x\right) \, ds.$$

We check this claim using the fact that

$$f(X_{t+h}) = f(X_t) + \int_t^{t+h} L_s(f)(X_s) dr + M_{t+h}(f) - M_t(f)$$

with $\mathbb{E}\left(M_{t+h}(f) - M_t(f) \mid \mathcal{F}_t\right) = 0$.

Applying theorem 14.2.1 to the function $(s,x) \mapsto g(s,x) := L_s(f)(x)$, with $t \le s$, we find that

$$L_s(f)(X_s) = L_t(f)(X_t) + \int_t^s \left[\partial_r + L_r\right](g)(r, X_r) \, dr + M_s(g) - M_t(g).$$

Using the fact that

$$\mathbb{E}\left(M_s(g) - M_t(g) \mid \mathcal{F}_t\right) = 0 \quad \text{and} \quad \int_t^s \left[\partial_r + L_r\right](g)(r, X_r) \, dr = \mathrm{O}(s-t)$$

we conclude that

$$P_{t,t+h}(f)(x) = f(x) + L_t(f)(x) \, h + \mathrm{O}(h^2). \tag{14.14}$$

Combining these expansions with theorem 12.3.1 we prove the following theorem.

Theorem 14.3.2 *For any $t \ge 0$, we have the first order operator expansion*

$$P_{t,t+h}^h = Id + h \, L_t + \mathrm{O}(h^{1+1/2}) = P_{t,t+h}. \tag{14.15}$$

In addition, for any time steps $t_p = h\lfloor s/h \rfloor \le t_n = h\lfloor t/h \rfloor$, we have the first order approximation formulae

$$P_{t_p,t_n}^h = P_{t_p,t_n} + \mathrm{O}\left(h^{1/2}\right).$$

For any finite sequence of times $s_1 < \ldots < s_n$ we also have the finite dimensional approximation

$$\prod_{1 \le i < n} P_{s_i,s_{i+1}}^h(x_i, dx_{i+1}) = \prod_{1 \le i < n} P_{s_i,s_{i+1}}(x_i, dx_{i+1}) + \mathrm{O}\left(h^{1/2}\right)$$

where P_{s_1,s_2}^h stands for the continuous time semigroup of the process X_t^h defined by interpolation:

$$\forall t \in [t_n, t_{n+1}] \quad X_t^h := X_{t_n}^h + b_t\left(X_{t_n}^h\right) \, (t - t_n) + \sigma_t\left(X_{t_n}^h\right) \, (W_t - W_{t_n}).$$

Notice that the interpolated model is coupled to the limit one by the same Brownian motion and the same initial condition. This allows us to derive $\mathbb{L}_p$-type bounds under natural Lipchitz conditions on (b_t, σ_t) using Gronwall's type techniques. These approximations are often called strong approximations. In other instances, it may be hard to couple a given Markov chain $X_{t_n}^h$ to its limiting diffusive process, and we prefer to use the càdlàg continuous time embedding (12.14).

In this context, following the Taylor's expansions discussed above, a natural and simple way to ensure (14.15) is to check that

$$\mathbb{E}\left((X_{t_n+h}^h - X_{t_n}^h) \mid X_{t_n}^h = x\right) = b_t(x)\,h + O(h^{1+1/2}) \qquad (14.16)$$

$$\mathbb{E}\left((X_{t_n+h}^h - X_{t_n}^h)^2 \mid X_{t_n}^h = x\right) = \sigma_t^2(x)\,h + O(h^{1+1/2})$$

$$\mathbb{E}\left(\left|X_{t_n+h}^h - X_{t_n}^h\right|^3 \mid X_{t_n}^h = x\right) = O(h^{1+1/2}) \quad \text{as soon as } t_n = n\lfloor t/h \rfloor.$$

For a more thorough discussion of weak approximation of diffusion processes and more general sufficient conditions we refer to [174].

14.3.3   A backward stochastic differential equation

Combining (14.14) with the decompositions (13.6) and (13.7), we prove the following theorem.

> **Theorem 14.3.3** *We have the forward and backward evolution equations*
>
> $$\partial_t P_{s,t}(f) = P_{s,t}(L_t(f)) \quad \text{and} \quad \partial_s P_{s,t}(f) = -L_s(P_{s,t}(f)).$$
>
> *In addition, for any fixed time horizon T and any smooth function $g(t,x)$ the stochastic process*
>
> $$t \in [0,T] \mapsto Y_t := P_{t,T}(g_T)(X_t) \qquad (14.17)$$
>
> *is a martingale ending at $Y_T = g_T(X_T)$.*

Proof :
Applying theorem 14.2.1 to the function $f(t,x) = P_{t,T}(g_T)(x)$ we find that

$$dY_t = df(t, X_t) = dP_{t,T}(g_T)(X_t) = dM_t(f)$$

with the martingale $M_t(f)$ defined in (14.11). We prove this claim using the fact that

$$[\partial_t + L_t]\,(f) = \partial_t P_{t,T}(g_T) + L_t P_{t,T}(g_T) = 0.$$

This ends the proof of the theorem.   ∎

> The proof of the above theorem also shows that for any $t \in [0,T]$ we have the backward stochastic differential equation
>
> $$dY_t = \sigma_t(X_t)\,U_t\,dW_t \quad \text{with} \quad U_t := \partial_x P_{t,T}(g_T)(X_t).$$
>
> By construction, the above backward stochastic differential equation (abbreviated BSDE) has the terminal condition $Y_T = f_T(X_T)$. More sophisticated BSDEs based on backward Feynman-Kac semigroups are discussed in section 15.7.2.

14.4 Multidimensional diffusions

14.4.1 Multidimensional stochastic differential equations

The description of d-dimensional diffusions follows the same line of arguments as in the one-dimensional case.

Definition 14.4.1 *A d-dimension Brownian motion $W_t = (W^i)_{1 \leq i \leq d}$ is a sequence of independent copies of a one-dimensional Brownian motion.*

A d-dimensional diffusion $X_t = (X_t^i)_{1 \leq i \leq d}$ is given by d stochastic differential equations

$$dX_t^i = b_t^i(X_t)\, dt + \sum_{1 \leq j \leq d} \sigma_{j,t}^i(X_t)\, dW_t^j \tag{14.18}$$

with some regular functions b_t^i and $\sigma_{j,t}^i$. In vector notation, we have

$$d \begin{bmatrix} X_t^1 \\ \vdots \\ X_t^d \end{bmatrix} = \begin{bmatrix} b_t^1(X_t) \\ \vdots \\ b_t^d(X_t) \end{bmatrix} dt + \begin{bmatrix} \sigma_{1,t}^1(X_t) & \cdots & \sigma_{d,t}^1(X_t) \\ \vdots & & \\ \sigma_{1,t}^d(X_t) & \cdots & \sigma_{d,t}^d(X_t) \end{bmatrix} d \begin{bmatrix} W_t^1 \\ \vdots \\ W_t^d \end{bmatrix}$$

or in a more synthetic formulation

$$dX_t = b_t(X_t)\, dt + \sigma_t(X_t)\, dW_t.$$

We apply the rules

$$dt \times dt = 0 \qquad dW_t^i \times dW_t^j = 1_{i=j}\, dt \quad \text{and} \quad dt \times dW_t^i = 0.$$

To simplify the presentation, we set

$$\partial_i := \partial_{x^i} = \frac{\partial}{\partial x^i} \quad \text{and} \quad \partial_{i,j} := \partial_{x^i, x^j} = \frac{\partial^2}{\partial x^i \partial x^j}.$$

In this notation, we have

$$df(X_t) = f(X_t + dX_t) - f(X_t) = \sum_{i=1}^d \partial_i f(X_t)\, dX_t^i + \frac{1}{2} \sum_{i,j=1}^d \partial_{i,j} f(X_t)\, dX_t^i dX_t^j.$$

Using the fact that

$$dX_t^i dX_t^j = \sum_{1 \leq k,l \leq d} \sigma_{k,t}^i(X_t)\, \sigma_{l,t}^j(X_t)\, dW_t^k dW_t^l$$

and applying the rules $dW_t^k \times dW_t^l = 1_{k=l}\, dt$, and $dt \times dW_t^i = 0$, we conclude that

$$dX_t^i dX_t^j = \sum_{1 \leq k \leq d} \sigma_{k,t}^i(X_t)\, \sigma_{k,t}^j(X_t)\, dt = a_t^{i,j}(X_t)\, dt$$

with the symmetric $d \times d$-matrix valued function

$$a_t = \sigma_t(\sigma_t)' \Leftrightarrow \forall i,j \qquad a_t^{i,j}(x) := (\sigma_t(\sigma_t)')^{i,j}(x) = \sum_{1 \leq k \leq d} \sigma_{k,t}^i(x)\sigma_{k,t}^j(x).$$

This yields the Doeblin-Itō differential formula

$$df(X_t) = L_t(f)(X_t)\, dt + dM_t(f)$$

with the operator L_t defined by the formula

$$L_t \; := \; \sum_{i=1}^d b_t^i\, \partial_i + \frac{1}{2}\sum_{i,j=1}^d a_t^{i,j}\, \partial_{i,j}. \qquad (14.19)$$

The martingale term is

$$dM_t(f) := \sum_{1 \leq j \leq d}\left[\sum_{1 \leq i \leq d} \partial_i f(X_t)\, \sigma_{j,t}^i(X_t)\right]\, dW_t^j$$

with the angle bracket

$$d\langle M(f)\rangle_t \; := \; \sum_{1 \leq j \leq d}\left[\sum_{1 \leq i \leq d} \sigma_{j,t}^i(X_t)\partial_i f(X_t)\right]^2\, dt = \Gamma_{L_t}(f,f)(X_t)\, dt.$$

The last assertion follows from the fact that

$$\mathbb{E}\left(d\left(M_t^2(f)\right) \mid \mathcal{F}_t\right)$$

$$= \mathbb{E}\left(M_{t+dt}^2(f) - M_t^2(f) \mid \mathcal{F}_t\right) = \mathbb{E}\left((M_t(f) + dM_t(f))^2 - M_t^2(f) \mid \mathcal{F}_t\right)$$

$$= 2M_t(f)\,\overbrace{\mathbb{E}\left(dM_t(f) \mid \mathcal{F}_t\right)}^{=0} + \mathbb{E}\left((dM_t(f))^2 \mid \mathcal{F}_t\right) = \mathbb{E}\left((dM_t(f))^2 \mid \mathcal{F}_t\right)$$

$$= \sum_{1 \leq j,k \leq d}\left[\sum_{1 \leq i \leq d} \sigma_{j,t}^i(X_t)\, \partial_i f(X_t)\right]\left[\sum_{1 \leq i' \leq d} \sigma_{k,t}^{i'}(X_t)\, \partial_{i'} f(X_t)\right]\, \underbrace{dW_t^j dW_t^k}_{1_{j=k}\; dt}$$

$$= \sum_{1 \leq j \leq d}\left[\sum_{1 \leq i \leq d} \sigma_{j,t}^i(X_t)\, \partial_i f(X_t)\right]^2\, dt$$

with $\mathcal{F}_t = \sigma(X_s, \ s \leq t)$. We also have the carré du champ decompositions

$$\Gamma_{L_t}(f,f)(x)$$

$$= L_t \left[(f(.) - f(x))^2 \right](x) = \sum_{1 \leq i,j \leq d} a_t^{i,j}(x) \ \partial_j \left[(f(.) - f(x)) \ \partial_i (f - f(x)) \right](x)$$

$$= \sum_{1 \leq i,j \leq d} \sum_{1 \leq k \leq d} \sigma_{k,t}^i(x) \sigma_{k,t}^j(x) \ \partial_j f(x) \ \partial_i f(x) = \sum_{1 \leq k \leq d} \left(\sum_{1 \leq i \leq d} \sigma_{k,t}^i(x) \partial_i f(x) \right)^2.$$

The infinitesimal generator (14.19) of the diffusion process (14.18) is sometimes rewritten in the following form

$$L_t(f) = b_t^T \ \nabla f + \frac{1}{2} \ \mathrm{Tr} \left(a_t \nabla^2 f \right)$$

with the row vector $b_t^T = \left(b_t^i \right)_{1 \leq i \leq d}$, the column gradient vector $\nabla f = (\partial_i f)_{1 \leq i \leq d}^T$, and the Hessian matrix $\nabla^2 f = (\partial_{i,j} f)_{1 \leq i,j \leq d}$.

Important remark : Using the same arguments as in the one-dimensional case, the semi-group of the multidimensional diffusion $P_{s,t}$ also satisfies the forward and backward evolution equations stated in theorem 14.3.3. In addition, the backward evolution stochastic process $M_t := P_{t,T}(f_T)(X_t)$, with $t \in [0,T]$, is a martingale ending at some given function $f_T(X_T)$.

14.4.2 An integration by parts formula

Applying the Doeblin-Itō differential rule to the function $f(X_t^1, X_t^2) = X_t^1 X_t^2$ we find that

$$\begin{aligned} d\left(X_t^1 X_t^2 \right) &= df(X_t^1, X_t^2) \\ &= \partial_1 f(X_t^1, X_t^2) \ dX_t^1 + \partial_2 f(X_t^1, X_t^2) \ dX_t^2 \\ &\qquad + \frac{1}{2} \left(\partial_{1,2} + \partial_{2,1} \right) f(X_t^1, X_t^2) \ dX_t^1 dX_t^2 \\ &= X_t^1 \ dX_t^2 + X_t^2 \ dX_t^1 + dX_t^1 dX_t^2. \end{aligned}$$

This yields the integration by parts formula

$$d\left(X_t^1 X_t^2 \right) = X_t^1 \ dX_t^2 + X_t^1 \ dX_t^1 + dX_t^1 dX_t^2. \tag{14.20}$$

This integration by parts formula is clearly valid for any diffusion processes X_t^1 and X_t^2 (since we can always write the pair process (X_t^1, X_t^2) in terms of a diffusion equation of the form (14.18)). We can also show that the stochastic process $\langle M^1, M^2 \rangle_t$ defined by

$$d\langle M^1, M^2 \rangle_t := \sum_{1 \leq k \leq d} \sigma_{k,t}^1(X_t) \ \sigma_{k,t}^2(X_t) \ dt := dX_t^1 dX_t^2$$

is the angle bracket of the product $M_t^1 M_t^2$ of the martingales

$$dM_t^i = \sum_{1 \leq j \leq d} \sigma_{j,t}^i(X_t) \, dW_t^j$$

in the sense that $M_t^1 M_t^2 - \langle M^1, M^2 \rangle_t$ is a martingale.

14.4.3 Laplacian and orthogonal transformations

Definition 14.4.2 *The infinitesimal generator* $L^W = \frac{1}{2} \sum_{1 \leq i \leq r} \partial_i^2$ *of an r-dimensional Brownian motion* $W_t = (W_t^i)_{1 \leq i \leq r} \in \mathbb{R}^r$ *is called the (Euclidian) Laplacian.*

Given some orthogonal matrix O on $\mathbb{R}^r$ we set $X_t = OW_t$. We recall that an orthogonal matrix $O = (O_1, \ldots, O_r)'$ is a square matrix with orthonormal row vectors $O_i = (Q_i^j)_{1 \leq j \leq r}$ w.r.t. the Euclidian scalar product $\langle x, y \rangle = \sum_{1 \leq i \leq r} x_i y_i$ on $\mathbb{R}^r$. In matrix form, this property reads $O'O = OO' = Id$, where Id stands for the identity matrix.

Using the Doeblin-Itō differential formula (14.19) we check that the generators L^W and $L^X = L^{OW}$ of W_t and X_t coincide

$$L^X = \frac{1}{2} \sum_{1 \leq i,j \leq r} (OO')_{i,j} \, \partial_{i,j} = \frac{1}{2} \sum_{1 \leq i \leq r} \partial_i^2 = L^W \iff L^{OW} = L^W.$$

In terms of stochastic increments, we have the matrix formula

$$dX_t = OdW_t \Rightarrow (dX_t)'dX_t = (dW_t)'O'OdW_t = (dW_t)'dW_t = Id \times dt.$$

If we set

$$g(x) = f(Ox) = f(O_1 x, \ldots, O_r x) = f(\langle O_1, x \rangle, \ldots, \langle O_r, x \rangle),$$

we have

$$\partial_i g(x) = \sum_{1 \leq k \leq r} (\partial_k f)(Ox) \, \partial_i(\langle O_k, x \rangle) = \sum_{1 \leq k \leq r} (\partial_k f)(Ox) \, O_k^i$$

$$\Longrightarrow \partial_i^2 g(x) = \sum_{1 \leq k,l \leq r} (\partial_{k,l} f)(Ox) \, O_k^i O_l^i$$

$$\Longrightarrow \sum_{1 \leq i \leq r} \partial_i^2 g(x) = \sum_{1 \leq k,l \leq r} (\partial_{k,l} f)(Ox) \underbrace{\langle O_k, O_l \rangle}_{=1_{k=l}} = \sum_{1 \leq k \leq r} (\partial_k^2 f)(Ox).$$

In summary we have proved the orthogonal invariance theorem.

Theorem 14.4.3 *We have the invariance property of the Brownian motion and the Laplacian under orthogonal transformations*

$$L^{OW} = L^W, \quad \sum_{1 \leq i \leq r} \partial_i^2 (f(Ox)) = \sum_{1 \leq k \leq r} (\partial_k^2 f)(Ox), \quad (OdW_t)' (OdW_t) = Id \times dt.$$

$$(14.21)$$

Illustrations of these invariance properties are provided in exercises 240 and 241.

14.4.4   Fokker-Planck equation

In multidimensional settings, the probability density

$$\mathbb{P}\left(X_t \in dx\right) = p_t(x) \; dx := \eta_t(dx)$$

of the random states X_t on $\mathbb{R}^d$ satisfies the Fokker-Planck equations

$$\partial_t \eta_t(f) = \eta_t(L_t(f)) \quad \text{and} \quad \partial_t p_t = L_t^{\star}(p_t)$$

with the dual operator

$$L_t^{\star}(p_t) = -\sum_{i=1}^{d} \partial_i \left(b_t^i \; p_t\right) + \frac{1}{2} \sum_{i,j=1}^{d} \partial_{i,j} \left(\left(\sigma_t(\sigma_t)^T\right)_{i,j} \; p_t\right).$$

14.5   Exercises

Exercise 220 (Covariance function) *Let W_t be a (standard) Brownian motion. Compute the covariance function $\mathrm{Cov}(W_s, W_t)$, for any $s \leq t$.*

Exercise 221 (Brownian mixtures) *Let $(W_t^i)_{i \in I}$ be a collection of independent Brownian motions indexed by some parameter i in some countable set I. Consider a collection of parameters $(a_i)_{i \in I}$. Prove that the process $W_t := \sum_{i \in I} a_i W_t^i$ is a Brownian motion if and only if $\sum_{i \in I} a_i^2 = 1$.*

Exercise 222 (Correlations) *Let W_t^1, W_t^2 be two independent Brownian motions. Consider the Brownian motion W_t defined by*

$$W_t = \epsilon W_t^1 + \sqrt{1 - \epsilon^2} \; W_t^2$$

for some given parameter $\epsilon \in [0,1]$ (cf. exercise 221 to check that W_t is a Brownian motion). Compute the covariance functions $\mathrm{Cov}(W_s^1, W_t)$.

Exercise 223 (Brownian paths given a terminal condition) *Consider the stochastic process*

$$X_t = X_0 + b \, t + \sigma \, W_t$$

for some X_0 and a Brownian motion W_t (starting at the origin). Let t_n be a given time mesh with time step $t_n - t_{n-1} = \epsilon$, with $t_0 = 0$. prove that the conditional densities of the sequence of state $(X_{t_1}, \ldots, X_{t_{n-1}})$ given $(X_0, X_{t_n}) = (x_0, x_n)$ are given by

$$p_{t_1, \ldots, t_{n-1}}(x_1, \ldots, x_{n-1} \mid x_0, x_n) \propto \exp\left[-\frac{1}{2\sigma^2} \sum_{1 \leq k \leq n} \left(\frac{x_k - x_{k-1}}{\sqrt{\epsilon}}\right)^2\right].$$

Deduce that

$$\mathrm{Law}\left(\left(X_s\right)_{s \in [0,t]} \mid X_0 = x, X_t = y\right) = \mathrm{Law}\left(\left(x + \sigma \, W_s\right)_{s \in [0,t]} \mid x + \sigma \, W_t = y\right).$$

In mathematical finance the diffusion X_t discussed above is sometimes called the Bachelier model.

Exercise 224 (Poisson approximations) *Let N_t and N_t' be two independent Poisson processes with intensity $\lambda > 0$. We let X_t be the real valued stochastic process starting at the origin $X_t = 0$ and defined by the stochastic differential equation*

$$dX_t = \frac{1}{\sqrt{2\lambda}} \ (dN_t - dN_t').$$

Describe the infinitesimal generator L of X_t and check that

$$L(f)(x) = \frac{1}{2} \ f''(x) + \mathrm{O}\left(\lambda^{-1/2}\right).$$

Check that $(X_t - X_s)$ is independent of X_s for any $s \leq t$ and we have

$$\mathbb{E}(X_t - X_s) = 0 \quad and \quad \mathbb{E}\left((X_t - X_s)^2\right) = (t - s).$$

For any $\alpha \in \mathbb{R}$ prove that
$$\lim_{\lambda \to \infty} \log \mathbb{E}(e^{\alpha X_t}) = \alpha^2 t/2.$$

Exercise 225 (Brownian averages) *Let W_t be a Brownian motion. Check that*

$$W_t = \frac{1}{t} \int_0^t W_s \ ds + \frac{1}{t} \int_0^t s \ dW_s.$$

Exercise 226 (Gaussian martingales - 1) *Let W_t be a Brownian motion. Check that*

$$M_t = \frac{1}{2} \ t^2 \ W_t - \int_0^t s \ W_s \ ds$$

is a martingale w.r.t. $\mathcal{F}_t = \sigma(W_s, \ s \leq t)$. Compute its angle bracket $\langle M \rangle_t$.

Exercise 227 (Gaussian martingales - 2) *Let W_t be a Brownian motion and f a smooth function. Check that*

$$M_t := f(t) \ W_t - \int_0^t f'(s) \ W_s \ ds = \int_0^t f(s) \ dW_s.$$

Deduce that M_t is a martingale w.r.t. $\mathcal{F}_t = \sigma(W_s, \ s \leq t)$. Compute its angle bracket $\langle M \rangle_t$.

Exercise 228 (Metropolis-Hastings samplers [133]) *Let $\pi(dx) \propto e^{-V(x)}dx$ be a probability measure on $\mathbb{R}$ associated with a smooth and bounded potential function V with first and second bounded derivatives. Let t_n be a given time mesh with a fixed time step $(t_{n+1} - t_n) = h > 0$. Consider the Metropolis-Hastings transition given by*

$$M_h(x, dy) = P_h(x, dy) \ a(x, y) + \left(1 - \int P_h(x, dz) \ a(x, z)\right) \delta_x(dy)$$

with the acceptance rate $a(x, y) = \min\left(1, e^{-(V(y)-V(x))}\right)$ and the (symmetric) Markov transition P_h of the Brownian motion W_t between between time t_n and t_{n+1} (cf. for instance (14.4)). We let $X_{t_n}^h$ the chain with transitions M_h, and $\overline{X}_{t_n}^h$ the chain with transitions $\overline{M}_h$ defined as M_h by replacing $a(x, y)$ by the acceptance rate

$$\overline{a}(x, y) = \min\left(1, e^{-\partial_x V(x)(y-x)}\right).$$

- *Check that for any $u, v \in \mathbb{R}$ we have*

$$\left| \min\left(1, e^u\right) - \min\left(1, e^v\right) \right| \leq 1 - e^{-|u-v|} \leq |u - v|$$

and deduce that

$$|a(x, y) - \overline{a}(x, y)| \leq |V(y) - V(x) - \partial_x V(x)(y - x)| \leq c \; |x - y|^2$$

for some finite constant $c < \infty$.

- *Prove that*

$$\sup_{f: osc(f) \leq 1} \left\| \left(M_h - \overline{M}_h\right)(f) \right\| \leq c \; h \qquad and \qquad \sup_{f: lip(f) \leq 1} \left\| \left(M_h - \overline{M}_h\right)(f) \right\| \leq c \; h^{1+1/2}.$$

Exercise 229 (Random direction and bouncy particle samplers) *We consider the random direction Monte Carlo sampler discussed in exercise 120. We assume that the potential function U and its first and second derivatives are bounded. Check that for any sufficiently regular function f we have*

$$h^{-1}\left[M_h(f)(v, x) - f(v, x)\right] = v \; \partial_x f(v, x) + \left(v \; \partial_x U(x)\right)_+ \; (f(-v, x) - f(v, x)) + O(h).$$

Exercise 230 (Metroplis-Hastings sampler and Langevin diffusion [133]) ✒ *We consider the Metropolis-Hastings models discussed in exercise 228.*

- *Check that for any $p \geq 1$ we have*

$$\sup_{x \in \mathbb{R}} \left| \int \left(M_h - \overline{M}_h\right)(x, dy) \, |y - x|^p \right| \leq c \; h^{1 + \frac{p}{2}} \quad \text{for some finite constant } c < \infty.$$

- *Suppose that $\partial_x V(x) \geq 0$. Check that*

$$\int (y - x) \; P_h(x, dy) \; \overline{a}(x, y) = \sqrt{h} \; \mathbb{E}\left(W_1 \; 1_{W_1 > 0} \; \left[e^{-\sqrt{h} \; \partial_x V(x) \; W_1} - 1\right]\right)$$

$$= -\frac{h}{2} \; \partial_x V(x) + O(h^{1+1/2}).$$

Prove that the last expansion is also met when $\partial_x V(x) \leq 0$.

- *Suppose that $\partial_x V(x) \geq 0$. Check that*

$$h^{-1} \int (y - x)^2 \; P_h(x, dy) \; \overline{a}(x, y) = 1 - \mathbb{E}\left(W_1^2 \; 1_{W_1 > 0} \; \left[1 - e^{-\sqrt{h} \; \partial_x V(x) \; W_1}\right]\right)$$

$$= 1 + O(h^{1/2}).$$

Prove that the last expansion is also true when $\partial_x V(x) \leq 0$.

- *Deduce that $X_{t_n}^h$ and $\overline{X}_{t_n}^h$ satisfy the regularity conditions (14.16) with $b_t = -2^{-1}\partial_x V$ and $\sigma_t = 1$. Deduce that the processes $X_{t_n}^h$ and $\overline{X}_{t_n}^h$ weakly converge, as $h \downarrow 0$, in the sense of finite distributions to the stochastic flow (a.k.a. Langevin diffusion) given by the equation*

$$dX_t = -2^{-1}\partial_x V(X_t)dt + dW_t.$$

Exercise 231 (Heat-bath Markov chain sampler [133]) *defined as the Markov transition M_h discussed in exercise 228 by replacing $a(x, y)$ by the acceptance rate*

$$b(x, y) = \frac{e^{V(x)-V(y)}}{1 + e^{V(x)-V(y)}} = e^{V(x)-V(y)} \frac{1}{1 + e^{V(x)-V(y)}} = \frac{1}{1 + e^{V(y)-V(x)}}.$$

We denote by $\overline{K}_h$ and $\overline{\overline{K}}_h$ the transitions defined as K_h by replacing b by the acceptance rates

$$\overline{b}(x, y) = \left[\frac{1}{2} + \frac{1 - e^{(V(y)-V(x))}}{4} \right] 1_{V(y)-V(x) \leq 0}$$

$$+ e^{-(V(y)-V(x))} \left[\frac{1}{2} + \frac{1 - e^{-(V(y)-V(x))}}{4} \right] 1_{V(y)-V(x)>0}$$

and $\overline{\overline{b}}(x, y)$ defined as $\overline{b}(x, y)$ by replacing $(V(y) - V(x))$ by $\partial_x V(x)(y - x)$ in the above expression. Check that

$$\left| b(x, y) - \widehat{b}(x, y) \right| \leq c \, |x - y|^2$$

with $\widehat{b} = \overline{b}$ or $\widehat{b} = \overline{\overline{b}}$, for some finite constant c. Deduce that

$$\sup_{f : osc(f) \leq 1} \left\| \left(K_h - \widehat{K}_h \right)(f) \right\| \leq c \, h \quad\quad and \quad\quad \sup_{f : lip(f) \leq 1} \left\| \left(K_h - \widehat{K}_h \right)(f) \right\| \leq c \, h^{1+1/2}$$

with $\widehat{K}_h = \overline{K}_h$ or $\widehat{K}_h = \overline{\overline{K}}_h$.

Exercise 232 (Heat-bath Markov chain sampler and diffusion limit [133]) ✏ *We consider the heat-bath Markov chain samplers $\left(Y_{t_n}^h, \overline{Y}_{t_n}^h, \overline{\overline{Y}}_{t_n}^h \right)$ with the Markov transitions $\left(K_h, \overline{K}_h, \overline{\overline{K}}_h \right)$ discussed in exercise 231. When $\partial_x V(x) \geq 0$, prove that*

$$\int (y - x) \, P_h(x, dy) \, \overline{\overline{b}}(x, y)$$

$$= \sqrt{h} \, \mathbb{E} \left(W_1 \, 1_{W_1 \geq 0} \left[e^{-\sqrt{h} \, \partial_x V(x) W_1} - 1 \right] \left[\frac{1}{2} - \frac{e^{-\sqrt{h} \, \partial_x V(x) W_1} - 1}{4} \right] \right).$$

Following the arguments provided in the solution of exercise 230, check that these three processes weakly converge, as $h \downarrow 0$, in the sense of finite distributions to the Langevin diffusion

$$dX_t = -4^{-1} \partial_x V(X_t) \, dt + 2^{-1/2} \, dW_t.$$

Exercise 233 (Exit times - 1) *Let W_t be a Brownian motion on $\mathbb{R}$ (starting at the origin). We let T_D be the first time it exits the interval $D = [-a, a]$, for some $a > 0$. Applying (12.23) check that*

$$\mathbb{E}(T_D) \leq a^2.$$

The Brownian motion starting at some $x \in D$ is given by $W_t^x = x + W_t$. We let T_D^x be the first time it exits the interval $D \ni x$. Check that

$$\mathbb{E}(T_D^x) \leq (a^2 - x^2).$$

Exercise 234 (Exit times - 2) *Let W_t be an r-dimensional Brownian motion (starting at the origin), and let D be some open and bounded subset of $\mathbb{R}^r$. We let T_x be the first time $W_t^x := x + W_t$ exits the set $D \ni x$. We denote by $\mathrm{diam}(D) := \sup_{(x,y) \in D} \|x - y\|$ the diameter of D. Check that*

$$\mathbb{P}(T_x < 1) \geq \mathbb{P}(\|W_1\| > \mathrm{diam}(D)) := \epsilon > 0.$$

Prove that

$$\sup_{x \in D} \mathbb{P}(T_x \geq n) \leq (1 - \epsilon)^n$$

by induction w.r.t. the parameter $n \geq 1$. Check that that for any $p \geq 1$ we have

$$\mathbb{E}(T_x^p) = p \int_0^\infty \mathbb{P}(T_x \geq s) \ s^{p-1} \ ds$$

and deduce that $\sup_{x \in D} \mathbb{E}(T_x^p) < \infty$.

Exercise 235 (Time-changed Brownian motion) ✏ *Let W_t be a (standard) Brownian motion and $a : t \mapsto a(t) \in \mathbb{R}$ be a function s.t. $b(t) := \int_0^t a^2(s)ds < \infty$ for any $t \geq 0$. Consider the diffusion $dX_t = a(t)dW_t$ starting at $X_0 = 0$. Check that X_t is a Gaussian process with independent increments. Compute the mean and the variance of $(X_t - X_s)$, for $0 \leq s \leq t$. Deduce that the diffusion X_t starting at $X_0 = 0$ has the same law as the time-changed Brownian motion $W_{\langle X \rangle_t}$.*

Exercise 236 (Non-uniqueness of solution) *Apply the Doeblin-Itō formula to the function $f(W_t) = (a + \frac{1}{3} W_t)^3$ and deduce the non-uniqueness of the solution of the following stochastic differential equation*

$$dX_t = \frac{1}{3} \ X_t^{1/3} \ dt + X_t^{2/3} \ dW_t.$$

Exercise 237 *Let W_t be a Brownian motion. We consider the diffusion process*

$$dX_t = t^{-1} \ X_t \ dt + t \ dW_t$$

with the initial condition $X_1 = x_1$, and with $t \in [1, \infty[$. Solve this equation by applying the Doeblin-Itō formula to the function $f(t, x) = x/t$.

Exercise 238 *Let W_t be a Brownian motion and $a : t \in [t_0, \infty[\mapsto]0, \infty[$ some positive and smooth function, for some parameter $t_0 \geq 0$. We consider the diffusion process*

$$dX_t = -(\log a)'(t) \ X_t \ dt + a(t)^{-1} \ dW_t$$

with some given initial condition $X_{t_0} = x_{t_0}$, and with $t \in [t_0, \infty[$. Solve this equation.

Exercise 239 (Levy's characterization of Brownian motion) *Let $(W_t^i)_{1 \leq i \leq n}$ be a collection of n independent Brownian motions. Let $a = (a_i)_{1 \leq i \leq n}$ be some given parameters in $\mathbb{R}$ s.t. $|a|_2^2 := \sum_{1 \leq i \leq n} a_i^2 \in]0, \infty[$. Check that*

$$W_t^a = |a|_2^{-1} \sum_{1 \leq i \leq n} a_i \ W_t^i \ \overset{in \ law}{=} \ W_t$$

where W_t stands for a standard Brownian motion on $\mathbb{R}$.

Exercise 240 (Rotational invariance of Brownian motion) *We let R_α be the 2d-rotation matrix of an angle $\alpha \in [0, 2\pi]$ given by*

$$R_\alpha = \left(\begin{array}{cc} \cos(\alpha) & -\sin(\alpha) \\ \sin(\alpha) & \cos(\alpha) \end{array} \right).$$

We let $W_t = \left(W_t^1, W_t^2 \right)'$ be a 2d-Brownian motion. Check that the stochastic process $X_t = R_\theta W_t$ has the same law as a 2d-Brownian motion. Show that the generators L^X and L^W of X_t and W_t coincide. Prove that

$$f_\alpha(x) = f(R_\alpha x) \Rightarrow L^W(f_\theta)(R_{-\alpha}x) = L^W(f)(x).$$

Exercise 241 (Reflection invariance of Brownian motion) *We consider a line l_α passing through the origin in $\mathbb{R}^2$ and making an angle α with the $(0, x)$ axis. We let $\overline{R}_\alpha$ be the 2d-reflexion matrix w.r.t. l_α given by*

$$\overline{R}_\alpha = \left(\begin{array}{cc} \cos(2\alpha) & \sin(2\alpha) \\ \sin(2\alpha) & -\cos(2\alpha) \end{array} \right).$$

We let $W_t = \left(W_t^1, W_t^2 \right)'$ be a 2d-Brownian motion. Check that the stochastic process $X_t = \overline{R}_\theta W_t$ has the same law as a 2d-Brownian motion. Show that the generators L^X and L^W of X_t and W_t coincide. Prove that

$$f_\alpha(x) = f(\overline{R}_\alpha x) \Rightarrow L^W(f_\theta)(\overline{R}_\alpha x) = L^W(f)(x).$$

Exercise 242 ((0,1)-valued diffusion) *Let W_t be a (standard) Brownian motion. We let X_t be defined by the stochastic differential equation starting at $x \in]0, 1[$ and given by*

$$dX_t = X_t(1 - X_t)\left(\frac{1}{2} - X_t \right) dt + X_t(1 - X_t)\, dW_t.$$

Check that $Y_t := \log(X_t/(1 - X_t))$ has the same law as a Brownian motion W_t starting at $W_0 = \log(x/(1 - x))$.

Exercise 243 (Brownian rotations) *Apply the Doeblin-Itō formula to the couple of functions $f(W_t) = a \cos(W_t)$ and $g(W_t) = b \sin(W_t)$, for given parameters a, b, and compute the solution of the stochastic differential equation*

$$\left\{ \begin{array}{rcl} dX_t & = & -\frac{1}{2} X_t\, dt - \frac{a}{b} Y_t\, dW_t \\ dY_t & = & -\frac{1}{2} Y_t\, dt + \frac{b}{a} X_t\, dW_t. \end{array} \right.$$

Solve the equation

$$dX_t = -\frac{1}{2} X_t\, dt - \frac{b}{a} \sqrt{(a - X_t)(a + X_t)}\, dW_t.$$

Exercise 244 (Hyperbolic Brownian motions) *Apply the Doeblin-Itō formula to the functions $f(W_t) = a \cosh(\alpha W_t)$ and $g(W_t) = b \sinh(\alpha W_t)$, for given parameters a, b, α, and compute the solution of the stochastic differential equation*

$$\left\{ \begin{array}{rcl} dX_t & = & 2^{-1}\alpha^2 X_t\, dt + \alpha\, b^{-1}a\, Y_t\, dW_t \\ dY_t & = & 2^{-1}\alpha^2 Y_t\, dt + \alpha\, a^{-1}b\, X_t\, dW_t. \end{array} \right.$$

Exercise 245 (Scaling properties) ✎ *Let W_t be a (standard) Brownian motion, and let $\alpha > 0$ be a given parameter. Show that the stochastic processes $W_t^\alpha := \frac{1}{\alpha} W_{\alpha^2 t}$ and $W_t^- = tW(1/t)$ are again (standard) Brownian motions.*

Exercise 246 (Reflection principle) ✐ *Let T be the first time a standard Brownian motion W_t (starting at 0) hits the set $A = [a, \infty[$, for some $a > 0$. Knowing that after time T the Brownian motion is equally likely to move up or down from $W_T = a$, check that*

$$\mathbb{P}(T \leq t) = 2\,\mathbb{P}(W_t > a) = \sqrt{\frac{2}{\pi t}} \int_a^\infty \exp\left(-\frac{x^2}{2t}\right)\,dx.$$

Exercise 247 (Coupling diffusions) *We consider an $\mathbb{R}^r$-valued diffusion process X_t with generator*

$$L_t(f)(x) = \sum_{1 \leq i \leq r} b_t^i(x)\,\partial_{x_i} f(x) + \frac{1}{2} \sum_{1 \leq i,j \leq r} (\sigma_t(x)\sigma_t'(x))^{i,j}\,\partial_{x_i,x_j} f(x)$$

for some regular vector fields b_t and some regular matrices σ_t. We let $\mathcal{Z}_t = (\mathcal{X}_t, \mathcal{Y}_t)$ be the $(\mathbb{R}^r \times \mathbb{R}^r)$-valued diffusion with generator

$$\mathcal{L}_t(F)(x, y) = L(F(\,\cdot\,, y))(x) + L(F(x, \,\cdot\,))(y) + \sum_{1 \leq i,j \leq r} \tau_t(x, y)^{i,j}\,\partial_{x_i, y_j} F(x, y)$$

for some symmetric and regular $(r \times r)$-matrix field τ. When $\tau_t(x, y) = 0$ check that $\mathcal{X}_t$ and $\mathcal{Y}_t$ have the same law as X_t. When $\tau_t(x, y) = 2^{-1}\left[\sigma_t(x)\sigma_t'(y) + \sigma_t(y)\sigma_t'(x)\right]$, check that $\mathcal{L}$ is the generator of the diffusion

$$\begin{cases} d\mathcal{X}_t = b_t(\mathcal{X}_t)\,dt + \sigma_t(\mathcal{X}_t)\,dW_t \\ d\mathcal{Y}_t = b_t(\mathcal{Y}_t)\,dt + \sigma_t(\mathcal{Y}_t)\,dW_t \end{cases}$$

with the same r-dimensional Brownian motion W_t. Deduce that $\mathcal{X}_t$ and $\mathcal{Y}_t$ have the same law as X_t.

Exercise 248 (Coupling diffusions by reflections) *We equip $\mathbb{R}^r$ with the scalar product $\langle x, y \rangle = \sum_{1 \leq i \leq r} x_i y_i$ and the Euclidian norm $\|x\|^2 = \langle x, x \rangle$. For any $x \neq 0$ we set $U(x) = x/\|x\|$ the column vector associated with the projection on the unit sphere. Consider the diffusion*

$$\begin{cases} d\mathcal{X}_t &= b_t(\mathcal{X}_t)\,dt + \sigma_t(\mathcal{X}_t)\,dW_t \\ d\mathcal{Y}_t &= b_t(\mathcal{Y}_t)\,dt + \sigma_t(\mathcal{Y}_t)\,\left[I - 2\,U\,(\mathcal{X}_t - \mathcal{Y}_t)\,U\,(\mathcal{X}_t - \mathcal{Y}_t)'\right]\,dW_t \end{cases}$$

for some regular r-column vector fields b_t, some regular $(r \times r)$-matrix fields σ_t, and some r-dimensional Brownian motion W_t. In the above display, I stands for the $(r \times r)$-matrix. Check that

$$V_t := \int_0^t \left[I - 2\,U\,(\mathcal{X}_t - \mathcal{Y}_t)\,U\,(\mathcal{X}_t - \mathcal{Y}_t)'\right]\,dW_t$$

is an r-dimensional Brownian motion. Deduce that $\mathcal{X}_t$ and $\mathcal{Y}_t$ have the same law as X_t, and compute the generator of $(\mathcal{X}_t, \mathcal{Y}_t)$.

Exercise 249 (Maximum of Brownian motion) *Consider a one-dimensional Brownian motion W_t (starting at the origin) and set $W_t^\star := \sup_{0 \leq s \leq t} W_t$. Let T_y be the first time it reaches the value y so that $\{T_y \leq t\} = \{W_t^\star \geq y\}$. Check that for any $\epsilon > 0$, $y \geq 0$ and $y > x + \epsilon$ we have*

$$\begin{aligned} \mathbb{P}\left(W_t^\star \geq y\,,\ W_t \in [x, x + \epsilon]\right) &= \mathbb{P}\left(W_t - y \in [x - y, x - y + \epsilon]\,,\ T_y \leq t\right) \\ &= \mathbb{P}\left(W_t \in [2y - x - \epsilon, 2y - x]\right). \end{aligned}$$

Deduce that

$$\mathbb{P}\left(W_t^\star \geq y \mid W_t \in [x, x+\epsilon]\right) = \frac{\mathbb{P}\left(W_t \in [2y - x - \epsilon, 2y - x]\right)}{\mathbb{P}\left(W_t \in [x, x+\epsilon]\right)}$$

and

$$\mathbb{P}\left(W_t^\star \geq y \mid W_t = x\right) = \exp\left[-2y(y - x)/t\right].$$

Exercise 250 (Supremum of Brownian motion with drift) *Consider a one-dimensional diffusion X_t discussed in exercise 223 and set $X_t^\star := \sup_{0 \leq s \leq t} X_t$. Let T_y the first time X_t reaches the value y so that $\{T_y \leq t\} = \{X_t^\star \geq y\}$. Using exercise 223 and exercise 249 check that for any $y \geq x$ we have*

$$\mathbb{P}\left(X_t^\star \geq y \mid X_t = x\right) = \exp\left[-2y(y - x)/(\sigma^2 t)\right].$$

Check that for any $y \geq 0$ we have

$$\mathbb{P}\left(X_t^\star \geq y \mid X_0 = 0\right) = \exp\left[2yb/\sigma^2\right] \mathbb{P}\left(W_1 \geq [y + bt]/(\sigma\sqrt{t})\right)$$
$$+ \mathbb{P}\left(W_1 \geq [y - bt]/(\sigma\sqrt{t})\right).$$

Exercise 251 (Infimum of a Brownian motion with drift) *Consider a 1-dimensional diffusion X_t discussed in exercise 250 and set $X_{\star,t} := \inf_{0 \leq s \leq t} X_t$. Check that for any $y \leq 0$ we have*

$$\mathbb{P}\left(X_{\star,t} \leq y \mid X_0 = 0\right) = \exp\left[2yb/\sigma^2\right] \mathbb{P}\left(W_1 \leq [y + bt]/(\sigma\sqrt{t})\right)$$
$$+ \mathbb{P}\left(W_1 \leq [y - bt]/(\sigma\sqrt{t})\right).$$

Exercise 252 *We let $U_t := \frac{2}{\pi} \arctan\left(a(W_t + b)\right) \in [-1, 1]$, for some parameters a, b and a Brownian motion W_t. Check that U_t satisfies the stochastic differential equation*

$$dU_t = -\frac{2a^2}{\pi} \cos^3\left(\frac{\pi}{2} U_t\right) \sin\left(\frac{\pi}{2} U_t\right) dt + \frac{2a}{\pi} \cos^2\left(\frac{\pi}{2} U_t\right) dW_t.$$

Exercise 253 (Correlated Brownian motions) ✏ *We let $\mathcal{F}_t$ be the filtration generated by three independent Brownian motions $W_t^{(i)}$, with $i = 1, 2, 3$. We let U_t be a $[-1, 1]$-valued diffusion defined by*

$$dU_t = b_t(U_t)dt + \sigma_t(U_t)dW_t^{(1)}$$

for some regular functions b_t, σ_t. We set

$$\overline{W}_t := \int_0^t U_s \, dW_s^{(2)} + \int_0^t \sqrt{1 - U_s^2} \, dW_s^{(3)}.$$

For any $s \leq t$, check that

$$\mathbb{E}(\overline{W}_t - \overline{W}_s \mid \mathcal{F}_s) = 0 \quad and \quad \mathbb{E}\left((\overline{W}_t - \overline{W}_s)^2\right) = t - s$$

and

$$\mathbb{E}\left(\overline{W}_t W_t^{(2)}\right) = \mathbb{E}\left(\int_0^t U_s \, ds\right).$$

By Lévy's characterization, these properties ensure that $\overline{W}_t$ is again a Brownian motion.

Exercise 254 (Squared Bessel process) 🖋 *We let $W_t = \left(W_t^i\right)_{1 \le i \le n}$ be n independent Brownian motions for some $n \ge 1$, and we set*

$$X_t = \|W_t\|^2 := \sum_{1 \le i \le n} (W_t^i)^2 \quad \text{and} \quad d\overline{W}_t := 1_{X_t \ne 0} \sum_{1 \le i \le n} \frac{W_t^i}{\sqrt{X_t}} dW_t^i.$$

- *For any $\lambda > 0$, and $s < t$, check that*

$$\mathbb{E}\left(e^{-\lambda X_t} \mid \mathcal{F}_s\right) = (1 + 2\lambda(t-s))^{-n/2} \exp\left(-\frac{\lambda X_s}{1 + 2\lambda(t-s)}\right)$$

with $\mathcal{F}_s = \sigma(W_r^i, 0 \le r \le s\ 1 \le i \le n)$, and deduce that $\mathbb{P}(X_t = 0 \mid \mathcal{F}_s) = 0$.

- *Check that $\overline{W}_t$ is a martingale with continuous trajectories and predictable angle bracket $\langle \overline{W} \rangle_t = t$ (by Lévy's characterization, these properties ensure that $\overline{W}_t$ is again a Brownian motion).*

- *Prove that*

$$dX_t = 2\sqrt{X_t}\, d\overline{W}_t + n\, dt.$$

Exercise 255 (Ornstein-Uhlenbeck process [214]) *We consider the Ornstein-Uhlenbeck process given by*

$$dX_t = a\,(b - X_t)\, dt + \sigma\, dW_t$$

where $a > 0$, b and $\sigma > 0$ are fixed parameters, and W_t is a Brownian process.

- *Check that the density $p_t(x)$ of the random states X_t satisfies the Fokker-Planck equation*

$$\partial_t p_t = a\, \partial_x \left((x - b)\, p_t(x)\right) + \frac{\sigma^2}{2}\, \partial_x^2\,(p_t).$$

- *Applying the Doeblin-Itō lemma to the function $f(t,x) = e^{at}\, x$, check that*

$$X_t = e^{-at}\, X_0 + b\,\left(1 - e^{-at}\right) + \sigma \int_0^t e^{-a(t-s)}\, dW_s.$$

- *Check that*

$$\mathbb{E}(X_t \mid X_0) = e^{-at}\, X_0 + b\,\left(1 - e^{-at}\right) \longrightarrow_{t \to \infty} b$$

$$\mathrm{Var}(X_t \mid X_0) = \frac{\sigma^2}{2a}\,\left(1 - e^{-2at}\right) \longrightarrow_{t \to \infty} \frac{\sigma^2}{2a}$$

and

$$\mathbb{P}(X_t \in dy \mid X_0 = x)$$
$$= \sqrt{\frac{a}{\pi \sigma^2 (1 - e^{-2at})}}\, \exp\left(-\frac{a}{\sigma^2\,(1 - e^{-2at})}\,\left[y - (e^{-at}\, x + b\,(1 - e^{-at}))\right]^2\right)\, dy.$$

In mathematical finance, X_t is also called the Vasicek model [254]. This stochastic process represents the evolution of the interest rates. The parameter b stands for the long term asymptotic level, a is the speed of reversion, and σ denotes the stochastic volatility. Notice that this interest rate model allows negative values.

Exercise 256 (Ornstein-Uhlenbeck process - Estimation of moments) *We consider the Ornstein-Uhlenbeck process discussed in exercise 255. Using exercise 235, check that*

$$X_t \overset{law}{=} e^{-at} X_0 + b \left(1 - e^{-at}\right) + e^{-at} W_{c(t)} \overset{law}{=} e^{-at} X_0 + b \left(1 - e^{-at}\right) + \sqrt{d(t)} W_1,$$

with the time change $c(t) = \frac{\sigma^2}{2a} \left(e^{2at} - 1\right)$ and $d(t) := \frac{\sigma^2}{2a} \left(1 - e^{-2at}\right)$. Using the Gaussian moments formula (5.10) check that

$$\mathbb{E}\left(\left[X_t - \left[e^{-at} X_0 + b \left(1 - e^{-at}\right)\right]\right]^{2n}\right) \leq \left(\frac{\sigma^2}{2a}\right)^n \frac{(2n)!}{n! 2^n} \quad and \quad \sup_{t \geq 0} \mathbb{E}\left(X_t^{2n}\right) < \infty.$$

$$(14.22)$$

Exercise 257 *We let A be an $(r_1 \times r_1)$-matrix with all eigenvalues having a negative real part, for some $r_1 \geq 1$. We let W_t be some r_2-dimensional Brownian motion and B some $(r_1 \times r_2)$-matrix. Consider the r_1-dimensional diffusion given in matrix form by the differential equation*

$$dX_t = AX_t \, dt + B \, dW_t$$

starting at some r_1-dimensional Gaussian random variable X_0 with mean and covariance matrix (m_0, P_0). Prove that X_t is an r_1-dimensional Gaussian random variable X_0 with mean and covariance matrix (m_t, P_t) given by

$$m_t = e^{At} m_0 \to_{t \to \infty} 0$$

and

$$P_t = e^{At} P_0 e^{A't} + \int_0^t e^{sA} BB' e^{sA'} \, ds \to_{t \to \infty} P_\infty := \int_0^\infty e^{sA} BB' e^{sA'} \, ds.$$

In the above display 0 stands for the origin in $\mathbb{R}^{r_1}$. Check that

$$\dot{P_t} = AP_t + P_t A' + BB' \quad and \quad AP_\infty + P_\infty A' + BB' = 0.$$

In the above display 0 stands for the null $(r_1 \times r_1)$-square matrix.

Exercise 258 (Poincaré inequality) ✒ *Consider the Ornstein-Uhlenbeck process X_t introduced in exercise 255 with $a = 1$, $b = 0$, and $\sigma^2 = 2$. We set $\epsilon_t = e^{-2t}$. We also let π be the centered Gaussian distribution with unit variance. Check that*

$$X_t \overset{law}{=} \sqrt{\epsilon_t} X_0 + \sqrt{1 - \epsilon_t} W_1$$

and describe the infinitesimal generator L of X_t. We let $P_t(f)(x) = \mathbb{E}(f(X_t) \mid X_0 = x)$ be the corresponding Markov semigroup. For any smooth functions f_1, f_2, check that

$$\pi \left(f_1 P_t(f_2)\right) = \pi \left(P_t(f_1) f_2\right)$$

and

$$\pi \left(f_1 L(f_2)\right) = \pi \left(L(f_1) f_2\right) = -\mathbb{E}\left(f_1'(W_1) f_2'(W_1)\right).$$

We consider the Dirichlet form $\mathcal{E}(f_1, f_2) = -\pi \left(f_1 L(f_2)\right)$ associated with L and π (see also exercise 187 in the context of finite state space models). Check that

$$\partial_x P_t(f)(x) = e^{-t} P_t \left(f'\right)(x) \quad and \quad \mathcal{E}\left(P_t(f), P_t(f)\right) = e^{-2t} \pi \left[\left(P_t \left(f'\right)\right)^2\right] \leq \|f'\|_{\mathbb{L}_2(\pi)}^2.$$

Check the formulae stated in exercise 187 and exercise 188. Deduce the Poincaré inequality

$$\mathrm{Var}_\pi(f) \leq \mathcal{E}(f, f) = \|f'\|_{\mathbb{L}_2(\pi)}^2 \quad and \; check \; that \quad \mathrm{Var}_\pi(P_t(f)) \leq e^{-2t} \, \mathrm{Var}_\pi(f).$$

Exercise 259 (Poincaré inequality - Exponential distribution) ✎
  We let
$$\pi(dx) = 1_{[0,\infty[}(x) \ e^{-x} \ dx$$
be the exponential distribution on $S := [0, \infty[$.

- *Check that for any function* f *with compact support on* S *we have*
$$\mathrm{Var}_\pi(f) := \pi[(f - \pi(f))^2] \leq \pi[(f - f(0))^2].$$

- *For any function* f *with compact support on* S *s.t.* $f(0) = 0$ *prove that*
$$\pi(f^2) \leq 2\pi(f f') \leq 2\pi(f^2)^{1/2} \ \pi((f')^2)^{1/2}.$$

- *Deduce the Poincaré inequality*
$$\mathrm{Var}_\pi(f) \leq 4 \, \|f'\|_{\mathbb{L}_2(\pi)} \,.$$

Exercise 260 (Diffusion - bi-Laplace invariant measure) *Consider the real-valued diffusion process*
$$dX_t = -\frac{\lambda}{2} \ \mathrm{sign}(X_t) \ dt + dW_t$$
where $\mathrm{sign}(x)$ *stands for the sign of* $x \in \mathbb{R}$, $\lambda > 0$ *and* W_t *a Brownian motion. Describe the generator* L *of* X_t *and check that its invariant measure is the Laplace distribution* $\pi(dx) = \frac{\lambda}{2} \ e^{-\lambda|x|} dx$ *(i.e.* $\pi L(f) = 0$ *for any smooth compactly supported function* l*).*

Exercise 261 (Square of Ornstein-Uhlenbeck processes) *We consider a couple of independent Ornstein-Uhlenbeck processes given by*
$$dU_t = -U_t \ dt + dB_t \quad and \quad dV_t = -V_t \ dt + dB_t'$$
where W_t, W_t' *are independent Brownian processes. Check that the process*
$$X_t = U_t^2 + V_t^2$$
has the same law as the Cox-Ingersoll-Ross diffusion process defined by the equation
$$dX_t = 2(1 - X_t) \ dt + 2\sqrt{X_t} \ dW_t.$$
(Here W_t *is a Brownian process.)*

Exercise 262 (Cox-Ingersoll-Ross diffusion) ✎ *We let* $\pi(dx) = 1_{[0,\infty[}(x) \ e^{-x} \ dx$ *be the exponential distribution on* $S := [0, \infty[$. *Consider the diffusion process*
$$dX_t = 2(1 - X_t) \ dt + 2 \ \sqrt{X_t} \ dW_t$$
where W_t *is a Brownian process.*

- *Describe the generator of* X_t *and check that for any smooth functions* f_1, f_2 *we have*
$$\pi \left(f_1 \ L(f_2) \right) = \pi \left(L(f_1) \ f_2 \right).$$

- *Prove that the Dirichlet form* $\mathcal{E}(f_1, f_2) := -\pi \left(f_1 \ L(f_2) \right)$ *is given by the formula*
$$\mathcal{E}(f_1, f_2) = 2 \ \pi(g f_1' f_2') \quad with \ the \ function \quad g(x) := x.$$

Exercise 263 ✏ *Let W_t be a (standard) Brownian motion. Consider the diffusion process defined by*

$$dY_t = (a_t + b_t \ Y_t) \ dt + \ (\tau_t + \sigma_t \ Y_t) \ dW_t$$

for some functions a_t, b_t, τ_t and σ_t. Consider the diffusion process X_t defined in (3.16) (or alternatively in (3.17)) and starting at $X_0 = 1$. Check that the stochastic process $Z_t := Y_t/X_t$ satisfies the diffusion equation

$$X_t \ dZ_t = (a_t - \tau_t \sigma_t) \ dt + \tau_t \ dW_t.$$

Deduce that

$$Y_t = e_{0,t} \ Y_0 + \int_0^t \ e_{s,t} \ (a_s - \sigma_s \tau_s) \ ds + \int_0^t \ e_{s,t} \ \tau_s \ dW_s$$

with the exponential process

$$e_{s,t} = \exp\left(\int_s^t \left(b_r - \frac{\sigma_r^2}{2} \right) dr + \int_s^t \sigma_r \ dW_r \right).$$

In mathematical finance, the process Y_t associated with the parameters $(a_t, b_t) = (a\mu, -a)$ and $(\tau_t, \sigma_t) = (0, \sigma)$ is also called the mean reverting geometric Brownian motion.

Exercise 264 (Brownian bridge-1) ✏ *Let W_t be a (standard) Brownian motion. For any $0 < s < t$, prove that $W_s - \frac{s}{t} \ W_t$ is independent of W_t. Compute the conditional expectation $\mathbb{E}(W_s \mid W_t)$ and the conditional variance $\mathrm{Var}(W_s \mid W_t)$. Find the distribution of W_s given $W_t = 0$. Check that the process $V_t = W_t - tW_1$, with $t \in [0,1]$, is independent of the terminal random variable W_1 of the Brownian motion. The process V_t is called a Brownian bridge with $V_1 = 0 = V_0$.*

Exercise 265 (Brownian bridge-2) *We fix a couple of parameters $a, b \in \mathbb{R}$, and a time horizon t. We consider the stochastic process given by*

$$\forall s \in [0,t] \qquad dX_s = \frac{b - X_s}{t - s} \ ds + dW_s.$$

Applying the Doeblin-Itō formula to the function $f(s, X_s) = \frac{X_s - b}{t - s}$, check that

$$\forall 0 \le r \le s \le t \qquad X_s = \frac{t - s}{t - r} \ X_r + \frac{s - r}{t - r} \ b \ + \int_r^s \frac{t - s}{t - u} \ dW_u.$$

Check that $X_t = b$ and compute the distribution of X_s given X_r.

Exercise 266 (Milstein scheme) ✏ *We consider the one-dimensional diffusion*

$$dX_t = b(X_t)dt + \sigma(X_t)dW_t$$

with some smooth functions b and σ with bounded derivatives at any order. Prove that

$$X_{t+h} - X_t = b(X_t) \ h + \sigma(X_t) \ (W_{t+h} - W_t) + \frac{1}{2} \ \sigma'(X_t)\sigma(X_t) \ \left[(W_{t+h} - W_t)^2 - h \right] + \mathcal{R}_{t,t+h}$$

with a second order remainder term $\mathcal{R}_{t,t+h}$ such that $\mathbb{E}\left(|\mathcal{R}_{t,t+h}|^2 \right)^{1/2} \le c \ h^{3/2}$ for some finite constant $c < \infty$.

15

Jump diffusion processes

Jump diffusion processes introduced in this chapter represent an extension of the piecewise deterministic processes introduced in chapter 13. Informally, the deterministic flows between the jumps of the piecewise deterministic process are replaced by the diffusion processes introduced in chapter 14. We discuss the Doeblin-Itō differential calculus for these processes and use it to derive the Fokker-Planck equation. Further on, we introduce jump diffusion processes with killing in terms of Feynman-Kac semigroups. At the end of the chapter, we show in some details the usefulness of these models for solving some classes of partial differential equations.

> *Mathematics is the art of giving the same name to different things.*
> Jules Henri Poincare (1854-1912).

15.1 Piecewise diffusion processes

As we mentioned in the introduction of chapter 14 dedicated to pure diffusion processes, it is beyond the scope of this chapter to describe in full details the construction of jump-diffusion processes and stochastic integrals w.r.t. to the induced random measure on the time axis. We have chosen to present these continuous time stochastic processes from the practitioner's point of view, using simple arguments based on their discrete approximations. For a more thorough and rigorous discussion on these probabilistic

models, we refer the reader to the seminal book by Stewart Ethier and Thomas Kurtz [122].

We also refer the reader to section 12.5.2 for the definitions of continuous time filtrations, predictable processes, martingales and their angle brackets.

Jump diffusion processes (*abbreviated JDP*) are defined as the piecewise deterministic jump models discussed in section 13.2, replacing the deterministic flow between the jumps by the diffusion process. More formally, we have the following definitions.

Definition 15.1.1 *For any $s \in \mathbb{R}_+$ and $x \in \mathbb{R}^d$ we denote by*

$$t \in [s, \infty[\ \mapsto \ \varphi_{s,t}(x) \in \mathbb{R}^d$$

the solution of the SDE (14.18) starting at x at time $t = s$. The mappings $\varphi_{s,t}(x)$ are called the stochastic flow of the diffusion process (14.18).

A d-dimensional jump diffusion process with jump intensity function $\lambda_t(x)$ and jump *amplitude transition $K_t(x, dy)$ is defined as a pure jump process introduced in chapter 12 by replacing the constant evolution model between the jumps (12.1) by the stochastic flow of the diffusion process*

$$\forall T_n \leq s < T_{n+1} \qquad X_s := \varphi_{T_n,s}(X_{T_n}). \tag{15.1}$$

Important remark : Jump diffusion processes represent the most general class of Markovian stochastic processes encountered in practice. For null diffusion functions $\sigma_t = 0$, JDP models reduce to PDMP processes. In addition, when the drift function $b_t = 0$ is also null, the JDP processes reduces to pure jump models with intensity $\lambda_t(x)$ and jump amplitude transition $K_t(x, dy)$.

15.2 Evolution semigroups

By construction, for any $s \geq t$ we have

$$\mathbb{P}\left(T_{n+1} \in dt, X_{T_{n+1}} \in dy \mid T_n = s, \varphi_{s,r}(x), r \geq s\right) \tag{15.2}$$

$$= \underbrace{\lambda_t(\varphi_{s,t}(x)) \exp\left(-\int_s^t \lambda_r(\varphi_{s,r}(x))\, dr\right) dt}_{\mathbb{P}(T_{n+1} \in dt \mid T_n = s, \varphi_{s,r}(x), r \geq s)}$$

$$\times \quad \underbrace{K_t(\varphi_{s,t}(x), dy)}_{\mathbb{P}(X_{T_{n+1}} \in dy \mid T_{n+1} = t, \varphi_{s,r}(x), r \geq s)}.$$

In the same vein, if $T^{(s)}$ is the first jump time after time s, we have

$$\mathbb{P}\left(T^{(s)} \in dt, X_{T^{(s)}} \in dy \mid X_s, \varphi_{s,r}(X_s), r \geq s\right)$$

$$= \underbrace{\lambda_t(\varphi_{s,t}(X_s)) \exp\left(-\int_s^t \lambda_r(\varphi_{s,r}(X_s))\, dr\right) 1_{s \leq t}\, dt}_{\mathbb{P}(T^{(s)} \in dt \mid X_s, \varphi_{s,r}(X_s), r \geq s)}$$

$$\times \quad \underbrace{K_t(\varphi_{s,t}(X_s), dy)}_{\mathbb{P}(X_{T^{(s)}} \in dy \mid T^{(s)} = t, \varphi_{s,r}(X_s), r \geq s)}. \tag{15.3}$$

This yields the formula

$$\mathbb{P}\left(T^{(t)} \in dt, X_{t+dt} \in dy \mid X_t\right) = \lambda_t(X_t)\, dt\, K_t(X_t, dy).$$

Last but not least, we also observe that for any $t \geq s$ we have

$$\mathbb{P}\left(T^{(s)} > t \mid X_s, \varphi_{s,r}(X_s), r \geq s\right) = \exp\left(-\int_s^t \lambda_r(\varphi_{s,r}(X_s))\, dr\right) 1_{s \leq t}. \tag{15.4}$$

We check this claim by using the fact that

$$\mathbb{P}\left(T^{(s)} > t \mid X_s, \ \varphi_{s,r}(X_s), r \geq s\right) = \int_t^\infty \mathbb{P}\left(T^{(s)} \in du \mid X_s, \ \varphi_{s,r}(X_s), r \geq s\right) du$$

$$= \int_t^\infty \lambda_u(\varphi_{s,u}(X_s)) \exp\left(-\int_s^u \lambda_r(\varphi_{s,r}(X_s)) \, dr\right) du$$

$$= -\int_t^\infty \frac{\partial}{\partial u} \exp\left(-\int_s^u \lambda_r(\varphi_{s,r}(X_s)) \, dr\right) du.$$

Definition 15.2.1 *For any intensity function λ, we let $P_{s,t}^\lambda$ and $P_{s,t}$ with $s \leq t$ denote the semigroup (sg) defined by*

$$P_{s,t}^\lambda(f)(x) \quad := \quad \mathbb{E}\left[f(\varphi_{s,t}(x)) \exp\left(-\int_s^t \lambda_u(\varphi_{s,u}(x))du\right)\right]$$

$$P_{s,t}(f)(x) \quad = \quad \mathbb{E}\left(f(X_t) \mid X_s = x\right). \tag{15.5}$$

For the null intensity function, the sg $P_{s,t}^0$ is the evolution sg of the d-dimensional diffusion $X_t = \left(X^i\right)_{1 \leq i \leq d}$ defined in (14.18)

$$P_{s,t}^0(f)(x) = \mathbb{E}\left[f(\varphi_{s,t}(x))\right].$$

If $T^{(s)}$ stands for the first time of a jump of the process with intensity λ_u after time s, we have the formula

$$P_{s,t}^\lambda(f)(x) = \mathbb{E}\left[f(\varphi_{s,t}(x)) \exp\left(-\int_s^t \lambda_u(\varphi_{s,u}(x))du\right)\right] = \mathbb{E}\left[f(\varphi_{s,t}(x)) \, 1_{T^s > t}\right]. \tag{15.6}$$

We check this claim by combining a simple conditioning argument with the fact that

$$(15.4) \implies \mathbb{E}\left[1_{T^s > t} \mid X_s, \ \varphi_{s,r}(X_s), r \geq s\right] = \exp\left(-\int_s^t \lambda_u(\varphi_{s,u}(x))du\right).$$

Extending the arguments to diffusion processes, all the results presented in section 13.2.2, section 13.2.3, and section 13.2.6 remain valid by replacing the generators L_t^λ and L_t defined in (13.21) and (13.23) by the generators

$$L_t^\lambda(f) = L_t^0(f) - \lambda_t \, f \qquad L_t = L_t^0 + \lambda_t \, [K_t - Id] := L_t^c + L_t^d \tag{15.7}$$

with the infinitesimal generator of the continuous d-dimensional diffusion

$$L_t^c := L_t^0 = \sum_{i=1}^d b_t^i \, \partial_i + \frac{1}{2} \sum_{i,j=1}^d \left(\sigma_t(\sigma_t)^T\right)_{i,j} \, \partial_{i,j}, \tag{15.8}$$

and the generator associated with the discontinuous jump process

$$L_t^d := \lambda_t \, [K_t - Id]. \tag{15.9}$$

Given a couple of JDP (X_t^1, X_t^2), we also have the easily checked integration by parts formula

$$
\begin{aligned}
d\left(X_t^1 X_t^2\right) &= X_{t+dt}^1 X_{t+dt}^2 - X_t^1 X_t^2 \\
&= \left(X_t^1 + dX_t^1\right)\left(X_t^2 + dX_t^2\right) - X_t^1 X_t^2 \\
&= X_t^1 \, dX_t^2 + X_t^2 \, dX_t^1 + dX_t^1 dX_t^2.
\end{aligned}
$$

On the other hand, for $i = 1, 2$ we have

$$
dX_t^i = \underbrace{b_t^i(X_t^i)dt + \sigma_t^i(X_t^i)\, dW_t^i}_{=dX_t^{c,i}} + \Delta X_t^i
$$

for some drift and diffusion functions with appropriate dimensions, and some d^i-dimensional Brownian motion W_t^i. The term $\Delta X_t^i = X_{t+dt}^i - X_t^i$ represents the jump of the process at time t.

Using the rules

$$
dt \times \Delta X_t^i = 0 = \Delta X_t^i \times dW_t^j
$$

implies that

$$
dX_t^1 dX_t^2 = dX_t^{c,1} dX_t^{c,2} + \Delta X_t^1 \Delta X_t^2.
$$

A discrete time approximation model on some time mesh sequence t_n, with time step h, is defined using the same constructions as the ones we used in section 13.2.5, by considering the semigroup

$$
P_{t_n,t_{n+1}}^{0,h}(f)(x) = \mathbb{E}\left(f\left(\varphi_{t_n,t_{n+1}}^h(x)\right)\right)
$$

with the discrete time approximation of the diffusion (14.18) on the time step $[t_n, t_{n+1}[$

$$
\varphi_{t_n,t_{n+1}}^h(x) = x + b_{t_n}(x)\, h + \sigma_t(x)\, \sqrt{h}\, W_1.
$$

The extension of the weak approximation theorem 14.3.2 to JDP processes is proved using the same arguments as the ones we used in the proof of theorem 13.2.8.

15.3 Doeblin-Itō formula

We denote by $\mathcal{F}_t$ the filtration generated by the d-dimensional Brownian motion W_t and the random times T_n and the jump-amplitudes $\Delta X_{T_n} := X_{T_n} - X_{T_n-}$ of the process X_t; that is, we have that

$$
\mathcal{F}_t = \sigma\left(X_s, s \le t\right) = \sigma\left(X_0,\, W_s - W_r,\, T_n,\, \Delta X_{T_n},\, r \le s \le t,\, T_n \le t\right).
$$

The definition of a martingale w.r.t. the increasing sequence of σ-fields $\mathcal{F}_t$ is provided in section 12.5.2. To simplify the presentation, we also set

$$
\partial_i := \partial_{x^i} = \frac{\partial}{\partial x^i} \quad \text{and} \quad \partial_{i,j} := \partial_{x^i, x^j} = \frac{\partial^2}{\partial x^i \partial x^j}.
$$

Expanding a Taylor series and applying the chain rules presented in section 14.4, for twice continuously differentiable functions f on $\mathbb{R}_+ \times \mathbb{R}^d$ we have

$$
\begin{aligned}
df(t, X_t) &= f(t + dt, X_t + dX_t) - f(t, X_t) \\
&= \partial_t f(t, X_t) dt + \sum_{i=1}^{d} \partial_i f(X_t) \, dX_t^i + \frac{1}{2} \sum_{i,j=1}^{d} \partial_{i,j} f(X_t) \, dX_t^i dX_t^j \\
&\quad + \Delta f(t, X_t) - \left[\sum_{i=1}^{d} \partial_i f(X_t) \, \Delta X_t^i + \frac{1}{2} \sum_{i,j=1}^{d} \partial_{i,j} f(X_t) \, \Delta X_t^i \Delta X_t^j \right].
\end{aligned}
$$

$$(15.10)$$

The r.h.s. term follows from the decomposition

$$
dX_t^i = dX_t^{c,i} + \Delta X_t^i
$$

with the jump term ΔX_t^i, and the increment

$$
dX_t^{c,i} := b_t^i(X_t) \, dt + \sum_{1 \le j \le d} \sigma_{j,t}^i(X_t) \, dW_t^j
$$

of the continuous process

$$
X_t^{c,i} := X_0^i + \int_0^t b_t^i(X_s) \, ds + \sum_{1 \le j \le d} \int_0^t \sigma_{j,s}^i(X_t) \, dW_s^j.
$$

In this context, the quadratic term is interpreted as the increment of the covariation process

$$
\begin{aligned}
dX_t^i dX_t^j &= \left(\sigma_t (\sigma_t)^T \right)_{i,j} (X_t) \, dt + \Delta X_t^i \Delta X_t^j \\
&:= d\left[X^i, X^j \right]_t = d\langle X^{c,i}, X^{c,j} \rangle_t + \Delta X_t^i \Delta X_t^j
\end{aligned}
$$

with

$$
d\langle X^{c,i}, X^{c,j} \rangle_t = dX_t^{c,i} dX_t^{c,j} = \left(\sigma_t (\sigma_t)^T \right)_{i,j} (X_t) \, dt.
$$

Rewritten in terms of the continuous and jump parts of the process, the Doeblin-Itō formula (15.10) takes the form

$$
\begin{aligned}
df(t, X_t) &= \partial_t f(t, X_t) dt + \sum_{i=1}^{d} \partial_i f(X_t) \, dX_t^{c,i} \\
&\quad + \frac{1}{2} \sum_{i,j=1}^{d} \partial_{i,j} f(X_t) \, dX_t^{c,i} dX_t^{c,j} + \Delta f(t, X_t). \quad (15.11)
\end{aligned}
$$

To take the final step, using (15.3) we check that

$$
\Delta f(t, X_t) = \mathbb{E}\left(\Delta f(t, X_t) \mid \mathcal{F}_t \right) + \underbrace{\Delta f(t, X_t) - \mathbb{E}\left(\Delta f(t, X_t) \mid \mathcal{F}_t \right)}_{:= d\mathcal{M}_t^d(f)}
$$

with the predictable jump amplitude

$$\mathbb{E}\left(\Delta f(t, X_t) \mid \mathcal{F}_t\right) = \lambda_t(X_t)\, dt \int \left(f(t, y) - f(t, X_t)\right)\, K_t(X_t, dy),$$

and the martingale increment $d\mathcal{M}_t^d(f)$.

In summary, we have proved the following result.

We have the Doeblin-Itō formula

$$df(t, X_t) = [\partial_t + L_t]\,(f)(t, X_t)\, dt + d\mathcal{M}_t(f) \tag{15.12}$$

with the infinitesimal generator

$$L_t = \underbrace{\sum_{i=1}^{d} b_t^i\, \partial_i + \frac{1}{2} \sum_{i,j=1}^{d} \left(\sigma_t(\sigma_t)^T\right)_{i,j}\, \partial_{i,j}}_{:=L_t^c} + \underbrace{\lambda_t\, [K_t - Id]}_{:=L_t^d} \tag{15.13}$$

acting on the set $D(L)$ of twice differentiable functions with bounded derivates. The martingale increment $d\mathcal{M}_t(f) = d\mathcal{M}_t^c(f) + d\mathcal{M}_t^d(f)$, with the discontinuous and the continuous parts is given by

$$\begin{aligned}
d\mathcal{M}_t^d(f) &= \Delta f(t, X_t) - \mathbb{E}\left(\Delta f(t, X_t) \mid \mathcal{F}_t\right) \\
d\mathcal{M}_t^c(f) &= \sum_{i,j=1}^{d} \partial_i f(X_t)\, \sigma_{j,t}^i(X_t)\, dW_t^j.
\end{aligned}$$

We recall that the angle bracket $\langle \mathcal{M} \rangle_t$ (a.k.a. the predictable quadratic variation) of a given martingale $\mathcal{M}_t$ w.r.t. some filtration $\mathcal{F}_t$ is the predictable stochastic process $\langle \mathcal{M} \rangle_t$ s.t. $\mathcal{M}_t^2 - \langle \mathcal{M} \rangle_t$ is a martingale. By construction, arguing as in the discrete time case, the angle bracket of $\mathcal{M}_t(f)$ (w.r.t. the filtration $\mathcal{F}_t$) is the sum

$$\langle \mathcal{M}(f) \rangle_t = \langle \mathcal{M}^c(f) \rangle_t + \langle \mathcal{M}^d(f) \rangle_t$$

of the angle brackets of the martingales $\mathcal{M}_t^c(f)$ and $\mathcal{M}_t^d(f)$ given by

$$\langle \mathcal{M}^c(f) \rangle_t = \sum_{i,j=1}^{d} \int_0^t \mathbb{E}\left((d\mathcal{M}_s^c(f))^2 \mid \mathcal{F}_s\right) = \sum_{j=1}^{d} \int_0^t \left(\sum_{i=1}^{d} \partial_i f(X_s)\, \sigma_{j,s}^i(X_s)\right)^2 ds \tag{15.14}$$

and

$$\begin{aligned}
\langle \mathcal{M}^d(f) \rangle_t &= \int_0^t \mathbb{E}\left((d\mathcal{M}_s^d(f))^2 \mid \mathcal{F}_s\right) = \int_0^t \mathbb{E}\left((\Delta \mathcal{M}_s^d(f))^2 \mid \mathcal{F}_s\right) \\
&= \int_0^t \mathbb{E}\left((\Delta f(s, X_s))^2 \mid \mathcal{F}_s\right) \\
&= \int_0^t \lambda_s(X_s)\, \left[\int \left(f(s, y) - f(s, X_s)\right)^2 K_s(X_s, dy)\right] ds.
\end{aligned} \tag{15.15}$$

These angle brackets can be rewritten in a more synthetic form

$$\langle \mathcal{M}(f) \rangle_t = \int_0^t \Gamma_{L_s}(f(s,.), f(s,.))(X_s)\, ds$$

in terms of the carré du champ operators

$$\begin{aligned}\Gamma_{L_t}(f,f)(x) &:= L_t((f - f(x))^2)(x) \\ &= L_t(f^2)(x) - 2f(x)L_t(f)(x) = \Gamma_{L_t^c}(f,f)(x) + \Gamma_{L_t^d}(f,f)(x).\end{aligned}$$

We can also show that

$$\mathcal{M}_t(f)\mathcal{M}_t(g) - \langle \mathcal{M}(f), \mathcal{M}(g) \rangle_t \quad \text{is a martingale (w.r.t. the filtration } \mathcal{F}_t)$$

with the angle bracket

$$d\langle \mathcal{M}(f), \mathcal{M}(g) \rangle_t = \mathbb{E}\left(d\mathcal{M}_t(f)d\mathcal{M}_t(g) \mid X_t\right) = \Gamma_L(f,g)(X_t)\, dt$$

defined in terms of the carré du champ operators

$$\Gamma_{L_t}(f,g)(x) = L_t((f - f(x))(g - g(x)))(x) \simeq_{h\downarrow 0} \frac{[P_{t,t+h} - Id][(f - f(x))(g - g(x))]}{h}(x).$$

We notice that the carré du champ is non-negative since we have

$$\Gamma_{L_t}(f,f)(x) \simeq_{h\downarrow 0} \frac{[P_{t,t+h} - Id][(f - f(x))^2]}{h}(x) = \frac{P_{t,t+h}[(f - f(x))^2]}{h}(x) \geq 0.$$

Using the same arguments as in the pure diffusion or as in the pure jump cases (see for instance the decompositions (13.7) and (13.8)), the semigroup of the multidimensional jump-diffusion $P_{s,t}$ also satisfies the forward and backward evolution equations stated in theorem 14.3.3; that is,

$$\partial_t P_{s,t}(f) = P_{s,t}(L_t(f)) \quad \text{and} \quad \partial_s P_{s,t}(f) = -L_s(P_{s,t}(f)). \tag{15.16}$$

In addition, the backward evolution stochastic process $M_t := P_{t,T}(f_T)(X_t)$, with $t \in [0, T]$, is a martingale ending at some given function $f_T(X_T)$.

It is also important to notice that the finite sum of a collection of jump generators $L_t^{d,i} = \lambda_t^{(i)} [K_t^{(i)} - Id]$, with $i \in I$, for some finite set I can be rewritten in terms of a single jump generator

$$\sum_{i \in I} L_t^{d,i} = \lambda_t [K_t - Id] := L_t^d \qquad (15.17)$$

with jump intensity and Markov jump amplitude transitions

$$\lambda_t = \sum_{i \in I} \lambda_t^{(i)} \quad \text{and} \quad K_t(x, dy) = \sum_{i \in I} \frac{\lambda_t^{(i)}(x)}{\sum_{j \in I} \lambda_t^{(j)}(x)} \, K_t^{(i)}(x, dy).$$

We check this claim using the elementary decomposition

$$\sum_{i \in I} L_t^{d,i}(f)(x) = \sum_{i \in I} \lambda_t^{(i)}(x) \int (f(y) - f(x)) \, K_t^{(i)}(x, dy)$$

$$:= \lambda_t(x) \int (f(y) - f(x)) \sum_{i \in I} \frac{\lambda_t^{(i)}(x)}{\sum_{i \in I} \lambda_t^{(i)}(x)} \, K_t^{(i)}(x, dy).$$

In this situation the jump diffusion process discussed in (15.1) is defined in terms of jump times T_n with intensity λ_t, that is,

$$T_{n+1} = \inf \left\{ t \geq T_n \; : \; \int_{T_n}^t \lambda_u(X_u) \, du \geq \mathcal{E}_{n+1} \right\}$$

with a sequence $\mathcal{E}_n$ of independent exponential random variables with unit parameters. At any jump time, say T_n, we select an index $\epsilon_n \in I$ with the discrete probability measure on I defined by

$$\sum_{i \in I} \frac{\lambda_{T_n-}^{(i)}(X_{T_n-})}{\sum_{j \in I} \lambda_{T_n-}^{(j)}(X_{T_n-})} \, 1_i$$

and we perform a jump amplitude $X_{T_n-} \rightsquigarrow X_{T_n}$ by sampling a random variable X_{T_n} with distribution $K_{T_n-}^{(\epsilon_n)}(X_{T_n-}, dy)$.

Following the discussion provided in section 12.4, the jump process discussed above can be interpreted as the superposition of $\text{Card}(I)$ jump processes with intensities $\left(\lambda_t^{(i)}\right)_{i \in I}$. To be more precise, for each $i \in I$ and any $n \geq 0$ we set

$$\tau_{n+1}^i := \inf \left\{ k > \tau_n^i \; : \; \epsilon_k = i \right\}$$

with $\tau_0^i = 1$. By the superposition principle of Poisson processes, the random times $T_{\tau_n^i}$ occur with the intensity $\lambda_t^{(i)}$, that is,

$$T_{\tau_{n+1}^i} \stackrel{law}{=} \inf \left\{ t \geq T_{\tau_n^i} \; : \; \int_{T_{\tau_n^i}}^t \lambda_u^{(i)}(X_u) \, du \geq \mathcal{E}_{n+1}^{(i)} \right\} \qquad (15.18)$$

with a sequence $\mathcal{E}_n^{(i)}$ of independent exponential random variables with unit parameter.

For smooth functions f we also have

$$\partial_t P_{s,t}(f) = P_{s,t}(L_t(f)) \Rightarrow \frac{d^2}{dt^2} P_{s,t}(f) = \partial_t P_{s,t}(L_t(f)) = P_{s,t}(L_t^2(f))$$

with $L_t^2(f) = L_t(L_t(f))$. Iterating this formula, we prove that

$$\frac{d^n}{dt^n} P_{s,t}(f) = P_{s,t}(L_t^n(f))$$

with the n-th composition operator

$$L_t^n = L_t^{n-1} \circ L_t = L_t \circ L_t^{n-1}.$$

Consequently, we can use Taylor's expansion to check that

$$P_{s,t}(f) \quad = \quad \sum_{n \geq 0} \frac{(t-s)^n}{n!} \frac{d^n}{dt^n} P_{s,t}(f)_{|t=s} = \sum_{n \geq 0} \frac{(t-s)^n}{n!} L_s^n(f) := e^{(t-s)L_s} f.$$

For time homogeneous models $L_t = L$, this formula is sometimes rewritten in terms of the exponential operator

$$P_{0,t}(f) := P_t(f) := \sum_{n \geq 0} \frac{t^n}{n!} L^n(f) := e^{tL} f. \tag{15.19}$$

We let $\eta_t = \mathrm{Law}(X_t)$. For any smooth function f on $\mathbb{R}^d$, we also notice that

$$(15.12) \quad \Rightarrow \quad \eta_t(f) \quad = \quad \mathbb{E}(f(X_t)) = \mathbb{E}(X_0)) + \int_0^t \mathbb{E}(L_s(f)(X_s)) \, ds$$

$$= \quad \eta_0(f) + \int_0^t \eta_s(L_s(f)) \, ds$$

$$\Leftrightarrow \quad \partial_t \eta_t(f) = \eta_t(L_t(f)).$$

In a more synthetic way, the above equation takes the form

$$\partial_t \eta_t = \eta_t L_t. \tag{15.20}$$

15.4 Fokker-Planck equation

We further assume that $K_t(x, dy) = m_t(x, y) \, dy$ and the law of the random states

$$\mathbb{P}(X_t \in dx) = p_t(y) \, dy$$

have smooth densities $m_t(x, y)$, and $p_t(y)$ w.r.t. the Lebesgue measure dy on $\mathbb{R}^d$. In this case, p_t satisfies the Fokker-Planck equation

$$\partial_t p_t(x) = L_t^\star(p_t)(x) = L_t^{c,\star}(p_t)(x) + L_t^{d,\star}(p_t)(x) \tag{15.21}$$

with the dual operators of the generators discussed in (15.13)

$$L_t^{c,\star}(p_t) = -\sum_{i=1}^{d} \partial_i \left(b_t^i \, p_t \right) + \frac{1}{2} \sum_{i,j=1}^{d} \partial_{i,j} \left(\left(\sigma_t(\sigma_t)^T \right)_{i,j} \, p_t \right)$$

and

$$L_t^{d,\star}(p_t) = \left(\int p_t(y) \, \lambda_t(y) \, m_t(y,x) \, dy \right) - p_t(x) \, \lambda_t(x).$$

The Fokker-Planck type integro-differential equation (15.21) is sometimes rewritten in the form

$$\partial_t p_t(x) + \text{div} \left(b_t \, p_t \right) - \frac{1}{2} \, \nabla^2 : \left(\sigma_t(\sigma_t)^T p_t \right) - \Theta_t(p_t) = 0,$$

with the operators

$$\text{div} \left(b_t \, p_t \right) \quad := \quad \sum_{i=1}^{d} \partial_i \left(b_{t,i} \, p_t \right)$$

$$\nabla^2 : \left(\sigma_t(\sigma_t)^T p_t \right) \quad = \quad \sum_{i,j=1}^{d} \partial_{i,j} \left(\left(\sigma_t(\sigma_t)^T \right)_{i,j} \, p_t \right)$$

$$\Theta_t(p_t) \quad := \quad \int p_t(y) \, \lambda_t(y) \, \left[m_t(y,x) \, dy - \delta_x(dy) \right].$$

15.5 An abstract class of stochastic processes

15.5.1 Generators and carré du champ operators

We let X_t be some stochastic process on some state space S. The state space S may be discrete, continuous, or can be the product of a continuous state space with a discrete one, and so on.

The stochastic process X_t can be homogeneous or not, it may contain a diffusion part and jumps, or it can be a pure jump process or a sum of jump processes.

In all cases, the generator and the corresponding carré du champ operators are defined in terms of the semigroup $P_{s,t}(f)(x) = \mathbb{E}(f(X_t) \mid X_s = x)$, $s \leq t$ by the formulae

$$L_t(f)(x) \simeq_{h\downarrow 0} \frac{[P_{t,t+h} - Id]f}{h}(x) = \frac{\mathbb{E} \left((f(X_{t+h}) - f(X_t)) \mid X_t = x \right)}{h}$$

and

$$\Gamma_{L_t}(f,g)(x) \quad = \quad L_t(fg)(x) - f L_t(g)(x) - g L_t(f)(x).$$

The carré du champ operator can be represented in many ways:

$$
\begin{aligned}
\Gamma_{L_t}(f,g)(x) \quad &= \quad L_t(fg)(x) - f L_t(g)(x) - g L_t(f)(x) \\[2mm]
&= \quad L_t((f - f(x))(g - g(x)))(x) \\[2mm]
&\simeq_{h\downarrow 0} \quad \frac{[P_{t,t+h} - Id][(f - f(x))(g - g(x))]}{h}(x) \\[2mm]
&= \quad \frac{P_{t,t+h}[(f - f(x))(g - g(x))]}{h}(x) \\[2mm]
&= \quad \frac{\mathbb{E}\left((f(X_{t+h}) - f(X_t))(g(X_{t+h}) - g(X_t)) \mid X_t = x\right)}{h}.
\end{aligned}
$$

Of course, the above formulae are valid for sufficiently regular functions f. For instance, for diffusion processes, L_t is a second order differential operator only defined on twice differentiable functions. For piecewise deterministic models, L_t is a first order differential operator defined on differentiable functions, and so on. The set of these regular functions $D(L)$ depends on the nature of the stochastic process, and it is called the domain of the generator.

For sufficiently regular functions $f(t,x)$ we have the Doeblin-Itō formula

$$df(t, X_t) = [\partial_t + L_t](f)(t, X_t)\, dt + d\mathcal{M}_t(f) \qquad (15.22)$$

for a collection of martingales $\mathcal{M}_t(f)$ (w.r.t. $\mathcal{F}_t = \sigma(X_s,\ s \leq t)$) with angle bracket defined for any $f(t,x)$ and $g(t,x)$ by the formulae

$$\langle \mathcal{M}(f), \mathcal{M}(g) \rangle_t = \int_0^t \Gamma_{L_s}(f(s,.), g(s,.))(X_s)\, ds.$$

Combining the definition of the carré du champ operator with the application of the Doeblin-Itō formula (15.22) to the product of functions $h(t,x) = f(t,x)g(t,x)$, we have

$$
\begin{aligned}
dh(t, X_t) &= \quad [\partial_t + L_t](fg)(t, X_t)\, dt + d\mathcal{M}_t(fg) \\[2mm]
&= \quad [f(t, X_t)\ [\partial_t + L_t](g)(t, X_t) + g(t, X_t)\ [\partial_t + L_t](f)(t, X_t)]\ dt \\[1mm]
&\qquad + \Gamma_{L_t}(f(t,.), g(t,.))(X_t)\, dt + d\mathcal{M}_t(fg). \qquad (15.23)
\end{aligned}
$$

In other words, for regular functions (f, g) we have the integration by part formula

$$L_t(fg) = f L_t(g) + g L_t(f) + \Gamma_{L_t}(f,g). \qquad (15.24)$$

We also have the integration by part formulae

$dh(t, X_t)$

$= f(t, X_t) \, dg(t, X_t) + g(t, X_t) \, df(t, X_t) + df(t, X_t) \, dg(t, X_t)$

$= \{ f(t, X_t) \, (\partial_t + L_t) g(t, X_t) + g(t, X_t) \, (\partial_t + L_t) f(t, X_t) + \Gamma_{L_t}(f(t, .), g(t, .))(X_t) \} \, dt$

$+ f(t, X_t) \, d\mathcal{M}_t(g) + g(t, X_t) \, d\mathcal{M}_t(f) + \{ d\mathcal{M}_t(f) d\mathcal{M}_t(g) - \mathbb{E} \left(df(t, X_t) \, dg(t, X_t) \mid X_t \right) \}.$

In the last assertion, we used the fact that

$$\mathbb{E} \left(df(t, X_t) \, dg(t, X_t) \mid X_t \right) \left(= \mathbb{E} \left(d\mathcal{M}_t(f) \, d\mathcal{M}_t(g) \mid X_t \right) \right)$$

$$= \mathbb{E} \left([f(t + dt, X_{t+dt}) - f(t, X_t)] \, [g(t + dt, X_{t+dt}) - g(t, X_t)] \mid X_t \right)$$

$$= \mathbb{E} \left([f(t, X_{t+dt}) - f(t, X_t)] \, [g(t, X_{t+dt}) - g(t, X_t)] \mid X_t \right) + \mathrm{O}((dt)^2)$$

$$= \Gamma_{L_t}(f(t, .), g(t, .))(X_t) \, dt \; \left(+ \mathrm{O}((dt)^2) \right).$$

In other words, we have the martingale increment formulae

$$\{ d\mathcal{M}_t(f) d\mathcal{M}_t(g) - \mathbb{E} \left(d\mathcal{M}_t(f) \, d\mathcal{M}_t(g) \mid X_t \right) \}$$

$$= d\mathcal{M}_t(fg) - f(t, X_t) \, d\mathcal{M}_t(g) - g(t, X_t) \, d\mathcal{M}_t(f).$$

Let $P_{s,t}$, $s \leq t$ be the Markov semigroup of X_t and we set $\eta_t = \mathrm{Law}(X_t)$. Applying the Doeblin-Itō formula to the function $f(s, x) = P_{s,t}(\varphi)(x)$ w.r.t. $s \in [0, t]$, we find that

$$dP_{s,t}(\varphi)(X_s) = - \left(\partial_s P_{s,t}(\varphi) \right) (X_s) \, ds + L_s(P_{s,t}(\varphi))(X_s) \, ds + dM_s(P_{.,t}(\varphi)) = dM_s(P_{.,t}(\varphi))$$

with a martingale $M_s(P_{.,t}(\varphi))$ with angle bracket

$$\forall s \in [0, t] \qquad \langle M(P_{.,t}(\varphi)), M(P_{.,t}(\varphi)) \rangle_s = \int_0^s \left(\Gamma_{L_\tau}(P_{\tau,t}(\varphi), P_{\tau,t}(\varphi)) \right) (X_\tau) \, d\tau.$$

This implies that for any $0 \leq s_1 \leq s_2 \leq t$ we have

$$\mathbb{E} \left([P_{s_2,t}(\varphi)(X_{s_2}) - P_{s_1,t}(\varphi)(X_{s_1})]^2 \mid X_{s_1} \right)$$

$$= \mathbb{E} \left((M_{s_2}(P_{.,t}(\varphi)) - M_{s_1}(P_{.,t}(\varphi)))^2 \mid X_{s_1} \right) = \int_{s_1}^{s_2} P_{s_1,\tau} \left[\Gamma_{L_\tau} \left(P_{\tau,t}(\varphi), P_{\tau,t}(\varphi) \right) \right] (X_{s_1}) \, d\tau.$$

Choosing $(s_1, s_2) = (0, t)$ and applying the expectations, we conclude that

$$\mathbb{E} \left([\varphi(X_t) - P_{0,t}(\varphi)(X_0)]^2 \right) = \int_0^t \eta_\tau \left[\Gamma_{L_\tau} \left(P_{\tau,t}(\varphi), P_{\tau,t}(\varphi) \right) \right] \, d\tau.$$

Using the fact that $\eta_0 \left[P_{0,t}(\varphi) \right] = \eta_t(\varphi)$ we also have the formula

$$\eta_t \left([\varphi - \eta_t(\varphi)]^2 \right) = \eta_0 \left([P_{0,t}(\varphi) - \eta_t(\varphi)]^2 \right) + \int_0^t \eta_\tau \left[\Gamma_{L_\tau} \left(P_{\tau,t}(\varphi), P_{\tau,t}(\varphi) \right) \right] \, d\tau.$$

15.5.2 Perturbation formulae

Let $X_t^{(1)}$ and $X_t^{(2)}$ be a couple of Markov processes evolving in some state space S with generators $L_t^{(1)}$ and $L_t^{(2)}$. We let $P_{s,t}^{(1)}$ and $P_{s,t}^{(2)}$ the Markov semigroup of $X_t^{(1)}$ and $X_t^{(2)}$.

We consider the Markov process X_t with generator $L_t = L_t^{(1)} + L_t^{(2)}$ and we let $P_{s,t}$ be its Markov semigroup. We further assume that all the semigroups $P_{s,t}$, $P_{s,t}^{(1)}$ and $P_{s,t}^{(2)}$ satisfy the forward and backward equations (15.16).

In this situation, for any $r \leq t$ we have the Gelfand-Pettis perturbation formulae:

$$P_{r,t} = P_{r,t}^{(1)} + \int_r^t P_{r,s} L_s^{(2)} P_{s,t}^{(1)} \, ds = P_{r,t}^{(1)} - \int_r^t P_{r,s}^{(1)} L_s^{(2)} P_{s,t} \, ds. \qquad (15.25)$$

Proof :
We check these formulae using the interpolating integral operators $P_{r,s} P_{s,t}^{(1)}$ and $P_{r,s}^{(1)} P_{s,t}$, with $r \leq s \leq t$. For instance, we have

$$
\begin{aligned}
\partial_s \left(\left[P_{r,s} P_{s,t}^{(1)} \right](f) \right) &= \left[\partial_s P_{r,s} \right] \left(P_{s,t}^{(1)}(f) \right) + P_{r,s} \left(\partial_s P_{s,t}^{(1)}(f) \right) \\
&= \left[P_{r,s} L_s \right] \left(P_{s,t}^{(1)}(f) \right) + P_{r,s} \left(-L_s^{(1)} P_{s,t}^{(1)}(f) \right) \\
&= P_{r,s} \left(L_s - L_s^{(1)} \right) P_{s,t}^{(1)}(f).
\end{aligned}
$$

These equations are of course valid for sufficient regular functions f on S depending on the form of the generators $L_t^{(1)}$ and $L_t^{(2)}$. In addition, it is implicitly assumed that $P_{s,t}^{(1)}(f)$ are sufficiently regular so that $L_s \left(P_{s,t}^{(1)}(f) \right)$ is well defined.

This implies that

$$P_{r,t}(f) - P_{r,t}^{(1)}(f) = \int_r^t \partial_s \left(\left[P_{r,s} P_{s,t}^{(1)} \right](f) \right) ds = \int_r^t P_{r,s} L_s^{(2)} P_{s,t}^{(1)}(f) \, ds.$$

The second assertion is proved by symmetry arguments. This ends the proof of (15.25). ∎

Suppose that $X_t^{(2)}$ is a pure jump process with generator

$$L_t^{(2)}(f)(x) = \lambda_t(x) \int [f(y) - f(x)] \, K_t(x, dy) \qquad (15.26)$$

for some jump rate $\lambda_t(x)$ and some Markov transition $K_t(x, dy)$. In this case (15.25) implies that

$$P_{r,t}(f) = P_{r,t}^{(1)}(f) + \int_r^t P_{r,s}^{(1)} \left(\lambda_s \left[P_{s,t}(f) - K_s \left(P_{s,t}(f) \right) \right] \right) ds.$$

Between jump times at rate λ_t the process X_t evolves as $X_t^{(1)}$. Following the analysis presented in section 15.2, we let T^s be the first jump time of X_t with rate λ_t, that is

$$T^s := \inf \left\{ r \geq s \; : \; \int_s^r \lambda_u(X_u) du \geq \mathcal{E} \right\}$$

where $\mathcal{E}$ stands for an exponential random variable with unit parameter. Arguing as in section 13.2.2 we have

$$
\begin{aligned}
P_{s,t}(f)(x) &= \mathbb{E}\left(f(X_t)\,1_{T^s>t} \mid X_s=x\right) + \mathbb{E}\left(f(X_t)\,1_{T^s\le t} \mid X_s=x\right) \\
&= \mathbb{E}\left(f(X_t^{(1)})\,\exp\left[-\int_s^t \lambda_r(X_r^{(1)})dr\right] \mid X_s^{(1)}=x\right) \\
&\qquad + \mathbb{E}\left(K_{T^s}\left(P_{T^s,t}(f)\right)(X_{T^s-})\,1_{T^s\le t} \mid X_s=x\right)
\end{aligned}
$$

and

$$
\mathbb{E}\left(g_{T^s}(X_{T^s-})\,1_{T^s\le t} \mid X_s=x\right)
$$

$$
= \mathbb{E}\left[\int_s^t g_r(X_r^{(1)})\,\underbrace{\lambda_r\left(X_r^{(1)}\right)\,\exp\left[-\int_s^r \lambda_u\left(X_u^{(1)}\right)du\right]dr}_{=\mathbb{P}\left(T^s\in dr \mid X_u^{(1)},\,s\le u\le r\right)} \mid X_s^{(1)}=x\right].
$$

This implies that

$$
P_{s,t}(f)(x)
$$

$$
= \mathbb{E}\left(f(X_t^{(1)})\,\exp\left[-\int_s^t \lambda_r(X_r^{(1)})dr\right] \mid X_s^{(1)}=x\right)
$$

$$
+ \mathbb{E}\left[\int_s^t K_r\left(P_{r,t}(f)\right)(X_r^{(1)})\,\lambda_r\left(X_r^{(1)}\right)\,\exp\left[-\int_s^r \lambda_u\left(X_u^{(1)}\right)du\right]dr \mid X_s^{(1)}=x\right].
$$

This yields the semigroup decomposition

$$
P_{s,t}(f) = Q_{s,t}(f) + \int_s^t Q_{s,r}\left[\lambda_r K_r\left(P_{r,t}(f)\right)\right]\,dr \tag{15.27}
$$

with the Feynman-Kac semigroup

$$
Q_{s,t}(f)(x) := \mathbb{E}\left(f(X_t^{(1)})\,\exp\left[-\int_s^t \lambda_r(X_r^{(1)})dr\right] \mid X_s^{(1)}=x\right).
$$

When $\lambda_t(x)=\lambda$ is a constant function, the perturbation formula (15.27) reverts to

$$
P_{s,t}(f) = e^{-\lambda(t-s)}\,P_{s,t}^{(1)}(f) + \int_s^t \lambda\,e^{-\lambda(r-s)}\,P_{s,r}^{(1)}\left[K_r\left(P_{r,t}(f)\right)\right]\,dr. \tag{15.28}
$$

15.6 Jump diffusion processes with killing

15.6.1 Feynman-Kac semigroups

We return to the definition of a jump diffusion process given in (15.1) but we replace the stochastic diffusion flow by a jump diffusion process $\varphi_{s,t}(x)$. Notice that in this situation the random paths $t \in [s, \infty[\mapsto \varphi_{s,t}(x)$ of the stochastic flow are càdlàg. To avoid confusion we denote by V_t the jump rate of this new jump diffusion process. In this situation, the developments of section 15.2 remain valid if we replace λ_t by V_t. In particular, (15.6) takes the form

$$P_{s,t}^V(f)(x) := \mathbb{E}\left[f(\varphi_{s,t}(x)) \exp\left(-\int_s^t V_u(\varphi_{s,u}(x))du\right)\right] = \mathbb{E}\left[f(\varphi_{s,t}(x)) \, 1_{T^s(x)>t}\right]$$

where $T^{(s)}$ stands for the first jump time of the process with the intensity V_u after time s (discarding the possible jump times of the càdlàg paths of the stochastic flow $t \in [s, \infty[\mapsto \varphi_{s,t}(x)$). More precisely, we have

$$T^{(s)}(x) = \inf\left\{t \geq s \; : \; \int_s^t V_u(\varphi_{s,u}(x)) \, du \geq \mathcal{E}\right\} \tag{15.29}$$

where $\mathcal{E}$ denotes some exponential random variable with unit parameter. We have not said anything about what happens when the jump occurs. Let us add a cemetery state c to the state space $S := \mathbb{R}^d$ and set $S^c := \mathbb{R}^d \cup \{c\}$. We extend any function f to S^c by setting $f(c) = 0$. This convention prevents the existence of the unit function on S^c.

When the jump with intensity V_t occurs we place the process in c. In other words, at the first jump time T with intensity V_t we kill the process by placing it to the cemetery state c, and we set $X_t^c = c$ for any $t \geq T$. This clearly amounts to choosing the amplitude transitions $K_t(x, dy) = \delta_c(dy)$. Notice that the Markov semigroup $P_{s,t}^V$ of X_t^c and the one $P_{s,t}$ of the jump diffusion process X_t associated with the stochastic flow $X_t := \varphi_{0,t}(x)$ are connected by the formula

$$1_S(x) \, P_{s,t}^V(x, dy) \, 1_S(y) \;=\; 1_S(x) \, P_{s,t}(x, dy) \, 1_S(y)$$
$$1_S(x) \, P_{s,t}^V(x, \{c\}) \;=\; 1_S(x) \, (1 - P_{s,t}(x, S))$$

and clearly $P_{s,t}^V(\{c\}, S) = 0$ and $P_{s,t}^V(\{c\}, \{c\}) = 1$.

By construction, the generator of the killed process X_t^c on S^c is defined by

$$\forall x \in S \qquad L_t^V(f)(x) = L_t(f)(x) + V_t(x) \, (f(c) - f(x)) = L_t(f)(x) - V_t(x) \, f(x) \tag{15.30}$$

and $L_t^c(f)(c) = 0$, where L_t stands for the generator of X_t.

As usual, we have the forward and backward equations

$$\partial_t P_{s,t}^V(f) = P_{s,t}^V(L_t^V(f)) \quad \text{and} \quad \partial_s P_{s,t}^V(f) = -L_s^V(P_{s,t}^V(f)). \tag{15.31}$$

The forward equation can be checked directly using the fact for any $s \leq t$ we have

$$Z_{s,t} \;:=\; e^{-\int_s^t V_r(X_r)dr} \, f(X_t) - f(X_s) - \int_s^t e^{-\int_s^r V_u(X_u)du} \, L_r^V(f)(X_r) \, dr$$

$$=\; \int_s^t e^{-\int_s^r V_u(X_u)du} \, dM_r(f) \tag{15.32}$$

with the martingale

$$M_t(f) = f(X_t) - f(X_0) - \int_0^t L_s(f)(X_s)ds.$$

By applying the Doeblin-Itō formula w.r.t. the time parameter τ we have

$$
\begin{aligned}
dZ_{s,\tau} &= e^{-\int_s^\tau V_r(X_r)dr}\ ([-V_\tau(X_\tau)f(X_\tau) + L_\tau(f)(X_\tau)]d\tau + dM_\tau(f)) \\
&\qquad\qquad -e^{-\int_s^\tau V_u(X_u)du}\ L_\tau^V(f)(X_\tau)\ d\tau \\
&= e^{-\int_s^\tau V_r(X_r)dr}\ dM_\tau(f).
\end{aligned}
$$

This implies that

$$Z_{s,t} = Z_{s,t} - Z_{s,s} = \int_s^t dZ_{s,\tau} = \int_s^t e^{-\int_s^\tau V_r(X_r)dr}\ dM_\tau(f).$$

We conclude that

$$\mathbb{E}(Z_{s,t} \mid X_s) = 0 \Rightarrow P_{s,t}^V(f)(x) = f + \int_s^t P_{s,r}^V\left(L_r^V(f)\right)dr \Rightarrow \partial_r P_{s,r}^V(f) = P_{s,r}^V\left(L_r^V(f)\right).$$

We refer to exercise 294 for more details on the time evolution equations of Feynman-Kac measures.

By construction, for any $x \in S$ we have the Feynman-Kac semigroup formulae

$$
\begin{aligned}
P_{s,t}^V(f)(x) &= \mathbb{E}\left(f(X_t^c) \mid X_s^c = x\right) = \mathbb{E}\left(f(X_t)\ 1_{T^{(s)} > t} \mid X_s = x\right) \\
&= \mathbb{E}\left(f(X_t)\ \exp\left(-\int_s^t V_u(X_u)du\right) \mid X_s = x\right)
\end{aligned}
\qquad (15.33)
$$

with the first jump time $T^{(s)} := T^{(s)}(X_s)$ after time s defined in (15.29).

Important remark : The same formulae remain valid if we consider an abstract Markov process X_t with some generator L_t on some state space S. This claim should be clear from the proof of the forward equation given above.

The semigroup property in (15.33) can be used to derive the backward evolution equation from the forward equations.

More precisely we have

$$-h^{-1}\left[P_{s-h,t}^V - P_{s,t}^V\right] = -h^{-1}\left[P_{s-h,s}^V - Id\right]P_{s,t}^V \to_{h \to 0}\ -L_s P_{s,t}^V$$

as well as

$$h^{-1}\left[P_{s+h,t}^V - P_{s,t}^V\right] = h^{-1}\left[Id - P_{s,s+h}^V\right]P_{s+h,t}^V \to_{h \to 0}\ -L_s P_{s,t}^V.$$

15.6.2 Cauchy problems with terminal conditions

Observe that

$$\exp\left(-\int_s^t V_u(X_u)du\right) = 1 - \int_s^t V_u(X_u)\ \exp\left(-\int_s^u V_v(X_v)dv\right)du.$$

This yields the integral decomposition

$$
\begin{aligned}
P_{s,t}^V(f) &= \mathbb{E}\left(f(X_t)\ \exp\left(-\int_s^t V_u(X_u)du\right) \mid X_s = x\right) \\
&= \mathbb{E}\left(f(X_t)\left[1 - \int_s^t V_u(X_u)\ \exp\left(-\int_s^u V_v(X_v)dv\right)\ du\right] \mid X_s = x\right) \\
&= P_{s,t}(f) - \int_s^t P_{s,u}^V\left(V_u P_{u,t}(f)\right)\ du.
\end{aligned} \tag{15.34}
$$

By (15.31), for any given t and some given terminal condition f_t, the function

$$
s \in [0,t] \mapsto u_s = P_{s,t}^V(f_t)
$$

satisfies the Cauchy problem with terminal condition

$$
\begin{cases}
\partial_s u_s + L_s(u_s) &= V_s\, u_s \qquad \forall s \in [0,t] \\
u_t &= f_t.
\end{cases} \tag{15.35}
$$

In the reverse direction, if (15.35) admits a solution u_s with $0 \le s \le t$, then $u_s = P_{s,t}^V(f_t)$. To check this claim, we fix the time s, and for any $t \ge s$ we set

$$
\forall s \ge t \qquad Y_t := u_t(X_t)\ e^{-\int_s^t V_r(X_r)dr}.
$$

By applying the Doeblin-Itō formula (w.r.t. the time parameter t, with s fixed) we get

$$
\begin{aligned}
dY_t &= e^{-\int_s^t V_r(X_r)dr} du_t(X_t) + u_t(X_t)\, de^{-\int_s^t V_r(X_r)dr} \\
&= e^{-\int_s^t V_r(X_r)dr} du_t(X_t) - V_t(X_t)\, u_t(X_t)\, e^{-\int_s^t V_r(X_r)dr}\, dt.
\end{aligned}
$$

Recalling that

$$
du_t(X_t) = (\partial_t + L_t)\, u_t(X_t)dt + dM_t(u) = V_t(X_t)\, u_t(X_t)\, dt + dM_t(u) \quad \Leftarrow (15.35)
$$

for some martingale $\mathcal{M}_t(u)$, we conclude that $dY_t = e^{-\int_s^t V_r(X_r)dr}\, d\mathcal{M}_t(u)$ is a martingale. This implies that

$$
\mathbb{E}\left(u_t(X_t)\ e^{-\int_s^t V_r(X_r)dr} \mid X_s = x\right) = \mathbb{E}(Y_t \mid X_s = x) = \mathbb{E}(Y_s \mid X_s = x) = u_s(x).
$$

Recalling that $u_t(X_t) = f_t(X_t)$, we conclude that $u_s = P_{s,t}^V(f_t)$.

We fix a time horizon t and we consider the integrated Feynman-Kac model defined for any $s \in [0,t]$ by

$$
\begin{aligned}
v_s(x) &:= P_{s,t}^V(f_t) + \int_s^t P_{s,r}^V(g_r)\ dr \\
&= \mathbb{E}\left(f_t(X_t)\ e^{-\int_s^t V_r(X_r)dr} + \int_s^t g_r(X_r)\ e^{-\int_s^r V_u(X_u)du} \mid X_s = x\right)
\end{aligned}
$$

for some given function $(s,x) \in ([0,t] \times S) \mapsto g_s(x) \in \mathbb{R}$.

In this case, we have

$$\frac{\partial v}{\partial s} = -L_s^V \left(P_{s,t}^V(f) \right) - \int_s^t L_s^V \left(P_{s,r}^V(g_r) \right) \, dr - g_s = L_s^V(v_s) - g_s.$$

This shows that $v : (s,x) \in ([0,t] \times S) \mapsto v_s(x)$ satisfies the Cauchy problem with terminal condition

$$\begin{cases} \partial_s v_s + L_s(v_s) + g_s &= V_s \, v_s \qquad \forall s \in [0,t] \\ v_t &= f_t. \end{cases} \tag{15.36}$$

For a more thorough discussion on these killed processes, Feynman-Kac semigroups and their applications in physics and more precisely in the spectral analysis of Schrödinger operators, we refer the reader to chapter 27 (see for instance section 27.2).

We end this section with a discrete time approximation of (15.36) based on the discrete time models discussed in (9.42). We consider a time mesh sequence t_p with some time step $\delta = t_n - t_{n-1}$, and we consider the discrete time model (9.42) with

$$G_p = e^{-V_{t_p}\delta} = 1 - V_{t_p}\delta + \mathrm{O}(\delta^2) \quad \text{and} \quad M_p = Id + L_{t_p}\,\delta + \mathrm{O}(\delta^2). \tag{15.37}$$

We also replace in (9.42) the functions g_p and f_n by the function $g_{t_p}\,\delta$ and f_{t_n}, and we let v_{t_p} the solution of (9.42). In this notation (9.42) takes the form

$$\begin{aligned} v_{t_p} &= \left(1 - V_{t_p}\delta\right) \left(v_{t_{p+1}} + L_{t_p}(v_{t_{p+1}})\delta\right) + g_{t_p}\,\delta + \mathrm{O}(\delta^2) \\ &= \left(1 - V_{t_p}\delta\right) \left(v_{t_{p+1}} + L_{t_p}(v_{t_p})\delta\right) + g_{t_p}\,\delta + \mathrm{O}(\delta^2) \\ &= \left(1 - V_{t_p}\delta\right) v_{t_{p+1}} + L_{t_p}(v_{t_p})\delta + g_{t_p}\,\delta + \mathrm{O}(\delta^2) \end{aligned}$$

with the terminal condition $v_{t_n} = f_{t_n}$. Rearranging the terms, the above equation takes the form

$$\left(v_{t_{p+1}} - v_{t_p}\right)/\delta + L_{t_p}(v_{t_p}) + g_{t_p} = V_{t_p} + \mathrm{O}(\delta^2)/\delta \rightarrow_{\delta \downarrow 0} \text{ equation (15.36)}.$$

15.6.3 Dirichlet-Poisson problems

We let D be some open subset of $S = \mathbb{R}^d$. We consider a couple of functions (g, V) and the generator L of some jump diffusion process on S, and some function h on $S - D$.

The Dirichlet-Poisson problem consists with finding a smooth continuous function v on S and satisfying the following equations:

$$\begin{cases} L(v)(x) + g(x) &= V(x)v(x) \quad \text{for any} \quad x \in D \\ v(x) &= h(x) \qquad\quad \text{for any} \quad x \notin D. \end{cases} \tag{15.38}$$

Without any boundary conditions we have $D = S = \mathbb{R}^d$ (and $(h, S - D) = (0, \emptyset)$). In this situation the Dirichlet-Poisson problem (15.38) reduces to the Poisson equation

$$L(v) + g = V \, v. \tag{15.39}$$

When L is the generator of a pure diffusion process X_t, the trajectories of X_t are continuous. In this situation the Dirichlet-Poisson problem (15.38) reduces to finding a smooth continuous function v on $\overline{D} = D \cup \partial D$, where ∂D stands for the boundary of the open set D, satisfying the equations

$$\begin{cases} L(v)(x) + g(x) &= V(x)v(x) \quad \text{for any} \quad x \in D \\ v(x) &= h(x) \qquad\quad \text{for any} \quad x \in \partial D. \end{cases} \qquad (15.40)$$

Arguing as in the end of section 15.6.2, the Dirichlet-Poisson problem (15.38) can be approximated by the discrete time model (9.43). To be more precise, we consider a time mesh sequence t_p with some time step $\delta = t_n - t_{n-1}$, and we consider the discrete time model (9.43) with g replaced by $g\delta$ and

$$G = e^{-V\delta} = 1 - V\delta + \mathrm{O}(\delta^2) \quad \text{and} \quad M = Id + L\,\delta + \mathrm{O}(\delta^2). \qquad (15.41)$$

With a slight abuse of notation we let v be the corresponding solution of (9.43). In this notation (9.43) takes the form

$$\begin{aligned} v(x) &= (1 - V(x)\delta)\,(v(x) + L(v)(x)\delta) + g(x)\delta + \mathrm{O}(\delta^2) \\ &= (1 - V(x)\delta)\,v(x) + L(v)(x)\delta + g(x)\delta + \mathrm{O}(\delta^2) \end{aligned}$$

for any $x \in D$. This clearly implies that

$$\forall x \in D \quad L(v)(x) + g(x) = V(x)\,v(x) + \mathrm{O}(\delta^2)/\delta \to_{\delta\downarrow 0} \text{ equation (15.40)}.$$

We further assume that V is a non-negative function on D. The solving of this problem is based on a stopping procedure of the martingale starting at $\mathcal{N}_0 = v(X_0)$ defined by

$$\begin{aligned} \mathcal{N}_t &:= e^{-\int_0^t V(X_s)ds}\,v(X_t) + \int_0^t e^{-\int_0^s V(X_r)dr}\,\{-L(v)(X_s) + V(X_s)\,v(X_s)\}\,ds \\ &= \int_0^t e^{-\int_0^s V(X_r)dr}\,d\mathcal{M}_s(v) \quad \left(\Longrightarrow d\mathcal{N}_t = e^{-\int_0^t V(X_r)dr}\,d\mathcal{M}_t(v)\right) \end{aligned}$$

with the martingale $d\mathcal{M}_t(v) := dv(X_t) - L(v)(X_t)\,dt$.

Let T_D be the (first) time X_t exits the set D. Whenever (15.38) is satisfied we have

$$v(x) = \mathbb{E}\left(e^{-\int_0^{T_D} V(X_s)ds}\,h(X_{T_D}) + \int_0^{T_D} e^{-\int_0^s V(X_r)dr}\,g(X_s)\,ds \mid X_0 = x\right) \qquad (15.42)$$

as soon as $\mathbb{E}(\mathcal{N}_{T_D} \mid \mathcal{F}_0) = \mathcal{N}_0 (= v(X_0))$. This formula provides an explicit description of the solution of (15.38) in terms of the functions (g, h) and T_D. For instance, if $(V, g) = (0, 0)$ and $h = 1_A$ with $A \subset D$, we have

$$v(x) = \mathbb{P}(X_{T_D} \in A \mid X_0 = x).$$

Without any boundary conditions we have $D = S = \mathbb{R}^d$ (and $(h, S - D) = (0, \emptyset)$) so that $T_D = \infty$ and the solution of the Poisson equation (15.39) is given by

$$v(x) = \mathbb{E}\left(\int_0^\infty e^{-\int_0^s V(X_r)dr}\,g(X_s)\,ds \mid X_0 = x\right). \qquad (15.43)$$

In the above display, we have implicitly assumed that T_D is a well defined stopping time (cf. section 12.5.2).

For instance we have

$$(15.38) \quad \text{with} \quad L = \frac{1}{2} \sum_{1 \leq i \leq d} \partial_{x_i}^2 \quad \text{and} \quad (V, g) = (0, 0) \implies v(x) = \mathbb{E}\left(h(x + W_{T_D})\right)$$

where W_t stands for a d-dimensional Brownian motion.

In addition, for any closed ball $B := B(x, \epsilon) \subset D$, with center $x \in D$ and radius $\epsilon > 0$,

$$\mathbb{E}\left(h(x + W_{T_B} + (W_{T_D} - W_{T_B})) \mid x + W_{T_B}\right) = v(x + W_{T_B}) \Rightarrow v(x) = \mathbb{E}\left(v(x + W_{T_B})\right).$$

Invoking the rotational invariance of the Brownian motion, we note that $x + W_{T_B}$ is uniformly distributed on the boundary ∂B of B. This shows that

$$\forall \rho \in [0, \epsilon] \quad v(x) = \int_{\partial B(x, \rho)} v(y) \, \mu_{\partial B(x, \rho)}(dy) \Rightarrow v(x) = \int_{B(x, \epsilon)} v(y) \, \mu_{B(x, \epsilon)}(dy) \quad (15.44)$$

where $\mu_{\partial B(x, \rho)}$, respectively $\mu_{B(x, \epsilon)}$, stands for the uniform measure on the sphere $S(x, \rho) := \partial B(x, \rho)$, respectively on the ball $B(x, \epsilon)$.

The r.h.s. formula in (15.44) is proved by integrating the radii $\rho \in [0, \epsilon]$ of the boundary spheres $\partial B(x, \rho)$ with the uniform measure on $[0, \epsilon]$. We can check this radial integration by simply recalling that a uniform point in $B(x, \epsilon)$ is obtained by sampling a uniform radius ρ between 0 and ϵ and sampling randomly a state in the corresponding sphere $S(x, \rho)$). This is called the mean value property.

A function v satisfying the mean value property for every ball inside the domain D is called a harmonic function on D.

We also notice that

$$|\mathcal{N}_{T_D}| \leq C(g, h)(1 + T_D) \quad \text{with} \quad C(g, h) = \sup_{x \in S - D} |h(x)| \vee \sup_{x \in D} |g(x)|. \quad (15.45)$$

In the above discussion, we have implicitly assumed the continuous time version of the Doob's stopping theorem 8.4.12 presented in section 8.4.3.

Theorem 15.6.1 (Doob's stopping theorem) *The stopped process $t \mapsto \mathcal{M}_{t \wedge T}$ of an $\mathcal{F}_t$-martingale $\mathcal{M}_t$ w.r.t. some stopping time T is also an $\mathcal{F}_t$-martingale. In addition, a stochastic process $\mathcal{M}_t$ is an $\mathcal{F}_t$-martingale if and only if $\mathbb{E}(\mathcal{M}_T) = \mathbb{E}(\mathcal{M}_0)$ for any finite stopping time s.t. $\mathbb{E}(|\mathcal{M}_T|) < \infty$.*

Proof :
The detailed proof of this theorem follows the same arguments as those of theorem 8.4.12 and theorem 8.4.16 thus it is only sketched. We only prove that the r.h.s. of the last assertion implies that $\mathcal{M}_t$ is a martingale. To this end, we choose $0 \leq s \leq t$ and some event $A \in \mathcal{F}_s$. We clearly have

$$T = s \, 1_A + t \, 1_{A^c} \Rightarrow \mathcal{M}_T = \mathcal{M}_s \, 1_A + \mathcal{M}_t \, 1_{A^c} = \mathcal{M}_t - 1_A (\mathcal{M}_t - \mathcal{M}_s).$$

Using the fact that $\mathbb{E}(\mathcal{M}_T) = \mathbb{E}(\mathcal{M}_0) = \mathbb{E}(\mathcal{M}_t)$ we conclude that

$$(\forall A \in \mathcal{F}_s \quad \mathbb{E}\left((\mathcal{M}_t - \mathcal{M}_s)\, 1_A\right) = 0) \iff \mathbb{E}(\mathcal{M}_t \mid \mathcal{F}_s) = \mathcal{M}_s.$$

This ends the proof of the theorem. ∎

By theorem 15.6.1, using (15.45) we have

$$\mathbb{E}\left(T_D \mid X_0 = x\right) < \infty \implies \mathbb{E}(\mathcal{N}_{T_D} \mid X_0 = x) = v(x).$$

In the reverse angle, by choosing $(g, h, V) = (1, 0, 0)$ the function (15.42) takes the form
$$v(x) = \mathbb{E}\left(T_D \mid X_0 = x\right).$$

In other words, the computation of mean exit times (a.k.a. mean confinement times) requires to solve the Dirichlet problem

$$\begin{cases} -L(v)(x) & = & 1 \quad \text{for any} \quad x \in D \\ v(x) & = & 0 \quad \text{for any} \quad x \in \partial D. \end{cases} \tag{15.46}$$

For instance, for a one-dimensional rescaled Brownian motion $X_t = x + \sqrt{2}\, W_t$ starting at some $x \in D =]a, b[$, we have $-L(v) = -v'' = 1$. In this situation, (15.46) reverts to

$$v'' = -1 \qquad \text{with} \quad v(a) = 0 = v(b).$$

The solution is clearly given by the quadratic polynomial

$$v(x) = \frac{1}{2}\, (a - x)\, (b - x)\, 1_{[a,b]}(x) = \mathbb{E}\left(T_{]a,b[} \mid X_0 = x\right).$$

Notice that this formula coincides with the estimate discussed in exercise 233. For more general one-dimensional diffusions with generator

$$L(f) = b\, f' + \frac{1}{2}\, \sigma^2\, f''$$

with a smooth drift and diffusion functions (b, σ), the strategy is to introduce the function

$$\alpha(x) = \int_a^x \frac{2b}{\sigma^2}(y) \implies \frac{1}{2}\, \sigma^2\, e^{-\alpha}\, (e^\alpha\, f')' = L(f).$$

This shows that we need to solve the equation

$$\frac{1}{2}\, \sigma^2\, e^{-\alpha}\, (e^\alpha\, v')' = -1.$$

The solution is given by

$$v'(y) = e^{-\alpha(y)}\, \left[c_1 - 2 \int_a^y \sigma^{-2}(z)\, e^{\alpha(z)}\, dz\right]$$

$$v(x) = 1_{[a,b]}(x) \left\{c_2 + \int_a^x e^{-\alpha(y)}\, \left[c_1 - 2 \int_a^y \sigma^{-2}(z)\, e^{\alpha(z)}\, dz\right]\, dy\right\}$$

with the constants c_1, c_2 given by the boundary conditions $v(a) = c_2 = 0$ and

$$v(b) = 0 = c_1 \int_a^b e^{-\alpha(y)} \, dy - 2 \int_a^b e^{-\alpha(y)} \int_a^y \sigma^{-2}(z) \, e^{\alpha(z)} \, dz \, dy$$

$$\implies c_1 = 2 \int_a^b \frac{e^{-\alpha(y)}}{\int_a^b e^{-\alpha(y')} \, dy'} \left[\int_a^y \sigma^{-2}(z) \, e^{\alpha(z)} \, dz \right] \, dy.$$

This implies that

$$\mathbb{E} \left(T_{]a,b[} \mid X_0 = x \right) = 2 \, 1_{]a,b[}(x)$$

$$\times \int_a^x e^{-\alpha(y)} \left[\int_a^b \frac{e^{-\alpha(z)}}{\int_a^b e^{-\alpha(z')} \, dz'} \left\{ \int_a^z \sigma^{-2}(s) \, e^{\alpha(s)} \, ds - \int_a^y \sigma^{-2}(t) \, e^{\alpha(t)} \, dt \right\} \, dz \right] \, dy.$$

Most of the time we only need to ensure that $\mathbb{E} \left(T_D \mid X_0 = x \right) < \infty$. In this situation, we can use the following simple criteria.

Lemma 15.6.2

$$\exists \, w \text{ s.t. } \sup_{x \in D} L(w)(x) \leq -1 \quad \text{and} \quad \mathrm{osc}_D(w) := \sup_{(x,y) \in D^2} |w(x) - w(y)| < \infty$$

$$\implies \sup_{x \in \mathbb{R}^d} \mathbb{E} \left(T_D \mid X_0 = x \right) \leq \mathrm{osc}_D(w).$$

$$\tag{15.47}$$

Proof :

To check this claim, it clearly suffices to assume that $X_0 \in D$. Using the fact that the stopped process

$$w(X_{t \wedge T_D}) - w(X_0) - \int_0^{t \wedge T_D} L(w)(X_s) ds \geq (t \wedge T_D) + w(X_{t \wedge T_D}) - w(X_0)$$

is a martingale starting at the origin, we check that

$$\mathbb{E} \left(t \wedge T_D \mid X_0 \right) \leq w(X_0) - \mathbb{E} \left(w(X_{t \wedge T_D}) \mid X_0 \right) \leq \mathrm{osc}_D(w).$$

Passing to the limit $t \uparrow \infty$ with Fatou's lemma ends the proof of the desired estimate. $\blacksquare$

For instance, for time homogeneous and pure diffusion processes with a generator $L = L^c$ given in (15.13) (with $\sigma_t = \sigma$ and $b_t = b$) we have

$$\exists 1 \leq i \leq d \quad \text{s.t.} \quad \min_{x \in \overline{D}} (\sigma \sigma^T)_{i,i}(x) > 0 \implies \sup_{x \in \mathbb{R}^d} \mathbb{E} \left(T_D \mid X_0 = x \right) < \infty.$$

To check this claim, we let $\alpha := \min_{x \in \overline{D}} (\sigma \sigma^T)_{i,i}(x) > 0$, $\beta := \max_{x \in \overline{D}} |b^i(x)|$, $\delta := \min_{x \in \overline{D}} x_i$, $\delta' = \max_{x \in \overline{D}} x_i$ and we set $w(x) := -c_1 \, e^{-c_2 x_i} \in [-c_1 \, e^{-c_2 \delta}, 0]$, for some non-negative parameters (c_1, c_2), and any $x \in D$. Observe that

$$
\begin{aligned}
-L(w) \;&=\; c_1 \, c_2 \, e^{-c_2 x_i} \left(\frac{1}{2} \, (\sigma\sigma^T)_{i,i} \, c_2 - b^i \right) \\[2mm]
&\geq\; c_1 \, e^{-c_2\delta'} \left(\frac{1}{2} \, \alpha c_2^2 - \beta c_2 \right) = c_1 \, e^{-c_2\delta'} \, \frac{\alpha}{2} \left[\left(c_2 - \frac{\beta}{\alpha} \right)^2 - \left(\frac{\beta}{\alpha} \right)^2 \right] \geq 1
\end{aligned}
$$

for well chosen parameters (c_1, c_2). For instance, this lower bound is satisfied for $c_2 = (1 + \sqrt{2}) \, \beta/\alpha$ and any $c_1 \geq 2\alpha \, \beta^{-2} \, e^{(1+\sqrt{2})\beta\delta/\alpha}$. This ends the proof of the desired estimate. ∎

A weaker condition is discussed in exercise 290.

It is important to observe that we have not discussed the uniqueness properties of the solution of (15.39). For instance when $(h, g, S - D) = (0, 0, \emptyset)$ the Feynman-Kac formula (15.43) only provides a null solution $v = 0$ to the equation $Lv = Vv$, while all constant functions v satisfy this equation. In exercise 299 we provide a non-trivial example of diffusion and quadratic potential functions V for which a solution o $Lv = Vv$ is given by Bessel functions of the first kind.

For diffusion generators L and constant potential functions $V(x) = \lambda$, with $(h, g) = (0, 0)$ but $\partial D \neq \emptyset$, the Poisson-Dirichlet equation (15.38) takes the form

$$
\begin{cases}
L(v)(x) &=\; \lambda \, v(x) \quad \text{for any} \quad x \in D \\
v(x) &=\; 0 \qquad\quad \text{for any} \quad x \in \partial D.
\end{cases}
$$

These Dirichlet equations are related to the spectrum of the operator L subject to a null boundary condition. Another important class of equations arising in physics is given by the same type of equation with Neuman boundary conditions

$$
\begin{cases}
L(v)(x) &=\; \lambda \, v(x) \quad \text{for any} \quad x \in D \\
\langle \nabla v(x), N^\perp(x) \rangle &=\; 0 \qquad\quad \text{for any} \quad x \in \partial D
\end{cases}
$$

where $N^\perp(x)$ stands for the outward unit normal to the boundary, and $\nabla v(x)$ stands for the gradient of the function v. These two problems are directly related to an integration by parts formula on manifolds discussed in (23.21). We refer to exercise 303 for a simple illustration of these models for a $1d$-Laplacian on an interval.

15.6.4 Cauchy-Dirichlet-Poisson problems

We let D be some open subset of $S = \mathbb{R}^d$. We consider some functions (f_t, g_t) and V_t on S, and some function h_t on $S - D$.

The Cauchy-Dirichlet-Poisson problem consists of finding a smooth functional mapping $v \; : \; (s, x) \in ([0, t] \times S) \mapsto v_s(x)$, with a fixed time horizon t satisfying the following equations:

$$
\begin{cases}
\partial_s v_s(x) + L_s(v_s)(x) + g_s(x) &=\; V_s(x) v_s(x) \quad \text{for any} \quad (s, x) \in ([0, t] \times D) \\
v_s(x) &=\; h_s(x) \qquad\quad\;\; \text{for any} \quad (s, x) \in ([0, t] \times S - D) \\
v_t(x) &=\; f_t(x) \qquad\quad\;\; \text{for any} \quad x \in D.
\end{cases}
\tag{15.48}
$$

As in (15.40), when L is the generator of a pure diffusion process X_t, the trajectories of X_t are continuous. In this situation, the Cauchy-Dirichlet-Poisson problem (15.48) reduces to find a smooth continuous function $v : (s, x) \in ([0, t] \times \overline{D}) \mapsto v_s(x)$, with $\overline{D} = D \cup \partial D$, where ∂D stands for the boundary of the set D, satisfying the equations

$$
\begin{cases}
\partial_s v_s(x) + L_s(v_s)(x) + g_s(x) &= V_s(x)v_s(x) & \text{for any} \quad (s, x) \in ([0, t] \times D) \\
v_s(x) &= h_s(x) & \text{for any} \quad (s, x) \in ([0, t] \times \partial D) \quad (15.49) \\
v_t(x) &= f_t(x) & \text{for any} \quad x \in D.
\end{cases}
$$

Arguing as in the end of section 15.6.2, the Cauchy-Dirichlet-Poisson problem (15.48) can be approximated by the discrete time model (9.48) using the potential functions and the Markov transitions (15.37) on some time mesh.

Suppose that (15.48) has a solution v_s, with $0 \leq s \leq t$. We fix the time s, and for any $t \geq s$ we set

$$
Y_t := v_t(X_t) \, e^{-\int_s^t V_r(X_r)dr} + \int_s^t g_r(X_r) \, e^{-\int_s^r V_u(X_u)du} \, dr.
$$

By applying the Doeblin-Itō formula (w.r.t. the time parameter t, with s fixed) we clearly have

$$
\begin{aligned}
dY_t &= e^{-\int_s^t V_r(X_r)dr} \, [dv_t(X_t) - v_t(X_t) \, V_t(X_t) + g_t(X_t)] \\
&= e^{-\int_s^t V_r(X_r)dr} \, \underbrace{[\partial_t v_t + L_t(v_t) + g_t - v_t V_t] (X_t)}_{=0} \, dt + e^{-\int_s^t V_r(X_r)dr} \, dM_t(v),
\end{aligned}
$$

with the martingale $dM_t(v) = dv_t(X_t) - [\partial_t v_t + L_t(v_t)] (X_t)dt$. This shows that $(Y_t)_{t \geq s}$ with s fixed is a martingale. We consider the stopped martingale $Y_{t \wedge T_D^{(s)}}$, with the first exit time of the domain D by $(X_t)_{t \geq s}$ (after time s). We implicitly assume that $T_D^{(s)}$ is a well defined stopping time (cf. section 12.5.2).

The martingale property implies that

$$
\mathbb{E}\left(Y_{t \wedge T_D^{(s)}} \mid X_s = x\right) = \mathbb{E}\left(Y_{s \wedge T_D^{(s)}} \mid X_s = x\right) = v_s(x).
$$

On the other hand, we have

$$
Y_{t \wedge T_D^{(s)}}
$$

$$
= v_{t \wedge T_D^{(s)}}(X_{t \wedge T_D^{(s)}}) \, e^{-\int_s^{t \wedge T_D^{(s)}} V_r(X_r)dr} + \int_s^{t \wedge T_D^{(s)}} g_r(X_r) \, e^{-\int_s^r V_u(X_u)du} \, dr
$$

$$
= 1_{T_D^{(s)} > t} \, \overbrace{v_t(X_t)}^{=f_t(X_t)} \, e^{-\int_s^t V_r(X_r)dr}
$$

$$
+ 1_{T_D^{(s)} \leq t} \, \overbrace{v_{T_D^{(s)}}(X_{T_D^{(s)}})}^{=h_{T_D^{(s)}}(X_{T_D^{(s)}})} \, e^{-\int_s^{T_D^{(s)}} V_r(X_r)dr} + \int_s^{t \wedge T_D^{(s)}} g_r(X_r) \, e^{-\int_s^r V_u(X_u)du} \, dr
$$

and

$$
\int_s^{t \wedge T_D^{(s)}} g_r(X_r) \, e^{-\int_s^r V_u(X_u)du} \, dr = \int_s^t 1_{T_D^{(s)} > r} \, g_r(X_r) \, e^{-\int_s^r V_u(X_u)du} \, dr.
$$

Whenever (15.48) is satisfied we have

$$v_s(x) = \mathbb{E}\left(1_{T_D^{(s)}>t}\, f_t(X_t)\, e^{-\int_s^t V_r(X_r)dr} \mid X_s = x\right)$$

$$+ \mathbb{E}\left(\int_s^t 1_{T_D^{(s)}>r}\, g_r(X_r)\, e^{-\int_s^r V_u(X_u)du}\, dr \mid X_s = x\right)$$

$$+ \mathbb{E}\left(1_{T_D^{(s)}\leq t}\, h_{T_D^{(s)}}(X_{T_D^{(s)}})\, e^{-\int_s^{T_D^{(s)}} V_r(X_r)dr} \mid X_s = x\right).$$

In particular for $(g, h) = (0, 0)$, the solution (whenever it exists) of (15.48) is given for any $(s, x) \in ([0, t] \times S)$ by the Feynman-Kac semigroup $Q_{s,t}$ defined by

$$v_s(x) = Q_{s,t}(f_t)(x) := \mathbb{E}\left(1_{T_D^{(s)}>t}\, f_t(X_t)\, e^{-\int_s^t V_r(X_r)dr} \mid X_s = x\right). \qquad (15.50)$$

In addition, for time homogeneous functions $f_t = f$ we have the forward and backward semigroup formulae

$$\partial_s Q_{s,t}(f)(x) = -L_s^V\left(Q_{s,t}(f)\right)(x) \quad \text{and} \quad \partial_t Q_{s,t}(f)(x) = Q_{s,t}\left(L_t^V(f)\right)(x) \qquad (15.51)$$

for any $x \in D$ and $s \leq t$, with the boundary conditions

$$Q_{s,s}(f)(x) = f(x) = Q_{t,t}(f)(x) \quad \text{and} \quad Q_{s,t}(f)(x) = 0 \text{ as soon as } x \notin D.$$

In the above display L_t^V stands for the Schrödinger operator defined in (15.30).

Proof :
The semigroup property in (15.50) follows from the fact that

$$\forall 0 \leq s \leq r \leq t \qquad Z_{s,t} := 1_{T_D^{(s)}>t}\, e^{-\int_s^t V_\tau(X_\tau)d\tau} = Z_{s,r} \times Z_{r,t}.$$

The backward equation in (15.51) is a reformulation of (15.49); the forward equation is checked using the semigroup property

$$h^{-1}\left[Q_{s,t+h} - Q_{s,t}\right] = Q_{s,t}\left\{h^{-1}\left[Q_{t+h-h,t+h} - Id\right]\right\} \to_{h \to 0} Q_{s,t}L_t^V.$$

This ends the proof of (15.51). ∎

Here again we have implicitly assumed the continuous time version of Doob's stopping theorem 8.4.12 presented in section 8.4.3. We leave the reader to find regularity conditions justifying the above developments following the discussion provided in section 15.6.3.

15.7   Some illustrations

15.7.1   One-dimensional Dirichlet-Poisson problems

This section provides explicit description of the solution of the Dirichlet-Poisson problem (15.38) for one-dimensional diffusion processes on some bounded open interval $D =]c_1, c_2[$ (with boundary $\partial D = \{c_1, c_2\}$), when $g = 0$. We further assume that the generator of the process has the form

$$L(f) = b \, f' + \frac{1}{2} \, \sigma^2 \, f''$$

for some functions (b, σ) on $\mathbb{R}$ with $\sigma(x) > 0$ for any $x \in [c_1, c_2]$.

We fix some constant $c \in \mathbb{R}$ and we set

$$s(x) := \exp\left[-\int_c^x \frac{2b(y)}{\sigma^2(y)} \, dy \right] \qquad \text{and} \qquad S(x) = \int_c^x s(y) dy.$$

In this situation, by (15.38) for any $x \in]c_1, c_2[$ we have

$$\sigma^2(x) \, s(x) \, \left(\frac{1}{s(x)} \right)' = b(x)$$

and

$$\begin{aligned}
\frac{1}{2} \, \sigma^2(x) \, s(x) \, \left(\frac{1}{s(x)} \, v'(x) \right)' &= \frac{1}{2} \, \sigma^2(x) \, s(x) \, \left(\frac{1}{s(x)} \right)' \, v'(x) + \frac{1}{2} \, \sigma^2(x) \, v''(x) \\
&= b(x) \, v'(x) + \frac{1}{2} \, \sigma^2(x) \, v''(x) = 0.
\end{aligned}$$

$$\left(\frac{1}{s(x)} \, v'(x) \right)' = 0 \implies \frac{1}{s(x)} \, v'(x) = \alpha$$

$$\implies v'(x) = \alpha \, s(x) \Rightarrow v(x) = \alpha \, S(x) + \beta$$

for some constants α, β defined by the boundary conditions

$$\left. \begin{aligned}
v(c_1) &= \alpha \, S(c_1) + \beta = h(c_1) \\
v(c_2) &= \alpha \, S(c_2) + \beta = h(c_2)
\end{aligned} \right\} \Rightarrow \begin{cases}
\alpha &= \frac{h(c_2) - h(c_1)}{S(c_2) - S(c_1)} \\
\beta &= h(c_1) - \frac{h(c_2) - h(c_1)}{S(c_2) - S(c_1)} \, S(c_1).
\end{cases}$$

We conclude that

$$v(x) = \mathbb{E} \left(h(X_{T_D}) \mid X_0 = x \right) = h(c_1) + \frac{S(x) - S(c_1)}{S(c_2) - S(c_1)} \, (h(c_2) - h(c_1)).$$

In particular, when $h = 1_{c_2}$ we find that

$$v(x) = \mathbb{P}\left(X_{T_D} = c_2 \mid X_0 = x\right) = \frac{S(x) - S(c_1)}{S(c_2) - S(c_1)}.$$

In addition, we have

$$\sigma = 1 \quad b = 0 = c \implies s(x) = 1 \Rightarrow S(x) = x \Rightarrow v(x) = \mathbb{P}\left(W_{T_D} = c_2 \mid X_0 = x\right) = \frac{x - c_1}{c_2 - c_1}$$

where W_t stands for a Brownian motion on $\mathbb{R}$.

15.7.2 A backward stochastic differential equation

Suppose that the reference process X_t of the Feynman-Kac semigroup (15.50) is a one-dimensional diffusion with generator (14.10). In this situation, applying theorem 14.2.1 to the function $v_s(x) = Q_{s,t}(f_t)(x)$ (with a fixed time horizon t) and recalling that

$$L_s^V(v_s) = L_s(v_s) - v_s V_s \Rightarrow (\partial_s + L_s)\, v_s = -L_s^V(v_s) + L_s(v_s) = v_s V_s,$$

we find that

$$\begin{aligned} dv_s(X_s) &= (\partial_s + L_s)\, v_s(X_s)\, ds + \partial_x v_s(X_s)\, \sigma_s(X_s)\, dW_s \\ &= V_s(X_s)\, v_s(X_s)\, ds + \partial_x v_s(X_s)\, \sigma_s(X_s)\, dW_s. \end{aligned}$$

Thus, if we set
$$Y_s = v_s(X_s) \quad \text{and} \quad U_s := \partial_x v_s(X_s),$$

we obtain the backward stochastic differential equation

$$\forall s \in [0, t] \qquad dY_s = V_s(X_s)\, Y_s\, ds + \sigma_s(X_s)\, U_s\, dW_s$$

with the terminal condition $Y_t = f_t(X_t)$.

15.8 Exercises

Exercise 267 (Levy processes) *We consider a Poisson process N_t with intensity $\lambda > 0$, a Brownian motion W_t starting at the origin, and a sequence of independent and identically distributed random variables $Y = (Y_n)_{n \geq 1}$ with common distribution μ on $\mathbb{R}$. We assume that (N_t, W_t, Y) are independent. Propose a simple way to sample the Levy process*

$$X_t = a\, t + b\, V_t + c\, W_t \quad \text{with} \quad V_t := \sum_{1 \leq n \leq N_t} Y_n$$

for some given parameters (a, b, c) on a time mesh $t_0 = 0 \leq t_1 < \ldots < t_m \leq t$ of a finite horizon interval $[0, t]$. Compute the characteristic function $\phi_{X_t}(u) := \mathbb{E}(e^{iuX_t})$ in terms of $\phi_Y(u) := \mathbb{E}(e^{iuY})$ and the parameters (a, b, c, t).

Exercise 268 (Exponential Levy processes) *We consider the Levy process X_t discussed in exercise 267 and we set $Z_t = Z_0 \, e^{X_t}$, for some random variable Z_0 independent of X_t. Check that*

$$dZ_t = Z_t \, \left(a + \frac{c^2}{2} \right) \, dt + c \, Z_t \, dW_t + Z_t \, dU_t \quad \text{with} \quad U_t = \sum_{1 \leq n \leq N_t} \left(e^{bY_n} - 1 \right).$$

Exercise 269 (Generators of Levy processes) *Describe the infinitesimal generators L_t^X and L_t^Z of the Levy processes X_t and Z_t discussed in exercises 267 and 268.*

Exercise 270 (Doeblin-Itō formula) *Consider the $\mathbb{R}$-valued jump-diffusion process given by*

$$dX_t \; = \; a_t(X_t) \, dt + b_t(X_t) \, dW_t + c_t(X_t) \, dN_t$$

with some regular functions a_t, b_t, c_t on $\mathbb{R}$. In the above display, N_t is a Poisson process with intensity $\lambda_t(X_t)$ and W_t is a Brownian motion (starting at the origin). We assume that these processes are independent. Write the Doeblin-Itō formula for a smooth function $f(t, X_t)$.

Exercise 271 (Generator of jump diffusion processes) ✐ *Consider the $\mathbb{R}^r$-valued jump diffusion process given by*

$$dX_t \; = \; a_t(X_t) \, dt + b_t(X_t) \, dW_t + c_t(X_t) \, dN_t$$

with some regular functions a_t on $\mathbb{R}^r$, and $(r \times r)$-matrices $b_t(x), c_t(x)$. In the above display, W_t is an r-dimensional Brownian motion (starting at the origin), and $N_t = (N_t^i)_{1 \leq i \leq r}$ is a column vector of independent Poisson processes N_t^i with intensity $\lambda_t^i(X_t)$, in the sense that the exponential random variables (15.18) governing their jumps are independent. We assume that these exponential random variables are independent of W_t. Write the generator of the process X_t.

Exercise 272 (Variance - Local carré du champ decompositions) *Let X_t be a Markov process with generator L_t on some state space S. We let $P_{s,t}$ the semigroup of X_t, with $s \leq t$, and we assume that $P_{s,t}$ satisfies the forward and backward evolution equations (15.16). We fix a time horizon $t \geq 0$. We let $\eta_t = \mathrm{Law}(X_t)$. Check that the process $s \in [0, t] \mapsto M_s := P_{s,t}(f)(X_s)$ is a martingale w.r.t. $\mathcal{F}_s = \sigma(X_r, \ r \leq s)$. Deduce that for any sufficiently regular functions f we have*

$$\eta_t \left[(f - \eta_t(f))^2 \right] = \eta_0 \left[(P_{0,t}(f) - \eta_0 \left[P_{0,t}(f) \right])^2 \right] + \int_0^t \eta_s \left(\Gamma_{L_s}(P_{s,t}(f), P_{s,t}(f)) \right) \, ds.$$

Exercise 273 (Law of large numbers - Carré du champ operators) ✐ *Consider a Markov process X_t with a generator L_t on some state space S. We let $P_{s,t}$, with $s \leq t$ be the Markov semigroup of X_t and we set $\eta_t = \mathrm{Law}(X_t)$. We let $\xi_t := (\xi_t^i)_{1 \leq i \leq N}$ be N independent copies of X_t and we set $\mathcal{X}_t := \sum_{1 \leq i \leq N} \delta_{\xi_t^i}$. Describe the Doeblin-Itō formula (15.22) to the Markov process $\xi_t \in S^N$ on functions of the form $f(t, \xi_t) = F(t, \mathcal{X}_t(\varphi))$, where φ is a regular function on S and $F(t, u)$ a regular function from $[0, \infty[\times \mathbb{R}$ into $\mathbb{R}$.*

- *Prove that*
$$\mathbb{E} \left(\left[N^{-1} \, \mathcal{X}_t(\varphi) - \eta_t(\varphi) \right]^2 \right) = N^{-1} \, \eta_t \left(\left[\varphi - \eta_t(\varphi) \right]^2 \right).$$

- *Check that*

$$d\mathcal{X}_t(\varphi) = \mathcal{X}_t(L_t(\varphi)) \, dt + dM_t^{(1)}(\varphi)$$

with a martingale $M_t^{(1)}(\varphi)$ w.r.t. $\mathcal{F}_t = \sigma(\xi_s, \ s \leq t)$ with angle bracket

$$\partial_t \left\langle M^{(1)}(\varphi), M^{(1)}(\varphi) \right\rangle_t = \mathcal{X}_t \left(\Gamma_{L_t}(\varphi, \varphi) \right).$$

- *Applying the Doeblin-Itō formula to the function $f(s, x) = \sum_{1 \leq i \leq N} P_{s,t}(\varphi)(x_i)$ w.r.t. $s \in [0, t]$ prove that $M_s^{(2)} := \mathcal{X}_s(P_{s,t}(\varphi))$ is a martingale w.r.t. $\mathcal{F}_s$, with angle bracket*

$$\partial_s \left\langle M^{(2)}(\varphi), M^{(2)}(\varphi) \right\rangle_s = \mathcal{X}_s \left(\Gamma_{L_s}(P_{s,t}(\varphi), P_{s,t}(\varphi)) \right).$$

Deduce that

$$\mathbb{E}\left([\mathcal{X}_t(\varphi) - \mathcal{X}_0(P_{0,t}(\varphi))]^2 \right) = N \, \mathbb{E}\left([\varphi(X_t) - P_{0,t}(\varphi)(X_0)]^2 \right)$$

and

$$\mathbb{E}\left([\mathcal{X}_t(\varphi) - N \, \eta_t(\varphi)]^2 \right) = N \, \eta_t \left([\varphi - \eta_t(\varphi)]^2 \right).$$

- *Check that*

$$d\left(\mathcal{X}_t(\varphi)\right)^2 = [2 \, \mathcal{X}_t(\varphi) \, \mathcal{X}_t(L_t(\varphi)) + \mathcal{X}_t \left(\Gamma_{L_t}(\varphi, \varphi) \right)] \, dt + dM_t^{(3)}(\varphi)$$

with a martingale $M_t^{(3)}(\varphi)$.

- *Applying the Doeblin-Itō formula to the function $f(s, x) = \left(\sum_{1 \leq i \leq N} P_{s,t}(\varphi)(x_i) \right)^2$ w.r.t. $s \in [0, t]$, prove that*

$$d\left(\mathcal{X}_s P_{s,t}(\varphi)\right)^2 = \mathcal{X}_s \left(\Gamma_{L_s}(P_{s,t}(\varphi), P_{s,t}(\varphi)) \, ds + dM_s^{(4)}(\varphi) \right)$$

with a martingale $M_t^{(4)}(\varphi)$. Deduce that

$$\mathbb{E}\left((\mathcal{X}_t(\varphi))^2 \right) - \mathbb{E}\left((\mathcal{X}_0 P_{0,t}(\varphi))^2 \right) = N \, \mathbb{E}\left([\varphi(X_t) - P_{0,t}(\varphi)(X_0)]^2 \right)$$

and

$$\mathbb{E}\left((\mathcal{X}_t(\varphi))^2 \right) = N^2 \, (\eta_t(\varphi))^2 + N \, \eta_t \left([\varphi - \eta_t(\varphi)]^2 \right).$$

Exercise 274 (Coupled jump diffusion processes - 1) *We consider a jump diffusion process $X_t := (X_t^1, X_t^2) \in (\mathbb{R}^{r_1} \times \mathbb{R}^{r_2})$ defined by the stochastic differential equations*

$$\begin{cases} dX_t^1 &= a_t^1(X_t^1) \, dt + b_t^1(X_t^1) \, dW_t^1 + c_t^1(X_t^1) \, dN_t^1 \\ dX_t^2 &= a_t^2(X_t) \, dt + b_t^2(X_t) \, dW_t^2 + c_t^2(X_t) \, dN_t^2 \end{cases}$$

with r_i-dimensional Brownian motions W_t^i and Poisson processes N_t^i with intensity $\lambda_t^1(X_t^1)$, respectively $\lambda_t^2(X_t)$. with some column vectors $a_t^1(x^1)$, $a_t^2(x)$, $c_t^1(x^1)$, $c_t^2(x)$ and matrices $b_t^1(x^1)$, $b_t^2(x)$ with appropriate dimensions. The Brownian motions and the exponential random variables (15.18) governing the Poisson jumps are assumed to be independent. Write the generator of the processes X_t^1 and X_t.

Exercise 275 (Coupled jump diffusion processes - 2) ✐ *We consider the jump diffusion process* $X_t := (X_t^1, X_t^2) \in (\mathbb{R}^{r_1} \times \mathbb{R}^{r_2})$ *discussed in exercise 274. We assume that* $N_t^1 = (N_t^{1,j})_{1 \le i \le r_1}$, *respectively* $N_t^2 = (N_t^{2,j})_{1 \le i \le r_2}$, *is a column vector with independent Poisson entries with intensities* $\lambda_t^{1,j}(X_t^1)$, *respectively* $\lambda_t^{2,j}(X_t)$. *In this situation,* $c_t^1(x^1)$, *and* $c_t^2(x)$ *are* $(r_1 \times r_1)$-matrices and respectively $(r_2 \times r_2)$-matrices. Write the generator of the processes X_t^1 and X_t.

Exercise 276 (Flashing diffusions - Generator) *We let* X_t *be a homogeneous pure jump process taking values in* $\{0,1\}$. *Given* X_t *we let* Y_t *be the* r-dimensional diffusion process

$$dY_t = b_t(Y_t) \; dt + X_t \; \sigma_t(Y_t) \; dW_t.$$

In the above display, $b_t(y)$ *and* σ_t *denote a smooth* r-column vector and a smooth $(r \times r)$-matrix, respectively, and W_t is an r-dimensional Brownian motion, independent of X_t. Write the generator of the processes X_t and $Z_t = (X_t, Y_t)$. As mentioned in [185, 235], these flashing diffusion models appear in transport phenomena in sponge-type structures.

Exercise 277 (Flashing diffusions - Fokker-Planck equation) ✐ *We consider the flashing diffusion process* $Z_t := (X_t, Y_t) \in (\{0,1\} \times \mathbb{R}^r)$ *discussed in exercise 276. Assume that* (X_t, Y_t) *has a density given by*

$$\forall x \in \{0,1\} \qquad \mathbb{P}(X_t = x \; , \; Y_t \in dy) = p_t(x,y) \; dy$$

where dy *stands for the Lebesgue measure on* $\mathbb{R}^r$. *Describe the (coupled) evolution equation of the densities* $p_t(0,y)$ *and* $p_t(1,y)$.

Exercise 278 (Merton's jump diffusion model) *We let* N_t *be a Poisson process with intensity* λ, *and* $\{Y_n\}$ *be a sequence of i.i.d. random copies of a real valued random variable* Y. *Consider the compound Poisson process* $Z_t = \sum_{1 \le n \le N_t}(Y_n - 1)$, *and the* $\mathbb{R}$-valued jump-diffusion process given by

$$dX_t \; = \; a \; X_t \; dt \; + \; b \; X_t \; dW_t + X_t \; dZ_t$$

with some parameters $a, b \in \mathbb{R}$. *We assume that* Y_n, N_t *and the Brownian motion* W_t *are independent. Compute the exact solution of this stochastic differential equation. We assume that* $Y_n = e^{V_n}$ *for a sequence* V_n *of i.i.d. random variables. Check that*

$$X_t = X_0 \; \exp\left(\left[a - \frac{b^2}{2}\right] \; t + b \; W_t + \sum_{1 \le n \le N_t} V_n\right).$$

Exercise 279 (Integration by parts) *We let* N_t *be a Poisson process with intensity* $\lambda > 0$, W_t *be Brownian motion, and* $a_t^{(i)}(x), b_t^{(i)}(x), c_t^{(i)}(x)$ *some given functions. We consider the stochastic process* $X_t = (X_t^{(1)}, X_t^{(2)})$ *defined for any* $i = 1, 2$ *by*

$$\left\{ \begin{array}{rcl} dX_t^{(i)} & = & a_t^{(i)}(X_t) \; dt + b_t^{(i)}(X_t) \; dN_t + c_t^{(i)}(X_t) \; dW_t \\ i & = & 1, 2. \end{array} \right.$$

Check that

$$X_t^{(1)} X_t^{(2)} = X_0^{(1)} X_0^{(2)} + \int_0^t X_s^{(1)} \; dX_s^{(2)} + \int_0^t X_s^{(2)} \; dX_s^{(1)} \; + \; \left[X^{(1)}, X^{(2)}\right]_t + \langle X^{(1)}, X^{(2)}\rangle_t$$

with the processes $\left[X^{(1)}, X^{(2)}\right]_t$ *and* $\langle X^{(1)}, X^{(2)}\rangle$ *defined by*

$$\left[X^{(1)}, X^{(2)}\right]_t = \int_0^t b_s^{(1)}(X_s) b_s^{(2)}(X_s) \; dN_s \quad and \quad \langle X^{(1)}, X^{(2)}\rangle = \int_0^t c_s^{(1)}(X_s) c_s^{(2)}(X_s) \; ds.$$

Exercise 280 (Diffusions on circles) *We consider a Poisson process N_t with intensity $\lambda > 0$ and a Brownian motion W_t starting at the origin. We let $Z_t := (X_t, Y_t)$ be the two-dimensional diffusion given by the stochastic differential equations*

$$\begin{cases} dX_t &= -X_t \, dt + \sqrt{2} \, Y_t \, dW_t + a(Z_t) \, X_t \, dN_t \\ dY_t &= -Y_t \, dt - \sqrt{2} \, X_t \, dW_t + a(Z_t) \, Y_t \, dN_t \end{cases}$$

for some function a from $\mathbb{R}^2$ into $\mathbb{R}$. We let $\|Z_t\| = \sqrt{X_t^2 + Y_t^2}$. Check that

$$d\|Z_t\|^2 = \|Z_t\|^2 \left[(1 + a(Z_t))^2 - 1 \right] dN_t.$$

When

$$a(Z_t) = -1 + \epsilon \, \sqrt{1 + b(Z_t)/\|Z_t\|^2}$$

for some $\epsilon \in \{-1, +1\}$ and some function b from $\mathbb{R}^2$ into $\mathbb{R}$, prove that

$$d\|Z_t\|^2 = b(Z_t) \, dN_t$$

for any $Z_t \neq (0,0)$. Deduce that that the process Z_t evolves on circles between the jump times of N_t.

Exercise 281 (Regularity properties) ✒ *We let $X_t = X_0 + \sigma \, W_t$, for some initial state X_0, some parameter $\sigma > 0$ and a one-dimensional Brownian motion. We consider a non-necessarily smooth function f_t on $\mathbb{R}$ and we fix some final time horizon t. Check that the function*

$$u_s(x) = \mathbb{E}\left[f_t(X_t) \mid X_s = x \right] = \mathbb{E}\left[f_t \left(x + \sigma\sqrt{t - s} \, W_1 \right) \right]$$

is smooth for any $s < t$ and it satisfies the backward equation

$$\partial_s u_s + \frac{1}{2}\sigma^2 \partial_x^2 u_s = 0$$

for any $s \in [0, t]$, with the terminal condition $u_t = f_t$.

Exercise 282 (Martingale design) *We let X_t be some stochastic process on some state space S with generator L_t. For ay regular functions $f(t, x)$ and $g(t, x)$ check that*

$$M_t \;:=\; f(t, X_t)g(t, X_t) - f(0, X_0)g(t, X_0)$$
$$- \int_0^t \left[f(s, \cdot) \left[\partial_s + L_s \right] g(s, \cdot) + g(s, \cdot) \left[\partial_s + L_s \right] f(s, \cdot) + \Gamma_{L_s}(f(s, \cdot), g(s, \cdot)) \right](X_s) \, ds$$

is a martingale (w.r.t. $\mathcal{F}_t = \sigma(X_s, \ s \leq t)$). Deduce that

$$f(t, X_t) \, e^{-\lambda t} - f(0, X_0) + \int_0^t e^{-\lambda s} \left[\lambda \, f(s, \cdot) - \left[\partial_s + L_s \right] f(s, \cdot) \right](X_s) \, ds$$

is a martingale for any $\lambda \in \mathbb{R}$.

Exercise 283 (Feynman-Kac martingales) *We consider the Feynman-Kac semigroup (15.33) associated with some Markov process X_t with a generator L_t on some state space S, and some bounded and non-negative functions V_t on S. We let $f(t, x)$ be some regular function on $([0, \infty[\times S$ (the required regularity depends on the nature of the generator L_t; for instance second order differential operators L_t act on twice differentiable functions on $S = \mathbb{R}^d$) and we set*

$$Z_{s,t}(f) := e^{-\int_s^t V_r(X_r)dr} \, f(t, X_t) - f(s, X_s) - \int_s^t e^{-\int_s^r V_u(X_u)du} \, \left(\partial_r + L_r^V \right) f(r, X_r) \, dr$$

for any $s \leq t$. Check that for any s the process $(Z_{s,t})_{t \in [s,\infty]}$ is a martingale w.r.t. $\mathcal{F}_t :=$
$\sigma(X_r, \; s \leq t)$. Deduce that

$$\partial_t \left(P_{s,t}^V(f(t, \cdot)) \right) = P_{s,t}^V((\partial_t + L_t^V)f(t, \cdot)).$$

Exercise 284 (Semigroup series expansions - 1) *Consider the perturbation model discussed in section 15.5.2. Using (15.25) check that for any $r \leq s_0$ we have*

$$P_{r,s_0} = \sum_{n \geq 0} \int_r^{s_0} \cdots \int_r^{s_{n-1}} P_{r,s_n}^{(1)} L_{s_n}^{(2)} P_{s_n,s_{n-1}}^{(1)} \cdots L_{s_1}^{(2)} P_{s_1,s_0}^{(1)} \; ds_n \; \cdots \; ds_1.$$

By convention, the term of order $n = 0$ in the above series is equal to $P_{r,s_0}^{(1)}$.

Exercise 285 (Semigroup series expansions - 2) *Consider the perturbation model discussed in section 15.5.2 with the jump generator (15.26). Check that*

$$P_{s,t} = Q_{s,t} + \int_s^t Q_{s,r} \overline{K}_r P_{r,t} \; dr$$

with the Feynman-Kac semigroup $Q_{s,t}$ defined in (15.27) and the integral operator $\overline{K}_r(f) = \lambda_r K_r(f)$. Check that for any $r \leq s_0$ we have

$$P_{s_0,t} = \sum_{n \geq 0} \int_{s_0}^t \cdots \int_{s_{n-1}}^t Q_{s_0,s_1} \overline{K}_{s_1} Q_{s_1,s_2} \cdots \overline{K}_{s_n} Q_{s_n,t} \; ds_n \; \cdots \; ds_1.$$

By convention, the term of order $n = 0$ in the above series is equal to $Q_{s_0,t}$.

Exercise 286 (Semigroup series expansions - 3) *Consider the perturbation model discussed in exercise 285 with a spatially homogeneous jump rate function $\lambda_t(x) = \lambda_t$. Prove that for any $r \leq s_0$ we have*

$$P_{s_0,t} = \sum_{n \geq 0} \lambda^n \; e^{-\lambda(t-s_0)} \int_{s_0}^t \cdots \int_{s_{n-1}}^t P_{s_0,s_1}^{(1)} K_{s_1} P_{s_1,s_2}^{(1)} \cdots K_{s_n} P_{s_n,t}^{(1)} \; ds_n \; \cdots \; ds_1.$$

Exercise 287 (Feynman-Kac semigroup series expansions) *Consider a Feynman-Kac semigroup $Q_{s,t}$ associated with some reference Markov process X_t with sg $P_{s,t}$, and potential functions W_t on some state space S; that is, for any $s \leq t$ we have*

$$Q_{s,t}(f)(x) = \mathbb{E}\left(f(X_t) \; \exp\left(\int_s^t W_u(X_u) du \right) \mid X_s = x \right).$$

We denote by $\overline{W}_t$ the multiplication operator $\overline{W}_t(f) = W_t f$. Arguing as in the proof of (15.34) check that

$$Q_{s,t} = P_{s,t} + \int_s^t Q_{s,s_1} \overline{W}_{s_1} P_{s_1,t} \; ds_1.$$

Prove that for any $r \leq s_0$ we have

$$Q_{r,s_0} = \sum_{n \geq 0} \int_r^{s_0} \cdots \int_r^{s_{n-1}} P_{r,s_n} \overline{W}_{s_n} P_{s_n,s_{n-1}} \cdots \overline{W}_{s_1} P_{s_1,s_0} \; ds_n \ldots ds_1.$$

By convention, the term of order $n = 0$ in the above series is equal to $P_{s_0,t}$.

Exercise 288 (Feynman-Kac martingales - Switching processes [121]) *We let $S_1 = \mathbb{Z}/m\mathbb{Z} = \{0, 1, \ldots, m-1\}$ be the cyclic group equipped with the addition $i+j \bmod(m)$, and S_2 some state space. We consider a potential function $v : i \in S_1 \mapsto v(i) \in \mathbb{R}$. We extend v to the product space $S = (S_1 \times S_2)$ by setting $V(i, y) = v(i)$, for any $x = (i, y) \in (S_1 \times S_2)$. Let $X_t = (I_t, Y_t)$ be a Markov process on the product space $S = (S_1 \times S_2)$ with infinitesimal generator defined for any regular functions f on $(S_1 \times S_2)$ and any $x = (i, y) \in (S_1 \times S_2)$ by*

$$L(f)(x) = \mathcal{L}_i(f(i, \cdot))(y) + w(i) \ (f(i+1, y) - f(i, y))$$

where $\mathcal{L}_i$ stands for a collection of generators indexed by $i \in S_1$, and $w(i) \geq 0$ some rate functions on S_1. We set

$$\langle v, \overline{I}_t \rangle := \sum_{i \in S_1} v(i) \ \overline{I}_t^i(i) \quad \text{with} \quad \overline{I}_t^i(i) := \int_0^t 1_{I_s = i} \ ds$$

and

$$\mathcal{L}_i^{v-w}\left(f(i, \cdot)\right) := \mathcal{L}_i(f(i, \cdot))(y) + (v(i) - w(i)) \ f(i, y).$$

Check that

$$\langle v, \overline{I}_t \rangle = \int_0^t V(X_s) \ ds$$

and deduce that the process

$$
\begin{aligned}
Z_t(f) = \ & e^{\langle v, \overline{I}_t \rangle} \ f(X_t) - f(X_0) - \sum_{i \in S_1} \int_0^t e^{\langle v, \overline{I}_s \rangle} \\
& \times \left[w(i) \ f(i+1, Y_s) + L^{v-w} f(i, Y_s) \right] 1_{I_s = i} \ ds
\end{aligned}
$$

is a martingale w.r.t. $\mathcal{F}_t := \sigma(X_r, \ s \leq t)$.

Exercise 289 (Feynman-Kac switching processes [121]) *We consider the Feynman-Kac switching process discussed in exercise 288. We let $f(i, y)$, $0 \leq i < m$ be the collection of functions defined by the induction*

$$
\begin{aligned}
f(0, \cdot) &:= g \\
w(0) \ f(1, \cdot) &:= -L^{v-w} f(0, \cdot) \\
&\vdots \quad = \quad \vdots \\
w(m-2) \ f(m-1, \cdot) &:= -L^{v-w} f(m-2, \cdot) = (-1)^{m-1} \left(\mathcal{L}_{m-2}^{v-w} \ldots \mathcal{L}_0^{v-w} \right) (g)
\end{aligned}
$$

for some initial condition g. We let h be the function defined by

$$h := -w(m-1) \ f(0, \cdot) - L^{v-w} f(m-1, \cdot) = (-1)^m \left(\mathcal{L}_{m-1}^{v-w} \ldots \mathcal{L}_0^{v-w} \right) (g) - w(m-1) \ g.$$

Check that

$$e^{\langle v, \overline{I}_t \rangle} \ f(I_t, Y_t) - f(I_0, Y_0) + \int_0^t e^{\langle v, \overline{I}_s \rangle} \ h(Y_s) \ 1_{I_s = m-1} \ ds$$

is a martingale (w.r.t. $\mathcal{F}_t := \sigma(X_r, \ s \leq t)$).

Exercise 290 (Diffusion exit times) *Consider the diffusion processes X_t with generator $L_t = L_t^c$ given in (15.13), starting at $X_0 \in D$ for some bounded open subset D of $\mathbb{R}^d$. We let T_D the first time X_t exits D. Assume that for any $x = (x_i)_{1 \leq i \leq d} \in D$, any $y = (y_i)_{1 \leq i \leq d} \in \mathbb{R}^d$ and any $t \geq 0$ we have*

$$\sum_{1 \leq i,j \leq d} \left(\sigma_t(x) \sigma_t^T(x) \right)_{i,j} y_i \, y_j \; \geq \; \rho \, \|y\|_1^2$$

for some ρ, with the $\mathbb{L}_1$-norm $\|y\|_1^2 = \sum_{1 \leq i \leq d} |y_i|$. Also assume that

$$\|b\|_D = \max_{1 \leq i \leq d} \sup_{t \geq 0} \sup_{x \in D} |b_t^i(x)| < \infty \quad \text{and set} \quad w_{c,y}(x) := -c \, \exp \left[-\sum_{1 \leq i \leq d} y_i \, x_i \right].$$

for some non-negative parameters c and $(y_i)_{1 \leq i \leq d}$, and for any $x \in D$

- *Check that*

$$-L_t(w_{c,y})(x) \; \geq \; c \, e^{-\|y\|_1 \delta} \, \|y\|_1 \, [\rho \|y\|_1 - \|b\|_D]$$

with $\delta = \max_{1 \leq i \leq d} \sup_{x \in D} x_i$.

- *Prove that*

$$\|y\|_1 = \rho^{-1} \left[1 + \|b\|_D \right] \quad \text{and} \quad c = e^{\|y\|_1 \delta} \, \|y\|_1^{-1} \Rightarrow -L_t(w_{c,y})(x) \geq 1.$$

Using lemma 15.6.2 deduce that $\sup_{x \in \mathbb{R}^d} \mathbb{E}(T_D \mid X_0 = x) \leq \mathrm{osc}_D(w_{c,y})$.

Exercise 291 (Martingale stopping) *Let M_t be a martingale w.r.t. some filtration $\mathcal{F}_t$ with angle bracket $\langle M, M \rangle_t$ s.t. $|\langle M, M \rangle_t - \langle M, M \rangle_s| \, 1_{s \vee t \leq T} \leq c \, |t - s|$, for some constant c, and some stopping time T s.t. $\mathbb{E}(T) < \infty$. Check that $M_{t \wedge T}$ is a Cauchy sequence in $\mathbb{L}_2(\mathbb{P})$ and deduce that $M_{t \wedge T}$ converges in $\mathbb{L}_2(\mathbb{P})$ to M_T and $\mathbb{E}(M_T) = 0$. Apply this result to the martingale*

$$M_t = \int_0^t X_s \, dW_s$$

where W_t is a one-dimensional Brownian motion and X_s some real valued stochastic process adapted to $\mathcal{F}_t = \sigma(W_s, \ s \leq t)$ (in the sense that X_t is an $\mathcal{F}_t$-measurable random variable) s.t. $|X_{t \wedge T}| \leq c$ for some finite constant c.

Exercise 292 (Cauchy problem with terminal condition) *We consider the semigroup $P_{s,t}$ of a jump diffusion process with generator L_t on $\mathbb{R}^r$, for some $r \geq 1$. We let $f \ : \ (t, x) \in ([0, \infty[\times \mathbb{R}^r) \mapsto f_t(x) \in \mathbb{R}$ and $g \ : \ (t, x) \in ([0, \infty[\times \mathbb{R}^r) \mapsto g_t(x) \in \mathbb{R}$ be some smooth and bounded functions. We fix the time horizon t and for any $s \in [0, t]$ we set*

$$v_s(x) = P_{s,t}(f_t)(x) + \int_s^t P_{s,u}(g_u)(x) \, du = \mathbb{E} \left(f(X_t) + \int_s^t g_u(X_u) \, du \mid X_s = x \right).$$

Using the backward equation (15.16) check that v_s satisfies the integro-differential equation

$$\forall s \in [0, t] \qquad \partial_s v_s + L_s v_s + g_s = 0$$

with terminal condition $v_t = f_t$.

Exercise 293 (Cauchy problem with initial condition) *We consider a time homogeneous Markov process X_t with generator L on some state space S. We let f and g be a couple of regular functions on S. Check that*

$$w_t(x) := \mathbb{E}\left(f(X_t) + \int_0^t g_s(X_{t-s})\, ds \mid X_0 = x \right)$$

satisfies the equation

$$\partial_t w_t = L w_t + g_t$$

with the initial condition $w_0 = f$.

Exercise 294 (Killed processes - Feynman-Kac semigroups) *We consider the Feynman-Kac semigroup $Q_{s,t} := P_{s,t}^V$ presented in (15.33), and we let $\eta_0 = \mathrm{Law}(X_0)$. We introduce the positive measures γ_t and their normalized version η_t defined for any bounded function f on $\mathbb{R}^d$ by*

$$\eta_t(f) = \gamma_t(f)/\gamma_t(1) \quad and \quad \gamma_t(f) := \mathbb{E}\left(f(X_t) \, \exp\left(-\int_0^t V_u(X_u)du \right) \right).$$

Check that

$$\forall s \leq t \qquad \gamma_t = \gamma_s Q_{s,t} \quad and \quad \gamma_t(1) := \exp\left(-\int_0^t \eta_s(V_s)\, ds \right).$$

Deduce that for any bounded and smooth function f on $\mathbb{R}^d$ we have

$$\partial_t \eta_t(f) = \eta_t(L_t(f)) + \eta_t(f)\eta_t(V_t) - \eta_t(fV_t).$$

Exercise 295 (Embedded Markov chains) *Consider the processes $(X_t, X_t^{(1)}, X_t^{(2)})$ discussed in section 15.5.2 with the generator of $X_t^{(2)}$ given by the jump generator (15.26). Between these jump times the process evolves (as $X_t^{(1)}$) according to the generator $L_t^{(1)}$. We let T_n be the jump times of the process X_t arriving at rate $\lambda_t(X_t)$. We assume that $T_n \uparrow \infty$ (a.k.a. no explosion). Describe the Markov transition $\mathcal{K}$ of the Markov chain $Y_n = (T_n, X_{T_n})$.*

For time homogeneous models check that $\mathcal{K}((s,x), d(t,y))$ does not depend on the parameter s. Describe the Markov transition $\mathcal{M}$ of the Markov chain $Z_n = X_{T_n}$. Examine the time homogeneous situation with a constant rate function $\lambda(x) = \lambda > 0$.

Exercise 296 (Embedded Markov chain - Stability) *Consider the time homogeneous embedded Markov chains discussed in exercise 295. Assume that $\lambda_\star \leq \lambda(x) \leq \lambda^\star$ for some positive and finite parameters $\lambda_\star$ and $\lambda^\star$. Also assume that $K(x, dy) \geq \epsilon \, \nu(dy)$ for some $\epsilon > 0$ and some probability measure ν on S. Check that the Markov transition $\mathcal{M}$ of the embedded chain $Z_n = X_{T_n}$ satisfies the minimization condition*

$$\mathcal{M}(x, dy) \geq (\lambda_\star/\lambda^\star)\, \nu(dy).$$

Deduce that the law of Z_n converges exponentially fast to a unique invariant measure $\pi = \pi \mathcal{M}$.

Exercise 297 (Embedded Markov chain - Invariant measures) *Consider the time homogeneous embedded Markov chains discussed in exercise 296. Consider the integral operator $\mathcal{Q}$ defined by*

$$\mathcal{Q} = \int_0^\infty Q_t \, dt \qquad \left(\Leftrightarrow \forall (f,x) \quad \mathcal{Q}(f)(x) = \int_0^\infty Q_t(f)(x) \, dt \right)$$

with the Feynman-Kac semigroup

$$Q_t(f)(x) = \mathbb{E}\left(f(X_t^{(1)}) \, \exp\left(-\int_0^t \lambda(X_s^{(1)}) \, ds \right) \mid X_0^{(1)} = x \right).$$

Check that $\mathcal{M}(f) = \mathcal{Q}(\lambda K(f))$. Using the evolution equations (15.31) check that $\mathcal{M} - Id = \mathcal{Q}L$ and deduce that the the invariant probability measure μ of X_t and the one π of X_{T_n} are connected by the formula $\mu(f) = \pi(\mathcal{Q}(f))/\pi(\mathcal{Q}(1))$.

Exercise 298 (Hit-and-run jump processes) *Let X_t be a Markov process evolving in some state space S. When X_t it some subset $D \subset S$ it jumps according to some Markov transition $K_t(x, dy)$ from D into $S - D$. We set $L_s^D = K_s - Id$. Between these jumps X_t evolves as a stochastic process with generator L_t. Let T_n be the n-th time the process X_t hits the set D, with the convention $T_0 = 0$, and we set $\mu^D := \sum_{n \geq 1} \delta_{T_n-}$. Prove that*

$$f(X_t) - f(X_0) - \int_0^t \left[L_s(f)(X_s) \, ds + L_s^D(f)(X_s) \, \mu^D(ds) \right]$$

is a martingale. For time homogeneous models $(L_t, K_t, L_t^D) = (L, K, L^D)$ and functions f s.t. $f = K(f)$, check that $f(X_t) - f(X_0) - \int_0^t L_s(f)(X_s) \, ds$ is a martingale.

Exercise 299 (Bessel functions) ✎ *We consider the generator L of the diffusion process*

$$dX_t = X_t \, dt + \sqrt{2} \, X_t \, dW_t$$

with a Brownian motion W_t. We consider the potential function $V(x) = (n - x)(n + x)$, for some integer $n \geq 0$. Check that a solution of the Poisson equation $L(v) = v \, V$ is given by the n-th Bessel function

$$B_n(x) := \left(\frac{x}{2} \right)^n \sum_{i \geq 0} \frac{(-1)^i}{i!(n+i)!} \left(\frac{x}{2} \right)^{2i}.$$

Exercise 300 (Average survival time) *We let W_t a one-dimensional Brownian motion (starting at the origin), and we set $X_t = X_0 + W_t$ for some initial condition X_0. Let D be an open and bounded subset of $\mathbb{R}$. We let T_D be the (first) time X_t exits the set D. Check that the solution of the Dirichlet-Poisson problem*

$$\begin{cases} \frac{1}{2}\partial_x^2 v(x) & = & -g(x) & \text{for any} \quad x \in D \\ v(x) & = & 0 & \text{for any} \quad x \in \partial D \end{cases}$$

is given by

$$v(x) = \mathbb{E}\left(\int_0^{T_D} g(X_s) \, ds \mid X_0 = x \right).$$

Deduce that the solution of

$$\begin{cases} \frac{1}{2}\partial_x^2 v(x) & = & -1 & \text{for any} \quad x \in D \\ v(x) & = & 0 & \text{for any} \quad x \in \partial D \end{cases}$$

is given by $v(x) = \mathbb{E}(T_D \mid X_0 = x)$.

Exercise 301 (Biharmonic functions) *Let D be an open and bounded subset of $\mathbb{R}^r$, for some $r \geq 1$, and (f, g) a couple of functions on the boundary ∂D. We also let W_t be an r-dimensional Brownian motion (starting at the origin), and we set $X_t = X_0 + W_t$ for some initial condition X_0. Check that*

$$N_t(u) = u(X_t) - u(X_0) - tL(u)(X_t) + \int_0^t s\, L^2(u)(X_s)\,ds \quad with \quad L := \frac{1}{2} \sum_{1 \leq i \leq r} \partial_{x_i}^2$$

is a martingale w.r.t. $\mathcal{F}_t = \sigma(X_s,\ s \leq t)$. We let T_D be the first time X_t exits D. Deduce that the solution of the problem

$$\begin{cases} L^2 u(x) &= 0 & if \quad x \in D \\ (u(x), Lu(x)) &= (f(x), g(x)) & if \quad x \in \partial D \end{cases}$$

is given by

$$u(x) = \mathbb{E}\left(f(X_{T_D}) - T_D\ g(X_{T_D}) \mid X_0 = x\right).$$

Exercise 302 (Brownian motion - Exit time distribution) *Let W_t be Brownian motion and $W_0 = 0 \in D =]-a, a[\subset \mathbb{R}$ for some $a > 0$. We let T_D^x be the first time $X_t := x + \sigma W_t$ exits the interval $D \ni x$. Using (15.50) check that the solution of the Cauchy problem*

$$\begin{cases} \partial_t v_t(x) &= 2^{-1}\sigma^2\, \partial_x^2 v_t(x) & for\ any \quad (t, x) \in ([0, \infty[\times]-a, a[) \\ v_0(x) &= 1 & for\ any \quad x \in]-a, a[\\ v_t(x) &= 0 & for\ any \quad x \in \{+a, -a\} \end{cases}$$

is given by

$$v_t(x) = \mathbb{P}(T_D > t \mid X_0 = x).$$

Check that the solution of the above problem is given by the formula

$$v_t(x) = \frac{4}{\pi} \sum_{n \geq 0} \frac{(-1)^n}{2n + 1}\ \cos\left((2n + 1)\ \frac{\pi}{2}\ \frac{x}{a}\right)\ \exp\left(-\frac{\sigma^2}{2}\left((2n + 1)\ \frac{\pi}{2}\ \frac{1}{a}\right)^2\ t\right).$$

Exercise 303 (Dirichlet and Neuman boundary conditions on an interval) *We consider the generator $L = \partial_x^2$ of a rescaled Brownian motion $X_t := \sqrt{2}\, W_t$. We let $D = [0, a]$ to be some compact interval for some $a > 0$.*

- *Check that*

$$v_n(x) := \sin(n\pi x/a) \Rightarrow L(v_n) = \lambda_n\, v_n \quad with \quad \lambda_n = -(n\pi/a)^2$$

and $v_n(x) = 0$ for any $x \in \partial D = \{0, a\}$, with $n \geq 1$.

- *Check that*

$$v_n(x) := \cos(n\pi x/a) \Rightarrow L(v_n) = \lambda_n\, v_n \quad with \quad \lambda_n = -(n\pi/a)^2$$

and $v_n'(x) = 0$ for any $x \in \partial D = \{0, a\}$, with $n \geq 1$.

Exercise 304 (Dirichlet and Neuman boundary conditions on a cell) *We consider the generator $L = \partial_{x_1}^2 + \partial_{x_2}^2$ of a rescaled Brownian motion $X_t := \sqrt{2}\, (W_t^1, W_t^2)$. We let $D = ([0, a_1] \times [0, a_2])$ be a cell in $\mathbb{R}^2$ for some $a_1, a_2 > 0$.*

- *Check that for any $n = (n_1, n_2) \in (\mathbb{N} - \{0\})^2$ and $x = (x_1, x_2) \in D$ we have*

$$v_n(x) := \sin(n_1 \pi x_1 / a_1) \ \sin(n_2 \pi x_2 / a_2)$$

$$\Rightarrow L(v_n) = \lambda_n \ v_n \quad \text{with} \quad \lambda_n = -\left[(n_1 \pi / a_1)^2 + (n_2 \pi / a_2)^2\right]$$

and $v_n(x) = 0$ for any $x \in \partial D$.

- *Check that for any $n = (n_1, n_2) \in (\mathbb{N} - \{0\})^2$ and $x = (x_1, x_2) \in D$ we have*

$$v_n(x) = \cos(n_1 \pi x_1 / a_1) \ \cos(n_2 \pi x_2 / a_2)$$

$$\Rightarrow L(v_n) = \lambda_n \ v_n \quad \text{with} \quad \lambda_n = -\left[(n_1 \pi / a_1)^2 + (n_2 \pi / a_2)^2\right]$$

and $\langle \nabla v(x), N^\perp(x) \rangle = 0$ for any $x \in \partial D$.

Exercise 305 (Heat equation - Boundary conditions - 1) *Let W_t be Brownian motion and $W_0 = 0 \in D =]0, a[\subset \mathbb{R}$ for some $a > 0$. We let T_D^x be the first time $X_t := x + \sigma \ W_t$ exits the interval $D \ni x$. Using (15.50) check that the solution of the Cauchy problem*

$$\begin{cases} \partial_t v_t(x) & = & 2^{-1}\sigma^2 \ \partial_x^2 v_t(x) & \text{for any} & (t, x) \in ([0, \infty[\times]0, a[) \\ v_0(x) & = & f & \text{for any} & x \in]0, a[\\ v_t(x) & = & 0 & \text{for any} & x \in \{0, a\}. \end{cases}$$

is given by

$$v_t(x) = \mathbb{E}\left(1_{T_D > t} \ f(X_t) \mid X_0 = x\right).$$

Check that the solution of the above problem is given by the formula

$$v_t(x) = \sum_{n \geq 1} \left[\frac{2}{a} \int_0^a f(x) \ \sin(n\pi x/a) \ dx\right] e^{\sigma^2 \lambda_n t / 2} \ \sin(n\pi x/a)$$

with $\lambda_n := -(n\pi/a)^2$. Examine the situations $f(x) = \sin(x)$ with $a = \pi$, and $f(x) = x(1 - x)$ with $a = 1$.

Exercise 306 (Heat equation - Boundary conditions - 2) *Consider the diffusion model discussed in exercise 305 with $a = 1$. We let Q_t be the Feynman-Kac semigroup*

$$Q_t(f)(x) = \mathbb{E}\left(f(X_t) \ 1_{T_D > t} \mid X_0 = x_0\right) = \int_0^1 q_t(x, y) \ f(y) \ dy$$

defined for any $x \in D$ and any function f. In the above display, we assumed implicitly that Q_t has a density function $q_t(x, y)$ w.r.t. the Lebesgue measure dy. Check that

$$\partial_t q_t(x, y) = \frac{1}{2}\partial_y^2 q_t(x, y)$$

with the boundary conditions $q_t(x, 0) = q_t(x, 1) = 0$. Check that

$$q_t(x, y) = \sum_{n \geq 1} \exp\left[-(n\pi\sigma)^2 \ t/2\right] \ \sin(n\pi x) \ \sin(n\pi y).$$

16

Nonlinear jump diffusion processes

The results from the previous chapter 15 are extended now to nonlinear Markov processes in section 16.1 and their mean field particle interpretations in section 16.2. The usefulness of these models is illustrated by some applications in systemic risk analysis as demonstrated in the description of the Fouque-Sun model and also in fluid mechanics and particle physics.

Great things are done by a series of small things brought together.
Vincent Van Gogh (1853-1890).

16.1   Nonlinear Markov processes

We let X_t be a jump diffusion process on $\mathbb{R}^d$, with the infinitesimal generator L_t defined in (15.13). We recall that $\eta_t = \mathrm{Law}(X_t)$ satisfies the equation (15.20). We further assume that the parameters

$$(b_t, \sigma_t, \lambda_t, K_t) = (\overline{b}_{t,\eta_t}, \overline{\sigma}_{t,\eta_t}, \overline{\lambda}_{t,\eta_t}, \overline{K}_{t,\eta_t})$$

depend on the distribution η_t. With a slight abuse of notation, we let L_{t,η_t} be the operator defined as L_t by replacing $(b_t, \sigma_t, \lambda_t, K_t)$ by $(\overline{b}_{t,\eta_t}, \overline{\sigma}_{t,\eta_t}, \overline{\lambda}_{t,\eta_t}, \overline{K}_{t,\eta_t})$. For instance, given some functions $\widetilde{b}_t$ and $\widetilde{\sigma}_t$ on $(\mathbb{R}^d)^2$, we can choose

$$\overline{b}_{t,\eta_t}(x) = \int \widetilde{b}_t(x,y)\, \eta_t(dy) \quad \text{and} \quad \overline{\sigma}_{t,\eta_t}(x) = \int \widetilde{\sigma}_t(x,y)\, \eta_t(dy). \tag{16.1}$$

We further assume that these functionals are sufficiently regular so that the process $\overline{X}_t$ with generator L_{t,η_t} is a well defined stochastic process on $\mathbb{R}^d$. In this context, $\eta_t = \mathrm{Law}(\overline{X}_t)$ satisfies the nonlinear evolution equation

$$\frac{d}{dt}\eta_t(f) = \eta_t(L_{t,\eta_t}(f)) \tag{16.2}$$

for any smooth function f on $\mathbb{R}^d$. The process $\overline{X}_t$ is often called a McKean interpretation of the nonlinear equation (16.2).

16.1.1   Pure diffusion models

For pure diffusion processes (that is when $\overline{\lambda}_{t,\eta_t} = 0$) with a drift and diffusion functionals $(\overline{b}_{t,\eta_t}, \overline{\sigma}_{t,\eta_t})$ given in (16.1), L_{t,η_t} is the infinitesimal generator of the nonlinear diffusion model $\overline{X}_t = \left(\overline{X}_t^i \right)_{1 \leq i \leq d}$ is given for any $1 \leq i \leq d$ by

$$d\overline{X}_t^i = \left[\int \widetilde{b}_t^i(\overline{X}_t, y) \; \eta_t(dy) \right] \; dt + \sum_{1 \leq j \leq d} \left[\int \widetilde{\sigma}_{j,t}^i(\overline{X}_t, y) \; \eta_t(dy) \right] \; dW_t^j. \qquad (16.3)$$

In the above formula, $\left(W^i \right)_{1 \leq i \leq d}$ is a d-dimension Brownian motion. In flip mechanics and applied probability literature, these models are often called McKean-Vlasov diffusions.

16.1.2 Burgers equation

We consider the nonlinear model (16.3) on $\mathbb{R}$ with

$$\widetilde{b}_t(x, y) \; = 1_{[x, \infty[}(y) \quad \text{and} \quad \widetilde{\sigma}_t(x, y) = \sigma > 0.$$

In this situation, the law of the random states

$$\mathbb{P}\left(\overline{X}_t \in dx \right) = p_t(y) \; dy$$

has a smooth density $p_t(y)$ w.r.t. the Lebesgue measure dy on $\mathbb{R}^d$. In addition, in this context p_t satisfies the Fokker-Planck equation (15.21) given by

$$\frac{\partial p_t}{\partial t} = -\partial_x \left(V_t \; p_t \right) + \frac{1}{2} \partial_{x,x} p_t$$

with

$$V_t(x) := \int \widetilde{b}_t(x, y) \; \eta_t(dy) = \int_x^\infty p_t(y) dy \implies \partial_x V_t = -p_t.$$

This implies that

$$
\begin{aligned}
\partial_t p_t &= -\partial_t \partial_x V_t = -\partial_x \; \partial_t V_t \\
&= -\partial_x \left(V_t \; p_t \right) + \frac{\sigma^2}{2} \partial_{x,x} p_t = \partial_x \left[V_t \; \partial_x V_t \; - \frac{\sigma^2}{2} \; \partial_{x,x} V_t \right].
\end{aligned}
$$

In other words, V_t satisfies the Burgers equation

$$\partial_t V_t = -V_t \; \partial_x V_t \; + \frac{\sigma^2}{2} \; \partial_{x,x} V_t. \qquad (16.4)$$

Our next objective is to provide a more explicit description of V_t. To this end, we recall that the solution of the heat equation

$$\partial_t q_t = \frac{\sigma^2}{2} \partial_{x,x} q_t$$

is given by the Gaussian density $q_t(x) = \frac{1}{\sqrt{2\pi\sigma^2 t}} \exp\left(-\frac{x^2}{2\sigma^2 t} \right)$. We denote by

$$U_t(x) := \int U_0(y) \; q_t(x - y) \; dy$$

the transport equation associated with the heat equation. Next, we show that

$$V_t = -\sigma^2 \partial_x \log(q_t)$$

is a solution of the Burgers equation (16.4). To check this claim, we observe that

$$V_t = -\sigma^2 \frac{\partial_x q_t}{q_t} \quad \Rightarrow \quad V_t \partial_x q_t + q_t \partial_x V_t = -\sigma^2 \partial_{x,x} q_t$$

$$\Rightarrow \quad V_t^2 - \sigma^2 \partial_x V_t = \sigma^4 \frac{\partial_{x,x} q_t}{q_t}$$

$$\Rightarrow \quad 2V_t \partial_x V_t - \sigma^2 \partial_{x,x} V_t = \sigma^4 \partial_x \left(\frac{\partial_{x,x} q_t}{q_t} \right).$$

We also notice that

$$\sigma^2 \partial_x \left(\frac{\partial_{x,x} q_t}{q_t} \right) = \sigma^2 \frac{\partial_{x,x,x} q_t}{q_t} - \sigma^2 \frac{\partial_x q_t}{q_t} \frac{\partial_{x,x} q_t}{q_t}$$

$$= \sigma^2 \frac{\partial_{x,x,x} q_t}{q_t} + V_t \frac{\partial_{x,x} q_t}{q_t} = -\frac{2}{\sigma^2} \partial_t V_t.$$

The last assertion follows from the fact that

$$\partial_t V_t = -\sigma^2 \partial_t \left(\frac{1}{q_t} \partial_x q_t \right) = \sigma^2 \frac{1}{q_t^2} \partial_t q_t \, \partial_x q_t - \sigma^2 \frac{1}{q_t} \partial_x \partial_t q_t$$

$$= \frac{\sigma^2}{2} \left(\frac{\partial_{x,x} q_t}{q_t} \sigma^2 \frac{\partial_x q_t}{q_t} - \sigma^2 \frac{\partial_{x,x,x} q_t}{q_t} \right)$$

$$= -\frac{\sigma^2}{2} \left(\frac{\partial_{x,x} q_t}{q_t} V_t + \sigma^2 \frac{\partial_{x,x,x} q_t}{q_t} \right).$$

This shows that $V_t = -\sigma^2 \partial_x \log(q_t)$ satisfies (16.4), so that $V_t = -\sigma^2 \partial_x \log(U_t)$ also satisfies the Burgers equation (16.4).

We also observe that

$$V_0 = -\sigma^2 \partial_x \log(U_0) \Rightarrow \partial_x U_0 = -\frac{1}{\sigma^2} V_0 U_0 \Rightarrow U_0(x) = U_0(0) \times \exp \left\{ -\frac{1}{\sigma^2} \int_0^x V_0(y) dy \right\}.$$

By a simple integration by part, this implies that

$$\partial_x U_t(x) = \int U_0(y) \, \partial_x q_t(x - y) \, dy$$

$$= -\frac{U_0(0)}{\sqrt{2\sigma^2 \pi t}} \int e^{-\frac{1}{\sigma^2} \int_0^y V_0(z) dz} \, \partial_y e^{-\frac{(x-y)^2}{2\sigma^2 t}} \, dy$$

$$= \frac{U_0(0)}{\sqrt{2\sigma^2 \pi t}} \int \partial_y \left(e^{-\frac{1}{\sigma^2} \int_0^y V_0(z) dz} \right) e^{-\frac{(x-y)^2}{2\sigma^2 t}} \, dy$$

and

$$-\sigma^2 \partial_x U_t(x) = \frac{U_0(0)}{\sqrt{2\sigma^2 \pi t}} \int V_0(y) \, e^{-\frac{1}{\sigma^2} \int_0^y V_0(z) dz - \frac{(x-y)^2}{2\sigma^2 t}} \, dy$$

from which we conclude that

$$V_t(x) = -\sigma^2 \partial_x \log(U_t)(x) = \frac{\int V_0(y) \, e^{-\frac{1}{\sigma^2} \int_0^y V_0(z) dz - \frac{(x-y)^2}{2\sigma^2 t}} \, dy}{\int e^{-\frac{1}{\sigma^2} \int_0^y V_0(z) dz - \frac{(x-y)^2}{2\sigma^2 t}} \, dy}.$$

This shows that

$$V_t(x) = \frac{\mathbb{E} \left(1_{]-\infty,0]}(x + \sigma W_t) \, e^{-\frac{1}{\sigma^2} \int_0^{x + \sigma W_t} 1_{]-\infty,0]}(y) dy} \right)}{\mathbb{E} \left(e^{-\frac{1}{\sigma^2} \int_0^{x + \sigma W_t} 1_{]-\infty,0]}(y) \, dy} \right)}.$$

16.1.3   Feynman-Kac jump type models

We consider the nonlinear jump model associated with drift-diffusion functions

$$(\widetilde{b}_t(x,y), \widetilde{\sigma}_t(x,y)) = (b_t(x), \sigma_t(x))$$

that only depend on the first coordinate, and a jump part given by

$$\overline{\lambda}_{t,\eta_t}(x) := V_t(x) \quad \text{and} \quad \overline{K}_{t,\eta_t}(x, dy) = \eta_t(dy).$$

In this situation, we have

$$L_{t,\eta_t}(f)(x) = L_t^c(f)(x) + V_t(x) \int \ (f(y) - f(x)) \ \eta_t(dy)$$

$$\Rightarrow \frac{d}{dt}\eta_t(f) = \eta_t\left(L_{t,\eta_t}(f)\right) = \eta_t\left[L_t^c(f) - V_t[f - \eta_t(f)]\right] = \eta_t\left[L_t^c(f) - [V_t - \eta_t(V_t)]f\right],$$

$$(16.5)$$

with the generator L_t^c defined in (15.13).

We consider the Feynman-Kac measures

$$\gamma_t(f) := \eta_0\left(P_{0,t}^V(f)\right) = \mathbb{E}\left(f(X_t) \ \exp\left\{-\int_0^t V_s(X_s)ds\right\}\right)$$

with the semigroup $P_{0,t}^V$ defined in (15.5) with $\lambda_t = V_t$. In the above formula, $X_t = \varphi_{0,t}(X_0)$ denotes a diffusion process with infinitesimal generator $L_t^c := L_t^0$ and initial distribution $\mathrm{Law}(X_0) = \eta_0$.

We let η_t be the normalized distribution defined for any bounded function f on $\mathbb{R}^d$ by

$$\eta_t(f) := \gamma_t(f)/\gamma_t(1).$$

Using the equation (13.20), which is valid for jump diffusion processes with the operator L_t^λ defined in (15.7) (following the remark we made on page 427), we have

$$\frac{d}{dt}\gamma_t(f) = \gamma_t\left[L_t^c(f) - V_t f\right]. \tag{16.6}$$

We also notice that

$$\frac{d}{dt}\log\gamma_t(1) = \frac{1}{\gamma_t(1)} \ \mathbb{E}\left(V_t(X_t) \ e^{-\int_0^t V_s(X_s)ds}\right) = \gamma_t(V_t)/\gamma_t(1) = \eta_t(V_t).$$

This implies that

$$\gamma_t(1) = \exp\left\{-\int_0^t \eta_s(V_s) \ ds\right\}$$

$$\Longrightarrow \eta_t(f) = \mathbb{E}\left(f(X_t) \ \exp\left\{-\int_0^t[V_s(X_s) - \eta_s(V_s)] \ ds\right\}\right)$$

as well as

$$\gamma_t(f) = \eta_t(f) \times \exp\left\{-\int_0^t \eta_s(V_s) \ ds\right\}. \tag{16.7}$$

By replacing V_t by $[V_t - \eta_t(V_t)]$ in (16.6), we prove that the normalized measure η_t satisfies (16.5). We can alternatively apply the forward evolution equations (15.31) to the measures $\gamma_t(f) := \eta_0 \left(P_{0,t}^V(f) \right)$ with the initial condition $\eta_0 = \delta_x$.

The flow of measures $\eta_t = \text{Law}(\overline{X}_t)$ can be interpreted as the distributions of the random states $\overline{X}_t$ of a jump type Markov process.

Between the jumps, $\overline{X}_t$ follows the diffusion X_t with generator L_t^c. At jump times T_n, occurring with the stochastic rate $V_t(X_t)$, the process $\overline{X}_{T_n-} \rightsquigarrow \overline{X}_{T_n}$ jumps to a new location, randomly chosen with the distribution $\eta_{T_n-}(dy)$.

The same formulae remain valid if we consider a Markov process X_t with some generator L_t on some state space S. In this situation, V_t is a non-negative potential function on S and we also have the nonlinear evolution equations

$$
\begin{aligned}
\partial_t \gamma_t(f) &= \gamma_t(L_t^V(f)) \\
\partial_t \eta_t(f) &= \eta_t(L_t(f)) + \eta_t(V_t)\eta_t(f) - \eta_t(V_t f) = \eta_t(L_{t,\eta_t}(f)) \quad (16.8)
\end{aligned}
$$

with the collection of generators

$$
L_t^V(f) = L_t(f) - V_t f \quad \text{and} \quad L_{t,\eta_t}(f)(x) = L_t(f)(x) + V(x) \int (f(y) - f(x)) \, \eta_t(dy).
$$

For a more detailed discussion on these models and their applications in physics and molecular chemistry, we refer to chapter 27. Section 15.6.1 also provides an interpretation of Feynman-Kac measures in terms of jump diffusion processes with killing. We also refer to exercises 307 and 308 for an analytic description of these measures for two-state models.

16.1.4 A jump type Langevin model

We consider a non-homogeneous overdamped Langevin diffusion on an energy landscape associated with a given energy function $V \in \mathcal{C}^2(\mathbb{R}^d, \mathbb{R}_+)$ on $S = \mathbb{R}^d$, for some $d \geq 1$. This model is defined by the following diffusion equation

$$
dX_t = -\beta_t \, \nabla V(X_t) + \sqrt{2} \, dB_t
$$

where ∇V denotes the gradient of V, β is an inverse temperature parameter, and B_t is a standard Brownian motion on $\mathbb{R}^d$. The infinitesimal generator associated with this continuous time process is given by the second order differential operator

$$
L_{\beta_t}^c = -\beta_t \, \nabla V \cdot \nabla + \triangle
$$

with the Laplacian operator $\triangle = \sum_{1 \leq i \leq d} \partial_{i,i}$.

Under some regularity conditions on V, for any fixed $\beta_t = \beta$, the diffusion X_t is geometrically ergodic with an invariant measure given by

$$
d\pi_\beta = \frac{1}{\mathcal{Z}_\beta} \, e^{-\beta V} \, d\lambda \quad (16.9)
$$

where λ denotes the Lebesgue measure on $\mathbb{R}^d$, and $\mathcal{Z}_\beta$ is a normalizing constant. When the inverse temperature parameter β_t depends on the time parameter t, the time inhomogeneous diffusion X_t has a time inhomogeneous generator L_{β_t}.

We further assume that $\pi_{\beta_0} = \mathrm{Law}(X_0)$, and we set $\beta_t' := \frac{d\beta_t}{dt}$. By construction, we have

$$\frac{d}{dt}\pi_{\beta_t}(f) = \beta_t' \left(\pi_{\beta_t}(V)\pi_{\beta_t}(f) - \pi_{\beta_t}(fV)\right) \quad \text{and} \quad \pi_{\beta_t} L_{\beta_t} = 0.$$

This implies that π_{β_t} satisfies the Feynman-Kac evolution equation in (16.5) (by replacing V_t by $\beta_t'V$). More formally,

$$\frac{d}{dt}\pi_{\beta_t}(f) = \pi_{\beta_t}\left(L_{\beta_t}(f)\right) + \beta_t'\left(\pi_{\beta_t}(V)\pi_{\beta_t}(f) - \pi_{\beta_t}(fV)\right)$$

from which we conclude that

$$\pi_{\beta_t}(f) = \eta_t(f) := \gamma_t(f)/\gamma_t(1) \quad \text{with} \quad \gamma_t(f) := \mathbb{E}\left(f(X_t)\,\exp\left(-\int_0^t \beta_s'\,V(X_s)ds\right)\right).$$
$$(16.10)$$

It is also easily checked that

$$\gamma_t(1) := \mathbb{E}\left(\exp\left(-\int_0^t \beta_s'\,V(X_s)ds\right)\right) = \exp\left(-\int_0^t \beta_s'\,\eta_s(V)ds\right) = \mathcal{Z}_{\beta_t}/\mathcal{Z}_{\beta_0}.$$

This formula is known as the Jarzinsky identity [48, 49, 157, 158]. In statistical physics, the weight functions

$$\mathcal{V}_t(X) = \int_0^t \beta_s'\,V(X_s)\,ds$$

represent the out-of-equilibrium virtual work of the system at the time horizon t.

In summary, we have described a McKean interpretation of Boltzmann-Gibbs measures (16.9) associated with some non-decreasing inverse cooling schedule. In this situation, the flow of measures

$$\eta_t := \pi_{\beta_t} = \mathrm{Law}(\overline{X}_t)$$

can be interpreted as the distributions of the random states $\overline{X}_t$ of a jump type Markov process. Between the jumps, X_t follows the Langevin diffusion equation (23.32). At jump times T_n, with the stochastic rate $\beta_t'V_t(\overline{X}_t)$, the process $\overline{X}_{T_n-} \rightsquigarrow \overline{X}_{T_n}$ jumps to a new site, chosen randomly according to the distribution $\eta_{T_n-}(dy)$.

16.2   Mean field particle models

The N-mean field particle interpretation associated with the nonlinear evolution equation (16.2) is of a Markov chain $\xi_t^{(N)} := (\xi_t^{(i,N)})_{1 \le i \le N}$ on the product state space $(\mathbb{R}^d)^N$, with infinitesimal generator defined, for sufficiently regular functions F on $(\mathbb{R}^d)^N$, by the formulae

$$\mathcal{L}_t(F)(x^1, \ldots, x^N) := \sum_{1 \le i \le N} L_{t,m(x)}^{(i)}(F)(x^1, \ldots, x^i, \ldots, x^N). \tag{16.11}$$

In the above display, $m(x) := \frac{1}{N}\sum_{1 \le i \le N} \delta_{x^i}$ denotes the occupation measure of the population $x = (x^i)_{1 \le i \le N} \in (\mathbb{R}^d)^N$; and for any probability measure η on $\mathbb{R}^d$, $L_{t,\eta}^{(i)}$ stands for the operator $L_{t,\eta}$ acting on the function $x^i \mapsto F(x^1, \ldots, x^i, \ldots, x^N)$ (the other coordinates $(x^j)_{j \ne i}$ being fixed). In other words, every individual ξ_t^i follows a jump diffusion model defined as the McKean process $\overline{X}_t$ by replacing the unknown measures η_t by their particle approximations $\eta_t^N = \frac{1}{N} \sum_{j=1}^N \delta_{\xi_t^{(j,N)}}$.

Assuming that the size of the population N is fixed, to simplify notation we often suppress the upper index N and write $\xi_t := (\xi_t^i)_{1 \leq i \leq N}$ instead of $\xi_t^{(N)} := (\xi_t^{(i,N)})_{1 \leq i \leq N}$.

Using (16.5), the generator associated with the Feynman-Kac jump process discussed in section 16.1.3 is given for any sufficiently regular function F by the formula

$$L_{t,m(x)}^{(i)}(F)(x^1, \ldots, x^N)$$

$$:= \sum_{1 \leq i \leq N} L_t^{c,(i)}(F)(x^1, \ldots, x^i, \ldots, x^N) + \sum_{1 \leq i \leq N} V_t(x^i)$$

$$\times \int \left[F(x^1, \ldots, x^{i-1}, y, x^{i+1}, \ldots, x^N) - F(x^1, \ldots, x^i, \ldots, x^N) \right] \, m(x)(dy)$$

$$(16.12)$$

with $m(x) := \frac{1}{N} \sum_{j=1}^N \delta_{x^j}$. By construction, between jumps the particles evolve independently as a diffusion process with generator L_t^c. At rate V_t, the particles jump to a new, randomly selected location in the current population.

Using Ito's formula we have

$$dF(\xi_t) = \mathcal{L}_t(F)(\xi_t) \, dt + d\mathcal{M}_t(F)$$

for some martingale $\mathcal{M}_t(F)$ with predictable increasing process defined by

$$\langle \mathcal{M}(F) \rangle_t := \int_0^t \Gamma_{\mathcal{L}_s}(F, F)(\xi_s) \, ds.$$

We recall that the carré du champ operator $\Gamma_{\mathcal{L}_s}$ associated to $\mathcal{L}_s$ is defined by

$$\Gamma_{\mathcal{L}_s}(F, F)(x) := \mathcal{L}_s \left[(F - F(x))^2 \right](x) = \mathcal{L}_s(F^2)(x) - 2F(x)\mathcal{L}_s(F)(x).$$

For empirical test functions of the form

$$F(x) = m(x)(f) = \frac{1}{N} \sum_{i=1}^N f(x^i)$$

with some sufficiently smooth function f, we find that

$$\mathcal{L}_s(F)(x) = m(x)(L_{s,m(x)}(f))$$
$$\Gamma_{\mathcal{L}_s}(F, F)(x) = \frac{1}{N} m(x) \left(\Gamma_{L_{s,m(x)}}(f, f) \right). \qquad (16.13)$$

From this discussion, it should be clear that

$$\eta_t^N := \frac{1}{N} \sum_{1 \leq i \leq N} \delta_{\xi_t^i} \implies d\eta_t^N(f) = \eta_t^N(L_{t,\eta_t^N}(f)) \, dt + \frac{1}{\sqrt{N}} dM_t^N(f) \qquad (16.14)$$

with the martingale

$$M_t^N(f) = \sqrt{N} \, \mathcal{M}_t(F). \qquad (16.15)$$

The predictable angle bracket is given by

$$\langle M^N(f) \rangle_t := \int_0^t \eta_s^N \left(\Gamma_{L_{s,\eta_s^N}}(f, f) \right) \, ds.$$

From the r.h.s. perturbation formulae (16.14), we conclude that η_t^N "almost solves," as $N \uparrow \infty$, the nonlinear evolution equation (16.2).

For instance, for the Feynman-Kac jump process discussed above and in section 16.1.3, using (16.7) we have

$$\eta_t^N(f) \quad \simeq_{N\uparrow\infty} \quad \eta_t(f)$$

$$\gamma_t^N(f) \quad := \quad \eta_t^N(f) \times \exp\left\{-\int_0^t \eta_s^N(V_s)\,ds\right\} \quad \simeq_{N\uparrow\infty} \quad \gamma_t(f). \quad (16.16)$$

We can also check that $\gamma_t^N(f)$ is unbiased (in the sense that $\mathbb{E}\left(\gamma_t^N(f)\right) = \gamma_t(f)$).

For a more thorough discussion on these continuous time models, we refer the reader to the review article [76], and the references therein.

16.3   Some application domains

16.3.1   Fouque-Sun systemic risk model

We consider the log-monetary reserves $(\xi_t^i)_{1\leq i\leq N}$ of N banks. The inter-bank exchanges (borrowing and lending) are represented by the diffusion equation

$$d\xi_t^i = \frac{\alpha}{N}\sum_{1\leq j\leq N}(\xi_t^j - \xi_t^i)\,dt + \sigma\,dW_t^i$$

where $(W_t^i)_{1\leq i\leq N}$ are N independent Brownian motions, and α and σ are fixed parameters. This model is the mean field approximation of the nonlinear process (16.3) associated with the parameters

$$\widetilde{b}_t(x,y) = \alpha \times (y-x) \qquad \widetilde{\sigma}_t(x,y) = \sigma \quad \text{and} \quad \lambda_t(x) = 0.$$

The model has been introduced by J.P. Fouque and L.H. Sun in [130] (see also [40] for a mean field game interpretation of this model). Simulations show that stability is created by increasing the parameter α. Nevertheless the systemic risk is also increased when α is large.

The following two graphs illustrate Euler type discrete time approximations of the model with 10 banks when $\alpha = 100$ and $\alpha = 0$, respectively.

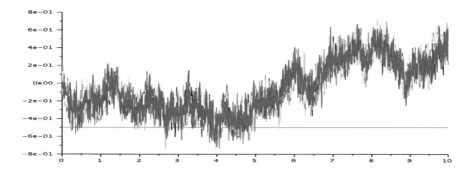

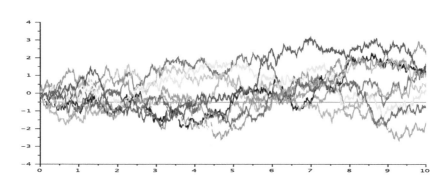

16.3.2 Burgers equation

The mean field particle interpretation of the Burgers model from section 16.1.2 is given for any $1 \leq i \leq N$ by the interacting diffusion equations:

$$d\xi_t^i = \frac{1}{N} \sum_{1 \leq j \leq N} 1_{[\xi_t^i, \infty[}(\xi_t^j) \ dt + \sigma \ dW_t^i$$

where again $(W_t^i)_{1 \leq i \leq N}$ are N independent Brownian motions. In this case, we have

$$V_t^N(x) := \frac{1}{N} \sum_{1 \leq j \leq N} 1_{[x, \infty[}(\xi_t^j) \ \simeq_{N\uparrow\infty} \ V(x).$$

The following graphs illustrate these three approximations. The top l.h.s graph compares the exact solution with the mean field particle estimate based on Euler type scheme with $N = 100$ particles and a $\Delta t = .01$ time step. The top r.h.s. represents the exact values on the simulated states, and the bottom graph compares the crude Monte Carlo method with the exact solution.

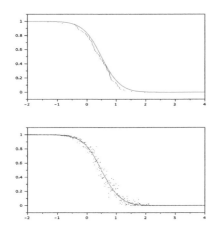

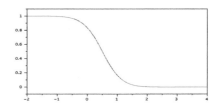

16.3.3   Langevin-McKean-Vlasov model

The Langevin-McKean-Vlasov model is a stochastic gradient process on $\mathbb{R}$ associated with some smooth energy function V, coupled with an attraction or repulsion force around ensemble averages. This model is defined by N interacting diffusion processes

$$d\xi^i(t) = -\beta \, \partial_x V(\xi_t^i)dt + \alpha \, \left(\frac{1}{N} \sum_{1 \leq j \leq N} \xi_t^j - \xi_t^i \right) \, dt + \sigma \, dW_t^i$$

with $(W_t^i)_{1 \leq i \leq N}$ being N independent Brownian motions, and α, β and σ as fixed parameters. This model was introduced by S. Herrmann, and J. Tugaut in [151] and coincides with the mean field particle interpretation of the nonlinear model (16.1) on $\mathbb{R}$ with

$$\widetilde{b}_t(x,y) \, = -\beta \, V'(x) + \alpha \times (y - x) \qquad \widetilde{\sigma}_t(x,y) = \sigma > 0 \quad \text{and} \quad \lambda_t(x) = 0.$$

The following graphs illustrate the time evolution Langevin-McKean-Vlasov model with a double well potential with $\alpha = -7$, and $N = 50$ particles.

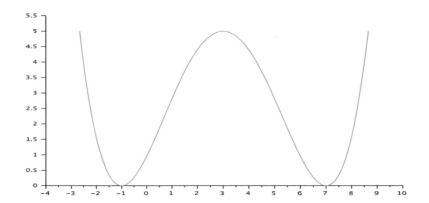

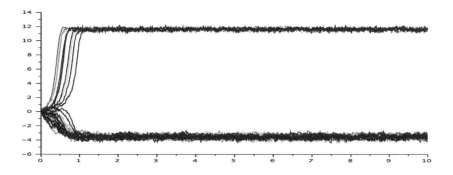

16.3.4 Dyson equation

In nuclear physics, the statistical properties of the spectrum of quantum systems can be analyzed using the nonlinear Dyson equations

$$
\begin{cases}
d\lambda_i(t) = \dfrac{1}{N} \sum_{j \neq i} \dfrac{1}{\lambda_i(t) - \lambda_j(t)} \, dt + \sqrt{\dfrac{2}{N}} \, dW_t^i \\[2mm]
1 \leq i \leq N
\end{cases}
$$

with some initial conditions $\lambda_1(0) < \ldots < \lambda_N(0)$ and N independent Brownian motions $(W_t^i)_{1 \leq i \leq N}$.

This mean field type particle model is slightly different from the nonlinear processes discussed in these lectures. One can show (cf. exercises 5.3.2 through 5.3.4 in [86]) that $\lambda_1(t) < \ldots < \lambda_N(t)$ coincide with the eigenvalues of the symmetric Gaussian matrices

$$
A_{i,i}(t) = W^i(t)/\sqrt{N/2} \quad \text{et} \quad A_{i,j}(t) = A_{j,i}(t) = W^{i,j}(t)/\sqrt{N}.
$$

In the above display $W_{i,j}$, $1 \leq i < j \leq N$ and W_t^i, $1 \leq i \leq N$, stand for $N(N+1)/2$ independent Brownian motions on the real line.

The following illustrates the time evolution of $N = 30$ eigenvalues of the matrices $A(t)$ on the interval $[0,1]$.

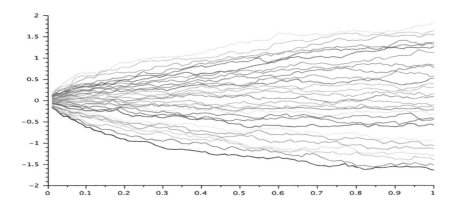

16.4   Exercises

Exercise 307 (Feynman-Kac particle models) *We let X_t be a Markov process on $S = \{0, 1\}$ with some infinitesimal generator L defined by*

$$L(f)(0) = \lambda(0) \ (f(1) - f(0)) \quad and \quad L(f)(1) = \lambda(1) \ (f(0) - f(1))$$

for some rate function λ on S. We consider a potential function $V \ : \ x \in \{0, 1\} \mapsto [0, \infty[$. We let γ_t, η_t be the Feynman-Kac measures defined by

$$\eta_t(f) = \gamma_t(f)/\gamma_t(1) \quad with \quad \gamma_t(f) = \mathbb{E}\left(f(X_t) \ \exp\left(-\int_0^t V(X_s)ds\right)\right).$$

Using (16.8), check the evolution equations

$$\partial_t \gamma_t(f) = \gamma_t \left(L^V(f)\right) \quad and \quad \partial_t \eta_t(f) = \eta_t \left(L_{\eta_t}(f)\right)$$

with the Schrödinger operator $L^V = L - V$, and the collection of jump type generators

$$\begin{aligned}
L_\eta(f)(0) &= L(f)(0) + V(0) \ (f(1) - f(0)) \ \eta(1) \\
L_\eta(f)(1) &= L(f)(1) + V(1) \ (f(0) - f(1)) \ \eta(0).
\end{aligned}$$

Describe the evolution of the nonlinear jump process $X_t \in S = \{0, 1\}$ with generator L_{η_t}. Describe the mean field particle model associated with this nonlinear process. Provide an interpretation of these models in terms of an epidemic process. The state 1 represents the infected individuals, while 0 represents the susceptible ones.

Exercise 308 (Feynman-Kac two-state models - Explicit formulae) ✎ *Consider the Feynman-Kac model discussed in exercise 307. Check that $\eta_t(0)$ satisfies the Riccati equation*

$$\partial_t \eta_t(0) = a \ \eta_t(0)^2 - b \ \eta_t(0) + c$$

with the parameters

$$a := [V(0) - V(1)] \quad and \quad b := a + c + \lambda(0) \quad with \quad c := \lambda(1).$$

We assume that $a \geq 0$ so that $a, b, c \geq 0$. Check that $b^2 \geq 4ac$.

- When $a = 0$ check that

$$\eta_t(0) = \mathbb{P}(X_t = 0) = e^{-bt}\, \eta_0(0) + \frac{c}{b}\,\left(1 - e^{-bt}\right).$$

- When $c = 0$, and $(b \geq)a > 0$, check that the solution is given by

$$\eta_t(0) = e^{-bt}\, \frac{\eta_0(0)}{1 - \frac{a}{b}\eta_0(0)\,(1 - e^{-bt})}.$$

Notice that we have two solutions: one $\eta_t(0) = 0 \Rightarrow \eta_t(1) = 1$ corresponding to the absorbing state 1 when $\eta_0(1) = 1$, and a non-null one when $\eta_0(1) < 1$.

- Assume that $a \wedge c > 0$ ($\Rightarrow b \geq a + c > 0$).

  - When $b^2 = 4ac$, check that $a = c = b/2$ and

  $$\eta_t(1) = 1 - \eta_t(0) = \frac{\eta_0(1)}{1 + a\eta_0(1)\,t}.$$

  - When $b^2 > 4ac$, check that

  $$\eta_t(0) + \frac{z_2}{a} = \left(\eta_0(0) + \frac{z_2}{a}\right)\underbrace{\frac{(z_2 - z_1)\,e^{-(z_2-z_1)t}}{(a\eta_0(0) + z_2)\,e^{-(z_2-z_1)t} - (a\eta_0(0) + z_1)}}_{>0}$$

  with

  $$z_1 = -\frac{1}{2}\left(b + \sqrt{b^2 - 4ac}\right) \leq z_2 = -\frac{1}{2}\left(b - \sqrt{b^2 - 4ac}\right) < 0.$$

  Prove that $-\frac{z_2}{a} \in [0,1]$ and $-\frac{z_1}{a} > 1$.

This exercise can be completed with the exercises *446* and *447* dedicated to application of these formulae to the computations of quasi-invariant measures.

Exercise 309 (Nonlinear switching models) *We consider a nonlinear evolution equation in the 2d-simplex*

$$\begin{cases} \partial_t u_t &= -a(u_t)\,u_t + b(u_t)\,v_t \\ \partial_t v_t &= a(u_t)\,u_t - b(u_t)\,v_t \end{cases}$$

for some positive functions a, b and some initial condition $u_0 + v_0 = 1$ with $u_0, v_0 \geq 0$. We let η_t be the probability distribution on $S = \{0,1\}$ defined by $\eta_t(0) = u_t$ and $\eta_t(1) = v_t$, and we set $\lambda(\eta_t, 0) := a(u_t)$ and $\lambda(\eta_t, 1) := b(u_t)$. Check that

$$\partial_t \eta_t(f) = \eta_t(L_{\eta_t}(f))$$

for any function f on S with the generator

$$L_{\eta_t}(f)(0) = \lambda(\eta_t, 0)\,(f(1) - f(0)) \quad and \quad L_{\eta_t}(f)(1) = \lambda(\eta_t, 1)\,(f(0) - f(1)).$$

Describe the evolution of the nonlinear jump process $X_t \in S = \{0,1\}$ with generator L_{η_t}. Describe the mean field particle model associated with this nonlinear process. Discuss the cases $(\lambda(\eta_t, 0), \lambda(\eta_t, 1)) = (\eta_t(1), \eta_t(0))$ and $(\lambda(\eta_t, 0), \lambda(\eta_t, 1)) = (\eta_t(0), \eta_t(1))$.

Exercise 310 (McKean-Vlasov equation) *We let $a(u)$ be some function on $\mathbb{R}$. We consider the solution $p_t : x \in \mathbb{R} \mapsto p_t(x) \in \mathbb{R}$ of the integro-differential evolution equation*

$$\partial_t p_t(x) = \partial_x^2 p_t - \partial_x (p_t \ (p_t \star a))$$

with some initial condition $p_0(x)$ given by some probability density on $\mathbb{R}$. In the above display, $(p_t \star a)$ stands for the convolution operation $(p_t \star a)(x) = \int a(x - y) \ p_t(y) \ dy$. We set $\eta_t(dx) = p_t(x)dx$. Check that

$$\partial_t \eta_t(f) = \eta_t(L_{\eta_t}(f))$$

for any smooth and compactly supported function f, with the collection of diffusion generators

$$L_{\eta_t}(f)(x) = \partial_x^2 f(x) + b(x, \eta_t) \ \partial_x f(x) \quad \text{with} \quad b(x, \eta_t) = \int \int a(x - y) \ \eta_t(dy).$$

Describe the evolution of the nonlinear diffusion process X_t with generator L_{t,η_t}. Describe the mean field particle model associated with this nonlinear process. Show that the Fouque-Sun systemic risk model discussed in section 16.3.1 corresponds to the case $\sigma = \sqrt{2}$, and $a(u) = \alpha u$, for some parameters α. Check that in this case the nonlinear model has the same form as the Ornstein-Uhlenbeck process discussed in exercise 255 with $a = -\alpha$ and $b = \mathbb{E}(X_0)$.

Exercise 311 (Nonlinear soliton-like jump process [154]) *We let $q(u)$ be some probability density of some random variable U on the positive axis $[0, \infty[$, and $h(.)$ be some function on $\mathbb{R}$. We consider the solution $p_t : x \in \mathbb{R} \mapsto p_t(x) \in \mathbb{R}$ of the evolution equation*

$$\partial_t p_t(x) = -H(x, p_t) \ p_t(x) + \int_{-\infty}^{x} q(x - y) \ H(y, p_t) \ p_t(y) \ dy$$

with some initial condition $p_0(x)$ given by some probability density on $\mathbb{R}$ and

$$H(x, p_t) := \int_x^{+\infty} h\left(y - \int_{-\infty}^{+\infty} z \ p_t(z) \ dz\right) \ p_t(y) \ dy.$$

We set $\eta_t(dx) = p_t(x)dx$ and we consider the Markov transitions and the jump intensities

$$M(x, dy) = 1_{[x, \infty[}(y) \ q(y - x) \ dy \quad \text{and} \quad \lambda(x, \eta_t) = \int_x^{+\infty} h\left(y - \int_{-\infty}^{+\infty} z \ p_t(z) \ dz\right) \ \eta_t(dy).$$

Check that

$$\partial_t \eta_t(f) = \eta_t(L_{\eta_t}(f))$$

with the collection of jump generators

$$L_{\eta_t}(f)(x) = \lambda(x, \eta_t) \int (f(y) - f(x)) \ M(x, dy).$$

Describe the evolution of the nonlinear jump diffusion process X_t with generator L_{t,η_t}. Describe the mean field particle model associated with this nonlinear process. Discuss the case $h = 1$. As underlined by Max-Olivier Hongler in [154], the mean field model can be interpreted as a pool of interacting agents subject to mutual imitation. Their interaction is measured by their relative position w.r.t. their empirical average. The interacting jump rate encapsulates two different features. The first one can be seen as imitation or go-with-the-winner type mechanism. Leader agents on the right of the state space attract laggards that try to imitate them. The jump rate also measures the barycentric range modulation between the agents.

Exercise 312 (Opinion dynamics [56]) *We consider the collection of generators L_η indexed by probability measures on $S = \mathbb{R}^d$ defined for any bounded function f on S and for any $x \in S$ by the formula*

$$L_\eta(f)(x) = \int \left(f((1-\epsilon)x + \epsilon y) - f(x) \right) \kappa(x,y) \; \eta(dy)$$

with confidence parameter $\epsilon \in [0,1]$ and some bounded interacting density $k(x,y) \geq 0$. We consider the weak solution η_t of the equation

$$\partial_t \eta_t(f) = \eta_t(L_{\eta_t}(f)).$$

Describe the evolution of the nonlinear jump diffusion process X_t with generator L_{η_t}.

Describe the mean field particle model associated with this nonlinear process. Examine the situations

$$\kappa(x,y) = 1_{[0,R]}(\|x-y\|) \quad and \quad \kappa(x,y) = \exp\left(-\|x-y\|^2/\sigma^2\right)$$

for some parameters $R, \sigma > 0$. The model corresponding to the l.h.s. function κ is called the Deffuant-Weisbuch model.

Exercise 313 (Interacting agents dynamics - Quantitative estimates [56]) *Consider the opinion dynamic model discussed in exercise 312 with symmetric densities $k(x,y) = k(y,x)$. We set $f_1(x) = x$ and $f_2(x) = \|x\|^2$. We assume that $\eta_0(f_1) = 0$ and $\eta_0(f_2) < \infty$. Check that $\eta_t(f_1) = 0$ and $\eta_t(f_2) \leq \eta_0(f_2)$, for any $t \geq 0$. When $\kappa = 1$ prove that*

$$\eta_t(f_2) = e^{-2\epsilon(1-\epsilon)t} \; \eta_0(f_2) \quad and \quad \|\eta_t(f)\| \leq \|\eta_0(f)\| + (1-\epsilon)\,(2\eta_0(f_2))^{1/2}$$

for any Lipschitz function $\|f(x) - f(y)\| \leq \|x - y\|$.

Exercise 314 (Rank-based interacting diffusions) *We consider the solution $p_t : x \in \mathbb{R} \mapsto p_t(x) \in \mathbb{R}$ of the evolution equation*

$$\partial_t p_t = -\partial_x \left(b_{p_t} \; p_t \right) + \frac{1}{2}\partial_x^2 \left(\sigma_{p_t}^2 \; p_t \right)$$

with some initial condition $p_0(x)$ given by some probability density on $\mathbb{R}$. In the above display, the drift and diffusion functions (b_{p_t}, σ_{p_t}) are defined by

$$b_{p_t}(x) = \alpha \left(\int_{-\infty}^x p_t(y) \; dy \right) \quad and \quad \sigma_{p_t}(x) = \beta \left(\int_{-\infty}^x p_t(y) \; dy \right)$$

for some smooth functions (α, β) on $\mathbb{R}$. We set $\eta_t(dx) = p_t(x)dx$ and we set

$$B_{\eta_t}(x) = \alpha \left(\int_{-\infty}^x \eta_t(dy) \right) \quad and \quad D_{\eta_t}(x) = \beta \left(\int_{-\infty}^x \eta_t(dy) \right).$$

Check that for smooth functions f with compact support we have

$$\partial_t \eta_t(f) = \eta_t(L_{\eta_t}(f))$$

with the collection of diffusion generators

$$L_{\eta_t} = B_{\eta_t} \; \partial_x + \frac{1}{2} \; D_{\eta_t}^2 \; \partial_x^2.$$

Describe the evolution of the nonlinear diffusion process X_t with generator L_{t,η_t}. Describe the mean field particle model associated with this nonlinear process.

Exercise 315 (Granular media equations) *Let* $V_i : \mathbb{R}^r \mapsto \mathbb{R}$, $i = 1, 2$ *be a couple of smooth potential functions on* $\mathbb{R}^r$, *for some* $r \geq 1$. *We also let* $W_t = (W_t^i)_{1 \leq i \leq r}$ *be* r-*dimensional Brownian motions and* σ *some* $(r \times r)$-*matrix. Describe the generator and the mean field approximation of the nonlinear diffusion equation*

$$dX_t = \left[\partial V_1(X_t) + \int \partial V_2(X_t - x) \, \mathbb{P}(X_t \in dx) \right] dt + \sigma(X) \, dW_t$$

where $\partial F = (\partial_{x_i} F)_{1 \leq i \leq r}$ *stands for the gradient column vector of a smooth function* F *on* $\mathbb{R}^r$. *In the above display,* W_t *stands for an* r-*dimensional Brownian motion and* $\sigma(x)$ *some* $(r \times r)$-*mtarix. We assume that* $\mathbb{P}(X_t \in dx) = p_t(x) \, dx$ *has a density* $p_t(x)$ *w.r.t. the Lebesgue measure* dx *on* $\mathbb{R}^r$. *Describe the evolution equation of* p_t.

Exercise 316 (Membrane potential of interacting neurons [87]) *The membrane potentials of* N-*interacting neurons* $\xi_t = (\xi_t^i)_{1 \leq i \leq N} \in [0, \infty[$ *are represented by a Markov process on* $[0, \infty[^N$ *with infinitesimal generator*

$$\mathcal{L}(f)(x)$$
$$= \sum_{1 \leq i \leq N} \lambda(x_i) \left(f(x + \rho_{1/N}^i(x)) - f(x) \right) + \sum_{1 \leq i \leq N} a \left(x_i - \frac{1}{N} \sum_{1 \leq j \leq N} x_j \right) \partial_{x_i} f(x)$$

for some parameter $a > 0$, *some rate function* $\lambda(u) \geq 0$ *and the jump amplitude functions* $\rho_{1/N}^i$ *defined by*

$$\rho_{1/N}^i(x) = (\rho_{j,1/N}^i(x))_{1 \leq j \leq N} \quad \text{with} \quad \rho_{j,1/N}^i(x) = 1_{i \neq j} \frac{1}{N} - 1_{i=j} \, x_i.$$

The coordinates ξ_t^i *represent the membrane potential of the* i-*th neuron. The jump term in the generator represents the random spikes of the neuron. At some rate* $\lambda(x_i)$ *the* i-*th neuron sets its energy to 0 and gives the other neurons a small quantity of energy* $1/N$. *The first order differential term represents the transport equation of the electrical synapses. The value of each neuron potential tends to reach the average potential of the neurons. We let* $\mathcal{G}$ *be the generator*

$$\mathcal{G}(f)(x) = \sum_{1 \leq i \leq N} \lambda(x_i) \left(f(x + \rho_0^i(x)) - f(x) \right)$$
$$+ \sum_{1 \leq i \leq N} \left[a \left(x_i - \frac{1}{N} \sum_{1 \leq j \leq N} x_j \right) + \frac{1}{N} \sum_{1 \leq i \leq N} \lambda(x_i) \right] \partial_{x_i} f(x).$$

For empirical functions $f(x) = \frac{1}{N} \sum_{1 \leq i \leq N} \varphi(x_i)$, *for some bounded and smooth functions* φ *check that*

$$\mathcal{L}(f)(x) = \mathcal{G}(f)(x) + \mathrm{O}\left(\frac{1}{N} \right) \quad \text{and} \quad \mathcal{G}(f)(x) = m(x) \left(L_{m(x)}(\varphi) \right)$$

with the collection of generators L_η *indexed by probability measures* η *on* $[0, \infty[$ *and defined by*

$$L_\eta(\varphi)(u) = \lambda(u) \left(\varphi(0) - \varphi(u) \right) + \left[a \left(u - \int v \, \eta(dv) \right) + \int \lambda(v) \, \eta(dv) \right] \varphi'(u).$$

We let η_t *be the solution of the nonlinear equation*

$$\partial_t \eta_t(\varphi) = \eta_t(L_{\eta_t}(\varphi))$$

starting from some probability measure η_0 on $S = [0, \infty[$, for any bounded and smooth functions φ. Describe a nonlinear Markov process X_t with distribution η_t on S.

Exercise 317 (Feynman-Kac particle models - 1) *We consider the Feynman-Kac measures η_t discussed in exercise 294. Check that*

$$\partial_t \eta_t(f) = \eta_t(L_{t,\eta_t}(f)) \quad with \quad L_{t,\eta_t}(f)(x) := L_t(f)(x) + V_t(x) \int (f(y) - f(x)) \, \eta_t(dy).$$

Describe the evolution of the nonlinear jump diffusion process X_t with generator L_{t,η_t}. Describe the mean field particle model associated with this nonlinear process.

Exercise 318 (Feynman-Kac particle models - 2) *We let η_t be the Feynman-Kac measures η_t defined as in exercise 294 by replacing V_t by $-V_t$. Check that*

$$\partial_t \eta_t(f) = \eta_t(L_{t,\eta_t}(f)) \quad with \quad L_{t,\eta_t}(f)(x) := L_t(f)(x) + \int (f(y) - f(x)) \, V_t(y) \, \eta_t(dy).$$

Describe the evolution of the nonlinear jump diffusion process X_t with generator L_{t,η_t}. Describe the mean field particle model associated with this nonlinear process.

Exercise 319 (Mean field particle models) *We consider the generator (16.11) of a mean field particle model. Check the formulae (16.13).*

Exercise 320 (Interacting jump Langevin model) *We consider the Feynman-Kac interpretation (16.10) of the jump Langevin process presented in section 16.1.4. Describe the mean field particle model of the sequence of non-homogeneous Boltzmann-Gibbs measures π_{β_t} defined in (16.9).*

17

Stochastic analysis toolbox

This chapter summarises some analytical tools that are frequently used when studying the asymptotic behaviour of continuous time stochastic processes as the time parameter tends to infinity. We start with the change of time technique. The time change method is used as a simple rule to transfer the invariance property of probability measures for jump diffusion processes. We also describe Foster-Lyapunov conditions and related Dobrushin contraction inequalities. Section 17.5 shows a series of application examples of Lyapunov functions for different classes of diffusion processes. Section 17.6 discusses more advanced functional and spectral analysis techniques, including a derivation of the Poincaré inequality. These techniques are applied to investigate the exponential decays to equilibrium.

> *Beauty is the first test: there is no permanent place in the world for ugly mathematics.*
> Godfrey H. Hardy (1877-1947).

17.1   Time changes

We let X_t be a jump diffusion process with generator L on some state space S. We consider a function a on S taking values in $[\epsilon, 1/\epsilon]$, for some $\epsilon > 0$. We set

$$A_t := \int_0^t \frac{1}{a(X_s)} \, ds$$

and denote its inverse map by
$$\tau_t := A^{-1}(t) \quad (\text{with } \tau_0 = 0).$$

By construction, we have

$$A'_t = 1/a(X_t) \quad \text{and} \quad A_{\tau_t} = t \Rightarrow \tau'_t = 1/A'(\tau_t) = a(X_{\tau_t}) \Rightarrow \tau_t = \int_0^t a(X_{\tau_s}) \, ds.$$

On the other hand, for any sufficiently regular function f we have

$$f(X_{\tau_t}) - f(X_0) = \int_0^{\tau_t} L(f)(X_s) \, ds + \mathcal{M}_{\tau_t}(f)$$

for some martingale $\mathcal{M}_s(f)$ (w.r.t. the filtration $\mathcal{F}_s$). Changing the time integration variable $s \rightsquigarrow u$ in the above display, we have

$$s = \tau_u = \int_0^u a(X_{\tau_r}) \, dr \Rightarrow ds = a(X_{\tau_u}) \, du \Rightarrow \int_0^{\tau_t} L(f)(X_s) \, ds = \int_0^t a(X_{\tau_u}) \, L(f)(X_{\tau_u}) \, du.$$

If we set $Y_t = X_{\tau_t}$ and $\mathcal{N}_t(f) = \mathcal{M}_{\tau_t}(f)$, this implies that

$$f(Y_t) - f(Y_0) = \int_0^t a(Y_u) \, L(f)(Y_u) \, du + \mathcal{N}_t(f)$$

for the martingale $\mathcal{N}_s(f)$ associated with the filtration $\mathcal{G}_s = \mathcal{F}_{\tau_s}$. In summary we have proved the following time change property.

$$\left.\begin{array}{l} X \qquad \text{with generator } L^X \\[3em] A_t \ := \ \displaystyle\int_0^t 1/a(X_s) \, ds \end{array}\right\} \Rightarrow Y_t = X_{A_t^{-1}} \quad \text{with generator } L^Y = a \times L^X.$$

$$\tag{17.1}$$

For instance, if $a(x) = a$ is a constant function, $A_t = t/a \Rightarrow A_t^{-1} = at$. In this case, $Y_t = X_{at}$ has a generator given by $L^Y = a \times L^X$.

17.2 Stability properties

A probability measure π on some state space S is invariant w.r.t. some time homogeneous Markov semigroup $P_t = P_{0,t}$ if we have $\pi P_t = \pi$, for any $t \geq 0$. The invariance property is also characterized in terms of the generator L of the semigroup:

$$\pi \text{ is } P_t\text{-invariant} \iff (\forall t \geq 0 \ \ \pi P_t = \pi) \iff \pi L = 0.$$

We also say that π is L-invariant.

We check this claim by using the formulae

$$\pi \left(\frac{P_t - Id}{t} \right) \ \to_{t \downarrow 0} \ \pi L \quad \text{and} \quad P_t = Id + \int_0^t LP_s \, ds.$$

The measure π is reversible w.r.t. P_t or w.r.t. L if we have

$$\pi(dx) \, P_t(x, dy) = \pi(dy) \, P_t(y, dx)$$

or equivalently, for any pair of sufficiently regular functions f, g

$$\forall t \geq 0 \quad \pi(fP_t(g)) = \pi(P_t(f)g) \iff \pi(fL(g)) = \pi(L(f)g). \tag{17.2}$$

We also say that π is L-reversible.

The invariant measures of pure jump processes with a generator $L^d = \lambda \, [M - Id]$ have been discussed in section 12.7.1. In (12.29) we characterized these limiting measures in terms of the invariant measure of the Markov transition M and the jump rate function λ.

By (17.1) the jump rate λ can be interpreted as a time change of the process with generator $L = [M - Id]$. We recall that this process is a continuous time embedding of the Markov chain with transition probability M.

The following proposition provides a simple rule to transfer the invariance property of given probability measures through a time change procedure.

Proposition 17.2.1 *Let L be the generator of some jump diffusion process and π some probability measure on some state space S. We let a be a positive function on S s.t. $\pi\left(a^{-1}\right) < \infty$, with $a^{-1} = 1/a$. We consider the generator L_a and the probability measure π_a defined for any sufficiently regular function f on S by the formulae*

$$L_a(f) = a\, L(f) \quad \text{and} \quad \pi_a(f) = \pi\left(a^{-1}f\right)/\pi\left(a^{-1}\right).$$

It is readily checked that

$$\pi_a(fL_a(g)) = \pi(fL(g)) \tag{17.3}$$

for any sufficiently regular functions f and g on S This shows that π is L-invariant, respectively L-reversible, if and only if π_a is L_a-invariant, respectively L_a-reversible.

We end this section with a simple property which can be used to design and combine stochastic processes with a prescribed invariant measure.

Proposition 17.2.2 *Let (L_1, L_2) be a couple of generators of some stochastic processes on some state space S and let π be some probability measure on S. If we set $L = L_1 + L_2$, then we have the following properties:*

$$\pi \quad L_1 - reversible \quad and \quad L_2 - reversible \quad \Rightarrow \quad \pi \quad L - reversible$$
$$\pi \quad L_1 - invariant \quad and \quad L_2 - invariant \quad \Rightarrow \quad \pi \quad L - invariant. \tag{17.4}$$

The proofs of these two elementary properties are left to the reader. Notice that (17.4) allows us to combine MCMC samplers with a given target probability measure. Furthermore, adding a generator to a given MCMC sampler may improve the stability properties of the simulation process.

For instance, using the perturbation formulae (15.27) discussed in section 15.5.2, the semigroup $P_{s,t}$ associated with the generator $L = L_1 + L_2$ satisfies the minorization property (8.15) as soon as the semigroup $P_{s,t}^{(1)}$ satisfies this property and L_2 is any jump generator with a bounded intensity.

17.3 Some illustrations

17.3.1 Gradient flow processes

We consider a diffusion generator on $\mathbb{R}^d$ of the following form

$$L(f) \;=\; e^V \sum_{1 \le i \le d} \partial_{x_i}\left(e^{-V}\partial_{x_i}f\right) = -\sum_{1 \le i \le d} \partial_{x_i}V\,\partial_{x_i}f + \sum_{1 \le i \le d} \partial_{x_i}^2 f$$

for some function V on $\mathbb{R}^d$ s.t. $\int e^{-V(x)} \, dx \in]0, \infty[$, where dx stands for the Lebesgue measure on $\mathbb{R}^d$. The first formulation of L in the above display is sometimes called the Sturm-Liouville formulation of the generator.

For any smooth functions f, g with compact support on $\mathbb{R}^d$ using a simple integration by parts procedure we have

$$\int e^{-V(x)} \, g(x) \, L(f)(x) \, dx = \sum_{1 \leq i \leq d} \int g(x) \, \partial_{x_i} \left(e^{-V(x)} \, \partial_{x_i} f \right)(x) \, dx$$

$$= - \sum_{1 \leq i \leq d} \int e^{-V(x)} \, \partial_{x_i} g(x) \, \partial_{x_i} f(x) \, dx.$$

This simple observation clearly implies that the probability measure

$$\pi(dx) \propto e^{-V(x)} \, dx \quad \text{is } L\text{-reversible.} \tag{17.5}$$

These gradient flow diffusions are discussed in further details in section 23.4, along with manifold valued models.

17.3.2 One-dimensional diffusions

The two properties discussed above allow us to characterize the invariant measures of one-dimensional diffusion processes with generators of the form

$$L(f) = b \, f' + \frac{1}{2} \, \sigma^2 \, f''$$

for some functions (b, σ) on $\mathbb{R}$ such that $\int_S 2\sigma^{-2}(x) \, e^{\int_c^x 2b(y)\sigma^{-2}(y) \, dy} dx \in]0, \infty[$, for some constant $c \in \mathbb{R}$ and $S = \sigma^{-1}(]0, \infty[)$. In this situation, the probability measure

$$\pi(dx) \propto 1_S(x) \, \sigma^{-2}(x) \, \exp\left[\int_c^x 2b(y)\sigma^{-2}(y) \, dy \right] dx \quad \text{is } L\text{-reversible.} \tag{17.6}$$

In addition, we have the Sturm-Liouville formula

$$L(f)(x) = \frac{1}{2} \, \sigma^2(x) \, e^{-\int_c^x \frac{2b(y)}{\sigma^2(y)} \, dy} \, \partial_x \left(e^{\int_c^x \frac{2b(y)}{\sigma^2(y)} \, dy} \, \partial_x f \right)(x). \tag{17.7}$$

The proof is rather elementary. By (17.3) it clearly suffices to consider the case where $\sigma = \sqrt{2}$ so that

$$L(f) = b \, f' + f'' = V' \, f' + f'' \quad \text{with} \quad V(x) = \int_0^x b(y) dy.$$

By (17.5) we conclude that π is L-reversible. The proof of (17.7) is immediate. Notice that for any functions (f, g) with compact support, by a simple integration by parts formula we find that

$$-2\pi(fL(g)) = \int \sigma^2(x) \, (\partial_x f)(x) \, (\partial_x g)(x) \, \pi(dx) = \pi \left(\Gamma_L(f, g) \right)$$

with the carré du champ operator $\Gamma_L(f,g) = L(fg) - fL(g) - gL(f)$ associated with the operator L.

Several illustrations of this formula for Pearson type diffusion processes are provided in exercises 321 to 328.

17.4 Foster-Lyapunov techniques

17.4.1 Contraction inequalities

We start with a general result that allows us to apply the V-norm contraction techniques we developed in section 8.2.7.

We consider the semigroup (sg) $P_t = P_{s,s+t}$, for any $s \geq 0$, of a time homogeneous Markov process on some state space S satisfying the evolution equations (13.22) for some infinitesimal generator L acting on a domain $D(L)$ of sufficiently regular functions.

Theorem 17.4.1 *We assume that there exists some non-negative function $W \in D(L)$ on S such that*

$$L(W) \leq -a\,W\,+\,c \qquad (17.8)$$

for some parameters $a > 0$, and $c \geq 0$. Then for any $t > 0$, the Markov transition P_t satisfies the Foster-Lyapunov condition (8.28)

$$P_t(W) \leq \epsilon_t\,W + c_t \quad \text{with} \quad \epsilon_t = \frac{1}{1+at} \quad \text{with} \quad c_t = c\,t.$$

In addition, if P_t satisfies the Dobrushin local contraction condition (8.27) for any $t > 0$, there exists an unique invariant measure $\pi = \pi M$. Furthermore, for any $h > 0$ there exists some positive function V s.t. $\beta_V(P_h) < 1$. In particular, we have the exponential contraction inequality

$$\|P_{h\lfloor t/h\rfloor}(x,\cdot) - \pi\|_V \leq \beta_V(P_h)^{\lfloor t/h\rfloor}\,(1+V(x)+\pi(V)).$$

Proof :
Using (13.22) we have

$$
\begin{aligned}
P_t(W) &= W + \int_0^t P_s(L(W))\,ds \\
&\leq W + \int_0^t [-a\,P_s(W) + c]\,ds = W + ct - a\int_0^t P_s(W)ds.
\end{aligned}
$$

On the other hand, through integration by parts we have

$$
\begin{aligned}
\int_0^t P_s(W)ds &= [s\,P_s(W)]_0^t - \int_0^t s\,\frac{d}{ds}P_s(W)\,ds \\
&= t\,P_t(W) - \int_0^t s\,P_s(\underbrace{L(W)}_{\leq c}))\,ds \geq t\,P_t(W) - ct^2/2.
\end{aligned}
$$

This implies that

$$P_t(W) \leq W + ct - a\left(t\,P_t(W) - ct^2/2\right)$$

from which we conclude that

$$P_t(W) \leq \frac{1}{1+at} W + ct \frac{1+at/2}{1+at} \leq \frac{1}{1+at} W + ct.$$

The last assertion is a direct consequence of the theorem 8.2.21 applied to the Markov transition $P_{nh} = P_h^n$. This ends the proof of the theorem. ∎

Remark : Replacing in (17.8) W by $W + b$ for some $b > 0$ we find that

$$L(W + b) = L(W) + b \underbrace{L(1)}_{=0} = L(W) \leq -a \ (W + b) \ + \ (c + ab).$$

This shows that without loss of generality we can assume that (17.8) is met for some function $W \geq b$, for some $b > 0$.

On the other hand, using the fact that

$$(17.8) \quad \text{with} \quad c > 0 \quad \Rightarrow \quad L\left(\frac{W}{c}\right) \leq -a \ \frac{W}{c} \ + \ 1,$$

we can also assume without loss of generality that (17.8) is met for some non-negative function $W \geq 0$, with $c = 1$.

Last, but not least it suffices to check that there exists a subset $A \subset S$ such that

$$\forall x \in S - A \quad W^{-1}(x)L(W)(x) \leq -a \quad \text{and} \quad \forall x \in A \quad L(W)(x) \leq c.$$

In this situation, it is readily checked that

$$L(W) = W^{-1}L(W)1_{S-A} \ W \ + \ L(W) \ 1_A \Rightarrow L(W) \leq -a \ W \ +c. \tag{17.9}$$

For instance, when S is equipped with some norm $\|.\|$, and $L(W)$ is continuous, it suffices to find a sufficiently large radius R such that

$$\forall \|x\| \geq R \qquad W^{-1}(x)L(W)(x) \leq -a. \tag{17.10}$$

17.4.2 Minorization properties

We mention that any sg P_t on $S = \mathbb{R}^d$ satisfies the Dobrushin local contraction condition (8.27) for any $t > 0$ as soon as the Markov transitions $P_t(x, dy) = p_t(x, y) \ dy$ have continuous densities $(x, y) \mapsto p_t(x, y) > 0$ w.r.t. the Lebesgue measure dy. Much more is true. Rephrasing proposition 8.2.18 we have:

Proposition 17.4.2 *We assume that the sg P_t satisfies the minorization property*

$$P_t(x, dy) \geq q_t(x, y) \ dy \tag{17.11}$$

for some function $q_t(x, y)$ that is lower semicontinuous w.r.t. the first variable, and upper semicontinuous w.r.t. the second. In this situation the sg P_t satisfies the Dobrushin local contraction condition (8.27) for any $t > 0$.

Proposition 17.4.3 *We consider a d-dimensional jump diffusion process with jump intensity function $\lambda(x)$, jump amplitude transition $M(x, dy)$ and a stochastic flow $\varphi_{s,t}(x)$. We assume that the sg P_t^0 of the stochastic flow satisfies the minorisation condition (17.11) for some $q_t^0(x, y)$. When the intensity function λ is bounded, the sg P_t of the jump diffusion model satisfies (17.11) with*

$$P_t(x, dy) \geq e^{-\|\lambda\|t} \ q_t^0(x, y) \ dy.$$

Proof :

The proof is a direct consequence of the integral formula (13.18) (which is valid for jump diffusion processes following the remark given on page 427) and the definition (15.5) of the sg P_t^λ. Indeed, by (13.18) we have

$$
\begin{aligned}
P_t(f) &= P_t^\lambda(f) + \int_0^t P_u^\lambda(\lambda M P_{t-u}(f))\, du \\
&\geq P_t^\lambda(f) \geq e^{-\|\lambda\| t}\, P_t^0(f)
\end{aligned}
$$

for any non-negative function f. This clearly ends the proof of the proposition. ∎

17.5 Some applications

It is far beyond the scope of these lectures to discuss in full details the absolute continuity properties of Markov semigroups. A brief discussion is provided in section 18.4.2. For a more detailed discussion we refer the reader to [57, 143, 161], and the references therein.

In this section we present examples of Lyapunov functions for some classes of diffusion processes.

17.5.1 Ornstein-Uhlenbeck processes

The infinitesimal generator of the Ornstein-Uhlenbeck process is given by

$$
L := -\sum_{i=1}^d a_i\, x_i\, \partial_i + \frac{1}{2} \sum_{i,j=1}^d \left(\sigma(\sigma)^T\right)_{i,j} \partial_{i,j}
$$

with some deterministic covariance matrix σ and a collection of parameters $a_i < \infty$. In this situation, the sg $P_t(x, dy) = p_t(x, y)\, dy$ has smooth densities $(x, y) \mapsto p_t(x, y) > 0$ w.r.t. the Lebesgue measure dy [143]. When $a_{min} = \wedge_{1 \leq i \leq d} a_i > 0$, we can choose the Lyapunov function

$$
W(x) = \frac{1}{2}\, \|x\|^2 := \frac{1}{2} \sum_{1 \leq i \leq d} x_i^2.
$$

$$
\begin{aligned}
L(W)(x) &= -\sum_{i=1}^d a_i\, x_i^2 + \frac{1}{2}\, \mathrm{Trace}\left(\sigma(\sigma)^T\right) \\
&\leq -2\, a_{min}\, W(x) + \mathrm{Trace}\left(\sigma\sigma^T\right).
\end{aligned}
$$

17.5.2 Stochastic gradient processes

The infinitesimal generator of the stochastic gradient process is given by

$$
L := -\sum_{i=1}^d \partial_i V\, \partial_i + \frac{1}{2} \sum_{i=1}^d \partial_{i,i}
$$

with a smooth function V behaving as $\|x\|^\alpha$ with $\alpha \geq 1$ at infinity; that is, there exists some sufficiently large radius R such that for any $\|x\| \geq R$ we have

$$\sum_{i=1}^{d} |\partial_{i,i} V(x)| \leq c_1 \|x\|^{\alpha-2} \quad \text{and} \quad \sum_{i=1}^{d} (\partial_i V(x))^2 \geq c_2 \|x\|^{2(\alpha-1)}$$

for some constants $c_1 < \infty$ and $c_2 > 0$. More generally, potentials s.t. $\lim_{\|x\| \to \infty} V(x) = \infty$ and $e^{-\beta V} \in \mathbb{L}_1(dx)$, for any $\beta > 0$ are called confining potentials. Under these conditions, the stochastic gradient process is ergodic with the unique invariant measure given by the Gibbs probability measure

$$\pi(dx) = Z^{-1} e^{-2^{-1} V(x)} \, dx.$$

The normalizing constant Z is often called the partition function. Further details on these processes are provided in section 23.4 dedicated to stochastic flows in Euclidian and Riemannian manifolds.

In this situation, the sg $P_t(x, dy) = p_t(x, y) \, dy$ has smooth densities $(x, y) \mapsto p_t(x, y) > 0$ w.r.t. the Lebesgue measure dy [143]. In addition, we can choose the Lyapunov function

$$W(x) = \exp(2\epsilon V)$$

for any $\epsilon \in]0, 1[$. To check this claim, we observe that

$$W^{-1} L(W)$$

$$= -2\epsilon \sum_{i=1}^{d} (\partial_i V)^2 + \frac{1}{2} \sum_{i=1}^{d} \left(4\epsilon^2 (\partial_i V)^2 + 2\epsilon \partial_{i,i} V\right) \leq -2\epsilon \left[(1-\epsilon) \|\nabla V\|^2 - \frac{1}{2} \triangle V\right].$$

Under our assumptions, for any $\|x\| \geq R > [c_1/(2c_2)]^{1/\alpha}$ we have

$$\begin{aligned}
(1-\epsilon) \|\nabla V\|^2 - \frac{1}{2} \triangle V &\geq (1-\epsilon) c_2 \|x\|^{2(\alpha-1)} - \frac{c_1}{2} \|x\|^{\alpha-2} \\
&= \|x\|^{\alpha-2} \left[c_2 \|x\|^\alpha - \frac{c_1}{2}\right] \\
&\geq R^{\alpha-2} \left[c_2 R^\alpha - \frac{c_1}{2}\right] > 0.
\end{aligned}$$

We conclude that (17.10) holds with $a := 2\epsilon R^{\alpha-2} \left[c_2 R^\alpha - \frac{c_1}{2}\right]$.

17.5.3   Langevin diffusions

We consider the $\mathbb{R}^2$-valued stochastic process $X_t = (q_t, p_t)$ defined by

$$\begin{cases}
dq_t &= \beta \frac{p_t}{m} \, dt \\
dp_t &= -\beta \left(\frac{\partial V}{\partial q}(q_t) + \frac{\sigma^2}{2} \frac{p_t}{m}\right) \, dt + \sigma \, dW_t
\end{cases} \tag{17.12}$$

with some positive constants β, m, σ, a Brownian motion W_t, and a smooth positive function V on $\mathbb{R}$ such that for sufficiently large R we have

$$|q| \geq R \qquad q \frac{\partial V}{\partial q}(q) \geq \delta \left(V(q) + q^2\right)$$

for some positive constant δ. This condition is clearly met when V behaves as $q^{2\alpha}$ for certain $\alpha \geq 1$ at infinity, that is, when there exists some sufficiently large radius R such that for any $|q| \geq R$ we have

$$q \frac{\partial V}{\partial q}(q) \geq c_1 q^{2\alpha} \quad \text{and} \quad c_2 q^{2\alpha} \geq V(q).$$

Here $c_1 < \infty$ and $c_2 > 0$ are suitably chosen constants.

The generator of the process (17.12) is defined by

$$L(f)(q,p) = \beta \, \frac{p}{m} \, \frac{\partial f}{\partial q} - \beta \left(\frac{\partial V}{\partial q} + \frac{\sigma^2}{2} \, \frac{p}{m} \right) \frac{\partial f}{\partial p} + \frac{\sigma^2}{2} \, \frac{\partial^2 f}{\partial p^2}.$$

We let $W(q,p)$ be the function on $\mathbb{R}^2$ defined by

$$W(q,p) = \frac{1}{2m} \, p^2 + V(q) + \frac{\epsilon}{2} \left(\frac{\sigma^2}{2} \, q^2 + 2pq \right) \quad \text{with} \quad \epsilon < \frac{\sigma^2}{2m}.$$

Recalling that $2pq \le p^2 + q^2$, we prove that

$$
\begin{aligned}
W(q,p) &\le \frac{1}{2} \left(\frac{1}{m} + \epsilon \right) p^2 + \frac{\epsilon}{2} \left(\frac{\sigma^2}{2} + 1 \right) q^2 + V(q) \\
&\le C^\star(\epsilon) \, (p^2 + q^2 + V(q))
\end{aligned}
$$

with

$$C^\star(\epsilon) := \max \left\{ \frac{1}{2} \left(\frac{1}{m} + \epsilon \right), \frac{\epsilon}{2} \left(\frac{\sigma^2}{2} + 1 \right), 1 \right\}.$$

On the other hand, we have

$$
\begin{aligned}
L(W) &= \beta \frac{p}{m} \left(\frac{\partial V}{\partial q} + \epsilon \frac{\sigma^2}{2} \, q + \epsilon \, p \right) \\
&\quad - \beta \left(\frac{\partial V}{\partial q} + \frac{\sigma^2}{2} \, \frac{p}{m} \right) \left(\frac{p}{m} + \epsilon \, q \right) + \frac{\sigma^2}{2m} \\
&= -\beta \left[\frac{1}{m} \left(\frac{\sigma^2}{2m} - \epsilon \right) p^2 + \epsilon \, q \, \frac{\partial V}{\partial q} \right] + \frac{\sigma^2}{2m}.
\end{aligned}
$$

Under our assumptions, this implies that for any $|q| \ge R$ we have

$$
\begin{aligned}
L(W) &\le -\beta \left[\frac{1}{m} \left(\frac{\sigma^2}{2m} - \epsilon \right) p^2 + \epsilon \, \delta \, (V(q) + q^2) \right] + \frac{\sigma^2}{2m} \\
&\le -C_\star(\epsilon, \delta) \, (p^2 + q^2 + V(q)) + \frac{\sigma^2}{2m}
\end{aligned}
$$

with

$$C_\star(\epsilon, \delta) := \beta \, \min \left\{ \left(\frac{1}{m} \left(\frac{\sigma^2}{2m} - \epsilon \right) \right), \epsilon \, \delta \right\}.$$

We conclude that for any $|q| \ge R$ we have

$$
\begin{aligned}
(W^{-1} L(W))(q,p) &\le -\frac{C_\star(\epsilon, \delta) \, (p^2 + q^2 + V(q)) - \frac{\sigma^2}{2m}}{W(q,p)} \\
&\le -\frac{C_\star(\epsilon, \delta) \, (p^2 + q^2 + V(q)) - \frac{\sigma^2}{2m}}{C^\star(\epsilon) \, (p^2 + q^2 + V(q))} \\
&= -\frac{C_\star(\epsilon, \delta)}{C^\star(\epsilon)} + \frac{\sigma^2}{2mC^\star(\epsilon)} \, \frac{1}{p^2 + q^2 + V(q)} \\
&\le -\left[\frac{C_\star(\epsilon, \delta)}{C^\star(\epsilon)} - \frac{\sigma^2}{2mC^\star(\epsilon)} \, \frac{1}{p^2 + q^2} \right] \\
&\le -\left[\frac{C_\star(\epsilon, \delta)}{C^\star(\epsilon)} - \frac{\sigma^2}{2mC^\star(\epsilon)} \, \frac{1}{p^2 + R^2} \right].
\end{aligned}
$$

Choosing $\overline{R}$ sufficiently large to satisfy

$$|p| \geq \overline{R} \quad \text{and} \quad |q| \geq R$$

$$\Rightarrow \frac{C_\star(\epsilon,\delta)}{C^\star(\epsilon)} - \frac{\sigma^2}{2mC^\star(\epsilon)} \, \frac{1}{p^2+q^2} \geq \frac{C_\star(\epsilon,\delta)}{C^\star(\epsilon)} - \frac{\sigma^2}{2mC^\star(\epsilon)} \, \frac{1}{\overline{R}^2+R^2} \geq \frac{C_\star(\epsilon,\delta)}{2C^\star(\epsilon)} > 0,$$

we conclude that (17.9) is met with the set

$$A = \{|p| \wedge |q| < \overline{R} \vee R\}$$

and the parameter $a = C_\star(\epsilon,\delta)/(2C^\star(\epsilon)) > 0$; that is,

$$|p| \wedge |q| \geq \overline{R} \vee R \Rightarrow (W^{-1}L(W))(q,p) \leq -a.$$

17.6 Spectral analysis

17.6.1 Hilbert spaces and Schauder bases

Suppose we are given a generator L of a jump diffusion process and probability measure π on some state space S which is L-reversible. We let P_t be the Markov semigroup of the stochastic process X_t with generator L.

We denote by $\mathcal{E}_L(f,g)$ the Dirichlet bilinear form defined on sufficiently smooth functions by the formula

$$\mathcal{E}_L(f,g) = \pi(f(-L)(g)) = \frac{1}{2} \, \pi \, (\Gamma_L(f,g))$$

with the carré du champ operator associated to L (the r.h.s. formula in the above display comes from the fact that $\pi L = 0$).

The required regularity properties on (f,g) depend on the state space S and the generator L. The prototype of one-dimensional reversible process is the Ornstein-Uhlenbeck diffusion on $S = \mathbb{R}$ given by

$$dX_t = -2 \, X_t \, dt + \sqrt{2} \, dW_t \tag{17.13}$$

with a Brownian motion W_t. By (17.5), the generator L of X_t and its reversible probability measure are given by

$$L(f)(x) = -2x \, \partial_x f(x) + \partial_x^2 f(x) = e^{x^2} \partial_x \left(e^{-x^2} \, \partial_x f \right)(x) \quad \text{and} \quad \pi(dx) = \frac{1}{\sqrt{\pi}} \, e^{-x^2} \, dx. \tag{17.14}$$

In this situation, for any smooth functions (f,g) with compact support by an integration by parts,

$$\mathcal{E}_L(f,g) = -\int \pi(dx) \, f(x) \, L(g)(x) \, dx \tag{17.15}$$

$$= -\frac{1}{\sqrt{\pi}} \int f(x) \, \partial_x \left(e^{-x^2} \, \partial_x g \right)(x) \, dx = \int \pi(dx) \, \partial_x(f)(x) \partial_x g(x) \, dx.$$

We further assume that the Hilbert space $\mathbb{L}_2(\pi)$ equipped with the scalar product $\langle f, g \rangle_\pi = \pi(fg)$, has a complete orthonormal basis of sufficiently smooth eigenfunctions $(\varphi_n)_{n \geq 0}$ associated with the eigenvalues $(\lambda_n)_{n \geq 0}$ of $(-L)$, with $0 \leq \lambda_n \leq \lambda_{n+1}$ for any $n \geq 0$, and $\sum_{n \geq 0} e^{-\lambda_n} < \infty$. In other words, for any $n \geq 0$ and any $f \in \mathbb{L}_2(\pi)$,

$$L(\varphi_n) = -\lambda_n \, \varphi_n \quad \text{and} \quad f = \sum_{n \geq 0} \langle f, \varphi_n \rangle_\pi \, \varphi_n \tag{17.16}$$

where the convergence of the r.h.s. series in the above display is in $\mathbb{L}_2(\pi)$. In other words, $(\varphi_n)_{n \geq 0}$ is dense in $\mathbb{L}_2(\pi)$. Complete orthogonal bases for an infinite dimensional Hilbert space are called Schauder bases.

It is important to notice that

$$L(1) = 0 \Rightarrow (\lambda_0 = 0 \quad \text{and} \quad \varphi_0(x) = 1) \Rightarrow f = \pi(f) + \sum_{n \geq 1} \langle f, \varphi_n \rangle_\pi \, \varphi_n. \tag{17.17}$$

This yields the variance formula

$$\begin{aligned} \mathrm{Var}_\pi(f) &:= \pi[(f - \pi(f))^2] \\ &= \sum_{n \geq 1} \langle f, \varphi_n \rangle_\pi^2 \leq \lambda_1^{-1} \sum_{n \geq 1} \lambda_n \, \langle f, \varphi_n \rangle_\pi^2 \quad (\Leftarrow \lambda_n \uparrow). \tag{17.18} \end{aligned}$$

We illustrate this rather abstract model with our prototype Gaussian model (17.13). We let $\mathbb{H}_n$ be the Hermite polynomials defined by the generating function

$$\mathbb{S}_t(x) = e^{2tx - t^2} = \sum_{n \geq 0} \frac{t^n}{n!} \, \mathbb{H}_n(x). \tag{17.19}$$

The orthogonality property of the functions $\varphi_n = \mathbb{H}_n$ is easily checked in terms of $\mathbb{S}_t$. Indeed, we clearly have the decompositions

$$\begin{aligned} \pi(\mathbb{S}_s \mathbb{S}_t) &= \frac{1}{\sqrt{\pi}} \int e^{2(s+t)x - s^2 - t^2} \, e^{-x^2} \, dx = e^{2st} \frac{1}{\sqrt{\pi}} \int e^{-(x-(t+s))^2} \, dx = e^{2st} \\ &= \sum_{n \geq 0} 2^n \frac{(st)^n}{n!} = \sum_{m, n \geq 0} \frac{t^n s^m}{n! m!} \, \pi(\mathbb{H}_n \mathbb{H}_m) \Rightarrow \pi(\mathbb{H}_n \mathbb{H}_m) = 1_{m=n} \, 2^n \, n! \end{aligned}$$

The claim that the normalized Hermite polynomials $2^{n/2}\sqrt{n!} \, \mathbb{H}_n$ form an orthonormal Schauder basis of $\mathbb{L}_2(\pi)$ can be found in any textbook on Hilbert spaces. For the convenience of the reader we sketch this elementary proof. The key ingredient is to prove that the only functions $f(x)$ that are orthogonal to all Hermite polynomials are the null functions (almost everywhere). Since any polynomial can be expressed as a finite linear combination of Hermite polynomials, it is equivalent to show that the functions that are orthogonal to all monomials $(x^n)_{n \geq 0}$ are the null functions. Notice that the entire function on the complex plane $z \in \mathbb{T}$ defined by

$$F(z) = \int f(x) \, e^{xz} \, \pi(dx) = \sum_{n \geq 0} \frac{z^n}{n!} \int f(x) \, x^n \, \pi(dx)$$

is holomorphic (i.e., differentiable in the complex domain in a neighborhood of every point z) and is null on the real line $z = y \in \mathbb{R}$ as soon as f is orthogonal to all monomials $(x^n)_{n \geq 0}$. The interchange of the summation and the integration in the above display is ensured by the monotone convergence theorem. This implies that F is also null on the whole complex plane. On the other hand $F(-iy) = 0$ is the Fourier transform of the function $e^{-x^2} f(x)$ evaluated as y. Taking the Fourier inverse we check that $e^{-x^2} f(x) = 0$ almost everywhere, from which we readily conclude that $f(x) = 0$ almost everywhere.

Using $\partial_t e^{-(t-x)^2} = -\partial_x e^{-(t-x)^2} \Rightarrow (\partial_t^n e^{-(t-x)^2})_{t=0} = (-1)^n \partial_x^n e^{-x^2}$, we also have

$$\mathbb{S}_t(x) = e^{x^2} \, e^{-(t-x)^2} = e^{x^2} \sum_{n \geq 0} (-1)^n \partial_x^n e^{-x^2} \, \frac{t^n}{n!} = \sum_{n \geq 0} \mathbb{H}_n(x) \, \frac{t^n}{n!}.$$

> This yields the Rodrigues formula
>
> $$\mathbb{H}_n(x) = (-1)^n \, e^{x^2} \, \partial_x^n e^{-x^2}.$$

We also notice that

$$
\begin{aligned}
\partial_t \mathbb{S}_t(x) &= 2(x - t) \, \mathbb{S}_t(x) = \sum_{n \geq 0} \frac{t^n}{n!} \, [2x \, \mathbb{H}_n(x) - 2n \, \mathbb{H}_{n-1}(x)] \\
&= \sum_{n \geq 0} \frac{t^n}{n!} \, \mathbb{H}_{n+1}(x) \Rightarrow \mathbb{H}_{n+1}(x) = 2x \, \mathbb{H}_n(x) - 2n \, \mathbb{H}_{n-1}(x)
\end{aligned}
$$

with the convention $\mathbb{H}_{-1} = 0$ and

$$
\begin{aligned}
\partial_x \mathbb{S}_t(x) &= 2t \, \mathbb{S}_t(x) = 2 \sum_{n \geq 1} \frac{t^n}{n!} \, n \mathbb{H}_{n-1}(x) \\
&= \sum_{n \geq 0} \frac{t^n}{n!} \, \partial_x \mathbb{H}_n(x) \Rightarrow \partial_x \mathbb{H}_n = 2n \mathbb{H}_{n-1}.
\end{aligned}
$$

This yields

$$\mathbb{H}_{n+1}(x) = 2x \, \mathbb{H}_n(x) - \partial_x \mathbb{H}_n(x) \quad \text{and} \quad \partial_x \mathbb{H}_{n+1} = 2(n+1)\mathbb{H}_n$$

from which we prove that

$$2(n+1)\mathbb{H}_n = \partial_x \mathbb{H}_{n+1} = \partial_x \left(2x \mathbb{H}_n - \partial_x \mathbb{H}_n\right) = 2\mathbb{H}_n + 2x\partial_x \mathbb{H}_n - \partial_x^2 \mathbb{H}_n.$$

Hence

$$\partial_x^2 \mathbb{H}_n - 2x\partial_x \mathbb{H}_n = L(\mathbb{H}_n) = -\lambda_n \, \mathbb{H}_n \quad \text{with} \quad \lambda_n = 2n. \tag{17.20}$$

This shows that (17.16) is satisfied with $\varphi_n = 2^{n/2}\sqrt{n!} \, \mathbb{H}_n$ and $\lambda_n = 2n$.

> Using the Sturm-Liouville formulation (17.14) of the generator L, we readily check the Sturm-Liouville formulation of the Hermite polynomials
>
> $$(17.20) \iff -\frac{1}{2} \, e^{x^2} \partial_x \left(e^{-x^2} \, \partial_x \mathbb{H}_n\right)(x) = n\mathbb{H}_n(x).$$

17.6.2 Spectral decompositions

We quote some direct consequences of (17.16). Firstly, we have

$$\frac{d}{dt} P_t(\varphi_n) = P_t(L(\varphi_n)) = -\lambda_n \, P_t(\varphi_n) \; \Rightarrow \; P_t(\varphi_n) = e^{-\lambda_n t} \varphi_n.$$

This implies that

$$P_t(f) = \sum_{n \geq 0} \langle f, \varphi_n \rangle_\pi \, P_t(\varphi_n) = \sum_{n \geq 0} e^{-\lambda_n t} \, \langle f, \varphi_n \rangle_\pi \, \varphi_n.$$

By (17.17) we obtain the spectral decomposition of the Markov semigroup

$$P_t(f) = \pi(f) + \sum_{n \geq 1} e^{-\lambda_n t} \, \langle f, \varphi_n \rangle_\pi \, \varphi_n.$$

The spectral decomposition of the generator L is obtained using the same types of arguments. By (17.16) and (17.17) we clearly have the spectral decompositions

$$L(f) = \sum_{n \geq 1} \overline{\lambda}_n \, \langle f, \varphi_n \rangle_\pi \, \varphi_n \quad \text{with} \quad \overline{\lambda}_n = -\lambda_n$$

and

$$\mathcal{E}_L(f, g) = \frac{1}{2} \, \pi \left(\Gamma_L(f, g) \right) = \langle f, (-L)(g) \rangle_\pi = \sum_{n \geq 1} \lambda_n \, \langle g, \varphi_n \rangle_\pi \, \langle f, \varphi_n \rangle_\pi. \quad (17.21)$$

The spectral decomposition formula of L is sometimes written in the form

$$L(f) = \sum_{n \geq 0} \overline{\lambda}_n \, E_{\overline{\lambda}_n}(f) \quad \text{with} \quad E_{\overline{\lambda}_n}(f) = \langle f, \varphi_n \rangle_\pi \, \varphi_n.$$

We introduce the projection valued measures $A \mapsto E_A(f)$ on the Borel subsets $A \subset \mathbb{R}$ defined by

$$E_A(f) = \sum_{n \geq 0} 1_A(\overline{\lambda}_n) \, \langle f, \varphi_n \rangle_\pi \, \varphi_n := \int 1_A(\lambda) \, dE_\lambda(f).$$

In this notation, we have

$$L(f) = \int \lambda \, dE_\lambda(f) \quad \text{or in a more synthetic form} \quad L = \int \lambda \, dE_\lambda.$$

Notice that

$$
\begin{aligned}
E_A(E_B(f)) &= \sum_{n \geq 0} 1_A(\overline{\lambda}_n) \, \langle E_B(f), \varphi_n \rangle_\pi \, \varphi_n \\
&= \sum_{m,n \geq 0} 1_A(\overline{\lambda}_n) \, 1_B(\overline{\lambda}_m) \, \underbrace{\langle \varphi_m, \varphi_n \rangle_\pi}_{=1_{m=n}} \, \langle f, \varphi_m \rangle_\pi \, \varphi_n = E_{A \cap B}(f).
\end{aligned}
$$

17.6.3   Poincaré inequality

Combining (17.21) with (17.18) we obtain the Poincaré inequality

$$\lambda_1 \, \mathrm{Var}_\pi(f) \; \leq \; \mathcal{E}_L(f, f) = \frac{1}{2} \, \pi \left(\Gamma_L(f, f) \right) \qquad\qquad (17.22)$$

for any smooth and compactly supported function f.

A simple semigroup derivation shows that

$$\frac{1}{2} \frac{d}{dt} \pi \left[(P_t(f))^2 \right] \;\; = \;\; \pi \left[P_t(f) \, \frac{\partial}{\partial t} P_t(f) \right] = \pi \left[P_t(f) \, L\left(P_t(f) \right) \right]$$

$$= \;\; -\frac{1}{2} \, \pi \left(\Gamma_L \left[P_t(f), P_t(f) \right] \right) \leq -\lambda_1 \, \pi \left(\left(P_t(f) \right)^2 \right)$$

for any function f such that $\pi(f) = 0$.

This clearly implies the exponential decay to equilibrium

$$\pi \left[\left(P_t(f) - \pi(f) \right)^2 \right] \leq e^{-2\lambda_1 t} \, \pi \left[\left(f - \pi(f) \right)^2 \right] . \qquad\qquad (17.23)$$

In the reverse direction, suppose that (17.23) holds for some constant λ_1. In this case, recalling that

$$P_\epsilon(f) = f + L(f) \, \epsilon \, + \mathrm{o}(\epsilon) \quad \text{and} \quad \leq e^{-2\lambda_1 \epsilon} = 1 - 2\lambda_1 \epsilon + \mathrm{o}(\epsilon)$$

for any centered function f s.t. $\pi(f) = 0$, by (17.23),

$$\pi \left(f^2 \right) + 2 \, \pi(f L(f)) \, \epsilon \, + \mathrm{o}(\epsilon) = \pi(f^2) - 2 \, \mathcal{E}_L(f, f) \, + \mathrm{o}(\epsilon) \leq (1 - 2\lambda_1 \epsilon) \, \pi \left(f^2 \right) + \mathrm{o}(\epsilon).$$

After some elementary manipulations, this implies that

$$\mathcal{E}_L(f, f) \geq \lambda_1 \, \pi \left(f^2 \right) .$$

In other words, the exponential decay to equilibrium (17.23) is equivalent to the Poincaré inequality (17.22) for some constant λ_1. The best constant is clearly the lowest eigenvalue λ_1 of the spectral decomposition of L presented in (17.16).

For instance, using (17.13), (17.15) and (17.20) we obtain without further work the Poincaré inequality for the Gaussian distribution

$$\pi(dx) = \frac{1}{\sqrt{\pi}} \, e^{-x^2} \, dx \Rightarrow 2 \, \mathrm{Var}_\pi(f) \; \leq \; \pi \left((f')^2 \right) .$$

We readily recover the Poincaré inequality discussed in exercise 258 for the rescaled Gaussian distribution

$$\overline{\pi}(dx) = \frac{1}{\sqrt{2\pi}} \, e^{-x^2/2} \, dx \implies \mathrm{Var}_{\overline{\pi}}(\overline{f}) \; \leq \; \overline{\pi}((\overline{f}')^2) \qquad\qquad (17.24)$$

for any smooth and compactly supported function $\overline{f}$. We check this claim using

$$\pi(f^2) = \int \overline{\pi}(dx) \; f^2(x/\sqrt{2}) \; dx \quad \text{and} \quad \pi\left((f')^2\right) = \int \overline{\pi}(dx) \; f'\left(x/\sqrt{2}\right)^2 \; dx$$

for any function f s.t. $\pi(f) = 0$. If we set $\overline{f}(x) = f(x/\sqrt{2})$ we clearly have $\sqrt{2} \; \overline{f}'(x) = f'(x/\sqrt{2})$ from which we conclude that

$$2 \int \overline{\pi}(dx) \; \overline{f}(x)^2 \; dx \leq 2 \int \overline{\pi}(dx) \; \overline{f}'(x)^2 \; dx \Longleftrightarrow (17.24).$$

For more general rescaling properties, we refer the reader to exercise 329.

17.7 Exercises

Exercise 321 (Gaussian processes) *Consider the diffusion process*

$$dX_t = -(\alpha + \beta \; X_t) \; dt + \tau \; dW_t \tag{17.25}$$

for some parameters $(\alpha, \beta, \tau) \in \mathbb{R} \times]0, +\infty[^2$ and a Brownian motion W_t. Using (17.6) find the reversible measure of X_t.

Exercise 322 (Landau-Stuart diffusions) *Consider the one-dimensional Landau-Stuart diffusion process*

$$dX_t = \alpha \; X_t \; (1 - X_t^2) \; dt + \sqrt{2} \; \tau \; X_t \; dW_t \tag{17.26}$$

with some reflecting and diffusion parameters α and $\tau \in [0, \infty[$ s.t. $\tau^2 < \alpha$, and a Brownian motion W_t. Using (17.6) find the stationary measure of the process X_t starting in $]0, \infty[$.

Exercise 323 (Gamma distributions) *Consider the diffusion process*

$$dX_t = -(\alpha + \beta \; X_t) \; dt + \sqrt{\tau^2 + \rho \; X_t} \; dW_t \tag{17.27}$$

with $X_0 > m := -\frac{\tau^2}{\rho}$, and some parameters $(\alpha, \beta, \tau, \rho)$, with $\alpha < 0$, and $\beta, \rho > 0$, and a Brownian motion W_t. Using (17.6) find the reversible measure of X_t. Compare this model with the diffusion process discussed in exercise 262.

Exercise 324 (Square root processes) *Consider the diffusion X_t defined in (17.27). Find the evolution equation of $Y_t := (X_t - m)$. We further assume that $\gamma = n\rho/4$, for some integer $n \geq 1$. We consider a sequence $U_t = \left(U_t^{(1)}, \ldots, U_t^{(n)}\right)$ of n independent Ornstein-Uhlenbeck processes of the following form*

$$\forall 1 \leq i \leq n \qquad dU_t^{(i)} = -\frac{\beta}{2} \; U_t^{(i)} \; dt + \frac{\sqrt{\rho}}{2} \; dW_t^{(i)}$$

with n independent Brownian motion $W_t^{(i)}$. Find the evolution equation of $Z_t := \|U_t\|^2 = \sum_{1 \leq i \leq n} (U_t^{(i)})^2$. Deduce that $Y \stackrel{law}{=} Z$ and $X \stackrel{law}{=} m + Z$.

Exercise 325 (Jacobi processes - Beta distributions) *Consider the diffusion process*

$$dX_t = -(\alpha + \beta\ X_t)\ dt + \sqrt{\tau^2\ (X_t - \gamma_1)\ (\gamma_2 - X_t)}\ dW_t \qquad (17.28)$$

with $X_0 \in S := [\gamma_1, \gamma_2]$, and some parameters $(\alpha, \beta, \tau, \gamma_1, \gamma_2)$, with $\gamma_1 < \gamma_2$ and $\alpha + \beta\gamma_1 < 0 < \alpha + \beta\gamma_2$. Using (17.6) find the reversible measure of X_t. Examine the situations:

$$(\gamma_1, \gamma_2) = (0,1) \qquad \beta > 0 \qquad -\alpha/\beta = m \in]0,1[\quad and \quad \tau^2 = 2\beta\nu \quad with \quad \nu > 0$$

as well as the models:

- *$(\alpha, \beta) = (0,1)$, $\tau^2 = 2$ and $(\gamma_1, \gamma_2) = (-1,1)$.*

- *$(\alpha, \beta) = (0,2)$, $\tau^2 = 2$ and $(\gamma_1, \gamma_2) = (-1,1)$.*

In these two cases, write the equations, the generator, and the reversible probability measure of the corresponding $[-1,1]$-valued diffusion processes.

Exercise 326 (Inverse gamma distributions) *Consider the diffusion process given by the equation*

$$dX_t = -(\alpha + \beta\ X_t)\ dt + \tau X_t\ dW_t \qquad (17.29)$$

with $X_0 \in S :=]0, \infty[$, and some parameters (α, β, τ) with $\tau > 0$ and $\alpha < 0$. Using (17.6) find the reversible measure of X_t.

Exercise 327 (Fisher distributions) *Consider the diffusion process given by the equation*

$$dX_t = -(\alpha + \beta\ X_t)\ dt + \sqrt{\tau^2\ (X_t + \gamma_1)\ (X_t + \gamma_2)}\ dW_t \qquad (17.30)$$

with $X_0 \in S :=] - \gamma_1, \infty[$. and some parameters $(\alpha, \beta, \tau, \gamma_1, \gamma_2)$, s.t. $\alpha/\beta < \gamma_1 < \gamma_2$ and $2\beta + \tau^2 > 0$. Using (17.6) find the reversible measure π of X_t. Let X be a random variable with distribution π. Find the distribution of the random variable

$$Y = \frac{d_2}{d_1}\ \left(\frac{X + \gamma_1}{\delta}\right) \quad with \quad d_1/2 = \frac{2\beta}{\tau^2}\ \frac{\gamma_1 - \frac{\alpha}{\beta}}{\gamma_2 - \gamma_1} > 0 \quad and \quad d_2/2 = 1 + 2\beta/\tau^2 > 0.$$

Examine the situation

$$(\gamma_1, \gamma_2) = (0,1) \qquad \beta > 0 \qquad -\alpha/\beta = m > 0 \quad and \quad \tau^2 = 2\beta\nu \quad with \quad \nu > 0.$$

Exercise 328 (Student distribution) *Consider the Student diffusion process given by the equation*

$$dX_t = -(\alpha + \beta\ X_t)\ dt + \sqrt{\tau^2\ ((\alpha + \beta\ X_t)^2 + \gamma^2)}\ dW_t \qquad (17.31)$$

with $X_0 \in S = \mathbb{R}$, and some parameters $(\alpha, \beta, \tau, \gamma))$, with $\beta > 0$. Using (17.6) find the reversible measure π of X_t.

Exercise 329 (Rescaled Hermite polynomials) *We consider a function h on $\mathbb{R}$ such that $h'(x) > 0$ for any $x \in \mathbb{R}$. We denote by π_h the probability measure $\pi_h(dx) \propto h'(x)\ e^{-h^2(x)}\ dx$. We consider the Ornstein-Uhlenbeck generators*

$$\overline{L}_h(f) := (h')^{-2}\ L_h(f) \quad with \quad L_h := f'' - \left(h^2 + \log h'\right)'\ f'.$$

We also denote by $\overline{\mathbb{H}}_n := \mathbb{H}_n \circ h$ the h-rescaling of the Hermite polynomials $\mathbb{H}_n$ defined in (17.19).

- *Check that the sequence of functions $2^{n/2}\sqrt{n!}\ \overline{\mathbb{H}}_n$ forms an orthonormal basis of $\mathbb{L}_2(\pi_h)$.*

- *Find the reversible probability measure of the generators L_h and $\overline{L}_h$, and check that*

$$\forall n \geq 1 \qquad \overline{L}_h(\overline{\mathbb{H}}_n) = -2n\ \overline{\mathbb{H}}_n.$$

- *Find the reversible probability measure π of the generator L defined by*

$$L(f) := -(ax+b)\ f' + \frac{1}{2}\ \sigma^2\ f''$$

for some parameters (a, b, σ^2) with $a > 0$. Prove that the functions

$$\widehat{\mathbb{H}}_n(x) := 2^{n/2}\sqrt{n!}\ \mathbb{H}_n\left(\sqrt{\frac{a}{\sigma^2}}\ \left(x + \frac{b}{a}\right)\right)$$

form an orthonormal basis of $\mathbb{L}_2(\pi)$ s.t. $\widehat{L}_h(\widehat{\mathbb{H}}_n) = -na\ \widehat{\mathbb{H}}_n$.

Exercise 330 (Laguerre polynomials) *We consider the generalized Laguerre polynomials $\mathbb{I}_n$ defined by the generating function*

$$\mathbb{S}_t(x) = \sum_{n \geq 0} \mathbb{I}_n(x)\ \frac{t^n}{n!} = (1-t)^{-(\alpha+1)}\ e^{-\frac{xt}{1-t}}$$

for any $1 + \alpha > 0$, $x \geq 0$ and $t \neq 1$. We also consider the gamma distribution

$$\pi(dx) = \frac{1}{\Gamma(\alpha+1)}\ 1_{[0,\infty[}(x)\ x^\alpha\ e^{-x}\ dx$$

and the differential operator

$$L(f)(x) = x\ f''(x) + ((\alpha+1) - x)\ f'(x)$$

of the $[0, \infty[$-valued diffusion process

$$dX_t = ((\alpha+1) - X_t)\ dt + \sqrt{2X_t}\ dW_t$$

with $X_0 > 0$. This diffusion process belongs to the class of square root processes discussed in exercise 323.

- *Check the Sturm-Liouville formula*

$$L(f) = x^{-\alpha}\ e^x\ \partial_x\left(x^{\alpha+1}\ e^{-x}\ \partial_x(f)\right).$$

- *We set $t_m := \frac{m}{m+1}$, for $m \geq 0$. Check that $e^{-mx} = (m+1)^{-(\alpha+1)}\sum_{n \geq 0} \mathbb{I}_n(x)\ \frac{t_m^n}{n!}$ and for any $u \neq 1$ we have*

$$\frac{1}{(1-u)^{\alpha+1}} = \sum_{n \geq 0} \frac{\Gamma(\alpha+n+1)}{\Gamma(\alpha+1)\Gamma(n+1)}\ u^n.$$

- *Prove that*

$$(1-t)^2\ \partial_t\mathbb{S}_t(x) + (x - (\alpha+1)(1-t))\ \mathbb{S}_t(x) = 0 = (1-t)\ \mathbb{S}_t'(x) + t\ \mathbb{S}_t(x).$$

- *Deduce that for any $n \geq 0$ we have*

$$((n+1)\, \mathbb{I}_n - \mathbb{I}_{n+1})' = (n+1)\, \mathbb{I}_n$$

$$x\, \mathbb{I}_n(x) + (n+\alpha)\,(n\, \mathbb{I}_{n-1}(x) - \mathbb{I}_n(x)) = (n+1)\, \mathbb{I}_n(x) - \mathbb{I}_{n+1}(x).$$

- *Prove that*

$$\mathbb{I}_0(x) = 1 \qquad \mathbb{I}_1(x) = ((\alpha+1) - x) \quad and \quad \mathbb{I}_2(x) = (x - (\alpha+2))^2 - (\alpha+2).$$

- *Check that*

$$\pi\,(\mathbb{S}_s\mathbb{S}_t) = \frac{1}{(1 - st)^{1+\alpha}} = \sum_{n \geq 0} \frac{\Gamma(\alpha+n+1)}{\Gamma(\alpha+1)\Gamma(n+1)}\,(st)^n$$

and

$$\pi\,(\mathbb{I}_n\mathbb{I}_m) = 1_{m=n}\,\frac{\Gamma(\alpha+n+1)\Gamma(n+1)}{\Gamma(\alpha+1)}.$$

Deduce that $\varphi_n := \sqrt{\frac{\Gamma(\alpha+1)}{\Gamma(\alpha+n+1)\Gamma(n+1)}}\, \mathbb{I}_n$ is a Schauder basis of the Hilbert space $\mathbb{L}_2(\pi)$.

- *Prove that φ_n are the eigenfunctions of $(-L)$ associated with the eigenvalues $\lambda_n = n$.*

- *Check the Sturm-Liouville and Rodrigues formula:*

$$\mathbb{I}_n(x) = e^x\, x^{-\alpha}\partial_x^n\left(e^{-x}\,x^{n+\alpha}\right) \quad and \quad -x^{-\alpha}\,e^x\,\partial_x\left(x^{\alpha+1}\,e^{-x}\,\partial_x\mathbb{I}_n\right)(x) = n\mathbb{I}_n(x).$$

Exercise 331 (Tchebyshev polynomials) *We consider the Tchebyshev polynomials $\mathbb{T}_n$ on $S = [-1,1]$ defined by the generating function*

$$\mathbb{S}_t(x) = \sum_{n \geq 0} \mathbb{T}_t(x)\, t^n = \frac{1 - tx}{1 - 2tx + t^2} \quad and\ we\ set \quad \pi(dx) := 1_{]-1,1[}(x)\,\frac{1}{\pi}\,\frac{1}{\sqrt{1 - x^2}}\, dx.$$

We also consider the generator L of the $S = [-1,1]$-valued diffusion X_t discussed at the end of exercise 325 and given

$$dX_t = -X_t\, dt + \sqrt{2(1 - X_t^2)}\, dW_t \;\Rightarrow\; L(f)(x) = -x\, f'(x) + (1 - x^2)\, f''(x).$$

- *Using the series $(1 - te^{i\theta})^{-1} = \sum_{n \geq 0}(te^{i\theta})^n$, with $t \in\,]0,1[$ check that $\mathbb{T}_n(\cos\theta) = \cos(n\theta)$, for any $\theta \in\,]0, \pi[$.*

- *Check that*

$$x\mathbb{T}_n(x) = \frac{1}{2}\,(\mathbb{T}_{n+1}(x) + \mathbb{T}_{n-1}(x)) \quad (and \quad x\mathbb{T}_0(x) = \mathbb{T}_1(x)).$$

By induction w.r.t. the degree n, check that $x^n = \sum_{0 \leq k \leq n} a_{k,n}\, \mathbb{T}_k(x)$, for some parameters $(a_{k,n})_{0 \leq k \leq n}$. Prove that $\sqrt{2}\,\mathbb{T}_n$ forms a Schauder basis of $\mathbb{L}_2(\pi)$.

- *Check that*

$$-L(\mathbb{T}_n)(x) = -\sqrt{1 - x^2}\,\partial_x\left(\sqrt{1 - x^2}\,\partial_x\mathbb{T}_n(x)\right) = \lambda_n\mathbb{T}_n(x) \quad with\ the\ eigenvalue\ \lambda_n = n^2.$$

Exercise 332 (Legendre polynomials) *We consider the Legendre polynomials on $S = [-1, 1]$ defined by the Rodrigues formula*

$$\mathbb{J}_n = \frac{(-1)^n}{2^n n!} \, \partial_x^n J_n \quad \text{with} \quad J_n(x) = (1 - x^2)^n.$$

We also let $\pi(dx) = 2^{-1} \, 1_{[-1,1]}(x)dx$ be the uniform probability distribution on the interval $[-1, 1]$. We let L be the generator of the diffusion process X_t on $S = [-1, 1]$ defined by

$$dX_t = -2X_t \, dt + \sqrt{2(1 - X_t^2)} \, dW_t.$$

This model belongs to the class of Jacobi processes discussed in exercise 325.

- *Check that $(1 - x^2) \, J_n'(x) + 2nx J_n(x) = 0$. Using the Leibniz formula*

$$\partial_x^n (fg) = \sum_{0 \leq m \leq n} \frac{n!}{(n - m)!m!} \, \partial_x^{n-m} f \, \partial_x^m g \tag{17.32}$$

  check the spectral properties and the Sturm-Liouville formulae

$$-L (\mathbb{J}_n) = - \, \partial_x \left((1 - x^2) \, \partial_x \mathbb{J}_n \right) = \lambda_n \mathbb{J}_n \quad \text{with the eigenvalue } \lambda_n = n(n + 1).$$

- *Prove that $\sqrt{2n + 1} \, \mathbb{J}_n$ forms a Schauder basis of $\mathbb{L}_2(\pi)$.*

- *Check that the generating function of the Legendre polynomials is given for any $x \in]0, 1[$ and $|t| < 1$ by the formula*

$$\frac{1}{\sqrt{(1 - 2xt + t^2)}} = \sum_{n \geq 0} \mathbb{J}_n(x) \, t^n.$$

18

Path space measures

Continuous time stochastic processes are defined in terms of a sequence of random variables X_t with a time index t taking values in the uncountable continuous time axis $\mathbb{R}_+$. As a result, proving the existence and the uniqueness of their distribution on the set of trajectories requires some sophisticated probabilistic and analytic tools. In contrast with discrete time stochastic processes, we have no explicit descriptions of these probability measures.

Nevertheless, apart from some mathematical technicalities, the analysis of these path space measures often follows the same construction as in the discrete time case. Furthermore, despite the informal derivation of our constructions, all the formulae presented in this section are mathematically correct.

Path space measures are not only of pure mathematical interest. They are commonly used in engineering sciences, in statistical machine learning, in reliability analysis, as well as in mathematical finance. Most of these applications are in Bayesian inference, maximum likelihood estimation, importance sampling techniques and Girsanov type change of probability measures.

We have chosen to guide the reader's intuition and for this reason alone we present a rather informal discussion of the path space models and their applications.

> *If you find a path with no obstacles, it probably doesn't lead anywhere.*
> Frank A. Clark (1860-1936).

18.1 Pure jump models

We start with a pure jump process X_t with intensity function $\lambda_t(x)$ and a jump amplitude transition $M_t(x, dy)$ on some state space S. In other words, X_t is a stochastic process with infinitesimal generator

$$L_t(f)(x) = \lambda_t(x) \int [f(y) - f(x)] \, M_t(x, dy).$$

We let $(T_k)_{k \geq 0}$ be the sequence of jump times of X_t, with the convention $T_0 = 0$, and we assume that $X_0 = \omega_0$ for some $\omega_0 \in S$.

The random trajectories of the process on some time interval $[0, t]$ are càdlàg paths

$$\omega \; : \; s \in [0, t] \;\; \mapsto \;\; \omega_s \in S$$

with a finite number, say n, of jump epochs $t_k \in [0, t]$, $k \leq n$, defined by the fact that $\Delta \omega_{t_k} \neq 0$, with $k \leq n$. In addition, between the jumps the trajectory remains constant, in the sense that $\omega_s = \omega_{t_k}$, for any $s \in [t_k, t_{k-1}[$. We let $D([0, t], S)$ be the set of all the càdlàg paths from $[0, t]$ into S, and $D_0([0, t], S)$ the subset of the càdlàg piecewise constant paths, with a finite number of jump times.

By construction, we have

$$\mathbb{P}\left((T_{k+1}, X_{T_{k+1}}) \in d(t_{k+1}, \omega_{t_{k+1}}) \mid (T_k, X_{T_k}) \in d(t_k, \omega_{t_k})\right)$$

$$= \lambda_{t_{k+1}}(\omega_{t_k}) \, \exp\left(-\int_{t_k}^{t_{k+1}} \lambda_s(\omega_{t_k}) \, ds\right) dt_{k+1} \times M_{t_{k+1}}(\omega_{t_k}, d\omega_{t_{k+1}})$$

and

$$\mathbb{P}\left(T_{n+1} \geq t \mid (T_n, X_{T_n}) \in d(t_n, \omega_{t_n})\right) = \exp\left(-\int_{t_n}^t \lambda_s(\omega_{t_n}) \, ds\right).$$

In the above formula, dt_k and $d\omega_{t_k}$ denote infinitesimal neighborhoods of the points $t_k \in [0, t]$ and $\omega_{t_k} \in S$.

Further on, we denote by $(t_k)_{1 \leq k \leq n} \in [0, t]^n$ the jump times of a trajectory $\omega = (\omega_s)_{s \leq t} \in D_0([0, t], S)$. Recalling that $\omega_s = \omega_{t_k}$, for any $s \in [t_k, t_{k+1}[$, we have the following construction.

> The path space distribution of $X = (X_s)_{s \leq t}$ is defined for any $\omega = (\omega_s)_{s \leq t} \in D_0([0, t], S)$ by
>
> $$\mathbb{P}(X \in d\omega)$$
>
> $$:= \mathbb{P}\left((T_1, X_{T_1}) \in d(t_1, \omega_{t_1}), \ldots, (T_n, X_{T_n}) \in d(t_n, \omega_{t_n}), T_{n+1} \geq t\right) \qquad (18.1)$$
>
> $$= \exp\left(-\int_0^t \lambda_s(\omega_{s-}) \, ds\right) \times \prod_{s \leq t \, : \, \Delta \omega_s \neq 0} \left[\lambda_s(\omega_{s-}) \, ds \, M_s(\omega_{s-}, d\omega_s)\right].$$

One direct consequence of this result is the conditional distribution formula

$$\mathbb{P}\left((X_r)_{s \leq r \leq t} \in d(\omega_s)_{s \leq r \leq t}) \mid (X_r)_{0 \leq r < s} = (\omega_r)_{0 \leq r < s}\right)$$

$$= \mathbb{P}\left((X_r)_{s \leq r \leq t} \in d(\omega_s)_{s \leq r \leq t}) \mid X_{s-} = \omega_{s-}\right)$$

$$= \exp\left(-\int_s^t \lambda_r(\omega_{r-}) \, dr\right) \times \prod_{s \leq r \leq t \, : \, \Delta \omega_r \neq 0} \left[\lambda_r(\omega_{r-}) \, dr \, M_r(\omega_{r-}, d\omega_r)\right].$$

Theorem 18.1.1 *There exists a unique path space measure $\mathbb{P}$ on $\Omega :=$ $D([0,\infty[,\mathbb{R})$ such that its restrictions to every time mesh sequence are given by (18.1). This path space measure is the distribution of the (canonical) pure jump process*

$$X = (X_s)_{s\in[0,t]} : \omega \in \Omega \mapsto X(\omega) = (X_s(\omega))_{s\in[0,t]} = (\omega_s)_{s\in[0,t]} \in D([0,\infty[,\mathbb{R}).$$

With a slight abuse of notation, it is defined by the path space measures $\mathbb{P}_t$ on $\Omega_t := D([0,t],\mathbb{R})$ given by

$$\mathbb{P}_t(d\omega) := \exp\left(-\int_s^t \lambda_r(\omega_{r-})\ dr\right) \times \prod_{s\leq r\leq t\ :\ \Delta\omega_r\neq 0} [\lambda_r(\omega_{r-})\ dr\ M_r(\omega_{r-}, d\omega_r)].$$

Proof :
One strategy is to consider the set of path sequences indexed by rationals and taking values in the compact space $\overline{\mathbb{R}} = \mathbb{R}\cup\{\infty\}$:

$$\Omega := D([0,\infty[\cap\mathbb{Q},\mathbb{R}) := \prod_{t\in\mathbb{Q}} \overline{\mathbb{R}} = \left\{(\omega_s)_{s\in\mathbb{Q}}\ :\ \omega_s\in\overline{\mathbb{R}}\ \forall s\in\mathbb{Q}\right\}.$$

By construction the product space Ω is a compact metrizable state space. Then, we interpret the measures $\mathbb{P}$ defined in (18.1) as positive linear functionals on the space $C(\Omega)$ of real valued continuous functions over $\Omega_{\mathbb{Q}}$ s.t. $\mathbb{P}(1) = 1$. For any $F\in C(\Omega)$, that depends only on the values of ω on some mesh sequence $t_n\in\mathbb{Q}$,

$$\mathbb{P}(F) = \int F(\omega)\ \mathbb{P}\left(X\in d\omega\right)$$

with the measure $\mathbb{P}$ defined in (18.1). This proves the existence and the consistency of these measures on the subspace $C_{finite}(\Omega)$ of functions that depend only on the values of ω on some finite mesh sequence $t_n\in\mathbb{Q}$. Since $C_{finite}(\Omega)$ is dense in $C(\Omega)$, invoking the Stone-Weierstrass theorem, we conclude that there exists a unique extension to the set $C(\Omega)$. This ends the proof of the theorem. ∎

The Poisson process with time non-homogeneous intensity λ_t corresponds to the situation where $M_t(x,dy) = \delta_{x+1}(dy)$ and $\lambda_t(x) = \lambda_t$.

In this case, for any $\omega\in D_0([0,t],S)$, such that $\Delta\omega_{t_k} = 1$, with $k\leq n$, we have

$$\begin{aligned}
\mathbb{P}\left(X\in d\omega\right) &= \exp\left(-\int_0^t \lambda_s\ ds\right) \times \prod_{s\leq t\ :\ \Delta\omega_s\neq 0} [\lambda_s\ ds] \\
&= \exp\left(-\int_0^t \lambda_s\ ds + \int_0^t \log\left(\lambda_s\right)\ d\omega_s\right)\ dt_1\ldots dt_n. \quad (18.2)
\end{aligned}$$

The corresponding path space measure on $\omega\in D_0([0,t],S)$ is called the Poisson measure. For time homogeneous models, this formula also reduces to

$$\begin{aligned}
\mathbb{P}\left(X\in d\omega\right) &= e^{-\lambda t}\ \lambda^n\ dt_1\ldots dt_n \\
&= e^{-\lambda t}\ \lambda^{\omega_t}\ dt_1\ldots dt_n.
\end{aligned}$$

18.1.1   Likelihood functionals

We let Λ be some positive valued random variable. Given $\Lambda = \lambda$, we let $X = (X_s)_{s \le t}$ be a Poisson process with intensity λ. Combining

$$\mathbb{P}\left(X \in d\omega \mid \Lambda = \lambda\right) = e^{-\lambda t} \ \lambda^{\omega_t} \ dt_1 \ldots dt_n$$

with the Bayes rule, we find that the conditional distribution of Λ given a realization

$$X = (X_s)_{s \le t} = (\omega_s)_{s \le t} = \omega$$

of the Poisson process is given by the formula

$$\mathbb{P}\left[\Lambda \in d\lambda \mid X = \omega\right] \propto \exp\left(L\left(\lambda \mid \omega_t\right)\right) \times \mathbb{P}\left(\Lambda \in d\lambda\right)$$

with the log likelihood function

$$L\left(\lambda \mid \omega_t\right) := -\lambda t \ + \ \omega_t \ \log \lambda.$$

We observe that the maximum value of the log-likelihood function is given by

$$\frac{\partial}{\partial \lambda} L\left(\lambda \mid \omega_t\right) = -t \ + \ \omega_t \frac{1}{\lambda} = 0 \Leftrightarrow \lambda = \omega_t / t.$$

We let $\Theta = (\Theta_1, \Theta_2)$ be some positive valued random variables. Given $\Theta = \theta = (\theta_1, \theta_2)$, we let $X = (X_s)_{s \le t}$ be a time non-homogeneous Poisson process with power law intensity function $\lambda_t = \theta_1 \theta_2 t^{\theta_2 - 1}$. Then we have

$$
\begin{aligned}
\mathbb{P}\left(X \in d\omega \mid \Lambda = \lambda\right) &= \exp\left(-\int_0^t \lambda_s \ ds\right) \times \left[\prod_{1 \le k \le n} \lambda_{t_k}\right] dt_1 \ldots dt_n \\
&= \exp\left(-\theta_1 t^{\theta_2}\right) \times (\theta_1 \theta_2)^n \left[\prod_{1 \le k \le n} t_k^{\theta_2 - 1}\right] dt_1 \ldots dt_n.
\end{aligned}
$$

Using the Bayes rule, we find that the conditional distribution of Θ given a realization $(X_s)_{s \le t} = (\omega_s)_{s \le t}$ of the Poisson process is given by

$$\mathbb{P}\left[\Theta \in d\theta \mid X = \omega\right] \propto \exp\left[L\left(\theta \mid \omega\right)\right] \times \mathbb{P}\left(\Theta \in d\theta\right)$$

---

with the log likelihood function

$$L\left(\theta \mid \omega\right) := -\theta_1 t^{\theta_2} + \omega_t \ \log\left(\theta_1 \theta_2\right) + (\theta_2 - 1) \sum_{1 \le k \le \omega_t} \log t_k.$$

Maximum value of the log-likelihood function is delivered by the point θ that represents a root of the equation system

$$
\begin{cases}
\dfrac{\partial}{\partial \theta_1} L\left(\theta \mid (\omega_s)_{s \le t}\right) &= -t^{\theta_2} + \dfrac{\omega_t}{\theta_1} = 0 \iff \omega_t = \theta_1 t^{\theta_2} \\[2mm]
&= \\[2mm]
\dfrac{\partial}{\partial \theta_2} L\left(\theta \mid (\omega_s)_{s \le t}\right) &= -\underbrace{\theta_1 \ t^{\theta_2}}_{=\omega_t} \ \log t + \dfrac{\omega_t}{\theta_2} + \log\left(\prod_{1 \le k \le n} t_k\right) = 0.
\end{cases}
$$

This implies that

$$\hat{\theta}_2 = \omega_t \left(\log \frac{t^{\omega_t}}{\prod_{1 \leq k \leq n} t_k} \right)^{-1} = \omega_t \left(\sum_{1 \leq k \leq n} \log \frac{t}{t_k} \right)^{-1}$$

and

$$\hat{\theta}_1 = \omega_t \ \exp \left(- \frac{1}{1 - \frac{1}{\omega_t} \sum_{1 \leq k \leq \omega_t} \frac{\log t_k}{\log t}} \right)$$

are the maximum likelihood estimators.

18.1.2 Girsanov's transformations

We let (X_t, X_t') be a pair of pure jump processes with *positive* intensity functions $(\lambda_t(x), \lambda_t'(x))$, and jump amplitude transitions $(M_t(x, dy), M_t'(x, dy))$ on some state space S.

We further assume that M_t and M_t' have a densities m_t and m_t' with respect to some measure ν on S; that is,

$$M_t(x, dy) = m_t(x, y) \ \nu(dy) \quad \text{and} \quad M_t'(x, dy) = m_t'(x, y) \ \nu(dy).$$

For instance, all the Gaussian transitions on $S = \mathbb{R}$ are absolutely continuous w.r.t. the Lebesgue measure $\nu(dx) = dx$.

In this situation, the path space measures of

$$X = (X_s)_{s \leq t} \quad \text{and} \quad X' = (X_s')_{s \leq t}$$

are defined for any $\omega = (\omega_s)_{s \leq t} \in D_0([0, t], S)$ (with jump times $(t_k)_{1 \leq k \leq n} \in [0, t]^n$) by the formulae

$$\mathbb{P}(X \in d\omega)$$

$$= \exp \left(- \int_0^t \lambda_s(\omega_{s-}) \ ds \right) \times \prod_{s \leq t \ : \ \Delta \omega_s \neq 0} [\lambda_s(\omega_{s-}) \ ds \ m_s(\omega_{s-}, \omega_s) \ \nu(d\omega_s)]$$

and

$$\mathbb{P}(X' \in d\omega)$$

$$= \exp \left(- \int_0^t \lambda_s'(\omega_{s-}) \ ds \right) \times \prod_{s \leq t \ : \ \Delta \omega_s \neq 0} [\lambda_s'(\omega_{s-}) \ ds \ m_s'(\omega_{s-}, \omega_s) \ \nu(d\omega_s)].$$

It is readily checked that

$$\mathbb{P}(X' \in d\omega) = \mathcal{Z}(\omega) \times \mathbb{P}(X \in d\omega) \tag{18.3}$$

with the function

$$\mathcal{Z}(\omega) = \exp \left(- \int_0^t [\lambda_s' - \lambda_s] (\omega_{s-}) \ ds \right)$$

$$\times \prod_{s \leq t \ : \ \Delta \omega_s \neq 0} \left[\frac{\lambda_s'(\omega_{s-})}{\lambda_s(\omega_{s-})} \ \frac{m_s'(\omega_{s-}, \omega_s)}{m_s(\omega_{s-}, \omega_s)} \right].$$

The above formula is called the Girsanov change of measure or Girsanov transformation. It is often written for any function F on the path space $D_0([0,t], S)$ as

$$\mathbb{E}\left(F\left((X'_s)_{s \leq t}\right)\right) = \mathbb{E}\left(F\left((X_s)_{s \leq t}\right) \, Z_t\right) \tag{18.4}$$

with the exponential stochastic processes Z_t defined by

$$Z_t = \exp\left(-\int_0^t [\lambda'_s - \lambda_s]\,(X_{s-})\,ds\right)$$

$$\times \prod_{0 \leq s \leq t \,:\, \Delta\omega_s \neq 0} \left[\frac{\lambda'_s(X_{s-})}{\lambda_s(X_{s-})} \, \frac{m'_s(X_{s-}, X_s)}{m_s(X_{s-}, X_s)}\right].$$

Writing $\mathbb{P}$ and $\mathbb{P}'$ the path space measures on $\Omega := D([0, \infty[, \mathbb{R})$ with time marginals $\mathbb{P}_t(d\omega) := \mathbb{P}((X_s)_{s \leq t} \in d\omega)$ and $\mathbb{P}'_t(d\omega) := \mathbb{P}((X'_s)_{s \leq t} \in d\omega)$, the Girsanov formula (18.4) takes the following form

$$\mathbb{P}'_t(d\omega) = \mathcal{Z}_t(\omega) \times \mathbb{P}_t(d\omega) \qquad \Longleftrightarrow \qquad \frac{d\mathbb{P}'_t}{d\mathbb{P}_t}(\omega) := \mathcal{Z}_t(\omega).$$

18.1.3 Exponential martingales

The Poisson processes $(X_t, X'_t) = (N_t, N'_t)$ with time non-homogeneous intensity (λ_t, λ'_t) correspond to the situation where

$$M_t(x, dy) = M'_t(x, dy) = \delta_{x+1}(dy) \quad \text{and} \quad (\lambda_t(x), \lambda'_t(x)) = (\lambda_t, \lambda'_t).$$

In this situation, for any $\omega \in D_0([0,t], S)$, such that $\Delta\omega_{t_k} = 1$, with $k \leq n$, we have

$$\mathcal{Z}(\omega) = \exp\left[-\int_0^t [\lambda'_s - \lambda_s]\,ds + \int_0^t \log\left(\lambda'_s/\lambda_s\right)\,d\omega_s\right].$$

Sometimes, we also rewrite this function in the following form

$$\mathcal{Z}(\omega) = \exp\left(-\int_0^t [\lambda'_s - \lambda_s]\,ds\right) \prod_{0 \leq s \leq t} \left(1 + \left(\frac{\lambda'_s}{\lambda_s} - 1\right)\Delta\omega_s\right).$$

We also have that

$$Z_t = \exp\left[-\int_0^t [\lambda'_s - \lambda_s]\,ds + \int_0^t \log\left(\lambda'_s/\lambda_s\right)\,dN_s\right]$$

$$= \exp\left(-\int_0^t [\lambda'_s - \lambda_s]\,ds\right) \prod_{0 \leq s \leq t} \left(1 + \left(\frac{\lambda'_s}{\lambda_s} - 1\right)\Delta N_s\right).$$

In terms of the martingale

$$dM_t^\lambda = \left(\frac{\lambda'_t}{\lambda_t} - 1\right)(dN_t - \lambda_t dt) = \underbrace{\left(\frac{\lambda'_t}{\lambda_t} - 1\right)dN_t}_{\Delta M_t^\lambda} - (\lambda'_t - \lambda_t)\,dt$$

we also have the exponential formula

$$Z_t = e^{M_t^\lambda} \prod_{0 \le s \le t} \left(\left(1 + \Delta M_s^\lambda\right) e^{-\Delta M_s^\lambda} \right).$$

We also notice that

$$
\begin{aligned}
Z_{t+dt} - Z_t &= Z_t \left(e^{\left(M_{t+dt}^\lambda - M_t^\lambda\right) - \Delta M_t^\lambda} \left(1 + \Delta M_t^\lambda\right) - 1 \right) \\
&= Z_t \left(e^{-(\lambda_t' - \lambda_t)dt} \left[1 + \Delta M_t^\lambda\right] - 1 \right) \\
&= Z_t \left(\left[1 - (\lambda_t' - \lambda_t) dt\right] \left[1 + \Delta M_t^\lambda\right] - 1 \right) \\
&= Z_t \left(\Delta M_t^\lambda - (\lambda_t' - \lambda_t) dt \right) = Z_t \, dM_t^\lambda.
\end{aligned}
$$

This implies that

$$dZ_t = Z_t \, dM_t^\lambda. \tag{18.5}$$

In this notation, the change of measure formula (18.4) takes the following form.

Theorem 18.1.2 (Girsanov's theorem) *For any $t \ge 0$, and any functional F on $D_0([0,t], \mathbb{R})$ we have*

$$\mathbb{E}\left(F\left((N_s)_{s \le t}\right) Z_t \right) = \mathbb{E}\left(F\left((N_s')_{s \le t}\right) \right) \tag{18.6}$$

with the martingale $dZ_t = Z_t \, dM_t^\lambda$ defined by the exponential formula

$$Z_t = \exp\left(M_t^\lambda\right) \prod_{0 \le s \le t} \left\{ \left(1 + \Delta M_s^\lambda\right) \exp\left(-\Delta M_s^\lambda\right) \right\}$$

and with the martingale increments

$$dM_t^\lambda = \left(\lambda_t' \lambda_t^{-1} - 1\right) \left(dN_t - \lambda_t \, dt\right).$$

Remark : If we interpret $dZ_t = Z_t - Z_{t-dt}$, it is preferable to write $dZ_t = Z_{t-} \, dM_t^\lambda$. In our notational system, we interpret $dZ_t = Z_{t+dt} - Z_t$ so that $dZ_t = Z_t \, dM_t^\lambda$ is a well defined infinitesimal increment.

For time homogeneous models $(\lambda_t, \lambda_t') = (\lambda, \lambda')$, this formula also reduces to

$$\mathcal{Z}(\omega) = e^{(\lambda - \lambda')t} \times (\lambda'/\lambda)^{\omega_t} \iff Z_t = e^{(\lambda - \lambda')t} \times (\lambda'/\lambda)^{N_t}.$$

18.2 Diffusion models

18.2.1 Wiener measure

The random trajectories of a Brownian process on some time interval $[0, t]$ are continuous paths

$$\omega : s \in [0, t] \mapsto \omega_s \in \mathbb{R}.$$

We let $C([0, t], S)$ be the set of continuous trajectories from $[0, t]$ into $\mathbb{R}$.

We consider a time mesh sequence $t_n = nh$, with $n \in \mathbb{N}$, and a given time step $\Delta t :=$

$h \in]0,1[$. By construction, the path space distribution of $W = (W_{t_k})_{0 \le k \le n}$ is defined for any $\omega = (\omega_s)_{s \le t_n} \in C([0, t_n], \mathbb{R})$ by

$$\mathbb{P}(W \in d\omega)$$

$$:= \mathbb{P}(W_{t_1} \in d\omega_{t_1}, \ldots, W_{t_n} \in d\omega_{t_n})$$

$$= \frac{1}{(2\pi \Delta t)^{n/2}} \exp\left[-\frac{1}{2} \sum_{1 \le k \le n} \left(\frac{\Delta \omega_{t_k}}{\Delta t}\right)^2 \Delta t\right] d\omega_{t_1} \ldots d\omega_{t_n}.$$

(18.7)

Here we denote $\Delta \omega_{t_k} = \omega_{t_k} - \omega_{t_{k-1}}, 1 \le k \le n$.

Since the path space measure of $W := (W_s)_{s \le t}$ is supported by the set of functions $\omega = (\omega_s)_{s \le t}$ that are nowhere differentiable, we cannot pass in the limit $\Delta t \to 0$ in the above formula. Nevertheless, using the same lines of arguments as in the proof of theorem 18.1.1, we have the following theorem.

Theorem 18.2.1 *There exists a unique path space measure $\mathbb{P}$ on $C([0, \infty[, \mathbb{R})$ such that its restrictions to every time mesh sequence are given by (18.7). This path space measure is the distribution of the Brownian motion $W = (W_s)_{s \ge 0}$, and it is called the Wiener measure on the set of continuous trajectories $\Omega := C([0, \infty[, \mathbb{R})$. The set Ω equipped with the Wiener measure $\mathbb{P}$ is called the Wiener space.*

18.2.2   Path space diffusions

To simplify the presentation, we restrict ourselves with a one-dimensional diffusion X_t defined by the SDE

$$dX_t = b(X_t)\, dt + dW_t \tag{18.8}$$

with some regular homogeneous function b.

We denote by $X_{t_n}^h$ the discrete approximation model on some time mesh defined in (14.7). We assume that $X_0^h = x_0 = X_0 = W_0$, for some given $x_0 \in \mathbb{R}$.

By construction, the the path space distribution of $X^h = (X_{t_k}^h)_{0 \le k \le n}$ is defined for any $\omega = (\omega_s)_{s \le t_n} \in C([0, t_n], \mathbb{R})$ s.t. $\omega_0 = x_0$ by

$$\mathbb{P}(X^h \in d\omega)$$

$$:= \mathbb{P}(X_{t_1}^h \in d\omega_{t_1}, \ldots, X_{t_n}^h \in d\omega_{t_n})$$

$$= \frac{1}{(2\pi \Delta t)^{n/2}} \exp\left(-\frac{1}{2} \sum_{1 \le k \le n} \left[\frac{\Delta \omega_{t_k}}{\Delta t} - b(\omega_{t_{k-1}})\right]^2 \Delta t\right) d\omega_{t_1} \ldots d\omega_{t_n}.$$

It is now readily checked that

$$\mathbb{P}(X^h \in d\omega) := \mathcal{Z}^h(\omega) \times \mathbb{P}(W \in d\omega)$$

with the density function

$$\mathcal{D}^h(\omega) = \exp\left(\sum_{1 \le k \le n} b(\omega_{t_{k-1}}) \Delta \omega_{t_k} - \frac{1}{2} \sum_{1 \le k \le n} b^2(\omega_{t_{k-1}}) \Delta t\right).$$

Choosing $t_n = h\lfloor t/h \rfloor$, and taking the limit $h = \Delta t \to 0$, we obtain the path space measure of the diffusion process $X = (X_s)_{s \leq t}$ on $C([0, t], \mathbb{R})$

$$\mathbb{P}(X \in d\omega) := D(\omega) \times \mathbb{P}(W \in d\omega) \tag{18.9}$$

with the density function

$$D(\omega) = \exp\left(\int_0^t b(\omega_s)\, d\omega_s - \frac{1}{2} \int_0^t b^2(\omega_s)\, ds \right)$$

called the Cameron Martin formula or Girsanov's theorem for diffusion processes.

If $\mathbb{P}'$ stands for the distribution of $X = (X_s)_{s \leq t}$ and $\mathbb{P}$ for $W = (W_s)_{s \leq t}$, the above result takes the form

$$\mathbb{P}'(d\omega) = D(\omega) \times \mathbb{P}(d\omega) \iff \frac{d\mathbb{P}'}{d\mathbb{P}}(\omega) := D(\omega).$$

Under $\mathbb{P}$, the trajectories are distributed like those of a Brownian motion $W = (W_s)_{s \leq t}$ (starting at x_0). Defining $\mathbb{P}'$ as above, the trajectories are distributed like the ones of the diffusion $X = (X_s)_{s \leq t}$. We often say say under $\mathbb{P}'$ the Brownian motion becomes a diffusion with a drift function b (starting at x_0).

18.2.3 Girsanov transformations

As in the pure jump case, this formula is often rewritten as follows.

For any function F on the path space $C([0, t], \mathbb{R})$ we have

$$\mathbb{E}\left(F\left((X_s)_{s \leq t} \right) \right) = \mathbb{E}\left(F\left((W_s)_{s \leq t} \right)\, D\left((W_s)_{s \leq t} \right) \right) \tag{18.10}$$

with a density function D on the path space $C([0, t], \mathbb{R})$ defined by

$$Z_t := D\left((W_s)_{s \leq t} \right) = \exp\left(\int_0^t b(W_s)\, dW_s - \frac{1}{2} \int_0^t b^2(W_s)\, ds \right).$$

Formula (18.10) is valid when the Brownian motion W_s and the diffusion X_s start at the same location $W_0 = X_0$.

Notice that

$$Z_t = e^{U_t} \quad \text{with} \quad U_t := \int_0^t b(W_s)\, dW_s - \frac{1}{2} \int_0^t b^2(W_s)\, ds.$$

Combining the Doeblin-Itō formula with

$$dU_t = b(W_t)\, dW_t - \frac{1}{2} b(W_t)^2\, dt \quad \text{and} \quad dU_t dU_t = b(W_t)^2\, dW_t^2 = b(W_t)^2\, dt,$$

> we prove that Z_t is a martingale with increments given by
>
> $$dZ_t = e^{U_t} \, dU_t + \frac{1}{2} \, e^{U_t} \, dU_t dU_t = Z_t \, b(W_t) \, dW_t.$$

Theorem 18.2.2 (Girsanov's theorem) *We let X_t be the diffusion process defined in (18.8). For any $t \geq 0$, and any functional F on $C([0, t], \mathbb{R})$ we have*

$$\mathbb{E}\left(F\left((X_s)_{s\leq t}\right) \, U_t\right) = \mathbb{E}\left(F\left((W_s)_{s\leq t}\right)\right) \tag{18.11}$$

with the martingale

$$dU_t = U_t \, b(X_t) \, dW_t$$

given by the exponential formula

$$U_t = \exp\left(-\int_0^t b(X_s) \, dW_s - \frac{1}{2} \int_0^t b^2(X_s) \, ds\right).$$

Formula (18.11) is valid when the Brownian motion W_s and the diffusion X_s start at the same location $W_0 = X_0$.

Proof :
We first observe that

$$-\int_0^t b(X_s) \, dW_s - \frac{1}{2} \int_0^t b^2(X_s) \, ds = -\int_0^t b(X_s) \, dX_s + \frac{1}{2} \int_0^t b^2(X_s) \, ds.$$

We check this claim using the fact that

$$dX_t = b(X_t) \, dt + dW_t \quad \Longrightarrow \quad dW_t = dX_t - b(X_t) \, dt$$
$$\Longrightarrow \quad b(X_t) \, dW_t = b(X_t) \, dX_t - b^2(X_t) \, dt.$$

Replacing in (18.10) the function $F\left((X_s)_{s\leq t}\right)$ by the function

$$F\left((X_s)_{s\leq t}\right) \, \exp\left(-\int_0^t b(X_s) \, dW_s - \int_0^t b^2(X_s) \, ds\right)$$

$$= F\left((X_s)_{s\leq t}\right) \exp\left(-\int_0^t b(X_s) \, dX_s + \frac{1}{2} \int_0^t b^2(X_s) \, ds\right)$$

we find that

$$\mathbb{E}\left(F\left((X_s)_{s\leq t}\right) \exp\left(-\int_0^t b(X_s) \, dW_s - \frac{1}{2} \int_0^t b^2(X_s) \, dr\right)\right)$$

$$= \mathbb{E}\left(F\left((X_s)_{s\leq t}\right) \exp\left(-\int_0^t b(X_s) \, dX_s + \frac{1}{2} \int_0^t b^2(X_s) \, ds\right)\right).$$

The end of the proof is now a direct consequence of (18.10). This ends the proof of the theorem. ∎

If $\mathbb{P}'$ stands for the distribution of $X = (X_s)_{s \leq t}$ and $\mathbb{P}_t$ for the distribution of $W = (W_s)_{s \leq t}$, the above result takes the form

$$\mathbb{P}\,(d\omega) = D^{-1}(\omega)\mathbb{P}'(d\omega) = \quad \Longleftrightarrow \quad \frac{d\mathbb{P}}{d\mathbb{P}'}(\omega) := D^{-1}(\omega).$$

Under $\mathbb{P}'$ the trajectories are distributed like those of the diffusion $X = (X_s)_{s \leq t}$. Defining $\mathbb{P}$ as above, the trajectories are distributed like those of a Brownian motion $W = (W_s)_{s \leq t}$. We often say say under $\mathbb{P}$ the diffusion becomes a Brownian motion (without any drift function). With a slight abuse of notation, we often write

$$\begin{aligned}
\frac{d\mathbb{P}}{d\mathbb{P}'} &= U_t = \exp\left(-\int_0^t b(X_s)\,dW_s - \frac{1}{2}\int_0^t b^2(X_s)\,ds\right) \\
&\left(= \exp\left(-\int_0^t b(X_s)\,dX_s + \frac{1}{2}\int_0^t b^2(X_s)\,ds\right)\right)
\end{aligned}$$

instead of

$$\frac{d\mathbb{P}}{d\mathbb{P}'}(\omega) = D^{-1}(\omega) = \exp\left(-\int_0^t b(\omega_s)\,d\omega_s + \frac{1}{2}\int_0^t b^2(\omega_s)\,ds\right).$$

Important remark : The theorem 18.2.2 and formula (18.9) can be extended to d-dimensional non-homogeneous diffusions

$$dX_t = b_t(X_t)\,dt + R_t^{1/2}\,dW_t \tag{18.12}$$

where W_t is a d-dimensional Brownian motion (i.e., $W_t = \left(W_t^i\right)_{1 \leq i \leq d}$ with d independent Brownian motions W_t^i, $1 \leq i \leq d$), and b_t is a function taking values in $\mathbb{R}^d$. We also assume that the diffusion matrix functional $t \mapsto R_t^{1/2}$ is smooth and invertible.

In this situation, the formulae (18.9), (18.10), and (18.11) are valid, replacing in the r.h.s. of (18.9), (18.10), and (18.11) the Brownian motion W_s by a Brownian motion with increments $R_s^{1/2}\,dW_s$, with the Radon-Nikodym derivatives

$$Z_t = D\left((W_s)_{0 \leq s \leq t}\right) = \exp\left(\int_0^t b_s(W_s)' R_t^{-1}\,dW_s - \frac{1}{2}\int_0^t \left\langle b_s(W_s), R_s^{-1}b_s(W_s)\right\rangle\,ds\right)$$

and

$$\begin{aligned}
U_t &= \exp\left(-\int_0^t b_s(X_s)' R_t^{-1/2}\,dW_s - \frac{1}{2}\int_0^t \left\langle R_s^{-1/2}b_s(X_s), R_s^{-1/2}b_s(X_s)\right\rangle\,ds\right) \\
&= \exp\left(-\int_0^t b_s(X_s)'\,R_s^{-1}\,dX_s + \frac{1}{2}\int_0^t \left\langle b_s(X_s), R_s^{-1}b_s(X_s)\right\rangle\,ds\right).
\end{aligned}$$

The last assertion is implied by the following decompositions

$$-b_t(X_t)'\,R_t^{-1}\,dX_t + \tfrac{1}{2}\left\langle b_t(X_t), R_t^{-1}b_t(X_t)\right\rangle\,dt$$

$$= -b_t(X_t)'\,R_t^{-1}\left(b_t(X_t)\,dt + R_t^{1/2}\,dW_t\right) + \tfrac{1}{2}\left\langle b_t(X_t), R_t^{-1}b_t(X_t)\right\rangle\,dt$$

$$= -b_t(X_t)'\,R_t^{-1/2}\,dW_t - \tfrac{1}{2}\underbrace{\left\langle b_t(X_t), R_t^{-1}b_t(X_t)\right\rangle}_{=\left\langle R_t^{-1/2}b_t(X_t), R_t^{-1/2}b_t(X_t)\right\rangle}\,dt.$$

18.3 Exponential change twisted measures

We let X_t be a Markov process on some state space S with infinitesimal generator L_t. We recall that for any sufficiently regular function $f(t,x)$ we have

$$df(t, X_t) = \left[\frac{\partial}{\partial t} + L_t\right](f)(t, X_t)dt + dM_t(f)$$

with some martingale $M_t(f)$ with an angle bracket

$$d\langle M(f)\rangle_t = \Gamma_{L_t}(f(t,.), f(t,.))(X_t)\, dt.$$

We also consider a collection of sufficiently smooth positive functions h_t, and we set

$$M_t^h := h_0^{-1}(X_0)h_t(X_t)\, \exp\left(-\int_0^t \left[h_s^{-1}(\partial_s + L_s)h_s\right](X_s)ds\right)$$

with $h_t^{-1} = 1/h_t$. Using Doeblin-Itō formula (15.12), we prove that M_t^h is a unit mean positive martingale.

More precisely, we have

$$
\begin{aligned}
dM_t^h &= M_t^h \left\{\left(h_t^{-1}\left(\partial_t h_t + L_t(h_t)\right)\right)(X_t)\, dt + h_t^{-1}(X_t)dM_t(h)\right\} \\
&\qquad - M_t^h \left[\partial_t h_t + h_t^{-1}L_t(h_t)\right](X_t)\, dt \\
&= h_t^{-1}(X_t)M_t^h dM_t(h).
\end{aligned}
\tag{18.13, 18.14}
$$

We let X^h be the process defined by the change of probability measure

$$\mathbb{E}\left(F((X_s^h)_{s\leq t})\right) = \mathbb{E}\left(F((X_s)_{s\leq t})\, M_t^h\right). \tag{18.15}$$

The conditional expectations w.r.t. $(X_r^h)_{r\leq s}$, with $s \leq t$ are given by the formula

$$\mathbb{E}\left(F_1((X_r^h)_{r\leq s})\, F_2((X_r^h)_{s\leq r\leq t})\right) = \mathbb{E}\left(F_1((X_r^h)_{r\leq s})\, \mathbb{E}\left(F_2((X_r^h)_{s\leq r\leq t}) \mid (X_r^h)_{r\leq s}\right)\right).$$

On the other hand, using the Markov property we have

$$\mathbb{E}\left(F_1((X_r^h)_{r\leq s})\, F_2((X_r^h)_{s\leq r\leq t})\right)$$

$$= \mathbb{E}\left(F_1((X_r)_{r\leq s})M_s^h\, \mathbb{E}\left(F_2((X_r)_{s\leq r\leq t})M_t^h/M_s^h \mid X_s\right)\right)$$

$$= \mathbb{E}\left(F_1((X_r)_{r\leq s})M_s^h\, \mathbb{P}_{s,X_s}^h(F_2)\right) = \mathbb{E}\left(F_1((X_r^h)_{r\leq s})\, \mathbb{P}_{s,X_s^h}^h(F_2)\right)$$

with the functional

$$
\begin{aligned}
\mathbb{P}_{s,x}^h(F_2) &:= \mathbb{E}\left(F_2((X_r)_{s\leq r\leq t})M_t^h/M_s^h \mid X_s = x\right) \\
&= h_s^{-1}(x)\, \mathbb{E}\left(F_2((X_r)_{s\leq r\leq t})\, h_t(X_t)\right. \\
&\qquad\qquad \left.\times \exp\left(-\int_s^t h_s^{-1}(X_s)(\partial_s + L_s)(h_s)(X_s)ds\right) \mid X_s = x\right).
\end{aligned}
$$

This implies that

$$\mathbb{E}\left(F_2((X_r^h)_{s\leq r\leq t}) \mid (X_r^h)_{r\leq s}\right) = \mathbb{E}\left(F_2((X_r^h)_{s\leq r\leq t}) \mid X_s^h\right) = \mathbb{P}_{s,X_s^h}^h(F_2).$$

To get one step further, for any sufficiently regular function f on S, combining (18.14) with Doeblin-Itō formula we prove that

$$d(f(X_t)M_t^h) = M_t^h \overbrace{(L_t(f)(X_t)dt + dM_t(f))}^{df(X_t)} + f(X_t) \overbrace{h_t^{-1}(X_t)M_t^h dM_t(h)}^{=dM_t^h}$$

$$+\mathbb{E}\left(df(X_t)dM_t^h \mid (X_s)_{s\leq t}\right) + \underbrace{df(X_t)dM_t^h - \mathbb{E}\left(df(X_t)dM_t^h \mid (X_s)_{s\leq t}\right)}_{:=\mathcal{M}_t}$$

for some martingale $\mathcal{M}_t$. On the other hand, we have

$$\mathbb{E}\left(df(X_t)dM_t^h \mid (X_s)_{s\leq t}\right) = \mathbb{E}\left(dM_t(f)dM_t^h \mid (X_s)_{s\leq t}\right)$$

$$= h_t^{-1}(X_t)M_t^h \,\mathbb{E}\left(dM_t(f)dM(h)_t \mid (X_s)_{s\leq t}\right)$$

$$= h_t^{-1}(X_t)M_t^h \, d\langle M(h), M(f)\rangle_t$$

$$= M_t^h \left(h_t^{-1}\Gamma_{L_t}(h_t, f)\right)(X_t) \, dt.$$

This yields

$$f(X_{t+dt}) \, M_{t+dt}^h/M_t^h - f(X_t) = \left(f(X_{t+dt})M_{t+dt}^h - f(X_t)M_t^h\right)/M_t^h$$

$$= \left[L_t(f) + h_t^{-1}\Gamma_{L_t}(h_t, f)\right](X_t)dt + d\widetilde{\mathcal{M}}_t$$

for some martingale $\widetilde{\mathcal{M}}_t$. In particular, this implies that

$$\frac{1}{dt}\left[\mathbb{E}\left(f(X_{t+dt}^h) \mid X_t^h = x\right) - f(x)\right]$$

$$= \frac{1}{dt}\left[\mathbb{E}\left(f(X_{t+dt}^h) \, M_{t+dt}^h/M_t^h \mid X_t = x\right) - f(x)\right]$$

$$\simeq_{dt\downarrow 0} L_t^{[h]}(f)(x) := L_t(f)(x) + h_t^{-1}(x)\Gamma_{L_t}(h_t, f)(x).$$

This shows that X_t^h has an infinitesimal generator

$$L_t^{[h]}(f) := L_t(f) + h_t^{-1}\Gamma_{L_t}(h_t, f). \tag{18.16}$$

18.3.1 Diffusion processes

For the diffusion generator $L_t = L_t^c$ defined in (15.8), using (15.14) we have

$$L_t^{[h]}(f) = \sum_{i=1}^d b_t^{h,i} \, \partial_i f + \frac{1}{2} \sum_{i,j=1}^d \left(\sigma_t(\sigma_t)^T\right)_{i,j} \, \partial_{i,j},$$

with the drift functions $b_t^{h,i}$ defined for any $1 \leq i \leq d$ by

$$b_t^{h,i} := b_t^i + \sum_{k=1}^d \left(\sigma_t\sigma_t^T\right)_i^k \, \partial_k \log h_t \qquad \left(\text{with} \quad \left(\sigma_t\sigma_t^T\right)_i^k = \sum_{j=1}^d \sigma_{j,t}^k\sigma_{j,t}^i\right).$$

18.3.2 Pure jump processes

For the jump generator $L_t = L_t^d$ defined in (15.9), using (15.15) we find that

$$
\begin{aligned}
L_t^{[h]}(f)(x) &= L_t(f)(x) + \lambda_t(x) \int (f(y) - f(x)) \left(\frac{h_t(y)}{h_t(x)} - 1 \right) S_t(x, dy) \\
&= \lambda_t^h(x) \int (f(y) - f(x)) \, M_t^h(x, dy)
\end{aligned}
$$

with the jump intensity and transition kernels

$$
\lambda_t^h := \lambda_t \, h_t^{-1} \, M_t(h) \quad \text{and} \quad M_t^h(x, dy) := \frac{M_t(x, dy) \, h_t(y)}{M_t(h)(x)}.
$$

18.4 Some illustrations

18.4.1 Risk neutral financial markets

Path space measures and Girsanov's theorem are some of the most useful results of stochastic analysis in financial engineering applications. The fundamental theorem of financial mathematics states that the market is tradeable (i.e., there are no arbitrage opportunities) if and only if the deflated risky assets are martingales w.r.t. some probability measure $\mathbb{Q}$. This $\mathbb{Q}$ is called the risk neutral measure.

18.4.1.1 Poisson markets

Suppose we are given the evolution of a risky asset price in a Poisson market in terms of an SDE of the form

$$
d\mathcal{S}_t = b_t \, \mathcal{S}_t \, dt + \sigma_t \, \mathcal{S}_t \, dN_t
$$

for some volatility type function $\sigma_t > 0$, and some return rate function $b_t < r_t$, where r_t stands for the exponential return of the riskless asset.

In the above display, N_t stands for a Poisson process with intensity function λ_t.

The Poisson process $(N_s)_{s \in [0,t]}$ and the corresponding stochastic process $(\mathcal{S}_s)_{s \in [0,t]}$ are defined on a probability space Ω equipped with some probability measure, say $\mathbb{P}$ (for instance, we can choose the Poisson measure defined in (18.2)). We define an equivalent measure on Ω by setting for any measurable subset $A \subset \Omega$

$$
\mathbb{Q}(A) = \mathbb{E}_{\mathbb{Q}}(1_A) := \mathbb{E}_{\mathbb{P}}(1_A \, Z_t)
$$

with the exponential martingale

$$
Z_t = \exp\left[-\int_0^t \left[\frac{r_s - b_s}{\sigma_s} - \lambda_s \right] ds + \int_0^t \log\left(\frac{r_s - b_s}{\sigma_s \lambda_s} \right) dN_s \right].
$$

By (18.6), for any functional F on $D_0([0,t], \mathbb{R})$ we have

$$
\mathbb{E}_{\mathbb{Q}}(F(N_s)_{s \le t}) = \mathbb{E}_{\mathbb{P}}(F((N_s)_{s \le t}) \, Z_t) = \mathbb{E}_{\mathbb{P}}(F((N_s')_{s \le t}))
$$

where N_s' stands for a Poisson process with intensity function $\lambda_s' = \frac{r_s - b_s}{\sigma_s}$.

Thus, if we set

$$
dM_t = dN_t - \left(\frac{r_t - b_t}{\sigma_t} \right) dt,
$$

we find that

$$
\begin{aligned}
d\mathcal{S}_t &= b_t \, \mathcal{S}_t \, dt + \sigma_t \, \mathcal{S}_t \, dN_t \\
&= b_t \, \mathcal{S}_t \, dt + \sigma_t \, \mathcal{S}_t \, (dM_t + \sigma_t^{-1} \, (r_t - b_t) \, dt) \\
&= r_t \, \mathcal{S}_t \, dt + \sigma_t \, \mathcal{S}_t \, dM_t.
\end{aligned}
$$

This shows that under $\mathbb{Q}$ the stock price satisfies the $\mathbb{Q}$-SDE evolution equation

$$
d\mathcal{S}_t = r_t \, \mathcal{S}_t \, dt + \sigma_t \, \mathcal{S}_t \, dM_t
$$

for some Poisson martingale process

$$
dM_t = dN_t - \left(\frac{r_t - b_t}{\sigma_t} \right) dt.
$$

As a direct consequence, the $\mathbb{Q}$-dynamics of the deflated risky asset price defined in (3.19) is given by the martingale increments

$$
d\overline{\mathcal{S}}_t = \sigma_t \, \overline{\mathcal{S}}_t \, dM_t.
$$

18.4.1.2 Diffusion markets

Suppose we are given the evolution of a risky asset price in terms of an SDE of the form

$$
d\mathcal{S}_t = b_t \, \mathcal{S}_t \, dt + \sigma_t \, \mathcal{S}_t \, dW_t
$$

for some return rate function b_t and volatility deterministic function σ_t. The Brownian motion $(W_s)_{s \in [0,t]}$ and the stochastic process $(\mathcal{S}_s)_{s \in [0,t]}$ are defined on a probability space Ω equipped with some probability measure, say $\mathbb{P}$. We define an equivalent measure on Ω by setting for any measurable subset $A \subset \Omega$

$$
\mathbb{Q}(A) = \mathbb{E}_{\mathbb{Q}} \left(1_A \right) := \mathbb{E}_{\mathbb{P}} \left(1_A \, U_t \right)
$$

with the exponential martingale

$$
U_t = \exp \left(-\int_0^t \left(\frac{b_s - r_s}{\sigma_s} \right) dW_s - \frac{1}{2} \int_0^t \left(\frac{b_s - r_s}{\sigma_s} \right)^2 ds \right).
$$

We set

$$
dM_s = \frac{1}{\sigma_s} \, (b_s - r_s) \, ds + dW_s.
$$

By theorem 18.2.2, for any functional F on $C([0,t], \mathbb{R})$ we have

$$
\mathbb{E}_{\mathbb{Q}} \left(F \left((M_s)_{s \le t} \right) \right) = \mathbb{E}_{\mathbb{P}} \left(F \left((M_s)_{s \le t} \right) \, U_t \right) = \mathbb{E}_{\mathbb{P}} \left(F \left((W_s)_{s \le t} \right) \right).
$$

This shows that $(X_s)_{s \le t}$ is a Brownian motion with variance parameter $\sigma = 1$ under the measure $\mathbb{Q}$. We end up with

$$
\begin{aligned}
d\mathcal{S}_t &= b_t \, \mathcal{S}_t \, dt + \sigma_t \, \mathcal{S}_t \, dW_t \\
&= b_t \, \mathcal{S}_t \, dt + \sigma_t \, \mathcal{S}_t \, (dM_t - \sigma_t^{-1} \, (b_t - r_t) \, dt) \\
&= r_t \, \mathcal{S}_t \, dt + \sigma_t \, \mathcal{S}_t \, dM_t.
\end{aligned}
$$

Here again, the $\mathbb{Q}$-dynamics of the deflated risky asset price defined in (3.19) is given by the martingale increments

$$
d\overline{\mathcal{S}}_t = \sigma_t \, \overline{\mathcal{S}}_t \, dM_t.
$$

18.4.2   Elliptic diffusions

We consider the d-dimensional diffusion process $\boldsymbol{X_t} = \left(\boldsymbol{X^i}\right)_{1 \leq i \leq d}$ defined for any $1 \leq i \leq d$ by the stochastic differential equations

$$dX_t^i = b_t^i(\boldsymbol{X_t})\, dt + \sum_{1 \leq j \leq d_1} \sigma_{j,t}^i(\boldsymbol{X_t})\, dW_t^j$$

or in vector notation

$$d\boldsymbol{X_t} = \boldsymbol{b_t}(\boldsymbol{X_t})\, dt + \sum_{1 \leq j \leq d_1} \boldsymbol{\sigma_{j,t}}(\boldsymbol{X_t})\, dW_t^j.$$

Notice that these d-dimensional equations have the same form as the one discussed in (14.18) except that the number d_1 of Brownian motions $W_t = \left(W_t^i\right)_{1 \leq i \leq d_1}$ may be less than d.

For sufficiently regular drift functions $\boldsymbol{b_t}$, and for a positive diffusion matrix-valued function $\boldsymbol{a_t} = \boldsymbol{\sigma_t \sigma_t'}$ satisfying the uniform ellipticity condition

$$\forall u = (u_i)_{1 \leq i \leq d} \qquad \sum_{1 \leq i,j \leq d} a_t^{i,j}(\boldsymbol{x})\, u_i\, u_j \geq \epsilon \, \|u\|^2 \tag{18.17}$$

for some ϵ, the random states $\boldsymbol{X_t}$ have a smooth density w.r.t. the Lebesgue measure on $\mathbb{R}^d$.

We illustrate these models with the $d = d_1 + d_2$-dimensional diffusion process $X_t = (X_t, Y_t) \in \mathbb{R}^{d_1 + d_2}$, given by

$$\begin{cases} dX_t &= b_t(X_t, Y_t)\, dt + dW_t \\ dY_t &= c_t\, X_t\, dt \quad \Leftrightarrow Y_t = y_0 + \int_0^t c_s\, X_s\, ds \end{cases} \tag{18.18}$$

with some initial value $y_0 \in \mathbb{R}^{d_2}$, some functions b_t taking values in $\mathbb{R}^{d_1}$ and some $d_2 \times d_1$-matrix $c_t = (c_t^{i,j})_{1 \leq i \leq d_2, 1 \leq j \leq d_1}$ mapping $\mathbb{R}^{d_1}$ into $\mathbb{R}^{d_2}$. In this situation, we have

$$\boldsymbol{X_t^i} = \begin{cases} X_t^i & \text{for} \quad i \in \{1, \ldots, d_1\} \\ Y_t^{i-d_1} & \text{for} \quad i \in \{d_1 + 1, \ldots, d_1 + d_2\} \end{cases}$$

and

$$\boldsymbol{b_t^i}(x, y) = \begin{cases} b_t^i(x, y) & \text{for} \quad i \in \{1, \ldots, d_1\} \\ \sum_{1 \leq j \leq d_1} c_t^{i-d_1, j} x^j & \text{for} \quad i \in \{d_1 + 1, \ldots, d_1 + d_2\} \end{cases}$$

and

$$\boldsymbol{\sigma_{t,j}^i}(x, y) = \begin{cases} 1_{i=j} & \text{for} & i, j \in \{1, \ldots, d_1\} \\ 0 & \text{for} & (i,j) \in \{d_1 + 1, \ldots, d_1 + d_2\} \times \{1, \ldots, d_1\}. \end{cases}$$

The form of the $\boldsymbol{\sigma_{t,j}^i}$ coefficients above implies

$$\boldsymbol{a_t^{i,j}}(\boldsymbol{x}, \boldsymbol{y}) = \begin{cases} 1_{i=j} & \text{for} & i, j \in \{1, \ldots, d_1\} \\ 0 & \text{for} & (i,j) \in \{d_1 + 1, \ldots, d_1 + d_2\} \times \{1, \ldots, d_1\}. \end{cases}$$

It is not difficult to check that the uniform ellipticity condition (18.17) fails for this class of multidimensional diffusions.

By applying Girsanov's theorem to the non-homogeneous diffusions (18.12), for any $t \geq 0$, and any functional F on $C([0,t], \mathbb{R}^{d_1}) \times C([0,t], \mathbb{R}^{d_2})$ we have

$$\mathbb{E}\left(F\left((X_s)_{s \leq t}, (Y_s)_{s \leq t}\right) U_t\right) = \mathbb{E}\left(F\left((W_s)_{s \leq t}, \left(y_0 + \int_0^s c_r\, W_r\, dr\right)_{0 \leq s \leq t}\right)\right)$$

and

$$\mathbb{E}\left(F\left((X_s)_{s\leq t},(Y_s)_{s\leq t}\right)\right) = \mathbb{E}\left(F\left((W_s)_{s\leq t},\left(y_0+\int_0^s c_r\ W_r\ dr\right)_{0\leq s\leq t}\right)Z_t\right)$$

with

$$\log Z_t = \int_0^t b_s\left(W_s, y_0+\int_0^s c_r\ W_r\ dr\right)dW_s$$

$$-\frac{1}{2}\int_0^t\left\|b_s\left(W_s, y_0+\int_0^s c_r\ W_r\ dr\right)\right\|^2\ ds$$

and

$$U_t = \exp\left(-\int_0^t b_s(X_s,Y_s)'\ dW_s - \frac{1}{2}\int_0^t\|b_s(X_s,Y_s)\|^2\ ds\right).$$

For instance, when $d_1 = d_2$, $y_0 = 0$ and $c_t = Id$, we have

$$\mathbb{E}\left(F\left((X_s)_{s\leq t},(Y_s)_{s\leq t}\right)\right) = \mathbb{E}\left(F\left((W_s)_{s\leq t},\left(\int_0^s W_r\ dr\right)_{0\leq s\leq t}\right)Z_t\right).$$

In this situation, the law of the random paths $[(X_s)_{s\leq t},(Y_s)_{s\leq t}]$ has a density Z_t w.r.t. the distribution of the Gaussian random paths

$$\mathrm{Law}\left[(W_s)_{s\leq t},\left(\int_0^s W_r\ dr\right)_{0\leq s\leq t}\right].$$

With a little extra work, this result implies that the law of the random states (X_t,Y_t) is absolutely continuous w.r.t. the law of the Gaussian r.v. $(W_t,\int_0^t W_s ds)$. This clearly implies that (X_t,Y_t) has a density $p_t(x,y)$ w.r.t. the Lebesgue measure on $\mathbb{R}^d$. For smooth drift functions b_t, we can also show that $p_t(x,y)$ is smooth.

18.5 Nonlinear filtering

18.5.1 Diffusion observations

We consider a two-dimensional diffusion process (X_t,Y_t) of the form

$$\begin{cases} dX_t &=& b(X_t)\ dt\ +\ \sigma(X_t)\ dW_t \\ dY_t &=& h(X_t)\ dt\ +\ dV_t \end{cases} \tag{18.19}$$

with some regular functions (b,h,σ) and a couple of independent Brownian motions (V_t,W_t). The Markov process X_t represents some signal partially observed by the process Y_t. The h is called the sensor function. The Brownian motion V_t represents the perturbations of the sensor measurements. To simplify the presentation, we assume that $Y_0 = 0$.

We fix some time horizon, say $t > 0$. The random paths $(X,Y) = ((X_s)_{0\leq s\leq t},(Y_s)_{0\leq s\leq t})$ are defined on the product space $\Omega := (C([0,t],\mathbb{R})\times C([0,t],\mathbb{R}))$. Given the random path $X = \omega_1 \in C([0,t],\mathbb{R})$, the observation trajectories Y are given by $Y_0 = 0$ and

$$dY_s = h(\omega_1(s))\ ds + dV_s.$$

The sensor function $h(\omega_1(t))$ is now purely deterministic. We let $V = (V_s)_{0 \leq s \leq t}$. In this notation, by the Cameron Martin density formula (18.9) the conditional distribution of Y given $X = \omega_1$ is defined by

$$\mathbb{P}\left(Y \in d\omega_2 \mid X = \omega_1\right) = Z(\omega) \times \mathbb{P}\left(V \in d\omega_2\right)$$

with the density function Z defined for any $\omega = (\omega_1, \omega_2) \in \Omega$ by the formula

$$Z(\omega) = \exp\left(\int_0^t h(\omega_1(s))\, d\omega_2(s) - \frac{1}{2} \int_0^t h^2(\omega_1(s))\, ds \right).$$

This yields the change of probability formula

$$\begin{aligned}
\mathbb{P}\left((X,Y) \in d(\omega_1, \omega_2)\right) &= \mathbb{P}\left(Y \in d\omega_2 \mid X = \omega_1\right) \times \mathbb{P}\left(X \in d\omega_1\right) \\
&= Z(\omega) \times \left[\mathbb{P}\left(X \in d\omega_1\right) \mathbb{P}\left(V \in d\omega_2\right)\right].
\end{aligned}$$

If we set

$$\mathbb{P}^0\left((X,Y) \in d(\omega_1, \omega_2)\right) := \mathbb{P}\left(X \in d\omega_1\right) \times \mathbb{P}\left(V \in d\omega_2\right)$$

the above formula takes the form

$$\mathbb{P}\left((X,Y) \in d\omega\right) = Z(\omega) \times \mathbb{P}^0\left((X,Y) \in d\omega\right).$$

Notice that under $\mathbb{P}^0$ the process X is unchanged but Y becomes a Brownian motion independent of X. In probability theory, this is often called the Kallianpur-Striebel formula.

Finally, using Bayes' rule we obtain the conditional distribution

$$\mathbb{P}\left(X \in d\omega_1 \mid Y = \omega_2\right) \propto Z(\omega_1, \omega_2)\, \mathbb{P}\left(X \in d\omega_1\right) \qquad (18.20)$$

or equivalently

$$\mathbb{P}\left(X \in d\omega_1 \mid Y\right) \propto \exp\left(\int_0^t h(\omega_1(s))\, dY_s - \frac{1}{2} \int_0^t h^2(\omega_1(s))\, ds \right) \mathbb{P}\left(X \in d\omega_1\right).$$

18.5.2 Duncan-Zakai equation

Taking the t-marginal of the conditional distribution (18.20) for any bounded function f on $\mathbb{R}$ we have

$$\eta_t(f) := \mathbb{E}\left(f(X_t) \mid (Y_s)_{0 \leq s \leq t}\right) = \gamma_t(f)/\gamma_t(1)$$

with the random non-negative measures γ_t defined by

$$\gamma_t(f) = \int f(\omega_1(t)) \exp\left(\int_0^t h(\omega_1(s))\, dY_s - \frac{1}{2} \int_0^t h^2(\omega_1(s))\, ds \right) \mathbb{P}\left(X \in d\omega_1\right).$$

Since we have assumed that $Y_0 = 0$, we have $\gamma_0 = \eta_0 = \mathrm{Law}(X_0)$.

This formula is often written in the following form

$$\gamma_t(f) = \mathbb{E}_X \left(f(X_t) \exp \left(\int_0^t h(X_s) \, dY_s - \frac{1}{2} \int_0^t h^2(X_s) \, ds \right) \right)$$

where the lower index X in the expectation $\mathbb{E}_X(.)$ underlines the fact that the observation trajectories $(Y_s)_{0 \leq s \leq t}$ are not integrated.

Applying the Doeblin-Itō formula to

$$Z_t := e^{H_t} \quad \text{with} \quad H_t = \int_0^t h(X_s) \, dY_s - \frac{1}{2} \int_0^t h^2(X_s) \, ds \qquad (18.21)$$

we find that

$$\begin{aligned}
dZ_t &= e^{H_t} \, dH_t + \frac{1}{2} e^{H_t} \, dH_t dH_t \\
&= Z_t \left(h(X_t) \, dY_t - \frac{1}{2} h^2(X_t) \, dt + \frac{1}{2} h^2(X_t) \, dt \right) = Z_t \, h(X_t) \, dY_t.
\end{aligned}$$

We let L be the generator of the diffusion X_t. Using the integration by part formula

$$\begin{aligned}
d(f(X_t)Z_t) &= Z_t \, df(X_t) + f(X_t) \, dZ_t + df(X_t) \, dZ_t \\
&= Z_t \, L(f)(X_t) \, dt + f(X_t) \, Z_t \, h(X_t) \, dY_t + Z_t \, dM_t(f)
\end{aligned}$$

with the martingale increment $dM_t(f) = df(X_t) - L(f)(X_t)dt = f'(X_t)\sigma(X_t)dW_t$. In the last assertion we used the fact that $dV_t \times dW_t = 0$.

We conclude that

$$\begin{aligned}
d\gamma_t(f) &= d\mathbb{E}_X(f(X_t)Z_t) \\
&= \mathbb{E}_X(Z_t \, L(f)(X_t)) \, dt + \mathbb{E}_X(f(X_t) \, Z_t \, h(X_t)) \, dY_t \\
&= \gamma_t(L(f)) \, dt + \gamma_t(fh) \, dY_t.
\end{aligned}$$

In summary, we have proved that γ_t satisfies the Zakai stochastic partial differential equation

$$d\gamma_t(f) = \gamma_t(L(f)) \, dt + \gamma_t(fh) \, dY_t \qquad (18.22)$$

with the initial condition $\gamma_0 = \eta_0 = \text{Law}(X_0)$.

Applying the Doeblin-Itō formula to $\log \gamma_t(1)$ we also have

$$\begin{aligned}
d \log \gamma_t(1) &= \frac{1}{\gamma_t(1)} \, d\gamma_t(1) - \frac{1}{2} \frac{1}{\gamma_t(1)^2} \, d\gamma_t(1)d\gamma_t(1) \\
&= \frac{\gamma_t(h)}{\gamma_t(1)} \, dY_t - \frac{1}{2} \frac{\gamma_t(h)^2}{\gamma_t(1)^2} \, dt = \eta_t(h) \, dY_t - \frac{1}{2} \eta_t(h)^2 \, dt.
\end{aligned}$$

In the second line of the above display, we used $dY_t dY_t = dV_t dV_t = dt$, and $L(1) = 0$.

This yields the exponential formula

$$\gamma_t(1) = \exp \left[\int_0^t \eta_s(h) \, dY_s - \frac{1}{2} \int_0^t \eta_s(h)^2 \, ds \right].$$

18.5.3   Kushner-Stratonovitch equation

We have proved that

$$\eta_t(f) = \mathbb{E}_X\left(f(X_t)\,\exp\left(\int_0^t\,(h(X_s)-\eta_s(h))\,\,dY_s - \frac{1}{2}\int_0^t\,\left(h^2(X_s)-\eta_s(h)^2\right)\,ds\right)\right).$$

On the other hand, we also have

$$(h(X_s)-\eta_s(h))\,\,dY_s - \tfrac{1}{2}\left(h^2(X_s)-\eta_s(h)^2\right)\,ds$$

$$= (h(X_s)-\eta_s(h))\,(dY_s-\eta_s(h)ds) - \tfrac{1}{2}\left[\left(h^2(X_s)-\eta_s(h)^2\right)-2\,(h(X_s)-\eta_s(h))\,\eta_s(h)\right]ds$$

$$= (h(X_s)-\eta_s(h))\,(dY_s-\eta_s(h)ds) - \tfrac{1}{2}\,(h(X_s)-\eta_s(h))^2\,ds.$$

This yields the implicit formula

$$\eta_t(f)$$
$$= \mathbb{E}_X\left(f(X_t)\exp\left[\int_0^t\,(h(X_s)-\eta_s(h))\,[dY_s-\eta_s(h)ds] - \tfrac{1}{2}\int_0^t\,(h(X_s)-\eta_s(h))^2\,ds\right]\right).$$

This formula shows that η_t is defined as γ_t by replacing h by $(h-\eta_t(h))$ and dY_t by the so-called innovation process $(dY_t - \eta_t(h)dt)$.

Using the same proof as the one of the Zakai equation, we prove that the normalized conditional distribution η_t (a.k.a. the optimal filter) satisfies the Kushner-Stratonovitch equation

$$d\eta_t(f) = \eta_t(L(f))\,dt + \eta_t(f(h-\eta_t(h))\,(dY_t-\eta_t(h)dt) \qquad (18.23)$$

with the initial condition $\eta_0 = \mathrm{Law}(X_0)$.

Important remark : The Bayes' formula (18.20), the Duncan-Zakai equation (18.22) as well as the Kushner-Stratonovitch equation (18.23) can be extended to Markov processes X_t with some generator L_t on some state space and to any r-dimensional non-homogeneous diffusion observation processes of the form

$$dY_t = h_t(X_t)\,dt + R_t^{1/2}dV_t. \qquad (18.24)$$

In the above display, V_t is an r-dimensional Brownian motion (i.e. $V_t = \left(V_t^i\right)_{1\le i\le r}$ with r independent Brownian motions V_t^i, $1\le i\le r$), and $h_t\,:\,x\in S\mapsto h_t(x) = (h_t^i(x))_{1\le i\le r}\in \mathbb{R}^r$ is a function taking values in $\mathbb{R}^r$. We also assume that the symmetric matrix functional $t\mapsto R_t^{1/2}$ is smooth and invertible, and $Y_0=0$.

In this situation, the Bayes' formula (18.20) remains valid with

$$Z(\omega_1,\omega_2) := \exp\left(\int_0^t\,h_s(\omega_1(s))'R_s^{-1}\,d\omega_2(s) - \frac{1}{2}\int_0^t\,\langle h_s(\omega_1(s)),R_s^{-1}h_s(\omega_1(s))\rangle\,ds\right)$$

with the transposition $h_s(\omega_1(s))'$ of the column vector $h_s(\omega_1(s))$ and the Euclidian norm $\|h_s(\omega_1(s))\|^2 = \sum_{1 \leq i \leq r} \left(h_s^i(\omega_1(s)) \right)^2$. In this situation, the change of probability measure Z_t defined in (18.21) takes the form

$$Z_t := \exp\left(\int_0^t h_s(X_s)' R_s^{-1} \, dY_s - \frac{1}{2} \int_0^t \langle h_s(X_s), R_s^{-1} h_s(X_s) \rangle \, ds \right)$$

and satisfies the evolution equation

$$dZ_t = Z_t \, h_t(X_t)' \, R_t^{-1} \, dY_t.$$

In much the same way, the Duncan-Zakai and the Kushner-Stratonovitch formula, are given by the evolution equations

$$\begin{aligned}
d\gamma_t(f) &= \gamma_t(L_t(f)) \, dt + \gamma_t(f h_t)' R_t^{-1} \, dY_t \\
d\eta_t(f) &= \eta_t(L_t(f)) \, dt + \eta_t(f(h_t - \eta_t(h_t)))' R_t^{-1} \, (dY_t - \eta_t(h_t) dt).
\end{aligned} \quad (18.25)$$

In the above formulae, $\eta_t(h_t)$, $\gamma_t(f h_t)$, and $\eta_t(f(h_t - \eta_t(h_t)))$ stand for the column vectors with entries $\eta_t(h_t^i)$, $\gamma_t(f h_t^i)$, and $\eta_t(f(h_t^i - \eta_t(h_t^i)))$, with $1 \leq i \leq r$.

We also have the exponential formula

$$\gamma_t(1) = \exp\left[\int_0^t \eta_s(h_s)' R_s^{-1} \, dY_s - \frac{1}{2} \int_0^t \langle \eta_s(h_s), R_s^{-1} \eta_s(h_s) \rangle \, ds \right].$$

18.5.4 Kalman-Bucy filters

Consider a linear Gaussian filtering model of the following form

$$\begin{cases}
dX_t &= (A_t \, X_t + a_t) \, dt + R_{1,t}^{1/2} \, dW_t \\
dY_t &= (C_t \, X_t + c_t) \, dt + R_{2,t}^{1/2} \, dV_t.
\end{cases} \quad (18.26)$$

In the above display, (W_t, V_t) is an $(r_1 + r_2)$-dimensional Brownian motion, X_0 is a r_1-valued Gaussian random vector with mean and covariance matrix $(\mathbb{E}(X_0), P_0)$ (independent of (W_t, V_t)), the symmetric matrix functionals $t \mapsto R_{1,t}^{1/2}$ and $t \mapsto R_{2,t}^{1/2}$ are smooth and invertible, A_t is a square $(r_1 \times r_1)$-matrix, C_t is an $(r_2 \times r_1)$-matrix, a_t is a given r_1-dimensional column vector and c_t is an r_2-dimensional column vector, and $Y_0 = 0$. The infinitesimal generator of X_t is given by the second order differential operator

$$L_t(f)(x) = \sum_{1 \leq i \leq r_1} \left(a_t(i) + \sum_{1 \leq j \leq r_1} A_t(i,j) x_j \right) \partial_{x_i} f(x) + \frac{1}{2} \sum_{1 \leq i,j \leq r_1} R_{1,t}(i,j) \, \partial_{x_i, x_j} f(x).$$

For each $1 \leq i \leq r_1$, we set $f_i(x) = x_i$. In this notation, we have

$$L_t(f_i) = a_t(i) + \sum_{1 \leq j \leq r_1} A_t(i,j) f_j.$$

The Kushner-Stratonovitch equation (18.25) applied to these monomial functions takes the form

$$\begin{aligned}
d\eta_t(f) &= (A_t \eta_t(f) + a_t) \, dt + P_{\eta_t} \, C_t' R_{2,t}^{-1} \, (dY_t - (C_t \eta_t(f) + c_t) \, dt) \\
&= (A_t \eta_t(f) + a_t) \, dt + P_{\eta_t} \, C_t' R_{2,t}^{-1} \left[C_t \, (f(X_t) - \eta_t(f)) \, dt + R_{2,t}^{1/2} \, dV_t \right]
\end{aligned}$$

with the column vectors $f = (f_i)_{1 \leq i \leq r_1}$ and $\eta_t(f) = (\eta_t(f_i))_{1 \leq i \leq r_1}$ and the covariance matrix

$$
\begin{aligned}
P_{\eta_t} &= \eta_t \left[(f - \eta_t(f))(f - \eta_t(f))' \right] = \left[\eta_t((f_i - \eta_t(f_i))(f_j - \eta_t(f_j))) \right]_{1 \leq i,j \leq r_1} \\
&= \mathbb{E} \left([f(X_t) - \eta_t(f)] \, [f(X_t) - \eta_t(f)]' \mid \sigma(Y_s, \ s \leq t) \right) \\
&= \mathbb{E} \left([f(X_t) - \eta_t(f)] \, [f(X_t) - \eta_t(f)]' \right).
\end{aligned}
$$

The last assertion comes from the fact that (X_t, Y_t) is a linear Gaussian process. On the other hand,

$$
f(X_t) = X_t \Rightarrow df(X_t) = (A_t f(X_t) + a_t) \ dt \ + \ R_{1,t}^{1/2} \ dW_t.
$$

We set $f_t = f - \eta_t(f)$ and $f_{t,i} = f_i - \eta_t(f_i)$. In this notation, we have the equation

$$
\begin{aligned}
df_t(X_t) &= A_t f_t(X_t) \ dt \ + \ R_{1,t}^{1/2} \ dW_t - P_{\eta_t} \ C_t' R_{2,t}^{-1} \left[C_t \ f_t(X_t) \ dt \ + \ R_{2,t}^{1/2} \ dV_t \right] \\
&= \left[A_t - P_{\eta_t} \ C_t' R_{2,t}^{-1} \ C_t \right] \ f_t(X_t) \ dt + \ R_{1,t}^{1/2} \ dW_t - P_{\eta_t} \ C_t' R_{2,t}^{-1/2} \ dV_t.
\end{aligned}
$$

Using the matrix differential formula

$$
d \left[f_t(X_t) f_t(X_t)' \right] = f_t(X_t) \ df_t(X_t)' + df_t(X_t) \ f_t(X_t)' + \underbrace{df_t(X_t) \ df_t(X_t)'}_{= [R_{1,t} + P_{\eta_t} C_t' R_{2,t}^{-1} C_t P_{\eta_t}] dt}
$$

we prove that

$$
d \left[f_t(X_t) f_t(X_t)' \right]
$$

$$
= \left\{ f_t(X_t) f_t(X_t)' \left[A_t - P_{\eta_t} \ C_t' R_{2,t}^{-1} \ C_t \right]' + \left[A_t - P_{\eta_t} \ C_t' R_{2,t}^{-1} \ C_t \right] \ f_t(X_t) f_t(X_t)' \right.
$$

$$
\left. + [R_{1,t} + P_{\eta_t} C_t' R_{2,t}^{-1} C_t P_{\eta_t}] \right\} \ dt
$$

$$
+ \ f_t(X_t) \left(R_{1,t}^{1/2} \ dW_t - P_{\eta_t} \ C_t' R_{2,t}^{-1/2} \ dV_t \right)' + \left(R_{1,t}^{1/2} \ dW_t - P_{\eta_t} \ C_t' R_{2,t}^{-1/2} \ dV_t \right) \ f_t(X_t)'.
$$

Taking the expectation we find that

$$
\begin{aligned}
\partial_t P_{\eta_t} &= P_{\eta_t} \left[A_t - P_{\eta_t} \ C_t' R_{2,t}^{-1} \ C_t \right]' + \left[A_t - P_{\eta_t} \ C_t' R_{2,t}^{-1} \ C_t \right] \ P_{\eta_t} + [R_{1,t} + P_{\eta_t} C_t' R_{2,t}^{-1} C_t P_{\eta_t}] \\
&= P_{\eta_t} A_t' + A_t P_{\eta_t} + R_{1,t} - P_{\eta_t} C_t' R_{2,t}^{-1} C_t P_{\eta_t}.
\end{aligned}
$$

If we set $\widehat{X}_t := \eta_t(f)$ and $P_t := P_{\eta_t}$, we obtain the Kalman-Bucy filter equations

$$
\begin{cases}
d\widehat{X}_t &= \left(A_t \ \widehat{X}_t + a_t \right) \ dt + P_t \ C_t' R_{2,t}^{-1} \left(dY_t - \left(C_t \widehat{X}_t + c_t \right) dt \right) \\
\partial_t P_t &= P_t A_t' + A_t P_t + R_{1,t} - P_t C_t' R_{2,t}^{-1} C_t P_t
\end{cases} \tag{18.27}
$$

with the initial conditions $(\widehat{X}_0, P_0)$ given by the mean and covariance matrix $(\mathbb{E}(X_0), P_0)$ of the initial condition X_0 of the signal (since we have assumed that $Y_0 = 0$; otherwise the initial condition is given by the regression formula).

The quadratic evolution equation of the covariance matrices P_t belong to the class of Riccati equations.

18.5.5 Nonlinear diffusion and ensemble Kalman-Bucy filters

We return to the linear Gaussian filtering model (18.26) discussed in section 18.5.4. We use the same notation as in section 18.5.4.

We fix an observation path $y = (y_t)_{t \geq 0}$ and we consider the nonlinear diffusion process

$$d\overline{X}_t = \left(A_t \, \overline{X}_t + a_t \right) \, dt \, + \, R_{1,t}^{1/2} \, d\overline{W}_t + P_{\eta_t} C_t' R_{2,t}^{-1} \left(dy_t - \left((C_t \overline{X}_t + c_t) dt + R_{2,t}^{1/2} \, d\overline{V}_t \right) \right)$$
(18.28)

where $(\overline{W}_t, \overline{V}_t)$ is an $(r_1 + r_2)$-dimensional Brownian motion and $\overline{X}_0$ is a r_1-valued Gaussian random vector with mean and covariance matrix $(\mathbb{E}(X_0), P_0)$ (independent of $(\overline{W}_t, \overline{V}_t)$). We also assume that $(\overline{X}_0, \overline{W}_t, \overline{V}_t)$ are independent of the observation path. In the above formula, P_{η_t} stands for the covariance matrix

$$P_{\eta_t} = \eta_t \left[(f - \eta_t(f))(f - \eta_t(f))' \right] \quad \text{with} \quad \eta_t := \text{Law}(\overline{X}_t).$$

By construction, we have

$$d\mathbb{E}\left(\overline{X}_t \right) = \left(A_t \, \mathbb{E}\left(\overline{X}_t \right) + a_t \right) \, dt + P_{\eta_t} \, C_t' R_{2,t}^{-1} \left(dy_t - \left(C_t \mathbb{E}\left(\overline{X}_t \right) + c_t \right) dt \right). \tag{18.29}$$

We set $\widetilde{X}_t := \overline{X}_t - \mathbb{E}\left(\overline{X}_t \right)$. In this notation we have

$$d\widetilde{X}_t = \left[A_t \, - P_{\eta_t} C_t' R_{2,t}^{-1} C_t \right] \, \widetilde{X}_t \, dt \, + \, R_{1,t}^{1/2} \, d\overline{W}_t - P_{\eta_t} \, C_t' R_{2,t}^{-1/2} d\overline{V}_t.$$

This implies that

$$\begin{aligned}
d\left(\widetilde{X}_t \widetilde{X}_t' \right) &= \left[A_t \, - P_{\eta_t} C_t' R_{2,t}^{-1} C_t \right] \, \widetilde{X}_t \widetilde{X}_t' \, dt \, + \left[\, R_{1,t}^{1/2} \, d\overline{W}_t - P_{\eta_t} \, C_t' R_{2,t}^{-1/2} d\overline{V}_t \right] \widetilde{X}_t' \\
&\quad + \widetilde{X}_t \widetilde{X}_t' \left[A_t \, - P_{\eta_t} C_t' R_{2,t}^{-1} C_t \right]' \, dt \, + \, \widetilde{X}_t \left[R_{1,t}^{1/2} \, d\overline{W}_t - P_{\eta_t} \, C_t' R_{2,t}^{-1/2} d\overline{V}_t \right]' \\
&\quad + \left[R_{1,t} + P_{\eta_t} \, C_t' R_{2,t}^{-1} C_t P_{\eta_t} \right] \, dt.
\end{aligned}$$

Taking the expectations, we find that

$$\partial_t P_{\eta_t} = A_t P_{\eta_t} + P_{\eta_t} A' + R_{1,t} - P_{\eta_t} C_t' R_{2,t}^{-1} C_t P_{\eta_t}. \tag{18.30}$$

The evolution equations (18.29) and (18.30) coincide with the Kalman-Bucy equations (18.27).

The ensemble Kalman filter coincides with the mean field particle interpretation of the nonlinear diffusion process (18.28). More precisely, we sample N copies $\xi_0 = (\xi_0^i)_{1 \leq i \leq N}$ of $\overline{X}_0$ and we consider the Mckean-Vlasov type interacting diffusion process

$$\begin{cases} d\xi_t^i = \left(A_t \, \xi_t^i + a_t \right) dt + R_{1,t}^{1/2} d\overline{W}_t^i + P_{\eta_t^N} C_t' R_{2,t}^{-1} \left(dY_t - \left((C_t \xi_t^i + c_t) dt + R_{2,t}^{1/2} \, d\overline{V}_t^i \right) \right) \\ i = 1, \ldots, N \end{cases}$$

with the empirical measures $\eta_t^N := \frac{1}{N} \sum_{1 \leq i \leq N} \delta_{\xi_t^i}$. In the above formula, $(\overline{W}_t^i, \overline{V}_t^i)_{1 \leq i \leq N}$ stands for N independent copies of $(\overline{W}_t, \overline{V}_t)$.

For a more detailed discussion on nonlinear diffusion processes and their mean field particle interpretations, we refer the reader to chapter 16 and section 7.10.2 dedicated to their discrete time version. The discrete time version of the ensemble-Kalman filters is presented in some details in section 9.9.6.3. We also refer to the articles [83, 84, 85] and the references therein for stability and uniform propagation of chaos properties of these mean field particle samplers.

18.5.6 Robust filtering equations

We return to the two-dimensional filtering model (X_t, Y_t) discussed in (18.19). Recall that

$$dh(X_t) = L(h)(X_t) \, dt + dM_t(h)$$

with some martingale $M_t(h)$ (w.r.t. $\sigma(X_s, s \leq t)$). Using the integration by part formula, we find that

$$
\begin{aligned}
h(X_t)dY_t &= d(h(X_t)Y_t) - Y_t dh(X_t) \\
&= d(h(X_t)Y_t) + \left[e^{Y_t h} L(e^{-Y_t h}) \right] (X_t) \, dt + d \log M_t^h(Y)
\end{aligned}
$$

with the exponential process $M_t^h(Y)$ defined by the following formula

$$d \log M_t^h(Y) := -Y_t dM_t(h) - \left[e^{Y_t h} L(e^{-Y_t h}) + Y_t L(h) \right] (X_t) dt.$$

Notice that the stochastic integration w.r.t. dY_t has been removed and $M_t^h(y)$ depends only on the observation path $y = (y_s)_{0 \leq s \leq t}$.

$$\log M_t^h(y) = -\int_0^t y_s \, dM_s(h) - \int_0^t \left[e^{y_s h} L(e^{-y_s h}) + y_s L(h) \right] (X_s) \, ds$$

For smooth observation paths, we have

$$d \left(y_t h(X_t) \right) = (\partial_t + L) \left(y_t h(.) \right) \, dt + y_t \, dM_t(h)$$

and

$$e^{y_t h} L(e^{-y_t h}) + y_t L(h) = e^{y_t h} (\partial_t + L) (e^{-y_t h}) + (\partial_t + L) (y_t h).$$

This yields

$$-y_t \, dM_t(h) = (\partial_t + L) \left(y_t h(.) \right)(X_t) \, dt - d \left(y_t h(X_t) \right)$$

and therefore

$$
\begin{aligned}
d \log M_t^h(y) &= -y_t \, dM_t(h) - \left[e^{y_t h} L(e^{-y_t h}) + y_t L(h) \right] (X_t) dt \\
&= - \left[e^{y_t h} (\partial_t + L) (e^{-y_t h}) \right] (X_t) dt - d \left(y_t h(X_t) \right).
\end{aligned}
$$

This shows that

$$
\begin{aligned}
M_t^h(y) &= \exp \left(-y_t h(X_t) + y_0 h(X_0) - \int_0^t \left[e^{y_s h} (\partial_s + L) (e^{-y_s h}) \right] (X_s) ds \right) \\
&= h_0^{-1}(X_0) h_t(X_t) \exp \left\{ - \int_0^t \left[h_s^{-1} (\partial_s + L) (h_s) \right] (X_s) ds \right\}
\end{aligned}
$$

with the function

$$h_t(x) := \exp \left(-y_t h(x) \right). \tag{18.31}$$

This implies that

$$\gamma_t(f) = \mathbb{E}_X \left(f(X_t) M_t^h(y) \, \exp\left(y_t h(X_t) - y_0 h(X_0) + \int_0^t V_s(X_s) ds \right) \right)$$

with the random potential function V_t defined by

$$V_t = e^{y_t h} L(e^{-y_t h}) - h^2/2. \tag{18.32}$$

Finally, using (18.15) and (18.16) we obtain the Feynman-Kac representation of γ_t in terms of the function h_t and the potential function V_t defined in (18.31) and (18.32); that is,

$$\gamma_t(f) = \eta_0(h_0) \, \mathbb{E}_X \left(f(X_t^h) h_t^{-1}(X_t^h) \, \exp\left(\int_0^t V_s(X_s^h) ds \right) \right)$$

where X_t^h has an infinitesimal generator

$$L_t^{[h]}(f) := L(f) + h_t^{-1} \Gamma_L(h_t, f)$$

and initial distribution $\text{Law}(X_0^h) = \Psi_{h_0}(\eta_0)$ with $\eta_0 = \text{Law}(X_0)$.

The Feynman-Kac formula stated above does not require any regularity property on the observation path and it does not involve any stochastic integration w.r.t. dY_t. We recall that continuously differentiable trajectories are dense in the set of all continuous paths equipped with the uniform norm. Thus, using smooth approximations (see for instance lemma 4.2 in [81]), the Feynman-Kac representation we obtained is also valid for any continuous observation paths.

18.5.7 Poisson observations

We consider a two-dimensional stochastic process (X_t, Y_t). The process X_t has some generator L_t and it evolves on some state space S. We fix the time horizon t, and we let Ω_1 be the set of random trajectories of $X = (X_s)_{0 \le s \le t}$.

Given a realization of the process $X = (X_s)_{0 \le s \le t} = \omega_1 \in \Omega_1$, the observation process Y_s is a Poisson process with intensity $\lambda_s(\omega_1(s))$, with $s \in [0, t]$. The random paths $(X, Y) = ((X_s)_{0 \le s \le t}, (Y_s)_{0 \le s \le t})$ are defined on the product space $\Omega := (\Omega_1 \times D([0, t], \mathbb{R}))$.

We let $N = (N_s)_{0 \le s \le t}$ be a Poisson process with unit intensity.

See the discussion on the Poisson processes provided in section 18.1.3, and Girsanov theorem 18.1.2. Using (18.3) and replacing (λ_s, λ_s') by $(1, \lambda_s(\omega_1(s)))$, we have the formula:

$$\mathbb{P}(Y \in d\omega_2 \mid X = \omega_1) = Z(\omega_1, \omega_2) \, \mathbb{P}(N \in d\omega_2)$$

with the density function Z defined for any $\omega = (\omega_1, \omega_2)$ by the formula

$$Z(\omega) = \exp\left[\int_0^t [1 - \lambda_s(\omega_1(s))] \, ds + \int_0^t \log(\lambda_s(\omega_1(s))) \, d\omega_2(s) \right].$$

This yields the change of probability formula

$$\begin{aligned} \mathbb{P}((X, Y) \in d(\omega_1, \omega_2)) &= \mathbb{P}(Y \in d\omega_2 \mid X = \omega_1) \times \mathbb{P}(X \in d\omega_1) \\ &= Z(\omega) \times [\mathbb{P}(X \in d\omega_1) \, \mathbb{P}(N \in d\omega_2)]. \end{aligned}$$

Finally, using the Bayes rule, we obtain the conditional distribution

$$\mathbb{P}\left(X \in d\omega_1 \mid Y = \omega_2\right) \propto Z(\omega_1, \omega_2) \, \mathbb{P}\left(X \in d\omega_1\right) \qquad (18.33)$$

or equivalently

$$\mathbb{P}\left(X \in d\omega_1 \mid Y\right)$$

$$\propto \exp\left[\int_0^t \left[1 - \lambda_s(\omega_1(s))\right] \, ds + \int_0^t \log\left(\lambda_s(\omega_1(s))\right) \, dY_s\right] \mathbb{P}\left(X \in d\omega_1\right).$$

Taking the t-marginal of the conditional distribution for any bounded function f on $\mathbb{R}$ we have

$$\eta_t(f) := \mathbb{E}\left(f(X_t) \mid (Y_s)_{0 \le s \le t}\right) = \gamma_t(f)/\gamma_t(1).$$

The random non-negative measures γ_t defined by this formula are often written in the form

$$\gamma_t(f) = \mathbb{E}_X\left(f(X_t)\, Z_t\right) \quad \text{with} \quad Z_t = \exp\left[\int_0^t \left[1 - \lambda_s(X_s)\right] \, ds + \int_0^t \log\left(\lambda_s(X_s)\right) \, dY_s\right]$$

where the lower index X in the expectation $\mathbb{E}_X\left(.\right)$ underlines the fact that the observation trajectories $(Y_s)_{0 \le s \le t}$ are not integrated. By (18.6) we have

$$dZ_t = Z_t \, \left(\lambda_t(X_t) - 1\right)\left(dY_t - dt\right).$$

This implies that

$$Z_t = 1 + \int_0^t Z_s \, \left(\lambda_s(X_s) - 1\right)\left(dY_s - ds\right)$$

from which we prove that

$$\begin{aligned} \gamma_t(f) &= \mathbb{E}_X\left(f(X_t) \left[1 + \int_0^t Z_s \, \left(\lambda_s(X_s) - 1\right)\left(dY_s - ds\right)\right]\right) \\ &= \mathbb{E}(f(X_t)) + \int_0^t \gamma_s \left(P_{s,t}(f)(\lambda_s - 1)\right)\left(dY_s - ds\right). \end{aligned}$$

Here $P_{s,t}$ stands for the semigroup of X_t. In differential form, we have proved that

$$\begin{aligned} d\gamma_t(f) &= \mathbb{E}(L_t(f)(X_t)) \, dt + \gamma_t \left(f(\lambda_t - 1)\right)\left(dY_t - dt\right) \\ &\qquad + \left[\int_0^t \gamma_s \left(P_{s,t}(L_t(f))(\lambda_s - 1)\right)\left(dY_s - ds\right)\right] dt \\ &= \gamma_t(L_t(f)) \, dt + \gamma_t \left(f(\lambda_t - 1)\right)\left(dY_t - dt\right). \end{aligned}$$

In summary, we obtain the Duncan-Zaikai equation

$$d\gamma_t(f) = \gamma_t(L_t(f)) \, dt + \gamma_t \left(f(\lambda_t - 1)\right)\left(dY_t - dt\right).$$

Notice that

$$d\gamma_t(1) = \gamma_t\left((\lambda_t - 1)\right)\left(dY_t - dt\right) = \gamma_t(1) \, \left(\eta_t(\lambda_t) - 1\right) \, \left(dY_t - dt\right).$$

This implies that

$$\gamma_t(1) = \exp\left[\int_0^t (1 - \eta_s(\lambda_s))\, ds + \int_0^t \log\left(\eta_s(\lambda_s)\right) dY_s\right].$$

This yields the normalized exponential formula

$$\begin{aligned}
\overline{Z}_t &:= Z_t/\gamma_t(1) \\
&= \exp\left[-\int_0^t (\lambda_s(X_s) - \eta_s(\lambda_s))\, ds + \int_0^t \log\left(\lambda_s(X_s)/\eta_s(\lambda_s)\right) dY_s\right].
\end{aligned}$$

Replacing in (18.5) (λ'_s, λ_s) by $(\lambda_s(X_s), \eta_s(\lambda_s))$ we check that

$$\begin{aligned}
d\overline{Z}_t &= \overline{Z}_t \left[(\overline{\lambda}_t(X_t) - 1)\, dY_t - (\lambda_t - \eta_t(\lambda_t))\, dt\right] \\
&= \overline{Z}_t \left(\overline{\lambda}_t(X_t) - 1\right)\, (dY_t - \eta_t(\lambda_t)\, dt) \quad \text{with} \quad \overline{\lambda}_t = \lambda_t/\eta_t(\lambda_t).
\end{aligned}$$

This shows that $\overline{Z}_t$ is defined as Z_t by replacing λ_t by the normalized intensities $\overline{\lambda}_t$ and $dY_t - dt$ by the innovation process $dY_t - \eta_t(\lambda_t)dt$.

> Using the same proof as the one for the Zakai equation, we prove that the normalized conditional distributions η_t (a.k.a. the optimal filter) satisfies the Kushner-Stratonovitch equation
>
> $$\begin{aligned}
d\eta_t(f) &= \eta_t(L_t(f))\, dt + \eta_t\left[f\left(\frac{\lambda_t}{\eta_t(\lambda_t)} - 1\right)\right] (dY_t - \eta_t(\lambda_t)\, dt) \quad (18.34) \\
&= \left[\eta_t(L_t(f)) + \eta_t\left(f\left(\eta_t(\lambda_t) - \lambda_t\right)\right)\right] dt + \eta_t\left[f\left(\frac{\lambda_t}{\eta_t(\lambda_t)} - 1\right)\right] dY_t.
\end{aligned}$$

18.6   Exercises

Exercise 333 *We consider a Brownian process W_t starting at the origin. For any $b \in \mathbb{R}$ and any time horizon t, check that*

$$\mathbb{E}\left[F\left((W_s)_{s\in[0,t]}\right) Z_t^{(b)}\right] = \mathbb{E}\left[F\left((W_s + bs)_{s\in[0,t]}\right)\right] \quad \text{with} \quad Z_t^{(b)} = \exp\left[b\, W_t - \frac{b^2}{2}\, t\right].$$

Prove that $Z_t^{(b)}$ is a martingale w.r.t. $\mathcal{F}_t = \sigma(W_s,\ s \le t)$.

Exercise 334 *Consider the one-dimensional diffusion X_t starting at the origin and discussed in exercise 223. Using exercise 223 check that*

$$\mathbb{E}\left[F\left((X_s)_{s\in[0,t]}\right)\right] = \exp\left[-\frac{t}{2}\left(\frac{b}{\sigma}\right)^2\right] \mathbb{E}\left\{F\left((\sigma\, W_s)_{s\in[0,t]}\right)\, \exp\left[\left(\frac{b}{\sigma}\right) W_t\right]\right\}.$$

Exercise 335 *Consider the one-dimensional diffusion X_t starting at the origin and discussed in exercise 334. Using exercise 334 check that*

$$\mathbb{E}\left[F\left((X_s)_{s\in[0,t]}\right) \exp\left\{-\left(\frac{b}{\sigma}\right) W_t - \frac{t}{2}\left(\frac{b}{\sigma}\right)^2\right\}\right] = \mathbb{E}\left[F\left((\sigma\, W_s)_{s\in[0,t]}\right)\right].$$

Exercise 336 *We consider the one-dimensional diffusion process*

$$dX_t = b_t \, dt + \sigma_t \, dW_t \quad \text{and} \quad d\overline{W}_t = \sigma_t \, dW_t$$

starting at $\overline{W}_0 = X_0$, with some Brownian motion W_t. Check that

$$\mathbb{E}\left(F\left((X_s)_{s \leq t}\right) \, U_t\right) = \mathbb{E}\left(F\left((\overline{W}_s)_{s \leq t}\right)\right)$$

with

$$U_t := \exp\left(-\int_0^t \left(\frac{b_s}{\sigma_s}\right) \, dW_s - \frac{1}{2} \int_0^t \left(\frac{b_s}{\sigma_s}\right)^2 \, ds\right).$$

Exercise 337 (Conditional expectations - Change of measures) *We let $\mathbb{P}$ and $\mathbb{P}'$ be a couple of probability measures on a measurable space $(\Omega, \mathcal{F})$ equipped with a σ-field $\mathcal{F}$ and such that*

$$\mathbb{E}'(X) = \int X(\omega) \, \mathbb{P}'(d\omega) = \int X(\omega) \, Z(\omega) \, \mathbb{P}(d\omega) = \mathbb{E}(XZ)$$

for any random variable X on $(\Omega, \mathcal{F})$, and for some positive random variable Z s.t. $\mathbb{E}(Z) = 1$. The random variable Z is sometimes written $Z = d\mathbb{P}'/d\mathbb{P}$ and it is called the Radon-Nikodym derivative of $\mathbb{P}'$ w.r.t. $\mathbb{P}$. We let $\mathcal{G} \subset \mathcal{F}$ be a smaller σ-field on Ω. (For instance $\mathcal{F} = \sigma(U, V)$ may be generated by a couple of random variables (U, V) on Ω and $\mathcal{G} = \sigma(U)$ may be the information contained in the first coordinate only.) We let $\mathbb{E}'(X|\mathcal{G})$ be the conditional expectation of X w.r.t. $\mathcal{G}$ on the probability space $(\Omega, \mathcal{F}, \mathbb{P}')$. Check that

$$\mathbb{E}'(X|\mathcal{G}) = \mathbb{E}(ZX|\mathcal{G})/\mathbb{E}(Z|\mathcal{G}).$$

Exercise 338 *We let $(V_s)_{s \in [0,t]}$ be a Brownian motion on $\mathbb{R}$, and h_s a given regular function. We set $dY_s := h_s ds + dV_s$, with $s \in [0, t]$ Check that*

$$\mathbb{E}\left(F\left((V_s)_{s \leq t}\right)\right) = \mathbb{E}\left(F\left((Y_s)_{s \leq t}\right) \, \exp\left(-\int_0^t h_s \, dY_s + \frac{1}{2} \int_0^t h_s^2 \, ds\right)\right)$$

and

$$\mathbb{E}\left(F\left((Y_s)_{s \leq t}\right)\right) = \mathbb{E}\left(F\left((V_s)_{s \leq t}\right) \, \exp\left(\int_0^t h_s \, dV_s - \frac{1}{2} \int_0^t h_s^2 \, ds\right)\right).$$

Exercise 339 (Wonham filter) *We let X_t be a Markov process on a finite space $S = \{1, \ldots, n\}$, for some $n \geq 1$, with generator*

$$L_t(f)(i) = \sum_{1 \leq j \leq n} (f(j) - f(i)) \, Q_t(i, j)$$

for some matrices Q_t with non-negative entries. Let Y_t be the observation process given by

$$dY_t = h_t(X_t) \, dt + \sigma_t \, dV_t$$

for some regular functions h_t from S into $\mathbb{R}$, a Brownian motion V_t, and some function $\sigma_t > 0$. We set $\eta_t(i) = \mathbb{P}(X_t = i \mid \mathcal{Y}_t)$, where $\mathcal{Y}_t = \sigma(Y_s, \, s \leq t)$. Describe the evolution equation of the random vector

$$\eta_t = [\eta_t(1), \ldots, \eta_t(n)].$$

Exercise 340 (Wonham filter - Telegraphic signal) *We let X_t be a Markov process on a finite space $S = \{0, 1\}$ switching from one state to another with a given intensity $\theta > 0$. Let Y_t be the observation process given by*

$$dY_t = X_t \, dt \, + \, dV_t$$

with a Brownian motion V_t. We set $\eta_t(1) := \mathbb{P}(X_t = 1 \mid \mathcal{Y}_t)$, where $\mathcal{Y}_t = \sigma(Y_s, \ s \le t)$. Using exercise 339, check that

$$d\eta_t(1) = \lambda \ (1 - 2\eta_t(1)) \ dt + \eta_t(1)(1 - \eta_t(1)) \ (dY_t - \eta_t(1) \ dt).$$

Exercise 341 (Kalman filter) *Consider a two-dimensional linear Gaussian signal/observation model of the form*

$$\begin{cases} dX_t & = & (A \ X_t + a) \ dt \ + \ \sigma_1 \ dW_t \\ dY_t & = & (C \ X_t + c) \ dt \ + \ \sigma_2 \ dV_t. \end{cases} \qquad (18.35)$$

In the above display, (W_t, V_t) is a two-dimensional Brownian motion, X_0 is a real valued Gaussian random vector with mean and covariance matrix $(\mathbb{E}(X_0), P_0)$ (independent of (W_t, V_t)), $\sigma_1, \sigma_2 > 0$, (A, a) and $(C, , c)$ are some given parameters, and $Y_0 = 0$. Describe the Kalman-Bucy filter associated with this estimation problem. Discuss the situation $C = 0$. When $C \ne 0$ check that the Kalman-Bucy filter is given by

$$\begin{cases} d\widehat{X}_t & = & \left(A \ \widehat{X}_t + a \right) \ dt + Q_t C^{-1} \left(dY_t - \left(C\widehat{X}_t + c \right) dt \right) \\ \dot{Q}_t & = & -Q_t^2 + 2AQ_t + B^2, \end{cases}$$

with $B := C(\sigma_1/\sigma_2) \ne 0$ and $Q_0 = (C/\sigma_2)^2 \, P_0$. Prove that

$$Q_t = z_2 - \frac{(z_2 - Q_0)e^{-(z_2 - z_1)t}(z_2 - z_1)}{(z_2 - Q_0) \ e^{-(z_2 - z_1)t} + (Q_0 - z_1)}$$

and

$$\int_0^t (A - Q_s) \, ds \le -\frac{(z_2 - z_1)t}{2} + \log \left(1 + |z_2/z_1| \right)$$

with

$$z_1 := A - \sqrt{A^2 + B^2} < 0 < z_2 := A + \sqrt{A^2 + B^2}.$$

Exercise 342 (Martingales - Change of measures) *We let $(M_s)_{s \in [0,t]}$ be a martingale w.r.t. some filtration $(\mathcal{F}_s)_{s \le t}$ on some probability space $(\Omega_t, \mathcal{F}_t, \mathbb{P}_t)$. we let $\mathbb{P}'_t$ be another probability measure on $(\Omega_t, \mathcal{F}_t)$ defined by $Z_t = d\mathbb{P}'_t/d\mathbb{P}_t$ for some non-negative martingale $(Z_s)_{s \le t}$ w.r.t. some filtration $(\mathcal{F}_s)_{s \le t}$ on $(\Omega_t, \mathcal{F}_t, \mathbb{P}_t)$. In other words, we have $\mathbb{E}(Z_t | \mathcal{F}_s) = Z_s$ for any $s \le t$ and*

$$\mathbb{E}'_t (X) := \int \ X(\omega) \ \mathbb{P}'_t(d\omega) = \int \ X(\omega) \ Z_t(\omega) \ \mathbb{P}_t(d\omega) := \mathbb{E}_t (X \ Z_t)$$

for any random variable X on $(\Omega_t, \mathcal{F}_t)$. Using exercise 337, check that for any $s \le t$

$$\mathbb{E}'(M_t \mid \mathcal{F}_s) = M_s \iff \mathbb{E}(M_t Z_t | \mathcal{F}_s) = Z_s M_s.$$

This shows that $(M_s)_{s \in [0,t]}$ is a martingale (w.r.t. $(\mathcal{F}_s)_{s \le t}$) on $(\Omega_t, \mathcal{F}_t, \mathbb{P}'_t)$ if and only if $(M_s Z_s)_{s \in [0,t]}$ is a martingale (w.r.t. $(\mathcal{F}_s)_{s \le t}$) on $(\Omega_t, \mathcal{F}_t, \mathbb{P}_t)$. In this situation check that

$$\mathbb{E}'(M_t \mid \mathcal{F}_s) = Z_s^{-1}\mathbb{E}(M_t Z_t | \mathcal{F}_s).$$

Exercise 343 (Martingale transformations) ✎ *We consider the martingale* $(M_s)_{s \leq t}$
*and the change of probability model discussed in exercise 342. Using exercise 342, check
that*

$$M'_s := M_s - \int_0^t Z_{s+}^{-1} \, d\,[Z, M]_s$$

is a martingale on $(\Omega_t, \mathcal{F}_t, \mathbb{P}'_t)$.

Exercise 344 (Martingale transformations - Diffusions) ✎ *Consider a Brownian
motion* $(W_s)_{s \leq t}$ *on* $(\Omega_t, \mathbb{P}_t)$ *and the change of probability* $Z_t = d\mathbb{P}'_t/d\mathbb{P}_t$ *with the non-
negative martingale* $(Z_s)_{s \leq t}$ *defined in (18.10). Using exercise 343, check that* $W'_s =
W_s - \int_0^s b(W_r) \, dr$, *with* $s \in [0, t]$, *is a martingale on* $(\Omega_t, \mathcal{F}_t, \mathbb{P}'_t)$.

Exercise 345 (Martingale transformations - Poisson process) *Consider a Poisson
process* $(N_s)_{s \leq t}$ *with unit intensity* $\lambda_s = 1$ *on* $(\Omega_t, \mathbb{P}_t)$ *and the change of probability* $Z_t =
d\mathbb{P}'_t/d\mathbb{P}_t$ *with the non-negative martingale* $(Z_s)_{s \leq t}$ *defined in (18.4) with* $m = m'$ *and some
regular intensity function* λ'_s. *Using exercise 343, check that* $M'_s = \int_0^s \frac{1}{\lambda'_r(N_r)} \, dN_r - s$,
with $s \in [0, t]$, *is a martingale on* $(\Omega_t, \mathcal{F}_t, \mathbb{P}'_t)$. *Deduce that* $N'_s = N_s - \int_0^s \lambda'_r(N_r) \, dr$ *is a
martingale on* $(\Omega_t, \mathcal{F}_t, \mathbb{P}'_t)$.

Exercise 346 (Exponential martingales) ✎ *We let* f *be a bounded function on* $\mathbb{R}$, g
be a bounded function taking values in $]-1, \infty[$ *and* N_t *be a Poisson process with intensity*
λ, *and* $\mathcal{F}_t = \sigma(N_s, \, s \leq t)$. *Check that the following processes are* $\mathcal{F}_t$-*martingales:*

$$
\begin{aligned}
M_t^{(1)} &:= \exp\left(\int_0^t f(s) \, dN_s - \lambda \int_0^t (e^{f(s)} - 1) ds \right) \\
M_t^{(2)} &:= \exp\left(\int_0^t \log(1 + g(s)) \, dN_s - \lambda \int_0^t g(s) ds \right).
\end{aligned}
$$

Exercise 347 (Girsanov's theorem) *We let* f *be a bounded function on* $\mathbb{R}$, W_t *be a
Brownian motion, and* $\mathcal{F}_t = \sigma(W_s, \, s \leq t)$. *Check that the following process is an* $\mathcal{F}_t$-
martingale:

$$M_t = \exp\left(\int_0^t f(s) dW_s - \frac{1}{2} \int_0^t f(s)^2 ds \right).$$

Exercise 348 (Importance sampling - Twisted processes) *We let* X_t *be a Markov
process on* S *with infinitesimal generator* L_t *defined on some domain* $D(L)$ *of functions
(for instance we can take* $S = \mathbb{R}^d$, *for some* $d \geq 1$, *with the infinitesimal generator (15.13)
acting on the set* $D(L)$ *of twice differentiable functions with bounded derivates). We let*
$\eta_0 = Law(X_0)$. *We consider some sufficiently smooth functions* $\varphi \in D(L)$. *For any bounded
function* F *on the set of paths* $(X_t)_{0 \leq s \leq t}$, *check the Feynman-Kac formula*

$$\mathbb{E}(F((X_s)_{s \leq t})) = \mathbb{E}(\varphi(X_0)) \, \mathbb{E}\left(F((X_s^\varphi)_{s \leq t}) \, \varphi(X_t^\varphi)^{-1} \, \exp\left(\int_0^t V_s^\varphi(X_s) ds \right) \right)$$

with the potential function $V_t^\varphi = \varphi^{-1} L_t(\varphi)$, *with a stochastic process* X_t^φ *with initial distribution* $\eta_0^\varphi = \Psi_\varphi(\eta_0)$, *and with the infinitesimal generator defined for any sufficiently regular
function* g *by*

$$L_t^{[\varphi]}(g) = L_t(g) + \varphi^{-1} \, \Gamma_{L_t}(\varphi, g). \tag{18.36}$$

The X_t^φ *is called the twisted process. Illustrations are provided in section 18.3.1 and in
section 18.3.2 in the context of diffusion and pure jump processes.*

Exercise 349 (Trial and guiding wave functions) *We consider the stochastic models described in exercise 348. We further assume that $L_t = L$ is homogeneous w.r.t. the time parameter. We let V be a regular energy function on S. Prove that for any bounded function f on S we have*

$$\mathbb{E}\left(f(X_t) \exp\left(-\int_0^t V(X_s)ds \right) \right) = \eta_0(\varphi)\ \mathbb{E}\left(f(X_t^\varphi)\ \varphi(X_t^\varphi)^{-1}\ \exp\left(-\int_0^t V^\varphi(X_s)ds \right) \right)$$

with the potential function

$$V^\varphi = \varphi^{-1}\mathcal{H}(\varphi) \quad \text{and the Hamiltonian operator} \quad \mathcal{H}(g) := -L(g) + Vg.$$

Exercise 350 (Twisted process reversibility) *We consider a generator L that is reversible w.r.t. some measure μ (in the sense of (17.2)). Check that the twisted generator $L^{[\varphi]}(g) = L(g) + \varphi^{-1}\ \Gamma_L(\varphi, g)$ is reversible w.r.t. $\Psi_{\varphi^2}(\mu)$.*

Part V

Processes on manifolds

19

A review of differential geometry

As its name indicates, `differential geometry` is a field of mathematics concerned with differential and integral calculus on differentiable manifolds This chapter summarises some basic tools that are frequently used to design and analyze the behavior of stochastic processes in constraint type manifolds, including parametric type Riemannian manifolds.

We start with brief discussion on projection operators and symmetric bilinear forms on finite dimensional vector spaces. The second part of the chapter is dedicated to first and second covariant derivatives of functions and vector fields. We also present more advanced operators such as the divergence, the Lie bracket, the Laplacian, and the Ricci curvature. The end of the chapter is concerned with the Bochner-Lichnerowicz formula and several change-of-variable formulae. The local expressions of these geometric objects and formulae in a given parametric space are discussed in chapter 21. The Bochner-Lichnerowicz formula is used in section 23.7 to analyze the stability properties of diffusions on manifolds.

> *And since geometry is the right foundation of all painting, I have decided to teach*
> *its rudiments, and principles to all youngsters eager for art.*
> `Albrecht Durer (1471-1528).`

19.1   Projection operators

We let $\mathcal{V} = \mathrm{Vect}\,(V_1, \ldots, V_p) \subset \mathbb{R}^r$ be a p-dimensional vector space with a (not necessarily orthonormal) basis $(V_1, \ldots, V_p) \in (\mathbb{R}^r)^p$, with the column vectors $V_i = \begin{bmatrix} V_i^1 \\ \vdots \\ V_i^r \end{bmatrix}$, with $1 \leq i \leq p \leq r$. We equip $\mathcal{V}$ with the Euclidian inner product

$$g_{i,j} := \langle V_i, V_j \rangle = V_i^T V_j = \sum_{1 \leq k \leq r} V_i^k V_j^k = \mathrm{tr}(V_i V_j^T)$$

and the corresponding Euclidian norm $\|W\|_2^2 = \langle W, W \rangle$, for any $W \in \mathcal{V}$. The matrix g is called the Gramian matrix or the Gram matrix associated with the p vectors $(V_1, \ldots, V_p)$. In geometry literature, g is sometimes written as

$$g = \mathrm{Gram}\,(V_1, \ldots, V_p)\,.$$

We let $g^{i,j}$ be the entries of the inverse g^{-1} of the matrix $g = (g_{i,j})_{1 \le i,j \le p}$ and we set

$$V = [V_1, \ldots, V_p] \quad \text{and} \quad V^T := \begin{bmatrix} V_1^T \\ \vdots \\ V_p^T \end{bmatrix} \implies g = V^T V. \qquad (19.1)$$

Proposition 19.1.1 *The orthogonal projection*

$$\pi_{\mathcal{V}} : W \in \mathbb{R}^r \mapsto \pi_{\mathcal{V}}(W) \in \mathcal{V}$$

on $\mathcal{V}$ is given by the matrix

$$\pi_{\mathcal{V}} = V g^{-1} V^T \implies \pi_{\mathcal{V}}(W) = \sum_{1 \le i \le p} \left\langle \sum_{1 \le j \le p} g^{i,j} V_j, W \right\rangle V_i.$$

The r.h.s. formula is implied by the fact that

$$g^{-1} V^T = \begin{bmatrix} \sum_{1 \le j \le p} g^{1,j} V_j^T \\ \vdots \\ \sum_{1 \le j \le p} g^{p,j} V_j^T \end{bmatrix} \implies g^{-1} V^T W = \begin{bmatrix} \sum_{1 \le j \le p} g^{1,j} V_j^T W \\ \vdots \\ \sum_{1 \le j \le p} g^{p,j} V_j^T W \end{bmatrix}.$$

For any vector W_1, W_2 we notice that

$$\langle \pi_{\mathcal{V}}(W_1), W_2 \rangle = \sum_{1 \le i,j \le r} g^{i,j} \langle W_1, V_i \rangle \langle V_j, W_2 \rangle.$$

In particular, for any $W_1, W_2 \in \mathcal{V}$ we have

$$\langle W_1, W_2 \rangle = \sum_{1 \le i,j \le r} g^{i,j} \langle W_1, V_i \rangle \langle V_j, W_2 \rangle. \qquad (19.2)$$

Definition 19.1.2 *Given a collection of vectors $(W_i)_{1 \le i \le k}$ we set*

$$\pi_{\mathcal{V}}([W_1, \ldots, W_k]) = [\pi_{\mathcal{V}}(W_1), \ldots, \pi_{\mathcal{V}}(W_k)].$$

In this notation, for any $W_1, W_2 \in \mathbb{R}^r$, we observe that

$$W_1 W_2^T := [W_2^1 W_1, \ldots W_2^r W_1] \implies \begin{aligned} \pi_{\mathcal{V}}(W_1 W_2^T) &= [W_2^1 \pi_{\mathcal{V}}(W_1), \ldots W_2^r \pi_{\mathcal{V}}(W_1)] \\ &= \pi_{\mathcal{V}}(W_1) W_2^T. \end{aligned}$$

In summary, we have proved that

$$\pi_{\mathcal{V}}(W_1 W_2^T) = \pi_{\mathcal{V}}(W_1) W_2^T.$$

If we choose an orthonormal basis $(U_1, \ldots, U_p) \in (\mathbb{R}^r)^p$ we have

$$V_i = \sum_{1 \leq j \leq p} \langle V_i, U_j \rangle U_j = \underbrace{[U_1, \ldots, U_p]}_{U} \underbrace{\begin{bmatrix} U_1^T \\ \vdots \\ U_p^T \end{bmatrix}}_{U^T} V_i := UU^T V_i \implies V = UP$$

with

$$P = U^T V = \begin{bmatrix} \langle U_1, V_1 \rangle & \cdots & \langle U_1, V_p \rangle \\ \vdots & & \\ \langle U_p, V_1 \rangle & \cdots & \langle U_p, V_p \rangle \end{bmatrix}$$

from which we conclude that

$$\begin{aligned} Vg^{-1}V^T &= V(V^TV)^{-1}V^T \\ &= UP((UP)^T UP)^{-1}(UP)^T = UP(P^T U^T UP)^{-1}(UP)^T \\ &= UP(P^T P)^{-1} P^T U^T = UPP^{-1}(P^T)^{-1}P^T U^T = UU^T. \end{aligned}$$

This shows that the projection matrix $\pi_{\mathcal{V}} = \left(\pi_{\mathcal{V},l}^k \right)_{1 \leq k,l \leq r}$ does not depend on the choice of the basis of the vector field $\mathcal{V}$, and we have

$$\pi_{\mathcal{V}} = \pi_{\mathcal{V}}^T \quad \text{and} \quad \pi_{\mathcal{V}} \pi_{\mathcal{V}} = \pi_{\mathcal{V}} \Rightarrow \forall 1 \leq k,l \leq r \quad \sum_{1 \leq i \leq r} \pi_{\mathcal{V},l}^i \pi_{\mathcal{V},k}^i = \pi_{\mathcal{V},l}^k.$$

The entries π_i^j of the matrix $\pi := \pi_{\mathcal{V}}$ can be expressed in terms of the basis vectors V_k by the formula

$$e_i := \begin{bmatrix} 0 \\ \vdots \\ 0 \\ 1 \\ 0 \\ \vdots \\ 0 \end{bmatrix} \text{ (i-th)} \Rightarrow \pi_i := \begin{bmatrix} \pi_i^1 \\ \vdots \\ \pi_i^r \end{bmatrix} = \pi(e_i) = \sum_{1 \leq l \leq p} \left[\sum_{1 \leq k \leq p} g^{k,l} V_k^i \right] V_l \in \mathcal{V}. \quad (19.3)$$

For any vector field $W \in \mathcal{V}$, we notice that

$$(19.3) \Rightarrow \sum_{1 \leq i \leq p} W^i \, \pi_i = \sum_{1 \leq l \leq p} \left[\sum_{1 \leq k \leq p} g^{l,k} \langle V_k, W \rangle \right] V_l = \pi(W) = W.$$

This yields the following decomposition.

For any $1 \leq i \leq r$ and vector field $W \in \mathcal{V}$ we have

$$W = \sum_{1 \leq i \leq p} W^i \, \pi_i \quad \text{and} \quad \pi_i = \sum_{1 \leq k,l \leq p} g^{k,l} V_k^i V_l. \quad (19.4)$$

Definition 19.1.3 *We let $B : \mathcal{V} \times \mathcal{V} \mapsto \mathbb{R}$ be a symmetric bilinear form. The trace of B is defined by*

$$\mathrm{tr}(B) = \sum_{1 \le i \le p} B(\overline{V}_i, \overline{V}_i)$$

where $\overline{V}_i$ stands for an orthonormal basis of $\mathcal{V}$ w.r.t. the Euclidian inner product.

The vectors $\overline{V}_i$ can be obtained from V_i using a Gram-Schmidt procedure. A symmetric bilinear form $B : \mathcal{V} \times \mathcal{V} \mapsto \mathbb{R}$ on $\mathcal{V}$ is expressed in terms of some linear form $b : \mathcal{V} \mapsto \mathcal{V}$ defined for any $W_1, W_2 \in \mathcal{V}$ by

$$B(W_1, W_2) = \langle b(W_1), W_2 \rangle = \langle W_1, b(W_2) \rangle.$$

The form b is a non necessarily symmetric matrix on $\mathbb{R}^r$ but its action on the vector space $\mathcal{V}$ is symmetric. It is said to be positive semi-definite when we have

$$\forall W \in \mathcal{V} \qquad B(W, W) = \langle b(W), W \rangle \ge 0. \tag{19.5}$$

Notice that $\mathcal{V}$ equipped with the inner product can be interpreted as a finite dimensional Hilbert space. In this context, the famous Riesz-Fischer theorem ensures that there is a one-to-one correspondence between bilinear forms B and linear operators b on $\mathcal{V}$. Thus, we often identify B with b; for instance we set $\mathrm{tr}(B) = \mathrm{tr}(b)$. Whenever B is symmetric, the linear form b is self adjoined and we can find an orthonormal basis $\overline{V}_i$ such that $b(\overline{V}_i) = \lambda_i \overline{V}_i$ for some collection of (possibly equal) eigenvalues $\lambda_1 \ge \ldots \ge \lambda_p$. For positive semi-definite forms (19.5) ensures that $\lambda_i \ge 0$, for any $1 \le i \le p$.

The spectral decomposition of b (a.k.a. Schmidt decomposition) is given by the formula

$$W = \sum_{1 \le i \le p} \langle \overline{V}_i, W \rangle \, \overline{V}_j \in \mathcal{V} \Rightarrow b(W) = \sum_{1 \le i \le p} \lambda_i \, \langle \overline{V}_i, W \rangle \, \overline{V}_j.$$

Observe that $b^2 = b \circ b \Rightarrow b^2(\overline{V}_i) = \lambda_i \, b(\overline{V}_i) = \lambda_i^2 \, \overline{V}_i$.

The trace of b is independent of the choice of the basis vectors and thus we have the formulae

$$\mathrm{tr}(b) = \sum_{1 \le i \le p} \lambda_i \quad \text{and} \quad \mathrm{tr}(b^2) = \sum_{1 \le i \le p} \lambda_i^2. \tag{19.6}$$

In this situation, we also have

$$V_i = \sum_{1 \le k \le p} \langle V_i, \overline{V}_k \rangle \, \overline{V}_k \quad \text{and} \quad g_{i,j} = \langle V_i, V_j \rangle = \sum_{1 \le k, l \le p} \langle V_i, \overline{V}_k \rangle \, \langle V_j, \overline{V}_l \rangle.$$

The trace of B expressed in the basis vectors V_i takes the form

$$\mathrm{tr}(B) = \sum_{1 \le i, j \le p} g^{i,j} B(V_i, V_j) = \sum_{1 \le i, j \le p} g^{i,j} \langle b(V_i), V_j \rangle. \tag{19.7}$$

We check this claim recalling that

$$B(V_i, V_j) = \sum_{1 \le k, l \le p} \langle V_i, \overline{V}_k \rangle \, \langle V_j, \overline{V}_l \rangle \, B\left(\overline{V}_k, \overline{V}_l\right)$$

and

$$\sum_{1 \le i, j \le p} g^{i,j} \, \langle V_i, \overline{V}_k \rangle \, V_j = \overline{V}_k$$

from which we prove that

$$
\begin{aligned}
\sum_{1 \leq i,j \leq p} g^{i,j} B(V_i, V_j) &= \sum_{1 \leq k,l \leq p} \left[\sum_{1 \leq i,j \leq p} g^{i,j} \langle V_i, \overline{V}_k \rangle \ \langle V_j, \overline{V}_l \rangle \right] B\left(\overline{V}_k, \overline{V}_l\right) \\
&= \sum_{1 \leq k,l \leq p} \langle \overline{V}_k, \overline{V}_l \rangle \ B\left(\overline{V}_k, \overline{V}_l\right) = \operatorname{tr}(B).
\end{aligned}
$$

This ends the proof of the assertion. ∎

We let B_1 and $B_2 : \mathcal{V} \times \mathcal{V} \mapsto \mathbb{R}$ be a pair of bilinear forms defined for any $W_1, W_2 \in \mathcal{V}$ by

$$\forall i = 1,2 \qquad B_i(W_1, W_2) = \langle b_i(W_1), W_2 \rangle \quad \text{for some linear mapping} \quad b_i : \mathcal{V} \mapsto \mathcal{V}.$$

Notice that

$$\langle b_1\left(\overline{V}_1\right), b_2\left(\overline{V}_2\right) \rangle = \langle \overline{V}_1, b_1 b_2\left(\overline{V}_2\right) \rangle = \langle \overline{V}_2, b_2 b_1\left(\overline{V}_1\right) \rangle.$$

This leads to the following definition of the trace type inner product.

> **Definition 19.1.4** *We equip the space of symmetric bilinear forms with the Hilbert-Schmidt inner product*
>
> $$\langle B_1, B_2 \rangle := \sum_{1 \leq i \leq p} \langle b_1\left(\overline{V}_i\right), b_2\left(\overline{V}_i\right) \rangle = \langle b_1, b_2 \rangle = \operatorname{tr}\left(b_1 b_2\right) \qquad (19.8)$$
>
> *and the corresponding norm* $\|B\| = \sqrt{\langle B, B \rangle} = \|b\| = \sqrt{\langle b, b \rangle} = \operatorname{tr}(b^2)$ *where* $\overline{V}_i$ *stands for an orthonormal basis of* $\mathcal{V}$ *w.r.t. the Euclidian inner product.*

We also recall the operator norm inequality

$$\|B\|_2 := \|b\|_2 := \sup \left\{ \|b(W)\|_2 : W \in \mathcal{V} \ s.t. \ \|W\|_2 = 1 \right\} \leq \|B\| \qquad (19.9)$$

for any bilinear form B on $\mathcal{V}$ associated with some linear form b on $\mathcal{V}$.

We check this claim using an orthonormal basis $\overline{V}_i$ of $\mathcal{V}$. We assume that B is associated with some linear mapping $b : \mathcal{V} \mapsto \mathcal{V}$, so that

$$W = \sum_{1 \leq i \leq p} \langle W, \overline{V}_i \rangle \ \overline{V}_j \Rightarrow b(W) \sum_{1 \leq i \leq p} \langle W, \overline{V}_i \rangle \ b(\overline{V}_j).$$

By the triangle inequality and the Cauchy-Schwartz inequality,

$$
\begin{aligned}
\|b(W)\|_2 &\leq \sum_{1 \leq i \leq p} |\langle W, \overline{V}_i \rangle| \ \| b\left(\overline{V}_i\right) \|_2 \\
&\leq \left(\sum_{1 \leq i \leq p} \|b\left(\overline{V}_i\right)\|_2^2 \right)^{1/2} = \left(\sum_{1 \leq i \leq p} \langle b\left(\overline{V}_i\right), b\left(\overline{V}_i\right) \rangle \right)^{1/2} = \|B\|
\end{aligned}
$$

since $\sum_{1 \leq i \leq p} |\langle W, \overline{V}_i \rangle|^2 = 1 = \|W\|_2$.

> The Hilbert-Schmidt inner product expressed in the basis vectors V_i takes the form
>
> $$\langle B_1, B_2 \rangle := \sum_{1 \leq i,j \leq p} g^{i,j} \langle b_1\left(V_i\right), b_2\left(V_j\right) \rangle = \operatorname{tr}(b_1 b_2). \qquad (19.10)$$

We check this claim recalling that

$$b_l\left(V_i\right) = \sum_{1 \le k \le p} \left\langle V_i, \overline{V}_k \right\rangle \, b_l\left(\overline{V}_k\right)$$

$$\left\langle b_1\left(V_i\right), b_2\left(V_j\right)\right\rangle = \sum_{1 \le k,l \le p} \left\langle V_i, \overline{V}_k \right\rangle \, \left\langle V_j, \overline{V}_l \right\rangle \, \left\langle b_1\left(\overline{V}_k\right), b_2\left(\overline{V}_l\right)\right\rangle$$

and

$$\sum_{1 \le i,j \le p} g^{i,j} \left\langle V_i, \overline{V}_k \right\rangle \, V_j = \overline{V}_k$$

from which we prove that

$$\sum_{1 \le i,j \le p} g^{i,j} \left\langle b_1\left(V_i\right), b_2\left(V_j\right)\right\rangle$$

$$= \sum_{1 \le k,l \le p} \left[\sum_{1 \le i,j \le p} g^{i,j} \sum_{1 \le k,l \le p} \left\langle V_i, \overline{V}_k \right\rangle \, \left\langle V_j, \overline{V}_l \right\rangle\right] \left\langle b_1\left(\overline{V}_k\right), b_2\left(\overline{V}_l\right)\right\rangle$$

$$= \sum_{1 \le k,l \le p} \left\langle \overline{V}_k, \overline{V}_l \right\rangle \left\langle b_1\left(\overline{V}_k\right), b_2\left(\overline{V}_l\right)\right\rangle = \left\langle B_1, B_2 \right\rangle.$$

This ends the proof of the assertion.   ∎

Notice that

$$(19.4) \Rightarrow \operatorname{tr}(B) = \sum_{1 \le k,l \le r} \sum_{1 \le i,j \le p} g^{i,j} \, V_i^k \, V_j^l \, B(\pi_k, \pi_l) = \sum_{1 \le k,l \le r} \pi_k^l \, B(\pi_k, \pi_l).$$

In much the same way we check that

$$\left\langle B_1, B_2 \right\rangle := \sum_{1 \le i,j \le r} \pi_i^j \, \left\langle b_1\left(\pi_i\right), b_2\left(\pi_j\right)\right\rangle \quad \text{and} \quad \operatorname{tr}(B) = \sum_{1 \le i,j \le r} \pi_i^j \, \left\langle b(\pi_i), \pi_j \right\rangle.$$

$$(19.11)$$

Suppose that b is associated with some matrix $b = (b_{i,j})_{1 \le i,j \le r}$ acting on the basis vectors V_i of $\mathcal{V}$. In this situation, we have

$$\sum_{1 \le k,l \le p} g^{k,l} \left\langle b(V_k), V_l \right\rangle = \sum_{1 \le i,j \le r} \left[\sum_{1 \le k,l \le p} g^{k,l} \, V_k^i V_l^j\right] b_{i,j}$$

$$= \sum_{1 \le i,j \le r} \pi_{i,j} \, b_{i,j}.$$

If b_k stands for some matrix $b_k = (b_k(i,j))_{1 \le i,j \le r}$ we have

$$\left\langle B_1, B_2 \right\rangle = \sum_{1 \le i,j \le r} b_1(i,j) \, \pi_j^i \, b_2^T(j,i).$$

In summary, we have proved that

$$\operatorname{tr}(b) = \operatorname{tr}_e(\pi b) \quad \text{and} \quad \langle B_1, B_2 \rangle = \operatorname{tr}_e(b_1 \pi b_2^T) = \operatorname{tr}_e(\pi b_1^T b_2) \tag{19.12}$$

where $\operatorname{tr}_e(A) = \sum_{1 \le i \le r} A_{i,i}$ stands for the trace of a matrix A on the Euclidian space $\mathbb{R}^r$. Notice that

$$\operatorname{tr}(\pi b) = \operatorname{tr}_e(\pi b) \quad \text{and} \quad \operatorname{tr}(b\pi) = \operatorname{tr}_e(b\pi)$$

and for symmetric matrices b_1, b_2 we have

$$\langle \pi b_1, \pi b_2 \rangle = \operatorname{tr}_e((\pi b_1)\,(\pi b_2)).$$

We end this section with some well known trace inequalities.

For any symmetric positive semi-definite bilinear form B on $\mathcal{V}$ associated with some linear form b on $\mathcal{V}$ we have the estimates

$$\frac{1}{p}\,(\operatorname{tr}(b))^2 \le \operatorname{tr}(b^2) = \langle b, b \rangle^2 = \|b\|^2 \le (\operatorname{tr}(b))^2. \tag{19.13}$$

These formulae are direct consequences of (19.6). For instance

$$\lambda_i \ge 0 \Rightarrow \left(\sum_{1 \le i \le p} \lambda_i^2 \right) \le \left(\sum_{1 \le i \le p} \lambda_i \right)^2 \quad \text{and} \quad \left(\frac{1}{p} \sum_{1 \le i \le p} \lambda_i \right)^2 \le \frac{1}{p} \sum_{1 \le i \le p} \lambda_i^2.$$

19.2   Covariant derivatives of vector fields

We further assume that we are given a collection of smooth vector functionals (a.k.a. vector fields) $V_i\ :\ x \in \mathbb{R}^r \mapsto V_i(x) \in \mathbb{R}^r$ and $V_j^\perp\ :\ x \in \mathbb{R}^r \mapsto V_j^\perp(x) \in \mathbb{R}^r$, with $1 \le i \le p$ and $1 \le j \le q = r - p$ such that

$$\mathbb{R}^{p+q} = \underbrace{\operatorname{Vect}(V_1, \ldots, V_p)}_{=\mathcal{V}} \overset{\perp}{+} \underbrace{\operatorname{Vect}(V_1^\perp, \ldots, V_q^\perp)}_{=\mathcal{V}^\perp}, \tag{19.14}$$

in the sense that

$$\forall x \in \mathbb{R}^r \qquad \mathbb{R}^{p+q} = \underbrace{\operatorname{Vect}(V_1(x), \ldots, V_p(x))}_{=\mathcal{V}(x)} \overset{\perp}{+} \underbrace{\operatorname{Vect}(V_1^\perp(x), \ldots, V_q^\perp(x))}_{=\mathcal{V}^\perp(x)}.$$

The vector fields V_i do not need to be defined on the whole Euclidian space $\mathbb{R}^r$. The analysis developed in the forthcoming section and chapter is based on local differential calculus only, so that the vector fields need only be defined on some open subset of $S \subset \mathbb{R}^r$ or on some manifold S embedded in $\mathbb{R}^r$. In these settings, the mapping $V_i\ : x \in S \subset \mathbb{R}^r \mapsto V_i(x)$ the velocity field of a particle moving on a manifold with a velocity vector $V_i(x)$ at some visited state x. We have chosen $S = \mathbb{R}^r$ to simplify the presentation and to avoid unnecessary technicalities. We slightly abuse notation and we let $\mathcal{V}$ be the set of these vector fields. In the further development of this chapter we shall assume without further mentioning that

$$\forall 1 \leq i, j \leq p \quad \forall 1 \leq k \leq r \qquad \sum_{1 \leq m \leq r} V_i^m \, \partial_{x_m} V_j^k = \sum_{1 \leq m \leq r} V_j^m \, \partial_{x_m} V_i^k. \qquad (19.15)$$

Two important examples where (19.15) holds follow below:

- The condition is clearly satisfied when $V_i = e_i$ (and $V_j^\perp = e_{p+j}$), where e_i stands for the unit vectors (19.3).

- When the vector fields V_i have the form $V_i = (\partial_{\theta_i} \psi)_\phi := (\partial_{\theta_i} \psi) \circ \phi$ for some smooth parametrization mapping $\psi \, : \, \theta \in \mathbb{R}^p \mapsto \psi(\theta) \in \mathbb{R}^r$ with some inverse $\phi \, : \, x \in \mathbb{R}^r \mapsto \phi(x) \in \mathbb{R}^p$ s.t. $(\phi \circ \psi)(x) = x$ we have the formula

$$\sum_{1 \leq m \leq r} V_i^m \, \partial_{x_m} V_j^k = \sum_{1 \leq m \leq r} (\partial_{\theta_i} \psi^m)_\phi \, \partial_{x_m} \left(\left[\partial_{\theta_j} \psi^k \right]_\phi \right) \qquad (19.16)$$

$$= \left[\partial_{\theta_i} \left(\left[(\partial_{\theta_j} \psi^k)_\phi \right] \circ \psi \right) \right]_\phi = \left[\partial_{\theta_i} \partial_{\theta_j} \psi^k \right]_\phi = \left[\partial_{\theta_j} \partial_{\theta_i} \psi^k \right]_\phi.$$

This shows that (19.15) is also met in this case.

  - When $p = r$ the mapping ψ can be interpreted as a change of coordinates.
  - When $r > p$ the mapping ψ represents a local parametrization of a manifold of dimension p embedded in $\mathbb{R}^r$.

We mention that the commutation condition (19.15) ensures that the vector fields $\mathcal{V}(x)$ form a Lie algebra. These important geometric properties are discussed in section 19.4. When (19.15) is satisfied, we can also define a symmetric and natural second order derivation on vector fields in $\mathcal{V}$. These second order models are introduced in section 19.2.2. The symmetry property is discussed in section 19.4.

In this case (19.15) is clearly satisfied. When working with the basis vector fields of the orthogonal space $\mathcal{V}^\perp$, we shall assume that the following condition is satisfied.

$$\forall 1 \leq i \leq q \quad \forall 1 \leq k, l \leq r \qquad \partial_{x_k} V_i^{\perp, l} = \partial_{x_l} V_i^{\perp, k}. \qquad (19.17)$$

This condition is clearly satisfied for gradient type vector field models $V_i^\perp = \partial \varphi_i$ associated with some smooth constraint type functions φ_i on $\mathbb{R}^r$. These models are discussed in section 20.1.

We also consider the coordinate projection mappings

$$\forall 1 \leq i \leq r \qquad \chi_i \, : \, x = (x_1, \ldots, x_r) \in \mathbb{R}^r \mapsto \chi_i(x) = x_i. \qquad (19.18)$$

In the further development of this chapter, to clarify the presentation, we suppress the dependency on the state x. In this simplified notation,

$$\langle V_i, V_j^\perp \rangle = 0 \quad \forall \, 1 \leq i \leq p \quad \text{and} \quad \forall \, 1 \leq j \leq q.$$

By construction, for any vector field we have

$$W = \pi(W) + \pi_{\mathcal{V}^\perp}(W) \qquad \text{with} \qquad \pi := \pi_{\mathcal{V}}. \qquad (19.19)$$

Notice that $\pi :=$ and $\pi_\perp := \pi_{\mathcal{V}^\perp}$ are smooth matrix functionals and

$$\pi_\perp(W) = \sum_{1 \leq i \leq q} \left\langle \sum_{1 \leq j \leq q} g_\perp^{i,j} V_j^\perp, W \right\rangle V_i^\perp \qquad (19.20)$$

with the entries $g_\perp^{i,j}$ of the inverse $g_\perp^{-1}$ of the matrix $g_\perp = (g_{\perp,i,j})_{1 \leq i, j \leq p}$ given by

$$g_{\perp,i,j} := \langle V_i^\perp, V_j^\perp \rangle.$$

19.2.1 First order derivatives

Definition 19.2.1 *Given a smooth function F and a smooth vector field $W = \begin{bmatrix} W^1 \\ \vdots \\ W^r \end{bmatrix}$ on $\mathbb{R}^r$, for any $1 \le i \le r$ we set*

$$\partial F := \begin{bmatrix} \partial_{x_1} F \\ \vdots \\ \partial_{x_r} F \end{bmatrix} \qquad \partial_{x_i} W := \begin{bmatrix} \partial_{x_i} W^1 \\ \vdots \\ \partial_{x_i} W^r \end{bmatrix} \quad \text{and} \quad \partial W := \begin{bmatrix} \partial W^1, \dots, \partial W^r \end{bmatrix}.$$

The Euclidean gradient operator $\partial \; : \; F \mapsto \partial F$ maps smooth functions to vector fields $\partial F \; : \; x \mapsto (\partial F)(x)$ that encapsulate information about the change of the function F w.r.t. infinitesimal variations of the individual coordinates x_i of the state $x = (x_1, \dots, x_r)^T$.

Definition 19.2.2 *We consider the operators*

$$\partial_W(F) := \sum_{1 \le k \le r} W^k \, \partial_{x_k}(F) = W^T \partial F = \langle W, \partial F \rangle \quad \text{and} \quad \nabla := \pi \partial.$$

and their extension to vector fields defined by

$$\partial_{W_1}(W_2) = \begin{bmatrix} \partial_{W_1} W_2^1 \\ \vdots \\ \partial_{W_1} W_2^r \end{bmatrix} \quad \text{and} \quad \nabla_{W_1}(W_2) := \pi \partial_{W_1}(W_2).$$

In this notation, the commutation condition (19.15) takes the form

$$\forall 1 \le i, j \le p \qquad \partial_{V_i}(V_j) = \partial_{V_j}(V_i).$$

Definition 19.2.3 *The Christoffel symbols $C_{i,j}^k$ are defined by the projection coordinates of $\partial_{V_i}(V_j)$ in the basis V_k, that is,*

$$\nabla_{V_i}(V_j) := \sum_{1 \le k \le r} C_{i,j}^k \, V_k \iff C_{i,j}^k := \sum_{1 \le l \le r} g^{k,l} \, \langle V_l, \partial_{V_i}(V_j) \rangle = C_{j,i}^k. \quad (19.21)$$

The symmetry property is a direct consequence of the commutation condition (19.15). In differential geometry, the symmetry of the Christoffel symbols $C_{i,j}^k$ ensures that the covariant derivative is without torsion.

We return to the model discussed in (19.16). Recall that the vector fields V_i have the form $V_i = (\partial_{\theta_i} \psi)_\phi$ for some smooth parametrization mapping $\psi \; : \; \theta \in \mathbb{R}^p \mapsto \psi(\theta) \in \mathbb{R}^r$ with an inverse $\phi \; : \; x \in \mathbb{R}^r \mapsto \phi(x) \in \mathbb{R}^p$. In this situation we have

$$\partial_{V_i}(V_j) = \left[\partial_{\theta_i} \partial_{\theta_j} \psi \right]_\phi \Rightarrow C_{i,j}^k := \sum_{1 \le l \le r} g^{k,l} \, \langle (\partial_{\theta_l} \psi)_\phi, \left[\partial_{\theta_i} \partial_{\theta_j} \psi \right]_\phi \rangle.$$

In the same way we have

$$\partial_{V_i}(F) = \sum_{1 \le m \le r} (\partial_{\theta_i}\psi^m)_\phi \ \partial_{x_m}F = (\partial_{\theta_i}f)_\phi \qquad (19.22)$$

with $f := F \circ \psi$ and

$$\nabla F = \sum_{1 \le i,j \le r} g^{i,j} \ \partial_{V_i}(F) \ V_j = \sum_{1 \le i,j \le r} g^{i,j} \ (\partial_{\theta_i}f)_\phi \ (\partial_{\theta_j}\psi)_\phi \ .$$

We clearly have the linearity property

$$\partial_{f_1 W_1 + f_2 W_2} = f_1 \ \partial_{W_1} + f_2 \ \partial_{W_2} \implies \nabla_{f_1 W_1 + f_2 W_2} = f_1 \ \nabla_{W_1} + f_2 \ \nabla_{W_2} \quad (19.23)$$

for any functions f_i and any vector fields W_i.

Given a smooth curve

$$C \ : \ t \in [0,1] \mapsto C(t) = \left(C^1(t), \ldots, C^r(t)\right)^T \in \mathbb{R}^r$$

starting at some state $C(0) = x \in \mathbb{R}^r$, with a velocity vector field W, we have

$$\frac{d}{dt}F(C(t)) = \sum_{1 \le k \le r} W^k(C(t)) \ \partial_{x_k}(F)(C(t)) = (\partial_W(F))(C(t)) = \langle W(C(t)), (\partial F)(C(t)) \rangle.$$

The function $\partial_W(F)$ is called the directional derivative of F w.r.t. the vector field W. The r.h.s. equation makes clear the dependency of the gradient on the inner product structure on $\mathbb{R}^r$. If $A((\partial F)(x), W(x))$ represents the angle between $(\partial F)(x)$ and $W(x)$ we have

$$\partial_W(F)(x) = \|(\partial F)(x)\| \ \|W(x)\| \ \cos\left(A((\partial F)(x), W(x))\right).$$

When $W(x)$ is perpendicular to $(\partial F)(x)$, the rate of change of $(\partial F)(x)$ in the direction $W(x)$ is null. In contrast, the rate of change of $(\partial F)(x)$ in the direction $W(x)$ is maximal when $W(x)$ is parallel to $(\partial F)(x)$.

The covariant derivative (w.r.t. the vector space $\mathcal{V}$)

$$\nabla F := \pi(\partial F) = \partial F - \pi_\perp(\partial F) \qquad (19.24)$$

expresses the changes of the function F w.r.t. vectors $W \in \mathcal{V}$:

$$\forall W \in \mathcal{V} \qquad \langle \partial F, W \rangle = \langle \pi(\partial F), W \rangle = \langle \nabla F, W \rangle.$$

In terms of the basis vector fields V_i of $\mathcal{V}$ we clearly have

$$\nabla F \;=\; \pi\,(\partial F) = \begin{bmatrix} \partial_{\pi_1}(F) \\ \vdots \\ \partial_{\pi_r}(F) \end{bmatrix} \tag{19.25}$$

$$=\; \sum_{1\leq i,j\leq p} g^{i,j}\,\langle V_i,\partial F\rangle\,V_j = \sum_{1\leq i,j\leq p} g^{i,j}\,\partial_{V_i}(F)\,V_j\ \in\mathcal{V}.$$

Choosing $F=\chi_i$ we have

$$\partial\chi_i = e_i \quad\Longrightarrow\quad \nabla\chi_i = \pi(e_i) = \pi_i \tag{19.26}$$

with the projection mappings χ_i and the unit vectors e_i defined in (19.18) and (19.3). In this notation, we have the formula

$$\nabla F = \sum_{1\leq i\leq r} \pi_i\,\partial_{x_i}F. \tag{19.27}$$

The formula (19.27) can be checked in various ways. For instance,

$$f(x) = f(\chi_1,\dots,\chi_r)(x) \Rightarrow \partial f = \sum_{1\leq m\leq r} \partial_{x_m}(f)\,\partial\chi_m \Rightarrow \nabla f = \sum_{1\leq m\leq r} \partial_{x_m}(f)\,\underbrace{\nabla\chi_m}_{=\pi_m}.$$

Using (19.25) and (19.27) we check the inner product formulae

$$\langle \nabla F_1,\nabla F_2\rangle = \sum_{1\leq i,j\leq r} \langle\pi_i,\pi_j\rangle\,\partial_{x_i}F_1\,\partial_{x_i}F_2 = \sum_{1\leq i,j\leq p} g^{i,j}\,\partial_{V_i}F_1\,\partial_{V_j}(F_2). \tag{19.28}$$

By construction, we also readily check that

$$\partial(FW) = \partial F\,W^T + F\,\partial W \tag{19.29}$$

and

$$\begin{aligned} \nabla(FW) &= \pi\partial(FW) \\ &= \pi\left(\partial F\,W^T\right) + \pi\left(F\,\partial W\right) = \pi\left(\partial F\right)\,W^T + F\,\pi\left(\partial W\right). \end{aligned}$$

In a similar way, for any vector fields W_1,W_2 we have

$$\partial_{W_1}(FW_2) = F\,\partial_{W_1}(W_2) + \partial_{W_1}(F)\,W_2 \Rightarrow \nabla_{W_1}(FW_2) = F\,\nabla_{W_1}(W_2) + \nabla_{W_1}(F)\,W_2.$$

We summarize this discussion with the following proposition.

Proposition 19.2.4 *For any vector fields W,W_1,W_2 on $\mathbb{R}^r$ and any smooth function we have*

$$\nabla(FW) = \nabla F\,W^T + F\,\nabla W \quad and \quad \nabla_{W_1}(FW_2) = F\,\nabla_{W_1}(W_2) + \nabla_{W_1}(F)\,W_2. \tag{19.30}$$

When $W \in \mathcal{V}$, ∇W is given by the matrix

$$\nabla W = \pi \left[\partial W^1, \ldots, \partial W^r \right] = \begin{bmatrix} \partial_{\pi_1} W^1, & \ldots & , \partial_{\pi_1} W^r \\ \vdots & \vdots & \vdots \\ \partial_{\pi_r} W^1, & \ldots & , \partial_{\pi_r} W^1 \end{bmatrix} = \begin{bmatrix} (\partial_{\pi_1} W)^T \\ \vdots \\ (\partial_{\pi_r} W)^T \end{bmatrix}. \tag{19.31}$$

For any $W_1, W_2 \in \mathcal{V}$, by the linearity property (19.23) we observe that

$$\pi_i = \sum_{1 \leq k, l \leq p} g^{k,l} \, V_k \, V_l^i$$

$$\Rightarrow \sum_{1 \leq k, l \leq p} g^{k,l} \, \langle \partial_{V_k} W_1, W_2 \rangle \, V_l = \begin{bmatrix} \langle \partial_{\pi_1} W_1, W_2 \rangle \\ \vdots \\ \langle \partial_{\pi_r} W_1, W_2 \rangle \end{bmatrix} = \begin{bmatrix} (\partial_{\pi_1} W_1)^T W_2 \\ \vdots \\ (\partial_{\pi_r} W_1)^T W_2 \end{bmatrix} = (\nabla W_1) W_2.$$

This yields for any $W_1, W_2 \in \mathcal{V}$ the matrix formula

$$\sum_{1 \leq k, l \leq p} g^{k,l} \, \langle \nabla_{V_k} W_1, W_2 \rangle \, V_l = (\nabla W_1) W_2. \tag{19.32}$$

19.2.2 Second order derivatives

We notice that

$$\nabla F = \pi \partial F = \pi^T \partial F = \begin{bmatrix} \pi_1^T \\ \vdots \\ \pi_r^T \end{bmatrix} \partial F = \begin{bmatrix} \partial_{\pi_1} F \\ \vdots \\ \partial_{\pi_r} F \end{bmatrix}.$$

This yields the following Hessian representation formula of the second order covariant derivative

$$\nabla^2 F = \nabla(\nabla F) = [\nabla \partial_{\pi_1} F, \ldots, \nabla \partial_{\pi_r} F] = \begin{pmatrix} \partial_{\pi_1} \partial_{\pi_1} F & \ldots & \partial_{\pi_1} \partial_{\pi_r} F \\ \vdots & \vdots & \vdots \\ \partial_{\pi_r} \partial_{\pi_1} F & \ldots & \partial_{\pi_r} \partial_{\pi_r} F \end{pmatrix}. \tag{19.33}$$

Lemma 19.2.5 *For any vector field* $W, W_1, W_2 \in \mathcal{V}$, *using (19.3) and (19.4) we check that*

$$W = \sum_{1 \leq i \leq p} W^i \, \pi_i \; \Rightarrow \; \sum_{1 \leq i \leq p} W^i \, \partial_{\pi_i} = \partial_W \quad \text{and} \quad \sum_{1 \leq i \leq p} W^i \, \nabla_{\pi_i} = \nabla_W$$

and

$$\sum_{1 \leq i \leq p} \partial_{W_1} \left(W_2^i \right) \, \partial_{\pi_i}(F) = \langle \nabla_{W_1}(W_2), \nabla F \rangle.$$

In addition, we have

$$\sum_{1 \leq i,j \leq r} W_1^i \, \partial_{\pi_i} \partial_{\pi_j}(F) \, W_2^j = \langle W_1, \nabla_{W_2} \nabla F \rangle. \tag{19.34}$$

Proof :
The first formula is a direct consequence of (19.4). To prove the second one we notice that

$$\sum_{1 \leq i \leq p} \partial_{W_1} \left(W_2^i \right) \, \partial_{\pi_i}(F) \;\; = \;\; \langle \partial_{W_1} W_2, \nabla F \rangle = \langle \nabla_{W_1} W_2, \nabla F \rangle.$$

By construction for any $1 \leq k, l \leq r$ we also have

$$\sum_{1 \leq j \leq r} \left(\sum_{1 \leq i \leq r} V_k^i \, \partial_{\pi_i} \right) \partial_{\pi_j}(F) \, V_l^j = \sum_{1 \leq j \leq r} \partial_{V_k}((\nabla F)^j) \, V_l^j = \langle \partial_{V_k}(\nabla F), V_l \rangle = \langle \nabla_{V_k}(\nabla F), V_l \rangle.$$

Recalling that $W_k^i = \sum_{1 \leq m,n \leq p} g^{m,n} \langle V_m, W_k \rangle \, V_n^i$ for any $k = 1, 2$, we check that

$$\sum_{1 \leq i,j \leq r} W_1^i \, \partial_{\pi_i} \partial_{\pi_j}(F) \, W_2^j \;\; = \;\; \langle W_1, \nabla_{W_2}(\nabla F) \rangle.$$

This ends the proof of the lemma.

∎

19.3 Divergence and mean curvature

We return to the setting of section 19.2. For $q = 1$, we have $g_\perp = \left\| V_1^\perp \right\|^2$, $g_\perp^{-1} = \left\| V_1^\perp \right\|^{-2}$ and

$$\pi_\perp(W) = \left\langle \overline{V}_1^\perp, W \right\rangle \overline{V}_1^\perp = \overline{V}_1^\perp \, \overline{V}_1^{\perp,T} W = \frac{V_1^\perp \, V_1^{\perp,T}}{V_1^{\perp,T} V_1^\perp} \, W \quad \text{with} \quad \overline{V}_1^\perp = \frac{V_1^\perp}{\left\| V_1^\perp \right\|}.$$

In this particular case, we have the following result.

Proposition 19.3.1 *We have the formula*

$$\mathrm{tr}\left(\nabla\pi_\perp(W)\right) = \langle \mathbb{H}, W \rangle \tag{19.35}$$

with the mean curvature vector $\mathbb{H}$ defined by

$$\mathbb{H} = \mathrm{div}_\perp\left(\overline{V}_1^\perp\right) \overline{V}_1^\perp$$

$$= -\sum_{1\leq k\leq r}\left(\sum_{1\leq l\leq r}\partial_{\pi_l}\pi_k^l\right) e_k = -\sum_{1\leq k\leq r}\mathrm{tr}\left(\nabla\pi_k\right) e_k \tag{19.36}$$

with

$$\mathrm{div}_\perp\left(\overline{V}_1^\perp\right) := \sum_{1\leq i\leq r}\partial_{x_i}\left(\overline{V}_1^{\perp,i}\right)$$

and the unit vectors e_i on $\subset \mathbb{R}^r$ defined in (19.3).

To check this assertion, we use the fact that

$$\partial\pi_\perp(W) = \partial\left(\left\langle\overline{V}_1^\perp,W\right\rangle\right) \overline{V}_1^{\perp,T} + \left\langle\overline{V}_1^\perp,W\right\rangle \partial\overline{V}_1^\perp$$

from which we prove that

$$\nabla\pi_\perp(W) = \pi\partial\pi_\perp(W) = \pi\left(\partial\left\langle\overline{V}_1^\perp,W\right\rangle\right) \overline{V}_1^{\perp,T} + \left\langle\overline{V}_1^\perp,W\right\rangle \pi\left(\partial\overline{V}_1^\perp\right).$$

To analyze the r.h.s. term, we observe that

$$\pi\left(\partial\overline{V}_1^\perp\right) = \left[\partial\overline{V}_1^{\perp,1},\ldots,\partial\overline{V}_1^{\perp,r}\right] - \left[\pi_\perp\left(\partial\overline{V}_1^{\perp,1}\right),\ldots,\pi_\perp\left(\partial\overline{V}_1^{\perp,r}\right)\right].$$

On the other hand,

$$\mathrm{tr}\left(\pi\left(\partial\left\langle\overline{V}_1^\perp,W\right\rangle\right) \overline{V}_1^{\perp,T}\right) = \left\langle\pi\left(\partial\left\langle\overline{V}_1^\perp,W\right\rangle\right),\overline{V}_1^\perp\right\rangle = 0$$

and

$$\mathrm{tr}\left(\pi\left(\partial\overline{V}_1^\perp\right)\right) = \sum_{1\leq i\leq r}\partial_{x_i}\overline{V}_1^{\perp,i} - \mathrm{tr}\left[\pi_\perp\left(\partial\overline{V}_1^{\perp,1}\right),\ldots,\pi_\perp\left(\partial\overline{V}_1^{\perp,r}\right)\right].$$

Finally, we check that

$$\mathrm{tr}\left[\pi_\perp\left(\partial\overline{V}_1^{\perp,1}\right),\ldots,\pi_\perp\left(\partial\overline{V}_1^{\perp,r}\right)\right] = \mathrm{tr}\left[\left\langle\overline{V}_1^\perp,\partial\overline{V}_1^{\perp,1}\right\rangle \overline{V}_1^\perp,\ldots,\left\langle\overline{V}_1^\perp,\partial\overline{V}_1^{\perp,r}\right\rangle \overline{V}_1^\perp\right]$$

$$= \sum_{1\leq i\leq r}\left\langle\overline{V}_1^\perp,\partial\overline{V}_1^{\perp,i}\right\rangle \overline{V}_1^{\perp,i}$$

$$= \sum_{1\leq j\leq r}\overline{V}_1^{\perp,j} \sum_{1\leq i\leq r}\partial_{x_j}\overline{V}_1^{\perp,i} \overline{V}_1^{\perp,i}$$

$$= \sum_{1\leq j\leq r}\overline{V}_1^{\perp,j} \left\langle\partial_{x_j}\overline{V}_1^\perp,\overline{V}_1^\perp\right\rangle$$

$$= \frac{1}{2}\sum_{1\leq j\leq r}\overline{V}_1^{\perp,j}\partial_{x_j}\left\langle\overline{V}_1^\perp,\overline{V}_1^\perp\right\rangle = 0.$$

This ends the proof of (19.35).

Our next objective is to extend this formula to any dimensional vector spaces $\mathcal{V}^\perp$ spanned by a given basis of vector fields $V_i^\perp$, $1 \leq i \leq q$. In this general situation, we have

$$\pi_\perp(W) = \sum_{1 \leq i \leq q} \left\langle \sum_{1 \leq j \leq q} g_\perp^{i,j} V_j^\perp, W \right\rangle V_i^\perp$$

$$\Rightarrow \partial\pi_\perp(W) = \sum_{1 \leq i \leq q} \left[\partial \left\langle \sum_{1 \leq j \leq q} g_\perp^{i,j} V_j^\perp, W \right\rangle \right] V_i^{\perp,T} + \sum_{1 \leq i \leq q} \left\langle \sum_{1 \leq j \leq q} g_\perp^{i,j} V_j^\perp, W \right\rangle \partial V_i^\perp$$

$$\Rightarrow \nabla\pi_\perp(W) = \sum_{1 \leq i \leq q} \pi \left[\partial \left\langle \sum_{1 \leq j \leq q} g_\perp^{i,j} V_j^\perp, W \right\rangle \right] V_i^{\perp,T} + \sum_{1 \leq i \leq q} \left\langle \sum_{1 \leq j \leq q} g_\perp^{i,j} V_j^\perp, W \right\rangle \nabla V_i^\perp.$$

Using the fact that

$$\text{tr}\left(\pi \left[\partial \left\langle \sum_{1 \leq j \leq q} g_\perp^{i,j} V_j^\perp, W \right\rangle \right] V_i^{\perp,T} \right) = \left\langle \pi \left[\partial \left\langle \sum_{1 \leq j \leq q} g_\perp^{i,j} V_j^\perp, W \right\rangle \right], V_i^\perp \right\rangle = 0$$

we prove the following result.

Proposition 19.3.2 *We have the formula*

$$\text{tr}\left(\nabla\pi_\perp(W) \right) = \sum_{1 \leq i \leq q} \left\langle \sum_{1 \leq j \leq q} g_\perp^{i,j} V_j^\perp, W \right\rangle \text{tr}\left(\nabla V_i^\perp \right)$$

$$= \left\langle \sum_{1 \leq j \leq q} \left[\sum_{1 \leq i \leq q} g_\perp^{i,j} \text{tr}\left(\nabla V_i^\perp \right) \right] V_j^\perp, W \right\rangle. \qquad (19.37)$$

We also observe that

$$\nabla V_i^\perp = \pi \partial V_i^\perp = \partial V_i^\perp - \pi_\perp \partial V_i^\perp$$

$$= \left[\partial V_i^{\perp,1}, \ldots, \partial V_i^{\perp,r} \right] - \left[\pi_\perp \partial V_i^{\perp,1}, \ldots, \pi_\perp \partial V_i^{\perp,r} \right]$$

with

$$\pi_\perp \partial V_i^{\perp,j} = \sum_{1 \leq k \leq q} \left\langle \sum_{1 \leq l \leq q} g_\perp^{k,l} V_l^\perp, \partial V_i^{\perp,j} \right\rangle V_k^\perp.$$

This yields

$$\text{tr}\left(\nabla V_i^\perp \right) = \sum_{1 \leq m \leq r} \partial_{x_m} V_i^{\perp,m} - \sum_{1 \leq m \leq r} \sum_{1 \leq k,l \leq q} g_\perp^{k,l} \left\langle V_l^\perp, \partial V_i^{\perp,m} \right\rangle V_k^{\perp,m}$$

and therefore

$$\sum_{1 \leq i \leq q} g_\perp^{i,j} \text{tr}\left(\nabla V_i^\perp \right) = \sum_{1 \leq i \leq q} g_\perp^{i,j} \left[\sum_{1 \leq m \leq r} \partial_{x_m} V_i^{\perp,m} - \sum_{1 \leq m \leq r} \sum_{1 \leq k,l \leq q} g_\perp^{k,l} \left\langle V_l^\perp, \partial V_i^{\perp,m} \right\rangle V_k^{\perp,m} \right].$$

Under the condition (19.17) we have

$$
\begin{aligned}
\left\langle \partial_{x_i} V_k^\perp, V_l^\perp \right\rangle &= \sum_{1 \le j \le r} \partial_{x_i} V_k^{\perp,j} V_l^{\perp,j} \\
&= \sum_{1 \le j \le r} V_l^{\perp,j} \, \partial_{x_j} V_k^{\perp,i} = \partial_{V_l^\perp} V_k^{\perp,i} = \left\langle \partial V_k^{\perp,i}, V_l^\perp \right\rangle
\end{aligned}
$$

and

$$
\begin{aligned}
\left\langle \partial_{V_j^\perp} V_k^\perp, V_l^\perp \right\rangle &= \sum_{1 \le i \le r} V_j^i \left\langle \partial_{x_i} V_k^\perp, V_l^\perp \right\rangle \\
&= \sum_{1 \le i, i' \le r} V_j^i \, \partial_{x_i} V_k^{\perp,i'} V_l^{\perp,i'} = \sum_{1 \le i, i' \le r} V_j^i \, \partial_{x_{i'}} V_k^{\perp,i} V_l^{\perp,i'} \\
&= \sum_{1 \le i, i' \le r} V_l^{\perp,i'} \, \partial_{x_i'} V_k^{\perp,i} \; V_j^i = \left\langle \partial_{V_l^\perp} V_k^\perp, V_j^\perp \right\rangle. \tag{19.38}
\end{aligned}
$$

Using the fact that

$$
\partial_{x_j} \left(g_{\perp,k,m} \right) = \partial_{x_j} \left\langle V_k^\perp, V_m^\perp \right\rangle = \left\langle \partial_{x_j} V_k^\perp, V_m^\perp \right\rangle + \left\langle V_k^\perp, \partial_{x_j} V_m^\perp \right\rangle
$$

we have

$$
\begin{aligned}
\partial_{V_j^\perp} \left(g_{\perp,k,m} \right) &= \left\langle \partial_{V_j^\perp} V_k^\perp, V_m^\perp \right\rangle + \left\langle \partial_{V_j^\perp} V_m^\perp, V_k^\perp \right\rangle \tag{19.39} \\
&= \left\langle \partial_{V_m^\perp} V_k^\perp, V_j^\perp \right\rangle + \left\langle \partial_{V_k^\perp} V_m^\perp, V_j^\perp \right\rangle = \left\langle \partial_{V_m^\perp} V_k^\perp + \partial_{V_k^\perp} V_m^\perp, V_j^\perp \right\rangle.
\end{aligned}
$$

In addition, for any $1 \le k, l, m \le q$ we have

$$
\left\langle \partial_{V_m^\perp} V_k^\perp, V_l^\perp \right\rangle = \sum_{1 \le i \le r} V_m^{\perp,i} \left\langle \partial_{x_i} V_k^\perp, V_l^\perp \right\rangle = \sum_{1 \le i \le r} V_m^{\perp,i} \left\langle \partial V_k^{\perp,i}, V_l^\perp \right\rangle. \tag{19.40}
$$

This yields

$$
\begin{aligned}
\sum_{1 \le i \le q} g_\perp^{i,j} \, \mathrm{tr} \left(\nabla V_i^\perp \right) &= \sum_{1 \le i \le q} \sum_{1 \le m \le r} g_\perp^{i,j} \, \partial_{x_m} V_i^{\perp,m} \\
&\qquad - \sum_{1 \le i,k,l \le q} g_\perp^{i,j} \, g_\perp^{k,l} \sum_{1 \le m \le r} V_k^{\perp,m} \left\langle \partial V_i^{\perp,m}, V_l^\perp \right\rangle \\
&= \sum_{1 \le i \le q} \sum_{1 \le m \le r} g_\perp^{i,j} \left[\partial_{x_m} V_i^{\perp,m} - \sum_{1 \le k,l \le q} g_\perp^{k,l} \left\langle \partial_{V_k^\perp} V_i^\perp, V_l^\perp \right\rangle \right].
\end{aligned}
$$
$$\tag{19.41}$$

To get one step further in our discussion, we need to recall some basic facts about the differentiation of the determinants of invertible matrices. We let

$$
\epsilon \mapsto A(\epsilon) = \begin{bmatrix} A_1^1(\epsilon) & \cdots & A_r^1(\epsilon) \\ \vdots & \vdots & \vdots \\ A_1^r(\epsilon) & \cdots & A_r^r(\epsilon) \end{bmatrix}
$$

be a smooth $(r \times r)$-invertible matrix functional. The co-factor expansion of the determinant of $A(\epsilon)$ along the i-th row is given by the formula

$$
\det(A(\epsilon)) = \sum_{1 \le j \le r} A_j^i(\epsilon) \, C_j^i(\epsilon) \implies C_j^i(\epsilon) = \frac{\partial \det(A)}{\partial A_j^i}(\epsilon)
$$

where $C_j^i(\epsilon)$ denotes the co-factor of the entry $A_j^i(\epsilon)$. This co-factor is obtained by multiplying by $(-1)^{i+j}$ the determinant of the minor of the entry in the i-th row and the j-th column. We recall that this (i, j)-minor is the determinant of the sub-matrix deduced from $A(\epsilon)$ by deleting the i-th row and j-th column.

The inverse of the matrix $A(\epsilon)$ is defined by

$$A^{-1}(\epsilon) = \frac{1}{\det(A(\epsilon))} \begin{bmatrix} C_1^1(\epsilon) & \cdots & C_1^r(\epsilon) \\ \vdots & \vdots & \vdots \\ C_r^1(\epsilon) & \cdots & C_r^r(\epsilon) \end{bmatrix} = \frac{1}{\det(A(\epsilon))} C^T(\epsilon).$$

This leads quickly to the following result.

Proposition 19.3.3 *The Jacobi formula for the derivative of the determinant is given by*

$$
\begin{aligned}
\frac{d}{d\epsilon}\left(\det(A(\epsilon))\right) &= \sum_{1 \leq i,j \leq r} \frac{\partial \det(A)}{\partial A_j^i}(\epsilon) \frac{dA_j^i(\epsilon)}{d\epsilon} = \sum_{1 \leq i \leq r} \sum_{1 \leq j \leq r} (C^T(\epsilon))_i^j \left(\frac{dA(\epsilon)}{d\epsilon}\right)_j^i \\
&= \operatorname{tr}_e\left(C^T(\epsilon) \frac{dA(\epsilon)}{d\epsilon}\right) \\
&= \det(A(\epsilon)) \operatorname{tr}_e\left(A^{-1}(\epsilon) \frac{dA(\epsilon)}{d\epsilon}\right).
\end{aligned}
$$

For any smooth vector field W on $\mathbb{R}^r$ we set

$$\operatorname{div}_\perp(W) = \frac{1}{\sqrt{\det(g_\perp)}} \sum_{1 \leq m \leq r} \partial_{x_m}\left(\sqrt{\det(g_\perp)} \, W^m\right). \qquad (19.42)$$

We have

$$
\begin{aligned}
\frac{1}{\sqrt{\det(g_\perp)}} \partial_{x_m}\left(\sqrt{\det(g_\perp)} \, W^m\right) &= \frac{1}{\sqrt{\det(g_\perp)}} \partial_{x_m}\left(\sqrt{\det(g_\perp)}\right) W^m + \partial_{x_m}(W^m) \\
&= \frac{1}{2\det(g_\perp)} \partial_{x_m}(\det(g_\perp)) W^m + \partial_{x_m}(W^m) \\
&= \frac{1}{2} \operatorname{tr}\left(g_\perp^{-1} \partial_{x_m} g_\perp\right) W^m + \partial_{x_m}(W^m) \\
&= \frac{1}{2} \sum_{1 \leq k,l \leq q} g_\perp^{k,l} \partial_{x_m} g_{\perp,k,l} W^m + \partial_{x_m}(W^m)
\end{aligned}
$$

from which we find that

$$\frac{1}{\sqrt{\det(g_\perp)}} \partial_{x_m}\left(\sqrt{\det(g_\perp)} \, W^m\right) = \sum_{1 \leq k,l \leq q} g_\perp^{k,l} \left\langle \partial_{x_m} V_k^\perp, V_l^\perp \right\rangle W^m + \partial_{x_m}(W^m).$$

On the other hand, we have

$$\sum_{1 \le j \le q} g_{\perp,i,j}\, g_{\perp}^{j,k} = 1_{i=k} \quad \Rightarrow \quad \sum_{1 \le j \le q} g_{\perp,i,j}\, \partial_{x_m} g_{\perp}^{j,k} = - \sum_{1 \le j \le q} (\partial_{x_m} g_{\perp,i,j})\, g_{\perp}^{j,k}$$

$$\Rightarrow \quad \sum_{1 \le i,j \le q} g_{\perp}^{l,i} g_{\perp,i,j}\, \partial_{x_m} g_{\perp}^{j,k}$$

$$= \partial_{x_m} g_{\perp}^{l,k} = - \sum_{1 \le i,j \le q} g_{\perp}^{l,i} g_{\perp}^{k,j}\, \partial_{x_m} g_{\perp,i,j}.$$

Applying these formulae to $W = \sum_{1 \le j \le q} g_{\perp}^{i,j} V_j^{\perp}$ we find that

$$\text{div}_{\perp}\left(\sum_{1 \le j \le q} g_{\perp}^{i,j} V_j^{\perp}\right)$$

$$= \sum_{1 \le j,k,l \le q} g_{\perp}^{i,j}\, g_{\perp}^{k,l} \sum_{1 \le m \le r} V_j^{\perp m} \left\langle \partial_{x_m} V_k^{\perp}, V_l^{\perp}\right\rangle$$

$$- \sum_{1 \le j \le q} \sum_{1 \le m \le r} V_j^{\perp,m} \sum_{1 \le k,l \le q} g_{\perp}^{i,k} g_{\perp}^{j,l}\, \partial_{x_m} g_{\perp,k,l} + \sum_{1 \le j \le q} g_{\perp}^{i,j} \sum_{1 \le m \le r} \partial_{x_m} V_j^{\perp,m}.$$

Using (19.38) we obtain

$$\text{div}_{\perp}\left(\sum_{1 \le j \le q} g_{\perp}^{i,j} V_j^{\perp}\right) = \sum_{1 \le j \le q} g_{\perp}^{i,j} \left\langle \sum_{1 \le k,l \le q} g_{\perp}^{k,l}\, \partial_{V_l^{\perp}} V_k^{\perp}, V_j^{\perp}\right\rangle$$

$$- \sum_{1 \le j \le q} g_{\perp}^{i,j} \sum_{1 \le k,l \le q} g_{\perp}^{k,l}\, \partial_{V_k^{\perp}} g_{\perp,j,l} + \sum_{1 \le j \le q} g_{\perp}^{i,j} \sum_{1 \le m \le r} \partial_{x_m} V_j^{\perp,m}.$$

Using (19.39) we have

$$\sum_{1 \le k,l \le q} g_{\perp}^{k,l}\, \partial_{V_k^{\perp}} (g_{\perp,j,l}) = \sum_{1 \le k,l \le q} g_{\perp}^{k,l} \left\langle \partial_{V_k^{\perp}} V_j^{\perp}, V_l^{\perp}\right\rangle + \left\langle \sum_{1 \le k,l \le q} g_{\perp}^{k,l} \partial_{V_k^{\perp}} V_l^{\perp}, V_j^{\perp},\right\rangle.$$

Combining this formula with (19.41) we conclude that

$$\text{div}_{\perp}\left(\sum_{1 \le j \le q} g_{\perp}^{i,j}\, V_j^{\perp}\right)$$

$$= \sum_{1 \le j \le q} g_{\perp}^{i,j}\left[\sum_{1 \le m \le r} \partial_{x_m} V_j^{\perp,m} - \sum_{1 \le k,l \le q} g_{\perp}^{k,l} \left\langle \partial_{V_k^{\perp}} V_j^{\perp}, V_l^{\perp}\right\rangle\right] = \sum_{1 \le j \le q} g_{\perp}^{i,j}\, \text{tr}\left(\nabla V_j^{\perp}\right).$$

$$(19.43)$$

Finally, using (19.37) we prove the following result.

Proposition 19.3.4 *We have the formula*

$$\mathrm{tr}\left(\nabla \pi_\perp(W)\right) = \mathrm{tr}\left((\pi \partial \pi_\perp)W\right) = \langle \mathbb{H}, W \rangle \qquad (19.44)$$

with the mean curvature vector

$$\mathbb{H} = \sum_{1 \leq i \leq q} \mathrm{div}_\perp \left(\sum_{1 \leq j \leq q} g_\perp^{i,j} \, V_j^\perp \right) \, V_i^\perp$$

$$= -\sum_{1 \leq k \leq r} \left(\sum_{1 \leq l \leq r} \partial_{\pi_l} \pi_k^l \right) \, e_k = -\sum_{1 \leq k \leq r} \mathrm{tr}\left(\nabla \pi_k\right) \, e_k. \qquad (19.45)$$

Definition 19.3.5 *We let* $\mathrm{div}(W)$ *be the divergence of a vector field* W *defined by*

$$\mathrm{div}(W) = \mathrm{tr}\left(\nabla W\right).$$

Using (19.64), we have

$$\mathrm{div}(W) = \mathrm{tr}\left(\nabla \pi(W)\right) + \mathrm{tr}\left(\nabla \pi_\perp(W)\right).$$

Choosing $W = \sum_{1 \leq j \leq q} g_\perp^{i,j} \, V_j^\perp$, for some $1 \leq i \leq q$, we have

$$\mathrm{div}\left(\sum_{1 \leq j \leq q} g_\perp^{i,j} \, V_j^\perp \right) = \mathrm{tr}\left(\nabla \left(\sum_{1 \leq j \leq q} g_\perp^{i,j} \, V_j^\perp \right) \right)$$

$$= \mathrm{div}_\perp \left(\sum_{1 \leq j \leq q} g_\perp^{i,j} \, V_j^\perp \right) = \sum_{1 \leq j \leq q} g_\perp^{i,j} \, \mathrm{div}\left(V_j^\perp\right).$$

We check this claim using the fact that

$$\nabla \left(\sum_{1 \leq j \leq q} g_\perp^{i,j} \, V_j^\perp \right) = \sum_{1 \leq j \leq q} \nabla \left(g_\perp^{i,j} \, V_j^\perp \right)$$

$$= \sum_{1 \leq j \leq q} \nabla \left(g_\perp^{i,j} \right) \, V_j^{\perp,T} + \sum_{1 \leq j \leq q} g_\perp^{i,j} \, \nabla V_j^\perp$$

so that

$$\mathrm{div}\left(\sum_{1 \leq j \leq q} g_\perp^{i,j} \, V_j^\perp \right) = \sum_{1 \leq j \leq q} \left\langle \nabla \left(g_\perp^{i,j} \right), V_j^\perp \right\rangle + \sum_{1 \leq j \leq q} g_\perp^{i,j} \, \mathrm{tr}\left(\nabla V_j^\perp\right)$$

$$= \sum_{1 \leq j \leq q} g_\perp^{i,j} \, \mathrm{tr}\left(\nabla V_j^\perp\right).$$

By (19.66) we have

$$
\mathrm{div}(W) \;=\; \mathrm{tr}\,(\nabla W) = \sum_{1 \le k \le r} \partial_{\pi_k} W^k \tag{19.46}
$$

$$
=\; \mathrm{tr}\,(\nabla \pi(W)) + \left\langle \sum_{1 \le i \le q} \mathrm{div}_\perp \left(\sum_{1 \le j \le q} g_\perp^{i,j}\, V_j^\perp \right) V_i^\perp, W \right\rangle. \tag{19.47}
$$

19.4   Lie brackets and commutation formulae

At this level of generality, it is important to notice that the bilinear form on $\mathcal{V} \times \mathcal{V}$ induced by the Hessian matrix with entries $\left(\partial_{\pi_i} \partial_{\pi_j}(F)\right)_{1 \le i \le j \le r}$ has no reason to be symmetric. Our next objective is show that the commutation property (19.15) ensures the symmetry of this bilinear form. To this end, we need to introduce some new mathematical objects.

Definition 19.4.1 *The Lie bracket $[W_1, W_2]$ of any two vector fields W_1 and W_2 is the vector field defined by*

$$
[W_1, W_2] = \partial_{W_1}(W_2) - \partial_{W_2}(W_1) = \begin{pmatrix} \partial_{W_1}(W_2^1) - \partial_{W_2}(W_1^1) \\ \vdots \\ \partial_{W_1}(W_2^r) - \partial_{W_2}(W_1^r) \end{pmatrix}
$$

or alternatively by the formula

$$
\left(\partial_{W_1}\partial_{W_2} - \partial_{W_2}\partial_{W_1}\right)(F) = \partial_{[W_1,W_2]}(F).
$$

The set $\mathcal{V}$ of vector fields $x \mapsto W(x) \in \mathcal{V}(x)$ is called a Lie algebra when $[W_1, W_2] \in \mathcal{V}$ for any W_1, W_2 (or equivalently $[V_i, V_j] \in \mathcal{V}$, for any $1 \le i, j \le p$).

Proposition 19.4.2 *The commutation condition (19.15) is equivalent to the fact that $[V_i, V_j] = 0$ for any $1 \le i, j \le p$. In this case, the set $\mathcal{V}$ of vector fields $x \mapsto W(x) \in \mathcal{V}(x)$ is a Lie algebra in the sense that $[W_1, W_2] \in \mathcal{V}$ for any W_1, W_2.*

Proof :
The first assertion is immediate. To check the second, we assume that $W_k = \sum_{1 \le i \le p} v_k^i\, V_i$, for some coordinate functions v_k^i. A simple calculation shows that

$$
[W_1, W_2] = \partial_{W_1}(W_2) - \partial_{W_2}(W_1)
$$

$$
= \sum_{1 \le i,j \le p} v_1^i v_2^j \underbrace{[\partial_{V_i}(V_j) - \partial_{V_j}(V_i)]}_{=0} + \sum_{1 \le i \le p} [\partial_{W_1}(v_2^i) - \partial_{W_2}(v_1^i)]\, V_i \;\in \mathcal{V}.
$$

■

Theorem 19.4.3 *For any function F and any vector fields $W_1, W_2 \in \mathcal{V}$ we have*

$$[W_1, W_2] = \partial_{W_1}(W_2) - \partial_{W_2}(W_1) = \nabla_{W_1}(W_2) - \nabla_{W_2}(W_1) \in \mathcal{V} \qquad (19.48)$$

and

$$\langle W_2, \nabla_{W_1}(\nabla F) \rangle = \langle W_1, \nabla_{W_2}(\nabla F) \rangle. \qquad (19.49)$$

The matrix $\left(\partial_{\pi_i}\partial_{\pi_j}(F)\right)_{1 \leq i \leq j \leq r}$ is symmetric; that is, for any $1 \leq k, l \leq p$ we have

$$\sum_{1 \leq i,j \leq r} V_k^i \, \partial_{\pi_i}\partial_{\pi_j}(F) \, V_l^j \;=\; \sum_{1 \leq i,j \leq r} V_l^i \, \partial_{\pi_i}\partial_{\pi_j}(F) \, V_k^j$$

$$= \; \langle V_k, \nabla_{V_l}\nabla F \rangle = \langle V_l, \nabla_{V_k}\nabla F \rangle. \qquad (19.50)$$

This yields for any $W_1, W_2 \in \mathcal{V}$ the symmetric bilinear formulation of the second covariant derivative

$$\nabla^2 F(W_1, W_2) \;=\; \sum_{1 \leq i,j \leq r} W_1^i \, \partial_{\pi_i}\partial_{\pi_j}(F) \, W_2^j$$

$$= \; W_1^T \nabla^2 F W_2 = W_2^T \nabla^2 F W_1$$

$$= \; \langle W_1, (\nabla^2 F)W_2 \rangle = \langle (\nabla^2 F)W_1, W_2 \rangle$$

$$= \; \langle W_1, \nabla_{W_2}\nabla F \rangle = \langle \nabla_{W_1}\nabla F, W_2 \rangle \qquad (19.51)$$

with the Hessian matrix $\nabla^2 F$ matrix introduced in (19.33). In addition, we have

$$\langle (\nabla^2 F_1)W_1, (\nabla^2 F_2)W_2 \rangle = \langle \nabla_{W_1}\nabla F_1, \nabla_{W_2}\nabla F_2 \rangle. \qquad (19.52)$$

Proof :

The first assertion is a direct consequence of proposition 19.4.2. To check (19.49) we use the decompositions

$$\partial_{W_1}\partial_{W_2}(F) \;=\; \partial_{W_1}\langle W_2, \nabla F \rangle = \sum_{1 \leq i \leq r} W_2^i \, \left(\partial_{W_1}(\nabla F)^i\right) + \left(\partial_{W_1} W_2^i\right) (\nabla F)^i$$

$$= \; \langle W_2, \partial_{W_1}(\nabla F) \rangle + \langle \nabla F, \partial_{W_1} W_2 \rangle = \langle W_2, \nabla_{W_1}(\nabla F) \rangle + \langle \nabla F, \partial_{W_1} W_2 \rangle.$$

This implies that

$$\left(\partial_{W_1}\partial_{W_2} - \partial_{W_2}\partial_{W_1}\right)(F) \;=\; \langle W_2, \nabla_{W_1}(\nabla F) \rangle - \langle W_1, \nabla_{W_2}(\nabla F) \rangle + \langle \partial_{W_1} W_2 - \partial_{W_2} W_1, \nabla F \rangle$$

$$= \; \langle W_2, \nabla_{W_1}(\nabla F) \rangle - \langle W_1, \nabla_{W_2}(\nabla F) \rangle + \langle [W_1, W_2], \nabla F \rangle.$$

Now (19.49) follows from the fact that

$$\left(\partial_{W_1}\partial_{W_2} - \partial_{W_2}\partial_{W_1}\right)(F) = \partial_{[W_1, W_2]}(F) = \langle [W_1, W_2], \partial F \rangle \langle [W_1, W_2], \nabla F \rangle.$$

Formula (19.50) is a direct consequence of (19.34) and (19.48). To check the last assertion we use

$$W_k = \sum_{1 \leq i,j \leq p} g^{i,j}\langle V_i, W_k \rangle \, V_j \Rightarrow \nabla_{W_k} = \sum_{1 \leq i,j \leq p} g^{i,j}\langle V_i, W_k \rangle \, \nabla_{V_j}.$$

This implies

$$\langle W_1, \nabla_{W_2}(\nabla F) \rangle = \sum_{1 \leq i,j,k,l \leq p} g^{i,j}g^{k,l}\langle V_i, W_1 \rangle\langle V_k, W_2 \rangle \overbrace{\langle V_j, \nabla_{V_l}(\nabla F) \rangle}^{=\langle V_l, \nabla_{V_j}(\nabla F) \rangle}.$$

The end of the proof of (19.51) is now clear. We check (19.52) using the decomposition

$$\langle \nabla_{W_1} \nabla F_1, \nabla_{W_2} \nabla F_2 \rangle = \sum_{1 \le i,j \le p} g^{i,j} \langle \nabla_{W_1} \nabla F_1, V_i \rangle \langle \nabla_{W_2} \nabla F_2, V_j \rangle.$$

Using (19.51) we find that

$$
\begin{aligned}
\langle \nabla_{W_1} \nabla F_1, \nabla_{W_2} \nabla F_2 \rangle &= \sum_{1 \le i,j \le p} g^{i,j} \langle \nabla_{V_i} \nabla F_1, W_1 \rangle \langle \nabla_{V_j} \nabla F_2, W_2 \rangle \\
&= \sum_{1 \le i,j \le p} g^{i,j} \langle (\nabla^2 F_1)(W_1), V_i \rangle \langle (\nabla^2 F_2)(W_2), V_j \rangle \\
&= \langle (\nabla^2 F_1) W_1, (\nabla^2 F_2) W_2 \rangle.
\end{aligned}
$$

This ends the proof of the theorem.   ∎

Corollary 19.4.4 *For any $W \in \mathcal{V}$ we have the commutation property*

$$[\nabla, \nabla_W](F) = (\nabla W) \nabla F \tag{19.53}$$

with the bracket $[\nabla, \nabla_W] = \nabla \nabla_W - \nabla_W \nabla$. In addition we have the formula

$$
\begin{aligned}
& \langle \nabla \nabla_W(F_1), \nabla F_2 \rangle + \langle \nabla \nabla_W(F_2), \nabla F_1 \rangle \\
& = \nabla_W \langle \nabla F_1, \nabla F_2 \rangle + (\nabla F_1)^T \left((\nabla W) + (\nabla W)^T \right) \nabla F_2.
\end{aligned}
\tag{19.54}
$$

Proof :
Using (19.32) and (19.51) we check that

$$
\begin{aligned}
\nabla \nabla_W F &= \nabla \langle W, \nabla F \rangle \\
&= \sum_{1 \le i,j \le p} g^{i,j} \, \partial_{V_i} \langle W, \nabla F \rangle \, V_j \\
&= \sum_{1 \le i,j \le p} g^{i,j} \left(\langle \nabla_{V_i} W, \nabla F \rangle + \langle \nabla_{V_i} \nabla F, W \rangle \right) V_j = (\nabla W) \nabla F + \nabla_W \nabla F.
\end{aligned}
$$

This ends the proof of the first assertion. The formula (19.53) is a direct consequence of the decomposition

$$\langle \nabla \nabla_W(F_1), \nabla F_2 \rangle = \langle \nabla_W \nabla F_1, \nabla F_2 \rangle + (\nabla F_2)^T (\nabla W) \nabla F_1.$$

This ends the proof of the corollary.   ∎

19.5   Inner product derivation formulae

For any vector fields $W_1, W_2, W_3 \in \mathcal{V}$ we have

$$
\begin{aligned}
\partial_{W_1} \langle W_2, W_3 \rangle &= \langle W_1, \nabla \langle W_2, W_3 \rangle \rangle \\
&= \langle W_3, \nabla_{W_1} W_2 \rangle + \langle W_2, \nabla_{W_1} W_3 \rangle.
\end{aligned}
\tag{19.55}
$$

We check this formula using the fact that

$$
\begin{aligned}
\partial_{W_1}\langle W_2, W_3\rangle &= \langle W_1, \partial\langle W_2, W_3\rangle\rangle = \langle W_1, \nabla\langle W_2, W_3\rangle\rangle \\
&= \langle \partial_{W_1} W_2, W_3\rangle + \langle W_2, \partial_{W_1} W_3\rangle \\
&= \langle \nabla_{W_1} W_2, W_3\rangle + \langle W_2, \nabla_{W_1} W_3\rangle.
\end{aligned}
$$

Using (19.25) and (19.51), we also notice that

$$
\begin{aligned}
\nabla\langle \nabla F_1, \nabla F_2\rangle &= \sum_{1\le i,j\le p} g^{i,j}\, \partial_{V_i}(\langle \nabla F_1, \nabla F_2\rangle)\, V_j \\
&= \sum_{1\le i,j\le p} g^{i,j}\, (\langle \nabla_{V_i}\nabla F_1, \nabla F_2\rangle + \langle \nabla F_1, \nabla_{V_i}\nabla F_2\rangle)\, V_j \\
&= \sum_{1\le i,j\le p} g^{i,j}\, (\langle \nabla_{\nabla F_2}\nabla F_1, V_i\rangle + \langle \nabla V_i, \nabla_{\nabla F_1}\nabla F_2\rangle)\, V_j \\
&= \nabla_{\nabla F_2}\nabla F_1 + \nabla_{\nabla F_1}\nabla F_2.
\end{aligned}
$$

In summary, for any $W \in \mathcal{V}$ and any smooth functions F_1, F_2 we have the formulae

$$
\nabla\langle \nabla F_1, \nabla F_2\rangle = \nabla_{\nabla F_2}\nabla F_1 + \nabla_{\nabla F_1}\nabla F_2 \tag{19.56}
$$

and

$$
\partial_W\langle \nabla F_1, \nabla F_2\rangle = \langle \nabla_{\nabla F_2}\nabla F_1 + \nabla_{\nabla F_1}\nabla F_2, W\rangle. \tag{19.57}
$$

Several algebraic and analytic formulae can be derived using (19.52) and (19.56). For instance,

$$
\left\|\nabla\langle \nabla F_1, \nabla F_2\rangle\right\|^2
$$

$$
= \langle \nabla\langle \nabla F_1, \nabla F_2\rangle, \nabla\langle \nabla F_1, \nabla F_2\rangle\rangle
$$

$$
= \langle \nabla_{\nabla F_2}\nabla F_1, \nabla_{\nabla F_2}\nabla F_1\rangle + 2\,\langle \nabla_{\nabla F_1}\nabla F_2, \nabla_{\nabla F_2}\nabla F_1\rangle\langle \nabla_{\nabla F_1}\nabla F_2, \nabla_{\nabla F_1}\nabla F_2\rangle
$$

$$
= \langle (\nabla^2 F_1)\nabla F_2, (\nabla^2 F_1)\nabla F_2\rangle + \langle (\nabla^2 F_2)\nabla F_1, (\nabla^2 F_2)\nabla F_1\rangle + 2\langle (\nabla^2 F_1)\nabla F_2, (\nabla^2 F_2)\nabla F_1\rangle
$$

$$
= \left\|(\nabla^2 F_1)\nabla F_2\right\|^2 + \left\|(\nabla^2 F_2)\nabla F_1\right\|^2 + 2\langle (\nabla^2 F_1)\nabla F_2, (\nabla^2 F_2)\nabla F_1\rangle.
$$

Using (19.9) we obtain the following estimates

$$
\begin{aligned}
\left\|\nabla\langle \nabla F, \nabla F\rangle\right\| &= 2\,\left\|(\nabla^2 F)\nabla F\right\| \\
&\le 2\,\|\nabla^2 F\|_2\,\|\nabla F\|_2 \le 2\,\|\nabla^2 F\|\,\|\nabla F\|_2. \tag{19.58}
\end{aligned}
$$

Using (19.26) the formulae (19.49) and (19.55) applied to the projection vector fields π_i take the following form.

Corollary 19.5.1 *For any $1 \leq i, j, k \leq r$ we have the formula*

$$\langle \pi_k, \nabla_{\pi_i} \pi_j \rangle = \langle \pi_i, \nabla_{\pi_k} \pi_j \rangle + \langle [\pi_i, \pi_k], \pi_\perp(e_j) \rangle \qquad (19.59)$$

with the unit vectors e_i defined in (19.3). For any $1 \leq i, j, k \leq k$, we also have the commutation formulae

$$\nabla \langle \pi_i, \pi_j \rangle = \nabla_{\pi_i} \pi_j + \nabla_{\pi_j} \pi_i \quad \text{and} \quad \langle \pi_k, \nabla_{\pi_i} \pi_j \rangle = \langle \pi_i, \nabla_{\pi_k} \pi_j \rangle. \qquad (19.60)$$

In addition, we have

$$\partial_{\pi_i} \langle \pi_j, \pi_k \rangle = \langle \pi_i, \nabla \langle \pi_j, \pi_k \rangle \rangle = \langle \pi_i, \nabla_{\pi_j} \pi_k + \nabla_{\pi_k} \pi_j \rangle \qquad (19.61)$$

as well as

$$\frac{1}{2} \partial_{\pi_i} \langle \pi_j, \pi_k \rangle = \langle \pi_i, \nabla_{\pi_j} \pi_k \rangle + \frac{1}{2} \left(\langle [\pi_k, \pi_j], \pi_i \rangle \right) \qquad (19.62)$$

and

$$\langle \pi_i, \nabla_{\pi_j} \pi_k \rangle + \frac{1}{2} \langle \pi_k, [\pi_j, \pi_i] \rangle = \partial_{\pi_j} \langle \pi_i, \pi_k \rangle - \frac{1}{2} \partial_{\pi_k} \langle \pi_i, \pi_j \rangle. \qquad (19.63)$$

Proof :
Applying (19.49) to $F = \chi_j$, $W_1 = \pi_i$, and $W_2 = \pi_k$ we check (19.59). Using (19.26) and (19.51) we find that

$$\langle \pi_k, \nabla_{\pi_i} \pi_j \rangle = \langle \pi_k, \nabla_{\pi_i} \nabla \chi_j \rangle = \langle \pi_i, \nabla_{\pi_k} \nabla \chi_j \rangle = \langle \pi_i, \nabla_{\pi_k} \pi_j \rangle.$$

On the other hand, recalling that $\pi_k \in \mathcal{V}$ for any k, we have

$$\begin{aligned}
\partial_{\pi_i} \left(\langle \pi_j, \pi_k \rangle \right) &= \langle \partial_{\pi_i} \pi_j, \pi_k \rangle + \langle \pi_j, \partial_{\pi_i} \pi_k \rangle \\
&= \langle \nabla_{\pi_i} \pi_j, \pi_k \rangle + \langle \pi_j, \nabla_{\pi_i} \pi_k \rangle = \langle \nabla_{\pi_k} \pi_j, \pi_i \rangle + \langle \pi_i, \nabla_{\pi_j} \pi_k \rangle.
\end{aligned}$$

The last assertions are direct consequences of (19.61). To check the first one, we use following decompositions

$$\frac{1}{2} \partial_{\pi_i} \langle \pi_j, \pi_k \rangle = \frac{1}{2} \langle \pi_i, \nabla_{\pi_j} \pi_k + \nabla_{\pi_k} \pi_j \rangle = \langle \pi_i, \nabla_{\pi_j} \pi_k \rangle + \frac{1}{2} \langle \pi_i, \nabla_{\pi_k} \pi_j - \nabla_{\pi_j} \pi_k \rangle.$$

The end of the proof follows now from

$$\langle \pi_i, \nabla_{\pi_k} \pi_j - \nabla_{\pi_j} \pi_k \rangle = \langle \pi_i, \partial_{\pi_k} \pi_j - \partial_{\pi_j} \pi_k \rangle \quad \text{and} \quad \partial_{\pi_k} \pi_j - \partial_{\pi_j} \pi_k = [\pi_k, \pi_j].$$

The second assertion is a consequence of (19.61) and the following decomposition

$$\begin{aligned}
\langle \pi_i, \nabla_{\pi_j} \pi_k \rangle &= \partial_{\pi_j} \langle \pi_i, \pi_k \rangle - \langle \pi_k, \nabla_{\pi_j} \pi_i \rangle \\
\frac{1}{2} \partial_{\pi_k} \langle \pi_i, \pi_j \rangle &= \frac{1}{2} \langle \pi_k, \nabla_{\pi_i} \pi_j + \nabla_{\pi_j} \pi_i \rangle \\
&= \langle \pi_k, \nabla_{\pi_j} \pi_i \rangle + \frac{1}{2} \langle \pi_k, \nabla_{\pi_i} \pi_j - \nabla_{\pi_j} \pi_i \rangle.
\end{aligned}$$

Recalling that

$$\langle \pi_i, \nabla_{\pi_j} \pi_k - \nabla_{\pi_k} \pi_j \rangle = \langle \pi_i, \partial_{\pi_j} \pi_k - \partial_{\pi_k} \pi_j \rangle \quad \text{and} \quad \partial_{\pi_j} \pi_k - \partial_{\pi_k} \pi_j = [\pi_i, \pi_j]$$

we find that

$$\langle \pi_k, \nabla_{\pi_j} \pi_i \rangle + \frac{1}{2} \langle \pi_k, [\pi_i, \pi_j] \rangle = \frac{1}{2} \partial_{\pi_k} \langle \pi_i, \pi_j \rangle.$$

This clearly implies that

$$\langle \pi_i, \nabla_{\pi_j} \pi_k \rangle = \left(\partial_{\pi_j} \langle \pi_i, \pi_k \rangle - \frac{1}{2} \partial_{\pi_k} \langle \pi_i, \pi_j \rangle \right) + \frac{1}{2} \langle [\pi_i, \pi_j], \pi_k \rangle.$$

This ends the proof of the corollary. ∎

19.6 Second order derivatives and some trace formulae

For any vector field W on $\mathbb{R}^r$ we have

$$(19.19) \implies \partial W = \partial \pi(W) + \partial \pi_\perp(W) \implies \nabla W = \nabla \pi(W) + \nabla \pi_\perp(W). \quad (19.64)$$

On the other hand, we have

$$\pi(W) = \begin{bmatrix} \pi_1^1 & \cdots & \pi_r^1 \\ \vdots & \vdots & \vdots \\ \pi_1^r & \cdots & \pi_r^r \end{bmatrix} \begin{bmatrix} W^1 \\ \vdots \\ W^r \end{bmatrix} = \begin{bmatrix} \sum_{1 \le k \le r} \pi_k^1 \, W^k \\ \vdots \\ \sum_{1 \le k \le r} \pi_k^r \, W^k \end{bmatrix}$$

and therefore

$$
\begin{aligned}
\partial \pi(W) &= \\
&= \left[\sum_{1 \le k \le r} \partial \left(\pi_k^1 \, W^k \right), \ldots, \sum_{1 \le k \le r} \partial \left(\pi_k^r \, W^k \right) \right] \\
&= \left[\sum_{1 \le k \le r} W^k \, \partial \pi_k^1, \ldots, \sum_{1 \le k \le r} W^k \, \partial \pi_k^r \right] + \left[\sum_{1 \le k \le r} \pi_k^1 \, \partial W^k, \ldots, \sum_{1 \le k \le r} \pi_k^r \, \partial W^k \right] \\
&= \sum_{1 \le k \le r} W^k \left[\partial \pi_k^1, \ldots, \partial \pi_k^r \right] + \begin{bmatrix} \partial_{x_1} W^1 & \cdots & \partial_{x_1} W^r \\ \vdots & \vdots & \vdots \\ \partial_{x_r} W^1 & \cdots & \partial_{x_r} W^r \end{bmatrix} \begin{bmatrix} \pi_1^1 & \cdots & \pi_r^1 \\ \vdots & \vdots & \vdots \\ \pi_r^1 & \cdots & \pi_r^r \end{bmatrix} \\
&= \sum_{1 \le k \le r} W^k \, \partial \pi_k + \left[\partial W^1, \ldots, \partial W^r \right] \pi = \sum_{1 \le k \le r} W^k \, \partial \pi_k + (\partial W) \pi.
\end{aligned}
$$

This yields the formula

$$\nabla \pi(W) \;=\; \pi \partial \pi(W) = \sum_{1 \le k \le r} W^k \, \nabla \pi_k + \pi \partial W \pi \qquad (19.65)$$

with the matrix

$$\pi \partial W \pi = \begin{bmatrix} \sum_{1 \leq l \leq r} \pi_l^1 \sum_{1 \leq k \leq r} \pi_k^1 \, \partial_{x_l} W^k, & \cdots & , \sum_{1 \leq l \leq r} \pi_l^1 \sum_{1 \leq k \leq r} \pi_k^r \, \partial_{x_l} W^k \\ \vdots & \vdots & \vdots \\ \sum_{1 \leq l \leq r} \pi_l^r \sum_{1 \leq k \leq r} \pi_k^1 \, \partial_{x_l} W^k, & \cdots & , \sum_{1 \leq l \leq r} \pi_l^r \sum_{1 \leq k \leq r} \pi_k^r \, \partial_{x_l} W^k \end{bmatrix}.$$

On the other hand, we have the equivalent formulations

$$\mathrm{tr}\,(\pi \partial W \pi) = \sum_{1 \leq k,l \leq r} \left(\sum_{1 \leq i \leq r} \pi_l^i \pi_k^i \right) \partial_{x_l} W^k \qquad (19.66)$$

$$= \sum_{1 \leq k,l \leq r} \pi_k^l \, \partial_{x_l} W^k = \mathrm{tr}\,(\pi \partial W) \qquad (19.67)$$

$$= \mathrm{tr}\,(\nabla W) = \sum_{1 \leq k \leq r} \partial_{\pi_k} W^k.$$

We recall that ∇W is given by the matrix

$$\begin{aligned} \nabla W &= \pi \partial W = [\pi(\partial W^1), \ldots, \pi(\partial W^r)] = [\nabla W^1, \ldots, \nabla W^r] \\ &= \pi^T \partial W = \begin{bmatrix} \pi_1^T \\ \vdots \\ \pi_r^T \end{bmatrix} [\partial W^1, \ldots, \partial W^r] = \begin{pmatrix} \partial_{\pi_1} W^1, & \cdots & , \partial_{\pi_1} W^r \\ \vdots & \vdots & \vdots \\ \partial_{\pi_r} W^1, & \cdots & , \partial_{\pi_r} W^r \end{pmatrix}. \end{aligned}$$

In much the same way, we have the formula

$$\begin{aligned} \nabla \pi_k &= \pi \partial \pi_k = \pi^T \partial \pi_k \\ &= \begin{bmatrix} \pi_1^T \\ \vdots \\ \pi_r^T \end{bmatrix} [\partial \pi_k^1, \ldots, \partial \pi_k^r] = \begin{bmatrix} \partial_{\pi_1} \pi_k^1, & \cdots, & \partial_{\pi_1} \pi_k^r \\ \vdots & \vdots & \vdots \\ \partial_{\pi_r} \pi_k^1, & \cdots, & \partial_{\pi_r} \pi_k^r \end{bmatrix}. \end{aligned}$$

Recalling that

$$\partial_{\pi_i} \chi_j = \sum_{1 \leq k \leq r} \pi_i^k \partial_{x_k} \chi_j = \pi_i^j = \pi_j^i$$

and using (19.51) and (19.63) we prove the following proposition.

Proposition 19.6.1 *For any $1 \leq k \leq r$ we have the Hessian formula*

$$
\nabla \pi_k = \nabla^2 \chi_k = \begin{bmatrix} \partial_{\pi_1} \pi_k^1, & \cdots, & \partial_{\pi_1} \pi_k^r \\ \vdots & \vdots & \vdots \\ \partial_{\pi_r} \pi_k^1, & \cdots, & \partial_{\pi_r} \pi_k^r \end{bmatrix} = \begin{bmatrix} \partial_{\pi_1} \partial_{\pi_1} \chi_k, & \cdots, & \partial_{\pi_1} \partial_{\pi_r} \chi_k \\ \vdots & \vdots & \vdots \\ \partial_{\pi_r} \partial_{\pi_1} \chi_k, & \cdots, & \partial_{\pi_r} \partial_{\pi_r} \chi_k \end{bmatrix}
$$

with $\nabla^2 := \nabla\nabla$ and the projection mappings χ_k defined in (19.18). In addition, we have

$$
\begin{aligned}
\nabla \pi_k(\pi_i, \pi_j) &:= \pi_i^T \, \nabla \pi_k \, \pi_j = \langle \pi_i, \nabla_{\pi_j} \pi_k \rangle = \langle \pi_j, \nabla_{\pi_i} \pi_k \rangle \\
&= \left(\partial_{\pi_j} \langle \pi_i, \pi_k \rangle - \frac{1}{2} \, \partial_{\pi_k} \langle \pi_i, \pi_j \rangle \right) + \frac{1}{2} \langle \, [\pi_i, \pi_j] \, , \pi_k \rangle
\end{aligned}
$$

$$(19.68)$$

and the trace formulae

$$
\begin{aligned}
\langle \nabla \pi_k, \nabla \pi_l \rangle &= \sum_{1 \leq i,j \leq r} \pi_i^j \, \langle \nabla \pi_k (\pi_i), \nabla \pi_l (\pi_j) \rangle \qquad (\Leftarrow (19.11)) \\
&= = \operatorname{tr}_e \left(\pi \, (\nabla \pi_l)^T \nabla \pi_k \right). \qquad (\Leftarrow (19.12))
\end{aligned}
$$

Using (19.65) we readily check that

$$
\begin{aligned}
\operatorname{tr} \left(\nabla \pi(W) \right) &= \sum_{1 \leq k \leq r} W^k \operatorname{tr} \left(\nabla \pi_k \right) + \sum_{1 \leq k \leq r} \partial_{\pi_k} W^k \\
&= \sum_{1 \leq k \leq r} W^k \operatorname{tr} \left(\nabla \pi_k \right) + \operatorname{tr} \left(\pi \partial W \right) = \sum_{1 \leq k \leq r} W^k \operatorname{tr} \left(\nabla \pi_k \right) + \operatorname{tr} \left(\nabla W \right)
\end{aligned}
$$

and

$$
\begin{aligned}
(19.64) \implies \operatorname{tr} \left(\nabla W \right) &= \operatorname{tr} \left(\nabla \pi(W) \right) + \operatorname{tr} \left(\nabla \pi_\perp (W) \right) \\
&= \sum_{1 \leq k \leq r} W^k \operatorname{tr} \left(\nabla \pi_k \right) + \operatorname{tr} \left(\nabla W \right) + \operatorname{tr} \left(\nabla \pi_\perp (W) \right).
\end{aligned}
$$

Using (19.47) we find the orthonormal divergence formula

$$
\begin{aligned}
\operatorname{tr} \left(\nabla \pi_\perp (W) \right) &= - \sum_{1 \leq k \leq r} W^k \operatorname{tr} \left(\nabla \pi_k \right) \\
&= \left\langle \sum_{1 \leq i \leq q} \operatorname{div}_\perp \left(\sum_{1 \leq j \leq q} g_\perp^{i,j} \, V_j^\perp \right) V_i^\perp, W \right\rangle.
\end{aligned}
$$

In other words, we recover the trace formula

$$
\operatorname{tr} \left(\nabla \pi_k \right) = - \sum_{1 \leq i \leq q} \operatorname{div}_\perp \left(\sum_{1 \leq j \leq q} g_\perp^{i,j} \, V_j^\perp \right) V_i^{\perp,k} = \mathbb{H}^k
$$

with the mean curvature vector $\mathbb{H}$ discussed in (19.36).

19.7   Laplacian operator

We return to the settings of section 19.2. Choosing $W = \partial F$ in (19.65) we prove the following result.

Proposition 19.7.1 *We have the second covariant derivative formula*

$$\nabla^2 F = \sum_{1 \le k \le r} \partial_{x_k}(F)\, \nabla \pi_k + \pi \partial^2 F \pi \qquad (19.69)$$

and the Laplacian formula

$$\Delta F := \operatorname{tr}\left(\nabla^2 F\right) = \operatorname{tr}\left(\pi \partial^2 F\right) + \sum_{1 \le k \le r} \partial_{x_k} F \, \operatorname{tr}\left(\nabla \pi_k\right) \qquad (19.70)$$

with $\nabla^2 F = \nabla(\nabla F)$, and the Hessian matrix $\partial^2 F = (\partial_{x_k, x_l} F)_{1 \le k, l \le r}$.

This shows that

$$\operatorname{tr}\left(\nabla^2 F\right) = \sum_{1 \le i \le r} \partial_{\pi_i}^2 F = \operatorname{tr}\left(\pi \partial^2 F\right) + \sum_{1 \le k \le r} \partial_{x_k} F \, \operatorname{tr}\left(\nabla \pi_k\right).$$

Using (19.44), we also have

$$
\begin{aligned}
\Delta F &= \operatorname{tr}\left(\nabla^2 F\right) \\
&= \sum_{1 \le i \le r} \partial_{\pi_i}^2 F = \operatorname{tr}\left(\pi \partial^2 F\right) - \langle \mathbb{H}, \partial F \rangle = \operatorname{tr}\left(\pi \partial^2 F\right) - \partial_{\mathbb{H}} F \\
&= \sum_{1 \le k,l \le p} g^{k,l} \langle V_k, \nabla_{V_l} \nabla F \rangle = \sum_{1 \le k,l \le p} g^{k,l} \nabla^2 F(V_k, V_l) \qquad (19.71)
\end{aligned}
$$

with the mean curvature vector $\mathbb{H} \in \mathcal{V}^\perp$ defined in (19.45), and the basis vector fields V_i of $\mathcal{V}$ defined in (19.14). In addition we have the formula

$$\langle \nabla^2 F_1, \nabla^2 F_2 \rangle = \sum_{1 \le k,l \le p} g^{k,l} \langle \nabla_{V_k} \nabla F_1, \nabla_{V_l} \nabla F_2 \rangle \qquad (19.72)$$

with the Hilbert-Schmidt inner product defined in (19.8).

The formulae (19.71) and (19.72) are consequences of the fact that $B_F = \nabla^2 F$ can be seen as the bilinear form on $\mathcal{V}$ defined by its action on the basis random fields

$$
\begin{aligned}
B_F(V_k, V_l) &= \nabla^2 F(V_k, V_l) := V_k^T \, \nabla^2 F \, V_l \\
&= \sum_{1 \le i,j \le r} V_k^i \, \partial_{\pi_i} \partial_{\pi_j}(F) \, V_l^j = \langle V_k, \nabla_{V_l}(\nabla F) \rangle \quad \Leftarrow (19.34) \\
&= \langle V_k, b_F(V_l) \rangle \qquad \text{with} \quad b_F(V_l) = \nabla_{V_l}(\nabla F) = \nabla^2(V_l).
\end{aligned}
$$

The result is now a direct consequence of the trace formulae (19.7) and (19.10). Notice that

$$(19.13) \implies \frac{1}{p} \left(\Delta F\right)^2 \leq \left\|\nabla F^2\right\|^2 = \langle \nabla^2 F, \nabla^2 F \rangle \leq \left(\Delta F\right)^2. \tag{19.73}$$

In terms of the projection mappings χ_i defined in (19.18) and (19.3) we have

$$\Delta \chi_i = \operatorname{tr}\left(\nabla \pi_i\right) = \operatorname{tr}\left(\nabla^2 \chi_i\right) = -\mathbb{H}^i. \tag{19.74}$$

For any function F on $\mathbb{R}^n$ and any sequence $f = (f_1, \ldots, f_n)$ of functions f_i on $\mathbb{R}^r$, with $1 \leq i \leq n$, we have the change of variable formula

$$\nabla(F(f)) = \sum_{1 \leq m \leq n} \left(\partial_{x_m} F\right)(f) \, \nabla f_m. \tag{19.75}$$

In addition, using (19.69) we have

$$\nabla^2(F(f)) = \sum_{1 \leq m \leq n} \left(\partial_{x_m} F\right)(f) \, \nabla^2 f_m + \sum_{1 \leq m, m' \leq n} \left(\partial_{x_m, x_{m'}} F\right)(F) \, \nabla f_m (\nabla f_{m'})^T \tag{19.76}$$

and therefore

$$\Delta(F(f)) = \sum_{1 \leq m \leq n} \left(\partial_{x_m} F\right)(f) \, \Delta f_m + \sum_{1 \leq l, m \leq n} \left(\partial_{x_l, x_m} F\right)(f) \, \langle \nabla f_l, \nabla f_m \rangle. \tag{19.77}$$

19.8 Ricci curvature

Definition 19.8.1 *The curvature operator/tensor R is defined for any vector fields W_1, W_2 by the formulae*

$$R\left(W_1, W_2\right) \quad := \quad [\nabla_{W_1}, \nabla_{W_2}] - \nabla_{[W_1, W_2]}. \tag{19.78}$$

The linearity properties of the curvature tensor are summarized in the following technical lemma.

Lemma 19.8.2 *For any smooth functions F_1, F_2, F_3 and any $W_1, W_2, W_3 \in \mathcal{V}$ we have*

$$R(F_1 W_1, F_2 W_2)(F_3 W_3) = F_1 F_2 F_3 \, R(W_1, W_2)(W_3) \tag{19.79}$$

and for any $W \in \mathcal{V}$ we have

$$\begin{aligned}
R(W_1 + W, W_2) &= R(W_1, W_2) + R(W, W_2) \\
R(W_1, W_2 + W) &= R(W_1, W_2) + R(W_1, W) \\
R(W_1, W_2)(W_3 + W) &= R(W_1, W_2)(W_3) + R(W_1, W_2)(W).
\end{aligned} \tag{19.80}$$

Proof :
Using (19.30) we have

$$\nabla_{W_2}(FW_3) = F \ \nabla_{W_2}(W_3) + \partial_{W_2}(F) \ W_3$$

and

$$\nabla_{W_1}\nabla_{W_2}(FW_3)$$

$$= F \ \nabla_{W_1}\nabla_{W_2}(W_3) + \partial_{W_1}(F)\nabla_{W_2}(W_3) + \partial_{W_2}(F) \ \nabla_{W_1}W_3 + \partial_{W_1}\partial_{W_2}(F) \ W_3.$$

This implies that

$$[\nabla_{W_1},\nabla_{W_2}](FW_3) = F \ [\nabla_{W_1},\nabla_{W_2}](W_3) + \nabla_{[W_1,W_2]}(F) \ W_3.$$

The r.h.s. term in the above display follows from the fact that $\partial_{[W_1,W_2]}(F) = \nabla_{[W_1,W_2]}(F)$. Recalling that

$$\nabla_{[W_1,W_2]}(FW_3) = F \ \nabla_{[W_1,W_2]}(W_3) + \partial_{[W_1,W_2]}(F) \ W_3$$

we conclude that

$$R(W_1,W_2)(FW_3) = F \ R(W_1,W_2)(W_3).$$

Using (19.30) we also check that

$$\begin{aligned}
\nabla_{FW_1}\nabla_{W_2}(W_3) &= F \ \nabla_{W_1}\nabla_{W_2}(W_3) \\
\nabla_{W_2}\nabla_{FW_1}(W_3) &= \nabla_{W_2}(F \ \nabla_{W_1}(W_3)) \\
&= F \ \nabla_{W_2}\nabla_{W_1}(W_3) + \partial_{W_2}(F) \ \nabla_{W_1}(W_3)
\end{aligned}$$

from which we find that

$$[\nabla_{FW_1},\nabla_{W_2}] = F \ [\nabla_{W_1},\nabla_{W_2}] - \partial_{W_2}(F) \ \nabla_{W_1}.$$

In a similar way, for any function G we have

$$\begin{aligned}
\partial_{FW_1}\partial_{W_2}(G) &= \partial_{FW_1}\partial_{W_2}(G) \\
\partial_{W_2}\partial_{FW_1}(G) &= \partial_{W_2}(F\partial_{W_1}(G)) \\
&= \partial_{W_2}\partial_{W_1}(G) + \partial_{W_2}(F) \ \partial_{W_1}(G).
\end{aligned}$$

Hence we find that

$$\begin{aligned}
\partial_{[FW_1,W_2]}(G) &= (\partial_{FW_1}\partial_{W_2} - \partial_{W_2}\partial_{FW_1})(G) \\
&= \partial_{F \ [W_1,W_2]}(G) - \partial_{W_1\partial_{W_2}(F)}(G).
\end{aligned}$$

Combining this with (19.23) we prove that

$$[FW_1,W_2] = F \ [W_1,W_2] - \partial_{W_2}(F) \ W_1 \Rightarrow \nabla_{[FW_1,W_2]} = F \ \nabla_{[W_1,W_2]} - \partial_{W_2}(F) \ \nabla_{W_1}$$

from which we conclude that $R(FW_1,W_2) = F \ R(W_1,W_2)$, and by symmetry arguments $R(W_1,FW_2) = F \ R(W_1,W_2)$. This ends the proof of (19.79). The proof of (19.80) is a direct consequence of the linearity properties (19.23). The last assertion is immediate. This completes the proof of the lemma. ∎

Definition 19.8.3 *The Ricci curvature is the trace* $\mathrm{Ric}\,(W_1, W_2)$ *of the linear forms* $W \mapsto$
$\mathrm{Ric}\,(W, W_1)\,(W_2) + \mathrm{Ric}\,(W, W_2)\,(W_1)$ *defined by*

$$\mathrm{Ric}\,(W_1, W_2) \quad := \quad 2^{-1} \sum_{1 \le k, l \le p} g^{k,l} \, \langle V_k, R\,(V_l, W_1)\,(W_2) + R\,(V_l, W_2)\,(W_1)\rangle. \ (19.81)$$

The linearity property is a direct consequence of lemma 19.8.2, that is,

$$\mathrm{Ric}\,(F_1 W_1 + F_2 W_2, W_3) = F_1 \, \mathrm{Ric}\,(W_1, W_3) + F_2 \, \mathrm{Ric}\,(W_2, W_3).$$

Combining this lemma with (19.26) and (19.27) we prove the formula

$$\mathrm{Ric}\,(\nabla F_1, \nabla F_2)$$

$$= \sum_{1 \le i, j \le p} \left[\sum_{1 \le k \le p} g^{i,k} \, \partial_{V_k} F_1 \right] \left[\sum_{1 \le l \le p} g^{j,l} \, \partial_{V_l} F_2 \right] \mathrm{Ric}\,(V_i, V_j) \qquad (19.82)$$

$$= \sum_{1 \le k, l \le p} \partial_{V_k} F_1 \, \partial_{V_l} F_2 \sum_{1 \le i, j \le p} \left(g^{k,i} \, \mathrm{Ric}\,(V_i, V_j)\, g^{j,l} \right).$$

In the same way, recalling that $V_i = \sum_{1 \le k \le r} V_i^k \, \pi_k$ and $\pi_i = \sum_{1 \le k, l \le p} g^{l,k} \, V_k^i \, V_l$ we
also have

$$\mathrm{Ric}\,(V_i, V_j) = \sum_{1 \le k, l \le r} V_i^k \, V_j^l \, \mathrm{Ric}\,(\pi_k, \pi_l)$$

$$\mathrm{Ric}\,(\pi_i, \pi_j) = \sum_{1 \le k, k' \le p} V_k^i \, V_{k'}^j \sum_{1 \le l, l' \le p} \left(g^{k,l} \mathrm{Ric}\,(V_l, V_{l'})\, g^{l',k'} \right). \qquad (19.83)$$

In the parametrization model discussed in (19.16), $\mathrm{Ric}\,(V_i, V_j)$ can be interpreted as the
expression of the Ricci curvature in the local coordinates induced by the mappings ψ from
$\mathbb{R}^p$ into some manifold S. One particular important class of manifolds arising in physics
satisfies the following property

$$\mathrm{Ric}\,(V_i, V_j) = \rho \, \langle V_i, V_j \rangle \qquad (19.84)$$

for some constant ρ. These are called Einstein manifolds. For instance, the 2-sphere
discussed in section 24.1.2 satisfies (19.84) with $\rho = 1$. The Euclidian space which is flat
(.e. $g_{i,j} = 1_{i=j}$) also satisfies (19.84) with $\rho = 1$. In this situation, using (19.83), we have

$$(19.84) \iff R(\pi_i, \pi_j) = \rho \, \pi_i^j \implies \mathrm{Ric}\,(\nabla F_1, \nabla F_2) = \rho \, \langle \nabla F_1, \nabla F_2 \rangle. \qquad (19.85)$$

Recalling that

$$\nabla F = \sum_{1 \le k \le p} g^{i,k} \, \partial_{V_k} F_1 \, V_l$$

we prove the following formula.

$$\mathrm{Ric}\,(\nabla F_1, \nabla F_2) = \sum_{1 \le i, j \le r} (\nabla F_1)^i \, (\nabla F_2)^j \, \mathrm{Ric}\,(\pi_i, \pi_j). \qquad (19.86)$$

Proposition 19.8.4 *For any* $1 \leq i, j, k \leq p$ *we have the first Bianchi identity (a.k.a. algebraic Bianchi identity)*

$$R(V_i, V_j)(V_k) + R(V_j, V_k)(V_i) + R(V_k, V_i)(V_j) = 0. \tag{19.87}$$

In addition we have

$$R(V_i, V_j)(V_k) = -R(V_j, V_i)(V_k) \quad and \quad \langle R(V_i, V_j)(V_k), V_l \rangle = -\langle V_k, R(V_i, V_j)(V_l) \rangle. \tag{19.88}$$

Proof :
We recall that

$$[V_i, V_j] = \nabla_{V_i}(V_j) - \nabla_{V_j}(V_i) = 0 \Rightarrow \nabla_{V_k}\nabla_{V_i}(V_j) - \nabla_{V_k}\nabla_{V_j}(V_i) = \nabla_{V_k}([V_i, V_j]) = 0$$

so that

$$\left[\nabla_{V_i}, \nabla_{V_j}\right](V_k) + \left[\nabla_{V_j}, \nabla_{V_k}\right](V_i) + \left[\nabla_{V_k}, \nabla_{V_i}\right](V_j)$$

$$= \left[\nabla_{V_i}\nabla_{V_j}(V_k) - \nabla_{V_j}\nabla_{V_i}(V_k)\right] + \left[\nabla_{V_j}\nabla_{V_k}(V_i) - \nabla_{V_k}\nabla_{V_j}(V_i)\right] + \left[\nabla_{V_k}\nabla_{V_i}(V_j) - \nabla_{V_i}\nabla_{V_k}(V_j)\right]$$

$$= \nabla_{V_i}([V_j, V_k]) + \nabla_{V_j}([V_k, V_i]) + \nabla_{V_k}([V_i, V_j]) = 0.$$

This ends the proof of the first assertion. The l.h.s. formula in (19.88) follows from

$$R(V_i, V_j)(V_k) = \left[\nabla_{V_i}, \nabla_{V_j}\right](V_k) = -R(V_j, V_i)(V_k).$$

We also have

$$\langle R(V_i, V_j)(V_k), V_l \rangle = \langle \nabla_{V_i}\nabla_{V_j}(V_k), V_l \rangle - \langle \nabla_{V_j}\nabla_{V_i}(V_k), V_l \rangle.$$

On the other hand

$$\langle \nabla_{V_i}\nabla_{V_j}(V_k), V_l \rangle = \partial_{V_i}\langle \nabla_{V_j}(V_k), V_l \rangle - \langle \nabla_{V_j}(V_k), \nabla_{V_i}V_l \rangle$$

and

$$\langle \nabla_{V_j}(V_k), V_l \rangle = \partial_{V_j}\langle V_k, V_l \rangle - \langle V_k, \nabla_{V_j}(V_l) \rangle.$$

This implies that

$$\langle \nabla_{V_i}\nabla_{V_j}(V_k), V_l \rangle = \partial_{V_i}\partial_{V_j}\langle V_k, V_l \rangle - \partial_{V_i}\langle V_k, \nabla_{V_j}V_l \rangle - \langle \nabla_{V_j}V_k, \nabla_{V_i}V_l \rangle$$

from which we prove that

$$\langle R(V_i, V_j)(V_k), V_l \rangle$$

$$= \partial_{V_j}\langle V_k, \nabla_{V_i}V_l \rangle + \langle \nabla_{V_i}V_k, \nabla_{V_j}V_l \rangle - \partial_{V_i}\langle V_k, \nabla_{V_j}V_l \rangle - \langle \nabla_{V_j}V_k, \nabla_{V_i}V_l \rangle$$

$$= \langle V_k, \nabla_{V_j}\nabla_{V_i}V_l \rangle - \langle V_k, \nabla_{V_i}\nabla_{V_j}V_l \rangle = \langle V_k, \left[\nabla_{V_j}, \nabla_{V_i}\right](V_l) \rangle = -\langle R(V_i, V_j)(V_l), V_k \rangle.$$

This ends the proof of the proposition. ∎

Proposition 19.8.5 *The Ricci curvature parameters* $\mathrm{Ric}\,(V_i, V_j) = R_{i,j} = R_{j,i}$ *are defined by the functions*

$$R_{i,j} = \sum_{1 \le m,n \le p} \left[C^n_{i,j}\, C^m_{m,n} - C^n_{i,m}\, C^m_{j,n} \right] + \sum_{1 \le m \le p} \left[\partial_{V_m} C^m_{i,j} - \partial_{V_j} C^m_{i,m} \right],$$

with the Christoffel symbols introduced in (19.21). In addition we have

$$R_{i,j} = \sum_{1 \le m \le p} R^m_{i,m,j} = \sum_{1 \le m \le p} R^m_{j,m,i}$$

with

$$R^m_{k,i,j} := \sum_{1 \le n \le p} \left[C^n_{j,k}\, C^m_{i,n} - C^n_{i,k}\, C^m_{j,n} \right] + \left[\partial_{V_i}(C^m_{j,k}) - \partial_{V_j}(C^m_{i,k}) \right]. \tag{19.89}$$

Proof :
Firstly, we observe that

$$\begin{aligned}
\nabla_{V_i} \nabla_{V_j}(V_k) &= \sum_{1 \le l \le p} \nabla_{V_i}\left(C^l_{j,k}\, V_l \right) = \sum_{1 \le l \le p} \left(C^l_{j,k}\, \nabla_{V_i}(V_l) + \partial_{V_i}(C^l_{j,k})\, V_l \right) \\
&= \sum_{1 \le m \le rp} \left[\sum_{1 \le l \le r} C^l_{j,k}\, C^m_{i,l} + \partial_{V_i}(C^m_{j,k}) \right] V_m.
\end{aligned}$$

This yields

$$\left[\nabla_{V_i}, \nabla_{V_j} \right](V_k)$$

$$= \sum_{1 \le m \le p} \left[\sum_{1 \le n \le r} \left[C^n_{j,k}\, C^m_{i,n} - C^n_{i,k}\, C^m_{j,n} \right] + \left[\partial_{V_i}(C^m_{j,k}) - \partial_{V_j}(C^m_{i,k}) \right] \right] V_m$$

from which we check that

$$\begin{aligned}
R(V_i, V_j)(V_k) &= \left[\nabla_{V_i}, \nabla_{V_j} \right](V_k) = \sum_{1 \le m \le p} R^m_{k,i,j}\, V_m \\
\langle R\,(V_i, V_j)\,(V_k), V_l \rangle &= \sum_{1 \le m \le p} g_{l,m}\, R^m_{k,i,j} := R_{l,k,i,j}. \Leftarrow (19.88)
\end{aligned}$$

Also notice that

$$\sum_{1 \le l \le p} g^{i,l}\, R_{l,k,i,j} = \sum_{1 \le l,m \le p} g^{i,l}\, g_{l,m}\, R^m_{k,i,j} = R^i_{k,i,j}.$$

On the other hand we have

$$(19.88) \;\Rightarrow\; R_{l,i,j,k} = R_{i,l,j,k} \Rightarrow \sum_{1 \le i,l \le p} g^{i,l} R_{l,i,j,k} = -\sum_{1 \le i,l \le p} g^{i,l} R_{i,l,j,k} = 0. \tag{19.90}$$

The Bianchi formula (19.87) ensures that

$$R_{l,k,i,j} + R_{l,i,j,k} + R_{l,j,k,i} = 0 = R_{l,k,i,j} + R_{l,i,j,k} - R_{l,j,i,k} \Leftarrow (19.88)$$

$$\sum_{1 \leq l \leq p} g^{i,l} \left(R_{l,k,i,j} + R_{l,i,j,k} - R_{l,j,i,k} \right) = 0 = R^i_{k,i,j} + \left(\sum_{1 \leq l \leq p} g^{i,l} R_{l,i,j,k} \right) - R^i_{j,i,k}.$$

By (19.90), this clearly implies the symmetric formula

$$\forall 1 \leq j, k \leq p \qquad \sum_{1 \leq i \leq p} R^i_{j,i,k} = \sum_{1 \leq i \leq p} R^i_{k,i,j}$$

from which we deduce that

$$\sum_{1 \leq i, l \leq p} g^{i,l} \langle V_l, R(V_i, V_j)(V_k) \rangle = \sum_{1 \leq i, l \leq p} g^{i,l} R_{l,k,i,j} = \sum_{1 \leq i \leq p} R^i_{k,i,j} = \sum_{1 \leq i \leq p} R^i_{j,i,k}$$

for any indices $1 \leq j, k \leq p$. We conclude that

$$\mathrm{Ric}\,(V_i, V_j) \; := \; 2^{-1} \sum_{1 \leq k, l \leq p} g^{k,l} \, \langle V_k, R(V_l, V_i)(V_j) + R(V_l, V_j)(V_i) \rangle$$

$$= \; \sum_{1 \leq k, l \leq p} g^{k,l} \, \langle V_k, R(V_l, V_i)(V_j) \rangle = \sum_{1 \leq m \leq p} R^m_{i,m,j} = \sum_{1 \leq m \leq p} R^m_{j,m,i}.$$

This ends the proof of the proposition.   ■

19.9   Bochner-Lichnerowicz formula

Definition 19.9.1 *Given some some functional operator* L, *we denote by* Γ_L *and* $\Gamma_{2,L}$ *the operators defined for any smooth functions* F_1, F_2 *on* $\mathbb{R}^r$ *by the formulae*

$$\Gamma_L(F_1, F_2) \; := \; L(F_1 F_2) - F_1 \, L(F_2) - F_2 \, L(F_1)$$
$$\Gamma_{2,L}(F_1, F_2) \; := \; L\left(\Gamma_L(F_1, F_2)\right) - \Gamma_L(F_1, L(F_2)) - \Gamma_L(L(F_1), F_2). \qquad (19.91)$$

The operators Γ_L and $\Gamma_{2,L}$ defined in (19.91) are often called the carré du champ and the gamma-two associated with the operator L.

As expected with second order differential operators, the following proposition shows that Γ_Δ is a symmetric bilinear form on smooth functions.

Theorem 19.9.2 (Bochner-Lichnerowicz formula) *For any smooth functions* F_1, F_2 *on* $\mathbb{R}^r$ *we have*

$$L := \frac{1}{2} \, \Delta \; \Longrightarrow \; \Gamma_L(F_1, F_2) = \langle \nabla F_1, \nabla F_2 \rangle.$$

In addition, we have the Bochner-Lichnerowicz formula

$$\Gamma_{2,L}(F_1, F_2) = \left\langle \nabla^2 F_1, \nabla^2 F_2 \right\rangle + \mathrm{Ric}\,(\nabla F_1, \nabla F_2). \qquad (19.92)$$

Proof :
We readily check that

$$\nabla(F_1 F_2) = F_1 \, \nabla F_2 + F_2 \, \nabla F_1.$$

Using (19.30) this implies that

$$
\begin{aligned}
\nabla_{V_i} \nabla (F_1 F_2) &= \nabla_{V_i} (F_1 \, \nabla F_2) + \nabla_{V_i} (F_2 \, \nabla F_1) \\
&= F_1 \, \nabla_{V_i} (\nabla F_2) + F_2 \, \nabla_{V_i} (\nabla F_1) + \partial_{V_i} (F_1) \, \nabla F_2 + \partial_{V_i} (F_2) \, \nabla F_1.
\end{aligned}
$$

Recalling $\nabla_{V_i} (\nabla F) = \langle V_l, \partial F \rangle = \langle V_l, \nabla F \rangle$ implies that

$$
\langle V_k, \nabla_{V_i} \nabla (F_1 F_2) \rangle
$$

$$
= F_1 \, \langle V_k, \nabla_{V_i} \nabla F_1 \rangle + F_2 \, \langle V_k, \nabla_{V_i} \nabla F_1 \rangle + (\langle V_l, \nabla F_1 \rangle \, \langle V_k, \nabla F_2 \rangle + \langle V_l, \nabla F_2 \rangle \, \langle V_k, \nabla F_1 \rangle) .
$$

Recalling (19.2) shows that

$$
\Gamma_\Delta (F_1, F_2) = 2 \sum_{1 \le k, l \le p} g^{k,l} \, \langle V_l, \nabla F_1 \rangle \, \langle V_k, \nabla F_2 \rangle \; = 2 \, \langle \nabla F_1, \nabla F_2 \rangle .
$$

Using (19.56) we have

$$
\begin{aligned}
\Delta \langle \nabla F_1, \nabla F_2 \rangle &= \sum_{1 \le k, l \le p} g^{k,l} \, \langle V_k, \nabla_{V_i} \nabla \langle \nabla F_1, \nabla F_2 \rangle \rangle \\
&= \sum_{1 \le k, l \le p} g^{k,l} \, \langle V_k, \nabla_{V_i} \nabla_{\nabla F_1} \nabla F_2 \rangle + \sum_{1 \le k, l \le p} g^{k,l} \, \langle V_k, \nabla_{V_i} \nabla_{\nabla F_2} \nabla F_1 \rangle .
\end{aligned}
$$

On the other hand, we have

$$
\begin{aligned}
\nabla_{V_i} \nabla_{\nabla F_1} &= \nabla_{\nabla F_1} \nabla_{V_i} + \nabla_{[V_i, \nabla F_1]} + \left([\nabla_{V_i}, \nabla_{\nabla F_1}] - \nabla_{[V_i, \nabla F_1]} \right) \\
&= \nabla_{\nabla F_1} \nabla_{V_i} + \nabla_{[V_i, \nabla F_1]} + R \left(V_l, \nabla F_1 \right) .
\end{aligned}
$$

This implies that

$$
\Delta \langle \nabla F_1, \nabla F_2 \rangle
$$

$$
= \sum_{1 \le k, l \le p} g^{k,l} \, \langle V_k, \left(\nabla_{\nabla F_1} \nabla_{V_i} + \nabla_{[V_i, \nabla F_1]} \right) \nabla F_2 \rangle
$$

$$
+ \sum_{1 \le k, l \le p} g^{k,l} \, \langle V_k, \left(\nabla_{\nabla F_2} \nabla_{V_i} + \nabla_{[V_i, \nabla F_2]} \right) \nabla F_1 \rangle + 2 \, \mathrm{Ric} \left(\nabla F_1, \nabla F_2 \right) .
$$

We observe that

$$
\begin{aligned}
\partial_{\nabla F_1} \langle V_k, \nabla_{V_i} \nabla F_2 \rangle &= \langle \nabla F_1, \nabla \langle V_k, \nabla_{V_i} \nabla F_2 \rangle \rangle \\
&= \langle V_k, \nabla_{\nabla F_1} \nabla_{V_i} \nabla F_2 \rangle + \langle \nabla_{\nabla F_1} V_k, \nabla_{V_i} \nabla F_2 \rangle .
\end{aligned}
$$

$$
\Rightarrow \langle V_k, \nabla_{\nabla F_1} \nabla_{V_i} \nabla F_2 \rangle = \langle \nabla F_1, \nabla \langle V_k, \nabla_{V_i} \nabla F_2 \rangle \rangle - \langle \nabla_{\nabla F_1} V_k, \nabla_{V_i} \nabla F_2 \rangle . \tag{19.93}
$$

On the other hand, we have

$$
\begin{aligned}
\partial_{\nabla F_1} \langle V_j, V_k \rangle &= \partial_{\nabla F_1} (g_{j,k}) = \langle \nabla F_1, \partial g_{j,k} \rangle = \langle \nabla F_1, \nabla g_{i,j} \rangle \\
&= \langle \nabla_{\nabla F_1} V_j, V_k \rangle + \langle V_j, \nabla_{\nabla F_1} V_k \rangle .
\end{aligned}
$$

By symmetry arguments, this clearly implies that for any $1 \le i, l \le r$

$$
\begin{aligned}
\sum_{1 \le j, k \le p} g^{i,j} \, g^{k,l} \, \langle V_j, \nabla_{\nabla F_1} V_k \rangle &= \frac{1}{2} \sum_{1 \le j, k \le p} g^{i,j} \, g^{k,l} \, \langle \nabla F_1, \nabla g_{j,k} \rangle \\
&= -\frac{1}{2} \sum_{1 \le j, k \le p} g_{j,k} \, g^{k,l} \, \langle \nabla F_1, \nabla g^{i,j} \rangle = -\frac{1}{2} \, \langle \nabla F_1, \nabla g^{i,l} \rangle .
\end{aligned}
$$

In the last assertion we used

$$\nabla \left(\sum_{1 \le j \le p} g^{i,j} g_{j,k} \right) = 0 = \sum_{1 \le j \le p} g^{i,j} \nabla g_{j,k} + \sum_{1 \le j \le p} g_{j,k} \nabla g^{i,j} \quad \text{and} \quad \sum_{1 \le k \le p} g_{j,k} \, g^{k,l} = 1_{j=l}.$$

Using (19.2), we have

$$\langle \nabla_{\nabla F_1} V_k, \nabla_{V_l} \nabla F_2 \rangle \quad = \quad \sum_{1 \le i,j \le p} g^{i,j} \, \langle \nabla_{V_l} \nabla F_2, V_i \rangle \, \langle \nabla_{\nabla F_1} V_k, V_j \rangle$$

from which we find the decomposition

$$\sum_{1 \le k,l \le p} g^{k,l} \langle \nabla_{\nabla F_1} V_k, \nabla_{V_l} \nabla F_2 \rangle \quad = \quad \sum_{1 \le i,l \le p} \left[\sum_{1 \le j,k \le p} g^{i,j} \, g^{k,l} \langle \nabla_{\nabla F_1} V_k, V_j \rangle \right] \langle \nabla_{V_l} \nabla F_2, V_i \rangle$$

$$= \quad -\frac{1}{2} \sum_{1 \le i,l \le p} \langle \nabla F_1, \nabla g^{i,l} \rangle \, \langle \nabla_{V_l} \nabla F_2, V_i \rangle. \qquad (19.94)$$

Combining this result with (19.93) we check that

$$\sum_{1 \le k,l \le p} g^{k,l} \langle V_k, \nabla_{\nabla F_1} \nabla_{V_l} \nabla F_2 \rangle$$

$$= \sum_{1 \le i,l \le p} g^{i,l} \langle \nabla F_1, \nabla \langle V_k, \nabla_{V_l} \nabla F_2 \rangle \rangle + \frac{1}{2} \sum_{1 \le i,l \le p} \langle \nabla F_1, \nabla g^{i,l} \rangle \, \langle \nabla_{V_l} \nabla F_2, V_i \rangle$$

$$= \left\langle \nabla F_1, \nabla \left[\sum_{1 \le i,l \le p} g^{i,l} \, \langle V_k, \nabla_{V_l} \nabla F_2 \rangle \right] \right\rangle - \frac{1}{2} \sum_{1 \le i,l \le p} \langle \nabla F_1, \nabla g^{i,l} \rangle \, \langle \nabla_{V_l} \nabla F_2, V_i \rangle.$$

We conclude that

$$\sum_{1 \le k,l \le p} g^{k,l} \langle V_k, \nabla_{\nabla F_1} \nabla_{V_l} \nabla F_2 \rangle = \langle \nabla F_1, \nabla \Delta F_2 \rangle - \frac{1}{2} \sum_{1 \le i,l \le p} \langle \nabla F_1, \nabla g^{i,l} \rangle \, \langle \nabla_{V_l} \nabla F_2, V_i \rangle.$$

$$(19.95)$$

Using the Hessian formula (19.51) we have the commutation property

$$\langle V_k, \nabla_{[V_l, \nabla F_1]} \nabla F_2 \rangle \quad = \quad \langle [V_l, \nabla F_1], \nabla_{V_k} \nabla F_2 \rangle$$

$$= \quad \langle \nabla_{V_l}(\nabla F_1) - \nabla_{\nabla F_1}(V_l), \nabla_{V_k} \nabla F_2 \rangle$$

$$= \quad \langle \nabla_{V_l} \nabla F_1, \nabla_{V_k} \nabla F_2 \rangle - \langle \nabla_{\nabla F_1} V_l, \nabla_{V_k} \nabla F_2 \rangle.$$

Using (19.94) and (19.72) we prove that

$$\sum_{1 \le k,l \le p} g^{k,l} \langle V_k, \nabla_{[V_l, \nabla F_1]} \nabla F_2 \rangle$$

$$= \sum_{1 \le k,l \le p} g^{k,l} \langle \nabla_{V_l} \nabla F_1, \nabla_{V_k} \nabla F_2 \rangle + \frac{1}{2} \sum_{1 \le i,l \le p} \langle \nabla F_1, \nabla g^{i,l} \rangle \, \langle \nabla_{V_l} \nabla F_2, V_i \rangle \qquad (19.96)$$

$$= \langle \nabla^2 F_1, \nabla^2 F_2 \rangle + \frac{1}{2} \sum_{1 \le i,l \le p} \langle \nabla F_1, \nabla g^{i,l} \rangle \, \langle \nabla_{V_l} \nabla F_2, V_i \rangle.$$

Combining (19.95) and (19.96) we conclude that

$$\sum_{1 \le k,l \le p} g^{k,l} \, \langle V_k, \left(\nabla_{\nabla F_1} \nabla_{V_l} + \nabla_{[V_l, \nabla F_1]} \right) \nabla F_2 \rangle = \langle \nabla F_1, \nabla \Delta F_2 \rangle + \langle \nabla^2 F_1, \nabla^2 F_2 \rangle.$$

This ends the proof of the theorem.   ∎

Corollary 19.9.3 *We let L be the second order operator defined by*

$$L = \frac{1}{2} \Delta - \nabla_W \tag{19.97}$$

for some vector field $W \in \mathcal{V}$. In this situation, we have $\Gamma_L = \Gamma_{\frac{1}{2}\Delta}$ and

$$\Gamma_{2,L}(F_1, F_2) = \left\langle \nabla^2 F_1, \nabla^2 F_2 \right\rangle + \mathrm{Ric}\left(\nabla F_1, \nabla F_2\right) + 2\left(\nabla W\right)_{sym}\left(\nabla F_1, \nabla F_2\right) \tag{19.98}$$

with the symmetric bilinear form $(\nabla W)_{sym}$ induced by the matrix

$$(\nabla W)_{sym} = \frac{1}{2}\left((\nabla W) + (\nabla W)^T\right).$$

In particular, for covariant gradient vector fields $W = \nabla V$, we have

$$L(F) = \frac{1}{2} \Delta(F) - \langle \nabla V, \nabla F \rangle$$

and

$$\Gamma_{2,L}(F_1, F_2) = \left\langle \nabla^2 F_1, \nabla^2 F_2 \right\rangle + \mathrm{Ric}\left(\nabla F_1, \nabla F_2\right) + 2\, \nabla^2 V\left(\nabla F_1, \nabla F_2\right). \tag{19.99}$$

Proof :
Since Γ_L measures the derivation rule defects, we clearly have $\Gamma_L = \Gamma_{\frac{1}{2}\Delta}$. To verify the second assertion we use the easily checked decomposition

$$\Gamma_{2,L}(F_1, F_2) - \Gamma_{2, \frac{1}{2}\Delta}(F_1, F_2) = \left\langle \nabla \nabla_W F_1, \nabla F_2 \right\rangle + \left\langle \nabla \nabla_W F_2, \nabla F_1 \right\rangle - \nabla_W \left\langle \nabla F_1, \nabla F_2 \right\rangle.$$

The end of the proof is now a direct consequence of the commutation property (19.54). The second assertion is a direct consequence of (19.98) since

$$(\nabla F_1)^T \left(\nabla^2 V + (\nabla^2 V)^T\right) \nabla F_2 = 2\, \nabla^2 V\, \left(\nabla F_1, \nabla F_2\right).$$

This ends the proof of the corollary. ∎

Using (19.77), we readily check the following change of variable formula.

Proposition 19.9.4 *Let L be the diffusion generator defined in (19.97). For any function F on $\mathbb{R}^n$ and any sequence $f = (f_1, \ldots, f_n)$ of functions f_i on $\mathbb{R}^r$ with $1 \leq i \leq n$, we set $\mathcal{F} = F(f)$. In this notation, we have*

$$L(\mathcal{F}) = \sum_{1 \leq m \leq n} \left(\partial_{x_m} F\right)(f)\, L(f_m) + \frac{1}{2} \sum_{1 \leq l, m \leq n} \left(\partial_{x_l, x_m} F\right)(f)\, \Gamma_L(f_l, f_m). \tag{19.100}$$

Applying the Bochner-Lichnerowicz formula (19.92) to the coordinate mappings $F_1 = \chi_i$ and $F_2 = \chi_j$ (or directly using (19.3)) and recalling that $\nabla \chi_i = \pi_i$ and $\Delta \chi_i = \mathrm{tr}\left(\nabla \pi_i\right) = -\mathbb{H}^i$ (cf. (19.74)), we check that

$$\mathrm{Ric}\left(\pi_i, \pi_j\right) + \left\langle \nabla \pi_i, \nabla \pi_j \right\rangle = \frac{1}{2}\left\{\Delta \langle \pi_i, \pi_j \rangle + \left\langle \pi_i, \nabla \mathbb{H}^j \right\rangle + \left\langle \pi_j, \nabla \mathbb{H}^i \right\rangle\right\}.$$

The next theorem provides a more explicit description of the operator $\Gamma_{2,L}$ in terms of the projection matrix π.

Theorem 19.9.5 *When $L := \frac{1}{2} \Delta$, for any $1 \le i, j \le r$ we have*

$$\Gamma_{2,L}(\chi_i, \chi_j) = \text{Ric}(\pi_i, \pi_j) + \langle \nabla \pi_i, \nabla \pi_j \rangle = \frac{1}{2} \left\{ \Delta \pi_i^j + \partial_{\pi_i}(\mathbb{H}^j) + \partial_{\pi_j}(\mathbb{H}^i) \right\}$$

and

$$\Gamma_{2,L}(F_1, F_2) = \text{tr}(\nabla \partial F_1 \, \nabla \partial F_2) + \sum_{1 \le i,j \le r} \Gamma_{2,L}(\chi_i, \chi_j) \, \partial_{x_i} F_1 \, \partial_{x_j} F_2 \qquad (19.101)$$

$$+ \sum_{1 \le i,j,k \le r} \nabla \pi_k(\pi_i, \pi_j) \left(\partial_{x_k} F_1 \, \partial_{x_i, x_j} F_2 + \partial_{x_k} F_2 \, \partial_{x_i, x_j} F_1 \right).$$

In addition, the same formula holds when we can replace the term $\nabla \pi_k(\pi_i, \pi_j)$ in the above display by $\left(\partial_{\pi_j} \pi_k^i - \frac{1}{2} \partial_{\pi_k} \pi_j^i \right)$.

Proof :
Firstly we observe that

$$\nabla F = \sum_{1 \le i \le r} \partial_{x_i}(F) \, \pi_i \Rightarrow \nabla_{V_k} \nabla F = \sum_{1 \le i \le r} \partial_{x_i}(F) \, \nabla_{V_k} \pi_i + \sum_{1 \le i \le r} \partial_{V_k}(\partial_{x_i} F) \, \pi_i.$$

Using (19.72) implies that

$$\langle \nabla^2 F_1, \nabla^2 F_2 \rangle$$

$$= \sum_{1 \le k,l \le p} g^{k,l} \langle \nabla_{V_l} \nabla F_1, \nabla_{V_l} \nabla F_2 \rangle$$

$$= \sum_{1 \le i,j \le r} \partial_{x_i} F_1 \, \partial_{x_j} F_2 \sum_{1 \le k,l \le p} g^{k,l} \langle \nabla_{V_k} \nabla \chi_i, \nabla_{V_l} \nabla \chi_j \rangle$$

$$+ \sum_{1 \le i,j \le r} \partial_{x_i} F_1 \sum_{1 \le k,l \le p} g^{k,l} \, \partial_{V_l}(\partial_{x_j} F_2) \, \langle \nabla_{V_k} \pi_i, \pi_j \rangle$$

$$+ \sum_{1 \le i,j \le r} \partial_{x_j} F_2 \sum_{1 \le k,l \le p} g^{k,l} \, \partial_{V_k}(\partial_{x_i} F_1) \, \langle \pi_i, \nabla_{V_l} \pi_j \rangle$$

$$+ \sum_{1 \le i,j \le r} \sum_{1 \le k,l \le p} g^{k,l} \, \partial_{V_k}(\partial_{x_i} F_1) \, \partial_{V_l}(\partial_{x_j} F_2) \, \langle \pi_i, \pi_j \rangle.$$

Recalling that

$$\sum_{1 \le k,l \le p} g^{k,l} V_k^m V_k^n = \pi_n^m \qquad V_k = \sum_{1 \le i \le r} V_k^i \, \pi_i \quad \text{and} \quad \langle \pi_i, \pi_j \rangle = \pi_j^i$$

we obtain the following decomposition

$$\langle \nabla^2 F_1, \nabla^2 F_2 \rangle$$

$$= \sum_{1 \le i,j \le r} \partial_{x_i} F_1 \, \partial_{x_j} F_2 \, \langle \nabla \pi_i, \nabla \pi_j \rangle + \sum_{1 \le i,j,m,n \le r} \pi_n^m \, \partial_{x_i} F_1 \, \partial_{x_n, x_j} F_2 \, \langle \nabla_{\pi_m} \pi_i, \pi_j \rangle$$

$$+ \sum_{1 \le i,j,m,n \le r} \pi_n^m \, \partial_{x_n, x_i} F_1 \, \partial_{x_j} F_2 \, \langle \pi_i, \nabla_{\pi_m} \pi_j \rangle + \sum_{1 \le i,j,m,n \le r} \pi_n^m \, \partial_{x_m, x_i} F_1 \, \pi_i^j \, \partial_{x_j, x_n} F_2.$$

Using $\sum_{1 \leq m \leq r} \pi_n^m \, \pi_m = \pi_n$, we also check that

$$
\begin{aligned}
\left\langle \nabla^2 F_1, \nabla^2 F_2 \right\rangle \;=\; & \sum_{1 \leq i,j \leq r} \partial_{x_i} F_1 \, \partial_{x_j} F_2 \, \left\langle \nabla \pi_i, \nabla \pi_j \right\rangle + \sum_{1 \leq i,j \leq r} \left(\pi \partial^2 F_1 \right)_{i,j} \, \left(\pi \partial^2 F_2 \right)_{j,i} \\
& + \sum_{1 \leq i,j,k \leq r} \left(\partial_{x_k} F_1 \, \partial_{x_i, x_j} F_2 + \partial_{x_k} F_2 \, \partial_{x_i, x_j} F_1 \right) \overbrace{\left\langle \nabla_{\pi_i} \pi_k, \pi_j \right\rangle}^{= \nabla \pi_k(\pi_i, \pi_j)}.
\end{aligned}
$$

By (19.68) and

$$
\nabla F = \sum_{1 \leq i \leq r} \partial_{x_i}(F) \, \pi_i \Rightarrow \mathrm{Ric}\,(\nabla F_1, \nabla F_2) = \sum_{1 \leq i,j \leq r} \partial_{x_i} F_1 \, \partial_{x_j} F_2 \, \mathrm{Ric}\,(\pi_i, \pi_j)
$$

we conclude that

$$
\begin{aligned}
\Gamma_{2,L}(F_1, F_2) \;=\; & \left\langle \nabla^2 F_1, \nabla^2 F_2 \right\rangle + \mathrm{Ric}\,(\nabla F_1, \nabla F_2) \\
=\; & \sum_{1 \leq i,j \leq r} \partial_{x_i} F_1 \, \partial_{x_j} F_2 \, \overbrace{\left(\left\langle \nabla \pi_i, \nabla \pi_j \right\rangle + \mathrm{Ric}\,(\pi_i, \pi_j) \right)}^{= \Gamma_{2,L}(\chi_i, \chi_j)} + \mathrm{tr}\,(\nabla \partial F_1 \, \nabla \partial F_2) \\
& + \sum_{1 \leq i,j,k \leq r} \left(\partial_{x_k} F_1 \, \partial_{x_i, x_j} F_2 + \partial_{x_k} F_2 \, \partial_{x_i, x_j} F_1 \right) \left(\partial_{\pi_j} \pi_k^i - \frac{1}{2} \, \partial_{\pi_k} \pi_j^i \right).
\end{aligned}
$$

This ends the proof of the theorem. ∎

Definition 19.9.6 *We consider the operator Υ_L defined for any smooth functions F_i on $\mathbb{R}^r$, with $i = 1,2,3$, by the formula*

$$
\Upsilon_L(F_3)(F_1, F_2) := \Gamma_{2,L}(F_1 F_2, F_3) - F_1 \, \Gamma_{2,L}(F_2, F_3) - F_2 \, \Gamma_{2,L}(F_1, F_3). \tag{19.102}
$$

Proposition 19.9.7 *When $L := \frac{1}{2} \, \Delta$, for any smooth functions F_i, with $i = 1,2,3$, we have*

$$
\begin{aligned}
\frac{1}{2} \, \Upsilon_L(F_3)(F_1, F_2) \;=\; & \left(\partial^2 F_3 + (\partial F_3)^T \nabla^2 \chi \right) (\nabla F_1, \nabla F_2) \\
=\; & \nabla^2 F_3 \, (\nabla F_1, \nabla F_2), \tag{19.103}
\end{aligned}
$$

with the bilinear form

$$
(\partial F_3)^T \nabla^2 \chi := \sum_{1 \leq i \leq r} \partial_{x_i} F_3 \, \nabla \pi_i.
$$

We consider a sequence of smooth functions $f_k = (f_{k,i})_{1 \leq i \leq n_k}$ on $\mathbb{R}^r$, and $F_k(x_1, \ldots, x_{n_k})$ on $\mathbb{R}^{n_k}$, and we set $\mathcal{F}_k = F_k(f_k) = F_k(f_{k,1}, \ldots, f_{k,n_k})$, with $k = 1,2$. In this situation, we have the change of variable formulae

$$
\Gamma_L(\mathcal{F}_1, \mathcal{F}_2) = \sum_{1 \leq i \leq n_1} \sum_{1 \leq k \leq n_2} (\partial_{x_i} F_1)(f_1) \, (\partial_{x_k} F_2)(f_2) \, \Gamma_L(f_{1,i}, f_{2,k})
$$

and

$$\Gamma_{2,L}(\mathcal{F}_1, \mathcal{F}_2) - \sum_{1 \le i \le n_1} \sum_{1 \le k \le n_2} (\partial_{x_i} F_1)(f_1)(\partial_{x_k} F_2)(f_2) \, \Gamma_{2,L}(f_{1,i}, f_{2,k})$$

$$= \sum_{1 \le i,j \le n_1} \sum_{1 \le k,l \le n_2} (\partial_{x_i, x_j} F_1)(f_1)(\partial_{x_k, x_l} F_2)(f_2) \; \Gamma_L(f_{1,i}, f_{2,l}) \; \Gamma_L(f_{1,j}, f_{2,k})$$

$$+ \frac{1}{2} \sum_{1 \le i \le n_1} \sum_{1 \le k,l \le n_2} (\partial_{x_i} F_1)(f_1)(\partial_{x_k, x_l} F_2)(f_2) \; \Upsilon_L(f_{1,i})(f_{2,k}, f_{2,l})$$

$$+ \frac{1}{2} \sum_{1 \le k \le n_2} \sum_{1 \le i,j \le n_1} (\partial_{x_k} F_2)(f_2)(\partial_{x_i, x_j} F_1)(f_1) \; \Upsilon_L(f_{2,k})(f_{1,i}, f_{1,j}).$$

Proof :
The first assertion is a direct consequence of (19.75). The proof of the second is based on the well known formula

$$\partial^2(F_1 F_2) = F_1 \, \partial^2 F_2 + F_2 \, \partial^2 F_1 + \partial F_1 (\partial F_2)^T + \partial F_2 (\partial F_1)^T.$$

This implies that

$$\begin{aligned}
\operatorname{tr}\left(\pi \partial^2(F_1 F_2) \, \pi \partial^2 F_3\right) &= F_1 \operatorname{tr}\left(\pi \partial^2 F_2 \, \pi \partial^2 F_3\right) + F_2 \operatorname{tr}\left(\pi \partial^2 F_1 \, \pi \partial^2 F_3\right) \\
&\quad + \operatorname{tr}\left(\pi \partial F_1 (\partial F_2)^T \, \pi \partial^2 F_3\right) + \operatorname{tr}\left(\pi \partial F_2 (\partial F_1)^T \, \pi \partial^2 F_3\right).
\end{aligned}$$

We readily check that

$$\begin{aligned}
\operatorname{tr}\left(\pi \partial F_1 (\partial F_2)^T \, \pi \partial^2 F_3\right) &= \operatorname{tr}\left(\pi \partial F_1 (\pi \partial F_2)^T \, \partial^2 F_3\right) \\
&= \langle \nabla F_1, (\partial^2 F_3) \, \nabla F_2 \rangle = \operatorname{tr}\left(\pi \partial F_2 (\partial F_1)^T \, \pi \partial^2 F_3\right).
\end{aligned}$$

On the other hand, recalling that $\sum_{1 \le i \le r}(\partial_{x_i} F) \, \pi_i = \nabla F$, we have

$$\sum_{1 \le i,j,k \le r} \partial_{x_k} F_3 \, \nabla \pi_k(\pi_i, \pi_j) \, \partial_{x_i} F_1 \, \partial_{x_j} F_2 = \sum_{1 \le k \le r} \partial_{x_k} F_3 \, \nabla \pi_k(\nabla F_1, \nabla F_2).$$

The end of the proof of the first assertion is now clear. The second covariant derivative formula (19.103) is a direct consequence of (19.69).

$$\mathcal{F}_k = F_k(f_k) = F_k(f_{k,1}, \ldots, f_{k,n_k}) \Rightarrow \partial \mathcal{F}_k = \sum_{1 \le i \le n_k} (\partial_{x_i} F_k)(f_k) \, \partial f_{k,i}.$$

Arguing as before, we have the matrix formula

$$\partial^2 \mathcal{F}_k = \sum_{1 \le i,j \le n_k} (\partial_{x_i, x_j} F_k)(f_k) \, \partial f_{k,i} \partial f_{k,j}^T + \sum_{1 \le i \le n_k} (\partial_{x_i} F_k)(f_k) \, \partial^2 f_{k,i}.$$

This implies that

$$\text{tr}\left(\nabla \partial \mathcal{F}_1 \, \nabla \partial \mathcal{F}_2\right)$$

$$= \sum_{1 \le i,j \le n_1} \sum_{1 \le k,l \le n_2} \left(\partial_{x_i,x_j} F_1\right)(f_1)\left(\partial_{x_k,x_l} F_2\right)(f_2) \,\, \text{tr}\left(\nabla f_{1,i}\partial f_{1,j}^T \, \nabla f_{2,k}\partial f_{2,l}^T\right)$$

$$+ \sum_{1 \le i,j \le n_1} \sum_{1 \le k \le n_2} \left(\partial_{x_i,x_j} F_1\right)(f_1)\left(\partial_{x_k} F_2\right)(f_2) \, \text{tr}\left(\nabla f_{1,i}\partial f_{1,j}^T \, \pi \partial^2 f_{2,k}\right)$$

$$+ \sum_{1 \le i \le n_1} \sum_{1 \le k,l \le n_2} \left(\partial_{x_i} F_1\right)(f_1)\left(\partial_{x_k,x_l} F_2\right)(f_2) \, \text{tr}\left(\nabla f_{2,k}\partial f_{2,l}^T \, \pi \partial^2 f_{1,i}\right)$$

$$+ \sum_{1 \le i \le n_1} \sum_{1 \le k \le n_2} \left(\partial_{x_i} F_1\right)(f_1)\left(\partial_{x_k} F_2\right)(f_2) \, \text{tr}\left(\nabla \partial f_{1,i} \, \nabla \partial f_{2,k}\right).$$

Notice that

$$
\begin{aligned}
\text{tr}\left(\nabla f_{1,i}\partial f_{1,j}^T \, \nabla f_{2,k}\partial f_{2,l}^T\right) &= \text{tr}\left(\nabla f_{2,k}\partial f_{2,l}^T \nabla f_{1,i}\partial f_{1,j}^T\right) \\
&= \langle \partial f_{2,l}, \nabla f_{1,i}\rangle \, \langle \partial f_{2,k}, \nabla f_{1,j}\rangle = \langle \nabla f_{2,l}, \nabla f_{1,i}\rangle \, \langle \nabla f_{2,k}, \nabla f_{1,j}\rangle \\
&= \Gamma_L\left(f_{1,i}, f_{2,l}\right) \, \Gamma_L\left(f_{1,j}, f_{2,k}\right)
\end{aligned}
$$

and

$$
\begin{aligned}
\text{tr}\left(\nabla f_{1,i}\partial f_{1,j}^T \, \pi \partial^2 f_{2,k}\right) &= \text{tr}\left(\nabla f_{1,i} \left(\pi \partial f_{1,j}^T \, \partial^2 f_{2,k}\pi\right)\right) \\
&= \langle \nabla f_{1,i}, \partial^2 f_{2,k}\nabla f_{1,j}\rangle = \partial^2 f_{2,k}\left(\nabla f_{1,i}, \nabla f_{1,j}\right).
\end{aligned}
$$

Arguing as above we check that

$$\Gamma_{2,L}(\mathcal{F}_1, \mathcal{F}_2) - \sum_{1 \le i \le n_1} \sum_{1 \le k \le n_2} \left(\partial_{x_i} F_1\right)(f_1)\left(\partial_{x_k} F_2\right)(f_2) \, \Gamma_{2,L}(f_{1,i}, f_{2,k})$$

$$= \sum_{1 \le i,j \le n_1} \sum_{1 \le k,l \le n_2} \left(\partial_{x_i,x_j} F_1\right)(f_1)\left(\partial_{x_k,x_l} F_2\right)(f_2)\Gamma_L\left(f_{1,i}, f_{2,l}\right) \, \Gamma_L\left(f_{1,j}, f_{2,k}\right)$$

$$+ \sum_{1 \le i \le n_1} \sum_{1 \le k,l \le n_2} \left(\partial_{x_i} F_1\right)(f_1)\left(\partial_{x_k,x_l} F_2\right)(f_2)\left[\partial^2 f_{1,i} + \sum_{1 \le m \le r} \partial_{x_m} f_{1,i} \, \nabla \pi_m\right]\left(\nabla f_{2,k}, \nabla f_{2,l}\right)$$

$$+ \sum_{1 \le k \le n_2} \sum_{1 \le i,j \le n_1} \left(\partial_{x_k} F_2\right)(f_2)\left(\partial_{x_i,x_j} F_1\right)(f_1)\left[\partial^2 f_{2,k} + \sum_{1 \le i \le r} \partial_{x_m} f_{2,k} \, \nabla \pi_m\right]\left(\nabla f_{1,i}, \nabla f_{1,j}\right).$$

This completes the proof of the proposition. ∎

19.10 Exercises

Exercise 351 (Euclidian space) *We consider the two-dimensional euclidian space* $\mathbb{R}^2 = \mathcal{V} = \mathrm{Vect}(e_1, e_2)$ *generated by the unit vectors* $e_1 = \begin{pmatrix} 1 \\ 0 \end{pmatrix}$ *and* $e_2 = \begin{pmatrix} 0 \\ 1 \end{pmatrix}$. *Describe the covariant derivatives* ∇F, $\nabla^2 F$ *and the Laplacian* ΔF *of a smooth function* F *on* $\mathbb{R}^2$.

Exercise 352 (Bochner-Lichnerowicz formula in Euclidian spaces) *We consider the two-dimensional euclidian space* $\mathbb{R}^2 = \mathcal{V}$ *discussed in exercise 351. For a smooth function* F *on* $\mathbb{R}^2$ *check that*

$$\frac{1}{2} \Delta \left(\|\nabla F\|^2 \right) = \mathrm{tr} \left(\nabla^2 F \, \nabla^2 F \right) + \frac{1}{2} \left\langle \nabla F, \nabla(\Delta F) \right\rangle.$$

Exercise 353 (Carré de champ operator) *We consider first and second order generators* L_1 *and* L_2 *defined (on smooth functions* f *on* $\mathbb{R}^r$) *by* $L = L_1 + L_2$, *with the first and second order operators*

$$L_1(f) := \sum_{1 \leq i \leq r} b^i \, \partial_i(f) \quad and \quad L_2(f) := \frac{1}{2} \sum_{1 \leq i,j \leq r} a^{i,j} \, \partial_{i,j}(f)$$

for some drift function $b = (b^i)_{1 \leq i \leq r}$ *and some symmetric matrix functional* $a = (a^{i,j})_{1 \leq i,j \leq r}$ *on* $\mathbb{R}^r$. *Check that*

$$L_1(fg) = f \, L_1(g) + g \, L_1(f)$$

and

$$L_2(fg) = f \, L_2(g) + g \, L_2(f) + \Gamma_{L_2}(f, g) \quad with \quad \Gamma_{L_2}(f, g) := \sum_{1 \leq i,j \leq r} a^{i,j} \, \partial_i f \, \partial_j g$$

for smooth functions (f, g). *Deduce that*

$$L(fg) = f \, L(g) + g \, L(f) + \Gamma_L(f, g) \quad with \quad \Gamma_L(f, g) = \Gamma_{L_2}(f, g).$$

To avoid complex summation formulae, we recommend Einstein notation and write $b^i \, \partial_i$ *and* $a^{i,j} \, \partial_{i,j}$ *instead of* $\sum_{1 \leq i \leq r} b^i \, \partial_i$ *and* $\sum_{1 \leq i,j \leq r} a^{i,j} \, \partial_{i,j}$.

Exercise 354 (Change of variable formulae) *We consider the differential operator* L *discussed in exercise 353. Check that* $(f, g) \mapsto \Gamma_L(f, g)$ *is a symmetric bilinear form satisfying the property*

$$\Gamma_L(f, gh) = g \, \Gamma_L(f, h) + h \, \Gamma_L(f, g).$$

We consider collections of smooth functions $f = (f^i)_{1 \leq i \leq n}$ *and* $g = (g^j)_{1 \leq j \leq m}$ *on* $\mathbb{R}^r$, *and some smooth functions* $F(x_1, \ldots, x_n)$ *and* $G(x_1, \ldots, x_m)$ *on* $\mathbb{R}^n$ *and* $\mathbb{R}^m$, *for some* $m, n \geq 1$. *We also set*

$$F(f) = F(f^1, \ldots, f^n) \quad and \quad G(g) = G(g^1, \ldots, g^m).$$

Check that

$$\Gamma_L \left(F(f), G(g) \right) = \sum_{1 \leq k \leq n} \sum_{1 \leq l \leq m} (\partial_k F)(f)(\partial_l G)(g) \, \Gamma_L(f^k, g^l)$$

and

$$L(F(f)) = \sum_{1 \leq k \leq n} (\partial_k F)(f) \, L(f^k) + \frac{1}{2} \sum_{1 \leq k,l \leq n} (\partial_{k,l} F)(f) \, \Gamma_L(f^k, f^l).$$

Exercise 355 (Gamma-two operator) *We consider the differential operator L discussed in exercise 353 and exercise 354. Check that*

$$\sum_{1 \le i,j \le r} a^{i,j} \; \partial_i f \; L(\partial_j g) = \sum_{1 \le k \le r} b^k \; \Gamma_L(f, \partial_k g) + \frac{1}{2} \sum_{1 \le k,l \le r} a^{k,l} \; \Gamma_L(f, \partial_{k,l} g).$$

Deduce that

$$\Gamma_L(f, L(g)) + \Gamma_L(g, L(f)) + \sum_{1 \le i,j \le r} a^{i,j} \; [\Gamma_L(\partial_i f, \partial_j g) - L(\partial_i f \partial_j g)]$$

$$= \sum_{1 \le i \le r} \left[\{ \Gamma_L(b^i, f) \; \partial_i g + \Gamma_L(b^i, g) \; \partial_i f \} + \frac{1}{2} \sum_{1 \le j \le r} \{ \Gamma_L(a^{i,j}, f) \; \partial_{i,j} g + \Gamma_L(a^{i,j}, g) \; \partial_{i,j} f \} \right].$$

Prove the formula

$$L\left(\Gamma_L(f,g) \right) - \sum_{1 \le i,j \le r} a^{i,j} \; L\left(\partial_i f \partial_j g \right)$$

$$= \sum_{1 \le i,j \le r} \left[L\left(a^{i,j} \right) \; \partial_i f \; \partial_j g + \Gamma_L\left(a^{i,j}, \partial_j g \right) \; \partial_i f + \Gamma_L\left(a^{i,j}, \partial_i f \right) \; \partial_j g \right]$$

and deduce that

$$\Gamma_{2,L}(f,g) = \sum_{1 \le i,j \le r} L\left(a^{i,j} \right) \; \partial_i f \; \partial_j g - \sum_{1 \le i \le r} \{ \Gamma_L(b^i, f) \; \partial_i g + \Gamma_L(b^i, g) \; \partial_i f \}$$

$$+ \sum_{1 \le i,j \le r} \left\{ \left[\Gamma_L\left(a^{i,j}, \partial_i g \right) \; \partial_j f + \Gamma_L\left(a^{i,j}, \partial_i f \right) \; \partial_j g \right] \right.$$

$$\left. - \frac{1}{2} \left[\Gamma_L(a^{i,j}, f) \; \partial_{i,j} g + \Gamma_L(a^{i,j}, g) \; \partial_{i,j} f \right] \right\} + \sum_{1 \le i,j \le r} a^{i,j} \; \Gamma_L(\partial_i f, \partial_j g).$$

Exercise 356 (Gamma-two operator - Explicit formula) *We consider the differential operator L discussed in the exercises 353 through 355. Check the explicit formula*

$$\Gamma_{2,L}(f,g) = \sum_{1 \le i,j \le r} \left(L\left(a^{i,j} \right) - \sum_{1 \le l \le r} \left[a^{i,l} \; \partial_l b^j + a^{j,l} \; \partial_l b^i \right] \right) \partial_i f \; \partial_j g$$

$$+ \sum_{1 \le i,j,k \le r} \sum_{1 \le l \le r} \left[a^{k,l} \; \partial_l a^{i,j} - \frac{1}{2} \; a^{i,l} \; \partial_l a^{k,j} \right] \; [\partial_i f \; \partial_{j,k} g + \partial_i g \; \partial_{j,k} f]$$

$$+ \sum_{1 \le i,j,k,l \le r} a^{i,j} \; a^{k,l} \; \partial_{i,k} f \; \partial_{j,l} g. \tag{19.104}$$

Exercise 357 (Hessian operator) *We consider the differential operator L discussed in the exercises 353 through 355. Check the formula*

$$\begin{aligned} \Upsilon_L(f)(g,h) \; &:= \; \Gamma_{2,L}(f, gh) - h \; \Gamma_{2,L}(f,g) - g \; \Gamma_{2,L}(f,h) \\ &= \; \Gamma_L\left(g, \Gamma_L(f,h) \right) + \Gamma_L\left(h, \Gamma_L(f,g) \right) - \Gamma_L\left(f, \Gamma_L(g,h) \right). \end{aligned}$$

Exercise 358 (Hessian operator - Explicit formula) *We consider the differential operator L discussed in the exercises 353 through 356. Check the explicit formula*

$$\Upsilon_L(f)(g,h)$$

$$= \sum_{1 \leq i,j,k,l \leq r} \left(\left(a^{k,l} \, \partial_l a^{i,j} - \frac{1}{2} \, a^{i,l} \, \partial_l a^{k,j} \right) \, \partial_i f + a^{i,j} \, a^{k,l} \, \partial_{i,l} f \right) \, (\partial_j g \, \partial_k h + \partial_j h \, \partial_k g).$$

Exercise 359 (Projection matrix) *We consider the plane $\mathcal{V} = \{(x_1, x_2, x_3) \in \mathbb{R}^3 \, : \, x_1 + x_2 + x_3 = 0\}$. Find an orthogormal basis (U_1, U_2) of the two-dimensional vector space $\mathcal{V}$ and compute the orthogonal projection matrix $\pi_{\mathcal{V}}$.*

Exercise 360 (Normal vector fields) *We consider a smooth level surface $S = \{x = (x_1, x_2, x_3) \in \mathbb{R}^3 \, : \, \varphi(x) = 0\}$, with a continuous gradient $x \in \mathbb{R}^3 \mapsto (\partial \varphi)(x)$ which is non-null at any point $x \in S$ of the surface. Prove that the vector $(\partial \varphi)(x)$ is orthogonal to each tangent vector at any state $x \in S$. Find the equations of the tangent planes to the following surfaces at some point $x = (x_1, x_2, x_3) \in S$:*

- *The hyperboloid $\varphi(x) = x_1^2 - x_2^2 - x_3^2 - 4 = 0$.*

- *The circular cone $\varphi(x) = x_1^2 + x_2^2 - x_3^2 = 0$.*

Exercise 361 (Parametric surfaces) *Find the equation of the tangent plane and the surface unit vector of the torus given by the parametrization*

$$\psi(\theta) = \begin{pmatrix} (R + r\cos(\theta_1))\cos(\theta_2) \\ (R + r\cos(\theta_1))\sin(\theta_2) \\ r\sin(\theta_1) \end{pmatrix} \quad \text{with} \quad r < R$$

at some point $\psi(\theta) = x = (x_1, x_2, x_3)$.

Exercise 362 (Monge parametrization) *We consider a surface*

$$S = \{x = (x_1, x_2, x_3) \in \mathbb{R}^3, \ \varphi(x) = 0 \in \mathbb{R}^2\}$$

with $\varphi(x) = h(x_1, x_2) - x_3$ and some height function $\theta = (\theta_1, \theta_2) \in \mathbb{R}^2 \mapsto h(\theta) \in \mathbb{R}$. Give two natural tangent vectors $V_1(x), V_2(x)$ and an orthogonal vector $V_1^\perp(x)$ at some point $x = (x_1, x_2, x_3) = (x_1, x_2, h(x_1, x_2)) \in S$. Describe the metrics $g(x) = (g_{i,j}(x))_{1 \leq i,j \leq 2}$ and $g_\perp(x)$ on the tangent space $T_x(S) = \text{Vect}(V_1(x), V_2(x))$ and the orthogonal space $T_x^\perp(S) = \text{Vect}(V_1^\perp(x))$.

Exercise 363 *We consider the surface S presented in the exercise 362. Describe the projection matrices $\pi(x)$ and $\pi_\perp(x)$ on $T_x(S)$ and on $T_x^\perp(S)$ in terms of derivatives of the height function and in terms of the vector $(\partial \varphi)_x$.*

Exercise 364 *We consider the surface S presented in the exercise 362. Describe the covariant derivative $(\nabla F)(x)$ of a smooth function F on S with respect to the tangent space $T_x(S)$. Compute the mean curvature vector $\mathbb{H}(x)$ and deduce a formula for the Laplacian ΔF in terms of $\mathbb{H}(x)$.*

20

Stochastic differential calculus on manifolds

This chapter is concerned with the construction of Brownian motion and more general diffusion processes evolving in constraint type manifolds. These stochastic processes are defined in terms of projection of Euclidian Brownian motions and multidimensional diffusions adjusted by mean curvature drift functions. We also discuss the Doeblin-Itō formula associated with these manifold valued diffusions in the ambient space. The chapter also provides a brief introduction to the Stratonovitch differential calculus. We illustrate these probabilistic models with a variety of concrete examples, including Brownian motion and diffusions on the sphere, the cylinder, the simplex or the more general orbifold. The expressions of these stochastic diffusions on local coordinates and parameter spaces are discussed in chapter 22.

> *A circle is a round straight line with a hole in the middle.*
> Mark Twain (1971-1910).

20.1 Embedded manifolds

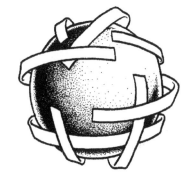

In this section we briefly recall some terminology that is used frequently in geometry and differential calculus. We only consider submanifolds S of dimension p which are "smooth subsets" of the ambient Euclidian space $\mathbb{R}^{p+q}$, for some $q \geq 1$. The case $q = 1$ corresponds to hypersurfaces (a.k.a. hypermanifolds). We let

$$\varphi : x \in \mathbb{R}^{r=p+q} \mapsto \varphi(x) = (\varphi_1(x), \ldots, \varphi_q(x))^T \in \mathbb{R}^q$$

be a smooth function with a non empty and connected null-level set $S := \varphi^{-1}(0)$ s.t.

$$\forall x \in S \qquad \text{rank}\,(\partial\varphi(x)) = \text{rank}\,(\partial\varphi_1(x), \ldots, \partial\varphi_q(x)) = q.$$

We consider a smooth curve $C : t \in [0,1] \mapsto C(t) = (C^1(t), \ldots, C^r(t))^T \in S$ starting at some state $C(0) = x \in S$, with a velocity vector field W, that is,

$$\frac{dC}{dt} = \left(\frac{dC^1}{dt}, \ldots, \frac{dC^r}{dt}\right)^T = W(C(t)) = (W^1(C(t)), \ldots, W^r(C(t)))^T.$$

By construction, we have

$$\forall 1 \leq i \leq q \qquad \frac{d}{dt}\varphi_i(C(t)) = \sum_{1 \leq j \leq r} (\partial_{x_j}\varphi_i)(C(t))\, W^j(C(t)) = 0.$$

For $t = 0$, this implies that

$$\langle \partial \varphi_i(x), W(x) \rangle = \sum_{1 \le j \le r} \left(\partial_{x_j} \varphi_i \right)(x) \, W^j(x) = 0 \quad \Longleftrightarrow \quad W(x) \in \ker \left(\partial \varphi_i(x) \right).$$

We let $T_x(S)$ be the vector space spanned by the kernels $\ker \left(\partial \varphi_i(x) \right)$ of the gradient vectors $\partial \varphi_i(x)$, with $1 \le i \le q$, that is,

$$T_x(S) = \text{Vect} \left(\cup_{1 \le i \le q} \ker \left(\partial \varphi_i(x) \right) \right).$$

Under our assumptions, we have

$$\mathbb{R}^r = T_x(S) \overset{\perp}{+} T_x^\perp(S) \quad \text{with} \quad T_x^\perp(S) = \text{Vect} \left(\partial \varphi_1(x), \ldots, \partial \varphi_q(x) \right).$$

This implies that $T_x(S)$ is a p-dimensional vector space, so that S is a p-dimensional manifold embedded in the ambient space $\mathbb{R}^r$.

Definition 20.1.1 *We let $\pi(x)$ be the orthogonal projection from $\mathbb{R}^r$ into $T_x(S)$, and $\mathbb{H}(x)$ be the mean curvature vector given by*

$$\mathbb{H} = \sum_{1 \le i \le q} \text{div}_\perp \left(\sum_{1 \le j \le q} g_\perp^{i,j} \, \partial \varphi_j \right) \partial \varphi_i.$$

Here

$$g_\perp^{-1} = \left(g_\perp^{i,j} \right)_{1 \le i,j \le q} \quad \text{and} \quad g_\perp = (g_{\perp,i,j})_{1 \le i,j \le q} = (\langle \partial \varphi_i, \partial \varphi_j \rangle)_{1 \le i,j \le q}$$

where $\text{div}_\perp(.)$ denotes the divergence operator defined in (19.42). We also recall from (19.45) that

$$\forall 1 \le k \le r \qquad \mathbb{H}^k = - \sum_{1 \le i,j \le r} \pi_j^i \partial_{x_i} \pi_k^j = - \sum_{1 \le j \le r} \partial_{\pi_j} \pi_j^k \, \Leftrightarrow \, \mathbb{H}^T = - \sum_{1 \le j \le r} \partial_{\pi_j} \pi_j$$

with the vector fields π_j on $\mathbb{R}^r$ defined by the column vectors

$$\forall 1 \le j \le r \qquad \pi_j := \begin{bmatrix} \pi_j^1 \\ \vdots \\ \pi_j^r \end{bmatrix}.$$

By construction, we also notice that

$$\forall 1 \le i \le q \qquad \partial \varphi_i \in T^\perp(S) \quad \Longrightarrow \quad \nabla \varphi_i = \pi(\partial \varphi_i) = 0. \qquad (20.1)$$

In the case of orthogonal constraints

$$\langle \partial \varphi_i, \partial \varphi_j \rangle = 1_{i=j} \, \|\partial \varphi_j\|^2 \Rightarrow g_\perp^{i,j} = 1_{i=j} \, \|\partial \varphi_j\|^{-2} \Rightarrow \mathbb{H} = \sum_{1 \le i \le q} \text{div}_\perp \left(\frac{\partial \varphi_i}{\|\partial \varphi_i\|^2} \right) \partial \varphi_i.$$

In addition, by (19.42), we find the computationally useful formula

$$q = 1 \Rightarrow \ \mathbb{H} \ = \ \operatorname{div}_\perp \left(\frac{\partial \varphi_1}{\|\partial \varphi_1\|^2} \right) \partial \varphi_1 = \left[\sum_{1 \leq m \leq r} \partial_{x_m} \left(\frac{\partial_{x_m} \varphi_1}{\|\partial \varphi_1\|} \right) \right] \frac{\partial \varphi_1}{\|\partial \varphi_1\|}. \qquad (20.2)$$

In the special case of the sphere $S = \mathbb{S}^p \subset \mathbb{R}^{p+1}$, with $p = r - 1$, we use

$$\varphi(x) = \|x\| - 1 \Rightarrow \partial \varphi(x) = x/\|x\| \quad \text{and} \quad \pi(x) = Id - \partial \varphi(x) \partial \varphi(x)^T = Id - \frac{x x^T}{x^T x}. \quad (20.3)$$

In this situation, for any $x \neq 0$ we have

$$\mathbb{H}(x) = \left[\sum_{1 \leq m \leq r} \partial_{x_m} \left(\frac{x_m}{\sqrt{x_1^2 + \ldots + x_r^2}} \right) \right] \frac{x}{\sqrt{x_1^2 + \ldots + x_r^2}} = (r - 1) \frac{x}{\|x\|^2} = (r - 1) \frac{x}{x^T x}.$$
$$(20.4)$$

We check this claim by using

$$\partial_{x_m} \left(\frac{x_m}{\sqrt{x_1^2 + \ldots + x_r^2}} \right) = \frac{1}{\sqrt{x_1^2 + \ldots + x_r^2}} \left[1 - \frac{x_m^2}{(x_1^2 + \ldots + x_r^2)} \right]$$

$$\Rightarrow \operatorname{div}_\perp \left(\frac{\partial \varphi}{\|\partial \varphi\|^2} \right) = \frac{1}{\|x\|} \sum_{1 \leq m \leq r} \partial_{x_m} \left(\frac{x_m}{\|x\|} \right) = \frac{1}{\|x\|^2} \sum_{1 \leq m \leq r} \left(1 - \frac{x_m^2}{\|x\|^2} \right) = \frac{r - 1}{\|x\|^2}.$$

For $p = 2$, the projection on the unit sphere can also be represented in terms of the cross product

$$\pi(x) W(x) = \frac{x}{\|x\|} \wedge W_x \stackrel{x \in \mathbb{S}^2}{=} x \wedge W_x = \begin{pmatrix} x_2 W^3(x) - x_3 W^2(x) \\ W^1(x) x_2 - W^2(x) x_1 \\ x_1 W^2(x) - x_2 W^1(x) \end{pmatrix}. \qquad (20.5)$$

In much the same way, the cylinder of unit radius on $\mathbb{R}^2$ is given $\varphi(x_1, x_2, x_3) = \sqrt{x_1^2 + x_2^2} - 1 = 0$. In this case, we have

$$\partial \varphi(x) = \frac{1}{\sqrt{x_1^2 + x_2^2}} \begin{pmatrix} x_1 \\ x_2 \\ 0 \end{pmatrix} \quad , \quad \pi(x) = Id - \partial \varphi(x) \partial \varphi(x)^T = \begin{pmatrix} \frac{x_2^2}{x_1^2 + x_2^2} & -\frac{x_1 x_2}{x_1^2 + x_2^2} & 0 \\ -\frac{x_1 x_2}{x_1^2 + x_2^2} & \frac{x_1^2}{x_1^2 + x_2^2} & 0 \\ 0 & 0 & 1 \end{pmatrix}$$
$$(20.6)$$

and

$$\mathbb{H}(x) = \left(\sum_{m=1,2} \partial_{x_m} (\partial_{x_m} \varphi)(x) \right) \partial \varphi(x) = \frac{1}{x_1^2 + x_2^2} \begin{pmatrix} x_1 \\ x_2 \\ 0 \end{pmatrix}.$$

20.2 Brownian motion on manifolds

20.2.1 A diffusion model in the ambient space

We consider the embedded manifold $S = \varphi^{-1}(0)$ described in section 20.1. We let π and $\mathbb{H}$ be the projection matrix and the mean curvature vector described in definition 20.1.1.

We let $X_t = \begin{pmatrix} X_t^1 \\ \vdots \\ X_t^r \end{pmatrix}$ be the $\mathbb{R}^r$-valued diffusion defined by

$$dX_t = -\frac{1}{2}\, \mathbb{H}(X_t)\, dt + \pi(X_t)\, dB_t \tag{20.7}$$

where B_t stands for a standard r-dimensional Brownian motion. The process X_t is called the Brownian motion on the manifold $S = \varphi^{-1}(0)$.

In the special case of the sphere $S = \mathbb{S}^p \subset \mathbb{R}^{r=p+1}$ we have

$$dX_t = -\frac{r-1}{2}\, \frac{X_t}{X_t^T X_t}\, dt + \left(Id - \frac{X_t X_t^T}{X_t^T X_t} \right)\, dB_t. \tag{20.8}$$

In terms of the cross product (20.5) we have

$$dX_t = -\frac{r-1}{2}\, \frac{X_t}{X_t^T X_t}\, dt + \frac{X_t}{\sqrt{X_t^T X_t}} \wedge dB_t.$$

The graph below illustrates a realization of a Brownian motion on a unit sphere.

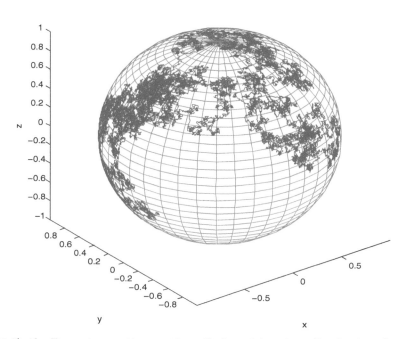

Using (20.6), the Brownian motion on the cylinder with unit radius is given by

$$\begin{cases} dX_t^1 &=& -\frac{1}{2}\, \frac{X_t^1}{(X_t^1)^2+(X_t^2)^2}\, dt + \left(\frac{(X_t^2)^2}{(X_t^1)^2+(X_t^2)^2}\, dB_t^1 - \frac{X_t^1 X_t^2}{(X_t^1)^2+(X_t^2)^2}\, dB_t^2 \right) \\ dX_t^2 &=& -\frac{1}{2}\, \frac{X_t^2}{(X_t^1)^2+(X_t^2)^2}\, dt + \left(-\frac{X_t^1 X_t^2}{(X_t^1)^2+(X_t^2)^2}\, dB_t^1 + \frac{(X_t^1)^2}{(X_t^1)^2+(X_t^2)^2}\, dB_t^2 \right) \\ dX_t^3 &=& dB_t^3. \end{cases} \tag{20.9}$$

The graph below illustrates a realization of a Brownian motion on a cylinder of unit radius.

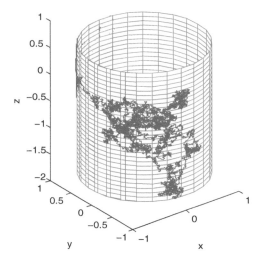

20.2.2 The infinitesimal generator

Recalling that $\pi = \pi^T$, for any $1 \leq k \leq r$ we have

$$dX_t^k = -\frac{1}{2}\,\mathbb{H}^k(X_t)\,dt + \sum_{1 \leq j \leq r} \pi_j^k(X_t)\,dB_t^j = \sum_{1 \leq j \leq r}\left[\frac{1}{2}\,\partial_{\pi_j}(\pi_j^k)(X_t)\,dt + \pi_j^k(X_t)\,dB_t^j\right].$$

Notice that

$$dX_t^k dX_t^l \sum_{1 \leq i,j \leq r} \pi_i^k(X_t)\pi_j^l(X_t)\,dB_t^i dB_t^j \simeq \underbrace{\sum_{1 \leq i \leq r}(\pi_i^k\pi_i^l)(X_t)\,dt}_{:=(\pi\pi^T)_t^k} = \pi_l^k(X_t)\,dt.$$

Using Ito's formula, for any smooth function F on $\mathbb{R}^r$ we have

$$\begin{aligned}
dF(X_t) &= \sum_{1 \leq k \leq r} \partial_{x_k}(F)(X_t)\,dX_t^k + \frac{1}{2}\sum_{1 \leq k,l \leq r} \partial_{x_k,x_l}(F)(X_t)\,dX_t^k dX_t^l \\
&= \langle \partial F(X_t), dX_t\rangle + \frac{1}{2}\mathrm{tr}\left(\partial^2 F(X_t)\,dX_t dX_t^T\right) \\
&= \langle \partial F(X_t), dX_t\rangle + \frac{1}{2}\mathrm{tr}\left(\pi(X_t)\partial^2 F(X_t)\right)\,dt = L(F)(X_t)\,dt + dM_t(F).
\end{aligned}$$

The infinitesimal generator L is defined by

$$\begin{aligned}
L(F) &= \frac{1}{2}\left[\mathrm{tr}\left(\pi\partial^2 F\right) - \partial_{\mathbb{H}}F\right] = \frac{1}{2}\sum_{1 \leq j \leq r}\partial_{\pi_j}^2 F \\
&= \frac{1}{2}\,\Delta(F) = \frac{1}{2}\,\mathrm{tr}\left(\nabla^2 F\right). \quad (\Longleftarrow (19.70) \text{ and } (19.71))
\end{aligned}$$

The martingale $M_t(F)$ is given by

$$
\begin{aligned}
dM_t(F) &= \langle (\partial F)(X_t), \pi(X_t) dB_t \rangle = \langle \pi(X_t)(\partial F)(X_t), dB_t \rangle = \langle (\nabla F)(X_t), dB_t \rangle \\
&= \sum_{1 \le j \le r} \left[\sum_{1 \le k \le r} \pi_j^k \, \partial_{x_k}(F) \right] (X_t) \, dB_t^j = \sum_{1 \le j \le r} \partial_{\pi_j}(F)(X_t) \, dB_t^j.
\end{aligned}
$$

Using (20.1), we check that $X_t \in S$ for any t, as soon as $X_0 \in S$. More precisely, we have

$$
\begin{aligned}
F = \varphi_i \quad &\Rightarrow \quad \nabla F = 0 \\
&\Rightarrow \quad dM_t(F) = \langle (\nabla F)(X_t), dB_t \rangle = 0 \ \ \& \ \ L(F) = \frac{1}{2} \, \mathrm{tr}\,(\nabla(\nabla F)) = 0 \\
&\Rightarrow \quad \varphi_i(X_t) = \varphi(X_0) = 0.
\end{aligned}
$$

Thus, (20.8) can be rewritten as follows

$$
dX_t = -\frac{r-1}{2} \, X_t \, dt + \left(Id - X_t X_t^T \right) \, dB_t.
$$

20.2.3   Monte Carlo simulation

In practice, the sampling of the diffusion process (20.7) requires a discrete time approximation of a sort. For instance, Euler type approximation on a time mesh $(t_n)_{n \ge 0}$ with $(t_n - t_{n-1}) = \epsilon \simeq 0$ is given by the equation

$$
X_{t_n}^\epsilon - X_{t_{n-1}}^\epsilon = -\frac{1}{2} \, \mathbb{H}(X_{t_{n-1}}^\epsilon) \, (t_n - t_{n-1}) + \pi(X_{t_{n-1}}^\epsilon) \, \sqrt{t_n - t_{n-1}} \, \overline{B}_n
$$

where $\overline{B}_n$ stands for a sequence of i.i.d. centered and normalized Gaussian r.v. on $\mathbb{R}^r$. Unfortunately these schemes do not ensure that $X_{t_n}^\epsilon$ stays in the manifold S. For deterministic dynamical systems, we often handle this issue by projecting each step on the manifold

$$
X_{t_n}^\epsilon = \mathrm{proj}_S \left(X_{t_{n-1}}^\epsilon - \frac{1}{2} \, \mathbb{H}(X_{t_{n-1}}^\epsilon) \, (t_n - t_{n-1}) + \pi(X_{t_{n-1}}^\epsilon) \, \sqrt{t_n - t_{n-1}} \, \overline{B}_n \right).
$$

Another strategy is to use a description of the stochastic process in some judicious chart space. Manifold parametrizations and chart spaces are discussed in chapter 21. We also refer the reader to chapter 22 for an overview of stochastic calculus on chart spaces.

20.3   Stratonovitch differential calculus

We recall that an r-dimensional stochastic differential equation of the form

$$
dX_t = b(X_t) \, dt + \sigma(X_t) \, dB_t
$$

can be rewritten as a Stratonovitch differential equation

$$
\partial X_t = \left[b - \frac{1}{2} \sum_{1 \le j \le r} \partial_{\sigma_j} (\sigma_j)^T \right] (X_t) \, \partial t + \sigma(X_t) \, \partial B_t \tag{20.10}
$$

with the vector fields σ_j on $\mathbb{R}^r$ defined by the column vectors

$$\forall 1 \leq j \leq r \qquad \sigma_j := \begin{bmatrix} \sigma_j^1 \\ \vdots \\ \sigma_j^r \end{bmatrix} \implies \partial_{\sigma_j} (\sigma_j)^T = \begin{bmatrix} \partial_{\sigma_j} \sigma_j^1 \\ \vdots \\ \partial_{\sigma_j} \sigma_j^r \end{bmatrix}.$$

In other words,

$$\forall 1 \leq k \leq r \qquad \partial X_t^k = \left[b^k - \frac{1}{2} \sum_{1 \leq j \leq r} \partial_{\sigma_j} \sigma_j^k \right] (X_t) \ \partial t + \sum_{1 \leq j \leq r} \sigma_j^k (X_t) \partial B_t^j.$$

An heuristic but constructive derivation of these formulae is given below. The Stratonovich and the Itō increments are connected by

$$b^k(X_t) \ \partial t = b^k \left(X_t + \frac{1}{2} \ dX_t \right) \times dt \quad \text{and} \quad \sigma_j^k(X_t) \ \partial B_t^j = \sigma_j^k \left(X_t + \frac{1}{2} \ dX_t \right) \times dB_t^j$$

with the middle state of the increment of X_t given by

$$\frac{X_t + X_{t+dt}}{2} := X_t + \frac{1}{2} \ dX_t.$$

Using this rule, we have

$$b^k(X_t) \ \partial t + \sum_{1 \leq j \leq r} \sigma_j^k(X_t) \ \partial B_t^j$$

$$\simeq b^k \left(X_t + \frac{1}{2} \ dX_t \right) \times dt + \sum_{1 \leq j \leq r} \sigma_j^k \left(X_t + \frac{1}{2} \ dX_t \right) \times dB_t^j$$

$$= b^k(X_t) \times dt + \sum_{1 \leq j \leq r} \sigma_j^k(X_t) \ dB_t^j + \frac{1}{2} \sum_{1 \leq j \leq r} \sum_{1 \leq i \leq r} \left(\partial_{x_i} \sigma_j^k \right)(X_t) \times dX_t^i dB_t^j + \ldots$$

$$= b^k(X_t) \times dt + \sum_{1 \leq j \leq r} \sigma_j^k(X_t) \ dB_t^j + \frac{1}{2} \sum_{1 \leq j \leq r} \left(\partial_{\sigma_j} \sigma_j^k \right)(X_t) \ dt.$$

The last assertion is a consequence of

$$dX_t^i dB_t^j = \sum_{1 \leq l \leq r} \sigma_l^i(X_t) \ dB_t^l dB_t^j = \sigma_j^i(X_t) \ dt.$$

Thus, the Stratonovitch formulation of (20.7) is given by

$$\partial X_t = \pi(X_t) \ \partial B_t.$$

In much the same way, using

$$\begin{aligned} \left(\partial_{\pi_j} F \right)(X_t) \ \partial B_t^j &= \left(\partial_{\pi_j} F \right) \left(X_t + \frac{1}{2} \ dX_t \right) dB_t^j \\ &= \left(\partial_{\pi_j} F \right)(X_t) \ dB_t^j + \frac{1}{2} \sum_{1 \leq l \leq r} \partial_{x_l} \left(\partial_{\pi_j} F \right)(X_t) \ dX_t^l dB_t^j \end{aligned}$$

and

$$
\begin{aligned}
\sum_{1 \leq l \leq r} \partial_{x_l} \left(\partial_{\pi_j} F \right) (X_t) \; dX_t^l dB_t^j &= \sum_{1 \leq l \leq r} \partial_{x_l} \left(\partial_{\pi_j} F \right) (X_t) \sum_{1 \leq m \leq r} \pi_m^l (X_t) \; dB_t^m dB_t^j \\
&= \left[\sum_{1 \leq l \leq r} \pi_l^j (X_t) \; \partial_{x_l} \left(\partial_{\pi_j} F \right) (X_t) \right] dt = \partial_{\pi_j}^2 F(X_t) \; dt
\end{aligned}
$$

we prove that

$$
\left(\partial_{\pi_j} F \right) (X_t) \; \partial B_t^j = \left(\partial_{\pi_j} F \right) (X_t) \; dB_t^j + \frac{1}{2} \left(\partial_{\pi_j}^2 F \right) (X_t) \; dt.
$$

Therefore

$$
\left(\partial_{\pi_j} F \right) (X_t) \; dB_t^j = \left(\partial_{\pi_j} F \right) (X_t) \; \partial B_t^j - \frac{1}{2} \left(\partial_{\pi_j}^2 F \right) (X_t) \; dt.
$$

Hence the Stratonovitch formulation of the equation

$$
dF(X_t) = L(F)(X_t) \; dt + \sum_{1 \leq j \leq r} \left(\partial_{\pi_j} F \right) (X_t) \; dB_t^j
$$

is given by

$$
\begin{aligned}
\partial F(X_t) &= \left[L(F) - \frac{1}{2} \sum_{1 \leq j \leq r} \partial_{\pi_j}^2 F \right] (X_t) \; dt + \sum_{1 \leq j \leq r} \left(\partial_{\pi_j} F \right) (X_t) \; \partial B_t^j \\
&= \sum_{1 \leq j \leq r} \left(\partial_{\pi_j} F \right) (X_t) \; \partial B_t^j = \sum_{1 \leq k \leq r} \left(\partial_{x_k} F \right) (X_t) \sum_{1 \leq j \leq r} \pi_j^k (X_t) \; \partial B_t^j \\
&= \sum_{1 \leq k \leq r} \partial_{x_k} F(X_t) \; \partial X_t^k = \langle (\partial F)(X_t), \partial X_t \rangle .
\end{aligned}
$$

20.4   Projected diffusions on manifolds

We consider the embedded manifold $S = \varphi^{-1}(0)$ described in section 20.1. We let π and $\mathbb{H}$ be the projection matrix and the mean curvature vector described in definition 20.1.1.

We let $X_t = \begin{pmatrix} X_t^1 \\ \vdots \\ X_t^r \end{pmatrix}$ be the $\mathbb{R}^r$-valued diffusion defined by

$$
\begin{aligned}
dX_t &= \pi(X_t) \; (b(X_t)dt + \sigma(X_t) \; dB_t) - \frac{1}{2} \, \mathbb{H}_\sigma(X_t) \; dt \qquad (20.11)\\
&= \overline{b}(X_t)dt + \left[\overline{\sigma}(X_t) \; dB_t - \frac{1}{2} \, \mathbb{H}_\sigma(X_t) \; dt \right]
\end{aligned}
$$

with

$$
\overline{\sigma}(x) = \pi(x)\sigma(x) \quad \text{and} \quad \overline{b}(x) = \pi(x)b(x)
$$

where B_t stands for a standard r-dimensional Brownian motion,

$$
\sigma = \begin{bmatrix} \sigma_1^1 & \cdots & \sigma_r^1 \\ \vdots & \vdots & \vdots \\ \sigma_1^r & \cdots & \sigma_r^r \end{bmatrix} \quad \text{and} \quad \mathbb{H}_\sigma(x) = - \begin{bmatrix} \sum_{1 \le j \le r} \partial_{\overline{\sigma}_j} \overline{\sigma}_j^1 \\ \vdots \\ \sum_{1 \le j \le r} \partial_{\overline{\sigma}_j} \overline{\sigma}_j^r \end{bmatrix}. \qquad (20.12)
$$

Using (20.10), the Stratonovitch formulation of the above equation is given by

$$
\partial X_t = \overline{b}(X_t) \; \partial t + \overline{\sigma}(X_t) \; \partial B_t.
$$

In this situation, we have

$$
dX_t dX_t^T = \overline{\sigma}(X_t) \; dB_t \; dB_t^T \overline{\sigma}(X_t)^T = \left(\overline{\sigma}\overline{\sigma}^T \right)(X_t).
$$

This yields the following result.

Theorem 20.4.1 *For any smooth function F on $\mathbb{R}^r$, we have the Ito formula:*

$$
\begin{aligned}
dF(X_t) &= \langle \partial F(X_t), dX_t \rangle + \frac{1}{2} \mathrm{tr} \left(\partial^2 F(X_t) \; dX_t dX_t^T \right)\\
&= \langle \partial F(X_t), dX_t \rangle + \frac{1}{2} \mathrm{tr} \left(\left(\overline{\sigma}\overline{\sigma}^T \right)(X_t)\partial^2 F(X_t) \right) \; dt\\
&= L(F)(X_t) \; dt + dM_t(F)
\end{aligned}
$$

with the infinitesimal generator

$$
L(F) = \partial_{\overline{b}} F + \frac{1}{2} \left[\mathrm{tr} \left(\left(\overline{\sigma}\overline{\sigma}^T \right) \partial^2 F \right) - \partial_{\mathbb{H}_\sigma} F \right]
$$

and the martingale

$$
dM_t(F) = \langle \partial F(X_t), \overline{\sigma}(X_t)dB_t \rangle = \langle \nabla F(X_t), \sigma(X_t)dB_t \rangle.
$$

To get one step further in our discussion, we notice that

$$\operatorname{tr}\left(\left(\overline{\sigma}\overline{\sigma}^T\right)(x)\partial^2 F(x)\right)$$

$$= \sum_{1\le k,l\le r} \partial_{x_k,x_l}(F)(x)\ \left(\overline{\sigma}(x)\overline{\sigma}^T(x)\right)_l^k = \sum_{1\le j,k,l\le r} \overline{\sigma}_j^k(x)\overline{\sigma}_j^l(x)\ \partial_{x_k,x_l}(F)(x)$$

$$= \sum_{1\le j\le r}\ \sum_{1\le k\le r} \overline{\sigma}_j^k(x)\ \partial_{x_k}\left(\sum_{1\le l\le r}\overline{\sigma}_j^l\ \partial_{x_l}F\right)(x)$$

$$- \sum_{1\le l\le r}\left\{\sum_{1\le j\le r}\left[\sum_{1\le k\le r}\overline{\sigma}_j^k(x)\ \partial_{x_k}\left(\overline{\sigma}_j^l\right)(x)\right]\right\}\partial_{x_l}F$$

$$= \sum_{1\le j\le r} \partial_{\overline{\sigma}_j}\left(\partial_{\overline{\sigma}_j}F\right)(x) + \partial_{\mathbb{H}_\sigma}F(x).$$

This implies that

$$L(F)\ =\ \partial_{\overline{b}}(F) + \frac{1}{2}\sum_{1\le j\le r}\partial_{\overline{\sigma}_j}^2(F) = \langle b,\nabla F\rangle + \frac{1}{2}\operatorname{tr}\left(\sigma^T\nabla\ \sigma^T\nabla F\right).\tag{20.13}$$

The r.h.s. formulation follows from the fact that

$$\overline{\sigma}^T\partial F = \begin{pmatrix}\sum_{1\le k\le r}\sigma_1^k\partial_{x_k}F\\ \vdots\\ \sum_{1\le k\le r}\sigma_r^k\partial_{x_k}F\end{pmatrix}$$

$$\Longrightarrow\ \partial\left(\overline{\sigma}^T\partial F\right)\ =\ \left[\sum_{1\le k\le r}\partial\left(\sigma_1^k\partial_{x_k}F\right),\ldots,\sum_{1\le k\le r}\partial\left(\sigma_r^k\partial_{x_k}F\right)\right]$$

$$=\ \begin{bmatrix}\sum_{1\le k\le r}\partial_{x_1}\left(\sigma_1^k\partial_{x_k}F\right) & \cdots & \sum_{1\le k\le r}\partial_{x_1}\left(\sigma_r^k\partial_{x_k}F\right)\\ \vdots & & \vdots\\ \sum_{1\le k\le r}\partial_{x_r}\left(\sigma_1^k\partial_{x_k}F\right) & \cdots & \sum_{1\le k\le r}\partial_{x_r}\left(\sigma_r^k\partial_{x_k}F\right)\end{bmatrix}$$

$$\Longrightarrow\overline{\sigma}^T\left(\partial\left(\overline{\sigma}^T\partial F\right)\right) = \begin{bmatrix}\sum_{1\le k,l\le r}\sigma_1^l\partial_{x_l}\left(\sigma_1^k\partial_{x_k}F\right) & \cdots & \sum_{1\le k,l\le r}\sigma_1^l\partial_{x_l}\left(\sigma_r^k\partial_{x_k}F\right)\\ \vdots & & \vdots\\ \sum_{1\le k,l\le r}\sigma_r^l\partial_{x_l}\left(\sigma_1^k\partial_{x_k}F\right) & \cdots & \sum_{1\le k,l\le r}\sigma_r^l\partial_{x_l}\left(\sigma_r^k\partial_{x_k}F\right)\end{bmatrix}.$$

This yields

$$\operatorname{tr}\left(\overline{\sigma}^T\partial\ \overline{\sigma}^T\partial F\right) = \operatorname{tr}\left(\sigma^T\nabla\ \sigma^T\nabla F\right) = \sum_{1\le j\le r}\sum_{1\le k,l\le r}\sigma_j^l\partial_{x_l}\left(\sigma_j^k\partial_{x_k}F\right) = \sum_{1\le j\le r}\partial_{\overline{\sigma}_i}^2(F).$$

Using (20.1), we check that $X_t\in S$ for any t, as soon as $X_0\in S$. More precisely, we have

$$F=\varphi_i\ \Rightarrow\ \nabla F = 0$$

$$\Rightarrow\ \begin{cases} dM_t(F)\ =\ \langle(\nabla F)(X_t),\sigma(X_t)dB_t\rangle = 0\\ L(F)\ =\ \langle b,\nabla F\rangle + \frac{1}{2}\operatorname{tr}\left(\sigma^T\nabla\ \sigma^T\nabla F\right) = 0\end{cases}$$

$$\Rightarrow\ \varphi_i(X_t) = \varphi(X_0) = 0.$$

20.5 Brownian motion on orbifolds

We let $S = \varphi^{-1}(0) \subset \mathbb{R}^{r=p+1}$ be some hypersurface, and we let π and $\mathbb{H}$ be the projection matrix and the mean curvature vector described in definition 20.1.1. We consider a subgroup $\mathcal{H}$ of the orthogonal group $\mathbb{O}(r)$ on $\mathbb{R}^r$, acting on S, such that

$$\forall h \in \mathcal{H} \quad \forall x \in S \quad hx \in S \quad (\Longrightarrow \varphi(x) = \varphi(hx)).$$

The prototype of a model we have in mind as a toy example is the unit sphere

$$S = \mathbb{S}^p = \left\{ x = (x_i)_{1 \leq i \leq r}^T \; : \; \varphi(x) := \sqrt{\sum_{1 \leq i \leq r} x_i^2} - 1 = 0 \right\} \subset \mathbb{R}^{p+1}$$

and the group action induced by the subgroup

$$\mathcal{O} := \left\{ h = \begin{pmatrix} \epsilon_1 & 0 & \cdots & 0 \\ 0 & \epsilon_2 & \cdots & 0 \\ \vdots & \vdots & & \vdots \\ 0 & 0 & \cdots & \epsilon_r \end{pmatrix} \; : \; \forall 1 \leq i \leq r \quad \epsilon_i \in \{-1, 1\} \right\}. \tag{20.14}$$

In the case of the sphere, the quotient manifold is isomorphic to the positive orthant $S/\mathcal{H} = S \cap \mathbb{R}_+^r$

$$\langle hx, hy \rangle = x^T h^T h y = x^T y = \langle x, y \rangle.$$

Quotient manifolds defined by the orbit space $S/\mathcal{H}$ are often called orbifolds.

By construction, we have

$$\varphi_h(x) := \varphi(hx) \implies \partial_{x_i}(\varphi_h)(x)$$

$$= \sum_{1 \leq k \leq r} (\partial_{x_k}\varphi)(hx) \, \partial_{x_i} \left(\sum_{1 \leq j \leq r} h_k^j \, x_j \right)$$

$$= \sum_{1 \leq k \leq r} h_k^j \, (\partial_{x_k}\varphi)(hx) = \sum_{1 \leq k \leq r} (h^T)_j^k \, (\partial_{x_k}\varphi)(hx)$$

$$\implies (\partial\varphi_h)(x) = h^{-1}(\partial\varphi)(hx) = h^T(\partial\varphi)(hx)$$

$$\implies \|(\partial\varphi_h)(x)\|^2 = \langle (\partial\varphi_h)(x), (\varphi_h)(x) \rangle$$

$$= (\partial\varphi)(hx)^T h h^T (\partial\varphi)(hx) = (\partial\varphi)(hx)^T (\partial\varphi)(hx)$$

$$= \|(\partial\varphi)(hx)\|^2.$$

This shows that the unit normal at hx is given by

$$h \frac{(\partial\varphi_h)(x)}{\|(\partial\varphi_h)(x)\|} = \frac{(\partial\varphi)(hx)}{\|(\partial\varphi)(hx)\|}.$$

On the other hand, under our assumptions we have

$$\varphi_h(x) := \varphi(hx) = \varphi(x) \implies \frac{(\partial\varphi_h)(x)}{\|(\partial\varphi_h)(x)\|} = \frac{(\partial\varphi)(x)}{\|(\partial\varphi)(x)\|}.$$

This implies that

$$h\,\frac{(\partial\varphi_h)(x)}{\|(\partial\varphi_h)(x)\|} = h\,\frac{(\partial\varphi)(x)}{\|(\partial\varphi)(x)\|} = \frac{(\partial\varphi)(hx)}{\|(\partial\varphi)(hx)\|}$$

from which we prove that

$$\pi(x) = Id - \frac{(\partial\varphi)(x)}{\|(\partial\varphi)(hx)\|}\,\frac{(\partial\varphi)(x)^T}{\|(\partial\varphi)(x)\|}$$

$$\implies h\pi(x)h^T = Id - h\frac{(\partial\varphi)(x)}{\|(\partial\varphi)(x)\|}\left(h\,\frac{(\partial\varphi)(x)}{\|(\partial\varphi)(x)\|}\right)^T \tag{20.15}$$

$$= Id - \frac{(\partial\varphi)(hx)}{\|(\partial\varphi)(hx)\|}\,\frac{(\partial\varphi)^T(hx)}{\|(\partial\varphi)(hx)\|} \implies h\pi(x)h^T = \pi(hx).$$

On the other hand, we have

$$\frac{(\partial\varphi_h)}{\|(\partial\varphi_h)\|}(x) = h^T\,\frac{(\partial\varphi)(hx)}{\|(\partial\varphi)(hx)\|} = \frac{\partial\varphi}{\|(\partial\varphi)\|}(x)$$

$$\Rightarrow \partial_{x_m}\left[\frac{(\partial_{x_m}\varphi_h)(x)}{\|(\partial\varphi_h)(x)\|}\right] = \sum_{1\le j\le r} h_j^m\,\partial_{x_m}\left[\frac{(\partial_{x_j}\varphi)(hx)}{\|(\partial\varphi)(hx)\|}\right] = \partial_{x_m}\left[\frac{(\partial_{x_m}\varphi)}{\|(\partial\varphi)\|}\right](x)$$

$$= \sum_{1\le j\le r} h_j^m \sum_{1\le k\le r}\partial_{x_k}\left[\frac{(\partial_{x_j}\varphi)}{\|(\partial\varphi)\|}\right](hx)\,\partial_{x_m}\left(\sum_{1\le i\le r} h_k^i x_i\right)$$

$$= \sum_{1\le j\le r} h_j^m h_k^m \sum_{1\le k\le r}\partial_{x_k}\left[\frac{(\partial_{x_j}\varphi)}{\|(\partial\varphi)\|}\right](hx).$$

Taking the sum over m, this implies that

$$\sum_{1\le j\le r}\overbrace{\left[\sum_{1\le m\le r} h_j^m (h^T)_m^k\right]}^{=1_{j=k}}\sum_{1\le k\le r}\partial_{x_k}\left[\frac{(\partial_{x_j}\varphi)}{\|(\partial\varphi)\|}\right](hx) = \sum_{1\le m\le r}\partial_{x_m}\left[\frac{(\partial_{x_m}\varphi)}{\|(\partial\varphi)\|}\right](x) \tag{20.16}$$

$$\implies \sum_{1\le m\le r}\partial_{x_m}\left[\frac{(\partial_{x_m}\varphi)}{\|(\partial\varphi)\|}\right](hx) = \sum_{1\le m\le r}\partial_{x_m}\left[\frac{(\partial_{x_m}\varphi)}{\|(\partial\varphi)\|}\right](x).$$

Combining this result with (20.2) we conclude that

$$h\,\mathbb{H}(x) = \left[\sum_{1\le m\le r}\partial_{x_m}\left(\frac{\partial_{x_m}\varphi}{\|\partial\varphi\|}\right)(x)\right] h\left[\frac{\partial\varphi}{\|\partial\varphi\|}(x)\right]$$

$$= \left[\sum_{1\le m\le r}\partial_{x_m}\left(\frac{\partial_{x_m}\varphi}{\|\partial\varphi\|}\right)(hx)\right]\frac{(\partial\varphi)(hx)}{\|(\partial\varphi)(hx)\|} \implies h\,\mathbb{H}(h^T x) = \mathbb{H}(x).$$

We let X_t be the Brownian motion on S defined in (20.7). Using (20.15) and (20.16), for any $h\in\mathcal{H}$ we have

$$Y_t = hX_t \Rightarrow dY_t = hdX_t = -\frac{1}{2}\,h\mathbb{H}(h^T(hX_t))\,dt + h\pi(X_t)h^T\,hdB_t$$

$$= -\frac{1}{2}\,\mathbb{H}(Y_t)\,dt + \pi(Y_t)\,dB_t^{(h)} \tag{20.17}$$

where $B_t^{(h)}$ is a standard r-dimensional Brownian motion. Roughly speaking, this result shows that all the stochastic processes in the same orbit

$$\text{Orb}_{\mathcal{H}}(X) = \{(hX_t)_{t\geq 0} \; : \; h \in \mathcal{H}\}$$

differ only by changing their driving Brownian motion. This coupling technique allows us to define in a unique way the Brownian motion on the quotient manifold $S/\mathcal{H}$.

20.6 Exercises

Exercise 365 (Brownian motion on the graph of a curve) *We consider the graph of a curve in the plane $\mathbb{R}^2$ defined by*

$$S = \left\{x = (x_1, x_2) \in \mathbb{R}^2 \; : \; x_2 = h(x_1)\right\}$$

for some smooth height function h. Describe the tangent space $T_x(S)$ and the orthogonal space $T_x^{\perp}(S)$ at a point $x \in S$. Compute the projection matrix $\pi(x)$ on $T_x(S)$ and the mean curvature vector $\mathbb{H}(x)$ at any state $x \in S$. Describe the diffusion equation of a Brownian motion on the graph of the function h.

Exercise 366 (Brownian motion on an ellipsoid) *We consider an ellipsoid S, embedded in $\mathbb{R}^3$, centered at the origin and defined by the equation*

$$\left(\frac{x_1}{a_1}\right)^2 + \left(\frac{x_2}{a_2}\right)^2 + \left(\frac{x_3}{a_3}\right)^2 = 1$$

for some parameters $a_i > 0$, with $i = 1, 2, 3$. Describe the equation of the Brownian motion on the surface of this ellipsoid.

Exercise 367 (Brownian motion - Monge parametrization) *Describe the Brownian motion on the surface S presented in the exercise 362.*

Exercise 368 (Stratonovitch calculus) *We consider the one-dimensional diffusion given by the stochastic differential equation*

$$dX_t = b(X_t) \; dt + \sigma(X_t) \; dW_t$$

where W_t stands for a Brownian motion on the real line. Write the Stratonovitch formulation of this equation.

Exercise 369 (Stratonovitch calculus for solving stochastic equations) *We consider the one-dimensional diffusion given by the stochastic differential equation*

$$dX_t = aX_t + bX_t \; dW_t$$

where W_t stands for a Brownian motion, a, b are parameters, and $X_0 > 0$. Check that the Stratonovitch formulation of this equation is given by

$$\partial \log X_t = \left(a - \frac{b^2}{2}\right) \; \partial t + b \; \partial W_t \; \Rightarrow \; \log\left(X_t/X_0\right) = \left(a - \frac{b^2}{2}\right) t + b \; W_t.$$

Exercise 370 (Stratonovitch Geometric Brownian motion) *We consider the one-dimensional Stratonovitch differential equation*

$$\partial X_t = a \ X_t \ \partial t + \ b \ X_t \ \partial W_t$$

where W_t stands for a Brownian motion on the real line, and $X_0 > 0$. Write the corresponding (Itō sense) stochastic diffusion equation and check that

$$\log\left(X_t/X_0\right) = \ a \ t + b \ W_t.$$

Exercise 371 (Projected pure diffusions) *We consider the $\mathbb{R}^3$-valued diffusion $dY_t = \sigma(Y_t)dB_t$ associated with an $\mathbb{R}^3$-dimensional Brownian motion B_t and with some regular function $\sigma \ : \ y \in \mathbb{R}^3 \mapsto \sigma(y) \in \mathbb{R}^{3\times3}$. Construct the projection X_t of Y_t on the surface S presented in the exercise 362. Describe the Itō formula for some function $F(X_t)$.*

Exercise 372 (Brownian motion on a p-sphere) *Describe the evolution of the Brownian motion on the sphere $S = \mathbb{S}^p \subset \mathbb{R}^{p+1}$.*

Exercise 373 (Stratonovitch diffusion) *Describe the Stratonovitch formulation of the $\mathbb{R}^3$-valued diffusion*

$$dX_t = -\frac{X_t}{X_t^T X_t} \ dt + \left(Id - \frac{X_t X_t^T}{X_t^T X_t}\right) \ dB_t \qquad (20.18)$$

on the unit sphere $S = \mathbb{S}^2$, with an $\mathbb{R}^3$-dimensional Brownian motion B_t.

Exercise 374 (Time discretisation) *Propose a discrete time numerical simulation on a time mesh $(t_n)_{n\geq0}$ with $(t_n - t_{n-1}) = \epsilon \simeq 0$ of the Brownian motion on the sphere $S = \mathbb{S}^2$ described in (20.18).*

21

Parametrizations and charts

This chapter is concerned with differentiable manifolds and with their parametrization. Most of the chapter is devoted to expressing the geometrical objects and formulae presented in chapter 19 to the parameter space. The geometry of the manifold in the ambient space is expressed in the space of parameters in terms of a Riemannian geometry. These mathematicical models are essential for designing diffusions on parameter spaces associated with a given constraints manifold. From the numerical viewpoint, these stochastic processes are easier to handle than the diffusion on ambient spaces as soon as we find a judicious parametrization of the manifold.

> *Geometry is not true, it is advantageous.*
> Jules Henri Poincare (1854-1912).

21.1   Differentiable manifolds and charts

Differentiable manifolds can be described *locally* by some smooth parametrization function on some open subset. The set of these parametrization functions forms a chart (a.k.a. atlas). We only consider embedded manifolds of some dimension p which are smooth subsets of the ambient Euclidian space $\mathbb{R}^{p+q}$, for some $q \geq 1$ (equipped with the cartesian coordinates) defined by a non-empty and connected null-level set $S := \varphi^{-1}(0)$ of some regular function φ from $\mathbb{R}^{p+q}$ into $\mathbb{R}^q$. We refer to section 20.1 for a detailed discussion on these geometrical objects.

Definition 21.1.1 *We denote by*

$$\psi \ : \ \theta \in S_\psi \subset \mathbb{R}^p \mapsto \psi(\theta) = \left(\psi^1(\theta), \ldots, \psi^r(\theta)\right)^T \in S \subset \mathbb{R}^r \tag{21.1}$$

a given smooth parametrization of S, with a well defined smooth inverse mapping

$$\phi = \psi^{-1} \ : \ x \in S \mapsto \phi(x) = \left(\phi^1(x), \ldots, \phi^p(x)\right)^T \in S_\phi \subset \mathbb{R}^p.$$

To clarify the presentation, we further assume that the manifold S can be parametrized by a single map ψ, and thus with a single chart coordinate. At the end of this section, we provide a discussion about more general situations.

By construction, we have

$$\forall 1 \leq l \leq q \quad \forall \theta \in S_\psi \qquad \varphi_l(\psi(\theta)) = 0$$

$$\Downarrow$$

$$\partial_{\theta^j}\left(\varphi_l\circ\psi\right)(\theta)=\sum_{1\leq k\leq r}\left(\partial_{x_k}\varphi_l\right)\left(\psi(\theta)\right)\partial_{\theta_i}\psi^l(\theta)=\langle(\partial\varphi_l)\left(\psi(\theta)\right),\partial_{\theta_i}\psi(\theta)\rangle=0$$

$$\Downarrow$$

$$\forall x\in S\qquad\langle\partial\varphi_l(x),\left(\partial_{\theta_i}\psi\right)(\phi(x))\rangle=0.$$

Further on, we set

$$\left(\partial_{\theta_i}\psi\right)_\phi\ :\ x\in S\mapsto\left(\partial_{\theta_i}\psi\right)_\phi(x):=\left(\partial_{\theta_i}\psi\right)(\phi(x))\in T_x(S)$$

and

$$\left(\partial\phi^i\right)_\psi\ :\ \theta\in S_\psi\mapsto\left(\partial\phi^i\right)_\psi(\theta)=\left(\partial\phi^i\right)(\psi(\theta))\in\mathbb{R}^r.$$

In this notation, we have shown that

$$T(S)=\mathrm{Vect}\left(\left(\partial_{\theta_1}\psi\right)_\phi,\ldots,\left(\partial_{\theta_p}\psi\right)_\phi\right)$$

in the sense that

$$\forall x\in S\qquad T_x(S)=\mathrm{Vect}\left(\left(\partial_{\theta_1}\psi\right)_\phi(x),\ldots,\left(\partial_{\theta_p}\psi\right)_\phi(x)\right).$$

We let e_i be the unit vectors on $S_\psi(\subset\mathbb{R}^p)$ defined by

$$\forall 1\leq i\leq p\qquad e_i=\begin{bmatrix}0\\\vdots\\0\\1\\0\\\vdots\\0\end{bmatrix}\leftarrow i\text{-th term.}\qquad(21.2)$$

We end this section with an alternative description of the vector fields $\left(\partial_{\theta_i}\psi\right)_\phi$. For each $1\leq i\leq p$, we let $c_i(t)=\phi(x)+t\,e_i$ be a curve in S_ψ with starting point $\phi(x)$ and velocity e_i and

$$C_i(t)=\psi\left(\phi(x)+t\,e_i\right)$$

is the pushed forward curve in the manifold S with velocity

$$\frac{dC_i}{dt}(0)=\sum_{1\leq j\leq p}\left(\partial_{\theta_j}\psi\right)_\phi(x)\,e_i^j=\left(\partial_{\theta_i}\psi\right)_\phi(x).$$

In differential geometry, these vector fields in $T_x(S)$ are often denoted using the somehow misleading notation

$$\frac{\partial}{\partial\theta_i}\mid_x:=\left(\partial_{\theta_i}\psi\right)_\phi(x).\qquad(21.3)$$

The main reason for this strange notation will become clear in section 21.3 (see formula (21.17)).

We end this section with some comments on the parametrization of manifolds.

It may happen that the manifold S cannot be parametrized by a single mapping ϕ (21.1). In this situation, we need to resort to a collection of parametrizations $(\phi_{i \in I}, S_{i \in I})_{i \in I}$ between some open subsets S_i covering $S = \cup_{i \in S_i}$ onto some open parametric subsets $S_{\psi_i} \subset \mathbb{R}^p$. The collection $(\phi_{i \in I}, S_{i \in I})_{i \in I}$ is called an atlas or a chart on the manifold. The mappings $\phi_j \circ \psi_i \; : \; \phi_i(S_i \cap S_j) \mapsto \phi_j(S_i \cap S_j)$ are called the transition maps of the atlas. The regularity of the transition maps characterizes the regularity of the manifold, for instance, a C^k-manifold is a manifold which can be equipped with an atlas whose transition maps are C^k-differentiable (i.e., k-times continuously differentiable). Of course, the existence of atlases requires a manifold equipped with some topology with well defined open subsets.

For instance, suppose we have parametrizations

$$\psi_1 \; : \; \theta \in \phi_1(S_1 \cap S_2) \subset \mathbb{R}^p \; \mapsto \; \psi_2(\theta) \in S_1 \cap S_2 \subset \mathbb{R}^r$$
$$\psi_2 \; : \; \alpha \in \phi_2(S_1 \cap S_2) \subset \mathbb{R}^p \; \mapsto \; \psi_2(\alpha) \in S_1 \cap S_2 \subset \mathbb{R}^r$$

with transition maps

$$\phi_2 \circ \psi_1 \; : \; \phi_1(S_1 \cap S_2) \mapsto \phi_2(S_1 \cap S_2).$$

In this situation, for any $\theta \in \phi_1(S_1 \cap S_2)$,

$$\psi_1 = \psi_2 \circ (\phi_2 \circ \psi_1) \Rightarrow \partial_{\theta_i} \psi_1 = \sum_{1 \leq j \leq p} \frac{\partial (\phi_2 \circ \psi_1)^j}{\partial \theta_i} \, \partial_{\alpha_j} \psi_2.$$

This formula is often written for any state $x \in (S_1 \cap S_2)$ in the synthetic form

$$\frac{\partial}{\partial \theta_i} \mid_x = \sum_{1 \leq j \leq p} \frac{\partial \alpha^j}{\partial \theta_i}(\phi_1(x)) \frac{\partial}{\partial \alpha_j} \mid_x. \tag{21.4}$$

The partition of unity patching technique:

For locally compact and completely separable manifolds (i.e., second-countable) every open neighborhood of a point $x \in S$ can be covered with a finite number of open sets S_i. The resulting chart forms a locally finite covering of the manifold. In addition, there exist some smooth mappings

$$\epsilon_i \; : \; x \in M \mapsto \epsilon_i(x) \in [0, 1] \quad \text{such that} \quad \epsilon_i^{-1}(]0, 1]) \subset S_i \quad \text{and} \quad \sum_{i \in I} \epsilon_i = 1.$$

The set of mappings $(\epsilon_i)_{i \in I}$ is called a partition of unity for the locally finite chart $(\phi_{i \in I}, S_{i \in I})_{i \in I}$ (cf. for instance [260], theorem 1.11). This clearly implies that for any function F on M we have

$$F = \sum_{i \in I} F_i \quad \text{with the functions} \quad F_i := \epsilon_i \, F \quad \text{with support} \subset S_i. \tag{21.5}$$

These decompositions allow us to reduce the analysis to manifolds $S = S_i$ equipped with a single parametrization $(\phi, \psi) = (\phi_i, \phi_i^{-1})$ and patch if need arises by using a partition of unity.

21.2   Orthogonal projection operators

Definition 21.2.1 *We let* $g = (g_{i,j})_{1 \leq i,j \leq p}$ *be the* $(p \times p)$*-matrix field on* S_ψ *defined by*

$$\forall 1 \leq i, j \leq p \qquad g_{i,j} := \langle \partial_{\theta_i} \psi, \partial_{\theta_j} \psi \rangle \qquad (21.6)$$

in the sense that

$$\forall \theta \in S_\psi \quad g_{i,j}(\theta) := \langle \partial_{\theta_i} \psi(\theta), \partial_{\theta_j} \psi(\theta) \rangle .$$

Observe that

$$\psi = \begin{pmatrix} \psi^1 \\ \vdots \\ \psi^r \end{pmatrix} \Rightarrow \partial \psi = (\partial \psi^1, \cdots, \partial \psi^r) = \begin{pmatrix} (\partial_{\theta_1} \psi)^T \\ \vdots \\ (\partial_{\theta_p} \psi)^T \end{pmatrix}$$

and

$$(\partial \psi)^T = (\partial_{\theta_1} \psi, \cdots, \partial_{\theta_p} \psi) .$$

This yields the Gram matrix formulae

$$g = (\partial \psi) \ (\partial \psi)^T \quad \text{and} \quad \sqrt{\det(g)} = \sqrt{\det \left((\partial \psi) \ (\partial \psi)^T \right)}. \qquad (21.7)$$

In geometry literature the determinant of the matrix $g = \text{Gram} \left(\partial_{\theta_1} \psi, \cdots, \partial_{\theta_p} \psi \right)$ is called the Gramian. When $p = r$ we have the formula

$$\sqrt{\det(g)} = |\det (\partial_{\theta_1} \psi, \cdots, \partial_{\theta_r} \psi)| .$$

In differential geometry, one often uses the notation for the metric g in local coordinate frames

$$g = g_{i,j} \ d\theta_i \otimes d\theta_j \quad \text{or simply as} \quad g = g_{i,j} \ d\theta_i d\theta_j,$$

with the Einstein summation convention and the dual forms $d\theta_i \in T^\star(S_\psi)$ of the vector fields $\frac{\partial}{\partial \theta_i} \in T(S_\psi)$ introduced in (21.3).

In this notation, the metric of the Euclidian space $S = \mathbb{R}^2$, equipped with the Euclidian metric and with the canonical parametrization

$$\psi \ : \ x = (x_1, x_2) \in \mathbb{R}^2 \mapsto \psi(x) = x \Rightarrow \partial_{x_1} \psi = \begin{pmatrix} 1 \\ 0 \end{pmatrix} \quad \text{and} \quad \partial_{x_2} \psi = \begin{pmatrix} 0 \\ 1 \end{pmatrix}$$

is given by $g = (dx_1)^2 + (dx_2)^2$. The manifold $\mathbb{R}^2 \cap \{x = (x_1, x_2) \; : \; x_2 > 0\}$ can be parametrized by the polar coordinates

$$\psi \; : \; \theta = (\theta_1, \theta_2) \in S_\psi = \mathbb{R} \times [0, 2\pi] \mapsto \psi(\theta) = \begin{cases} \psi^1(\theta) & = & \theta_1 \cos(\theta_2) \\ \psi^2(\theta) & = & \theta_1 \sin(\theta_2). \end{cases} \qquad (21.8)$$

The change of coordinates happens by the differential rules

$$\begin{aligned} d\psi^1(\theta) &= \partial_{\theta_1}\psi^1 \, d\theta_1 + \partial_{\theta_2}\psi^1 \, d\theta_2 = \cos(\theta_2) \, d\theta_1 - \theta_1 \sin(\theta_2) \, d\theta_2 \\ d\psi^2(\theta) &= \partial_{\theta_1}\psi^2 \, d\theta_1 + \partial_{\theta_2}\psi^2 \, d\theta_2 = \sin(\theta_2) \, d\theta_1 + \theta_1 \cos(\theta_2) \, d\theta_2. \end{aligned}$$

These formulae are often written in the following synthetic form

$$dx_i := d\psi^i(\theta) \quad \text{and} \quad \frac{\partial x_i}{\partial \theta_j} := \partial_{\theta_j}\psi^i \Rightarrow dx_i := \frac{dx_i}{\partial \theta_j} \, d\theta_j$$

with the Einstein summation convention. In this notation, we have

$$\begin{aligned} (dx_1)^2 + (dx_2)^2 &= [\cos(\theta_2) \, d\theta_1 - \theta_1 \sin(\theta_2) \, d\theta_2]^2 + [\sin(\theta_2) \, d\theta_1 + \theta_1 \cos(\theta_2) \, d\theta_2]^2 \\ &= (d\theta_1)^2 + \theta_1^2 \, (d\theta_2)^2. \end{aligned}$$

Definition 21.2.2 *The push forward of the scalar product g on $T(S)$ is the matrix field on S given by*

$$g_\phi = (g_{\phi,i,j})_{1 \le i,j \le p} = (g_{i,j} \circ \phi)_{1 \le i,j \le p}.$$

Observe that

$$g_{\phi,i,j}(x) := \left\langle \left(\partial_{\theta_i}\psi(\theta)\right)_{\phi(x)}, \left(\partial_{\theta_j}\psi(\theta)\right)_{\phi(x)} \right\rangle$$

with the tangent vector fields

$$x \in S \;\mapsto\; \left(\partial_{\theta_i}\psi(\theta)\right)_{\phi(x)} := \left(\left(\partial_{\theta_i}\psi(\theta)\right) \circ \phi\right)(x) \in T_x(S).$$

By construction, the projection of any vector field W on $\mathbb{R}^r$ onto $T(S)$ is given by

$$\pi(W) = \sum_{1 \le i \le p} \left\langle \sum_{1 \le i \le p} g_\phi^{i,j} \left(\partial_{\theta_j}\psi\right)_\phi, W \right\rangle \left(\partial_{\theta_i}\psi\right)_\phi \quad \text{with} \quad g^{-1} = (g^{i,j})_{1 \le i,j \le p}, \qquad (21.9)$$

---

in the sense that

$$\pi(x)(W(x)) = \sum_{1 \le i \le p} \left\langle \sum_{1 \le i \le p} g_\phi^{i,j}(x) \left(\partial_{\theta_j}\psi\right)_\phi(x), W(x) \right\rangle \left(\partial_{\theta_i}\psi\right)_\phi(x).$$

We let $W_\psi = W \circ \psi$ be the pull back vector field on the parameter space, and denote by $\pi_\psi(\theta) = \pi(\psi(\theta)) = \pi(x)$ the orthonormal projection functional onto $T_x(S)$ with $x = \psi(\theta)$. In this notation,

$$\pi_\psi(\theta)(W_\psi(\theta)) = \sum_{1 \le i \le p} \left\langle \sum_{1 \le i \le p} g^{i,j}(\theta) \left(\partial_{\theta_j}\psi\right)(\theta), W_\psi(\theta) \right\rangle \left(\partial_{\theta_i}\psi\right)(\theta)$$

or in a more synthetic form

$$\pi_\psi(W_\psi) = \sum_{1 \le i \le p} \left\langle \sum_{1 \le i \le p} g^{i,j} \, \partial_{\theta_j} \psi, W_\psi \right\rangle \partial_{\theta_i} \psi.$$

By construction, for any $1 \le i, j \le p$,

$$\phi^i(\psi(\theta)) = \theta^i \quad \Rightarrow \quad \partial_{\theta^j}\left(\phi^i \circ \psi)\right)(\theta) = \sum_{1 \le k \le r} \left(\partial_{x_k}\phi^i\right)_\psi (\theta) \, \partial_{\theta^j}\psi^k(\theta)$$
$$= \left\langle \left(\partial\phi^i\right)_\psi (\theta), \partial_{\theta^j}\psi(\theta) \right\rangle = 1_{i=j}$$

so that

$$\forall x \in S \qquad \left\langle \left(\partial\phi^i\right)(x), \left(\partial_{\theta^j}\psi\right)_\phi (x) \right\rangle = 1_{i=j}.$$

This implies that

$$\nabla\phi^i = \pi\left(\partial\phi^i\right) = \sum_{1 \le k \le p} \left\langle \sum_{1 \le l \le p} g_\phi^{k,l} \left(\partial_{\theta^l}\psi\right)_\phi, \partial\phi^i \right\rangle \left(\partial_{\theta^k}\psi\right)_\phi$$
$$= \sum_{1 \le k \le p} g_\phi^{i,k} \left(\partial_{\theta^k}\psi\right)_\phi \qquad (21.10)$$

and

$$\sum_{1 \le i \le p} g_{\phi,j,i} \, \nabla\phi^i = \sum_{1 \le k \le p} \sum_{1 \le i \le p} g_{\phi,j,i} \, g_\phi^{i,k} \left(\partial_{\theta^k}\psi\right)_\phi = \left(\partial_{\theta^j}\psi\right)_\phi.$$

By construction, we have

$$\left\langle \left(\nabla\phi^i\right), \left(\partial_{\theta^j}\psi\right)_\phi \right\rangle = 1_{i=j} \quad \text{and} \quad \left\langle \nabla\phi^i, \nabla\phi^j \right\rangle = g_\phi^{i,j}. \qquad (21.11)$$

We check these claims using

$$\left\langle \left(\nabla\phi^i\right), \left(\partial_{\theta^j}\psi\right)_\phi \right\rangle = \sum_{1 \le k \le p} g_\phi^{i,k} \, g_{\phi,k,j} = 1_{i=j} \qquad (21.12)$$

and

$$\left\langle \nabla\phi^i, \nabla\phi^j \right\rangle = \sum_{1 \le k,l \le p} g_\phi^{i,k} \, g_\phi^{j,l} \left\langle \left(\partial_{\theta^k}\psi\right)_\phi, \left(\partial_{\theta^l}\psi\right)_\phi \right\rangle$$
$$= \sum_{1 \le k \le p} g_\phi^{i,k} \sum_{1 \le l \le p} g_\phi^{j,l} \, g_{\phi,l,k} = g_\phi^{i,j}.$$

In summary, we have the following result.

Proposition 21.2.3 *The vector fields* $\nabla\phi^i$ *form a new basis of* $T(S)$

$$T(S) = \text{Vect}\left(\nabla\phi^1, \ldots, \nabla\phi^p\right) \quad \text{with the scalar product} \quad \left\langle\nabla\phi^i, \nabla\phi^j\right\rangle = g_\phi^{i,j}$$

and the change of basis formulae are given by

$$\left(\partial_{\theta^i}\psi\right)_\phi = \sum_{1\leq j\leq p} g_{\phi,i,j}\,\nabla\phi^j \quad \text{and} \quad \nabla\phi^i = \sum_{1\leq j\leq p} g_\phi^{i,j}\,\left(\partial_{\theta^j}\psi\right)_\phi. \tag{21.13}$$

Expressed in these new basis vector fields, the orthogonal projection operators π *take the form*

$$\pi(W) = \sum_{1\leq i\leq p}\left\langle\sum_{1\leq j\leq p} g_{\phi,i,j}\,\nabla\phi^j, W\right\rangle\,\nabla\phi^i.$$

The property (21.12) ensures that $\left(\nabla\phi^i\right)_{1\leq i\leq p}$ and $\left(\left(\partial_{\theta_i}\psi\right)_\phi\right)_{1\leq i\leq p}$ form a biorthogonal system. These constructions together with the differential formula (21.44) can be used to define the dual forms of vector fields.

Notice that

$$W = \sum_{1\leq i\leq p} V_\phi^i\,\left(\partial_{\theta^i}\psi\right)_\phi \implies \forall 1\leq i\leq p \quad V_\phi^i = \left\langle W, \nabla\phi^i\right\rangle.$$

Rewritten in a slightly different form we have

$$\begin{aligned}
\pi(W) &= \sum_{1\leq j\leq p}\left\langle\nabla\phi^j, W\right\rangle\sum_{1\leq i\leq p} g_{\phi,j,i}\,\nabla\phi^i \\
&= \sum_{1\leq j\leq p}\left\langle\nabla\phi^j, W\right\rangle\,\left(\partial_{\theta^j}\psi\right)_\phi.
\end{aligned} \tag{21.14}$$

21.3 Riemannian structures

We consider a smooth curve in the parameter space

$$c\,:\,t\in[0,1]\mapsto c(t) = \left(c^1(t), \ldots, c^p(t)\right)^T\in S_\psi$$

starting at some parameter state $c(0) = \theta\in S_\psi$, with a velocity vector field $V(\in\mathbb{R}^p)$. That is,

$$\frac{dc}{dt} = \left(\frac{dc^1}{dt}, \ldots, \frac{dc^p}{dt}\right)^T = V(c(t)) = \left(V^1(c(t)), \ldots, V^p(c(t))\right)^T.$$

The function c is called an integral curve of V (a.k.a. V-integral curve). By construction, $C(t) := \psi(c(t))$ is a smooth curve on S and we have

$$\forall 1\leq i\leq q \quad \frac{dC}{dt} = \frac{d}{dt}\psi(c(t)) = \sum_{1\leq i\leq p}\left(\partial_{\theta_i}\psi\right)(c(t))\,V^i(c(t)).$$

For $t = 0$, this implies that

$$\frac{dC}{dt}(0) = \sum_{1 \leq i \leq p} V_\phi^i(x) \ (\partial_{\theta_i} \psi)_{\phi(x)}$$

with

$$x = \psi(\theta) \Leftrightarrow \theta = \phi(x) \quad \text{and} \quad V_\phi^i(x) = V^i(\phi(x)).$$

In other words, C is an integral curve of the vector field

$$W(x) = \sum_{1 \leq i \leq p} V_\phi^i(x) \ (\partial_{\theta_i} \psi)_{\phi(x)} \ .$$

For any smooth function $F = f \circ \phi$ on S we have

$$
\begin{aligned}
\frac{d}{dt} F(C(t)) &= \sum_{1 \leq k \leq r} W^k(C(t)) \ (\partial_{x_k} F)(C(t)) = \partial_W(F)(C(t)) \\
&= \sum_{1 \leq j \leq p} V^j(c(t)) \sum_{1 \leq k \leq r} (\partial_{x_k} F)(\psi(c(t))) \ (\partial_{\theta_j} \psi^k) \ (c(t)) \\
&= \sum_{1 \leq j \leq p} V^j(c(t)) \ (\partial_{\theta_j} (F \circ \psi)) \ (c(t)) \\
&= \sum_{1 \leq j \leq p} V^j(c(t)) \ (\partial_{\theta_j} f) \ (c(t)) = \partial_V(f)(c(t)) = \frac{d}{dt} f(c(t)). \quad (21.15)
\end{aligned}
$$

This shows that

$$\frac{d}{dt} F(C(t)) = \partial_W(F)(C(t)) = \partial_V(f)(c(t)) = \frac{d}{dt} f(c(t)) \quad \overset{t=0}{\Longrightarrow} \quad \partial_W F = (\partial_V f) \circ \phi.$$
$$(21.16)$$

Vector fields can also be interpreted as differential operators

$$W \ : \ F \ \mapsto \ W(F) = \partial_W(F) = W^T \partial F = \langle W, \partial F \rangle \ .$$

In this interpretation, rewritten in terms of (21.3), we have

$$(\partial_{\theta_j} (F \circ \psi))_\phi = \sum_{1 \leq k \leq r} \left(\frac{\partial}{\partial \theta_i} \right)^k (\partial_{x_k} F) = \left\langle \frac{\partial}{\partial \theta_i}, \partial F \right\rangle = \frac{\partial}{\partial \theta_i}(F)$$

and therefore

$$\forall x \in S \qquad W(x) = \sum_{1 \leq i \leq p} V_\phi^i(x) \left(\frac{\partial}{\partial \theta_i} \right)_{|x} \Leftrightarrow W = \sum_{1 \leq i \leq p} V_\phi^i \frac{\partial}{\partial \theta_i} . \qquad (21.17)$$

In this synthetic notation,

$$W(F) = \sum_{1 \leq i \leq p} V_\phi^i \frac{\partial}{\partial \theta_i}(F) = \sum_{1 \leq i \leq p} V_\phi^i \partial_{\theta_i}(f) \quad \text{with} \quad f = F \circ \psi \quad \Leftrightarrow F = f \circ \phi.$$

This induces a one-to-one *linear* mapping between the tangent spaces $T_\theta(S_\psi)$ of the parameter space and the tangent space $T_{\psi(\theta)}(S)$ on the manifold S. This mapping is called the push forward of the vector fields on $T_\theta(S_\psi)$ into $T_{\psi(\theta)}(S)$, and it is given by

$$(d\psi) \ : \ V \in T(S_\psi) \mapsto (d\psi)(V) := \sum_{1 \le i \le p} V^i \ (\partial_{\theta_i} \psi) \ \in T_\psi(S),$$

in the sense that

$$(d\psi)_\theta \ : \ V(\theta) \in T(S_\psi) \mapsto (d\psi)_\theta (V(\theta)) := \sum_{1 \le i \le p} V^i(\theta) \ (\partial_{\theta_i} \psi) (\theta) \ \in T_{\psi(\theta)}(S).$$

Alternatively, we have

$$(d\psi)_\phi \ : \ V_\phi \in T_\phi(S_\psi) \mapsto (d\psi)_\phi (V_\phi) := \sum_{1 \le i \le p} V_\phi^i \ (\partial_{\theta_i} \psi)_\phi \ \in T(S),$$

in the sense that

$$(d\psi)_{\phi(x)} \ : \ V(\phi(x)) \ = \ V_\phi(x) \ \mapsto \ (d\psi)_{\phi(x)} (V_\phi(x)) \ := \ \sum_{1 \le i \le p} V_\phi^i(x) \ (\partial_{\theta_i} \psi)_\phi (x)$$

$$\in \ T_{\phi(x)}(S_\psi) \qquad\qquad \in \ T_x(S).$$

Finally, we notice that

$$W = (d\psi)_\phi(V_\phi) = \sum_{1 \le i \le p} V_\phi^i \ (\partial_{\theta i} \psi)_\phi \Rightarrow \langle \nabla \phi^j, W \rangle = \sum_{1 \le i \le p} V_\phi^i \left\langle \nabla \phi^j, (\partial_{\theta i} \psi)_\phi \right\rangle = V_\phi^j.$$

Thus, for any $W \in T(S)$ we have

$$V_\phi = (d\psi)_\phi^{-1}(W) = \begin{bmatrix} (\nabla \phi^1)^T \\ \vdots \\ (\nabla \phi^p)^T \end{bmatrix} W.$$

The parameter space $S_\psi \subset \mathbb{R}^p$ is free of any constraints and we have

$$T_\theta(S_\psi) = \text{Vect}\,(e_1, \ldots, e_p)$$

with the unit vectors defined in (21.2).

In this notation, $(d\psi)$ maps the basis functions e_i of $T_\theta(S_\psi)$ into the basis functions $(\partial_{\theta_i} \psi)$ of $T_{\psi(\theta)}(S)$, that is,

$$(d\psi) (e_i) = (\partial_{\theta_i} \psi) \quad \text{and} \quad (d\psi)_\phi^{-1} (\partial_{\theta_i} \psi) = e_i.$$

It is also essential to notice that

$$\langle (d\psi) (V_1), (d\psi) (V_2) \rangle = \sum_{1 \le i \le p} V_1^i \ g_{i,j} \ V_2^j = V_1^T g \ V_2.$$

Thus, if we equip the tangent space $T_\theta(S_\psi)$ with the scalar product

$$\langle V_1, V_2 \rangle_g = \sum_{1 \leq i \leq p} g_{i,j} \, V_1^i V_2^j,$$

the description of $T(S)$ in the chart ϕ is given by

$$(W_k)_\psi = (d\psi)(V_k) \Rightarrow \langle (d\psi)(V_1), (d\psi)(V_2) \rangle = \sum_{1 \leq i \leq p} V_1^i V_2^j \, g_{i,j}.$$

In summary,
$$\langle V_1, V_2 \rangle_g = \langle (W_1)_\psi, (W_2)_\psi \rangle.$$

More formally, the (linear) pushed forward mappings $(d\psi)_\theta$ are smooth isomorphisms between the inner product spaces $(T_\theta(S_\psi), \langle ., . \rangle_{g(\theta)})$ and $(T_{\psi(\theta)}(S), \langle ., . \rangle)$. The scalar product induced by g on the tangent space $T(S_\psi)$ of the parameter space S_ψ is called the Riemannian scalar product. This construction equips the tangent space

$$T(S_\psi) = \text{Vect}(e_1, \ldots, e_p) = \mathbb{R}^p$$

with the inner product

$$\langle e_i, e_j \rangle_g = \langle \partial_{\theta_i} \psi, \partial_{\theta_j} \psi \rangle. \tag{21.18}$$

In the above display, $e_i \, : \, \theta \in \mathbb{R}^p \mapsto e_i(\theta)$ stands for the unit basis vector fields on $\mathbb{R}^p$ defined in (21.2); where $e_i(\theta)$ stands for the unit vector attached to the state $\theta \in \mathbb{R}^p$.

21.4   First order covariant derivatives

21.4.1   Pushed forward functions

Smooth functions F on S are the push forwards of functions f on S_ψ, and inversely functions f on the parameter space are the pull backs of functions F on S using the relations

$$F = f \circ \phi \quad \text{and} \quad f = F \circ \psi.$$

As a rule, we use the letters F and W to denote functions, and vector fields on S and f, V to denote functions and vector fields on the parameter space S_ψ.
We also denote by $F_\psi = F \circ \psi$, resp. $W_\psi = W \circ \psi$, and $f_\phi = f \circ \phi$, resp. $V_\phi = V \circ \phi$, the pull back of W, resp. F, and V, resp. f, w.r.t. ψ and ϕ.

In this notation, differentials of push forward functions are given by

$$\partial_{\theta_i}(F_\psi) = \sum_{1 \leq j \leq r} \left(\partial_{x_j} F \right)_\psi \, \partial_{\theta_i} \psi^j = (\partial_{\theta_i} \psi)^T \, (\partial F)_\psi$$

in terms of (21.3).

In much the same way, differentials of pull back functions are given by the formula

$$\partial_{x_i}(f_\phi) = \sum_{1 \le j \le p} \left(\partial_{\theta_j} f\right)_\phi \, \partial_{x_i} \phi^j \quad \Leftrightarrow \quad (\partial f_\phi)_\psi = \sum_{1 \le j \le p} \partial_{\theta_j} f \, \left(\partial \phi^j\right)_\psi.$$

Therefore, we have the following result.

Proposition 21.4.1 *We have the formulae*

$$\begin{aligned}
(\nabla F)_\psi &= \pi_\psi (\partial F)_\psi = \sum_{1 \le j \le p} \left(\partial_{\theta_j} f\right) \, \pi_\psi \left(\partial \phi^j\right)_\psi \\
&= \sum_{1 \le j \le p} \left(\partial_{\theta_j} f\right) \, (\nabla \phi^j)_\psi = \sum_{1 \le i \le p} \left[\sum_{1 \le j \le p} g^{i,j} \left(\partial_{\theta_j} f\right) \right] \partial_{\theta^i} \psi \quad (21.19) \\
&= \sum_{1 \le i \le p} (\nabla_g f)^i \, \partial_{\theta^i} \psi = d\psi (\nabla_g f) \quad\quad\quad\quad (21.20)
\end{aligned}$$

with the vector field $\nabla_g f$ *on* S_ψ *given by*

$$\nabla_g f := \begin{bmatrix} \sum_{1 \le j \le p} g^{1,j} \left(\partial_{\theta_j} f\right) \\ \vdots \\ \sum_{1 \le j \le p} g^{p,j} \left(\partial_{\theta_j} f\right) \end{bmatrix} = g^{-1} \partial f. \quad\quad (21.21)$$

The last assertion follows from the change of basis formula (21.13). It is also instructive to observe that

$$\nabla_g f = \sum_{1 \le i \le p} \langle \sum_{1 \le j \le p} g^{i,j} \, e_j, \partial f \rangle \, e_i = \sum_{1 \le i \le p} \left[\sum_{1 \le i \le p} g^{i,j} \, \partial_{\theta^j} f \right] \, e_i \quad \text{and} \quad (\nabla \phi^i)^j_\psi = (\nabla_g \psi^j)^i.$$

Furthermore, using (21.11) for any

$$F = f \circ \phi \quad \text{and} \quad W = (d\psi)_\phi (V_\phi)$$

we have

$$\begin{aligned}
\langle \nabla F, W \rangle &= \sum_{1 \le i,j \le p} V_\phi^i \left(\partial_{v_j} f\right)_\phi \left\langle \nabla \phi^j, (\partial_{\theta_i} \psi)_\phi \right\rangle \\
&= \sum_{1 \le i \le p} V_\phi^i \left(\partial_{v_i} f\right)_\phi = \langle (\partial f)_\phi, V_\phi \rangle.
\end{aligned}$$

In much the same way, we have

$$\begin{aligned}
\langle (\nabla F)_\psi, W_\psi \rangle &= \sum_{1 \le i,j \le p} (\nabla_g(f))^i \, V^j \langle \partial_{\theta^i} \psi, \partial_{\theta^j} \psi \rangle \\
&= \sum_{1 \le i,j \le p} g_{i,j} \, (\nabla_g(f))^i \, V^j = \langle \nabla_g f, V \rangle_g = \langle \partial f, V \rangle.
\end{aligned}$$

In terms of directional derivatives, we have

$$(\partial_W(F)) \circ \psi = \langle (\nabla F)_\psi, W_\psi \rangle = \langle \nabla_g f, V \rangle_g = \langle \partial f, V \rangle = \partial_V(f). \qquad (21.22)$$

In particular, for functions $F_1 = f_1 \circ \phi$ and $F_2 = f_2 \circ \phi$ we have

$$\langle \nabla F_1, \nabla F_2 \rangle = \Big\langle (\nabla_g f_1)_\phi, (\nabla_g f_2)_\phi \Big\rangle_{g_\phi} \quad \Big(= \Big\langle (\partial f_1)_\phi, (\partial f_2)_\phi \Big\rangle_{g_\phi^{-1}} \Big). \qquad (21.23)$$

We consider the coordinate projection mappings

$$\chi_i := \psi^i \circ \phi \; : \; x \in \mathbb{R}^r \mapsto \chi^i(x) = (\psi^i \circ \phi)(x) = x_i \in \mathbb{R}$$
$$\epsilon_i := \phi^i \circ \psi \; : \; \theta \in \mathbb{R}^p \mapsto \epsilon_i(\theta) = (\phi^i \circ \psi)(\theta) = \theta_i \in \mathbb{R}. \qquad (21.24)$$

Notice that $\phi^i = \epsilon^i \circ \phi$ and $\psi^i = \chi_i \circ \psi$, Applying (21.19) to the functions $f = \epsilon_i = \phi^i \circ \psi$ and $F = \chi_i = \psi^i \circ \phi$ using (21.20) we find that

$$\left(\nabla \phi^i\right)_\psi = d\psi\left(\nabla_g \epsilon^i\right) \quad \text{and} \quad \left(\nabla \chi_i\right)_\psi = d\psi\left(\nabla_g \psi^i\right) \quad \text{with} \quad \nabla_g \epsilon^i = g^i = \begin{pmatrix} g^{i,1} \\ \vdots \\ g^{i,p} \end{pmatrix}.$$

It is also readily checked that

$$\nabla_g f = \sum_{1 \le i \le p} \partial_{\theta_i} f \; \nabla_g \epsilon_i \implies \forall 1 \le j \le r \quad \nabla_g \psi^j = \sum_{1 \le i \le p} \partial_{\theta_i} \psi^j \; \nabla_g \epsilon_i.$$

We also have the differential product rule

$$\nabla_g(f_1 f_2) = f_1 \nabla_g f_2 + f_2 \nabla_g f_1. \qquad (21.25)$$

By (21.14) we have

$$\pi(W) = \sum_{1 \le j \le p} \langle \nabla \phi^j, W \rangle \; (\partial_{\theta^j} \psi)_\phi \qquad (21.26)$$

$$= \begin{bmatrix} \sum_{1 \le j \le p} \langle \nabla \phi^j, W \rangle \; (\partial_{\theta^j} \psi^1)_\phi \\ \vdots \\ \sum_{1 \le j \le p} \langle \nabla \phi^j, W \rangle \; (\partial_{\theta^j} \psi^r)_\phi \end{bmatrix} = [\nabla \chi_1, \dots, \nabla \chi_r] \; W.$$

21.4.2 Pushed forward vector fields

We consider the push forward W_x on $T_x(S)$ of a vector field V on $T(S_\psi)$ given by the formula

$$W(x) = (d\psi)_{\phi(x)} (V_\phi(x)) := \sum_{1 \le j \le p} V_\phi^j(x) \; (\partial_{\theta_j} \psi)_\phi(x).$$

We have

$$V_\phi^j = V^j \circ \phi \implies \partial_{x_k}\left(V_\phi^j\right)(x) = \sum_{1 \le l \le p}\left(\partial_{\theta_l}V^j\right)(\phi(x))\ \left(\partial_{x_k}\phi^l\right)(x)$$

and

$$\partial_{x_k}\left(\left(\partial_{\theta_j}\psi^i\right)_\phi\right)(x) = \sum_{1 \le l \le p}\left(\partial_{\theta_l,\theta_j}\psi^i\right)(\phi(x))\ \left(\partial_{x_k}\phi^l\right)(x).$$

Rewritten in a more synthetic way,

$$\partial_{x_k}V_\phi^j = \sum_{1 \le l \le p}\left(\partial_{\theta_l}V^j\right)_\phi\ \partial_{x_k}\phi^l$$

$$\partial_{x_k}\left(\partial_{\theta_j}\psi\right)_\phi = \sum_{1 \le l \le p}\left(\partial_{\theta_l,\theta_j}\psi\right)_\phi\ \partial_{x_k}\phi^l.$$

This implies that

$$\partial_{x_k}W^i = \sum_{1 \le j \le p}\left[\partial_{x_k}(V_\phi^j)\ \left(\partial_{\theta_j}\psi^i\right)_\phi + V_\phi^j\ \partial_{x_k}\left(\partial_{\theta_j}\psi^i\right)_\phi\right]$$

$$= \sum_{1 \le j,l \le p}\left[\left(\partial_{\theta_l}V^j\right)_\phi\ \left(\partial_{\theta_j}\psi^i\right)_\phi + V_\phi^j\ \left(\partial_{\theta_l,\theta_j}\psi^i\right)_\phi\right]\partial_{x_k}\phi^l.$$

In vector form, we have

$$\partial W^i = \sum_{1 \le j,l \le p}\left[\left(\partial_{\theta_l}V^j\right)_\phi\ \left(\partial_{\theta_j}\psi^i\right)_\phi + V_\phi^j\ \left(\partial_{\theta_l,\theta_j}\psi^i\right)_\phi\right]\partial\phi^l$$

and

$$\partial W = \sum_{1 \le j,l \le p}\left[\left(\partial_{\theta_l}V^j\right)_\phi\ \partial\phi^l\ \left(\partial_{\theta_j}\psi\right)_\phi^T + V_\phi^j\ \partial\phi^l\ \left(\partial_{\theta_l,\theta_j}\psi\right)_\phi^T\right].$$

This implies that

$$\nabla W = \pi(\partial W)$$

$$= \sum_{1 \le j,l \le p}\left[\left(\partial_{\theta_l}V^j\right)_\phi\ \nabla\phi^l\ \left(\partial_{\theta_j}\psi\right)_\phi^T + V_\phi^j\ \nabla\phi^l\ \left(\partial_{\theta_l,\theta_j}\psi\right)_\phi^T\right]$$

$$= \sum_{1 \le j,k,l \le p}g_\phi^{l,k}\left[\left(\partial_{\theta_l}V^j\right)_\phi\ \left(\partial_{\theta_k}\psi\right)_\phi\ \left(\partial_{\theta_j}\psi\right)_\phi^T + V_\phi^j\ \left(\partial_{\theta_k}\psi\right)_\phi\ \left(\partial_{\theta_l,\theta_j}\psi\right)_\phi^T\right].$$

Taking the trace, we obtain

$$\mathrm{tr}\left(\nabla W\right) = \sum_{1 \le j,k,l \le p}g_\phi^{l,k}\left[\left(\partial_{\theta_l}V^j\right)_\phi\ g_{\phi,k,j} + V_\phi^j\ \left\langle\left(\partial_{\theta_k}\psi\right)_\phi,\left(\partial_{\theta_l,\theta_j}\psi\right)_\phi\right\rangle\right]$$

$$= \sum_{1 \le j \le p}\left(\partial_{\theta_j}V^j\right)_\phi + \frac{1}{2}\sum_{1 \le j,k,l \le p}g_\phi^{l,k}\ V_\phi^j\ \left(\partial_{\theta_j}\left\langle\partial_{\theta_k}\psi,\partial_{\theta_l}\psi\right\rangle\right)_\phi.$$

This yields

$$\mathrm{tr}\left(\nabla W\right)_\psi = \sum_{1 \le j \le p}\partial_{\theta_j}V^j + \sum_{1 \le j \le p}V^j\ \frac{1}{2}\sum_{1 \le k,l \le p}g^{l,k}\ \partial_{\theta_j}g_{k,l}.$$

We also have the formula

$$\sum_{1 \le k,l \le p} g^{l,k} \, \partial_{\theta_j} g_{k,l} = \operatorname{tr}\left(g^{-1}\partial_{\theta_j}g\right) = \frac{1}{\det(g)} \, \partial_{\theta_j}\left(\det(g)\right) = \frac{2}{\sqrt{\det(g)}} \, \partial_{\theta_j}\left(\sqrt{\det(g)}\right) \tag{21.27}$$

from which we prove the following proposition.

Proposition 21.4.2 *We have the formula*

$$\operatorname{div}\left(W\right)_\psi := \operatorname{tr}\left(\nabla W\right)_\psi = \sum_{1 \le j \le p} \frac{1}{\sqrt{\det(g)}} \, \partial_{\theta_j}\left(\sqrt{\det(g)} \, V^j\right) := \operatorname{div}_g(V). \tag{21.28}$$

In particular, choosing $W = \nabla F = (d\psi)_\phi \left(\nabla_g(f)\right)_\phi$, *we have*

$$
\begin{aligned}
(\Delta F)_\psi &= \operatorname{div}\left(\nabla F\right)_\psi := \operatorname{tr}\left(\nabla^2 F\right)_\psi \\
&= \sum_{1 \le j \le p} \frac{1}{\sqrt{\det(g)}} \, \partial_{\theta_j}\left(\sqrt{\det(g)} \sum_{1 \le i \le p} g^{j,i} \, \partial_{\theta^i} f\right) := \operatorname{div}_g(\nabla_g f).
\end{aligned}
\tag{21.29}
$$

21.4.3 Directional derivatives

We let W_1, W_2 be a couple of vector fields in $T(S)$. We let V_1, V_2 be their pull back vector fields so that

$$W_k \circ \psi = (d\psi)\left(V_k\right) := \sum_{1 \le j \le p} V_k^j \, \partial_{\theta_j}\psi$$

for any $k = 1, 2$. We let C_1 be a W_1-integral curve, that is,

$$\frac{dC_1}{dt}(t) = W_1\left(C_1(t)\right) \Rightarrow \frac{d}{dt}F(C_1(t)) = \sum_{1 \le k \le r} W_1^k\left(C_1(t)\right)\left(\partial_{x_k}F\right)(C_1(t)) = \partial_{W_1}(F)(C_1(t)).$$

We recall from (21.15) that

$$\partial_{W_1}(F) \circ \psi = \sum_{1 \le j \le p} V_1^j \, \partial_{\theta_j}\left(F \circ \psi\right) = \partial_{V_1}(F \circ \psi) = \left\langle \nabla_g(F \circ \psi), V_1 \right\rangle_g. \tag{21.30}$$

Notice that

$$
\begin{aligned}
\left(\partial_{W_1}\left(\partial_{W_2}(F)\right)\right) \circ \psi &= \left(\partial_{W_1}(F_2)\right) \circ \psi \quad \text{with} \quad F_2 \circ \psi = \sum_{1 \le j \le p} V_2^j \, \partial_{\theta_j}\left(F \circ \psi\right) \\
&= \partial_{V_2}(F \circ \psi) \\
&= \partial_{V_1}(F_2 \circ \psi) = \partial_{V_1}\left(\partial_{V_2}(F \circ \psi)\right) \\
&= \sum_{1 \le i,j \le p} V_1^i \, \partial_{\theta_i}(V_2^j) \, \partial_{\theta_j}\left(F \circ \psi\right) + \sum_{1 \le i,j \le p} V_1^i V_2^j \, \partial_{\theta_i,\theta_j}\left(F \circ \psi\right).
\end{aligned}
\tag{21.31}
$$

Definition 21.4.3 *The directional derivative of the vector field* W_2 *along the curve* C_1 *is given by*

$$\frac{d}{dt}W_2(C_1(t)) = \begin{bmatrix} \frac{d}{dt}W_2^1(C_1(t)) \\ \vdots \\ \frac{d}{dt}W_2^r(C_1(t)) \end{bmatrix} = \begin{bmatrix} (\partial_{W_1}W_2^1)(C_1(t)) \\ \vdots \\ (\partial_{W_1}W_2^r)(C_1(t)) \end{bmatrix} := \partial_{W_1}(W_2)(C_1(t))$$

with

$$\partial_{W_1}(W_2) = \begin{bmatrix} \partial_{W_1}W_2^1 \\ \vdots \\ \partial_{W_1}W_2^r \end{bmatrix} \Rightarrow \partial_{W_1}(W_2) \circ \psi = \begin{bmatrix} \partial_{V_1}(W_2^1 \circ \psi) \\ \vdots \\ \partial_{V_1}(W_2^r \circ \psi) \end{bmatrix}.$$

Using

$$\begin{aligned}
\partial_{V_1}(W_2^k \circ \psi) &= \sum_{1 \le i \le p} V_1^i \, \partial_{\theta_i}\left[\sum_{1 \le j \le p} V_2^j \, \partial_{\theta_j}\psi^k\right] \\
&= \sum_{1 \le i,j \le p} V_1^i \, \partial_{\theta_i}(V_2^j) \, \partial_{\theta_j}\psi^k + \sum_{1 \le i \le p} V_1^i \, V_2^j \, \partial_{\theta_i,\theta_j}\psi^k
\end{aligned}$$

we conclude that

$$\partial_{W_1}(W_2) \circ \psi = \sum_{1 \le i,j \le p} V_1^i \, \partial_{\theta_i}(V_2^j) \, \partial_{\theta_j}\psi + \sum_{1 \le i \le p} V_1^i \, V_2^j \, \partial_{\theta_i,\theta_j}\psi.$$

The directional covariant derivative is defined by taking the projection on the tangent space $T(S)$

$$\nabla_{W_1}W_2 = \pi\left(\partial_{W_1}(W_2)\right) = \sum_{1 \le i,j \le p} V_{1,\phi}^i \left(\partial_{\theta_i}V_2^j\right)_\phi \left(\partial_{\theta_j}\psi\right)_\phi + \sum_{1 \le i \le p} V_{\phi,1}^i \, V_{\phi,2}^j \, \pi\left(\left(\partial_{\theta_i,\theta_j}\psi\right)_\phi\right)$$

or equivalently

$$(\nabla_{W_1}W_2) \circ \psi = \pi_\psi\left(\left(\partial_{W_1}(W_2)\right)_\psi\right) = \sum_{1 \le i,j \le p} V_1^i \left(\partial_{\theta_i}V_2^j\right) \partial_{\theta_j}\psi + \sum_{1 \le i \le p} V_1^i \, V_2^j \, \pi_\psi\left(\left(\partial_{\theta_i,\theta_j}\psi\right)\right).$$

Definition 21.4.4 *The Christoffel symbols are the coordinate functions* $C_{i,j}^k$ *defined by*

$$\pi_\psi\left(\left(\partial_{\theta_i,\theta_j}\psi\right)\right) = \sum_{1 \le k \le p} \underbrace{\left\langle \sum_{1 \le l \le p} g^{k,l} \, \partial_{\theta_l}\psi, \partial_{\theta_i,\theta_j}\psi \right\rangle}_{:=C_{i,j}^k} \partial_{\theta_k}\psi. \qquad (21.32)$$

In this notation, we have

$$
\begin{aligned}
(\nabla_{W_1} W_2) \circ \psi &= \sum_{1 \le k \le p} \left[\sum_{1 \le i \le p} V_1^i \left(\partial_{\theta_i} V_2^k \right) + \sum_{1 \le i,j \le p} C_{i,j}^k \, V_1^i \, V_2^j \right] \partial_{\theta_k} \psi \\
&\qquad\qquad\qquad\qquad\qquad\qquad\qquad\qquad\qquad\qquad (21.33)
\end{aligned}
$$

$$
= \sum_{1 \le k \le p} \left[\sum_{1 \le i \le p} V_1^i \left\{ \left(\partial_{\theta_i} V_2^k \right) + \sum_{1 \le j \le p} C_{i,j}^k \, V_2^j \right\} \right] \partial_{\theta_k} \psi
$$

$$
= (d\psi) \left(\nabla_{g, V_1} V_2 \right)
$$

with the Riemannian directional derivative

$$
\nabla_{g, V_1} V_2 = \begin{bmatrix} \sum_{1 \le i \le p} V_1^i \left\{ \left(\partial_{\theta_i} V_2^1 \right) + \sum_{1 \le j \le p} C_{i,j}^1 \, V_2^j \right\} \\ \vdots \\ \sum_{1 \le i \le p} V_1^i \left\{ \left(\partial_{\theta_i} V_2^p \right) + \sum_{1 \le j \le p} C_{i,j}^p \, V_2^j \right\} \end{bmatrix}. \qquad (21.34)
$$

In terms of the unit vectors e_i on $\mathbb{R}^p$ we have

$$
\nabla_{g, e_i} e_j = C_{i,j} := \begin{bmatrix} C_{i,j}^1 \\ \vdots \\ C_{i,j}^p \end{bmatrix}. \qquad (21.35)
$$

This also shows that

$$
\left(\nabla_{W_1} \left(\partial_{\theta_l} \psi \right)_\phi \right) \circ \psi = \sum_{1 \le i,k \le p} C_{i,l}^k \, V_1^i \, \partial_{\theta_k} \psi \implies \left(\nabla_{\left(\partial_{\theta_i} \psi \right)_\phi} \left(\partial_{\theta_j} \psi \right)_\phi \right) \circ \psi = \sum_{1 \le k \le p} C_{i,j}^k \, \partial_{\theta_k} \psi.
$$

On the other hand, we have

$$
\begin{aligned}
c_1^i(t) = \phi^i(C_1(t)) \Rightarrow \dot{c}_1^i(t) &= \partial_{W_1}(\phi^i)(C_1(t)) = \sum_{1 \le j \le p} V_1^j(c_1(t)) \underbrace{\partial_{\theta_j} \left(\phi^i \circ \psi \right)(c_1(t))}_{=1_{i=j}} \\
&= V_1^i(c_1(t)).
\end{aligned}
$$

Thus, using (21.33), we find that

$$
(\nabla_{W_1} W_2)(C_1(t))
$$

$$
= \sum_{1 \le k \le p} \left[\sum_{1 \le i \le p} \dot{c}_1^i(t) \left(\partial_{\theta_i} V_2^k \right)(c_1(t)) + \sum_{1 \le i,j \le p} C_{i,j}^k(c_1(t)) \, \dot{c}_1^i(t) \, V_2^j(c_1(t)) \right] \left(\partial_{\theta_k} \psi \right)(c_1(t))
$$

$$
= \sum_{1 \le k \le p} \left[\frac{d}{dt} (V_2^k(c_1(t))) + \sum_{1 \le i,j \le p} C_{i,j}^k(c_1(t)) \, \dot{c}_1^i(t) \, V_2^j(c_1(t)) \right] \left(\partial_{\theta_k} \psi \right)(c_1(t)).
$$

In differential geometry, the above formula is sometimes expressed in terms of the linear differential operator

$$\frac{DW_2}{dt}(t) := \left(\nabla_{\dot{C}_1(t)} W_2\right)(C_1(t))$$

$$= \sum_{1 \leq k \leq p} \left[\frac{d}{dt}(V_2^k(c_1(t)) + \sum_{1 \leq i,j \leq p} C_{i,j}^k(c_1(t))\ \dot{c}_1^i(t)\ V_2^j(c_1(t))\right]\ (\partial_{\theta_k}\psi)(c_1(t))$$

$$= \sum_{1 \leq j \leq p} \left[\frac{d}{dt}(V_2^j(c_1(t))\ (\partial_{\theta_j}\psi)(c_1(t))\right.$$

$$\left. + V_2^j(c_1(t)) \sum_{1 \leq i,k \leq p} C_{i,j}^k(c_1(t))\ \dot{c}_1^i(t)\ (\partial_{\theta_k}\psi)(c_1(t))\right]$$

$$= \sum_{1 \leq j \leq p} \left[\frac{d}{dt}(V_2^j(c_1(t))\ (\partial_{\theta_j}\psi)(c_1(t)) + V_2^j(c_1(t))\ \left(\nabla_{\dot{C}_1(t)}(\partial_{\theta_l}\psi)_\phi\right)(C_1(t))\right]$$

$$= (d\psi)_{c_1(t)}\left(\nabla_{g,\dot{c}_1} V_2\right)(c_1(t)).$$

We say that the vector field $V(t) = V_2(c_1(t))$ is parallel along the curve $C_1(t) = C(t) = \psi(c(t))$, with $c(t) = c_1(t)$ if we have

$$\frac{DV}{dt}(t) := \left(\nabla_{g,\dot{c}} V_2\right)(c(t)) = \dot{V}(t) + \sum_{1 \leq i,j \leq p} C_{i,j}(c(t))\ \dot{c}^i(t)\ V^j(t) = 0\ , \qquad (21.36)$$

with the column vector function $C_{i,j} = \begin{pmatrix} C_{i,j}^1 \\ \vdots \\ C_{i,j}^p \end{pmatrix}$. In other words, we have the linear

ordinary equation w.r.t. the coordinates $V^j(t)$ given for any $1 \leq k \leq p$ by

$$\dot{V}^k(t) + \sum_{1 \leq i,j \leq p} C_{i,j}^k(c(t))\ \dot{c}^i(t)\ V^j(t) = 0.$$

Note that for any fixed initial vector field V', there always exists a vector field curve $V : t \in [0,1] \mapsto V(t) \in \mathbb{R}^p$ parallel to $c(t)$ s.t. $V(0) = V'$. In this case, we also say that $V(1) = V''$ is obtained from $V(0) = V'$ by parallel transport along the curve c. Replacing $[0,1]$ by $[s,t]$, we obtain the following definition:

$$\mathrm{parall}_{c,s,t} : V' \in T_{c(s)}S_\phi \mapsto \mathrm{parall}_{c,s,t}(V') = V(c(t)) \in T_{c(s)}S_\phi \qquad (21.37)$$

where $V(c(\tau))$, $\tau \in [s,t]$, is a unique vector field on $(c(\tau))$, $\tau \in [s,t]$, such that

$$V(c(s)) = V' \quad \text{and} \quad \left(\nabla_{g,\dot{c}} V_2\right)(c(\tau)) = 0.$$

21.5 Second order covariant derivative

21.5.1 Tangent basis functions

We recall from (21.10) that

$$(\nabla\phi^i)_\psi = \sum_{1 \leq k \leq p} g^{i,k}\ \partial_{\theta_k}\psi.$$

Notice that

$$\partial_{\theta_m} \left((\nabla \phi^i)_\psi \right) = \sum_{1 \le k, l \le p} \left(\partial_{\theta_m} (g^{i,k}) \, \partial_{\theta^k} \psi + g^{i,k} \, \partial_{\theta_m, \theta_k} \psi \right). \qquad (21.38)$$

Using the differential rule (19.30) we also prove that

$$\nabla^2 \phi^i = \sum_{1 \le k \le p} \left[\nabla \left(g_\phi^{i,k} \right) \, (\partial_{\theta^k} \psi)_\phi^T + g_\phi^{i,k} \, \nabla \left((\partial_{\theta^k} \psi)_\phi \right) \right].$$

Notice that

$$\partial_{x_l} \left(g_\phi^{i,k} \right) = \partial_{x_l} \left(g^{i,k} \circ \phi \right) = \sum_{1 \le m \le p} \left(\partial_{\theta_m} g^{i,k} \right)_\phi \, \partial_{x_l} \phi^m \Rightarrow \partial \left(g_\phi^{i,k} \right) = \sum_{1 \le m \le p} \left(\partial_{\theta_m} g^{i,k} \right)_\phi \, \partial \phi^m$$

from which we derive that

$$\nabla \left(g_\phi^{i,k} \right) \;=\; \pi \left(\partial \left(g_\phi^{i,k} \right) \right) = \sum_{1 \le m \le p} \left(\partial_{\theta_m} g^{i,k} \right)_\phi \, \nabla \phi^m.$$

On the other hand, we have

$$\partial_{x_m} \left(\left(\partial_{\theta_k} \psi^l \right)_\phi \right) = \sum_{1 \le i \le p} \left(\partial_{\theta_i, \theta_k} \psi^l \right)_\phi \, \partial_{x_m} \phi^i$$

$$\Rightarrow \partial \left(\left(\partial_{\theta_k} \psi^l \right)_\phi \right) = \sum_{1 \le i \le p} \left(\partial_{\theta_k, \theta_i} \psi^l \right)_\phi \, \partial \phi^i$$

$$\Rightarrow \nabla \left(\left(\partial_{\theta_k} \psi^l \right)_\phi \right) = \sum_{1 \le i \le p} \left(\partial_{\theta_k, \theta_i} \psi^l \right)_\phi \, \nabla \phi^i.$$

This implies that

$$\partial \left((\partial_{\theta_k} \psi)_\phi \right) = \left[\partial \left(\left(\partial_{\theta_k} \psi^1 \right)_\phi \right), \dots, \partial \left(\left(\partial_{\theta_k} \psi^r \right)_\phi \right) \right]$$

$$\Rightarrow \nabla \left((\partial_{\theta_k} \psi)_\phi \right) = \left[\nabla \left(\left(\partial_{\theta_k} \psi^1 \right)_\phi \right), \dots, \nabla \left(\left(\partial_{\theta^k} \psi^r \right)_\phi \right) \right] = \sum_{1 \le m \le p} \nabla \phi^m \, (\partial_{\theta_k, \theta_m} \psi)_\phi^T.$$

Using (21.38), we conclude that

$$\nabla^2 \phi^i \;=\; \sum_{1 \le m \le p} \nabla \phi^m \, \partial_{\theta_m} \left((\nabla \phi^i)_\psi^T \right) \qquad (21.39)$$

$$=\; \sum_{1 \le k, m \le p} \left[\left(\partial_{\theta_m} g^{i,k} \right)_\phi \, \nabla \phi^m \, (\partial_{\theta_k} \psi)_\phi^T + g_\phi^{i,k} \, \nabla \phi^m \, (\partial_{\theta_k, \theta_m} \psi)_\phi^T \right]$$

and

$$\mathrm{tr} \left(\nabla^2 \phi^i \right) \;=\; \sum_{1 \le k, m \le p} \left[\left(\partial_{\theta_m} g^{i,k} \right)_\phi \, \left\langle \nabla \phi^m, (\partial_{\theta_k} \psi)_\phi \right\rangle + g_\phi^{i,k} \, \left\langle \nabla \phi^m, (\partial_{\theta_k, \theta_m} \psi)_\phi \right\rangle \right]$$

$$=\; \sum_{1 \le m \le p} \left[\left\langle \nabla \phi^m, \sum_{1 \le k \le p} \left(\partial_{\theta_m} g^{i,k} \right)_\phi (\partial_{\theta_k} \psi)_\phi + \sum_{1 \le k \le p} g_\phi^{i,k} \, (\partial_{\theta_k, \theta_m} \psi)_\phi \right\rangle \right].$$

Using (21.38) this formula can be rewritten as follows

$$\mathrm{tr} \left(\nabla^2 \phi^i \right)_\psi = \sum_{1 \le m \le p} \left\langle (\nabla \phi^m)_\psi, \partial_{\theta_m} \left((\nabla \phi^i)_\psi \right) \right\rangle.$$

Using

$$\nabla \phi^m = \sum_{1 \leq l \leq p} g_\phi^{m,l} \, (\partial_{\theta_l} \psi)_\phi \quad \Longrightarrow \quad \left\langle \nabla \phi^m, (\partial_{\theta_k} \psi)_\phi \right\rangle = 1_{m=k}$$

we also have the following formulae.

$$
\begin{aligned}
(\Delta \phi^i)_\psi &= \operatorname{tr} \left(\nabla^2 \phi^i \right)_\psi \\
&= \sum_{1 \leq m \leq p} \partial_{\theta_m} g^{i,m} + \sum_{1 \leq k \leq p} g^{i,k} \sum_{1 \leq m,l \leq p} g^{m,l} \, \left\langle \partial_{\theta_l} \psi, \partial_{\theta_k,\theta_m} \psi \right\rangle \\
&= \sum_{1 \leq j \leq p} \partial_{\theta_j} g^{i,j} + \frac{1}{2} \sum_{1 \leq j \leq p} g^{i,j} \sum_{1 \leq k,l \leq p} g^{k,l} \, \partial_{\theta_j} g_{k,l} \\
&= \sum_{1 \leq j \leq p} \frac{1}{\sqrt{\det(g)}} \, \partial_{\theta_j} \left(\sqrt{\det(g)} \, g^{i,j} \right) = \operatorname{div} \left(\nabla \phi^i \right)_\psi . \quad (21.40)
\end{aligned}
$$

The last assertion is a direct consequence of (21.28) when applied to the vector field

$$W = \nabla \phi^i = \sum_{1 \leq j \leq p} g_\phi^{i,j} \, (\partial_{\theta_j} \psi)_\phi$$

$$\Longrightarrow \left\langle (\nabla \phi^i)_\psi, \partial_{\theta_m \theta_l} \psi \right\rangle = \sum_{1 \leq j \leq p} g^{i,j} \, \left\langle \partial_{\theta_j} \psi, \partial_{\theta_m \theta_l} \psi \right\rangle = C^i_{m,l}. \quad (21.41)$$

We end this section with a formula relating $\Delta \phi^i$ to the Christoffel symbols introduced in (21.32). Firstly, we observe that

$$\sum_{1 \leq m \leq m} \partial_{\theta_p} \left\langle (\nabla \phi^m)_\psi, (\nabla \phi^i)_\psi \right\rangle = \sum_{1 \leq m \leq p} \partial_{\theta_m} g^{i,m}. \quad (21.42)$$

On the other hand, we have

$$\sum_{1 \leq m \leq p} \partial_{\theta_m} \left\langle (\nabla \phi^m)_\psi, (\nabla \phi^i)_\psi \right\rangle$$

$$= \sum_{1 \leq m \leq p} \left\langle \partial_{\theta_m} \left((\nabla \phi^m)_\psi \right), (\nabla \phi^i)_\psi \right\rangle + \sum_{1 \leq m \leq p} \left\langle (\nabla \phi^m)_\psi, \partial_{\theta_m} \left((\nabla \phi^i)_\psi \right) \right\rangle$$

$$= \sum_{1 \leq m,l \leq p} \left\langle \partial_{\theta_m} \left(g^{m,l} \, \partial_{\theta_l} \psi \right), (\nabla \phi^i)_\psi \right\rangle + (\Delta \phi^i)_\psi$$

$$= \sum_{1 \leq m,l \leq p} \partial_{\theta_m} g^{m,l} \, \left\langle \partial_{\theta_l} \psi, (\nabla \phi^i)_\psi \right\rangle + \sum_{1 \leq m,l \leq p} g^{m,l} \, \left\langle \partial_{\theta_m \theta_l} \psi, (\nabla \phi^i)_\psi \right\rangle + (\Delta \phi^i)_\psi .$$

Combined with (21.41) and (21.42), this implies that

$$\sum_{1 \leq m \leq p} \partial_{\theta_m} g^{i,m} = \sum_{1 \leq m \leq p} \partial_{\theta_m} g^{m,i} + \sum_{1 \leq m,l \leq p} g^{m,l} \underbrace{\left\langle \partial_{\theta_m \theta_l} \psi, (\nabla \phi^i)_\psi \right\rangle}_{=C^i_{m,l}} + (\Delta \phi^i)_\psi .$$

Using (21.40), we conclude that

$$(\Delta \phi^i)_\psi = - \sum_{1 \leq m,l \leq p} g^{m,l} \, C^i_{m,l} = \operatorname{div} \left(\nabla \phi^i \right)_\psi = \sum_{1 \leq j \leq p} \frac{1}{\sqrt{\det(g)}} \, \partial_{\theta_j} \left(\sqrt{\det(g)} \, g^{i,j} \right).$$

$$(21.43)$$

21.5.2   Composition formulae

Suppose we are given a function $F = f \circ \phi$ on S. By (21.20) we have

$$\nabla F = \nabla (f \circ \phi) = \sum_{1 \leq j \leq p} \left(\partial_{\theta_j} f \right)_\phi \nabla \phi^j \qquad (21.44)$$

and

$$F = \left(\partial_{\theta_j} f \right)_\phi = \left(\partial_{\theta_j} f \right) \circ \phi \implies \nabla \left(\left(\partial_{\theta_j} f \right)_\phi \right) = \sum_{1 \leq i \leq p} \left(\partial_{\theta_i, \theta_j} f \right)_\phi \nabla \phi^i.$$

Using the differential rule (19.30) we find that

$$\nabla^2 F = \sum_{1 \leq j \leq p} \left[\nabla \left(\left(\partial_{\theta_j} f \right)_\phi \right) \left(\nabla \phi^j \right)^T + \left(\partial_{\theta_j} f \right)_\phi \nabla^2 \phi^j \right].$$

By (21.39), this yields the second covariant derivative formula

$$\begin{aligned}
\nabla^2 (f \circ \phi) &= \sum_{1 \leq i,j \leq p} \left(\partial_{\theta_i, \theta_j} f \right)_\phi \nabla \phi^i \left(\nabla \phi^j \right)^T + \sum_{1 \leq j \leq p} \left(\partial_{\theta_j} f \right)_\phi \nabla^2 \phi^j \\
&= \sum_{1 \leq i,j \leq p} \left[\left(\partial_{\theta_i, \theta_j} f \right)_\phi \nabla \phi^i \left(\nabla \phi^j \right)^T + \left(\partial_{\theta_j} f \right)_\phi \nabla \phi^i \left[\partial_{\theta_i} \left(\left(\nabla \phi^j \right)^T_\psi \right) \right]_\phi \right]. (21.45)
\end{aligned}$$

We also readily check that

$$\begin{aligned}
\mathrm{tr} \left(\nabla^2 F \right) &= \sum_{1 \leq i,j \leq p} \left(\partial_{\theta_i, \theta_j} f \right)_\phi \left\langle \nabla \phi^i, \nabla \phi^j \right\rangle + \sum_{1 \leq j \leq p} \left(\partial_{\theta_j} f \right)_\phi \mathrm{tr} \left(\nabla^2 \phi^j \right) \\
&= \sum_{1 \leq i,j \leq p} g_\phi^{i,j} \left(\partial_{\theta_i, \theta_j} f \right)_\phi + \sum_{1 \leq j \leq p} \left(\partial_{\theta_j} f \right)_\phi \mathrm{tr} \left(\nabla^2 \phi^j \right).
\end{aligned}$$

This yields the Laplacian formula

$$\begin{aligned}
\Delta (f \circ \phi) &:= \mathrm{tr} \left(\nabla^2 (f \circ \phi) \right) \\
&= \sum_{1 \leq i,j \leq p} g_\phi^{i,j} \left(\partial_{\theta_i, \theta_j} f \right)_\phi + \sum_{1 \leq j \leq p} \left(\partial_{\theta_j} f \right)_\phi \Delta \phi^j. \qquad (21.46)
\end{aligned}$$

Using (21.43) we also have

$$\begin{aligned}
\Delta (f \circ \phi) &:= \mathrm{tr} \left(\nabla^2 (f \circ \phi) \right) \\
&= \sum_{1 \leq l,m \leq p} g_\phi^{l,m} \left[\left(\partial_{\theta_l, \theta_m} f \right)_\phi - \sum_{1 \leq j \leq p} C_{\phi,l,m}^j \left(\partial_{\theta_j} f \right)_\phi \right] \qquad (21.47)
\end{aligned}$$

with the Christoffel symbols $C_{\phi,l,m}^j = C_{l,m}^j \circ \phi$.

Hence we prove the following result.

Proposition 21.5.1 *We have a divergence formulation of the Riemannian Laplacian:*

$$(\Delta(f \circ \phi))_\psi$$

$$= \sum_{1 \le i \le p} \left[\sum_{1 \le j \le p} g^{i,j} \partial_{\theta_i}(\partial_{\theta_j} f) + \frac{1}{\sqrt{det(g)}} \sum_{1 \le j \le p} \partial_{\theta_i}\left(\sqrt{det(g)} \ g^{i,j}\right) \partial_{\theta_j} f \right]$$

$$= \sum_{1 \le i \le p} \frac{1}{\sqrt{det(g)}} \partial_{\theta_i} \left(\sqrt{det(g)} \sum_{1 \le j \le p} g^{i,j} \partial_{\theta_j} f \right) := \mathrm{div}_g\left(\nabla_g(f)\right) := \Delta_g(f).$$

$$(21.48)$$

In terms of the coordinate mappings ϵ_i introduced in (21.24), we have $\chi_i \circ \phi = \phi^i$ and

$$\Delta_g \epsilon_i = (\Delta \phi^i)_\psi \quad \text{and} \quad (\Delta \chi_i)_\psi = \Delta_g \psi^i.$$

For any couple of functions f_1 and f_2, we quote the following formula

$$\mathrm{div}_g\left(f_1 \nabla_g f_2\right) = f_1 \mathrm{div}_g\left(\nabla_g f_2\right) + \langle \nabla_g f_1, \nabla_g f_2 \rangle_g. \qquad (21.49)$$

We check this claim using

$$\partial_{\theta_i}\left(\sqrt{\det(g)} \ f_1 \sum_{1 \le j \le p} g^{i,j} \partial_{\theta_j} f_2\right)$$

$$= f_1 \partial_{\theta_i}\left(\sqrt{\det(g)} \sum_{1 \le j \le p} g^{i,j} \partial_{\theta_j} f_2\right) + \partial_{\theta_i}(f_1) \times \sqrt{\det(g)} \sum_{1 \le j \le p} g^{i,j} \partial_{\theta_j} f_2$$

and

$$\sum_{1 \le i \le p} \partial_{\theta_i} f_1 \sum_{1 \le i,j \le p} g^{i,j} \partial_{\theta_j} f_2 = \sum_{1 \le i,j \le p} g^{i,j} \partial_{\theta_i} f_1 \partial_{\theta_j} f_2 = \langle \partial f_1, \partial f_2 \rangle_{g^{-1}} = \langle \nabla_g f_1, \nabla_g f_2 \rangle_g.$$

21.5.3 Hessian operators

We end this section with an Hessian interpretation of the second covariant derivative $\nabla^2 F$. We let W_1, W_2 be a couple of vector fields in $T(S)$. We let V_1, V_2 be their pull back vector fields so that

$$W_k \circ \psi = (d\psi)(V_k) := \sum_{1 \le m \le p} V_k^m \partial_{\theta_m} \psi$$

for any $k = 1, 2$. Using (21.45) we prove that

$$W_1^T \, \nabla^2 F \, W_2$$

$$= \sum_{1 \leq m, m' \leq p} V_{\phi,1}^m V_{\phi,2}^{m'} \sum_{1 \leq i,j \leq p} (\partial_{\theta_i, \theta_j} f)_\phi \overbrace{(\partial_{\theta_m} \psi)_\phi^T \nabla \phi^i}^{=1_{m=i}} \overbrace{(\nabla \phi^j)^T (\partial_{\theta_{m'}} \psi)_\phi}^{=1_{m'=j}}$$

$$+ \sum_{1 \leq m, m' \leq p} V_{\phi,1}^m V_{\phi,2}^{m'} \sum_{1 \leq i,j \leq p} (\partial_{\theta_j} f)_\phi \underbrace{(\partial_{\theta_m} \psi)_\phi^T \nabla \phi^i}_{=1_{m=i}} \underbrace{\left[\partial_{\theta_i} \left((\nabla \phi^j)^T\right)\right]_\phi (\partial_{\theta_{m'}} \psi)_\phi}_{=-C_{\phi,i,m'}^j}$$

The assertion in the r.h.s. follows from

$$\partial_{\theta_i} \left(\underbrace{(\nabla \phi^j)_\psi^T (\partial_{\theta_m} \psi)}_{=1_{m=j}} \right) = 0 \Rightarrow \partial_{\theta_i} \left((\nabla \phi^j)_\psi^T\right) (\partial_{\theta_m} \psi) = -(\nabla \phi^j)_\psi^T (\partial_{\theta_i, \theta_m} \psi) = -C_{i,m}^j.$$

This yields

$$W_1^T \nabla^2 F \, W_2 = \sum_{1 \leq m, m' \leq p} V_{\phi,1}^m V_{\phi,2}^{m'} \left[(\partial_{\theta_m, \theta_{m'}} f)_\phi - \sum_{1 \leq j \leq p} C_{\phi,m,m'}^j (\partial_{\theta_j} f)_\phi \right]$$

$$= V_{\phi,1}^T \, (\text{Hess}_g(f))_\phi \, V_{\phi,2}.$$

Equivalently, we have the following result.

Proposition 21.5.2 *We have the formula*

$$\left(W_1^T \nabla^2 F \, W_2\right)_\psi = \langle W_1, \nabla^2 F \, W_2 \rangle = V_1^T \, \text{Hess}_g(f) \, V_2$$

$$= \langle V_1, \text{Hess}_g(f) \, V_2 \rangle = \langle V_1, \nabla_g^2 f \, V_2 \rangle_g,$$

with the Hessian matrix field $\text{Hess}_g(f) = ((\text{Hess}_g(f))_{m,m'})_{1 \leq m, m' \leq p}$ *on* S_ψ *with entries*

$$(\text{Hess}_g(f))_{m,m'} = \partial_{\theta_m, \theta_{m'}} f - \sum_{1 \leq j \leq p} C_{m,m'}^j \, \partial_{\theta_j} f$$

and the Riemannian second covariant derivative

$$\nabla_g^2(f) = g^{-1} \text{Hess}_g(f)$$

$$\left[\Rightarrow \langle V_1, \nabla_g^2 f \, V_2 \rangle_g = V_1^T \, g \, (g^{-1} \text{Hess}_g(f)) \, V_2 = V_1^T \, \text{Hess}_g(f) \, V_2\right]. \tag{21.50}$$

By (21.47), if we set $F = f \circ \phi$, *we also have the Laplacian formula*

$$\Delta(F) \circ \psi = \text{tr}\left(\nabla^2 F\right) \circ \psi = \text{div}\left(\nabla F\right) \circ \psi$$

$$= \text{tr}\left(\nabla_g^2 f\right) = \text{div}_g\left(\nabla_g(f)\right) = \Delta_g(f).$$

In differential calculus literature, the above formulae are sometimes written in the following form

$$\left(\nabla^2 F\right) (W_1, W_2) := \langle W_1, \nabla^2 F \, W_2 \rangle = \langle V_1, \nabla_g^2 f \, V_2 \rangle_g := \left((\nabla_g^2 f) (V_1, V_2)\right) \circ \phi$$

or using the Hessian symbol

$$\text{Hess}(F)\,(W_1, W_2) := \left(\nabla^2 F\right)\,(W_1, W_2) = \left(\left(\nabla_g^2 f\right)(V_1, V_2)\right) \circ \phi := (\text{Hess}_g(f)(V_1, V_2)) \circ \phi.$$

The bilinear form induced by the matrix $g^{-1}\text{Hess}_g(f)$ acts on the tangent space $T(S_\psi) = \text{Vect}(e_1, \ldots, e_p) = \mathbb{R}^p$ equipped with the inner product (21.18) with the unit basis vectors e_i on $S_\psi(\subset \mathbb{R}^p)$ defined in (21.2).

In this situation, we have the Hilbert-Schmidt inner product formula

$$f_1 = F_1 \circ \psi \quad f_2 = F_2 \circ \psi$$

$$\Rightarrow \quad \langle \nabla^2 F_1, \nabla^2 F_2 \rangle = \sum_{1 \le k,l \le p} g^{k,l}\, \langle \nabla_g^2 f_1(e_k), \nabla_g^2 f_2(e_l) \rangle_g \qquad (21.51)$$
$$= \text{tr}\left(\nabla_g^2 f_1\, \nabla_g^2 f_2\right) := \langle \nabla_g^2 f_1, \nabla_g^2 f_2 \rangle_g$$

with the Hilbert-Schmidt inner product defined in (19.8).

In view of (21.31) and (21.33) we also have

$$(\partial_{W_1}(\partial_{W_2}(F))) = \sum_{1 \le m,m' \le p} V_{\phi,1}^m \left(\partial_{\theta_m} V_2^{m'}\right)_\phi \left(\partial_{\theta_{m'}} f\right)_\phi$$
$$+ \sum_{1 \le m,m' \le p} V_{\phi,1}^m V_{\phi,2}^{m'} \left(\partial_{\theta_m, \theta'_m} f\right)_\phi,$$

$$\nabla_{W_1} W_2 = \sum_{1 \le m' \le p} \left[\sum_{1 \le m \le p} V_{\phi,1}^m \left(\partial_{\theta_m} V_2^{m'}\right)_\phi \right.$$
$$\left. + \sum_{1 \le m,j \le p} C_{\phi,m,j}^{m'}\, V_{\phi,1}^m\, V_{\phi,2}^j \right] \left(\partial_{\theta_{m'}} \psi\right)_\phi.$$

Since

$$\left(\partial_{\theta_k} \psi\right)^T (\partial F)_\psi = \sum_{0 \le l \le r} (\partial_{x_l} F)_\psi\, \partial_{\theta_k} \psi^l = \partial_{\theta_k}\,(F \circ \psi) = \partial_{\theta_k} f$$

we find that

$$\left(\nabla_{W_1} W_2\right)^T \partial F$$

$$= \sum_{1 \le m,m' \le p} V_{\phi,1}^m \left(\partial_{\theta_m} V_2^{m'}\right)_\phi \left(\partial_{\theta_{m'}} f\right)_\phi + \sum_{1 \le m,m' \le p} V_{\phi,1}^m\, V_{\phi,2}^{m'} \sum_{1 \le j \le p} C_{\phi,m,m'}^j \left(\partial_{\theta_j} f\right)_\phi.$$

We recover the fact that

$$W_1^T\, \nabla^2 F\, W_2 = (\partial_{W_1}(\partial_{W_2}(F))) - (\nabla_{W_1} W_2)^T \partial F = (\partial_{W_1}(\partial_{W_2}(F))) - \langle \nabla_{W_1} W_2, \partial F \rangle$$
$$= (\partial_{W_1}(\partial_{W_2}(F))) - \langle \nabla_{W_1} W_2, \nabla F \rangle = \langle W_2, \nabla_{W_1} \nabla F \rangle.$$

By (21.30), for and $F = f \circ \phi$ we have

$$\partial_{W_1}(F) \circ \psi = \langle \nabla_g f, V_1 \rangle_g = \partial_{V_1}(f).$$

On the other hand,

$$\langle W_2, W_3 \rangle = \langle V_2, V_3 \rangle_g \circ \phi \Rightarrow (\partial_{W_1} \langle W_2, W_3 \rangle) \circ \psi = \partial_{V_1} \langle V_2, V_3 \rangle_g$$

and by (21.34) we prove that

$$\langle \nabla_{W_1} W_2, W_3 \rangle \circ \psi = \langle (d\psi)(\nabla_{g,V_1} V_2), (d\psi)(V_3) \rangle = \langle \nabla_{g,V_1} V_2, V_3 \rangle_g.$$

We conclude that

$$\partial_{V_1} \langle V_2, V_3 \rangle_g = \langle \nabla_{g,V_1} V_2, V_3 \rangle_g + \langle V_2, \nabla_{g,V_1} V_3 \rangle_g.$$

We give some comments on the parallel transport technique introduced in (21.36). We let $c_1(t)$ be a given curve in S_ϕ with $\dot{c}_1(t) = V_1(c_1(t))$, and $U_i : t \in [0,1] \mapsto U_i(t) = V_i(c_1(t)) \in \mathbb{R}^p$ two parallel vectors to $c(t)$ s.t. $U_i(0) = V_i(c_1(0))$, with $i = 2,3$. In this situation, using (21.36) we have

$$\frac{d}{dt} \langle V_2(c_1(t)), V_3(c_1(t)) \rangle_{g(c_1(t))} = \left(\partial_{V_1} \left(\langle V_2, V_3 \rangle_g \right) \right) (c_1(t))$$

$$= \left(\langle \nabla_{g,V_1} V_2, V_3 \rangle_g + \langle V_2, \nabla_{g,V_1} V_3 \rangle_g \right) (c_1(t)) = 0.$$

This shows that the parallel transport is an isometry

$$\langle V_2(c_1(0)), V_3(c_1(0)) \rangle_{g(c_1(0))} = \langle V_2(c_1(1)), V_3(c_1(1)) \rangle_{g(c_1(1))}. \tag{21.52}$$

Using (21.35) we readily check that

$$\partial_{e_i} \langle e_j, e_k \rangle_g = \partial_{\theta_i} g_{j,k}$$

$$= \langle C_{i,j}, e_k \rangle_g + \langle e_j, C_{i,k} \rangle_g = \sum_{1 \leq l \leq p} g_{k,l}\, C_{i,j}^l + \sum_{1 \leq l \leq p} g_{j,l}\, C_{i,k}^l$$

the unit vectors e_i on $\mathbb{R}^p$. This implies that

$$\partial_{\theta_j} g_{i,k} = \sum_{1 \leq l \leq p} g_{k,l}\, C_{i,j}^l + \sum_{1 \leq l \leq p} g_{i,l}\, C_{j,k}^l$$

$$\partial_{\theta_k} g_{i,j} = \sum_{1 \leq l \leq p} g_{j,l}\, C_{i,k}^l + \sum_{1 \leq l \leq p} g_{i,l}\, C_{j,k}^l.$$

Hence after simple addition, we get

$$\partial_{\theta_k} g_{i,j} + \partial_{\theta_j} g_{i,k} - \partial_{\theta_i} g_{j,k} = 2 \sum_{1 \leq l \leq p} g_{i,l}\, C_{j,k}^l.$$

This yields the formula

$$C_{j,k}^m = \frac{1}{2} \sum_{1 \leq i \leq m} g^{m,i} \left(\partial_{\theta_k} g_{i,j} + \partial_{\theta_j} g_{i,k} - \partial_{\theta_i} g_{j,k} \right). \tag{21.53}$$

21.6 Bochner-Lichnerowicz formula

Using (19.22) and (19.82), the Ricci curvature is given in terms of the basis vector fields $(\partial_{\theta_i}\psi)_\phi$ by the formula

$$f_1 = F_1 \circ \psi \quad f_2 = F_2 \circ \psi$$

$$\begin{aligned} \implies \operatorname{Ric}(\nabla F_1, \nabla F_2) &= \sum_{1 \le i,j \le p} (\nabla_g f_1)^i_\phi (\nabla_g f_2)^j_\phi \operatorname{Ric}((\partial_{\theta_i}\psi)_\phi, (\partial_{\theta_i}\psi)_\phi) \\ &:= \operatorname{Ric}_g(\nabla_g f_1, \nabla_g f_2) \circ \phi \end{aligned}$$

with

$$\operatorname{Ric}(\partial_{\theta_i}\psi, \partial_{\theta_i}\psi) = R_{i,j} \circ \psi$$

$$= \sum_{1 \le m,n \le p} \left[C^n_{i,j} C^m_{m,n} - C^n_{i,m} C^m_{j,n} \right] + \sum_{1 \le m \le p} \left[\partial_{\theta_m} C^m_{i,j} - \partial_{\theta_j} C^m_{i,m} \right]$$

and the Christoffel symbols $C^n_{i,j}$ defined in (21.32). (If we use the definition (19.21) with the tangent vector fields $V_i = (\partial_{\theta_i}\psi)_\phi \in T(S)$, the above formula is satisfied by replacing $C^n_{i,j}$ by the expression $C^n_{i,j} \circ \psi$ in the coordinate system ψ.)

The computation of the matrix R in terms of the Christoffel symbols is rather elementary but can be tedious to derive. Next, we provide a natural matrix decomposition to compute R. We consider the matrices

$$C_k := \begin{pmatrix} C^1_{k,1} & \cdots & C^1_{k,p} \\ \vdots & \vdots & \vdots \\ C^p_{k,1} & \cdots & C^p_{k,p} \end{pmatrix} \quad \text{and} \quad E := \begin{pmatrix} \operatorname{tr}_e(C_1 C_1) & \cdots & \operatorname{tr}_e(C_1 C_p) \\ \vdots & \vdots & \vdots \\ \operatorname{tr}_e(C_p C_1) & \cdots & \operatorname{tr}_e(C_p C_p) \end{pmatrix}.$$

We also set

$$t(C) := \begin{pmatrix} \operatorname{tr}_e(C_1) \\ \vdots \\ \operatorname{tr}_e(C_p) \end{pmatrix} \quad \text{and} \quad B := \begin{pmatrix} \langle C_{1,1}, t(C) \rangle & \cdots & \langle C_{1,p}, t(C) \rangle \\ \vdots & \vdots & \vdots \\ \langle C_{p,1}, t(C) \rangle & \cdots & \langle C_{p,p}, t(C) \rangle \end{pmatrix}$$

and finally the differential matrices

$$T := \begin{pmatrix} \partial_{\theta_1} \operatorname{tr}_e(C_1 & \cdots & \partial_{\theta_p} \operatorname{tr}_e(C_1) \\ \vdots & \vdots & \vdots \\ \partial_{\theta_1} \operatorname{tr}_e(C_p) & \cdots & \partial_{\theta_p} \operatorname{tr}_e(C_p) \end{pmatrix} \quad \text{and} \quad S = \begin{pmatrix} \operatorname{div}(C_{1,1}) & \cdots & \operatorname{div}(C_{1,p}) \\ \vdots & \vdots & \vdots \\ \operatorname{div}(C_{p,1}) & \cdots & \operatorname{div}(C_{p,p}) \end{pmatrix}$$

with

$$C_{i,j} = \begin{pmatrix} C^1_{i,j} \\ \vdots \\ C^p_{i,j} \end{pmatrix} \quad \text{and} \quad \operatorname{div}(C_{i,j}) = \sum_{1 \le k \le p} \partial_{\theta_k}(C^k_{i,j}).$$

In this notation, we have the matrix formulation of the Ricci curvature

$$R \circ \psi = B - E + S - T \tag{21.54}$$

with the matrices (B, E, S, T) defined above.

To check this decomposition we simply observe that

$$\sum_{1\leq n\leq p} C_{i,j}^n \sum_{1\leq m\leq p} C_{n,m}^m = \sum_{1\leq n\leq p} C_{i,j}^n \, \mathrm{tr}(C_n) = \langle C_{i,j}, t(C)\rangle$$

$$\sum_{1\leq n\leq p}\sum_{1\leq m\leq p} C_{i,m}^n C_{j,n}^m = \sum_{1\leq n\leq p} (C_i C_j)_{n,n} = \mathrm{tr}_e\,(C_i C_j)$$

$$\sum_{1\leq m\leq p} \partial_{\theta_m} C_{i,j}^m = \mathrm{div}(C_{i,j}) \quad \text{and} \quad \sum_{1\leq m\leq p} \partial_{\theta_j}(C_{i,m}^m) = \partial_{\theta_j}\mathrm{tr}_e(C_i) = T_{i,j}.$$

A detailed analysis of the Ricci curvature of the sphere and the torus is provided in section 24.1.2 and in section 24.1.3.

We also notice that

$$\mathrm{Ric}_g(\nabla_g\epsilon_i, \nabla_g\epsilon_i) = \sum_{1\leq k,l\leq p} \overbrace{(\nabla_g\epsilon_i)^k}^{=g^{i,k}} (\nabla_g\epsilon_j)^l \, \mathrm{Ric}\,(\partial_{\theta_k}\psi, \partial_{\theta_i}\psi) = (g^{-1}Rg^{-1})_{i,j}$$

with the coordinate projection mappings ϵ_i introduced in (21.24). This yields

$$\nabla_g f = \sum_{1\leq i\leq p} \partial_{\theta_i} f \, \nabla_g\epsilon_i$$

$$\mathrm{Ric}_g\,(\nabla_g f_1, \nabla_g f_2) = \sum_{1\leq i,j\leq p} \partial_{\theta_i} f_1 \, \partial_{\theta_j} f_2 \, \mathrm{Ric}_g(\nabla_g\epsilon_i, \nabla_g\epsilon_i) = \langle \nabla_g f_1, R\, \nabla_g f_2\rangle.$$

Rephrasing theorem 19.9.2 on the Riemannian manifold we obtain the following theorem.

Theorem 21.6.1 (Bochner-Lichnerowicz formula) *For any smooth functions f_1, f_2 on $\mathbb{R}^p$ of we have*

$$L_g = \frac{1}{2}\,\Delta_g \Longrightarrow \Gamma_{L_g}(f_1, f_2) = \langle \nabla_g f_1, \nabla_g f_2\rangle_g.$$

In addition, we have

$$\begin{aligned} L_g\langle \nabla_g f_1, \nabla_g f_2\rangle_g &= \langle \nabla_g f_1, \nabla_g L_g f_2\rangle_g + \langle \nabla_g f_2, \nabla_g L_g f_1\rangle_g \\ &\quad + \left[\langle \nabla_g^2 f_1, \nabla_g^2 f_2\rangle_g + \mathrm{Ric}_g\,(\nabla_g f_1, \nabla_g f_2)\right]. \end{aligned}$$

$$(21.55)$$

Rewritten in terms of Γ_{2,L_g}, we have the synthetic formula

$$\Gamma_{2,L_g}(f_1, f_2) = \langle \nabla_g^2 f_1, \nabla_g^2 f_2\rangle_g + \mathrm{Ric}_g\,(\nabla_g f_1, \nabla_g f_2). \qquad (21.56)$$

Our next objective is to provide a more explicit description of Γ_{2,L_g} in terms of the inverse g^{-1} of the metric.

To this end, we first notice that

$$C_{j,k}^m = -\frac{1}{2} \sum_{1\leq i\leq m} \left(g_{i,j}\,\partial_{\theta_k} g^{m,i} + g_{i,k}\,\partial_{\theta_j} g^{m,i} + g^{m,i}\,\partial_{\theta_i} g_{j,k}\right).$$

This implies that

$$\sum_{1 \le j \le p} g^{\alpha,j} C_{j,k}^m = -\frac{1}{2} \left(\partial_{\theta_k} g^{m,\alpha} + \sum_{1 \le i,j \le p} \left(g_{k,i} \, g^{\alpha,j} \, \partial_{\theta_j} g^{m,i} - g^{m,i} \, g_{k,j} \, \partial_{\theta_i} g^{\alpha,j} \right) \right).$$

These decompositions readily imply the following formula.

$$
\begin{aligned}
\left(g^{-1} C^m g^{-1} \right)_{\alpha,\beta} &= \sum_{1 \le j,k \le p} g^{\alpha,j} \, g^{\beta,k} \, C_{j,k}^m \\
&= -\frac{1}{2} \left(\partial_{g^\beta} g^{m,\alpha} + \partial_{g^\alpha} g^{m,\beta} - \partial_{g^m} g^{\alpha,\beta} \right), \quad (21.57)
\end{aligned}
$$

with the Christofell symbol matrix $C^m := \left(C_{i,j}^m \right)_{1 \le i,j \le p}$ and the operators ∂_{g^i} defined by

$$\forall 1 \le i \le p \qquad \partial_{g^i} f := \sum_{1 \le j \le p} g^{i,j} \, \partial_{\theta_j} f \left(= (\nabla_g f)^i \right).$$

We also have the matrix decomposition formula

$$\nabla_g^2 f = g^{-1} \partial^2 f - \sum_{1 \le m \le p} g^{-1} C^m \, \partial_{\theta_m} f = \nabla_g \partial f - \sum_{1 \le m \le p} \nabla_g^2 \epsilon_m \, \partial_{\theta_m} f.$$

In this notation we have a more explicit form of the Hilbert-Schmidt inner product formulae

$$\langle \nabla_g^2 f_1, \nabla_g^2 f_2 \rangle_g$$

$$= \langle \nabla_g \partial f_1 \, \nabla_g \partial f_2 \rangle_g + \sum_{1 \le m,n \le p} \langle \nabla_g^2 \epsilon_m, \nabla_g^2 \epsilon_n \rangle_g \, \partial_{\theta_m} f_1 \, \partial_{\theta_n} f_2$$

$$+ \sum_{1 \le \alpha,\beta,m \le p} \left(\partial_{g^\beta} g^{m,\alpha} - \frac{1}{2} \partial_{g^m} g^{\alpha,\beta} \right) \left[\partial_{\theta_m} f_1 \, \partial_{\theta_\beta,\theta_\alpha} f_2 + \partial_{\theta_m} f_2 \, \partial_{\theta_\beta,\theta_\alpha} f_1 \right]$$

$$(21.58)$$

and

$$\langle \nabla_g \partial f_1 \, \nabla_g \partial f_2 \rangle_g = \sum_{1 \le i,j \le p} \partial_{g^i} \partial_{\theta_j} f_1 \, \partial_{g^j} \partial_{\theta_i} f_2 = \sum_{1 \le i,j \le p} g^{i,j} \, \Gamma_{L_g} \left(\partial_{\theta_i} f_1, \partial_{\theta_j} f_2 \right).$$

To check (21.58) we use (21.51) to obtain the following decomposition

$$\langle \nabla_g^2 f_1, \nabla_g^2 f_2 \rangle_g$$

$$= \langle \nabla_g \partial f_1 \, \nabla_g \partial f_2 \rangle_g + \sum_{1 \le m,n \le p} \overbrace{\mathrm{tr} \left(g^{-1} C^m g^{-1} C^n \right)}^{\langle \nabla_g^2 \epsilon_n, \nabla_g^2 \epsilon_n \rangle_g} \, \partial_{\theta_m} f_1 \, \partial_{\theta_n} f_2$$

$$- \sum_{1 \le m \le p} \mathrm{tr} \left(g^{-1} C^m g^{-1} \, \partial^2 f_2 \right) \, \partial_{\theta_m} f_1 - \sum_{1 \le m \le p} \mathrm{tr} \left(g^{-1} C^m g^{-1} \, \partial^2 f_1 \right) \, \partial_{\theta_m} f_2.$$

By (21.57) we conclude that

$$
\begin{aligned}
\mathrm{tr}\left(g^{-1}C^m g^{-1}\,\partial^2 f_2\right) &= \sum_{1\le \alpha,\beta\le p} \left(g^{-1}C^m g^{-1}\right)_{\alpha,\beta}\,\partial_{\theta_\beta,\theta_\alpha} f_2 \\
&= -\frac{1}{2}\sum_{1\le\alpha,\beta\le p}\left(\partial_{g^\beta}g^{m,\alpha} + \partial_{g^\alpha}g^{m,\beta} - \partial_{g^m}g^{\alpha,\beta}\right)\,\partial_{\theta_\beta,\theta_\alpha} f_2 \\
&= -\sum_{1\le\alpha,\beta\le p}\left(\partial_{g^\beta}g^{m,\alpha} - \frac{1}{2}\partial_{g^m}g^{\alpha,\beta}\right)\,\partial_{\theta_\beta,\theta_\alpha} f_2.
\end{aligned}
$$

The end of the proof of (21.58) is now clear.   ∎

Theorem 21.6.2 *For any $1\le i,j\le p$ we have*

$$
\Gamma_{2,L_g}(\epsilon_i,\epsilon_j) = \left\langle \nabla_g^2\epsilon_i, \nabla_g^2\epsilon_j\right\rangle_g + \mathrm{Ric}_g(\nabla_g\epsilon_i,\nabla_g\epsilon_j) = L_g(g^{i,j}) - \partial_{g^j}b^i - \partial_{g^i}b^j
$$

with the drift vector field

$$
b^i := \frac{1}{\sqrt{det(g)}}\sum_{1\le i\le p}\partial_{\theta_i}\left(\sqrt{det(g)}\,g^{i,j}\right).
$$

For any smooth functions f_1, f_2 we have the formula

$$
\Gamma_{2,L_g}(f_1,f_2)
$$

$$
= \Gamma_{2,L_g}(\epsilon_i,\epsilon_j)\,\partial_{\theta_i}f_1\,\partial_{\theta_j}f_2 + \left\langle \nabla_g\partial f_1\ \nabla_g\partial f_2\right\rangle_g
$$

$$
+ \sum_{1\le i,j,k\le p}\left(\partial_{g^k}g^{i,j} - \frac{1}{2}\,\partial_{g^i}g^{j,k}\right)\left[\partial_{\theta_i}f_1\,\partial_{\theta_j,\theta_k}f_2 + \partial_{\theta_i}f_2\,\partial_{\theta_j,\theta_k}f_1\right].
$$

$$
(21.59)
$$

Proof :
We have

$$
\nabla_g f = \sum_{1\le i\le p}\partial_{\theta_i}f\,\nabla_g\epsilon_i
$$

$$
\Rightarrow \mathrm{Ric}_g\left(\nabla_g f_1,\nabla_g f_2\right) = \sum_{1\le i,j\le p}\partial_{\theta_i}f_1\,\partial_{\theta_j}f_2\,\mathrm{Ric}_g(\nabla_g\epsilon_i,\nabla_g\epsilon_i).
$$

Using (21.56) and (21.58) we check (21.59). On the other hand, we have

$$
\begin{aligned}
\Gamma_{2,L_g}(\epsilon_i,\epsilon_j) &= L_g\langle\nabla_g\epsilon_i,\nabla_g\epsilon_j\rangle_g - \langle\nabla_g\epsilon_i,\nabla_g L_g\epsilon_j\rangle_g - \langle\nabla_g\epsilon_j,\nabla_g L_g\epsilon_i\rangle_g \\
&= \left\langle\nabla_g^2\epsilon_i,\nabla_g^2\epsilon_j\right\rangle_g + \mathrm{Ric}_g(\nabla_g\epsilon_i,\nabla_g\epsilon_j).
\end{aligned}
$$

Using (21.48) we check that
$$
\Delta_g f = \mathrm{tr}\left(g^{-1}\partial^2 f\right) + \partial_b f
$$

with the differential operators

$$
\mathrm{tr}\left(g^{-1}\partial^2 f\right) := \sum_{1\le i,j\le p} g^{i,j}\,\partial_{\theta_i,\theta_j}f \quad\text{and}\quad \partial_b f := \sum_{1\le i\le p} b^i\,\partial_{\theta_i}f.
$$

This implies that

$$\langle \nabla_g \epsilon_i, \nabla_g \epsilon_j \rangle_g = g^{i,j} \Rightarrow L_g \langle \nabla_g \epsilon_i, \nabla_g \epsilon_j \rangle_g = \frac{1}{2} \mathrm{tr} \left(g^{-1} \partial^2 g^{i,j} \right) + \partial_b g^{i,j}$$

and

$$L_g \epsilon_i = b^i \Rightarrow \langle \nabla_g \epsilon_j, \nabla_g L_g \epsilon_i \rangle_g = \partial_{g^j} b^i.$$

The proof of the theorem is now completed. ∎

The reader may have noticed that the Riemannian derivatives and the Bochner formula have the same form as discussed in chapter 19. We end this section with a discussion on the correspondence principles between these geometrical objects. Suppose we are given some positive definite and symmetric matrix $g = (g_{i,j})_{1 \le i,j \le p}$ with functional entries $g_{i,j} : \theta \in \mathbb{R}^p \mapsto \mathbb{R}$. Firstly, we recall that parameter space $S_\psi \subset \mathbb{R}^p$ discussed in this chapter is free of any constraints and we have

$$T_\theta(S_\psi) = \mathrm{Vect}\,(e_1, \ldots, e_p)$$

with the unit vectors fields discussed in (21.18). We recall that we distinguish $e_i(\theta)$ from $e_i(\theta')$ for different states θ and θ'. We equip this vector space with the scalar product $\langle e_i(\theta), e_j(\theta) \rangle_{g(\theta)} = g_{i,j}(\theta)$. In this context, the Riemannian versions of the r projection vector fields $\pi_i = \nabla \chi_i$ discussed in (19.3), (21.24) and (19.26) are given by

$$\forall 1 \le i \le p \qquad \sum_{1 \le k,l \le p} g^{k,l} e_k^i\, e_l = g^i = \nabla_g \epsilon_i$$

with the p coordinate mappings ϵ_i defined in (21.24). The Riemannian version of the second covariant derivative $\nabla^2 F$ is given by $\nabla_g^2 f$, and its action is defined by

$$\nabla_g^2 f(V_1, V_2) = \langle V_1, \nabla_g^2 f\, V_2 \rangle_g = \langle V_1,\ \mathrm{Hess}_g(f)\, V_2 \rangle.$$

The Riemannian versions of the r derivatives $\nabla^2 \chi_i$ are defined by the p derivatives $\nabla_g^2 \epsilon_j$, and so on. In this notation, we observe that

$$\nabla_g^2 \epsilon_j = g^j \Rightarrow \partial_{\nabla_g^2 \epsilon_j} = \sum_{1 \le k \le p} g^{j,k}\, \partial_{\theta_k} \quad \text{and} \quad \langle \nabla_g \epsilon_i, \nabla_g \epsilon_j \rangle_g = g^{i,j}.$$

This yields the formulae

$$\partial_{g^k} g^{i,j} - \frac{1}{2}\, \partial_{g^i} g^{j,k} \quad = \quad \partial_{\nabla_g \epsilon_k} \langle \nabla_g \epsilon_i, \nabla_g \epsilon_j \rangle_g - \frac{1}{2}\, \partial_{\nabla_g \epsilon_i} \langle \nabla_g \epsilon_j, \nabla_g \epsilon_k \rangle_g$$

and

$$\nabla_g^2 \epsilon_i \left(\nabla_g \epsilon_j, \nabla_g \epsilon_k \right) = \langle g^j, \mathrm{Hess}_g(\epsilon_i)\, g^k \rangle = - \left(g^{-1} C^i g^{-1} \right)_{j,k} = \frac{1}{2} \left(\partial_{g^k} g^{i,j} + \partial_{g^j} g^{i,k} - \partial_{g^i} g^{j,k} \right).$$

The above formulae are the Riemannian versions of (19.63) and (19.68). This shows that the formula (21.59) is the Riemannian version of (19.101).

Definition 21.6.3 *We consider the operator* Υ_{L_g} *defined for any smooth functions* f_i *on* $\mathbb{R}^p$, *with* $i = 1, 2, 3$, *by the formula*

$$\Upsilon_{L_g}(f_3)(f_1, f_2) := \Gamma_{2,L_g}(f_1 f_2, f_3) - f_1\, \Gamma_{2,L_g}(f_2, f_3) - f_2\, \Gamma_{2,L_g}(f_1, f_3). \qquad (21.60)$$

Proposition 21.6.4 *For any smooth functions f_i, with $i = 1, 2, 3$, we have the Hessian formula*

$$\frac{1}{2} \, \Upsilon_{L_g}(f_3)(f_1, f_2) = \left(\partial^2 f_3 + (\partial f_3)^T \nabla_g^2 \epsilon \right) (\nabla_g f_1, \nabla_g f_2) \tag{21.61}$$

$$= \mathrm{Hess}_g(f_3)(\nabla_g f_1, \nabla_g f_2) \tag{21.62}$$

with the bilinear form

$$(\partial f_3)^T \nabla_g^2 \epsilon := \sum_{1 \leq i \leq p} \partial_{\theta_i} f_3 \, \nabla_g^2 \epsilon_i.$$

We consider a sequence of smooth functions $f_k = (f_{k,i})_{1 \leq i \leq n_k}$ on $\mathbb{R}^p$, and $F_k(x_1, \ldots, x_{n_k})$, and we set $\mathcal{F}_k = F_k(f_k) = F_k(f_{k,1}, \ldots, f_{k,n_k})$, with $k = 1, 2$. In this situation, we have the change of variable formula

$$\Gamma_{2,L_g}(\mathcal{F}_1, \mathcal{F}_2) - \sum_{1 \leq i \leq n_1} \sum_{1 \leq k \leq n_2} (\partial_{x_i} F_1)(f_1)(\partial_{x_k} F_2)(f_2) \, \Gamma_{2,L_g}(f_{1,i}, f_{2,k})$$

$$= \sum_{1 \leq i,j \leq n_1} \sum_{1 \leq k,l \leq n_2} (\partial_{x_i, x_j} F_1)(f_1)(\partial_{x_k, x_l} F_2)(f_2) \, \Gamma_{L_g}(f_{1,i}, f_{2,l}) \, \Gamma_{L_g}(f_{1,j}, f_{2,k})$$

$$+ \frac{1}{2} \sum_{1 \leq i \leq n_1} \sum_{1 \leq k,l \leq n_2} (\partial_{x_i} F_1)(f_1)(\partial_{x_k, x_l} F_2)(f_2) \, \Upsilon_{L_g}(f_{1,i})(f_{2,k}, f_{2,l})$$

$$+ \frac{1}{2} \sum_{1 \leq k \leq n_2} \sum_{1 \leq i,j \leq n_1} (\partial_{x_k} F_2)(f_2)(\partial_{x_i, x_j} F_1)(f_1) \, \Upsilon_{L_g}(f_{2,k})(f_{1,i}, f_{1,j}).$$

Proof :

The proof follows the same lines of arguments as the arguments in proposition 19.9.7, and thus it is only sketched below. The formula (21.61) is based on

$$\nabla_g f = \sum_{1 \leq i \leq p} \partial_{\theta_i} f \, \nabla_g \epsilon_i$$

$$\Rightarrow \nabla_g^2 \epsilon_i (\nabla_g f_1, \nabla_g f_2) = \sum_{1 \leq j,j \leq p} \nabla_g^2 \epsilon_i (\nabla_g \epsilon_j, \nabla_g \epsilon_k) \, \partial_{\theta_j} f_1 \, \partial_{\theta_k} f_2$$

$$= \frac{1}{2} \sum_{1 \leq j,k \leq p} \left(\partial_{g^k} g^{i,j} + \partial_{g^j} g^{i,k} - \partial_{g^i} g^{j,k} \right) \partial_{\theta_j} f_1 \, \partial_{\theta_k} f_1$$

and

$$\sum_{1 \leq i,j,k,l \leq p} g^{i,k} \, g^{j,l} \, \partial_{\theta_k, \theta_l} f_3 \, \partial_{\theta_i} f_1 \partial_{\theta_j} f_2 = \sum_{1 \leq k,l \leq p} \partial_{\theta_k, \theta_l} f_3 \, (\nabla_g f_1)^k \, (\nabla_g f_2)^l$$

$$= \left(\partial^2 f_3 \right) (\nabla_g f_1, \nabla_g f_2).$$

To check (21.62) we use

$$\left(\partial^2 f_3 + \sum_{1 \leq i \leq p} \partial_{\theta_i} f_3 \, \nabla_g^2 \epsilon_i \right) (\nabla_g f_1, \nabla_g f_2)$$

$$= \sum_{1 \leq j,k \leq p} \partial_{\theta_j} f_1 \partial_{\theta_k} f_2 \left[\partial^2 f_3 (\nabla_g \epsilon_j, \nabla_g \epsilon_k) + \sum_{1 \leq i \leq p} \partial_{\theta_i} f_3 \, \nabla_g^2 \epsilon_i (\nabla_g \epsilon_j, \nabla_g \epsilon_k) \right]$$

$$= \sum_{1 \leq j,k \leq p} \partial_{\theta_j} f_1 \left(g^{-1} \left(\partial^2 f_3 - C^i \right) g^{-1} \right)_{j,k} \partial_{\theta_k} f_2 = (\nabla_g f_1)^T \mathrm{Hess}_g(f_3) \nabla_g f_2.$$

The proof of the last assertion follows exactly the same arguments as the ones provided in the proof of proposition 19.9.7 and thus it is omitted. This ends the proof of the proposition.

∎

21.7 Exercises

Exercise 375 (Product derivative formula) *We consider the Laplacian Δ_g associated with some Riemannian metric g and defined in (21.46). Check the formula*

$$\Gamma_{\Delta_g}(f_1, f_2 f_3) = f_2 \, \Gamma_{\Delta_g}(f_1, f_3) + f_3 \, \Gamma_{\Delta_g}(f_1, f_2).$$

Exercise 376 (Hessian operator) *We consider the Laplacian Δ_g associated with some Riemannian metric g and defined in (21.46). Check the formula*

$$\begin{aligned} \Upsilon_{\Delta_g}(f_1)(f_2, f_3) \; &:= \; \Gamma_{2,\Delta_g}(f_1, f_2 f_3) - f_2 \, \Gamma_{2,\Delta_g}(f_1, f_3) - f_3 \, \Gamma_{2,\Delta_g}(f_1, f_2) \\ &= \; \Gamma_{\Delta_g}\left(f_2, \Gamma_{\Delta_g}(f_1, f_3)\right) + \Gamma_{\Delta_g}\left(f_3, \Gamma_{\Delta_g}(f_1, f_2)\right) - \Gamma_{\Delta_g}\left(f_1, \Gamma_{\Delta_g}(f_2, f_3)\right). \end{aligned}$$

Exercise 377 (Parametric surfaces) *Find a parametrization ψ of the surfaces S defined by:*

- *The elliptic paraboloid $\left(\dfrac{x_1}{a}\right)^2 + \left(\dfrac{x_2}{b}\right)^2 = \dfrac{x_3}{c}$, for some given parameters a, b, c.*

- *The hyperbolic paraboloid $\left(\dfrac{x_2}{a}\right)^2 - \left(\dfrac{x_1}{b}\right)^2 = \dfrac{x_3}{c}$, for some given parameters a, b, c.*

- *The sphere $x_1^2 + x_2^2 + x_3^2 = r^2$, for some radius r.*

- *The p-sphere in $\mathbb{R}^{p+1} : x_1^2 + \ldots + x_{p+1}^2 = r^2$, for some radius r.*

- *The cylinder $x_1^2 + x_2^2 = r^2$ and $x_3 \in \mathbb{R}$, for some radius r.*

Describe the tangent spaces $T_x(S)$ at the states $x \in S$.

Exercise 378 (Orthogonal projections) *Describe the normal unit vectors and the orthogonal projections on the tangent spaces $T_x(S)$ of the surfaces discussed in exercise 377.*

Exercise 379 (Riemannian structures) *Describe a Riemannian scalar product on the tangent spaces $T(S_\psi)$ associated with the parametrization ψ of the surfaces S discussed in exercise 377.*

Exercise 380 (Projection matrices) *Find the orthogonal projection matrices on the tangent spaces $T(S)$ of the surfaces S discussed in exercise 377.*

Exercise 381 (Riemannian gradient) *Compute the Riemannian gradient $\nabla_g f$ of a smooth function f on the parametrization space S_ψ associated with the parametrization ψ of the surfaces S discussed in exercise 377.*

Exercise 382 (Second covariant derivative) *Compute the Riemannian second covariant derivative* $\nabla_g^2 f$ *of a smooth function f on the unit sphere* $\mathbb{S}^2 = \{(x_1, x_2, x_3) \in \mathbb{R}^3 : x_1^2 + x_2^2 + x_3^2 = 1\}$ *equipped with the spherical coordinates. Check that the corresponding Riemannian Laplacian on the unit sphere is defined by*

$$\Delta_g(f) = \frac{1}{\sin(\theta_1)} \, \partial_{\theta_1} \left(\sin(\theta_1) \, \partial_{\theta_1} f \right) + \frac{1}{\sin^2(\theta_1)} \, \partial_{\theta_2}^2 (f).$$

Exercise 383 *We consider the cone $S \in \mathbb{R}^3$ defined by the equation*

$$S := \left\{ x = (x_1, x_2, x_3) \in \mathbb{R}^3 \; : \; x_3 \geq 0 \quad \text{and} \quad a \, x_3 = \sqrt{x_1^2 + x_2^2} \right\}$$

for some given parameter $a > 0$. We consider polar parametrization is given by the function

$$\psi \; : \; \theta = (\theta_1, \theta_2) \in [0, \infty[\times [0, 2\pi] \mapsto \psi(\theta_1, \theta_2) = \begin{pmatrix} a \, \theta_1 \cos(\theta_2) \\ a \, \theta_1 \sin(\theta_2) \\ \theta_1 \end{pmatrix} \in S.$$

Compute the Riemannian gradient ∇_g and the Hessian ∇_g^2. Deduce that the corresponding Riemannian Laplacian on the cone is defined by

$$\Delta_g = (1 + a^2)^{-1} \partial_{\theta_1}^2 f + a^{-2} \theta_1^{-2} \partial_{\theta_2}^2 f + (\theta_1 (1 + a^2))^{-1} \, \partial_{\theta_1} f.$$

Exercise 384 (Ellipsoid) *We consider the ellipsoid S embedded in $\mathbb{R}^3$, centered at the origin and defined by the equation*

$$\left(\frac{x_1}{a_1} \right)^2 + \left(\frac{x_2}{a_2} \right)^2 + \left(\frac{x_3}{a_3} \right)^2 = 1$$

for some parameters $a_1 \geq a_2 \geq a_3 > 0$. We equip S with the spherical parametrization defined by

$$\psi(\theta_1, \theta_2) = \begin{pmatrix} a_1 \, \sin(\theta_1) \cos(\theta_2) \\ a_2 \, \sin(\theta_1) \sin(\theta_2) \\ a_3 \, \cos(\theta_1) \end{pmatrix}$$

with the restrictions $S_\psi = \{(\theta_1, \theta_2) \; : \; \theta_1 \in [0, \pi], \, \theta_2 \in [0, 2\pi]\}$. Compute the Riemannian gradient ∇_g and the Hessian ∇_g^2.

Exercise 385 (Surface of revolution) *We consider the surface of revolution S resulting from rotating the graph of a smooth positive function $z \mapsto y = u(z)$ around the $(0, z)$ axis. We parametrize S using the coordinate system*

$$\psi(\theta_1, \theta_2) \in S_\psi := ([a, b] \times [0, 2\pi]) \mapsto \psi(\theta_1, \theta_2) = \begin{pmatrix} u(\theta_1) \cos(\theta_2) \\ u(\theta_1) \sin(\theta_2) \\ \theta_1 \end{pmatrix} \in S$$

for some given parameters $a < b$. Check that the Riemannian gradient operator on the surface is given by the formula

$$\nabla_g \;=\; \left(1 + (\partial_{\theta_1} u(\theta_1))^2 \right)^{-1} \, \partial_{\theta_1} + u(\theta_1)^{-2} \, \partial_{\theta_2}.$$

Compute the Hessian ∇_g^2 and deduce that the Laplacian operator Δ_g is given by the formula

$$\Delta_g = \frac{1}{1 + (\partial_{\theta_1} u)^2} \, \partial_{\theta_1}^2 + \frac{1}{u^2} \, \partial_{\theta_2}^2 + \frac{1}{1 + (\partial_{\theta_1} u)^2} \, \partial_{\theta_1} \log \left(u / \sqrt{1 + (\partial_{\theta_1} u)^2} \right) \, \partial_{\theta_1}.$$

Exercise 386 (Surface of revolution - Rotating a cosinus) *Apply exercise 385 to the function $u(z) = c + \cos z$, for some given parameter $c \in]-1, \infty[$. Check that the Riemannian gradient operator is given by the formula*

$$\nabla_g = \left(1 + \sin^2(\theta_1)\right)^{-1} \partial_{\theta_1} + (c + \cos(\theta_1))^{-2} \partial_{\theta_2}.$$

Prove that the Riemannian Laplacian on this surface is defined by

$$\Delta_g = \frac{1}{1 + \sin^2(\theta_1)} \partial_{\theta_1}^2 + \frac{1}{(c + \cos(\theta_1))^2} \partial_{\theta_2}^2$$
$$- \frac{\sin(\theta_1)}{1 + \sin^2(\theta_1)} \left(\frac{\cos(\theta_1)}{1 + \sin^2(\theta_1)} + \frac{1}{c + \cos(\theta_1)}\right) \partial_{\theta_1}.$$

Exercise 387 (Catenoid - Rotating a catenary) *Apply exercise 385 to the function $u(z) = \cosh z \, (:= (e^z + e^{-z})/2)$. Check that the corresponding Riemannian gradient and Laplacian operators on the surface are given by the formulae*

$$\nabla_g = \cosh^{-2}(\theta_1) \, (\partial_{\theta_1} + \partial_{\theta_2}) \quad and \quad \Delta_g = \cosh^{-2}(\theta_1) \, \left(\partial_{\theta_1}^2 + \partial_{\theta_2}^2\right).$$

Exercise 388 (Tangential and normal accelerations) *We consider some parametrization $\psi : \theta = (\theta_1, \theta_2) \in S_\psi \mapsto \psi(\theta) \in S = \varphi^{-1}(0)$ of some surface $S \subset \mathbb{R}^3$ associated with the null level set of a function $\varphi : \mathbb{R}^3 \mapsto \mathbb{R}$. To simplify notation, we write $C_{i,j}^k$ instead of $C_{i,j}^k \circ \psi$ the Christoffel symbols expressed in the parameter space S_ψ. We let the $N^\perp$ be the unit normal field to the surface $N^\perp = \frac{\partial \varphi}{\|\partial \varphi\|}$.*

- *Check the Gauss equations*

$$\partial_{\theta_i, \theta_j} \psi = \sum_{k=1,2} C_{i,j}^k \, \partial_{\theta_k} \psi + \Omega^{i,j} \, n^\perp$$

with the matrix $\Omega = \left(\Omega^{i,j}\right)_{1 \leq i,j \leq 2}$ of the orthogonal component entries

$$\Omega^{i,j} = \langle \partial_{\theta_i, \theta_j} \psi, n^\perp \rangle \quad and \quad n^\perp = N^\perp \circ \psi = (+/-) \times \frac{\partial_{\theta_1} \psi \wedge \partial_{\theta_2} \psi}{\|\partial_{\theta_1} \psi \wedge \partial_{\theta_2} \psi\|}.$$

- *We let $t \in [0,1] \mapsto \alpha(t) = (\alpha_1(t), \alpha_2(t)) \in S_\psi$ be some curve in the parameter space and we set $c(t) = \psi(\alpha(t))$ the corresponding curve on the surface S. Prove that*

$$c'(t) = \alpha_1'(t) \, (\partial_{\theta_1} \psi)_{\alpha(t)} + \alpha_2'(t) \, (\partial_{\theta_2} \psi)_{\alpha(t)} \quad and \quad c''(t) = c_{tan}''(t) + c_\perp''(t)$$

with the tangential and the normal acceleration

$$c_{tan}''(t) := \sum_{k=1,2} \left(\alpha_k''(t) + \sum_{1 \leq i,j \leq 2} C_{i,j}^k(\alpha(t)) \, \alpha_i'(t) \, \alpha_j'(t)\right) (\partial_{\theta_k} \psi)_{\alpha(t)} \in T_{c(t)}(S)$$

$$c_\perp''(t) := \left(\sum_{1 \leq i,j \leq 2} \Omega^{i,j}(\alpha(t)) \, \alpha_i'(t) \, \alpha_j'(t)\right) n^\perp(\alpha(t)) \in T_{c(t)}^\perp(S).$$

Exercise 389 (Tangential and normal curvature) *We consider the surface model discussed in exercise 388. We introduce the arc length*

$$\omega(t) = \int_0^t \|c'(u)\| \, du \quad and \; its \; inverse \quad \tau = \omega^{-1}$$

and the time changed curve $\bar{c}(s) = c(\tau(s))$.

- *Check that*

$$\overline{c}'(s) = \frac{c'(\tau(s))}{\|c'(\tau(s))\|} \quad and \quad T_{\overline{c}(s)}(S) = \text{Vect}\left(\overline{c}'(s), \overline{c}'(s) \wedge N_{\overline{c}(s)}^{\perp}\right).$$

- *Prove that*

$$\overline{c}''(s) = \frac{c''(\tau(s))}{\|c'(\tau(s))\|^2} - \frac{1}{\|c'(\tau(s))\|^2} \ \langle c''(\tau(s)), \overline{c}'(s)\rangle \ \overline{c}'(s).$$

- *Deduce that*

$$\|\overline{c}''(s)\|^2 \quad = \quad \kappa_{tan}^2(\overline{c}(s)) + \kappa_{\perp}^2(\overline{c}(s)) := \kappa^2(\overline{c}(s))$$

with the tangential and the normal curvature

$$\kappa_{tan}(\overline{c}(s)) := \left\|\frac{c''_{tan}(\tau(s))}{\|c'(\tau(s))\|^2}\right\| \quad and \quad \kappa_{\perp}(\overline{c}(s)) := \left\|\frac{c''_{\perp}(\tau(s))}{\|c'(\tau(s))\|^2}\right\|.$$

Exercise 390 (Principal, Gaussian and mean curvature) *We consider the surface model discussed in exercises 388 and 389, and we set $\phi = \psi^{-1}$. Check that the normal curvature at a given point of the curve $\overline{c}(s)$ in the surface is given by the formula*

$$\kappa_{\perp}(\overline{c}(s)) := \left\|\frac{c''_{\perp}(\tau(s))}{\|c'(\tau(s))\|^2}\right\| = R_{\overline{c}(s)}(\alpha_1'(\tau(s)), \alpha_2'(\tau(s)))$$

with the polynomial ratio

$$R_x(v_1, v_2) = \frac{\sum_{1 \leq i,j \leq 2} \Omega^{i,j}(\phi(x)) \ v_i \ v_j}{\sum_{1 \leq i,j \leq 2} g_{i,j}(\phi(x)) \ v_i \ v_j}.$$

Find the minimal and maximal curvatures k_1 and k_2 at a given point x of the surface S (w.r.t. the velocities parameters (v_1, v_2)). Show that the mean and Gaussian curvature are given by

$$\kappa_{mean} := \frac{k_1 + k_2}{2} = \frac{1}{2} \ \text{tr}(S) \quad and \quad \kappa_{Gauss} := k_1 k_2 = \det(S)$$

with the shape matrix $S = g^{-1}\Omega$ (a.k.a. the Weingarten map).

Exercise 391 (Weingarten's equations [261]) *We consider the surface model discussed in exercises 388 through 390. Check that*

$$\forall i, j = 1, 2 \qquad \Omega^{i,j} = -\left\langle \partial_{\theta_j}\psi, \partial_{\theta_i}n^{\perp}\right\rangle$$

and deduce the Weingarten's equations

$$\forall i = 1, 2 \qquad \partial_{\theta_i}n^{\perp} = -\sum_{l=1,2} S_{l,i} \ \partial_{\theta_l}\psi$$

with the shape matrix S defined in exercise 390.

Exercise 392 (Peterson-Codazzi-Mainardi equations) *We consider the surface model discussed in exercises 388 through 391. To simplify notation, we slightly abuse through it* *and write $C_{i,j}^k$, $R_{i,j,k}^m$ and $R_{i,k}$ instead of $C_{i,j}^k \circ \psi$, $R_{i,j,k}^m \circ \psi$ and $R_{i,k} \circ \psi$ in the chart co-ordinate ψ of the Christoffel symbols and in the Ricci curvature coefficients $R_{i,j,k}^m$ and $R_{i,k}$ introduced in (19.89). Prove that*

$$\partial_{\theta_k, \theta_i, \theta_j} \psi$$

$$= \sum_{m=1,2} \left[\partial_{\theta_k} C_{i,j}^m + \sum_{l=1,2} C_{i,j}^l C_{l,k}^m - \Omega^{i,j} S_{m,k} \right] \partial_{\theta_m} \psi + \left[\partial_{\theta_k} \Omega^{i,j} + \sum_{l=1,2} C_{i,j}^l \Omega^{l,k} \right] n^\perp.$$

Deduce the Peterson-Codazzi-Mainardi equations

$$R_{i,j,k}^m = \Omega^{i,k} S_{m,j} - \Omega^{i,j} S_{m,k} \quad and \quad \partial_{\theta_j} \Omega^{i,k} - \partial_{\theta_k} \Omega^{i,j} = \sum_{l=1,2} \left[C_{i,j}^l \Omega^{l,k} - C_{i,k}^l \Omega^{l,j} \right].$$

Prove the Theorema Egregium

$$\kappa_{scalar} := \operatorname{tr}(g^{-1} R) = 2 \, \kappa_{Gauss} = \frac{1}{\det(g)} \sum_{i \neq j} \sum_{m=1,2} g_{i,m} \, R_{j,i,j}^m.$$

22

Stochastic calculus in chart spaces

This chapter is concerned with the design of Brownian motion and diffusion processes in local coordinate systems. The expressions of these processes in ambient spaces are discussed in chapter 20. We also discuss the Doeblin-Itō formula associated with these diffusion processes in Riemannian manifolds. Several illustrations are discussed including Brownian motion on the sphere, on the torus, and on the simplex.

> *If only I had the theorems! Then I should find the proofs easily enough.*
> Georg Friedrich Bernhard Riemann (1826-1866).

22.1   Brownian motion on Riemannian manifolds

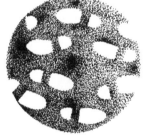

We let $\Theta_t = \begin{pmatrix} \Theta_t^1 \\ \vdots \\ \Theta_t^p \end{pmatrix}$ be the $\mathbb{R}^p$-diffusion on the parameter space S_ϕ defined for any $1 \leq i \leq p$ by the diffusion equation

$$
\begin{aligned}
d\Theta_t^i &= \frac{1}{2} \left(\Delta\phi^i\right)_\psi (\Theta_t) \, dt + \left(\nabla\phi^i\right)_\psi^T (\Theta_t) \, dB_t \\
&= -\sum_{1 \leq j,k \leq p} g^{j,k}(\Theta_t) \, C_{j,k}^i(\Theta_t) \, dt + \left(\nabla\phi^i\right)_\psi^T (\Theta_t) \, dB_t \quad (\Leftarrow (21.43))
\end{aligned}
$$

$$(22.1)$$

where B_t stands for a standard r-dimensional Brownian motion. In the above display, $C_{i,j}^n$ stands for the Christoffel symbols defined in (21.32). In this situation, we notice that

$$
\begin{aligned}
d\Theta_t d\Theta_t^T &= \begin{pmatrix} \left(\nabla\phi^1\right)_\psi^T \\ \vdots \\ \left(\nabla\phi^p\right)_\psi^T \end{pmatrix} (\Theta_t) \, dB_t dB_t^T \left(\left(\nabla\phi^1\right)_\psi, \ldots, \left(\nabla\phi^p\right)_\psi \right) (\Theta_t) \\
&= \begin{pmatrix} \left(\nabla\phi^1\right)_\psi^T \left(\nabla\phi^1\right)_\psi & \cdots & \left(\nabla\phi^1\right)_\psi^T \left(\nabla\phi^r\right)_\psi \\ \vdots & \vdots & \\ \left(\nabla\phi^r\right)_\psi^T \left(\nabla\phi^1\right)_\psi & \cdots & \left(\nabla\phi^r\right)_\psi^T \left(\nabla\phi^r\right)_\psi \end{pmatrix} (\Theta_t) \, dt.
\end{aligned}
$$

This yields

$$d\Theta_t d\Theta_t^T = \begin{pmatrix} \left\langle (\nabla\phi^1)_\psi, (\nabla\phi^1)_\psi \right\rangle & \cdots & \left\langle (\nabla\phi^1)_\psi, (\nabla\phi^r)_\psi \right\rangle \\ \vdots & & \vdots \\ \left\langle (\nabla\phi^r)_\psi, (\nabla\phi^1)_\psi \right\rangle & \cdots & \left\langle (\nabla\phi^r)_\psi, (\nabla\phi^r)_\psi \right\rangle \end{pmatrix} (\Theta_t)\, dt = g^{-1}(\Theta_t)\, dt.$$

Thus, by using the Ito formula we prove the following theorem.

Theorem 22.1.1 *For any smooth function f on $\mathbb{R}^p$ we have*

$$df(\Theta_t) = \sum_{1 \le i \le p} \partial_{\theta_i}(f)(\Theta_t)\, d\Theta_t^i + \frac{1}{2} \sum_{1 \le i,j \le p} \partial_{\theta_i,\theta_j}(f)(\Theta_t)\, d\Theta_t^i d\Theta_t^j$$

$$= \mathcal{L}(f)(\Theta_t)dt + d\mathcal{M}_t(f).$$

The generator $\mathcal{L}$ associated with the diffusion process Θ_t is given by

$$\mathcal{L}(f) = \frac{1}{2}\left[\sum_{1 \le i \le p} \partial_{\theta_i} f\, (\Delta\phi^i)_\psi + \sum_{1 \le i,j \le p} \partial_{\theta_i,\theta_j}(f) \left\langle (\nabla\phi^i)_\psi, (\nabla\phi^j)_\psi \right\rangle \right]$$

$$= \frac{1}{2}\left[\sum_{1 \le i \le p} \partial_{\theta_i} f\, (\Delta\phi^i)_\psi + \sum_{1 \le i,j \le p} g^{i,j}\, \partial_{\theta_i,\theta_j} f \right]$$

$$= \frac{1}{2}\,\Delta_g(f) = \frac{1}{2}\,\text{div}_g\left(\nabla_g(f)\right). \quad (\Leftarrow (21.48)) \tag{22.2}$$

The martingale term $\mathcal{M}_t(f)$ is defined by

$$d\mathcal{M}_t(f) = \sum_{1 \le i \le p} \partial_{\theta_i}(f)(\Theta_t)\, \left(\nabla\phi^i\right)_\psi^T(\Theta_t)\, dB_t.$$

Using (21.44) and (21.46) we find that

$$\sum_{1 \le i \le p} (\partial_{\theta_i} f)_\phi\, \left(\nabla\phi^i\right)^T = \nabla(f \circ \phi) = \nabla F$$

$$\mathcal{L}(f) \circ \phi = \frac{1}{2}\,\Delta(f \circ \phi) = \Delta(F) \quad \text{with} \quad F = f \circ \phi.$$

Therefore, if we set $X_t = \psi(\Theta_t) \Rightarrow \Theta_t = \phi(X_t)$ we find that

$$df(\Theta_t) = d(f \circ \phi)(X_t) = dF(X_t) = \frac{1}{2}\,\Delta(F)(X_t)dt + dM_t(F)$$

with the martingale

$$dM_t(F) = (\nabla F)^T(X_t)\, dB_t.$$

Choosing $f = \psi^k \Rightarrow F = \psi^k \circ \phi = \chi^k$ (cf. (21.24)) we find that

$$\pi(X_t)\, dB_t = \begin{bmatrix} (\nabla \chi^1)^T \\ \vdots \\ (\nabla \chi^r)^T \end{bmatrix} (X_t)\, dB_t.$$

This yields

$$\begin{aligned}
dX_t^k &= d\psi^k(\Theta_t) = \frac{1}{2}\,\Delta(\chi^k)(X_t)\, dt + (\nabla \chi^k)^T(X_t)\, dB_t \\
&= -\frac{1}{2}\,\mathbb{H}^k(X_t)\, dt + \sum_{1 \le j \le r} \pi_j^k(X_t)\, dB_t^j. \quad (\Leftarrow (21.26))
\end{aligned} \qquad (22.3)$$

22.2 Diffusions on chart spaces

Starting from the equation (22.3), if we set $\Theta_t^i = \phi^i(X_t)$, we find that

$$(21.19) \quad \Rightarrow \quad \nabla \chi^k = \nabla(\psi^k \circ \phi) = \sum_{1 \le j \le p} \left(\partial_{\theta_j} \psi^k\right)_\phi \nabla \phi^j$$

$$\Rightarrow \quad (\nabla \chi^k)^T(X_t)\, dB_t = \sum_{1 \le j \le p} \left(\partial_{\theta_j} \psi^k\right)_\phi (\nabla \phi^j)^T(X_t)\, dB_t$$

$$= \sum_{1 \le j \le p} \left(\partial_{\theta_j} \psi^k\right)_\phi \langle \nabla \phi^j(X_t), dB_t \rangle$$

and

$$\begin{aligned}
dX_t^k dX_t^l &= \sum_{1 \le i,j \le p} \left(\partial_{\theta_i} \psi^k\right)_\phi \left(\partial_{\theta_j} \psi^l\right)_\phi \langle \nabla \phi^i(X_t), dB_t \rangle \langle \nabla \phi^j(X_t), dB_t \rangle \\
&= \sum_{1 \le i,j \le p} \left(\partial_{\theta_i} \psi^k\right)_\phi \left(\partial_{\theta_j} \psi^l\right)_\phi \langle \nabla \phi^i(X_t), \nabla \phi^j(X_t) \rangle\, dt.
\end{aligned}$$

Therefore, by using Ito's formula we have

$$d\phi^i(X_t)$$

$$= \sum_{1 \le k \le r} \left(\partial_{x_k} \phi^i\right)(X_t) \left[\frac{1}{2}\,\Delta(\chi^k)(X_t)\, dt + (\nabla \chi^k)^T(X_t)\, dB_t\right]$$

$$+ \frac{1}{2} \sum_{1 \le k,l \le r} \left(\partial_{x_l, x_k} \phi^i\right)(X_t) \sum_{1 \le m,n \le p} \left(\partial_{\theta_m} \psi^k\right)_\phi \left(\partial_{\theta_n} \psi^l\right)_\phi \langle \nabla \phi^m(X_t), \nabla \phi^n(X_t) \rangle\, dt.$$

Notice that

$$\begin{aligned}
\sum_{1 \le k \le r} \left(\partial_{x_k} \phi^i\right) (\nabla \chi^k)^T &= \sum_{1 \le j \le p} \left[\sum_{1 \le k \le r} \left(\partial_{x_k} \phi^i\right)\left(\partial_{\theta_j} \psi^k\right)_\phi\right] (\nabla \phi^j)^T \qquad (22.4) \\
&= \sum_{1 \le j \le p} \left(\partial_{\theta_j}(\phi^i \circ \psi)\right)_\phi (\nabla \phi^j)^T = (\nabla \phi^i)^T \quad \left(\Leftarrow (\phi^i \circ \psi)(v) = v^i\right)
\end{aligned}$$

and using (21.46) we have

$$\Delta(\chi^k) = \Delta(\psi^k \circ \phi) = \sum_{1 \le m,n \le p} \langle \nabla \phi^m, \nabla \phi^n \rangle \ \left(\partial_{\theta_m, \theta_n} \psi^k \right)_{\phi} + \sum_{1 \le j \le p} \left(\partial_{\theta_j} \psi^k \right)_{\phi} \Delta \phi^j.$$

This implies that

$$\sum_{1 \le k \le r} \left(\partial_{x_k} \phi^i \right) \Delta(\chi^k) + \sum_{1 \le k,l \le r} \left(\partial_{x_l, x_k} \phi^i \right) \sum_{1 \le m,n \le p} \left(\partial_{\theta_m} \psi^k \right)_{\phi} \left(\partial_{\theta_n} \psi^l \right)_{\phi} \ \langle \nabla \phi^m, \nabla \phi^n \rangle$$

$$= \sum_{1 \le m,n \le p} \langle \nabla \phi^m, \nabla \phi^n \rangle$$

$$\times \left[\sum_{1 \le k \le r} \left(\partial_{x_k} \phi^i \right) \left(\partial_{\theta_m, \theta_n} \psi^k \right)_{\phi} + \sum_{1 \le k,l \le r} \left(\partial_{x_l, x_k} \phi^i \right) \ \left(\partial_{\theta_m} \psi^k \right)_{\phi} \ \left(\partial_{\theta_n} \psi^l \right)_{\phi} \right]$$

$$+ \sum_{1 \le j \le p} \sum_{1 \le k \le r} \left(\partial_{x_k} \phi^i \right) \left(\partial_{\theta_j} \psi^k \right)_{\phi} \Delta \phi^j$$

$$= \sum_{1 \le m,n \le p} \langle \nabla \phi^m, \nabla \phi^n \rangle \ \left(\partial_{\theta_m, \theta_n} (\phi^i \circ \psi) \right)_{\phi} + \sum_{1 \le j \le p} \left(\partial_{\theta_j} (\phi^i \circ \psi) \right)_{\phi} \Delta \phi^j = \Delta \phi^i.$$

Therefore

$$\forall 1 \le i \le p \qquad d\phi^i(X_t) = \frac{1}{2} \left(\Delta \phi^i \right)(X_t) dt + \left(\nabla \phi^i \right)^T (X_t) dB_t. \tag{22.5}$$

Letting $\Theta_t := \phi(X_t) \ \Rightarrow \ X_t = \psi(\Theta_t)$, we arrive at the equation of the Brownian motion on the Riemannian manifold:

$$d\Theta_t^i = \frac{1}{2} \left(\Delta \phi^i \right)_{\psi} (\Theta_t) \ dt + \left(\nabla \phi^i \right)_{\psi}^T (\Theta_t) \ dB_t, \tag{22.6}$$

where B_t stands for a standard r-dimensional Brownian motion on $\mathbb{R}^r$.

22.3   Brownian motion on spheres

22.3.1   The unit circle $S = \mathbb{S}^1 \subset \mathbb{R}^2$

The unit circle can be described in terms of the polar coordinates mapping

$$\psi(\theta) = \begin{pmatrix} \cos(\theta) \\ \sin(\theta) \end{pmatrix}.$$

It can be easily checked (cf. (24.1) and (24.2)) that

$$(\nabla \phi)_{\psi} = \partial_{\theta} \psi = \begin{bmatrix} -\sin(\theta) \\ \cos(\theta) \end{bmatrix} \quad \text{and} \quad (\Delta \phi)_{\psi} = 0.$$

Hence (22.1) is equivalent to

$$d\Theta_t = (\nabla\phi)^T_\psi(\Theta_t) \, dB_t = -\sin(\Theta_t) \, dB^1_t + \cos(\Theta_t) \, dB^1_t := d\overline{B}_t.$$

Notice that $\overline{B}_t$ is itself a standard Brownian motion

$$d\overline{B}_t d\overline{B}_t = \left(\cos^2(\Theta_t) + \sin^2(\Theta_t)\right) \, dt = dt.$$

22.3.2 The unit sphere $S = \mathbb{S}^2 \subset \mathbb{R}^3$

The 2-sphere can be parametrized by the spherical coordinates mapping

$$\psi(\theta) = \begin{pmatrix} \sin(\theta_1)\cos(\theta_2) \\ \sin(\theta_1)\sin(\theta_2) \\ \cos(\theta_1) \end{pmatrix}.$$

In this situation, we can check (cf. also (24.3) and (24.4))

$$(\nabla\phi^1)_\psi(\theta) = \begin{pmatrix} \cos(\theta_1)\cos(\theta_2) \\ \cos(\theta_1)\sin(\theta_2) \\ -\sin(\theta_1) \end{pmatrix} \qquad (\nabla\phi^2)_\psi(\theta) = \frac{1}{\sin(\theta_1)} \begin{pmatrix} -\sin(\theta_2) \\ \cos(\theta_2) \\ 0 \end{pmatrix}$$

$$(\Delta\phi^1)_\psi(\theta) = \cot(\theta_1) \qquad\qquad (\Delta\phi^2)_\psi(\theta) = 0.$$

As a consequence, we have:

(22.1)

$$\Longleftrightarrow \begin{cases} d\Theta^1_t &= \dfrac{1}{2}\cot(\Theta^1_t) \, dt \\ &\qquad + \left[\cos(\Theta^1_t)\left(\cos(\Theta^2_t)dB^1_t + \sin(\Theta^2_t)dB^2_t\right) - \sin(\Theta^1_t) \, dB^3_t\right] \\ &:= \dfrac{1}{2}\cot(\Theta^1_t) \, dt + d\overline{B}^1_t \\ d\Theta^2_t &= \dfrac{1}{\sin(\Theta^1_t)}\left[-\sin(\Theta^2_t) \, dB^1_t + \cos(\Theta^2_t) \, dB^2_t\right] := \dfrac{1}{\sin(\Theta^1_t)} \, d\overline{B}^2_t. \end{cases}$$

Notice that

$$d\overline{B}^1_t d\overline{B}^2_t = 0 \quad\text{and}\quad d\overline{B}^1_t d\overline{B}^1_t = dt = d\overline{B}^2_t d\overline{B}^2_t$$

so that (22.1) can be rewritten as

$$\begin{cases} d\Theta^1_t &= \dfrac{1}{2}\cot(\Theta^1_t) \, dt + dB^1_t \\ d\Theta^2_t &= \dfrac{1}{\sin(\Theta^1_t)} \, dB^2_t. \end{cases}$$

The computation of the generator $\frac{1}{2}\Delta_g$ of this diffusion readily yields the formula of the Laplacian on the unit 2-sphere

$$\begin{aligned} \Delta_g(f) &= \cot(\theta_1) \, \partial_{\theta_1}(f) + \partial^2_{\theta_1}(f) + \frac{1}{\sin^2(\theta_1)} \, \partial^2_{\theta_2}(f) \\ &= \frac{1}{\sin(\theta_1)} \, \partial_{\theta_1}\left(\sin(\theta_1) \, \partial_{\theta_1}f\right) + \frac{1}{\sin^2(\theta_1)} \, \partial^2_{\theta_2}(f). \end{aligned}$$

22.4   Brownian motion on the torus

The 2-torus is the null level set of the function

$$\varphi(x) = \left(R - \sqrt{x_1^2 + x_2^2}\right)^2 + x_3^2 - r^2$$

with $r < R$. It can be parametrized by the spherical coordinates mapping

$$\psi(\theta) = \begin{pmatrix} (R + r\cos(\theta_1))\cos(\theta_2) \\ (R + r\cos(\theta_1))\sin(\theta_2) \\ r\sin(\theta_1) \end{pmatrix}.$$

In this situation, we can check (cf. section 24.1.3)

$$(\nabla\phi^1)_\psi(\theta) = r^{-1} \begin{pmatrix} -\sin(\theta_1)\cos(\theta_2) \\ -\sin(\theta_1)\sin(\theta_2) \\ \cos(\theta_1) \end{pmatrix}$$

$$(\nabla\phi^2)_\psi(\theta) = (R + r\cos(\theta_1))^{-1} \begin{pmatrix} -\sin(\theta_2) \\ \cos(\theta_2) \\ 0 \end{pmatrix}$$

$$(\Delta\phi^1)_\psi(\theta) = -\frac{\sin(\theta_1)}{r(R + r\cos(\theta_1))} \quad \text{and} \quad (\Delta\phi^2)_\psi(\theta) = 0.$$

Hence (22.1) is equivalent to

$$\begin{cases} d\Theta_t^1 = -\dfrac{\sin(\Theta_t^1)}{2r(R + r\cos(\Theta_t^1))}\, dt \\[2mm] \qquad\quad + \frac{1}{r}\left[-\sin\left(\Theta_t^1\right)\left(\cos\left(\Theta_t^2\right) dB_t^1 + \sin\left(\Theta_t^2\right) dB_t^2\right) + \cos\left(\Theta_t^1\right) dB_t^3\right] \\[3mm] d\Theta_t^2 = \dfrac{1}{(R + r\cos(\Theta_t^1))}\left[-\sin\left(\Theta_t^2\right) dB_t^1 + \cos\left(\Theta_t^2\right) dB_t^2\right]. \end{cases}$$

We set

$$d\overline{B}_t^1 = -\sin\left(\Theta_t^1\right)\left(\cos\left(\Theta_t^2\right) dB_t^1 + \sin\left(\Theta_t^2\right) dB_t^2\right) + \cos\left(\Theta_t^1\right) dB_t^3$$
$$d\overline{B}_t^2 = -\sin\left(\Theta_t^2\right) dB_t^1 + \cos\left(\Theta_t^2\right) dB_t^2.$$

It is readily checked that

$$d\overline{B}_t^1 d\overline{B}_t^2 = 0 \quad \text{and} \quad d\overline{B}_t^1 d\overline{B}_t^1 = dt = d\overline{B}_t^2 d\overline{B}_t^2$$

so that (22.1) can be rewritten as

$$(22.1) \iff \begin{cases} d\Theta_t^1 := -\dfrac{\sin(\Theta_t^1)}{2r(R + r\cos(\Theta_t^1))}\, dt + \dfrac{1}{r}\, dB_t^1 \\[3mm] d\Theta_t^2 = \dfrac{1}{(R + r\cos(\Theta_t^1))}\, dB_t^2. \end{cases}$$

An illustration of a realization of a Brownian motion on the torus is provided in the illustration below.

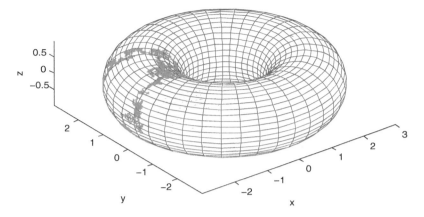

22.5 Diffusions on the simplex

We return to the Brownian motion on the orbifold $S/\mathcal{H} = \mathbb{S}^p \cap \mathbb{R}_+^{r=p+1}$ discussed in section 20.5. The positive orthant of the sphere is in bijection with the p-simplex

$$\text{Simplex}(p) = \{\theta = (\theta_i)_{1 \le i \le r} \in \mathbb{R}_+^r \ : \ \sum_{1 \le i \le r} \theta_i = 1\}.$$

One diffeomorphism is given by the square mapping

$$x = (x_i)_{1 \le i \le r}^T \in \mathbb{S}^p \cap \mathbb{R}_+^{r=p+1} \ \xrightarrow{\ \Xi\ } \ \Xi(x) = \left(x_1^2, \ldots, x_r^2\right)^T \in \text{Simplex}(p)$$

$$\left(\sqrt{\theta_1}, \ldots, \sqrt{\theta_r}\right)^T = \Xi^{-1}(\theta) \in \mathbb{S}^p \cap \mathbb{R}_+^{r=p+1} \ \xleftarrow{\ \Xi^{-1}\ } \ \theta = (\theta_i)_{1 \le i \le r}^T \in \text{Simplex}(p).$$

Notice that

$$\partial_{x_k} \Xi^i = 2 \, 1_{k=i} \, x^i \quad \Longrightarrow \quad \frac{1}{2} \, \partial_{x_k, x_l} \Xi^i = 1_{k,l=i}. \tag{22.7}$$

Using (20.3), the projection $\pi(x)$ onto $T_x(\mathbb{S}^p \cap \mathbb{R}_+^{r=p+1})$ is given by

$$x = \Xi^{-1}(\theta)$$

$$\begin{aligned}
\Rightarrow \ \pi(x) \ &= \ Id - \frac{(\partial \varphi)(x)}{\|(\partial \varphi)(x)\|} \frac{(\partial \varphi)(x)^T}{\|(\partial \varphi)(x)\|} \\
&= \begin{pmatrix}
1 - \frac{x_1^2}{\|x\|^2} & -\frac{x_1 x_2}{\|x\|^2} & -\frac{x_1 x_3}{\|x\|} & \cdots & -\frac{x_1 x_r}{\|x\|} \\
-\frac{x_2 x_1}{\|x\|} & 1 - \frac{x_2^2}{\|x\|^2} & -\frac{x_2 x_3}{\|x\|} & \cdots & -\frac{x_2 x_r}{\|x\|} \\
\vdots & \vdots & \vdots & \vdots & \vdots \\
-\frac{x_r x_1}{\|x\|} & -\frac{x_r x_2}{\|x\|} & \cdots & \cdots & 1 - \frac{x_r^2}{\|x\|}
\end{pmatrix} \\
&= \begin{pmatrix}
1 - \frac{\theta_1}{\sum_{1 \le j \le r} \theta_j} & -\frac{\sqrt{\theta_1}\sqrt{\theta_2}}{\sum_{1 \le j \le r} \theta_j} & -\frac{\sqrt{\theta_1}\sqrt{\theta_3}}{\sum_{1 \le j \le r} \theta_j} & \cdots & -\frac{\sqrt{\theta_1}\sqrt{\theta_r}}{\sum_{1 \le j \le r} \theta_j} \\
-\frac{\sqrt{\theta_2}\sqrt{\theta_1}}{\sum_{1 \le j \le r} \theta_j} & 1 - \frac{\theta_2}{\sum_{1 \le j \le r} \theta_j} & -\frac{\sqrt{\theta_2}\sqrt{\theta_3}}{\sum_{1 \le j \le r} \theta_j} & \cdots & -\frac{\sqrt{\theta_2}\sqrt{\theta_r}}{\sum_{1 \le j \le r} \theta_j} \\
\vdots & \vdots & \vdots & \vdots & \vdots \\
-\frac{\sqrt{\theta_r}\sqrt{\theta_1}}{\sum_{1 \le j \le r} \theta_j} & -\frac{\sqrt{\theta_r}\sqrt{\theta_2}}{\sum_{1 \le j \le r} \theta_j} & \cdots & \cdots & 1 - \frac{\theta_r}{\sum_{1 \le j \le r} \theta_j}
\end{pmatrix}.
\end{aligned}$$

In addition, using (20.4), the mean curvature vector $\mathbb{H}$ on the sphere is given by

$$x = \Xi^{-1}(\theta) \Rightarrow \mathbb{H}^i(x) = p\frac{x_i}{x^T x} = p\,\frac{\sqrt{\theta_i}}{\sum_{1 \leq l \leq r}\theta_l}.$$

We let X_t be the Brownian motion on the positive orthant defined in (20.17). We recall that

$$dX_t^k = -\frac{1}{2}\,\mathbb{H}^k(X_t)\,dt + \sum_{1 \leq j \leq r}\pi_j^k(X_t)\,dB_t^j \Rightarrow dX_t^k dX_t^k = \left[\sum_{1 \leq j \leq r}\pi_j^k\pi_j^l\right](X_t)\,dt.$$

Applying the Ito formula to

$$\Theta_t = \Xi(X_t) \quad \Leftrightarrow \quad \Xi^{-1}(\Theta_t) = X_t = \left(\sqrt{\Theta_t^1}, \ldots, \sqrt{\Theta_t^r}\right)$$

we find that

$$
\begin{aligned}
d\Theta_t^i &= \sum_{1 \leq k \leq r}\left(\partial_{x_k}\Xi^i\right)\left(\Xi^{-1}(\Theta_t)\right)\left[-\frac{1}{2}\,\mathbb{H}^k(\Xi^{-1}(\Theta_t))\,dt + \sum_{1 \leq j \leq r}\pi_j^k(\Xi^{-1}(\Theta_t))\,dB_t^j\right]\\
&\quad +\frac{1}{2}\sum_{1 \leq k,l \leq r}\left(\partial_{x_k,x_l}\Xi^i\right)\left(\Xi^{-1}(\Theta_t)\right)\left[\sum_{1 \leq j \leq r}\pi_j^k\pi_j^l\right](\Xi^{-1}(\Theta_t))\,dt\\
&= 2\sqrt{\Theta_t^i}\left[-\frac{1}{2}\,\mathbb{H}^i(\Xi^{-1}(\Theta_t))\,dt + \sum_{1 \leq j \leq r}\pi_j^i(\Xi^{-1}(\Theta_t))\,dB_t^j\right]\\
&\quad +\left[\sum_{1 \leq k \leq r}\pi_k^i\pi_k^i\right](\Xi^{-1}(\Theta_t))\,dt.
\end{aligned}
$$

The last assertion is a direct consequence of (22.7). To take the final step, we observe that

$$
\begin{aligned}
\sum_{1 \leq k \leq r}\left(\pi_k^i(\Xi^{-1}(\theta))\right)^2 &= \left(1 - \frac{\theta_i}{\sum_{1 \leq j \leq r}\theta_j}\right)^2 + \frac{\theta_i}{\sum_{1 \leq j \leq r}\theta_j}\sum_{1 \leq k \leq r,\ k \neq i}\frac{\theta_k}{\sum_{1 \leq j \leq r}\theta_j}\\
&= 1 - 2\frac{\theta_i}{\sum_{1 \leq j \leq r}\theta_j} + \frac{\theta_i}{\sum_{1 \leq j \leq r}\theta_j}\sum_{1 \leq k \leq r}\frac{\theta_k}{\sum_{1 \leq j \leq r}\theta_j}\\
&= 1 - \frac{\theta_i}{\sum_{1 \leq j \leq r}\theta_j}
\end{aligned}
$$

and

$$
\begin{aligned}
\sum_{1 \leq j \leq r}\pi_j^i(\Xi^{-1}(\Theta_t))\,dB_t^j &= \left(1 - \frac{\Theta_t^i}{\sum_{1 \leq j \leq r}\Theta_t^j}\right)dB_t^i - \sum_{1 \leq j \leq r,\ j \neq i}\frac{\sqrt{\Theta_t^i}\sqrt{\Theta_t^j}}{\sum_{1 \leq j \leq r}\Theta_t^j}\,dB_t^j\\
&= dB_t^i - \sum_{1 \leq j \leq r}\frac{\sqrt{\Theta_t^i}\sqrt{\Theta_t^j}}{\sum_{1 \leq j \leq r}\Theta_t^j}\,dB_t^j.
\end{aligned}
$$

We conclude that

$$
d\Theta_t^i = \left(-p\, \frac{\Theta_t^i}{\sum_{1 \le l \le r} \Theta_t^l} + \left(1 - \frac{\Theta_t^i}{\sum_{1 \le j \le r} \Theta_t^j} \right) \right) dt
$$

$$
+ 2\sqrt{\Theta_t^i} \left[dB_t^i - \sum_{1 \le j \le r} \frac{\sqrt{\Theta_t^i}\,\sqrt{\Theta_t^j}}{\sum_{1 \le j \le r} \Theta_t^j}\, dB_t^j \right]
$$

$$
= \left(1 - r\, \frac{\Theta_t^i}{\sum_{1 \le j \le r} \Theta_t^j} \right) dt + 2\sqrt{\Theta_t^i} \left[dB_t^i - \sum_{1 \le j \le r} \frac{\sqrt{\Theta_t^i}\,\sqrt{\Theta_t^j}}{\sum_{1 \le j \le r} \Theta_t^j}\, dB_t^j \right].
$$

It is instructive to notice that

$$
d\Theta_t^i d\Theta_t^j
$$

$$
= 4\sqrt{\Theta_t^i \Theta_t^i} \left[\left(1_{i=j} - \frac{\sqrt{\Theta_t^i}\,\sqrt{\Theta_t^j}}{\sum_{1 \le l \le r} \Theta_t^l} \right) + \left(-\frac{\sqrt{\Theta_t^i}\,\sqrt{\Theta_t^j}}{\sum_{1 \le j \le r} \Theta_t^j} + \sum_{1 \le k \le r} \frac{\Theta_t^k}{\sum_{1 \le l \le r} \Theta_t^l} \frac{\sqrt{\Theta_t^i}\,\sqrt{\Theta_t^j}}{\sum_{1 \le l \le r} \Theta_t^l} \right) \right] dt
$$

$$
= 4\sqrt{\Theta_t^i \Theta_t^i} \left[\left(1_{i=j} - \frac{\sqrt{\Theta_t^i}\,\sqrt{\Theta_t^j}}{\sum_{1 \le l \le r} \Theta_t^l} \right) \right] dt = 4\Theta_t^i \left(1_{i=j} - \frac{\Theta_t^j}{\sum_{1 \le l \le r} \Theta_t^l} \right) dt.
$$

22.6 Exercises

Exercise 393 (The Laplacian in polar coordinates) *We consider the polar coordinates on $S = \mathbb{R}^2$ discussed in (21.8). Check the formulation of the Laplacian in polar coordinates*

$$
\Delta_g = \frac{1}{\theta_1}\, \partial_{\theta_1} + \partial_{\theta_1}^2 + \frac{1}{\theta_1^2}\, \partial_{\theta_2}^2.
$$

Exercise 394 (The Laplacian in spherical coordinates) *We consider the spherical coordinates on $S = \mathbb{R}^3$ given by*

$$
\psi(\theta) = \begin{cases} \psi^1(\theta) = \theta_1 \sin(\theta_2)\cos(\theta_3) \\ \psi^2(\theta) = \theta_1 \sin(\theta_2)\sin(\theta_3) \\ \psi^3(\theta) = \theta_1 \cos(\theta_2). \end{cases}
$$

Check the formulation of the Laplacian in spherical coordinates

$$
\Delta_g = \frac{2}{\theta_1}\, \partial_{\theta_1} + \frac{1}{\theta_1^2}\, \cot(\theta_2)\, \partial_{\theta_2} + \partial_{\theta_1}^2 + \frac{1}{\theta_1^2}\, \partial_{\theta_2}^2 + \frac{1}{\theta_1^2 \sin^2(\theta_2)}\, \partial_{\theta_3}^2.
$$

Exercise 395 (Brownian motion on the ellipsoid) *We consider the ellipsoid S embedded in $\mathbb{R}^3$ and equipped with the spherical coordinates discussed in exercise 384. Describe the Brownian motion on S.*

Exercise 396 (Brownian motion on a cone) *We consider the cone S embedded in $\mathbb{R}^3$ and equipped with the coordinate system discussed in exercise 383. Describe the Brownian motion on S. Check that the generator L_g of the above diffusion coincides with half of the Laplacian operator presented in exercise 383.*

Exercise 397 (Brownian motion on a surface of revolution - 1) *We consider the surface of revolution S embedded in $\mathbb{R}^3$ and equipped with the coordinate system discussed in exercise 385. Describe the Brownian motion on S.*

Exercise 398 (Brownian motion on a surface of revolution - 2) *We consider the surface of revolution S associated with the function $u(z) = c + \cos z$, for some given parameter $c \in]-1, \infty[$. We equip the surface with the coordinate system discussed in exercise 385. Describe the Brownian motion on S.*

Exercise 399 (Brownian motion on the Riemannian helicoid) *We consider the helicoid S parametrized by the function*

$$\psi \; : \; (\theta_1, \theta_2) \in S_\psi :=]a, b[\times [0, 2\pi] \mapsto \psi(\theta_1, \theta_2) = \begin{pmatrix} \theta_1 \, \cos(\theta_2) \\ \theta_1 \, \sin(\theta_2) \\ \theta_2 \end{pmatrix} \in S = \psi(S_\psi)$$

for some given parameters $-\infty \leq a < b \leq +\infty$. Describe the Brownian motion on S in the coordinate chart defined by ψ. Check that the Laplacian on the helicoid is given in this chart by the operator

$$\Delta_g = \partial_{\theta_1}^2 + \frac{1}{1 + \theta_1^2} \, \partial_{\theta_2}^2 + \frac{\theta_1}{1 + \theta_1^2} \, \partial_{\theta_1}.$$

Exercise 400 (Brownian motion on the helicoid - ambient space) *Check that the parametrization discussed in exercise 399 provides a chart coordinate of the helicoid in the ambient space defined as the null level set $S = \varphi^{-1}(0)$ of the smooth function*

$$\varphi(x_1, x_2, x_2) = x_2 \cos(x_3) - x_1 \sin(x_3).$$

Describe the Brownian motion in the helicoid in the ambient space $\mathbb{R}^3$.

Exercise 401 (Brownian motion on the catenoid) *We consider the catenoid revolution surface discussed in exercise 387. Check that the Brownian motion on this surface is given by the diffusion equation*

$$d\Theta_t^1 = \frac{1}{\cosh(\Theta_t^1)} \, dB_t^1 \quad \text{and} \quad d\Theta_t^2 = \frac{1}{\cosh(\Theta_t^1)} \, dB_t^2,$$

where (B_t^1, B_t^2) stands for a couple of independent Brownian motions on $\mathbb{R}$.

Exercise 402 (Brownian motion on the unit circle) *Describe the Brownian motion on the unit circle equipped with the polar coordinates.*

Exercise 403 (Brownian motion on the unit 2-sphere) *Describe the Brownian motion on the unit 2-sphere equipped with the spherical coordinates.*

Exercise 404 (Brownian motion on the unit p-sphere) *Find a direct description on the ambient space of the Brownian motion on the unit p-sphere.*

Exercise 405 (Brownian motion on cylinder) *Find a direct description on the ambient space of the Brownian motion on the cylinder.*

Exercise 406 (Brownian motion on the Torus) *Describe the Brownian motion on the unit 2-torus equipped with the spherical coordinates.*

Exercise 407 (Brownian motion on the Simplex) *Describe the Brownian motion on the p-simplex.*

23

Some analytical aspects

The chapter presents some key analytical tools used to study the behaviors of stochastic processes on manifolds. The first part is concerned with the notion of a geodesic and with the construction of a distance on a manifold. We use these mathematical objects to derive a Taylor type expansion of functions around two states related by a geodesic. The second part of the chapter provides a review of the integration theory on manifolds. We present some pivotal integration formulae such as the divergence theorem and Green's identities. These integration formulae allow us to derive the weak formulation of the time evolution of the distribution of random states of stochastic processes evolving on a manifold.

The third part of the chapter is dedicated to gradient flows and to Langevin type diffusions in Euclidian and Riemannian manifolds. The final part of the chapter is concerned with the stability properties of diffusions on manifolds in terms of the Ricci curvature using the Bochner-Lichnerowicz formulae presented earlier in chapter 19 and in chapter 21.

A great deal of my work is just playing with equations and seeing what they give.
Paul A. M. Dirac (1902-1984).

23.1 Geodesics and the exponential map

Definition 23.1.1 *The distance between two states $x, y \in S$ on a manifold equipped with some parametrization ψ is defined in a chart coordinate by the formula*

$$d(x, y) = \inf \int_a^b \left\| \dot{c}(t) \right\|_{g(c(t))} dt$$

where the infimum is taken over all parametric curves

$$c : t \in [a, b] \mapsto c(t) \in S_\psi$$

s.t. $\phi(c(a)) = x$, and $\phi(c(b)) = y$, and

$$\left\| \dot{c}(t) \right\|_{g(c(t))}^2 := \left\langle \dot{c}(t), \dot{c}(t) \right\rangle_{g(c(t))} = \sum_{1 \leq i,j \leq p} g_{i,j}(c(t)) \; \dot{c}^i(t) \; \dot{c}^j(t).$$

We notice that

$$\left\| \dot{c}(t) \right\|_{g(c(t))}^2 = \left\langle \sum_{1 \leq i \leq p} \dot{c}^i(t) \; (\partial_{\theta_i} \psi)_{c(t)}, \sum_{1 \leq j \leq p} \dot{c}^j(t) \; (\partial_{\theta_i} \psi)_{c(t)} \right\rangle$$

$$= \left\langle \dot{C}(t), \dot{C}(t) \right\rangle = \left\| \dot{C}(t) \right\|^2 \quad \text{with} \quad C(t) = \psi(c(t)).$$

Definition 23.1.2 *The length of some curve* $C : t \in [a,b] \mapsto C(t) = \psi(c(t)) \in S$
is given by the integral

$$\int_a^b \|C(t)\| \ dt = \int_a^b \|\dot{c}(t)\|_{g(c(t))} \ dt.$$

To understand this definition better, we simply notice that

$$\|c(t+dt) - c(t)\|_{g(c(t))} \simeq \|\dot{c}(t)\|_g \ dt$$

$$\Rightarrow \sum_{a \leq t \leq b} \|c(t+dt) - c(t)\|_{g(c(t))} \ dt \simeq \int_a^b \|\dot{c}(t)\|_{g(c(t))} \ dt.$$

Definition 23.1.3 *The energy of a given curve* c *is given by*

$$\mathcal{E}(c) = \frac{1}{2} \int_a^b \|\dot{c}(t)\|_{g(c(t))}^2 \ dt = \int_a^b L\left(c(t) \ \dot{c}(t)\right) \ dt$$

with the Lagrangian

$$L(c, \dot{c}) := \frac{1}{2} \|\dot{c}\|_{g(c)}^2 = \sum_{1 \leq i,j \leq p} g_{i,j}(c) \ \dot{c}^i \dot{c}^j .$$

To find the extremal curves, we let $c_\epsilon(t) = c(t) + \epsilon c'(t)$ be an ϵ-perturbation of c, with some
curve $c'(t)$ s.t. $c'(a) = 0 = c'(b)$. For any $t \in [a,b]$, we have

$$\frac{d}{d\epsilon} L(c_\epsilon(t), \dot{c}_\epsilon(t))_{|\epsilon=0} = \langle (\partial_c L)(c(t), \dot{c}(t)), c'(t) \rangle + \left\langle (\partial_{\dot{c}} L)(c(t), \dot{c}(t)), \dot{c}'(t) \right\rangle$$

with the gradients

$$(\partial_c L) = \begin{bmatrix} (\partial_{c^1} L) \\ \vdots \\ (\partial_{c^p} L) \end{bmatrix} \quad \text{and} \quad (\partial_{\dot{c}} L) = \begin{bmatrix} (\partial_{\dot{c}^1} L) \\ \vdots \\ (\partial_{\dot{c}^p} L) \end{bmatrix}.$$

An integration by part w.r.t. the time parameter yields

$$\int_a^b \left\langle (\partial_{\dot{c}} L)(c(t), \dot{c}(t)), \dot{c}'(t) \right\rangle dt$$

$$= \left[\langle (\partial_{\dot{c}} L)(c(t), \dot{c}(t)), c'(t) \rangle \right]_a^b - \int_a^b \left\langle \frac{d}{dt} \left[(\partial_{\dot{c}} L)(c(t), \dot{c}(t)) \right], c'(t) \right\rangle dt.$$

This implies that for any perturbation c' we have

$$\frac{d}{d\epsilon} \mathcal{E}(c_\epsilon)_{|\epsilon=0} = \int_a^b \left\langle (\partial_c L)(c(t), \dot{c}(t)) - \frac{d}{dt} \left[(\partial_{\dot{c}} L)(c(t), \dot{c}(t)) \right], c'(t) \right\rangle dt.$$

This implies that the extremal curves must satisfy the Euler-Lagrange differential equation
of calculus of variations

$$(\partial_c L)(c(t), \dot{c}(t)) = \frac{d}{dt} \left[(\partial_{\dot{c}} L)(c(t), \dot{c}(t)) \right]. \tag{23.1}$$

In our context, we have

$$\partial_{c^k} L = \sum_{1 \leq i,j \leq p} (\partial_{\theta_k} g_{i,j})(c)\, \overset{\cdot}{c}^i\, \overset{\cdot}{c}^j$$

and

$$\partial_{\overset{\cdot}{c}^k} L = 2 \sum_{1 \leq i \leq p} g_{k,i}\, \overset{\cdot}{c}^i$$

$$\Rightarrow \frac{d}{dt}\left[(\partial_{\overset{\cdot}{c}^k} L)(c(t),\dot{c}(t))\right]$$

$$= 2 \sum_{1 \leq i,j \leq p} (\partial_{\theta_j} g_{k,i})(c(t))\, \overset{\cdot}{c}^i(t)\, \overset{\cdot}{c}^j(t) + 2 \sum_{1 \leq i \leq p} g_{k,i}(c(t))\, \overset{\cdot\cdot}{c}^i(t).$$

Thus, the Euler-Lagrange equations take the form

$$(23.1) \Leftrightarrow \forall 1 \leq k \leq p \quad \sum_{1 \leq i \leq p} g_{k,i}(c(t))\, \overset{\cdot\cdot}{c}^i(t)$$

$$= \sum_{1 \leq i,j \leq p} \left[\frac{1}{2}(\partial_{\theta_k} g_{i,j})(c(t)) - (\partial_{\theta_j} g_{k,i})(c(t)) \right] \overset{\cdot}{c}^i(t)\, \overset{\cdot}{c}^j(t).$$

This yields the following result.

Proposition 23.1.4 *For any $1 \leq p \leq m$, we have the formula*

$$\overset{\cdot\cdot}{c}^m(t) = \sum_{1 \leq i,k \leq p} g^{m,k}(c(t))\, g_{k,i}(c(t))\, \overset{\cdot\cdot}{c}^i(t)$$

$$= \sum_{1 \leq i,j \leq p} \left(\sum_{1 \leq k \leq p} g^{m,k} \left[\frac{1}{2}\, \partial_{\theta_k} g_{i,j} - \partial_{\theta_j} g_{k,i} \right] \right)(c(t))\, \overset{\cdot}{c}^i(t)\, \overset{\cdot}{c}^j(t).$$

Next, we express this formula in terms of the Christoffel symbols $C^k_{i,j}$ introduced in (21.32). Firstly, we notice that

$$\langle \partial_{\theta_l}\psi, \partial_{\theta_i,\theta_j}\psi \rangle = \partial_{\theta_i}\langle \delta_{\theta_l}\psi, \delta_{\theta_j}\psi \rangle - \langle \partial_{\theta_i\theta_l}\psi, \partial_{\theta_j}\psi \rangle = \partial_{\theta_i} g_{l,j} - \langle \partial_{\theta_i\theta_l}\psi, \partial_{\theta_j}\psi \rangle$$

$$\Rightarrow \quad C^k_{i,j} = C^k_{j,i}$$

$$= \sum_{1 \leq l \leq p} g^{k,l}\, \langle \partial_{\theta_l}\psi, \partial_{\theta_i,\theta_j}\psi \rangle$$

$$= \sum_{1 \leq l \leq p} g^{k,l}\, \partial_{\theta_i} g_{l,j} - \sum_{1 \leq l \leq p} g^{k,l}\, \langle \partial_{\theta_i\theta_l}\psi, \partial_{\theta_j}\psi \rangle.$$

Thus, for any symmetric functionals $f^{i,j} = f^{j,i}$, $1 \leq i,j \leq p$, on S_ψ we have

$$\sum_{1 \leq i,j \leq p} C_{i,j}^m \, f^{i,j}$$

$$= \sum_{1 \leq i,j \leq p} f^{i,j} \sum_{1 \leq k \leq p} g^{m,k} \, \partial_{\theta_i} g_{k,j} - \sum_{1 \leq i,j \leq p} f^{i,j} \sum_{1 \leq k \leq p} g^{m,k} \, \langle \partial_{\theta_k \theta_i} \psi, \partial_{\theta_j} \psi \rangle$$

$$= - \sum_{1 \leq i,j \leq p} \sum_{1 \leq k \leq p} g^{m,k} \left[\frac{1}{2} \partial_{\theta_k} g_{i,j} - \partial_{\theta_i} g_{k,j} \right] f^{i,j}.$$

We conclude that

$$\forall 1 \leq m \leq p \qquad \ddot{c}^m(t) = - \sum_{1 \leq i,j \leq p} C_{i,j}^m \, \dot{c}^i(t) \, \dot{c}^j(t).$$

The solution of these equations gives a curve that minimizes the distances between two states $\phi(x)$ and $\phi(y)$ in the parameter space. These curves $c(t)$ and their mapping $C(t) = \psi(c(t))$ into the manifold S are called the geodesics.

It is instructive to observe that the velocity vector $C(t) = \psi(c(t))$ of a given curve on S is given by the formula

$$\frac{dC}{dt}(t) = \sum_{1 \leq i \leq p} (\partial_{\theta_i} \psi)(c(t)) \, \dot{c}_t^i.$$

Thus, its acceleration takes the form

$$\frac{d^2 C}{dt^2}(t) = \sum_{1 \leq i \leq p} (\partial_{\theta_i} \psi)(c(t)) \, \ddot{c}_t^i + \sum_{1 \leq i,j \leq p} (\partial_{\theta_j, \theta_i} \psi)(c(t)) \, \dot{c}_t^i \, \dot{c}_t^j.$$

Taking the orthogonal projection on the tangent plane $T_{C(t)}(S)$, we have

$$\pi(C(t)) \left(\frac{d^2 C}{dt^2}(t) \right) = \sum_{1 \leq i \leq p} (\partial_{\theta_i} \psi)(c(t)) \, \ddot{c}_t^i + \sum_{1 \leq i,j \leq p} \pi \left[(\partial_{\theta_j, \theta_i} \psi)(c(t)) \right] \dot{c}_t^i \, \dot{c}_t^j$$

$$= \sum_{1 \leq m \leq p} \left[\ddot{c}_t^m + \sum_{1 \leq i,j \leq p} C_{i,j}^m(c(t)) \, \dot{c}_t^i \, \dot{c}_t^j \right] (\partial_{\theta_m} \psi)(c(t)) = 0$$

from which we find that

$$\forall 1 \leq m \leq p \qquad \ddot{c}_t^m + \sum_{1 \leq i,j \leq p} C_{i,j}^m(c(t)) \, \dot{c}_t^i \, \dot{c}_t^j = 0. \qquad (23.2)$$

This shows that the acceleration vector of the geodesics is orthogonal to the tangent place $T(S)$. In other words, the speed geodesic vector $\dot{c}(t)$ is parallel to the curve $c(t)$ (cf. (21.36)).

By the existence and uniqueness theorem of solutions of ordinary differential equations, given a tangent vector field $W(x) \in T_x(S)$, for any $x = \psi(\theta) \in S$ there exists unique geodesics $C_x(t)$ with velocity vector $W(x) = \frac{dC_x}{dt}(0)$ at the origin and starting at $x = C_x(0)$. The geodesics $c_{\phi(x)}(t) = \phi(C_x(t))$ and $C_x(t)$ associated with a given velocity vector $V_\phi(x)$ and $W_x = (d\psi)_{\phi(x)}(V_{\phi(x)})$ are denoted in terms of the exponential maps

$$C_x(t) := Exp_x(tW) \quad \text{and} \quad c_\theta(t) := Exp_\theta(tV).$$

23.2 Taylor expansion

Given a smooth function f on the parameter space S_ψ, we have

$$
\begin{aligned}
\frac{d}{dt} f\left(c_\theta(t)\right) &= \sum_{1 \leq i \leq p} (\partial_{\theta_i} f)\left(c_\theta(t)\right) \; \overset{\cdot i}{c_\theta}(t) \\
&= \sum_{1 \leq i \leq p} (\partial_{\theta_i} f)\left(c_\theta(t)\right) \; \overset{\cdot i}{c_\theta}(t) := \partial_{\overset{\cdot}{c_\theta}(t)}(f)\left(c_\theta(t)\right) \\
&= \left\langle \overset{\cdot}{c_\theta}(t), (\partial f)\left(c_\theta(t)\right) \right\rangle = \left\langle \overset{\cdot}{c_\theta}(t), \nabla_g f\left(c_\theta(t)\right) \right\rangle_{g(c_\theta(t))}. \quad (23.3)
\end{aligned}
$$

Notice that this first order formula is valid for any curve $c_\theta(t)$ with velocity vector

$$
\overset{\cdot}{c_\theta}(t) = \left(\overset{\cdot i}{c_\theta}(t) \right)_{1 \leq i \leq p} \quad \text{with} \quad \overset{\cdot i}{c_\theta}(t) := \frac{dc_\theta^i}{dt}(t).
$$

Whenever $c_\theta(t)$ is a geodesic for $t = 0$ we have

$$
\overset{\cdot}{c_\theta}(0) = V(\theta) \Rightarrow \frac{d}{dt} f\left(Exp_\theta(tV)\right)_{|t=0} = \left\langle V(\theta), \nabla_g f(\theta) \right\rangle_{g(\theta)}.
$$

In much the same way, for a geodesic curve we have

$$
\begin{aligned}
\frac{d^2}{dt^2} f\left(c_\theta(t)\right) &= \sum_{1 \leq i,j \leq p} (\partial_{\theta_i, \theta_j} f)\left(c_\theta(t)\right) \; \overset{\cdot i}{c_\theta}(t) \overset{\cdot j}{c_\theta}(t) + \sum_{1 \leq i \leq p} (\partial_{\theta_i} f)\left(c_\theta(t)\right) \; \overset{\cdot\cdot i}{c_\theta}(t) \\
&= \sum_{1 \leq k,l \leq p} \left[(\partial_{\theta_k, \theta_l} f)\left(c_\theta(t)\right) - \sum_{1 \leq i \leq p} C_{k,l}^i(c(t)) \; (\partial_{\theta_i} f)\left(c_\theta(t)\right) \right] \overset{\cdot k}{c_\theta}(t) \overset{\cdot l}{c_\theta}(t) \\
&= \left\langle \overset{\cdot}{c_\theta}(t), \nabla_g^2 f(c_\theta(t)) \; \overset{\cdot}{c_\theta}(t) \right\rangle_{g(c_\theta(t))}. \quad (23.4)
\end{aligned}
$$

Thus, for $t = 0$ we have

$$
\frac{d^2}{dt^2} f\left(Exp_\theta(tV)\right)_{|t=0} = V^T(\theta) \; g \; \nabla_g^2 f(\theta) \; V^T(\theta) = \left\langle V(\theta), \nabla_g^2 f(\theta) \; V(\theta) \right\rangle_{g(\theta)}.
$$

Theorem 23.2.1 *For regular vector fields V this yields the Taylor expansion*

$$f\left(Exp_\theta(tV)\right) = f(\theta) + t\,\langle V(\theta), \nabla_g f(\theta)\rangle_{g(\theta)} + \frac{t^2}{2}\,\langle V(\theta), \nabla_g^2 f(\theta)\,V(\theta)\rangle_{g(\theta)} + \mathrm{O}(t^3)$$

or equivalently

$$f\left(Exp_\theta(V)\right) = f(\theta) + \langle V(\theta), \nabla_g f(\theta)\rangle_{g(\theta)} + \frac{1}{2}\,\langle V(\theta), \nabla_g^2 f(\theta)\,V(\theta)\rangle_{g(\theta)} + \mathrm{O}\left(\|V\|^3\right).$$

Letting $F = f \circ \phi$, and using the fact that $Exp_\theta(V) = \phi\left(Exp_x(W)\right)$ when $\theta = \phi(x)$ and $W = (d\psi)_\phi(V_\phi)$, the above formula takes the form

$$F\left(Exp_x(W)\right) = F(x) + \langle W(x), \nabla F(x)\rangle + \frac{1}{2}\,\langle W(x), (\nabla^2 F)(x)W(x)\rangle + \mathrm{O}\left(\|W\|^3\right).$$

We end this section with some Taylor expansions with integral remainders.

Using (23.3) we have the first order expansion

$$f\left(c_\theta(t)\right) - f(\theta) = \int_0^t \langle \dot{c}_\theta(t), \nabla_g f(c_\theta(s))\rangle_{g(c_\theta(s))} \, ds.$$

Applying (21.22) to the pushed forward velocity vectors $\dot{C}_x(s) = (d\psi)_{c_\theta(t)}\left(\dot{c}_\theta(t)\right)$ implies that

$$F\left(C_x(t)\right) - F(x) = \int_0^t \left\langle \dot{C}_x(s), \nabla F\left(C_x(s)\right)\right\rangle \, ds. \tag{23.5}$$

As mentioned above, these first order formulae are valid for any curves $C_x(t) = \psi\left(c_\theta(t)\right)$ (not necessarily geodesics).

By (23.4) we also have the first order Taylor expansion

$$\frac{d}{ds} f\left(c_\theta(s)\right) = \langle V(\theta), \nabla_g f(\theta)\rangle_{g(\theta)} + \int_0^s \langle \dot{c}_\theta(r), \nabla_g^2 f(c_\theta(r))\,\dot{c}_\theta(r)\rangle_{g(c_\theta(r))} \, dr \tag{23.6}$$

from which we find the second order expansion

$$\begin{aligned}
f\left(c_\theta(t)\right) &= f(\theta) + \int_0^t \frac{d}{ds} f\left(c_\theta(s)\right) ds \\
&= f(\theta) + t\,\langle V(\theta), \nabla_g f(\theta)\rangle_{g(\theta)} \\
&\quad + \int_0^t \left[\int_0^s \langle \dot{c}_\theta(r), \nabla_g^2 f(c_\theta(r))\,\dot{c}_\theta(r)\rangle_{g(c_\theta(r))} \, dr\right] ds.
\end{aligned}$$

Recalling that

$$\int_0^t \left(\int_0^s a(r)\, dr\right) ds = t \int_0^t a(s)\, ds - \int_0^t a(s)\, ds = \int_0^t a(s)\,(t-s)\, ds$$

we prove the second order Taylor expansion with integral remainder for geodesics with prescribed initial velocity vectors

$$f\left(c_\theta(t)\right) = f\left(\theta\right) + t\,\left\langle V\left(\theta\right), \nabla_g f\left(\theta\right)\right\rangle_{g(\theta)} \tag{23.7}$$
$$+ \int_0^t \left\langle \dot{c}_\theta\left(s\right), \nabla_g^2 f(c_\theta(s))\,\dot{c}_\theta\left(s\right)\right\rangle_{g(c_\theta(s))}\,(t-s)\,ds.$$

Combining (21.22) with proposition 21.5.2 implies that

$$F\left(C_x(t)\right) = F\left(x\right) + t\,\left\langle W(x), \nabla F(x)\right\rangle \tag{23.8}$$
$$+ \int_0^t \left\langle \dot{C}_x\left(s\right), \nabla^2 F(C_x(s))\,\dot{C}_x\left(s\right)\right\rangle\,(t-s)\,ds.$$

23.3 Integration on manifolds

23.3.1 The volume measure on the manifold

Heuristically, for manifolds S with dimension

$$p = 1 = \dim(T_x(S)) = \dim(\mathrm{Vect}((\partial_{\theta_i}\psi)_\phi(\theta)))$$

the volume element $\mu_S(dx)$ at some state $x = \psi(\theta)$ reduces the length $\mathrm{length}_S(\psi(\delta\theta))$ of the ψ-image curve $\psi(\delta\theta)$ of an infinitesimal interval

$$\delta\theta := [\theta, \theta + d\theta] \in S_\psi \subset \mathbb{R},$$

that is,

$$\psi(\delta\theta) \simeq \psi(\theta + d\theta) - \psi(\theta) \simeq (\partial_\theta\psi)(\theta)\,d\theta$$

so that

$$\mu_S(dx) = \mathrm{length}_S(\psi(\delta\theta)) \simeq \|\psi(\theta + d\theta) - \psi(\theta)\|$$
$$\simeq \|(\partial_\theta\psi)(\theta)\|\,d\theta = \sqrt{\langle(\partial_\theta\psi)(\theta), (\partial_\theta\psi)(\theta)\rangle}\,d\theta.$$

More rigorously, for any function F with compact support we have

$$\int_S F(x)\,\mu_S(dx) = \int_{S_\psi} f(\theta)\,\sqrt{\langle(\partial_\theta\psi)(\theta), (\partial_\theta\psi)(\theta)\rangle}\,d\theta \quad \text{with} \quad f = F \circ \psi.$$

In larger dimensions, the ψ-image $\psi(\delta\theta)$ of a cell $\delta\theta = \prod_{1 \leq i \leq p}[\theta_i, \theta_i + d\theta_i] \in S_\psi \subset \mathbb{R}^p$ is given by

$$\psi(\delta\theta) \simeq \psi(\theta + d\theta) - \psi(\theta) \simeq \sum_{1 \leq i \leq p}(\partial_{\theta_i}\psi)(\theta)\,d\theta_i$$
$$= (d\psi)\left(\left\{\sum_{1 \leq i \leq p}\epsilon_i\,e_i\,:\,\epsilon_i \in [0, d\theta_i]\right\}\right) = (d\psi)\left(\prod_{1 \leq i \leq p}[0, d\theta_i]\right)$$

with the unit vectors e_i in $\mathbb{R}^p$, $1 \le i \le p$. We recall that $(d\psi)(e_i) = \partial_{\theta_i}$, for any $1 \le i \le p$. On the other hand, by the change of variables formula

$$\text{Vol}\left[(d\psi)\left(\prod_{1 \le i \le p}[0, d\theta_i]\right)\right] \simeq \det((d\psi)(\theta)) \underbrace{\prod_{1 \le i \le p} d\theta_i}_{=d\theta} = \det((d\psi)(\theta))\ d\theta.$$

Recalling that $\sqrt{\det(A^T A)} = \det(A)$, and $A_{i,j} = \langle Ae_i, e_j \rangle$ for any $(p \times p)$-matrix A, we have

$$\begin{aligned}
\det((d\psi)(\theta)) &= \sqrt{\det((d\psi)(\theta)^T (d\psi)(\theta))} \\
&= \sqrt{\det(\langle(d\psi)(\theta)^T (d\psi)(\theta)e_i, e_j\rangle)_{1 \le i,j \le p}} \\
&= \sqrt{\det(\langle(d\psi)(\theta)e_i, (d\psi)(\theta)e_j\rangle)_{1 \le i,j \le p}} \\
&= \sqrt{\det(\langle(\partial_{\theta_i}\psi)(\theta), (\partial_{\theta_j}\psi)(\theta)\rangle)_{1 \le i,j \le p}} = \sqrt{\det(g(\theta))}.
\end{aligned}$$

In summary, for any function F with compact support,

$$\int_S F\ d\mu_S := \int_S F(x)\ \mu_S(dx) = \int_{S_\psi} f(\theta)\ \sqrt{\det(g(\theta))}\ d\theta \quad \text{with} \quad f = F \circ \psi.$$

If we set
$$\mu_g(d\theta) = \sqrt{\det(g(\theta))}\ d\theta$$

the above formulae can be rewritten in a more synthetic form

$$\mu_S(F) := \int_S F\ d\mu_S = \int_{S_\psi} f(\theta)\ \mu_g(d\theta) := \mu_g(f). \tag{23.9}$$

For instance, the arc length on the circle $S = \mathbb{S}^1 := \{(x, y) \in \mathbb{R}^2 : x^2 + y^2 = 1\}$ equipped with the polar coordinates $\psi(\theta) = \begin{pmatrix} \cos(\theta) \\ \sin(\theta) \end{pmatrix}$, $\theta \in [0, 2\pi[$, is given by

$$\partial_\theta \psi(\theta) = \begin{pmatrix} -\sin(\theta) \\ \cos(\theta) \end{pmatrix} \Rightarrow \|\partial_\theta \psi(\theta)\| = 1 \Rightarrow \mu_g(d\theta) = d\theta.$$

In cartesian coordinates, using the change of variables

$$\theta = \arccos x \Rightarrow \partial_x \arccos x = -\frac{1}{\sqrt{1 - x^2}}$$

we have

$f = F \circ \psi$

$$\begin{aligned}
\Rightarrow \int_0^{2\pi} f(\theta)\ d\theta &= \int_0^\pi f(\theta)d\theta + \int_\pi^{2\pi} f(\theta)d\theta \\
&= \int_{-1}^1 F(x, \sqrt{1-x^2})\ \frac{dx}{\sqrt{1-x^2}} + \int_{-1}^1 F(x, -\sqrt{1-x^2})\ \frac{dx}{\sqrt{1-x^2}}.
\end{aligned}$$

Notice that in cartesian coordinates the uniform distribution $\eta(d(x,y))$ of a random point (X,Y) in $S = \mathbb{S}^1$ takes the form

$$\eta(d(x,y)) \;=\; \underbrace{\frac{1}{\pi}\,\frac{1}{\sqrt{1-x^2}}\,1_{]-1,1[}(x)\,dx}_{\mathbb{P}(X \in dx)} \;\times\; \underbrace{\frac{1}{2}\left[\delta_{-\sqrt{1-x^2}} + \delta_{\sqrt{1-x^2}}\right](dy)}_{\mathbb{P}(Y \in dy \mid X = x)}.$$

The arc length of a curve of the form

$$\mathcal{C} = \left\{ \psi(\theta) = \left(\begin{array}{c} r(\theta)\cos(\theta) \\ r(\theta)\sin(\theta) \end{array} \right) \in \mathbb{R}^2 \;:\; \theta \in [\theta_1, \theta_2] \right\}$$

is given by

$$\partial_\theta \psi(\theta) = \left(\begin{array}{c} \cos(\theta)\partial_\theta r(\theta) - r(\theta)\sin(\theta) \\ \sin(\theta)\partial_\theta r(\theta) + r(\theta)\cos(\theta) \end{array} \right)$$

$$\Rightarrow \|\partial_\theta \psi(\theta)\|^2 = (\partial_\theta r(\theta))^2 + (r(\theta))^2 \Rightarrow \mu_g(d\theta) = \sqrt{(\partial_\theta r(\theta))^2 + (r(\theta))^2}\, d\theta.$$

In much the same way, the arc length of the graph of a function $h \;:\; \mathbb{R} \mapsto \mathbb{R}$

$$\mathcal{C} = \left\{ \psi(\theta) = \left(\begin{array}{c} \theta \\ h(\theta) \end{array} \right) \in \mathbb{R}^2 \;:\; \theta \in [\theta_1, \theta_2] \right\}$$

is given by

$$\partial_\theta \psi(\theta) = \left(\begin{array}{c} 1 \\ \partial_\theta h(\theta) \end{array} \right)$$

$$\Rightarrow \|\partial_\theta \psi(\theta)\|^2 = 1 + (\partial_\theta f(\theta))^2 \Rightarrow \mu_g(d\theta) = \sqrt{1 + (\partial_\theta h(\theta))^2}\, d\theta.$$

In terms of cartesian coordinates, the uniform distribution $\eta(d(x,y))$ of a random point $(X,Y) \in \mathcal{C} = \left\{ \psi(\theta) = \left(\begin{array}{c} x \\ h(x) \end{array} \right) \in \mathbb{R}^2 \;:\; x \in [x_1, x_2] \right\}$ takes the form

$$\eta(d(x,y)) \;\propto\; \underbrace{\sqrt{1 + (\partial_x h(x))^2}\, 1_{[x_1,x_2]}(x)\, dx}_{\mathbb{P}(X \in dx)} \;\times\; \underbrace{\delta_{h(x)}(dy)}_{\mathbb{P}(Y \in dy \mid X = x)}.$$

Line segments correspond to linear height functions $h(\theta) = a\theta + b$, for some $a,b \in \mathbb{R}$. In this case, we find that $\mu_g(d\theta) = \sqrt{1 + a^2}\, d\theta$. It is instructive to see that the arc length computation resumes to the Pythagoras theorem

$$\left[\int_{\theta_1}^{\theta_2} \mu_g(d\theta) \right]^2 \;=\; (1 + a^2)(\theta_2 - \theta_1)^2 = (\theta_2 - \theta_1)^2 + (a(\theta_2 - \theta_1))^2.$$

In terms of cartesian coordinates, the uniform distribution $\eta(d(x,y))$ of a random point $(X,Y) \in \{(x,y) \;:\; y = ax + b \;,\; x \in [x_1, x_2]\}$ takes the form

$$\eta(d(x,y)) \;=\; \underbrace{\frac{1}{(x_2 - x_1)}\, 1_{[x_1,x_2]}(x)\, dx}_{\mathbb{P}(X \in dx)} \;\times\; \underbrace{\delta_{ax+b}(dy)}_{\mathbb{P}(Y \in dy \mid X = x)}.$$

We end this section with a couple of important observations.

Firstly, in the above construction, we have implicitly assumed that the manifold S can be parametrized by a single map ψ, and thus with a single chart coordinate. In a more general situation, using the patching technique presented in (21.5),

$$\mu_S(F) := \int_S F \, d\mu_S = \sum_{i \in I} \int_{S_i} F_i \, d\mu_{S_i} = \sum_{i \in I} \int_{S_{\psi_i}} f_i(\theta) \, \mu_{g_{\psi_i}}(d\theta) := \mu_g(f) \qquad (23.10)$$

with $F_i \circ \psi_i := f_i$ and with the Riemannian metric g_{ψ_i} on the parameter space S_{ψ_i} defined by the formula

$$g_{\psi_i, k; l} = \langle \partial_{\theta_k} \psi_i, \partial_{\theta_l} \psi_i \rangle.$$

Last, but not least, we fix some state $x = \phi(\theta) \in S$ and we consider the parallelotope manifold formed by the vectors $(\partial_{\theta_i} \psi)_{\phi(x)}$ given by

$$\mathcal{P}_x = \left\{ \sum_{1 \leq i \leq p} \epsilon_i \, (\partial_{\theta_i} \psi)_{\phi(x)} \; : \; \epsilon := (\epsilon_i)_{1 \leq i \leq p} \in [0,1]^p \right\}.$$

The manifold P is clearly parametrized by the mapping

$$\overline{\psi}(\epsilon) = \sum_{1 \leq i \leq p} \epsilon_i \, (\partial_{\theta_i} \psi)_{\phi(x)} \Rightarrow \partial_{\epsilon_i} \overline{\psi} = (\partial_{\theta_i} \psi)_{\phi(x)} \quad \text{and} \quad \langle \partial_{\epsilon_i} \overline{\psi}, \partial_{\epsilon_j} \overline{\psi} \rangle = g_{i,j}(\phi(x)).$$

Applying (23.9) to $(S, \psi, S_\psi) = (\mathcal{P}_x, \overline{\psi}, [0,1]^p)$ and using (21.7) we find that

$$\mathrm{Vol}\,(\mathcal{P}_x) = \sqrt{\det\,(g(\phi(x)))} = \sqrt{\det\left((\partial \psi)\,(\partial \psi)^T \right)} \qquad (23.11)$$

with the matrix formed by the generating column vectors

$$(\partial \psi)^T = \left(\partial_{\theta_1} \psi, \cdots, \partial_{\theta_p} \psi \right). \qquad (23.12)$$

In other words, the Gramian $\det\,(g(\phi(x)))$ is the square of the volume of the parallelotope $\mathcal{P}(W)$ formed by the vectors $W_i := (\partial_{\theta_i} \psi)_{\phi(x)}$, with $1 \leq i \leq p$. When $p = r$ we have the formula

$$\mathrm{Vol}\,(\mathcal{P}_x) = \sqrt{\det\,(g(\phi(x)))} = \left| \det\left((\partial_{\theta_1} \psi)_{\phi(x)}, \cdots, (\partial_{\theta_r} \psi)_{\phi(x)} \right) \right|.$$

23.3.2   Wedge product and volume forms

This section provides an algebraic interpretation of volume elements in terms of wedge products, also called exterior products.

Wedge products can also be used to compute determinants and volumes of parallelotopes (a.k.a. p-dimensional parallelopipeds) generated by some independent vectors. We let

$$\mathcal{P}_x(W_1, \ldots, W_p)) = \left\{ x + \sum_{1 \leq i \leq p} w_p \, W_p \; : \; w = (w_i)_{1 \leq i \leq p} \in [0,1]^p \right\}$$

be the parallelotope with a basis edge at x and formed by a collection of of p column independent vectors $(W_i)_{1 \le i \le p}$ in $\mathbb{R}^r$. When $x = 0$, we simplify notation and we write $\mathcal{P}(W)$ instead of $\mathcal{P}_0(W)$.

We equip the space of p-exterior products with the inner product defined by the duality formula

$$\ll W_1 \wedge \ldots \wedge W_p, V_1 \wedge \ldots \wedge V_p \gg := \det \left[(\langle W_i, V_j \rangle)_{1 \le i,j \le p} \right]$$

between the p-exterior products $(W_1 \wedge \ldots \wedge W_p)$ and $(V_1 \wedge \ldots \wedge V_p)$ of p vectors in any vector space equipped with some inner product $\langle ., . \rangle$. We also consider the norm

$$\||W_1 \wedge \ldots \wedge W_p\||^2 = \ll W_1 \wedge \ldots \wedge W_p, W_1 \wedge \ldots \wedge W_p \gg .$$

With the introduced notation, we can also check the volume formula

$$\||W_1 \wedge \ldots \wedge W_p\|| = \sqrt{\det \left[(\langle W_i, W_j \rangle)_{1 \le i,j \le p} \right]} = \text{Volume} \left(\mathcal{P}(W_1, \ldots, W_p) \right) \quad (23.13)$$

using simple induction w.r.t. the parameter p. For $p = 1$ the result is clear. Suppose the result has been proved for some p. The new vector W_{p+1} can be written as

$$W_{p+1} = W_{p+1}^{\perp} + \sum_{1 \le i \le p} a_i \, W_i \quad \text{for some } a_i \in \mathbb{R} \text{ and for any } i \text{ we have } W_{p+1}^{\perp} \perp W_i.$$

By construction, we have

$$
\begin{aligned}
\||W_1 \wedge \ldots \wedge W_p \wedge W_{p+1}^{\perp}\||^2 &= \det \left((\langle W_i, W_j \rangle)_{1 \le i,j \le p} \right) \langle W_{p+1}^{\perp}, W_{p+1}^{\perp} \rangle \\
&= \text{Volume} \left(\mathcal{P}(W_1, \ldots, W_p) \right) \times \|W_{p+1}^{\perp}\| \\
&= \text{Volume} \left(\mathcal{P}(W_1, \ldots, W_p, W_{p+1}) \right).
\end{aligned}
$$

In the above display, we used the easily checkable Lagrange indentity

$$\|U \wedge V\|^2 + |\langle U, V \rangle|^2 = \|U\|^2 \times \|V\|^2$$

for $U = W_1 \wedge \ldots \wedge W_p$ and $V = W_{p+1}^{\perp}$. For a proof of this formula, we refer to exercise 409. This ends the proof of (23.13).

We let e_i the r unit vector in $\mathbb{R}^r$. Arguing as above, it is also readily checked that

$$W_1 \wedge \ldots \wedge W_p = \sum_{1 \le i_1 < \ldots < i_p \le r} w_{i_1, \ldots, i_p} \left(e_{i_1} \wedge \ldots \wedge e_{i_p} \right)$$

$$\implies \||W_1 \wedge \ldots \wedge W_p\||^2 = \sum_{1 \le i_1 < \ldots < i_p \le r} w_{i_1, \ldots, i_p}^2$$

for some parameters $w_{i_1, \ldots, i_p}$.

Observe that $W_i = W(e_i)$ with the matrix $W = (W_1, \ldots, W_p)$ generated by the column vectors W_i. In this notation, we have

$$\mathcal{P}(W_1, \ldots, W_p) = W\left(\mathcal{P}(e_1, \ldots, e_p)\right)$$

as well as the volume formulae

$$
\begin{aligned}
\|W(e_1) \wedge \ldots \wedge W(e_p)\| &= \sqrt{\det\left[(\langle W(e_i), W(e_j)\rangle)_{1 \le i, j \le p}\right]} \\
&= \sqrt{\det(W^T W)} = \text{Volume}\left(\mathcal{P}(W(e_1), \ldots, W(e_p))\right)
\end{aligned}
$$

and therefore

$$
\begin{aligned}
\text{Volume}\left(W(\mathcal{P}(e_1, \ldots, e_p))\right) &= \sqrt{\det(W^T W)} \times \text{Volume}\left(\mathcal{P}(e_1, \ldots, e_p)\right) \\
&\overset{\text{if } p = r}{=} |\det(W)| \times \text{Volume}\left(\mathcal{P}(e_1, \ldots, e_p)\right). \quad (23.14)
\end{aligned}
$$

We let $\varphi : \theta \in S_\psi \mapsto \varphi(\theta) \in S = \psi(S_\psi)$ be some differentiable function on some open parameter set $S_\psi \subset \mathbb{R}^p$. The change of variables formula discussed in section 23.3.1 takes the form

$$\text{Volume}\left(\psi(\mathcal{P}_\theta(\epsilon_1\, e_1, \ldots, \epsilon_p\, e_p))\right)$$

$$= \sqrt{\det((\partial\psi)(\theta)(\partial\psi)(\theta)^T)} \times \text{Volume}\left(\mathcal{P}_{\psi(\theta)}(\epsilon_1\, e_1, \ldots, \epsilon_p\, e_p)\right) + \text{o}(\epsilon)$$

with $\epsilon = \min_i \epsilon_i$ and the matrix $(\partial\psi)(\theta)$ formed by the generating column vectors $(\partial_{\theta_i}\psi)(\theta)$ defined in (23.12). In terms of integrals, formula (23.9) is sometimes rewritten in the following form

$$\int_S F(x)\, dx_1 \wedge \ldots \wedge dx_r = \int_{S_\psi} F(\psi(\theta)) \sqrt{\det((\partial\psi)(\theta)(\partial\psi)(\theta)^T)}\, d\theta_1 \wedge \ldots \wedge d\theta_r.$$

23.3.3 The divergence theorem

We consider the push forward $W = (d\psi)_\phi(V_\phi) \in T(S)$ and $F = f \circ \psi$ of a smooth vector field V and a smooth function f on some open parametric space S_ψ (so that S is also an open subset of $\mathbb{R}^p$). We further assume that either f or V has compact support so that f or V is null at the boundary, whenever it exists, of the parametric space S_ψ. In other words, F or W is compactly supported in the chart $\phi = \psi^{-1} : S \mapsto S_\psi$.

By construction (cf. (21.16)),

$$W \in T(S) \Rightarrow \langle W, \nabla F\rangle = \langle W, \partial F\rangle = \partial_W F = (\partial_V f) \circ \phi = \langle V, \partial f\rangle \circ \phi$$

and

$$
\begin{aligned}
\int_S \langle W, \nabla F\rangle\, d\mu_S &= \int_{S_\psi} \langle V(\theta), (\partial f)(\theta)\rangle \sqrt{\det(g(\theta))}\, d\theta \\
&= \int_{S_\psi} \langle V(\theta), (\nabla_g f)(\theta)\rangle_{g(\theta)}\, \mu_g(d\theta)
\end{aligned}
$$

with

$$\mu_g(d\theta) = \sqrt{\det(g(\theta))}\, d\theta.$$

Applying a simple integration by part formula (recalling that V and f have compact support on the open set S_ψ), we prove that

$$\int_{S_\psi} \langle V(\theta), (\partial f)(\theta) \rangle \ \sqrt{\det(g(\theta))} \ d\theta = \sum_{1 \le i \le p} \int_{S_\psi} V^i(\theta) \ \partial_{\theta_i}(f)(\theta) \ \sqrt{\det(g(\theta))} \ d\theta$$

$$= - \sum_{1 \le i \le p} \int_{S_\psi} f(\theta) \ \partial_{\theta_i} \left[\sqrt{\det(g(\theta))} \ V^i(\theta) \right] \ d\theta$$

$$= - \int_{S_\psi} f(\theta) \ \underbrace{\frac{1}{\sqrt{\det(g(\theta))}} \sum_{1 \le i \le p} \partial_{\theta_i} \left[\sqrt{\det(g(\theta))} \ V^i(\theta) \right]}_{=\mathrm{div}_g(V)(\theta)} \ \sqrt{\det(g(\theta))} \ d\theta.$$

This implies that

$$\int_S \langle W, \nabla F \rangle \ d\mu_S \ = \ - \int_{S_\psi} f(\theta) \ \mathrm{div}_g(V)(\theta) \ \sqrt{\det(g(\theta))} \ d\theta$$

$$= \ - \int_{S_\psi} f(\theta) \ \mathrm{div}_g(V)(\theta) \ \mu_g(d\theta).$$

On the other hand, using (21.28),

$$\mathrm{div}_g(V)(\theta) = \mathrm{tr}(\nabla W)_\psi = \mathrm{div}(W) \circ \psi.$$

Here again we have implicitly assumed that the manifold S can be parametrized by a single map ψ. In a more general situation, using the patching technique presented in (21.5) and (23.10), we have

$$\int_S \langle W, \nabla F \rangle \ d\mu_S \ = \ \sum_{i \in I} \int_{S_i} \langle W_i, \nabla F \rangle \ d\mu_{S_i}$$

$$= \ - \sum_{i \in I} \int_{S_{\psi_i}} f_i(\theta) \ \mathrm{div}_{g_{\psi_i}}(V_i)(\theta) \ \mu_{g_{\psi_i}}(d\theta)$$

with

$$f_i = (\epsilon_i F) \circ \psi_i \quad \text{and} \quad W_i = \epsilon_i W = (d\psi_i)_{\phi_i}(\epsilon_i \ V_{i,\phi_i}) \quad \text{with} \quad V_{i,\phi_i} = \epsilon_i \ V_{\phi_i}.$$

Hence we proved the result:

Theorem 23.3.1 *If either F or W is compactly supported on S, we have the divergence formula*

$$\int_S F \ \mathrm{div}(W) \ d\mu_S = - \int_S \langle W, \nabla F \rangle \ d\mu_S. \qquad (23.15)$$

We quote a series of direct consequences of this integration-by-part formula:

- On closed manifolds (i.e., compact without boundaries), choosing $F = 1$, we have

$$\int_S \text{div}\,(W)\ d\mu_S = 0.$$

- Choosing $W = F_1\,\nabla F_2$ and $F = F_3$, with either W or F with compact support we have

$$
\begin{aligned}
\int_S \text{div}\,(F_1\,\nabla F_2)\ F_3\ d\mu_S &= -\int_S \langle F_1\,\nabla F_2, \partial F_3\rangle\ d\mu_S\\
&= -\int_S F_1\ \langle \nabla F_2, \partial F_3\rangle\ d\mu_S = -\int_S F_1\ \langle \nabla F_2, \nabla F_3\rangle\ d\mu_S.
\end{aligned}
$$

The r.h.s. assertion arises from

$$\nabla F_2 \in T(S) \implies \langle \nabla F_2, \partial F_3\rangle = \langle \nabla F_2, \pi\,(\partial F_3)\rangle = \langle \nabla F_2, \nabla F_3\rangle.$$

- Choosing $F_1 = 1$ in the above formula, and combining (21.23) with (21.29), we prove Green's first identity; that is, for any smooth functions (F_1, F_2) with (at least one with) compact support,

$$\int_S \Delta(F_1)\ F_2\ d\mu_S = \int_S \text{div}\,(\nabla F_1)\ F_2\ d\mu_S = -\int_S \langle \nabla F_1, \nabla F_2\rangle\ d\mu_S = \int_S F_1\ \Delta(F_2)\ d\mu_S$$

$$=$$

$$
\begin{aligned}
\int_{S_\psi} \Delta_g(f_1)\ f_2\ d\mu_g &= \int_S \text{div}_g\,(\nabla_g f_1)\ f_2\ d\mu_g\\
&= -\int_{S_\psi} \langle \nabla_g f_1, \nabla_g f_2\rangle_g\ d\mu_g = \int_{S_\psi} f_1\ \Delta_g(f_2)\ d\mu_g.
\end{aligned}
$$

- On closed manifolds, choosing $F_2 = 1$ in the above formula, we find that for any smooth function F

$$\int_S \Delta(F)\ d\mu_S = 0.$$

We further assume that S is an open manifold with smooth boundaries ∂S s.t. $\overline{S} = S \cup \partial S$ is compact. We also assume that these manifolds are oriented in the sense that there is a consistent choice of oriented tangent vectors, or equivalently all changes of coordinates associated with the transition maps have positive Jacobian determinants. These properties ensure a consistent choice of unit outward-pointing normal vector fields (a.k.a. unit exterior normal fields) at every point of the boundary manifold. For surfaces this property is equivalent to the well-known right-hand rule to define the three-dimensional orientation.

Arguing as above, we have

$$
\begin{aligned}
\int_S \langle W, \nabla F\rangle\ d\mu_S &= \int_{S_\psi} \langle V(\theta), (\partial f)(\theta)\rangle\ \sqrt{\det\,(g(\theta))}\ d\theta\\
&= -\int_{S_\psi} f(\theta)\ \text{div}_g(V)(\theta)\ \mu_g(d\theta)\\
&\qquad + \int_{S_\psi} \sum_{1\le i\le p} \partial_{\theta_i}\left[f(\theta)\ \sqrt{\det\,(g(\theta))}\ V^i(\theta)\right]\ d\theta.
\end{aligned}
$$

Stokes' theorem implies that

$$
\int_{S_\psi} \sum_{1 \leq i \leq p} \partial_{\theta_i} \left[f(\theta) \, \sqrt{\det\left(g(\theta)\right)} \, V^i(\theta) \right] \, d\theta
$$

$$
= \int_{S_\psi} \operatorname{div}_g(fV)(\theta) \, \mu_g(d\theta) = \int_S \operatorname{div}\left(F \, W\right) \, d\mu_S
$$

$$
\stackrel{\text{Stokes' theo}}{=} \int_{\partial S} \langle FW, N^\perp \rangle \, d\mu_{\partial S} = \int_{\partial S} F \, \langle W, N^\perp \rangle \, d\mu_{\partial S}
$$

where $N^\perp$ stands for the outward unit normal field to ∂S. In summary, we have the following integration formula.

Theorem 23.3.2 *The divergence formula, sometimes also called Gauss' theorem, is given by*

$$
\int_S F \operatorname{div}\left(W\right) \, d\mu_S = -\int_S \langle W, \nabla F \rangle \, d\mu_S + \int_{\partial S} F \, \langle W, N^\perp \rangle \, d\mu_{\partial S}, \qquad (23.16)
$$

with the outward unit normal field $N^\perp$ to the boundary ∂S.

The detailed proof of Stokes' theorem can be found in any textbook on differential geometry. Most of these books are based on advanced differential calculus including differential forms techniques. For the convenience of the reader, we provide a simple and short proof in dimension 3 based on standard integration calculus that can be easily extended to larger dimensions. We consider three-dimensional vector fields $W \in \mathbb{R}^3$ and three-dimensional simple manifolds of the form $S = S_1 = S_2 = S_3(= S_1 \cap S_2 \cap S_3)$ with

$$
\begin{aligned}
S_1 &= \left\{ x \in \mathbb{R}^3 \, : \, (x_2, x_3) \in D_1 \quad \text{and} \quad h_1^+(x_2, x_3) \leq x_1 \leq h_1^+(x_2, x_3) \right\} \\
S_2 &= \left\{ x \in \mathbb{R}^3 \, : \, (x_1, x_3) \in D_2 \quad \text{and} \quad h_2^-(x_1, x_3) \leq x_2 \leq h_2^+(x_1, x_2) \right\} \\
S_3 &= \left\{ x \in \mathbb{R}^3 \, : \, (x_1, x_2) \in D_3 \quad \text{and} \quad h_3^-(x_1, x_2) \leq x_3 \leq h_3^+(x_1, x_2) \right\}.
\end{aligned}
$$

In the above display, the regions D_1, D_2 and D_3 stand respectively for the projection of our solid region S on the $(0, x_2, x_3)$, $(0, x_1, x_3)$ and respectively $(0, x_1, x_2)$; and $h_i^{+/-}$ stands for some smooth functions. In the further development, we use the decomposition $W = \sum_{1 \leq i \leq 3} W^i \, e_i$ of W on the unit basis vector fields e_i of $\mathbb{R}^3$.

We observe that

$$
\int_S \partial_{x_1} W^1 \, d\mu_S = \int_{D_1} \left[W^1\left(h_1^+(x_2, x_3), x_2, x_3\right) - W^1\left(h_1^-(x_2, x_3), x_2, x_3\right) \right] \, d\mu_{D_1}.
$$

On the other hand, we have the boundary decomposition

$$
\partial S = \partial_1^- S \cup \partial_1^0 S \cup \partial_1^+ S
$$

where $\partial_1^{+/-} S$ stands for the graphs of the functions $(x_2, x_3) \in D_1 \mapsto h_1^+(x_2, x_3)$ and $\partial_1^0 S = \partial S - \left(\partial_1^- S \cup \partial_1^+ S\right)$ the remaining surface (whenever it is not empty) that connects the front and back surfaces $\partial_1^+ S$ and $\partial_1^- S$ of the boundary. By construction, $\partial_1^0 S \subset (\partial D_1 \times \mathbb{R})$ so that the the outward pointing orthogonal vector fields $N_1^{0,\perp} \perp e_1$ on the boundary $\partial_1^+ S$.

The parametrization of $\partial_1^{+/-} S$ is clearly given by the functions

$$\psi_1^{-/-} \; : \; (x_2, x_3) \in D_1 \mapsto \psi_1^{-/+}(x_2, x_3) := \begin{pmatrix} h_1^{-/+}(x_2, x_3) \\ x_2 \\ x_3 \end{pmatrix}.$$

The outward pointing orthogonal vector fields on the boundary $\partial_1^- S$ are given by the usual formula

$$N_1^{-,\perp} \; = \; -\partial_{x_2}\psi_1^- \wedge \partial_{x_3}\psi_1^- = - \begin{pmatrix} \partial_{x_2} h_1^- \\ 1 \\ 0 \end{pmatrix} \wedge \begin{pmatrix} \partial_{x_3} h_1^- \\ 0 \\ 1 \end{pmatrix} = \begin{pmatrix} -1 \\ \partial_{x_2} h_1^- \\ \partial_{x_3} h_1^- \end{pmatrix}.$$

In much the same way, the outward pointing orthogonal vector fields on the boundary $\partial_1^+ S$ are given by the usual formula

$$N_1^{+,\perp} \; = \; \partial_{x_2}\psi_1^+ \wedge \partial_{x_3}\psi_1^+ = \begin{pmatrix} \partial_{x_2} h_1^+ \\ 1 \\ 0 \end{pmatrix} \wedge \begin{pmatrix} \partial_{x_3} h_1^+ \\ 0 \\ 1 \end{pmatrix} = \begin{pmatrix} 1 \\ -\partial_{x_2} h_1^+ \\ -\partial_{x_3} h_1^+ \end{pmatrix}.$$

We let $N^\perp = 1_{\partial_1^- S} N_1^{-,\perp} + 1_{\partial_1^0 S} N_1^{0,\perp} + 1_{\partial_1^+ S} N_1^{+,\perp}$ be the the outward pointing orthogonal vector fields on ∂S. Recalling that $N_1^{0,\perp} \perp e_1$ we find that

$$\int_{\partial S} W^1 \left\langle e_1, N^\perp \right\rangle \, d\mu_{\partial S}$$

$$= \int_{\partial_1^+ S} W^1 \left\langle e_1, N_1^{+,\perp} \right\rangle \, d\mu_{\partial_1^+ S} + \int_{\partial_1^- S} W^1 \left\langle e_1, N_1^{-,\perp} \right\rangle \, d\mu_{\partial_1^- S}$$

$$= \int_{D_1} \left[W^1 \left(h_1^+(x_2, x_3), x_2, x_3 \right) - W^1 \left(h_1^-(x_2, x_3), x_2, x_3 \right) \right] \, \mu_{D_1}(d(x_2, x_3)).$$

This yields

$$\int_{\partial S} W^1 \left\langle e_1, N^\perp \right\rangle \, d\mu_{\partial S}$$

$$= \int_{D_1} \int_{h_1^-(x_2, x_3)}^{h_1^+(x_2, x_3)} (\partial_{x_1} W^1)(x_1, x_2, x_3) \, dx_1 \, \mu_{D_1}(d(x_2, x_3)) = \int_S (\partial_{x_1} W^1) \, d\mu_S.$$

We prove in the same way that

$$\int_{\partial S} W^i \left\langle e_i, N^\perp \right\rangle \, d\mu_{\partial S} = \int_S (\partial_{x_i} W^i) \, d\mu_S$$

for any $i = 1, 2, 3$. This ends the proof of (23.16) when $F = 1$. ∎

We illustrate the divergence theorem in terms of parametrizations of the manifold and its boundary. We consider the open disk of radius $R > 0$ given by

$$S = \{(x_1, x_2) \in \mathbb{R}^2 \; : \; x_1^2 + x_2^2 < R\} \implies \partial S = C := \{(x_1, x_2) \in \mathbb{R}^2 \; : \; x_1^2 + x_2^2 = R\},$$

equipped respectively (when we extract the state $(0, R)$ to S and ∂S) with the polar coordinates ψ and ψ_∂ defined by

$$\psi(\theta_1, \theta_2) = \begin{pmatrix} R(1 - \theta_1) \cos(\theta_2) \\ R(1 - \theta_1) \sin(\theta_2) \end{pmatrix} \quad \text{and} \quad \psi_\partial(\theta_2) = \psi(0, \theta_2) \tag{23.17}$$

with $(\theta_1, \theta_2) \in S_\psi := (]0, 1[\times \in]0, 2\pi[)$. In this situation, we observe that $\psi(\{0\} \times C_{\psi_\partial}) = C = \partial S$ with the parametric space $C_{\psi_\partial} =]0, 2\pi[$. Also notice that

$$\partial_{\theta_1} \psi = - \begin{pmatrix} R \cos(\theta_2) \\ R \sin(\theta_2) \end{pmatrix} \perp \partial_{\theta_2} \psi = - \begin{pmatrix} -R(1 - \theta_1) \sin(\theta_2) \\ R(1 - \theta_1) \cos(\theta_2) \end{pmatrix}$$

yields the volume measure on the disk

$$g = \begin{pmatrix} \langle \partial_{\theta_1} \psi, \partial_{\theta_1} \psi \rangle & \langle \partial_{\theta_1} \psi, \partial_{\theta_2} \psi \rangle \\ \langle \partial_{\theta_2} \psi, \partial_{\theta_1} \psi \rangle & \langle \partial_{\theta_2} \psi, \partial_{\theta_2} \psi \rangle \end{pmatrix} = \begin{pmatrix} R^2 & 0 \\ 0 & R^2(1 - \theta_1)^2 \end{pmatrix}.$$

The surface measure on the circle reduces to

$$g_\partial = \langle \partial_{\theta_2} \psi_\partial, \partial_{\theta_2} \psi_\partial \rangle = R^2.$$

For $F = 1$ and $W(x) = \frac{1}{2} \begin{pmatrix} x_1 \\ x_2 \end{pmatrix} \Rightarrow \mathrm{div}(W) = 1$ we readily find the surface of the disk

$$\begin{aligned} \int_S d\mu_S &= \int_{\partial S} \langle W, N^\perp \rangle \, d\mu_{\partial S} \\ &= \frac{R^2}{2} \int_0^{2\pi} \left\langle \begin{pmatrix} \cos(\theta_2) \\ \sin(\theta_2) \end{pmatrix}, n^\perp(\theta_2) \right\rangle d\theta_2 = \pi R^2 \end{aligned} \tag{23.18}$$

with the outward pointing unit normal $n^\perp(\theta_2) = N^\perp(\psi_\partial(\theta_2)) = \begin{pmatrix} \cos(\theta_2) \\ \sin(\theta_2) \end{pmatrix}$.

Notice that any vector fields W on S can be expressed in terms of the local coordinates ψ by the formula

$$W \circ \psi = V^1 \, \partial_{\theta_1} \psi + V^2 \, \partial_{\theta_2} \psi.$$

We clearly have

$$\partial_{\theta_1} \psi = -R \, N^\perp \circ \psi \implies \langle W \circ \psi, N^\perp \circ \psi \rangle = -R \, V^1. \tag{23.19}$$

The computation of the vector fields V^i can be done using the formula (21.14). Also observe that for any $\theta_1 \in [0, 1[$ we have

$$\psi(\theta_1, 0) = R(1 - \theta_1) \begin{pmatrix} 1 \\ 0 \end{pmatrix} = \psi(\theta_1, 2\pi) \implies W \circ \psi(\theta_1, 0) = W \circ \psi(\theta_1, 2\pi) \tag{23.20}$$

$$\implies \forall i = 1, 2 \quad V^i(\theta_1, 0) = V^i(\theta_1, 2\pi).$$

In this situation, using (23.19) and (23.20),

$$\begin{aligned} \int_S \mathrm{div}_g(V) \, d\mu_S &= R^2 \int_0^\pi \int_0^{2\pi} \left[\partial_{\theta_1} \left((1 - \theta_1) V^1 \right) + \partial_{\theta_2} \left((1 - \theta_1) V^2 \right) \right] d\theta_1 d\theta_2 \\ &= \int_0^{2\pi} \underbrace{(-R \, V^1(0, \theta_2))}_{=\langle W \circ \psi_\partial, N^\perp \circ \psi_\partial \rangle(\theta_2)} \underbrace{R d\theta_2}_{\mu_{g_\partial}(d\theta_2)} \\ &\qquad + \int_0^1 (1 - \theta_1) \underbrace{\left[V^2(\theta_1, 2\pi) - V^2(\theta_1, 0) \right]}_{=0} d\theta_2 \end{aligned}$$

from which we conclude that

$$
\int_S \operatorname{div}(W) \, d\mu_S = \int_{S_\psi} \operatorname{div}_g(V) \, d\mu_g
$$

$$
= \int_{C_{\psi_\partial}} \langle W \circ \psi_\partial, N^\perp \circ \psi_\partial \rangle \, d\mu_{g_\partial} = \int_{\partial S} \langle W, N^\perp \rangle \, d\mu_{\partial S}.
$$

Arguing as above, we prove the integration by part formula

$$
\int_S \Delta(F_1) \, F_2 \, d\mu_S = \int_S \operatorname{div}(\nabla F_1) \, F_2 \, d\mu_S \qquad (23.21)
$$

$$
= -\int_S \langle \nabla F_1, \nabla F_2 \rangle \, d\mu_S + \int_{\partial S} F_2 \, \langle \nabla F_1, N^\perp \rangle \, d\mu_{\partial S}.
$$

For a given smooth function F, the quantity $\langle \nabla F, N^\perp \rangle$ is called the normal derivative. In differential geometry, with some abuse of notation, this derivative is often denoted by

$$
\nabla_{N^\perp} F := \langle \nabla F, N^\perp \rangle \quad \text{or as} \quad \frac{\partial F}{\partial N^\perp} := \langle \nabla F, N^\perp \rangle. \qquad (23.22)
$$

We also have Green's identity

$$
\int_S \left(\Delta(F_1) \, F_2 - \Delta(F_2) \, F_1 \right) \, d\mu_S = \int_{\partial S} \langle (F_2 \, \nabla F_1 - F_1 \, \nabla F_2), N^\perp \rangle \, d\mu_{\partial S}.
$$

In the Riemannian manifold (a.k.a. in local coordinates), these formulae take the following form.

Theorem 23.3.3 *For any smooth functions f_1, f_2, we have the divergence formula*

$$
\int_{S_\psi} \Delta_g(f_1) \, f_2 \, d\mu_g
$$

$$
= -\int_{S_\psi} \langle \nabla_g f_1, \nabla_g f_2 \rangle_g \, d\mu_g + \int_{(\partial S)_{\psi_\partial}} (F_2)_{\psi_\partial} \, \langle (\nabla F_1)_{\psi_\partial}, N^\perp_{\psi_\partial} \rangle \, d\mu_{g_\partial}.
$$

In particular, this yields Green's identity

$$\int_{S_\psi} \left(\Delta_g(f_1) \, f_2 - \Delta_g(f_2) \, f_1 \right) \, d\mu_g$$

$$= \int_{(\partial S)_{\psi_\partial}} \langle \left((F_2)_{\psi_\partial} \, (\nabla F_1)_{\psi_\partial} - (F_1)_{\psi_\partial} (\nabla F_2)_{\psi_\partial} \right), N^\perp_{\psi_\partial} \rangle \, d\mu_{g_\partial}.$$

In the above display, μ_{g_∂} stands for the volume measure on the parameter space $(\partial S)_{\psi_\partial}$ of the boundary ∂S equipped with some chart or parametrization $\psi_\partial \; : \; (\partial S)_{\psi_\partial} \mapsto \partial S$. We have also used the synthetic notation $N^\perp_{\psi_\partial} = (N^\perp \circ \psi_\partial)$, $(\nabla F_i)_{\psi_\partial} = (\nabla F_i) \circ \psi_\partial$, and $(F_i)_{\psi_\partial} = F_i \circ \psi_\partial$.

Once again we implicitly assumed that the manifold S and its boundary ∂S can be parametrized by a single map ψ and ψ_∂. More general situations can be handled using the patching technique presented in (21.5) and (23.10).

23.4 Gradient flow models

23.4.1 Steepest descent model

As their name indicates, stochastic gradient flow models are the stochastic versions of the well known steepest descent dynamical systems. Suppose we are given a manifold S. The steepest descent evolution equation in a chart

$$\phi \; : \; x \in S \mapsto \phi(x) \in S_\phi \subset \mathbb{R}^p$$

is given by the dynamical system

$$\dot{\theta}_t = -(\nabla_g V)(\theta_t)$$

with the Riemannian vector field gradient defined in (21.21). For Euclidian state spaces $S = \mathbb{R}^p$, the chart reduces to the identity mapping $\phi(x) = \theta = x = \phi^{-1}(\theta)$, and $\nabla_g V = \partial V$ is the traditional gradient.

We further assume that the manifold $S = \varphi^{-1}(0)$ is the null level set of a smooth function $\varphi \; : \; x \in \mathbb{R}^{r=p+q} \mapsto \mathbb{R}^q$ s.t. $\text{rank}(\partial \varphi(x)) = q$, for any $x \in \mathbb{R}^r$. Then for any $1 \le l \le q$ we have

$$\begin{aligned}
\frac{d}{dt} \varphi_l(\psi(\theta_t)) &= \sum_{1 \le i \le p} (\partial_{\theta_i}(\varphi_l \circ \psi)) \, (\theta_t) \, \dot{\theta}_t^i \\
&= - \sum_{1 \le i,j \le p} g^{i,j}(\theta_t) \sum_{1 \le k \le r} (\partial_{x_k} \varphi_l) \, (\psi(\theta_t)) \, (\partial_{\theta_i} \psi^k) \, (\theta_t) \, (\partial_{\theta_j} V) \, (\theta_t) \\
&= - \sum_{1 \le i,j \le p} g^{i,j}(\theta_t) \underbrace{\langle (\partial \varphi_l) \, (\psi(\theta_t)), (\partial_{\theta_i} \psi) \, (\theta_t) \rangle}_{=0} \, (\partial_{\theta_j} V) \, (\theta_t).
\end{aligned}$$

This shows that the gradient flow keeps the state x_t in the constraint manifold at any time t as soon as we start in the desired manifold. More formally, we have

$$x_0 = \psi(\theta_0) \in S \Rightarrow \forall t \ge 0 \qquad x_t = \psi(\theta_t) \in S.$$

We also notice that

$$\frac{d}{dt} V(\theta_t) = -\sum_{1 \le i \le p} g^{i,j}(\theta_t) \ (\partial_{\theta_i} V)(\theta_t) \ (\partial_{\theta_j} V)(\theta_t)$$

$$= -\langle (\nabla_g V)(\theta_t), (\nabla_g V)(\theta_t) \rangle_{g(\theta_t)} = -\langle \dot{\theta}_t, \dot{\theta}_t \rangle_{g(\theta_t)} = -\left\| \dot{\theta}_t \right\|_{g(\theta_t)} \Rightarrow V(\theta_t) \downarrow.$$

23.4.2   Euclidian state spaces

We start with an elementary example. The distribution

$$\pi(dx) \propto e^{-\frac{x^2}{2\sigma^2}} \ dx$$

on $\mathbb{R}$ is reversible w.r.t. the Ornstein-Uhlenbeck semigroup P_t associated with the generator

$$L(f)(x) = -\frac{x}{\sigma^2} \ \partial_x f(x) + \partial_x^2 f(x).$$

A simple way to check this claim is to rewrite the generator as

$$L(f)(x) = e^{\frac{x^2}{2\sigma^2}} \ \partial_x \left(e^{-\frac{x^2}{2\sigma^2}} \ \partial_x f \right)(x).$$

By a simple integration by parts, we have

$$\int e^{-\frac{x^2}{2\sigma^2}} \ f(x) \ L(g)(x) \ dx = \int f(x) \ \partial_x \left(e^{-\frac{x^2}{2\sigma^2}} \ \partial_x g \right)(x) \ dx$$

$$= -\int \partial_x f(x) \ e^{-\frac{x^2}{2\sigma^2}} \ \partial_x g(x) \ dx$$

$$= \int \partial_x \left(e^{-\frac{x^2}{2\sigma^2}} \partial_x f \right)(x) \ g(x) \ dx$$

$$= \int e^{-\frac{x^2}{2\sigma^2}} \ L(f)(x) \ g(x) \ dx.$$

More generally, we let S be a finite set,

$$V \ : \ x = (x_i)_{i \in S} \in E = \mathbb{R}^S \mapsto V(x) \in [0, \infty[$$

be a sufficiently smooth function that tends to infinity sufficiently fast when one of the coordinates of x tends to infinity and $\sigma, \beta \in]0, \infty[$ be some given parameters.

The Boltzmann Gibbs measure

$$\pi(dx) \propto e^{-\frac{2\beta}{\sigma^2} \ V(x)} \ \lambda(dx) \tag{23.23}$$

where λ stands for the Lebesgue measure on E, is reversible w.r.t. the semigroup P_t associated with the generator

$$L = -\beta \ \nabla V \cdot \nabla + \tfrac{1}{2} \ \sigma^2 \ \triangle$$

$$\Longleftrightarrow L(f)(x) = \tfrac{1}{2} \ \sigma^2 \sum_{i \in S} \partial_{x_i}^2 f(x) - \beta \sum_{i \in S} \partial_{x_i} V(x) \ \partial_{x_i} f(x). \tag{23.24}$$

Here again, a natural way to check this claim is to rewrite the generator as

$$L(f)(x) = \frac{1}{2}\,\sigma^2\,e^{2\beta/\sigma^2\,V(x)} \sum_{i \in S} \partial_{x_i}\left(e^{-2\beta/\sigma^2\,V(x)}\,\partial_{x_i}f\right)(x) \qquad (23.25)$$

and use an integration by part to check the desired reversibility property. The stochastic gradient diffusion with generator L is given by

$$dX_t = -\beta\,\nabla V(X_t)\,dt + \sigma\,dB_t \qquad (23.26)$$

where $B_t = (B_t^i)_{i \in S}$ stands for a sequence of independent Brownian motions on $\mathbb{R}$. The stochastic process X_t is reversible w.r.t. the invariant probability measure $\pi(dx)$.

The reversibility property is clearly satisfied if we replace L by αL, for some given parameter $\alpha > 0$. The corresponding process X_t (with generator αL) is defined as in (23.26) by replacing $(\beta, \sigma^2/2)$ by $(\alpha\beta, \alpha\sigma^2/2)$. In particular, the reversibility property w.r.t. the measure (23.23) is preserved if we replace $(\sigma^2/2, \beta)$ by $\left(1, \frac{2\beta}{\sigma^2}\right)$ in (23.24). Notice that in this case the proportionality factor is given by $\alpha^{-1} = \frac{1}{2}\,\sigma^2$ and the corresponding diffusion is given by

$$dX_t = -2\beta\,\sigma^{-2}\,\nabla V(X_t)\,dt + \sqrt{2}\,dB_t.$$

23.5 Drift changes and irreversible Langevin diffusions

This section is concerned with drift changes that preserve the target measure but not the reversibility property. We let $S = \cup_{1 \leq i \leq n} S_i$ be a partition of some finite set S. We consider the potential function

$$V \;:\; x = (x_i)_{i \in S_i} \in E = (\mathbb{R})^S \mapsto V(x) = \sum_{1 \leq i \leq n} \beta_i\,V_i(x_i) \in [0, \infty[$$

associated with a collection of parameters $\beta_i \geq 0$ and some sufficiently smooth functions

$$V_i \;:\; x_i = (x_{i,j})_{j \in S_i} \in E_i = (\mathbb{R})^{S_i} \mapsto V_i(x_i) \in [0, \infty[.$$

We let $L = \sum_{1 \leq i \leq n} (L_i + L_{-i})$ with the couple of generators (L_i, L_{-i}) defined by

$$\begin{aligned}
L_i(f)(x) : &= \frac{1}{2}\,\sigma_i^2\,e^{\frac{2\beta_i}{\sigma_i^2}\,V_i(x_i)} \sum_{j \in S_i} \partial_{x_{i,j}}\left(e^{-\frac{2\beta_i}{\sigma_i^2}\,V_i(x_i)}\,\partial_{x_{i,j}}f\right)(x) \\
&= \frac{1}{2}\,\sigma_i^2 \sum_{j \in S_i} \partial_{x_{i,j}}^2 f(x) - \beta_i \sum_{j \in S_i} \partial_{x_{i,j}}V_i(x_i)\,\partial_{x_{i,j}}f(x)
\end{aligned}$$

and

$$L_{-i}(f)(x) = \left[\sum_{k \in \{1,\dots,n\}-\{i\}} \alpha_{k,i}(x_k) \sum_{j \in S_k} \partial_{x_{k,j}}V_k(x_k)\right] \sum_{j \in S_i} \partial_{x_{i,j}}f(x)$$

for some functions $\alpha_{k,j}(x_k)$. We set

$$\partial_{x_i}f(x) := \sum_{j \in S_i} \partial_{x_{i,j}}f(x) \qquad \pi_i(dx_i) \propto e^{-\sum_{1 \leq i \leq n} \beta_i V_i(x_i)}\,\lambda(dx_i)$$

and

$$\pi(dx) \propto e^{-\sum_{1 \le i \le n} \beta_i V_i(x_i)} \lambda(dx).$$

It is now easy to check that for any $k \ne i$ and any function f with compact support, we have

$$\int e^{-\beta_i V_i(x_i)} \alpha_{k,i}(x_k) \, \partial_{x_k} V_k(x_k) \, \partial_{x_i} f(x) \, dx_i$$

$$= \int \pi_i(dx_i) \, f(x) \, \alpha_{k,i}(x_k) \, \partial_{x_k} V_k(x_k) \, \partial_{x_i} V_i(x_i)$$

$$\Rightarrow \int \pi(dx) \, L_{-i}(f)(x) = \int \pi(dx) \, f(x) \left[\sum_{k \in \{1,\dots,n\}-\{i\}} \alpha_{k,i}(x_k) \, \partial_{x_k} V_k(x_k) \right] \partial_{x_i} V_i(x_i).$$

We conclude that

$$\sum_{1 \le i \le n} \left[\sum_{k \in \{1,\dots,n\}-\{i\}} \alpha_{k,i}(x_k) \, \partial_{x_k} V_k(x_k) \right] \partial_{x_i} V_i(x_i) = 0 \Rightarrow \pi L = 0. \qquad (23.27)$$

Let us examine some consequence of these results. We let $E = E_1 \times E_2$ with $E_1 = \mathbb{R}^S$ and $E_2 = \mathbb{R}^S$ for some finite set S. A generic state is defined by $(x,y) \in (E_1 \times E_2)$ with $x = (x_i)_{i \in S}$ and $y = (y_i)_{i \in S}$. We consider the generators

$$L_1 := \frac{1}{2} \sigma_1^2 \sum_{i \in S} \partial_{x_i}^2 - \sum_{i \in S} \left[\alpha_{(1,1)}(x,y) \partial_{x_i} V_1(x) + \alpha_{(1,2)}(y) \partial_{y_i} V_2(y) \right] \partial_{x_i}$$

$$L_2 := \frac{1}{2} \sigma_2^2 \sum_{i \in S} \partial_{y_i}^2 - \sum_{i \in S} \left[\alpha_{(2,1)}(x) \partial_{x_i} V_1(x) + \alpha_{(2,2)}(x,y) \partial_{y_i} V_2(y) \right] \partial_{y_i}.$$

These generators rewritten in a slightly different form read as

$$L_1 := \left[\frac{1}{2} \sigma_1^2 \sum_{i \in S} \partial_{x_i}^2 - \sum_{i \in S} \alpha_{(1,1)}(x,y) \partial_{x_i} V_1(x) \; \partial_{x_i} \right] - \alpha_{(1,2)}(y) \sum_{i \in S} \partial_{y_i} V_2(y) \, \partial_{x_i}$$

$$L_2 := \left[\frac{1}{2} \sigma_2^2 \sum_{i \in S} \partial_{y_i}^2 - \sum_{i \in S} \alpha_{(2,2)}(x,y) \partial_{y_i} V_2(y) \; \partial_{y_i} \right] - \alpha_{(2,1)}(x) \sum_{i \in S} \partial_{x_i} V_1(x) \, \partial_{y_i}.$$

In this situation, applying (23.27) to $n = 2$, we prove the following result.

Proposition 23.5.1 *The Boltzmann Gibbs measure*

$$\pi(d(x_1, x_2)) \propto e^{-\beta_1 V_1(x_1) - \beta_2 V_1(x_2)} \, \lambda(dx_1) \, \lambda(dx_2)$$

is an invariant measure of the semigroup associated with the generator $L = L_1 + L_2$ as soon as the following conditions are satisfied:

$$\forall i = 1, 2 \qquad \alpha_{(i,i)} = \frac{1}{2} \sigma_i^2 \beta_i \quad \text{and} \quad \alpha_{(1,2)} + \alpha_{(2,1)} = 0.$$

23.5.1 Langevin diffusions on closed manifolds

We consider the projected diffusion model (20.11) associated with a gradient $b = \partial V$ of some smooth function $V : \mathbb{R}^r \mapsto \mathbb{R}$. In the further development of this section, we assume that the manifold S is smooth and closed (such as the sphere or the torus).

When $\sigma = Id$, we have $\mathbb{H}_\sigma = \mathbb{H}$ and $\pi(x)\partial V(x) = \nabla V(x)$ so that (20.11) can be interpreted as the projection on the manifold of the Langevin diffusion; that is, we have

$$
\begin{aligned}
dX_t &= \pi(X_t) \ (-\partial V(X_t)dt + dB_t) - \frac{1}{2} \ \mathbb{H}(X_t) \ dt \\
&= -\nabla V(X_t) \ dt + \left[\pi(X_t) \ dB_t - \frac{1}{2} \ \mathbb{H}(X_t) \ dt \right].
\end{aligned}
\tag{23.28}
$$

In this particular situation, the generator (20.13) of X_t is given by

$$
\begin{aligned}
L(F) &= \frac{1}{2} \ \Delta F - \langle \nabla V, \nabla F \rangle \quad (\text{ with } \ \Delta F = \mathrm{tr} \left(\nabla^2 F \right)) \\
&= \frac{1}{2} \ e^{2V} \ \mathrm{div} \left(e^{-2V} \ \nabla F \right).
\end{aligned}
$$

The divergence formulation given above is checked using the fact that

$$
\nabla \left[e^{-2V} \ \nabla F \right] = \nabla \left[e^{-2V} \right] \ [\nabla F]^T + e^{-2V} \ \nabla [\nabla F]
$$

and

$$
\nabla \left[e^{-2V} \right] = -2 \ e^{-2V} \ \nabla V
$$

so that

$$
\begin{aligned}
\mathrm{div} \left(e^{-2V} \ \nabla F \right) &= \mathrm{tr} \left(\nabla \left[e^{-2V} \ \nabla F \right] \right) \\
&= -2 \ e^{-2V} \ \mathrm{tr} \left(\nabla V \ [\nabla F]^T \right) + e^{-2V} \ \mathrm{tr} \left(\ \nabla [\nabla F] \right) \\
&= -2 \ e^{-2V} \ \langle \nabla V, \nabla F \rangle + e^{-2V} \ \Delta F.
\end{aligned}
$$

The end of the proof is now clear.

For any smooth function F and any smooth vector field $W \in T(S)$, we have

$$
\int_S F \ \mathrm{div} \left(W \right) \ d\mu_S = - \int_S \ \langle W, \nabla F \rangle \ d\mu_S
$$

where μ_S denotes the volume measure on S. This integration by part divergence theorem (23.16) is proved in section 23.3.

We let η be the Boltzmann-Gibbs measure on S defined by

$$
d\eta = \frac{1}{\mathcal{Z}} \ e^{-2V} \ d\mu_S,
$$

with $\mathcal{Z}$ being a normalizing constant.

Here we implicitly assumed that $e^{-2V} \, d\mu_S \in]0, \infty[$. In this notation,

$$
\begin{aligned}
2 \, \mathcal{Z} \int_S F_1 \, L(F_2) \, d\eta &= \int_S F_1 \, e^{2V} \, \mathrm{div}\left(e^{-2V} \, \nabla F_2\right) \, e^{-2V} \, d\mu_S \\
&= \int_S F_1 \, \mathrm{div}\left(e^{-2V} \, \nabla F_2\right) \, d\mu_S \\
&= -\int_S \left\langle e^{-2V} \, \nabla F_2, \nabla F_1 \right\rangle \, d\mu_S = -\int_S \left\langle \nabla F_1, \nabla F_2 \right\rangle \, e^{-2V} \, d\mu_S.
\end{aligned}
$$

> This implies the reversibility property of L w.r.t. η. More precisely, we have the formula
> $$
> \int_S F_1 \, L(F_2) \, d\eta = -\frac{1}{2} \int_S \left\langle \nabla F_1, \nabla F_2 \right\rangle \, d\eta = \int_S L(F_1) \, F_2 \, d\eta.
> $$

23.5.2   Riemannian Langevin diffusions

We consider the projected Langevin diffusion model (23.28), and we set $\phi(X_t) = \Theta_t$. In this situation,

$$
dX_t^k = -(\nabla \chi^k)^T (X_t) \partial V(X_t) \, dt + \frac{1}{2} \, \Delta(\chi^k)(X_t) \, dt + (\nabla \chi^k)^T (X_t) \, dB_t.
$$

Arguing as in the proof of (22.5) and using Ito's formula we have

$$
\begin{aligned}
d\phi^i(X_t) &= -\sum_{1 \leq k \leq r} \left(\partial_{x_k} \phi^i\right)(X_t)(\nabla \chi^k)^T(X_t) \, \partial V(X_t) \, dt \\
&\qquad\qquad\qquad\qquad + \frac{1}{2} \left(\Delta \phi^i\right)(X_t) dt + \left(\nabla \phi^i\right)^T (X_t) dB_t \\
&= -\left(\nabla \phi^i\right)^T (X_t)(\partial V)(X_t) \, dt + \frac{1}{2} \left(\Delta \phi^i\right)(X_t) dt + \left(\nabla \phi^i\right)^T (X_t) \, dB_t. \quad (\Leftarrow (22.4))
\end{aligned}
$$

Using the fact that $X_t = \psi(\Theta_t)$, we arrive at the equation

$$
\forall 1 \leq i \leq q \quad d\Theta_t^i = \left[-\left(\nabla \phi^i\right)_\psi^T (\Theta_t) \, (\partial V)_\psi(\Theta_t) + \frac{1}{2} \left(\Delta \phi^i\right)_\psi (\Theta_t) \right] dt + \left(\nabla \phi^i\right)_\psi^T (\Theta_t) \, dB_t.
$$

Using (21.40) and (21.41), we can express these Riemannian Langevin equations in terms of the Riemannian inner product g as follows:

$$
\begin{aligned}
\left(\Delta \phi^i\right)_\psi &= \sum_{1 \leq j \leq p} \frac{1}{\sqrt{\det(g)}} \, \partial_{\theta_j} \left(\sqrt{\det(g)} \, g^{i,j}\right) \\
\left(\nabla \phi^i\right)_\psi^T (\partial V)_\psi &= \sum_{1 \leq j \leq p} g^{i,j} \left\langle \left(\partial_{\theta_j} \psi\right), (\partial V)_\psi \right\rangle \\
d\Theta_t^i d\Theta_t^j &= \left(\nabla \phi^i\right)_\psi^T (\Theta_t) \, dB_t dB_t^T \left(\nabla \phi^j\right)_\psi (\Theta_t) = \left\langle \left(\nabla \phi^i\right)_\psi (\Theta_t), \left(\nabla \phi^j\right)_\psi (\Theta_t) \right\rangle dt \\
&= g^{i,j}(\Theta_t) \, dt = \left\langle \sum_{1 \leq k \leq p} \sqrt{g^{-1}}_k^i (\Theta_t) \, dB_t^k, \sum_{1 \leq l \leq p} \sqrt{g^{-1}}_l^j (\Theta_t) \, dB_t^l \right\rangle
\end{aligned}
$$

with

$$\left\langle \left(\partial_{\theta_j} \psi \right), (\partial V)_\psi(\theta) \right\rangle = \sum_{1 \leq k \leq r} \left(\partial_{\theta_j} \psi^k \right)(\theta) \, (\partial_{x_k} V)_\psi(\theta)$$

$$= \partial_{\theta_j} (V \circ \psi)(\theta) = \left(\partial_{\theta_j} U \right)(\theta), \quad \text{where} \quad U = V \circ \psi.$$

This yields

$$d\Theta_t^i = -\sum_{1 \leq j \leq p} g^{i,j} \left(\partial_{\theta_j} U \right)(\Theta_t) \, dt$$

$$+ \left[\sum_{1 \leq k \leq p} \sqrt{g^{-1}}^i_k(\Theta_t) \, dB_t^k + \frac{1}{2} \sum_{1 \leq j \leq p} \frac{1}{\sqrt{\det(g(\Theta_t))}} \, \partial_{\theta_j} \left(\sqrt{\det(g)} \, g^{i,j} \right)(\Theta_t) \, dt \right]$$

$$= - (\nabla_g U)^i (\Theta_t) \, dt + d\overline{B}_t^i$$

with the d-dimensional Brownian motion $\overline{B}_t$ on the Riemannian manifold (a.k.a. in local coordinates) defined for any $1 \leq i \leq p$ by

$$d\overline{B}_t^i = \sum_{1 \leq k \leq p} \sqrt{g^{-1}}^i_k(\Theta_t) \, dB_t^k + \frac{1}{2} \sum_{1 \leq j \leq p} \frac{1}{\sqrt{\det(g(\Theta_t))}} \, \partial_{\theta_j} \left(\sqrt{\det(g)} \, g^{i,j} \right)(\Theta_t) \, dt. \quad (23.29)$$

Here $\sqrt{g^{-1}}^i_k$ denotes the (i,k)-entry of the square root matrix $\sqrt{g^{-1}}$ of $g^{-1} \left(= \sqrt{g^{-1}} \sqrt{g^{-1}} \right)$.

Notice that

$$\frac{1}{2} \sum_{1 \leq j \leq p} \frac{1}{\sqrt{\det(g)}} \, \partial_{\theta_j} \left(\sqrt{\det(g)} \, g^{i,j} \right)$$

$$= \frac{1}{2} \sum_{1 \leq j \leq p} \partial_{\theta_j} \left(g^{i,j} \right) + \frac{1}{4} \sum_{1 \leq j \leq p} g^{i,j} \, \text{tr} \left(g^{-1} \partial_{\theta_j} g \right).$$

This shows that

$$d\Theta_t = - (\nabla_g U)(\Theta_t) \, dt + d\overline{B}_t. \quad (23.30)$$

Alternatively, in terms of the potential function U' defined by

$$e^{-2U'} := e^{-2U} \sqrt{\det(g)} \Longleftrightarrow U' = U - \frac{1}{4} \log \det(g(\theta))$$

we have

$$d\Theta_t = \overbrace{\left[- (\nabla_g U) + \frac{1}{4} \sum_{1 \leq j \leq p} g^{i,j} \, \text{tr} \left(g^{-1} \partial_{\theta_j} g \right) \right]}^{= -(\nabla_g U')} (\Theta_t) dt$$

$$+ \frac{1}{2} \sum_{1 \leq j \leq p} \partial_{\theta_j} \left(g^{i,j} \right)(\Theta_t) \, dt + \sum_{1 \leq k \leq p} \sqrt{g^{-1}}^i_k(\Theta_t) \, dB_t^k.$$

$$(23.31)$$

Arguing as in (22.2), and recalling that

$$\langle \partial f, \nabla_g U \rangle = \langle \nabla_g f, \nabla_g U \rangle_g ,$$

we check that the generator $\mathcal{L}$ of Θ_t is given by

$$
\begin{aligned}
\mathcal{L}(f) &= -\langle \nabla_g f, \nabla_g U \rangle_g + \frac{1}{2} \Delta_g(f) = -\langle \nabla_g f, \nabla_g U \rangle_g + \frac{1}{2} \operatorname{div}_g (\nabla_g(f)) \\
&= -\langle \nabla_g f, \nabla_g U \rangle_g + \frac{1}{2} \sum_{1 \le i \le p} \frac{1}{\sqrt{\det(g)}} \partial_{\theta_i} \left(\sqrt{\det(g)} \, (\nabla_g f)^i \right) \quad (\Leftarrow (21.48)) \\
&= \frac{1}{2} e^{2U} \sum_{1 \le i \le p} \frac{1}{\sqrt{\det(g)}} \partial_{\theta_i} \left(e^{-2U} \sqrt{\det(g)} \, (\nabla_g f)^i \right) .
\end{aligned}
$$

The above Sturm-Liouville formula can be rewritten in terms of the divergence operator

$$\mathcal{L}(f) = \frac{1}{2} e^{2U} \operatorname{div}_g \left(e^{-2U} \nabla_g(f) \right) .$$

We consider the Riemannian volume measure μ_g and the Boltzmann-Gibbs measure η on S_ψ defined by

$$\mu_g(d\theta) = \sqrt{\det(g(\theta))} \, d\theta \quad \text{and} \quad \eta(d\theta) = \frac{1}{\mathcal{Z}} e^{-2U(\theta)} \mu_g(d\theta) = \frac{1}{\mathcal{Z}} e^{-2U'(\theta)} \, d\theta,$$

with the normalizing constant

$$\mathcal{Z} = \int e^{-2U(\theta)} \mu_g(d\theta) = \int e^{-2U'(\theta)} \, d\theta.$$

For any smooth functions f_1, f_2 with compact support, using a simple integration by parts we have

$$
\begin{aligned}
\int f_1(\theta) \, \mathcal{L}(f_2)(\theta) \, e^{-2U(\theta)} \mu_g(d\theta) &= -\frac{1}{2} \sum_{1 \le i \le p} \int \partial_{\theta_i}(f_1)(\theta) \, (\nabla_g f_2)^i(\theta) \, e^{-2U(\theta)} \mu_g(d\theta) \\
&= -\frac{1}{2} \int \langle (\nabla_g f_1)(\theta), (\nabla_g f_2)(\theta) \rangle_{g(\theta)} \, e^{-2U(\theta)} \mu_g(d\theta).
\end{aligned}
$$

This shows that $\mathcal{L}$ is reversible w.r.t. η, that is,

$$
\begin{aligned}
\int f_1(\theta) \, \mathcal{L}(f_2)(\theta) \, \eta(d\theta) &= -\frac{1}{2} \sum_{1 \le i \le p} \int \langle (\nabla_g f_1)(\theta), (\nabla_g f_2)(\theta) \rangle_{g(\theta)} \, \eta(d\theta) \\
&= \int \mathcal{L}(f_1)(\theta) \, f_2(\theta) \, \eta(d\theta).
\end{aligned}
$$

23.6 Metropolis-adjusted Langevin models

The choice of the time discretization schemes is extremely important. For instance, a simple Euler type discretization model may fail to transfer the desired regularity properties of the continuous model to the discrete time model.

We illustrate this assertion with a discussion on an overdamped Langevin diffusion on an energy landscape associated with a given energy function $V \in \mathcal{C}^2(\mathbb{R}^d, \mathbb{R}_+)$ on $E = \mathbb{R}^d$ for some $d \geq 1$. This model is defined by the diffusion equation

$$dX_t = -\beta \, \nabla V(X_t) + \sqrt{2} \, dW_t \tag{23.32}$$

where ∇V denotes the gradient of V, β is an inverse temperature parameter, and W_t is a standard Brownian motion on $\mathbb{R}^d$. The infinitesimal generator associated with this continuous time process is given by the second order differential operator

$$L_\beta = -\beta \, \nabla V \cdot \nabla + \triangle.$$

Under some regularity conditions on V, the diffusion X_t' is geometrically ergodic with an invariant measure given by

$$d\pi_\beta = \frac{1}{\mathcal{Z}_\beta} \, e^{-\beta V} \, d\lambda,$$

where λ stands for the Lebesgue measure on $\mathbb{R}^d$, and $\mathcal{Z}_\beta$ is a normalizing constant.

As usual, in the continuous time framework, to find some feasible solution we need to introduce a time discretization scheme. To this end, we let $\mathcal{W}_{n+1}$ be a sequence of centered and standardized Gaussian variables on $\mathbb{R}^d$.

Firstly, starting from some random state $\mathcal{X}_n$, we propose a random state $\mathcal{Y}_{n+1}$ using the Euler scheme

$$\mathcal{Y}_{n+1} = \mathcal{X}_n - \beta \, \nabla V(\mathcal{X}_n)/m + \sqrt{2/m} \, \mathcal{W}_{n+1}. \tag{23.33}$$

Then, we accept this state $\mathcal{X}_{n+1} = \mathcal{Y}_{n+1}$ with probability

$$1 \wedge \left(e^{-\beta(V(\mathcal{Y}_{n+1}) - V(\mathcal{X}_n))} \times \frac{p_m(\mathcal{Y}_{n+1}, \mathcal{X}_n)}{p_m(\mathcal{X}_n, \mathcal{Y}_{n+1})} \right).$$

Otherwise, we stay in the same location $\mathcal{X}_{n+1} = \mathcal{X}_n$.

In the above display, the function p_m denotes the probability density of the Euler scheme proposition

$$p_m(x, y) = \frac{1}{(4\pi/m)^{d/2}} \, \exp\left(-\frac{m}{4} \, \|y - x + \beta \, \nabla V(x)/m\|^2 \right).$$

The resulting Markov chain model X_n is often referred to as the Metropolis-adjusted Langevin algorithm (*abbreviated (MALA)*). One of the main advantages of the above construction is that the Markov chain X_n is reversible w.r.t. to π_β, and it has the same fixed point π_β as the continuous time model. Without the acceptance-rejection rate, the Markov chain (23.33) reduces to the standard Euler approximation of the Langevin diffusion model (23.32). In this situation, the Markov chain may even fail to be ergodic, when the vector field ∇V is not globally Lipschitz [232, 191]. We refer the reader to [31, 135, 232, 233, 262] for further details on the stochastic analysis of these Langevin diffusion models. We also mention that the Euler scheme diverges in many cases, even for uniformly convex functions V. At the cost of some additional computational effort, a better idea is to replace (23.33) by the implicit backward Euler scheme given by

$$\mathcal{Y}_{n+1} + \beta \, \nabla V(\mathcal{Y}_{n+1})/m = \mathcal{X}_n + \sqrt{2/m} \, W_{n+1}.$$

23.7   Stability and some functional inequalities

We let $S = \varphi^{-1}(0) \subset \mathbb{R}^{r=p+1}$ be some hypersurface, and we let π and $\mathbb{H}$ be the projection matrix and the mean curvature vector described in definition 20.1.1. We assume that $T(S) = \text{Vect}(V_1, \ldots, V_p)$ for some basis vector fields satisfying the commutation property (19.15). We let L be a second order operator of the form

$$L = \frac{1}{2} \, \Delta \, - \nabla_W$$

for some vector field W on $T(S)$. In this situation, we recall that

$$\Gamma_L(F, F) = \langle \nabla F, \nabla F \rangle.$$

We let P_t be the transition semigroup associated with the Markov process on S with generator L, that is,

$$P_t(F)(x) = \mathbb{E}(F(X_t) \mid X_0 = x) \quad \text{and} \quad \frac{\partial}{\partial t} P_t(F) = L P_t(F).$$

This implies that

$$\begin{aligned}
\frac{\partial}{\partial t} \Gamma_L \left[P_t(F), P_t(F) \right] &= \frac{\partial}{\partial t} \langle \nabla P_t(F), \nabla P_t(F) \rangle \\
&= 2 \, \langle \nabla P_t(F), \nabla L(P_t(F)) \rangle = 2 \, \Gamma_L \left[P_t(F), L(P_t(F)) \right]
\end{aligned}$$

and therefore

$$L \left(\Gamma_L \left[P_t(F), P_t(F) \right] \right) - \frac{\partial}{\partial t} \Gamma_L \left[P_t(F), P_t(F) \right] = \Gamma_{2,L} \left[P_t(F), P_t(F) \right]. \tag{23.34}$$

We consider the interpolating function

$$s \in [0, t] \mapsto a(s) := P_s \Gamma_L \left[P_{t-s}(F), P_{t-s}(F) \right]$$

between

$$a(0) = \Gamma_L \left[P_t(F), P_t(F) \right] = \| \nabla P_t(F) \|^2 \quad \text{and} \quad a(t) = P_t \Gamma_L \left[F, F \right] = P_t \left(\| \nabla F \|^2 \right).$$

By (23.34) we have

$$a'(s) = P_s \left(\Gamma_{2,L} \left[P_{t-s}(F), P_{t-s}(F) \right] \right).$$

Using the Bochner-Lichnerowicz formula presented in (19.98) we have

$$\Gamma_{2,L}(F, F) = \left\| \nabla^2 F \right\|^2 + \text{Ric} \left(\nabla F, \nabla F \right) + 2 \, (\nabla W)_{\text{sym}} \left(\nabla F, \nabla F \right) \tag{23.35}$$

with the symmetric bilinear form $(\nabla W)_{\mathrm{sym}}$ defined in (19.98) and the Ricci curvature of the manifold discussed in section 19.8.

> Assuming that
>
> $$\exists \rho > 0 \; : \; \Gamma_{2,L}(F,F) \geq \rho \, \Gamma_L(F,F) \quad \Rightarrow \quad a'(s) \geq \rho \, a(s) \quad \Rightarrow \quad a(0) \leq e^{-\rho t} \, a(t) \tag{23.36}$$
>
> we readily conclude that
>
> $$\|\nabla P_t(F)\|^2 \leq e^{-\rho t} \, P_t \left(\|\nabla F\|^2 \right).$$

The l.h.s. condition in (23.36) allows us to apply Grownwall's inequality based on the key Bochner-Lichnerowicz formula. In probability and statistic literature, this condition is often called the Bakry-Emery criterion [9].

Rewritten in terms of Γ_L we have proved that

$$\Gamma_L \left(P_t(F), P_t(F) \right) \leq e^{-\rho t} \, P_t \left(\Gamma_L (F,F) \right). \tag{23.37}$$

We consider the interpolating sequence

$$t \in [0,t] \mapsto \widehat{a}(s) := P_s \left((P_{t-s}(F))^2 \right)$$

between

$$\widehat{a}(0) := (P_t(F))^2 \quad \text{and} \quad \widehat{a}(t) := P_t \left(F^2 \right).$$

It is readily checked that

$$
\begin{aligned}
\widehat{a}'(s) &= \frac{\partial P_s}{\partial s} \left((P_{t-s}(F))^2 \right) + P_s \left(\frac{\partial}{\partial s} (P_{t-s}(F))^2 \right) \\
&= P_s \left(L \left[(P_{t-s}(F))^2 \right] \right) - 2 P_s \left(P_{t-s}(F) \, L \left[P_{t-s}(F) \right] \right) \\
&= P_s \left(\Gamma_L \left[P_{t-s}(F), P_{t-s}(F) \right] \right) \leq e^{-\rho(t-s)} \, P_t \left(\Gamma_L (F,F) \right). \qquad \Leftarrow (23.37)
\end{aligned}
$$

> This yields the Poincaré inequality for the Markov semigroup
>
> $$P_t \left(F^2 \right) - (P_t(F))^2 \leq \frac{1}{\rho} \left[1 - e^{-\rho t} \right] \, P_t \left(\Gamma_L (F,F) \right).$$

Suppose that $P_t(F) \to_{t \to \infty} \pi(F)$ for some invariant measure $\pi = \pi P_t \; (\Rightarrow \pi L = 0)$. In this situation, we check that π satisfies the Poincaré inequality

$$\pi \left(F^2 \right) - (\pi(F))^2 \leq \frac{1}{\rho} \, \pi \left[\|\nabla F\|^2 \right] \quad \left(= \frac{1}{\rho} \, \pi \left(\Gamma_L (F,F) \right) \right). \tag{23.38}$$

Following the development provided in section 17.6, let us examine some more or less direct consequences of (23.38). A simple derivation shows that

$$
\begin{aligned}
\frac{d}{dt} \pi \left[(P_t(F))^2 \right] &= 2 \pi \left[P_t(F) \, \frac{\partial}{\partial t} P_t(F) \right] = 2 \pi \left[P_t(F) \, L \left(P_t(F) \right) \right] \\
&= -\pi \left(\Gamma_L \left[P_t(F), P_t(F) \right] \right) \quad (\Leftarrow \pi L = 0) \\
&\leq -\rho \, \pi \left((P_t(F))^2 \right) \quad \Leftarrow (23.38)
\end{aligned}
$$

for any function F such that $\pi(F) = 0$.

> This clearly implies the exponential decay to equilibrium
>
> $$\pi\left[(P_t(F) - \pi(F))^2\right] \leq e^{-\rho t}\, \pi\left[(F - \pi(F))^2\right].$$

For a further discussion on Poincaré inequalities and their applications in the stability analysis of Markov processes, we refer the reader to exercise 258 in the context of the Ornstein-Uhlenbeck process, and to exercise 259 for exponential limiting distributions.

> For instance, the regularity condition in the l.h.s. of (23.35) is met when the bilinear forms induced by Ric and $(\nabla W)_{\mathrm{sym}}$ satisfy the minorization condition
>
> $$\mathrm{Ric}\,(\nabla F, \nabla F) + 2\,(\nabla W)_{\mathrm{sym}}\,(\nabla F, \nabla F) \geq\ \rho\,\langle \nabla F, \nabla F\rangle = \rho\,\Gamma_L(F, F). \quad (23.39)$$
>
> In view of (19.86), this condition is also met when the smallest eigenvalues $\lambda(x)$ of the matrices $\mathrm{Ric}\,(\pi_i(x), \pi_j(x)) + (\nabla W)_{\mathrm{sym}}(x)$ are lower bounded by some parameter ρ. Using (19.85) this minorization condition is also satisfied for any Einstein manifold. For gradient vector fields $W = \nabla V$, the regularity condition (23.39) is expressed in terms of the bilinear form induced by second covariant derivative of the function V, that is,
>
> $$\mathrm{Ric}\,(\nabla F, \nabla F) + 2\,\nabla^2 V\,(\nabla F, \nabla F) \geq\ \rho\,\langle \nabla F, \nabla F\rangle.$$

Notice that

$$(23.39) \Rightarrow \Gamma_{2,L}(F, F) \geq \rho\,\Gamma_L(F, F) + \left\|\nabla^2 F\right\|^2$$

which by (19.58) yields

$$\Gamma_{2,L}(F, F) - \rho\,\Gamma_L(F, F) \geq \left\|\nabla^2 F\right\|^2 \geq \frac{1}{4}\,\frac{\left\|\nabla\langle \nabla F, \nabla F\rangle\right\|^2}{\|\nabla F\|_2^2}$$

or equivalently

$$\Gamma_{2,L}(F, F) - \frac{1}{4}\,\frac{\left\|\nabla\langle \nabla F, \nabla F\rangle\right\|^2}{\|\nabla F\|_2^2} \geq \rho\,\Gamma_L(F, F). \quad (23.40)$$

In probability theory, the norm $\|\nabla\langle \nabla F, \nabla F\rangle\|$ is often rewritten in terms of the Γ_L bilinear form

$$\begin{aligned}
\left\|\nabla\langle \nabla F, \nabla F\rangle\right\|^2 &=\ \langle \nabla\langle \nabla F, \nabla F\rangle, \nabla\langle \nabla F, \nabla F\rangle\rangle = \Gamma_L(\langle \nabla F, \nabla F\rangle, \langle \nabla F, \nabla F\rangle) \\
&=\ \Gamma_L\left(\Gamma_L\left(F, F\right), \Gamma_L\left(F, F\right)\right).
\end{aligned}$$

In this notation, we have the equivalent inequality

$$\Gamma_{2,L}(F, F) - \frac{1}{4}\,\frac{\Gamma_L\left(\Gamma_L\left(F, F\right), \Gamma_L\left(F, F\right)\right)}{\Gamma_L\left(F, F\right)} \geq \rho\,\Gamma_L(F, F).$$

By applying the formula (19.100) to the function $h(u) = \sqrt{u}$, for any $0 \leq s \leq t$ we find that

$$L\left\{h\left[\Gamma_L\left(P_{t-s}(F), P_{t-s}(F)\right)\right]\right\}$$

$$= \frac{1}{2}\,\frac{1}{\sqrt{\Gamma_L(P_{t-s}(F), P_{t-s}(F))}}\,L\left(\Gamma_L\left(P_{t-s}(F), P_{t-s}(F)\right)\right)$$

$$-\frac{1}{\sqrt{\Gamma_L(P_{t-s}(F), P_{t-s}(F))}}\,\frac{1}{4}\frac{\Gamma_L[\Gamma_L[P_{t-s}(F), P_{t-s}(F)], \Gamma_L[P_{t-s}(F), P_{t-s}(F)]]}{\Gamma_L(P_{t-s}(F), P_{t-s}(F))}.$$

Using the interpolating function

$$\overline{a}(s) := P_s\left(h\left[\Gamma_L\left(P_{t-s}(F), P_{t-s}(F)\right)\right]\right)$$

between

$$\overline{a}(0) = \sqrt{\Gamma_L\left[P_t(F), P_t(F)\right]} = \|\nabla P_t(F)\| \quad \text{and} \quad \overline{a}(t) = P_t\sqrt{\Gamma_L\left[F, F\right]} = P_t\left(\|\nabla F\|\right)$$

we find that

$$a'(s) = P_s\left(L\left\{h\left[\Gamma_L\left(P_{t-s}(F), P_{t-s}(F)\right)\right]\right\}\right)$$

$$-\frac{1}{2}\,P_s\left(\frac{1}{\sqrt{\Gamma_L(P_{t-s}(F), P_{t-s}(F))}}\,\Gamma_L\left[P_{t-s}(F), L(P_{t-s}(F))\right]\right)$$

$$= \frac{1}{2}\,P_s\left(\frac{1}{\sqrt{\Gamma_L(P_{t-s}(F), P_{t-s}(F))}}\right.$$

$$\left.\times\left\{\Gamma_{2,L}\left[P_{t-s}(F), P_{t-s}(F)\right] - \frac{1}{4}\frac{\Gamma_L[\Gamma_L[P_{t-s}(F), P_{t-s}(F)], \Gamma_L[P_{t-s}(F), P_{t-s}(F)]]}{\Gamma_L(P_{t-s}(F), P_{t-s}(F))}\right\}\right).$$

Using (23.40) we find that $2\,a'(s) \geq \rho\,a(s)$ from which we deduce that

$$(23.39) \implies \|\nabla P_t(F)\| \leq e^{-\rho\,t/2}\,P_t\left(\|\nabla F\|\right).$$

23.8 Exercises

Exercise 408 (Gradient flows - Gradient estimates) *We let $W \in T(S)$ be a vector field on the tangent space of some manifold S embedded in $\mathbb{R}^r$, for some $r \geq 1$. We let $t \in [0, \infty[\mapsto C_x(t)$ be some curve in S starting at some location $C_x(0) = x$ and such that $\dot{C}_x(t) = W(C_x(t))$. For any smooth function F on $\mathbb{R}^r$ check that*

$$\frac{1}{2}\partial_t\|\nabla F(C_x(t))\|^2 = \langle\nabla F(C_x(t)), (\nabla^2 F)(C_x(t))\,W(C_x(t))\rangle.$$

We further assume that $W = -\nabla F$ for some smooth function s.t. $\nabla^2 F \geq \lambda\,Id$ (in the sense that $\langle W, \nabla^2 F(W)\rangle \geq \lambda\,\langle W, W\rangle$, for any $W \in T(S)$). In this situation, check that

$$\|\nabla F(C_x(t))\| \leq e^{-\lambda t}\,\|\nabla F(x)\|.$$

Deduce that

$$0 \leq F(x) - F(C_x(t)) \leq \frac{1}{2\lambda}\,\|\nabla F(x)\|^2.$$

Exercise 409 (Lagrange indentity) *We consider r-dimensional vectors $U = \left(U^i\right)_{1 \le i \le r}$ and $V = \left(V^i\right)_{1 \le i \le r} \in \mathbb{R}^r$. Prove that*

$$U \wedge V = \sum_{1 \le i < j \le r} \left(U^i V^j - U^j V^i\right) \ (e_i \wedge e_j)$$

with the r unit basis vectors $(e_i)_{1 \le i \le r}$ of $\mathbb{R}^r$. Deduce the Lagrange identity

$$\|U \wedge V\|^2 + |\langle U, V \rangle|^2 = \|U\|^2 \times \|V\|^2 \,.$$

Exercise 410 (Volume of parallepiped) *We consider the parallepiped $\mathcal{P}(W_1, W_2, W_3)$ in $\mathbb{R}^3$ formed by three independent vectors $W_1, W_2, W_3 \in \mathbb{R}^3$. Check the volume formula*

$$\text{Volume}\left(\mathcal{P}(W_1, W_2, W_3)\right) = |\langle W_1 \wedge W_2, W_3 \rangle| = |\det\left(W_1, W_2, W_3\right)| \,.$$

Exercise 411 (Geodesics on the sphere) *Describe the geodesics on the unit 2-sphere (we can use the Christoffell symbols associated with spherical coordinates derived in section 24.1.2).*

Exercise 412 (Arc length of a curve) *We consider a surface $S \subset \mathbb{R}^3$ parametrized by a function $\theta = (\theta_1, \theta_2) \in S_\psi \mapsto \psi(\theta) = \left(\psi^i(\theta)\right)_{1 \le i \le 3}$. Compute the arc length $s = \int_a^b \|C(t)\| \ dt$ of some curve $C : t \in [a, b] \mapsto C(t) \in S$ in terms of the Riemannian scalar product g associated with the parametrization ψ. Compute the lengths $\mathcal{L}(C, [a, b])$ of a curve $C : t \in [a, b] \mapsto C(t) = \psi(c(t)) \in S$ on the cylinder S defined by $x_1^2 + x_2^2 = r$ and $x_3 \in \mathbb{R}$, equipped with the polar parametrization*

$$\theta = (\theta_1, \theta_2) \mapsto \psi(\theta) = \begin{cases} \psi^1(\theta) &=& r\cos(\theta_1) \\ \psi^2(\theta) &=& r\sin(\theta_1) \\ \psi^3(\theta) &=& \theta_2 \end{cases}$$

in terms of the velocity vector of a curve $c(t)$ in S_ψ. Compute the length of the curves $C_0(t) = \psi(\beta, r\alpha t)$, $C_1(t) = \psi(\alpha t, \beta)$ and $C_2(t) = \psi(\alpha t^2, \beta)$ for given parameters α, β.

Exercise 413 (Surface of the disk) *We consider the disk S and its polar coordinates ψ on $S_\psi =]0, 1[\times]0, 2\pi[$ discussed in (23.17). We let $W \in T(S)$. Using (21.14) compute the vector field $V \in T(S_\psi)$ such that*

$$W \circ \psi = V^1 \ \partial_{\theta_1} \psi + V^2 \ \partial_{\theta_1} \psi.$$

When $W(x) = \frac{1}{2} \begin{pmatrix} x_1 \\ x_2 \end{pmatrix}$ deduce that $\int_{S_\psi} \text{div}_g(V) \ d\mu_g = \pi \, R^2$.

Exercise 414 (Volume of 3-ball) *We consider the 3-dimensional ball and the 2-sphere boundary*

$$\mathbb{B} = \{x = (x_1, x_2, x_3) : x_1^2 + x_2^2 + x_3^3 \le r^2\} \Rightarrow \partial\mathbb{B} = \mathbb{S} = \{x = (x_1, x_2, x_3) : x_1^2 + x_2^2 + x_3^3 = r^2\}$$

for some radius $r > 0$. Applying the divergence theorem (23.16) to the vector field $W(x) = \frac{1}{3}(x_1, x_2, x_3)^T$ compute the volume of the 3-ball given by

$$\mu_{\mathbb{B}}\left(\mathbb{B}\right) = \int_{\mathbb{S}} \langle W, N^\perp \rangle \ d\mu_{\mathbb{S}} = \frac{4r^3\pi}{3}.$$

Check that $\mu_{\mathbb{B}}\left(\mathbb{B}\right) = \frac{r}{3} \, \mu_{\mathbb{S}}\left(\mathbb{S}\right)$, and $\mu_{\mathbb{S}}\left(\mathbb{S}\right) = 4r^2\pi$.

Exercise 415 (Volume and surface measures) *We parametrize the 3-ball $\mathbb{B}$ discussed in exercise 414 with the spherical coordinates*

$$\psi_0(\theta_0, \theta_1, \theta_2) = \begin{pmatrix} r(1-\theta_0) \ \sin(\theta_1)\cos(\theta_2) \\ r(1-\theta_0) \ \sin(\theta_1)\sin(\theta_2) \\ r(1-\theta_0) \ \cos(\theta_1) \end{pmatrix}$$

with $(\theta_0, \theta_1, \theta_2) \in ([0,1] \times [0,\pi] \times [0,2\pi[)$. We also denote by $\psi(\theta_1,\theta_2) = \psi_0(0,\theta_1,\theta_2)$ the spherical parametrization of the sphere $\mathbb{S} = \partial\mathbb{B}$. We denote by g and g_∂ the corresponding Riemannian metric on $\mathbb{B}$ and its boundary $\partial\mathbb{B}$. Check that

$$\mu_g(d(\theta_0,\theta_1,\theta_2)) = r \ (1-\theta_0)^2 \ d\theta_0 \ \mu_{g_\partial}(d(\theta_1,\theta_2)) \ \text{ with } \ \mu_{g_\partial}(d(\theta_1,\theta_2)) := r^2 \ \sin(\theta_1)d\theta_1 d\theta_2.$$

We let $n^\perp(\theta_1,\theta_2)$ be the unit outward pointing normal to the sphere at the point $\psi(\theta_1,\theta_2)$. Check that

$$\mu_g(d(\theta_0,\theta_1,\theta_2)) = \left\langle -\partial_{\theta_0}\psi_0, n^\perp \right\rangle \ d\theta_0 \times \|(\partial_{\theta_1}\psi_0 \wedge \partial_{\theta_2}\psi_0)\| \ d\theta_1 d\theta_2.$$

Exercise 416 (Gauss theorem for the ball) *We consider the 3-ball $\mathbb{B}$ discussed in exercise 414 with the spherical coordinates presented in exercise 415. We use the same notation as in exercise 415. Using (21.14) check that any vector field W on $\mathbb{B}$ takes the form*

$$W \circ \psi_0 = V^0 \ \partial_{\theta_0}\psi_0 + V^1 \ \partial_{\theta_1}\psi_0 + V^2 \ \partial_{\theta_2}\psi_0$$

for some vector field $V = (V^i)_{0 \leq i \leq 2}$ on the product space $([0,1] \times [0,\pi] \times [0,2\pi])$. Check that

$$\left\langle (W \circ \psi)(\theta_1,\theta_2), (N^\perp \circ \psi)(\theta_1,\theta_2) \right\rangle = -r \ V^0(0,\theta_1,\theta_2)$$

where stands for the unit outward-pointing normal vector field to the sphere $\mathbb{S} = \partial\mathbb{B}$. Prove that

$$\forall i = 0,1,2 \qquad V^i(\theta_0,\theta_1,0) = V^i(\theta_0,\theta_1,2\pi).$$

Without using Gauss' theorem, show that

$$\int_0^{2\pi} \left[\int_0^\pi \left[\int_0^1 \partial_{\theta_0}\left(V^0\sqrt{\det(g)} \right) \ d\theta_0 \right] d\theta_1 \right] d\theta_2$$

$$= -r^3 \ \int_0^{2\pi} \left[\int_0^\pi V^0(0,\theta_1,\theta_2) \ \sin(\theta_1)d\theta_1 \right] d\theta_2.$$

Deduce Gauss' theorem

$$\int_{\mathbb{B}_{\psi_0}} \mathrm{div}_g(V) \ d\mu_g = \int_{\partial\mathbb{B}} \langle W, N^\perp \rangle \ d\mu_{\partial\mathbb{B}}.$$

Exercise 417 (Langevin equation) *The velocity of a one-dimensional particle X_t of mass m in a viscous medium is represented by the Langevin diffusion process*

$$m \ dX_t = -b \ X_t \ dt + \sigma \ dW_t$$

where W_t stands for the standard Brownian motion. The parameter b is the friction coefficient. The latter depends on the geometry of the particle and the viscosity of the medium. The Brownian motion represents the random force.

- *Find an explicit representation of X_t in terms of t and $(W_s)_{s \leq t}$.*

- *Describe the invariant probability measure of the diffusion X_t.*

- *We let $\mathbb{W}$ be the Wasserstein distance associated with the metric $d(x, y) = |x - y|$ (cf. definition 8.3.8). Check that*

$$\mathbb{W}\left(\mathrm{Law}(X_t), \pi\right) \le e^{-t(b/m)} \left(\mathbb{E}(X_0) + \frac{\sigma^2}{2mb} \, e^{-(b/m)t}\right).$$

Exercise 418 (Projected Langevin equation) *We let μ_S be the volume measure on the unit sphere $\mathbb{S}^2 = \{(x_1, x_2, x_3) \in \mathbb{R}^3, \ x_1^2 + x_2^2 + x_3^2 = 1\}$ and π be the Boltzmann-Gibbs measure on S defined by*

$$d\pi = \frac{1}{\mathcal{Z}} \, e^{-2x^T A x} \, d\mu_S$$

for some normalizing constant $\mathcal{Z} < \infty$ and some matrix A. Find a diffusion equation on $\mathbb{S}^2$ with reversible measure π.

Exercise 419 (Riemannian Langevin equation) *We let μ_g be the Riemannian volume measure on the unit sphere $\mathbb{S}^2$ equipped with the spherical coordinates*

$$\psi \ : \ \theta = (\theta_1, \theta_2) \in S_\psi = ([0, \pi] \times [0, 2\pi])$$

$$\mapsto \psi(\theta_1, \theta_2) = (\sin(\theta_1)\cos(\theta_2), \sin(\theta_1)\sin(\theta_2), \cos(\theta_1)).$$

We let η be the Boltzmann-Gibbs measure on S_ψ defined by

$$\eta(d(\theta_1, \theta_2)) = \frac{1}{\mathcal{Z}} \, e^{-2\psi(\theta)^T A \psi(\theta)} \, \sin(\theta_1) \, d\theta_1 d\theta_2$$

for some normalizing constant $\mathcal{Z} < \infty$ and some matrix A. Find a diffusion equation on S_ψ with reversible measure η.

Exercise 420 (Kinetic Langevin diffusion) *Let U^X and U^V be some non-negative and smooth potential functions on $\mathbb{R}$ s.t. $\int e^{-U^X(x)} \, dx$ and $\int e^{-U^V(v)} \, dv \in]0, \infty[$. We let (π^X, π^V) be the Boltzmann-Gibbs probability measures given by $\pi^X(dx) \propto e^{-U^X(x)}dx$ and $\pi^V(dv) \propto e^{-U^V(v)}dv$. For any given $\epsilon \in \{-1, +1\}$, we let $\mathcal{X}_t^\epsilon := (X_t, V_t)$ be the diffusion given by*

$$\begin{cases} dX_t &= \ \epsilon \, \partial_v U^V(V_t) \, dt \\ dV_t &= \ -\left[\partial_v U^V(V_t) + \epsilon \, \partial_x U^X(X_t)\right] \, dt + \sqrt{2} \, dW_t \end{cases}$$

where W_t stands for a Brownian motion. Compute the generator L^ϵ of $\mathcal{X}_t^\epsilon$. Check that $\pi(gL^\epsilon(f)) = \pi(fL^{-\epsilon}(g))$ and deduce that $\pi(d(x, v)) = \pi^X(dx)\pi^V(dv)$ is L^ϵ-invariant. Applications of these kinetic Langevin samplers in computational physics are discussed in section 27.1.2 (see also section 23.4.2 for a more detailed discussion on these non-reversible kinetic samplers).

24

Some illustrations

The first part of this chapter presents some illustrations of the main mathematical objects and geometrical models discussed in earlier chapters. We provide worked out and detailed examples of chart and parametric spaces for some classical manifolds such as the circle, the sphere and the torus. In each case, we present a detailed derivation of the mean curvature vectors, the Riemannian metrics, the geodesics, the Christoffel symbols, and the Ricci curvature. The second part of the chapter presents selected applications of Riemannian geometry to statistics and physics.

It is always more easy to discover and proclaim general principles than to apply them.
Winston S. Churchill (1874-1965).

24.1 Prototype manifolds

24.1.1 The circle

The prototype of hypermanifold is the unit circle $S = \mathbb{S}^1 \subset \mathbb{R}^2 \ni x = \begin{pmatrix} x_1 \\ x_2 \end{pmatrix}$ described as the null level set $S = \varphi^{-1}(0)$ of the function

$$\varphi(x) = (x_1^2 + x_2^2 - 1)/2 \Rightarrow (\partial\varphi)(x) = x = \begin{pmatrix} x_1 \\ x_2 \end{pmatrix}.$$

The orthogonal projection $\pi_\perp$ onto the normal axis $T_x^\perp(S) = \mathrm{Vect}\,((\partial\varphi)(x))$ at $x \in S$ is given by the formula

$$\pi_\perp(x) = \frac{(\partial\varphi)\,(x)\,(\partial\varphi)^T\,(x)}{(\partial\varphi)\,(x)^T\,(\partial\varphi)\,(x)} = \frac{xx^T}{x^Tx} = \frac{1}{x_1^2 + x_2^2}\begin{pmatrix} x_1^2 & x_1x_2 \\ x_2x_1 & x_2^2 \end{pmatrix},$$

and the orthogonal projection on $T_x(S)$ is defined by $\pi(x) = Id - \pi_\perp(x)$. The (mean) curvature vector $\mathbb{H}$ defined by (20.2) on the circle is simply

$$\forall x \neq 0 \qquad \mathbb{H}(x) = \left[\sum_{1 \leq m \leq 2} \partial_{x_m}\left(\frac{x_m}{\sqrt{x_1^2 + x_2^2}}\right)\right]\frac{x}{\sqrt{x_1^2 + x_2^2}} = \frac{x}{\|x\|^2}.$$

We check this claim using

$$\partial_{x_1}\left(\frac{x_1}{\sqrt{x_1^2+x_2^2}}\right) = \frac{1}{\sqrt{x_1^2+x_2^2}}\left[1-\frac{x_1^2}{(x_1^2+x_2^2)}\right]$$

$$\Rightarrow \operatorname{div}_\perp\left(\frac{\partial\varphi}{\|\partial\varphi\|^2}\right) = \sum_{1\le m\le 2}\partial_{x_m}\left(\frac{x_m}{\sqrt{x_1^2+x_2^2}}\right) = \frac{1}{\sqrt{x_1^2+x_2^2}}.$$

The circle $S-\{(1,0)\}$ can be parametrized by the polar angle mapping $\psi : \theta \in]0,2\pi[\mapsto S-\{(1,0)\}$

$$\psi(\theta) = \begin{pmatrix} \cos(\theta) \\ \sin(\theta) \end{pmatrix} \implies (\partial_\theta\psi)(\theta) = \begin{pmatrix} -\sin(\theta) \\ \cos(\theta) \end{pmatrix}$$

so that

$$T_x(S) = \operatorname{Vect}\left((\partial_\theta\psi)_{\phi(x)}\right) \quad \text{with} \quad (\partial_\theta\psi)_{\phi(x)} = \begin{pmatrix} -\sin(\theta) \\ \cos(\theta) \end{pmatrix}_{\theta=\phi(x)} = \begin{pmatrix} -x_2 \\ x_1 \end{pmatrix}.$$

The Riemannian metric on $S_\psi =]0,2\pi[\Rightarrow T(S_\psi) = \mathbb{R} = \operatorname{Vect}(1)$ reduces to

$$g(\theta) = \langle(\partial_\theta\psi)(\theta),(\partial_\theta\psi)(\theta)\rangle = 1 = g(\theta)^{-1} \Rightarrow (\nabla\phi)_\psi = \partial_\theta\psi. \qquad (24.1)$$

Using (19.71) and (19.42) for any smooth function F on $S \ni x$ we have

$$\operatorname{div}_\perp(\partial\varphi) = \partial_{x_1}(\partial_{x_1}\varphi) + \partial_{x_2}(\partial_{x_2}\varphi) = 2$$

and therefore

$$\frac{1}{2}\Delta F = \operatorname{tr}\left(\pi\partial^2 F\right) - \langle\partial\varphi,\partial F\rangle.$$

In addition, we have

$$(\partial_{\theta,\theta}\psi)(\theta) = -\begin{pmatrix} \cos(\theta) \\ \sin(\theta) \end{pmatrix} \in T^\perp(S) \Rightarrow C_{1,1}^1 = 0 \quad \text{and} \quad (\Delta\phi)_\psi = 0. \qquad (24.2)$$

The geodesics $c_\theta(t) := Exp_\theta(tV)$, with $V(\theta) \in \mathbb{R}$ are given by

$$\ddot{c}_\theta(t) = 0 \Rightarrow \dot{c}_\theta(t) = V(\theta)$$

$$\Rightarrow c_\theta(t) = t\,V(\theta) + \theta \Rightarrow C_x(t) = \psi(c_\theta(t)) = \begin{pmatrix} \cos(t\,V(\theta)+\theta) \\ \sin(t\,V(\theta)+\theta) \end{pmatrix}.$$

24.1.2   The 2-sphere

The unit sphere $S = \mathbb{S}^2 \subset \mathbb{R}^3 \ni x = \begin{pmatrix} x_1 \\ x_2 \\ x_3 \end{pmatrix}$ is described as the null level set $S = \varphi^{-1}(0)$ of the function $\varphi(x) = (x_1^2+x_2^2+x_3^2-1)/2$. Now

$$(\partial\varphi)(x) = x = \begin{pmatrix} x_1 \\ x_2 \\ x_3 \end{pmatrix}$$

and we notice that $(\partial\varphi)(x)$ is the unit normal at any state $x \in S$. Thus, the orthogonal projection $\pi_\perp$ onto the normal axis $T_x^\perp(S) = \operatorname{Vect}((\partial\varphi)(x))$ at $x \in S$ is given by the formula

$$\pi_\perp(x) = (\partial\varphi)(x)(\partial\varphi)^T(x) = xx^T = \begin{pmatrix} x_1^2 & x_1x_2 & x_1x_3 \\ x_2x_1 & x_2^2 & x_2x_3 \\ x_3x_1 & x_2x_2 & x_3^2 \end{pmatrix}$$

and the orthogonal projection on $T_x(S)$ is defined by

$$\pi(x) = Id - \pi_\perp(x) = \begin{pmatrix} 1 - x_1^2 & -x_1 x_2 & -x_1 x_3 \\ -x_2 x_1 & 1 - x_2^2 & -x_2 x_3 \\ -x_3 x_1 & -x_2 x_2 & 1 - x_3^2 \end{pmatrix}.$$

The sphere S can be parametrized by the spherical coordinates mapping $\psi : \theta = (\theta_1, \theta_2) \in ([0, \pi] \times [0, 2\pi[) \mapsto S$

$$\psi(\theta) = \begin{pmatrix} \sin(\theta_1)\cos(\theta_2) \\ \sin(\theta_1)\sin(\theta_2) \\ \cos(\theta_1) \end{pmatrix} = (\partial\varphi)_\psi(\theta).$$

The first coordinate θ_1 is called the colatitude angle (a.k.a. zenith or normal angle or inclination, the latitude is the angle $(\frac{\pi}{2} - \theta_1)$), and the second θ_2 is called the azimuthal angle. We have

$$\partial_{\theta_1}\psi(\theta) = \begin{pmatrix} \cos(\theta_1)\cos(\theta_2) \\ \cos(\theta_1)\sin(\theta_2) \\ -\sin(\theta_1) \end{pmatrix}$$

and

$$\partial_{\theta_2}\psi(\theta) = \begin{pmatrix} -\sin(\theta_1)\sin(\theta_2) \\ \sin(\theta_1)\cos(\theta_2) \\ 0 \end{pmatrix} = -\sin(\theta_1) \begin{pmatrix} \sin(\theta_2) \\ -\cos(\theta_2) \\ 0 \end{pmatrix}$$

so that

$$\partial_{\theta_1}\psi(\theta) \wedge \partial_{\theta_2}\psi(\theta) = \sin(\theta_1) \begin{pmatrix} \sin(\theta_1)\cos(\theta_2) \\ \sin(\theta_1)\sin(\theta_2) \\ \cos(\theta_1) \end{pmatrix} = \sin(\theta_1)\, (\partial\varphi)_\psi(\theta) \in T^\perp(S).$$

This implies that

$$T_x(S) = \text{Vect}\left((\partial_{\theta_1}\psi)_{\phi(x)}, (\partial_{\theta_2}\psi)_{\phi(x)}\right) \quad \text{and} \quad T_x^\perp(S) = \text{Vect}\left((\partial\varphi)(x)\right).$$

The Riemannian metric on $S_\psi = ([0, \pi] \times [0, 2\pi[)$ is given by

$$\begin{aligned} g_{1,1}(\theta) &= \langle (\partial_{\theta_1}\psi)(\theta), (\partial_{\theta_1}\psi)(\theta) \rangle = 1 \\ g_{2,2}(\theta) &= \langle (\partial_{\theta_2}\psi)(\theta), (\partial_{\theta_2}\psi)(\theta) \rangle = \sin^2(\theta_1) \\ g_{1,2}(\theta) &= g_{2,1}(\theta) = \langle (\partial_{\theta_1}\psi)(\theta), (\partial_{\theta_2}\psi)(\theta) \rangle = 0. \end{aligned}$$

Up to the top and bottom points ($\theta_1 \in \{0, \pi\}$), we have

$$g^{-1}(\theta) = \begin{pmatrix} 1 & 0 \\ 0 & \sin^{-2}(\theta_1) \end{pmatrix}.$$

Our next objective is to compute the Christoffel symbols $C_{i,j}^n$ introduced in (21.32). To this end, we notice that

$$\partial_{\theta_1,\theta_2}\psi(\theta) = \begin{pmatrix} -\cos(\theta_1)\sin(\theta_2) \\ \cos(\theta_1)\cos(\theta_2) \\ 0 \end{pmatrix} = \frac{\cos(\theta_1)}{\sin(\theta_1)} \times \partial_{\theta_2}\psi(\theta)$$

$$\Rightarrow \quad C_{1,2}^1 = 0 = C_{2,1}^1 \quad \text{and} \quad C_{1,2}^2(\theta) = C_{2,1}^2(\theta) = \frac{\cos(\theta_1)}{\sin(\theta_1)}.$$

In much the same way, we have

$$\partial_{\theta_1,\theta_1}\psi(\theta) = \begin{pmatrix} -\sin(\theta_1)\cos(\theta_2) \\ -\sin(\theta_1)\sin(\theta_2) \\ -\cos(\theta_1) \end{pmatrix} = -(\partial\varphi)_\psi(\theta) \quad\Rightarrow\quad \forall k \in \{1,2\}\ \ C_{1,1}^k = 0$$

$$\partial_{\theta_2,\theta_2}\psi(\theta) = \begin{pmatrix} -\sin(\theta_1)\cos(\theta_2) \\ -\sin(\theta_1)\sin(\theta_2) \\ 0 \end{pmatrix} = -\sin(\theta_1)\begin{pmatrix} \cos(\theta_2) \\ \sin(\theta_2) \\ 0 \end{pmatrix}.$$

In addition, it is readily checked that

$$\partial_{\theta_2,\theta_2}\psi(\theta) \perp \partial_{\theta_2}\psi(\theta) \iff \langle\partial_{\theta_2,\theta_2}\psi(\theta), \partial_{\theta_2}\psi(\theta)\rangle = 0 \quad\Rightarrow\quad C_{2,2}^2 = 0$$

and

$$C_{2,2}^1(\theta) = \langle\partial_{\theta_2,\theta_2}\psi(\theta), \partial_{\theta_1}\psi(\theta)\rangle = -\sin(\theta_1)\cos(\theta_1) = -\frac{1}{2}\sin(2\theta_1).$$

To compute the Ricci curvature R of the sphere, we use the matrix decomposition (21.54). In this situation,

$$C_1 = \begin{pmatrix} 0 & 0 \\ 0 & \cot(\theta_1) \end{pmatrix} \quad\text{and}\quad C_2 = \begin{pmatrix} 0 & -\frac{1}{2}\sin(2\theta_1) \\ \cot(\theta_1) & 0 \end{pmatrix} \Rightarrow t(C) = \begin{pmatrix} \cot(\theta_1) \\ 0 \end{pmatrix}.$$

This implies that

$$T = \begin{pmatrix} -\frac{1}{\sin^2(\theta_1)} & 0 \\ 0 & 0 \end{pmatrix}.$$

We also readily check that

$$C_1 C_1 = \begin{pmatrix} 0 & 0 \\ 0 & \cot^2(\theta_1) \end{pmatrix}, \quad C_1 C_2 = \begin{pmatrix} 0 & 0 \\ \cot^2(\theta_1) & 0 \end{pmatrix}, \quad C_2 C_1 = \begin{pmatrix} 0 & -\cos^2(\theta_1) \\ 0 & 0 \end{pmatrix}$$

and $C_2 C_2 = -\begin{pmatrix} \cos^2(\theta_1) & 0 \\ 0 & \cos^2(\theta_1) \end{pmatrix}$. Taking the traces of each of these matrices we find that

$$E = \begin{pmatrix} \cot^2(\theta_1) & 0 \\ 0 & -2\cos^2(\theta_1) \end{pmatrix}.$$

On the other hand, we have

$$C_{1,1} = \begin{pmatrix} 0 \\ 0 \end{pmatrix}, \quad C_{1,2} = C_{2,1} = \begin{pmatrix} 0 \\ \cot(\theta_1) \end{pmatrix} \quad\text{and}\quad C_{2,2} = \begin{pmatrix} -\frac{1}{2}\sin(2\theta_1) \\ 0 \end{pmatrix}.$$

This implies that

$$B = \begin{pmatrix} 0 & 0 \\ 0 & -\cos^2(\theta_1) \end{pmatrix}, \quad \text{div}(C_{1,1}) = 0 = \text{div}(C_{1,1}) \quad\text{and}\quad \text{div}(C_{2,2}) = 1 - 2\cos^2(\theta_1)$$

from which we find

$$S = \begin{pmatrix} 0 & 0 \\ 0 & 1 - 2\cos^2(\theta_1) \end{pmatrix}.$$

Combining together, we conclude that

$$R \circ \psi = B - E + S - T = \begin{pmatrix} 1 & 0 \\ 0 & \sin^2(\theta_1) \end{pmatrix}.$$

Notice that $R = g$. Riemannian manifolds with Ricci curvatures proportional to the metric are called Einstein spaces or Einstein manifolds.

Using (21.43),

$$(\Delta\phi^1)_\psi(\theta) = -\sum_{1 \le i,j \le 2} C^1_{i,j}(\theta)\, g^{i,j}(\theta) = -\frac{C^1_{2,2}(\theta)}{\sin^2(\theta_1)} = \frac{\cos(\theta_1)}{\sin(\theta_1)} = \cot(\theta_1),$$

$$(\Delta\phi^2)_\psi(\theta) = 0 \qquad\qquad (24.3)$$

and by (21.10) we have

$$(\nabla\phi^1)_\psi(\theta) = \sum_{1 \le j \le 2} g^{1,j}(\theta)\, (\partial_{\theta_j}\psi)(\theta) = (\partial_{\theta_1}\psi)(\theta)$$

$$(\nabla\phi^2)_\psi(\theta) = \sum_{1 \le j \le 2} g^{2,j}(\theta)\, (\partial_{\theta_j}\psi)(\theta)$$

$$= \frac{1}{\sin^2(\theta_1)}\, (\partial_{\theta_2}\psi)(\theta) = \frac{1}{\sin(\theta_1)} \begin{pmatrix} -\sin(\theta_2) \\ \cos(\theta_2) \\ 0 \end{pmatrix}. \qquad (24.4)$$

The geodesics $c_\theta(t) := Exp_\theta(tV) = \begin{pmatrix} c^1_\theta(t) \\ c^2_\theta(t) \end{pmatrix}$, with $V(\theta) \in \mathbb{R}^2$ satisfy the differential equations

$$\begin{cases} \ddot{c}^1_\theta(t) = \sin(c^1_\theta(t))\cos(c^1_\theta(t))\, \dot{c}^2_\theta(t)\, \dot{c}^2_\theta(t) \\[2ex] \ddot{c}^2_\theta(t) = -2\, \dfrac{\cos(c^1_\theta(t))}{\sin(c^1_\theta(t))}\, \dot{c}^1_\theta(t)\, \dot{c}^2_\theta(t) \end{cases}$$

with initial conditions

$$c_\theta(0) = \theta \quad \text{and} \quad \dot{c}_\theta(0) = V(\theta).$$

These equations cannot be solved explicitly, and we need to resort to some numerical approximation. The second equation can be rewritten as

$$\frac{d}{dt}\left(\dot{c}^2_\theta(t)\, \sin^2(c^1_\theta(t))\right) = \ddot{c}^2_\theta(t)\, \sin^2(c^1_\theta(t)) + 2\sin(c^1_\theta(t))\cos(c^1_\theta(t))\, \dot{c}^1_\theta(t)\, \dot{c}^2_\theta(t)$$

$$= -2\, \frac{\cos(c^1_\theta(t))}{\sin(c^1_\theta(t))}\, \sin^2(c^1_\theta(t))\, \dot{c}^1_\theta(t)\, \dot{c}^2_\theta(t)$$

$$+ 2\sin(c^1_\theta(t))\cos(c^1_\theta(t))\, \dot{c}^1_\theta(t)\, \dot{c}^2_\theta(t) = 0.$$

This shows that

$$\dot{c}^2_\theta(t)\, \sin^2(c^1_\theta(t)) = \dot{c}^2_\theta(0)\, \sin^2(c^1_\theta(0)).$$

The geodesics $C_x(t) := Exp_x(tW)$ have a more explicit description given by the equations

$$Exp_x(tW) = \cos\left(t\|W(x)\|\right) x + \sin\left(t\|W(x)\|\right) \frac{W(x)}{\|W(x)\|}.$$

We readily check that $C_x(t)$ satisfies the required conditions

$$\dot{C}_x(t) = \left[-\sin\left(t\|W(x)\|\right) x + \cos\left(t\|W(x)\|\right) \frac{W(x)}{\|W(x)\|} \right] \|W(x)\| \overset{t=0}{=} W(x)$$

and

$$C_x(t) \in \mathbb{S}^2 \Longrightarrow \ddot{C}_x(t) = -\|W(x)\|^2\, C_x(t) \in T^\perp(\mathbb{S}^2) \Rightarrow \pi\left(\ddot{C}_x(t)\right) = 0.$$

24.1.3   The torus

The torus $\mathcal{T}$ can be seen as a surface of revolution obtained by revolving a circle

$$\mathcal{C}(R,r) = \left\{ \begin{pmatrix} R \\ 0 \\ 0 \end{pmatrix} + \begin{pmatrix} r\cos(\theta_1) \\ 0 \\ r\sin(\theta_1) \end{pmatrix} : \theta_1 \in \mathbb{R} \right\}$$

of radius r and center $x_1 = R > r$ about the symmetry x_3-axis. The Cartesian coordinates of the torus are parametrized by the function

$$\psi : \theta = \begin{pmatrix} \theta_1 \\ \theta_2 \end{pmatrix} \in \mathbb{R}^2 \mapsto \psi(\theta) = \begin{pmatrix} (R + r\cos(\theta_1))\cos(\theta_2) \\ (R + r\cos(\theta_1))\sin(\theta_2) \\ r\sin(\theta_1) \end{pmatrix}.$$

Alternatively, $\mathcal{T} = \varphi^{-1}(0)$ can be represented as the null level set of the function

$$\varphi : x = \begin{pmatrix} x_1 \\ x_2 \\ x_3 \end{pmatrix} \in \mathbb{R}^3 \mapsto \varphi(x) = \left(R - \sqrt{x_1^2 + x_2^2} \right)^2 + x_3^2 - r^2.$$

After some elementary manipulations, we find that

$$\partial_{\theta_1}\psi(\theta) = \begin{pmatrix} -r\sin(\theta_1)\cos(\theta_2) \\ -r\sin(\theta_1)\sin(\theta_2) \\ r\cos(\theta_1) \end{pmatrix} , \quad \partial_{\theta_2}\psi(\theta) = (R + r\cos(\theta_1)) \begin{pmatrix} -\sin(\theta_2) \\ \cos(\theta_2) \\ 0 \end{pmatrix}$$

and

$$\frac{R - \sqrt{x_1^2 + x_2^2}}{\sqrt{x_1^2 + x_2^2}} x_1 \stackrel{x=\psi(\theta)}{=} \frac{-r\cos(\theta_1)}{(R + r\cos(\theta_1))} (R + r\cos(\theta_1))\cos(\theta_2) = -r\cos(\theta_1)\cos(\theta_2)$$

$$\Longrightarrow \partial\varphi(x) = 2 \begin{pmatrix} -x_1 \frac{R - \sqrt{x_1^2 + x_2^2}}{\sqrt{x_1^2 + x_2^2}} \\ -x_2 \frac{R - \sqrt{x_1^2 + x_2^2}}{\sqrt{x_1^2 + x_2^2}} \\ x_3 \end{pmatrix} \stackrel{x=\psi(\theta)}{=} 2r \begin{pmatrix} \cos(\theta_1)\cos(\theta_2) \\ \cos(\theta_1)\sin(\theta_2) \\ \sin(\theta_1) \end{pmatrix}.$$

In addition, we have

$$\partial_{\theta_1}\psi(\theta) \perp \partial_{\theta_2}\psi(\theta) \quad (\Leftrightarrow \langle \partial_{\theta_1}\psi(\theta), \partial_{\theta_2}\psi(\theta)\rangle = 0)$$

$$\|\partial_{\theta_1}\psi(\theta)\| = r^2 \quad \text{and} \quad \|\partial_{\theta_2}\psi(\theta)\| = (R + r\cos(\theta_1))^2$$

and

$$\partial_{\theta_2}\psi(\theta) \wedge \partial_{\theta_1}\psi(\theta) = \begin{pmatrix} -(R + r\cos(\theta_1))\sin(\theta_2) \\ (R + r\cos(\theta_1))\cos(\theta_2) \\ 0 \end{pmatrix} \wedge \begin{pmatrix} -r\sin(\theta_1)\cos(\theta_2) \\ -r\sin(\theta_1)\sin(\theta_2) \\ r\cos(\theta_1) \end{pmatrix}$$

$$= r(R + r\cos(\theta_1)) \begin{pmatrix} \cos(\theta_1)\cos(\theta_2) \\ \cos(\theta_1)\sin(\theta_2) \\ \sin(\theta_1) \end{pmatrix}.$$

This shows that

$$g(\theta) = \begin{pmatrix} r^2 & 0 \\ 0 & (R + r\cos(\theta_1))^2 \end{pmatrix} \quad \text{and} \quad g^{-1}(\theta) = \begin{pmatrix} r^{-2} & 0 \\ 0 & (R + r\cos(\theta_1))^{-2} \end{pmatrix}$$

$$\Longrightarrow \sqrt{\det(g(\theta))} = r(R + r\cos(\theta_1)).$$

Using (21.10) we have

$$
\begin{aligned}
(\nabla \phi^1)_\psi(\theta) &= r^{-2} \, (\partial_{\theta_1} \psi)(\theta) \\
(\nabla \phi^2)_\psi(\theta) &= (R + r \cos(\theta_1))^{-2} \, (\partial_{\theta_2} \psi)(\theta)
\end{aligned}
$$

and

$$
\begin{aligned}
\langle (\nabla \phi^1)_\psi(\theta), (\nabla \phi^2)_\psi(\theta) \rangle &= 0 \\
\langle (\nabla \phi^1)_\psi(\theta), (\nabla \phi^1)_\psi(\theta) \rangle &= r^{-2} \quad \text{and} \quad \langle (\nabla \phi^2)_\psi(\theta), (\nabla \phi^2)_\psi(\theta) \rangle = (R + r \cos(\theta_1))^{-2}.
\end{aligned}
$$

By (21.40), we also find that

$$
\begin{aligned}
(\Delta \phi^1)_\psi(\theta) &= \frac{1}{r(R + r \cos(\theta_1))} \, \partial_{\theta_1} \left(r(R + r \cos(\theta_1)) \, r^{-2} \right) = -\frac{\sin(\theta_1)}{r(R + r \cos(\theta_1))} \\
(\Delta \phi^2)_\psi(\theta) &= \frac{1}{r(R + r \cos(\theta_1))} \, \partial_{\theta_2} \left(r(R + r \cos(\theta_1)) \, (R + r \cos(\theta_1))^{-2} \right) = 0.
\end{aligned}
$$

Our next objective is to compute the Christofell symbols (21.32). In our situation, we have

$$
\begin{aligned}
C^1_{i,j} &= g^{1,1} \, \langle \partial_{\theta_1} \psi, \partial_{\theta_i, \theta_j} \psi \rangle \\
C^2_{i,j} &= g^{2,2} \, \langle \partial_{\theta_2} \psi, \partial_{\theta_i, \theta_j} \psi \rangle .
\end{aligned}
$$

Firstly, we observe that

$$
\partial_{\theta_1, \theta_1} \psi(\theta) = - \begin{pmatrix} r \cos(\theta_1) \cos(\theta_2) \\ r \cos(\theta_1) \sin(\theta_2) \\ r \sin(\theta_1) \end{pmatrix} \quad \text{and} \quad \partial_{\theta_2, \theta_2} \psi(\theta) = -(R + r \cos(\theta_1)) \begin{pmatrix} \cos(\theta_2) \\ \sin(\theta_2) \\ 0 \end{pmatrix} .
$$

In much the same way, we find that

$$
\partial_{\theta_1, \theta_2} \psi(\theta) = \partial_{\theta_2, \theta_1} \psi(\theta) = r \sin(\theta_1) \begin{pmatrix} \sin(\theta_2) \\ -\cos(\theta_2) \\ 0 \end{pmatrix} .
$$

Using elementary calculations, we find that

$$
\begin{aligned}
\langle \partial_{\theta_1, \theta_1} \psi, \partial_{\theta_1} \psi \rangle &= \left\langle \begin{pmatrix} r \cos(\theta_1) \cos(\theta_2) \\ r \cos(\theta_1) \sin(\theta_2) \\ r \sin(\theta_1) \end{pmatrix}, \begin{pmatrix} r \sin(\theta_1) \cos(\theta_2) \\ r \sin(\theta_1) \sin(\theta_2) \\ -r \cos(\theta_1) \end{pmatrix} \right\rangle = 0 \\
\langle \partial_{\theta_1, \theta_1} \psi, \partial_{\theta_2} \psi \rangle &= -(R + r \cos(\theta_1)) \left\langle \begin{pmatrix} r \cos(\theta_1) \cos(\theta_2) \\ r \cos(\theta_1) \sin(\theta_2) \\ r \sin(\theta_1) \end{pmatrix}, \begin{pmatrix} -\sin(\theta_2) \\ \cos(\theta_2) \\ 0 \end{pmatrix} \right\rangle = 0.
\end{aligned}
$$

This implies that $C^1_{1,1} = 0 = C^2_{1,1}$. We also have

$$
\begin{aligned}
\langle \partial_{\theta_2, \theta_2} \psi, \partial_{\theta_1} \psi \rangle &= (R + r \cos(\theta_1)) \left\langle \begin{pmatrix} \cos(\theta_2) \\ \sin(\theta_2) \\ 0 \end{pmatrix}, \begin{pmatrix} r \sin(\theta_1) \cos(\theta_2) \\ r \sin(\theta_1) \sin(\theta_2) \\ -r \cos(\theta_1) \end{pmatrix} \right\rangle \\
&= r(R + r \cos(\theta_1)) \sin(\theta_1)
\end{aligned}
$$

and

$$\langle \partial_{\theta_2,\theta_2}\psi, \partial_{\theta_2}\psi\rangle \;=\; -(R+r\cos(\theta_1))^2 \left\langle \begin{pmatrix} \cos(\theta_2) \\ \sin(\theta_2) \\ 0 \end{pmatrix}, \begin{pmatrix} -\sin(\theta_2) \\ \cos(\theta_2) \\ 0 \end{pmatrix} \right\rangle = 0.$$

This implies that

$$C^2_{2,2} = 0 \quad \text{and} \quad C^1_{2,2} = r^{-1}\,(R+r\cos(\theta_1))\sin(\theta_1).$$

On the other hand, we have

$$\langle \partial_{\theta_1,\theta_2}\psi, \partial_{\theta_1}\psi\rangle \;=\; -r\sin(\theta_1) \left\langle \begin{pmatrix} \sin(\theta_2) \\ -\cos(\theta_2) \\ 0 \end{pmatrix}, \begin{pmatrix} r\sin(\theta_1)\cos(\theta_2) \\ r\sin(\theta_1)\sin(\theta_2) \\ -r\cos(\theta_1) \end{pmatrix} \right\rangle = 0$$

and

$$\langle \partial_{\theta_1,\theta_2}\psi, \partial_{\theta_2}\psi\rangle \;=\; r(R+r\cos(\theta_1))\sin(\theta_1) \left\langle \begin{pmatrix} \sin(\theta_2) \\ -\cos(\theta_2) \\ 0 \end{pmatrix}, \begin{pmatrix} -\sin(\theta_2) \\ \cos(\theta_2) \\ 0 \end{pmatrix} \right\rangle$$
$$=\; -r(R+r\cos(\theta_1))\sin(\theta_1).$$

This shows that

$$C^2_{1,2} \;=\; g^{2,2}\,\langle \partial_{\theta_2}\psi, \partial_{\theta_1,\theta_2}\psi\rangle = -\frac{r\sin(\theta_1)}{R+r\cos(\theta_1)} \quad \text{and} \quad C^1_{1,2} = 0.$$

We conclude that the only nonzero Christoffell symbols are given by

$$C^1_{2,2} = r^{-1}\,(R+r\cos(\theta_1))\sin(\theta_1) \quad \text{and} \quad C^2_{1,2} = C^2_{2,1} = -\frac{r\sin(\theta_1)}{R+r\cos(\theta_1)}.$$

To compute the Ricci curvature R of the sphere, we use the matrix decomposition (21.54). In this situation, we have

$$C_1 = \begin{pmatrix} 0 & 0 \\ 0 & -\frac{r\sin(\theta_1)}{R+r\cos(\theta_1)} \end{pmatrix} \quad \text{and} \quad C_2 = \begin{pmatrix} 0 & r^{-1}\sin(\theta_1)(R+r\cos(\theta_1)) \\ -\frac{r\sin(\theta_1)}{R+r\cos(\theta_1)} & 0 \end{pmatrix}.$$

This implies that

$$t(C) = \begin{pmatrix} -\frac{r\sin(\theta_1)}{R+r\cos(\theta_1)} \\ 0 \end{pmatrix} \quad \text{and} \quad T = \begin{pmatrix} -\frac{r\cos(\theta_1)}{R+r\cos(\theta_1)} - \left(\frac{r\sin(\theta_1)}{R+r\cos(\theta_1)}\right)^2 & 0 \\ 0 & 0 \end{pmatrix}.$$

We also readily check that

$$C_1 C_1 = \begin{pmatrix} 0 & 0 \\ 0 & \left(\frac{r\sin(\theta_1)}{R+r\cos(\theta_1)}\right)^2 \end{pmatrix}, \quad C_1 C_2 = \begin{pmatrix} 0 & 0 \\ \left(\frac{r\sin(\theta_1)}{R+r\cos(\theta_1)}\right)^2 & 0 \end{pmatrix}, \quad C_2 C_1 = \begin{pmatrix} 0 & -\sin^2(\theta_1) \\ 0 & 0 \end{pmatrix}$$

and $C_2 C_2 = \begin{pmatrix} -\sin^2(\theta_1) & 0 \\ 0 & -\sin^2(\theta_1) \end{pmatrix}$.Taking the traces of each of these matrices we find that

$$E = \begin{pmatrix} \left(\frac{r\sin(\theta_1)}{R+r\cos(\theta_1)}\right)^2 & 0 \\ 0 & -2\sin^2(\theta_1) \end{pmatrix}.$$

On the other hand, we have

$$C_{1,1} = \begin{pmatrix} 0 \\ 0 \end{pmatrix}, \qquad C_{1,2} = C_{2,1} = \begin{pmatrix} 0 \\ -\frac{r\sin(\theta_1)}{R+r\cos(\theta_1)} \end{pmatrix}, \qquad C_{2,2} = \begin{pmatrix} r^{-1}\sin(\theta_1)(R+r\cos(\theta_1)) \\ 0 \end{pmatrix}.$$

This implies that

$$B = \begin{pmatrix} 0 & 0 \\ 0 & -\sin^2(\theta_1) \end{pmatrix} \qquad \mathrm{div}(C_{1,1}) = 0 = \mathrm{div}(C_{1,1})$$

and

$$\mathrm{div}(C_{2,2}) = \frac{1}{r}\,\cos(\theta_1)\,(R+r\cos(\theta_1)) - \sin^2(\theta_1)$$

from which we find

$$S = \begin{pmatrix} 0 & 0 \\ 0 & \frac{1}{r}\,\cos(\theta_1)\,(R+r\cos(\theta_1)) - \sin^2(\theta_1) \end{pmatrix}.$$

Combining together, we conclude that

$$\begin{aligned} R \circ \psi = B - E + S - T &= \begin{pmatrix} \frac{r\cos(\theta_1)}{R+r\cos(\theta_1)} & 0 \\ 0 & \frac{1}{r}\,\cos(\theta_1)\,(R+r\cos(\theta_1)) \end{pmatrix} \\ &= \frac{\cos(\theta_1)}{r(R+r\cos(\theta_1))}\,g. \end{aligned}$$

This shows that the curvature is null on the top and bottom circles of the torus. The curvature is positive outside the torus and negative inside.

24.2   Information theory

24.2.1   Nash embedding theorem

In differential geometry, a (smooth) Riemannian manifold $(\mathcal{S}, g)$ is a real state space $\mathcal{S}$ equipped with a smooth inner product g on the tangent space $T(S)$; that is, for any $\theta \in \mathcal{S}$, and any vector fields $\theta \mapsto V_i(\theta) \in T_\theta(S)$ the mapping

$$\theta \mapsto \langle V_1(\theta), V_2(\theta) \rangle_{g(\theta)}$$

is a smooth function. This geometric Riemannian structure allows us to define various geometric notions such as angles, lengths of curves, volumes, curvatures, gradients of functions and divergences of vector fields.

The Nash embedding theorem states that every Riemannian manifold with dimension p can be (locally) isometrically embedded into some ambient Euclidean space with sufficiently high dimension r (but $r \le 2p + 1$). The isometric embedding problem amounts to finding a function $\psi : \theta \in \mathcal{S} \mapsto \psi(\theta) \in \mathbb{R}^r$ such that

$$g_{i,j}(\theta) := \langle \partial_{\theta_i}\psi, \partial_{\theta_j}\psi \rangle = \sum_{1 \le k \le r} \partial_{\theta_i}\psi^k(\theta)\partial_{\theta_j}\psi^k(\theta).$$

24.2.2   Distribution manifolds

When $\mathcal{S} = S_\psi = \phi(S)$ is the parameter space of a given manifold S as discussed in (21.1) the natural Riemannian inner product is given by the matrix field (21.6).

The space of discrete distributions $S := \mathcal{P}(E)$ on a finite set $E = \{1, \dots, r\}$ is represented by the $p = (r-1)$-dimensional simplex

$$\text{Simplex}(p) = \{z = (z_i)_{1 \le i \le r} \in \mathbb{R}_+^r \; : \; \varphi(z) := \sum_{1 \le i \le r} z_i - 1 = 0\}.$$

The tangent space $T_z(S)$ at each point z is given by

$$\partial\varphi = \begin{pmatrix} 1 \\ \vdots \\ 1 \end{pmatrix} \Rightarrow T_z(S) = \Big\{W(z) \in \mathbb{R}^r \; : \; z_k = 0 \Rightarrow W^k(z) = 0$$

$$\text{and} \quad \langle W(z), \partial\varphi(z)\rangle = \sum_{1 \le k \le r} W^k(z) = 0\Big\}.$$

The Fisher information metric on $T_z(S)$ is defined by the inner product

$$\forall W_1(z), W_2(z) \in T_z(S) \quad \langle W_1(z), W_2(z)\rangle_{h(z)} = \sum_{1 \le k \le r} \frac{W_1^k(z)}{z_k} \frac{W_2^k(z)}{z_k} \, z_k.$$

For instance, for $r = 3$ we have

$$T_z(S) := \text{Vect}\left(e_1(z) := \begin{pmatrix} 1 \\ 0 \\ -1 \end{pmatrix}, e_2(z) := \begin{pmatrix} 0 \\ 1 \\ -1 \end{pmatrix}\right).$$

In this case, for any $z = (z_k)_{1 \le k \le 3}$ s.t. $z_k > 0$ for any $k = 1, 2, 3$, we have

$$h_{1,1}(z) = \langle e_1(z), e_1(z)\rangle_{h(z)} = \frac{1}{z_1} + \frac{1}{z_3}, \quad h_{1,2}(z) = h_{2,1}(z) = \langle e_1(z), e_2(z)\rangle_{h(z)} = \frac{1}{z_3}$$

$$h_{2,2}(z) = \langle e_2(z), e_2(z)\rangle_{h(z)} = \frac{1}{z_2} + \frac{1}{z_3}.$$

More generally, let E be some measurable space equipped with some reference measure λ. The tangent space $T_\mu\left(\mathcal{P}^\lambda(E)\right)$ of the set of probability measures

$$\mathcal{P}^\lambda(E) := \{\mu \in \mathcal{P}(E) \; : \; \mu \ll \lambda\} \ni \mu$$

given by

$$T_\mu\left(\mathcal{P}^\lambda(E)\right) = \left\{\nu \in \mathcal{P}(E) \; : \; \nu \ll \mu \text{ s.t. } \int \left(\frac{d\nu}{d\mu}\right)^2 d\mu < \infty \quad \text{and} \quad \nu(1) = 0\right\}$$

is equipped with the Fisher inner product

$$\forall W_1(\mu), W_2(\mu) \in T_\mu(\mathcal{P}^\lambda(E)) \quad \langle W_1(\mu), W_2(\mu)\rangle_{h(\mu)} := \int \frac{dW_1(\mu)}{d\mu} \frac{dW_2(\mu)}{d\mu} \, d\mu.$$
$$\tag{24.5}$$

24.2.3 Bayesian statistical manifolds

Riemannian manifolds also arise in a natural way in Bayesian statistics and in information theory. To describe with some precision these statistical models, we let

$$\mu_\theta(dy) := P_\theta(y)\,\lambda(dy) \tag{24.6}$$

be a collection of distributions on some state space E, equipped with a reference measure $\lambda(dy)$, and indexed by some parameter θ on some space $\mathcal{S} \subset \mathbb{R}^p$ of dimension p. We assume that $\mathcal{S}$ is equipped with some probability measure of the form

$$\mu'(d\theta) = P'(\theta)\,\lambda'(d\theta)$$

where $\lambda'(d\theta)$ is some reference measure on $\mathcal{S}$. The probability measure μ' can be seen as the prior distribution of some unknown random parameter Θ. Given $\Theta = \theta$, $\mu_\theta(dy)$ denotes the distribution of some partial and noisy random observation Y of the parameter Θ. In this interpretation, the function

$$P(\theta, y) = P'(\theta)\,P_\theta(y)$$

represents the density of the random vector (Θ, Y) w.r.t. the reference measure $\lambda \otimes \lambda'$ on $\mathcal{S} \times E$. We consider the parametrization mapping

$$\psi \;:\; \theta \in \mathcal{S} \mapsto \psi(\theta) = \mu_\theta \in \mathcal{P}_\mathcal{S}(E) = \{\mu_\theta \in \mathcal{P}(E) \;:\; \theta \in \mathcal{S}\} \subset \mathcal{P}^\lambda(E)$$

and we equip $\mathcal{P}_\mathcal{S}(E)$ with the Fisher metric (24.5) induced by $\mathcal{P}^\lambda(E)$. Notice that for any $1 \le i \le p$ we have

$$(\partial_{\theta_i}\psi)(\theta) = \partial_{\theta_i}\mu_\theta$$

with the signed measure

$$\partial_{\theta_i}\mu_\theta(dy) := \partial_{\theta_i}P_\theta(y)\,\lambda(dy)$$

on E with null mass:

$$\int P_\theta(y)\,\lambda(dy) = 1 \Rightarrow \forall 1 \le i \le p \quad \int \partial_{\theta_i}P_\theta(y)\,\lambda(dy) = 0.$$

The tangent space

$$T_\theta(\mathcal{S}) = \mathrm{Vect}\,(e_i,\ i=1,\ldots,p) \quad \text{with} \quad e_i = \begin{pmatrix} 0 \\ \vdots \\ 0 \\ 1 \\ 0 \\ \vdots \\ 0 \end{pmatrix} \longleftarrow i-\text{th coordinate}$$

is mapped on the tangent space $T_{\mu_\theta}(\mathcal{P}_\mathcal{S}(E))$ by using the push forward mapping

$$(d\psi)_\theta \;:\; V(\theta) \;=\; \sum_{1 \le i \le p} V^i(\theta)\,e_i(\theta) \;\mapsto\; (d\psi)_\theta(V(\theta)) \;=\; \sum_{1 \le i \le p} V^i(\theta)\,(\partial_{\theta_i}\psi)(\theta).$$

$$\in \; T_\theta(\mathcal{S}) \qquad\qquad\qquad\qquad \in \; T_{\mu_\theta}(\mathcal{P}_\mathcal{S}(E)).$$

The Fisher information metric g on the parameter space $\mathcal{S}$ induced by the metric h on $\mathcal{P}_\mathcal{S}(E)$ is defined for any $1 \leq i, j \leq p$ by

$$
\begin{aligned}
g_{i,j}(\theta) &:= \left\langle (\partial_{\theta_i} \psi)(\theta), (\partial_{\theta_j} \psi)(\theta) \right\rangle_{h(\mu_\theta)} = \int \frac{\partial_{\theta_i} P_\theta(y)}{P_\theta(y)} \frac{\partial_{\theta_j} P_\theta(y)}{P_\theta(y)} \, P_\theta(y) \, \lambda(dy) \\
&= \int \partial_{\theta_i} \log P_\theta(y) \, \partial_{\theta_j} \log P_\theta(y) \, P_\theta(y) \, \lambda(dy) \\
&= \mathbb{E} \left(\partial_{\theta_i} \log P_\Theta(Y) \, \partial_{\theta_j} \log P_\Theta(Y) \mid \Theta = \theta \right).
\end{aligned}
$$

The Fisher metric can alternatively be defined by

$$
\int P_\theta(y) \, \lambda(dy) = 1 \quad \Rightarrow \quad \int \partial_{\theta_i} \log P_\theta(y) \, P_\theta(y) \, \lambda(dy) = 0
$$

$$
\begin{aligned}
\Rightarrow \quad g_{i,j}(\theta) &= \int \partial_{\theta_i} \log P_\theta(y) \, \partial_{\theta_j} \log P_\theta(y) \, P_\theta(y) \, \lambda(dy) \\
&= -\int \left(\partial_{\theta_j, \theta_i} \log P_\theta(y) \right) \, P_\theta(y) \, \lambda(dy) \\
&= -\mathbb{E} \left(\partial_{\theta_j, \theta_i} \log P_\Theta(Y) \mid \Theta = \theta \right).
\end{aligned}
$$

We end this section with a connection between the Fisher metric and the relative Boltzmann entropy. We fix a parameter $\theta^\star \in \mathcal{S}$ and we consider the Boltzmann entropy

$$
\mathcal{B}_{\theta^\star}(\theta) = \operatorname{Ent}(\mu_{\theta^\star} \mid \mu_\theta) = -\int \log \frac{P_\theta(y)}{P_{\theta^\star}(y)} \, P_{\theta^\star}(y) \, \lambda(dy).
$$

We have

$$
\begin{aligned}
\partial_{\theta_i} \log P_\theta(y) &= \frac{1}{P_\theta(y)} \, \partial_{\theta_i} P_\theta(y) \\
\partial_{\theta_j, \theta_i} \log P_\theta(y) &= -\frac{1}{P_\theta(y)^2} \, \partial_{\theta_j} P_\theta(y) \partial_{\theta_i} P_\theta(y) + \frac{1}{P_\theta(y)} \, \partial_{\theta_j, \theta_i} P_\theta(y)
\end{aligned}
$$

from which we conclude that

$$
(\partial_{\theta_i} \mathcal{B}_{\theta^\star})(\theta^\star) = -\int \frac{1}{P_{\theta^\star}(y)} \, \partial_{\theta_i} P_{\theta^\star}(y) \, P_{\theta^\star}(y) \, \lambda(dy) = -\int \partial_{\theta_i} P_{\theta^\star}(y) \, \lambda(dy) = 0
$$

and

$$
\begin{aligned}
(\partial_{\theta_j, \theta_i} \mathcal{B}_{\theta^\star})(\theta^\star) &= \int \frac{1}{P_{\theta^\star}(y)^2} \, \partial_{\theta_j} P_{\theta^\star}(y) \partial_{\theta_i} P_{\theta^\star}(y) \, P_{\theta^\star}(y) \lambda(dy) - \int \partial_{\theta_j, \theta_i} P_{\theta^\star}(y) \lambda(dy) \\
&= \int \frac{\partial_{\theta_j} P_{\theta^\star}(y)}{P_{\theta^\star}(y)} \frac{\partial_{\theta_j} P_{\theta^\star}(y)}{P_{\theta^\star}(y)} \, P_{\theta^\star}(y) \lambda(dy) = g_{i,j}(\theta^\star).
\end{aligned}
$$

This shows that

$$\text{Ent}\,(\mu_{\theta^\star} \mid \mu_\theta) = \frac{1}{2} \sum_{1 \le i,j \le p} g_{i,j}(\theta^\star)\,(\theta_i - \theta_i^\star)(\theta_j - \theta_j^\star) + \text{O}(\|(\theta - \theta^\star)\|^3)$$

$$= \frac{1}{2}\,(\theta - \theta^\star)^T g(\theta^\star)(\theta - \theta^\star) + \text{O}(\|(\theta - \theta^\star)\|^3).$$

The above formula shows that the Fisher matrix $g(\theta)$ encapsulates the infinitesimal changes of the model distribution μ_θ w.r.t. an infinitesimal fluctuation of the model parameter θ.

24.2.4 Cramer-Rao lower bound

Suppose we are given an unbiased estimator $\widehat{\Theta} = (\varphi^i(Y))_{1 \le i \le p}$ of the parameter $\theta = (\theta^i)_{1 \le i \le p}$ associated with an observation r.v. Y with distribution (24.6), that is,

$$\forall 1 \le i \le p \qquad \mathbb{E}\left(\varphi^i(Y)\right) = \theta^i.$$

The score function is defined by the gradient function

$$\mathbf{Score}_\theta^i(Y) := \partial_{\theta_i} \log P_\theta(Y).$$

Recalling that

$$\mathbb{E}\left(\mathbf{Score}_\theta^i(Y)\right) = 0$$

and using the Cauchy-Schwartz inequality, we find that

$$\mathbb{E}\left(\left[\mathbf{Score}_\theta^i(Y)\right]^2\right)^{1/2} \times \text{Var}(\varphi^j(Y))^{1/2}$$

$$\ge \mathbb{E}\left(\left[\mathbf{Score}_\theta^i(Y) - \mathbb{E}\left(\mathbf{Score}_\theta^i(Y)\right)\right]\left[\varphi^j(Y) - \mathbb{E}\left(\varphi^j(Y)\right)\right]\right)$$

$$= \mathbb{E}\left(\mathbf{Score}_\theta^i(Y)\varphi^j(Y)\right) = \int \varphi^j(y)\,\partial_{\theta_i} \log P_\theta(y)\,P_\theta(y)\,\lambda(dy)$$

$$= \int \varphi^j(y)\,\partial_{\theta_i} P_\theta(y)\,\lambda(dy) = \partial_{\theta_i}\mathbb{E}(\varphi^j(Y)) = \partial_{\theta_i}\theta^j = 1_{i=j}.$$

This implies that

$$\text{Var}(\varphi^j(Y)) \ge 1/g_{j,j}(\theta).$$

The quantity $1/g_{j,j}(\theta)$ gives the Cramer-Rao lower bound for the variance of an unbiased estimator of θ^i.

24.2.5 Some illustrations

24.2.5.1 Boltzmann-Gibbs measures

We consider a collection of Boltzmann-Gibbs measures associated with a potential function V on some state space E, and indexed by some real valued parameter θ:

$$\mu_\theta(dy) = \frac{1}{\mathcal{Z}_\theta}\,e^{-\theta V(y)}\,\lambda(dy).$$

In this situation, we have

$$\partial_\theta P_\theta(y) = -\frac{1}{\mathcal{Z}_\theta^2}\,\partial_\theta(\mathcal{Z}_\theta)\,e^{-\theta V(y)} + \frac{1}{\mathcal{Z}_\theta}\,\partial_\theta(e^{-\theta V(y)})$$

$$= (\mu_\theta(V) - V(y))\,P_\theta(y) \Longrightarrow \partial_\theta \log P_\theta = (\mu_\theta(V) - V).$$

Hence

$$g(\theta) = g_{1,1}(\theta) = \int \left[\mu_\theta(V) - V(y)\right]^2\,\mu_\theta(dy) = \mu_\theta(V^2) - \mu_\theta(V)^2.$$

24.2.5.2 Multivariate normal distributions

We consider the collection of distributions μ_θ indexed by a parameter $\theta \in \mathcal{S} \subset \mathbb{R}^p$ and given by

$$\mu_\theta(dy) = \frac{1}{\sqrt{2\pi}^{d^Y}\,\sqrt{\det(C(\theta))}}\,\exp\left(-\frac{1}{2}\,(y - m(\theta))^T C(\theta)^{-1}(y - m(\theta))\right)\,dy,$$

where $dy = \prod_{1 \leq i \leq d^Y} dy_i$ stands for an infinitesimal neighborhood of $y = (y_i)_{1 \leq i \leq d^Y} \in \mathbb{R}^{d^Y}$.

In this situation, we have

$$g_{i,j}(\theta^\star) \overset{\theta=\theta^\star}{=} \partial_{\theta_i,\theta_j}\mathrm{Ent}\,(\mu_{\theta^\star} \mid \mu_\theta)$$

$$\overset{\theta=\theta^\star}{=} \partial_{\theta_i}m(\theta^\star)^T C(\theta^\star)^{-1}\partial_{\theta_j}m(\theta^\star) + \frac{1}{2}\,\mathrm{tr}\left(\partial_{\theta_j}C(\theta^\star)\,C(\theta^\star)^{-1}\,(\partial_{\theta_i}C(\theta^\star))\,C(\theta^\star)^{-1}\right).$$

$$(24.7)$$

In particular, for $d^Y = 1$, and $p = 2$ with $m(\theta) = \theta_1 \in \mathbb{R}$ and $C(\theta) = \theta_2 \in\,]0, \infty[$ we have

$$g_{1,1}(\theta) = \frac{1}{\theta_2}$$

$$g_{1,2}(\theta) = g_{2,1}(\theta) = 0 \quad \text{and} \quad g_{2,2}(\theta) = \frac{1}{2\theta_2^2}.$$

The corresponding Riemannian gradient compensates for the fact that an infinitesimal change of a parameter in a Gaussian model μ_θ with small variance θ_2 has more pronounced effects:

$$(\nabla_g f)(\theta) = \theta_2\,\partial_{\theta_1}f + 2\theta_2^2\,\partial_{\theta_2}f.$$

To check (24.7), we observe that

$$\mathrm{Ent}\,(\mu_{\theta^\star} \mid \mu_\theta) = \frac{1}{2}\left[\log\det\left(C(\theta^\star)^{-1}C(\theta)\right)\right.$$

$$\left. + \int \left\{((y - m(\theta))^T C(\theta)^{-1}(y - m(\theta))) - (y - m(\theta^\star))^T C(\theta^\star)^{-1}(y - m(\theta^\star))\right\}\mu_{\theta^\star}(dy)\right]$$

$$= \tfrac{1}{2}\left[\log\det\left(C(\theta^\star)^{-1}C(\theta)\right)\right.$$

$$\left. + \mathbb{E}\left\{((Y - m(\theta))^T C(\theta)^{-1}(Y - m(\theta))) - (Y - m(\theta^\star))^T C(\theta^\star)^{-1}(Y - m(\theta^\star))\right\}\right]$$

with

$$Y = m(\theta^\star) + C(\theta^\star)^{1/2} Z \quad \text{where} \quad Z \sim N(0, Id_{d^Y \times d^Y})$$

$$\mathbb{E}\left(ZAZ^T\right) = \sum_{1 \le i,j \le d^Y} \mathbb{E}(Z^i A_{i,j} Z^j) = \sum_{1 \le i,j \le d^Y} A_{i,i} = \text{tr}(A).$$

We recall that for any invertible symmetric positive definite matrix A, its spectral decomposition $A = UDU^T$ holds with an orthogonal diagolizing matrix U and a diagonal matrix D. Given this spectral decomposition, the square root

$$A^{1/2} = U\sqrt{D}U^T$$

of A can be defined. This square root is symmetric and invertible, and we have

$$A^{-1/2}A^{1/2} = Id \qquad AA^{1/2} = A^{1/2}A \quad \text{and} \quad A^{1/2}A^{-1}A^{1/2} = Id.$$

We also recall that for any couple of matrices A and B with suitable numbers of rows and columns such that both $tr(AB)$ and $tr(BA)$ exist, we have

$$\text{tr}(AB) = \text{tr}(BA).$$

Hence

$$\text{tr}(A^{1/2}(BA^{1/2})) = \text{tr}(BA) = \text{tr}(AB).$$

Using these formulae, it is readily checked that

$$\begin{aligned}
\mathbb{E}\left((Y - m(\theta^\star))^T C(\theta^\star)^{-1}(Y - m(\theta^\star)))\right) &= \mathbb{E}\left(Z^T \; C(\theta^\star)^{1/2} \; C(\theta^\star)^{-1} C(\theta^\star)^{1/2} \; Z\right) \\
&= \mathbb{E}\left(Z^T Z\right) = d^Y
\end{aligned}$$

and

$$\mathbb{E}\left((Y - m(\theta))^T C(\theta)^{-1}(Y - m(\theta)))\right)$$

$$= \mathbb{E}\left(\left[(m(\theta^\star) - m(\theta)) + C(\theta^\star)^{1/2}Z\right]^T C(\theta)^{-1}\left[(m(\theta^\star) - m(\theta)) + C(\theta^\star)^{1/2}Z\right]\right)$$

$$= (m(\theta^\star) - m(\theta))^T C(\theta)^{-1}(m(\theta^\star) - m(\theta)) + \mathbb{E}\left(Z^T C(\theta^\star)^{1/2}C(\theta)^{-1}C(\theta^\star)^{1/2}Z\right)$$

$$= (m(\theta^\star) - m(\theta))^T C(\theta)^{-1}(m(\theta^\star) - m(\theta)) + \text{tr}\left(C(\theta^\star)^{1/2}C(\theta)^{-1}C(\theta^\star)^{1/2}\right)$$

$$= (m(\theta^\star) - m(\theta))^T C(\theta)^{-1}(m(\theta^\star) - m(\theta)) + \text{tr}\left(C(\theta^\star)C(\theta)^{-1}\right).$$

We conclude that

$$\begin{aligned}
2\,\text{Ent}\left(\mu_{\theta^\star} \mid \mu_\theta\right) = \; &\log \det\left(C(\theta^\star)^{-1}C(\theta)\right) + (m(\theta^\star) - m(\theta))^T C(\theta)^{-1}(m(\theta^\star) - m(\theta)) \\
&+ \left[\text{tr}\left(C(\theta^\star)C(\theta)^{-1}\right) - d^Y\right].
\end{aligned}$$

For any index $1 \le i \le p$, we have

$$\begin{aligned}
\partial_{\theta_i} \log \det\left(C(\theta^\star)^{-1}C(\theta)\right) &= \frac{1}{\det\left(C(\theta^\star)^{-1}C(\theta)\right)} \; \partial_{\theta_i} \log \det\left(C(\theta^\star)^{-1}C(\theta)\right) \\
&= \text{tr}\left(\left(C(\theta^\star)^{-1}C(\theta)\right)^{-1} \partial_{\theta_i} C(\theta^\star)^{-1}C(\theta)\right) \\
&= \text{tr}\left(C(\theta)^{-1}\partial_{\theta_i} C(\theta)\right)
\end{aligned}$$

so that

$$\partial_{\theta_i} \left(\log \det \left(C(\theta^\star)^{-1} C(\theta) \right) + \left[\operatorname{tr} \left(C(\theta^\star) C(\theta)^{-1} \right) - d^Y \right] \right)$$

$$= \operatorname{tr} \left(C(\theta)^{-1} \partial_{\theta_i} C(\theta) - C(\theta^\star) C(\theta)^{-1} \left(\partial_{\theta_i} C(\theta) \right) C(\theta)^{-1} \right) \overset{\theta = \theta^\star}{=} 0.$$

The second term in the trace formula is obtained using

$$\partial_\epsilon A(\epsilon)^{-1} = -A(\epsilon)^{-1} \ \left(\partial_\epsilon A(\epsilon) \right) \ A(\epsilon)^{-1}$$

for any smooth functional $\epsilon \mapsto A(\epsilon)$ in the space of invertible matrices. We check this claim by:

$$\partial_\epsilon \sum_j A_{i,j}(\epsilon) A^{j,k}(\epsilon) = 0 \ \Rightarrow \ \sum_j A_{i,j}(\epsilon) \partial_\epsilon A^{j,k}(\epsilon) = - \sum_j \partial_\epsilon A_{i,j}(\epsilon) A^{j,k}(\epsilon)$$

$$= \sum_{i,j} A^{l,i}(\epsilon) A_{i,j}(\epsilon) \partial_\epsilon A^{j,k}(\epsilon) = - \sum_{i,j} A^{l,i}(\epsilon) \partial_\epsilon A_{i,j}(\epsilon) A^{j,k}(\epsilon)$$

where $A(\epsilon)^{-1} = \left(A^{i,j}(\epsilon) \right)_{i,j}$ and $A(\epsilon) = (A_{i,j}(\epsilon))_{i,j}$.

We also have that

$$\partial_{\theta_i, \theta_j} \left(\log \det \left(C(\theta^\star)^{-1} C(\theta) \right) + \left[\operatorname{tr} \left(C(\theta^\star) C(\theta)^{-1} \right) - d^Y \right] \right)$$

$$= \operatorname{tr} \left\{ \left[\left(\partial_{\theta_j} C(\theta)^{-1} \right) \ \partial_{\theta_i} C(\theta) + C(\theta)^{-1} \partial_{\theta_i, \theta_j} C(\theta) \right] \right.$$

$$- C(\theta^\star) \left[\left(\partial_{\theta_j} C(\theta)^{-1} \right) \left(\partial_{\theta_i} C(\theta) \right) C(\theta)^{-1} \right.$$

$$\left. \left. + C(\theta)^{-1} \left(\partial_{\theta_i, \theta_j} C(\theta) \right) C(\theta)^{-1} + C(\theta)^{-1} \left(\partial_{\theta_i} C(\theta) \right) \ \partial_{\theta_j} C(\theta)^{-1} \right] \right\}$$

$$\overset{\theta = \theta^\star}{=} -\operatorname{tr} \left(C(\theta^\star) \left(\partial_{\theta_j} C(\theta^\star)^{-1} \right) \left(\partial_{\theta_i} C(\theta^\star) \right) C(\theta^\star)^{-1} \right)$$

$$= \operatorname{tr} \left(\partial_{\theta_j} C(\theta^\star) C(\theta^\star)^{-1} \left(\partial_{\theta_i} C(\theta^\star) \right) C(\theta^\star)^{-1} \right).$$

In much the same way,

$$\partial_{\theta_i} \left((m(\theta^\star) - m(\theta))^T C(\theta)^{-1} (m(\theta^\star) - m(\theta)) \right)$$

$$= -2 \partial_{\theta_i} m(\theta)^T C(\theta)^{-1} (m(\theta^\star) - m(\theta)) + \left((m(\theta^\star) - m(\theta))^T \partial_{\theta_i} C(\theta)^{-1} (m(\theta^\star) - m(\theta)) \right)$$

$$\overset{\theta = \theta^\star}{=} 0.$$

Therefore

$$\partial_{\theta_i, \theta_j} \left((m(\theta^\star) - m(\theta))^T C(\theta)^{-1} (m(\theta^\star) - m(\theta)) \right)$$

$$= 2 \ \partial_{\theta_i} m(\theta)^T C(\theta)^{-1} \partial_{\theta_j} m(\theta)$$

$$-2 \left[\left(\partial_{\theta_i, \theta_j} m(\theta) \right)^T C(\theta)^{-1} + \left(\partial_{\theta_i} m(\theta) \right)^T \partial_{\theta_j} C(\theta)^{-1} + \partial_{\theta_j} m(\theta)^T \partial_{\theta_i} C(\theta)^{-1} \right.$$

$$\left. -\frac{1}{2} (m(\theta^\star) - m(\theta))^T \partial_{\theta_i, \theta_j} C(\theta)^{-1} \right] (m(\theta^\star) - m(\theta))$$

$$\overset{\theta = \theta^\star}{=} 2 \ \partial_{\theta_i} m(\theta^\star)^T C(\theta^\star)^{-1} \partial_{\theta_j} m(\theta^\star).$$

We conclude that

$$\partial_{\theta_j} \mathrm{Ent}\left(\mu_{\theta^\star} \mid \mu_\theta\right) \overset{\theta=\theta^\star}{=} 0.$$

This ends the proof of the desired formula.

Part VI

Some application areas

25

Simple random walks

This chapter discusses two important classes of random walk type processes which are frequently used in applied probability. The first class of models relates to random walks on lattices and graphs. The second one is related to urn processes, such as the Ehrenfest and the Polya urn models.

> *All knowledge degenerates into probability.*
> David Hume (1711-1776).

25.1 Random walk on lattices

25.1.1 Description

We let S be the d-dimensional lattice $S = \mathbb{Z}^d$, equipped with the equivalence relation $x \sim y$ iff x and y are nearest neighbors. Notice that any state $x \in \mathbb{Z}^d$ has $2d$ nearest neighbors denoted by $\mathcal{N}(x)$. The simple random walk (abbreviated SRW) on the d-dimensional lattice $S = \mathbb{Z}^d$ is defined by the Markov transition

$$M(x,y) = \frac{1}{2d}\, 1_{\mathcal{N}(x)}(y).$$

The SRW on $\mathbb{Z}^d$ is sometimes called the drunkard's walk to emphasize that the evolution of the chain resembles the movement of a drunkard who stumbles at random from a bar to another bar. We can think of the origin as the home of the drunkard. In what follows we show that the drunkard returns infinitely often to each bar and is at home when $d = 1, 2$. However, in dimension $d = 3$, he will return home only a finite (random) number of times and then wander off in the universe of bars (visiting each one a finite number of times only).

25.1.2 Dimension 1

In dimension $d = 1$, the chain X_n moves at each step randomly to the right or to the left with a probability $1/2$. Starting from any state, say $X_0 = 0$, we need an even number of steps, say $(2n)$, to come back to the initial location, so that

$$\mathbb{P}\left(X_{2n} = 0 \mid X_0 = 0\right) = \binom{2n}{n} \left(\frac{1}{2}\right)^{2n} = \frac{2n!}{n!^2}\, 2^{-2n}.$$

Using Stirling formula $n! \simeq \sqrt{2\pi n}\, n^n e^{-n}$ we find that

$$\mathbb{P}\left(X_{2n} = 0 \mid X_0 = 0\right) \simeq \frac{\sqrt{4\pi n}\,(2n)^{2n}e^{-2n}}{2\pi n\, n^{2n}e^{-2n}}\, 2^{-2n} = \frac{1}{\sqrt{\pi n}}. \qquad (25.1)$$

We let $N_0 = \sum_{n \geq 1} 1_0(X_n)$ be the number of returns to 0. The expected number of returns to the origin is infinite:

$$\begin{aligned}
\mathbb{E}(N_0 \mid X_0 = 0) &= \sum_{n \geq 1} \mathbb{P}(X_n = 0 \mid X_0 = 0) \\
&= \sum_{n \geq 1} \mathbb{P}(X_{2n} = 0 \mid X_0 = 0) = \sum_{n \geq 1} \frac{1}{\sqrt{\pi n}} = \infty.
\end{aligned}$$

By symmetry arguments, the analysis remains the same if we change the initial state. Since the mean returns are infinite, all states are recurrent.

25.1.3 Dimension 2

In the planar lattice $\mathbb{Z}^2$, both coordinates of the excursion starting at $(0,0)$ must be null at the same time in order to return to the origin. This implies that

$$\mathbb{P}\left(X_{2n} = (0,0) \mid X_0 = (0,0)\right) \simeq \frac{1}{\sqrt{\pi n}} \times \frac{1}{\sqrt{\pi n}} = \frac{1}{\pi n}.$$

For a detailed proof of this result, we refer to exercise 426. Denote by T_1 the duration of the first return to the origin. In this situation, we have

$$\mathbb{E}(N_0 \mid X_0 = 0) = \infty \quad \text{and} \quad \mathbb{P}\left(T_1 < \infty\right) = 1.$$

Hence for $d = 2$ we say that the SRW is recurrent.

25.1.4 Dimension d ≥ 3

In larger dimensions $d \geq 3$, we have

$$\mathbb{P}\left(X_{2n} = (0,\ldots,0) \mid X_0 = (0,\ldots,0)\right) \simeq \frac{1}{(\pi n)^{d/2}}$$

so that

$$\mathbb{E}(N_0 \mid X_0 = 0) < \infty \quad \text{and} \quad \mathbb{P}\left(T_1 < \infty\right) < 1.$$

For a detailed proof of this result for $d = 3$, we refer to exercise 426. In this situation, we say that the SRW is transient.

A more detailed discussion on the long time behavior of the SRW in terms of the dimension is provided in section 8.5 dedicated to the analysis of Markov chains on countable state spaces.

25.2 Random walks on graphs

Random walks on graphs extend the simple random walk on the lattice $\mathbb{Z}^d$. To simplify the presentation, we consider only random walks on finite graphs in the sense that the number of vertices is finite.

We consider a finite graph $\mathcal{G} = (\mathcal{E}, \mathcal{V})$, where $\mathcal{V}$ stands for the set of vertices and $\mathcal{E} \subset (\mathcal{V} \times \mathcal{V})$ a set of edges. We further assume that $\mathcal{G}$ is non-oriented (i.e. $\mathcal{E}$ is symmetric), and there are no self-loops (i.e. $\mathcal{E}$ is anti-reflexive). We let $d(x)$ be the degree of a vertex $x \in \mathcal{V}$ (i.e. $d(x) = \mathrm{Card}\{y \in \mathcal{V} : (x, y) \in \mathcal{E}\}$). The $d(x)$ can be interpreted as the number of neighbors of the site x. We also assume that there are no isolated vertices, in the sense that $d(x) > 0$ for any $x \in \mathcal{V}$.

Random walks X_n on $\mathcal{G}$ are defined by the Markov transitions

$$M(x, y) = \frac{1}{d(x)} \, a_{x,y}$$

where $a_{x,y}$ stands for the adjacency matrix (i.e. $a_{x,y} = 1$ iff $(x, y) \in \mathcal{E}$). Since the graph is unoriented, the matrix $a_{x,y}$ is symmetric so that

$$d(x) \, M(x, y) = a_{x,y} = a_{y,x} = d(y) \, M(y, x).$$

This shows that M is reversible w.r.t. the probability measure

$$\pi(x) = \frac{d(x)}{\sum_{y \in \mathcal{V}} d(y)}.$$

25.3 Simple exclusion process

The simple exclusion process can be represented by a random walk on the set $S = \{0, 1\}^{\mathcal{V}}$, where $\mathcal{V}$ stands for some set of vertices of some finite graph $\mathcal{G} = (\mathcal{V}, \mathcal{E})$ of the same form as the one discussed in section 25.2. The value of the state 1 at some site $i \in \mathcal{V}$ is interpreted as the existence of a particle at site i, while the null value 0 encodes the fact that there are no particles at that site. Let $x = (x(i))_{i \in \mathcal{V}}$. The Markov transition of the chain X_n is defined as follows. Given some state X_n we pick an edge $(i, j) \in \mathcal{E}$ at random, and we interchange the values of the end points. More formally, we have

$$M(x, y) = \sum_{(i,j) \in \mathcal{E}} \frac{1}{\mathrm{Card}(\mathrm{E})} \, 1_{\mathcal{E}}((i, j)) \, 1_{x^{(i,j)}}(y)$$

with the mapping $x \mapsto x^{(i,j)}$ defined below:

$$\forall k \in \mathcal{V} - \{i, j\} \quad x^{i,j}(k) = x(k) \quad \text{and} \quad \left(x^{i,j}(i), x^{i,j}(j) \right) = (x(j), x(i)).$$

25.4 Random walks on the circle

25.4.1 Markov chain on cycle

The random walk on the cycle $S = \mathbb{Z}/m\mathbb{Z}$ is defined by the Markov transitions

$$M(x, y) = \begin{cases} \frac{1}{2} & \text{if} \quad y = x + 1 \ \mathrm{mod}(m) \\ \frac{1}{2} & \text{if} \quad y = x - 1 \ \mathrm{mod}(m) \\ 0 & \text{otherwise.} \end{cases}$$

For instance for $m = 4$, we have $S = \mathbb{Z}/4\mathbb{Z} = \{0, 1, 2, 3\}$ and M alternatively represented by the stochastic matrix

$$M = \begin{pmatrix} 0 & .5 & 0 & .5 \\ .5 & 0 & .5 & 0 \\ 0 & .5 & 0 & .5 \\ .5 & 0 & .5 & 0 \end{pmatrix}.$$

It is easy to check that

$$\pi = [\pi(0), \pi(1), \pi(2), \pi(3)]] = \left[\frac{1}{4}, \frac{1}{4}, \frac{1}{4}, \frac{1}{4} \right]$$

is the invariant measure of the chain.

We can alternatively define this walk by the recursion

$$X_n = X_{n-1} + U_n \quad \mathrm{mod}(m)$$

where U_n stands for a sequence of independent uniform r.v. on $\{+1, -1\}$.

25.4.2   Markov chain on circle

We can also identify the states $\mathbb{Z}/m\mathbb{Z}$ with the m-roots of unity $\{u^p : 0 \leq p < m\}$, with $u = e^{2i\pi/m}$; more formally

$$\forall 0 \leq k < m \quad k \mapsto u^k$$

is a bijection between these sets.

In this interpretation, the Markov chain X_n explores randomly the set $S = \{u^p : 0 \leq p < m\}$, with $u = e^{2i\pi/m}$ with the elementary Markov transition defined for any $x \in S$ by

$$\mathbb{P}(X_n = x \mid X_{n-1} = y) = \frac{1}{2} 1_{ux}(y) + \frac{1}{2} 1_{u^{-1}x}(y) := M(x, y).$$

Since $\pi(x)M(x, y) = \pi(y)M(y, x)$ is reversible w.r.t. the uniform distribution $\pi(x) = 1/m$, we conclude that $\pi M = \pi$. Our next objective to derive a complete description of the spectral representation of the random walk on the circle (cf. theorem 8.2.2).

25.4.3   Spectral decomposition

We consider the collection of power functions

$$\forall 0 \leq k < m \qquad \varphi_k : x \in S \mapsto \varphi_k(x) = x^k,$$

$$\begin{aligned} M(\varphi_k)(x) &= \frac{1}{2} \varphi_k(ux) + \frac{1}{2} \varphi_k(u^{-1}x) \\ &= \frac{u^k x^k + u^{-k} x^k}{2} = \frac{u^k + u^{-k}}{2} x^k. \end{aligned}$$

This clearly implies that

$$\forall 0 \leq k < m \qquad M(\varphi_k) = \lambda_k \, \varphi_k \quad \text{with} \quad \lambda_k := \cos(2\pi k/m).$$

Since the eigenvalues λ_k are real, the real parts and the imaginary parts of φ_k defined by

$$\begin{aligned} \forall 0 \leq l < m \quad \psi_k(u^l) &:= \frac{\varphi_k + \overline{\varphi}_k}{2}(u^l) = \cos(2\pi k l/m) \\ \phi_k(u^l) &:= \frac{\varphi_k - \overline{\varphi}_k}{2i}(u^l) = \sin(2\pi k l/m) \end{aligned}$$

are real valued eigenfunctions. We check this claim using

$$M(\psi_k) = \frac{1}{2}\left[M(\varphi_k) + \underbrace{M(\overline{\varphi}_k)}_{=\overline{M(\varphi_k)}=\lambda_k\overline{\varphi}_k}\right] = \lambda_k\psi_k$$

$$M(\phi_k) = \frac{1}{2i}\left[M(\varphi_k) \underbrace{-M(\overline{\varphi}_k)}_{=-\overline{M(\varphi_k)}=-\lambda_k\overline{\varphi}_k}\right] = \lambda_k\phi_k.$$

For $k = 0$, we notice that $\lambda_0 = 1$, and $\varphi_0 = 1 = \psi_0$. Note that

$$m = 2m' \Rightarrow \lambda_{m'} := \cos(\pi) = -1.$$

This shows that when m is even the chain is periodic, and we cannot expect to have convergence to the invariant measure (otherwise, the entries of M^m would be positive and by theorem 8.2.2 -1 would not be an eigenvalue). This is a consequence of the fact that for even numbers m the chain is periodic. According to the comments provided in section 8.5.1, we have $M^n(x,x) > 0$ only for $n \in 2m'\mathbb{N}$ so that the chain starting at x will never visit the states at odd distances. On the contrary, when m is odd, the chain is aperiodic and irreducible, and the spectral gap is positive.

Last, but not least, using the orthonormal properties

$$\langle \varphi_{k_1}, \varphi_{k_2} \rangle_\pi = \pi(\varphi_{k_1}\overline{\varphi}_{k_2}) = \sum_{0 \le l < m} \pi(u^l)\, e^{\frac{2i\pi lk_1}{m}} e^{-\frac{2i\pi lk_2}{m}}$$

$$= \frac{1}{m}\sum_{0 \le l < m} e^{\frac{2i\pi l(k_1-k_2)}{m}} = 1_{k_1=k_2}$$

we check that the real valued eigenfunctions ψ_k form an orthonormal basis of $l_2(\pi)$.

25.5 Random walk on hypercubes

25.5.1 Description

The random walk $X_n = (X_n^i)_{1 \le i \le d}$ on the hypercube $S = \{0,1\}^d$ is defined by the Markov transition

$$M(x,y) = \sum_{1 \le i \le d} \mu(i)\, 1_{x^{(i)}}(y)$$

for some probability measure μ on $\{1,\dots,d\}$, with the mapping $x \mapsto x^{(i)}$ defined below:

$$\forall k \ne i \quad x^{(i)}(k) = x(k) \quad \text{and} \quad x^{(i)}(i) = 1 - x(i).$$

We can check that M is reversible w.r.t. the product of the Bernoulli distributions on $\{0,1\}$

$$\pi(x) = \prod_{1 \le i \le d}\left(\frac{1}{2}1_0(x(i)) + \frac{1}{2}\,1_1(x(i))\right).$$

We check this claim using the fact that this law is the distribution of independent Bernoulli random variables $X := (X^1, \dots, X^d)$, and for any $1 \le i \le d$, the sequence $X^{(i)}$ has the same law as X. This can can be interpreted as the location of d molecules in two adjacent boxes. The number 1 indicates that the molecule is in the first box, and the number 0 indicates that it is not in the first box (i.e., it is in the second box).

25.5.2   A macroscopic model

Before adding more details about the spectral properties of the random walk on the hypercube, let us discuss the process that tracks the number of 1's in the sequence X_n:

$$\overline{X}_n := \sum_{1 \leq i \leq d} X_n^i. \tag{25.2}$$

By construction,

$$\mathbb{E}\left(f(\overline{X}_{n+1}) \mid X_n\right) = \left(\sum_{i:\ X_n^i = 1} \mu(i)\right) f(\overline{X}_n - 1) + \left(\sum_{i:\ X_n^i = 0} \mu(i)\right) f(\overline{X}_n + 1).$$

When μ is the uniform probability, for any function f on $\{1, \dots, d\}$ we find that

$$\mathbb{E}\left(f(\overline{X}_{n+1}) \mid X_n\right) = \frac{\overline{X}_n}{d}\, f(\overline{X}_n - 1) + \frac{d - \overline{X}_n}{d}\, f(\overline{X}_n + 1).$$

This shows that $\overline{X}_n$ is a Markov chain with transitions

$$\overline{M}(k, l) = \frac{k}{d}\, 1_{k-1}(l) + \frac{d - k}{d}\, 1_{k+1}(l) \tag{25.3}$$

associated with the transition diagram with reflection boundaries

$$0 \underset{1/d}{\overset{1}{\rightleftarrows}} 1 \underset{2/d}{\overset{(d-1)/d}{\rightleftarrows}} 2 \underset{3/d}{\overset{(d-2)/d}{\rightleftarrows}} 3 \quad \dots \quad (d-1) \underset{1}{\overset{1/d}{\rightleftarrows}} d.$$

We notice that $\frac{d-k}{d} > \frac{k}{d}$ when $k < d/2$. This shows that the chain is more likely to move to the right up to the state $k < d/2$. Inversely, when the chain is on the r.h.s. of $d/2 > k$ it is more likely to move back. We also see that the chain is irreducible, but periodic with period 2, so that we cannot expect good convergence properties. The mid-point acts like a central force attracting the random states of the chain. The eigenvalues $\lambda_k = 1 - \frac{2k}{d}$, with $0 \leq k \leq d$ and the corresponding Krawtchouk orthonormal polynomial w.r.t. the binomial distribution can be found using sophisticated combinatorial analysis (see for instance [175]). Notice that $\lambda_d = -1$ indicates that the chain is periodic.

25.5.3   A lazy random walk

We consider the $1/2$-lazy version of the random walk on the hypercube $S = \{0, 1\}^d$ defined by

$$M'(x, y) \;=\; \sum_{1 \leq i \leq d} \mu(i) \left(\frac{1}{2} 1_{x^{(i)}} + \frac{1}{2} 1_x\right)(y) = \sum_{1 \leq i \leq d} \mu(i) \frac{1}{2} \sum_{\epsilon \in \{0,1\}} 1_{x_\epsilon^{(i)}}(y)$$

with the mapping $x \mapsto x^{(i)}$ defined as:

$$\forall k \neq i \quad x_\epsilon^{(i)}(k) = x(k) \quad \text{and} \quad x_\epsilon^{(i)}(i) = \epsilon\,(1 - x(i)) + (1 - \epsilon)\, x(i).$$

Therefore, we can rewrite this Markov transition as

$$M'(x, y) = \sum_{1 \leq i \leq d} \mu(i) \left\{ \prod_{j \neq i} 1_{x(j)}(y(j)) \right\} M_i(x(i), y(i)) \tag{25.4}$$

with the Bernoulli transition

$$M_i(u, v) = \nu(v) = \frac{1}{2} \, 1_0(v) + \frac{1}{2} \, 1_1(v).$$

Using example 8.2.5, for any $J \subset \{1, \ldots, d\}$, the functions $\varphi_J(x) := \prod_{j \in J}(x(j) - .5)$ are eigenfunctions associated with the eigenvalues $\lambda_J = 1 - \mu(J)$. In particular, when $J = \{j\}$, $\varphi_J(x) := (x(j) - .5)$ are eigenfunctions associated with the eigenvalues $\lambda_J = 1 - \mu(j)$, and we have the partial order property

$$J_1 \subset J_2 \Rightarrow \lambda_{J_1} \geq \lambda_{J_2}$$

so that the absolute spectral gap defined in corollary 8.2.3 is given by

$$\rho^\star(M) = 1 - \sup_{1 < i \leq d} (1 - \mu(i)) = \inf_{1 \leq i \leq d} \mu(i).$$

We refer the reader to exercise 425 for the computation of the lazy version of the chain.

25.6   Urn processes

25.6.1   Ehrenfest model

A collection of d balls is distributed in a couple of urns $\mathcal{U}, \mathcal{V}$. We select a ball randomly from a given urn and we place it in the other urn. The number of balls B_n in $\mathcal{U}$ is a Markov chain taking values in $S = \{1, \ldots, d\}$ with transition probabilities defined for any $k \in S$ by

$$\mathbb{P}\left(B_n = l \mid B_{n-1} = k\right) = M(k, l) := \frac{d - k}{d} \, 1_{k+1}(l) + \frac{k}{d} \, 1_{k-1}(l).$$

The associated transition diagram is given on page 698. We consider the ϵ-lazy version associated with the parameter $\epsilon = \frac{1}{d+1}$ and the Markov transitions

$$M_\epsilon(k, l) := \frac{1}{d+1} \, 1_k(l) + \frac{d-k}{d+1} \, 1_{k+1}(l) + \frac{k}{d+1} \, 1_{k-1}(l)$$

and the corresponding transition diagram with reflecting boundaries

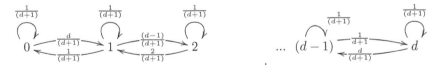

We also observe that the invariant measure $\pi M = \pi = \pi M_\epsilon$ of the chain is given by the binomial distribution

$$\pi(k) = \binom{d}{k} 2^{-d}. \tag{25.5}$$

To check this claim, we notice that

$$\binom{d}{k} M(k, l) = (d-1)! \left(\frac{1}{(d-l)!(l-1)!} \, 1_k(l-1) + \frac{1}{(d-(l+1))!l!} \, 1_k(l+1) \right).$$

This yields

$$\sum_{0 \le k \le b} \binom{d}{k} M(k, l) = (d-1)! \, \frac{l + (d-l)}{(d-l)! l!} = \frac{d!}{(d-l)! l!} = \binom{d}{l}.$$

Notice that the chain B_n can be interpreted as a projection of the random walk on the hypercube using (25.3). It can also be seen as a birth and death chain with a death rate B_n/b that depends on the size of the population. In the early days of statistical mechanics, Paul and Tatiana Ehrenfest introduced in 1907 this rather elementary model to analyze the thermodynamic equilibrium of gas molecules passing from one container into another through a hole [116]. In these settings, the number of molecules may be of order $d \simeq 10^{23}$, also known as Avogadro's number. The Ehrenfest model can be interpreted as the macroscopic version of the microscopic model associated with the random walk on the hypercube discussed in section 25.5. Notice that the microscopic model is reversible, but the macroscopic model is not reversible.

Another interpretation is to count the number of fleas jumping from one dog to another. In this case, B_n stands for the number of fleas on the first dog. In this model, we count one jump per unit of time from one of the dogs.

25.6.2 Pólya urn model

We consider an urn that contains b black balls and w white balls. We choose a ball from this this urn randomly and return it with c additional balls of the color drawn. More formally, if (B_n, W_n) counts the number of black and white balls in the urn, the transition probabilities of this chain are given by

$$\mathbb{P}\left((B_{n+1}, W_{n+1}) = (k, l) \mid (B_n, W_n) = (b, w)\right) = \frac{b}{b+w} \, 1_{(b+c, w)}(k, l) \; + \; \frac{w}{b+w} \, 1_{(b, w+c)}(k, l).$$

Notice that the total number of balls in the urn $N_n := B_n + W_n$ at time n is given by

$$N_n = N_{n-1} + c = \ldots = N_0 + n\, c.$$

We have $\binom{n}{k}$ possible ways of drawing k black balls in n different trials. Starting with (b, w) black and white balls, at each of these times, say $p_l \in \{1, \ldots, n\}$, with $l = 1, \ldots, k$, we have $(b + (l-1)c)$ black balls in an urn with $((b+w) + p_l c)$ balls. This shows that the the chance to pick a black ball at a given time p_l is $(b + (l-1)c)/((b+w) + p_l c)$.

It is also important to notice that any event with a specific ordering of the k black and l white balls in the $k + l = n$ trials has the same probability

$$\frac{\left[\prod_{0 \le l < k}(b + lc)\right] \left[\prod_{0 \le l < n-k}(w + lc)\right]}{\prod_{0 \le l < n}((b+w) + lc)}.$$

This implies that

$$\mathbb{P}\left(B_n = k \mid (B_0, W_0) = (b, w)\right) = \binom{n}{k} \frac{\left[\prod_{0 \le l < k}\left(\frac{b}{c} + l\right)\right] \left[\prod_{0 \le l < n-k}\left(\frac{w}{c} + l\right)\right]}{\prod_{0 \le l < n}\left(\frac{(b+w)}{c} + l\right)}.$$

Recalling that $\Gamma(z+1) = z\Gamma(z)$, this implies that

$$\mathbb{P}\left(B_n = k \mid (B_0, W_0) = (b, w)\right)$$

$$= \binom{n}{k} \frac{\Gamma\left(\frac{b}{c} + k\right) \times \Gamma\left(\frac{w}{c} + (n-k)\right)}{\Gamma\left(\frac{b}{c}\right)\Gamma\left(\frac{w}{c}\right)} \times \left(\frac{\Gamma\left(\frac{b+w}{c} + n\right)}{\Gamma\left(\frac{b+w}{c}\right)}\right)^{-1}$$

$$= \binom{n}{k} \frac{\Gamma\left(\frac{b}{c} + k\right)\Gamma\left(\frac{w}{c} + (n-k)\right)}{\Gamma\left(\frac{b+w}{c} + n\right)} \times \left(\frac{\Gamma\left(\frac{b}{c}\right)\Gamma\left(\frac{w}{c}\right)}{\Gamma\left(\frac{b+w}{c}\right)}\right)^{-1}.$$

This can be written as

$$\mathbb{P}\left(B_n = k \mid (B_0, W_0) = (b, w)\right) = \int_0^1 \binom{n}{k} u^k (1-u)^{(n-k)} p_{\left(\frac{b}{c}, \frac{w}{c}\right)}(u) \, du$$

with the density $p_{(\alpha, \beta)}$ of the beta(α, β) distribution defined in (4.6). We conclude that

$$\mathbb{P}\left(B_n = k \mid (B_0, W_0) = (b, w)\right) = \int_0^1 q_{n,u}(k) \, p_{\left(\frac{b}{c}, \frac{w}{c}\right)}(u) \, du$$

with the binomial distribution with parameters (n, u) defined by

$$q_{n,u}(k) = \binom{n}{k} u^k (1-u)^{(n-k)} = \mathbb{P}\left(\sum_{1 \leq i \leq n} 1_1(Y_i) = k \mid U_{\left(\frac{b}{c}, \frac{w}{c}\right)} = u\right).$$

Here $U_{\left(\frac{b}{c}, \frac{w}{c}\right)} \sim \text{beta}\left(\frac{b}{c}, \frac{w}{c}\right)$; and given $U_{\left(\frac{b}{c}, \frac{w}{c}\right)} = u$, $(Y_i)_{i \geq 1}$ is a collection of independent Bernoulli $\{0, 1\}$-valued r.v. with

$$\mathbb{P}\left(Y_i = 1 \mid U_{\left(\frac{b}{c}, \frac{w}{c}\right)} = u\right) = 1 - \mathbb{P}\left(Y_i = 0 \mid U_{\left(\frac{b}{c}, \frac{w}{c}\right)} = u\right) = u.$$

In summary, we have proved that

$$\mathbb{P}\left(B_n = k \mid (B_0, W_0) = (b, w)\right) = \mathbb{E}\left(\mathbb{P}\left(\sum_{1 \leq i \leq n} 1_1(Y_i) = k \mid U_{\left(\frac{b}{c}, \frac{w}{c}\right)}\right)\right).$$

Further details on Pólya urn models can be found in the book of H. Mahmoud [189]; see also exercice 427.

25.7 Exercises

Exercise 421 (Goldstein-Kac process [136, 159]) *We let ϵ be a $\{0, 1\}$-valued, and U be $\{-1, 1\}$-valued, Bernoulli random variables*

$$\mathbb{P}(\epsilon = 0) = e^{-a} = 1 - \mathbb{P}(\epsilon = 1) \quad \text{and} \quad \mathbb{P}(U = 1) = e^{-a} = 1 - \mathbb{P}(U = -1)$$

for some $a > 0$. We let ϵ_n and U_n be sequences of independent copies of ϵ and U. We set $N_n = \sum_{1 \leq k \leq n} \epsilon_k$, $X_n = U_1 \ldots U_n$ and $\Delta Y_n := Y_n - Y_{n-1} = v \, (-1)^{N_n}$ with some *given velocity parameter* $v \in \mathbb{R}$. *Check that* (X_n, Y_n) *is a Markov chain taking values in* $S = (\{-1, 1\} \times \mathbb{R})$ *and describe its Markov transition* K. *When* $a = \log 2$ *and* $v = v_0 t$, *check that*

$$w_t(y) := \mathbb{E}(g(Y_{n+1}) \mid (X_n, Y_n) = (1, y))$$

satisfies the wave equation

$$\partial_t^2 w = v_0^2 \partial_y^2 w \quad \text{with the initial conditions } w_0 = g \text{ and } \partial_t w_{|t=0} = 0.$$

We denote by K_h *the Markov transition associated with the parameters* $v = bh$ *and* $a = \lambda h$. *Check that*

$$K_h(f) = f + L(f)h + \mathrm{O}(h^2)$$

with the jump generator

$$L(f)(x, y) := b \; \partial_y f(x, y) + \lambda \; [f(-x, y) - f(x, y)].$$

Describe the evolution of the Markov process $(\mathcal{X}_t, \mathcal{Y}_t)$ *with generator* L. *Compare this stochastic process with the random 2-velocity process discussed in exercise 209.*

Exercise 422 (Telegraph equation [136, 159]) *Consider the Goldstein-Kac process discussed in exercise 421. Assume that* $(\mathcal{X}_t, \mathcal{Y}_t)$ *has a density given by*

$$\forall x \in \{-1, 1\} \qquad \mathbb{P}(\mathcal{X}_t = x \; , \; \mathcal{Y}_t \in dy) = p_t(x, y) \; dy.$$

We set

$$q_t^+(y) \;\; = \;\; p_t(1, y) + p_t(-1, y) \qquad q_t^-(y) \;\; = \;\; p_t(1, y) - p_t(-1, y).$$

The function $b \times q_t^-(y)$ *is sometimes called the current and it is denoted by* $j_t(y) = b \; q_t^-(y)$. *Describe the evolution equations of the functions* $(p_t(1, y), p_t(-1, y))$ *and* $(q_t^+(y), q_t^-(y))$. *Check that*

$$\partial_t q^+ = -\partial_y j \quad \text{and} \quad \partial_t j + 2\lambda \; j = -b^2 \; \partial_y q^+.$$

Prove that $q_t^+(y)$ *satisfies the telegraph equation*

$$\partial_t^2 q^+ + 2\lambda \; \partial_t q^+ = b^2 \; \partial_y^2 q^+$$

and compute $\mathbb{E}(\mathcal{Y}_t^2)$. *In the fluid mechanics and biology literature, this equation is sometimes called the macroscopic telegrapher's equation. For a detailed study on more general models we refer to exercise 209.*

Exercise 423 (The Walk on Sphere method [205]) *We let* W_t *be an* r-*dimensional brownian motion starting at the origin* $0 \in \mathbb{R}^r$, *we also set* $X_t := X_0 + W_t$ *and* $L := 2^{-1} \sum_{1 \leq i \leq r} \partial_{x_i}^2$. *Consider a bounded open domain* $D \subset \mathbb{R}^r$ *with a regular boundary* ∂D, *and let* T_D *be the first time* X_t *hits the set* ∂D *starting from* $X_0 \in D$. *Check that the solution of the Dirichlet-Poisson problem*

$$\begin{cases} L(v)(x) & = & 0 & \text{for any} \quad x \in D \\ v(x) & = & h(x) & \text{for any} \quad x \in \partial D \end{cases}$$

is given by

$$v(x) = \mathbb{E}\left(h(X_{T_D}) \mid X_0 = x\right).$$

Let $\mathbb{S}(x, \rho) \subset \mathbb{R}^r$ *be an* $(r-1)$-*dimensional sphere centered at some state* $x \in \mathbb{R}^r$ *with radius*

$\rho > 0$. We let $T_{\mathbb{S}(x,\rho)}$ be the first time the process X_t hits the sphere $\mathbb{S}(x,\rho)$. Check that $X_{T_{\mathbb{S}(x,\rho)}}$ is uniformly distributed on the sphere $\mathbb{S}(x,\rho)$.

Starting from a given state $x \in D$, we let $\mathbb{S}(x) \subset D$ be the maximal sphere included in D. We also let $M(x, dy)$ be the distribution of an uniform state on the sphere $\mathbb{S}(x)$. The random walk X_n starting at some $X_0 = x \in D$ with Markov transitions M is called the walk-on-sphere algorithm. When the random state X_n gets too close to ∂D we project X_n on D. The resulting random state has approximatively the same law as X_{T_D}.

Exercise 424 (Markov chain Polya urn model) *We consider the Pólya urn model from section 25.6.2, with $c = 1$. Write the Markov chain evolution of the number of black balls B_n in terms of i.i.d. uniform random variables U_n on $[0,1]$.*

Exercise 425 (Lazy processes) ✏ *Describe the Markov transitions of the counting process (25.2) associated with the lazy version (25.4) of the random walk on the hypercube. Check that the probability measure π defined in (25.5) is the unique invariant measure of the chain and that for any $k \in S$ we have*

$$\mathbb{E}(T_k \mid X_0 = k) = \frac{k!(b-k)!}{b!} \, 2^k.$$

Here T_k denotes the first return time to k. Compute the expected return time to the empty urn when we have $b = 100$ balls.

Exercise 426 (Return to origin) ✏ *We consider the random walk in the lattice $\mathbb{Z}^d$ as presented in section 25.1.1.*

- *When $d = 2$, prove that*

$$\mathbb{P}(X_{2n} = (0,0) \mid X_0 = (0,0)) = \sum_{0 \le k \le n} \frac{(2n)!}{k!^2(n-k)!^2} \left(\frac{1}{4}\right)^{2n}$$

  and deduce that

$$\mathbb{P}(X_{2n} = (0,0) \mid X_0 = (0,0)) \simeq \frac{1}{\pi n}.$$

- *When $d = 3$, prove that*

$$\mathbb{P}(X_{2n} = (0,0,0) \mid X_0 = (0,0,0)) = \sum_{0 \le k+l \le n} \frac{2n!}{k!^2 l!^2 (n-k-l)!^2} \left(\frac{1}{6}\right)^{2n}$$

  and deduce that

$$\mathbb{P}(X_{2n} = (0,0,0) \mid X_0 = (0,0,0)) \le c/n^{3/2}$$

  with some positive constant c.

Exercise 427 (Polya urn model) ✏ *We consider the Pólya urn model as discussed in section 25.6.2. We let $(X_i)_{1 \le i \le n}$ be the $\{0,1\}$-valued random variables defined by $X_i = 1$ iff a black ball is drawn at the i-th trial. Prove that for any function f on the set of outcomes $\{0,1\}^n$ we have*

$$\mathbb{E}(f(X_1, \ldots, X_n) \mid (B_0, W_0) = (b,w)) = \mathbb{E}\left(\mathbb{E}\left(f(Y_1, \ldots, Y_n) \mid U_{\left(\frac{b}{c}, \frac{w}{c}\right)}\right)\right).$$

Check that

$$\mathbb{E}\left(\frac{B_n}{B_n + W_n} \mid \mathcal{F}_{n-1}\right) = \frac{B_{n-1}}{B_{n-1} + W_{n-1}}$$

with $\mathcal{F}_{n-1} = \sigma\left((B_0, W_0), (X_1, \ldots, X_{n-1})\right)$. *Also, prove that for any $t \in \mathbb{R}$,*

$$\lim_{n \to \infty} \mathbb{E}\left(e^{\frac{itB_n}{B_n + W_n}} \mid (B_0, W_0) = (b, w) \right) = \mathbb{E}\left(e^{itU_{(b/c, w/c)}} \right).$$

Exercise 428 (Diffusion approximation) *We consider the Pólya urn model as discussed in section 25.6.2 with $c = 1$, and we let X_t be rescaled continuous process $X_t = \frac{B_{\lfloor Nt \rfloor}}{N + \lfloor Nt \rfloor}$, and we set $\Delta_h X_t := X_{t+h} - X_t$, for some time step h. For $h = 1/N$ and N sufficiently large, check that*

$$\mathbb{E}\left(\Delta_h X_t \mid X_t\right) = 0,$$
$$\mathbb{E}\left((\Delta_h X_t)^2 \mid X_t\right) = \frac{1}{N} \frac{X_t(1 - X_t)}{(1+t)^2} h \ (1 - \epsilon_t(N))$$

for some function $\epsilon_t(N) \in [0, 2/N]$. Compare X_t with the non-homogeneous neutral Wright-Fisher diffusion process

$$dY_t = \frac{1}{1+t} \sqrt{\frac{Y_t(1 - Y_t)}{N}} \ dV_t,$$

with V_t denoting a standard Brownian motion.

Exercise 429 (Coupon collector problem) ✎ *We sample with replacement from an urn with d numbered balls $S = \{1, \ldots, d\}$. We denote by X_n the S-valued sequence representing the label of the ball drawn at the n-th trial. For each $1 \le i \le d$, we let $A_n(i) = \{\forall 1 \le p \le n : X_p \in S - \{i\}\}$ denote the event "the ball i has not been drawn after n trials". Prove that*

$$\forall 1 \le i \le d \qquad \mathbb{P}(A_n(i)) = \left(1 - \frac{1}{d}\right)^n.$$

We let T be the first time n all the balls have been drawn. Prove that

$$\mathbb{P}(T > n) \le d \left(1 - \frac{1}{d}\right)^n \quad \text{and} \quad \mathbb{P}(T > d\log(d) + md) \le e^{-m}.$$

Exercise 430 (Brownian motion on torus) *The Brownian motion on the one-dimensional torus $\mathbb{T} := \mathbb{R}/(2\pi\mathbb{Z})$ can be defined by the equivalence class $\overline{W}_t = \{W_t + 2k\pi : k \in \mathbb{Z}\}$ of a one-dimensional Brownian motion W_t. We let $\mu(dw) = \frac{1}{2\pi} 1_{[0, 2\pi]}(w) \ dw$ be the uniform distribution on $\mathbb{T}$. We consider the mapping*

$$\psi : \theta \in \mathbb{T} \mapsto \psi(\theta) = e^{iW_t} \in \mathcal{C} = \{x = (x_1, x_2) \in \mathbb{R}^2 : x_1^2 + x_2^2 = 1\}.$$

For any function F on $\mathcal{C}$ with $\|F\| \le 1$ we set $f = F \circ \psi$. Check that

$$\mathbb{E}\left(F(\psi(W_t)) \mid W_0 = w_0\right) = \int_0^{2\pi} f(v) \ p_t(v - w_0) \ dv := P_t(f)(w_0)$$

with the probability density

$$p_t(v) := \frac{1}{\sqrt{2\pi t}} \sum_{n \in \mathbb{Z}} e^{-\frac{1}{2t}(v + 2n\pi)^2} = \frac{1}{2\pi} \sum_{n \in \mathbb{Z}} \mathbb{E}\left(e^{-inW_t}\right) e^{inv}.$$

The r.h.s. equality follows from the Poisson summation formula. Deduce that

$$\|P_t(f) - \mu(f)\| \le \sum_{n \ge 1} e^{-n^2 t/2} \to_{t \to \infty} 0.$$

26

Iterated random functions

This chapter is dedicated to stochastic processes defined in terms of iterations of random functions. We illustrate these models in biology with ancestral type evolution processes, as well as in combinatorics and group theory with card shuffling techniques. The last part of the chapter is concerned with a series of iterated random processes arising in the construction and the analysis of fractal images.

> *Trust everybody, but cut the cards.*
> Finley Peter Dunne (1867-1936).

26.1 Description

We consider a finite collection $\mathcal{A}$ of functions f from some Banach state space S into S equipped with some probability measure Γ. In other words, Γ is the law of some random variable F taking values in the finite set $\mathcal{F}$

$$\forall f \in \mathcal{A} \qquad \mathbb{P}\left(F = f\right) = \Gamma(f)$$

with a *finite* collection of numbers $\Gamma(f) \in [0, 1]$, indexed by $S \ni f$ and such that (s.t.) $\Gamma(S) := \sum_{f \in \mathcal{A}} \Gamma(f) = 1$.

Remark : Without loss of generality, we assume that $\mathcal{A} = \{f_i \ : \ i \in \mathcal{I}\}$ is a family of functions indexed by some finite set $\mathcal{I}$ and

$$\mathbb{P}\left(F = f_i\right) = \Gamma(f_i) = \nu(i) \tag{26.1}$$

where ν is a probability measure on the index set $\mathcal{I}$.

We further assume that $\mathcal{A} \subset \mathcal{F}$ is a subset of a Banach space $\mathcal{F}$ of functions from S into S which is *stable by the composition* of functions $f \circ g \in \mathcal{F}$, for any $f, g \in \mathcal{F}$ (i.e., $\mathcal{F}$ has a semiring structure w.r.t. the operations $(+, \times, \circ)$).

Definition 26.1.1 *Iterated random functions are defined by an $\mathcal{F}$-valued Markov chain using the forward formula*

$$X_n = X_{n-1} \circ F_n = F_0 \circ F_1 \circ \ldots \circ F_n \tag{26.2}$$

or the backward formula

$$X_n = F_n \circ X_{n-1} = F_n \circ F_{n-1} \circ \ldots \circ F_0. \tag{26.3}$$

In the above display, $(F_n)_{n \geq 1}$ denotes a sequence of independent random copies of F, and $X_0 = F_0$ stands for some initial function $F_0 \in \mathcal{F}$.

Notice that

$$\text{Law}\,(F_n \circ F_{n-1} \circ \ldots \circ F_0) = \text{Law}\,(F_0 \circ F_1 \circ \ldots \circ F_n).$$

In the example (26.1), the Markov chain $X_n(x)$ defined by (26.3) and starting at the identity function $F_0(x) = x$ can be written as follows:

$$X_0(x) = x \rightsquigarrow X_1(x) = f_{I_1}(x) \rightsquigarrow X_2(x) = f_{I_2}\left(f_{I_1}(x)\right)$$

$$\rightsquigarrow \ldots \rightsquigarrow X_n(x) = f_{I_n}\left(\ldots\left(f_{I_1}(x)\right)\ldots\right),$$

where $I_1, I_2, \ldots$ are independent draws from ν. The Markov transitions of the chain $X_n(x)$ are given by

$$M(x, dy) = \sum_{i \in \mathcal{I}} \nu(i)\,\delta_{f_i(x)}(dy).$$

Important remark : The Markov property for the function-valued Markov chains $X_n = X_{n-1} \circ f_{I_n}$ or $X_n = f_{I_n} \circ X_{n-1}$ is clear. In the second case, it is also true for the chain $X_n(x) = f_{I_n}(X_{n-1}(x))$ taking values on S and starting at some state x. *In this situation, sometimes we slightly abuse the notation and we denote by X_n instead of $X_n(x)$ the random states in S of the chain starting at the constant function $F_0(y) := x$. When there is no confusion, we often denote by X_n the random states in S of the chain starting at the random constant function $F_0(y) := X_0$. Here X_0 is a random variable taking values in S.*

As usual, we are interested in the limiting behavior (whenever it exists) of these Markov chains. More precisely, we would like to know whether there are limiting measures such that for any starting state x, and for any $B \subset S$,

$$\eta_n(B) := \mathbb{P}(X_n(x) \in B) \longrightarrow \eta_\infty(B)$$

and for any starting state $X_0 = f_0$ and for any $\mathcal{B} \subset \mathcal{F}$

$$\Gamma_n(\mathcal{B}) = \mathbb{P}(X_n \in \mathcal{B}) \longrightarrow \Gamma_\infty(\mathcal{B}).$$

For the convergence of the chain X_n in the set of random functions, we refer the reader to exercise 438.

Remark :
The set $\mathcal{F} = \mathcal{C}(\mathbb{R}^d, \mathbb{R}^d)$ can be equipped with a Banach space norm

$$\|f\|_{\mathcal{F}} = \sum_{k \geq 1} 2^{-k}\,\|f\|_{K_k} \quad \text{with} \quad \|f\|_{K_k} := \sup_{x \in K_k} |f(x)| \tag{26.4}$$

where $K_k \subset K_{k+1}$ is a sequence of strictly increasing compact sets covering $\mathbb{R}^d = \cup_{k \geq 1} K_k$. In this situation, we can extend the measure Γ to the state space $\mathcal{F}$ of the Markov chain X_n by considering the discrete measure

$$\Gamma(df) = \sum_{1 \leq i \leq d} \nu(i)\,\delta_{f_i}(df)$$

in the sense that for any bounded measurable function H on $\mathcal{F}$, we have the Lebesgue integral

$$\mathbb{E}\,(H(F)) = \int H(f)\,\Gamma(df) = \sum_{i \in \mathcal{I}} \nu(i)\,H(f_i).$$

When $\mathcal{I} \subset \mathbb{R}$ is not necessarily finite and is equipped with some probability measure ν, we have

$$\mathbb{E}\,(H(F)) = \int H(f)\,\Gamma(df) = \int H(f_u)\,\nu(du).$$

26.2 A motivating example

Suppose that $S = [-2, 2]$ and $\mathcal{A} = \{f_{-1}, f_{+1}\}$ consists of two functions

$$f_{-1}(x) = \frac{1}{2}\, x - 1 \quad \text{and} \quad f_{+1}(x) = \frac{1}{2}\, x + 1.$$

We equip the set $\mathcal{I} = \{-1, +1\}$ with the uniform measure $\nu(-1) = \nu(+1) = 1/2$. We can see that the Markov chain (26.3) is defined by

$$X_n(x) = f_{I_n}(X_{n-1}(x)) = \frac{1}{2}\, X_{n-1}(x) + I_n$$

where I_n is a $\{-1, +1\}$-Bernoulli r.v. with parameter $1/2$. We denote by M the Markov transition of the chain. Note that

$$f_{-1}([-2, 2]) = [-2, 0] \quad \text{and} \quad f_{+1}([-2, 2]) = [0, 2] \ \Rightarrow \ \forall x \in [-2, 2] \quad X_n(x) \in [-2, 2].$$

Elementary computations show that

$$\begin{aligned}
X_n(x) &= \left(\frac{1}{2}\right)^2 X_{n-2}(x) + \frac{1}{2}\, I_{n-1} + I_n \\
&= \ldots = \left(\frac{1}{2}\right)^n x + \sum_{0 \le p < n} \frac{1}{2^p}\, I_{n-p}.
\end{aligned}$$

On the other hand, since $(I_1, I_2, \ldots, I_n) \stackrel{Law}{=} (I_n, \ldots, I_2, I_1)$, we have

$$\sum_{0 \le p < n} \frac{1}{2^p}\, I_{n-p} \stackrel{Law}{=} \sum_{0 \le p < n} \frac{1}{2^p}\, I_{p+1} \to_{n \uparrow \infty} \sum_{p \ge 0} \frac{I_{p+1}}{2^p} := X_\infty.$$

From these observations, we check that $X_\infty \in [-2, 2]$ and

$$\text{Law}(X_n(x)) \to_{n \uparrow \infty} \text{Law}(X_\infty) = \pi. \tag{26.5}$$

We also notice that for any uniform r.v. I_0 on $\mathcal{I} = \{-1, +1\}$ we have

$$\frac{1}{2}\, X_\infty + I_0 = \sum_{p \ge 0} \frac{I_{p+1}}{2^{p+1}} + I_0 \stackrel{law}{=} \sum_{p \ge 0} \frac{I_{p+2}}{2^{p+1}} + I_1 = \sum_{p \ge 1} \frac{I_{p+1}}{2^p} + I_1 = X_\infty.$$

This implies that $\pi = \pi M$.

In addition, using the fact that $\text{lip}(f_i) = 1/2$, we prove that

$$\mathbb{W}(\text{Law}(X_n(x)), \text{Law}(X_n(y))) \le 2^{-n}\, |x - y|.$$

Using proposition 8.3.13 we also readily check that

$$\mathbb{W}(\text{Law}(X_n(x)), \pi) = \mathbb{W}(\delta_x M^n, \pi M^n) \le 2^{-n} \int \pi(dy)\, |x - y|.$$

Hence

$$\sup_{x \in [-2, 2]} \mathbb{W}(\text{Law}(X_{n+2}(x)), \pi) \le 2^{-n}.$$

Since S has a bounded diameter $\mathrm{diam}(S) := \sup_{x,y} |x - y| = 4$, using proposition 8.3.9 and proposition 8.3.13 we readily check that

$$\sup_{\mu \in \mathcal{P}(S)} \left\| \mu M^{n+4} - \pi \right\|_{tv} \leq 2^{-n}.$$

This shows that distribution of the random states X_n converges exponentially fast to the unique invariant measure π. We let $U_{a,b}$ be a uniform r.v. on $[a,b]$. Recalling that

$$U_{[a,b]} \overset{law}{=} a + (b - a)\, U_{0,1},$$

we check that

$$f_{-1}(U_{-2,2}) \overset{law}{=} \frac{1}{2}(-2 + 4\,U_{0,1}) - 1 = -2 + 2U_{0,1} \overset{law}{=} U_{-2,0}$$

$$f_{+1}(U_{-2,2}) \overset{law}{=} \frac{1}{2}(-2 + 4\,U_{0,1}) + 1 = -2U_{0,1} \overset{law}{=} U_{0,2}$$

from which we conclude that

$$\pi = \mathrm{Law}(U_{-2,2}).$$

Another illustration of these Bernoulli type iterated functions is provided in exercise 439.

In the next sections, we discuss three more typical examples arising in the literature on random iterated functions.

26.3   Uniform selection

26.3.1   An ancestral type evolution model

We discuss an elementary Markov chain that represents the ancestral genetic type evolution of d individuals $S = \{1, \ldots, d\}$. At each time n, the i-th individual chooses independently its parent type in the previous generation.

More formally, at the first generation, we sample d uniform random variables $A_1(i)$, $1 \leq i \leq d$, with common distribution

$$\forall 1 \leq j \leq d \qquad \mathbb{P}\left(A_1(1) = j\right) = \ldots = \mathbb{P}\left(A_1(d) = j\right) = \frac{1}{d} \sum_{1 \leq i \leq d} 1_i(j) = 1/d.$$

The index $A_1(i)$ can be interpreted as the ancestral type of the individual i at the first level, or equivalently $A_1(i)$ can be seen as the parent of the individual i. This transition can be summarized by the following diagram:

$$\begin{array}{ccc}
\text{Selected types at level } n = 0 & \longleftarrow & \text{Individual at } n = 1. \\
A_1(i) & \longleftarrow & i
\end{array}$$

At the second iteration, the i-th individual of type changes independently and randomly its type by choosing a new type in S. The ancestral tree of the individuals at levels $n = 0, 1, 2$ are encoded in the following diagram:

$$\begin{array}{ccccc}
n = 0 & \longleftarrow & n = 1 & \longleftarrow & n = 2. \\
A_1(A_2(i)) & \longleftarrow & A_2(i) & \longleftarrow & i
\end{array}$$

Notice that $A_2(i)$ can be interpreted as the ancestral type at level $n = 1$ of the individual i (at the current level $n = 2$); and $A_1(A_2(i))$ is the ancestral type at level $n = 0$ of the i-th individual.

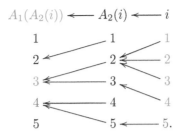

$$A_1(A_2(i)) \longleftarrow A_2(i) \longleftarrow i$$

The evolution of the genealogical tree based structure of the ancestral types of the individuals can be summarized by the following synthetic diagrams:

$$[A_1(i), i] \rightsquigarrow [A_1(A_2(i)), A_2(i), i] \rightsquigarrow [A_1(A_2(A_3(i))), A_2(A_3(i)), A_3(i), i].$$

Notice that the sequence of random variables $A_n := (A_n(1), \ldots, A_n(d))$ is a r.v. with a law given for any $a = (a(1), \ldots, a(d)) \in S^S$ by

$$\mathbb{P}(A_n = (a(1), \ldots, a(d)))$$

$$= \mathbb{P}(A_n(1) = a(1)) \times \ldots \times \mathbb{P}(A_n(1) = a(d))$$

$$= \tfrac{1}{d} \sum_{1 \leq b(1) \leq d} 1_{b(1)}(a(1)) \times \ldots \times \tfrac{1}{d} \sum_{1 \leq b(d) \leq d} 1_{b(d)}(a(d))$$

$$= \tfrac{1}{d^d} \sum_{(b(1), \ldots, b(d)) \in S^S} 1_{(b(1), \ldots, b(d))}(a(1), \ldots, a(d))$$

$$= \tfrac{1}{\mathrm{Card}(S^S)} \sum_{b \in S^S} 1_b(a) := \Gamma(a)$$

where Γ represents the the uniform probability measure on the set $\mathcal{A} = \mathcal{F} = S^S$ given by

$$\forall a \in \mathcal{F} \quad \Gamma(a) = \frac{1}{\mathrm{Card}(S^S)} = \frac{1}{d^d}.$$

The sampling of A_n consists in choosing randomly (with replacement) d indices $I_1, \ldots, I_d$ in the set $\{1, \ldots, d\}$ and then setting $A(j) = I_j$, with $1 \leq j \leq d$. In other words, we sample d random variables $I_1, \ldots, I_d$ with the uniform distribution $\tfrac{1}{d} \sum_{1 \leq i \leq d} 1_i$ on $\{1, \ldots, d\}$.

We also recall that we can equivalently use d uniform random variables U_i on $[0, 1]$ and then set $I_i = 1 + \lfloor d \, U_i \rfloor$, for each $1 \leq i \leq d$.

26.3.2 An absorbed Markov chain

We let $|a|$ be the cardinality of the set $a(S)$, and we consider the partial order relation

$$b_2 \leq b_1 \Leftrightarrow \exists a \in \mathcal{A} \ : \ b_2 = a \circ b_1.$$

We will use the term b_2 *is below* b_1 to refer to this relation.

We also denote by $S(q, p)$ the number of ways of partitioning q states into p blocks, and by $(d)_p = d!/(d-p)!$ the number of one-to-one mappings from p states into d states.

Remark : In the combinatorics literature, $S(q, p)$ is called the Stirling number of the second kind. For instance $S(q, q) = 1$, and $S(q, q-1) = q(q-1)/2$ (dividing q elements into $q-1$ sets amounts of dividing q into $q-2$ sets of size 1, and one set of size 2).

Suppose that $|b_1| = p_1$. By construction, the following assertions are satisfied:

- $|b_2| = p_2 \leq p_1$, so that $b_1(S)$ contains p_1 states $i_1, \ldots, i_{p_1}$, and $b_2(S)$ contains p_2 states $j_1, \ldots, j_{p_2}$

- The mappings b_2 below b_1 are defined by mapping the states $i_1, \ldots, i_{p_1}$ to the set $\{1, \ldots, d\}$, so that there are exactly d^{p_1} mappings b_2 below b_1. We check this claim using the fact that any choice of mapping from $j \notin \{i_1, \ldots, i_{p_1}\}$ to $\{1, \ldots, d\}$ does not change the construction of a given b_2 associated with some mapping from $i_1, \ldots, i_{p_1}$ to $\{1, \ldots, d\}$.

- Given some b_1 s.t. $|b_1| = p_1$, there are $S(p_1, p_2) \times (d)_{p_2}$ mappings b_2 with $|b_2| = p_2 \leq p_1$ such that $b_2 \leq b_1$.

  Each partition $\pi = \{\pi(1), \ldots, \pi(p_2)\}$ of the p_1 states in $b_1(S)$ into p_2 blocks is combined with one of the $(d)_{p_2}$ one-to-one mappings from the partition indexes $\{1, \ldots, p_2\}$ into $\{1, \ldots, d\}$.

The following diagram illustrates these observations when $p_1 = 3 \geq p_2 = 2$, and $\pi(1) = \{2\}$ $\pi(2) = \{4, 5\}$.

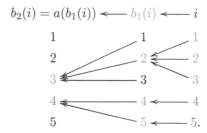

A useful general observation for iterated random functions is that the time reversed chain has the same law as the forward chain. More formally, for any $n \geq 1$,

$$\text{Law} \left(A_1 \circ A_2 \circ \ldots \circ A_n \right) = \text{Law} \left(A_n \circ A_{n-1} \circ \ldots \circ A_1 \right).$$

This observation allows us to analyze the genealogical structure of the process backward in time in terms of the Markov chain

$$B_n = A_n \circ A_{n-1} \circ \ldots \circ A_1 \Rightarrow B_n = A_n \circ B_{n-1} \leq B_{n-1}$$

with initial condition $B_0 = Id \in \mathcal{A}$. In terms of genealogy, this Markov chain represents the ancestral branching process from the present generation running backward into the past. In this interpretation, the mapping A_n represents the way the individuals choose their parents in the previous ancestral generation. The range of A_n represents successful parents with direct descendants, whereas the range of B_n represents successful ancestors with descendants in all the generations through terminal time.

In view of previous observations, eventually the sequential composition of random mappings becomes constant after some sufficiently large time. To see why, we consider the Markov chain defined for any $p \leq q \leq d$ by

$$\mathbb{P} \left(|B_n| = p \mid |B_{n-1}| = q \right) = \frac{1}{d^q} \, S \left(q, p \right) \, (d)_p.$$

We also observe that 1 is an absorbing state, i.e. $\mathbb{P} \left(|B_n| = 1 \mid |B_{n-1}| = 1 \right) = 1$.

We let $\partial \mathcal{A}$ the set of the d constants mappings, that is

$$\partial \mathcal{A} = \{a \in \mathcal{A} \ : \ |a| = 1\}.$$

We are interested in the random variable T that represents the time to most recent common ancestor of an initial population with $|B_0|$ individuals.

$$T = \inf \{ n \geq 1 \ : \ B_n \in \partial\mathcal{A} \} = \inf \{ n \geq 1 \ : \ |B_n| = 1 \ \}.$$

Of course, if one of the $A_p \in \partial\mathcal{A}$, for some $p \leq n$, we have $B_n \in \partial\mathcal{A}$. This yields the rather crude estimate

$$\begin{aligned}
\mathbb{P}\,(T > n) \quad &\leq \quad \mathbb{P}\,(\forall 1 \leq p \leq n \ A_p \notin \partial\mathcal{A}) \\
&= \quad \prod_{1 \leq p \leq n} (1 - \mathbb{P}\,(A_p \in \partial\mathcal{A})) = \left(1 - \frac{d}{d^d} \right)^n \leq e^{-n/d^{d-1}}.
\end{aligned}$$

This implies that $\mathbb{P}\,(T < \infty) = 1$. In addition, by symmetry arguments B_T is uniformly distributed in the set $\partial\mathcal{A}$.

Using remark 26.3.2, we have

$$\begin{aligned}
\mathbb{P}\,(|B_n| \neq q \mid |B_{n-1}| = q) \quad &= \quad 1 - \frac{(d)_q}{d^q} \\
&= \quad 1 - \frac{d(d-1)\ldots(d-(q-1))}{d^q} \\
&= \quad 1 - \prod_{1 \leq k < q} \left(1 - \frac{k}{d} \right) \\
&\leq \quad 1 - \left(1 - \frac{q-1}{d} \right)^{q-1} \leq \frac{(q-1)^2}{d}.
\end{aligned}$$

In the last assertion we used $1 - a^n = (1-a)\,(1 + a + a^2 + \ldots + a^{n-1}) \leq n(1-a)$, which is valid for any $0 \leq a \leq 1$, and any $n \in \mathbb{N}$. Recalling that $\log\,(1-x) \leq -x$, for any $x \in [0, 1[$, we also have the estimate

$$\begin{aligned}
\mathbb{P}\,(|B_n| \neq q \mid |B_{n-1}| = q) \quad &= \quad 1 - \prod_{1 \leq k < q} \left(1 - \frac{k}{d} \right) \\
&= \quad 1 - e^{\sum_{1 \leq k < q} \log\left(1 - \frac{k}{d} \right)} \\
&\geq \quad 1 - e^{-\frac{1}{d} \sum_{1 \leq k < q} k} = 1 - e^{-\frac{q(q-1)}{2d}}.
\end{aligned}$$

We consider the Markov chain

$$\mathbb{P}\,(R_n = p \mid R_{n-1} = q) = \frac{(d)_q}{d^q} \, 1_{p=q} + \left(1 - \frac{(d)_q}{d^q} \right) \, 1_{p=q-1}. \tag{26.6}$$

This chain has the same chance to stay in its position, and whenever it changes its values it decreases only by a unit. This indicates that the first time T' it reaches the absorbing state 1, is larger than T in the sense that

$$\mathbb{P}\,(T > n) \leq \mathbb{P}\,(T' > n) \leq \frac{n}{d} \, \exp\left[-\left(\frac{n}{d} - \frac{7}{2} \right) \right]. \tag{26.7}$$

A detailed proof of the r.h.s. estimate is provided in exercise 431. For a more thorough discussion on the long time behavior of these Markov chain models, we refer the reader to [80].

26.4   Shuffling cards

26.4.1   Introduction

Most card shuffling can be interpreted as a Markov chain taking values in the symmetric group $\mathcal{G}_d$, that is, in the set of all the $d!$ permutations of d cards $S := \{1, \ldots, d\}$, for some $d \geq 1$.

This section is mainly taken from the seminal article of D. Aldous and P. Diaconis [2] (see also the book by D. A. Levin and Y. Peres [182], for a more recent and updated treatment on Markov chains and mixing times). The forthcoming analysis also originates from this article. We also refer the reader to section 8.3.1 on p. 173 for a presentation of the key notion of strong stationary times and their use to derive quantitative estimates for the convergence to equilibrium of a given Markov chain.

26.4.2   The top-in-at-random shuffle

As its name indicates, the top-in-at-random shuffle takes the first card of a deck and inserts it at a uniformly chosen random position. One way to model this shuffle by iterated random functions is to consider the set $\mathcal{F} = \mathcal{G}_d$, and the set $\mathcal{A}$ of the d cycles $c_i = (i \; i-1 \; \ldots \; 2 \; 1)$, with $1 \leq i \leq d$ defined by

$$c_i(i) = i - 1 \quad c_i(i-1) = i - 2 \quad \ldots c_i(2) = 1 \quad c_i(1) = i$$

and $c_i(k) = k$ for any $k > i$, equipped with the uniform distribution

$$\Gamma(c_i) = 1/d.$$

The cycle permutations c_i are described in a synthetic diagram below.

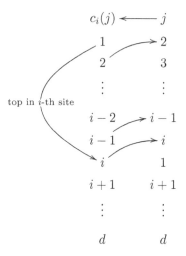

We let γ be a random variable with distribution Γ. The top-in-at-random shuffle is represented as the $\mathcal{G}_d$-valued Markov chain

$$X_n = X_{n-1} \circ \gamma_n = \gamma_1 \circ \ldots \circ \gamma_n = c_{I_1} \circ \ldots \circ c_{I_n}. \tag{26.8}$$

Here $\gamma_n = c_{I_n}$ stands for a sequence of independent random copies of $\gamma = c_I$, and $X_0 = Id$ denotes the identity permutation.

We let $T = 1 + \tau$, where τ is the first time the original bottom card reaches the top of the deck.

$$T = T_1 + \ldots + T_{d-1} + T_d \tag{26.9}$$

where $T_d = 1$, and for any $1 \leq i < d$, T_i stands for the number of times required for the bottom card to rise from position i to position $i+1$. We start counting from the bottom of the deck. When the bottom card is at position i, the chance for the randomly inserted top card to be inserted below i is i/d. This shows that $T_i \sim \text{Geo}(\frac{i}{d})$, that is,

$$\mathbb{P}\left(T_i = k\right) = \left(1 - \frac{i}{d}\right)^{k-1} \frac{i}{d}.$$

In addition the random variables T_i are clearly independent.

Remark : This model is intimately related to the coupon collector problem discussed in exercise 429. Assuming that the collector gets one out of d distinct coupons with equal probability $1/d$, the time it takes to obtain the first coupon is $T_d = 1$, the time to see the second T_{d-1} is a geometric random variable with parameter $(d-1)/d$, the time to see the third T_{d-2} is a geometric random variable with parameter $(d-2)/d$, and by induction the time $T_{d-(i-1)}$ to see the i-th coupon is a geometric random variable with parameter $(d-(i-1))/d$, for each $1 \leq i \leq d$. The total time T coincides with the sum defined above.

In this interpretation, we have

$$\mathbb{P}(T > n)$$

$$\leq \mathbb{P}\left(\cup_{1 \leq i \leq d} \{\text{the } i\text{-th coupon does not appear in the first } n \text{ trials}\}\right)$$

$$= \sum_{1 \leq i \leq d} \mathbb{P}\left(\{\text{the } i\text{-th coupon does not appear in the first } n \text{ trials}\}\right)$$

$$= \sum_{1 \leq i \leq d} \left(\frac{d-1}{d}\right)^n \leq d\, e^{-n/d}.$$

This implies that

$$\mathbb{P}(T > d\log(d) + n\,d) \leq d\, e^{-\log(d)) - n} = e^{-n}$$

and by theorem $8.3.18$ we conclude that

$$\|\text{Law}(X_{d\log d + nd}) - \pi\|_{tv} \leq e^{-n}.$$

Here π denotes the invariant uniform distribution $\pi(\sigma) = 1/d!$ of the shuffling Markov chain (26.8) on $\mathcal{G}_d$. We refer the reader to exercise 432 for the analysis of the mean and variance of T.

26.4.3 The random transposition shuffle

In the random transposition shuffle, we randomly pick two cards and exchange their positions. One way to model this shuffle by iterated random functions is to consider the set $\mathcal{A} = \mathcal{T} := \{\tau_{i,j} \in \mathcal{G}_d : 1 \leq i < j \leq d\}$, the subset of transpositions $\tau_{i,j}$ of the indexes i, j; that is, $\tau_{i,j}(i) = j$, $\tau_{i,j}(j) = i$, and $\tau_{i,j}(k) = k$ for any $k \in S - \{i,j\}$, equipped with the uniform distribution

$$\Gamma(\tau_{i,j}) = \frac{2}{d(d-1)}.$$

In this case, Γ is the distribution of the random variable γ defined by

$$\mathbb{P}\left(\gamma = \tau_{i,j}\right) = \Gamma(\tau_{i,j}) = \frac{2}{d(d-1)}$$

for any $1 \leq i < j \leq d$. The sampling of γ consists of choosing randomly (without replacement) two indices I, J in the set $\{1, \ldots, d\}$ and then setting $\gamma = \tau_{I,J}$. Then we can define the $\mathcal{G}_d$-valued Markov chain

$$X_n = X_{n-1} \circ \gamma_n = \gamma_1 \circ \ldots \circ \gamma_n = \tau_{I_1, J_1} \circ \ldots \tau_{I_n, J_n} \tag{26.10}$$

where $\gamma_n = \tau_{I_n, J_n}$ denotes a sequence of independent random copies of $\gamma = \tau_{I,J}$, and $X_0 = Id$ denotes the identity permutation.

Using $\tau_{i,j}^2 = Id$, we see that the Markov transition is clearly reversible

$$
\begin{aligned}
M(\sigma, \sigma \circ \tau_{i,j}) &= \mathbb{P}\left(X_n = \sigma \circ \tau_{i,j} \mid X_{n-1} = \sigma\right) \\
&= \mathbb{P}\left(X_n = \sigma \mid X_{n-1} = \sigma \circ \tau_{i,j}\right) = M(\sigma \circ \tau_{i,j}, \sigma).
\end{aligned}
$$

This shows that the invariant probability measure of the chain is the uniform distribution $\pi(\sigma) = 1/d!$ on $\mathcal{G}_d$. Recalling that any permutation can be written as a composition of elementary transpositions, this Markov chain X_n is irreducible.

We consider the lazy (and aperiodic) chain associated with the random variable γ defined by

$$\forall i \neq j \qquad \mathbb{P}\left(\gamma = \tau_{i,j}\right) = \Gamma(\tau_{i,j}) = \frac{2}{d^2}$$

and $\mathbb{P}\left(\gamma = Id\right) = 1/d$ (recall that $(d(d-1)/2)\, 2/d^2 + 1/d = 1$).

When cards are numbered $\{c_1, \ldots, c_d\} := \{1, \ldots, d\}$, an equivalent way of shuffling the cards is to choose at any time n a card label $L_n := i$ with the left hand, and a position $R_n = j$ with the right hand, and then switch the card i with the card in position j, so that

$$X_n = X_{n-1} \circ \tau_{(L_n, R_n)}.$$

Next, we consider two decks X_n and $X_n' = X_{n-1}' \circ \tau_{(L_n', R_n')}$ that start from a different initial configuration $X_n = x_0 \neq x_0' = X_0'$, and evolve with the same random transposition shuffles $(L_n, R_n) = (L_n', R_n')$, up to the first time T they coincide $X_T = X_T'$. After that time, we have $X_n = X_n'$.

We let N_n be the number of alignments, that is, the number of cards that occupy the same location in the decks X_n and X_n'. We can prove that

$$N_{n+1} \geq N_n + \epsilon_n \tag{26.11}$$

with a $\{0,1\}$-value random variable ϵ_n with distribution given by

$$\mathbb{P}\left(\epsilon_{n+1} = 1 \mid N_n\right) \geq \underbrace{\frac{d - N_n}{d}}_{\substack{\text{chance to choose} \\ \text{a non aligned card}}} \times \underbrace{\frac{d - N_n}{d}}_{\substack{\text{chance to choose} \\ \text{a non aligned position}}}.$$

This implies that

$$d \overset{n \uparrow \infty}{\longleftarrow} d - \left(1 - \frac{1}{d}\right)^n (d - N_0) \leq \mathbb{E}\left(N_n \mid N_0\right) \leq d. \tag{26.12}$$

A detailed proof of these claims can be found in exercise 434.

The initial state N_0 depends on the initial configuration of the deck. Obviously the largest coupling time occurs when the initial configurations are such that $N_0 = 0$.

The coupling time of the chains is decomposed as follows

$$T = T_1 + \ldots + T_d$$

where T_1 is the first time n s.t. $N_n = 1$, T_2 is the first time n s.t. $N_n = 2$, and T_i is the first time n s.t. $N_n = i$. In this situation, T_i ($\Rightarrow N_{T_i} = i$), is a geometric random variable with success parameter $((d - (i-1))/d)^2$; This implies that

$$\mathbb{E}\left(T_{i+1}\right) = (d/(d-i))^2 \Rightarrow \mathbb{E}(T) = d^2 \sum_{0 \leq i < d} \left(\frac{1}{(d-i)}\right)^2.$$

Hence we conclude that

$$\mathbb{E}(T) = d^2 \sum_{1 \leq i \leq d} \frac{1}{i^2} \;\leq\; d^2 \, \pi^2/6.$$

Using proposition 8.3.16, we conclude that

$$\left\|\mathrm{Law}(X_{n(\pi d)^2/6}) - \pi\right\|_{tv} \leq \frac{1}{n}$$

where π denotes the invariant uniform distribution $\pi(\sigma) = 1/d!$ of the lazy version of the Markov chain (26.10) on $\mathcal{G}_d$.

Our next objective is to design a judicious strong stationary time to improve this estimate. We start with a deck of d unmarked cards, and we mark them following the next step at each random transposition shuffle. At each step, we assume that the left and right hands independently pick uniformly a card, say L_n and R_n, and we mark the card R_n when

- L_n *is marked* and R_n *is unmarked*

- or when $L_n = R_n$ is unmarked.

Notice that R_n remains unmarked if and only if the left hand card L_n is marked, the right hand R_n is unmarked and $L_n \neq R_n$.

Notice that each time a card is marked, it is uniformly ordered in the set of the cards we have marked so far. Thus, the first time T all the cards are marked is a strong stationary time. A more detailed and formal proof of this assertion is provided in exercise 433.

We let $T_0 = 1$. For any $1 \leq i < d$, we let T_i be the number of transpositions after the i-th card is marked, until the $(i+1)$ card is marked (including the time of marking the second card). We can check that the random variables $(T_i)_{1 \leq i < d}$ are independent, geometrically distributed with success parameters

$$p_i = \frac{i \times (d - (i-1))}{d^2} \tag{26.13}$$

and

$$T = \sum_{0 \leq i < d} T_i \implies \mathbb{E}(T) \leq 2d(1 + \log d). \tag{26.14}$$

A detailed proof of this result is provided in exercise 433; see also exercise 435 for the estimation of $\mathbb{E}(T)$.

By theorem 8.3.18 we conclude that

$$\left\|\mathrm{Law}(X_{2d(1+\log d)n}) - \pi\right\|_{tv} \leq \frac{1}{n}.$$

Here π denotes the invariant uniform distribution $\pi(\sigma) = 1/d!$ of the lazy version of the Markov chain (26.10) on $\mathcal{G}_d$.

26.4.4   The riffle shuffle

This real shuffle consists of dividing the deck into two packs of roughly $d/2$ cards each. For instance, we can cut the pack of d cards according to the binomial distribution with parameter $1/2$. In other words, the probability to get a pack of $d = k + (d - k)$ cards to be cut just after the k-th card is $\binom{d}{k} \times 2^{-d}$.

Then these two packs are interwoven randomly. (We mention that a perfect interwoven card-by-card shuffle, when done by a good magician, is not random at all.)

The randomness of the shuffle is characterized by a sequence $(\epsilon_1, \ldots, \epsilon_d)$ of $\{0, 1\}$-valued independent Bernoulli random variables with probability $1/2$. The number of $L := \sum_{1 \le k \le d}(1 - \epsilon_k)$ of 0's represents the number of cards that go to the l.h.s. stack, and the number $R = \sum_{1 \le k \le d} \epsilon_k$ of 1's shows how many cards go to the r.h.s stack before performing the riffle. Notice that the decomposition $d = L + R$ represents the binomial cut of the deck with parameter $1/2$.

The order of the 0's and 1's in the sequence $(\epsilon_1, \ldots, \epsilon_d)$ characterizes the order the cards fall when riffling the two stacks. The 0's corresponds to cards in the left hand stack, and the 1's to cards the right hand stack. Each of these sequences $(\epsilon_1, \ldots, \epsilon_d)$ corresponds to a unique riffle shuffle. Thus, given the binomial numbers $(L, R) = (k, d - k)$, there are $\binom{d}{k}$ possible riffles, one of which $(0, \ldots, 0, 1, \ldots, 1)$ is the identity. Thus, the total number of possible riffle shuffles is given by

$$1 + \sum_{0 \le k \le d} \left(\binom{d}{k} - 1 \right) = 2^d - d.$$

In addition, the probability of any cut followed by any possible riffle is 2^{-d}.

For instance, the following schematic picture shows a riffle associated with the sequence $(0, 1, 1, 0, 0, 1, 1, 0)$

Assuming that the original sequence of ordered 8 cards is $(c_1, \ldots, c_8)$, this "real" riffle of the two stacks starts from the bottom, dropping first the card c_4 from the left hand stack, then the card c_8, c_7 from the right hand stack, then c_3, c_2 from the left hand stack, then c_6, c_5 from the right hand stack, and finally the remaining card c_1 from the left hand stack.

c_1	c_2	c_3	c_4	c_5	c_6	c_7	c_8
0	1	1	0	0	1	1	0
c_1	c_5	c_6	c_2	c_3	c_7	c_8	c_4
$c_{\sigma^{-1}(1)}$	$c_{\sigma^{-1}(2)}$	$c_{\sigma^{-1}(3)}$	$c_{\sigma^{-1}(4)}$	$c_{\sigma^{-1}(5)}$	$c_{\sigma^{-1}(6)}$	$c_{\sigma^{-1}(7)}$	$c_{\sigma^{-1}(8)}$

The new order of the cards $(1, 4, 5, 8, 2, 3, 6, 7)$ after the riffle is described by the following permutation

$$\sigma = \Sigma\,(0, 1, 1, 0, 0, 1, 1, 0) := \begin{pmatrix} 1 & 2 & 3 & 4 & 5 & 6 & 7 & 8 \\ 1 & 4 & 5 & 8 & 2 & 3 & 6 & 7 \end{pmatrix}$$

with the usual position and value encoding

$$j = \sigma(i) \quad = \quad \text{position of the card } c_i \text{ (originally in position } i)$$
$$c_{\sigma^{-1}(j)} \quad = \quad \text{value of the card in position } j \text{ (originally with value } c_j).$$

Notice that these permutations have two rising sequences

$$\sigma(1) < \sigma(2) < \sigma(3) < \sigma(4) \quad \text{and} \quad \sigma(5) < \sigma(6) < \sigma(7) < \sigma(8)$$

except for the identity $(0, \ldots, 0, 1, \ldots, 1)$ which has only one.

We also observe that the mapping $\Sigma : \epsilon \in \{0,1\}^d \mapsto \Sigma(\epsilon) \in \mathcal{G}_d$ is bijective. We consider a uniform random variable γ in the set of 2^d permutations $\mathcal{R} = \Sigma(\{0,1\}^d) \in \mathcal{G}_d$. The iteration of the riffle shuffle discussed above is described by the Markov chain model

$$X_n = X_{n-1} \circ \gamma_n = \gamma_1 \circ \ldots \circ \gamma_n$$

where γ_n denotes a sequence of independent random copies of γ, and $X_0 = Id$ is the identity permutation.

Notice that at the forth step of the above interleaving,

c_1	c_2	c_3	c_4	c_5	c_6	c_7	c_8
0	1	1	0	0	1	1	0
c_1	c_5	c_6	c_2	c_3	c_7	c_8	c_4
$c_{\sigma^{-1}(1)}$	$c_{\sigma^{-1}(2)}$	$c_{\sigma^{-1}(3)}$	$c_{\sigma^{-1}(4)}$	$c_{\sigma^{-1}(5)}$	$c_{\sigma^{-1}(6)}$	$c_{\sigma^{-1}(7)}$	$c_{\sigma^{-1}(8)}$

we have 3 cards $\{c_1, c_2, c_3\}$ on the left hand stack, and 2 cards $\{c_5, c_6\}$ in the right hand pack (since the cards c_7, c_8, c_9 have already fallen). Since the Bernoulli random variables that model the riffle are independent, the chance that the next card is dropped from the right hand pack to the right hand stack is $2/(2+3)$

$$\mathbb{P}\left(\epsilon_5 = 1 \mid \sum_{1 \le i \le 5} \epsilon_i = 2\right) = \frac{\mathbb{P}\left(\epsilon_5 = 1, \sum_{1 \le i \le 4} \epsilon_i = 1\right)}{\mathbb{P}\left(\sum_{1 \le i \le 5} \epsilon_i = 2\right)}$$

$$= \frac{1}{2} \frac{\binom{4}{1} 2^{-4}}{\binom{5}{2} 2^{-5}} = \frac{4!}{3!1!} \times \frac{3!2!}{5!} = 2/5.$$

The time *reversed shuffle* consists in labeling at random all cards with 0's and 1's using independent Bernoulli random variables with parameter $1/2$. Then, we pull out the cards with label 0's with their relative order (maintaining the cards with label 1's with their relative order). We place the cards with label 0 on the top of the stack of cards with label 1, keeping them in their relative order. In other words, the inverse shuffle takes cards with 0's, respectively 1's, in the top deck, respectively bottom deck, without changing their order. For instance, the reversed riffle associated with the sequence defined above $(0, 1, 1, 0, 0, 1, 1, 0)$ is given by

c'_1	c'_2	c'_3	c'_4	c'_5	c'_6	c'_7	c'_8
c_1	c_5	c_6	c_2	c_3	c_7	c_8	c_4
0	1	1	0	0	1	1	0
c_1	c_2	c_3	c_4	c_2	c_3	c_6	c_7
c'_1	c'_4	c'_5	c'_8	c'_2	c'_3	c'_6	c'_7
$c'_{\sigma(1)}$	$c'_{\sigma(2)}$	$c'_{\sigma(3)}$	$c'_{\sigma(4)}$	$c'_{\sigma(5)}$	$c'_{\sigma(6)}$	$c'_{\sigma(7)}$	$c'_{\sigma(8)}$

Notice that the new order of the cards $(1,6,5,2,3,7,8,4)$ after the inverse riffle $(0,1,1,0,0,1,1,0)$ is described by:

$$\sigma' = \left(\begin{array}{cccccccc} 1 & 2 & 3 & 4 & 5 & 6 & 7 & 8 \\ 1 & 5 & 6 & 2 & 3 & 7 & 8 & 4 \end{array} \right) = \sigma^{-1}.$$

Repeating inverse shuffles from a given configuration, we retain and mark the sequence of $0-1$ labels at the back of each card, so that after n inverse shuffles each card has a n-digit binary number. Notice that the labels at each step are assigned uniformly at random. Cards with distinct labels are in uniform random relative order, while cards with the same label are in the same order as the origin. An example of three successive inverse shuffles is provided below

0	c_1		01	c_1		001	c_5		0100	
1	c_2		01	c_4		101	c_3		0101	
1	c_3		00	c_5		101	c_6		0100	
0	c_4		01	c_8		010	c_1		0011	
0	c_5	$\leadsto$	11	c_2	$\leadsto$	010	c_4	$\leadsto$	1010	$\ldots/\ldots$
1	c_6		10	c_3		010	c_8		1011	
1	c_7		10	c_6		111	c_2		1110	
0	c_8		11	c_7		111	c_7		1110	

From previous observations, we notice that when all cards have distinct labels, their relative order is uniform. This shows that

$$T = \inf\{n \geq 1 \ : \ \text{all cards have distinct label}\} \tag{26.15}$$

is a strong stationary time.

At each step n, we have a collection of d digit binary numbers of size n. In addition, the labels of cards are independent. Since there are 2^n possible n-digit numbers, the chance that $T \leq n$ is the same as the probability, when putting randomly d balls into 2^n boxes, to get at most one ball in each box. We can also interpret the balls as students in a classroom where each box represents a different birthday:

$$\mathbb{P}(T > n) = 1 - \frac{2^n}{2^n} \times \frac{2^n - 1}{2^n} \times \frac{2^n - 2}{2^n} \cdots \times \frac{2^n - (d-1)}{2^n} = 1 - \prod_{1 \leq i \leq d} \left(1 - \frac{i}{2^n} \right).$$

Recalling that $-x \geq \log(1-x) \geq -x - x^2$, for any $0 \leq x \leq 1/2$ (since $1 + 2x \geq 1/(1-x) = 1 + x/(1-x) \geq 1$ for such x), we have

$$\sum_{1 \leq i \leq d} \log\left(1 - \frac{i}{2^n} \right) \geq -\frac{1}{2^n} \sum_{1 \leq i \leq d} i - \frac{1}{4^n} \sum_{1 \leq i \leq d} i^2$$

$$= -\frac{d(d+1)}{2^{n+1}} - \frac{1}{4^n} \frac{d(d+1)(2d+1)}{6}$$

for any $n \geq 1$ such that $(i/2^n \leq) \, d/2^n \leq 1/2$. This implies that for any n such that $2^n \geq 2d$, we have

$$\mathbb{P}(T > n) \leq 1 - \exp\left\{ -\frac{d(d+1)}{2^{n+1}} \left[1 + \frac{1}{2^n} \left(\frac{2}{3} d + \frac{1}{3} \right) \right] \right\}.$$

Using theorem 8.3.18, we conclude that

$$\|\text{Law}(X_n) - \pi\|_{tv} \leq 1 - \exp\left\{ -\frac{d(d+1)}{2^{n+1}} \left[1 + \frac{1}{2^n} \left(\frac{2}{3} d + \frac{1}{3} \right) \right] \right\}. \tag{26.16}$$

If we choose n such that $2^n \geq a\,(d+1)^2$ with a such that $1 - e^{-\frac{1}{a}} \leq \frac{1}{b}$ for a given $b \geq 1$,

$$\frac{d(d+1)}{2^{n+1}}\left[1 + \frac{1}{2^n}\left(\frac{2}{3}d + \frac{1}{3}\right)\right] \leq \frac{1}{2a}\left[1 + \frac{d}{2^n}\right]$$
$$\leq \frac{1}{2a}\left[1 + \frac{1}{a}\right] \leq \frac{1}{a}.$$

We find that $\mathbb{P}\,(T > n) \leq 1/b$. Using

$$d/2^n \leq 1/2 \Rightarrow 1 + \frac{1}{2^n}\left(\frac{2}{3}d + \frac{1}{3}\right) \leq 1 + \frac{d}{2^n} \leq 3/2$$

and the inequality $e^{-x} \geq 1 - x$, for $x \geq 0$, we obtain the rather crude estimates

$$\|\mathrm{Law}(X_n) - \pi\|_{tv} \leq \frac{3d(d+1)}{4\,2^n} \leq \frac{(d+1)^2}{2^n} \leq \frac{1}{a}.$$

These estimates lead us to the conclusion that

$$\left\|\mathrm{Law}(X_{1+\lfloor \log_2 a + 2\log_2 (d+1)\rfloor}) - \pi\right\|_{tv} \leq 1/a$$

with $\log_2 a := \log a / \log 2$ (so that $2^{\log_2 a} = e^{\log_2 (a) \times \log 2} = e^{\log a} = a$, for any $a > 0$).

The computation of the mean $\mathbb{E}(T)$ is presented in exercise 437.

26.5 Fractal models

Fractal forms are highly symmetric objects. They can be described in terms of random iterations of affine type functions on the Euclidian space $\mathbb{R}^d$, for a fixed dimension $d \geq 1$.

In the further development of this section, we let $\mathcal{F} := \mathcal{C}(S, S)$ be the set of continuous functions from some subset $S \subset \mathbb{R}^d$ into itself. We consider a collection of $d \times d$ matrices A_i and d dimensional vectors b_i, indexed by a finite set $\mathcal{I}$ and we set

$$\mathcal{A} = \{f_i \ : \ i \in \mathcal{I}\} \quad \text{with} \quad \forall i \in \mathcal{I} \quad f_i(x) := A_i x + b_i.$$

We equip $\mathcal{I}$ with a distribution ν and we set

$$\forall i \in \mathcal{I} \quad \Gamma(f_i) = \nu(i) \quad \text{and} \quad \Gamma\,(\mathcal{F} - \mathcal{A}) = 0.$$

In this situation, the sampling of a random variable F with distribution Γ on $\mathcal{A}$ consists of randomly choosing an index I in the set $\mathcal{I}$ with probability $\nu(I)$, and then setting $F = f_I$. The corresponding iterated random functions are defined by an $\mathcal{F}$-valued Markov chain

$$X_n = F_{I_n} \circ X_{n-1} = F_{I_n} \circ F_{I_{n-1}} \circ \ldots \circ F_{I_1} \circ F_0$$

where $(I_n)_{n \geq 1}$ is a sequence of independent random copies of I, and $X_0 = F_0$ is an initial function $F_0 \in \mathcal{F}$.

For instance, if $\mathcal{I} = \{0, 1\}$ then $\nu(1) = 1 - \nu(0)$. In this case, we flip a coin with probability of a head $\nu(1)$ to randomly choose a function in $\mathcal{A} = \{f_0, f_1\}$. More formally, we sample a Bernoulli random variable

$$\mathbb{P}\,(I = 1) = 1 - \mathbb{P}(I = 0) = \nu(1) \quad \text{and we set} \quad F = f_I.$$

In this situation, we have

$$X_n = \left(1_{[0,\nu(1)[}(U_n)\ f_1(X_{n-1}) + 1_{[\nu(1),1]}(U_n)\ f_0(X_{n-1})\right)$$

where $(U_n)_{n\geq 1}$ is a sequence of independent uniform random variables on $[0,1]$.

26.5.1 Exploration of Cantor's discontinuum

A Cantor discontinuum subset of $[0,1]$ is a closed and nowhere dense and non-empty subset of the unit interval. Thus, the Lebesgue integral of this set is equal to zero. These "almost" empty fractal sets were introduced in 1883 by the German mathematician Georg Cantor [38].

The Cantor ternary discontinuum is defined in terms of $1/3$-Lipschitz transformations on $S = [0,1]$

$$f_0(x) = \frac{x}{3} \quad \text{and} \quad f_1(x) = \frac{x+2}{3} = \frac{2}{3} + f_0(x).$$

We notice that

$$f_0(S) = \left[0, \frac{1}{3}\right] \quad \text{and} \quad f_1(S) = \frac{2}{3} + f_0(S) = \left[\frac{2}{3}, 1\right]$$

$$\Downarrow$$

$$f_0(S) \cup f_1(S) = \left[0, \frac{1}{3}\right] \cup \left[\frac{2}{3}, 1\right] \neq [0,1]$$

so that the set $[0,1]$ is not a fixed point of the set mapping

$$f \ : \ A \subset S \mapsto f(A) = f_0(A) \cup f_1(A).$$

A graphical description of the iterated mapping f^n is provided in figure 26.1.

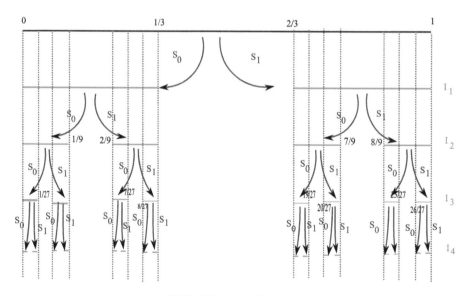

FIGURE 26.1: Cantor discontinuum

For instance,

$$
\begin{aligned}
S_1 &= f(S) = \left[0, \frac{1}{3}\right] \cup \left[\frac{2}{3}, 1\right] \\
S_2 &= f^2(S) = f(S_1) = \left[0, \frac{1}{9}\right] \cup \left[\frac{2}{9}, \frac{1}{3}\right] \cup \left[\frac{6}{9}, \frac{7}{9}\right] \cup \left[\frac{8}{9}, 1\right] \\
S_3 &= f^3(S) = f(S_2) \\
&= \left[0, \frac{1}{27}\right] \cup \left[\frac{2}{27}, \frac{1}{9}\right] \cup \left[\frac{6}{27}, \frac{7}{27}\right] \cup \left[\frac{8}{27}, \frac{1}{3}\right] \cup \left[\frac{2}{3}, \frac{19}{27}\right] \\
&\qquad \cup \left[\frac{20}{27}, \frac{21}{27}\right] \cup \left[\frac{8}{9}, \frac{25}{27}\right] \cup \left[\frac{26}{27}, 1\right].
\end{aligned}
$$

This shows that the subsets

$$
S = S_0 = [0, 1] \supset S_{n-1} \supset f(S_{n-1}) = S_n = f^n(S_0) = \cup_{k=1}^{2^n} S_{k,n}
$$

are decomposed into 2^n intervals $(S_{k,n})_{k=1,\dots,2^n}$ with length 3^{-n}.

When $n \uparrow \infty$, this decreasing sequence converges to the Cantor ternary discontinuum non-empty set

$$
S_n = f^n(S_0) \downarrow I_\infty = \cap_{n \geq 1} S_n \neq \emptyset.
$$

This set is the fixed point of the transformation f

$$
f(S_\infty) = \cap_{n \geq 1} f(S_n) = \cap_{n \geq 1} S_{n+1} = \cap_{n \geq 2} S_n = S_\infty.
$$

The complementary subset

$$
C_n = S_n^c = \cup_{k=1}^n \left(\cup_{l=1}^{2^{k-1}} C_{k,l} \right)
$$

is defined in terms of 2^{k-1} *disjoint intervals* $(J_{k,l})_{l=1,\dots,2^{k-1}}$ *with length* 3^{-k}, *with* $k = 1, \dots, n$. A J_n and $J_{k,l}$ are depicted in figure 26.2.

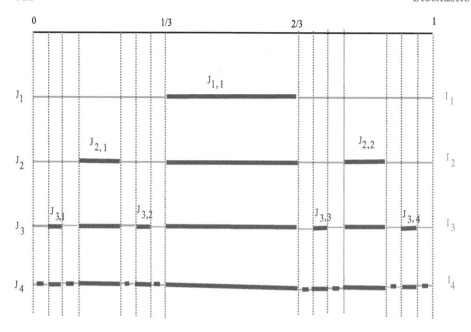

FIGURE 26.2: Complement of Cantor discontinuum

This implies that

$$\text{Full-length}(J_n) \;=\; 1 - \left(\frac{2}{3}\right)^n \uparrow \text{Full-length}(J_\infty) = 1 \quad \text{when } n \uparrow \infty$$

$$\implies \text{Full-length}(I_n) \downarrow \text{Full-length}(I_\infty) = 0. \quad (26.17)$$

A proof of this result is provided in exercise 440.

Our random walker moves sequentially in the set $S_n \subset S = [0,1]$

$$X_n = f_{\epsilon_n}(X_{n-1}) = \frac{1}{3} X_{n-1} + \frac{2}{3} \epsilon_n$$

starting from some state $X_0 \in [0,1]$, where ϵ_n is a sequence of independent random variables with distribution

$$\mathbb{P}(\epsilon_n = 0) = \mathbb{P}(\epsilon_n = 1) = \frac{1}{2}.$$

An elementary computation shows that

$$X_n \;=\; \frac{1}{3^n} X_0 + \frac{2}{3} \sum_{k=0}^{n-1} \frac{\epsilon_{n-k}}{3^k} \quad (26.18)$$

$$\stackrel{law}{=} \frac{1}{3^n} X_0 + \frac{2}{3} \sum_{k=0}^{n-1} \frac{\epsilon_k}{3^k} \rightarrow_{n \uparrow \infty} X_\infty := \frac{2}{3} \sum_{n \geq 0} \frac{\epsilon_n}{3^n}.$$

From these observations, we check that

$$X_\infty \in S_\infty = \left\{ \frac{2}{3} \sum_{k \geq 0} \frac{\alpha_k}{3^k} \;:\; \alpha_k \in \{0,1\} \right\}$$

(cf. exercise 440) and

$$\text{Law}(X_n(x)) \to_{n\uparrow\infty} \text{Law}(X_\infty) := \pi.$$

In addition, using $\text{lip}(f_i) = 1/3$ we prove that

$$\mathbb{W}(\text{Law}(X_n(x)), \text{Law}(X_n(y))) \le 3^{-n} |x - y|.$$

Using proposition 8.3.13 we also readily check that

$$\mathbb{W}(\text{Law}(X_n(x)), \pi) = \mathbb{W}(\delta_x M^n, \pi M^n) \le 3^{-n} \int \pi(dy) |x - y|$$

from which we conclude that

$$\sup_{x \in [0,1]} \mathbb{W}(\text{Law}(X_n(x)), \pi) \le 3^{-n}.$$

Since S has a bounded diameter $\text{diam}(S) := \sup_{x,y} |x - y| = 1$, by using proposition 8.3.9 and proposition 8.3.13 we readily check that

$$\sup_{\mu \in \mathcal{P}(S)} \|\mu M^n - \pi\|_{tv} \le 3^{-n}.$$

By the ergodic theorem,

$$\frac{1}{n} \sum_{p=1}^{n} \varphi(X_p) \longrightarrow \int_{S_\infty} \varphi(x) \, \pi(dx)$$

for any function φ on the Cantor continuum set S_∞. For instance, for any $A \subset [0,1]$

$$\frac{1}{n} \sum_{p=1}^{n} \varphi(X_p) = \frac{\text{Card}\{1 \le p \le n : X_p \in A\}}{n} \simeq_{n\uparrow\infty} \pi(A) = \mathbb{P}(X_\infty \in A).$$

26.5.2 Some fractal images

The analysis of iterated random functions on the plane follows essentially the same lines of arguments as those used in the one-dimensional case. In this section, we merely present some illustrations of fractal images without any convergence analysis.

An example of a fractal leaf.
We consider a sequence of independent Bernoulli random variables

$$\mathbb{P}(\epsilon_n = 1) = 1 - \mathbb{P}(\epsilon_n = 0) = 0.2993.$$

For each $i \in \{0,1\}$, choose the affine functions $f_i(x) = A_i.x + b_i$ with the matrices and the vectors

$$A_0 = \begin{pmatrix} +0.4000 & -0.3733 \\ +0.0600 & +0.6000 \end{pmatrix} \qquad b_0 = \begin{pmatrix} +0.3533 \\ +0.0000 \end{pmatrix}$$

and

$$A_1 = \begin{pmatrix} -0.8000 & -0.1867 \\ +0.1371 & +0.8000 \end{pmatrix} \qquad b_1 = \begin{pmatrix} +1.1000 \\ +0.1000 \end{pmatrix}.$$

Running the Markov chain with 10^5 iterations we obtain the fractal image in figure 26.3.

FIGURE 26.3: Second example of fractal leaf

A fractal tree We consider a sequence of independent uniform random variables ϵ_n on the set $I = \{1, 2, 3\}$. For each $i \in \{1, 2, 3\}$, we choose the affine functions $f_i(x) = A_i.x + b_i$ with the matrices and the vectors defined below

$$A_1 = \begin{pmatrix} 0 & 0 \\ 0 & c \end{pmatrix} \qquad b_1 = \begin{pmatrix} 1/2 \\ 0 \end{pmatrix}$$

$$A_2 = \begin{pmatrix} r\,\cos(\varphi) & -r\,\sin(\varphi) \\ r\,\sin(\varphi) & r\,\cos(\varphi) \end{pmatrix} \qquad b_2 = \begin{pmatrix} \frac{1}{2} - \frac{r}{2}\cos(\varphi) \\ c - \frac{r}{2}\sin(\varphi) \end{pmatrix}$$

and

$$A_3 = \begin{pmatrix} q\,\cos(\psi) & -r\,\sin(\psi) \\ q\,\sin(\psi) & r\,\cos(\psi) \end{pmatrix} \qquad b_3 = \begin{pmatrix} \frac{1}{2} - \frac{q}{2}\cos(\psi) \\ \frac{3c}{5} - \frac{q}{2}\sin(\psi) \end{pmatrix}$$

with the parameters

$$c = 0.255, \qquad r = 0.75, \qquad q = 0.625$$
$$\varphi = -\frac{\pi}{8}, \qquad \psi = \frac{\pi}{5}, \qquad |X_0| \le 1.$$

Running the Markov chain with 10^5 iterations we obtain the fractal image 26.4.

FIGURE 26.4: Fractal tree

Sierpinski carpet

We consider a sequence of independent uniform random variables ϵ_n on the set $I = \{1, 2, 3\}$. For each $i \in \{1, 2, 3\}$, we choose the affine functions $f_i(x) = A_i.x + b_i$ with the matrices and vectors

$$A_1 = A_2 = A_3 = \begin{pmatrix} 1/2 & 0 \\ 0 & 1/2 \end{pmatrix}$$

and

$$b_1 = \begin{pmatrix} 0 \\ 0 \end{pmatrix} \quad b_2 = \begin{pmatrix} 1/2 \\ 0 \end{pmatrix} \quad \text{and} \quad b_3 = \begin{pmatrix} 1/4 \\ \sqrt{3}/4 \end{pmatrix} \simeq \begin{pmatrix} 0,250 \\ 0,433 \end{pmatrix}.$$

Running the Markov chain with 10^5 iterations we obtain the fractal image in figure 26.5 given below.

FIGURE 26.5: Sierpinski carpet

Heighways dragons

We consider a sequence of independent uniform random variables ϵ_n on the set $I = \{1, 2\}$. For each $i \in \{1, 2\}$, we choose the affine functions $f_i(x) = A_i.x + b_i$ with the matrices and the vectors defined below

$$A_1 = \begin{pmatrix} 1/2 & -1/2 \\ 1/2 & 1/2 \end{pmatrix}, \quad A_2 = \begin{pmatrix} -1/2 & -1/2 \\ 1/2 & -1/2 \end{pmatrix}, \quad b_1 = \begin{pmatrix} 0 \\ 0 \end{pmatrix}, \quad b_2 = \begin{pmatrix} 1 \\ 0 \end{pmatrix}.$$

Running the Markov chain with 10^5 iterations we obtain the fractal image in figure 26.6.

26.6 Exercises

Exercise 431 (Adsorbed Markov chain) /// *We consider the Markov chain presented in (26.6), and we let $T_{q \to (q-1)}$ be the random time to move from state q to $q - 1$, for any $2 < q \leq d$.*

- *We let N be any geometric random variable N with success probability $\alpha \in]0, 1[$. Check*

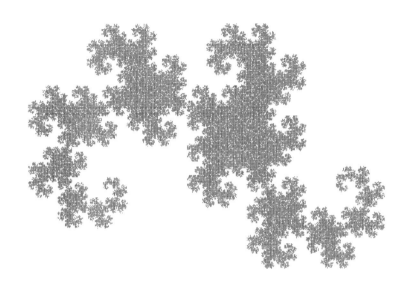

FIGURE 26.6: Heighways dragons

that for any $0 \leq t < -d \log(1 - \alpha)$ *we have*

$$\mathbb{E}(e^{t(N-1)/d}) = \frac{\alpha}{1 - (1 - \alpha)e^{t/d}}.$$

- *Check that* $T_{q \to (q-1)}$ *is a geometric random variable with success probability* $\left(1 - \frac{(d)_q}{d^q}\right)$. *Prove that*

$$\mathbb{E}\left(e^{t \, \frac{T' - (R_0 - 1)}{d}} \mid R_0\right) = \prod_{1 \leq q < R_0} \mathbb{E}\left(e^{t \, \frac{T_{q+1 \to q} - 1}{d}}\right)$$

and for any $0 \leq t < \frac{q(q-1)}{2}$

$$\mathbb{E}\left(e^{t \, \frac{T_{q \to (q-1)} - 1}{d}}\right) = \frac{1 - \beta_q}{1 - \beta_q \, e^{t/d}} \quad \text{with} \quad \beta_q := \frac{(d)_q}{d^q} \leq e^{-\frac{q(q-1)}{2d}}.$$

- *Consider for any $x > 0$ the function*

$$f \; : \; x \in \left[0, e^{-t/d}\right] \mapsto \frac{1-x}{1-x\; e^{t/d}}.$$

Check that f is increasing and deduce that

$$\frac{1-\beta_q}{1-\beta_q\; e^{t/d}} \le \frac{1 - e^{-\frac{q(q-1)}{2d}}}{1 - e^{-\frac{q(q-1)}{2d}}\; e^{t/d}}.$$

- *Consider for any $x > 0$ the function*

$$g \; : \; y \in [0, x] \mapsto g(y) = x\left(1 - e^{y-x}\right) + (y-x)\left(1 - e^{-x}\right).$$

Check that $g(y) \ge 0$, for any $y \in [0, x]$ and deduce the convexity property

$$e^x \le 1 + x\, \frac{e^x - e^y}{x - y}.$$

Applying this inequality to $x = \frac{q(q-1)}{2d} \ge y = t/d$, check that on the event $R_0 > 1$

$$\forall t \in [0, 1[\quad \mathbb{E}\left(e^{t\,\frac{T' - (R_0 - 1)}{d}} \mid R_0\right) = \prod_{1 \le q < R_0} \frac{1}{1 - \frac{t}{q(q+1)/2}}.$$

- *Deduce from the above that for any $0 \le t < 1$*

$$\mathbb{E}\left(e^{t\,\frac{T' - (R_0 - 1)}{d}} \mid R_0\right)$$
$$\le h(t) := \mathbb{E}\left(e^{t\sum_{1 \le q < \infty} \frac{2}{q(q+1)}\, \mathcal{E}_q}\right) = \prod_{q \ge 1} \frac{1}{1 - \frac{t}{q(q+1)/2}}$$

where $\mathcal{E}_q$ is a sequence of independent exponential random variables with unit parameter.

- *Prove that*
$$\mathbb{P}\left(\frac{T' - (R_0 - 1)}{d} \ge n \mid R_0\right) \le e^{-tn}\, h(t)$$

and deduce that

$$\mathbb{P}\left(T' \ge m \mid R_0\right) \le ae \inf_{t \in [0, 1[} \frac{e^{-mt/d}}{1 - t} = a\, \frac{m}{d}\, e^{-(m/d - 1)}. \qquad (26.19)$$

Here $a := \prod_{q \ge 2} \frac{1}{1 - \frac{1}{q(q+1)/2}}$.

- *Using $(-x \ge) \log(1 - x) \ge -x - x^2$, for any $0 \le x \le 1/2$ (cf. page 718), check that*

$$\mathbb{P}\left(T' \ge m \mid R_0 = q\right) \le \frac{m}{d}\, \exp\left[-\left(\frac{m}{d} - \frac{5}{2}\right)\right].$$

Exercise 432 (The top-in-at-random shuffle) ✎ *We consider the top-in-at-random card shuffle discussed on page 173 and in section 26.4. We let $T = 1 + \tau$, where τ is the first time the original bottom card reaches the top of the deck. Using the decomposition (26.9), prove that*

$$d\log(d + 1) \le \mathbb{E}(T) \le d(1 + \log d) \quad and \quad \mathrm{Var}(T) \le 2d^2.$$

Exercise 433 (Random transposition shuffle I) ///

The following exercise is taken from the book of D. A. Levin and Y. Peres [182]. We let T be the stopping time associated to the first time all cards are marked in the transposition shuffle introduced on page 715. Prove that T is a strong stationary time.

Exercise 434 // *We consider the coupled Markov chains described on page 714. Check (26.11) and (26.12).*

Exercise 435 / *We consider the strong stationary time T defined in (26.14). Prove that*

$$\mathbb{E}(T) \leq 2d \log d + \alpha d + \beta \quad \text{and} \quad \|\text{Law}(X_n) - \pi\|_{tv} \leq \frac{2d \log d + d + 1}{n}$$

for some finite constants α and β. Here π is the invariant uniform distribution $\pi(\sigma) = 1/d!$ of the shuffling Markov chain (26.10) on $\mathcal{G}_d$.

Exercise 436 (Random transposition shuffle II) / *We let T be the stopping time*

associated with the first time all cards are marked in the transposition shuffle introduced on page 715. Check the decomposition (26.14) of T, and prove that the random variables $(T_i)_{1 \leq i < d}$ are independent geometrically distributed with success parameters defined in (26.13).

Exercise 437 / *Compute the mean $\mathbb{E}(T)$ of the strong stationary time T introduced in (26.15).*

Exercise 438 / *We consider the example (26.5). We equip the set $\mathcal{F}$ of all continuous functions from $\mathbb{R}$ into itself with the norm $\|f\|_{\mathcal{F}}$ defined in (26.4) with the compact intervals $K_k = [-k, k]$. Check that*

$$\|X_n - Y_n\|_{\mathcal{F}} = 2^{-(n-1)}$$

with the constant (random) function

$$Y_n(x) = \sum_{0 \leq p < n} \frac{1}{2^p} I_{n-p} \stackrel{Law}{=} \sum_{0 \leq p < n} \frac{1}{2^p} I_{p+1} = Z_n \to_{n \to \infty} Z_\infty.$$

Prove that for any $F \in \mathcal{F}$ and any $\epsilon > 0$, we have

$$\lim_{n \to \infty} \mathbb{P}\left(\|X_{n+1} - F\|_{\mathcal{F}} \leq \epsilon\right) = \mathbb{P}\left(\|Z_\infty - F\|_{\mathcal{F}} \leq \epsilon\right).$$

Exercise 439 (A uniform excursion) *We consider the couple of transformations (f_0, f_1)*

of the unit interval $S = [0, 1]$, which associate to a given $x \in S$ the states at mid-distance to the boundaries:

$$f_0(x) = x + \frac{0 - x}{2} = \frac{x}{2} \quad \text{and} \quad f_1(x) = x + \frac{1 - x}{2} = f_0(x) + \frac{1}{2}.$$

- *Compute the Lipschitz constants $\text{lip}(f_i)$, and the sets $f_i(S)$ for $i = 0, 1$.*

- *The corresponding iterated random functions are defined by*

$$X_n = f_{\epsilon_n}(X_{n-1}) = \frac{1}{2} X_{n-1}(x) + \frac{\epsilon_n}{2}$$

where X_0 is some random variable on S, and ϵ_n are independent Bernoulli random variables with $\mathbb{P}(\epsilon_n = 0) = \mathbb{P}(\epsilon_n = 1) = \frac{1}{2}$. Check that

$$X_n(x) \to_{n \uparrow \infty} X_\infty := \sum_{n \geq 1} \frac{\epsilon_n}{2^n} \stackrel{law}{=} \frac{1}{2} X_\infty + \frac{\epsilon_0}{2}.$$

- *Prove that X_∞ is a conversion to base 2 of a uniform random number U on $[0,1]$, and $U \overset{law}{=} f_{\epsilon_1}(U)$.*

- *We let M be the Markov transition of the iterated random process X_n. Check that*

$$\sup_{\mu \in \mathcal{P}(S)} \mathbb{W}(\mu M^n, \pi) \leq 2^{-n}.$$

Exercise 440 (Cantor discontinuum set) *Check formulae (26.17) and (26.18). Deduce from (26.18) that*

$$S_n = \left\{ \frac{1}{3^n} \, x_0 + \frac{2}{3} \sum_{k=0}^{n-1} \frac{\alpha_k}{3^k} \; : \; x_0 \in [0,1] \quad \text{and} \quad \alpha_k \in \{0,1\} \right\}$$

and

$$S_\infty = \left\{ \frac{2}{3} \sum_{k \geq 0} \frac{\alpha_k}{3^k} \; : \; \alpha_k \in \{0,1\} \right\}.$$

We further assume that X_0 is uniformly chosen in $S_0 = [0,1]$. Check that

$$\mathbb{P}(X_n \in dx) = \left(\frac{3}{2} \right)^n \, 1_{S_n}(x) \, dx.$$

27

Computational and statistical physics

This chapter is dedicated to some applications of stochastic processes in computational and statistical physics. The first part is concerned with molecular dynamics simulation and Langevin type diffusion processes. The second part of the chapter presents some applications of the Feynman-Kac path integration theory to quantum mechanics and the computation of the ground states of Schrödinger operators. The last part of the chapter is dedicated to interacting particle systems.

> *If quantum mechanics hasn't profoundly shocked you,*
> *you haven't understood it yet.*
> Niels Bohr (1885-1962).

27.1 Molecular dynamics simulation

27.1.1 Newton's second law of motion

Molecular dynamics simulation is concerned with the analysis of the fluctuations and with the conformal changes of proteins and nucleic acids in biological molecules. The central problem is understanding the macroscopic properties of a molecule through the simulation of a microscopic system of atomic interacting particles in a given force field model. More formally, we consider the microscopic evolution of a many-body system formed by k atomic particles in the Euclidian space $E =$

$\mathbb{R}^3$ with possibly k different masses $m = (m_i)_{1 \leq i \leq k}$. Their spatial positions and their velocities are denoted by the letters $q = (q_i)_{1 \leq i \leq k}$, and $p = (p_i)_{1 \leq i \leq k}$. These particles move under the influence of some external forces $F_i(q)$ according to the Newton's second law

$$m_i \, \frac{d^2 q_i}{dt^2} = F_i(q). \tag{27.1}$$

The velocity vector

$$p_i = \begin{pmatrix} p_i^1 \\ p_i^2 \\ p_i^3 \end{pmatrix} = m_i \, \frac{dq_i}{dt} = m_i \begin{pmatrix} \dfrac{dq_i^1}{dt} \\ \dfrac{dq_i^2}{dt} \\ \dfrac{dq_i^3}{dt} \end{pmatrix}$$

is called the particle moment of the system, and the couple $x = (q, p)$ is called the phase vector.

We further assume that the force field is conservative in the sense that

$$F(q) = -\nabla_q V(q) = \left(-\frac{\partial V}{\partial q_i}(q) \right)_{1 \le i \le k}$$

for some interparticle potential function $V : E^k \to \mathbb{R}$. In this situation, we can reformulate the evolution equations (27.1) in terms of the Hamiltonian or energy functional

$$H(q, p) = \sum_{i=1}^{k} \frac{\|p_i\|^2}{2m_i} + V(q_1, \ldots, q_k) \tag{27.2}$$

with

$$\begin{cases} \dfrac{dq_i}{dt} & = \dfrac{p_i}{m_i} = \dfrac{\partial H}{\partial p_i}(q, p) \\[2mm] \dfrac{dp_i}{dt} & = F_i(q) = -\dfrac{\partial V}{\partial q_i}(q) = -\dfrac{\partial H}{\partial q_i}(q, p). \end{cases} \tag{27.3}$$

We notice that these evolution equations are time reversible in the sense that they have the same form if we consider the time transformation $\tau(t) = -t$. In other words, the microscopic physics does not depend on the direction of the time flow. We also notice the conservation property

$$\frac{d}{dt} H(q, p) = \sum_{i=1}^{k} \left[\frac{\partial H}{\partial q_i}(q, p) \frac{dq_i}{dt} + \frac{\partial H}{\partial p_i}(q, p) \frac{dp_i}{dt} \right] = 0. \tag{27.4}$$

In the above display we have used the conventions

$$\frac{\partial H}{\partial q_i} = \left(\frac{\partial H}{\partial q_i^1}, \frac{\partial H}{\partial q_i^2}, \frac{\partial H}{\partial q_i^3} \right) \quad \text{and} \quad \frac{dq_i}{dt} = \begin{pmatrix} \frac{dq_i^1}{dt} \\ \frac{dq_i^2}{dt} \\ \frac{dq_i^3}{dt} \end{pmatrix}, \tag{27.5}$$

and

$$\frac{\partial H}{\partial p_i} = \left(\frac{\partial H}{\partial p_i^1}, \frac{\partial H}{\partial p_i^2}, \frac{\partial H}{\partial p_i^3} \right) \quad \text{and} \quad \frac{dp_i}{dt} = \begin{pmatrix} \frac{dp_i^1}{dt} \\ \frac{dp_i^2}{dt} \\ \frac{dp_i^3}{dt} \end{pmatrix}. \tag{27.6}$$

We also mention that for $k = 1$ and if $V(q) = \frac{l}{2} q^2$, for some $l \ge 0$, the system (27.3) reduces to the linearized pendulum

$$\left. \begin{array}{ll} \dfrac{dq}{dt} & =: \ q' = \dfrac{p}{m} \\[2mm] \dfrac{dp}{dt} & =: \ p' = -kq \end{array} \right\} \Rightarrow \frac{d^2 q}{dt^2} + \omega^2 q = 0 \quad \text{with} \quad \omega = \sqrt{\frac{l}{m}}.$$

The solution of this system takes the form

$$q(t) = q(0) \cos(\omega t) + \frac{q'(0)}{\omega} \sin(\omega t).$$

Solid and liquid states of rare-gas elements with closed shell configurations only involve particles interacting with weak van der Waals bonds in terms of the pair-potential function

$$V(q_1, \ldots, q_k) = \sum_{1 \le i < j \le k} V_{LJ}(\|q_j - q_i\|)$$

with the Lennard-Jones potential functions

$$V_{LJ}(r) = 4\epsilon \left[\left(\frac{\tau}{r} \right)^{12} - \left(\frac{\tau}{r} \right)^6 \right]. \tag{27.7}$$

The parameter ϵ represents the depth of the potential well, and τ is the finite distance at which the interaction potential becomes null. Notice that

$$\inf_r V_{LJ}(r) = V_{LJ}(2^{1/6}\tau) = -\epsilon.$$

We say that for $r \ge 2^{1/6}\tau$ the potential is attractive, and it is repulsive for $r < 2^{1/6}\tau$.

The term $(\tau/r)^{12}$ describes the short range Pauli repulsion forces due to overlapping electron orbitals, while the term $(\tau/r)^6$ represents the attraction and the van der Waals dispersion forces at long range distances.

The repulsion term has no real theoretical foundations and it is sometimes replaced by the Buckingham exponential-6 potential $\exp(-r/\tau)$. To avoid the degeneracy of the Lennard-Jones potential at short range distances, we often use cut-off techniques. For instance, we can replace $V_{LJ}(r)$ by

$$\overline{V}_{LJ}(r) = (V_{LJ}(r) - V_{LJ}(r_c)) \; 1_{r<r_c}$$

or by

$$\overline{V}_{LJ}(r) = (V_{LJ}(r) - V_{LJ}(r_c) - V'_{LJ}(r_c)(r - r_c)) \; 1_{r<r_c}$$

for some well chosen cut-off radius r_c. The so-called Wayne-Chandler-Anderson potential is given by

$$r_c = 2^{1/6}\tau \implies \overline{V}_{WCA}(r) = \overline{V}_{LJ}(r) = (V_{LJ}(r) + \epsilon) \; 1_{r<2^{1/6}\tau}.$$

A very nice molecular dynamics simulation of supercritical water using the flexible simple-point-charge water model (a.k.a. the SPC water model) by MDSimulator can be found on YouTube.

The potential energy is a complicated function of the k atomic particles and we cannot expect to find an analytic solution of the system of equations (27.3). All the discrete integration schemes are based on the fact that the positions $q_i(t)$ and the velocities $v_i(t) = (p_i(t)/m_i)$ of the particle can be approximated by a Taylor expansion

$$\begin{aligned} q_i(t + dt) &= q_i(t) + v_i(t) \, dt + \frac{1}{2} \, a_i(t) \, dt^2 + \ldots \\ v_i(t + dt) &= v_i(t) + a_i(t) \, dt + \ldots \end{aligned}$$

The accelerations are always given by

$$a_i(t) = m_i^{-1} \, F_i(q(t)) = -m_i^{-1} \frac{\partial V}{\partial q_i}(q) \quad \text{with} \quad q(t) = (q_i(t))_{1 \le i \le k}.$$

- *The Verlet algorithm [257]* is based on the approximations

$$q_i(t + dt) = q_i(t) + v_i(t) \, dt + \frac{1}{2} \, a_i(t) \, dt^2$$

$$q_i(t - dt) = q_i(t) - v_i(t) \, dt + \frac{1}{2} \, a_i(t) \, dt^2.$$

Summing these two equations, we find that

$$q_i(t + dt) = 2q_i(t) - q_i(t - dt) + a_i(t) \, dt^2.$$

- *The Leap-frog algorithm* uses the approximations

$$q_i(t + dt) = q_i(t) + v_i\left(t + \frac{1}{2} \, dt\right) dt$$

$$v_i\left(t + \frac{1}{2} \, dt\right) = v_i\left(t - \frac{1}{2} \, dt\right) + a_i(t) \, dt.$$

In this algorithm, we first compute the velocities at time $t + \frac{1}{2} \, dt$ to calculate $q_i(t + dt)$. The velocities at time t are approximated by

$$v_i(t) = \frac{1}{2} \left[v_i\left(t + \frac{1}{2} \, dt\right) + v_i\left(t - \frac{1}{2} \, dt\right) \right].$$

- *The velocity Verlet algorithm [250]* uses the approximations

$$q_i(t + dt) = q_i(t) + v_i(t) \, dt + \frac{1}{2} \, a_i(t) \, dt^2$$

$$v_i(t + dt) = v_i(t) + \frac{1}{2} \left(a_i(t) + a_i(t + dt) \right) \, dt.$$

- *The Beeman's algorithm [14]* uses the approximations

$$q_i(t + dt) = q_i(t) + v_i(t) \, dt + \frac{1}{2} \left(\frac{4}{3} \, a_i(t) - \frac{1}{3} \, a_i(t - dt) \right) dt^2$$

$$v_i(t + dt) = v_i(t) + \frac{1}{2} \left(\frac{2}{3} \, a_i(t + dt) + \frac{5}{3} \, a_i(t) - \frac{1}{3} \, a_i(t - dt) \right) dt$$

or

$$v_i(t + dt) = v_i(t) + \frac{1}{2} \left(\frac{5}{6} \, a_i(t + dt) + \frac{8}{6} \, a_i(t) - \frac{1}{3} \, a_i(t - dt) \right) dt.$$

27.1.2 Langevin diffusion processes

We associate with the Hamiltonian function (27.2) the canonical measures on the phase space

$$\mu_\beta(dx) = \frac{1}{\mathcal{Z}_\beta} \, e^{-\beta H(x)} \, dx. \tag{27.8}$$

Here $\mathcal{Z}_\beta$ is a normalizing constant, $dx = d(q, p) = dq \times dp$ denotes the Lebesgue measure on $\mathbb{R}^{3k+3k}$, and $x = (q, p)$ denotes a given point in the phase space. We also consider the q-marginal measures

$$\overline{\mu}_\beta(dq) = \frac{1}{\overline{\mathcal{Z}}_\beta} \, e^{-\beta V(q)} \, dq \tag{27.9}$$

where $\mathcal{Z}_\beta$ is a normalizing constant, and dq is the Lebesgue measure on the position space $\mathbb{R}^{3k}$. In this notation, the measure μ_β is given by the product formula

$$\mu_\beta(d(q,p)) = \left[\prod_{1 \le i \le k} \frac{1}{\sqrt{2\pi m_i/\beta}} \, e^{-\beta \frac{p_i^2}{2m_i}} \, dp_i \right] \overline{\mu}_\beta(dq).$$

The Boltzmann-Gibbs measures μ_β and $\overline{\mu}_\beta$ can be interpreted as the invariant measures of the Langevin type stochastic dynamics

$$\begin{cases} dq_i = \beta \overbrace{\frac{\partial H}{\partial p_i}(q,p)}^{p_i/m_i} \, dt \\[2mm] dp_i = -\beta \underbrace{\left[\frac{\partial H}{\partial q_i}(q,p) + \sigma^2 \frac{\partial H}{\partial p_i}(q,p) \right]}_{=\frac{\partial V}{\partial q_i}(q) + \sigma^2 \, p_i/m_i} \, dt + \sigma \sqrt{2} \, dW_t^i \end{cases} \tag{27.10}$$

and respectively of

$$dq_i = -\beta \frac{\partial V}{\partial q_i}(q) \, dt + \sqrt{2} \, dW_t^i. \tag{27.11}$$

Here $(W_t^i)_{1 \le i \le k}$ represent k independent standard Brownian motions $W_t^i = \left(W_t^{i,j} \right)_{1 \le j \le 3}$ on $\mathbb{R}^3$.

The additional external Brownian forces represent the fluctuations of the many-body system, balanced by dissipative and viscous damping forces. In both cases, using the arguments presented in section 18.4.2 we can show that the Markov evolution semigroups of these diffusions have a density w.r.t. the Lebesgue measure on $\mathbb{R}^{3k}$ or on $\mathbb{R}^{3k+3k}$. The fact that these densities are smooth relies on more sophisticated stochastic analysis tools, including Malliavin calculus and differential geometry [57, 143, 161].

We check that μ_β, and $\overline{\mu}_\beta$ are the invariant measures of these diffusion models using the infinitesimal generators of the diffusion processes (27.10) and (27.11), given respectively on the set of smooth function f on $\mathbb{R}^{3k+3k}$ by the formulae

$$L_\beta(f) = \beta \sum_{i=1}^k \left[\frac{\partial H}{\partial p_i} \frac{\partial f}{\partial q_i} - \left(\frac{\partial H}{\partial q_i} + \sigma^2 \frac{\partial H}{\partial p_i} \right) \frac{\partial f}{\partial p_i} \right] + \sigma^2 \sum_{i=1}^k \frac{\partial^2 f}{\partial p_i^2}$$

and for any smooth function g on $\mathbb{R}^{3k}$ by

$$\overline{L}_\beta(g) = -\beta \sum_{i=1}^k \frac{\partial V}{\partial q_i} \frac{\partial g}{\partial q_i} + \sum_{i=1}^k \frac{\partial^2 g}{\partial q_i^2} = e^{\beta V} \sum_{i=1}^k \frac{\partial}{\partial q_i} \left(e^{-\beta V} \frac{\partial g}{\partial q_i} \right).$$

In the above display, we slightly abuse the notation by dropping the transposition operator $(.)'$ in the differential of the functions f and g. For instance, by using the conventions (27.5) and (27.6),

$$\frac{\partial H}{\partial p_i} \frac{\partial f}{\partial q_i} := \frac{\partial H}{\partial p_i} \left(\frac{\partial f}{\partial q_i} \right)' = \frac{\partial H}{\partial p_i^1} \frac{\partial f}{\partial q_i^1} + \frac{\partial H}{\partial p_i^2} \frac{\partial f}{\partial q_i^2} + \frac{\partial H}{\partial p_i^3} \frac{\partial f}{\partial q_i^3}.$$

To simplify the presentation, we denoted by $\frac{\partial^2}{\partial p_i^2}$ the Laplacian operator on $\mathbb{R}^3$; that is,

$$\frac{\partial^2 f}{\partial p_i^2} = \frac{\partial^2 f}{\partial (p_i^1)^2} + \frac{\partial^2 f}{\partial (p_i^2)^2} + \frac{\partial^2 f}{\partial (p_i^3)^2}$$

and

$$\frac{\partial}{\partial q_i}\left(e^{-\beta V}\frac{\partial g}{\partial q_i}\right) := \frac{\partial}{\partial q_i^1}\left(e^{-\beta V}\frac{\partial g}{\partial q_i^1}\right) + \frac{\partial}{\partial q_i^2}\left(e^{-\beta V}\frac{\partial g}{\partial q_i^2}\right) + \frac{\partial}{\partial q_i^3}\left(e^{-\beta V}\frac{\partial g}{\partial q_i^3}\right).$$

In this stochastic framework, the conservation properties (27.4) take the following form.

Lemma 27.1.1 *For any $\beta \in \mathbb{R}$, we have*

$$\mu_\beta L_\beta = 0 \quad \text{and} \quad \overline{\mu}_\beta \overline{L}_\beta = 0.$$

In addition $\overline{\mu}_\beta$ is $\overline{L}_\beta$-reversible, in the sense that for any smooth couple of functions (g, h) with compact support on $\mathbb{R}^{3k}$ we have

$$\overline{\mu}_\beta\left(g\,\overline{L}_\beta(h)\right) = \overline{\mu}_\beta\left(\overline{L}_\beta(g)\,h\right).$$

Proof :
By a simple integration by parts formula, for any smooth function f with compact support on $\mathbb{R}^{3k+3k}$ we check that

$$\int e^{-\beta H(x)}\,L_\beta(f)(x)\,dx$$

$$= -\beta\sum_{i=1}^k\,\int f(x)\,\frac{\partial}{\partial q_i}\left(e^{-\beta H}\,\frac{\partial H}{\partial p_i}\right)(x)\,dx$$

$$+\beta\sum_{i=1}^k\,\int f(x)\,\frac{\partial}{\partial p_i}\left(e^{-\beta H}\,\left(\frac{\partial H}{\partial q_i}+\sigma^2\frac{\partial H}{\partial p_i}\right)\right)(x)\,dx$$

$$+\sigma^2\sum_{i=1}^k\int f(x)\frac{\partial^2}{\partial p_i^2}\left(e^{-\beta H}\right)(x)\,dx.$$

This implies that

$$\mu_\beta\left(L_\beta(f)\right)$$

$$= \sum_{i=1}^k\,\mu_\beta\left\{f\,\left[\left(\beta^2\frac{\partial H}{\partial p_i}\frac{\partial H}{\partial q_i}-\beta\frac{\partial^2 H}{\partial q_i\partial p_i}\right)-\beta^2\frac{\partial H}{\partial p_i}\left(\frac{\partial H}{\partial q_i}+\sigma^2\frac{\partial H}{\partial p_i}\right)\right]\right\}$$

$$+\sum_{i=1}^k\,\mu_\beta\left\{f\,\left[\beta\left(\frac{\partial^2 H}{\partial q_i\partial p_i}+\sigma^2\frac{\partial^2 H}{\partial p_i^2}\right)-\sigma^2\beta\frac{\partial^2 H}{\partial p_i^2}+\sigma^2\beta^2\left(\frac{\partial H}{\partial p_i}\right)^2\right]\right\} = 0.$$

In much the same way, for any smooth functions (g, h) with compact support on $\mathbb{R}^{3k}$, we find that

$$\int e^{-\beta V(q)}g(q)\,\overline{L}_\beta(h)(q)\,dq = \sum_{i=1}^k\int g(q)\,\frac{\partial}{\partial q_i}\left(e^{-\beta V}\,\frac{\partial h}{\partial q_i}\right)(q)\,dq$$

$$= -\sum_{i=1}^k\int e^{-\beta V(q)}\,\frac{\partial g}{\partial q_i}(q)\frac{\partial h}{\partial q_i}(q)\,dq$$

$$= \sum_{i=1}^k\int h(q)\,\frac{\partial}{\partial q_i}\left(e^{-\beta V}\,\frac{\partial g}{\partial q_i}\right)(q)\,dq.$$

This clearly ends the proof of the lemma. ∎

Important remark : The stability properties of the Langevin models (27.10) and (27.11) can be analyzed using the tools developed in section 17.2, section 17.5, and section 18.4.2. More precisely, we first check that the semigroup P_t of these diffusion models has a smooth density w.r.t. the Lebesgue measure. This will ensure that P_t satisfies the Dobrushin local contraction condition (8.27) for any $t > 0$ (notice that (27.10) has the same form as the diffusion model (18.18)). The second step is to find a judicious Lyapunov function satisfying the condition (17.8). By theorem 17.4.1, these two properties imply that the laws of the random states of the Langevin models (27.10) and (27.11) converge exponentially fast, as the time parameter tends to infinity, to the invariant measures (27.8) and (27.9). Nevertheless, to the best of our knowledge, the Lyapunov functions developed in the literature on Langevin diffusions require that the potential functions behave as polynomials at infinity. These techniques cannot be used to analyze the Lennard-Jones potential functions presented in (27.7). The only work in this direction seems to be the article by B. Cooke, J.C. Mattingly, S.A. McKinley, and S.C. Schmidler [57] on a reduced two-dimensional Langevin diffusion model.

27.2 Schrödinger equation

27.2.1 A physical derivation

The Schrödinger equation is the quantum mechanics version of the Newton's second law of motion of classical mechanics (the mass times the acceleration equals the sum of the forces). This equation represents the wave function (a.k.a. the quantum state) evolution of a physical system, including molecular, atomic, subatomic, macroscopic systems like the universe [238].

The following physical derivation of the Schrödinger equation is taken from the lecture notes of James Cresser from the Department of Physics and Astronomy of Macquarie University, Sydney. In 1924 de Broglie made the hypothesis that if light waves of frequency ω behave as a population of particles of energy $E = \hbar\omega$, massive particles with energy E can also behave like waves of frequency $\omega = E/\hbar$. Here $\hbar$ denotes the Planck constant. More precisely, the wave function of a free particle of momentum $p = \hbar k$ and energy

$$E = \frac{p^2}{2m} = \hbar\omega \ \Rightarrow E = \frac{k^2\hbar^2}{2m} = \frac{p^2}{2m}$$

has the form

$$\psi(t,x) = \psi_0 \ e^{i(kx - \omega t)}.$$

This wave function is the result of two waves traveling in the x and t directions.

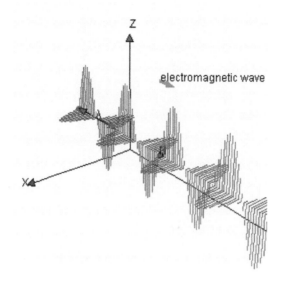

An elementary computation shows that

$$\partial_x \psi = ik\ \psi \Rightarrow -\frac{\hbar^2}{2m}\ \partial_x^2 \psi = k^2\ \frac{\hbar^2}{2m}\ \psi = \frac{p^2}{2m}\ \psi = E\ \psi$$

and

$$i\hbar\partial_t \psi = \hbar\omega\ \psi = E\ \psi$$

from which we conclude that

$$-\frac{\hbar^2}{2m}\partial_x^2 \psi = E\ \psi = i\hbar\partial_t \psi$$

and

$$i\hbar\ \partial_t \psi = -\frac{\hbar^2}{2m}\partial_x^2 \psi.$$

Extending these wave functions to particle motions in a potential energy $V(x)$, the energy E is the sum of the kinetic and the potential energies

$$E = \frac{p^2}{2m} + V(x).$$

Assuming that the above equations are valid in this case, we obtain the time dependent Schrödinger wave equation

$$i\hbar\partial_t \psi = E\ \psi = \frac{p^2}{2m}\psi + V\psi = -\frac{\hbar^2}{2m}\ \partial_x^2 \psi +\ V\psi.$$

Inversely, the solutions of the form $\psi(t,x) = \alpha(t)\ \psi_0(x)$ of the time dependent Schrödinger wave equation

$$i\hbar\partial_t \psi = -\frac{\hbar^2}{2m}\ \partial_x^2 \psi +\ V\psi$$

satisfy

$$i\hbar\ \partial_t \log \alpha(t) = \frac{1}{\psi_0(x)}\left[-\frac{\hbar^2}{2m}\ \partial_x^2\ \psi_0(x) +\ V\psi_0(x)\right].$$

Since the r.h.s. and l.h.s. do not depend on the parameters t and x, there is a constant E such that

$$\left. \begin{array}{rcl} i\hbar\,\partial_t\alpha(t) &=& E\,\alpha(t) \\ \frac{\hbar^2}{2m}\,\partial_x^2\,\psi_0(x) - V(x)\psi_0(x) &=& -E \end{array} \right\} \Leftrightarrow \left\{ \begin{array}{rcl} \alpha(t) &=& \exp\left(-iEt/\hbar\right)\alpha(0) \\ \partial_x^2\psi_0(x) &=& \frac{2m}{\hbar^2}\left[V(x) - E\right]\psi_0(x). \end{array} \right.$$

The second formula in the above equation is called the stationary or time independent Schrödinger equation.

27.2.2 Feynman-Kac formulation

Rewritten in a slightly different form, the Schrödinger wave equation takes the form

$$i\hbar\partial_t\psi = -L^V(\psi)$$

with the Schrödinger operator

$$L^V = \frac{\hbar^2}{2m}\,\partial_x^2 - V.$$

A formal change of time coordinate $t = i\tau/\hbar$ and $u(\tau, x) = \psi(i\tau/\hbar, x)$ transforms the above into a heat type equation

$$\partial_\tau u = L^V(u). \tag{27.12}$$

In physics, this change of coordinate is sometimes called a Wick rotation of the time axis, and the resulting equation is often referred to as the Schrödinger equation in imaginary time.

The equation (27.12) is sometimes written in terms of the Hamiltonian operator $\mathcal{H} = -L^V$, that is,

$$\partial_\tau u = L^V(u) \Leftrightarrow \partial_\tau u = -\mathcal{H}(u). \tag{27.13}$$

Definition 27.2.1 *We consider a time homogeneous stochastic process X_τ on a state space S with infinitesimal generator L acting on some domain of functions $D(L)$. We denote by Q_τ the integral operator defined for any bounded function f on S by the formula*

$$Q_\tau(f)(x) := \mathbb{E}\left(f(X_\tau)\,e^{-\int_0^\tau V(X_s)ds} \mid X_0 = x\right). \tag{27.14}$$

In the further development of this section we implicitly assume that $Q_\tau(D(L))$ and $L(D(L))$ are subsets of $D(L)$. This condition depends on the regularity property of the generator L. For jump type infinitesimal generators, this condition holds for any bounded potential function with $D(L) = \mathcal{B}(S)$. For diffusion type infinitesimal generators on $S = \mathbb{R}^d$, this condition holds for twice differentiable functions $D(L) = C_b^2(S)$ and for bounded smooth potential functions.

Proposition 27.2.2 *We have the semigroup (sg) property*

$$\forall s, t \geq 0 \qquad Q_{s+t} = Q_s Q_t \quad and \quad Q_0 = Id. \tag{27.15}$$

In addition, the following evolution is satisfied for any $f \in D(L)$

$$\partial_\tau Q_\tau(f) := Q_\tau(L^V(f)) = L^V(Q_\tau(f)). \tag{27.16}$$

In particular, the function $u(\tau, x) := Q_\tau(f)(x)$ satisfies the equation

$$\partial_\tau u = L^V(u) = -\mathcal{H}(u) \quad with \quad L^V = L - V = -\mathcal{H}. \tag{27.17}$$

The operator Q_t is sometimes termed a Feynman-Kac propagator and it is often written in the exponential form $Q_t = e^{-t\mathcal{H}}$.

Proof :

For any $s \leq \tau$, using the Markov property, we prove that

$$Q_\tau(f)(x) = \mathbb{E}\left(\underbrace{\mathbb{E}\left(f(X_\tau)\, e^{-\int_s^\tau V(X_r)dr} \mid X_s \right)}_{=Q_{\tau-s}(f)(X_s)} e^{-\int_0^s V(X_r)dr} \mid X_0 = x \right).$$

This yields the sg property (27.15). Now we turn to the proof of (27.16). We use the decomposition

$$Q_{\tau+d\tau}(f)(x) - Q_\tau(f)(x)$$

$$= \mathbb{E}\left(f(X_{\tau+d\tau}) \left(e^{-\int_0^{\tau+d\tau} V(X_s)ds} - e^{-\int_0^\tau V(X_s)ds} \right) \mid X_0 = x \right)$$

$$+ \mathbb{E}\left((f(X_{\tau+d\tau}) - f(X_\tau))\, e^{-\int_0^\tau V(X_s)ds} \mid X_0 = x \right).$$

The first term on the r.h.s. is given by

$$\mathbb{E}\left(f(X_{\tau+d\tau}) \left(e^{-\int_0^{\tau+d\tau} V(X_s)ds} - e^{-\int_0^\tau V(X_s)ds} \right) \mid X_0 = x \right)$$

$$= \mathbb{E}\left(f(X_{\tau+d\tau})\, e^{-\int_0^\tau V(X_s)ds} \left(e^{-\int_\tau^{\tau+d\tau} V(X_s)ds} - 1 \right) \mid X_0 = x \right)$$

$$\simeq \mathbb{E}\left((-V)(X_\tau)\, f(X_\tau)\, e^{-\int_0^\tau V(X_s)ds} \mid X_0 = x \right)\, d\tau = Q_\tau((-V)f)\, d\tau$$

and the second one is given by

$$\mathbb{E}\left((f(X_{\tau+d\tau}) - f(X_\tau))\, e^{-\int_0^\tau V(X_s)ds} \right)$$

$$= \mathbb{E}\left((\mathbb{E}\left(f(X_{\tau+d\tau}) \mid X_\tau \right) - f(X_\tau))\, e^{-\int_0^\tau V(X_s)ds} \right)$$

$$= \mathbb{E}\left(L(f)(X_\tau)\, e^{-\int_0^\tau V(X_s)ds} \right)\, d\tau = Q_\tau(L(f))\, d\tau.$$

This ends the proof of the first assertion. The r.h.s. of formula (27.16) follows from the fact that

$$Q_{\tau+d\tau} = Q_{d\tau} Q_\tau \implies Q_{\tau+d\tau} - Q_\tau = \underbrace{[Q_{d\tau} - Id]}_{\simeq L^V\, d\tau} Q_\tau.$$

This ends the proof of the proposition. ∎

We end this section with a series of important comments related to the Feynman-Kac models discussed above.

- The integral operators Q_τ can be made more explicit by using the following formulae

$$\mathbb{E}\left(f(X_\tau)\, e^{-\int_0^\tau V(X_s)ds} \mid X_0 = x \right)$$

$$= \mathbb{E}\left(\mathbb{E}\left(e^{-\int_0^\tau V(X_s)ds} \mid X_0, X_\tau \right) f(X_\tau) \mid X_0 = x \right)$$

$$= \int \underbrace{\mathbb{E}\left(e^{-\int_0^\tau V(X_s)ds} \mid X_0 = x,\ X_\tau = y \right) \mathbb{P}\left(X_\tau \in dy \mid X_0 = x \right)}_{:=Q_\tau(x,dy)} f(y).$$

- Using the sg property (27.15), the function $u(t,x) := Q_t(f)(x)$ satisfies the transport equation

$$u(s+t,x) := \int Q_s(x,dy)\, u(t,y).$$

In physics, the Feynman-Kac sg is also called the Green function or the Feynman-Kac propagator.

- The extension to time inhomogeneous model is given for any $s \leq t$ by the formula

$$Q_{s,t}(f)(x) = \mathbb{E}\left(f(X_t)\, e^{-\int_s^t V(X_r)dr} \mid X_s = x \right).$$

For time homogeneous models, we have $Q_{s,t} = Q_{0,t-s} := Q_{t-s}$ with the integral operator Q_τ defined in (27.14).

- The path space version of these models is given by the Boltzmann-Gibbs measures

$$d\mathbb{Q}_t := \frac{1}{\mathcal{Z}_t}\, e^{-\int_0^t V(X_s)ds}\, d\mathbb{P}_t \tag{27.18}$$

where $d\mathbb{P}_t$ denotes the distribution of the paths of the reference Markov process $(X_s)_{0 \leq s \leq t}$. When $X_0 = x$, for any function F_t of the paths $(X_s)_{0 \leq s \leq t}$ we have

$$
\begin{aligned}
\mathbb{Q}_t(F_t) &= \int F_t\left((x_s)_{0 \leq s \leq t}\right)\, \mathbb{Q}_t\left(d(x_s)_{0 \leq s \leq t}\right) \\
&\propto \int F_t\left((x_s)_{0 \leq s \leq t}\right)\, e^{-\int_0^t V(x_s)ds}\, \mathbb{P}_t\left(d(x_s)_{0 \leq s \leq t}\right).
\end{aligned}
$$

We end this section with a brief discussion on the description of the Hamiltonian operator (27.13) associated with a molecule in quantum physics. In this context, a state $x = ((x_{a,i})_{1 \leq i \leq N_a}, (x_{e,j})_{1 \leq j \leq N_e})$ represents the locations x_a^i of N_a atom nuclei, and the locations x_e^j of N_e electrons (we assume that each atom has the same number of electrons) w.r.t. a Cartesian reference frame. The (exact non-relativistic, time-independent molecular) Hamiltonian (27.13) is now given by

$$
\mathcal{H} = -L + \hbar^{-1}V \quad \text{with} \quad L := \underbrace{\frac{\hbar}{2} \sum_{1 \leq i \leq N_a} \frac{1}{m_{a,j}} \partial^2_{x_{a,i}}}_{:= L^{(a)} \text{ nuclear kinetic energy}} + \underbrace{\frac{\hbar}{2} \sum_{1 \leq i \leq N_e} \frac{1}{m_{e,j}} \partial^2_{x_{e,i}}}_{:= L^{(e)} \text{ electronic kinetic energy}}
$$

where $m_{a,j}$ stands for the mass of the j-th nuclei, $m_{e,i}$ stands for the mass of the i-th electron, and $\hbar$ is the Planck constant. The potential function is defined in terms of repulsive or attractive Coulomb forces

$$
V(x) := \underbrace{\frac{1}{2} \sum_{1 \leq i < j \leq N_a} \frac{z_{a,i} z_{a,j}}{\|x_{a,i} - x_{a,j}\|}}_{\text{nuclear repulsion}} + \underbrace{\sum_{1 \leq i < j \leq N_e} \frac{e^2}{\|x_{e,i} - x_{e,j}\|}}_{\text{electronic repulsion}} - \underbrace{\frac{1}{2} \sum_{1 \leq i \leq N_a} \sum_{1 \leq j \leq N_e} \frac{z_{a,i} e^2}{\|x_{a,i} - x_{e,j}\|}}_{\text{electron-nuclear attraction}}
$$

for some non-negative atomic numbers $z_{a,i}$. The nuclei are much heavier than electrons (for instance, the proton mass ($1.67 \ 10^{-27}$ kg) is 1800 times larger than the mass of the electron ($9.31 \ 10^{-31}$ kg)); in the Born-Oppenheimer approximation [28] the nuclei $(x_{a,i})_{1 \leq i \leq N_a}$ are fixed parameters, and we reduce the problem to the electronic configuration $x = ((x_{e,j})_{1 \leq j \leq N_e})$ associated with the Hamiltonian operator $\mathcal{H} = -L^{(e)} + V$. In physics, the Schrödinger (imaginary time) equation is often written as

$$\hbar\, \partial_t u_t(x) = \hbar\, L(u_t)(x) - V(x)u_t(x).$$

In this situation, the Hamiltonian operator is defined as above by replacing $\mathcal{H}$ by $\hbar\mathcal{H} = -\hbar L + V$.

27.2.3 Bra-kets and path integral formalism

In theoretical and computational physics, the state space S is generally the Euclidian space $S = \mathbb{R}^d$. We further assume that

$$\eta_0(dx) = \mathbb{P}(X_0 \in dx) = \mu_f(dx) := f(x)\, dx$$

for density function f on $\mathbb{R}^d$. We consider the Feynman-Kac measures

$$\gamma_t(\varphi) = \eta_0(Q_t(\varphi)) = \mathbb{E}\left[\varphi(X_t) \exp\left(-\int_0^t V(X_s) ds\right)\right] \quad \text{with} \quad \eta_0 = \text{Law}(X_0)$$

and the left and right actions of the integral operator Q_t defined on measures and functions

$$\mu \mapsto \mu Q_t \quad \text{and} \quad g \mapsto Q_t(g).$$

When $\mathbb{P}(X_t \in dy \mid X_0 = x)$ has a density $p_t(x, y)$ w.r.t. the Lebesgue measure dy, we have

$$Q_t(x, dy) = q_t(x, y) dy \quad \text{with} \quad q_t(x, y) = \mathbb{E}\left(\exp\left\{-\int_0^t V(X_s) ds\right\} \mid X_t = y\right) p_t(x, y).$$
$$(27.19)$$

In computational and theoretical physics, these functional operations are often written in terms of bra-kets

$$\eta_0(dx) = \mathbb{P}(X_0 \in dx) = \mu_f(dx) := f(x)\, dx \Rightarrow \gamma_t = \eta_0 Q_t = \mu_f Q_t =\prec f|e^{-t\mathcal{H}}$$

as well as

$$Q_t(g) = e^{-t\mathcal{H}} |g \succ \quad \text{so that} \quad \gamma_t(g) = \eta_0 Q_t(g) = \mu_f Q_t(g) =\prec f|e^{-t\mathcal{H}}|g \succ .$$

As in (27.19), we further assume that the semigroup $Q_t(x, dy)$ of the Hamiltonian operator

$$\mathcal{H} = -L^V = -L + V$$

has a density $q_t(x, y)$ w.r.t. the Lebesgue measure dy. In this context, the density $q_t(x, y)$ is written as

$$q_t(x, y) =\prec x|e^{-t\mathcal{H}}|y \succ \quad \text{or} \quad q_t(x, y) =\prec \delta_x|e^{-t\mathcal{H}}|\delta_y \succ$$

so that

$$\int dx\, f(x)\, q_t(x, y)\, g(y)\, dy = \int \prec x|e^{-t\mathcal{H}}|y \succ f(x)\, g(y)\, dxdy.$$

Representing functions formally on the basis of delta functions

$$"f(.) = \int f(x)\, \delta_x(.)\, dx" \quad \text{in the sense that} \quad \forall y \in \mathbb{R}^d \quad " \int f(x)\, \underbrace{\delta_x(y)}_{=1_{x=y}}\, dx = f(y)"$$

and using the linearity of the brackets, we arrive at the formal expression

$$\prec f|e^{-t\mathcal{H}}|g \succ \ = \ \prec \left(\int f(x)\, \delta_x(.)\, dx\right) |e^{-t\mathcal{H}}| \left(\int g(y)\, \delta_y(.)\, dy\right) \succ$$

$$= \int f(x)\, g(y) \prec \delta_x|e^{-t\mathcal{H}}|\delta_y \succ dxdy = \int \prec x|e^{-t\mathcal{H}}|y \succ f(x)\, g(y)\, dxdy.$$

In this notation, we have

$$\prec f_1 | e^{-t\mathcal{H}} | f_2 \succ \; = \; \int f_1(x) \; \prec x | e^{-t\mathcal{H}} | y \succ \; f_2(y) \; dx dy$$

$$= \int (\mu_{f_1} Q_t)(dy) \, f_2(y) = \int \mu_{f_1}(dx) \, Q_t(f_2)(x) = \mu_{f_1} Q_t(f_2).$$

Dividing $[0, n\Delta t] = [0, t_n]$ into n intervals $([0, \Delta t] \cup \ldots \cup [(n-1)\Delta t, n\Delta t])$ of length Δt, the semigroup property $Q_{t+s} = Q_t Q_s$ implies that

$$Q_{n\Delta t}(x_0, dx_n) \; = \; \overbrace{(Q_{\Delta t} \ldots Q_{\Delta t})}^{n \text{ times}}(x_0, dx_n)$$

$$= \left[\int q_{\Delta t}(x_0, x_1) \ldots q_{\Delta t}(x_{n-1}, x_n) \, dx_1 \ldots dx_{n-1} \right] dx_n$$

$$= \left[\int \left\{ \prod_{0 \le k < n} \prec x_k | e^{-\Delta t \, \mathcal{H}} | x_{k+1} \succ \right\} dx_1 \ldots dx_{n-1} \right] dx_n.$$

Whenever $X_t = \sqrt{\frac{\hbar}{m}} \, W_t$ with some Brownian motion W_t and some parameter $\hbar, m > 0$, replacing V by $\frac{1}{\hbar} V$ in (27.19) we have

$$\prec x_k | e^{-\Delta t \, \mathcal{H}} | x_{k+1} \succ \; \simeq_{\Delta t \downarrow 0} \; e^{-\frac{V(x_k)}{\hbar} \, \Delta t} \, p_{\Delta t}(x_k, x_{k+1}) = \sqrt{\frac{m}{2\pi\hbar\Delta t}} \, e^{-\frac{1}{\hbar} \left[\frac{m}{2} \left(\frac{x_{k+1} - x_k}{\Delta t} \right)^2 + V(x_k) \right] \Delta t}.$$

In this situation, the discrete time approximation of the integral operator Q_t is given by the formula

$$Q_{n\Delta t}(x_0, dx_n) \simeq_{\Delta t \downarrow 0} \left[\int \left(\sqrt{\frac{m}{2\pi\hbar\Delta t}} \right)^n e^{-\frac{1}{\hbar} S_n(x_0, \ldots, x_n)\Delta t} \, dx_1 \ldots dx_{n-1} \right] dx_n$$

with the so-called Euclidian action functional

$$S_n(x_0, \ldots, x_n) = \sum_{0 \le k < n} \left[\frac{m}{2} \left(\frac{x_{k+1} - x_k}{\Delta t} \right)^2 + V(x_k) \right].$$

Taking formally the limit $\Delta t \downarrow 0$, the density $q_t(x, y)$ is often written in the physics literature as a path integral

$$q_t(x, y) = \int_{x_0 = x}^{x_t = y} e^{-\frac{1}{\hbar} S_t(x)} \, \mathcal{D}x \quad \text{with} \quad S_t(x) = \int_0^t \left\{ \frac{m}{2} \, \dot{x}_s^2 + V(x_s) \right\} \, ds.$$

27.2.4   Spectral decompositions

We further assume that L is a self adjoint operator on $\mathbb{L}_2(\mathbb{R}^d)$ (equipped with the scalar product $\langle f, g \rangle = \int f(x)g(x)dx$), defined on some proper domain of functions $D(L)$, that is,

$$\langle f, L(g) \rangle = \langle L(f), g \rangle$$

for any $f, g \in D(L)$. In this situation, the Schrödinger operator $L^V = L - V$ is again a self adjoint operator on $\mathbb{L}_2(\mathbb{R}^d)$ (under appropriate regularity conditions on V, for instance by assuming that V is a bounded function). An important consequence of the self-adjoint property is that there exists a sequence of non-negative eigenvalues $0 \leq E_0 \leq E_1 \leq \ldots$ and a corresponding set of orthonormal eigenfunctions φ_i, $i \geq 0$, such that the integral operator Q_t has the spectral representation

$$Q_t(x, dy) = \sum_{i \geq 0} e^{-tE_i} \, \varphi_i(x) \varphi_i(y) \, dy. \qquad (27.20)$$

Therefore, by expanding the initial condition in the basis functions φ_i

$$f = \sum_{i \geq 0} \langle f, \varphi_i \rangle \, \varphi_i(x)$$

we find that

$$
\begin{aligned}
Q_t(f)(x) &= \int Q_t(x, dy) \, f(y) \\
&= \sum_{i,j \geq 0} e^{-tE_i} \, \langle f, \varphi_j \rangle \, \varphi_i(x) \underbrace{\langle \varphi_i, \varphi_j \rangle}_{=1_{i=j}} = \sum_{i \geq 0} e^{-tE_i} \, \langle f, \varphi_i \rangle \, \varphi_i(x).
\end{aligned}
$$

By (27.16), we also have

$$
\begin{aligned}
\frac{\partial}{\partial t} Q_t(f) &= -\sum_{i \geq 0} E_i \, e^{-tE_i} \, \langle f, \varphi_i \rangle \, \varphi_i \\
&= \sum_{i \geq 0} e^{-tE_i} \, \langle f, \varphi_i \rangle \, L^V(\varphi_i) = L^V(Q_t(f)).
\end{aligned}
$$

Choosing $f = \varphi_i$, we conclude that for any $i \geq 0$

$$\mathcal{H}(\varphi_i) = E_i \, \varphi \stackrel{\mathcal{H}:=-L^V}{\Longleftrightarrow} L^V(\varphi_i) = -E_i \, \varphi_i.$$

and therefore

$$\langle \varphi_i, L^V(\varphi_i) \rangle = -E_i \Longleftrightarrow \langle \varphi_i, \mathcal{H}(\varphi_i) \rangle = E_i.$$

For simplicity, we further assume that $E_0 < E_1$. In this case we have

$$Q_t(f)(x) \simeq_{t\uparrow\infty} e^{-tE_0} \, \langle f, \varphi_0 \rangle \, \varphi_0(x).$$

This implies that for any starting point x we have

$$-\frac{1}{t} \log Q_t(1)(x) \longrightarrow_{t\uparrow\infty} E_0 \quad \text{and} \quad \frac{Q_t(f)(x)}{Q_t(1)(x)} \simeq_{t\uparrow\infty} \frac{\langle f, \varphi_0 \rangle}{\langle 1, \varphi_0 \rangle}. \qquad (27.21)$$

For a more detailed discussion on the rate of convergence, we refer the reader to exercises 443 and 444.

Similar spectral decompositions can be derived when we replace $\mathbb{L}_2(\mathbb{R}^d)$ by some separable Hilbert space $\mathbb{L}_2(\mu)$ equipped with the scalar product $\langle f, g \rangle = \int f(x)g(x)\mu(dx)$, where μ stands for some measure on some state space S. In this situation L is the infinitesimal generator of some process X_t evolving in S and V is a potential function on S. Under some appropriate regularity conditions (compact and Hilbert-Schmidt) the operator $(-L^V)$ has a set of eigenvalues E_i and $\mathbb{L}_2(\mu)$ is equipped with an orthonormal basis of functions φ_i. The eigenvalues E_i may be complex but the Perron-Frobenious theorem (for finite spaces) or the Krein-Rutman theorem (for infinite dimensional spaces) ensure that $0 \leq E_0 \leq E_i$.

27.2.5 The harmonic oscillator

The harmonic oscillator is the prototype of physical model for which we can compute explicitly the spectrum of the operator $\mathcal{H} := -L^V$ acting on smooth functions of $\mathbb{L}_2(\mathbb{R}^p)$.

The harmonic oscillator is defined by choosing

$$V(x) = k\,x^2/2 \quad L := \frac{\hbar^2}{2m}\frac{\partial^2}{\partial x} \Rightarrow \begin{aligned} L^V &= \frac{\hbar^2}{2m}\frac{\partial^2}{\partial x} - k\,x^2/2 \\ &= \frac{\hbar^2}{2m}\frac{\partial^2}{\partial x} - \frac{1}{2}\,m\omega^2\,x^2 \quad \text{with} \quad \omega = \sqrt{\frac{k}{m}}, \end{aligned}$$

for some non-negative k.

We let φ_n be the eigenfunctions of L^V associated with the respective eigenvalues

$$E_n = \hbar\left(n + \frac{1}{2}\right)\omega \quad \Longleftrightarrow \quad \frac{2mE_n}{\hbar^2}\left(\frac{\hbar^2}{mk}\right)^{1/2} = \frac{2E_n}{\hbar\omega} = (2n+1)$$

$$L^V(\varphi_n) = -E_n\varphi_n \Longleftrightarrow \varphi_n''(x) = \left(\frac{mk}{\hbar^2}x^2 - \frac{2mE_n}{\hbar^2}\right)\varphi_n(x).$$

Notice that

$$\left(\frac{mk}{\hbar^2}\right)^{1/4} = \frac{1}{\sqrt{\hbar}}\left(m\left(\frac{k}{m}\right)^{1/2}\right)^{1/2} = \sqrt{\frac{m\omega}{\hbar}}.$$

We set

$$\psi_n(x) := \varphi_n\left(\left(\frac{mk}{\hbar^2}\right)^{-1/4}x\right) \quad \Longleftrightarrow \quad \varphi_n(x) = \psi_n\left(\sqrt{\frac{m\omega}{\hbar}}x\right).$$

We have

$$\begin{aligned} \psi_n'(x) &= \left(\frac{mk}{\hbar^2}\right)^{-1/4}\varphi_n'\left(\left(\frac{mk}{\hbar^2}\right)^{-1/4}x\right) \\ \psi_n''(x) &= \left(\frac{mk}{\hbar^2}\right)^{-1/2}\varphi_n''\left(\left(\frac{mk}{\hbar^2}\right)^{-1/4}x\right) \\ &= \left(\frac{\hbar^2}{mk}\right)^{1/2}\left(\frac{mk}{\hbar^2}\left(\left(\frac{mk}{\hbar^2}\right)^{-1/4}x\right)^2 - \frac{2mE_n}{\hbar^2}\right)\psi_n(x). \end{aligned}$$

This shows that

$$\psi_n''(x) = (x^2 - (2n + 1)) \, \psi_n(x).$$

Our next objective is to express these eigenfunctions in terms of the Hermite polynomials.

We recall that the Hermite polynomials can be defined using the Rodrigues formula

$$\mathbb{H}_n(x) = (-1)^n \, e^{x^2} \, \frac{d^n}{dx^n} e^{-x^2}. \tag{27.22}$$

Notice that

$$
\begin{aligned}
\mathbb{H}_n'(x) &:= \frac{d\mathbb{H}_n}{dx}(x) = 2x(-1)^n \, e^{x^2} \, \frac{d^n}{dx^n} e^{-x^2} + (-1)^n \, e^{x^2} \, \frac{d^{n+1}}{dx^{n+1}} e^{-x^2} \\
&= 2x\mathbb{H}_n(x) - \mathbb{H}_{n+1}(x) \Leftrightarrow \mathbb{H}_{n+1}(x) = 2x\mathbb{H}_n(x) - \mathbb{H}_n'(x). \tag{27.23}
\end{aligned}
$$

This formula shows that $\mathbb{H}_n$ is a polynomial of degree n with a leading coefficient 2^n so that $\frac{d^n \mathbb{H}_n}{dx^n}(x) = 2^n \, n!$. In addition, combining (27.22) with an integration by part, for any $m < n$ we find that

$$\int e^{-x^2} \mathbb{H}_m(x)\mathbb{H}_n(x) \, dx = (-1)^n \int \mathbb{H}_m(x) \, \frac{d^n}{dx^n} e^{-x^2} dx = \int \underbrace{\frac{d^n}{dx^n} \mathbb{H}_m(x)}_{=0} \, e^{-x^2} dx = 0$$

and for $m = n$

$$\int e^{-x^2} \mathbb{H}_n^2(x) \, dx = \int \underbrace{\frac{d^n}{dx^n} \mathbb{H}_n(x)}_{=2^n n!} \, e^{-x^2} dx = 2^n n! \int e^{-x^2} dx = 2^n n! \sqrt{\pi}.$$

Working a little more, one deduces that the set of functions

$$\widetilde{\mathbb{H}}_n(x) := (2^n n! \sqrt{\pi})^{-1/2} \, e^{-x^2/2} \mathbb{H}_n(x)$$

forms an orthonormal basis of $\mathbb{L}^2(\mathbb{R})$.

We recall the Leibniz formula

$$\frac{d^n}{dx^n}(fg) := (fg)^{(n)} = \sum_{0 \le k \le n} \binom{n}{k} \, f^{(k)} g^{(n-k)}.$$

This formula is proved by induction w.r.t. the parameter n as follows:

$$
\begin{aligned}
(fg)^{(n+1)} &= \sum_{0 \le k \le n} \binom{n}{k} f^{(k+1)} g^{((n+1)-(k+1))} + \sum_{0 \le l \le n} \binom{n}{l} f^{(l)} g^{(n+1-l)} \\
&= \sum_{1 \le l \le n} \underbrace{\left[\binom{n}{l-1} + \binom{n}{l} \right]}_{=\binom{n+1}{l}} f^{(l)} g^{(n+1-l)} + \binom{n}{n} f^{(n+1)} + \binom{n}{0} g^{(n+1)}.
\end{aligned}
$$

Applying this formula to $f(x) = -2x$ and $g(x) = e^{-x^2}$ we find that

$$
\begin{aligned}
\frac{d^{n+1}}{dx^{n+1}} e^{-x^2} &= \frac{d^n}{dx^n} (-2x e^{-x^2}) \\
&= \binom{n}{0} (-2x) \frac{d^n}{dx^n} e^{-x^2} + \binom{n}{1} (-2) \frac{d^{n-1}}{dx^{n-1}} e^{-x^2} \\
&= -2x \frac{d^n}{dx^n} e^{-x^2} - 2n \frac{d^{n-1}}{dx^{n-1}} e^{-x^2}
\end{aligned}
$$

and therefore

$$
\mathbb{H}_{n+1}(x) = 2x \mathbb{H}_n(x) - 2n \mathbb{H}_{n-1}(x). \tag{27.24}
$$

Combining (27.23) and (27.24), we have

$$
2x \mathbb{H}_n(x) - 2n \mathbb{H}_{n-1}(x) = 2x \mathbb{H}_n(x) - \mathbb{H}'_n(x) \Rightarrow \mathbb{H}'_n = 2n \mathbb{H}_{n-1} \Rightarrow \mathbb{H}''_n = 2n \mathbb{H}'_{n-1}
$$

and therefore

$$
\mathbb{H}'_n(x) = 2x \mathbb{H}_n(x) - \mathbb{H}_{n+1}(x) \Rightarrow \begin{aligned} \mathbb{H}''_n(x) &= 2\mathbb{H}_n(x) + 2x\mathbb{H}'_n(x) - \mathbb{H}'_{n+1}(x) \\ &= 2\mathbb{H}_n(x) + 2x\mathbb{H}'_n(x) - 2(n+1)\mathbb{H}_n(x) \\ &= 2x\mathbb{H}'_n(x) - 2n\mathbb{H}_n(x). \end{aligned}
$$
$$\tag{27.25}$$

We set
$$
\overline{\mathbb{H}}_n(x) := e^{-x^2/2} \mathbb{H}_n(x) \quad \Longleftrightarrow \quad \mathbb{H}_n(x) = e^{x^2/2} \overline{\mathbb{H}}_n(x).
$$

Using

$$
\begin{aligned}
\mathbb{H}'_n(x) &= e^{x^2/2} \left[x\overline{\mathbb{H}}_n(x) + \overline{\mathbb{H}}'_n(x) \right] \\
\mathbb{H}''_n(x) &= e^{x^2/2} \left(\left[\overline{\mathbb{H}}_n(x) + x\overline{\mathbb{H}}'_n(x) + \overline{\mathbb{H}}''_n(x) \right] + x\left[x\overline{\mathbb{H}}_n(x) + \overline{\mathbb{H}}'_n(x) \right] \right) \\
&= e^{x^2/2} \left[(1+x^2)\overline{\mathbb{H}}_n(x) + 2x\overline{\mathbb{H}}'_n(x) + \overline{\mathbb{H}}''_n(x) \right]
\end{aligned}
$$

we check that

$$
\begin{aligned}
(27.25) \Leftrightarrow 0 &= \mathbb{H}''_n(x) - 2x\mathbb{H}'_n(x) + 2n\mathbb{H}_n(x) \\
&= e^{x^2/2} \left[\left[(1+x^2)\overline{\mathbb{H}}_n(x) + 2x\overline{\mathbb{H}}'_n(x) + \overline{\mathbb{H}}''_n(x) \right] \right. \\
&\qquad\qquad \left. -2x\left[x\overline{\mathbb{H}}_n(x) + \overline{\mathbb{H}}'_n(x) \right] + 2n\overline{\mathbb{H}}_n(x) \right] \\
&= e^{x^2/2} \left[\overline{\mathbb{H}}''_n(x) + ((2n+1) - x^2)\overline{\mathbb{H}}_n(x) \right]
\end{aligned}
$$

and

$$\overline{\overline{\mathbb{H}}}_n''(x) = \left(x^2 - (2n+1)\right) \overline{\overline{\mathbb{H}}}_n(x).$$

This shows that $\psi_n(x) = \overline{\overline{\mathbb{H}}}_n(x)$ and therefore

$$\varphi_n(x) \propto \psi_n\left(\sqrt{\frac{m\omega}{\hbar}}x\right) = \overline{\overline{\mathbb{H}}}_n\left(\sqrt{\frac{m\omega}{\hbar}}x\right) = e^{-\frac{x^2}{2}\frac{m\omega}{\hbar}}\,\mathbb{H}_n\left(\sqrt{\frac{m\omega}{\hbar}}x\right).$$

Finally, we obtain the orthornormal basis of eigenfunctions by setting

$$\varphi_n(x) := \sqrt{\frac{1}{2^n n! \sqrt{\pi}}} \left(\frac{m\omega}{\hbar}\right)^{1/4} \exp\left[-\frac{x^2}{2}\frac{m\omega}{\hbar}\right]\mathbb{H}_n\left(\sqrt{\frac{m\omega}{\hbar}}x\right).$$

The numerical solving of the eigenvalues problem for general potential functions $V(x)$ can be done using the software `Maltalb Chebfun software` (based on one-dimensional dynamical system integrations). The figure below illustrates the first 10 eigenstates.

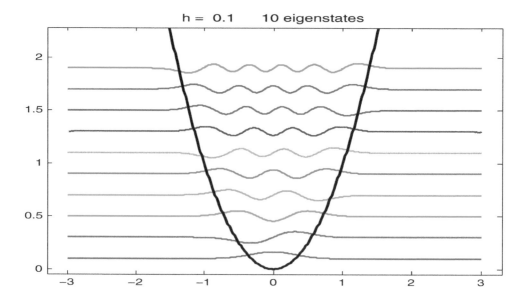

27.2.6 Diffusion Monte Carlo models

The discrete time version of the Feynman-Kac model (27.18) is given by the formula

$$
\begin{aligned}
d\mathbb{Q}_n &:= \frac{1}{\mathcal{Z}_n}\, e^{-\sum_{0\le p<n} V_p(X_p)}\, d\mathbb{P}_n \\
&= \frac{1}{\mathcal{Z}_n}\left\{\prod_{0\le p<n} G_p(X_p)\right\} d\mathbb{P}_n \quad \text{with} \quad G_p = e^{-V_p} \qquad (27.26)
\end{aligned}
$$

where $\mathbb{P}_n$ is the probability distribution of the paths $(X_0, X_1, \ldots, X_n)$ of a discrete time Markov chain model X_n, with $n \in \mathbb{N}$. The discrete time approximations of the continuous time models on a time mesh sequence t_n are defined by the Feynman-Kac formulae (27.26). These approximations are obtained by utilizing the approximation

$$\int_0^{t_n} V(X_s)ds = \sum_{0 \le p < n} \int_{t_p}^{t_{p+1}} V(X_s)ds \simeq \sum_{0 \le p < n} V(X_{t_p}) \, (t_{p+1} - t_p).$$

In this notation, for any function F_{t_n} of the discretized paths $(X_{t_p})_{0 \le p < n}$ we have

$$\mathbb{Z}_{t_n} \times \mathbb{Q}_{t_n}(F_{t_n}) \simeq \int F_{t_n}(x_{t_0}, \dots, x_{t_n})$$

$$\times \, e^{-\sum_{0 \le p < n} V(X_{t_p}) \, (t_{p+1} - t_p)} \, \mathbb{P}\left((X_{t_0}, \dots, X_{t_n}) \in d(x_{t_0}, \dots, x_{t_n})\right).$$

For reversible models on $S = \mathbb{R}^d$, the discrete time version of (27.21) is given by

$$-\frac{1}{n} \log \mathcal{Z}_n \longrightarrow_{t \uparrow \infty} E_0 \quad \text{and} \quad \eta_n(f) \simeq_{n \uparrow \infty} \frac{\langle f, \varphi_0 \rangle}{\langle 1, \varphi_0 \rangle}$$

where η_n denotes the n-th marginal of the probability measure $\mathbb{Q}_n$ from (27.26).

The Feynman-Kac distributions (27.26) and their normalizing constants $\mathcal{Z}_n$ (a.k.a. partition functions) can be computed using the mean field particle models developed in section 9.6. In computational physics these particle methods are called diffusion Monte Carlo methods. For a more detailed discussion on these models, we refer the reader to [66, 67].

27.3 Interacting particle systems

27.3.1 Introduction

Interacting particle systems are diffusion type or jump type stochastic processes that describe the evolution of a population of interacting individuals. The mean field processes from section 7.10.2 are specific examples of interacting particle systems. In this situation, each individual interacts with the occupation measure of the whole population. Another important class of interacting particle systems is defined in terms of jump type processes that describe the evolution of a collection of individuals in a configuration space of the form E^Λ, where Λ is a countable or a finite graph and E is a compact metric space.

More formally, we consider a finite set E, and a regular lattice Λ equipped with some equivalence relation $p \sim q$. For instance, if $\Lambda = (\mathcal{V}, \mathcal{E})$ is an undirected graph, we can choose the neighborhood distance between vertices $p, q \in \mathcal{V}$ given by the edge connections:

$$p \sim q \iff (p, q) \in \mathcal{E}.$$

We can also choose a finite subset of $\mathbb{Z}^d$, or $\Lambda = (\mathbb{Z}/m\mathbb{Z})^d$, for some dimension $d \ge 1$ and some cycle integer m, equipped with the distance $\rho(p, q)$. In this case, we can choose

$$p \sim q \iff \rho(p, q) \le a$$

for a given parameter $a \in \mathbb{R}_+$. For instance, for $\Gamma = [-n, n] \cap \mathbb{Z}$ we can choose the distance $\rho(p, q) = |p - q|$, and set $a = 1$. Then we will have

$$p \sim q \iff |p - q| \le 1.$$

The state space $S = E^\Lambda$ is often called the configuration space. A given state or configuration is a mapping

$$x \ : \ p \in \Lambda \ \mapsto x(p) \in E.$$

The set E depends on the application model. We can choose $\{0, 1\}$ to label vacant or occupied vertices or sites; or $E = \{-1, +1\}$ to label the spin of a site in a magnetic model. We can choose $E = \{0, s_1, \ldots, s_k\}$ to label the vacant site and the k different species in a biological system.

With a slight abuse of notation, we write $\sim$ for the equivalence relation on S induced by the one on Λ. This equivalence is given by

$$y \sim x \iff \exists (p, e) \in (\Lambda \times E) \ \text{s.t.} \ y = x^{p,e}$$

with the configuration $x^{p,e} \in S = E^\Lambda$ deduced from x by changing the value $x(p)$ by e, that is,

$$x^{p,e}(q) = \left\{ \begin{array}{cc} x(q) & \text{if} \quad q \neq p \\ e & \text{if} \quad q = p. \end{array} \right.$$

We associate with a given matrix $Q(x, y)$, $x, y \in S$, such that $Q(x, S) > 0$, the generator of a Markov process

$$
\begin{aligned}
L(f)(x) &= \sum_{y \sim x} Q(x, y) \ (f(y) - f(x)) \\
&= \lambda(x) \sum_{y \in S} [f(y) - f(x)] \ M(x, y) \qquad (27.27)
\end{aligned}
$$

with

$$\lambda(x) = \sum_{y \sim x} Q(x, y) \quad \text{and} \quad M(x, y) = Q(x, y)/Q(x, S).$$

By construction, given a configuration $x \in S$ the matrix $Q(x, y)$ only charges the configurations y in the vicinity of x. These are the configurations y that only differ from x at one given site $p \in \Lambda$ taking some possibly different value $e \in E$.

Identifying x with a vector $x = (x(p))_{p \in \Lambda}$, we see that

$$L(f)(x) = \sum_{(p,e) \in (\Lambda \times E)} \lambda_e(p, x) \ (f(x^{p,e}) - f(x))$$

with the intensity function

$$\lambda_e(p, x) = Q(x, x^{e,p}).$$

The Markov process evolves as follows.

Starting from a configuration x, all the possible neighboring configurations $y \sim x$ (that only differ from x at one site) start an exponential alarm with parameter

$$Q(x, y) = Q(x, x^{e,p}) = \lambda_e(p, x)$$

for each y of the form $x^{e,p}$, for some $p \in \Lambda$ and some $e \in E$. When the first alarm goes off for some configuration $y = x^{e,p}$, we change the value $x(p)$ by e. We refer to section 11.3.2, for a more thorough discussion of these jump processes.

Further details about this class of interacting particle systems can be found in the seminal work of T. Liggett [183, 184]. We also refer the reader to [51, 153], dedicated to invasion models and to the long-time behavior of these particle models.

27.3.2 Contact process

The contact process is defined on the state space $S = \{0, 1\}^\Lambda$, with a finite subset $\Lambda \subset \mathbb{Z}^d$

$$\lambda_0(p, x) = x(p)\, r_d \quad \text{and} \quad \lambda_1(p, x) = (1 - x(p))\, r_b \sum_{q \sim p} x(q).$$

Each particle at site p (i.e., $x(p) = 1$) dies at rate r_d

$$x(p) = 1 \xrightarrow{\text{rate } r_d} x^{0,p}(p) = 0.$$

On the contrary, every hole p (i.e. $x(p) = 0$) produces an offspring at a rate $r_b \sum_{q \sim p} x(q)$ that depends on the number of particles $x(q) = 1$ around its vicinity (i.e. with $q \sim p$)

$$x(p) = 0 \xrightarrow{\text{rate } r_b \sum_{q \sim p} x(q)} x^{1,p}(p) = 1.$$

The contact process can also be interpreted as a model of the spread of an infection. The 0 codes a healthy individual and 1 is a code for an infected individual. The infected individuals $x(p) = 1$ recover at a rate r_d, and the healthy ones become infected at a rate that depends on the number of infected neighbors. In other words, infected individuals infect their neighbors independently at rate r_d. We also notice that the increase of the number of infections increases the infection rate (since a given non-infected individual has more infected neighbors). Nevertheless the fully healthy configuration $x = 0$ (i.e., s.t. $x(p) = 0$, for any p) is stable.

The contact process is also of considerable interest in physics, where it provides simplified models for the analysis of directed percolation and Reggeon field theory. We refer the reader to [16, 139] and the references therein.

27.3.3 Voter process

The voter model is defined in the same way on the state space $S = \{1, \dots, r\}^\Lambda$ with the following rate function

$$\forall i \in E = \{1, \dots, r\} \qquad \lambda_i(p, x) = 1_{x(p) \neq i}\, c_i \sum_{q \sim p} 1_i(x(q)).$$

In words, the state space E is interpreted as a set of opinions. Every individual changes his opinion to $x(p) = i$, with a rate $\lambda_i(p, x)$ that depends on the number of individuals at his vicinity with opinion i.

When $r = 2$, we can also choose the rate functions such that

$$Q(x, x^p) = \sum_{q \sim p} P(p, q)\, 1_{\neq x(p)}(x(q))$$

for some Markov transition $P(p, q)$ on Λ, and with the change of opinion mapping

$$x^p(q) = \begin{cases} x(q) & \text{if } q \neq p \\ 3 - x(p) & \text{if } q = p. \end{cases}$$

Then the generator L takes the form

$$L(f)(x) = \sum_{p \sim q} P(p, q)\, 1_{\neq x(p)}(x(q))\, (f(x^p) - f(x)).$$

In this situation, at a unit rate, every individual p chooses a site q with probability $P(p, q)$, then he adopts his opinion. At any time, the configuration state x of the process can be partitioned into two sets: the set of individuals p with the first opinion, and the set of all the others:

$$x^{-1}(\{1\}) := \{p \in \Lambda \ : \ x(p) = 1\} \quad \text{and} \quad x^{-1}(\{2\}) := \{p \in \Lambda \ : \ x(p) = 2\}.$$

Any jump $x \mapsto x^p$ of the process flips the individual $p \in x^{-1}(\{x(p)\})$ into the set $x^{-1}(\{3 - x(p)\})$, that is,

$$x(p) = 1 \rightsquigarrow x^p(p) = 3 - x(p) = 3 - 1 = 2$$

and

$$x(p) = 2 \rightsquigarrow x^p(p) = 3 - x(p) = 3 - 2 = 1.$$

The resulting process can be interpreted in many ways: invasions between countries, immigration rates, and others.

27.3.4   Exclusion process

The exclusion process was introduced by Frank Spitzer in 1970 in [245]. We have $E = \{0, 1\}$. The 1's are interpreted as particles and the 0's as holes. The generator of the exclusion process is defined as in (27.27) with the equivalence relation

$$y \sim x \iff \exists (p, q) \in \Lambda^2 \ : \ y = x^{p,q}.$$

In the above display, $x^{p,q}$ stands for the (p, q)-transposition mapping associated with x. This mapping is given by

$$S = \{0, 1\}^\Lambda \ni x^{p,q} \ : \ r \in \{0, 1\} \mapsto x^{p,q}(r) = \begin{cases} x(p) & \text{if} \quad r = q \\ x(q) & \text{if} \quad r = p \\ x(r) & \text{if} \quad r \notin \{p, q\}. \end{cases}$$

We further assume that

$$x = y \Rightarrow Q(x, y) = 0.$$

This implies that jumps $x \rightsquigarrow y = x^{p,q}$ only occur when

$$(y(p), y(q)) = (x(q), x(p)) \in \{(1, 0), (0, 1)\}.$$

Since 0's are interpreted as non-occupied sites, we also add the condition

$$x(p) = 0 \Rightarrow Q(x, x^{p,q}) = 0.$$

Then the generator L is given by

$$L(f)(x) = \sum_{p,q \ : \ (x(p), x(q)) = (1,0)} Q(x, x^{p,q}) \ (f(x^{p,q}) - f(x)).$$

When the rate function $Q(x, x^{p,q}) = P(p, q)$ is associated with a Markov transition $P(p, q)$ on the sites space Λ, we have

$$L(f)(x) = \sum_{x \sim y} M(x, y) \ (f(y) - f(x))$$

with the Markov transition

$$M(x, y) := \sum_{p,q \ : \ (x(p), x(q)) = (1,0)} P(p, q) \ 1_{x^{p,q}}(y).$$

By construction, the number of particles does not change during the evolution of the particle system. The interaction between the particles is the result of the exclusion process that permits only jumps to holes.

When the initial configuration has no holes, the system is stopped. Also, if the system is empty, there are no particle births. This shows that the states $x = 1$ and $x = 0$ are stable, in the sense that the algorithm is stopped as soon as it enters one of these configurations.

27.4 Exercises

Exercise 441 (Klein-Kramers equation) *The evolution of a Brownian particle with mass m in a presence of an external gravity potential $U_t(x)$ is given by the diffusion*

$$
\begin{cases}
dX_t &= V_t \, dt \\
dV_t &= -\left(\alpha \, V_t + m^{-1} \, \partial_x U_t(X_t)\right) \, dt + \sigma \, dW_t
\end{cases}
$$

where W_t stands for a one-dimensional Brownian motion. In the above display, the column one vector $Y_t := (X_t, V_t)'$ denotes the position and the velocity of the particle, and $\alpha, \sigma > 0$ are some parameters. Assume that Y_t has a density $p_t(y)$ w.r.t. the Lebesgue measure $dy = dx dv$ around a state $y = (y_1, y_2) = (x, v)$. Check the Klein-Kramers equation

$$
m \partial_t p_t = -mv \, \partial_x p_t + \partial_x U_t \, \partial_v p_t + \alpha \, \partial_v \left[m \, v \, p_t + \kappa T \, \partial_v p_t\right]
$$

with $\sigma^2/\alpha := 2\kappa T/m$. In physics, the diffusion discussed above is sometimes written in a somewhat abusive Newtonian form

$$
m \, \frac{d^2 X_t}{dt^2} = -m\alpha \, \frac{dX_t}{dt} - \partial_x U_t(X_t) + m \, \sigma \, \frac{dW_t}{dt}.
$$

Exercise 442 (Smoluchowski equation) *The overdamped limit of the Newton-Langevin equation discussed in exercise 441 is given by the diffusion*

$$
\alpha \, dX_t = -m^{-1} \, \partial_x U_t(X_t) \, dt + \sigma \, dW_t
$$

where W_t is a one-dimensional Brownian motion. Assume that X_t has a density $p_t(x)$ w.r.t. the Lebesgue measure dx. Check the Smoluchowski equation

$$
m\alpha \, \partial_t p_t = \partial_x \left(p_t \, \partial_x U_t\right) + \kappa \, T \, \partial_x^2 p_t
$$

with $\alpha\sigma^2 = 2\kappa T/m$. In physics, this equation is sometimes rewritten as

$$
\partial_t p_t = -\partial_x J \quad \text{with the probability current} \quad -\alpha J := m^{-1} p_t \, \partial_x U_t + \kappa \, m^{-1} \, T \, \partial_x p_t.
$$

Exercise 443 (Normalized Schrödinger semigroups) *We consider the Schrödinger operator $L^V = L - V$ and the integral operator Q_t discussed in section 27.2.4. We assume that $E_1 > E_0$. Using the spectral decomposition (27.20) check that for any $f \in \mathbb{L}_2(\mathbb{R}^d)$ and for any $x \in \mathbb{R}^d$ we have*

$$
\frac{Q_t(f)(x)}{Q_t(1)(x)} - \frac{\langle f, \varphi_0 \rangle}{\langle 1, \varphi_0 \rangle} = O\left(e^{-t(E_1 - E_0)}\right).
$$

Exercise 444 (Unnormalized Schrödinger semigroups) *Let L be the infinitesimal generator of some process X_t evolving in some state space S equipped with some probability measure μ, and let V be some potential function on S. Assume that $(-L^V) := -L + V$ has a countable set of eigenvalues E_i and $\mathbb{L}_2(\mu)$ is equipped with an orthonormal basis of eigenfunctions φ_i. Check that for any $f \in \mathbb{L}_2(\mu)$ we have*

$$Q_t(f) = \sum_{i \geq 0} e^{-E_i t} \langle f, \varphi_i \rangle \varphi_i \quad and \quad \mu \left(Q_t(f)^2 \right)^{1/2} \leq e^{-E_0 t} \mu(f^2)^{1/2}.$$

For a finite space S, check that

$$\sup_{\|f\| \leq 1} \|Q_t(f)\| \leq c \, e^{-E_0 t} \quad and \quad e^{-\|V\| t} \leq \|Q_t(1)\| \leq c \, e^{-E_0 t} \quad for \ some \ c < \infty.$$

Exercise 445 (φ_0-processes) *We let X_t be a stochastic process on $\mathbb{R}^d$ with a self-adjoint infinitesimal generator L on $\mathbb{L}_2(\mathbb{R}^d)$ and an initial distribution $\eta_0 = \text{Law}(X_0)$. Let V be some bounded function on $\mathbb{R}^d$. We associate with these objects the Feynman-Kac measures (γ_t, η_t) defined for any bounded function f by the formulae*

$$\eta_t(f) := \frac{\gamma_t(f)}{\gamma_t(1)} \quad with \quad \gamma_t(f) = \mathbb{E}\left(f(X_t) \, \exp\left\{ -\int_0^t V(X_s) \, ds \right\} \right). \qquad (27.28)$$

We let $-E_0$ be the top eigenvalue of the Hamiltonian operator $\mathcal{H} = -L^V$ and φ_0 be the corresponding eigenfunction (see section 27.2.4). We assume that φ_0 is a smooth function s.t. $\varphi_0(x) > 0$ for any $x \in \mathbb{R}^d$ and $\eta_0(\varphi_0) < \infty$. Check that

$$\gamma_t(f) = e^{E_0 t} \, \eta_0(\varphi_0) \, \mathbb{E}\left(\varphi_0^{-1}(X_t^{\varphi_0}) \, f(X_t^{\varphi_0}) \right) \quad and \quad \eta_t(f) = \frac{\mathbb{E}\left(\varphi_0^{-1}(X_t^{\varphi_0}) \, f(X_t^{\varphi_0}) \right)}{\mathbb{E}\left(\varphi_0^{-1}(X_t^{\varphi_0}) \right)}$$

with the Markov process $X_t^{\varphi_0}$ having initial distribution $\eta_0^{[\varphi_0]} = \Psi_{\varphi_0}(\eta_0)$ and infinitesimal generator

$$L^{[\varphi_0]}(f) = L(f) + \varphi_0^{-1} \Gamma_L(\varphi_0, f). \qquad (27.29)$$

For $d = 1$ and $L = \frac{1}{2} \partial_x^2$, check that $X_t^{\varphi_0}$ satisfies the Langevin diffusion equation

$$dX_t^{\varphi_0} = (\partial_x \log \varphi_0) (X_t^{\varphi_0}) \, dt + dW_t$$

where W_t is a Brownian motion on the real line.

Exercise 446 (Feynman-Kac quasi-invariant measures) *We consider the Feynman-Kac models (27.28) associated with a stochastic process X_t on some state space S with a self-adjoint infinitesimal generator L on $\mathbb{L}_2(\mu)$, for some reference measure μ on S. Check that*

$$\frac{d}{dt} \eta_t(f) = \eta_t(L_{\eta_t}(f)) \quad with \quad L_{\eta_t}(f)(x) = L(f)(x) + V(x) \int [f(y) - f(x)] \, \eta_t(dy).$$

We consider the Schrödinger operator $L^V := L - V$ and the Hamiltonian $\mathcal{H} := -L^V$. We assume there exists some positive energy E_0 and some function φ_0 such that $\mathcal{H}(\varphi_0) = E_0 \varphi_0$ and $\mu(\varphi_0) < \infty$. Check that $\mathcal{H}$ is reversible w.r.t. μ, and for any sufficiently smooth function (depending on the differential order of the generator L) we have

$$\eta_\infty := \Psi_{\varphi_0}(\mu) \Longrightarrow \eta_\infty(L_{\eta_\infty}(f)) = 0 \quad and \quad E_0 = \eta_\infty(V).$$

Exercise 447 (Quasi-invariant measures - 2-state-model) ✒ *Consider the Feynman-Kac model discussed in exercises 307 and 308. Check that* $\lim_{t \to \infty} \mathbb{P}(X_t = x) = \mu(x)$ *with the L-reversible probability measure*

$$\mu(0) = 1 - \mu(1) = \lambda(1)/\lambda(0) + \lambda(1).$$

Also check that in all cases $\lim_{t \to \infty} \eta_t(x) = \eta_\infty(x)$ *for some probability measure on* $S = \{0,1\}$. *We let* φ_0 *be some function on* S *such that*

$$\mathcal{H}(\varphi_0) := (-L^V)(\varphi_0) = \eta_\infty(V)\, \varphi_0 \quad \text{with} \quad E_0 = \eta_\infty(V).$$

We also consider the probability $\pi_0 = \Psi_{\varphi_0}(\mu)$ *on* $S = \{0,1\}$ *as soon as* $\mu(\varphi_0) > 0$. *Check that the eigenvalues* $L^V(\varphi) = -E\,\varphi$ *of* L^V *are given by*

$$E_1 = V(1) - z_1 \ge E_0 = V(1) - z_2$$

with the parameters (z_1, z_2) *defined in exercise 308. Prove that*

$$\begin{aligned}
\lambda(0)\lambda(1) + E_0 E_1 &= (\lambda(0) + V(0)) \times (\lambda(1) + V(1)) \\
E_0 + E_1 &= (\lambda(0) + V(0)) + (\lambda(1) + V(1)).
\end{aligned}$$

- *When* $\lambda(0) > \lambda(1) = 0$, *and* $V(0) > V(1)$, *check that*

$$E_0 = V(1) < E_1 = V(0) + \lambda(0), \quad \lambda(0)\,\varphi_0(1) = (E_1 - E_0)\,\varphi_0(0), \quad \eta_\infty = \mu = \pi_0 = 1_{\{1\}}$$

as well as $(\varphi_1(0), \varphi_1(1)) \in (\mathbb{R} \times \{0\})$ *and*

$$\frac{1}{t} \log \gamma_t(1) = \eta_\infty(V) + \frac{1}{t} \log \left[1 + \frac{V(1) - V(0)}{E_1 - E_0}\, \eta_0(0) \left(1 - e^{-(E_1 - E_0)t} \right) \right].$$

Check that $\pi_0 L_{\pi_0} = 0$.

- *Assume that* $V(0) > V(1)$ *and* $\lambda(1) > 0$.

  - *When* $\lambda(1) = V(0) - V(1) > 0$ *and* $\lambda(0) = 0$ *check that*

$$E_0 = E_1 = V(0) \quad and \quad \eta_\infty = \mu = 1_{\{0\}}$$

as well as

$$\frac{1}{t} \log \gamma_t(1) = \eta_\infty(V) - \frac{1}{t}\, \log\left(1 + (V(0) - V(1))\,\eta_0(1)\, t \right).$$

  Check that $\eta_\infty L_{\eta_\infty} = 0$ *and* $(\varphi_0(0), \varphi_0(1)) \in (\{0\} \times \mathbb{R}) \Rightarrow \mu(\varphi_0) = 0$.

  - *When* $V(0) > V(1)$ *and* $\lambda(0), \lambda(1) > 0$ *check that*

$$\eta_\infty(0) = \frac{1}{2}\left[1 + \frac{\lambda(0) + \lambda(1)}{V(0) - V(1)} \right] - \sqrt{\left(\frac{1}{2}\left[1 + \frac{\lambda(0) + \lambda(1)}{V(0) - V(1)} \right] \right)^2 - \frac{\lambda(1)}{V(0) - V(1)}}$$

and

$$\frac{1}{t} \int_0^t \eta_s(V) ds$$

$$= \eta_\infty(V) + \frac{1}{t} \log \left[\frac{(E_1 - E_0)/(V(1) - V(0))}{(\eta_0(0) - \eta_\infty(0))\left(1 - e^{-(E_1 - E_0)t} \right) + (E_1 - E_0)/(V(1) - V(0))} \right].$$

Check that

$$\frac{\varphi_0(1)}{\varphi_0(0)} = \frac{\lambda(0) + (V(0) - V(1)) + z_2}{\lambda(0)} = \frac{\lambda(1)}{\lambda(1) + z_2} > 0$$

and

$$\frac{\varphi_1(1)}{\varphi_1(0)} = \frac{\lambda(0) + (V(0) - V(1)) + z_1}{\lambda(0)} = \frac{\lambda(1)}{\lambda(1) + z_1} < 0$$

as well as $\pi_0\left(L_{\pi_0}(f)\right) = 0$ *and* $\pi_0 = \eta_\infty$, *for any function* f *on* S. *Prove the orthogonality formula* $\mu(\varphi_0\varphi_1) = 0$.

Exercise 448 (Schrödinger equation with quadratic potential) *Check that*

$$\psi(t,x) = \alpha(t)\,\psi_0(x) \quad \text{with} \quad \alpha(t) = \exp\left(-\frac{i}{2}\sqrt{\frac{k}{m}}\,t\right) \quad \text{and} \quad \psi_0(x) = c\,\exp\left(-\frac{x^2}{2}\sqrt{\frac{km}{\hbar^2}}\right)$$

satisfies the Schrödinger equation

$$i\hbar\partial_t\psi(t,x) = -\frac{\hbar^2}{2m}\partial_x^2\psi(t,x) + V(x)\psi(t,x) \quad \text{with} \quad V(x) = \frac{km}{2}\,x^2.$$

Exercise 449 (Twisted guiding waves) *Consider the Feynman-Kac model* (γ_t, η_t) *discussed in exercise 446. We let* $X_t^{\varphi_T}$ *be the* φ_T *process associated with a trial energy function (a.k.a. guiding or trial wave function) denoted by* φ_T *and the initial distribution* $\eta_0^{[\varphi_T]} = \Psi_{\varphi_T}(\eta_0)$. *We assume that* ψ_T *is chosen so that the process* $X_t^{\varphi_T}$ *is well defined. Prove that*

$$\gamma_t(f) = \eta_0(\varphi_T)\,\mathbb{E}\left(\varphi_T^{-1}(X_t^{\varphi_T})\,f(X_t^{\varphi_T})\,\exp\left(-\int_0^t V_T(X_s^{\varphi_T})ds\right)\right)$$

with the trial ground state energy (a.k.a. local energy) V_T *given by*

$$V_T := V - \varphi_T^{-1}L(\varphi_T) = \varphi_T^{-1}\mathcal{H}(\varphi_T).$$

Check that $L^{[\varphi_T]}$ *is reversible w.r.t.* $\Psi_{\varphi_T^2}(\mu)$. *As underlined in [37], "the role of the trial function* φ_T *is to guide the stochastic walkers (a.k.a. particles) in the important regions (regions corresponding to an important contribution to the averages)." For a more detailed discussion on the choice of the trial waves functions in quantum systems, we also refer the reader to the review article by M. D. Towler [253].*

Exercise 450 (Variational principle) *Consider the Feynman-Kac twisted models discussed in exercise 449. We assume that* $\mathbb{L}_2(\mu)$ *is equipped with an orthonormal basis of eigenfunctions* $(\varphi_i)_{i\geq 0}$ *corresponding to the eigenvalues* $0 \leq E_0 \leq E_1 \leq \ldots$ *of the Hamiltonian operator* $\mathcal{H}$. *We let* $\langle f, g\rangle = \mu(fg)$ *be the inner product on* $\mathbb{L}_2(\mu)$. *Check the variational principles*

$$\Psi_{\varphi_T^2}(\mu)(V_T) \geq \frac{\langle\varphi_0, \mathcal{H}(\varphi_0)\rangle}{\langle\varphi_0, \varphi_0\rangle} = E_0 = \inf_{\varphi \in \mathbb{L}_2(\mu)} \frac{\langle\varphi, \mathcal{H}(\varphi)\rangle}{\langle\varphi, \varphi\rangle} = \Psi_{\varphi_0^2}(\mu)(V_0) \Leftarrow V_0 := \varphi_0^{-1}\mathcal{H}(\varphi_0).$$

Exercise 451 (Contact process) *Consider the contact process* $\xi_t \in S := \{0,1\}^\Lambda$ *discussed in section 27.3.2. For any given state* $q \in \Lambda$ *we set*

$$f_q \ : \ x \in \{0,1\}^\Lambda \mapsto f_q(x) = x(q).$$

We let $\eta_t(dx)$ be the distribution of the interacting particle model ξ_t. Check that

$$\partial_t \eta_t(f_q) = \eta_t(L(f_q)) = -r_d \ \eta_t(f_q) + r_b \sum_{p \sim q} \int \eta_t(dx) \ f_p(x) \ (1 - f_q(x)).$$

We set $\mu_t(q) = \eta_t(f_q)$. Prove that

$$\partial_t \mu_t(q) = -r_d \ \mu_t(q) + r_b \sum_{p \sim q} \mu_t(p)(1 - \mu_t(q)) - \sum_{p \sim q} \mathrm{Cov}_t(p, q)$$

with the covariance function

$$\mathrm{Cov}_t(p, q) := \sum_{p \sim q} \int \eta_t(dx) \ (f_p(x) - \eta_t(f_p)) \ (f_q(x) - \eta_t(f_q)).$$

For regular homogeneous lattices,

$$\forall p, q \in \Lambda \quad \eta_t(f_q) = \eta_t(f_p) := z_t \quad and \quad \|\{s \in \Lambda \ : \ s \sim p\}\| = \|\{s \in \Lambda \ : \ s \sim q\}\| := n(\geq 1).$$

Check that

$$\dot{z}_t = -r_d \ z_t + r_b \ n \ z_t(1 - z_t) - n\mathrm{Var}_t(p) \quad with \ the \ variance \ function \quad \mathrm{Var}_t(p) := \mathrm{Cov}_t(p, p).$$

Exercise 452 (Contact process - Mean field approximations) *We consider the contact process process discussed in exercise 451. We further assume that* $\mathrm{Var}_t(p) = 0$. *In this situation, check the quadratic mean field rate equation*

$$\dot{z}_t = -r_d \ z_t + r_b \ n \ z_t(1 - z_t).$$

Solve this equation when $\lambda := r_b/r_d \in \ ]\frac{1}{2n}, \frac{1}{n}[$. *Show that* $\lim_{t \to \infty} z_t = \frac{1}{n\lambda} - 1 \ (\in [0, 1])$.

Exercise 453 (Ising model - Interacting particle system) *Consider a configuration space of the form* $S = \{-1, 1\}^\Lambda$ *with a finite regular lattice* $\Lambda \subset \mathbb{R}^d$, *for some* $d \geq 1$. *We equip* S *with the Boltzmann-Gibbs measure*

$$\pi(x) = \frac{1}{\mathcal{Z}} \ \exp -\beta \ H(x)$$

with some normalizing constant $\mathcal{Z}$, *some inverse temperature parameter* β *and a Hamiltonian function*

$$H \ : \ x = (x(p))_{p \in \Lambda} \in S = \{-1, 1\}^\Lambda \mapsto H(x) = -\frac{1}{2} \sum_{(p,q) \in \Lambda^2} j(p-q) \ x(p)x(q) - \sum_{p \in \Lambda} h(p) \ x(p).$$

In the above display, $j \ : \ \mathbb{R}^d \mapsto \mathbb{R}$ *stands for some symmetric interacting potential function s.t.* $j(0) = 0$ *and the function* $h \ : \ \mathbb{R}^d \mapsto \mathbb{R}$ *represents some external field. We consider a jump type interacting particle system* X_t *on* S *with generator* L *defined for any function* f *on* S *by*

$$L(f)(x) = \sum_{p \in \Lambda} \lambda(p, x) \ (f(x^p) - f(x)) \quad with \quad x^p(q) = \begin{cases} x(q) & if \quad q \neq p \\ -x(q) & if \quad q = p. \end{cases}$$

Describe the evolution equations of $P_t(x, y) = \mathbb{P}(X_t = y \mid X_0 = x)$. *We choose*

$$\lambda(p, x) = \frac{e^{-\beta v(p,x)x(p)}}{e^{\beta v(p,x)} + e^{-\beta v(p,x)}} \quad with \quad v(p, x) = h(p) + \sum_{q \in \Lambda - \{p\}} j(p-q) \ x(q).$$

Check that π *is* L-*reversible.*

28

Dynamic population models

This chapter is dedicated to applications of stochastic processes to biology and more particularly to dynamic populations and evolutionary processes. We start with a presentation of discrete generation birth and death type processes. The second part of the chapter provides a discussion on the connections between deterministic type models (such as logistic or Lotka-Volterra type dynamical systems) and their stochastic versions expressed in terms of individual based models. The last part of the chapter is dedicated to discrete and continuous time branching and interacting processes.

> *Evolution is the fundamental idea in all of life science, in all of biology.*
> Bill Nye (1955-).

28.1   Discrete time birth and death models

In discrete time settings, a birth and death process is a Markov chain in the set of integers $S = \mathbb{N}$ with Markov transitions given for any $x \in \mathbb{N}$ by

$$M(x,y) = p(x) \, 1_{x+1}(y) \; + \; q(x) \, 1_{x-1}(y)$$
$$+ \; (1 - (p(x) + q(x))) \, 1_x(y).$$

Here p and q are non-negative functions from $\mathbb{N}$ in $[0,1]$ such that $p(x) + q(x) \leq 1$, for any $x \in \mathbb{N}$, with the boundary condition $q(0) = 0$. The condition $p(0) > 0$ can be interpreted as the probability of a restart of a population after extinction. When $p(x) + q(x) < 1$ for some $x > 0$, we have $M(x,x) > 0$ and the chain is aperiodic. If $p(0) = 0$, the chain is absorbed at the state 0; in other words it stays with a null population at any time after extinction.

In the further development of this section, we assume that $p(0) > 0$ and

$$\mathcal{Z} := \sum_{x \geq 0} \left\{ \prod_{y=1}^{x} \frac{p(y-1)}{q(y)} \right\} < \infty.$$

For time homogeneous models $p(x) = p$ and $q(x) = q$ and this condition is equivalent to a larger death rate:

$$p < q \iff \sum_{x \geq 0} (p/q)^x = \frac{q}{q-p} < \infty. \tag{28.1}$$

In addition, using

$$\left(\frac{p}{q}\right)^x M(x, x+1) = \left(\frac{p}{q}\right)^x p = \left(\frac{p}{q}\right)^{x+1} \times q = \left(\frac{p}{q}\right)^{x+1} M(x+1, x)$$

we see that M is reversible w.r.t. the (invariant) geometric distribution

$$\forall x \in \mathbb{N} \qquad \pi(x) = \left(\frac{p}{q}\right)^x \left(1 - \frac{p}{q}\right) > 0.$$

Let us discuss the existence of an invariant measure $\pi M = \pi$ that charges all states in the general situation. Notice that π solves the fixed point equation if and only if

$$\begin{aligned}
\pi(0) &= (\pi M)(0) = \pi(0) \, M(0,0) + \pi(1) M(1,0) \\
&= \pi(0)(1 - p(0)) + \pi(1)q(0) \iff \pi(0)p(0) = \pi(1)q(0)
\end{aligned}$$

and for any $x \geq 1$

$$\begin{aligned}
\pi(x) &= (\pi M)(x) \\
&= \pi(x-1) \, M(x-1, x) + \pi(x) \, M(x, x) + \pi(x+1) \, M(x+1, x) \\
&= \pi(x-1) \, p(x-1) + \pi(x+1) \, q(x+1) + \pi(x) \, (1 - p(x) - q(x)).
\end{aligned}$$

Rewritten in a slightly different form, this means

$$\pi(x+1) \, q(x+1) + \pi(x-1) \, p(x-1) = \pi(x) \, (p(x) + q(x)). \tag{28.2}$$

This helps us to prove that

$$\begin{aligned}
\pi(x+1) \, q(x+1) - \pi(x) \, p(x) &= \pi(x) \, q(x) - \pi(x-1) \, p(x-1) \\
&= \ldots = \pi(1) \, q(1) - \pi(0) \, p(0) = 0.
\end{aligned}$$

This relationship clearly implies that

$$\pi(x+1) = \frac{p(x)}{q(x+1)} \, \pi(x) = \ldots = \left\{ \prod_{y=0}^{x} \frac{p(y)}{q(y+1)} \right\} \pi(0). \tag{28.3}$$

Recalling that

$$\sum_{x \geq 0} \pi(x) = 1 = \sum_{x \geq 1} \left\{ \prod_{y=1}^{x} \frac{p(y-1)}{q(y)} \right\} \pi(0) + \pi(0) = \sum_{x \geq 0} \left\{ \prod_{y=1}^{x} \frac{p(y-1)}{q(y)} \right\} \pi(0),$$

we find that the fixed point is necessarily given by

$$\mathcal{Z} < \infty \Leftrightarrow \forall x \in \mathbb{N} \quad \pi(x) = \frac{1}{\mathcal{Z}} \left\{ \prod_{y=1}^{x} \frac{p(y-1)}{q(y)} \right\} > 0.$$

Conversely, using (28.3) we can also check that M is reversible w.r.t. π, that is,

$$\pi(x) M(x, x+1) = \pi(x) p(x) = \pi(x+1) q(x) = \pi(x+1) M(x+1, x).$$

This ensures that $\pi = \pi M$. By the remark 8.5.2, this condition ensures that all the states are null recurrent and using Kac's theorem 8.5.9,

$$\mathcal{Z} / \left\{ \prod_{y=1}^{x} \frac{p(y-1)}{q(y)} \right\} = \mathbb{E}\left(T_x \mid X_0 = x\right).$$

We denoted by T_x the hitting time of a state x. For the time homogeneous models discussed in (28.1), we have

$$p < q \iff \mathbb{E}\left(T_x \mid X_0 = x\right) = (q/p)^{x+1} \frac{1}{(q/p) - 1}.$$

As shown in the figure below, the coupling/merging time of two independent birth and death chains X_n and X'_n

$$T = \inf\left\{ n \ : \ X_n = X'_n \right\}$$

starting at two different locations $X_0 = x_0 \neq x'_0 = X'_0$ arises before both chains hit the origin

$$T \leq T_0 \vee T'_0$$

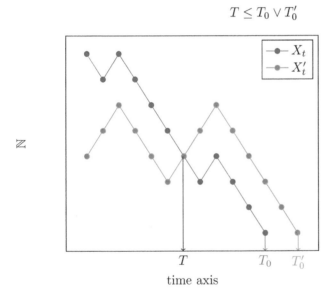

We conclude that

$$\mathbb{P}\left(T \geq n\right) \leq \mathbb{P}\left(T_0 \vee T'_0 \geq n\right) \leq \mathbb{P}\left(T_0 \geq n\right) + \mathbb{P}\left(T'_0 \geq n\right).$$

Using the proposition 8.3.16 we readily see that

$$\lim_{n \to \infty} \|\text{Law}(X_n) - \text{Law}(X'_n)\|_{tv} = 0.$$

For a more detailed discussion on this chain, we refer to exercise 462.

28.2   Continuous time models

28.2.1   Birth and death generators

These generators are Markov on $S := \mathbb{Z}$, with

$$L(f)(x) = \lambda_b(x)[f(x+1) - f(x)] + \lambda_d(x)[f(x-1) - f(x)]$$

for some positive functions λ_b and λ_d on $\mathbb{R}$. We can rewrite this generator as a pure jump model:

$$L(f)(x) = \lambda(x) \int (f(y) - f(x))\, M(x, dy)$$

with

$$\lambda(x) = \lambda_b(x) + \lambda_d(x)$$

and

$$M(x, dy) := \frac{\lambda_b(x)}{\lambda(x)}\, \delta_{x+1}(dy) + \frac{\lambda_d(x)}{\lambda(x)}\, \delta_{x-1}(dy).$$

28.2.2   Logistic processes

> The deterministic logistic model, also called the Verhulst-Pearl model, is given by the equation
>
> $$\frac{dx}{dt}(t) = \lambda\, x(t)\, \left(1 - \frac{x(t)}{K}\right). \tag{28.4}$$
>
> The state variable $x(t)$ can be interpreted as the population size. The parameter K stands for the carrying capacity and the size of the system, and λ represents the intrinsic growth rate.

Using the fraction decomposition

$$\frac{1}{x(1 - x/K)} = \frac{1}{x} + \frac{1/K}{(1 - x/K)}$$

for any $x \notin \{0, K\}$, we find that

$$(28.4) \iff \frac{dx}{x(1 - x/K)} = \lambda dt$$

$$\iff \frac{dx}{x} + \frac{1}{K}\frac{dx}{(1 - x/K)} = d(\lambda t)$$

$$\iff d(\log x) - d\log(1 - x/K) = d\log\left(\frac{x}{1 - x/K}\right) = d(\lambda t)$$

$$\iff -\log\left(\frac{x(t)}{1 - \frac{x(t)}{K}}\right) + \log\left(\frac{x(0)}{1 - \frac{x(0)}{K}}\right) = \lambda t.$$

This implies that

$$(28.4) \iff \log\left(\frac{1}{x(t)} - \frac{1}{K}\right) = -\lambda t + \log\left(\frac{1}{x(0)} - \frac{1}{K}\right)$$

$$\iff \frac{1}{x(t)} = \frac{1}{K} + e^{-\lambda t}\left(\frac{1}{x(0)} - \frac{1}{K}\right).$$

We conclude that the solution of (28.4), starting at $x(0) \notin \{0, K\}$, is given by the formula

$$\forall t \in \mathbb{R} \qquad x(t) = \frac{Kx(0)}{x(0) + [K - x(0)] \, e^{-\lambda t}}.$$

> The mathematical model (28.4) encapsulates dynamical systems of the form
>
> $$\frac{dx}{dt}(t) = \lambda_1 \, x(t) \left(1 - \alpha_1 \frac{x(t)}{N}\right) - \lambda_2 \, x(t) \left(1 + \alpha_2 \frac{x(t)}{N}\right) \qquad (28.5)$$
>
> with $\lambda_1 > \lambda_2 \geq 0$, and $\alpha_1, \alpha_2 \geq 0$, such that $\alpha_1 \leq 1$, and $\lambda_1 \alpha_1 + \lambda_2 \alpha_2 \geq (\lambda_1 - \lambda_2)$.

Indeed, using the fact that

$$\lambda_1 \, x \left(1 - \alpha_1 \frac{x}{N}\right) - \lambda_2 \, x \left(1 + \alpha_2 \frac{x}{N}\right)$$
$$= x \left([\lambda_1 - \lambda_2] - [\lambda_1 \alpha_1 + \lambda_2 \alpha_2] \frac{x}{N}\right) = [\lambda_1 - \lambda_2] \, x \left(1 - \frac{\lambda_1 \alpha_1 + \lambda_2 \alpha_2}{\lambda_1 - \lambda_2} \frac{x}{N}\right)$$

we find that (28.5) reduces to (28.4) with

$$\lambda = \lambda_1 - \lambda_2 > 0 \quad \text{and} \quad 0 \leq K = N \frac{\lambda_1 - \lambda_2}{\lambda_1 \alpha_1 + \lambda_2 \alpha_2} \leq N.$$

One way to incorporate the interaction with the environment and the model uncertainties is to consider the stochastic diffusion model

$$dX_t = \lambda \, X_t \left(1 - \frac{X_t}{K}\right) \, dt + \sigma \, dW_t$$

for some diffusion coefficient $\sigma \geq 0$ and some standard Brownian motion W_t on the real line.

Another strategy is to describe the evolution of each individual in the system.

> The individual-based stochastic version of the logistic model is defined in terms of a birth and death jump type model X_t with a state space $\mathbb{N}$. The generator is defined for any function f on S and for any $x \in S$ by
>
> $$L(f)(x) = \lambda_{\text{birth}}(x) \, (f(x + 1) - f(x)) + \lambda_{\text{death}}(x) \, (f(x - 1) - f(x)) \qquad (28.6)$$
>
> with
>
> $$\lambda_{\text{birth}}(x) = \lambda_1 \, x \left(1 - \alpha_1 \frac{x}{N}\right)_+ \quad \text{and} \quad \lambda_{\text{death}}(x) = \lambda_2 \, x \left(1 + \alpha_2 \frac{x}{N}\right).$$
>
> Notice that 0 is the absorbing state 0 and N is reflecting boundary at N when $\alpha_1 = 1$.

By construction, when $\alpha_1 = 1$, we have $\lambda_{\text{birth}}(x) \geq 0$ for any $x \in S := \{0, \dots, N\}$, and $\lambda_{\text{birth}}(0) = \lambda_{\text{birth}}(N) = 0$. In this situation, starting with some $X_0 \in S$ for any function f

on $\mathbb{R}$, we have

$$\frac{d}{dt}\mathbb{E}(f(X_t))$$

$$= \mathbb{E}(L(f)(X_t)) = \lambda_1 \, \mathbb{E}\left(X(t)\left(1 - \alpha_1 \frac{X(t)}{N}\right)[f(X_t + 1) - f(X_t)]\right)$$

$$+ \lambda_2 \, \mathbb{E}\left(X_t \left(1 + \alpha_2 \frac{X_t}{N}\right)[f(X_t - 1) - f(X_t)]\right).$$

Choosing $f(X_t) = X_t \Rightarrow [f(X_t + 1) - f(X_t)] = 1 = -[f(X_t - 1) - f(X_t)]$, we find that

$$\frac{d}{dt}\mathbb{E}(X_t) \;=\; \lambda_1 \, \mathbb{E}\left(X(t)\left(1 - \alpha_1 \frac{X(t)}{N}\right)\right) - \lambda_2 \, \mathbb{E}\left(X_t \left(1 + \alpha_2 \frac{X_t}{N}\right)\right)$$

$$\simeq\; \lambda_1 \, \mathbb{E}\left(X(t)\right)\left(1 - \alpha_1 \frac{\mathbb{E}(X(t))}{N}\right) - \lambda_2 \, \mathbb{E}\left(X_t\right)\left(1 + \alpha_2 \frac{\mathbb{E}(X_t)}{N}\right)$$

$$-\frac{1}{N}(\lambda_1 \alpha_1 + \lambda_2 \alpha_2) \, \mathrm{Var}(X_t).$$

This shows that the deterministic dynamical system (28.7) does not take into account the variance of X_t.

We end this section with an illustration of the stochastic logistic model in epidemiology. In this context X_t represents the number of infected individuals. It is a Markov model on a finite set $S = \{0, 1, \ldots, N\}$, where N denotes the total number of individuals. Its generator is given by

$$L(f)(x) = \lambda_{infect}(x)\,[f(x + 1) - f(x)] + \lambda_{recover}(x)\,[f(x - 1) - f(x)]$$

with the infection and the recovery rates

$$\lambda_{infect}(x) = \alpha \, N^2 \, \frac{x}{N}\left(1 - \frac{x}{N}\right) \quad \text{and} \quad \lambda_{recover}(x) = \beta \, N \, \frac{x}{N}.$$

The rates are associated with some non-negative constants α, β. The infection rate $\lambda_{infect}(x)$ depends on the proportion $\frac{x}{N}$ of infected individuals and on the proportion $\left(1 - \frac{x}{N}\right)$ of non-infected ones. Every infected individual (in a pool of x) remains infected with an exponential rate β.

28.2.3 Epidemic model with immunity

This model is Markov on a finite set $S = \{0, 1, \ldots, d\}$, where d denotes the total number of individuals. The state of the Markov chain $X_t = (Y_t, Z_t)$ represents the number Y_t of infected individuals and the number Z_t of immune individuals. The evolution of X_t is given by the jump generator

$$L(f)(y, z) \;=\; \alpha_1 \, d^2 \, \frac{y}{d}\left(1 - \frac{y}{d} - \frac{z}{d}\right)[f(y + 1, z) - f(y, z)]$$

$$+ \alpha_2 \, d \, \frac{y}{d}\,[f(y - 1, z + 1) - f(y, z)]$$

$$+ \alpha_3 \, d \, \frac{z}{d}\,[f(y, z - 1) - f(y, z)].$$

28.2.4   Lotka-Volterra predator-prey stochastic model

The deterministic Lotka-Volterra dynamical system is given by

$$\begin{cases} x' & = & a\,x - bxy \\ y' & = & -c\,y + dxy \end{cases} \tag{28.7}$$

for some parameters $a, b, c, d > 0$. The first coordinate x_t represents the number of preys and y_t the number of predators (such as rabbits and foxes, respectively). In the above system, $bx_t y_t$ indicates the prey death rate and ax_t indicates the prey birth rate. Similarly, the terms $dx_t y_t$ and cy_t represent the predator birth and death rates, respectively. The rates of change of the populations are proportional to their sizes. The food supply of the predators depends on the size of the prey population. In this model, the environmental interaction is not taken into account. One way to enter the interaction is to add noise to the model. The noise perturbations are also used to model the uncertainties of the mathematical model. One natural strategy is to consider the following two-dimensional diffusion:

$$\begin{cases} dX_t & = & [a\,X_t - b\,X_t Y_t]\,dt + \sigma_{1,1}X_t^2 dW_t^1 + \sigma_{1,2}X_t Y_t\,dW_t^2 \\ dY_t & = & [-c\,Y_t + d\,X_t\,Y_t]\,dt + \sigma_{2,2}Y_t^2 dW_t^2 + \sigma_{2,2}X_t Y_t\,dW_t^1, \end{cases}$$

with some independent Brownian motions W_t^1 and W_t^2.

The above models do not provide any information about the evolution of each individual in the system. The individual-based stochastic model is defined by a pure jump Markov process (X_t, Y_t) on $\mathbb{R}^2$ with an infinitesimal generator

$$L(f)(x, y)$$

$$= \lambda_{\text{prey-death}}(x, y)\,(f(x - 1, y) - f(x, y)) + \lambda_{\text{prey-birth}}(x)\,(f(x + 1, y) - f(x, y))$$

$$+ \lambda_{\text{pred-death}}(y)\,(f(x, y - 1) - f(x, y)) + \lambda_{\text{pred-birth}}(x, y)\,(f(x, y + 1) - f(x, y))$$

with the jump intensities

$$\begin{array}{ll} \lambda_{\text{prey-death}}(x, y) & = & bxy \qquad \lambda_{\text{prey-birth}}(x) & = & ax \\ \lambda_{\text{pred-death}}(y) & = & cy \qquad \lambda_{\text{pred-birth}}(x, y) & = & dxy. \end{array}$$

By construction, for any function f on $\mathbb{R}^2$, we have

$$\frac{d}{dt}\mathbb{E}(f(X_t, Y_t))$$

$$= \mathbb{E}(L(f)(X_t, Y_t))$$

$$= b\,\mathbb{E}\,(X_t Y_t[f(X_t - 1, Y_t) - f(X_t, Y_t)]) + a\,\mathbb{E}\,(X_t[f(X_t + 1, Y_t) - f(X_t, Y_t)])$$

$$+ c\,\mathbb{E}\,(Y_t[f(X_t, Y_t - 1) - f(X_t, Y_t)]) + d\,\mathbb{E}\,(X_t Y_t[f(X_t, Y_t + 1) - f(X_t, Y_t)])\,.$$

Choosing $f(X_t, Y_t) = X_t \Rightarrow [f(X_t + 1, Y_t) - f(X_t, Y_t)] = 1 = -[f(X_t - 1, Y_t) - f(X_t, Y_t)]$, we find that

$$\begin{array}{lll} \frac{d}{dt}\mathbb{E}(X_t) & = & a\,\mathbb{E}\,(X_t) - b\,\mathbb{E}\,(X_t Y_t) \\ & = & a\,\mathbb{E}\,(X_t) - b\,\mathbb{E}\,(X_t)\,\mathbb{E}\,(Y_t) - b\,\text{Cov}(X_t, Y_t). \end{array}$$

Choosing $f(X_t, Y_t) = Y_t \Rightarrow [f(X_t, Y_t + 1) - f(X_t, Y_t)] = 1 = -[f(X_t, Y_t - 1) - f(X_t, Y_t)]$, we find that

$$\frac{d}{dt}\mathbb{E}(Y_t) = d\,\mathbb{E}(X_t Y_t) - c\,\mathbb{E}(Y_t)$$
$$= d\,\mathbb{E}(X_t)\,\mathbb{E}(Y_t) - c\,\mathbb{E}(Y_t) + d\,\mathrm{Cov}(X_t, Y_t).$$

This shows that the deterministic dynamical system (28.7) does not take into account the covariance between X_t and Y_t.

To understand better the behavior of the Lotka-Volterra model, we end this section with a discussion on the phase diagram of the dynamical system (28.7). Firstly, using the change of variables

$$\tau = a\,t, \quad u = \frac{d}{c}\,x \quad \text{and} \quad v = \frac{b}{a}\,y$$

we find that

$$d\tau = a\,dt \qquad du = \frac{d}{c}\,dx \quad \text{and} \quad dv = \frac{b}{a}\,dy.$$

This allows us to reduce the system to the following equations

$$\frac{du}{d\tau} = \frac{d}{ac}\frac{dx}{dt} = \frac{d}{ac}\,(a\,x - b x y) = \frac{d}{c}\,x - \left(\frac{b}{a}\,y\right)\left(\frac{d}{c}\,x\right) = u\,(1 - v).$$

In much the same way, we have

$$\frac{dv}{d\tau} = \frac{b}{a^2}\frac{dy}{dt} = \frac{b}{a^2}\,(-c\,y + d x y) = -\frac{c}{a}\left(\frac{b}{a}\,y\right) + \frac{c}{a}\left(\frac{b}{a}\,y\right)\left(\frac{d}{c}\,x\right) = -\frac{c}{a}\,v\,(1 - u)$$

from which we conclude that

$$\frac{dv}{d\tau} = -\alpha\,v\,(1 - u) \quad \text{with} \quad \alpha = \frac{c}{a}.$$

The corresponding dynamical system is given by

$$\begin{cases} u' &= u\,(1 - v) \\ v' &= -\alpha\,v\,(1 - u). \end{cases}$$

The stationary states are given by

$$u\,(1 - v) = 0 \quad \text{and} \quad v\,(1 - u) = 0$$

$$\Updownarrow$$

$$(u, v) = (u_0, v_0) := (0, 0) \quad \text{and} \quad (u, v) = (u_1, v_1) := (1, 1).$$

To understand the behavior of this system, we notice that the integral curves are defined by the equations:

$$\frac{dv}{du} = \frac{dv/d\tau}{du/d\tau} = \frac{g(u, v)}{f(u, v)}$$

with

$$g(u, v) = -\alpha\,v\,(1 - u) \quad \text{and} \quad f(u, v) = u\,(1 - v).$$

The nullclines, sometimes called zero-growth isoclines, are defined by

$$g(u, v) = -\alpha\,v\,(1 - u) = 0 \iff u = 1 \quad \text{or} \quad v = 0$$

and

$$f(u, v) = u\,(1 - v) = 0 \iff u = 0 \quad \text{or} \quad v = 1.$$

At the intersecting states of these lines, the trajectory of the system has a vertical tangent ($f(u, v) = 0$) or a horizontal tangent ($g(u, v) = 0$).

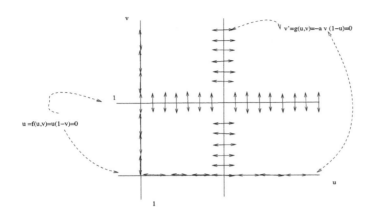

To be more precise, we have

$$\frac{dv}{du} = \frac{-\alpha \, v \, (1-u)}{u \, (1-v)} = \frac{-\alpha \, (1/u - 1)}{(1/v - 1)}.$$

Using an elementary analysis of the functions

$$u \mapsto -\alpha \, (1/u - 1) \quad \text{and} \quad v \mapsto \frac{1}{(1/v - 1)}$$

we obtain the following sign diagram

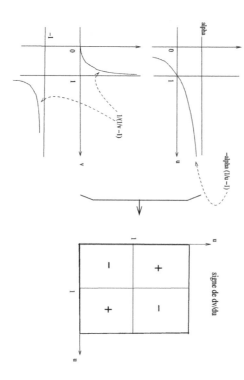

Following the tangent vector field at every state of the system trajectory, we obtain the following phase diagram.

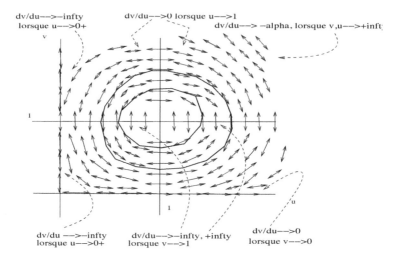

28.2.5 Moran genetic model

This model is Markov on a finite set $S = \{0/d, 1/d, \ldots, d/d\}$, where d is the total number of individuals. The state X_t of the Markov chain represents the proportion $x \in [0,1]$ of alleles of type A in a genetic model with alleles A and B. The evolution of X_t is given by the jump generator

$$L(f)(x) = \binom{d}{2} x(1-x) \left[f(x + 1/d) - f(x) \right]$$

$$+ \binom{d}{2} x(1-x) \left[f(x - 1/d) - f(x) \right].$$

Notice that this generator can be rewritten as

$$L(f)(x) = \lambda \int [f(y) - f(x)] \, M(x, dy)$$

with the jump rate $\lambda = \binom{d}{2}$ and the Markov transition

$$M(x, dy)$$

$$= x(1-x) \, \delta_{x+1/d}(dy) + d \, x(1-x) \, \delta_{x-1/d}(dy) + (1 - [x^2 + (1-x)^2]) \, \delta_x(dy).$$

The probability to stay in the same proportion is determined from the fact that $1 = (x + (1-x))^2 = x^2 + (1-x)^2 + 2x(1-x)$.

Using an elementary Taylor second order decomposition for large population sizes d and for a twice continuously differentiable function f, we find that

$$f(x + 1/d) - f(x) = f'(x)/d + \frac{1}{2} \, f''(x)/d^2 + O(1/d^3).$$

This implies that

$$L(f)(x) = \frac{d(d-1)}{2} \, x(1-x) \, \left([f(x + 1/d) - f(x)] + [f(x - 1/d) - f(x)] \right)$$

$$= \frac{1}{2} \, x(1-x) \, f''(x) + O(1/d).$$

28.3 Genetic evolution models

The cells of most organisms contain a specific hereditary material called DNA (deoxyribonucleic acid). The DNA is a double helix formed with base pairs attached to a sugar-phosphate backbone. These base pair sequences encode information in terms of four chemical bases: A (adenine), C (cytosine), G (guanine), and T (thymine). For instance, the human DNA has 3 billion bases, and 99% of these codes are common to all of us.

Proteins are sophisticated and complex molecules in cells that serve different functionalities. For instance, messenger proteins transmit signals to coordinate biological processes between cells, tissues, or organs.

Most of the genes in the cell produce these functional molecules. This chemical process converting DNA to RNA (ribonucleic acid) to proteins is decomposed into two steps: transcription and translation. The process is often called "central dogma" to reflect that it is one of the fundamental principles of molecular biology.

During the transcription process, the information stored in the DNA of the gene is transferred to a messenger molecule in the cell nucleus called mRNA . We assume that a given gene produces mRNA molecules at a constant rate λ_1 w.r.t. some time unit, e.g., hours.

During the translation process, the mRNA interacts with a ribosome which reads the sequence of mRNA bases. A transfer RNA called tRNA assembles the protein. We further assume that the molecules fabricate proteins P at rate λ_2. The mRNA and the proteins are degraded at rate λ_3, and resp. λ_4. The evolution of the process is summarized by the following diagram

$$
\begin{aligned}
\text{Gene} &\xrightarrow{\lambda_1} \text{Gene} + \text{mRNA} \\
\text{mRNA} &\xrightarrow{\lambda_2} \text{mRNA} + \text{Proteins} \\
\text{mRNA} &\xrightarrow{\lambda_3} \emptyset \quad \text{and} \quad \text{Proteins} \xrightarrow{\lambda_4} \emptyset.
\end{aligned}
$$

We assume that the number of genes in the system is constant over time, and we write $X_t = (Y_t, Z_t)$ for the $\mathbb{N}^2$-valued process of the number Y_t of mRNA molecules and the number Z_t of proteins.

$$
\begin{aligned}
L(f)(y, z) = \ & \lambda_1 \left[f(y+1, z) - f(y, z) \right] + \lambda_2 y \left[f(y, z+1) - f(y, z) \right] \\
& + \lambda_3 y \left[f(y-1, z) - f(y, z) \right] + \lambda_4 z \left[f(y, z-1) - f(y, z) \right].
\end{aligned}
$$

We observe that the last three rates are null as soon as both the number the molecules and the number of proteins vanish. This ensures that the process stays in $\mathbb{N}^2$.

28.4   Branching processes

28.4.1   Birth and death models with linear rates

We consider a population of individuals who die at a rate of r_d, and divide into two offsprings at a rate of r_b. We assume that these mechanisms are independent from one individual to another. The total size of the population X_t is a Markov process X_t on $S = \mathbb{N}$ with the generator given by

$$
L(f)(x) = \lambda_b(x)[f(x+1) - f(x)] + \lambda_d(x)[f(x-1) - f(x)]
$$

with $\lambda_b(x) = r_b x$ and $\lambda_d(x) = r_d x$.

$$
\begin{aligned}
P_t(x, y) &= \mathbb{P}\left(X_t = y \mid X_0 = x \right) \\
&= \mathbb{E}\left(f_y(X_t) \mid X_0 = x \right) = P_t(f_y)(x) \quad \text{with} \quad f_y = 1_y.
\end{aligned}
$$

Notice that the extinction probability, starting with a single individual, is given by

$$
e(t) := P_t(1, 0) = P_t(f_0)(1)
$$

and its evolution is given by the forward equation

$$
\begin{aligned}
\frac{d}{dt}P_t(1,0) &= \frac{d}{dt}P_t(f_0)(1) = L_t(P_t(f_0))(1) \\
&= r_b[P_t(f_0)(2) - P_t(f_0)(1)] + r_d\,[P_t(f_0)(0) - P_t(f_0)(1)] \\
&= r_b P_t(2,0) + r_d\,\underbrace{P_t(0,0)}_{=1} - [r_b + r_d]\,P_t(1,0).
\end{aligned}
$$

On the other hand, for any $x \in \mathbb{N}$ we have

$$
P_t(x,0) = P_t(1,0)^x.
$$

For $x = 0$, the result is obvious. This claim is a consequence of the independence between the individual branching processes. Starting the branching process from x individuals is equivalent to starting the branching process x times with a single individual. This implies that

$$
\begin{aligned}
e'(t) &= r_b e(t)^2 + r_d - [r_b + r_d]\,e(t) \\
&= (e(t) - 1)\,(r_b e(t) - r_d) = r_b\,(e(t) - 1)(e(t) - \tau)\ \ \text{with } \tau = r_d/r_b
\end{aligned}
$$

and with the initial condition $e(0) = P_0(1,0) = 0$.

When $r_d = r_b = r$, we set

$$
e_r(t) := e(t/r).
$$

The main simplification of this scaling follows from

$$
\begin{aligned}
e_r'(t) &= e'(t/r)/r = 1 + e(t/r)\,[e(t/r) - 2] \\
&= 1 + e_r(t)(e_r(t) - 2) \\
&= (e_r(t) - 1)^2 \qquad \text{with} \quad e_r(0) = 0.
\end{aligned}
$$

This implies that

$$
\left(-\frac{1}{e_r - 1}\right)' = \frac{e_r'}{(e_r - 1)^2} = 1 \ \Rightarrow\ \frac{1}{e_r(0) - 1} - \frac{1}{e_r(t) - 1} = t.
$$

It follows that

$$
\frac{1}{e_r(t) - 1} = -(1 + t) \Rightarrow e_r(t) = 1 - \frac{1}{1 + t}.
$$

We conclude that

$$
e(t) = 1 - \frac{1}{1 + rt}.
$$

When $r_d \neq r_b$, we set

$$
\bar{e}(t) := e(t/r_b) \Rightarrow d\bar{e}' = (\bar{e} - 1)(\bar{e} - \tau)\,dt.
$$

Now we use

$$
\frac{1}{\tau - 1}\left[\frac{1}{x - \tau} - \frac{1}{x - 1}\right] = \frac{1}{(x - \tau)(x - 1)}
$$

to prove that

$$
\begin{aligned}
dt &= \frac{d\bar{e}'}{(\bar{e} - 1)(\bar{e} - \tau)} = \frac{1}{\tau - 1}\left[\frac{d\bar{e}'}{\bar{e} - \tau} - \frac{d\bar{e}'}{\bar{e} - 1}\right] \\
&= \frac{1}{\tau - 1}\,d\left(\log\frac{\bar{e} - \tau}{\bar{e} - 1}\right).
\end{aligned}
$$

This implies that

$$\log \frac{\overline{e}(t) - \tau}{\overline{e}(t) - 1} - \underbrace{\log \frac{\overline{e}(0) - \tau}{\overline{e}(0) - 1}}_{=\log(\tau)} = (\tau - 1)(t - 0)$$

from which we prove that

$$\frac{\overline{e}(t) - \tau}{\overline{e}(t) - 1} = \tau \, \exp\left((\tau - 1)t\right) = 1 + \frac{1 - \tau}{\overline{e}(t) - 1}.$$

We conclude that

$$\overline{e}(t) = 1 - \frac{1 - \tau}{1 - \tau \, \exp\left((\tau - 1)t\right)}$$

and

$$e(t) = 1 - \frac{1 - \tau}{1 - \tau \, \exp\left((r_d - r_b)\, t\right)}.$$

When $r_d > r_b$, we have $\tau > 1$ and conclude that the system collapses exponentially fast

$$e(t) = 1 - \frac{\tau - 1}{\tau \, \exp\left((r_d - r_b)\, t\right) - 1} \simeq_{t\uparrow\infty} 1 - \exp\left(-(r_d - r_b)\, t\right) \uparrow_{t\uparrow\infty} 1.$$

When, on the contrary, $r_d < r_b$ holds, we have $\tau < 1$ and

$$
\begin{aligned}
e(t) &= 1 - \frac{1 - \tau}{1 - \tau \, e^{-(r_b - r_d)t}} \\
&= 1 - \frac{\left[1 - \tau e^{-(r_b - r_d)t}\right] - \tau(1 - e^{-(r_b - r_d)t})}{1 - \tau \, e^{-(r_b - r_d)\, t}} \\
&= \tau \, \frac{e^{(r_b - r_d)t} - \tau - (1 - \tau)}{e^{(r_b - r_d)t} - \tau} = \tau \left(1 - \frac{1 - \tau}{e^{(r_b - r_d)t} - \tau}\right).
\end{aligned}
$$

Hence in this case we conclude that the survival of the system increases exponentially fast to τ

$$0 \leq \tau - e(t) = \frac{\tau \, (1 - \tau)}{e^{(r_b - r_d)t} - \tau} \simeq_{t\uparrow\infty} \tau \, (1 - \tau)e^{-(r_b - r_d)t} \downarrow_{t\uparrow\infty} 0.$$

28.4.2 Discrete time branching processes

We denote by $\mathbf{S} = \cup_{p\geq 0}S^p$ the state space of a branching process with individuals taking values on some measurable state space S. The integer $p \geq 0$ represents the size of the population. It is implicitly assumed that the product spaces S^p are the p-symmetric product spaces (i.e., the orders of the states $(x_1, \ldots, x_p) \in S^p$ are not important), and test functions on S^p are symmetric. For $p = 0$ we use the convention $S^0 := \{c\}$, where c stands for an auxiliary cemetery state. The state c is an isolated point and functions f on S are extended to $S \cup c$ by setting $f(c) = 0$.

We let $M_n(x, dy)$, with $n \geq 1$, be a sequence of Markov transitions from S into inself, and we denote by $(g_n^i(x))_{i\geq 1, x\in S, n\geq 0}$ a collection of integer number-valued random variables with uniformly finite first moments. We further assume that for any $x \in S$ and any $n \geq 0$, $(g_n^i(x))_{i\geq 1}$ are identically distributed, and we set

$$G_n(x) := \mathbb{E}(g_n^i(x)).$$

Our branching process is defined as follows. We start at some point x_0 with a single particle, that is $p_0 = 1$ and $\zeta_0 = \zeta_0^1 = x_0 \in S^{p_0} = S$. This particle branches into $\widehat{p}_0$ offsprings $\widehat{\zeta}_0 = (\widehat{\zeta}_0^1, \ldots, \widehat{\zeta}_0^{\widehat{p}_0}) \in S^{\widehat{p}_0}$, with $\widehat{p}_0 = g_0^1(\zeta_0^1)$.

Each of these individuals explores randomly the state space S, according to the transition M_1. At the end of this mutation step, we have a population of $p_1 = \widehat{p}_0$ particles $\zeta_1^i \in S$ with distribution $M_1(\widehat{\zeta}_0^i, .)$, $i = 1, \ldots, p_1$. Then each of these particles ζ_1^i branches into $g_1^1(\zeta_1^i)$ offsprings. At the end of this transition, we have $\widehat{p}_1$ particles $\widehat{\zeta}_1 = (\widehat{\zeta}_1^1, \ldots, \widehat{\zeta}_1^{\widehat{p}_1}) \in S_1^{\widehat{p}_1}$, with $\widehat{p}_1 = \sum_{i=1}^{p_1} g_1^i(\zeta_1^i)$.

Then each of these individuals explores randomly the state space S, according to the transition M_2, and so on.

Whenever the system dies, $\widehat{p}_n = 0$ at a given time n, we set $\widehat{\zeta}_q = \zeta_{q+1} = c$, and $\widehat{p}_q = p_{q+1} = 0$, for any $q \geq n$.

By construction, we have $p_{n+1} = \widehat{p}_n$, and $\sum_{i=1}^{\widehat{p}_n} f(\widehat{\zeta}_n^i) = \sum_{i=1}^{p_n} g_n^i(\zeta_n^i) \, f(\zeta_n^i)$, for any function $f \in \mathcal{B}_b(S)$. If we consider the random measures

$$\mathcal{X}_n = \sum_{i=1}^{p_n} \delta_{\zeta_n^i} \quad \text{and} \quad \widehat{\mathcal{X}}_n = \sum_{i=1}^{\widehat{p}_n} \delta_{\widehat{\zeta}_n^i},$$

we find that

$$\mathbb{E}(\widehat{\mathcal{X}}_n(f) \mid \zeta_n) = \mathcal{X}_n(G_n \, f) \quad \text{and} \quad \mathbb{E}(\mathcal{X}_{n+1}(f) \mid \widehat{\zeta}_n) = \widehat{\mathcal{X}}_n(M_n(f)).$$

This clearly implies that

$$\mathbb{E}(\mathcal{X}_{n+1}(f) \mid \zeta_n) = \mathcal{X}_n(G_n M_{n+1}(f)).$$

We readily conclude that the first moments of the branching distributions $\mathcal{X}_n$ are given by the Feynman-Kac model

$$\mathbb{E}(\mathcal{X}_n(f)) = \mathbb{E}_{x_0}\left(f(X_n) \prod_{0 \leq k < n} G_k(X_k) \right) := \gamma_n(f).$$

Here X_n is a Markov chain on S with Markov transitions M_n. In this interpretation, the mean number of individuals in the current population is given by $\mathbb{E}(\mathcal{X}_n(1)) = \gamma_n(1)$.

In probability theory, the stochastic process $\mathcal{X}_n$ is called a branching Markov chain. The long-time behavior of these branching models, their connections with particle absorption models, and their applications in physics and biology are rapidly developing subjects in probability theory. We refer the reader to a series of articles [6, 5, 7, 23, 33, 156, 167, 168, 206], to more recent studies [1, 12, 18, 36, 88, 89, 146, 242], and to the references therein.

28.4.3 Continuous time branching processes

As their discrete time analogue, continuous time branching processes ξ_t take values in the state space $\mathbf{S} = \cup_{p \geq 0} S^p$ presented in section 28.4.2. The generator of a branching process is defined for any sufficiently regular function F on $\mathcal{S}$ and any $x = (x_1, \ldots, x_p) \in S^p$ by the a jump type generator

$$\mathcal{L}_t(F)(x) = \mathcal{L}_t^m(F)(x) + \mathcal{L}_t^b(F)(x). \tag{28.8}$$

The mutation generator

$$\mathcal{L}_t^m(F)(x) = \sum_{1 \leq i \leq p} L_t^{(i)}(F)(x)$$

represents the independent evolution of the particles between the branching transitions with some generator L_t. The upper index $L_t^{(i)}$ means that the generator L_t only acts on the i-th coordinate $x_i \mapsto F(x_1, \ldots, x_i, \ldots, x_p)$. The branching generator is a jump process in $\mathcal{S}$ given by

$$\mathcal{L}_t^b(F)(x) = \lambda_t(x) \int_{\mathcal{S}} (F(y) - F(x)) \, \mathcal{K}_t(x, dy)$$

for some branching rate function $\lambda_t(x)$ and some Markov transitions $\mathcal{K}_t(x, dy)$ on the set $\mathcal{S}$.

The state c is an absorbing state so that $\mathcal{K}_t(c, dy) = \delta_c(dy)$ and $\mathcal{L}_t^b(F)(c) = 0 = \mathcal{L}_t^m(F)(c)$ (recall the convention $\sum_\emptyset = 0$).

Let ξ_t is be the Markov process with generator $\mathcal{L}_t$. By the Doeblin-Itō formula, for sufficiently regular functions $F(t, x)$ on $([0, \infty[\times \mathcal{S})$ we have

$$dF(t, \xi_t) = (\partial_t + \mathcal{L}_t) \, F(t, \xi_t) \, dt + d\mathcal{M}_t(F) \tag{28.9}$$

with a collection of martingales $\mathcal{M}_t(F)$ with angle-bracket defined in terms of the carré du champ operators $\Gamma_{\mathcal{L}_t}$ associated with the generators $\mathcal{L}_t$ and given by the formulae

$$\langle \mathcal{M}(F), \mathcal{M}(F) \rangle_t = \int_0^t \Gamma_{\mathcal{L}_s}(F(s, .), F(s, .))(\xi_s) \, ds. \tag{28.10}$$

Some important classes of functions F are given by

$$F(t, \xi_t) = \mathcal{X}_t(f) \quad \text{with the occupation measures} \quad \mathcal{X}_t := \sum_{1 \leq i \leq N_t} \delta_{\xi_t^i} \tag{28.11}$$

where N_t stands for the size of the system at time t.

The branching transitions $\mathcal{K}_t(x, dy)$ often take the form

$$\mathcal{K}_t(x, dy) = \sum_{1 \leq i \leq p} \alpha_t^{(i)}(x) \sum_{q \geq 0} \rho_t^{(i)}(x, q) \int_{\mathcal{S} \cup \{c\}} K_t^{(i)}((x, q), du) \, \delta_{x_q^i(u)}(dy)$$

with

$$x_q^i(u) = \left(x_1, \ldots, x_{i-1}, \underbrace{u, \ldots, u}_{q-times}, x_{i+1}, \ldots, x_p \right) \in \mathcal{S}^{(p-1)+q}$$

for any $x = (x_1, \ldots, x_p) \in S^p$, for some $p \geq 1$. In the above display, $\alpha_t^{(i)}(x)$ is a probability measure on the index $i \in \{1, \ldots, p\}$; $\rho_t^{(i)}(x, q)$ is a probability measure on the index $q \in \mathbb{N}$, and $K_t^{(i)}((x, q), du)$ is a Markov transition from $(\mathcal{S} \times \mathbb{N})$ into $(\mathcal{S} \cup \{c\})$. In this construction when $q = 0$, the function $\rho_t^{(i)}(x, 0)$ stands for the probability of extinction of the i-th individual. In this situation, we take the convention $K_t^{(i)}((x, 0), du) = \delta_c(du)$.

At a rate $\lambda_t(x)$, an individual with label i is selected in the pool $x = (x_1, \ldots, x_p)$ with probability $\alpha_t^{(i)}(x)$. Then we chose randomly a number of individuals q with probability $\rho_t^{(i)}(x, q)$. If $q = 0$, the i-th individual dies. Otherwise, we select a state u with probability $K_t^{(i)}((x, q), du)$; and we replace the i-th individual by q copies of u.

In the next sections, we illustrate these models with two important examples. Further examples can be found in the article [79].

28.4.3.1 Absorption-death process

In the further development of this section we discuss the branching generator $\mathcal{L}_t^{b,-}$ associated with the parameters:

$$\lambda_t(x) = \sum_{1 \leq i \leq p} U_t(x_i) \qquad \alpha_t^{(i)}(x) = \frac{U_t(x_i)}{\sum_{1 \leq j \leq p} U_t(x_j)} \quad \text{and} \quad \rho_t^{(i)}(x,q) = 1_{q=0}$$

for some positive function U_t on S and for any $x = (x_1, \ldots, x_p) \in S^p$, for some $p \geq 1$. In this situation, each individual x_i dies at rate $U_t(x_i)$. We denote by $\mathcal{L}^{b,-}$ the corresponding branching generator.

For functions of the form

$$F(x) = p\, m(x)(\varphi) = \sum_{1 \leq j \leq p} f(x_j) \quad \text{with the empirical measures} \quad m(x) = \frac{1}{p} \sum_{1 \leq i \leq p} \delta_{x_i}$$

for any $x = (x_1, \ldots, x_p) \in S^p$, for some $p \geq 1$, we have

$$K_t^{(i)}(F)(x,0) - F(x) = -f(x_i)$$

$$\implies \mathcal{L}_t^{b,-}(F)(x) = \lambda_t(x)\, [K_t(F)(x) - F(x)] = -\sum_{1 \leq i \leq p} U_t(x_i)\, f(x_i) = -p\, m(x)(U_t f).$$

Using

$$L_t^{(i)}(F)(x) = L_t(f)(x_i) \implies \mathcal{L}_t^m(F)(x) = \sum_{1 \leq i \leq p} L_t(f)(x_i) = p\, m(x)(L_t(f))$$

we conclude that

$$F(x) = p\, m(x)(\varphi) \implies \mathcal{L}_t(F)(x) = p\, m(x)(L_t(f)) - p\, m(x)(U_t f).$$

Recalling that

$$\Gamma_{\mathcal{L}_t^{b,-}}(F,F)(x) = \lambda_t(x) \int_S (F(y) - F(x))^2\, K_t(x,dy)$$

and using

$$\int K_t^{(i)}((x,0),dy)[F(y) - F(x)]^2 = f(x_i)^2$$

we also find that

$$\Gamma_{\mathcal{L}_t^{b,-}}(F,F)(x) = p\, m(x)(U_t f^2).$$

Recalling that

$$\Gamma_{\mathcal{L}_t^m}(F,F)(x) = p\, m(x)\, (\Gamma_{L_t}(f,f))$$

we conclude that

$$\Gamma_{\mathcal{L}_t}(F,F)(x) = \Gamma_{\mathcal{L}_t^m}(F,F)(x) + \Gamma_{\mathcal{L}_t^{b,-}}(F,F)(x) = p\, m(x)\, (\Gamma_{L_t}(f,f)) + p\, m(x)(U_t f^2).$$

By applying (28.9) to the functions (28.11), we find that

$$d\mathcal{X}_t(f) = [\mathcal{X}_t(L_t(f)) - \mathcal{X}_t(U_t f)]\; dt + dM_t(f)$$

with a collection of martingales $M_t(f)$ with angle bracket given by the formulae

$$\langle M(f), M(f) \rangle_t = \int_0^t \left[\mathcal{X}_s \left(\Gamma_{L_t}(f, f) \right) + \mathcal{X}_s(U_s f^2) \right] \, ds.$$

We quote a direct consequence of this result. Note

$$\gamma_t(f) := \mathbb{E} \left(\mathcal{X}_t(f) \right) \Rightarrow \partial_t \gamma_t(f) = \gamma_t(L_t(f)) - \gamma_t(U_t f).$$

We further assume that the initial configuration $(\xi_0^i)_{1 \leq i \leq N_0}$ is given by N_0 independent random variables with some common law η_0 on S, for some given non-random $N_0 \geq 1$. In this situation, we have

$$\gamma_0(f) := \mathbb{E} \left(\mathcal{X}_0(f) \right) = N_0 \, \mathbb{E}(f(X_0)) = N_0 \, \eta_0(f).$$

These first moment measures coincide with the Feynman-Kac measures discussed in section 16.1.3 (see also section 15.6.1, exercise 294, exercises 317 and 318, and section 27.2.2). We have

$$\gamma_t(f) := \mathbb{E} \left(\mathcal{X}_t(f) \right) = N_0 \, \mathbb{E} \left[f(X_t) \, \exp \left\{ - \int_0^t U_s(X_s) \, ds \right\} \right]$$

where X_t is a Markov process with generator L_t on S with initial distribution $\eta_0 = \mathrm{Law}(X_0)$.

28.4.3.2   Birth type branching process

The simplest example of birth type branching process ξ_t is given by the spontaneous birth type generator

$$\mathcal{L}_t^{b,0}(F)(x) = \mu(1) \int_S \left(F(x, y) - F(x) \right) \overline{\mu}(dy)$$

where μ stands for some non-negative measure on S and $\overline{\mu}(dy) = \mu(dy)/\mu(1)$ the normalized probability measure. At rate $\mu(1)$ a new individual is added to the population. Thus at any time t the number of individuals is given by a Poisson process N_t with rate $\mu(1)$. Given $N_t = n$, the population $\xi_t = \left(\xi_t^1, \ldots, \xi_t^n \right)$ is formed by n independent random variables with common law $\overline{\mu}$. This spatial Poisson process is the continuous time version of the discrete generation model discussed in section 4.6. Observe that for any $x = (x_1, \ldots, x_p) \in S^p$, we have

$$F(x) = p \, m(x)(\varphi) \implies \mathcal{L}_t^{b,0}(F)(x) = \mu(f) \quad \text{and} \quad \Gamma_{\mathcal{L}_t^{b,0}}(F, F)(x) = \mu(f^2).$$

By applying (28.9) to the functions (28.11), we find that

$$d\mathcal{X}_t(f) = [\mathcal{X}_t(L_t(f)) + \mu(f)] \, dt + dM_t(f)$$

with a collection of martingales $M_t(f)$ with angle-bracket given by the formulae

$$\langle M(f), M(f) \rangle_t = \int_0^t \left[\mathcal{X}_s \left(\Gamma_{L_t}(f, f) \right) + \mu(f^2) \right] \, ds.$$

In the further development of this section we discuss the branching generator $\mathcal{L}_t^{b,+}$ associated with the parameters:

$$
\begin{aligned}
\lambda_t(x) &= \sum_{1 \leq i \leq p} V_t(x_i) & \alpha_t^{(i)}(x) &= \frac{V_t(x_i)}{\sum_{1 \leq j \leq p} V_t(x_j)} \\
\rho_t^{(i)}(x, q) &= 1_{q=2} & K_t^{(i)}((x, 2), du) &= \delta_{x_i}(du)
\end{aligned}
$$

for some positive function V_t on S, and for any $x = (x_1, \ldots, x_p) \in S^p$, for some $p \geq 1$. In this situation, each individual x_i duplicates at rate $V_t(x_i)$. We denote by $\mathcal{L}^{b,+}$ the corresponding branching generator.

For functions of the form

$$F(x) = p \, m(x)(\varphi) = \sum_{1 \leq j \leq p} f(x_j)$$

we have

$$K_t^{(i)}(F)(x, 2) - F(x) = f(x_i)$$

$$\implies \mathcal{L}_t^{b,+}(F)(x) = \lambda_t(x) \left[\mathcal{K}_t(F)(x) - F(x) \right] = \sum_{1 \leq i \leq p} V_t(x_i) \, f(x_i) = p \, m(x)(V_t f).$$

This implies that

$$F(x) = p \, m(x)(\varphi) \implies \mathcal{L}_t(F)(x) = p \, m(x)(L_t(f)) + p \, m(x)(V_t f).$$

Arguing as above, we also find that

$$\Gamma_{\mathcal{L}_t^b}(F, F)(x) \quad = \quad p \, m(x)(V_t f^2)$$

and

$$\Gamma_{\mathcal{L}_t}(F, F)(x) = \Gamma_{\mathcal{L}_t^m}(F, F)(x) + \Gamma_{\mathcal{L}_t^{b,+}}(F, F)(x) = p \, m(x) \, (\Gamma_{L_t}(f, f)) + p \, m(x)(V_t f^2).$$

After applying (28.9) to the functions (28.11), we find that

$$d\mathcal{X}_t(f) = \left[\mathcal{X}_t(L_t(f)) + \mathcal{X}_t(V_t f) \right] \, dt + dM_t(f)$$

with a collection of martingales $M_t(f)$ with angle bracket given by the formulae

$$\langle M(f), M(f) \rangle_t = \int_0^t \left[\mathcal{X}_s \left(\Gamma_{L_s}(f, f) \right) + \mathcal{X}_s(V_s f^2) \right] \, ds.$$

Arguing as above, we have the Feynman-Kac formulae

$$\gamma_t(f) = \mathbb{E}\left(\mathcal{X}_t(f) \right) = N_0 \, \mathbb{E}\left[f(X_t) \, \exp \left\{ \int_0^t V_s(X_s) \, ds \right\} \right]$$

where X_t is a Markov process with generator L_t on S with initial distribution $\eta_0 = \text{Law}(X_0)$.

28.4.3.3 Birth and death branching processes

Consider the generator defined by

$$\mathcal{L}_t = \mathcal{L}_t^m + \mathcal{L}_t^{b,-} + \mathcal{L}_t^{b,0} + \mathcal{L}_t^{b,+}$$

with the generators $\mathcal{L}_t^{b,-}$, $\mathcal{L}_t^{b,0}$ and $\mathcal{L}_t^{b,+}$ introduced in section 28.4.3.1 and section 28.4.3.2.

The state space S can be interpreted as a trait space. The first generator $\mathcal{L}_t^m$ represents the mutation transitions of the traits of the individuals. The function $u \in S \mapsto U_t(u) \in [0, \infty[$ represents the rate of natural death of an individual with traits u, while $u \in S \mapsto V_t(u) \in [0, \infty[$ represents the rate of natural birth of an individual with traits u.

The applications of these branching processes to model the microscopic behavior of individual trait-based models have been studied intensively by Nicolas Champagnat and his co-authors [45, 46, 47]. We also refer the reader to the earlier developments [60, 61, 64, 66, 79, 81] (and references therein) discussing applications branching processes including fixed population size processes in evolutionary computing, physics and chemistry, stochastic optimization, and nonlinear filtering.

The Doeblin-Itō formula presented in (28.9) and (28.10) is satisfied with the carré du champ operators $\Gamma_{\mathcal{L}_t}$

$$\Gamma_{\mathcal{L}_t}(F, F)(x) = \Gamma_{\mathcal{L}_t^m}(F, F)(x) + \Gamma_{\mathcal{L}_t^{b,-}}(F, F)(x) + \Gamma_{\mathcal{L}_t^{b,0}}(F, F)(x) + \Gamma_{\mathcal{L}_t^{b,+}}(F, F)(x).$$

Arguing as above, we find that

$$d\mathcal{X}_t(f) = [\mathcal{X}_t(L_t(f)) + \mathcal{X}_t((V_t - U_t)f) + \mu(f)] \ dt + dM_t(f)$$

with a collection of martingales $M_t(f)$ with angle bracket given by the formulae

$$\langle M(f), M(f) \rangle_t = \int_0^t \left[\mathcal{X}_s\left(\Gamma_{L_t}(f, f)\right) + \mathcal{X}_s((U_t + V_t)f^2) + \mu(f^2) \right] \ ds.$$

When $\mu = 0$, we have

$$\gamma_t(f) := \mathbb{E}\left(\mathcal{X}_t(f)\right) = N_0 \ \mathbb{E}\left[f(X_t) \ \exp\left\{ \int_0^t (V_s - U_s)(X_s) \ ds \right\} \right] \qquad (28.12)$$

where X_t is a Markov process with generator L_t on S with initial distribution $\eta_0 = \mathrm{Law}(X_0)$. We also notice that

$$\partial_t \mathbb{E}\left(\langle M(f), M(f) \rangle_t\right) = \gamma_t\left(\Gamma_{L_t}(f, f) + (U_t + V_t)f^2\right).$$

From the computational viewpoint, the simulation of these branching processes often involves extinctions and explosions of the number of particles when the time horizon increases. For instance, starting from a given state $X_0 = x_0$, with time homogeneous birth and death rate functions $(U_t, V_t) = (U, V)$, and without mutations, the mean population size is given by

$$\gamma_t(1) = \mathbb{E}(N_t) = N_0 \ \exp\left\{ (V - U)(x_0) \ t \right\} \rightarrow_{t \to \infty} \begin{cases} \infty & \text{if } V(x_0) > U(x_0) \\ N_0 & \text{if } V(x_0) = U(x_0) \\ 0 & \text{if } V(x_0) < U(x_0). \end{cases}$$

When $\mu \neq 0$ we have

$$\gamma_t(f) := \mathbb{E}\left(\mathcal{X}_t(f)\right) \Rightarrow \partial_t \gamma_t(f) = \gamma_t(L_t(f)) + \gamma_t((V_t - U_t)f) + \mu(f).$$

The solution is given by

$$\gamma_t(f) = \gamma_0 Q_{0,t}(f) + \int_0^t \mu Q_{s,t}(f) \ ds$$

with the Feynman-Kac semigroup

$$Q_{s,t}(f)(x) = \mathbb{E}\left[f(X_t) \ \exp\left\{ \int_s^t (V_r - U_r)(X_r) \ dr \right\} \mid X_s = x \right].$$

Using the mean field particle approximations (16.16) or preferably (for computational purposes) their discrete time versions discussed in section 16.2 and section 9.6, we can estimate these first moment distributions γ_t (a.k.a. intensity measures) using a particle model with *fixed* population sizes $N_t = N_0 := N$. For a more detailed study on these spatial branching processes, we refer the reader to the exercises 463 through 465.

28.4.3.4 Kolmogorov-Petrovskii-Piskunov equation

The Kolmogorov-Petrovskii-Piskunov equation [173], also known as the Fisher's equation was introduced in 1937 by A.N. Kolmogorov, I.G. Petrovskii, N.S. Piskunov in [173] and by R.A. Fisher in [127]. This dynamic population model describes the spatial spread of an advantageous allele. In this section we provide an interpretation of these equations in terms of branching processes discussed in section 28.4.3, following the ideas of H.P. McKean [187, 188].

Consider the branching process generators (28.8) with a jump intensity λ_t defined for any $x = (x_1, \ldots, x_p) \in S^p$ by

$$\lambda_t(x) = p\,\lambda \quad \text{and} \quad \mathcal{K}_t(x, dy) := \frac{1}{p} \sum_{1 \leq i \leq p} \sum_{k \geq 1} \alpha_{t,k}\, \delta_{\theta_i^k(x)}(dy)$$

for some parameter $\lambda \geq 0$, some probability $\alpha = (\alpha_{t,k})_{k \geq 1}$ on $\mathbb{N} - \{0\}$ and with the functions

$$\theta_i^k(x) = \left(x_1, \ldots, x_{i-1}, \underbrace{x_i, \ldots, x_i}_{k\text{-times}}, x_{i+1}, \ldots, x_p \right).$$

The process ξ_t evolves as follows. At a rate λ, each individual ξ_t^i dies and instantly gives birth to k offsprings with a probability p_k. Between the branching times, each individual evolves independently as a Markov process with generator L_t.

We let $\xi_t^x = \left(\xi_t^{x,i} \right)_{1 \leq i \leq N_t^x}$ be the branching process starting from p individuals $(x_1, \ldots, x_p)$. Since there are no interactions between the individuals we have

$$\xi_t^x \stackrel{law}{=} \left(\xi_t^{x_1}, \ldots, \xi_t^{x_p} \right).$$

Thus, recalling that we are working with the p-symmetric product spaces and symmetric test functions, the Markov semigroup $\mathcal{P}_{s,t}$ of the branching process ξ_t satisfies the product formula

$$\mathcal{P}_{s,t}(f)(x_1, \ldots, x_p) = \int_{S^p} \mathcal{P}_{s,t}(x_1, dy_1) \times \ldots \times \mathcal{P}_{s,t}(x_p, dy_p)\, f(y_1, \ldots, y_p).$$

In particular, for regular polynomial/product test functions of the form

$$\forall p \geq 0 \quad \forall x = (x_1, \ldots, x_p) \in S^p \quad f(x) = \prod_{0 \leq i \leq p} g(x_i) \left(= \prod_{0 \leq i \leq p} f(x_i) \right)$$

for some $[0, 1]$-valued function g on S, we have

$$\forall 1 \leq i \leq p \quad y_i = (y_{i,1}, \ldots, y_{i,p_i}) \in S^{p_i} \Rightarrow f(y_1, \ldots, y_p) = \prod_{1 \leq i \leq p} \prod_{0 \leq j \leq p_i} g(y_{i,j}) = \prod_{1 \leq i \leq p} f(y_i)$$

from which we prove that

$$\mathcal{P}_{s,t}(f)(x_1, \ldots, x_p) = \prod_{1 \leq i \leq p} \mathcal{P}_{s,t}(f)(x_i)$$

as soon as $\mathcal{P}_{s,t}(f)(x) < \infty$, for any $x \in S$. We also notice that

$$\mathcal{K}_r(f)(x) = \frac{1}{p} \sum_{1 \leq i \leq p} \sum_{k \geq 1} \alpha_{r,k}\, f(\theta_i^k(x)) = \frac{1}{p} \sum_{1 \leq i \leq p} \left[\sum_{k \geq 1} \alpha_{r,k}\, f(x_i)^k \right] \prod_{1 \leq j \neq i \leq p} f(x_j).$$

By the perturbation formula (15.28):

$$\mathcal{P}_{s,t}(f)(x) = e^{-\lambda(t-s)} \, \mathcal{P}_{s,t}^{(1)}(f)(x) + \int_s^t \lambda \, e^{-\lambda(s-r)} \, \mathcal{P}_{s,r}^{(1)} \left[\mathcal{K}_r \left(\mathcal{P}_{r,t}(f) \right) \right](x) \, dr \qquad (28.13)$$

with the Markov semigroup $\mathcal{P}_{s,t}^{(1)}$ of the process $\xi_t^{(1)}$ with generator $\mathcal{L}_t^m$ (corresponding to independent motions with generator L_t). By construction, for polynomial/product test functions we have

$$\mathcal{P}_{s,t}^{(1)}(f)(x_1, \ldots, x_p) = \prod_{1 \le i \le p} P_{s,t}^{(1)}(f)(x_i)$$

where $P_{s,t}^{(1)}$ stands for the semigroup of L_t. Choosing $p = 1$, this implies that for any $x \in S$ we have

$$\mathcal{P}_{s,t}(f)(x) = e^{-\lambda(t-s)} \left[P_{s,t}^{(1)}(f)(x) + \int_s^t \lambda \, e^{\lambda(t-r)} \left[\sum_{k \ge 1} \alpha_{r,k} \, P_{s,r}^{(1)}([\mathcal{P}_{r,t}(f)]^k)(x) \right] dr \right].$$

Taking the derivative w.r.t. s we find that

$$\begin{aligned}
\partial_s \mathcal{P}_{s,t}(f)(x) &= \lambda \, \mathcal{P}_{s,t}(f)(x) - L_s \mathcal{P}_{s,t}(f)(x) - \lambda \sum_{k \ge 1} \alpha_{s,k} \, [\mathcal{P}_{s,t}(f)](x)^k \\
&= -\left[L_s \mathcal{P}_{s,t}(f)(x) + \lambda \sum_{k \ge 1} \alpha_{s,k} \left([\mathcal{P}_{s,t}(f)](x)^k - \mathcal{P}_{s,t}(f)(x) \right) \right].
\end{aligned}$$

We fix the final time horizon t and some $[0,1]$-valued function f s.t. $\mathcal{P}_{s,t}(f)(x) < \infty$ exists for any $x \in S$ and $s \in [0,t]$. We also set $u : (s,x) \in ([0,t] \times S) \mapsto u_s(x) = \mathcal{P}_{s,t}(f)(x)$. In this notation, we have proved that

$$\partial_s u_s + L_s u_s + \lambda \sum_{k \ge 1} \alpha_{s,k} \left(u_s^k - u_s \right) = 0$$

for any $s \le t$, with a regular terminal condition $u_t = f$. For time homogeneous models $(L_s, \alpha_s) = (L, \alpha)$, we have $\mathcal{P}_{s,t}(f)(x) = \mathcal{P}_{0,t-s}(f)(x) := v_{t-s}(x)$ with

$$\partial_t v_t = L v_t + \lambda \sum_{k \ge 1} \alpha_k \left(v_t^k - v_t \right)$$

for any $t \ge 0$, with the initial condition $v_0 = f$.

The case $\alpha_k = 1_{k=2}$, $\lambda = 1$ and $L = \frac{1}{2} \partial_x^2$ yields the Kolmogorov-Petrovskii-Piskunov equations (a.k.a. KPP equations)

$$\partial_t v_t = \frac{1}{2} \, \partial_x^2 v_t + \left(v_t^2 - v_t \right). \qquad (28.14)$$

Further details on these equations can be found in exercises 467 through 469.

28.5 Exercises

Exercise 454 (Logistic diffusion process 1) *We consider the one-dimensional diffusion*

$$dX_t = X_t \ (\lambda - X_t) \ dt + \sigma \ X_t \ dW_t$$

where W_t stands for a Brownian motion, λ and σ are positive parameters. Show that the solution has the form

$$X_t = \frac{Y_t}{1 + \int_0^t Y_s ds} \tag{28.15}$$

for some geometric Brownian motion Y_t. Deduce that $X_0 = Y_0 \geq 0 \Rightarrow Y_t \geq X_t \geq 0$, for any time $t \geq 0$.

Exercise 455 (Logistic diffusion process 2) *We consider the one-dimensional diffusion*

$$dX_t = a \ X_t \ \left(1 - \frac{X_t}{b}\right) \ dt + \sigma \ X_t \ dW_t$$

where W_t stands for a Brownian motion and (a, b, σ) are positive parameters. Show that the solution has the form

$$X_t = \frac{Y_t}{1 + \frac{a}{b} \int_0^t Y_s ds} \quad \text{with} \quad Y_t = X_0 \ \exp\left(\left(a - \frac{\sigma^2}{2}\right) t + \sigma \ W_t\right). \tag{28.16}$$

Exercise 456 (Logistic model - Birth and death process) *Consider the deterministic logistic model (28.5) with $\alpha_1 = 1$. Check that*

$$x(t) = x(0) + (+1) \int_0^t \lambda_1 \ x(s) \ \left(1 - \frac{x(s)}{N}\right) \ ds + (-1) \int_0^t \lambda_2 \ x(s) \ \left(1 + \alpha_2 \ \frac{x(s)}{N}\right) \ ds.$$

We let N_t^1 and N^2 be two independent Poisson processes with unit intensity. Describe the generator of the process

$$\begin{aligned}
X_t \ = \ & X(0) + (+1) \ N^1 \left(\int_0^t \lambda_1 \ X(s) \ \left(1 - \frac{X(s)}{N}\right) \ ds\right) \\
& + (-1) N^2 \left(\int_0^t \lambda_2 \ X(s) \ \left(1 + \alpha_2 \ \frac{X(s)}{N}\right) \ ds\right).
\end{aligned}$$

Exercise 457 (Bimodal growth diffusion) *We consider the one-dimensional diffusion*

$$dX_t = a \ X_t \ \left(1 - \frac{X_t^2}{b}\right) \ dt + \sigma \ X_t \ dW_t$$

where W_t is a Brownian motion and (a, b, σ) are positive parameters. Find an explicit solution by applying the Doeblin-Itō formula to $f(X_t) = X_t^2$.

Exercise 458 (Facultative mutualism systems) *We consider the mutualism system between two species*

$$\begin{cases} x_t' \ = \ x_t \left((a_1 + b_{1,2} y_t) - b_{1,1} x_t\right) \\ y_t' \ = \ y_t \left((a_2 + b_{2,1} x_t) - b_{2,2} y_t\right). \end{cases} \tag{28.17}$$

These equations cannot be solved explicitly. Propose a stochastic version and a numerical simulation of this model.

Exercise 459 (SIS model) *We consider the SIS infection model between susceptible individuals x_t and infected individuals y_t for a population of size N, defined by*

$$\begin{cases} x_t' &= -\frac{\lambda_c}{N}\, xy + (\lambda_b + \lambda_r)\, y \\ y_t' &= \frac{\lambda_c}{N}\, xy - (\lambda_b + \lambda_r)\, y. \end{cases} \tag{28.18}$$

The parameter λ_c stands for the contact rate between the two populations, λ_r stands for the recovery rate of the infected individuals, and λ_b is the birth rate. The initial state is chosen so that $x_0 + y_0 = N$. These equations cannot be solved explicitly. Propose a stochastic version and a numerical simulation of this model.

Exercise 460 (Lotka-Volterra predator-prey - Birth and death process) *Consider the deterministic Lotka-Volterra predator-prey model (28.7). Check that*

$$\begin{pmatrix} x(t) \\ y(t) \end{pmatrix} = \begin{pmatrix} x(0) \\ y(0) \end{pmatrix} + \left[\int_0^t a\, x(s)\, ds \right] \begin{pmatrix} 1 \\ 0 \end{pmatrix} + \left[\int_0^t b\, x(s)y(s)\, ds \right] \begin{pmatrix} -1 \\ 0 \end{pmatrix}$$
$$+ \left[\int_0^t c\, y(s)\, ds \right] \begin{pmatrix} 0 \\ -1 \end{pmatrix} + \left[\int_0^t d\, x(s)y(s)\, ds \right] \begin{pmatrix} 0 \\ 1 \end{pmatrix}.$$

We let $(N_t^i)_{1 \le i \le 4}$ be four independent Poisson processes with unit intensity. Describe the generator of the process

$$\begin{pmatrix} X(t) \\ Y(t) \end{pmatrix}$$

$$= \begin{pmatrix} X(0) \\ Y(0) \end{pmatrix} + N^1 \left[\int_0^t a\, X(s)\, ds \right] \begin{pmatrix} 1 \\ 0 \end{pmatrix} + N^2 \left[\int_0^t b\, X(s)Y(s)\, ds \right] \begin{pmatrix} -1 \\ 0 \end{pmatrix}$$

$$+ N^3 \left[\int_0^t c\, Y(s)\, ds \right] \begin{pmatrix} 0 \\ -1 \end{pmatrix} + N^4 \left[\int_0^t d\, X(s)Y(s)\, ds \right] \begin{pmatrix} 0 \\ 1 \end{pmatrix}.$$

Exercise 461 (Kolmogorov equation) *We consider the Kolmogorov equation defined for any $x \in \mathbb{N} - \{0\}$ by*

$$\frac{d}{dt} p_t(x) = (a_1\,(x-1) + b)\, p_t(x-1) + a_2\,(x+1)\, p_t(x+1) - [b + (a_1 + a_2)\, x]\, p_t(x),$$

and for $x = 0$ by

$$\frac{d}{dt} p_t(0) = -b\, p_t(0) + a_2\, p_t(1),$$

with some positive parameters a_1, a_2 and $b \ge 0$. Find a stochastic process X_t such that $\mathrm{Law}(X_t) = p_t$. When $a_1 < a_2$, describe the invariant measure of this process.

Exercise 462 (Birth and death chain) *The aim of this exercise is to analyze the chance for $\alpha(x)$ to hit the origin when starting from certain state $X_0 = x \in \mathbb{N} = S$, that is, to find the probability of extinction of the population starting with x individuals. Let T be the first time the chain hits the origin. Check that*

$$\alpha(x) = p(x)\, \alpha(x+1) + q(x)\alpha(x-1) + \alpha(x)(1 - p(x) - q(x)) \tag{28.19}$$

and deduce that

$$\alpha(y+1) = 1 - (1 - \alpha(1)) \sum_{x=0}^{y} \left\{ \prod_{y=1}^{x} \frac{q(y)}{p(y)} \right\}. \tag{28.20}$$

We let $W := \sum_{x \ge 0} \left\{ \prod_{y=1}^{x} \frac{q(y)}{p(y)} \right\} \in\,]0, \infty]$.

- *Prove that $W = \infty \Rightarrow \alpha(y) = 1$ for any $y \in \mathbb{N}$.*

- *If $W < \infty$, check that for any choice of $\alpha(1) \geq 1 - W^{-1}$, (28.20) is a $[0,1]$-valued solution of (28.19).*

- *For any other solution $\beta(x)$ to (28.19) s.t. $\beta(0) = 1$, check that for any $x > 0$,*

$$\beta(x) = M(x,0) + \sum_{y_1 \geq 0} M(x,y_1) 1_{\neq 0}(y_1) \beta(y_1)$$

and deduce that

$$\begin{aligned} \beta(x) &= \mathbb{P}(T \leq n \mid X_0 = x) + \mathbb{E}(1_{T>n} \beta(X_n) \mid X_0 = x) \\ &\geq \mathbb{P}(T \leq n \mid X_0 = x) \uparrow_{n \uparrow \infty} \alpha(x). \end{aligned}$$

Conclude that the extinction probability $\alpha(x)$ coincides with the minimal solution.

Exercise 463 (Branching intensity measures) *Consider the intensity measures γ_t of the birth and death branching process defined in (28.12) without spontaneous branching ($\mu = 0$). We set $W_t := V_t - U_t$. We denote by η_t the normalized probability measures defined by $\eta_t(f) = \gamma_t(f)/\gamma_t(1)$.*

- *Check that*

$$\gamma_t(1) = \exp\left\{ \int_0^t \eta_s(W_s) \, ds \right\}$$

and

$$\eta_t(f) = \mathbb{E}\left(f(X_t) \exp\left\{ \int_0^t [W_s(X_s) - \eta_s(W_s)] \, ds \right\} \right).$$

- *Check that for any $s \leq t$ we have $\gamma_t = \gamma_s Q_{s,t}$, with the Feynman-Kac semigroup*

$$Q_{s,t}(f)(x) = \mathbb{E}\left(f(X_t) \exp\left\{ \int_s^t W_r(X_r) \, dr \right\} \mid X_s = x \right).$$

- *Prove that*

$$\partial_s Q_{s,t}(f) = -L_s^W(Q_{s,t}(f)) \quad \text{with the Schrödinger operator } L_s^W(f) = L_s(f) + W_s f.$$

- *Prove that*

$$\partial_t \eta_t(f) = \eta_t(L_{t,\eta_t}(f))$$

with the collection of generators

$$L_{t,\eta_t}(f) = L_t(f)(x) + U_t(x) \int (f(y) - f(x)) \, \eta_t(dy) + \int (f(y) - f(x)) \, V_t(y) \, \eta_t(dy).$$

- *Describe the evolution of the nonlinear jump diffusion process X_t with generator L_{t,η_t}. Describe the mean field particle model associated with this nonlinear process.*

Exercise 464 (Spontaneous branching - Renormalized measures [67]) *Consider the intensity measures γ_t of the birth and death branching process defined in (28.12) with non-necessarily null spontaneous branching. We denote by η_t the normalized probability measures defined by $\eta_t(f) = \gamma_t(f)/\gamma_t(1)$. Check that*

$$\begin{cases} \partial_t \gamma_t(1) &= \gamma_t(1) \, \eta_t(W_t) + \mu(1) \\ \partial_t \eta_t(f) &= \eta_t\left(\left(L_{t,\eta_t} + L_{t,\gamma_t(1),\eta_t}^0 \right)(f) \right) \end{cases}$$

with the jump type generator

$$L^0_{t,\gamma_t(1),\eta_t}(f)(x) = (\mu(1)/\gamma_t(1)) \int (f(y) - f(x)) \, \overline{\mu}(dy)$$

and the collection of generators L_{t,η_t} *presented in exercise 463. Describe the mean field particle model associated with this nonlinear process.*

Exercise 465 (Law of large numbers - Branching processes) ✎ *Consider the intensity measures of the birth and death branching process defined in (28.12) and further discussed in exercise 463. We let* $\xi_t^{(i)} := (\xi_t^{i,j})_{1 \leq i \leq N_t}$ *be the branching process starting at* ξ_0^i, *for each* $1 \leq i \leq N_0$ *(or equivalently the particles with ancestor* ξ_0^i *at the origin), and we set*

$$\forall 1 \leq i \leq N_0 \qquad \mathcal{X}_t^i := \sum_{1 \leq i \leq N_t} \delta_{\xi_t^{i,j}} \quad and \quad \overline{\mathcal{X}}_t = \frac{1}{N_0} \sum_{1 \leq i \leq N_0} \mathcal{X}_t^i.$$

We assume that $\xi_0 = (\xi_0^i)_{1 \leq i \leq N_0}$ *are* N_0 *independent random copies of* X_0.

- *Check that*

$$\mathbb{E}\left([\overline{\mathcal{X}}_t(\varphi) - \gamma_t(\varphi)]^2 \right) = \frac{1}{N_0} \, \mathbb{E}\left([\mathcal{X}_t^1(\varphi) - \gamma_t(\varphi)]^2 \right).$$

- *We further assume* $N_0 = 1$.

  - *Prove that*

  $$d\mathcal{X}_t(\varphi) = \mathcal{X}_t\left(L_t^W(\varphi)\right) \, dt + dM_t^{(1)}(\varphi)$$

  with a martingale $M_t^{(1)}(\varphi)$ *w.r.t.* $\mathcal{F}_t = \sigma(\xi_s, \, s \leq t)$ *with angle bracket*

  $$\partial_t \left\langle M^{(1)}(\varphi), M^{(1)}(\varphi) \right\rangle_t = \mathcal{X}_t \left[\Gamma_{L_t}(\varphi, \varphi) + (U_t + V_t) \, \varphi^2 \right].$$

  - *Deduce that*

  $$\begin{array}{llll}
  \mathbb{E}(N_t) & = & 1 & when \quad U_t = V_t \\
  \mathbb{E}(N_t) & \geq & e^{\epsilon t} & when \quad V_t \geq U_t + \epsilon \\
  \mathbb{E}(N_t) & \leq & e^{-\epsilon t} & when \quad V_t \leq U_t - \epsilon \qquad for \; some \; \epsilon \geq 0.
  \end{array}$$

  - *For any fixed time horizon* t, *prove that* $M_s^{(2)}(\varphi) := \mathcal{X}_s(Q_{s,t}(\varphi))$ *is a martingale for* $s \in [0, t]$ *with angle bracket*

  $$\partial_s \left\langle M^{(2)}(\varphi), M^{(2)}(\varphi) \right\rangle_s = \mathcal{X}_s \left[\Gamma_{L_s}(Q_{s,t}(\varphi), Q_{s,t}(\varphi)) + (U_s + V_s) \, (Q_{s,t}(\varphi))^2 \right].$$

  - *Deduce that*

  $$\mathbb{E}\left([\mathcal{X}_t(\varphi) - \mathcal{X}_0(Q_{0,t}(\varphi))]^2 \right)$$

  $$= \int_0^t \gamma_s \left[\Gamma_{L_s}(Q_{s,t}(\varphi), Q_{s,t}(\varphi)) + (U_s + V_s) \, (Q_{s,t}(\varphi))^2 \right] \, ds.$$

  - *Compute the derivative of the function* $s \mapsto \left(\gamma_s \left([Q_{s,t}(\varphi)]^2 \right) \right)$ *and check that*

  $$\mathbb{E}\left([\mathcal{X}_t(\varphi) - \gamma_t(\varphi)]^2 \right) = \gamma_t\left(\varphi^2\right) - \gamma_t\left(\varphi\right)^2 + 2 \int_0^t \gamma_s \left(V_s \left(Q_{s,t}(\varphi) \right)^2 \right) \, ds.$$

• *Deduce that*

$$N_0 \; \mathbb{E}\left(\left[\overline{\mathcal{X}}_t(\varphi) - \gamma_t(\varphi)\right]^2\right) = \gamma_t\left(\varphi^2\right) - \gamma_t\left(\varphi\right)^2 + 2\int_0^t \; \gamma_s\left(V_s\left(Q_{s,t}(\varphi)\right)^2\right) \; ds.$$

Exercise 466 (First moments of KPP equation) *We consider the spatial branching process $\xi_t = \left(\xi_t^i\right)_{1 \le i \le N_t}$ discussed in section 28.4.3 with the branching generator presented in section 28.4.3.4. We assume that the model is time homogeneous $(\mathcal{L}_t^m, L_t, \lambda_t, \alpha_t) = (\mathcal{L}^m, L_t, \lambda, \alpha)$ and the branching distribution is chosen so that $\overline{\alpha} := \sum_{k \ge 1} \alpha_k \; k < \infty$ and $\sigma^2 := \lambda \sum_{k \ge 1} \alpha_k \; (k-1)^2 < \infty$. We let $\mathcal{X}_t$ be the occupation measures defined in (28.11), and for any regular function f on S we set $\gamma_t(f) := \mathbb{E}\left(\mathcal{X}_t(f)\right)$. Check that*

$$d\mathcal{X}_t(f) = \mathcal{X}_t(L(f)) + \lambda \; \mathcal{X}_t(f) \; (\overline{\alpha} - 1) \; dt + dM_t(f)$$

with a collection of martingales $M_t(f)$ with angle bracket given by the formulae

$$\partial_t \left\langle M(f), M(f)\right\rangle_t = \left[\mathcal{X}_t\left(\Gamma_L(f, f)\right) + \sigma^2 \; \mathcal{X}_t(f^2)\right].$$

Deduce that

$$dN_t = \lambda \; N_t \; (\overline{\alpha} - 1) \; dt + dM_t$$

with a martingale M_t with angle bracket given by the formulae

$$\partial_t \left\langle M_t, M_t\right\rangle_t = \sigma^2 \; N_t.$$

Check that

$$\gamma_t(f) := N_0 \; e^{\lambda(\overline{\alpha}-1)t} \; \mathbb{E}\left[f(X_t)\right] \quad and \quad \frac{1}{t} \log \mathbb{E}(N_t/N_0) = \lambda\left(\overline{\alpha} - 1\right).$$

Examine the Kolmogorov-Petrovskii-Piskunov model discussed in (28.14).

Exercise 467 (Brownian branching process - KPP equation) *Consider a branching process $\xi_t = \left(\xi_t^i\right)_{1 \le i \le N_t}$ starting with a single Brownian motion starting at the origin. At unit rate each particle splits into two particles. Between these branching times, the particles evolve independently according to Brownian motions. For any function f on $\mathbb{R}$, using the developments of section 28.4.3.4, check that*

$$v_t(x) = \mathbb{E}\left(\prod_{1 \le i \le N_t} f\left(x + \xi_t^i\right)\right)$$

satisfies the Kolmogorov-Petrovskii-Piskunov equations (28.14) with the initial condition $v_0 = f$, for any given regular $[0,1]$-valued function f.

Exercise 468 (Brownian branching process - Right-most particle) *Consider the Brownian branching process discussed in exercise 467. Check that*

$$v_t(x) = \mathbb{P}\left(\sup_{1 \le i \le N_t} \xi_t^i \le x\right)$$

satisfies the the Kolmogorov-Petrovskii-Piskunov equation (28.14), with the initial condition $v_0 = 1_{[0,\infty[}$. Deduce that the function $v_t \; : \; x \in \mathbb{R} \mapsto [0,1]$ is strictly increasing from 0 to 1. Check that for any $\epsilon \in]0,1[$ there exists some $x_\epsilon(t)$ such that $v_t(x_\epsilon(t)) = \epsilon$.

Exercise 469 (Binary fission process - Generating function) *Consider the Brownian branching process discussed in exercise 467. Check that the function* $g_t(r) := \mathbb{E}\left(r^{N_t}\right)$, *with* $r \in]0, 1[$ *satisfies the equation*

$$\partial_t g_t(r) = g_t(r)\left(g_t(r) - 1\right).$$

Deduce that

$$\mathbb{E}\left(r^{N_t}\right) = \frac{re^{-t}}{(1 - r) + re^{-t}}.$$

When $r \geq 1$, *check that the solution blows up at the time* $t_b = -\log\left(1 - 1/r\right)$.

29

Gambling, ranking and control

This chapter is dedicated to some applications of stochastic processes in gambling theory and in operations research. We start with a brief discussion of the Google page rank algorithm and of ranking techniques based on the limiting behavior of Markov chains. The second part of the chapter is concerned with gambling betting systems and with martingale theory. We present and analyse a series of famous martingales such as the St. Petersburg technique, the grand martingale, and the D'Alembert and the Whittacker martingales. The last part of the chapter is dedicated to stochastic optimal control. We discuss discrete and continuous time control problems, as well as optimal stopping techniques.

> *If you must play, decide upon three things at the start:*
> *the rules of the game, the stakes, and the quitting time.*
> `Chinese Proverb.`

29.1   Google page rank

Internet search engines are based on ranking algorithms. They rank the pages of the Web and display them on the screen according to the order of their relative importance. The Google page rank model is based on two natural principles. The first one states that the more pages link to a given page, the more important the page. The second one states that a given page is more important when more important pages link to it.

More formally, we let $d \simeq 15 \times 10^9$ be the estimated number of Web pages in January 2015, and we set $S = \{1, \dots, d\}$. We equip S with the uniform counting distribution $\mu(i) = 1/d$, for any $i \in S$. Google uses a Web spider bot called Googlebot to count these links.

For any given Web page $i \in S$, we let d_i be the number of outgoing links from i. We denote by $P(i, j)$, with $i, j \in S$, the normalized hyperlink

$$P(i,j) = \begin{cases} \frac{1}{d_i} & \text{if} \quad j \text{ is one of the } d_i \text{ outgoing links} \\ 0 & \text{if} \quad d_i = 0 \text{ (a.k.a. a dangling node).} \end{cases}$$

By construction, $P = (P(i,j))_{i,j}$ is a stochastic matrix. Thus, its elements can be interpreted as the probabilities for elementary transitions of a Markov chain evolving on the set of all Web pages. P is a sparse matrix and the resulting Markov chain is not irreducible nor aperiodic. One way to turn this model into a regular Markov chain model is to consider the Markov transitions

$$M(i,j) = \epsilon\, P(i,j) + (1 - \epsilon)\, \mu(j)$$

for some parameter $\epsilon \in [0,1[$. The parameter ϵ reflects how likely is a Web navigator to restart a search by following the ranks of the pages. The matrix M is called the Google matrix associated with the damping factor ϵ.

It is an elementary exercise to check that M leads to an irreducible and aperiodic Markov chain as soon as $\epsilon \in [0,1[$. In addition, we have the minorization condition

$$M(i,j) \geq (1 - \epsilon)\, \mu(j).$$

This implies that the Dobrushin ergodic coefficient $\beta(M)$ of M satisfies the condition $\beta(M) \leq \epsilon$. By theorem 8.2.13, we conclude that

$$\|\mu_1 M^n - \mu_2 M^n\|_{tv} \leq \epsilon^n\, \|\mu_1 - \mu_2\|_{tv}$$

for any probability measures μ_1, μ_2 on S.

We let X_n be the Markov chain with elementary transition M starting at some random state X_0 with a distribution η_0. By the fixed point theorem, this implies that there exists a unique probability measure $\pi = \pi M$ and

$$\|\mathrm{Law}(X_n) - \pi\|_{tv} \leq \epsilon^n\, \|\eta_0 - \pi\|_{tv}.$$

The page rank is now defined by the order induced by the invariant measure $\pi = (\pi(i))_{i \in S}$

$$i \prec j \iff \pi(i) \leq \pi(j).$$

The rationale behind these constructions is that an individual wandering for a long time around the Web according to this Markov chain, after a long time, is not directly affected by the start of the navigation. The proportions of visits of the Web pages i are almost equal to $\pi(i)$. Thus, the higher this probability, the higher the rank of the page [218].

29.2   Gambling betting systems

29.2.1   Martingale systems

In probability theory, martingales refer to fair games in the sense that the average gain given the past information is always null for all gambling strategies we use. This terminology was introduced in 1939 by Jean Ville, but its literal origin comes from the early ages of casino gambling in the 18th century in France. In these early times, the terminology martingale was used to refer to the winning gambling style strategies. Most of the time, the idea was to increase the stake so that when we win eventually we would recover all our losses and would even gain profit. Martingales are also used by traders in financial markets, as well as by governments to define and control (when possible) some political and economical strategies. The interpretations of these stochastic processes depend on the application domain but they follow the same mathematical rules. We refer the reader to section 8.4. It is dedicated to the mathematical theory of martingale stochastic processes. In this section, we review some classical martingale gambling systems. For a more thorough discussion on this subject, we refer the reader to the book of S.N. Ethier [122] which is dedicated to the probabilistic aspects of gambling.

We recall the gambling model discussed in section $8.4.4$. We assume that X_n are independent $\{-1, 1\}$-valued random variables with common distribution

$$\mathbb{P}(X_n = -1) = p \quad \text{and} \quad \mathbb{P}(X_n = 1) = q = 1 - p \in [0, 1].$$

The fortune of the gambler presented in (8.45) is given by the discrete time stochastic process

$$M_n = M_0 + \sum_{1 \leq k \leq n} H_{k-1} X_k. \tag{29.1}$$

In this formula, $H_{k-1} \in \mathcal{F}_{k-1} := \sigma(X_1, \ldots, X_{k-1})$ denotes the betting strategy of the player at any time k. We recall that M_n is a martingale, respectively super-martingale, respectively sub-martingale, when $p = q$ (fair game), respectively $p > q$ (unfair), respectively $p < q$ (superfair). In each of these cases, we have

$$
\begin{aligned}
\mathbb{E}(\Delta M_n \mid \mathcal{F}_{n-1}) \quad &:= \quad H_{n-1} \, \mathbb{E}(X_n) \\
&= \begin{cases} 0 & \text{if } q = 1/2 \text{ (martingale = fair game)} \\ -H_{n-1} \, |q - p| & \text{if } 1/2 > q \text{ (super-martingale = unfair)} \\ +H_{n-1} \, |q - p| & \text{if } q > 1/2 \text{ (sub-martingale = superfair).} \end{cases}
\end{aligned}
$$

The initial random variable M_0 represents the gambler's fortune at the beginning of the game in given units, e.g., the number of dollars.

29.2.2 St. Petersburg martingales

In a classic martingale, gamblers always increase their bets after each loss. One way to recover all the previous losses is of course to bet the double of all losses until we win, and then leave the game. When using this strategy, the first win would recover all previous losses and win an extra profit that equals the amount of the previous losses.

This strategy is known as the St. Petersburg paradox.

In this section we formalize and analyze the above martingale. If we let T be the first time we win the game,

$$T := \inf\{m \geq 1 \; : \; X_m = 1\} > n \iff X_1 = X_2 = \ldots = X_n = -1,$$

then on the event $(T > n)$, the betting strategy at the $(n+1)$-th bet can be written as

$$
\begin{aligned}
H_n &= 2 \, (H_0 + H_1 + \ldots + H_{n-1}) \, 1_{T > n} \\
&= 3 H_{n-1} \, 1_{T > n} = 3^{n-1} \, H_1 \, 1_{T > n} = 2 \times 3^{n-1} \, H_0 \, 1_{T > n} \tag{29.2}
\end{aligned}
$$

for any $n \geq 1$. The last assertion is obvious:

$$H_{n-1} = 2 \, (H_0 + H_2 + \ldots + H_{n-2})$$

$$\Rightarrow H_n = 2 \, ((H_0 + H_2 + \ldots + H_{n-2}) + H_{n-1}) = 3 H_{n-1} = \ldots = 3^{n-1} \, H_1$$

and $H_1 = 2 H_0$ on the event $(T > n)$. We notice that the accumulated bets are given by

$$H_0 + \ldots + H_{n-1} = H_0 \left(1 + 2 \underbrace{[1 + 3 + \ldots + 3^{n-2}]}_{=(3^{n-1}-1)/2} \right) = 3^{n-1} \, H_0.$$

Therefore the fortune of the player M_T at the end of the game is given by the formula

$$
\begin{aligned}
M_T - M_0 &= (H_{T-1} \times 1 + (H_0 + \ldots + H_{T-2}) \times (-1)) \, 1_{T \geq 2} + H_0 \, 1_{T=1} \\
&= [2(H_0 + \ldots + H_{T-2}) - (H_0 + \ldots + H_{T-2})] \, 1_{T \geq 2} + H_0 \, 1_{T=1} \\
&= [H_0 + \ldots + H_{T-2}] \, 1_{T \geq 2} + H_0 \, 1_{T=1} \\
&= 3^{T-2} \, H_0 \, 1_{T \geq 2} + H_0 \, 1_{T=1}.
\end{aligned}
$$

This shows that

$$
M_T = M_0 + 3^{(T-2)^+} \, H_0.
$$

The duration of this game is a geometric random variable with success parameter q

$$
\mathbb{P}\,(T = n) = \mathbb{P}\,(X_1 = X_2 = \ldots = X_{n-1} = -1, \; X_n = 1) = p^{n-1} \, q.
$$

We see that the duration of the game is quite short

$$
\mathbb{P}\,(T \leq n) = q \sum_{1 \leq k \leq n} p^{k-1} = q \, (1 - p^n) / (1 - p) = 1 - p^n.
$$

Notice that

$$
\mathbb{E}(T) = \sum_{n \geq 1} \mathbb{P}(T \geq n) = \sum_{n \geq 1} p^n = 1/(1 - p) = 1/q.
$$

This clearly implies that T is almost surely finite. For instance, for fair games $p = 1/2$, the mean duration is of two bets only, and the probability that $T > 3$ is less than $(1/2)^3 = 12.5\%$.

For $p \neq 1/3$, we have

$$
\begin{aligned}
\mathbb{E}\left(3^{(T-2)^+} \, 1_{T \leq n}\right) &= \mathbb{P}\,(T = 1) + \sum_{2 \leq k \leq n} 3^{k-2} \, \mathbb{P}(T = k) \\
&= q\left(1 + p \sum_{0 \leq k \leq n-2} (3p)^k\right) = q\left(1 + p \, \frac{(3p)^{n-1} - 1}{3p - 1}\right)
\end{aligned}
$$

and

$$
\mathbb{E}\left(3^{(T-2)^+} \mid T \leq n\right) = \frac{1 - p}{1 - p^n}\left(1 + p \, \frac{(3p)^{n-1} - 1}{3p - 1}\right).
$$

For $p = 1/3$, we find that

$$
\mathbb{E}\left(3^{(T-2)^+} \mid T \leq n\right) = \frac{2}{3(1 - (1/3)^n)}\left(1 + \frac{(n-1)}{3}\right).
$$

This shows that for super-fair games

$$
p < 1/3 \Rightarrow \mathbb{E}\left(3^{(T-2)^+}\right) = \lim_{n \uparrow \infty} \mathbb{E}\left(3^{(T-2)^+} \, 1_{T \leq n}\right) = q\left(1 + \frac{p}{1 - 3p}\right)
$$

but for most of the fair games

$$
p \geq 1/3 \Rightarrow \mathbb{E}\left(3^{(T-2)^+}\right) = \lim_{n \uparrow \infty} \mathbb{E}\left(3^{(T-2)^+} \, 1_{T \leq n}\right) = \infty = \mathbb{E}(M_T).
$$

29.2.3 Conditional gains and losses

29.2.3.1 Conditional gains

In all situations we have

$$\mathbb{E}\left(M_T \mid M_0,\, H_0,\, T \le n\right) \;=\; M_0 \;+\; H_0\, \mathbb{E}\left(3^{(T-2)+} \mid T \le n\right)$$

$$= \begin{cases} M_0 \;+\; H_0\, \frac{1-p}{1-p^n}\left(1 + p\,\frac{(3p)^{n-1}-1}{3p-1}\right) & \text{if} \quad p \ne 1/3 \\[2mm] M_0 \;+\; H_0\, \frac{2/3}{1-(1/3)^n}\left(1 + \frac{(n-1)}{3}\right) & \text{for} \quad p = 1/3. \end{cases}$$

This shows for for most of the fair games (i.e., $p > 1/3$) the average conditional profit can be huge whenever T occurs before a given period

$$\mathbb{E}\left(M_T \mid M_0,\, H_0,\, T \le n\right) = M_0 \;+\; H_0\, \frac{1-p}{1-p^n}\left(1 + p\,\frac{(3p)^{n-1}-1}{3p-1}\right).$$

For instance, for the US roulette *red color* bet game discussed in section 2.3, we have $p = 20/38 \simeq .526$. The curve $n \mapsto \mathbb{E}\left(M_T \mid M_0,\, H_0,\, T \le n\right)$ of the expected profits when $M_0 = 1$ and $H_0 = 1$ for the first 20 bets is given below:

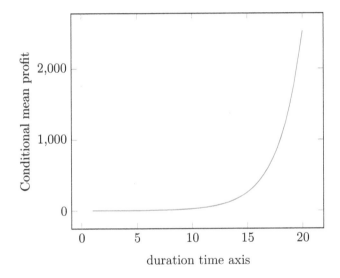

29.2.3.2 Conditional losses

Unfortunately, we need a large amount of money in the beginning of the game. More precisely, in the *rather unlikely* event $T > n$ (with probability p^n), we have

$$M_n\, 1_{T>n} = \left(M_0 - (H_0 + \ldots + H_{n-1})\right) 1_{T>n} = \left(M_0 - 3^{n-1}\, H_0\right) 1_{T>n}$$

so that the average losses are given by

$$\mathbb{E}\left(M_n \mid T > n\right) = \left(M_0 - 3^{n-1}\, H_0\right).$$

The following curve shows $n \mapsto \mathbb{E}\left(M_T \mid M_0,\, H_0,\, T > n\right)$ the expected losses when $M_0 = \$10,000$ and $H_0 = 1$ for the first 10 *red color* bets in US roulette:

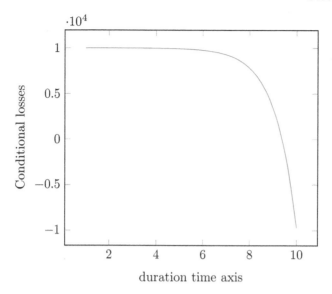

duration time axis

This shows that the player cannot gamble more than 8 or 9 steps without requiring a huge loan.

29.2.4   Bankroll management

We have not discussed the view of the casino. Like any bookmaker, a casino knows that bankruptcy is a possible outcome even if a game was not really fair for the gambler. To avoid martingale betting systems, a casino often uses table limits to control the maximum and the minimum bets a player can play. Major strip casinos in Las Vegas usually offer some tables with a $10,000 maximum (except three tables at Caesar's Palace which permit bets up to $50,000).

In this connection, it is interesting to notice that there are still no such table limits in financial markets which follow the same stochastic evolution models. Rogue traders can become extremely rich, but they can also provoke huge bankruptcy orders such as the largest bankruptcy filing in U.S. history, with Lehman holding over $600 billion in assets. Of course, governmental reserve banks will contribute to stabilising these random effects by printing money resources or more indirectly through general inflation.

We now come back to the gambling model's interpretation. Even if there are no table limits, any casino has a fixed amount of resources, say $K = \$3^{l-2}$, for some possibly large $l > 2$.

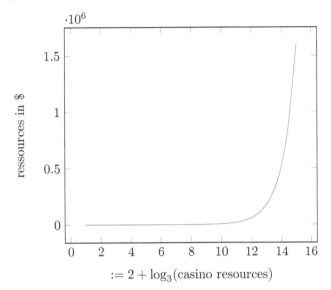

$$:= 2 + \log_3(\text{casino resources})$$

In this context, even when $H_0 = 1$, the maximum number of the times $T_c \geq 2$ the casino can play is given by

$$3^{(T_c-2)} \, H_0 = 3^{l-2} \Rightarrow T_c = l.$$

The expected gain of the game is given by

$$\mathbb{E}\left(\min\left(3^{(T-2)^+}, K\right)\right)$$

$$= q\left(1 + \sum_{2 \leq k \leq l} \, \min\left(3^{(k-2)}, 3^{l-2}\right) p^{k-1}\right) + 3^{l-2} \, \mathbb{P}\left(T > l\right)$$

$$= q\left(1 + p\sum_{0 \leq k \leq l-2} \, (3p)^k\right) = q\left(1 + p\,\frac{(3p)^{l-1}-1}{3p-1}\right) + 3^{l-2} \, p^l.$$

The curve below describes the expected value for the US roulette game discussed above.

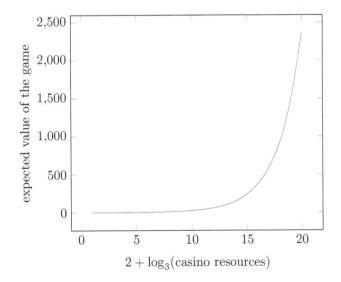

$$2 + \log_3(\text{casino resources})$$

29.2.5 Grand martingale

The grand martingale is defined similarly to the St Petersburg martingale except that (29.2) is replaced by:

$$
\begin{aligned}
H_n &= (2\,H_{n-1} + H_0)\,1_{T>n} \\
&= (2\,(2\,H_{n-2} + H_0) + H_0)\,1_{T>n} = \left(2^2\,H_{n-2} + 2H_0 + H_0\right)\,1_{T>n} \\
&= \ldots \\
&= \left(2^{n-1}\,H_1 + (2^{n-2} + 2^{n-2} + \ldots + 2 + 1)H_0\right)\,1_{T>n} \\
&= \left(2^{n-1}\,(2H_0 + H_0) + H_0\,(2^{n-1} - 1)\right)\,1_{T>n} \\
&= \left(2^{n+1} - 1\right)\,H_0\,1_{T>n}.
\end{aligned}
$$

In the event $T > n$, we have

$$
H_0 + H_1 + \ldots + H_{n-1}
$$

$$
= H_0\left[\left(2^1 - 1\right) + \left(2^2 - 1\right) + \left(2^3 - 1\right) + \ldots + \left(2^n - 1\right)\right]
$$

$$
= H_0\left((2^{n+1} - 1) - (n+1)\right).
$$

This implies that

$$
M_T - M_0
$$

$$
= \left[H_{T-1} - (H_0 + \ldots + H_{T-2})\right]\,1_{T\geq 2} + H_0\,1_{T=1}
$$

$$
= \left[H_0\,(2^T - 1) - H_0\,((2^T - 1) - T)\right]\,1_{T\geq 2} + H_0\,1_{T=1}
$$

$$
= H_0\,T\,1_{T\geq 1}
$$

and

$$
M_n\,1_{T>n} = (M_0 - (H_0 + \ldots + H_{n-1}))\,1_{T>n}
$$

$$
= (M_0 - [(2^{n+1} - 1) - (n+1)]\,H_0)\,1_{T>n}.
$$

In this situation, the expected profits are given by

$$
\mathbb{E}\left(M_T \mid M_0, H_0\right) = M_0 + H_0\,\mathbb{E}\left(T\right) = M_0 + H_0/q
$$

and the expected conditional losses are given by

$$
\mathbb{E}\left(M_T \mid M_0, H_0, T > n\right) = M_0 - \left[(2^{n+1} - 1) - (n+1)\right]\,H_0.
$$

29.2.6 D'Alembert martingale

The D'Alembert martingale is sometimes called the pyramid betting system. The name originates from the famous French 18th century mathematician Jean le Rond D'Alembert. The betting strategy is very simple. After a loss, we increase the bet by one unit (we can use the initial bet as an unit). After a win, we decrease the bet by one unit. More formally, on the event $(T > n)$, D'Alembert bets at the $(n+1)$-th epoch are given by

$$
H_n = H_{n-1} - X_n\,H_0
$$

for any $n \geq 1$. This shows that

$$H_n = [H_{n-1} - X_n \, H_0] = [1 - (X_1 + X_2 + \ldots + X_n)] \, H_0.$$

As a result, the relative fortune of the player is given by

$$R_n := (M_n - M_0)/H_0 = \sum_{l=1}^{n} \left[1 - \left(\sum_{1 \leq k < l} X_k \right) \right] X_l = \sum_{k=1}^{n} X_k - \sum_{1 \leq k < l \leq n} X_k X_l.$$

Hence we obtain for the expected value:

$$\begin{aligned}
\mathbb{E}\,(R_n) &= (q-p)n \left[1 - \frac{n-1}{2}\,(q-p) \right] \\
&= -2n \left(\frac{1}{2} - q \right) \left[1 + (n-1) \left(\frac{1}{2} - q \right) \right].
\end{aligned}$$

For instance, for the US roulette *red color* bet game discussed in section 2.3, we have $1/2 - q = 1/2 - 18/38 \geq 0.0263$. The expected relative losses are depicted below.

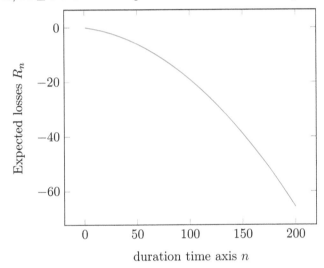

In the event $(T > n)$, this formula shows that $H_n = (1+n)\,H_0$. To simplify the presentation, we further assume that $H_0 = 1$. Then, in the event $(T > n)$, the losses are given by

$$M_n - M_0 = \sum_{p=1}^{n} H_{p-1}\, X_p = -\sum_{p=1}^{n} p = -n(n+1)/2.$$

This yields

$$\mathbb{E}\,(M_n - M_0 \mid T > n) = -n(n+1)/2.$$

In addition, at the time of winning we have

$$\begin{aligned}
M_T - M_0 &= (H_{T-1} - (H_0 + \ldots + H_{T-2}))\, 1_{T \geq 2} + 1_{T=1} \\
&= (T - (1 + 2 + \ldots + (T-1)))\, 1_{T \geq 2} + 1_{T=1} \\
&= T[1 - (T-1)/2].
\end{aligned}$$

This shows that

$$\mathbb{E}\,(M_T - M_0) = \mathbb{E}(T) - \frac{1}{2}\,\mathbb{E}(T(T-1)) = -\frac{1}{q^2}\,(1 + (1-q)) < 0.$$

29.2.7   Whittacker martingale

The Whittacker martingale refers to a progressive betting system whereby you bet the two latest bets until you win. The objective is to pull back the losses faster when you win. This betting system is often played on even money bets on roulette (even or odd or red or black, or other even strategies).

More formally, starting with two initial bets H_0 and H_1, the $(n+1)$-th bet is given by

$$\forall n \geq 2 \qquad H_n = H_{n-1} + H_{n-2}.$$

To get a more explicit expression for the cumulative bet, we recall that the solution of the above recurrence equation has the form

$$H_n = a_1 \, t_1^n + a_2 \, t_2^n$$

with (t_1, t_2) being the solutions of the characteristic polynomial function

$$t^n = t^{n-1} + t^{n-2} \quad \Leftrightarrow \quad t^2 - t - 1 = (t - 1/2)^2 - 5/4 = 0$$

$$\Leftrightarrow \quad (t_1, t_2) = \left(\frac{1 - \sqrt{5}}{2}, \frac{1 + \sqrt{5}}{2} \right).$$

We observe that $t_1 t_2 = -1,$ $\quad t_1 + t_2 = 1$ and $t_2 - t_1 = \sqrt{5}$. We can easily check directly that these functions satisfy the desired equation:

$$\left(a_1 \, t_1^{n-1} + a_2 \, t_2^{n-1} \right) + \left(a_1 \, t_1^{n-2} + a_2 \, t_2^{n-2} \right)$$

$$= a_1 \underbrace{\left(t_1^{n-1} + t_1^{n-2} \right)}_{=t_1^n} + a_2 \underbrace{\left(t_2^{n-1} + t_2^{n-2} \right)}_{=t_2^n}.$$

The constants (a_1, a_2) are computed in terms of the initial condition (H_0, H_1)

$$\left\{ \begin{array}{rcl} a_1 + a_2 & = & H_0 \\ a_1 t_1 + a_2 t_2 & = & H_1 \end{array} \right\}$$

$$\Leftrightarrow \left\{ \begin{array}{rcl} H_0 & = & a_1 + a_2 \\ H_1 & = & (a_1 + a_2)t_1 + a_2(t_2 - t_1) = H_0 t_1 + a_2(t_2 - t_1) \end{array} \right\}.$$

This implies that

$$a_2 = \frac{H_1 - H_0 t_1}{t_2 - t_1} \geq 0 \quad \text{and} \quad a_1 = H_0 - \frac{H_1 - H_0 t_1}{t_2 - t_1} = \frac{H_0 t_2 - H_1}{t_2 - t_1}.$$

Hence we get

$$\begin{aligned} H_n & = \frac{H_0 t_2 - H_1}{t_2 - t_1} \, t_1^n + \frac{H_1 - H_0 t_1}{t_2 - t_1} \, t_2^n \\ & = H_1 \frac{t_2^n - t_1^n}{t_2 - t_1} - H_0 t_1 t_2 \frac{t_2^{n-1} - t_1^{n-1}}{t_2 - t_1} \end{aligned}$$

and since $t_1 t_2 = -1$ holds we get

$$H_n = H_0 \frac{t_2^{n-1} - t_1^{n-1}}{t_2 - t_1} + H_1 \frac{t_2^n - t_1^n}{t_2 - t_1}.$$

When we start with the same bets $H_0 = H_1 = 1$, the sequence $H_n = H_{n-1} + H_{n-2}$ is given by the Fibonacci numbers

$$H_0 = 1, \; H_1 = 1 \; H_2 = 2, \; H_3 = 3, \; H_4 = 5, \; H_5 = 8, \; H_6 = 13, \; H_7 = 21 \ldots$$

A more synthetic form is given by Binet's formula

$$H_n = \frac{(t_2^{n-1} + t_2^n) - (t_1^{n-1} + t_1^n)}{t_2 - t_1} = \frac{t_2^{n+1} - t_1^{n+1}}{\sqrt{5}} \; \simeq_{n\uparrow\infty} \; \frac{(1+\sqrt{5})^{n+1}}{2^{n+1}\sqrt{5}}.$$

By construction, in the event $T > n$ we have

$$H_0 + H_1 + [H_2 + \ldots + H_{n-1}]$$

$$= (a_1 + a_2) + (a_1 t_1 + a_2 t_2)$$

$$+ \left[(a_1 t_1^2 + a_2 t_2^2) + \ldots + (a_1 t_1^{n-1} + a_2 t_2^{n-1}) \right]$$

$$= a_1 \frac{1 - t_1^n}{1 - t_1} + a_2 \frac{t_2^n - 1}{t_2 - 1}.$$

Using the fact that

$$(1 - t_1)(t_2 - 1) = -t_1 t_2 + (t_1 + t_2) - 1 = 1$$

and $(t_2 - 1) = -t_1 (\Leftrightarrow (1 - t_1) = t_2)$ we conclude that

$$
\begin{aligned}
H_0 + H_1 + [H_2 + \ldots + H_{n-1}] &= -a_1 t_1 (1 - t_1^n) + a_2 (t_2^n - 1) \\
&= a_2 t_2^{n+1} + a_1 t_1^{n+1} - (a_1 t_1 + a_2 t_2) \\
&= a_2 t_2^{n+1} + a_1 t_1^{n+1} - H_1.
\end{aligned}
$$

This shows that

$$M_T - M_0$$

$$= \left(\underbrace{H_{T-1}}_{=H_{T-3}+H_{T-2}} - (H_0 + \ldots + H_{T-2}) \right) 1_{T \geq 4}$$

$$+ (H_0 + H_1) 1_{T=3} + H_1 \, 1_{T=2} + H_0 \, 1_{T=1}$$

$$= -(H_0 + \ldots + H_{T-4}) \, 1_{T \geq 4} + (H_0 + H_1) 1_{T=3} + H_1 \, 1_{T=2} + H_0 \, 1_{T=1}.$$

In particular, this calculation demonstrates that the player will always lose if the success happens after the fourth bet. The strategy is then to follow this betting system up to the first time we hopefully recover some positive amount of money.

29.3 Stochastic optimal control

29.3.1 Bellman equations

A controlled Markov chain X_n evolving in some state spaces S_n is associated with a collection of Markov transitions $M_{u_n, n+1}(x_n, dx_{n+1})$ that depend on a control parameter $u_n \in U_n$

on some given control space U_n, that is,

$$M_{u_n,n+1}(X_n, dx_{n+1}) \quad := \quad \mathbb{P}_{u_n}(X_{n+1} \in dx_{n+1} \mid X_n)$$

for some probability measure $\mathbb{P}_{u_n}$ that depends on the parameter $u_n \in U_n$. We assume that X_0 is a random variable with a given distribution η_0 on S_0. In engineering sciences, controlled Markov chains are also called Markov decision processes and the control space u_n is known as the set of feasible or admissible actions or inputs. This model can be extended to situations where the control spaces $U_n = U_n(x_n)$ depend on the given state of the chain $X_n = x_n$. Deterministic discrete time dynamical systems are particular examples of controlled Markov chain models.

We illustrate these rather abstract models with the gambling model presented by Sheldon M. Ross in [234]. Let ϵ_n be a sequence on independent Bernoulli $\{0,1\}$-valued random variables with success probability $\mathbb{P}(\epsilon_n = 1) = 1 - \mathbb{P}(\epsilon_n = 0) = p_n$ representing the winning probability at the n-th game. At every time $(n+1)$ the gambler chooses to bet a proportion $u_n \in U_n := [0,1]$ of his cumulated earnings denoted by X_n. When $\epsilon_{n+1} = 1$, he gets back $\alpha_{n+1} \times (u_n X_n)$ units; otherwise he loses his bet.

After the $(n+1)$-th bet, his fortune is given by the controlled Markov chain

$$X_{n+1} = X_n + \epsilon_{n+1}\, \alpha_{n+1}\, u_n X_n - (1 - \epsilon_{n+1})\, u_n X_n = [(1 - u_n) + (1 + \alpha_{n+1})\, \epsilon_{n+1}\, u_n\,]\, X_n$$

with some initial fortune $X_0 = x_0$. In this situation, the Markov transitions are given by

$$M_{u_n,n+1}(X_n, dx) \quad = \quad p_{n+1}\, \delta_{(1+\alpha_{n+1}u_n)X_n}(dx) + (1 - p_{n+1})\, \delta_{(1-u_n)X_n}(dx).$$

Given the fortune $X_n = x_n$, the n-th proportion $u_n = v_n(x_n) \in [0,1]$ of the bet by the player may also depend on the given fortune x_n.

A regular function

$$v_n \;:\; x_n \in S_n \mapsto v_n(x_n) \in U_n$$

is called a control chart, a feedback control, or sometimes a Markov policy. We let $\mathcal{V}_n$ be the set of control charts. For any $k \le n$ we set $\mathcal{V}_{k,n}$ the set of feedback controls $v = (v_l)_{k \le l \le n} \in \prod_{k \le l \le n} \mathcal{V}_l$. This abstract framework allows us to consider historical Markov processes

$$X_n = (X'_0, \dots, X'_n) \in S_n = (S'_0 \times \dots \times S'_n) \tag{29.3}$$

associated with a Markov chain X'_n evolving in some state spaces S'_n. In this situation, the feedback control $v_n(x_n)$ may depend on the historical trajectory $x_n = (x'_0, \dots, x'_n)$.

The distribution of the trajectories of the Markov chain starting at $X_0 = x_0$ associated with a collection of controls $v = (v_k)_{0 \le k \le n} \in \mathcal{V}_{0,n}$ is defined by

$$\mathbb{P}_v\left((X_1, \dots, X_{n+1}) \in d(x_1, \dots, x_{n+1}) \mid X_0 = x_0\right)$$

$$= M_{v_0(x_0),1}(x_0, dx_1) M_{v_1(x_1),2}(x_1, dx_2) \dots M_{v_n(x_n),n+1}(x_n, dx_{n+1}). \tag{29.4}$$

Sometimes we use the superscript $\left(X_l^{(v)}\right)_{0 \le l \le n}$ to emphasize that the controlled Markov process is associated with some given $v = (v_l)_{0 \le l < n} \in \mathcal{V}_{0,n-1}$.

With a slight abuse of notation, for any feedback control $v_n \in \mathcal{V}_n$ we also write $M_{v_n,n+1}(x_n, dx_{n+1})$ for the Markov transition $M_{v_n(x_n),n}(x_n, dx_{n+1})$. In this notation, using (29.4) for any $k < l \leq (n+1)$ we have

$$\mathbb{P}_v\left(X_l \in dx_l \mid X_k = x_k\right)$$

$$= \left(M_{v_k,k+1} M_{v_{k+1},k+2} \cdots M_{v_{l-1},l}\right)(x_k, dx_l)$$

$$:= \int_{S_{k+1} \times \cdots \times S_{l-1}} M_{v_k(x_k),k+1}(x_k, dx_{k+1}) \cdots M_{v_{l-1}(x_{l-1}),l}(x_{l-1}, dx_l).$$

At any time k, given the value of the chain, say $X_k = x_k$, the objective is to maximize the average of some function $f_n(X_n)$ at a given time horizon, say $n \geq 1$. More formally, the objective is to maximize at every step k the value function

$$V_k(x_k) := \sup_{v \in \mathcal{V}_{k,n-1}} \mathbb{E}_v(f_n(X_n) \mid X_k = x_k). \qquad (29.5)$$

Depending on the application domains, the function f_n is also called the payoff (financial mathematics), the reward (reinforcement learning and operations research), the performance or the utility function (engineering sciences) or the gain criterion. The function $(-f_n)$ can also be thought of as a cost or some type of energy consumption function.

The function V_k is also called the optimal or the expected $(n - k)$ steps to go value or the return function.

We use backward reasoning w.r.t. the time parameter $k \leq n$:

At the terminal time $k = n$, we have the boundary condition $V_n = f_n$. When $k = (n-1)$ we have

$$V_{n-1}(x_{n-1}) = \sup_{v \in \mathcal{V}_{n-1}} \mathbb{E}_v(f_n(X_n) \mid X_{n-1} = x_{n-1})$$

$$= \sup_{v \in \mathcal{V}_{n-1}} M_{v,n}(f_n)(x_{n-1}) = \sup_{u \in U_{n-1}} M_{u,n}(f_n)(x_{n-1}).$$

In the last assertion, we implicitly assumed that supremums in the formulae are attained, that is, there exists some optimal feedback control $v_{n-1}^\star \in \mathcal{V}_{n-1}$, or equivalently some optimal control value $v_{n-1}^\star(x_{n-1}) = u_{n-1}^\star \in U_{n-1}$ for every given state $x_{n-1} \in S_{n-1}$. In this case,

$$V_{n-1}(x_{n-1}) = M_{v_{n-1}^\star,n}(f_n)(x_{n-1}) = \int_{S_n} M_{v_{n-1}^\star(x_{n-1}),n}(x_{n-1}, dx_n)\, f_n(x_n). \qquad (29.6)$$

When $k = (n-2)$ we have

$$V_{n-2}(x_{n-2}) = \sup_{v \in \mathcal{V}_{n-2,n-1}} \mathbb{E}_v(f_n(X_n) \mid X_{n-2} = x_{n-2})$$

$$= \sup_{u \in U_{n-2}} \sup_{v \in \mathcal{V}_{n-1}} M_{u,n-1}(M_{v,n}(f_n))(x_{n-2}).$$

It is important to observe that $u \in U_{n-2}$ is a control parameter in the set U_{n-2}, while

$v \in \mathcal{V}_{n-1}$ is a feedback control $v : x_{n-1} \in S_{n-1} \mapsto v(x_{n-1}) \in U_{n-1}$. This yields the formula

$$M_{u,n-1}(M_{v,n}(f_n))(x_{n-2})$$

$$= \int M_{u,n-1}(x_{n-2}, dx_{n-1}) \, M_{v(x_{n-1}),n}(x_{n-1}, dx_n) \, f_n(x_n).$$

If we choose the optimal feedback $v = v_{n-1}^\star$ defined in (29.6) we have

$$\sup_{w \in \mathcal{V}_{n-1}} M_{u,n-1}(M_{w,n}(f_n))(x_{n-2})$$

$$\geq M_{u,n-1}(M_{v_{n-1}^\star,n}(f_n))(x_{n-2}) = M_{u,n-1}(V_{n-1})(x_{n-2}). \quad \Longleftarrow (29.6)$$

In the reverse angle, we have

$$\sup_{w \in \mathcal{V}_{n-1}} M_{u,n-1}(M_{w,n}(f_n))(x_{n-2}) \leq M_{u,n-1}\left(\sup_{w \in \mathcal{V}_{n-1}} M_{w,n}(f_n) \right)(x_{n-2})$$

$$= M_{u,n-1}(V_{n-1})(x_{n-2}).$$

This implies that

$$V_{n-2}(x_{n-2}) = \sup_{u \in U_{n-2}} M_{u,n-1}(V_{n-1})(x_{n-2}) = \sup_{v \in \mathcal{V}_{n-2}} M_{v,n-1}(V_{n-1})(x_{n-2}).$$

In the r.h.s. formula we implicitly assumed that supremum is attained, that is, there exists some optimal feedback control $v_{n-2}^\star \in \mathcal{V}_{n-2}$, or equivalently some optimal control value $v_{n-2}^\star(x_{n-2}) = u_{n-2}^\star \in U_{n-1}$ for every given state $x_{n-2} \in S_{n-2}$.

Iterating this reasoning backward in time, we obtain the Bellman (optimality) backward equation

$$V_l(x_l) = \sup_{u \in U_l} \mathbb{E}_u\left(V_{l+1}(X_{l+1}) \mid X_l = x_l\right) = \sup_{u \in U_l} M_{u,l+1}(V_{l+1})(x_l) \qquad (29.7)$$

with $0 \leq l < n$ and the terminal (a.k.a. boundary) condition $V_n = f_n$. In addition, the optimal strategy in (29.5) is obtained by applying sequentially from time $l = k$ to the final time $l = (n-1)$ the optimal control charts $x_l \mapsto v_l(x_l)$ computed in the one-step backward recursion (29.7).

Definition 29.3.1 *We consider the collection of sub-martingales* $(\mathbb{V}_k(v))_{0 \leq k \leq n}$ *w.r.t. to* $\mathcal{F}_k^{(v)} = \sigma\left(X_l^{(v)}, \, l \leq k\right)$ *indexed by* $v = (v_l)_{0 \leq l < n} \in \mathcal{V}_{0,n-1}$ *and defined by*

$$\forall 0 \leq k \leq n \qquad \mathbb{V}_k(v) := V_k\left(X_k^{(v)}\right). \qquad (29.8)$$

The sub-martingale property is a direct consequence of (29.7). We have

$$\mathbb{E}_{v_l}\left(\mathbb{V}_{l+1}(v) \mid \mathcal{F}_l^{(v)}\right) = \mathbb{E}_{v_l}\left(V_{l+1}(X_{l+1}^{(v)}) \mid X_l^{(v)}\right) \leq V_l\left(X_l^{(v)}\right) = \mathbb{V}_l(v)$$

with the equality if and only if v_l is the the optimal control chart $x_l \mapsto v_l(x_l)$ computed in the one-step backward recursion (29.7).

In other words we have the equivalent martingale optimality principle by which

$$\mathbb{V}_k(v) \text{ is an } \mathcal{F}_k^{(v)} - \text{martingale} \iff v \text{ is an optimal control chart.} \qquad (29.9)$$

The derivation of the Bellman equation presented above is intuitive and natural but rather formal. A more rigorous mathematical formulation of abstract stochastic control problems on measurable state and control spaces including the Bellman equation requires a sophisticated measure-theoretic framework to ensure that the cost functions V_k and the optimal control charts are appropriately measurable. In this book we implicitly assume that the supremum in the Bellman equation (29.7) is achieved for any given state by some well defined optimal feedback control.

In much the same way, taking the supremum inside an expectation over an uncountable state space may require the use of outer expectations and probabilities. It is clearly not within the scope of this book to enter into these technical considerations. For more details on these intricate measure theoretic issues we refer the reader to the seminal book [21] by D. P. Bertsekas and S. E. Shreve, the one by O. Hernandez-Lerma and J.B. Lasserre [179], the more recent articles by N. El Karoui and X. Tan [119, 120], and the research monograph by N. Touzi [252].

We return to the Ross gambling model discussed earlier in this chapter. We assume that terminal returns are time homogeneous $V_n(x) = f_n(x) = f(x) := \log x$ and for any $u \in U_n = U = [0, 1]$ we have

$$
\begin{aligned}
M_{u,n}(V_n)(x) &= M_{u,n}(f)(x) \\
&= p_n \log((1 + \alpha_n u)x) + (1 - p_n) \log((1 - u)x) = f(x) + g_n(u)
\end{aligned}
$$

with

$$g_n(u) = p_n \log(1 + \alpha_n u) + (1 - p_n) \log(1 - u).$$

Notice that

$$\partial_u g_n(u) = p_n \frac{\alpha_n}{1 + \alpha_n u} - \frac{1 - p_n}{1 - u} = \frac{(p_n - u)\alpha - (1 - p_n)}{(1 - u)(1 + \alpha_n u)} = 0$$

as soon as

$$p_n \alpha_n (1 - u) = (1 - p_n)(1 + \alpha_n u) \quad \Leftrightarrow \quad u = r_n := p_n - \alpha_n^{-1}(1 - p_n) \ (\leq 1).$$

We set $q_n = 1 - p_n$. Whenever $\alpha_n \leq q_n/p_n$, the function $u \in [0, 1] \mapsto g_n(u)$ is decreasing and its maximum on $[0, 1]$ is attained when $u = 0$. Whenever $\alpha_n > q_n/p_n$, the function $u \in [0, 1] \mapsto g_n(u)$ is increasing on $[0, r_n]$ and decreasing on $[r_n, 1]$. In this situation, its maximum on $[0, 1]$ is attained when $u = r_n$. This implies that

$$V_{n-1}(x) = \sup_{u \in [0,1]} M_{u,n}(f)(x) = f(x) + g_n(u_n^\star) \quad \text{with} \quad u_{n-1}^\star = 1_{\alpha_n > q_n/p_n} \ r_n.$$

Iterating this reasoning, we check that

$$\forall 0 \leq k \leq n \qquad V_k(x) = f(x) + \sum_{k < l \leq n} g_l(u_l^\star).$$

29.3.2 Control dependent value functions

We extend the framework discussed in section 29.3.1 to control dependent value functions
of the form

$$F_n(X, v) := \sum_{0 \leq l < n} g_l(X_l, v_l(X_l)) + f_n(X_n) \tag{29.10}$$

for some functions g_l on $(S_l \times U_l)$ and for any $v = (v_l)_{0 \leq l < n} \in \mathcal{V}_{0,n-1}$. We consider the
backward interpolation functions

$$\forall 0 \leq k \leq n \quad \forall v \in \mathcal{V}_{k,n-1} \quad F_{k,n}(X, v) := \sum_{k \leq l < n} g_l(X_l, v_l(X_l)) + f_n(X_n)$$

from $F_{0,n}(X, v) = F_n(X, v)$ to the terminal payoff $F_{n,n}(X, v) = f_n(X_n)$. Sometimes we
slightly abuse the notation and write $F_{k,n}(X, v)$ instead of $F_{k,n}(X, v_{k,n-1})$ for any $v = (v_l)_{0 \leq l < n} \in \mathcal{V}_{0,n-1}$ and $v_{k,n-1} = (v_l)_{k \leq l < n}$.

In this situation, at any time k, given the value of the chain, say $X_k = x_k$, the
objective is to maximize at every step k the value function

$$V_k(x_k) := \sup_{v \in \mathcal{V}_{k,n-1}} \mathbb{E}_v \left(F_{k,n}(X, v) \mid X_k = x_k \right). \tag{29.11}$$

Arguing as in (29.7) we obtain the Bellman (optimality) backward equation

$$\begin{aligned}
V_l(x_l) &= \sup_{u \in U_l} \left[g_l(x_l, u) + \mathbb{E}_u(V_{l+1}(X_{l+1}) \mid X_l = x_l) \right] \\
&= \sup_{u \in U_l} \left[g_l(x_l, u) + M_{u,l+1}(V_{l+1})(x_l) \right]
\end{aligned} \tag{29.12}$$

with $0 \leq l < n$ and the terminal (a.k.a. boundary) condition $V_n = f_n$. In addition,
the optimal strategy in (29.5) is obtained by applying sequentially from time $l = k$
to the final time $l = (n-1)$ the optimal control charts $x_l \mapsto v_l(x_l)$ computed in
the one-step backward recursion (29.12) (see for instance exercise 471).

Proof :

We check this claim using a backward induction. For $l = n$ we clearly have $V_n = f_n$. For
$l = (n-1)$, we have

$$\begin{aligned}
V_{n-1}(x_{n-1}) &:= \sup_{v \in \mathcal{V}_{n-1}} \mathbb{E}_u(F_{n-1,n}(X, v) \mid X_{n-1} = x_{n-1}) \\
&= \sup_{u \in U_{n-1}} \left[g_{n-1}(x_{n-1}, u) + \mathbb{E}_u(f_n(X_n) \mid X_{n-1} = x_{n-1}) \right] \\
&= \sup_{u \in U_{n-1}} \left[g_{n-1}(x_{n-1}, u) + M_{u,n}(f_n)(x_{n-1}) \right].
\end{aligned}$$

Here again we have implicitly assumed the existence of some optimal feedback control
$v_{n-1}^\star \in \mathcal{V}_{n-1}$, or equivalently some optimal control value $v_{n-1}^\star(x_{n-1}) = u_{n-1}^\star \in U_{n-1}$ for
every given state $x_{n-1} \in S_{n-1}$. In this case, we have

$$V_{n-1}(x_{n-1}) = g_{n-1}(x_{n-1}, v_{n-1}^\star(x_{n-1})) + M_{v_{n-1}^\star, n}(f_n)(x_{n-1}).$$

When $k = (n-2)$ we have

$$
\begin{aligned}
V_{n-2}(x_{n-2}) &= \sup_{(u,v)\in\mathcal{V}_{n-2,n-1}} \left[g_{n-2}(X_{n-2}, u) + \mathbb{E}_{(u,v)}(F_{n-1,n}(X,v) \mid X_{n-2} = x_{n-2}) \right] \\
&= \sup_{u\in U_{n-2}} \left[g_{n-2}(X_{n-2}, u) + \sup_{v\in\mathcal{V}_{n-1}} \mathbb{E}_{(u,v)}(F_{n-1,n}(X,v) \mid X_{n-2} = x_{n-2}) \right].
\end{aligned}
$$

This yields the formula

$$
V_{n-2}(x_{n-2}) = \sup_{u\in U_{n-2}} \left[g_{n-2}(x_{n-2}, u) + \sup_{v\in\mathcal{V}_{n-1}} \mathbb{E}_{(u,v)}(F_{n-1,n}(X,v) \mid X_{n-2} = x_{n-2}) \right].
$$

Arguing as in section 29.3.1 we have

$$
\sup_{v\in\mathcal{V}_{n-1}} \mathbb{E}_{(u,v)}(F_{n-1,n}(X,v) \mid X_{n-2} = x_{n-2})
$$

$$
= \sup_{v\in\mathcal{V}_{n-1}} \mathbb{E}_u\big(\mathbb{E}_v \left[F_{n-1,n}(X,v) \mid X_{n-1} \right] \mid X_{n-2} = x_{n-2} \big)
$$

$$
\geq \mathbb{E}_u \left(\mathbb{E}_{v_{n-1}^*} \left[F_{n-1,n}(X, v_{n-1}^*) \mid X_{n-1} \right] \mid X_{n-2} = x_{n-2} \right)
$$

$$
= \mathbb{E}_u \left(V_{n-1}(X_{n-1}) \mid X_{n-2} = x_{n-2} \right).
$$

On the other hand, we have

$$
\mathbb{E}_u\big(\mathbb{E}_v \left[F_{n-1,n}(X,v) \mid X_{n-1} \right] \mid X_{n-2} = x_{n-2} \big)
$$

$$
\leq \mathbb{E}_u\big(\sup_{w\in\mathcal{V}_{n-1}} \mathbb{E}_w \left[F_{n-1,n}(X,w) \mid X_{n-1} \right] \mid X_{n-2} = x_{n-2} \big)
$$

$$
= \mathbb{E}_u\big(V_{n-1}(X_{n-1}) \mid X_{n-2} = x_{n-2} \big).
$$

We conclude that

$$
\sup_{v\in\mathcal{V}_{n-1}} \mathbb{E}_{u,v}(F_{n-1,n}(X,v) \mid X_{n-2} = x_{n-2}) = \mathbb{E}_u(V_{n-1}(X_{n-1}) \mid X_{n-2} = x_{n-2}).
$$

This implies that

$$
V_{n-2}(x_{n-2}) = \sup_{u\in U_{n-2}} \left[g_{n-2}(x_{n-2}, u) + \mathbb{E}_u(V_{n-1}(X_{n-1}) \mid X_{n-2} = x_{n-2}) \right].
$$

Iterating this reasoning backward in time we readily check (29.12). This ends the proof of (29.12). ∎

Using the same arguments as in the proof of (29.9), we check that the collection of processes $(\mathbb{V}_k(v))_{0\leq k\leq n}$ indexed by $v \in \mathcal{V}_{0,n-1}$ and defined by

$$
\forall 0 \leq k \leq n \qquad \mathbb{V}_k(v) = \sum_{0\leq l<k} g_l(X_l^{(v)}, v_l) + V_k\left(X_k^{(v)}\right), \tag{29.13}
$$

forms a sub-martingale w.r.t. to $\mathcal{F}_k^{(v)} = \sigma\left(X_l^{(v)}, \ l \leq k\right)$ and satisfy the martingale optimality principle stated in (29.9).

29.3.3   Continuous time models

In continuous time, a controlled Markov process X_t on some state space S is defined in terms of a collection of infinitesimal generators $L_{u,t}$ indexed by a control parameter $u \in U$ on some given control space U. For instance, the collection of partial differential operators

$$L_{u,t}(\varphi)(x) = b_t(x,u) \ \partial_x \varphi(x) + \frac{1}{2} \ \sigma_t^2(x,u) \ \partial_x^2 \varphi(x) + \lambda_t(x,u) \int \ (\varphi(y) - \varphi(x)) \ K_{u,t}(x,dy) \tag{29.14}$$

indexed by $u \in U = \mathbb{R}$ is associated to the infinitesimal generators L_{t,u_t} of a one-dimensional jump diffusion X_t, for some control function u_t.

At rate $\lambda_t(X_t, u_t)$ the process jumps from X_{t-} to X_t randomly chosen with some given probability $K_{u_t,t}(X_{t-}, dy)$. Between the jumps, X_t evolves according to the stochastic differential equation

$$dX_t = b_t(X_t, u_t) \ dt + \sigma_t(X_t, u_t) \ dW_t$$

where W_t stands for a Brownian motion and b_t and σ_t stand for some regular drift and diffusion functions. We assume that X_0 is a random variable with a given distribution η_0 on S.

> We let $\mathcal{V}_{s,t}$ be the set of feedback controls $v \ : \ (r,x) \in ([s,t] \times S) \mapsto v_r(x) \in U$, with $s \leq r \leq t$. Sometimes we slightly abuse the notation and write $L_{v,r}(\varphi)(x)$ instead of $L_{v(x),r}(\varphi)(x)$. We also use the superscript $X_s^{(v)}$ to emphasize that the controlled Markov process is associated with some given $v \in \mathcal{V}_{0,t}$.

We consider control dependent value functions of the form

$$F_{s,t}(X,u) := \int_s^t \ g_r(X_r, u_r) \ dr + f_t(X_t) \tag{29.15}$$

for any control mapping $u = (u_r)_{r \in [s,t]}$ from $[s,t]$ into U. For $s = 0$ we set $F_{0,t}(X,u) = F_t(X,u)$.

We assume the existence of a discrete time approximation $X_{t_n}^h$ of the process X_t on some time mesh $0 \leq t_n \leq t_{n+1}$ with time step $(t_{n+1} - t_n) = h$, and we let

$$P_{u,t_n}^h \left(X_{t_n}^h, dx \right) := \mathbb{P}_u \left(X_{t_{n+1}}^h \in dx \mid X_{t_n}^h \right) \tag{29.16}$$

be the one-step probability transitions of the process $X_{t_n}^h$ indexed by some control parameter $u \in U$. Also assume that

$$h^{-1} \left[P_{u,t_n}^h - Id \right] = L_{u,t_n} + O(h)$$

or equivalently

$$P_{u,t_n}^h (\varphi) = \varphi + L_{u,t_n}(\varphi) \ h + O(h^2) \tag{29.17}$$

for some sufficiently regular functions φ. The discrete time approximations of the value functions (29.15) are given by

$$F_{t_n}^h(X^h, u) := \sum_{0 \leq k < n} g_{t_k}(X_{t_k}^h, u_{t_k}) \ h + f_{t_n}(X_{t_n}^h). \tag{29.18}$$

By (29.12) the Bellman equation associated with the discrete time model is given by

$$V_{t_l}^h(x) = \sup_{u \in U} \left[g_{t_l}(x,u) \ h + P_{u,t_{l+1}}^h \left(V_{t_{l+1}}^h \right)(x) \right]$$

with the terminal condition $V_{t_n}^h = f_{t_n}$. By (29.17) we have

$$-h^{-1}\left[V_{t_{l+1}}^h(x) - V_{t_l}^h(x)\right] = \sup_{u \in U}\left[g_{t_l}(x,u) + L_{t_{l+1},u}\left(V_{t_{l+1}}^h\right)(x) + \mathrm{O}(h)\right].$$

Taking formally the limit as $h \downarrow 0$ with $t_l \downarrow s$ we find that the value function

$$\lim_{h \downarrow 0} V_{t_l}^h(x) = V_s(x) = \sup_{v \in \mathcal{V}_{s,t}} \mathbb{E}_v\left(F_{s,t}(X,v) \mid X_s = x\right)$$

satisfies the equation

$$-\partial_s V_s(x) = \sup_{u \in U}\left[g_s(x,u) + L_{u,s}(V_s)(x)\right] \tag{29.19}$$

with terminal condition $V_t = f_t$. As in the discrete time case, the optimal strategy is obtained by applying the optimal control charts $x \mapsto v_s(x)$ computed in the one-step backward recursion (29.19) (see for instance exercise 473 and exercise 476).

We also readily check the optimality principle:

$$
\begin{aligned}
V_{s_1}(x) &= \sup_{v=(v_1,v_2)\in\mathcal{V}_{s_1,s_2}\times\mathcal{V}_{s_2,t}} \mathbb{E}_{v_1}\left(\int_{s_1}^{s_2} g_s(X_s, v_{1,s}(X_s))\, ds \right. \\
&\qquad\qquad\qquad\qquad \left. + \mathbb{E}_{v_2}\left(F_{s_2,t}(X,v_2) \mid X_{s_2}\right) \mid X_{s_1} = x\right) \\
&= \sup_{v\in\mathcal{V}_{s_1,s_2}} \mathbb{E}_v\left(\int_{s_1}^{s_2} g_s(X_s, v_s(X_s))\, ds + V_{s_2}(X_{s_2}) \mid X_{s_1} = x\right). \tag{29.20}
\end{aligned}
$$

Using the same arguments as in the proof of (29.9) and (29.13), we check that the collection of processes $(\mathbb{V}_s(v))_{0 \le s \le t}$ defined by

$$\mathbb{V}_s(v) = \int_0^s g_r(X_r^{(v)}, v_r(X_r^{(v)}))\, dr + V_s(X_s^{(v)}) \tag{29.21}$$

is a sub-martingale w.r.t. to $\mathcal{F}_s^{(v)} = \sigma\left(X_r^{(v)}, r \le s\right)$ and satisfies the martingale optimality principle

$$\mathbb{V}_s(v) \text{ is an } \mathcal{F}_s^{(v)} - \text{martingale} \iff v \text{ is an optimal control chart.} \tag{29.22}$$

For the jump diffusion model (29.14), the Bellman equation (29.19) takes the form of a Hamilton-Jacobi-Bellman equation

$$\partial_s V_s(x) + \mathcal{H}_s\left(x, V_s, \partial_x V_s, \partial_x^2 V_s\right) = 0 \qquad (29.23)$$

(abbreviated HJB equation) with the terminal condition $V_t = f_t$ and the Hamiltonian functional

$$\mathcal{H}_s\left(x, h_0, h_1, h_2\right)$$

$$= \sup_{u \in U} \left\{ g_s(x, u) + b_s(x, u)\, h_1(x) + \frac{1}{2}\, \sigma_s^2(x, u)\, h_2(x) \right.$$

$$\left. + \lambda_s(x, u)\, \left[K_{u,s}(h_0)(x) - h_0(x)\right] \right\}.$$

As in the discrete time case, it is clearly beyond the scope of this section to provide a detailed discussion on the fully rigorous mathematical derivation of these limiting operations but ample discussions can be found in the stochastic control theory textbooks, for instance, the research monograph by N. Touzi [252]. The existence and the analysis of the solutions of Hamilton-Jacobi-Bellman equations requires study of the highly technical theory of viscosity solutions.

Despite its mathematical elegance and its usefulness in analyzing the Hamilton-Jacobi-Bellman equations, this theory is not really used to solve concrete optimization problems. In practice, when possible, we solve or approximate the solution of the Hamilton-Jacobi-Bellman equation and check that the resulting feedback controls are optimal. This technique is sometimes called *the verification argument*.

We also mention that deterministic and continuous time dynamical systems are particular examples of continuous time stochastic control problems.

We illustrate these rather abstract models using a pure diffusion control process with a generator

$$L_{u,s}(f)(x) = (b_s(x) + u)\ \partial_x f(x) + \frac{1}{2}\ \sigma^2\ \partial_x^2 f(x).$$

We consider the value function (29.15) with $g_s(x, u) = h_s(x) - u^2/2$, with $u \in U := \mathbb{R}$, and some regular negative functions h_s and $f_t(x)$. In this situation, the Hamilton-Jacobi-Bellman equation (29.23) takes the form

$$\partial_s V_s(x) + \sup_{u \in \mathbb{R}} \left(h_s(x) - \frac{u^2}{2} + (b_s(x) + u)\ \partial_x V_s(x) + \frac{1}{2}\ \sigma^2\ \partial_x^2 V_s(x) \right) = 0.$$

The optimal feedback control is clearly given by

$$u = v_s(x) := \partial_x V_s(x).$$

It remains to solve the Hamilton-Jacobi-Bellman equation. To this end, we observe that

$$\partial_s V_s + h_s + b_s\ \partial_x V_s + \frac{1}{2}\ \left[\sigma^2\ \partial_x^2 V_s + (\partial_x V_s)^2\right] = 0.$$

We use the Cole-Hopf transformation

$$q_s := \exp\left(V_s / \sigma^2\right) \iff V_s = \sigma^2\ \log q_s$$

$$\implies \partial_s V_s = \sigma^2 q_s^{-1} \partial_s q_s \quad \text{and} \quad \partial_x V_x = \sigma^2 q_s^{-1} \partial_x q_x$$

as well as

$$\sigma^2 \partial_x^2 V_s = \sigma^4 q_s^{-1} \partial_x^2 q_s - \sigma^4 q_s^{-2} (\partial_x q_x)^2 = \sigma^4 q_s^{-1} \partial_x^2 q_s - (\partial_x V_x)^2.$$

This implies that

$$
\begin{aligned}
-\sigma^2 q_s^{-1} \partial_s q_s &= -\partial_s V_s = h_s + b_s \; \partial_x V_s + \frac{1}{2} \left[\sigma^2 \; \partial_x^2 V_s - (\partial_x V_s)^2 \right] \\
&= h_s + b_s \; \sigma^2 \; q_s^{-1} \partial_x q_x + \frac{1}{2} \; \sigma^4 \; q_s^{-1} \partial_x^2 q_s.
\end{aligned}
$$

We conclude that

$$-\partial_s q_s = b_s \; \partial_x q_x + \frac{1}{2} \; \sigma^2 \; \partial_x^2 q_s + \sigma^{-2} h_s \; q_s$$

for any $0 \le s \le t$, with the terminal condition $q_t := \exp\left(V_t / \sigma^2\right) = \exp\left(f_t / \sigma^2\right)$. This equation can be rewritten as

$$-\partial_s q_s = \mathcal{L}(q_s) + \overline{h}_s \; q_s$$

with the potential function $\overline{h}_s = \sigma^{-2} h_s$ and the infinitesimal generator $\mathcal{L}$ of the diffusion process

$$dY_s = b_s(Y_s) \; ds + \sigma \; dW_s.$$

The solution of this equation is given by the Feynman-Kac formula

$$q_s(y) = Q_{s,t} \left[e^{\overline{f}_t} \right](y) := \mathbb{E}\left[\exp\left(\overline{f}_t(Y_t)\right) \; \exp\left(\int_s^t \overline{h}_r(Y_r) dr \right) \mid Y_s = y \right]$$

with $\overline{f}_t = \sigma^{-2} f_t$. We refer the reader to section 15.6.1, section 9.6 and section 16.1.3 for more thorough discussions on these Feynman-Kac semigroups, including their discrete time and particle interpretations. When $h_s(x) = 0$, the solution of the Hamilton-Jacobi-Bellman equation (29.23) reduces to the conditional expectation

$$q_s(y) = P_{s,t} \left[\exp\left(\overline{f}_t\right) \right](y) := \mathbb{E}\left[\exp\left(\overline{f}_t(Y_t)\right) \mid Y_s = y \right]$$

with the Markov semigroup $P_{s,t}$ of Y_s. For linear diffusion processes and quadratic functions f_t and h_s, the Feynman-Kac integration formulae can be solved explicitly. See exercises 473 through 476 For more general models we often need to resort to some additional approximation schemes. Further details on these Feynman-Kac values functions in multidimensional settings can be found in exercise 480.

29.4 Optimal stopping

29.4.1 Games with fixed terminal condition

We consider a real valued martingale Y_n equipped with a σ-field $\mathcal{F}_n := \sigma\left(Y_0, \ldots, Y_n\right)$. We assume that Y_n can take only two possibly random values a_n and b_n given $\mathcal{F}_{n-1}$, that is,

$$\mathbb{P}(Y_n = a_n \mid \mathcal{F}_{n-1}) = 1 - \mathbb{P}(Y_n = b_n \mid \mathcal{F}_{n-1}).$$

We illustrate this model with two examples. The first is

$$Y_n = \epsilon_0 + \ldots + \epsilon_n = Y_{n-1} + \epsilon_n$$

where ϵ_n denotes a sequence of centered independent random variables taking two possible values u_n and v_n. We readily check that

$$\sigma\left(Y_0,\ldots,Y_n\right) = \sigma\left(\epsilon_0,\ldots,\epsilon_n\right) \quad \text{and} \quad (a_n, b_n) = (Y_{n-1}+u_n, Y_{n-1}+v_n).$$

The second example is given by the random product model

$$Y_n = (1+\epsilon_0) \times \ldots \times (1+\epsilon_n) = Y_{n-1} \times (1+\epsilon_n)$$

with $1+u_n \le 1+v_n$. In this situation, $(a_n, b_n) = (Y_{n-1}(1+u_n), Y_{n-1}(1+v_n))$.

As in (29.1), we consider the $\mathcal{F}_k$-martingale

$$\forall 0 \le k \le n \qquad M_k := M_0 + \sum_{1 \le l \le k} H_{l-1}\, X_l \quad \text{with} \quad X_l = \Delta Y_l = Y_l - Y_{l-1}.$$

Our next objective is to find a player betting strategy H_k that ends at a fixed terminal condition

$$M_n = f_n(Y_0,\ldots,Y_n)$$

with some given function f_n on $\mathbb{R}^{n+1}$.

One natural way to solve this martingale control problem is to consider the sequence of functions V_k on $\mathbb{R}^{k+1}$ defined by the backward induction

$$\forall 0 \le k < n \qquad V_k(y_0,\ldots,y_k) = \mathbb{E}\left(V_{k+1}(y_0,\ldots,y_k,Y_{k+1}) \mid (Y_0,\ldots,Y_k) = (y_0,\ldots,y_k)\right)$$

with the terminal condition $V_n(y_0,\ldots,y_n) = f_n(y_0,\ldots,y_n)$. It is seen immediately that

$$V_k(Y_0,\ldots,Y_k) = \mathbb{E}\left(V_{k+1}(Y_0,\ldots,Y_{k+1}) \mid (Y_0,\ldots,Y_k)\right)$$

is a martingale. Therefore, if we can find a a betting strategy H_k such that

$$V_k(Y_0,\ldots,Y_k) = M_k \Leftrightarrow V_k(Y_0,\ldots,Y_k) - V_{k-1}(Y_0,\ldots,Y_{k-1}) = M_k - M_{k-1} = H_{k-1}(Y_k - Y_{k-1})$$

with the initial condition $V_0(Y_0) = M_0$, the problem will be solved. Notice that the r.h.s. in the above formula is satisfied as soon as

$$\begin{cases} H_{k-1}(a_k - Y_{k-1}) &= V_k(Y_0,\ldots,Y_{k-1},a_k) - V_{k-1}(Y_0,\ldots,Y_{k-1}) \\ H_{k-1}(b_k - Y_{k-1}) &= V_k(Y_0,\ldots,Y_{k-1},b_k) - V_{k-1}(Y_0,\ldots,Y_{k-1}). \end{cases}$$

Subtracting the two equations, we find

$$H_{k-1} = \frac{V_k(Y_0,\ldots,Y_{k-1},b_k) - V_k(Y_0,\ldots,Y_{k-1},a_k)}{b_k - a_k}.$$

29.4.2 Snell envelope

We let Z_k be the successive gains of the player in a game equipped with a filtration $\mathcal{F}_k$ on some fixed horizon $k \leq n$. For instance, we can choose f_k to be a collection of non-negative functions on a given space S, and a Markov chain X_k on the state space S with $\mathcal{F}_k = \sigma(X_0, \ldots, X_n)$. Then we define

$$Z_k = f_k(X_k). \tag{29.24}$$

We can extend these models to Markov chains X_k evolving in some state spaces S_k that may depend on the time parameter. In this situation, the payoff functions f_k are defined on the state spaces S_k. This general framework allows us to consider without further work the historical processes $X_n = (X'_0, \ldots, X'_n) \in S_n = (S')^{n+1}$ and payoff functions $f_n(X_n) = f_n(X'_0, \ldots, X'_n)$ that depend on the random trajectories of a given controlled Markov chain X'_k evolving in some state spaces S'. The optimal stopping problems associated with the history-dependent payoff functions

$$f_n(X_n) := \sum_{0 \leq l < n} g'_l(X'_l) + f'_n(X'_n) \tag{29.25}$$

and

$$f_n(X_n) := \left[\prod_{0 \leq l < n} g'_l(X'_l) \right] \times f'_n(X'_n) \tag{29.26}$$

for some non-negative functions f'_l, g'_l on S' are discussed in exercises 488 and 489.

At any time $k \in \{0, \ldots, n\}$ the player tries to stop the game at some $T_k \geq k$ using the information $\mathcal{F}_k$ he has at that time with the goal to maximize the expected gain given by the Snell envelope

$$U_k := \sup_{T \in \mathcal{T}_k} \mathbb{E}\left(Z_T \mid \mathcal{F}_k\right). \tag{29.27}$$

In the above display, $\mathcal{T}_k$ stands for the set of stopping times $(n \geq)T \geq k$ adapted to the filtration $\mathcal{F}_k$.

One way to solve this problem is to consider the sequence of stopping times defined using the backward induction

$$T_k = k \, 1_{Z_k \geq \mathbb{E}(Z_{T_{k+1}} \mid \mathcal{F}_k)} + T_{k+1} \, 1_{Z_k < \mathbb{E}(Z_{T_{k+1}} \mid \mathcal{F}_k)}$$

with the terminal condition $T_n = n$.

We use backward induction to check that these stopping times satisfy the optimal stopping problem. Since $\mathcal{T}_n$ is reduced to $T = n$, we clearly have at the final horizon n :

$$Z_{T_n} = Z_n = \mathbb{E}(Z_{T_n} \mid \mathcal{F}_n) = \sup_{T \in \mathcal{T}_n} \mathbb{E}\left(Z_T \mid \mathcal{F}_n\right).$$

Suppose that

$$\mathbb{E}\left(Z_{T_{k+1}} \mid \mathcal{F}_{k+1}\right) = \sup_{T \in \mathcal{T}_{k+1}} \mathbb{E}\left(Z_T \mid \mathcal{F}_{k+1}\right)$$

at some $k < n$. By definition of T_k we have

$$Z_{T_k} = Z_k \, 1_{Z_k \geq \mathbb{E}(Z_{T_{k+1}} \mid \mathcal{F}_k)} + Z_{T_{k+1}} \, 1_{Z_k \leq \mathbb{E}(Z_{T_{k+1}} \mid \mathcal{F}_k)}.$$

Taking the conditional expectation w.r.t. $\mathcal{F}_k$, this implies that

$$\mathbb{E}(Z_{T_k} \mid \mathcal{F}_k) \;=\; Z_k \, 1_{Z_k \geq \mathbb{E}(Z_{T_{k+1}} \mid \mathcal{F}_k)} + \mathbb{E}(Z_{T_{k+1}} \mid \mathcal{F}_k) \, 1_{Z_k \leq \mathbb{E}(Z_{T_{k+1}} \mid \mathcal{F}_k)}.$$

Hence we get

$$Z_k \vee \mathbb{E}(Z_{T_{k+1}} \mid \mathcal{F}_k) = \mathbb{E}(Z_{T_k} \mid \mathcal{F}_k) \leq \sup_{T \in \mathcal{T}_k} \mathbb{E}(Z_T \mid \mathcal{F}_k). \tag{29.28}$$

Finally, for any $T \in \mathcal{T}_k$ we observe that

$$T = k \, 1_{T=k} + (T \vee (k+1)) \, 1_{T \geq (k+1)}$$

with

$$\{T = k\} \quad , \{T \geq (k+1)\} \in \mathcal{F}_k \quad \text{and} \quad (T \vee (k+1)) \in \mathcal{T}_{k+1} \subset \mathcal{T}_k.$$

Thus, taking the expectation w.r.t. $\mathcal{F}_k$ we have

$$
\begin{aligned}
\mathbb{E}(Z_T \mid \mathcal{F}_k) &= Z_k \, 1_{T=k} + \mathbb{E}\left(Z_{T \vee (k+1)} \mid \mathcal{F}_k\right) \, 1_{T \geq (k+1)} \\
&= Z_k \, 1_{T=k} + \mathbb{E}\left(\mathbb{E}\left[Z_{T \vee (k+1)} \mid \mathcal{F}_{k+1}\right] \mid \mathcal{F}_k\right) \, 1_{T \geq (k+1)} \\
&\leq Z_k \, 1_{T=k} + \mathbb{E}\left(\sup_{S \in \mathcal{T}_{k+1}} \mathbb{E}\left[Z_S \mid \mathcal{F}_{k+1}\right] \mid \mathcal{F}_k\right) \, 1_{T \geq (k+1)} \\
&= Z_k \, 1_{T=k} + \mathbb{E}\left(\mathbb{E}(Z_{T_{k+1}} \mid \mathcal{F}_{k+1}) \mid \mathcal{F}_k\right) \, 1_{T \geq (k+1)}.
\end{aligned}
$$

This yields the upper bound

$$\mathbb{E}(Z_T \mid \mathcal{F}_k) \leq Z_k \vee \mathbb{E}\left(Z_{T_{k+1}} \mid \mathcal{F}_k\right) \Rightarrow \sup_{T \in \mathcal{T}_k} \mathbb{E}(Z_T \mid \mathcal{F}_k) \leq Z_k \vee \mathbb{E}\left(Z_{T_{k+1}} \mid \mathcal{F}_k\right). \tag{29.29}$$

Combining (29.28) and (29.29) we conclude that

$$
\begin{aligned}
U_k := \sup_{T \in \mathcal{T}_k} \mathbb{E}(Z_T \mid \mathcal{F}_k) &= \mathbb{E}(Z_{T_k} \mid \mathcal{F}_k) \\
&= Z_k \vee \mathbb{E}\left(Z_{T_{k+1}} \mid \mathcal{F}_k\right) = Z_k \vee \mathbb{E}\left(\mathbb{E}(Z_{T_{k+1}} \mid \mathcal{F}_{k+1}) \mid \mathcal{F}_k\right) \\
&= Z_k \vee \mathbb{E}\left(U_{k+1} \mid \mathcal{F}_k\right) \geq Z_k.
\end{aligned}
$$

In summary, the Snell envelope U_k is defined by the backward induction

$$U_k = Z_k \vee \mathbb{E}\left(U_{k+1} \mid \mathcal{F}_k\right) \tag{29.30}$$

starting at $U_n = Z_n$.

The optimal stopping time strategy is now given by

$$T_k := \inf\left\{l \in \{k, k+1, \ldots, n\} \,:\, U_l = Z_l\right\}. \tag{29.31}$$

We check this claim using

$$T_k = k \, 1_{\underbrace{Z_k \, \geq \, \mathbb{E}(U_{k+1} \mid \mathcal{F}_k)}_{U_k = Z_k}} + T_{k+1} \, 1_{\underbrace{Z_k \, < \, \mathbb{E}(U_{k+1} \mid \mathcal{F}_k)}_{\substack{U_k > Z_k \\ U_k = \mathbb{E}(U_{k+1} \mid \mathcal{F}_k)}}}.$$

Applying (29.30) to the stochastic model (29.24), the Snell envelope $U_k = V_k(X_k)$ is solved using the functions V_k defined by the backward induction

$$V_k(x_k) = f_k(x_k) \vee \mathbb{E}(V_{k+1}(X_{k+1}) \mid X_k = x_k) \qquad (29.32)$$

with the terminal condition $V_n = f_n$.

29.4.3 Continuous time models

We consider a Markov process X_t on some state space S associated with some infinitesimal generators L_t. As in (29.33), the goal is to maximize the expected gain given by the Snell envelope

$$V_s(x) := \sup_{T \in \mathcal{T}_s} \mathbb{E}\left(f_T(X_T) \mid X_s = x\right). \qquad (29.33)$$

In the above display, $\mathcal{T}_t$ stands for the set of admissible stopping times $(t \geq)T \geq s$, and f_t stands for some payoff function. As in section 29.3.3, we further assume the existence of a discrete time approximation $X_{t_n}^h$ of the process X_t on some time mesh $0 \leq t_n \leq t_{n+1}$ with time step $(t_{n+1} - t_n) = h$, and we let $P_{t_n}^h$ be the discrete time semigroup (29.16)

$$P_{t_n}^h\left(X_{t_n}^h, dx\right) := \mathbb{P}\left(X_{t_{n+1}}^h \in dx \mid X_{t_n}^h\right).$$

We assume that

$$P_{t_n}^h(\varphi) = \varphi + L_{t_n}(\varphi)\, h \, + \mathrm{O}\left(h^2\right).$$

By (29.32) the Snell envelope V_s is solved by the backward induction

$$
\begin{aligned}
V_{t_k}(x) &= f_{t_k}(x) \vee P_{t_{k+1}}^h\left(V_{t_{k+1}}\right)(x) \\
&= f_{t_k}(x) \vee \left[V_{t_{k+1}}(x) + L_{t_{k+1}}\left(V_{t_{k+1}}\right)(x)\, h \, + \mathrm{O}\left(h^2\right)\right].
\end{aligned}
$$

This implies that

$$
\begin{aligned}
0 &= \left(V_{t_k} - f_{t_k}\right) \wedge \left\{\left[h^{-1}\left(V_{t_k} - V_{t_k+h}\right) - L_{t_{k+1}}\left(V_{t_{k+1}}\right) + \mathrm{O}(h)\right]\, h \,\right\} \\
&\simeq \underbrace{\left(V_s - f_s\right)}_{\geq 0} \wedge \left\{-\left[\partial_s + L_s\right](V_s)\, h \,\right\}
\end{aligned}
$$

as soon as $t_k \to s$ as $h \downarrow 0$. Since $V_s \geq f_s$ we have

$$V_s > f_s \;\Rightarrow\; -\left[\partial_s + L_s\right](V_s) = 0 < (V_s - f_s)$$

as well as

$$V_s = f_s \Rightarrow -\left[\partial_s + L_s\right](V_s) \geq 0 = (V_s - f_s)$$

(otherwise we arrive at the contradiction

$$0 = 0 \wedge \left\{-\left[\partial_s + L_s\right](V_s)\, h \,\right\} < -\left[\partial_s + L_s\right](V_s)\, h < 0).$$

This yields the backward dynamic programming equation

$$(V_s - f_s) \wedge \left\{-\left[\partial_s + L_s\right](V_s)\right\} = 0$$

with the terminal condition $V_t = f_t$.

29.5   Exercises

Exercise 470 (Optimal allocation portfolio) *Self financing portfolios can be described by a one-dimensional stochastic differential equation of the form*

$$dX_t = (X_t - u_t)\, a_t\, dt\ +\ u_t\, [b_t\, dt + \sigma_t\, dW_t]$$

for some regular non-negative functions $t \mapsto (a_t, b_t, \sigma_t)$. *The functions* a_t *and* b_t *represent the return rates of the non-risky asset and of the risky one. The function* σ_t *stands for the volatility of the market. The control* u_t *represents the amount of risky asset invested at any time* t. *The objective is to maximize at every time* $s \in [0, t]$ *the expected power utility function*

$$V_s(x) := \sup_{v \in \mathcal{V}_{s,t}} \mathbb{E}\left(X_t^\alpha\, 1_{X_t \geq 0} \mid X_s = x\right)$$

for some given final time horizon t, *some given relative risk aversion coefficient* $\alpha \in\,]0, 1[$, *and the set* $\mathcal{V}_{s,t}$ *of feedback controls* $v\ :\ (r, x) \in ([s, t] \times [0, \infty[) \mapsto v_r(x) \in [0, \infty[$, *with* $s \leq r \leq t$. *Describe the HJB equation (29.19) associated with this stochastic control problem. Check that* $V_s(x) = \beta_s\, x^\alpha\, 1_{x \geq 0}$ *for some functions* $s \in [0, t] \mapsto \beta_s$ *and find the optimal feedback control.*

Exercise 471 (Linear quadratic control - Discrete time) *Check the square completion formula*

$$u'Ru + 2\, u'Sx = \left[u + R^{-1}Sx\right]'\, R\left[u + R^{-1}Sx\right] - x'S'R^{-1}Sx$$

which is valid for any column vectors $u \in \mathbb{R}^r$, $x \in \mathbb{R}^q$, *any symmetric and invertible* $(r \times r)$-*matrix* R, *and any* $(r \times q)$-*matrix* S. *In addition, when* R *is negative definite, deduce that*

$$\sup_{u \in U} [u'Ru + 2\, u'Sx] = -x'S'R^{-1}Sx$$

and the supremum is attained for $u = v(x) = -R^{-1}Sx$. *Let* $W_n = (W_n^i)_{1 \leq i \leq p}$ *be a sequence of independent* $\mathbb{R}^p$-*valued random variables with* $\mathbb{E}(W_n^i) = 0$ *and* $\mathbb{E}(W_n^i W_n^j) = 1_{i=j}$, *for any* $1 \leq i, j \leq p$. *Consider the linear controlled* $\mathbb{R}^q$-*valued Markov chain*

$$X_n = A_n X_{n-1} + B_n u_{n-1} + C_n W_n$$

with $u_n \in U_n = U := \mathbb{R}^r$, *and matrices* (A_n, B_n, C_n) *with appropriate dimensions. Consider the stochastic control problem (29.10) with*

$$f_n(x) = x'P_n x \quad and \quad g_l(x, u) = x'Q_l x + u'R_l u$$

for some definite negative and symmetric square matrices (P_n, Q_l, R_l) *with appropriate dimensions. Check that*

$$\forall 0 \leq k \leq n \qquad V_k(x) := x'P_k x + \alpha_k$$

for some symmetric and negative definite $(q \times q)$-*matrices* P_k, *and a sequence of parameters* α_k *with null terminal condition* $\alpha_n = 0$ *(so that* $V_n = f_n$) *satisfying the backward equations*

$$\begin{aligned}
P_k &= Q_k + A'_{k+1}P_{k+1}\left[Id - B_{k+1}\left[R_k + B'_{k+1}P_{k+1}B_{k+1}\right]^{-1} B'_{k+1}P_{k+1}\right] A_{k+1} \\
\alpha_k &= \alpha_{k+1} + \mathrm{tr}\left(C'_{k+1}P_{k+1}C_{k+1}\right).
\end{aligned}$$

In the above display, Id is a $(q \times q)$-identity matrix and $\mathrm{tr}(A)$ stands for the trace of a matrix A (we also recall that $\mathrm{tr}(AB) = \mathrm{tr}(BA)$). Prove that the optimal policy is given by the feedback control

$$v_k(x) := - \left[R_k + \left(B_{k+1}' P_{k+1} B_{k+1} \right) \right]^{-1} \left(B_{k+1}' P_{k+1} A_{k+1} \right) x.$$

Exercise 472 (Martingale optimality principle - Discrete time) *Consider the stochastic control problem discussed in exercise 471. Check that the stochastic process*

$$\forall 0 \leq k \leq n \qquad \mathbb{V}_k(v) := \alpha_k + X_k' P_k X_k + \sum_{0 \leq l < k} \left[X_l' Q_l X_l + v_l(X_l)' R_l v_l(X_l) \right]$$

defined in terms of the chain

$$X_{l+1} = A_{l+1} X_l + B_{l+1} v_l(X_l) + C_{l+1} W_{l+1}$$

is a martingale w.r.t. to $\mathcal{F}_k = \sigma(X_l, \ l \leq k)$ ending at $\mathbb{V}_n(v) = X_n' P_n X_n$, as soon as v_l is the optimal control feedback computed in exercise 471.

Exercise 473 (Linear quadratic control - Continuous time) *Let $W_t = (W_t^i)_{1 \leq i \leq p}$ be a p-dimensional Brownian motion. Consider the linear controlled $\mathbb{R}^q$-valued diffusion*

$$dX_t = (A_t X_t + B_t u_t) \ dt + C_t dW_t$$

with $u_t \in U := \mathbb{R}^r$, and matrices (A_t, B_t, C_t) with appropriate dimensions. Consider the stochastic control problem (29.15) with

$$f_t(x) = x' P_t x \quad and \quad g_s(x, u) = x' Q_s x + u' R_s u$$

for some negative definite and symmetric square matrices (P_t, Q_s, R_s) with appropriate dimensions and with an invertible matrix R_s. Prove that the solution of the Bellman equation (29.19) has the form

$$\forall 0 \leq s \leq t \qquad V_s(x) := x' P_s x + \alpha_s$$

for some symmetric and negative definite $(q \times q)$-matrices P_s and some parameters α_s satisfying the backward equations

$$
\begin{aligned}
-\partial_s P_s &= Q_s + A_s' P_s + P_s A_s - P_s B_s R_s^{-1} B_s' P_s \\
-\partial_s \alpha_s &= \mathrm{tr}(C_s' P_s C_s)
\end{aligned}
$$

with the boundary terminal condition $\alpha_t = 0$. Prove that the optimal policy is given by the feedback control

$$v_s(x) = -R_s^{-1} (B_s' P_s) x.$$

Exercise 474 (Linear quadratic control - Examples) *We consider the linear quadratic control problem discussed in exercise 473. Examine the one-dimensional case*

1. *$A_s = B_s = C_s = Q_s = R_s = 1$ and a null terminal condition $P_t = 0$.*
2. *$A_s = Q_s = 0$, $B_s = C_s = 1$, $P_t = P(< 0)$, $R_s = R(< 0)$. Check that*

$$\mathbb{V}_s(v) = -R \ \log \left(\frac{R}{R + P(t-s)} \right) + \frac{RP \ X_s^2}{R + P(t-s)} + R \int_0^s \left[\frac{PX_r}{R + P(t-r)} \right]^2 dr$$

is a martingale w.r.t. to $\mathcal{F}_k = \sigma(X_l, \ l \leq k)$ ending at $\mathbb{V}_t(v) = P_t X_t^2$.

Exercise 475 (Martingale optimality principle - Continuous time) *Consider the stochastic control problem discussed in exercise 473. Check that the stochastic process*

$$\forall 0 \leq k \leq n \qquad \mathbb{V}_s(v) := \alpha_s + X_s' P_s X_s + \int_0^s \left[X_r' Q_r X_r + v_r \left(X_r \right)' R_r v_r \left(X_r \right) \right] \, dr$$

defined in terms of the chain

$$dX_s = (A_s X_s + B_s v_s(X_s)) \, dt + C_s dW_s$$

is a martingale w.r.t. to $\mathcal{F}_s = \sigma \left(X_r, \ r \leq s \right)$ ending at $\mathbb{V}_t(v) = X_t' P_t X_t$ as soon as v_s is the optimal control feedback computed in exercise 473.

Exercise 476 (Bilinear quadratic control) *Let W_t be a one-dimensional Brownian motion. Consider the linear controlled one-dimensional diffusion*

$$dX_t = (A_t X_t + B_t u_t + C_t) \, dt + (a_t X_t + b_t u_t + c_t) \, dW_t$$

with $u_t \in U := \mathbb{R}$, and some parameters (A_t, B_t, C_t) and (a_t, b_t, c_t). We consider the stochastic control problem (29.15) with

$$f_t(x) = P_t x^2 \quad and \quad g_s(x, u) = Q_s x^2 + R_s u^2 + S_s \, x \, u$$

for some negative parameters (P_t, Q_s, R_s) with $R_s < 0$ and some $S_s \in \mathbb{R}$. Prove that the solution of the Bellman equation (29.19) has the form

$$\forall 0 \leq s \leq t \qquad V_s(x) := P_s x^2 + \beta_s x + \alpha_s$$

for parameters (P_s, β_s, α_s) satisfying some backward equation. Check that the optimal policy is given by the feedback control

$$v_s(x) = -x \, \frac{(a_s b_s + B_s) P_s + S_s/2}{R_s + b_s^2 P_s} - \frac{b_s c_s P_s + B_s \beta_s/2}{R_s + b_s^2 P_s}.$$

Exercise 477 (Path integral optimization - Discrete time) *We extend the framework discussed in section 29.3.2 to value functions of the form*

$$F_n(X, v) := \sum_{0 \leq l < n} Z_l(X, \nu) \, g_l(X_l, v_l) + Z_n(X, \nu) \, f_n(X_n) \quad with \quad Z_l(X, \nu) := \prod_{0 \leq k < l} z_k(v_k, X_k)$$

for some non-negative functions z_k on $(U_k \times S_k)$. Check that $F_n = F_{0,n}$ with the interpolating functions

$$F_{k,n}(X, v) \quad := \quad \sum_{k \leq l < n} Z_{k,l}(X, \nu) \, g_l(X_l, v_l) + Z_{k,n}(X, \nu) \, f_n(X_n)$$

and the Radon-Nikodym derivatives $Z_{k,n}(X, \nu) := Z_n(X, \nu)/Z_k(X, \nu)$. Prove that these interpolating functions satisfy the backward equation

$$F_{k,n}(X, v) = g_k(X_k, v_k) + z_k(v_k, X_k) \, F_{k+1,n}(X, v)$$

with the terminal condition $F_{n,n} = f_n$. Prove that

$$
\begin{aligned}
V_k(x) \quad &:= \quad \sup_{v \in \mathcal{V}_{k,n-1}} \mathbb{E}_v \left(F_{k,n}(X, v) \mid X_k = x \right) \\
&= \quad \sup_{u \in U_k} \left[g_k(x, u) + z_k(u, x) \, M_{u,k+1} \left(V_{k+1} \right)(x) \right].
\end{aligned}
$$

Exercise 478 (Path integral optimization - Continuous time) ✎ *We extend the frame-work discussed in section 29.3.3 to value functions of the form*

$$F_t(X, u) := \int_0^t Z_s(X, u) \, g_s(X_s, u_s) \, ds + Z_t(X, u) \, f_t(X_t) \quad with \quad Z_s(X, u) := e^{\int_0^s H_r(u_r, X_r) dr}$$

for some functions H_s on $(U \times S)$ and for any control mapping $u = (u_s)_{s \in [0,t]}$ from $[0, t]$ into U. We fix a time horizon t and consider the value functions

$$V_s(x) := \sup_{v \in \mathcal{V}_{s,t}} \mathbb{E}_v \left(\int_s^t Z_{s,r}(X, u) \, g_r(X_r, v_r(X_r)) \, dr + Z_{s,t}(X, u) \, f_t(X_t) \mid X_s = x \right)$$

with $Z_{s,t}(X, u) = Z_t(X, u)/Z_s(X, u)$, and any $s \leq t$. Following the derivation of the Bellman equation (29.19) presented in section 29.3.3 and exercise 477, check that V_s satisfies the backward equation

$$-\partial_s V_s(x) = \sup_{u \in U} \left[g_s(x, u) + L_{s,u}^H (V_s) (x) \right]$$

with terminal condition $V_t = f_t$ and the Schrödinger operator

$$L_{s,u}^H(\varphi)(x) = L_{s,u}(\varphi)(x) + H_s(u, x) \, \varphi(x).$$

Exercise 479 (Optimality Bellman equation) *We let V_s be some smooth function satisfying the optimality principle (29.20). Using the Doeblin-Itō formula check that*

$$\mathbb{E}_v \left(\int_{s_1}^{s_2} [\partial_s V_s(X_s) + g_s(X_s, v_s(X_s)) + L_{v_s, s}(V_s)(X_s)] \, ds \mid X_{s_1} = x \right) \leq 0$$

for any $0 \leq s_1 \leq s_2 \leq t$ and any control $v \in \mathcal{V}_{s_1, s_2}$, with the equality on optimal control policies. Deduce the Bellman equation (29.19).

Exercise 480 (Feynman-Kac value functions) ✎✎ *Let $W_t = (W_t^i)_{1 \leq i \leq r}$ be an r-dimensional Brownian motion. Consider the r-dimensional diffusion*

$$dX_t = b_t(X_t) \, dt + \sigma_t(X_t) \, (u_t \, dt + dW_t)$$

for some regular drift and diffusion functionals b_t and σ_t with appropriate dimensions, and a control function $u_t \in \mathbb{R}^r$. We consider the value function (29.15) with $g_s(x, u) = h_s(x) + \frac{1}{2} u' R_s u$, for some invertible and negative definite matrix R_s and some regular negative functions h_s and $f_t(x)$. Check that the value function V_s satisfies the Hamilton-Jacobi-Bellman equation

$$-\partial_s V_s = h_s + (\partial V_s)' \, b_s - \frac{1}{2}(\partial V_s)' \sigma_s R_s^{-1} \sigma_s'(\partial V_s) + \frac{1}{2} \, \mathrm{tr} \left(\sigma_s' \partial^2 V_s \sigma_s \right)$$

for any $s \leq t$ (with the terminal condition $V_t = f_t$), and the optimal feedback control is given by

$$v_s(x) = -R_s^{-1} \sigma_s(x)' (\partial V_s)(x).$$

Assume that $R_s = \lambda \, Id$, for some $\lambda < 0$, where Id stands for the $(r \times r)$-identity matrix. Following the argments given at the end of section 29.3.3, check that

$$V_s = -\lambda \log q_s$$

with q_s defined by the Feynman-Kac formula

$$q_s(y) = \mathbb{E} \left[\exp \left(\overline{f}_t(Y_t) \right) \exp \left(\int_s^t \overline{h}_r(Y_r) dr \right) \mid Y_s = y \right]$$

with the functions $\overline{f}_t = -\lambda^{-1} f_t$ and $\overline{h}_s = -\lambda^{-1} h_s$, and the r-dimensional diffusion process

$$dY_s = b_s(Y_s) \, ds + \sigma_s(Y_s) \, dW_s.$$

Exercise 481 (Optimal stopping - Minimization problems) *Consider the optimal stopping problem defined in section 29.4.2 by replacing the maximization problem (29.33) by minimization problems*

$$U_k := \inf_{T \in \mathcal{T}_k} \mathbb{E}\left(f_T(X_T) \mid \mathcal{F}_k\right)$$

for some given cost functions $f_k(X_k)$. Describe the Snell envelope and the optimal stopping rules associated with this problem.

Exercise 482 (Cayley-Moser optimal stopping problem [204]) 🖊 *We consider the optimal stopping problem discussed in section 29.4.2. We assume that $(X_k)_{0 \le k \le n}$ is a sequence of independent and identically distributed real valued random variables with some distribution μ and the payoff or reward functions are given by $f_k(x) = x$, with $0 \le k \le n$. Check that the Snell envelope (29.32) is given by $V_{n-k}(x) = x \vee m_k$, for any $0 \le k \le n$, with the sequence of numbers $m = (m_k)_{0 \le k \le n}$ given by*

$$m_{k+1} \quad = \quad \mathbb{E}\left(X \vee m_k\right)$$

with the initial condition $m_0 = -\infty$ (so that $m_1 = \mathbb{E}(X)$). Describe the optimal stopping strategy. Examine the situations where μ is the uniform distribution on $[0,1]$ and where $(X_k)_{0 \le k \le n}$ have exponential distribution with some parameter $\lambda > 0$.

Exercise 483 (Parking problem [193]) 🖊 *We drive in a street where parking places are indexed by the integer lattice $\mathbb{N}$. We let $(X_k)_{k \ge 0}$ be a sequence of independent copies of a Bernoulli random variable X defined by*

$$\mathbb{P}(X = 1) = 1 - \mathbb{P}(X = 0) = p$$

for some $p \in [0,1[$. The events $\{X_k = 0\}$, respectively $\{X_k = 1\}$, mean that the k-th parking place is available, respectively occupied by some car. We fix a given time horizon n representing a given target destination, say a pub. If we arrive to n-th parking place and it is filled, we choose the next first available place. Parking at the k-th place (if available) requires us to walk $(n-k)$-units to the pub. We cannot park if the place is filled; in this situation the cost is $(+\infty)$ as the car will be towed away. Consider the cost functions $(f_k)_{0 \le k \le n}$ defined by

$$\forall 0 \le k < n \qquad f_k(X_k) = (n-k) \; 1_{X_k=0} + \infty \; 1_{X_k=1}$$

and the terminal average cost

$$f_n(X_n) = 0 \; 1_{X_n=0} + \left[\sum_{k \ge 1} k \; \mathbb{P}\left[\cap_{1 \le l < k}(X_{n+l} = 1) \cap (X_{n+k} = 0)\right] \right] 1_{X_n=1}.$$

Describe the optimal stopping rules at any rank $0 \le k \le n$.

Exercise 484 (Myopic stopping rules - Monotone problems) 🖊 *We consider the optimal stopping problem defined in section 29.4.2. The one-stage look-ahead stopping rule (a.k.a. myopic rule) is defined by the non-optimal stopping time S given by*

$$S = \inf\{n \ge 0 \; : \; f_n(X_n) \ge \mathbb{E}\left(f_{n+1}(X_{n+1}) \mid X_n\right)\}.$$

We further assume that the optimal stopping problem is monotone in the sense that

$$f_{n-1}(X_{n-1}) \geq \mathbb{E}\left(f_n(X_n) \mid X_{n-1}\right) \Rightarrow f_n(X_n) \geq \mathbb{E}\left(f_{n+1}(X_{n+1}) \mid X_n\right)$$

for any time $n \geq 1$. Several examples of monotone optimal stopping problems are discussed in exercise 485, exercise 486, and exercise 487. We fix a given time horizon, say $n \geq 0$. Check that $S = T_0$, with the optimal stopping time T_0 defined in (29.31). In other words, check that the myopic rule is an optimal stopping time.

Exercise 485 (Burglar problem) *At each of its attempts, a burglar has a probability $p \in [0, 1]$ to be caught by police and lose his accumulated earnings; otherwise he enters a house and brings back an amount of money represented by a non-negative random variable W with a finite expectation $\mathbb{E}(W) = w$. Let ϵ be a $\{0, 1\}$-valued Bernoulli random variable with success probability $\mathbb{P}(\epsilon = 1) = p \in [0, 1[$ (the event $\{\epsilon = 1\}$ represents a successful burglary). We assume that ϵ is independent of X. We consider the Markov chain*

$$X_n = \left(X_n^1, X_n^2\right),$$

starting at $X_0 = (1, 0)$ and defined by

$$X_n^1 := \epsilon_n \, X_{n-1}^1 \quad and \quad X_n^2 := X_{n-1}^2 + W_n$$

where (ϵ_n, W_n) represent independent copies of (ϵ, W). We let $\mathcal{F}_n = \sigma(X_k, \ k \leq n)$. The payoff function is defined by

$$f_n(X_n) = X_n^1 X_n^2.$$

Check that $f_n(X_n) \geq \epsilon_n \, f_{n-1}(X_{n-1})$ and

$$f_n(X_n) - \mathbb{E}\left(f_{n+1}(X_{n+1}) \mid \mathcal{F}_n\right) = (1-p) \, f_n(X_n) - p \, w \, X_n^1.$$

Deduce that the optimal stopping problem is monotone and check that the best strategy for the burglar is to stop as soon as the accumulated earnings are at least $pw/(1-p)$.

Exercise 486 (Asset-selling problem) ✍ *The offers to buy a given asset are represented by a sequence of independent copies W_n of a non-negative random variable W with some given distribution μ on $\mathbb{R}_+$. We let $\mathcal{F}_n = \sigma(W_k, \ k \leq n)$. The trader is allowed to recall at any time n past offers, and pays amount of a units at each time a new offer is made before he decides to stop and sell the asset. The maximum of all offers at each time n and the payoff functions are given by*

$$X_n = X_{n-1} \vee W_n \quad and \quad f_n(X_n) = X_n - na$$

with the initial condition $X_0 = -\infty (= f_0(X_0))$. Check that the optimal stopping problem is monotone and find the best stopping strategy. Examine the situations where W is an exponential random variable with parameter $\lambda > 0$, or a uniform random variable on $[w_1, w_2]$, for some $w_1 \leq w_2$.

Exercise 487 (Proofreading problem) *The number of misprints in a given book follows a Poisson random variable N with a parameter $\lambda > 0$. The number of detected misprints at the $(n+1)$-th proofreading is a binomial random variable W_{n+1} with sample size $N_n = \left(N - \sum_{1 \leq k \leq n} W_k\right)$ and a given detection probability $p \in]0, 1[$. Check that the conditional distribution of N_n given $\mathcal{F}_n = \sigma(W_1, \ldots, W_n)$ is a Poisson distribution with parameter $\lambda(1-p)^n$. We let $X_n = (W_n^1, \ldots, W_n^2)$. The cost of each proofreading and each misprint* *left are given respectively by some non-negative parameters a_1 and a_2. Check that the total cost after the n-th proofreading is given by*

$$f_n(X_n) := \mathbb{E}\left(na_1 + (N - X_n)\, a_2 \mid X_n\right) = na_1 + \lambda(1 - p)^n a_2.$$

Check that the optimal stopping problem is monotone. Prove that the optimal stopping time on any finite horizon problem is given by

$$S = \inf\left\{n \geq 0 \ : \ \lambda p(1 - p)^n \ \leq a_1/a_2\right\}.$$

Exercise 488 (Path dependent payoff - 1) *We consider the optimal stopping problem in path space (29.25). Check that the Snell envelope V_k defined in (29.32) has the form*

$$V_k(x_k) = V_k(x'_0, \ldots, x'_k) = \sum_{0 \leq l < k} g'_l(x'_l) + V'_k(x')$$

for any historical path $x_k = (x'_0, \ldots, x'_k) \in S_k$, with the functions V'_k on S' defined by the backward induction

$$V'_k(x'_k) = f'_k(x'_k) \vee \left[g'_k(x'_k) + \mathbb{E}(V'_{k+1}(X'_{k+1}) \mid X'_k = x'_k)\right] \tag{29.34}$$

with the terminal condition $V'_n = f'_n$.

Exercise 489 (Path dependent payoff - 2) *We consider the optimal stopping problem in path space (29.26). Check that the Snell envelope V_k defined in (29.32) has the form*

$$V_k(x_k) = V_k(x'_0, \ldots, x'_k) = \left[\prod_{0 \leq l < k} g'_l(x'_l)\right] \times V'_k(x')$$

for any historical path $x_k = (x'_0, \ldots, x'_k) \in S_k$, with the functions V'_k on S' defined by the backward induction

$$V'_k(x'_k) = f'_k(x'_k) \vee \left[g'_k(x'_k) \times \mathbb{E}(V'_{k+1}(X'_{k+1}) \mid X'_k = x'_k)\right] \tag{29.35}$$

with the terminal condition $V'_n = f'_n$.

Exercise 490 (Kelly criterion [166]) ✎ *Consider the gambling model discussed in (29.1) with $q \geq p$ (fair or favorable game). We assume that the betting strategy $H_n = \alpha M_n$ is a proportion $\alpha \in [0, 1]$ of the fortune M_n of the gambler. Check that*

$$M_n = M_0 \prod_{1 \leq k \leq n} (1 + \alpha\, X_k)$$

and compute the growth rate $L_n(\alpha) = \frac{1}{n} \log(M_n/M_0)$ and its limit $L_\infty(\alpha) = \lim_{n \to \infty} L_n(\alpha)$. Check that $L_\infty(\alpha)$ is maximized for $\alpha = \alpha^\star = \mathbb{E}(X)$ and compute the maximal winning rate $L_\infty(\alpha^\star)$. Answer the same questions when the fortune of the gambler is given by

$$M_{n+1} = M_n + (1 - \alpha)\, M_n\, r + \alpha\, M_n\, X_{n+1}$$

where $r \in]0, 1[$ stands for the return rate of some riskless asset (discuss the two cases $(1 - r)/(1 + r) \leq p/q$, resp. $(1 - r)/(1 + r) \geq p/q$ corresponding to a more favorable return on the riskless asset, resp. on the game).

Exercise 491 (Monty Hall game show) *In the Monty Hall show, we present three doors with a new car behind one only of them (and nothing (or a goat) behind the other ones). The contestant chooses a door without opening it. Monty opens one door and shows that there is nothing behind it. The contestant has the opportunity to keep the door selected initially or switch. What is the best strategy?*

Suppose now that we have 1000 doors with a new car behind one of them, and the host opens 998 doors with nothing behind them. Compute the probability that the car is behind the door selected by the contestant and the probability that the car is behind the other door. What is the best strategy?

Exercise 492 (Gambler's ruin) *We consider a random walk X_n on $\mathbb{Z}$ that increases or decreases randomly by c units each time. We interpret X_n as the capital of a gambler, we choose two integers $a, b \in \mathbb{N}$, and assume that $X_0 = 0$. The gambler is ruined if he loses a dollars and wins if he earns b dollars. We let T represent the random time when the player is ruined or wins. Check that $Y_n := X_n^2 - c^2 n$ is a martingale with respect to the filtration $\mathcal{F}_n := \sigma(X_0, \ldots, X_n)$, and find an upper bound on $\mathbb{E}(T)$.*

Exercise 493 (Parrondo's game [221]) ✎ *In a game (G_1) we win $+1$ with some probability $\frac{1}{2} - \epsilon$ and lose -1 with a probability $\frac{1}{2} + \epsilon$, for some $\epsilon \in [0, 1/10[$ The game (G_2) depends on whether capital is a multiple of some number m, say $m = 3$. If it is the case, you win $+1$ with some probability $p_m = \frac{1}{10} - \epsilon$ and lose -1 with a probability $q_m = \frac{9}{10} + \epsilon$. Otherwise, you win $+1$ with some probability $p = \frac{3}{4} - \epsilon$ and lose -1 with a probability $q = \frac{1}{4} + \epsilon$.*

- *We consider the game (G_2). We let X_n be the fortune of the player at time n and set $Y_n = X_n \bmod(3) \in \{0, 1, 2\}$. Express the probability $P(n)$ to win game (G_2) at a time n in terms of the law of the random states X_n. Describe the invariant measure π of the chain and check that the game (G_2) is asymptotically fair for $\epsilon = 0$ and unfair for $\epsilon > 0$.*

- *In the game (G_3) we flip a coin and play (G_1) or (G_3). Check that for a sufficiently small ϵ the probability to win after long runs is strictly larger than 50%.*

Exercise 494 (Bold play strategy) *Consider a player in a red-and-black betting system who wants to achieve a given fortune a. We let p and q represent the corresponding probabilities to win and to lose in a red-and-black bet. Using the bold play strategy, at each time n the gambler bets either his fortune Y_n if $Y_n \leq a/2$ or the amount needed to achieve the target gain $a - Y_n$ if $Y_n \geq a/2$. For unfair games $q > p$, this bold strategy is optimal in the sense that it gives the maximal probability to win a given fortune with varying bet sizes. Rewrite this strategy in terms of the rescaled fortune of the gambler $X_n = Y_n/a$ on $[0, 1]$. Show that the probability $P(y) = Q(y/a)$ to reach a after starting with some initial fortune $y (\in [0, a])$ is defined in terms of the function Q on $[0, 1]$ given by*

$$Q(x) := \begin{cases} p\, Q(2x) & \text{if } x \leq 1/2 \\ p + q\, Q(2x - 1) & \text{if } 1/2 \leq x \leq 1, \end{cases}$$

with the boundary conditions $(Q(0), Q(1)) = (0, 1)$. Compute $Q(i/2^n)$ for $i \leq 2^n$ and $n = 1, 2, 3$.

Exercise 495 (Ballot problem) ✎ *Two candidates A and B received a and b votes in an election with $a \geq nb$ for some integer $n \geq 1$. The objective is to compute the number of ways $N_n(a, b)$ the ballots can be ordered so that $a \geq nb$. A permutation is said to be "good" if A is ahead of B by a factor of at least n, and it is said to be "bad" otherwise.*

- We let X_k the difference of A votes and B votes after counting $0 \le k \le (a+b)$ votes, and set $M_k := \frac{X_{n-k}}{n-k}$. Check that M_k is a martingale with respect to the filtration $\sigma(M_0, \ldots, M_k)$. Find the probability that candidate A is always ahead while processing $(a+b)$-votes.

- We interpret the ballot permutation as a path in the lattice $(\mathbb{N} \times \mathbb{Z})$ starting at $(0,0)$, with votes for A expressed by up-steps $(1,1)$ and votes for B as down steps $(1,-n)$. The axis $(0,x) = \mathbb{N}$ is interpreted as the time, and $(0,y) = \mathbb{Z}$ the state space of the states of the path. Paths in the $(0,x)$-axis are "good" while paths going below $(0,x)$ (ending or crossing) are "bad". We let $\mathcal{P}$ be the number of good paths. Down-steps starting above $(0,x)$ and ending on or below are called "bad" steps. We let $\mathcal{P}_k$ be the set of trajectories with the first bad step ending k units below $(0,x)$, with $0 \le k \le n$. Find $\mathrm{Card}(\mathcal{P}_n)$ and prove that $\mathrm{Card}(\mathcal{P}_n) = \mathrm{Card}(\mathcal{P}_k)$ holds for any k. Deduce that

$$N_n(a,b) = \mathrm{Card}(\mathcal{P}) = \binom{a+b}{a} - (n+1) \binom{a+b-1}{a} = \frac{a-nb}{a+b} \binom{a+b}{a}.$$

- Notice that $N_n(a,0) = 1$ and $N_n(nb,b) = 0$ for any $a, b > 0$, and check that they satisfy the above formula. For $a > nb > 0$, find a recurrence relation between $N_n(a,b)$ and the quantities $N_n(a-1,b)$ and $N_n(a,b-1)$. Prove the formula using induction.

Exercise 496 (Secretary problem) ✎ *Consider n candidates whose qualifications are ranked by some indices $q_1 > \ldots > q_n$ (the best being q_1), applying for one only available position of a secretary in some company. They are interviewed sequentially in a random order and we denote by $Y_1, \ldots, Y_n$ the corresponding qualifications (note that $(Y_1, \ldots, Y_n)$ is a random permutation of their qualifications $(q_1, \ldots, q_n)$). The selection committee accepts or rejects the applicants based on their rank, with the objective to select the best one. We let $X_k := \sum_{1 \le l \le k} 1_{Y_k < Y_l}$ be the (random) relative rank of the k-th applicant among the first k interviewed applicants. We let T be the stopping time of the selection process. We want to maximize the probability $\mathbb{P}(Y_T = q_1)$ w.r.t. the information generated sequentially by the observations $(X_1, \ldots, X_n)$.*

- Find the distribution of the random variables X_k.

- Check that $\mathbb{P}(Y_k = q_1 \mid X_1, \ldots, X_k) = f_k(X_k)$ with the function $f_k : x \in \{1, \ldots, n\} \mapsto f_k(x) = 1_{x=1} (k/n)$, and deduce that $\mathbb{P}(Y_T = q_1) = \mathbb{E}(f_T(X_T))$.

- Check that

$$\sup_{T \in \mathcal{T}_n} \mathbb{P}(Y_T = q_1) = V_1(1) = \frac{m_n - 1}{n} \left(\frac{1}{n-1} + \ldots + \frac{1}{m_n} + \frac{1}{m_n - 1} \right)$$

where $m_n \sim e^{-1} n$ is the first and unique value $2 \le k \le n$ such that $\frac{1}{k} + \frac{1}{k+1} + \ldots + \frac{1}{n-1} \le 1$.

- Describe the corresponding Snell envelope and solve the optimal stopping time problem.

30

Mathematical finance

This chapter is dedicated to applications of stochastic processes in mathematical finance, and more particularly to option pricing problems. We discuss several models such as the binomial, the Cox-Ross-Rubinstein and the celebrated Black-Scholes-Merton model. The last part of the chapter is concerned with European option pricing techniques and the design of self-financing and replicating portfolios.

> *The way to make money is to buy when blood is running in the streets.*
> John D. Rockefeller (1839-1937).

30.1 Stock price models

30.1.1 Up and down martingales

The up and down evolution of the price of a risky asset in a neutral financial market is defined in terms of a martingale $\overline{\mathcal{S}}_n$ with respect to some filtration $\mathcal{F}_n$ such that

$$
\begin{aligned}
P_n &:= \mathbb{P}\left(\overline{\mathcal{S}}_n = D_n \mid \mathcal{F}_{n-1}\right) \\
Q_n &:= \mathbb{P}\left(\overline{\mathcal{S}}_n = U_n \mid \mathcal{F}_{n-1}\right)
\end{aligned}
$$

for two random up and down variables (U_n, D_n) s.t. $D_n < \overline{\mathcal{S}}_{n-1} < U_n$. The filtration $\mathcal{F}_n$ represents the information available on the financial market. This discrete time model represents the evolution of the price between time units such as weeks, days, hours, or milliseconds. The figure below demonstrates the possible paths in the evolution of the price driven by this simplistic model.

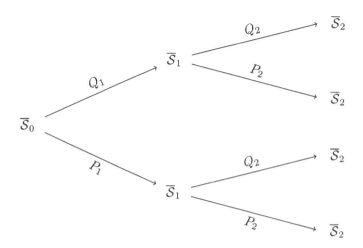

The random variables D_n and U_n represent the evolution up or down of the price of the underlying asset. From the martingale property

$$\mathbb{E}\left(\overline{S}_n \mid \mathcal{F}_{n-1}\right) = P_n\, D_n + Q_n\, U_n = \overline{S}_{n-1}$$

we see that the random probabilities (P_n, Q_n) need to be chosen so that

$$P_n\, D_n + (1 - P_n)\, U_n = U_n - P_n(U_n - D_n) = \overline{S}_{n-1}.$$

Therefore, they are necessarily given by

$$P_n = \frac{U_n - \overline{S}_{n-1}}{U_n - D_n} \quad \text{and} \quad Q_n = \frac{\overline{S}_{n-1} - D_n}{U_n - D_n}.$$

The stock price is usually defined as an exponential random walk

$$\overline{S}_n = \overline{S}_0 \times \mathcal{E}_1 \times \ldots \times \mathcal{E}_n = \overline{S}_{n-1}\, \mathcal{E}_n. \tag{30.1}$$

The walk starts at some strictly positive constant $\overline{S}_0$, and is governed by a sequence of independent random variables $\mathcal{E}_n$ with a unit mean taking two possible values $0 < d_n < 1 < u_n$. These deterministic coefficients are called the up or down factors.

In this context, the martingale $\overline{S}_n$ is also a Markov chain. In addition, the information available on the financial market at any time n is encapsulated in the the filtration $\mathcal{F}_n$ associated to this Markov chain

$$\mathcal{F}_n = \sigma(\overline{S}_1, \ldots, \overline{S}_n) = \sigma(\overline{S}_0, \mathcal{E}_1, \ldots, \mathcal{E}_n). \tag{30.2}$$

On the other hand, it is readily checked that

$$D_n = \overline{S}_{n-1} \times d_n \quad \text{and} \quad U_n = \overline{S}_{n-1} \times u_n$$

with the respective deterministic probabilities $P_n = p_n$ and $Q_n = q_n$ s.t

$$P_n = \frac{U_n - \overline{S}_{n-1}}{U_n - D_n} = \frac{u_n\, \overline{S}_{n-1} - \overline{S}_{n-1}}{u_n\, \overline{S}_{n-1} - d_n\, \overline{S}_{n-1}} = \frac{u_n - 1}{u_n - d_n}$$

and

$$Q_n = \frac{\overline{S}_{n-1} - D_n}{U_n - D_n} = \frac{\overline{S}_{n-1} - d_n\, \overline{S}_{n-1}}{u_n\, \overline{S}_{n-1} - d_n\, \overline{S}_{n-1}} = \frac{1 - d_n}{u_n - d_n}.$$

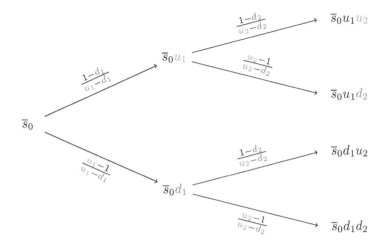

Example 30.1.1 *For instance, if $d_n = 1 - 10^{-2} < 1 < u_n = 1 + 10^{-2}$ for all n, then $P_n = Q_n = 1/2$. When the current asset price $\overline{S}_{n-1} = \$100$, there are 50% chances that the price will go up to $\overline{S}_n = \$100 \times (1 + 10^{-2}) = \$101$ and 50% chances that it will go down to $\overline{S}_n = \$100 \times (1 - 10^{-2}) = \$99$.*

In practice the random variables $\mathcal{E}_n$ are often expressed as the ratio between the returns of the $\{d'_n, u'_n\}$-valued risky asset ΔW_n and the returns of a riskless bond. The latter returns are known with certainty in advance (e.g. from a money market account) and are called risk free returns:

$$\mathcal{E}_n = \frac{1 + \Delta W_n}{1 + r_n}.$$

In this situation, the filtration (30.2) is also given by

$$\mathcal{F}_n = \sigma(\Delta W_1, \ldots, \Delta W_n) = \sigma(\mathcal{S}_1, \ldots, \mathcal{S}_n).$$

The martingale condition now takes the form

$$\mathbb{E}(\mathcal{E}_n \mid \mathcal{F}_n) = \frac{1 + \mathbb{E}(\Delta W_n \mid \mathcal{F}_n)}{1 + r_n} = 1 \iff \mathbb{E}(\Delta W_n \mid \mathcal{F}_n) = r_n.$$

This shows that

$$0 < d_n = \frac{1 + d'_n}{1 + r_n} < 1 < u_n = \frac{1 + u'_n}{1 + r_n} \iff d'_n < r_n < u'_n$$

as soon as $-1 < d'_n < u'_n$. The l.h.s. is called the "no-arbitrage" condition. It is meant to underline that we cannot choose between the returns of the risky or the riskless asset to achieve arbitrage. In this notation, we have

$$P_n = \frac{u_n - 1}{u_n - d_n} = \frac{\frac{1 + u'_n}{1 + r_n} - 1}{\frac{1 + u'_n}{1 + r_n} - \frac{1 + d'_n}{1 + r_n}} = \frac{u'_n - r_n}{u'_n - d'_n} \quad \text{and} \quad Q_n = \frac{r_n - d'_n}{u'_n - d'_n}.$$

By definition of the returns of the assets, the evolution of the riskless asset $\mathcal{S}_n^{(0)}$ is given by the deterministic model

$$\mathcal{S}_n^{(0)} = \mathcal{S}_{n-1}^{(0)} (1 + r_n) \iff \Delta \mathcal{S}_n^{(0)} := \mathcal{S}_n^{(0)} - \mathcal{S}_{n-1}^{(0)} = \mathcal{S}_{n-1}^{(0)} r_n$$

while the evolution of the risky asset is given by a Markov chain

$$\mathcal{S}_n = \mathcal{S}_{n-1}\left(1 + \Delta W_n\right) \Leftrightarrow \Delta\mathcal{S}_n := \mathcal{S}_n - \mathcal{S}_{n-1} = \mathcal{S}_{n-1}\,\Delta W_n. \tag{30.3}$$

In this context, the filtration (30.2) coincides with the one associated with the random returns of the risky assets, or equivalently with the one associated with the prices $\mathcal{S}_n$, that is,

$$\mathcal{F}_n = \sigma(\Delta W_1, \dots, \Delta W_n) = \sigma(\mathcal{S}_1, \dots, \mathcal{S}_n).$$

The stock price is given by the deflated price formula

$$\overline{\mathcal{S}}_n = \overline{\mathcal{S}}_{n-1}\,\frac{1 + \Delta W_n}{1 + r_n} = \mathcal{S}_n / \mathcal{S}_n^{(0)} \tag{30.4}$$

$$\overline{\mathcal{S}}_n = \prod_{1 \leq k \leq n} \left(\frac{1 + d_k'}{1 + r_k}\right)^{\epsilon_k} \left(\frac{1 + u_k'}{1 + r_k}\right)^{1 - \epsilon_k} \tag{30.5}$$

with a sequence of independent $\{0, 1\}$-valued Bernoulli random variables with a law

$$\mathbb{P}\left(\epsilon_k = 1\right) = P_k \quad \text{and} \quad \mathbb{P}\left(\epsilon_k = 0\right) = Q_k.$$

30.1.2   Cox-Ross-Rubinstein model

The evolution of the price of an asset is often described on some time mesh sequence t_n with a small time step $(t_n - t_{n-1}) = h$. In this case, the returns of the riskless asset are often given by the exponential functions

$$1 + r_{n+1} = e^{r_{t_n} h}.$$

The random returns of the risky asset are also given by Bernoulli type models

$$1 + \Delta W_{n+1} = e^{\epsilon_n \sigma_{t_n} \sqrt{h}}$$

for some $\{-1, +1\}$-valued random variables ϵ_n. The positive parameter $\sigma_{t_n}\sqrt{h}$ is called the standard deviation of returns in a time step h. The parameter σ_{t_n} is called the volatility of the market.

We clearly have the formula

$$\overline{\mathcal{S}}_n = \overline{\mathcal{S}}_0\,\exp\left(-\sum_{1 \leq k \leq n} r_{t_{k-1}} h + \sum_{1 \leq k \leq n} \epsilon_k\,\sigma_{t_{k-1}}\sqrt{h}\right). \tag{30.6}$$

By construction, we also have that

$$P_n = \mathbb{P}\left(\epsilon_n = -1 \mid \mathcal{F}_n\right) \quad \text{and} \quad Q_n = \mathbb{P}\left(\epsilon_n = +1 \mid \mathcal{F}_n\right).$$

Furthermore, in this context, the martingale

$$\mathbb{E}\left(1 + \Delta W_{n+1} \mid \mathcal{F}_n\right) = 1 + r_{n+1}$$

takes the form:

$$P_n\,e^{-\sigma_{t_n}\sqrt{h}} + (1 - P_n)\,e^{+\sigma_{t_n}\sqrt{h}} = e^{r_{t_n} h}.$$

This shows that

$$P_n = \frac{e^{\sigma_{t_n}\sqrt{h}} - e^{r_{t_n}h}}{e^{\sigma_{t_n}\sqrt{h}} - e^{-\sigma_{t_n}\sqrt{h}}} \quad \text{and} \quad Q_n = \frac{e^{r_{t_n}h} - e^{-\sigma_{t_n}\sqrt{h}}}{e^{\sigma_{t_n}\sqrt{h}} - e^{-\sigma_{t_n}\sqrt{h}}}$$

as soon as we have

$$\sigma_{t_n} > r_{t_n}\sqrt{h}.$$

Expanding these exponentials and neglecting the terms of smaller order, we find that

$$Q_n \simeq \frac{\left(r_{t_n} - \frac{\sigma_{t_n}^2}{2}\right)h + \sigma_{t_n}\sqrt{h}}{2\sigma_{t_n}\sqrt{h}} = \frac{1}{2}\left(1 + \left(r_{t_n} - \frac{\sigma_{t_n}^2}{2}\right)\frac{\sqrt{h}}{\sigma_{t_n}}\right).$$

Therefore $P_n := \frac{1}{2}\left(1 - \left(r_{t_n} - \frac{\sigma_{t_n}^2}{2}\right)\frac{\sqrt{h}}{\sigma_{t_n}}\right)$.

For constant parameters $(r_{t_n}, \sigma_{t_n}) = (r, \sigma)$, the formula (30.6) reduces to the Cox-Ross-Rubinstein binomial model

$$\overline{S}_n = \overline{S}_0 \exp\left(-nrh + \sigma\sqrt{h}\,\overline{\epsilon}_n\right) \quad \text{with} \quad \overline{\epsilon}_n = \sum_{1 \le k \le n} \epsilon_k. \tag{30.7}$$

With the same constant parameters we also have $(P_n, Q_n) = (P, Q)$ with

$$P := \frac{1}{2}\left(1 - \left(r - \frac{\sigma^2}{2}\right)\frac{\sqrt{h}}{\sigma}\right) := 1 - Q.$$

30.1.3 Black-Scholes-Merton model

The objective of this section is to analyze the limiting Cox-Ross-Rubinstein model when the time step h tends to 0. To clarify the presentation, we set

$$P = \frac{1}{2}(1-\alpha) \quad Q = \frac{1}{2}(1+\alpha) \quad \text{with} \quad \alpha := \left(r - \frac{\sigma^2}{2}\right)\frac{\sqrt{h}}{\sigma}.$$

In this notation, we have

$$\mathbb{E}(\overline{\epsilon}_n) = \sum_{1 \le k \le n} \mathbb{E}(\epsilon_k) = n\alpha$$

as well as

$$\begin{aligned}
\mathrm{Var}(\overline{\epsilon}_n) &= \sum_{1 \le k \le n} \mathbb{E}([\epsilon_k - \alpha]^2) \\
&= \frac{n}{2}\left[(-1-\alpha)^2(1-\alpha) + (1-\alpha)^2(1+\alpha)\right] \\
&= \frac{n}{2}(1-\alpha)(1+\alpha)\left[(1+\alpha)+(1-\alpha)\right] = n\,(1-\alpha^2).
\end{aligned}$$

Invoking the central limit theorem, as $n \uparrow \infty$ we have the following weak approximation

$$\sigma\sqrt{h}\,\frac{(\overline{\epsilon}_n - n\alpha)}{\sqrt{n(1-\alpha^2)}} \simeq \sigma\sqrt{h}\,W_1.$$

Here W_1 denotes a standard normal random variable. This implies that

$$\sigma\sqrt{h}\,\bar{\epsilon}_n \simeq \sigma\sqrt{h}\,n\,\alpha + \sigma\,\sqrt{(1-\alpha^2)}\sqrt{nh}\,W_1$$

$$= h\,n\left(r - \frac{\sigma^2}{2}\right) + \sigma\,\sqrt{1 - \left(\left(r - \frac{\sigma^2}{2}\right)\frac{\sqrt{h}}{\sigma}\right)^2}\,\sqrt{nh}\,W_1$$

$$\simeq h\,n\left(r - \frac{\sigma^2}{2}\right) + \sigma\,\sqrt{nh}\,W_1.$$

When $n = \lfloor\frac{t}{h}\rfloor$, we find the weak approximation

$$\sigma\sqrt{h}\,\bar{\epsilon}_n \xrightarrow{\ h\to 0\ } t\left(r - \frac{\sigma^2}{2}\right) + \sigma\,\underbrace{\sqrt{t}\,W_1}_{\stackrel{law}{=}\,W_t}$$

where W_t is the Brownian motion at time t starting at the origin (compare (14.1), section 14.1).

We write $\overline{\mathcal{S}}^h_{t_n} = \overline{\mathcal{S}}_n$ for the martingale defined in (30.7) on some time mesh sequence t_n, with time step h. From previous approximations, the limiting evolution of the stock prices is given by

$$\overline{\mathcal{S}}^h_{h\lfloor\frac{t}{h}\rfloor} \xrightarrow{\ h\to 0\ } \overline{\mathcal{S}}_t := \overline{\mathcal{S}}_0\,\exp\left(\sigma W_t - \frac{\sigma^2 t}{2}\right)$$

$$= e^{-rt}\left(\overline{\mathcal{S}}_0\,\exp\left(\sigma W_t + t\left(r - \frac{\sigma^2}{2}\right)\right)\right)$$

with the initial condition $\overline{\mathcal{S}}^h_0 = \overline{\mathcal{S}}_0$. The r.h.s. formula is the continuous time version of the deflated discrete time formula (30.4).

The process

$$S_t := S_0\,\exp\left(\sigma W_t + t\left(r - \frac{\sigma^2}{2}\right)\right) \tag{30.8}$$

is called the Black-Scholes model or the geometric Brownian motion. In the mathematical finance literature, it is also called Black-Scholes-Merton model, in reference to their pioneering articles [25, 194]. We refer the reader to the example of SDE (14.6) discussed in section 14.1.

30.2 European option pricing

30.2.1 Call and put options

The payoff function of a call with strike K and an expiration date n is given by

$$f_n(\overline{\mathcal{S}}_n) := \left(\overline{\mathcal{S}}_n - K_n\right)^+ \tag{30.9}$$

with the deflated strike

$$K_n := K \prod_{1 \leq k \leq n} (1 + r_k)^{-1}.$$

In the above display, r_k denotes the risk free returns of a secure and riskless reference asset or financial bond, and $\overline{S}_n$ is the price of the underlying risky asset. As its name indicates, *the payoff function represents the cost of the contract for the seller* if the owner exercises the option. For instance,

$$\overline{S}_n \leq K_n \implies f_n(\overline{S}_n) = 0.$$

This means that the owner of the option will prefer to buy the shares at price $\overline{S}_n$ directly in the market. If the opposite inequality holds, the prices are higher in the market, and the cost of the option will be the difference between the deflated market price $\overline{S}_n$ and the deflated strike $K \prod_{1 \leq k \leq n} (1 + r_k)^{-1}$.

The non-deflated payoff function is given by

$$f_n(\overline{S}_n) = \left[\prod_{1 \leq k \leq n} (1 + r_k) \right]^{-1} (S_n - K)^+$$

with the stock price presented in (30.3) and defined by

$$S_n := \left[\prod_{1 \leq k \leq n} (1 + r_k) \right] \overline{S}_n.$$

Example 30.2.1 *The payoff function associated with a $S_n = \$30$ stock price, with risk free rate $r = .05\%$ per year, over a 10-year term $n = 10$, and exercise/strike price $K = \$28$ is given by*

$$f_n(\overline{S}_n) = (1.05)^{-10} (30 - 28)^+ \simeq \$1.2278.$$

In much the same way, the payoff function of a put with strike K and an expiration date n is

$$f_n(\overline{S}_n) := (\overline{S}_n - K_n)^-. \tag{30.10}$$

In this situation,

$$\overline{S}_n \geq K_n \implies f_n(\overline{S}_n) = 0.$$

This means that the owner of the option will prefer to sell the shares at price $\overline{S}_n$ directly in the market.

30.2.2 Self-financing portfolios

Of course, the seller of a contract (a.k.a. the option writer) will ask for some fee or premium in exchange for his financial option. The portfolio of the seller has some number b_k of updated risky assets $\overline{S}_k$, and some number of (updated) risk free assets (with unit value) at any time k.

Thus, at any time k, portfolio value is given by the formula

$$\mathcal{P}_k(b) := c_{k-1} + b_{k-1} \overline{S}_k.$$

A negative b_k corresponds to short (uncovered) sales of $|b_k|$ assets, while a positive b_k corresponds to a purchase of a number b_k of assets.

Before the new market price is announced, the investor selects a new self-financing (c_k, b_k) (using all the information $\mathcal{F}_k$ he has on the past) such that

$$\mathcal{P}_k(b) := c_{k-1} + b_{k-1}\,\overline{\mathcal{S}}_k = c_k + b_k\,\overline{\mathcal{S}}_k.$$

Of course, it suffices to choose some b_k and set

$$c_k = \mathcal{P}_k(b) - b_k\,\overline{\mathcal{S}}_k.$$

This shows that the value of such self-financing portfolio depends only on the number of shares of the risky asset.

When the market prices $\overline{\mathcal{S}}_{k+1}$ are finally announced, the value of the updated portfolio is now given by

$$\mathcal{P}_{k+1}(b) := c_k + b_k\,\overline{\mathcal{S}}_{k+1}.$$

From the self-financing property, it is essential to observe that the increments of the portfolio are given by

$$
\begin{aligned}
\Delta\mathcal{P}_{k+1}(b) &= \mathcal{P}_{k+1}(b) - \mathcal{P}_k(b) \\
&= b_k\,\Delta\overline{\mathcal{S}}_{k+1} \quad \text{with} \quad \Delta\overline{\mathcal{S}}_{k+1} = \overline{\mathcal{S}}_{k+1} - \overline{\mathcal{S}}_k.
\end{aligned}
$$

We proved that a self-financing portfolio associated with a management strategy of $b = (b_k)_{0 \le k < n}$ risky assets at any time k is given by the martingale

$$\forall 0 \le l \le n \qquad \mathcal{P}_l(b) := \mathcal{P}_0(b) + \sum_{1 \le k \le l} b_{k-1}\,\Delta\overline{\mathcal{S}}_k. \tag{30.11}$$

30.2.3 Binomial pricing technique

The strategy of the option writer is to find some initial portfolio value $\mathcal{P}_0(b)$ such that there exists a self-financing strategy b for which the terminal value of the portfolio is precisely the payoff function, that is, $\mathcal{P}_n(b) = f_n(\overline{\mathcal{S}}_n)$. This minimal initial value is called the price of the option.

As shown in (7.2), without loss of generality, we can assume that $\overline{\mathcal{S}}_n$ is also a Markov process. For instance, we can consider the exponential random walk model in (30.1) or the time homogeneous Cox-Ross-Rubinstein binomial model in (30.7). To simplify the presentation, we consider the exponential model (30.4), and suppose that the initial value of the price is a known deterministic value $\overline{\mathcal{S}}_0 = x_0$.

In this situation, from theorem 8.4.7, we know that the unique martingale $\mathcal{M}_p^{(n)}$ that ends at $f_n(\overline{\mathcal{S}}_n)$ is given by

$$\forall 0 \le p \le n \qquad \mathcal{M}_p^{(n)} = \mathbb{E}\left(f_n(\overline{\mathcal{S}}_n) \mid \mathcal{F}_p\right).$$

Thus, we already know that the price C_n of the option is given by

$$C_n = \mathbb{E}\left(f_n(\overline{\mathcal{S}}_n) \mid \overline{\mathcal{S}}_0 = x_0\right).$$

Under our assumptions $\overline{\mathcal{S}}_k$ is a Markov chain, we thus have

$$\mathcal{M}_k^{(n)} = \mathbb{E}\left(f_n(\overline{\mathcal{S}}_n) \mid \overline{\mathcal{S}}_k\right) := \mathcal{V}_k^{(n)}(\overline{\mathcal{S}}_k)$$

with the collection of functions

$$\forall 0 \le k \le n \qquad \mathcal{V}_k^{(n)}(x) := \mathbb{E}\left(f_n(\overline{\mathcal{S}}_n) \mid \overline{\mathcal{S}}_k = x\right).$$

By construction, we have $\mathcal{V}_n^{(n)} = f_n$. The possible values of $\overline{\mathcal{S}}_n$ at time n can be interpreted as the branches of a binomial type of tree, taking into consideration all the possible states (30.5) when we vary at each time step $\epsilon_k = 0$ and $\epsilon_k = 1$, for $k \le n$.

Using the martingale property, we compute the values of $\mathcal{V}_{n-1}^{(n)}$ for all possible values of $\overline{\mathcal{S}}_{n-1}$ using the formula

$$
\begin{aligned}
\mathcal{V}_{n-1}^{(n)}(\overline{\mathcal{S}}_{n-1}) &:= \mathbb{E}\left(f_n(\overline{\mathcal{S}}_n) \mid \overline{\mathcal{S}}_{n-1}\right) \\
&= f_n\left(\overline{\mathcal{S}}_{n-1}\frac{1+d_n'}{1+r_n}\right)\frac{u_n'-r_n}{u_n'-d_n'} \\
&\qquad + f_n\left(\overline{\mathcal{S}}_{n-1}\frac{1+u_n'}{1+r_n}\right)\frac{r_n-d_n'}{u_n'-d_n'}.
\end{aligned}
$$

Iterating backward in time, these functions can be computed using the formula

$$
\begin{aligned}
\mathcal{V}_k^{(n)}(\overline{\mathcal{S}}_k) &:= \mathbb{E}\left(\mathcal{V}_{k+1}^{(n)}(\overline{\mathcal{S}}_{k+1}) \mid \overline{\mathcal{S}}_k\right) \\
&= \mathcal{V}_{k+1}^{(n)}\left(\overline{\mathcal{S}}_k\frac{1+d_{k+1}'}{1+r_{k+1}}\right)\frac{u_{k+1}'-r_{k+1}}{u_{k+1}'-d_{k+1}'} \\
&\qquad + \mathcal{V}_{k+1}^{(n)}\left(\overline{\mathcal{S}}_k\frac{1+u_{k+1}'}{1+r_{k+1}}\right)\frac{r_{k+1}-d_{k+1}'}{u_{k+1}'-d_{k+1}'}.
\end{aligned}
$$

Coming back to the martingale portfolio of the option writer (30.11), and recalling that the martingale with the fixed terminal condition $f_n(\overline{\mathcal{S}}_n)$ is unique, we find that the price of the option is given by

$$\mathcal{V}_0^{(n)}(\overline{\mathcal{S}}_0) = \mathcal{V}_0^{(n)}(x_0) = \mathcal{P}_0(b).$$

In addition, we have

$$\Delta \mathcal{P}_{k+1}(b) = b_k\,\Delta\overline{\mathcal{S}}_{k+1} = \Delta\mathcal{V}_{k+1}^{(n)}(\overline{\mathcal{S}}_{k+1}) = \mathcal{V}_{k+1}^{(n)}(\overline{\mathcal{S}}_{k+1}) - \mathcal{V}_k^{(n)}(\overline{\mathcal{S}}_k).$$

Recalling that

$$\overline{\mathcal{S}}_{k+1} \in \left\{\overline{\mathcal{S}}_k\frac{1+d_{k+1}'}{1+r_{k+1}}\,,\ \overline{\mathcal{S}}_k\frac{1+u_{k+1}'}{1+r_{k+1}}\right\}$$

we conclude that the management strategy b_k satisfies the equations

$$
\begin{aligned}
b_k\,\overline{\mathcal{S}}_k\left(\frac{1+d_{k+1}'}{1+r_{k+1}}-1\right) &= \mathcal{V}_{k+1}^{(n)}\left(\overline{\mathcal{S}}_k\frac{1+d_{k+1}'}{1+r_{k+1}}\right) - \mathcal{V}_k^{(n)}(\overline{\mathcal{S}}_k) \\
b_k\,\overline{\mathcal{S}}_k\left(\frac{1+u_{k+1}'}{1+r_{k+1}}-1\right) &= \mathcal{V}_{k+1}^{(n)}\left(\overline{\mathcal{S}}_k\frac{1+u_{k+1}'}{1+r_{k+1}}\right) - \mathcal{V}_k^{(n)}(\overline{\mathcal{S}}_k).
\end{aligned}
$$

Subtracting these two equations we find that

$$b_k = \frac{\mathcal{V}_{k+1}^{(n)}\left(\overline{\mathcal{S}}_k\frac{1+u_{k+1}'}{1+r_{k+1}}\right) - \mathcal{V}_{k+1}^{(n)}\left(\overline{\mathcal{S}}_k\frac{1+d_{k+1}'}{1+r_{k+1}}\right)}{\overline{\mathcal{S}}_k\left(\frac{1+u_{k+1}'}{1+r_{k+1}} - \frac{1+d_{k+1}'}{1+r_{k+1}}\right)}.$$

For the time homogeneous Cox-Ross-Rubinstein model (30.7) from section 30.1.2, these equations take the following form :

$$b_k = \frac{\mathcal{V}_{k+1}^{(n)}\left(\overline{S}_k \exp\left(-rh + \sigma\sqrt{h}\right)\right) - \mathcal{V}_{k+1}^{(n)}\left(\overline{S}_k \exp\left(-rh - \sigma\sqrt{h}\right)\right)}{\overline{S}_k\left(\exp\left(-rh + \sigma\sqrt{h}\right) - \exp\left(-rh - \sigma\sqrt{h}\right)\right)}.$$

30.2.4   Black-Scholes-Merton pricing model

In section 30.1.3 we introduced the Black-Scholes-Merton model as the continuous time version of a discrete binomial pricing model. In these continuous time settings, the risky asset $\overline{S}_t$ is given by the geometric Brownian motion

$$\overline{S}_t := \overline{S}_0 \exp\left(\sigma W_t - \frac{\sigma^2 t}{2}\right), \tag{30.12}$$

where W_t is the Brownian motion at time t. We have seen in (14.6) that $\overline{S}_t$ is the solution of the SDE

$$d\overline{S}_t = \sigma \, \overline{S}_t \, dW_t. \tag{30.13}$$

We let $\overline{\mathcal{P}}_{s,t}$, with $0 \leq s \leq t$ be the transition semigroup (sg) of $\overline{S}_t$ defined for any bounded function φ by the formula

$$\overline{\mathcal{P}}_{s,t}(\varphi)(x) := \mathbb{E}\left(\varphi(\overline{S}_t) \mid \overline{S}_s = x\right).$$

We recall that the infinitesimal generator L of $\overline{S}_t$ is given for any smooth function φ by

$$L(f)(x) = \frac{\sigma^2}{2} \, x^2 \, f''(x).$$

Recall that

$$\frac{\partial}{\partial s}\overline{\mathcal{P}}_{s,t}(\varphi)(x) = -L\left(\overline{\mathcal{P}}_{s,t}(\varphi)\right)(x).$$

Using the Doeblin-Itō formula

$$d\varphi(s, \overline{S}_s) = \left(\frac{\partial}{\partial s} + L\right)\varphi(s, \overline{S}_s)\, ds + \frac{\partial\varphi}{\partial x}(s, \overline{S}_s)\, d\overline{S}_s.$$

Applying this differential formula to

$$\varphi(s, \overline{S}_s) = \overline{\mathcal{P}}_{s,T}(f_T)(\overline{S}_s)$$

for some function f_T with $s \leq T$ (and a fixed time horizon T), for any s ($\leq T$) we have

$$d\overline{\mathcal{P}}_{s,T}(f_T)(\overline{S}_s) = \frac{\partial \overline{\mathcal{P}}_{s,T}(f_T)}{\partial x}\, d\overline{S}_s.$$

Integrating from $s = 0$ to $s = t$, we conclude that

$$\overline{\mathcal{P}}_{t,T}(f_T)(\overline{S}_t) = \overline{\mathcal{P}}_{0,T}(f_T)(\overline{S}_0) + \int_0^t \frac{\partial \overline{\mathcal{P}}_{s,T}(f_T)}{\partial x}(\overline{S}_s)\, d\overline{S}_s. \qquad (30.14)$$

For $t = T$, we find the self-financing portfolio formula

$$f_T(\overline{S}_T) = \overline{\mathcal{P}}_{0,T}(f_T)(\overline{S}_0) + \int_0^T \frac{\partial \overline{\mathcal{P}}_{s,T}(f_T)}{\partial x}(\overline{S}_s)\, d\overline{S}_s.$$

30.2.5   Black-Scholes partial differential equation

In this section we provide a brief discussion on the connections between the deflated semi-group formulae (30.14) and the more classical Black-Scholes partial differential equation. As in section 30.2.2, the strategy is to find a self-financing portfolio on some time mesh sequence

$$\mathcal{P}_{t_k}(b) := c_{t_{k-1}} + b_{t_{k-1}}\,\overline{S}_{t_k} = c_{t_k} + b_{t_k}\,\overline{S}_{t_k} \rightsquigarrow \mathcal{P}_{t_{k+1}}(b) := c_{t_k} + b_{t_k}\,\overline{S}_{t_{k+1}}.$$

The management strategy of $b = (b_{t_k})_{0 \le k < n}$ risky assets at any time t_k is given by the martingale

$$\forall 0 \le t_l \le t_n \qquad \mathcal{P}_{t_l}(b) := \mathcal{P}_0(b) + \sum_{0 \le t_k \le t_l} b_{t_{k-1}}\,\Delta\overline{S}_{t_k}. \qquad (30.15)$$

The undeflated version of these portfolios is given by

$$V(t, S_t) = b_t\,S_t + c_t\,e^{rt}$$

with

$$S_t := e^{rt}\overline{S}_t \quad \text{and} \quad c_t = e^{-rt}[V(t, S_t) - b_t\,S_t].$$

In this case we have

$$\mathcal{P}_t(b) = \overline{\mathcal{P}}_{t,T}(f_T)(\overline{S}_t) = e^{-rt}\,V(t, S_t) = e^{-rt}\,V(t, e^{rt}\overline{S}_t).$$

In other words, the undeflated portfolio function $V(t, x)$ is given by

$$e^{-rt}\,V(t, e^{rt}x) = \overline{\mathcal{P}}_{t,T}(f_T)(x) \iff V(t, x) = e^{rt}\,\overline{\mathcal{P}}_{t,T}(f_T)(e^{-rt}\,x).$$

By the l.h.s. formula, we have

$$\frac{\partial V}{\partial x}(t, x) = e^{rt}\,\frac{\partial \overline{\mathcal{P}}_{t,T}(f_T)}{\partial x}(e^{-rt}\,x)\,e^{-rt} = \frac{\partial \overline{\mathcal{P}}_{t,T}(f_T)}{\partial x}(e^{-rt}\,x)$$

and

$$\frac{\partial^2 V}{\partial x^2}(t, x) = e^{-rt}\,\frac{\partial^2 \overline{\mathcal{P}}_{t,T}(f_T)}{\partial x^2}(e^{-rt}\,x).$$

Recalling that

$$\frac{\partial \overline{\mathcal{P}}_{t,T}(f_T)}{\partial t}(y) = -L(\overline{\mathcal{P}}_{t,T}(f_T))(y) = -\frac{\sigma^2}{2}\,y^2\,\frac{\partial^2 \overline{\mathcal{P}}_{t,T}(f_T)}{\partial y^2}(y)$$

we prove that

$$
\begin{aligned}
\frac{\partial}{\partial t} V(t,x) &= r\, V(t,x) + e^{rt}\, \frac{\partial \overline{\mathcal{P}}_{t,T}(f_T)}{\partial t}(e^{-rt}\, x) \\
&\qquad + e^{rt}\, \frac{\partial \overline{\mathcal{P}}_{t,T}(f_T)}{\partial x}(e^{-rt}\, x) \times \frac{\partial (e^{-rt}\, x)}{\partial t} \\
&= r\, V(t,x) - \frac{\sigma^2}{2}\, x^2\, e^{rt} e^{-2rt}\, \frac{\partial^2 \overline{\mathcal{P}}_{t,T}(f_T)}{\partial x^2}(e^{-rt}\, x) \\
&\qquad\qquad\qquad\qquad - r\, x\, \frac{\partial \overline{\mathcal{P}}_{t,T}(f_T)}{\partial x}(e^{-rt}\, x) \\
&= r\, V(t,x) - \frac{\sigma^2}{2}\, x^2\, \frac{\partial^2 V}{\partial x^2}(t,x) - r\, x\, \frac{\partial V}{\partial x}(t,x).
\end{aligned}
$$

We conclude that V satisfies the so-called Black-Scholes partial differential equation

$$
\frac{\partial V}{\partial t}(t,x) + \frac{\sigma^2}{2}\, x^2\, \frac{\partial^2 V}{\partial x^2}(t,x) + r\, x\, \frac{\partial V}{\partial x}(t,x) = r\, V(t,x)
$$

for any $0 \le t \le T$, with the terminal condition $V_T = f_T$.

30.2.6 Replicating portfolios

The continuous time version of the payoff function (30.9) of a call with a strike K and an expiration date T is given by

$$
f_T(\overline{\mathcal{S}}_T) := \left(\overline{\mathcal{S}}_T - K_T\right)^+ \quad \text{with the updated strike } K_T = Ke^{-rT}. \tag{30.16}
$$

The term e^{-rT} is sometimes called the deflator. The portfolio of the writer of the call option replicating the evolution of the option price is now given by the martingale

$$
\forall 0 \le t \le T \qquad \mathcal{P}_t(b) := \mathcal{P}_0(b) + \int_0^t b_s\, d\overline{\mathcal{S}}_s \tag{30.17}
$$

with the initial price $\mathcal{P}_0(b) = C_T$ of the call given by

$$
\begin{aligned}
C_T(x_0, K) &= \overline{\mathcal{P}}_{0,T}(f_T)(x_0) = \mathbb{E}\left(f_T(\overline{\mathcal{S}}_T) \mid \overline{\mathcal{S}}_0 = x_0\right) \\
&= \mathbb{E}\left(\left(x_0\, e^{\sigma W_T - \frac{\sigma^2 T}{2}} - Ke^{-rT}\right)^+\right).
\end{aligned}
$$

Using

$$
\overline{\mathcal{S}}_T := \overline{\mathcal{S}}_t\, e^{\sigma(W_T - W_t) - \frac{\sigma^2(T-t)}{2}} \quad \text{and} \quad K_T = K_t\, e^{-r(T-t)}
$$

we also find that the price of the option with maturity T, given the value of the stock at time t, is given by the semigroup formula

$$
\mathcal{P}_t(b) = \overline{\mathcal{P}}_{t,T}(f_T)(x) = \mathbb{E}\left(f_T(\overline{\mathcal{S}}_T) \mid \overline{\mathcal{S}}_t = x\right) = C_{T-t}(x, K_t).
$$

Comparing (30.17) with (30.14), we conclude that the hedging strategy b_s is given by

$$
b_s = \frac{\partial \overline{\mathcal{P}}_{s,T}(f_T)}{\partial x}(\overline{\mathcal{S}}_s) = \frac{\partial}{\partial x}\, C_{T-s}(x, K_s)\big|_{x = \overline{\mathcal{S}}_s}.
$$

30.2.7 Option price and hedging computations

The computation of a price relies on elementary manipulations of Gaussian random variables.

We recall that

$$V_T = \sigma W_T - \frac{\sigma^2 T}{2}$$

is a Gaussian random variable with mean $m_T = -\frac{\sigma^2 T}{2}$ and variance $\tau_T^2 = \sigma^2 \, T$.

Thus, if we set $v_T = \frac{K e^{-rT}}{x_0}$, using (3.28), we have

$$
\begin{aligned}
C_T(x_0, K)/x_0 &= \mathbb{E}\left((e^{V_T} - v_T) \, 1_{V_T \geq \log(v_T)} \right) \\
&= \mathbb{E}\left(e^{V_T} 1_{V_T \geq \log(v_T)} \right) - v_T \, \mathbb{P}\left(V_T \geq \log(v_T) \right) \\
&= e^{m_T + \frac{\tau_T^2}{2}} \, \mathbb{P}\left(V_T + \tau_T^2 \geq \log(v_T) \right) - v_T \, \mathbb{P}\left(V_T \geq \log(v_T) \right).
\end{aligned}
$$

Notice that

$$m_T + \tau_T^2/2 = 0 \quad \text{and} \quad \log v_T = -rT + \log(K/x_0).$$

Using the survival function $F(z) := \mathbb{P}(Z \geq z) = \mathbb{P}(-Z \geq z) = \mathbb{P}(Z \leq -z) := G(-z)$ of a standard normal random variable, we can express

$$
\begin{aligned}
C_T(x_0, K) &= x_0 \, e^{m_T + \frac{\tau_T^2}{2}} F\left(\frac{1}{\tau_T} \left[-(m_T + \tau_T^2) + \log v_T \right] \right) \\
&\quad - x_0 v_T \, F\left(\frac{1}{\tau_T} \left[-m_T + \log v_T \right] \right) \\
&= x_0 \, F\left(-\frac{1}{\sigma\sqrt{T}} \left[\log\left(\frac{x_0}{K}\right) + \left(r + \frac{\sigma^2}{2}\right)T \right] \right) \\
&\quad - e^{-rT} K \, F\left(-\frac{1}{\sigma\sqrt{T}} \left[\log\left(\frac{x_0}{K}\right) + \left(r - \frac{\sigma^2}{2}\right)T \right] \right).
\end{aligned}
$$

Finally, we have proved that

$$C_T(x, K) = x \, G\left(d_{T,K}^{(1)}(x) \right) - e^{-rT} K \, G\left(d_{T,K}^{(2)}(x) \right)$$

with

$$
\begin{aligned}
d_{T,K}^{(1)}(x) &= \frac{1}{\sigma\sqrt{T}} \left[\log\left(\frac{x}{K}\right) + \left(r + \frac{\sigma^2}{2}\right)T \right] \\
d_{T,K}^{(2)}(x)) &= d_{T,K}^{(1)}(x) - \sigma\sqrt{T}.
\end{aligned}
$$

In addition, the hedging strategies are easily computed using the differential formula

$$
\begin{aligned}
\frac{\partial}{\partial x} C_T(x, K) &= G\left(d_{T,K}^{(1)}(x) \right) \\
&\quad + \frac{1}{\sigma\sqrt{T}} \left[g\left(d_{T,K}^{(1)}(x) \right) - \frac{e^{-rT} K}{x} \, g\left(d_{T,K}^{(2)}(x) \right) \right],
\end{aligned}
$$

with the Gaussian density $g(x) = G'(x) = \frac{1}{\sqrt{2\pi}} \, e^{-x^2/2}$.

30.2.8 A numerical illustration

We assume that the current price of shares of a company is $S_0 = \$100$, and you would like to get a call option that allows you to purchase one share of this stock for $K = \$90$. The standard deviation of the daily logarithmic stocks return is $\sigma = 1\%$, and the annual return r_{annual} of the risk free stock is 4% (i.e., a daily return $r = ((1.04)^{1/365} - 1)$, so that $(1 + r)^{365} = 1 + \frac{4}{100} = r_{annual}$). The annualized volatility σ_{annual} is computed with the formula $\sigma_{annual} = \sigma \times \sqrt{365} \simeq 19.1\%$, or sometimes with the formula $\sigma_{annual} = \sigma \times \sqrt{252} = 15.87\%$ when we consider only the 252 working days of the year.

The current stock price is defined by the geometric Brownian motion

$$S_t = e^{rt}\,\overline{S}_t \quad \text{with} \quad \overline{S}_t := \overline{S}_0\,\exp\left(\sigma W_t - \frac{\sigma^2 t}{2}\right)$$

and the option price with maturity T is given by

$$
\begin{aligned}
C_T(x, K) &= e^{-rT}\,\mathbb{E}\left(\left(x\,e^{\sigma W_T + T\left(r - \frac{\sigma^2}{2}\right)} - K\right)^+\right)\\
&= x\,G\left(d^{(1)}_{T,K}(x)\right) - e^{-rT}K\,G\left(d^{(2)}_{T,K}(x)\right)
\end{aligned}
$$

with the functions

$$
\begin{aligned}
d^{(1)}_{T,K}(x) &= \frac{1}{\sigma\sqrt{T}}\left[\log\left(\frac{x}{K}\right) + \left(r + \frac{\sigma^2}{2}\right)T\right]\\
d^{(2)}_{T,K}(x)) &= d^{(1)}_{T,K}(x) - \sigma\sqrt{T}.
\end{aligned}
$$

A graphical description of the call option prices for different values of the strike and the maturity is provided on page 63.

The wealth of the replicating portfolio is given by

$$
\begin{aligned}
\mathcal{P}_t(b) &= e^{-rt}\,V(t, S_t)\\
&= C_{T-t}(\overline{S}_t, K_t)\\
&= \overline{S}_t\,G\left(d^{(1)}_{T-t, K_t}(\overline{S}_t)\right) - e^{-r(T-t)}K_t\,G\left(d^{(2)}_{T-t, K_t}(\overline{S}_t)\right).
\end{aligned}
$$

This shows that the wealth of a replicating portfolio is given by

$$V(t, S_t) = S_t\,G\left(d^{(1)}_{T-t, K}(S_t)\right) - e^{-r(T-t)}K\,G\left(d^{(2)}_{T-t, K}(S_t)\right).$$

The last assertion follows from $d^{(1)}_{T-t, K_t}(\overline{S}_t) = d^{(1)}_{T-t, K}(S_t)$.

Our next objective is to compute the number of risk free shares and the number of shares of risky stock. To this end, we recall that the replicating portfolio is defined by

$$V(t, S_t) = b_t\,S_t + c_t\,e^{rt}$$

with the number of risk free bonds

$$c_t = e^{-rt}[V(t, S_t) - b_t\,S_t]$$

and the number of shares of the risky asset

$$
\begin{aligned}
b_t &= \frac{\partial C_{T-t}}{\partial x}(\overline{S}_t, K_t) = \frac{\partial C_{T-t}}{\partial x}(S_t, K)\\
&= G\left(d^{(1)}_{T-t, K}(S_t)\right)\\
&\quad + \frac{1}{\sigma\sqrt{T-t}}\left[g\left(d^{(1)}_{T-t, K}(S_t)\right) - \frac{e^{-r(T-t)}K}{S_t}\,g\left(d^{(2)}_{T-t, K}(S_t)\right)\right].
\end{aligned}
$$

A graphical description of a replicating portfolio is provided on page 65.

30.3   Exercises

Exercise 497 (Neutralization of market) *We consider a deflated risky stock price* $(\overline{S}_k)_{0 \leq k \leq n}$ *in a market model with two periods* $n = 2$. *Assume that the initial price* $\overline{S}_0 = s_0$ *is given and the prices* $\overline{S}_1 \in \{s_{0,1}, s_{0,2}\}$ *and* $\overline{S}_2 \in \cup_{(i,j) \in \{(0,1),(0,2)\}} \{s_{(i,j),1}, s_{(i,j),2}\}$ *can take only two possible values with*

$$s_{(0,1)} < s_0 < s_{(0,2)} \qquad s_{(0,1),1} < s_{0,1} < s_{(0,1),2} \quad and \quad s_{(0,2),1} < s_{0,2} < s_{(0,2),2}.$$

Design the Markov transitions of the chain $\overline{S}_k$ *such that* $\overline{S}_k$ *is a martingale with respect to the natural filtration* $\mathcal{F}_k = \sigma(\overline{S}_l,\ 0 \leq l \leq k)$. *The probability measure under which* $\overline{S}_k$ *is a martingale is called the neutral probability.*

Exercise 498 (Black-Scholes model - 1) *Consider the Black-Scholes model* S_t *discussed in (30.8). Check that*

$$dS_t = r S_t dt + \sigma S_t dW_t.$$

Exercise 499 (Black-Scholes model - 2) *Consider the Black-Scholes model* S_t *discussed in (30.8). We let* $\alpha \geq 1$, $\beta > 0$, *and we assume that* $S_0 = 1$. *Compute the probability of the event* $\{S_{\alpha t} \geq S_t^{\beta}\}$ *in terms of a Gaussian distribution function.*

Exercise 500 (Black-Scholes model - 3) *Consider the Black-Scholes model* S_t *discussed in (30.8). Check that for any* $s \leq t$ *we have*

$$\mathbb{E}(S_t \mid \mathcal{F}_s) = e^{r(t-s)}\ S_s$$

with the σ*-field* $\mathcal{F}_s = \sigma(W_u,\ u \leq s)$.

Exercise 501 (Call-put parity formula) *The prices of a European call and put options with maturity* T *and strike* K *are given by the formulae*

$$P_T^{call} := \mathbb{E}\left((\overline{S}_T - K_T)^{+} \mid \overline{S}_0\right) \quad and \quad P_T^{put} := \mathbb{E}\left((\overline{S}_T - K_T)^{-} \mid \overline{S}_0\right)$$

with the geometric Brownian motion $\overline{S}_t$ *discussed in section 30.2.4 and the updated strike* $K_T = e^{-rT}K$, *for some* $r \geq 0$. *Check the call-put parity formula*

$$P_T^{call} - P_T^{put} = \overline{S}_0 - K_T.$$

Exercise 502 (Replicating portfolios) *We consider the neutral 2-periods market discussed in exercise 497. We consider a call option with the payoff function*

$$f(\overline{S}_2) = \left(\overline{S}_2 - \overline{K}\right)^{+} = \max\left(0, \left(\overline{S}_2 - \overline{K}\right)\right)$$

with some deflated strike $\overline{K} = (1+r)^{-2}K$ *and some* $K > 0$. *Find the functions*

$$\mathcal{V}_k(s) := \mathbb{E}\left(f(\overline{S}_2) \mid \overline{S}_k = s\right)$$

for $k = 0, 1, 2$ *using a backward recursion. We let* $\mathcal{P}_k(b)$ *be a self-financing portfolio with a management strategy* $(b_k)_{k=0,1}$ *defined for any* $k \leq 2$ *by*

$$\mathcal{P}_k(b) = \mathcal{P}_0(b) + \sum_{1 \leq l \leq k} b_{l-1}\,\Delta\overline{S}_l \quad with \quad \Delta\overline{S}_l = \overline{S}_l - \overline{S}_{l-1}.$$

Find the initial value $\mathcal{P}_0(b)$ *of the portfolio and the management strategy* (b_0, b_1) *such that* $\mathcal{P}_2(b) = f(\overline{S}_2)$. *Discuss the price of the call option.*

Exercise 503 (Wilkie inflation model) *We let $L_n = \log R_n$ be the logarithm of some retail price index R_n at year n. The force $I_n = \log(X_n/X_{n-1})$ of the inflation over the year n is given by the stochastic equation*

$$I_n = a + b\,(I_{n-1} - a) + \sigma\,W_n$$

with a sequence of i.i.d. centered Gaussian random variables with unit variance and parameters $a, b, \sigma \in [0, 1[$.

Exercise 504 (Life function martingales) *We let L_n be the number of people alive, relative to an original cohort with no new entrants, at age n. As age increases, the number of people alive decreases. We let l_n be the expected number of lives at age n. Assuming independent of lives, given L_n, L_{n+1} is a binomial random variable with parameters L_n and l_{n+1}/l_n. For some given parameter $\alpha > 0$, check that $M_n = \left(1 + \frac{\alpha}{l_n}\right)^{L_n}$ is a martingale w.r.t. the filtration $\mathcal{F}_n$ generated by the variables $(L_0, \ldots, L_n)$.*

Exercise 505 (Cox-Ross-Rubinstein model) *In time homogeneous settings, the formula (30.6) can be rewritten as*

$$\overline{S}^h_{t_n+h} = \overline{S}^h_{t_n}\,\exp\left(-rh + \epsilon_n \sigma \sqrt{h}\right)$$

with a collection of independent $\{-1, +1\}$-valued Bernoulli random variables with a common law

$$
\begin{aligned}
p_h &= \mathbb{P}(\epsilon_n = -1) = \frac{e^{\sigma\sqrt{h}} - e^{rh}}{e^{\sigma\sqrt{h}} - e^{-\sigma\sqrt{h}}} \\
q_h &= \mathbb{P}(\epsilon_n = +1) = \frac{e^{rh} - e^{-\sigma\sqrt{h}}}{e^{\sigma\sqrt{h}} - e^{\sigma-\sqrt{h}}}.
\end{aligned}
$$

We let T_h be the Markov transition of the chain $\overline{S}^h_{t_n}$ on a time step h. For any bounded function f we have

$$T_h(f)(x) = f(x\,y_h)\,p_h + f(x\,z_h)\,q_h$$

with

$$y_h = e^{-rh - \sigma\sqrt{h}} \quad \text{and} \quad z_n = x\,e^{-rh + \sigma\sqrt{h}}.$$

Prove that

$$\lim_{h \to 0} \frac{T_h(f)(x) - f(x)}{h} = \frac{\sigma^2}{2}\,f''(x)\,x^2 := L(f)(x),$$

with the infinitesimal generator L of the geometric Brownian motion (14.6).

Exercise 506 (Pricing zero-coupon bonds) *Consider a spot interest rate X_t given by some diffusion*

$$dX_t = b_t(X_t)\,dt + \sigma_t(X_t)\,dW_t$$

for some regular functions (b_t, σ_t), some Brownian motion W_t, and some initial condition $X_0 = x_0$. The price of a zero-coupon bond with maturity time t given the spot rate $X_s = x$ at some time $s \in [0, t]$ is given by the Feynman-Kac formula

$$q_s(x) := \mathbb{E}\left(\exp\left[-\int_s^t V(X_r)\,dr\right] \,\Big|\, X_s = x\right) \quad \text{with} \quad V(x) = x.$$

Check that $(s,x) \in ([0,t] \times \mathbb{R}) \mapsto q_s(x)$ *satisfies the backward equation*

$$\partial_s q_s = -(L_s(q_s) - Vq_s) = -L_s^V(q_s) \quad \text{with the Schrödinger operator} \quad L_s^V = L_s - V$$

with the terminal boundary condition $q_t = 1$ *and the generator* L_s *of the process* X_s. *Using the exercise 255, provide an explicit formula in the following situations*

$$1) \quad dX_t = b \; dt + \sigma \; dW_t \quad \text{and} \quad 2) \quad dX_t = a \; (b - X_t) \; dt + \sigma \; dW_t$$

for some parameters (a, b, σ).

Exercise 507 (Dupire's formula) *Consider a stock price given by the one-dimensional diffusion*

$$dX_t = r_t(X_t) \; X_t \; dt + \sigma_t(X_t) \; X_t \; dW_t$$

with some Brownian motion W_t, *some regular interest rate drift function* r_t, *and some regular volatility diffusion function* $\sigma_t(> 0)$. *Consider a final time horizon* t *and a payoff function* f_t. *Describe the backward evolution equation of the price of the option given by*

$$\forall s \in [0,t] \qquad u_s(x) := Q_{s,t}(f_t)(x)$$

with the Feynman-Kac semigroup $Q_{s,t}$ *defined for any regular function* f *by*

$$Q_{s,t}(f)(x) = \mathbb{E}\left(\exp\left(-\int_s^t r_\tau(X_\tau) \; d\tau \right) f_t(X_t) \mid X_s = x \right).$$

We further assume that the integral operator $Q_{s,t}$ *has a density* $q_{s,t}(x, y)$ *w.r.t. the Lebesgue measure, that is,*

$$Q_{s,t}(f)(x) = \int q_{s,t}(x, y) \; f(y) \; dy.$$

Describe the forward evolution equation of the function $(t, y) \in ([s, \infty[\times \mathbb{R}) \mapsto q_{s,t}(x, y)$ *(in the weak sense). Consider the payoff function* $f_t(x) = (x - K)^+$ *associated with a European call option with a given strike* K, *and set*

$$v_{s,t}(x, y) := \mathbb{E}\left(\exp\left(-\int_s^t r_\tau(X_\tau) \; d\tau \right) (X_t - y)^+ \mid X_s = x \right).$$

Check that $\partial_y^2 v_{s,t}(x, y) = q_{s,t}(x, y)$.

Bibliography

[1] E. Aïdékon, B. Jaffuel. Survival of branching random walks with absorption. *Stochastic Processes and Their Applications*, vol. 121, pp. 1901–1937 (2011).

[2] D. Aldous, P. Diaconis. Shuffling cards and stopping times. *Amer. Math. Monthly*, vol. 93, pp. 333–348 (1986).

[3] H.C. Andersen, P. Diaconis. Hit and run as a unifying device. *Journal de la société Franqise de statistique*, vol. 148, no.4, pp. 5–28 (2007)

[4] C. Andrieu, A. Doucet and R. Holenstein. Particle Markov chain Monte Carlo methods. *J. R. Statist. Soc. B*, 72, Part 3, pp. 269–342 (2010).

[5] S. Asmussen, P.W. Glynn. *Stochastic Simulation. Algorithms and Analysis.* Springer Series: Stochastic Modelling and Applied Probability, vol. 57 (2007).

[6] S. Asmussen, H. Hering. *Branching Processes.* Birkhäuser, Boston (1983).

[7] K. B. Athreya, P.E. Ney. *Branching Processes*, Springer, NewYork (1972).

[8] L. Bachelier. The theory of speculation. *Annales scientifiques de l'École Normale Supérieure*, vol. 3, no. 17, pp. 21–86 (1900).

[9] D. Bakry, M. Emery. Diffusions hypercontractives, Séminaire de probabilités XIX. Strasbourg University, Springer (1983).

[10] A. R. Bansal, V. P. Dimri and K. K. Babu. Epidemic type aftershock sequence (ETAS) modeling of northeastern Himalayan seismicity. *Journal of Seismology*, vol. 17, no. 2, pp. 255–264 (2013).

[11] M. Barnsley, *Fractals Everywhere*. San Diego, Academic Press (1988).

[12] J. Barral, R. Rhodes and V. Vargas. Limiting laws of supercritical branching random walks. *Comptes Rendus de l'Académie des Sciences - Series I - Mathematics*, vol. 350, no 9-10, pp. 535–538 (2012).

[13] D. Bayer, P. Diaconis. Trailing the dovetail shuffle to its lair. *Annals of Applied Probability*, vol. 2, no. 2, pp. 294–313 (1992).

[14] D. Beeman. Some multistep methods for use in molecular dynamics calculations.*Journal of Computational Physics*, vol. 20, no. 2, pp. 130–139 (1976).

[15] C. J. P. Bélisle, H. E. Romeijn, and R. L. Smith. Hit-and-run algorithms for generating multivariate distributions. *Math. Oper. Res.*, vol. 18, no. 2, pp. 255–266 (1993).

[16] E. Ben-Naim, P.L. Krapivsky. Cluster approximation for the contact process. *J. Phys. A: Math. Gen.*, vol. 27 pp. 481–487 (1994).

[17] M. Benaïm, S. Le Borgne, F. Malrieu and P.A. Zitt Quantitative ergodicity for some switched dynamical systems. *Electronic Communications in Probability*, vol. 17, no. 56, pp. 1–14 (2012).

[18] J. Berestycki, N. Berestycki, and J. Schweinsberg. Critical branching Brownian motion with absorption: survival probability. *ArXiv:1212.3821v2 [math.PR]* (2012).

[19] J. M. Bernardo, A. F. M. Smith. *Bayesian Theory.* Wiley, Chichester (1994).

[20] J. Bertoin. *Random fragmentation and coagulation processes.* Cambridge Studies in Advanced Mathematics, Cambridge University Press (2006).

[21] D. P. Bertsekas, S. E. Shreve. *Stochastic Optimal Control: The Discrete-Time Case.* Academic Press (1978).

[22] J. Besag, P.J. Diggle. Simple Monte Carlo tests for spatial pattern. *Journal of the Royal Statistical Society*, vol. 26, No. 3, pp. 327–333 (1977).

[23] J. D. Biggins. Uniform convergence of martingales in the branching random walk. *Ann. Probab.* vol. 20, pp. 137–151 (1992).

[24] Z. W. Birnbaum. An inequality for Mill's ratio. *The Annals of Mathematical Statistics*, vol. 13, pp. 245–246 (1942).

[25] F. Black, M. Scholes. The pricing of options and corporate liabilities. *Journal of Political Economy*, vol. 81, no. 3, pp. 637–654 (1973).

[26] S. Bobzien. *Determinism and Freedom in Stoic Philosophy.* Oxford University Press (2001).

[27] C. G. E. Boender, R. J. Caron, J. F. McDonald, A. H. G. Rinnooy Kan, H. E. Romeijn, R. L. Smith, J. Telgen and A. C. F. Vorst. Shake-and-bake algorithms for generating uniform points on the boundary of bounded polyhedra. *Operations Research*, vol. 39, no.6. pp. 945–954 (1991).

[28] M. Born, J.R. Oppenheimer. On the quantum theory of molecules. *Annalen der Physik* (in German), vol. 389, no. 20, pp. 457–484 (1927).

[29] L. von Bortkiewicz. *Das Gesetz der kleinen Zahlen*, Leipzig: B.G. Teubner (1898).

[30] A. Bouchard-Côté, S.J. Vollmer and A. Doucet. *The bouncy particle sampler: a non-reversible rejection-free Markov chain Monte Carlo method.* ArXiv:1510.02451 (2015).

[31] N. Bou-Rabee, M. Hairer. Nonasymptotic mixing of the MALA algorithm. *IMA Journal of Numerical Analysis* (2012).

[32] C. G. Bowsher. Modelling security market events in continuous time: intensity based, multivariate point process models. *Journal of Econometrics*, vol. 141, pp. 876–912 (2007).

[33] M. D. Bramson. Maximal displacement of branching Brownian motion. *Comm. Pure Appl. Math.*, vol 31, pp. 531–581 (1978).

[34] J. K. Brooks. Representations of weak and strong integrals in Banach spaces. *Proc. Nat. Acad. Sci. U.S.A.*, vol. 63, pp. 266–270 (1969).

[35] S. BROWNE, K. SIGMAN. Work-modulated queues with applications to storage processes. *Journal of Applied Probability*, vol. 29, no. 3, pp. 699–712 (1992).

[36] B. E. BRUNET, B. DERRIDA, A. H. MULLER, AND S. MUNIER. Effect of selection on ancestry: an exactly soluble case and its phenomenological generalization. *Phys. Review E*, vol. 76, 041104 (2007).

[37] M. CAFFAREL, R. ASSARAF. A pedagogical introduction to quantum Monte Carlo. In *Mathematical Models and Methods for Ab Initio Quantum Chemistry* in Lecture Notes in Chemistry, eds. M. Defranceschi and C.Le Bris, Springer p. 45 (2000).

[38] G. CANTOR. Über unendliche, lineare Punktmannigfaltigkeiten, *Mathematische Annalen*, vol. 21, pp. 545–591 (1883).

[39] O. CAPPÉ, E. MOULINES AND T. RYDÈN. *Inference in Hidden Markov Models*, Springer (2005).

[40] R. CARMONA, J.-P. FOUQUE AND L.-H. SUN. Mean Field Games and Systemic Risk, *Communications in Mathematical Sciences*, vol. 13, no. 4, 911-933 (2015).

[41] F. CARON, P. DEL MORAL, A. DOUCET AND M. PACE. On the conditional distributions of spatial point processes *Adv. in Appl. Probab.*, vol. 43, no. 2, pp. 301–307 (2011).

[42] B. CASCALES, M. RAJA. Measurable selectors for the metric projection. *Mathematische Nachrichten*, vol. 254-255, 1, pp. 27–34 (2003).

[43] C. CASSANDRAS, J. LYGEROS (Editors). *Stochastic Hybrid Systems*, CRC Press (2007).

[44] D. G. CATCHSIDE, D. E. LEA AND J. M. THODAY. Types of chromosomal structural change induced by the irradiation of Tradescantia microspores. *Journal of Genetics*, vol. 47, pp. 113–136 (1945).

[45] N. CHAMPAGNAT. A microscopic interpretation for adaptive dynamics trait substitution sequence models. *Stochastic Process. Appl.*, vol. 116, no. 8, pp. 1127–1160 (2006).

[46] N. CHAMPAGNAT, R. FERRIÈRE AND S. MÉLÉARD. From individual stochastic processes to macroscopic models in adaptive evolution. *Stoch. Models*, vol. 24, no. 1, pp. 2–44 (2008).

[47] N. CHAMPAGNAT, S. MÉLÉARD. Polymorphic evolution sequence and evolutionary branching. *Probab. Theory Related Fields*, vol. 151, no. 1-2, pp. 45–94 (2011).

[48] R. CHELLI, S. MARSILI, A. BARDUCI AND P. PROCACCI. Generalization of the Jarzynski and Crooks nonequilibrium work theorems in molecular dynamics simulations *Phys. Rev. E*, vol. 75, no. 5, 050101(R) (2007).

[49] L. Y. CHEN. On the Crooks fluctuation theorem and the Jarzynski equality. *J. Chem. Phys.*, vol. 129, no. 9, 091101 (2008).

[50] M.H. CHEN, B. SCHMEISER. Performance of the Gibbs, hit-and-run, and Metropolis samplers. *Journal of Computational and Graphical Statistics*, vol. 2, no. 3, pp. 251–272 (1993).

[51] P. CLIFFORD, A. SUDBURY. A model for spatial conflict. *Biometrika*, vol. 60, no. 3, pp. 581–588 (1973).

[52] R. D. CLARKE. An application of the Poisson distribution. *Journal of the Institute of Actuaries*, vol. 72, p. 481 (1946).

[53] G. COLATA. In shuffling cards, 7 is winning number. Section C; Page 1, January 9, *New York Times* (1990).

[54] F. COMETS, T. SHIGA AND N. YOSHIDA. Probabilistic analysis of directed polymers in a random environment: a review. *Adv. Stud. in Pure Math.*, vol. 39. Mathematical Society of Japan (2004).

[55] F. COMETS, S. POPOV, G. M. SCHÜTZ AND M. VACHKOVSKAIA. Billiards in a general domain with random reflections. *Archive for Rational Mechanics and Analysis*, vol. 191, no. 3, pp. 497–537 (2008).

[56] G. COMO, F. FAGNANI. Scaling limits for continuous opinion dynamics systems. *The Annals of Applied Probability*, vol. 21, no. 4, pp. 1537–156 (2011).

[57] B. COOKE, J.C. MATTINGLY, S.A. McKINLEY AND S.C. SCHMIDLER. Geometric ergodicity of the two-dimensional Hamiltonian systems with a Lennard-Jones-like repulsive potential. eprint arXiv:1104.3842 (2011).

[58] D. R. COX. Some statistical methods connected with series of events. *Journal of the Royal Statistical Society*, vol. 17, no. 2, pp. 129–164 (1955).

[59] H. CRAMÉR. *Mathematical methods of statistics*. New Jersey: Princeton University Press. Princeton (1946). Reprinted (1974).

[60] CRISAN D., DEL MORAL P. AND LYONS T. Discrete filtering using branching and interacting particle systems. *Markov Processes and Related Fields*, vol. 5, no. 3, pp. 293–318 (1999).

[61] CRISAN D., DEL MORAL P. AND LYONS T. Interacting particle systems approximations of the Kushner Stratonovitch equation. *Advances in Applied Probability*, vol.31, no.3, pp. 819–838 (1999).

[62] A. CZUMAJ, P. KANAREK, M. KUTYLOWSKI AND K. LORYS. Delayed path coupling and generating random permutations via distributed stochastic processes. *Proceedings of the Tenth Annual ACM-SIAM Symposium on Discrete Algorithms*, pp. 271-280. Society for Industrial and Applied Mathematics, Philadelphia (1999).

[63] M.H.A. DAVIS. Piecewise-deterministic Markov processes: a general class of non-diffusion stochastic models. *Journal of the Royal Statistical Society. Series B*, vol. 46, no. 3, pp. 353–388 (1984).

[64] P. DEL MORAL. Nonlinear filtering: interacting particle solution. *Markov Processes and Related Fields*, vol. 2, no. 4, pp. 555–580 (1996).

[65] P. DEL MORAL. Measure valued processes and interacting particle systems. Application to nonlinear filtering problems. *Annals of Applied Probability*, vol. 8, no. 2, pp. 438–495 (1998).

[66] P. DEL MORAL. *Feynman-Kac Formulae. Genealogical and Interacting Particle Systems*, Springer, (2004).

[67] P. DEL MORAL. *Mean Field Simulation for Monte Carlo Integration*. Chapman & Hall/CRC Press. Monographs on Statistics and Applied Probability (2013).

[68] P. DEL MORAL, A. DOUCET AND A. JASRA. *Sequential Monte Carlo samplers.* *J. R. Statist. Soc. B*, vol. 68, pp. 411–436 (2006).

[69] P. DEL MORAL AND A. DOUCET. Particle motions in absorbing medium with hard and soft obstacles. *Stochastic Anal. Appl.*, vol. 22, pp. 1175–1207 (2004).

[70] P. DEL MORAL, A. DOUCET. Interacting Markov chain Monte Carlo methods for solving nonlinear measure-valued equations. *Annals of Applied Probability*, vol. 20, no. 2, pp. 593-639 (2010).

[71] P. DEL MORAL, A. DOUCET AND S. S. SINGH. A backward particle interpretation of Feynman-Kac formulae. *M2AN ESAIM.* vol 44, no. 5, pp. 947–976 (Sept. 2010).

[72] P. DEL MORAL, A. DOUCET AND S. S. SINGH. Forward smoothing using sequential Monte Carlo. Technical Report CUED/F-INFENG/TR 638. Cambridge University Engineering Department (2009).

[73] P. DEL MORAL, A. DOUCET AND S. S. SINGH. Computing the filter derivative using Sequential Monte Carlo. Technical Report. Cambridge University Engineering Department (2011).

[74] P. DEL MORAL, A. GUIONNET. On the stability of measure valued processes with applications to filtering. *C. R. Acad. Sci. Paris Sér. I Math.*, vol. 329, pp. 429–434 (1999).

[75] P. DEL MORAL, A. GUIONNET. On the stability of interacting processes with applications to filtering and genetic algorithms. *Ann. Inst. Henri Poincaré*, vol. 37, no. 2, pp. 155–194 (2001).

[76] P. DEL MORAL, N. HADJICONSTANTINOU. An introduction to probabilistic methods, with applications. *M2AN ESAIM*, vol. 44, no. 5, pp. 805–830 (2010).

[77] P. DEL MORAL, L. MICLO Self interacting Markov chains. *Stochastic Analysis and Applications*, vol. 24, no. 3, pp. 615–660 (2006).

[78] P. DEL MORAL, L. MICLO On convergence of chains with time empirical self-interactions. *Proc. Royal Soc. Lond. A.*, vol. 460, pp. 325–346 (2003).

[79] P. DEL MORAL, L. MICLO *Asymptotic Results for Genetic Algorithms with Applications to Nonlinear Estimation.* In KALLEL, L. ET AL. (EDS). *Theoretical Aspects of Evolutionary Computing.* Springer Berlin Heidelberg. Natural Computing Series, pp. 439–493 (2001).

[80] P. DEL MORAL, L. MICLO, F. PATRAS AND S. RUBENTHALER The convergence to equilibrium of neutral genetic models. *Stochastic Analysis and Applications*, vol. 28, no. 1, pp. 123–143 (2009).

[81] P. DEL MORAL, L. MICLO. *Branching and Interacting Particle Systems Approximations of Feynman-Kac Formulae with Applications to Non-Linear Filtering.* Seminaire de Probabilits XXXIV, Lecture Notes in Mathematics, Springer, vol. 1729, pp. 1–145 (2000).

[82] P. DEL MORAL, L. MICLO. Particle approximations of Lyapunov exponents connected to Schrdinger operators and Feynman-Kac semigroups. *ESAIM: Probability and Statistics*, no. 7, pp. 171–208 (2003).

[83] P. DEL MORAL AND J. TUGAUT. On the stability and the uniform propagation of chaos properties of Ensemble Kalman-Bucy filters Arxiv:1605.09329 (2016).

[84] P. DEL MORAL, A. KURTZMANN AND J. TUGAUT. On the Stability and the Exponential Concentration of Extended Kalman-Bucy filters. ArXiv:1606.08251 (2016).

[85] P. DEL MORAL, A. KURTZMANN AND J. TUGAUT. On the stability and the uniform propagation of chaos of Extended Ensemble Kalman-Bucy filters. ArXiv:1606.08256 (2016).

[86] P. DEL MORAL, CH. VERGÉ. *Stochastic models and methods. An introduction with applications (in French).* Springer Series on Maths and Applications (SMAI), vol. 75 (2014).

[87] A. DE MASI, A. GALVES,E. LÖCHERBACH AND E. PRESUTTI. Hydrodynamic limit for interacting neurons. `ArXiv preprint arXiv:1401.4264` (2014).

[88] B. DERRIDA, D. SIMON. The survival probability of a branching random walk in the presence of an absorbing wall. *Europhys. Lett. EPL*, vol. 78, Art. 60006 (2007).

[89] B. DERRIDA, D. SIMON. Quasi-stationary regime of a branching random walk in presence of an absorbing wall. *J. Statist. Phys.*, vol. 131, pp. 203–233 (2008).

[90] B. DERRIDA, M. R. EVANS AND E. R. SPEER. Mean field theory of directed polymers with random complex weights. *Commun. Math. Phys.*, vol. 156, pp. 221–244 (1993).

[91] B. DERRIDA, H. SPOHN. Polymers on disordered trees, spin glasses, and traveling waves. *Journal of Statistical Physics*, vol. 51, nos. 5/6, pp. 817–840 (1988).

[92] L. DEVROYE. *Nonuniform Random Variate Generation.* Springer (1986).

[93] P. DIACONIS. Some things we've learned (about Markov chain Monte Carlo). *Bernoulli*, vol. 19, no. 4, pp. 1294–1305 (2013).

[94] P. DIACONIS, G. LEBEAU AND L. MICHEL. Gibbs/Metropolis algorithm on a convex polytope. *Math Zeitschrift*, vol. 272, no. 1, 109–129 (2012).

[95] P. DIACONIS, G. LEBEAU AND L. MICHEL. Geometric analysis for the Metropolis algorithm on Lipschitz domains. *Invent. Math.*, vol. 185, no. 2, pp. 239–281 (2010).

[96] P. DIACONIS, L. MICLO. On Characterizations of Metropolis type algorithms in continuous time. *Alea*, vol. 6, pp. 199-238 (2009).

[97] P. DIACONIS, G. LEBEAU. Micro-local analysis for the Metropolis algorithm. *Mathematische Zeitschrift*, vol. 262, no. 2, pp. 441-447 (2009).

[98] P. DIACONIS. The Markov chain Monte Carlo revolution. *Bull. Amer. Math. Soc.*, Nov. (2008).

[99] P. DIACONIS, F. BASSETTI. Examples Comparing Importance Sampling and the Metropolis Algorithm. *Illinois Journal of Mathematics*, vol. 50, no. 1-4, pp. 67–91 (2006).

[100] P. DIACONIS J. NEUBERGER. Numerical results for the Metropolis algorithm. *Experimental Math.*, vol. 13, no. 2, pp. 207–214 (2004).

[101] P. DIACONIS, L. BILLERA. A geometric interpretation of the Metropolis-Hastings algorithm. *Statis. Sci.*, vol. 16, no. 4, pp. 335–339 (2001).

[102] P. DIACONIS, A. RAM. Analysis of systematic scan Metropolis algorithms using Iwahori-Hecke algebra techniques. *Michigan Journal of Mathematics*, vol. 48, no. 1, pp. 157–190 (2000).

[103] P. DIACONIS, L. SALOFF-COSTE. What do we know about the Metropolis algorithm? *Jour. Comp. System Sciences*, vol. 57, pp. 20–36 (1998).

[104] P. DIACONIS, P. HANLON. Eigen-analysis for some examples of the Metropolis algorithm. *Contemporary Math.*, vol. 138, pp. 99–117 (1992).

[105] P. DIACONIS, D. FREEDMAN. Iterated random functions. *SIAM Rev.*, vol. 41, pp. 45–76 (1999).

[106] P. DIACONIS, R. GRAHAM. *Magical Mathematics*. Princeton University Press (2012).

[107] P.J. DIGGLE *Statistical Analysis of Spatial Point Patterns* (2nd ed.). Academic Press (2003).

[108] P. DIRAC. A new notation for quantum mechanics. *Mathematical Proceedings of the Cambridge Philosophical Society*, vol. 35, no. 3, pp. 416–418 (1939).

[109] W. DOEBLIN. Exposé de la théorie des chaînes simples constantes de Markoff à un nombre fini d'états. *Revue Mathématique de l'Union Interbalkanique*, vol. 2, pp. 77–105 (1938).

[110] R. DOUC, A. GARIVIER, E. MOULINES AND J. OLSSON. On the forward filtering backward smoothing particle approximations of the smoothing distribution in general state spaces models. *Annals of Applied Probab.*, vol. 21, no. 6, pp. 2109–2145 (2011).

[111] A. DOUCET, N. DE FREITAS AND N. GORDON, EDITORS. *Sequential Monte Carlo Methods in Practice*. Statistics for Engineering and Information Science series. Springer (2001).

[112] V. DRAKOPOULOS, N.P. NIKOLAOU. Efficient computation of the Hutchinson metric between digitized images. *IEEE Transactions on Image Processing*, vol. 13, no. 12 (2004).

[113] R. ECKHARDT. Stam Ulam, John Von Neumann and the Monte Carlo method. *Los Alamos Science*, Special Issue: 131–136 (1987).

[114] A. ECONOMOU. Generalized product-form stationary distributions for Markov chains in random environments with queueing applications. *Adv. in Appl. Probab.*, vol. 37, no. 1, pp. 185–211 (2005).

[115] M. EGESDAL, C. FATHAUER, K. LOUIE AND J. NEUMAN. Statistical modeling of gang violence in Los Angeles, *SIAM Undergraduate Research Online*, 3 (2010).

[116] P. EHRENFEST, T. EHERENFEST. Uber zwei bekannte Einwande gegen das Boltzmannsche H-Theorem, *Phys. Z.*, vol. 8, pp. 311–314 (1907).

[117] P. EMBRECHTS, H. SCHMIDLI. Ruin estimation for a general insurance risk model. *Advances in Applied Probability*, vol. 26, no. 2, pp. 404–422 (1994).

[118] A. EINSTEIN. *Investigations on the Theory of the Brownian Movement*, Dover, New York (1956).

[119] N. EL KAROUI, X. TAN. Capacities, Measurable Selection and Dynamic Programming Part I: Abstract framework. `ArXiv:1310.3363` (2014).

[120] N. EL KAROUI, X. TAN. Capacities, Measurable Selection and Dynamic Programming Part II: Application in Stochastic Control Problems. `ArXiv:1310.3364` (2014).

[121] N. ENRIQUEZ. Correlated Processes and the Composition of Generators. *Séminaire de Probabilités XL*, Springer, vol. 1899, pp. 329–342 (2007).

[122] S. N. ETHIER. *The doctrine of chances. Probabilistic aspects of gambling.* Probability and its Applications series, Springer (2010).

[123] S. N. ETHIER, T. G. KURTZ. *Markov Processes: Characterization and Convergence*, Wiley Series on Probability & Statistics (1986).

[124] S.N. EVANS. Stochastic billiards on general tables. *Ann. Appl. Probab.*, vol. 11, no. 2, pp. 419–437 (2001).

[125] P. A. FERRARI, N. MARIC. Quasi-stationary distributions and Fleming-Viot processes in countable spaces. *Electron. J. Probab.*, vol. 12, no. 24, pp. 684–702 (2007).

[126] M. FIELD, M. GOLUBITSKY. *La symétrie du chaos, à la recherche des liens entre mathématiques et nature*, Inter-Éditions, Paris (1993).

[127] R.A. FISHER. The wave of advance of advantageous genes. *Annals of Eugenics*, vol. 7, pp. 355–369 (1937).

[128] R. FILLIGER, M.O. HONGLER. Supersymmetry in random two-velocity processes. *Physica A: Statistical Mechanics and its Applications*, vol. 332, no. 1. pp. 141–150 (2004).

[129] J. FONTBONA, H. GUÉRIN AND F. MALRIEU. Quantitative estimates for the long-time behavior of an ergodic variant of the telegraph process. *Adv. in Appl. Probab.* vol. 44, no. 4, pp. 977–994 (2012).

[130] J.-P. FOUQUE, L.-H. SUN. *Systemic Risk Illustrated: In Handbook on Systemic Risk*, Cambridge University Press (2013).

[131] C.F. GAUSS. Theoria motus corporum coelestium in sectionibus conicis solem ambientum (1809).

[132] C.F. GAUSS.Theoria combinationis observationum erroribus minimis obnoxiae (1821/1823).

[133] S.B. GELFAND, S.K. MITTER. Weak convergence of Markov chain sampling methods and annealing algorithms to diffusions. *Journal of Optimization Theory and Applications*, vol. 68, no.3, pp. 483–498 (1991).

[134] W. R. GILKS, S. RICHARDSON AND D. J. SPIEGELHALTER. *Markov Chain Monte Carlo in Practice.* Chapman-Hall/CRC Press (1996).

[135] M. GIROLAMI, B. CALDERHEAD. Riemann manifold Langevin and Hamiltonian Monte Carlo methods. *Journal of the Royal Statistical Society: Series B*, vol. 73, no. 2, pp. 123–214 (2011).

[136] S. GOLDSTEIN. On diffusion by discontinuous movements, and on the telegraph equation. *Quart. Journ. Mech. and Applied Math.*, vol. 4, no. 2, pp. 129–156 (1950).

[137] T. GONZALES. *EVE: The Empyrean Age.* Orion books (2008).

[138] F. GOSSELIN. Asymptotic behavior of absorbing Markov chains conditional on non-absorption for applications in conservation biology. *Ann. Appl. Probab.*, vol. 11, pp. 261–284 (2001).

[139] P. GRASSBERGER, A. DE LA TORRE. Reggeon field theory (Schlögl's first model) on a lattice: Monte Carlo calculation of critical behavior. *Ann. Phys.*, vol. 122, pp. 373–396 (1979).

[140] J. GRIME. Kruskal's count. (Unpublished manuscript).

[141] T.H. GRONWALL. Note on the derivatives with respect to a parameter of the solutions of a system of differential equations, *Ann. Math.*, vol. 20, no. 2, pp. 293–296 (1910).

[142] HAIGHT F.A., *Handbook of the Poisson Distribution*, Wiley (1967).

[143] M. HAIRER. Convergence of Markov processes. Lecture Notes, Warwick University (2010).

[144] P.R. HALMOS. *Measure Theory.* Van Nostrand (1950).

[145] T. HARA, G. SLADE. Self-avoiding walks in five or more dimensions: I. The critical behaviour. *Communications in Mathematical Physics*, vol. 147, pp. 101-136 (1992).

[146] J. W. HARRIS, S.C HARRIS. Survival probability for branching Brownian motion with absorption. *Electron. Comm. Probab.*, vol. 12, pp. 89–100 (2007).

[147] T. E. HARRIS, H. KAHN. Estimation of particle transmission by random sampling. *Natl. Bur. Stand. Appl. Math. Ser.*, vol. 12, pp. 27–30 (1951).

[148] B. HAYES. Statistics of deadly quarrels. *American Scientist*, vol. 90: pp. 10–14, (2002).

[149] A. HAWKES, L. ADAMOPOULOS. Cluster models for earthquakes: regional comparisons. *Bull. Int. Statist. Inst.*, vol. 45, no. 3, pp. 454–461 (1973).

[150] R. L. HERMAN. *A Course in Mathematical Methods for Physicists.* CRC Press (2013).

[151] S. HERRMANN, J. TUGAUT. Non-uniqueness of stationary measures for self-stabilizing diffusions. *Stochastic Processes & Their Applications*, Vol. 120, no. 7, pp. 1215–1246 (2010).

[152] A. HEUER, C. MÜLLER AND O. RUBNER. Soccer: Is scoring goals a predictable Poissonian process. *Europhysics Letters*, vol. 89, no. 3 (2010).

[153] R.A. HOLLEY, T.M. LIGGETT. Ergodic theorems for weakly interacting infinite systems and the voter model. *Annals of Probability*, vol. 3, no. 4, pp. 643–663 (1975).

[154] M.O. HONGLER. Exact soliton-like probability measures for interacting jump processes. ArXiv:1501.07061 (2015).

[155] V. JACOBSON Congestion avoidance and control. *Computer Communications Review*, vol. 18, no. 4, pp. 314–329 (1988).

[156] P. JAGERS. *Branching Processes with Biological Applications.* Wiley (1975).

[157] C. JARZYNSKI. Nonequilibrium equality for free energy differences. *Phys. Rev. Lett.*, vol. 78, 2690 (1997).

[158] C. JARZYNSKI. Equilibrium free-energy differences from nonequilibrium measurements: A master-equation approach. *Phys. Rev. E*, vol. 56, 5018 (1997).

[159] M. KAC. A stochastic model related to the telegrapher's equation. *Rocky Mountain Journal of Mathematics*, vol. 4, no. 3, pp. 497–509, (1974).

[160] R. E. KALMAN. A new approach to linear filtering and prediction problems. *Trans. ASME Ser. D. J. Basic Engrg*, vol. 82, pp. 35–45 (1960).

[161] I. KARATZAS, S. E. SHREVE, *Brownian Motion and Stochastic Calculus*, Graduate Texts in Mathematics, Springer (2004).

[162] S. KARLIN, H.M. TAYLOR. *A Second Course in Stochastic Processes*. Academic Press (1981).

[163] E. HUTCHINSON. Fractals and self similarity. *Indiana University Mathematics Journal*, vol. 30, pp. 713–747 (1981).

[164] R. E. KALMAN, R. S. BUCY. New results in linear filtering and prediction theory. *Trans. ASME Ser. D. J. Basic Engrg*, vol. 83, pp. 95–108 (1961).

[165] L. V. KANTOROVICH, G. S. RUBINSHTEIN. On a space of totally additive functions. *Vestnik Leningradskogo Universiteta*, vol. 13, no. 7, pp. 52–59, (1958).

[166] J. KELLY. A new interpretation of information rate. *Bell Sys. Tech. J.*, vol. 35, pp. 917–926 (1956)

[167] H. KESTEN. Branching Brownian motion with absorption. *Stochastic Process. and Appl.*, vol. 37, pp. 9–47 (1978).

[168] J. F. C. KINGMAN. The first birth problem for age dependent branching processes. *Ann. Probab.*, vol. 3, pp. 790–801 (1975).

[169] P. E. KLOEDEN, E. PLATEN AND N. HOFMANN. Extrapolation methods for the weak approximation of Itô diffusions. *SIAM Journal on Numerical Analysis*, vol. 32, no. 5, pp. 1519–1534 (1995).

[170] P. E. KLOEDEN, E. PLATEN. *Numerical Solution of Stochastic Differential Equations*, vol. 23, Springer (2011).

[171] P. E. KLOEDEN, E. PLATEN. A survey of numerical methods for stochastic differential equations. *Stochastic Hydrology and Hydraulics*, vol. 3, no. 5, pp. 155-178 (1989).

[172] P. R. KILLEEN, J. G. FETTERMAN. A behavioral theory of timing. *Psychological Review*, vol. .95, no. 2, pp. 274–295 (1988).

[173] A.N. KOLMOGOROV, I.G. PETROVSKII AND N.S. PISKUNOV. A study of the diffusion equation with increase in the amount of substance, and its application to a biological problem, *Bull. Moscow Univ. Math. Mech*, vol. 1, no. 6, pp. 1–26 (1937).

[174] H.J. KUSHNER. On the weak convergence of interpolated Markov chains to a diffusion, *Annals of Probability*, vol. 2, pp. 40–50 (1974).

[175] O. KRAFFT, M. SCHAEFER. Mean passage times for tridiagonal transition matrices and a two-parameter Ehrenfest urn model. *J. Appl. Prob.*, vol. 30: 964-970 (1993).

[176] J. F. LAGARIAS, E. BAINS AND R. J. VANDERBEI. The Kruskal count. ArXiv:math/0110143v1 (2001).

[177] Y.F. LEE, W.K. CHING. On the convergent probabilities of a random walk. *International Journal of Mathematical Education in Science and Technology*, vol. 37, no. 7 (2006).

[178] A.M. LEGENDRE. *Nouvelles méthodes pour la détermination des orbites des comètes.* Firmin Didot, Paris (1805).

[179] O. HERNANDEZ-LERMA, J.B. LASSERRE *Further topics on discrete-time Markov control processes.* Springer, Application of Mathematics. Stochastic modeling and Applied Probability series, vol. 42 (1999).

[180] C. LIANG, G. CHENG, D. L. WIXON AND T. C. BALSER. An absorbing Markov chain approach to understanding the microbial role in soil carbon stabilization. *Biogeochemistry*, vol. 106, pp. 303–309 (2011).

[181] S. LEE, J. R. WILSON AND M. CRAWFORD. Modeling and simulation of a non-homogeneous Poisson process having cyclic behavior. *Communications in Statistics, Simulation*, vol. 20, pp. 777–809 (1991).

[182] D. A. LEVIN, Y. PERES. *Markov Chains and Mixing Times.* American Mathematical Society (2008).

[183] T. M. LIGGETT. *Interacting Particle Systems.* Springer (1985).

[184] T. M. LIGGETT. *Stochastic Interacting Systems: Contact, Voter and Exclusion Processes.* Springer (1999).

[185] J. LUCZKA, R. RUDNICKI. Randomly flashing diffusion: asymptotic properties, *J. Statist. Phys.*, vol. 83, pp. 1149–1164 (1996).

[186] F. MALRIEU. Some simple but challenging Markov processes. `ArXiv:1412.7516`, dec. (2014).

[187] H. P. MCKEAN. Application of Brownian motion to the equation of Kolmogorov-Petrovskii- Piskunov, *Communications on Pure and Applied Mathematics*, vol. 28, pp. 323–331 (1975).

[188] H. P. MCKEAN. H.P. McKean, A correction to "Application of Brownian motion to the equation of Kolmogorov-Petrovskii-Piskunov", *Comm. Pure Appl. Math.*, vol. 29, pp. 553–554 (1976).

[189] H. MAHMOUD. *Polya Urn Models.* Chapman & Hall/CRC. Texts in Statistical Science (2009).

[190] B. MANDELBROT, *The Fractal Geometry of Nature.* W.H. Freeman & Co. (1982).

[191] J. C. MATTINGLY, A. M. STUART, AND D. J. HIGHAM. Ergodicity for SDEs and approximations: locally Lipschitz vector fields and degenerate noise. *Stochastic Process. Appl.*, vol. 101, no. 2, pp. 185–232 (2002).

[192] E. LEWIS, G. MOHLER, P. J. BRANTINGHAM AND A. BERTOZZI. Self-exciting point process models of civilian deaths in Iraq. *Security Journal*, vol. 25, pp. 244–264 (2011).

[193] J. MACQUEEN, R. G. MILLER JR. Optimal persistence policies, *Oper. Res.*, vol. 8, pp. 362–380 (1960).

[194] R.C. MERTON. Theory of rational option pricing. *Bell Journal of Economics and Management Science*, vol. 4, no. 1, pp. 141–183 (1973).

[195] A. R. MESQUITA. Exploiting Stochasticity in Multi-agent Systems, PhD thesis, University of California, Santa Barbara (2010).

[196] N. METROPOLIS, S. ULAM. The Monte Carlo Method. *Journal of the American Statistical Association*, Vol. 44, No. 247, pp. 335–341 (1949).

[197] N. METROPOLIS. The beginning of the Monte Carlo method. *Los Alamos Science*, no. 15, pp. 125–130 (1987).

[198] N. METROPOLIS, A.W. ROSENBLUTH, M.N. ROSENBLUTH, A.H. TELLER AND E. TELLER. Equation of state calculations by fast computing machines. *Journal of Chemical Physics*, vol. 21, no. 6, pp. 1087–1092 (1953).

[199] S. P. MEYN, R.L. TWEEDIE. *Markov Chains and Stochastic Stability.* Springer (1993).

[200] B. MEZRICH. *Bringing Down the House: The Inside Story of Six M.I.T. Students Who Took Vegas for Millions*, Free Press (2002).

[201] J. MICHAEL. Positive and negative reinforcement, a distinction that is no longer necessary; or a better way to talk about bad things. *Behaviorism*, vol. 3, no. 1, pp. 33–44 (1975).

[202] L. MLODINOW. *The Drunkard's Walk: How Randomness Rules Our Lives.* Edition Pantheon (2008).

[203] G. O. MOHLERA, M. B. SHORTA, P. J. BRANTINGHAMA, F. P. SCHOENBERGA AND G. E. TITA. Self-exciting point process modeling of crime. *Journal of the American Statistical Association*, vol. 106, no. 493 (2011).

[204] L. MOSER. On a problem of Cayley, *Scripta Math.*, vol. 22, pp. 289–292 (1956).

[205] M.E. MÜLLER. Some continuous Monte Carlo methods for the Dirichlet problem. *Ann. Math. Stat.*, vol. 27, pp. 569–589 (1956).

[206] J. NEVEU. Multiplicative martingales for spatial branching processes. In *Seminar on Stochastic Processes, 1987. Prog. Probab. Statist.* vol. 15, pp. 223–241. Birkhäuser, Boston (1988).

[207] Y. NISHIYAMA. The Kruskal principle. *Osaka Keidai Ronshu*, vol. 63, no. 3 (2012).

[208] F. POHL, J. BERKEY. *Drunkard's Walk.* Ballantine Books (1973).

[209] I. M. TOKE, F. POMPONIO. Modelling trades-through in a limit order book using Hawkes Processes. *Economics*, vol. 6, no. 2012-22, (2012).

[210] J. NORRIS. *Markov Chains.* Cambridge University Press (1998).

[211] Y. OGATA, D. VERE-JONES. Inference for earthquake models: A self-correcting model. *Stochastic Processes and their Applications*, vol. 17, no. 2, pp. 337-347 (1984).

[212] Y. OGATA. Statistical models for earthquake occurrences and residual analysis for point processes. *J. Amer. Statist. Assoc.*, vol. 83, pp. 9–27 (1988).

[213] B. K. ØKSENDAL. *Stochastic Differential Equations: An Introduction with Applications*, 6th ed. Springer (2003).

[214] L.S. ORNSTEIN, G.E. UHLENBECK. On the theory of Brownian motion. *Phys. Rev.*, vol. 36, pp. 823–841 (1930).

[215] T. J. OTT, J. H. B. KEMPERMAN. Transient behavior of processes in the TCP paradigm. *Probab. Engrg. Inform. Sci.*, vol. 22, no.3, pp. 431–471 (2008),

[216] T. J. OTT, J. H. B. KEMPERMAN AND M. MATHIS. The stationary behavior of ideal TCP congestion avoidance. Unpublished manuscript : available at http://www.teunisott.com/, 1996.

[217] M. PACE, P. DEL MORAL. Mean-field PHD filters based on generalized Feynman-Kac flow. *IEEE Journal of Selected Topics in Signal Processing*, vol. 7, no. 3, pp. 484-495 (2013).

[218] L. PAGE, S. BRIN. The anatomy of a large-scale hypertextual web search engine. *Computer Networks and ISDN Systems*, vol. 30, pp. 107–117 (1998).

[219] LARRY PAGE AND SERGEY BRIN. Encyclopedia of World Biography. Advameg, Inc. 2013.

[220] L. PAGE. PageRank: Bringing Order to the Web, Stanford Digital Library Project talk. August 18, 1997.

[221] J.M.R. PARRONDO, L. DINIS. Brownian motion and gambling: from ratchets to paradoxical games. *Contemporary Physics*, vol. 45, no. 2, pp. 147–157 (2004).

[222] E. A. J. F. PETERS, G. DE WITH. Rejection-free Monte-Carlo sampling for general potentials. *Physical Review E* 85.026703 (2012).

[223] V. M. PILLAI, M. P. CHANDRASEKHARAN. An absorbing Markov chain model for production systems with rework and scrapping. *Computers and Industrial Engineering*, vol. 55, no. 3, pp. 695–706 (2008).

[224] V. PLACHOURAS, I. OUNIS, AND G. AMATI. The static absorbing model for hyperlink analysis on the web. *Journal of Web Engineering*, vol. 4, no. 2, pp. 165–186 (2005).

[225] S. PINKER. *How The Mind Works*. W. W. Norton & Company, pp. 54–55 (2009).

[226] E. PLATEN, K. KUBILIUS. Rate of weak convergence of the Euler approximation for diffusion processes with jumps. School of Finance and Economics, University of Technology, Sydney (2001).

[227] J. POSTEL. Transmission Control Protocol, September 1981, RFC 793 (1981).

[228] N. PRIVAULT. *Notes on Markov chains*. Nanyang Technological University (2012).

[229] T. RAINSFORD, A. BENDER. Markov approach to percolation theory based propagation in random media. *IEEE Transactions on Antennas and Propagation*, vol. 56, no. 3, pp. 1402–1412, (2008).

[230] N. RATANOV. Double telegraph processes and complete market models. *Stoch. Anal. Appl.*, vol. 32, no. 4, pp. 555–574 (2014).

[231] D. REVUZ, M. YOR. *Continuous Martingales and Brownian Motion*, Springer (1991).

[232] G. O. ROBERTS, R. L. TWEEDIE. Exponential convergence of Langevin distributions and their discrete approximations. *Bernoulli*, vol. 2, pp. 341–363 (1996).

[233] G. O. ROBERTS, R. L. TWEEDIE. Geometric convergence and central limit theorems for multi-dimensional Hastings and Metropolis algorithms. *Biometrika*, vol. 83, no. 1, pp. 95–110 (1996).

[234] S. M. ROSS. *Introduction to stochastic dynamic programming*. Academic Press. Probability and Mathematical Statistics Series (1983).

[235] R. RUDNICKI. Markov operators: applications to diffusion processes and population dynamics. *Applicationes Mathematicae*, vol. 27, no. 1, pp. 67-79 (2000).

[236] L. SALOFF-COSTE. *Lectures on Finite Markov Chains*. École d'été de Saint Flour, Springer (1996).

[237] J. SANDEFUR. The gunfight at the OK corral. *Mathematics Magazine*, vol. 62, no. 2, pp. 119–124 (1989).

[238] E. SCHRÖDINGER. An undulatory theory of the mechanics of atoms and molecules. *Physical Review*, vol. 28, no. 6, pp. 1049–1070 (1926).

[239] E. SENETA. *Non-negative Matrices and Markov Chains*. Springer (2006).

[240] SINGER, R.A. Estimating optimal tracking filter performance for manned maneuvering targets. *IEEE Transactions on Aerospace and Electronic Systems*, vol. 6, no. 4, pp. 473–483 (1970).

[241] D. SHERRINGTON, S. KIRKPATRICK. Solvable model of a spin-glass, *Physics Review Letters*, vol. 35, no. 26, pp. 1792-1796 (1975).

[242] Z. SHI. *Branching Random Walks*. École d'été de Probabilités de Saint Flour, Springer (2012).

[243] M.B. SHORT, M. R. D'ORSOGNA, P.J. BRANTINGHAM AND G.E. TITA. Measuring and modeling repeat and near-repeat burglary effects, *J. Quant. Criminol.*, vol. 25, no. 3, 325–339 (2009).

[244] M. SMOLUCHOWSKI. Zur kinetischen Theorie der Brownschen Molekularbewegung und der Suspensionen. *Annalen der Physik*, vol. 21, no. 14, pp. 756–780 (1906).

[245] F. SPITZER. Interaction of Markov processes. *Advances in Mathematics*, vol. 5, pp. 246–290 (1970).

[246] R. L. STRATONOVICH. Conditional Markov processes. *Theory of Probability and Its Applications*, vol. 5, pp. 156–178 (1960).

[247] D. STEINSALTZ, S. N. EVANS. Markov mortality models: implications of quasi-stationarity and varying initial distributions. *Theor. Pop. Biol.*, vol. 65, pp. 319–337 (2004).

[248] L. M. SURHONE, M. T. TENNOE AND S. F. HENSSONOW. *Hutchinson Metric*. Betascript Publishing (2010).

[249] R.S. SUTTON, A.G. BARTO. *Reinforcement Learning: An Introduction*. MIT Press (1998).

[250] W.C. SWOPE, H.C. ANDERSEN, P.H. BERENS AND K. WILSON. A computer simulation method for the calculation of equilibrium constants for the formation of physical clusters of molecules: applications to small water clusters. *J. Chem. Phys.*, vol. 76, pp. 637–649 (1982).

[251] F. THORNDIKE. Applications of Poisson's probability summation. *Bell System Technical Journal*, vol. 5, no. 4, pp. 604–624 (1926).

[252] N. TOUZI. *Optimal Stochastic Control, Stochastic Target Problems and Backward SDE*. The Fields Institute for Research Springer (2013).

[253] M. D. TOWLER. Quantum Monte Carlo and the CASINO program: highly accurate total energy calculations for finite and periodic systems. *Psi-k Newsletter* December (2003).

[254] O. VASICEK. An equilibrium characterisation of the term structure. *Journal of Financial Economics*, vol. 5, no. 2, pp. 177–188 (1977).

[255] P.G.C. VASSILIOU. The evolution of stochastic mathematics that changed the financial world. *Linear Algebra and its applications*, vol. 203-204, pp. 1-66 (1994).

[256] C. VERGE, C. DUBARRY, P. DEL MORAL AND E. MOULINES. On parallel implementation of sequential Monte Carlo methods: the island particle model. *Statistics and Computing*, vol. 25, no. 2, pp. 243–260 (2015).

[257] L. VERLET. Computer experiments on classical fluids. I. Thermodynamical properties of Lennard-Jones Molecules. *Physical Review*, vol. 159, pp. 98–103. (1967).

[258] C. VILLANI. *Topics in Optimal Transportation*. Volume 58, Graduate Studies in Mathematics. American Mathematical Society (2003).

[259] J. VILLE. *Étude critique de la notion de collectif*, Monographies des Probabilités 3 (in French), Paris: Gauthier-Villars (1939).

[260] F.W. WARNER. *Foundations of Differentiable Manifolds and Lie Groups*. Scott, Foresman and Co. (1971).

[261] J. WEINGARTEN. Ueber eine Klasse auf einander abwickelbarer Flächen. *Journal für die reine und angewandte Mathematik*, vol. 59, pp. 382–393 (1861).

[262] K. Wolny. Geometric ergodicity of heterogeneously scaled Metropolis-adjusted Langevin algorithms (MALA). Seventh Workshop on Bayesian Inference in Stochastic Processes (BISP 2011), Madrid (2011).

[263] J. R. YOUNG. The Magical Mind of Persi Diaconis. Chronicle of Higher Education, October 16th (2011).

[264] L. ZHANG. Virtual-Clock: A new traffic control algorithm for packet switching networks. *ACM SIGCOMM Proc. Computer Communication Review*, vol. 20, no. 4, pp. 19–29 (1990).

[265] L. ZHANG AND S. DAI. Application of Markov model to environment fate of phenanthrene in Lanzhou reach of Yellow river. *Chemosphere*, vol. 67, pp. 1296–1299 (2007).

Index

3-Ball, 670

Absolute spectral gap, 147
Absorbed Markov chain, 709
Absorption models, 242, 243
Absorption rate, 242
Acceptance-rejection ratio, 227
Affine transformations, 14, 723
Aggregation, 77
Ambient space, 581
Ancestral line, 258
Ancestry evolution model, 708
Angle bracket, 181, 431
Annealed measures, 267
Aperiodicity, 203, 697
Arc length, 625, 647, 670
Arrival time, 21
Asset-selling problem, 817
Atlas, 595
Autoregressive models, 138
Average survival time, 460
Azimuthal angle, 675

Bachelier model, 414
Bachelier, Louis, 50, 60
Bachmann-Landau notation, xlviii
Backward filters, 281
Backward integration, 255
Backward stochastic differential equation
        (BSDE), 408, 451
Bakry-Emery criterion, 667
Ball walk Metropolis sampler, 288
Ballot problem, 819
Bankroll management, 792
Bayes' formula, xlvii, 46, 80
Bayesian inference, 77
Bayesian statistical manifolds, 683
Bayesian statistics, 84
Beeman's algorithm, 734
Bellman equation, 800, 802
Bernoulli process, 20, 304, 375, 380
Bernoulli random variable, 73
Bessel functions, 460

Best linear estimator, 82
Beta distribution, 78
Beta random variable, 77, 95
Betting system, 190
Biharmonic function, 461
Bimodal growth diffusion, 781
Binet formula, 797
Binomial pricing technique, 828
Binomial random variable, 73
Biochemical reaction network, 389
Biorthogonal system, 599
Birth and death process, 27, 759
Black, Fischer, 60
Black-box model, 108
Black-Scholes formula, 60
Black-Scholes-Merton formula, 60, 825
Bochner-Lichnerowicz formula, 568, 618
Bold play strategy, 819
Boltzmann entropy, 684
Boltzmann-Gibbs distribution, 44, 110, 111,
        226, 232, 236, 271, 467, 672, 685,
        734, 741
Boltzmann-Gibbs mapping, xlvi, 82, 102,
        134, 241
Bootstrapping, 253
Born-Oppenheimer approximation, 741
Bouncy particle samplers, 390
Bounded increment, 201
Bounded subset, 160
Box-Muller formula, 75
Bra-ket formalism, xlv, 742
Branching Markov chain, 773
Branching process, 27, 37, 770
Brownian bridge, 424
Brownian motion, 51, 393–395, 467, 665
        invariance properties, 412
        Monge parametrization, 591
        on p-sphere, 592
        on catenoid, 638
        on circle, 632
        on cone, 637
        on cylinder, 582
        on helicoid, 638

on simplex, 635
on sphere, 582
on torus, 634, 704
on ellipsoid, 591, 637
on graph, 591
on manifolds, 581
on orbifolds, 589
on quotient manifolds, 591
on revolution surfaces, 638
on Riemannian manifolds, 629
on sphere, 633
orthonormal transformation, 412
reflexion invariance, 418
rotational invariance, 418
Buckingham exponential-6 potential, 733
Buffon's needle, 99
Burgers equation, 471
Burglar problems, 817

Càdlàg process, 314, 502
Càdlàg property, 314
Call option, 62, 826
Canonical measure, 734
Canonical parametrization, 597
Cantor's discontinuum, 720
Carré du champ operator, 56, 185, 344, 403,
        431, 469, 568
Carrying capacity, 762
Casino roulette, 32
Cauchy problem, 247, 248, 441, 458
Cauchy random variable, 92
Cauchy-Dirichlet-Poisson problem, 250,
        251, 447, 449
Cayley-Moser problem, 816
Cemetery state, 242, 772
Central limit theorem, 100, 202
Change of basis formulae, 599
Change of coordinates, 542
Change of variable formula, 142
Change of variables, 74
Chapman-Kolmogorov equation, 124
Characteristic polynomial, 143
Chernov estimates, 105
Chinese restaurant process, 39
Christoffel symbols, 543, 607, 641
Circular cone, 578
Cloning, 253
Closed manifold, 652
Closed subset, 160
Co-factor expansion, 550
Coagulation, 77, 95

Codazzi-Mainardi equations, 626
Coincidences, 8
Colatitude angle, 675
Cole-Hopf transformation, 806
Compact set, 160
Compound Poisson process, 334, 356
Conditional distribution, 80, 96
Conditional expectation, xlvii
Confining potential, 488
Congestion, 36
Congestion window, 380
Conjugate distributions, 82, 97
Conjugate numbers, 195
Conjugate priors, 84
Connectivity constant, 278
Contact process, 751
Continuous time embedding, 313, 314, 329
Contraction, 152
Contraction theorem, 221
Control chart, 798
Coordinate projections, 542
Cosine law of reflection, 292
Cost function, 799
Coulomb forces, 741
Counting process, 299
Counting random variable, 73
Coupling, 6, 45, 166, 172, 176, 355
        diffusions, 419
        jump processes, 357
Coupling time, 12, 27
Coupon collector problem, 713
Covariance functional, 100
Covariance matrix, 101
Covariant derivative, 544
        first order, 541, 602
        second order, 562, 609
Covariation process, 429
Cox process, 307
Cox-Ingersoll-Ross diffusion, 423
Cox-Ross-Rubinstein model, 824
Cramer-Rao lower bound, 685
Critical fugacy, 278
Cross product, 581
Cryptography, 8
Curvature, 568
Cylinder, 581

D'Alembert martingale, 794
D'Alembert, Jean le Rond, 794
Data association, 8
Deflated asset, 62

Deflated risky asset, 55
Deflated stochastic differential equation, 55
Deflator, 832
Degrowth-production models, 385
Desintegration property , 229
Diaconis, Persi Warren, 7
Diagonalisation, 141
Diffeomorphism, 74
Differential geometry, 535
Differential operator, 53
    first order, 365
    second order, 396, 403, 467
Diffusion markets, 515
Diffusion Monte Carlo methods, 748, 749
Diffusion operator, 58
Diffusion process, 54, 393
Diffusions on chart spaces, 631
Diffusions on manifolds, 535
Diffusions on the simplex, 635
Dirac measure, xlii
Directed polymer, 279
Directional derivative, 544, 606
Dirichlet boundary conditions, 447
Dirichlet form, 360, 422
Dirichlet problem, 247, 248
Dirichlet random variable, 78, 96, 97
Dirichlet-Poisson problem, 248, 442
Discrete Laplacian, 320, 321
Discrete random variable, 73
Disordered models, 111
Distribution manifolds, 682
Divergence of vector fields, 553
Divergence operator, 580
Divergence theorem, 650
Dividends, 55
Dobrushin ergodic coefficient, 152, 229, 353
Dobrushin local contraction coefficient, 160
Dobrushin local contraction condition, 160,
    353, 485
Doeblin, Wolfgang, 6, 50
Doeblin-Itō formula, 53, 57, 185, 303, 350,
    376, 398, 402, 428, 583
Doléan-Dade exponential formula, 400
Domain of generators, 435
Dominating density, 77
Doob h-process, 244
Doob's convergence theorem, 197
Doob's upcrossing lemma, 197
Doob-Meyer decomposition, 346
Double telegraph process, 390
Doubly stochastic Poisson process, 307

Drunkard's walk, 693
Dual forms, 596
Dual mean field model, 264
Duncan-Zakai equation, 518
Dupire formula, 837
Dynamic population models, 759
Dynamical system, 36, 53, 363
Dyson equation, 473

Edging strategies, 63
Eigenmeasure, 144
Eigenvalue, 141
Eigenvector, 141
Einstein manifold, 565, 677
Einstein notation, 576
Einstein summation convention, 596
Einstein, Albert, 50
Electron-nuclear attraction, 741
Electronic repulsion, 741
Elementary transition, 122
Ellipsoid, 591
Elliptic diffusions, 516
Embedded manifolds, 579
Embedded Markov chain, 314, 329, 459
Empirical average, 99
Energy functional, 732
Enrichment, 253
Ensemble Kalman filters, 283, 523
Epidemic model with immunity, 764
Equilibrium measure, 124
Ergodic theorem, 221
Erlang random variable, 93
Euclidian distance, 13
Euclidian gradient operator, 543
Euclidian space, xli, 160
    inner product, 535
    metric, 597
Euler approximation, 366
European option, 56, 184
European option pricing, 826
European roulette, 32
Evolution equations, 240, 314, 338
Exclusion process, 752
Excursions, 11
Exit time, 172
Expectation of matrices, xlvii
Expectation operator, 314
Expiration date, 62
Exploration, 242
Exploration of the unit disk, 226
Exponential change of measures, 512

Exponential clock, 332
Exponential map, 639
Exponential martingale, 530
Exponential random variable, 20, 72, 92
Exterior product, 648
Extinction probability, 783
Extremal curve, 640

Facultative mutualism system, 781
Failure time, 21
Fair game, 32, 192
Feedback control, 798
Feynman-Kac
    semigroup, 439, 459
    semigroup series expansions, 456
Feynman-Kac formula, 530
Feynman-Kac jump process, 466, 469
Feynman-Kac measures, 239, 252, 254, 267,
        279, 739, 748
    branching processes, 773, 783
    Embedded Markov chains, 332
    intensity measures, 783
    normalizing constants, 239
    particle models, 474, 479
    Product formulae, 245
    two state models, 474
Feynman-Kac propagator, 741
Filtering, 113, 517
Filtering equation, 47
Finite dimensional approximation, 342, 375,
        407
Finite graph, 695
Finite state space model, 328
First order deviation, 100
First-in-first-out queue, 335
Fisher distribution, 496
Fisher information metric, 684
Fisher's equation, 780
Fixed point, 14, 125, 133
Flashing diffusions, 454
Flow map, 363
Fluctuation of martingales, 181
Fokker-Planck equation, 54, 405, 406, 413,
        433, 434
Forecasting, 253
Forward and backward equations, 408, 431,
        439
Forward filters, 280
Foster-Lyapunov condition, 160, 485
Fouque-Sun systemic risk model, 470
Fréchet, Maurice, 50

Fractal image, 13
Fractal leaf, 723
Fractal models, 719
Fractal tree, 724
Fragmentation, 77
Fredholm integral equations, 248
French tarot, 8
Frozen path, 264

Gain parameter, 82
Gambler's ruin, 819
Gambling betting systems, 788
Gambling model, 191
Games with fixed terminal condition, 807
Gamma random variable, 93, 95
Gamma-two operator, 568
Gauss' equations, 625
Gauss, Carl Friedrich, 46
Gaussian curvature, 626
Gaussian distribution, 54
Gaussian integration by part, 97
Gaussian process, 495
Gaussian random variable, 75
Gaussian subset shaker, 225
Gaussian updates, 82
Gelfand-Pettis equations, 316, 437
Gene expression, 389
Genealogical tree, 49, 258
Genealogical tree evolution, 254
Genealogy, 16
Generalized inverse, 71
Generator, 58
Genetic algorithm, 48, 137, 252, 267, 278
Genetic evolution models, 769
Geodesics, 639, 677
    on sphere, 670
Geometric Brownian motion, 55, 400, 826,
        834
Geometric Brownian-Poisson process, 59
Geometric clock, 332
Geometric drift condition, 160
Geometric ergodicity, 467
Geometric random variable, 21, 73, 76
geometric random variable, 7
Gibbs sampler, 229, 230, 266
Gibbs-Glauber dynamics, 229
Gillespie algorithm, 323
Girsanov transformation, 505, 506, 509
Glauber dynamics, 230
Global optimization, 274
Global positioning system, 46

Go with the winner, 253
Goldstein-Kac process, 701
Google matrix, 34, 788
Google PageRank, 34, 787
Gradient flow, 657
Gradient vector, 411
Gram matrices, 535, 596, 648
Gramian, 535, 596, 648
Grand martingale system, 794
Granular media equations, 478
Graph coloring, 8
Graph coloring model, 230, 232
Green formula (first type), 652
Green function, 741
Growth rate, 762
Growth-fragmentation models, 385

Hahn-Jordan decomposition, 156
Hamiltonian, 732, 806
Hamiltonian operator, 739
Hamiltonian systems, 731
Hard core model, 232
Harmonic function, 444
Harmonic oscillator, 745
Harris recurrent set, 219
Hawkes process, 308
Heat equation, 397
    boundary conditions, 462
Heat-bath Markov chain sampler, 227, 416
Hedging, 833
Heighways dragons, 726
Hermite polynomials, 496, 746
Hessian matrix, 411
Hessian operator, 613
Hilbert space, 145
Hilbert-Schmidt inner product, 539
Historical process, 133, 254
Historical processes, 245
Hit-and-run samplers, 289, 460
Hitting time, 11, 172, 207
Holding time, 322
Hyperboloid, 578
Hypermanifolds, 579
Hyperparameters, 84
Hypersurfaces, 579

Implicit backward Euler scheme, 666
Importance sampling, 102, 530
Increment, 58
Independent Metropolis-Hastings sampler, 227

Indicator function, xlii
Individual based model, 26, 778
Individuals, 134
Infinitesimal generator, 316, 371, 403, 430, 463, 467, 468, 487, 513, 525, 583, 630, 735
Input-Ouput models, 109
Instrumental distribution, 77
Instrumental transition, 66
Integral operator, 244
Integration by part formula
    divergence theorem, 661
Integration by parts formula, 411, 428, 454
Integration on manifolds, 645
Integro-differential equation, 56, 317
Integro-differential operator, 58
Intensity function, 307, 338, 351, 370
Intensity measure, 86
Inter-bank exchanges, 470
Interacting agents dynamics, 477
Interacting jump Langevin model, 479
Interacting jump process, 136
Interacting MCMC algorithm, 272
Interacting neurons process, 478
Interacting particle systems, 749
Invariant measures, 124, 351, 482
Inverse gamma distribution, 496
Inverse shuffle, 9
Inverse temperature parameter, 237
Inversion technique, 71
Irreducibility, 203, 697
Irreducible, 204
Ising model, 110, 232, 236, 361
Itō, Kiyoshi, 12, 50
Iterated random functions, 13, 705

Jacobi formula, 551
Jacobi process, 496
Jacobian, 74
Jarzinsky identity, 468
Jump diffusion process, 56, 57, 425
Jump integral operator, 58
Jump Langevin model, 467
Jump rate, 57, 322
Jump term, 58
Jump time, 57

Kac's model of gases, 360
Kac's random walk, 701
Kac's theorem, 208
Kallianpur-Striebel formula, 518

Kalman filter, 45
Kalman, Rudolf, 45
Kalman-Bucy filters, 280, 522
Kantorovich-Monge-Rubinstein metric, 169
Kantorovich-Rubinstein duality, 169
Kelly Criterion, 818
Kepler-22b, 15
Killed process, 440, 459
Killing transition, 242, 244
Kinetic parameters, 108
Klein-Kramers equation, 753
Knudsen random walk, 292
Knudsen stochastic billiard, 292
Kolmogorov equation, 782
Kolmogorov-Petrovskii-Piskunov equation,
    780
Kramer-Moyal expansion, 358, 359
Kriging interpolation, 117
Kruskal's count, 10, 177
Kushner-Stratonovitch equation, 520

Ladder chain, 235
Lagrange equations, 641
Lagrange identity, 649
Laguerre polynomials, 497
Landau-Stuart diffusions, 495
Langevin diffusion, 488, 665, 731, 734
    kinetic model, 672
    Metropolis-Hasting sampler, 416
    Metropolis-Hastings sampler, 414, 415
    on manifolds, 661
    on Riemannian manifolds, 662
Langevin equation, 671
Langevin-McKean-Vlasov model, 472
Laplace's rule of successions, 79
Laplace, Pierre-Simon, 79
Laplacian, 412, 562, 612, 614
    Euclidian space, 412
    invariance properties, 412
    on catenoid, 625
    on cone, 624
    on ellipsoid, 624, 637
    on helicoid, 638
    on circle, 638
    on cylinder, 638
    on integer lattices, 321
    on revolution surfaces, 624, 638
    on simplex, 638
    on sphere, 624
    orthonormal transformation, 412
    polar coordinates, 637

    reflexion invariance, 418
    Riemannian operator, 612
    rotational invariance, 418
    spectrum, 447
    spherical coordinates, 637
Latitude, 675
Law of large numbers, 100
Lazy Markov chain, 204, 714
Lazy random walk, 698
Leap-frog algorithm, 734
Lebesgue integral, xli
Lebesgue measure, 82
Legendre polynomials, 499
Legendre, Adrien Marie, 46
Leibniz formula, 746
Lennard-Jones potential, 733
Levy's characterization, 400
Lie algebra, 554
Lie bracket, 554, 560
Life function martingale, 836
Likelihood distribution, 80
Likelihood function, 84, 504
Limiting distribution, 124, 383
Linear Gaussian model, 45, 82, 90
Linear Markov chain model, 126
Linear quadratic control, 812
Linear quadratic optimization, 813
Lipschitz function, xliv, 170, 210
Local rate, 328
Log-Sobolev inequalities, 228
Logistic jump process, 781
Logistic process, 762, 781
Lotka-Volterra jump process, 782
Lotka-Volterra model, 765
Lyapunov function, 160, 487

M/M/1 queue, 360
M/M/m queue, 360
Macroscopic model, 698
Macroscopic telegrapher's equation, 702
Mandelbrot, Benoit, 13
Many-body Feynman-Kac measures, 260,
    261
Markov chain, 107, 121
    embedding, 313
    on circle, 696
    on cycle, 695
    on excursion spaces, 367
    on graphs, 6, 176
    restrictions, 276
Markov chain Monte Carlo, 110, 226

Markov integral operator, 123
Markov policy, 798
Markov semigroup, 61, 130
Markov semigroups
    series expansions, 456
Markov transition, 123
Markov, Anatoli, 121
Martingale, 28, 53, 56, 61, 178, 179, 346, 347
    convergence, 197
Martingale decompositions, 184, 345
Martingale limit theorem, 196
Martingale optimality principle, 801, 803, 805
Martingale part, 346
Martingale systems, 788
Martingale with fixed terminal value, 183
Master equation, 44
Matching problem, 8
Mathematical finance, 60, 821
Matrix representation, xliii
Maximal coupling, 167
Maximal inequalities, 194
Maximum principle, 360
McKean model, 135, 469
McKean transitions, 136
McKean-Vlasov diffusion, 135, 464, 476, 523
MCMC within Gibbs, 292
Mean curvature, 626
Mean curvature vector, 547, 548, 580, 636, 673, 678
Mean field games, 470
Mean field particle models, 102, 134, 254, 278, 468, 479, 523
Mean reverting Brownian motion, 424
Mean value property, 444
Meeting time, 27
Memoryless property, 301
Merton jump diffusion model, 454
Merton, Robert, 60
Mesquita bacterial chemotaxis process, 389
Mesquita MCMC samplers, 390
Metamodels, 117
Metropolis, Nicholas, 44
Metropolis-adjusted Langevin model, 665
Metropolis-Hasting algorithm
    diffusion approximation, 416
Metropolis-Hastings acceptance rate, 44
Metropolis-Hastings algorithm, 44, 66, 227, 262
    diffusion approximation, 414, 415

Metropolis-Hastings ratio, 227
Metropolis-Hastings transition, 44, 237
Micro-local analysis, 228
Midas equation, 60
Migration process, 25
Mill's inequalities, 103, 105
Milstein scheme, 424
Minorization condition, 160, 386
Molecular dynamics, 731
Monge parametrization, 578
Monotone stopping problem, 817
Monte Carlo integration, 99
Monte Carlo method, xxii, 44, 54, 584
Moran genetic model, 769
Multi-server queue, 335
Multidimensional diffusions, 409
Multidimensional function, xlii
Multidimensional Gaussian model, 82
Multilevel splitting, 253
Multinomial random variable, 74
Multiple target tracking, 85
Multivariate normal distributions, 686
Mutation, 253
Mutation transition, 137
Myopic stopping rules, 816

NASA Space Shuttle, 45
Nash embedding theorem, 681
Neuman boundary conditions, 447
Neutral financial market, 821
Newton's second law, 731, 737
Nicholas Metropolis, 43
Nonlinear filtering, 276
Nonlinear jump diffusion process, 463
Nonlinear Markov chain, 132
Nonlinear Markov process, 463
Normal angle, 675
Normal curvature, 625
Normal derivative, 656
Normal vector fields, 578
Novikov theorem, 105
Nuclear repulsion, 741

Observations, 113
Occupation measure, 185, 221, 227, 257
Offsprings, 37, 773
OK Corral process, 139
One stage look-ahead stopping rule, 816
One step optimal predictor, 46, 280
Opinion dynamics, 477
Optimal k-steps to go value function, 799

Optimal filter, 46, 520, 527
Optimal portfolio allocation, 812
Optimal stopping, 807
Optional stopping theorems, 187, 190, 349
Orbifolds, 589
Order statistics, 94
Ornstein-Uhlenbeck process, 421, 487, 658
Orthogonal group, 589
Orthogonal projection, 580, 596, 599, 623,
        673, 678
Orthonormal, 146
Orthonormal basis, 537
Out of equilibrium virtual work, 468

Packet sniffer, 36
Packets, 36, 380
Page rank, 34
Pairings, 101
Parallel transport, 609, 615
Parameter estimation, 224
Parametric surface, 578, 623
Parametrizations and charts, 593
Pareto random variable, 92
Parking problem, 816
Parrondos' paradox, 819
Partial differential equation, 56
Particle absorption models, 242
Particle filters, 49, 102
Particle Gibbs sampler, 269
Particle Markov chain Monte Carlo, 260
Particle Metropolis-Hastings algorithm, 271
Particle Metropolis-Hastings model, 261
Particle techniques, 110
Particles, 134
Partition function, 110, 488, 749
Partition of unity, 595
Path dependent payoff, 818
Path space models, 211, 254, 501, 742
Payoff function, 799, 826, 827
Pearson diffusions, 485
Perfect sampling algorithm, 237
Performance function, 799
Permutation, 8
Perron-Frobenius theorem, 143
Phase space, 734
Phase vector, 731
Piecewise deterministic processes, 363
Ping program, 36
Poincaré inequality, 361, 422, 667
Poincaré, Henri, 60
Poisson equation, 156, 185, 250, 443

Poisson point process, 85
Poisson process, 20, 299, 300, 352, 380
Poisson random variable, 85, 94, 97
Poisson thinning simulation, 309, 331
Poisson time rescaling, 308
Poisson, Siméon Denis, 17
Poisson-Gaussian clusters, 91
Polar coordinates, 597
Polygon, 176
Population dynamics, 45, 134
Portfolio, 62
Positive recurrent set, 219
Posterior distribution, 80
Power method, 35
Power utility function, 812
Predator-prey model, 765
Predictable part, 346
Predictable quadratic variation, 181
Prediction, 253
Principal curvature, 626
Prior distribution, 80
Projected diffusions on manifolds, 586
Projected Langevin equation, 672
Projection matrix, 537
Projection operators, 535
Proofreading problem, 817
Proportionality, xli
Propp and Wilson sampler, 233
Proton mass, 741
Pruning, 253
Pseudo random numbers, 8
Pull back
        functions, 603
        vector field, 597
Pure jump process, 337, 367
Push forward
        mappings, 683
Pushed forward
        curve, 594
        functions, 602
        mappings, 602
        scalar product, 597
        vector field, 601, 604
Put option, 62, 826
Pyramid betting system, 794

Quadratic variation, 181
Quantum Monte Carlo, 253
Quantum state, 737
Quantum systems, 473
Quantum teleportation, 253

Quasi-invariant measure, 245, 246, 754
Quenched measures, 267
Queueing system, 359
Quotient manifold, 589

R.c.l.l. property, 314
Radar, 46
Radial integration, 444
Random direction sampler, 286
Random dynamical system, 126
Random particle matrix model, 257
Random process, xxi
Random transposition shuffle, 713
Random two-velocity model, 388
Random walk
    in random environment, 138
    on circle, 695
    on graphs, 694
    on hypercubes, 697
    on lattices, 693
    on weighted graphs, 116
Rank-based interacting diffusions, 477
Ranking algorithm, 34, 787
Rare event simulation, 103, 274
Rayleigh random variable, 92
Reaction rate, 389
Recurrent state, 206, 694
Recycling mechanism, 48
Reflected Markov chain, 138
Reflecting boundaries, 699
Reflection principle, 22, 139, 419
Regenerative processes, 388
Regression formula, 46, 81, 82
Reinforcement process, 28, 132
Reinforcement rate, 29
Rejection technique, 75
Rejection-free Monte Carlo, 252, 273, 390
Relative risk aversion coeeficient, 812
Renewal process, 300
Replenish, 253
Replicating portfolio, 63, 184, 832
Reproducing kernel Hilbert space, 117
Resampling, 253
Return function, 799
Return of assets, 824
Return times, 5
Reversed shuffle, 717
Reversibility property, 44, 125, 131, 228, 353
Reward function, 799
Riccati equation, 474, 522

Ricci curvature, 563
Riemannian directional derivative, 608
Riemannian gradient, 623
    on cylinder, 623
    on ellipsoid, 624
    on paraboloid, 623
    on catenoid, 625
    on cone, 624
    on revolution surfaces, 624
    on sphere, 623
Riemannian Langevin equation, 672
Riemannian Laplacian, 613
Riemannian metric, 673, 678
Riemannian scalar product, 602
Riemannian second covariant derivative, 614
Riemannian structures, 599
Riemannian volume measure, 664
Riffle shuffle, 9, 716
Risk neutral financial markets, 514
Risk neutral measure, 61, 514
Riskless return rate, 55
Risky asset, 55
Robust filtering equation, 524
Rodrigues formula, 746
Rosenbluth, Arianna, 44
Rosenbluth, Marshall, 44
Rotation group MCMC, 290
Ruin process, 33, 41, 178

Scaling properties, 418
Schauder basis, 491
Scholes, Myron, 60
Schrödinger equation, 737
Schrödinger equation in imaginary time, 739
Schrödinger operator, 739, 743
Second covariant derivative, 612
Secretary problem, 820
Segments, 36, 380
Selection, 253
Selection transition, 137
Self avoiding walk, 277
Self interacting process, 132
Self-financing portfolio, 55, 62, 828
Semigroups, 61, 314, 338, 363, 395, 406, 426
    regularity properties, 455
    series expansions, 456
    telescoping sum, 367
Sensor, 517
Sensor function, 517

Sequential Monte Carlo, 102, 110
Shake-and-bake samplers, 291
Sherrington-Kirkpatrick model, 111
Shot noise process, 384
Shuffling cards, 7, 215, 712
Sierpinski carpet, 725
Sierpinski fractal, 14
Simple exclusion process, 695
Simple random walk, 4, 51, 139, 277, 393, 693
Simplex, xliii
Simulated annealing, 236
Singer radar model, 114
SIS model, 782
Slice sampling, 292
Small sets, 160
Smoluchowski equation, 753
Smoluchowski, Marian, 50
Snell envelope, 809
Soliton-like jump process, 333, 476
Sonar, 46
Spatial branching processes, 243
    continuous time processes, 773
    discrete time processes, 772
Spatial Poisson processes, 85
Spawning, 253
Spectral decomposition, 145, 146, 696, 743
Spectral gap, 146
Speed geodesic vector, 642
Sphere, 581
Spherical coordinates, 634, 672
Square field vector, 56, 344
Square of Ornstein-Uhlenbeck processes, 423
Square process, 56
Square root matrix, 663, 687
Square root process, 495
Squared Bessel process, 421
St. Petersburg martingales, 789
Stabilizing populations, 25
State space model, 129
Stationary population, 26
Stationary Schrödinger equation, 739
Steepest descent, 657
Stirling number of second kind, 709
Stirling's approximation, 5
Stochastic approximation, 216
Stochastic billards, 291
Stochastic calculus, 50
Stochastic differential equation, 52, 401
Stochastic flow, 426

Stochastic gradient process, 487
Stochastic matrix, xliii, 125
Stochastic partial differential equation, 519
Stochastic process, xxi
Stochastic simulation, xxii, 44, 226
Stock option market, 50
Stock price, 55
Stock price models, 821
Stoichiometric coefficients, 389
Stopping times, 172, 346
Storage model, 385
Stratonovitch differential calculus, 584
Strike, 62, 826
Strong Markov property, 172
Strong stationary time, 45, 173
Student diffusion, 496
Student distribution, 496
Sturm-Liouville formulation, 484
Sub-martingale, 179, 347
Submanifold, 579
Subset sampling, 231
Sum of generators, 342
Super-martingale, 179, 347
Superposition of Poisson processes, 87
Surface measure, 647
Switching processes, 377, 387, 475

Tail distributions, 103
Tangent basis functions, 609
Tangent space, 579, 594, 599, 602, 673, 678
Tangential curvature, 625
Target measure, 226, 236
Taylor expansion, 52, 58
Taylor expansion on manifolds, 644
Tchebyshev polynomials, 498
Telegraph equation, 702
Telegraph signal, 22
Teller, Augusta, 44
Teller, Edward, 44
Temperature parameter, 467
Theory of speculation, 60
Thinning Poisson processes, 87
Time discretization schemes, 340, 361, 366, 373, 394, 401, 405, 428, 584, 665
Time independent Schrödinger equation, 739
Time mesh, 366, 373
Timeout, 36
Top-in-at-random shuffle, 712
Topological vector space, 160
Torsion, 543, 568

Torus, 678
Total variation distance, 151, 166
Tradeable market, 61
Trading strategy, 66
Trait space, 777
Transient state, 206, 694
Transition diagram, 128, 699
Transmission control protocol, 36, 379
Transmission unit, 380
Traveling salesman problem, 111
Traveling waves, 737
Tree of outcomes, 128
Trial and guiding wave functions, 531
Triangular random variable, 93
Twisted distributions, 102
Twisted processes, 530
Twisted-guiding waves, 756
Two-state Markov model, 127, 141, 317

Ulam, Stanislaw, 43
Unbiased, 260
Uncertainties propagation, 109
Unfair game, 32, 193
Uniform distribution, 724
Uniform ellipticity condition, 516
Uniform Markov jump process, 314
Uniform random variable, 71
Uniform variable on the circle, 75
Unit vector representation, 356
Up and down martingales, 821
Updating, 253
Urn process
    diffusion approximation, 704
    Ehrenfest model, 699
    Polya model, 700
USA Roulette, 32

V-norm contraction, 164
V-norms, 156
V-oscillation, 157
van der Waals bonds, 733
Variational principle, 756
Vasicek model, 421
Velocity vector field, 579
Verhulst-Pearl model, 762
Verification argument, 806
Verlet algorithm, 734
Vertex, 6
Volatility, 55, 824
Volume forms, 648
Volume measure, 647

von Neumann, John, 43
Voter process, 751

Wald's identity, 189, 349
Walk on Sphere method, 702
Walkers, 253
Wasserstein metric, 169
Wealth increment, 62
Web page, 34
Web surfer, 34
Wedge product, 648
Weibull random variable, 93
Weighted Dirac measure, xlii
Weighted empirical measures, 102
Weingarten map, 626
Weingarten's equations, 626
Whittacker martingale, 796
Wick rotation of the time, 739
Wick's formula, 101, 105
Wiener measure, 507, 508
Wiener process, 395
Wiener space, 508
Wilkie inflation model, 836
Wonham filter, 528
Wright-Fisher diffusion, 704

Yaglom measure, 246
Yule process, 335

Zakai equation, 518
Zenith, 675